Bias Temperature Instability for Devices and Circuits

Tibor Grasser
Editor

Bias Temperature Instability for Devices and Circuits

Springer

Editor
Tibor Grasser
Institute for Microelectronics
Technische Universität Wien
Wien, Austria

ISBN 978-1-4614-7908-6 ISBN 978-1-4614-7909-3 (eBook)
DOI 10.1007/978-1-4614-7909-3
Springer New York Heidelberg Dordrecht London

Library of Congress Control Number: 2013948914

Printed on acid-free paper

Springer is part of Springer Science+Business Media (www.springer.com)

Editorial

First observations of a threshold voltage instability in MOS transistors which was found to be very sensitive to bias and temperature were made in the 1960s. However, this *bias temperature instability* (BTI) has remained a relatively obscure and unimportant phenomenon until the routine introduction of nitrogen into the oxide around the year 2000. Since then, particularly the *negative* BTI (NBTI) in pMOSFETs has received a considerable amount of attention, both from industry and from academia. With the advent of high-k gate stacks, the presumably related phenomenon of the *positive* BTI (PBTI) in nMOSFETs has been attracting a comparable amount of interest.

Despite its nearly 50-year-long history, BTI has remained a highly controversial issue, and many fundamental questions may be considered unresolved. The plethora of observations, explanations, as well as possible physical models found in literature is often highly confusing. One aspect of the confusion is related to the exponentially growing number of publications on the topic, which contain numerous contradictory claims. Furthermore, viewpoints within the same research group evolve as new aspects of the phenomenon are revealed. In order to resolve at least this part of the confusion, I have attempted to bring together world-leading groups working on that topic to review, define, and summarize their current understanding of a particular aspect of the phenomenon more clearly and in greater detail than is possible in regular journal and conference publications.

The book is structured in four chapters and encompasses characterization, defect/device modeling, technological impact, and circuit aspects. The opening chapter looks into the primary challenges we face in our understanding of BTI, namely characterization issues. The most prominent example is given by the fact that the degradation induced by BTI recovers once the stress bias is removed, an issue already discussed in the earliest papers. However, it was only realized about 10 years ago that this recovery has a dramatic impact on all characterization attempts. The first contribution by *Kerber and Cartier* (GF/IBM) discusses the most frequently used characterization methods, starting from standard stress and sense schemes, over on-the-fly techniques, to fast voltage ramp stresses, which are particularly useful during technology qualification. Since BTI is a strongly temperature-dependent phenomenon, a lot can be learned from fast and well-controlled switches of the device temperature. This is possible by using local polysilicon wires placed around the devices as discussed and exploited by *Aichinger, Pobegen, and Nelhiebel* (Infineon/KAI) in the next contribution.

As devices are scaled down into the nanometer regime, discrete charge capture and emission events can be monitored and studied. Just like in random telegraph noise (RTN), it was observed that the discrete changes in the threshold voltage caused by single defect discharging events show a very wide distribution. *Wang, Chiu, and Liu* (National Chiao Tung University) discuss this phenomenon in detail using statistics of threshold voltage steps and emission times and give an interpretation of their data based on a thermally assisted dispersive charge tunneling model. Along similar lines, *Reisinger* (Infineon) discusses a recently suggested analysis method named the "time-dependent defect spectroscopy," pointing out various experimental pitfalls as well as providing a glimpse at the numerous results obtained from this technique. Recently, it has been suggested that the traps responsible for RTN may also play an important role in BTI. The research on RTN has been going on for several decades and the state of the art is summarized in the contribution of *Frank and Miki* (IBM/Hitachi).

One of the most important consequences of device miniaturization is that future devices will only contain a countable number of defects, resulting in considerable variability during degradation. *Rauch* (IBM) gives a detailed study of intrinsic and extrinsic variability, properties of the distributions, scaling issues, as well as the impact of this variability on analog and digital circuits such as SRAM cells. In the next contribution, *Kaczer, Toledano-Luque, Franco, and Weckx* (IMEC) discuss their defect-centric view on this time-dependent variability, with a particular focus on the single defect threshold voltage distribution, its impact on the total threshold voltage shift, and its connection with time-zero variability.

An extremely important aspect in our understanding of BTI is the correct identification of the defects underlying the degradation. One of the few techniques which can reveal the chemical nature of those defects utilizes some form of electron spin resonance (ESR). *Campbell and Lenahan* (NIST/Penn State University) review their work in this field and discuss their observation of P_b, E', and K centers in response to negative bias temperature stress in SiO_2, SiON, and high-k gate stacks. In the subsequent contribution, *Afanas'ev, Houssa, and Stesmans* (University of Leuven) give their perspective on the topic but put a strong emphasis on the importance of hydrogen-related defects in a wide range of technologies. *Zhang* (Liverpool John Moores University), on the other hand, summarizes his efforts in identifying a wide range of possible defects in SiO_2, SiON, and high-k dielectrics using electrical measurements. Finally *Ang* (Nanyang Technological University) discusses observations which indicate the importance of hole trapping and trap transformations during cyclic bias temperature stress, which are inconsistent with the conventional reaction–diffusion formalism.

The second chapter is focused on theoretical and modeling aspects. In the opening contribution by *El-Sayed and Shluger* (UC London), theoretical properties of the most commonly observed oxide defects such as E' centers are discussed using density functional theory. When such defects are charged, they interact with the discrete dopands inside the device, leading to potentially enormous changes of the device characteristics, as discussed in detail by *Amoroso, Gerrer, and Asenov* (University of Glasgow). One of the first popular models for NBTI is the reaction–diffusion model, originally suggested in 1977, but continuously refined using more recent findings. The latest developments along these lines of thought are summarized by *Mahapatra* (IIT Bombay). Expressing a contrary view, the foundations and the applicability of reaction–diffusion theory are then examined using stochastic chemical kinetics by *Schanovsky and Grasser* (TU Wien). In the next contribution, *Goes, Schanovsky, and Grasser* (TU Wien) develop a rigorous formulation of charge trapping based on nonradiative multiphonon theory using additional metastable states in order to explain experimental data obtained from time-dependent defect spectroscopy. Finally, *Grasser* (TU Wien) discusses a recently suggested formalism for an approximate description of BTI as a collection of independent first-order reactions using capture/emission time maps.

The third chapter focuses on the impact of various technological processing steps on BTI. One of the most prominent actors frequently linked to the phenomenon in various theories is hydrogen and the wealth of literature together with recent findings are summarized by *Pobegen, Aichinger, and Nelhiebel* (KAI/Infineon). Next, the impact of various processing steps is discussed in detail by *Mahapatra* (IIT Bombay), including the impact of nitrogen, fluoride, and hydrogen versus deuterium in nitrided as well as high-k gate stacks. While most modern technologies employ very thin dielectric layers, power MOS transistors typically require thick layers of SiO_2. Experimental challenges together with other peculiarities related to such devices are discussed in the contribution of *Stojadinović* et al. (University of Niš). In order to suppress excessive leakage through ultrascaled insulating layers, high-k materials have to be used in these technologies. Unfortunately, in these materials PBTI in nMOSFETs becomes a critical issue. Differences and similarities between NBTI and PBTI, including processing issues and static versus dynamic stress, are discussed by *Zhao, Krishnan, Linder, and Stathis* (IBM) in the first high-k contribution. The following contribution by *Young and Bersuker* (UT Dallas/SEMATECH) is particularly devoted to experimental issues, fast transient charging, electron trapping, and concurrent defect generation. Finally, *Toledano-Luque and Kaczer* (IMEC) discuss recent experimental results obtained on nanoscaled high-k devices, with a focus on stochastic charge trapping and statistical lifetime prediction.

In order to overcome various scaling issues, alternative channel materials and device topologies have been suggested to replace conventional planar silicon technology. *Franco and Kaczer* (IMEC) present recent results obtained on SiGe channel devices, which show considerably improved reliability compared to their Si channel counterparts. *Huang, Wang, and Li* (Peking University) discuss the reliability of Si nanowires in comparison with planar technologies, with a particular focus on stochastic charge capture and emission in nanoscaled devices. Finally, *Fleetwood* et al. (Vanderbilt University) discuss bias temperature instabilities in 4H-SiC devices and suggest BTI to be of a different origin in these devices.

The final chapter is focused on the impact of BTI on circuits. First, *Keane, Wang, Jain, and Kim* (Intel/University of Minnesota) discuss odometers for the direct on-chip assessment of BTI on circuits for a better measurement and timing control. A bottom-to-top approach of reliability analysis from device level to system level aging is described by *Sutaria, Velamala, Ravi, and Cao* (Arizona State University). Then, *Wirth* et al. (UFRGS) summarize latest attempts in statistical modeling of charge trapping in the context of RTN and BTI from a circuit perspective. Finally, *Martin-Martinez, Rodriguez, and Nafria* (UA Barcelona) address different alternatives to translate the effects of NBTI degradation on the electrical properties of devices into circuit performance and reliability.

I sincerely hope that the information provided in these chapters proves useful to scientists and engineers working in this rapidly changing field by accurately capturing the state of the art and that it triggers further research into this elusive phenomenon.

Wien, Austria Tibor Grasser

Contents

Part I

1 **Bias Temperature Instability Characterization Methods** 3
Andreas Kerber and Eduard Cartier

2 **Application of On-Chip Device Heating for BTI Investigations** 33
Thomas Aichinger, Gregor Pobegen, and Michael Nelhiebel

3 **Statistical Characterization of BTI-Induced High-k Dielectric Traps in Nanoscale Transistors** 53
Tahui Wang, Jung-Piao Chiu, and Yu-Heng Liu

4 **The Time-Dependent Defect Spectroscopy** 75
Hans Reisinger

5 **Analysis of Oxide Traps in Nanoscale MOSFETs using Random Telegraph Noise** 111
David J. Frank and Hiroshi Miki

6 **BTI-Induced Statistical Variations** 135
Stewart E. Rauch III

7 **Statistical Distribution of Defect Parameters** 161
B. Kaczer, M. Toledano-Luque, J. Franco, and P. Weckx

8 **Atomic-Scale Defects Associated with the Negative Bias Temperature Instability** 177
Jason P. Campbell and Patrick M. Lenahan

9 **Charge Properties of Paramagnetic Defects in Semiconductor/Oxide Structures** 229
V.V. Afanas'ev, M. Houssa, and A. Stesmans

10 **Oxide Defects** 253
Jian F. Zhang

11 **Understanding Negative-Bias Temperature Instability from Dynamic Stress Experiments** 287
Diing Shenp Ang

Part II

12 **Atomistic Modeling of Defects Implicated in the Bias Temperature Instability** 305
Al-Moatasem El-Sayed and Alexander L. Shluger

13 **Statistical Study of Bias Temperature Instabilities by Means of 3D "Atomistic" Simulation** 323
Salvatore Maria Amoroso, Louis Gerrer, Fikru Adamu-Lema, Stanislav Markov, and Asen Asenov

14 **A Comprehensive Modeling Framework for DC and AC NBTI** 349
Souvik Mahapatra

15 **On the Microscopic Limit of the RD Model** 379
Franz Schanovsky and Tibor Grasser

16 **Advanced Modeling of Oxide Defects** 409
Wolfgang Goes, Franz Schanovsky, and Tibor Grasser

17 **The Capture/Emission Time Map Approach to the Bias Temperature Instability** 447
Tibor Grasser

Part III

18 **Impact of Hydrogen on the Bias Temperature Instability** 485
Gregor Pobegen, Thomas Aichinger, and Michael Nelhiebel

19 **FEOL and BEOL Process Dependence of NBTI** 507
Souvik Mahapatra

20 **Negative Bias Temperature Instability in Thick Gate Oxides for Power MOS Transistors** 533
Ninoslav Stojadinović, Ivica Manić, Danijel Danković, Snežana Djorić-Veljković, Vojkan Davidović, Aneta Prijić, Snežana Golubović, and Zoran Prijić

21 **NBTI and PBTI in HKMG** 561
Kai Zhao, Siddarth Krishnan, Barry Linder, and James H. Stathis

22 **PBTI in High-k Oxides** 585
Chadwin D. Young and Gennadi Bersuker

23 Characterization of Individual Traps in High-κ Oxides 597
M. Toledano-Luque and B. Kaczer

24 NBTI in (Si)Ge Channel Devices 615
Jacopo Franco and Ben Kaczer

25 Characteristics of NBTI in Multi-gate FETs for Highly Scaled CMOS Technology 643
Ru Huang, Runsheng Wang, and Ming Li

26 Bias-Temperature Instabilities in Silicon Carbide MOS Devices 661
D.M. Fleetwood, E.X. Zhang, X. Shen, C.X. Zhang, R.D. Schrimpf, and S.T. Pantelides

Part IV

27 On-Chip Silicon Odometers for Circuit Aging Characterization 679
John Keane, Xiaofei Wang, Pulkit Jain, and Chris H. Kim

28 Multilevel Reliability Simulation for IC Design 719
Ketul B. Sutaria, Jyothi B. Velamala, Venkatesa Ravi, Gilson Wirth, Takashi Sato, and Yu Cao

29 Charge Trapping in MOSFETS: BTI and RTN Modeling for Circuits 751
Gilson Wirth, Yu Cao, Jyothi B. Velamala, Ketul B. Sutaria, and Takashi Sato

30 Simulation of BTI-Related Time-Dependent Variability in CMOS Circuits 783
Javier Martin-Martinez, Rosana Rodriguez, and Montse Nafria

Part I

Chapter 1
Bias Temperature Instability Characterization Methods

Andreas Kerber and Eduard Cartier

Abstract Bias temperature instability (BTI) is one of the most critical device degradation mechanisms in conventional poly-Si/SiON and MG/HK CMOS technologies and is characterized with a variety of electrical measurement procedures. In this section, the most commonly used characterization procedures are introduced, and their advantages and possible pitfalls are discussed. Since the trapping and detrapping processes responsible for BTI cover a wide range of time constants, time-resolved characterization techniques are essential to capture the physics of the phenomena. Therefore, recent development efforts to this end are summarized, including fast reliability screening procedures for material and process evaluation.

1.1 Introduction

The electrical stability of metal oxide semiconductor (MOS) devices has been the subject of many scientific studies. In the early days of MOS device fabrication, research focused on ionic contamination [1, 2] and on the passivation of electrically active surface states at the silicon/silicon dioxide (Si/SiO_2) interface [3, 4]. During these early days, bias temperature instability (BTI) was hardly mentioned as a source of reliability concern [5]. However, over time, BTI has become one of the most frequently discussed degradation mechanisms in complementary metal

A. Kerber (✉)
Technology Reliability Development, GLOBALFOUNDRIES,
Yorktown Heights, NY, USA
e-mail: andreas.kerber@globalfoundries.com

E. Cartier
IBM Research Division, T. J. Watson Research Center,
Yorktown Heights, NY, USA
e-mail: ecartier@us.ibm.com

T. Grasser (ed.), *Bias Temperature Instability for Devices and Circuits*,
DOI 10.1007/978-1-4614-7909-3_1, © Springer Science+Business Media New York 2014

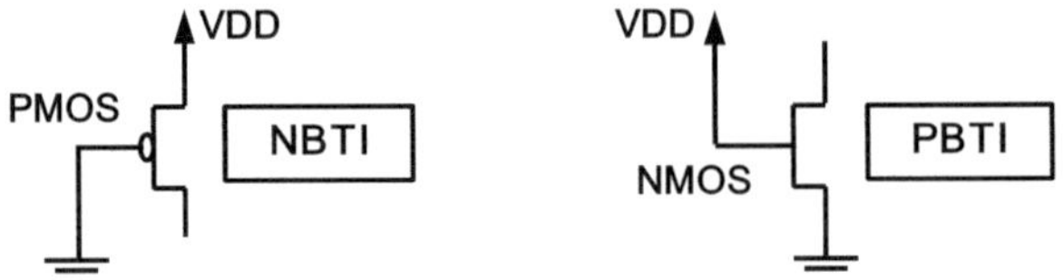

Fig. 1.1 Bias configuration for PMOS and NMOS devices in CMOS circuits. In inversion mode operation, for PMOS transistors, the gate is forced to ground with drain terminal high, and for NMOS transistors, the gate is biased at the supply voltage, VDD, with the source at ground

oxide semiconductor (CMOS) devices. In this chapter, we introduce the different characterization methods that are developed to capture the important aspects of BTI in advanced CMOS technologies.

As the name suggests, BTI refers to a time-dependent instability in transistors that is accelerated with increasing bias and temperature. Specifically, during a BTI test, the absolute threshold voltage of the metal oxide semiconductor field effect transistor (MOSFET) increases, while the device is biased in inversion mode. The threshold voltage shift leads to a decrease in drain current in the on-state of the transistor [6] and ultimately to a speed reduction of CMOS circuits. Degradation due to the BTI occurs during normal transistor operation, as shown in Fig. 1.1. For p-channel MOSFETs the term negative bias temperature instability (NBTI) is used, whereas for n-channel MOSFETs the degradation is called positive bias temperature instability (PBTI) since the corresponding gate bias conditions are negative and positive, respectively. Instead of transistors, capacitor structures can also be used to study NBTI (p-type Si) and PBTI (n-type Si). In this case, the devices are biased in accumulation mode, and flatband voltage shifts are monitored instead of threshold voltage shifts.

For conventional poly-Si electrodes, BTI is typically only observed in p-channel MOSFETs, and BTI is known to be more severe with SiON gate dielectrics as compared to SiO_2. With the introduction of high-k (HK) dielectrics into the CMOS manufacturing process [7, 8], both p-channel and n-channel MOSFETs are now exhibiting BTI.

The same characterization methods are used to study NBTI and PBTI, despite distinct differences in the physical degradation mechanisms, largely due to the defect structure of the dielectrics, the asymmetry in the band structure of the Si/HK/metal gate stack, and the opposite polarity of the gate bias. In Fig. 1.3, the schematic band structures of the p-channel and n-channel MOSFETs with MG/HK stacks during operation are compared, and the dominant degradation mechanisms are indicated.

During NBTI (PMOS, Fig. 1.3), both positive charge trapping and interface-state generation have been reported [10]. The basic physical degradation process for conventional poly-Si/SiON devices is very similar to the MG/HK stacks invoking positive charge trapping in the SiON layer and interface-state generation at the Si/SiON interface.

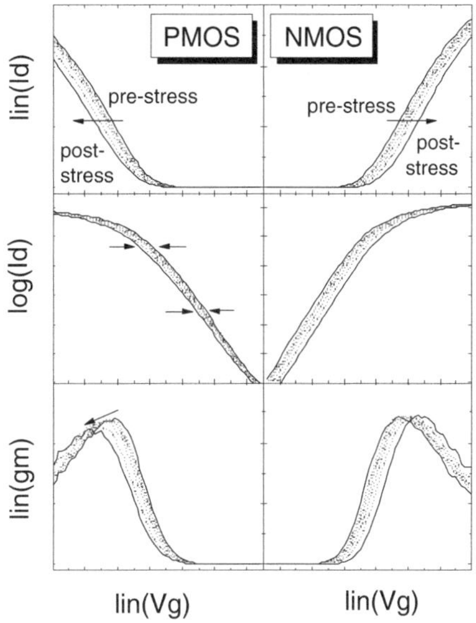

Fig. 1.2 Transistor characteristics taken prior (pre-stress), during (*dashed lines*), and after (post-stress) BTI stress. Note that BTI stress leads to an increase in the absolute threshold voltage for both PMOS and NMOS devices (*arrows top panels*). For NMOS devices the characteristic is typically only shifted, whereas for PMOS devices subthreshold and gm degradation occur (see *arrows*)

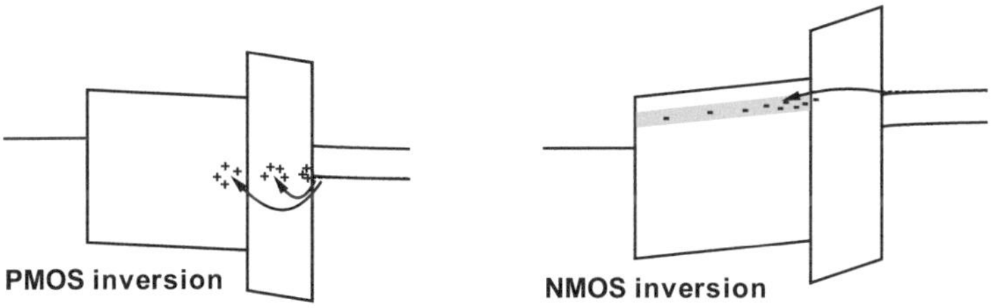

Fig. 1.3 Band diagram of MG/HK PMOS (*left*) and NMOS (*right*) biased in inversion mode using parameters from [9]. The trapping of positive and negative charge (BTI mechanism) for PMOS and NMOS devices, respectively, is indicated

During PBTI (NMOS, Fig. 1.3) in MG/HK devices, negative charge trapping in the HK layer or in the region between the HK layer and the interfacial oxide layer has been mainly observed [11]. Electron trapping in SiO_2 and SiON is small, and PBTI has not been a concern for NMOS devices in these technologies.

The difference in the charge location for NBTI and PBTI does lead to different degradation features in the transistor characteristic. The modifications of the device characteristics by BTI are summarized in Fig. 1.2. For PBTI (NMOS in Figs. 1.2 and 1.3), the trapped charge is separated from the inversion channel by the interfacial oxide layer. Therefore, Coulomb scattering is weak, and the channel carrier mobility remains constant, independent of the amount of trapped charge. As a result, the transistor characteristic is only horizontally shifted on the voltage scale, and the voltage shift throughout the entire range of transistor operation is determined by

$$\Delta V = \Delta Q/Cox.$$

For NBTI (PMOS in Figs. 1.2 and 1.3), the situation is different since positive charges can be trapped at interface-states, in near-interface defects (border traps), and in the bulk of the oxide. The trapped charges in close proximity of the inversion layer lead to strong additional Coulomb scattering and degrade the channel carrier mobility. Such degradation is frequently reported for NBTI in both conventional poly-Si/SiON [12] and MG/HK [13] PMOS devices. As a result of the mobility degradation, the peak transconductance of the device in the linear regime decreases with increasing voltage shift. At the same time the subthreshold characteristic during NBTI stress also degrades compared to the fresh characteristic as indicated by the arrows in Fig. 1.2. Since transconductance degradation and degradation of the subthreshold characteristic are observed during NBTI, the induced voltage shift depends on the drain current regime at which the voltage shift is extracted. In the subthreshold regime, due to the degradation of the subthreshold characteristic by interface-states, the extracted voltage shift is typically smaller than the values extracted near the device threshold voltage. When extracting the voltage shift above the device threshold, the values are increased due to transconductance degradation. These aspects of NBTI degradation are of importance when comparing measured NBTI degradation values from the literature.

Both NBTI and PBTI exhibit a strong voltage dependence which is frequently modeled using either the exponential or the power-law voltage acceleration model. The exponential BTI voltage acceleration model can be written as

$$\Delta V_T\left(V_g\right) = A\left(V_{gref}, t_{ref}\right) \cdot e^{M\left(V_g - V_{gref}\right)},$$

and the power-law BTI voltage acceleration model is given by

$$\Delta V_T\left(V_g\right) = A\left(V_{gref}, t_{ref}\right) \cdot \left(\frac{V_g}{V_{gref}}\right)^m,$$

where V_g is the gate voltage, t is the stress time, and the subscript "ref" refers to a reference condition. Frequently, $V_{gref} = 1$ V and $t_{ref} = 1$ s are chosen. The exponential voltage acceleration factor M is given in units of [V^{-1}], and the power-law acceleration factor, m, is dimensionless. The voltage acceleration factors at the reference condition are related by

$$m = M \cdot V_{gref}.$$

Typical examples for the voltage dependence of NBTI and PBTI are shown in Fig. 1.4. Threshold voltage shifts during NBTI are observed to be of comparable magnitude for conventional poly-Si/SiON and MG/HK stacks. However, during PBTI, conventional poly-Si/SiON NMOS devices show negligible threshold voltages shifts close to operation condition. Only when stressed at high voltage close to dielectric breakdown can the instability be observed at short stress times. This is typically attributed to bulk defect generation in the gate oxide. MG/HK NMOS on the other hand exhibits PBTI which is similar in magnitude to NBTI. It can clearly be observed close to the operation condition of the device even for short stress times of 100 s as shown in Fig. 1.4. Therefore, PBTI is of considerable concern with HK/MG technologies.

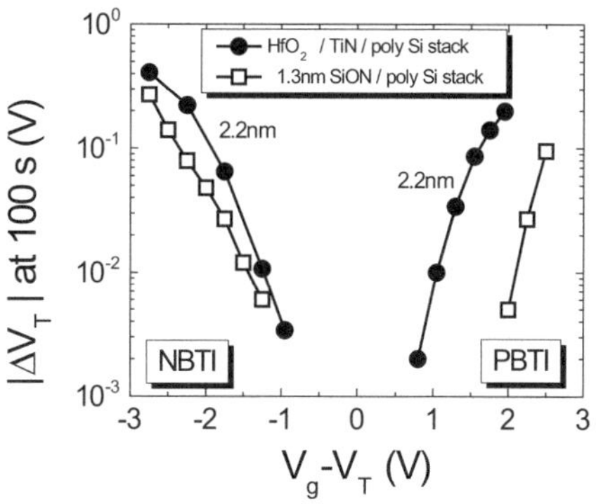

Fig. 1.4 Comparison of the stress voltage dependence of BTI-induced voltage shifts at a fixed stress time of 100 s, for conventional poly-Si/SiON and MG/HK MOS devices. Note the strong increase in the PBTI instability in n-channel devices for MG/HK MOSFETs (after [14])

Since BTI is largely caused by trapping and defect generation in the gate oxide, NBTI and PBTI are strongly dependent on gate stack processing of conventional and MG/HK technologies. The choice of the interlayer type (SiO_2 or SiON), its thickness, and the type and thickness of the high-k layer strongly impact the BTI characteristics. Additionally, thermal treatments, nitridation steps, and passivation anneals are important steps in the fabrication process which affect the instability for a given gate stack. While these knobs can greatly influence the magnitude of the instability, it is not expected that such optimizations can lead to the elimination of the BTI all together because of the underlying degradation physics.

For technology reliability qualifications, an accurate description of the BTI is paramount as BTI tests have to be conducted under accelerated degradation conditions. To predict the end-of-life (typically 5–10 years) shift, the BTI degradation is projected from typical test times ($t < 10^6$) using the time-evolution model. A voltage acceleration model is needed to project from high test voltages to operation conditions. To describe the time evolution of the BTI, various models can be found in the literature. Some models use logarithmic time dependence, some power-law model in time, and others include saturation like the stretched exponential model given in [15]. Combinations of these different models are also used to describe and combine contributions from individual physical processes to the instability. For illustration purpose, we use the power-law equation for the time evolution,

$$\Delta V_T\left(V_g,t\right) = A\left(V_{gref},t_{ref}\right)\cdot\left(\frac{t}{t_{ref}}\right)^n,$$

where t_{ref} is the reference time (typically set to $t_{ref} = 1$ unless specified differently) and n is the time-evolution exponent. The exponent, n, is typically measured to be in the range, $0.1 < n < 0.25$. As evident from the data in Fig. 1.5, the power-law time evolution is a reasonable approximation for modeling the threshold voltage shift as a function of time. However, caution needs to be exercised when modeling very small and very large voltage shifts where "turn-on" and saturation effects can lead to significant deviations from a power law.

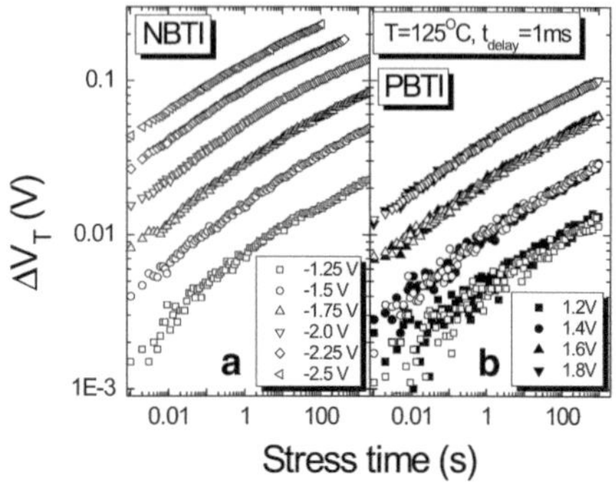

Fig. 1.5 Measured ΔV_T versus stress time for different stress voltages in HKMG PMOS transistors [NBTI (**a**)] and NMOS transistors [PBTI (**b**)] (after [16])

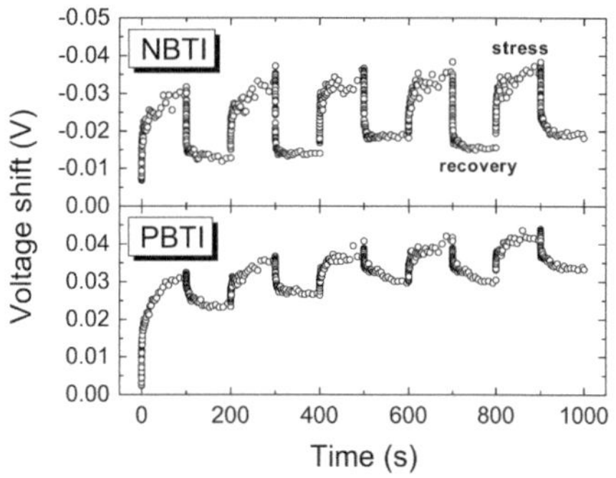

Fig. 1.6 Threshold voltage evolution of NMOS and PMOS devices during a test in which the gate bias is altered between inversion mode stress and transistor off-state recovery. When the gate stress is removed, the BTI-induced voltage shift recovers for both NMOS and PMOS devices

The observation of recovery effects during NBTI measurements in conventional poly-Si/SiON PMOS [17, 18] and during PBTI in MG/HK NMOS [19] stimulated a substantial effort in the development of fast time-resolved characterization techniques. Analysis based on slow characterization methods usually leads to an underestimation of the magnitude of the instability. This discovery led to a controversy about the time evolution of the instability. The BTI recovery effects can be easily illustrated with a test in which the voltage is cycled as shown in Fig. 1.6. First, the device is stressed for 100 s, and then the gate bias is reduced to a lower voltage near the transistor threshold voltage for 100 s and the threshold voltage is continuously monitored. The process is then repeated several times [17]. As can be seen, the BTI-induced voltage shift rapidly recovers when the stress voltage is removed. NBTI and PBTI show qualitatively similar recovery features, a fact which has been discussed in many reports in the literature [14, 20, 21]. The magnitude and the time dependence of the recovery are different for NBTI and PBTI due to the differences in the physical processes causing the instability. The NBTI is caused by interface or near-interface processes, leading to faster recovery than for PBTI, where trapping centers situated further away from the interface are involved. PMOS devices used to collect the data in Fig. 1.6 showed roughly 50% NBTI recovery, whereas NMOS devices showed only ~25% PBTI recovery.

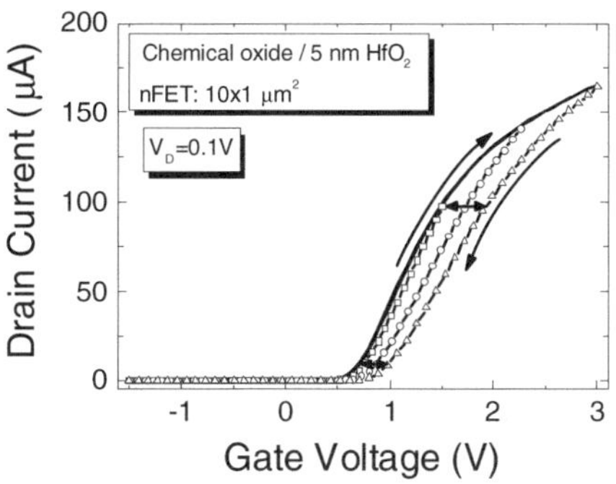

Fig. 1.7 Id–Vg hysteresis measurement taken on n-channel MOSFET with poly-Si/HfO_2/SiO_2 gate stack. When the device is swept from strong inversion to accumulation, a transient instability is observed, leading to full recovery when an accumulation bias of −1.5 V is applied. (The "up" traces (*solid line*) for three sequential double sweeps from −1.5 to +1.5 V, +2 V and +3 V are found to be identical.) (after [22])

The recovery of BTI-induced shifts plays an important role during cycling operation of CMOS circuit in advanced transistor technologies with conventional poly-Si/SiON and MG/HK stacks.

During the early exploration phase of alternative gate dielectrics for CMOS applications, hysteresis measurements were used to track BTI shifts during gate stack development. Hysteresis measurements performed on capacitors typically employ a high-frequency capacitance–voltage (HF C–V) sweep from depletion to accumulation and back to depletion. The shift between the two sweeps is used as an indicator of the stack stability. Evidently, sweep range and ramp rate are critical input parameters for the hysteresis measurements, and close attention needs to be paid to their selection when using a dual-ramp test for material selection or process optimization.

When transistors are available, the transfer characteristic of the transistor (Id–Vg) can be used to quantify the threshold voltage shift, as illustrated in Fig. 1.7. With transistors, the hysteresis is measured by sweeping the gate bias from accumulation to inversion and back to accumulation. Again, the magnitude of the shift in MG/HK NMOS depends on the electron injection current or gate bias, the duration of the stress, and the properties of the gate dielectric. As the inversion stress is gradually removed during the reverse sweep into accumulation, the dynamic aspects of the trapping process can be observed in a double-sweep experiment. As can be seen in Fig. 1.7, recovery already occurs during the down sweep, resulting in significant recovery near threshold. These types of double-sweep tests lead to an early postulation of reversible electron trapping in shallow defects (likely oxygen vacancies in the HK layer of MG/HK NMOS transistor) [9]. It has been demonstrated that the shallow defects are emptied by electron back-tunneling to the substrate at negative gate bias, while biasing in inversion mode leads to the filling of HK defects by tunneling and to a positive voltage shift of the transfer characteristics due to this negative charge.

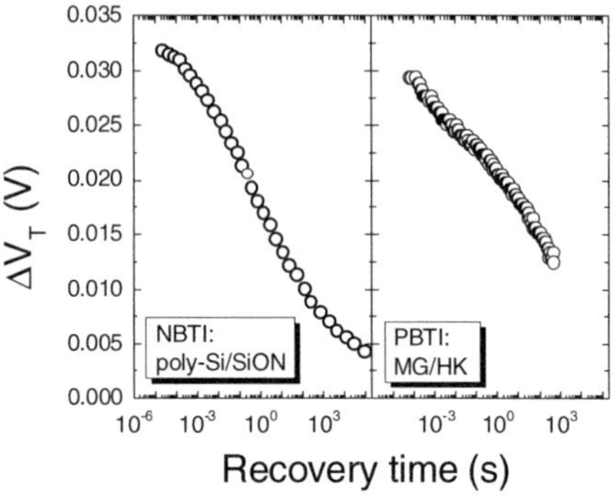

Fig. 1.8 Typical BTI recovery traces for poly-Si/SiON (NBTI, *left panel*) [20] and MG/HK (PBTI, *right panel*) [14] devices

The reported post-stress NBTI recovery for conventional poly-Si/SiON PMOS devices and transient charge trapping effects in NMOS devices with high-k dielectrics have led to large efforts to develop time-resolved BTI characterization techniques to more accurately capture the transient nature of the instabilities. The recovery phenomenon continuous to be studied in greater detail, utilizing the various time-resolved characterization methods proposed in the literature [14, 20, 21, 23, 24]. Typical recovery traces for conventional poly-Si/SiON NBTI and MG/HK PBTI are shown in Fig. 1.8, illustrating that recovery occurs over many orders of magnitude in time and to first order follows a logarithmic time dependence. Therefore, the threshold voltage recovery can be written as

$$\Delta V_T = \Delta V_T\,(1s) - \alpha \log(t),$$

where α is a constant and $\Delta V_T(1s)$ is the BTI degradation remaining after 1 s of recovery. The applicability of this logarithmic time dependence typically covers a time window from μs to hours. A more general function describing the recovery also at very short and very long times can be found in [21].

The recovery rate is given by

$$\frac{d\,(\Delta Vt)}{dt} = \frac{d\,(-\alpha \log(t))}{dt} = \frac{-\alpha}{t}.$$

When plotting the recovery rate on a log–log plot as shown in Fig. 1.9, a straight line with a slope close to −1 is obtained. The 1/t dependence has been reported in the literature for exchange between channel charge carriers and preexisting oxide traps [25, 26] invoking elastic or inelastic tunneling as reported in [27]. An alternative explanation for the $1/t$ dependence has been given in [28] attributing it to relaxation of the dielectric polarization. Independent of the physical origin of the $1/t$ dependence, the experimental data imply that recovery effects occur already at very short recovery times and extend over many orders of magnitude in time. Therefore, to accurately quantify these instabilities, fast measurement techniques are essential.

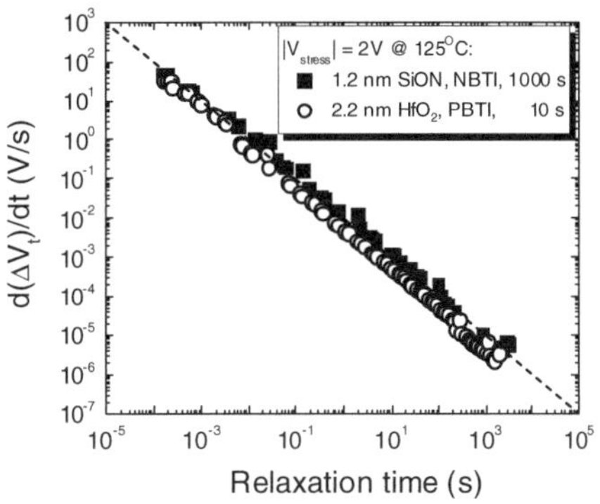

Fig. 1.9 BTI recovery rate versus relaxation time for conventional poly-Si/SiON p-channel and MG/HK n-channel MOS devices. Note that for both mechanisms the recovery rate follows to first order a 1/t dependence

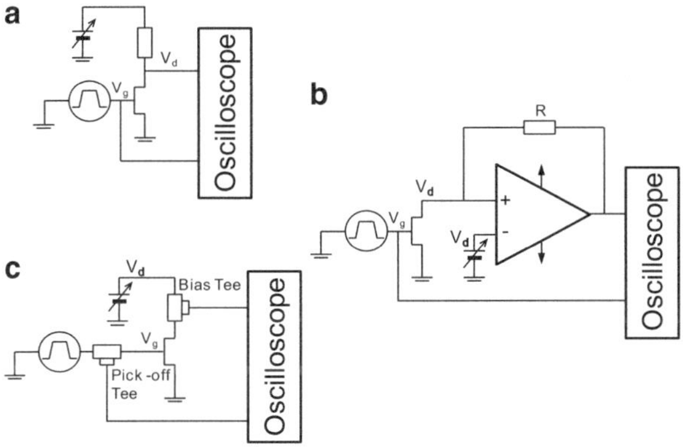

Fig. 1.10 Pulsed Id–Vg characterization setups used to capture transient instability effects in MOS devices. In setup (**a**) the MOS device is configured as an inverter circuit with resistive load, and in (**b**) the passive load is replaced by an active amplifier circuit. Setup (**c**) utilizes a pick-off tee and bias tee to enable ultrafast pulsed measurements

A first attempt to capture the magnitude of the transient charge trapping effect in NMOS devices comprising a high-k gate dielectric was made by using an inverter circuit with a resistive load, as illustrated in Fig. 1.10a [19]. In this setup, the gate was biased with a commercial pulse generator unit (PGU) using rise and fall times, t_r and t_f, respectively, in the range of 100 μs, and the drain was connected to a power supply through a resistor (R_L). A bias is applied in the linear regime (e.g. $V = 100$ mV). With a digital storage oscilloscope, the gate and drain voltages are recorded, and from these traces the Id–Vg characteristics are then constructed. The drain current in the linear regime is given by

$$I_D = \frac{100mV}{V_D} \cdot \left(\frac{100mV - V_D}{R_L} \right),$$

where V_D is the measured drain voltage and R_L is the load resistance of the inverter circuit. The power supply voltage is connected to the load resistor and is set to 100 mV. In this circuit configuration when the transistor is turned on, current flows from the power supply to ground and the voltage at the drain node drops. The first term in the equation accounts for the potential drop at the drain node. In order to obtain a measurable voltage drop across the load resistor, the resistance value has to be comparable to the channel resistance of the MOSFET. To properly capture the Id–Vg characteristic during the rise and fall portions of the pulse, the RC delay of the inverter circuit, τ, which is given by the load resistance and the parasitic capacitance, including the device under test (DUT), has to be much shorter than the rise and fall time,

$$\tau = R_L \cdot C_{par} \ll t_r \text{ and } t_f.$$

To overcome the RC delay limitation, the linear load in Fig. 1.10a was replaced with an active amplifier circuit as shown in Fig. 1.10b [29], enabling pulsed Id–Vg measurements with significantly shorter rise and fall times (transition times in the μs range).

A second modification to the setup was made as shown in Fig. 1.10c. In this setup, the pulse generator is connected to the gate using a "pick-off tee" connector for forwarding the gate signal to the storage oscilloscope. On the drain side, a "bias tee" is used for the DC connection to the power supply unit via an inductor and to the digital storage oscilloscope via a capacitor. The drain current was determined by the voltage drop across the 50 Ω input resistance of the digital oscilloscope. With the use of the "bias tee," as shown in Fig. 1.10c, pulse width of less than 100 ns can be implemented. The setup, however, only supports a pulsed measurement mode [30]. To maximize the system bandwidth, it is also recommended to use specific transistor designs which allow the use of ground-signal-ground probes.

Since all three setups illustrated in Fig. 1.10 employ digital storage oscilloscopes to measure the drain current, the accuracy of these setups is limited by the resolution of the analog-to-digital converter (ADC) and the operation mode of the oscilloscope. As oscilloscopes are optimized for high-frequency measurements, thus typically 8-bit ADCs providing 256 digitized levels over the dynamic range are employed. A higher resolution can be obtained when the signal is averaged using the "high-resolution" setting.

A further challenge in the use of these setups arises from the handling of the large amounts of data accumulated by the digital storage oscilloscope via general purpose interface bus (GPIB) communications.

In summary, the results obtained by these pulsed characterization methods highlighted the need for time-resolved measurement techniques, triggering a strong effort toward the development of improved commercial tools. With these improved tools, threshold voltage instability measurements with reduced delay times (in the ms to sub-μs range) are possible today.

1.2 Stress-and-Sense Characterization

One of the most frequently used characterization methods to study BTI in MOS devices is the stress-and-sense technique or measure-stress-measure technique. This technique can be applied to MOSFET and MOSCAP structures, utilizing drain current or capacitance measurements, respectively.

A generic flowchart of the stress-and-sense characterization method is shown in Fig. 1.11. Prior to applying the stress, a pre-stress, reference characteristics (Id–Vg or Cg–Vg) is measured. Additional pre-stress characterization may be useful, such as Ig–Vg, Id–Vd characterization, extended C–V characterization, or characterization with charge-pumping (CP) methods and DC-IV measurements.

After the pre-stress characterization is completed, the DUT is subjected to the stress for a specified period of time. The time period of the stress interval can either be determined using a linear or a logarithmic time base. Since many decades in time have to be covered with these characterization procedures a logarithmic time base is more convenient and thus frequently employed. During the stress period, various stress-related parameters may be monitored such as gate, drain, source, or substrate currents to collect additional useful information for the development of degradation models.

When the stress cycle is finished, the intermittent characterization is triggered which can consist of Id–Vg, Ig–Vg, Id–Vd, HF C–V, CP, DC-IV, or any other characterization technique including spot measurements of the various methods. After completion of the intermittent characterization, either another stress cycle is initiated or the final post-stress characterization is triggered.

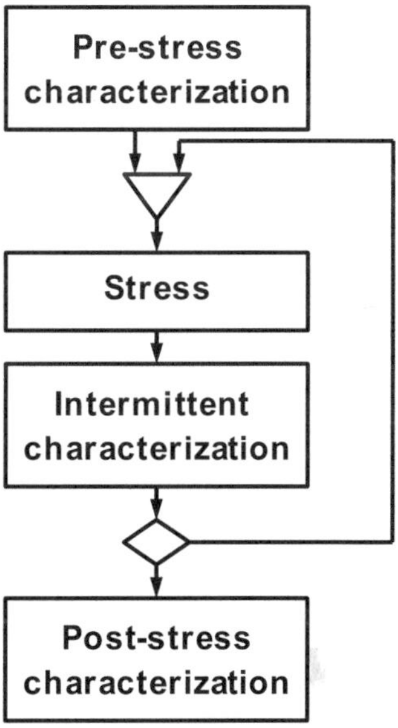

Fig. 1.11 Flowchart of a stress-and-sense characterization procedure

The post-stress characterization methods are typically identical to the pre-stress characterization methods. By comparing the post-stress characteristic with the pre-stress characteristic, important device degradation aspects, such as transconductance degradation, subthreshold degradation, and degradation of other device characteristics, can be determined.

1.2.1 Full Intermittent Id–Vg Characterization

The BTI characterization procedure including full intermittent characterization like Id–Vg or Id–Vd has been the primary choice for a long time since all relevant transistor degradation characteristics like V_T-degradation in the linear regime, V_T-degradation in the saturation regime, transconductance degradation, subthreshold degradation, off-state leakage degradation, saturation drain current degradation, and others can be evaluated. When full characterization is performed, not only the pre- and post-stress characterization sequence may be identical, but also the intermittent characterization may consist of the same test sequence. An Id–Vg sweep is an example of a full characterization sequence. The data analysis may be performed during stress or post-stress providing access to the various degradation parameters.

Since full characterization cycles are time consuming and may take several minutes, the interest in applying this method has diminished for scaled conventional poly-Si/SiON PMOS devices and MG/HK N- and PMOS devices because of the recovery effects. With full characterization, the magnitude of the degradation is typically underestimated, and it leads to a steeper time evolution. However, to obtain a generic understanding of various degradation features for new devices or materials used in gate stack processing, the full intermittent Id–Vg characterization procedure remains a very relevant characterization methodology.

1.2.2 Spot-Id Sense Measurement

To minimize BTI recovery, it is of great interest to reduce the measurement delay and measurement time during the intermittent characterization step. This can be achieved by performing just a single drain current measurement at a reference gate voltage instead of a full Id–Vg characterization, as introduced in [30]. A schematic flowchart of a spot-Id sense measurement procedure is illustrated in Fig. 1.12. Prior to the BTI stress, a pre-stress Id–Vg sweep is performed with the device biased either in the linear regime ($|Vd| < 100$ mV) or in the saturation regime ($|Vd| = VDD$). This Id–Vg characteristic serves as reference characteristic for the determination of the threshold voltage shift (ΔV_T). Then the DUT is subjected to BTI stress for a specified period of time. If the characteristic at high Vds (saturation) is used for ΔV_T extraction, then the drain bias has to be removed during the stress to avoid hot carrier degradation. If a characteristic with low Vds is used to determine

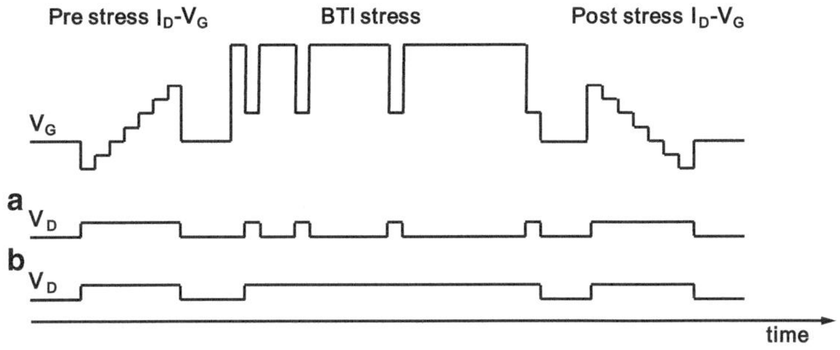

Fig. 1.12 Typical voltage time trace of a BTI characterization procedure with spot-Id sense measurement. Note that the drain can either be biased in the linear regime during the sense measurement only (**a**) or kept biased throughout the stress (**b**)

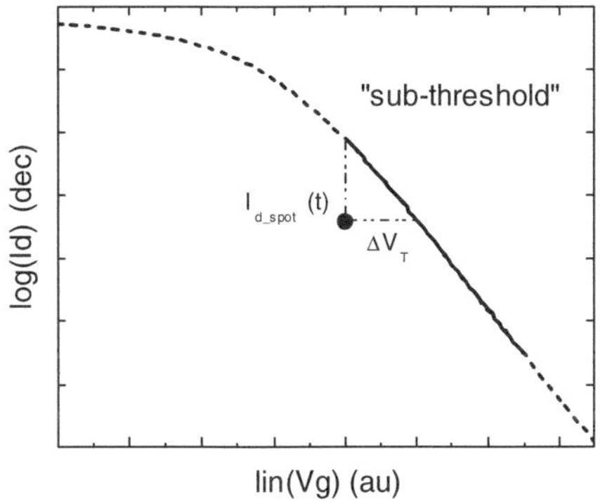

Fig. 1.13 Extraction procedure of the threshold voltage shift when the gate sense voltage is chosen in the subthreshold regime of the device. Pre-stress Id–Vg characteristic is used to determine the subthreshold slope

ΔV_T, then the low drain bias may or may not be applied during the stress. Using characterization in the saturation regime requires synchronization of the switching events and needs special attention.

After each stress cycle, the intermittent drain current measurement is performed using the specified sense voltage and delay time. With state-of-the-art commercial instruments, sense delays in the millisecond range can be achieved quite easily, and sub-μs delays are feasible nowadays with the most advanced instrumentation. After the sense cycle is completed, either the stress is continued or completed, which then triggers a post-stress characterization step.

The threshold voltage shift during the spot-Id sense measurement can be extracted in several ways. When the subthreshold characteristic is used to determine ΔV_T, then the first step is to determine the range where the logarithm of the drain current can be described by a linear expression as illustrated in Fig. 1.13. After the range is determined, the subthreshold slope at time zero (SS(0)) is calculated using the following relation:

$$SS(0) = \frac{1}{\frac{d\log_{10}(I_d)}{dV_g}} \qquad [\mathrm{V/dec}]\ .$$

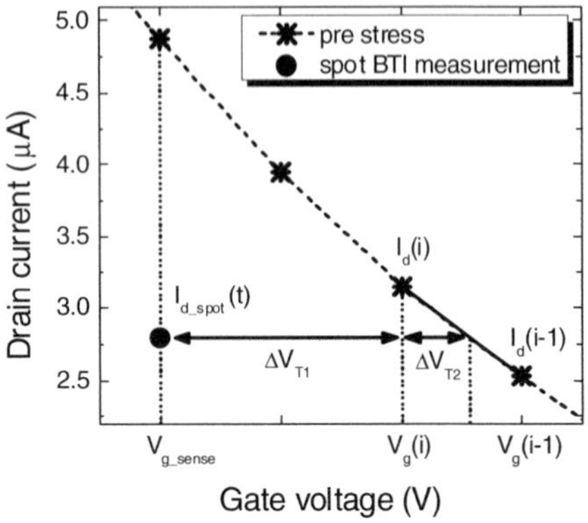

Fig. 1.14 Extraction procedure of the threshold voltage shift when the sense gate voltage is close to the time-zero threshold voltage of the device. Pre-stress Id–Vg characteristic is used to determine the shift using a linear interpolation scheme

The upper end of the subthreshold swing is the maximum gate bias which can be used for the sense measurement in this extraction procedure and needs to be considered when setting up the stress test. The intermittent drain currents can now simply be translated into voltage shifts following

$$\Delta V_T = \log_{10}\left(\frac{I_d(0)}{I_d\left(t_{stress}\right)}\right)\cdot SS(0),$$

where $I_d(0)$ is the pre-stress drain current at the sense condition, $SS(0)$ the subthreshold swing at time zero, and $I_d(t_{stress})$ the intermittent drain current during the stress. It is recommended to verify whether the subthreshold characteristic has degraded during the BTI stress by comparing pre- and post-stress characteristics.

The second method to extract the threshold voltage shift, discussed hereafter, utilizes the transfer characteristic around the device threshold as shown in Fig. 1.14 and described in [14, 31]. The drain current during the spot sense measurement is taken at a specified sense voltage, V_{g_sense}, near the device threshold (V_T) and is converted into a voltage shift using

$$\Delta V_T = \frac{\Delta I_d}{gm},$$

where ΔI_d is the drain current degradation and gm the transconductance. In short, the pre-stress device characteristic measured prior to the BTI stress is used as the reference characteristic. Then the local transconductance $\Delta V/\Delta I$ of the pre-stress characteristic in the vicinity of the measured sense current,

$$I_d(i-1) < I_{d\_spot}(t) < I_d(i),$$

is determined to extract the pre-stress gate voltage at the same current level by interpolation. Finally, the threshold voltage shift is calculated as

$$\Delta V_T = \Delta V_{T1} + \Delta V_{T2},$$

where

$$\Delta V_{T1} = V_{g\_sense} - V_g(i)$$

and

$$\Delta V_{T2} = \left(I_d(i) - I_{d\_spot}(t)\right) \cdot \frac{V_g(i) - V_g(i-1)}{I_d(i) - I_d(i-1)}.$$

To minimize interpolation errors, a small voltage step needs to be chosen for the pre-stress Id–Vg characterization ($Vg(i)$–$Vg(i-1) \leq \pm 20$ mV). Furthermore, if the current is measured in the subthreshold regime, a log-linear interpolation procedure should be applied.

Theoretically, both extraction procedures should yield the same result if subthreshold and transconductance degradation are negligible and the observed recovery effects are insensitive to the gate bias during the spot measurement. These conditions are reasonably well satisfied for the PBTI stress but not necessarily for the NBTI stress where transconductance degradation is typically reported. Therefore, it is recommended to assess the impact of the extraction procedure by comparing the pre-stress and post-stress characterization. This is sufficient since it has been shown that the magnitude of the instability is sensitive to the measurement delay but the correlation between the different degradation parameters is independent of the delay [32].

From a testing viewpoint, it should be mentioned that it is of interest to measure at large current levels since there is an intrinsic correlation between measurement current, measurement accuracy, and measurement time. A higher current can be measured with the same accuracy in shorter times. This is particularly important when measuring currents towards the μs time domain. Within limits, a higher drain current can either be achieved by increasing the gate sense voltage, the drain bias during the sense measurement, or by increasing the width of the DUT.

In summary, the spot-Id measurement procedure is a very popular choice for the characterization of NBTI and PBTI in MOSFETs using sense delays around ~1 ms since they can be achieved rather easily with commercial measurement instrumentation. Further efforts are being made to push the spot-Id measurement towards the sub-μs time domain, but since these fast sense measurements require tool upgrades, adaptation of these procedures by the semiconductor industry may be delayed.

1.2.3 Spot-CV Sense Measurement

An attractive alternative to reduce measurement delay for BTI measurements on capacitor structures is the use of spot-CV sense measurement analogues to the spot-Id measurement for the transistors. The spot-CV sense characterization procedure

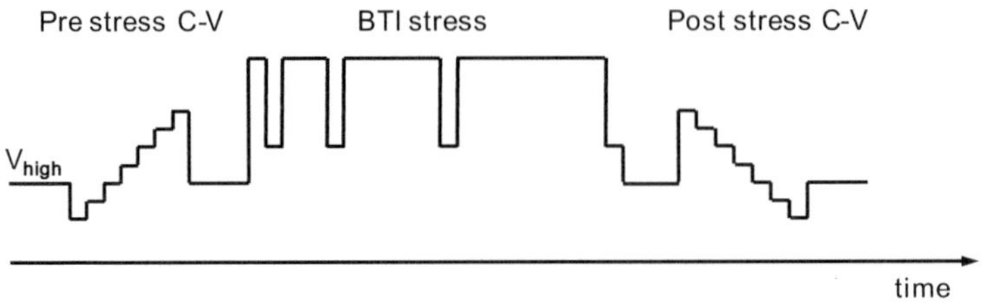

Fig. 1.15 Typical voltage time trace of a C–V BTI characterization procedure with spot capacitance sense measurement

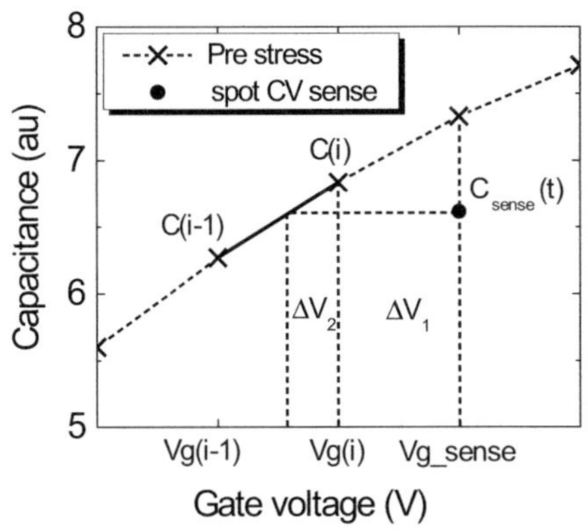

Fig. 1.16 Extraction procedure for the voltages shift from a spot-CV measurement on a capacitor structure. A pre-stress CV characteristic is used to determine the shift using a linear interpolation scheme

shown in Fig. 1.15 follows essentially the same sequence as the spot-Id technique [33]. First, a pre-stress CV characteristic is measured using a commercially available high-frequency LCR meter which serves as a reference in the determination of the BTI-induced voltage shift. Next, the DUT is subjected to the stress for a specified period of time. Since the measurement delay has become a critical issue in the BTI characterization, it may be necessary to stress and sense the device using the same instrument to avoid additional delays caused by switching between different instruments. After the stress cycle is completed, the gate bias is reduced to the sense condition, and a spot capacitance measurement is performed with minimum delay. It is recommended to choose a sense voltage such that $\Delta C/\Delta V$ is maximized for highest sensitivity. State-of-the-art LCR meters can yield accurate capacitance readings in the pF range within ~20 ms which is reasonably close to the spot-Id measurement. After the sense cycle is completed, either another stress cycle is initiated or the post-stress CV characterization is triggered. Comparison of the pre- and post-stress CV characteristic provides necessary information on bulk trapping versus interface-state degradation. The latter typically leads to CV *stretch-out* and an increase in the conductance signal. It is important to assess the impact of the BTI degradation on the CV characteristic when choosing the spot-CV sense voltage.

It should also be noted that the CV method for BTI characterization has the same limitations as for the standard capacitor measurements which may limit

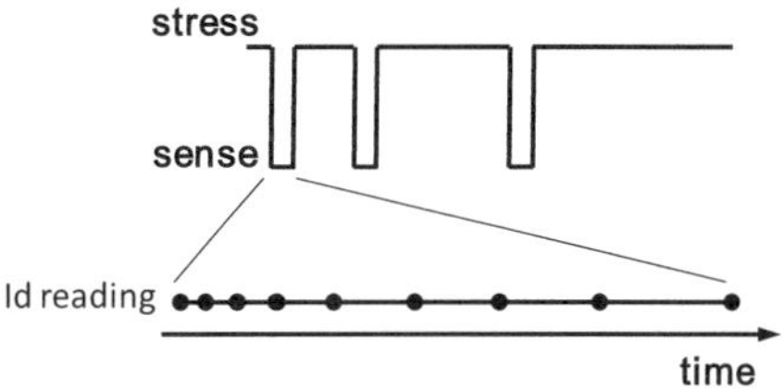

Fig. 1.17 Schematic illustration of the extended Id sense measurement proposed in [21] to monitor recovery after each stress cycle

its applicability for aggressively scaled gate stack materials. However, for early material screening, this technique can provide valuable insights regarding the stability of MOS structures.

The extraction of the voltage shift is outlined in Fig. 1.15. The pre-stress CV characteristic is used as reference, and the local CV swing (ΔC/ΔV) in the vicinity of the measured sense capacitance,

$$C(i-1) < C_{sense}(t) < C(i),$$

is approximated by a linear relation. Then the voltage shift can be calculated using

$$\Delta V = \Delta V_1 + \Delta V_2,$$

where

$$\Delta V_1 = V_{g\_sense} - V_g(i)$$

and

$$\Delta V_2 = (C(i) - C_{sense}(t)) \cdot \frac{V_g(i) - V_g(i-1)}{C(i) - C(i-1)}.$$

To minimize interpolation-related errors, small voltage steps need to be chosen for the pre-stress CV characterization ($Vg(i)$–$Vg(i-1) \leq \pm 20$ mV).

1.2.4 Extended Id Sense Measurement

To gain additional insights into the BTI process, an extended Id sense measurement methodology was proposed [21]. This method is particularly suited to study the BTI recovery effects. A schematic illustration of the test sequence of the extended Id sense measurement procedure is shown in Fig. 1.17. The procedure is basically identical to the spot-Id measurement except that the drain current during the sense cycle is measured as a function of time to enable BTI recovery modeling. After the bias is changed from stress to sense, the drain current is typically monitored on

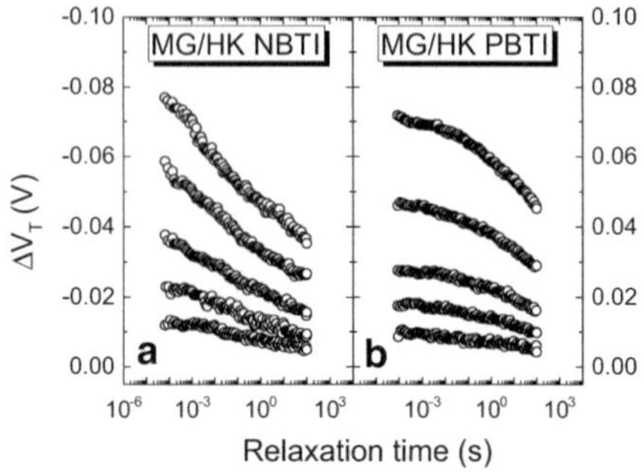

Fig. 1.18 NBTI (**a**) and PBTI (**b**) recovery traces for MG/HK devices measured with extended Id sense measurement at 125 °C after a voltage stress ranging from ~1.5× to ~2× of the nominal use voltage and after stress times up to 1,000 s

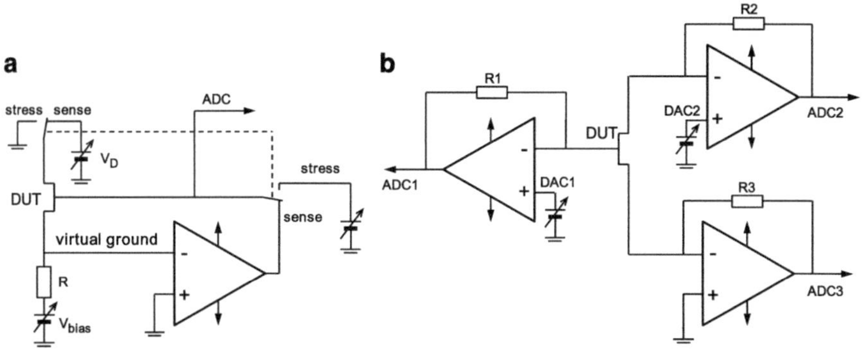

Fig. 1.19 Experimental circuit configurations enabling sub-ms sense measurements utilizing constant current sensing (**a**) and fast current reading at determined gate sense voltage (**b**)

a logarithmic time basis ranging from sub-ms to a few seconds, providing several readouts per decade in time (e.g., 10–20 data points per decade). Again, the current readings are converted to voltage shifts following the same procedure as for the spot-Id measurement. Typical extended recovery traces using Id-spot measurements for NBTI and PBTI in MG/HK CMOS devices are shown in Fig. 1.18. Such data are frequently used to derive recoverable and permanent degradation components [21].

1.2.5 *Ultrafast Sense Measurement*

Since it is important to understand the instantaneous BTI degradation during the on-state of the device, we discuss two ultrafast characterization techniques, enabling μs BTI measurements. The experimental circuits are shown in Fig. 1.19.

The first technique was proposed in [20] and utilizes an operational amplifier circuit forcing a constant current threshold condition. During the stress, the feedback loop is opened, and the stress source is directly connected to the gate. On the drain side the bias can be removed during the stress or can be retained if the bias is small.

When the stress is removed and the feedback loop is closed, the gate of the DUT is forced by the output of the amplifier. The amplifier sets the output to keep the inverting input at virtual ground forcing the DUT to maintain the threshold current. The value of the threshold current, I_{Th}, is set by the bias voltage, V_{bias}, and the resistor, R, following

$$I_{Th} = \frac{V_{bias}}{R}.$$

Therefore, the output voltage of the amplifier is equal to the threshold voltage of the devices and can be recorded directly by an ADC using a linear or logarithmic *time base*. The voltage resolution is determined by the accuracy of the ADC. When the parasitic capacitance is minimized, ultrafast V_T measurements down to a few μs are possible. No special requirements for either the source units or the switching devices are mentioned in [20].

The second technique proposed in [14] is based on peripheral component interconnect (PCI) card characterization methodology where a digital-to-analog converter (DAC) is used as voltage source and an ADC as meter in combination with a linear current voltage converter (IVC). In this setup, the DUT is directly biased by the DAC through the linear IVC. DAC calls of ~2 μs and ADC calls ~20 μs and single sense measurements within <30 μs have been demonstrated. The resistance values R_1, R_2, and R_3 set the amplification of the IVC converter and have values in the range of 100 Ω to 100 kΩ to enable maximum accuracy for different drain currents for transistors with varying geometries. The voltage resolution of the DAC and the ADC typically ranges from 12 to 16 bit. Since DAC and ADC are accessed through software calls, arbitrary waveforms can be programmed rather easily, enabling DC and various AC characterization tests up to a frequency of ~20 kHz.

Note that fast switching of large signals can introduce ringing in both setups which needs to be avoided. Filter elements may be required to eliminate undesired high-frequency components.

1.3 "On-the-Fly" Characterization Method

The "on-the-fly" (OTF) characterization method was introduced in [34, 35] to study BTI degradation directly at the stress condition without interrupting the stress. In principle, *the total, maximum* BTI shift can be obtained as recovery is eliminated. In this methodology, the BTI degradation is estimated using the local transconductance at the stress condition using

$$gm = \Delta I_d / \Delta V_g .$$

To obtain the local transconductance, the gate bias is modulated in small steps during the stress, and the drain current is recorded as shown in Fig. 1.20. In case

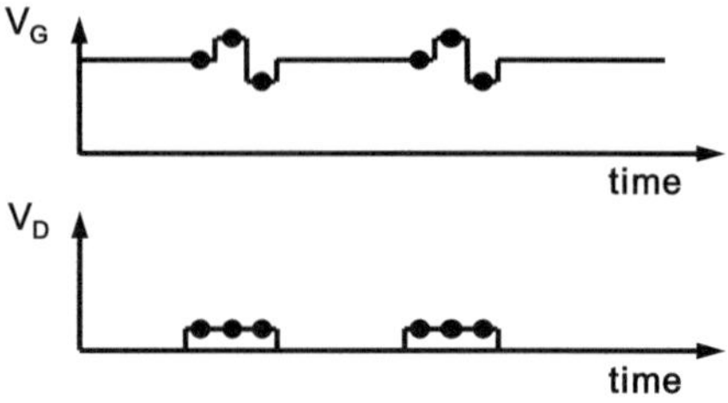

Fig. 1.20 Schematic voltage time traces of the "on-the-fly" characterization method proposed in [35]. The gate voltage is modulated to determine the transconductance for ΔV_T determination

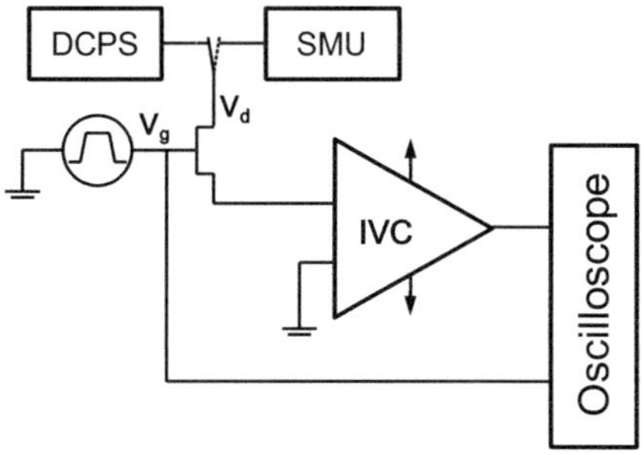

Fig. 1.21 Schematic diagram of the ultrafast on-the-fly characterization setup proposed in [23]. The drain current at stress is first monitored using the current voltage converter (IVC) and a digital oscilloscope. At longer times, the system is switched to a conventional source measurement unit (SMU)

there is no transconductance degradation during the stress condition, as typically observed for PBTI, the threshold voltage shift can be directly calculated using the relation $\Delta V_T = \Delta I_d / gm$. If, however, transconductance degradation is observed during the stress, the transformation of the measured current degradation into a voltage shift becomes more difficult and a *gm* degradation term needs to be included [36].

Accurate *time-zero* drain current determination is an issue with the OTF characterization technique [37]. To overcome this, an ultrafast version was proposed in [23] and is shown in Fig. 1.21. A pulse generator unit is used to drive the gate, and on the drain side a switch connects first a DC power supply (DCPS) during the initial part of the stress and then uses a conventional source measurement unit (SMU). This method enables first current readings within a few μs. The initial part of the trace up to, e.g., 100 ms is captured with the digital storage oscilloscope, and the drain current is determined by the oscilloscope reading (Vosc) divided by the amplifier gain. The second part of the trace is recorded using the SMU in trigger mode.

The sensitivity of the OTF characterization method is determined by the transfer characteristic of the DUT given by $\Delta V = \Delta I_d / gm(V_g)$. Since the maximum transconductance is typically obtained right above the device threshold, corresponding to the peak sensitivity for voltage shifts extractions, voltage shift extractions at higher gate biases will be less sensitive. The example given in Fig. 1.22 illustrates that the peak transconductance of 200 μA/V around 0.5 V gate bias drops by a factor of ~10 at 1.5 V and much less at 2.0 V stress conditions. This drop in transconductance with increasing gate bias must be taken into consideration when designing OTF BTI stress experiments.

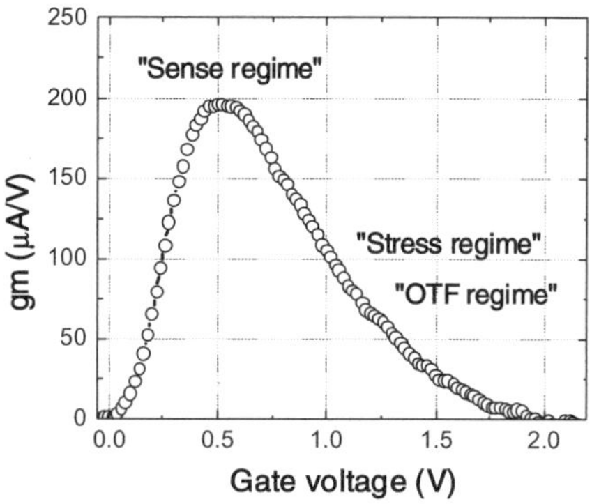

Fig. 1.22 Transconductance (gm) versus gate voltage of a NMOS device biased in the linear regime. The peak in gm occurs above device threshold and rapidly diminishes towards higher voltages where the on-the-fly characterization is typically performed

1.4 AC Characterization Method

The observation of recovery effects during stress-and-sense BTI characterization experiments initiated a discussion on its impact on digital circuit operation up to GHz frequencies [38–40]. For AC operation, stress and relaxation occur sequentially at the applied frequency, and recovery may provide relief in BTI degradation in circuits.

Many of the challenges noted for discrete device characterization such as time-zero characterization, recovery effects, and competing degradation mechanisms like hot carrier injection occurring during switching events also apply to CMOS circuits [41]. Customized circuits are often used to study BTI-induced degradation at CMOS relevant operation frequencies [40, 42] using either DC or AC stress mode.

The stress modes can be summarized in three different categories [41] spanning static to dynamic operation as shown in Fig. 1.23. The first operation mode is the DC equivalent stress mode which may occur in some circuits with low utilization, e.g., latches or possibly some parts of the SRAM circuit. For the second mode, AC switching is only applied to the gate. This mode can occur in CMOS logic where the gate terminal is altered between VDD and ground while the drain terminal is not. The third mode of operation is an inverter type stress where the gate is altered between stress and ground and the drain node altered between ground and stress simultaneously. The on-state represents the BTI stress mode, and the high drain terminal during the off-state likely enhances BTI recovery effects.

These operation modes can either be realized using customized circuits on-chip or by using off-chip pulse generator units enabling synchronized switching events. In general, higher frequency testing can be achieved with the on-chip circuits due to reduced parasitic capacitance. For accurate modeling of circuit BTI, large frequency and duty cycle domains need to be explored.

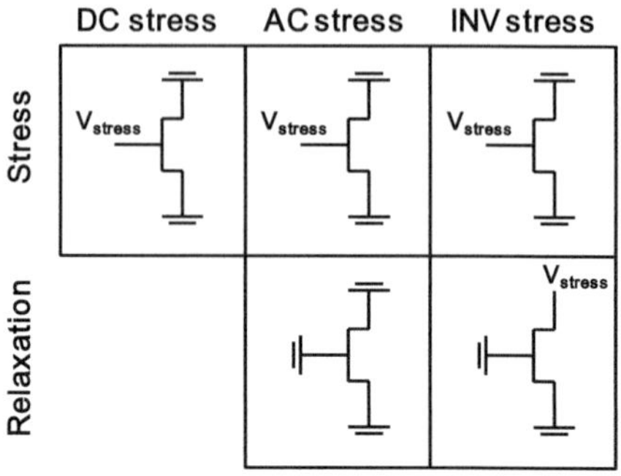

Fig. 1.23 Schematic illustration of the stress and relaxation cycle of transistors in circuits during a DC (*left*), AC (*middle*), and inverter type stress (*right*) (after [41])

1.5 Voltage Ramp Stress Characterization

Since the introduction of MG/HK into CMOS process technology, fast reliability screening during process development and reliability monitoring during manufacturing have gained importance as the use of several new materials has resulted in greater process complexity. The voltage ramp stress (VRS) characterization method, originally used for monitoring dielectric breakdown [43, 44], has been adapted to study BTI and has shown excellent quantitative agreement with conventional constant voltage stress (CVS) procedure [45, 46].

The flowchart for VRS characterization is given in Fig. 1.24. Similar to CVS tests, first, a pre-characterization is performed using either Id–Vg characteristics for transistors or CV characteristics for capacitors. Then, the device is subjected to a stress cycle for a specified stress period (t_{stress}). After the stress is completed, an intermittent characterization step is triggered. Either spot-Id or spot-CV can be used depending on the device structure. As in the CVS method, minimizing measurement delays is of great interest for the VRS test especially when correlating the results to the CVS test. After the intermittent characterization is completed, the next stress cycle is applied with increased stress bias given by

$$V_{stress} = V_{start} + i \cdot V_{step},$$

where V_{start} is the start voltage, V_{step} the step voltage ,and i the cycle index. The step voltage and stress period (t_{stress}) determine the ramp rate (RR):

$$RR = \frac{V_{step}}{t_{stress}}.$$

After the maximum stress voltage is reached and the final intermittent characterization is completed, the values of the intermittent spot-Id or spot-CV measurements are converted into voltage shifts by one of the methods described earlier. It is typically observed that the voltage dependence of the VRS tests is well described by a power law.

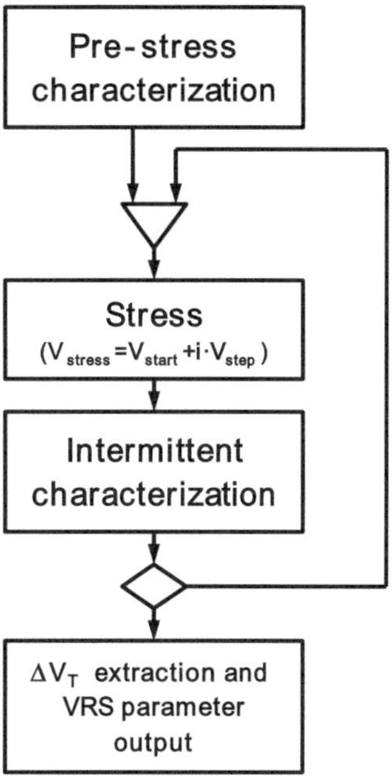

Fig. 1.24 Flowchart of a voltage ramp stress characterization procedure with intermittent device characterization for ΔV_T extraction

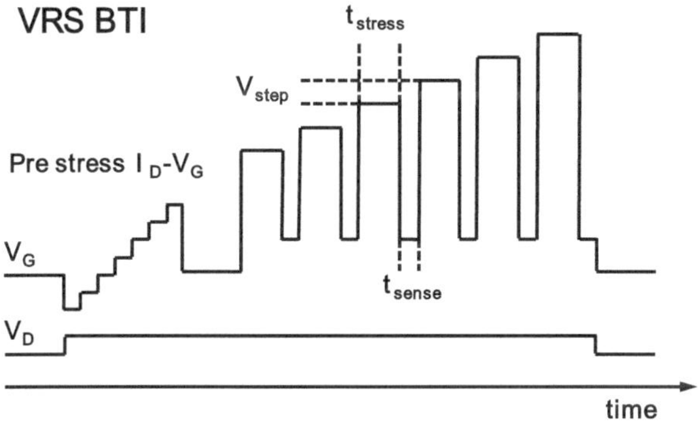

Fig. 1.25 Typical voltage time trace of a VRS BTI characterization procedure with spot-Id sense measurement. The pre-stress Id–Vg characteristic is used to extract the VRS-induced voltage shift

A schematic of the voltage time trace during VRS test for transistor structures employing the spot-Id method is shown in Fig. 1.25. Analogous to the spot-Id measurement in the CVS method, first, a pre-stress characterization is performed biasing the device in the linear regime. Then, a sequence of stress and intermittent characterization cycles is carried out with sequentially increasing stress bias. Note that when the device is biased in the linear regime, a small drain bias can be applied throughout the entire VRS test without compromising the results. After completing the VRS test, the drain current readings are converted into voltage shifts using procedures outlined in section 1.2.2. Under the assumption that a power-law voltage

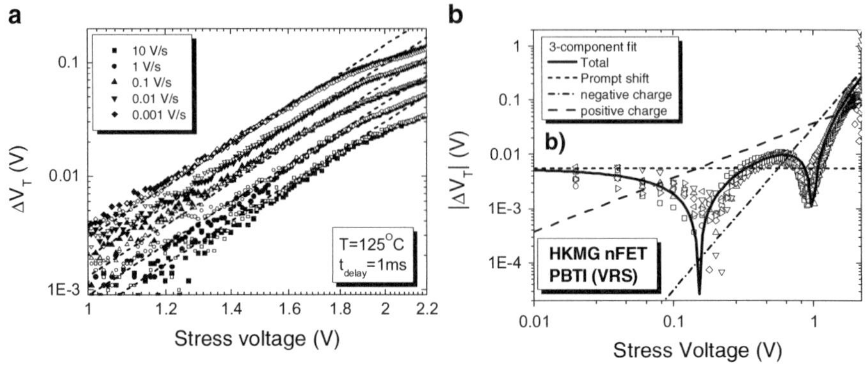

Fig. 1.26 (**a**) Typical VRS BTI traces measured at ramp rates of 10, 1, 0.1, 0.01, and 0.001 V/s. For each ramp rate, two devices are shown. Independent of the ramp rate, identical power-law exponents $(m+n)$ are measured with this method yielding a value of 5.5. Characterization was done at 125 °C with a sense delay of ~1 ms [45]. (**b**) A VRS PBTI trace measured on a gate stack during early MG/HK development; the complex trace can be separated into a prompt shift and two power-law components of opposite polarity [46]. With improved processing, the atypical positive charge trapping component could be eliminated from the gate stack

dependence and a power-law time dependence with acceleration factors, m and n, respectively, describe the CVS BTI characteristics, the voltage shift during the VRS test can be described by

$$\Delta V_T\left(RR_{VRS}, V_{VRS}\right) = \frac{A}{\left(\frac{m+n}{n}\right)^n}\frac{{V_{VRS}}^{m+n}}{{RR_{VRS}}^n},$$

where V_{VRS} is the gate voltage during the VRS test, RR is the ramp rate, and $A/((m+n)/n)^n$ is the pre-factor in the power-law expression [45]. This equation can be rewritten as

$$\begin{aligned}\log_{10}\left(\Delta V_T\left(RR_{VRS}, V_{VRS}\right)\right) &= \log_{10}\left(A\big/\left(\tfrac{m+n}{n}\right)^n\right)\\ &\quad + (m+n)\cdot\log_{10}\left(V_{VRS}\right) - n\cdot\log_{10}\left(RR_{VRS}\right),\end{aligned}$$

and the parameters $(m+n)$ and $A/((m+n)/n)^n$ can be determined by linear regression to the data in a log–log plot.

In Fig. 1.26a, typical VRS data for five different ramp rates for a MG/HK gate stack are compared. The data was measured at 125 °C with a sense delay of 1 ms. As can be seen, the threshold voltage shift exhibits a power-law behavior, showing that the power exponent, $m+n$, is well defined. Since $n << m$, the slope of the VRS data in a log–log plot directly measures the voltage acceleration. Since fast ramps of 1 V/s can be easily realized, the method provides voltage acceleration parameters in extremely short times and with high accuracy on a single device. With CVS stress, substantially longer test times and the characterization of multiple devices are typically used to obtain the voltage acceleration.

The data in Fig. 1.26b shows that a power-law model provides an accurate description for projection to operation voltage. At large stress voltages, a saturation behavior is observed. It is interesting to note that the saturation is not related to the absolute voltage shift and is therefore not caused by a limited precursor site density.

Finally, it is also of interest to notice that the VRS method can provide insights into multiple degradation phenomena, like concurrent electron and hole trapping within the same gate stack. In the example shown in Fig. 1.26b, a VRS trace, measured on a gate stack during early development, reveals multiple degradation phenomena, which disappeared later in an optimized gate stack process. In this specific case, the total degradation could be modeled as the sum of three different components: a constant prompt shift of $\sim$4 mV, which is already present by the time the first data point is measured, plus a power-law component due to positive charge trapping plus a power-law component due to electron trapping as typically observed for PBTI. With CVS, the presence of multiple component or physical processes is not easy to detect, and one may come to erroneous conclusions on the stack stability, if an unfortunate gate voltage is selected, where negative and positive charge trapping compensate each other in the measured time window.

1.6 Summary

In this chapter, a short overview on the experimental methods which were developed over the years to characterize the BTI phenomena in poly-Si/SiON and MG/HK devices was presented. The development of increasingly advanced experimental methods is largely driven by the need to develop accurate model for the predictions of the device lifetime during circuit operation with respect to the BTI instability. To accomplish this goal, a detailed physical understanding of the phenomena is important.

From a physics perspective the BTI phenomena are quite complex. A dominant contribution to the BTI instability is known to arise from trapping/detrapping of electrons (PBTI) and holes (NBTI) into/from preexisting defects in the gate dielectric. These processes exhibit a wide range of trapping and detrapping time constants and thus require time-resolved measurement methods for an accurate characterization of NBTI in conventional poly-Si/SiON and of NBTI as well as of PBTI in MG/HK devices. Other important contributors to the BTI instability are the generation of new defects, such as interface-states, which need to be comprehended in the modeling.

The successful introduction of fast characterization procedures employing pulse generator units and digital storage oscilloscopes in a research environment has led to the improvement of commercial measurement tools resulting in a significant reduction in measurement delays. State-of-the-art instruments are capable to provide BTI readouts with only millisecond delay, and it is expected that further improvements will be made to facilitate μs or sub-μs BTI characterization.

In addition to summarizing the developments in BTI characterization methods, the advantages and possible pitfalls of various test methods, such as stress-and-sense and OTF methods, were discussed.

Looking at recent developments, in addition to the DC stress and recovery testing, AC characterization will become increasingly important in order to gain better insight into the correlation between the degradation of discrete devices and CMOS circuit aging.

Finally, the VRS method was introduced as a fast reliability screening procedure for the development of novel, highly scaled MG/HK gate stacks for use in advanced CMOS devices and circuits.

References

1. E. H. SNOW, A. S. GROVE, B. E. DEAL, AND C. T. SAH, "Ion Transport Phenomena in Insulating Films", JOURNAL OF APPLIED PHYSICS, Vol. 36, No. 5, pp-1664-1673, 1965.
2. M. Kuhn and D. J. Silversmith, "Ionic Contamination and Transport of Mobile Ions in MOS Structures" J. Electrochem. Soc., Volume 118, Issue 6, Pages 966–970, 1971.
3. P. Balk, "Effects of Hydrogen Annealing on Silicon Surfaces", Electrochemical Society Extended Abstracts of Electronics Division, **14**, No. 1, Abst. 109, 237–240 (May, 1965)
4. A. S. Grove, B. E. Deal, E. H. Snow and C. T. Sah, "Investigation of Thermally Oxidised Silicon Surface using Metal-Oxide-Semiconductor Structures", Solid-State Electronics, Vol. 8, pp. 145–163, 1965.
5. Yoshio Miura and Yasuo Matukura, "Investigation of Silicon-Silicon Dioxide Interface Using MOS Structure", Japan. J. Appl. Phys., Vol. 5, pp. 180, 1966.
6. Anand T. Krishnan, Vijay Reddy, Srinivasan Chakravarthi, John Rodriguez, Soji John, Srikanth Krishnan, "NBTI Impact on Transistor & Circuit: Models, Mechanisms & Scaling Effects", in IEDM Tech. Digest, pp. 349–352, 2003.
7. K. Mistry, C. Allen, C. Auth, B. Beattie, D. Bergstrom, M. Bost, M. Brazier, M. Buehler, A. Cappellani, R. Chau, C.-H. Choi, G. Ding, K. Fischer, T. Ghani, R. Grover, W. Han, D. Hanken, M. Hattendorf, J. He, J. Hicks, R. Huessner, D. Ingerly, P. Jain, R. James, L. Jong, S. Joshi, C. Kenyon, K. Kuhn, K. Lee, H. Liu, J. Maiz, B. McIntyre, P. Moon, J. Neirynck, S. Pae, C. Parker, D. Parsons, C. Prasad, L. Pipes, M. Prince, P. Ranade, T. Reynolds, J. Sandford, L. Shifren, J. Sebastian, J. Seiple, D. Simon, S. Sivakumar, P. Smith, C. Thomas, T. Troeger, P. Vandervoorn, S. Williams, K. Zawadzki, "A 45nm Logic Technology with High-k + Metal Gate Transistors, Strained Silicon, 9 Cu Interconnect Layers, 193nm Dry Patterning, and 100% Pb-free Packaging," in IEDM Tech. Dig., pg. 247–250, 2007.
8. M. Chudzik, B. Doris, R. Mo, J. Sleight, E. Cartier, C. Dewan, D. Park, H. Bu, W. Natzle, W. Yan, C. Ouyang, K. Henson, D. Boyd, S. Callegari, R. Carter, D. Casarotto, M. Gribelyuk, M. Hargrove, W. He, Y. Kim, B. Linder, N. Moumen, V. K. Paruchuri, J. Stathis, M. Steen, A. Vayshenker, X. Wang, S. Zafar, T. Ando, R. Iijima, M. Takayanagi, V. Narayanan, R. Wise, Y. Zhang, R. Divakaruni, M. Khare, and T. C. Chen, "High-performance high-k/metal gates for 45 nm CMOS and beyond with gate-first processing," in *Symp. VLSI Technol.*, pp. 194–195, 2007.
9. R. G. Southwick III, A. Sup, A. Jain, and W. B. Knowlton, "An Interactive Simulation Tool For Complex Multilayer Dielectric Devices," IEEE Transactions on Device and Materials Reliability, vol. 11, pp. 236–243, 2011.
10. Dieter K. Schroder and Jeff A. Babcock, "Negative bias temperature instability: Road to cross in deep submicron silicon semiconductor manufacturing", J. Appl. Phys. Vol. 94, No.1, pp. 1–18, 2003.

11. A. Kerber, E. Cartier, L. Pantisano, R. Degraeve, T. Kauerauf, Y. Kim, A. Hou, G. Groeseneken, H.E. Maes, and U. Schwalke, "Origin of the threshold voltage instability in SiO_2/HfO_2 dual layer gate dielectrics", IEEE Electron Device Letters, Vol. 24, No. 2, pp. 87–89, 2003.
12. N. Kimizuka, K. Yamaguchi, K. Imai, T. Iizuka, C.T. Liu, R.C. Keller and T. Horiuchi "NBTI enhancement by nitrogen incorporation into ultrathin gate oxide for 0.10 μm gate CMOS generation" VLSI, pg. 92–93. 2000.
13. S. Pae, M. Agostinelli, M. Brazier, R. Chau, G. Dewey, T. Ghani, M. Hattendorf, J. Hicks, J. Kavalieros, K. Kuhn, M. Kuhn, J. Maiz, M. Metz, K. Mistry, C. Prasad, S. Ramey, A. Roskowski, J. Sandford, C. Thomas, J. Thomas, C. Wiegand, and J. Wiedemer, "BTI Reliability of 45 nm High-k + Metal-Gate Process Technology", IRPS, pg 352–357, 2008.
14. Andreas Kerber, Kingsuk Maitra, Amlan Majumdar, Mike Hargrove, Rick J. Carter, and Eduard Albert Cartier, "Characterization of Fast Relaxation During BTI Stress in Conventional and Advanced CMOS Devices with HfO_2/TiN Gate Stacks", IEEE Transaction on Electron Devices, Vol. 55, No. 11, pp. 3175, 2008.
15. Sufi Zafar, Byoung H. Lee, and James Stathis, "Evaluation of NBTI in HfO_2 Gate-Dielectric Stacks With Tungsten Gates", IEEE Electron Device Letters, Vol. 25, No. 3, pp. 153–155, 2004.
16. Andreas Kerber, Siddarth A. Krishnan, Eduard Albert Cartier, "Voltage Ramp Stress for Bias Temperature Instability testing of Metal-Gate/High-k Stacks", IEEE Electron Device Letters, vol. 30, Issue 12, pp. 1347–1349, 2009.
17. G. Chen, M. F. Li, C. H. Ang, J. Z. Zheng, and D. L. Kwong, "Dynamic NBTI of p-MOS Transistors and Its Impact on MOSFET Scaling" IEEE Electron Device Letters, Vol. 23, No. 12, pp. 734–736, 2002.
18. S. Rangan, N. Mielke, and E. C. C. Yeh, "Universal recovery behavior of negative bias temperature instability," in *IEDM Tech. Dig.*, 2003, pp. 341–344.
19. A. Kerber, E. Cartier, L. Pantisano, M. Rosmeulen, R. Degraeve, T. Kauerauf, G. Groeseneken, H.E. Maes, U. Schwalke, "Characterization of the V_T-instability in SiO_2 / HfO_2 gate dielectrics", in Proc. IRPS, pp. 41–45, 2003.
20. H. Reisinger, O. Blank, W. Heinrigs, A. Mühlhoff, W. Gustin, and C. Schlünder, "Analysis of NBTI degradation- and recovery-behavior based on ultra fast V_T-measurements," in Proc. IRPS, pp. 448–453, 2006.
21. B. Kaczer, T. Grasser, Ph. J. Roussel, J. Martin-Martinez2, R. O'Connor, B. J. O'Sullivan, G. Groeseneken, "Ubiquitous Relaxation in BTI stressing – New Evaluation and Insights", inProc. IRPS, pp. 20–27, 2008.
22. A. Kerber, E. Cartier, L. Pantisano, R. Degraeve, G. Groeseneken, H.E. Maes, U. Schwalke, "Charge trapping in SiO2/HfO2 gate dielectrics: Comparison between charge-pumping and pulsed I_D–V_G", Microelectronic Engineering, Vol. 72, pp. 267–272, 2004.
23. E. N. Kumar, V. D. Maheta, S. Purawat, A. E. Islam, C. Olsen, K. Ahmed, M. A. Alam and S. Mahapatra, "Material Dependence of NBTI Physical Mechanism in Silicon Oxynitride (SiON) p-MOSFETs: A Comprehensive Study by Ultra-Fast On-The-Fly (UF-OTF) IDLIN Technique", Technical Digest. International Electron Devices Meeting, (IEDM), pp. 809–812, 2008.
24. K. Zhao, J. H. Stathis, A. Kerber and E. Cartier, "PBTI Relaxation Dynamics after AC vs. DC Stress in High-k/Metal Gate Stacks" inProc. IRPS, pp. 50–54, 2010.
25. T.L. III Tewksbury andHae-Seung Lee,"Characterization, modeling, and minimization of transient threshold voltage shifts in MOSFETs", IEEE Journal of Solid-State Circuits, **29**, pp. 239–252, 1994.
26. Eduard Cartier, Rishikesh Krishnan, Andreas Kerber, Sandip De, Rajan Pandey, Takashi Ando, Marinus Hopstaken, Joseph F. Shepard Jr., Michael D. Sullivan, Kota Murali, Vijay Naraianan and Michael P. Chudzik "Characterization and Optimization of Charge Trapping in High-k Dielectrics", to be presented at IRPS 2013.
27. Tibor Grasser, Ben Kaczer, Wolfgang Goes, Hans Reisinger, Thomas Aichinger, Philipp Hehenberger, Paul-Jürgen Wagner, Franz Schanovsky, Jacopo Franco, María Toledano Luque,

and Michael Nelhiebel, "The Paradigm Shift in Understanding the Bias Temperature Instability: From Reaction–Diffusion to Switching Oxide Traps", IEEE Transaction on Electron Devices, Vol. 55, No. 11, pp. 3652, 2011.
28. H. Reisinger, G. Steinlesberger, S. Jakschik, M. Gutsche, T. Hecht, M. Leonhard, U. Schroder, H. Seidl, D. Schumann, "A comparative study of dielectric relaxation losses in alternative dielectrics", Technical Digest. International Electron Devices Meeting, (IEDM), pp. 12.2.1–12.2.4, 2001.
29. C. Shen, M. F. Li, X. P. Wang, H. Y. Yu, Y. P. Feng, A. T.-L. Lim, Y. C. Yeo, D. S. H. Chan, and D. L. Kwong, "Negative U traps in HfO_2 gate dielectrics and frequency dependence of dynamic BTI in MOSFETs," in IEDM Tech. Dig., pp. 733–736, 2004.
30. Chadwin D. Young, Yuegang Zhao, Michael Pendley, Byoung Hun Lee, Kenneth Matthews, Jang HoanSim, Rino Choi, George A. Brown, Robert W. Murto, and GennadiBersuker "Ultra-Short Pulse Current–Voltage Characterization of the Intrinsic Characteristics of High-κ Devices", Jpn. J. Appl. Phys. Vol. 44 pp. 2437–2440, 2005.
31. B. Kaczer, V. Arkhipov, R. Degraeve, N. Collaert, G. Groeseneken and M. Goodwin, "Disorder-Controlled-Kinetics Model for Negative Bias Temperature Instability and its Experimental Verification", in Proc. IRPS, pp. 381–387, 2005.
32. Andreas Kerber and Eduard Cartier, "A Fast Four-Point Sense Methodology for Extraction of Circuit-Relevant Degradation Parameters", IEEE Electron Device Letters, Vol. 31, No. 9, pp. 912–914, 2010.
33. M. Toledano-Luque, R. Degraeve, M. B. Zahid, B. Kaczer, J. Kittl, M. Jurczak, G. Groeseneken, and J. Van Houdt, "Resolving Fast V_{TH}Transients After Program/Erase of Flash Memory Stacks and Their Relation to Electron and Hole Defects", Technical Digest. International Electron Devices Meeting, (IEDM), pp. 749–752, 2009.
34. Sanjay Rangan, Neal Mielke, Everett C.C. Yeh, "Universal Recovery Behavior of Negative Bias Temperature Instability", Technical Digest.International Electron Devices Meeting, (IEDM), pp. 14.3.1–14.3.4, 2003.
35. M. Denais, A. Bravaix, V. Huard, C. Parthasarathy, G. Ribes, F. Perrier, Y. Rey-Tauriac, N. Revil, "On-the-fly characterization of NBTI in ultra-thin gate oxide PMOSFET's", Technical Digest. International Electron Devices Meeting, (IEDM), pp. 109–112, 2004.
36. V. Huard, M. Denais, C. Parthasarathy, "NBTI degradation: From physical mechanisms to modeling", Microelectronics Reliability Vol. 46, pp. 1–23, 2006.
37. Hans Reisinger, Ulrich Brunner, Wolfgang Heinrigs, Wolfgang Gustin, and Christian Schlünder, "A Comparison of Fast Methods for Measuring NBTI Degradation", IEEE Transactions on Device and Materials Reliability, vol. 7, pp. 531–539, 2007.
38. V.K. Reddy, A.T. Krishnan, A. Marshall, J. Rodriguez, S. Natarajan, T.A. Rost, and S. Krishnan, "Impact of Negative Bias Temperature Instability on Digital Circuit Reliability" inProc. IRPS, pp. 248–254, 2002.
39. T. Nigam and E. B. Harris, "Lifetime Enhancement under High Frequency NBTI measured on RingOscillators", inProc. IRPS, pp. 289–293, 2006.
40. R. Fernández, B. Kaczer, A. Nackaerts, S. Demuynck, R. Rodríguez, M. Nafría, G. Groeseneken, "AC NBTI studied in the 1 Hz – 2 GHz range on dedicated on-chip CMOS circuits" in IEDM Tech. Digest, pp. 337–240, 2006.
41. Barry P. Linder, Jae-Joon Kim, Rahul Rao, Keith Jenkins, Aditya Bansal, "Separating NBTI and PBTI effects on the Degradation of Ring Oscillator Frequency", IEEE International Integrated Reliability Workshop Final Report (IIRW), pp. 1–6, 2011
42. Jae-Joon Kim, Barry P. Linder, Rahul M. Rao, Tae-Hyoung Kim, Pong-Fei Lu, Keith A Jenkins, "Reliability Monitoring Ring Oscillator Structures for Isolated/Combined NBTI and PBTI Measurement in High-K Metal Gate Technologies, "in the Proceedings of the International Reliability Physics Symposium, pp. 47–48, 2011.
43. E.S. Snyder, J. Suehle, "Detecting Breakdown in Ultra-thin Dielectrics Using a Fast Voltage Ramp", Int. Integrated Rel. Workshop Final Report, pp. 118–123, 1999.

44. A. Kerber, L. Pantisano, A. Veloso, G. Groeseneken, and M. Kerber, "Reliability screening of high-k dielectrics based on voltage ramp stress," *Microelectron.Reliab.*, vol. 47, no. 4/5, pp. 513–517, Apr./May 2007.
45. Andreas Kerber, Siddarth A. Krishnan, and Eduard Albert Cartier, "Voltage Ramp Stress for Bias Temperature Instability Testing of Metal-Gate/High-k Stacks", IEEE Electron Device Letters, Vol. 30, No. 12, pp. 1347–1349, 2009.
46. E. Cartier, A. Kerber, S. Krishnan, B. Linder, T. Ando, M. M. Frank, K. Choi, and V. Narayanan, "Voltage Ramp Stress (VRS) based stress-and-sense test method for reliability characterization of Hf-base high-k/metal gate stacks for CMOS technologies", ECS Transaction, 41, (3), pp. 337–348, 2011.

Chapter 2
Application of On-Chip Device Heating for BTI Investigations

Thomas Aichinger, Gregor Pobegen, and Michael Nelhiebel

Abstract This chapter introduces a new experimental approach allowing to switch the temperature of a device in a very fast and defined way. The new hardware tool, which we will herein refer to as polycrystalline silicon heater or simply poly-heater, allows overcoming previously strict experimental limitations regarding the speed of temperature variation and the accessibility of temperature range. Having broadened one's mind to the possibility of switching the temperature very fast at arbitrary points in time, the poly-heater technique opens up unprecedented experimental capabilities for bias temperature instability (BTI) characterization. For instance, one can achieve decoupling of stress and characterization temperature by making use of degradation quenching. Such or similar experiments can probe our understanding of the BTI physics in a novel manner.

2.1 In Situ Device Heating and Cooling Using On-Chip Poly(-crystalline Silicon) Heaters

In this paragraph we are going to address the calibration procedure of the poly-heater tool, investigate its capabilities, and demonstrate dedicated experimental setups which become feasible once having a well-calibrated and stable temperature switching tool at hand.

Having embedded the poly-heater in the measurement software environment, the technique allows to run stress-recovery experiments with variable temperature. From these experiments new fundamental features of Negative bias temperature

T. Aichinger (✉)
Infineon Technologies Austria AG, Siemensstraße 2, Villach, Austria
e-mail: thomas.aichinger@infineon.com

G. Pobegen • M. Nelhiebel
KAI Kompetenzzentrum für Automobil- und Industrielektronik, Europastraße 8, Villach, Austria
e-mail: gregor.pobegen@k-ai.at; michael.nelhiebel@k-ai.at

T. Grasser (ed.), *Bias Temperature Instability for Devices and Circuits*,
DOI 10.1007/978-1-4614-7909-3_2, © Springer Science+Business Media New York 2014

instability (NBTI) arise, helping to clarify conflicting issues in literature and allowing to draw new conclusions which lead to a more consistent microscopic picture of the degradation and the recovery mechanism.

2.1.1 Hardware Assembly and Poly-heater Design

Polycrystalline silicon resistors surrounding an active device can be used to perform fast in situ heating on a single device on wafer level which is commonly applied in time-critical Fast wafer level reliability (fWLR) monitoring [1–5]. The main advantage of the poly-heater is that it provides an elevated stress temperature without the use and the limitations associated with a conventional heating system like a thermo chuck. Although this alone is already a valuable feature, it does not even begin to explore the multitude of possibilities one finds in a more scientific use of the tool. By correct calibration and automation of the poly-heater–device system, the temperature becomes a quasi arbitrary experimental parameter which can be switched easily by more than ± 200 K within a couple of seconds. For such an application conventional thermo chuck systems are unsuitable because they are slow and their heating/cooling durations depend strongly on the difference to the target temperature. Also, when attempting to keep the junction biases applied during the temperature switch, it is a mandatory requirement not to lose probe needle contact. Since heating and cooling of a wafer on a thermo chuck involves considerable thermal expansion of wafer, needles, and pads, this request cannot be fulfilled over a wide temperature range without continuous manual readjustments. As opposed to heating the whole wafer on the thermo chuck, in situ heating by poly-heater wires is very local. Hence, there is no thermal expansion of needles and pads, which saves us from losing mechanical contact during the temperature switch.

Once calibrated and implemented in the software, the poly-heater feature opens up unprecedented possibilities for device characterization and reliability testing. Its application potential is thereby way beyond the scope of NBTI characterization. The concept of in situ heating by poly-heater wires has been taken up also for instance for thermo cycling [6] or for performing on chip high temperature annealing of irradiated p-channel metal oxide semiconductor (PMOS) dosimeters [7].

A common challenge of in situ heating can be found in the fact that the device which we want to heat is in most cases a distance away from the actual heating source. Consequently, the poly-heater wires are always hotter than the tested structure itself, which results in a temperature gradient between heater, device, and ambient. In our following experiments, the ambient temperature is always controlled by the programmable temperature of a conventional thermo chuck. From a design point of view, the distance between the poly-heater wires and the active areas of the device must be large enough and well isolated to prevent leakage current flow between the individual components. However, a larger distance between heater and

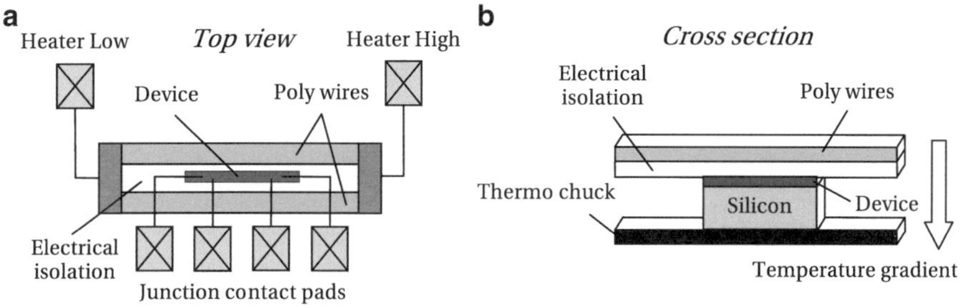

Fig. 2.1 (**a**) A schematic illustration of a poly-heater–device system placed on a thermo chuck. Electrically isolated poly-heater wires surround the silicon device. When a voltage is applied to the wires, a current flows and the dissociated heat elevates the temperature within the subjacent electrically active device regions. A schematic cross section of the poly-heater–device system is given in (**b**). Due to the lower chuck temperature, a temperature gradient arises between heater and wafer bottom

device reduces the accessible temperature range because more power has to be applied to the wires in order to adjust a certain elevated device temperature. Last but not least, the time delay for restoring thermal equilibrium must also be taken into account as we switch the heater on or off abruptly. The farther the heater is away from the device, the longer it takes to restore thermal equilibrium. Thus, in order to find an optimal design of the heating source and the active device for a particular application, the above-specified issues have to be considered carefully.

Figure 2.1 schematically illustrates a simple design proposal of an active transistor surrounded by a poly-heater. To adjust a well-defined ambient temperature, the wafer is placed on a thermo chuck which is held at a fixed programmed temperature. Depending on the thermo chuck temperature, a certain power has to be applied to the heater in order to reach a certain elevated device temperature. From a layout point of view the heater should overlap the device considerably in order to guarantee a homogeneous temperature distribution over the whole active transistor area. A calculation of the poly-heater and the device temperature as a function of the applied heater power needs some extra work, which will be treated in more detail in Sect. 2.1.3, but is not needed if the target device temperature is within the temperature range which is accessible by the thermo chuck system. As such, we present in the following subsections a procedure allowing to experimentally determine the power supply which is necessary to bring an active transistor to a certain device temperature. We will evaluate both the heater and the device temperature which enables us to estimate also the temperature gradient between the actively heated poly-heater wires and the active device. Furthermore, we will extract the transient heating and cooling characteristics for different target temperatures and discuss effects that may delay restoration of thermal equilibrium and destabilize the adjusted target temperature.

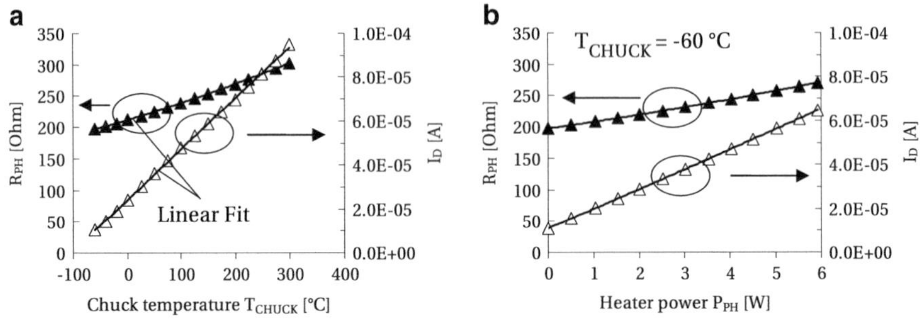

Fig. 2.2 (**a**) The poly-heater resistance R_{PH} (*full triangles*) and the drain current I_D (*open triangles*) as a function of the chuck temperature T_{CHUCK}. R_{PH} is characteristic for the poly temperature while I_D reflects the interface temperature of the active device. (**b**) The poly-heater resistance R_{PH} (*full triangles*) and the drain current I_D (*open triangles*) as a function of the heater power P_{PH} ($T_{CHUCK} = -60\,^\circ C$). The increase in the heater power causes a linear increase of R_{PH} and I_D

2.1.2 Calibration of Heater and Device Temperature

In order to extract the poly-heater and device temperature as a function of the applied power, we call on an appropriate temperature-dependent parameter of the material. In the case of the highly doped poly-heater wire, the temperature dependent electric resistance (R_{PH}) would be such a parameter. In the case of the Metal oxide semiconductor (MOS) transistor the forward current of the source/bulk diode or the source/drain current (around the threshold voltage of the device) can be used as an appropriate thermometer. For reliability issues, the source/drain current (I_D) is the preferred reference since it directly reflects the temperature of the interface between the silicon substrate and the gate oxide. This interface is of major interest because most studies suggest this region to be the location of concern for NBTI.

In the following, we demonstrate the temperature calibration using a lateral PMOS transistor embedded into two poly-heater wires similar as illustrated in Fig. 2.1. To record reference values for the poly-heater resistance (R_{PH}) and the drain current (I_D), we heat the wafer on the thermo chuck from -60 to $300\,^\circ C$ and measure R_{PH} and I_D at different temperatures, cf. Fig. 2.2a. The sense currents and voltages (which represent certain temperatures) must be chosen carefully in order to prevent self-heating during the measurement. Within the scanned temperature range (-60 to $300\,^\circ C$) the increase of the poly-heater resistance and the drain current can be fitted very well by a polynomial of first (linear) or second order. From a physical point of view, the increase in the poly resistance is due to a reduction of the carrier mobility due to enhanced lattice scattering at higher temperatures, while the increase in the drain current is due to an enhancement of the concentration of inversion carriers at higher temperatures. From the coefficients of the polynomial fit, we can interpolate the poly-heater and device temperature for arbitrary heating powers. Figure 2.2b illustrates the poly-heater resistance and the drain current as a function of the power supply (P_{PH}). The chuck temperature was −60 °C. Note that the relation between the measured poly-heater resistance and the applied heating power is not perfectly linear. The reasons will be elaborated in more detail in the next section. When applying 0 W (P_{PH} = 0 W), the poly resistance and the drain current correspond to their values extracted for −60 °C in Fig. 2.2a. As we increase the power supply from 0 W toward 6 W, the poly-heater resistance and the drain current grow simultaneously.

2.1.3 Maximum Accessible Temperature Range

Another remarkable application of the poly-heater technique can be found in the ability of reaching device temperatures far beyond the scope of conventional thermo chuck systems. This allows, for instance, probing a much wider temperature range to study Arrhenius-type processes. Extremely high temperatures can be reached when providing additional heating power at the maximum temperature range of the thermo chuck. The calibration procedure of the poly-heater–device system in such high temperature regimes requires, however, somewhat more effort because the above-described method for device temperature determination is only applicable for target temperatures within the maximum temperature range of the thermo chuck system. To determine the drain current for the experimentally not accessible range one would need to rely on TCAD device simulations. But the simulation itself would be based on material parameters in the high temperature regime which are rather difficult to characterize. Additionally, at high temperatures also the low biasing required to measure the drain current could already lead to degradation of the device, which makes an accurate calibration impossible. To avoid these problems we developed a method which extrapolates the functional dependence of the drain current *on the power supplied to the poly-heater* rather than on the temperature.

To illustrate this, we measured the increase of the device temperature T_{DV} (determined from the drain current increase) with poly-heater power P_{PH} at several different chuck temperatures and depicted the result in Fig. 2.3. It becomes evident from Fig. 2.3 that the device temperature does not depend linearly on the poly-heater power. A linear dependence is suggested for simple Joule heating where dissipated power is directly converted into a proportional temperature increase. The reason for the stronger than linear increase of the device temperature with power is the simultaneous increase of the thermal resistivity R^{th} of the materials which surround the heater and the device. This effect allows the device to become even hotter than what is expected from simple Joule heating. The functional dependence of the device temperature increase is, however, not a simple low-ordered polynomial and can therefore not be straightforwardly extrapolated.

In order to find an appropriate analytical expression for the observed increase of the device temperature, we exploit the definition of the thermal resistance and use Joule's first law that the electrical power is transferred into an equivalent amount of heat flow $\dot{Q}$, to find

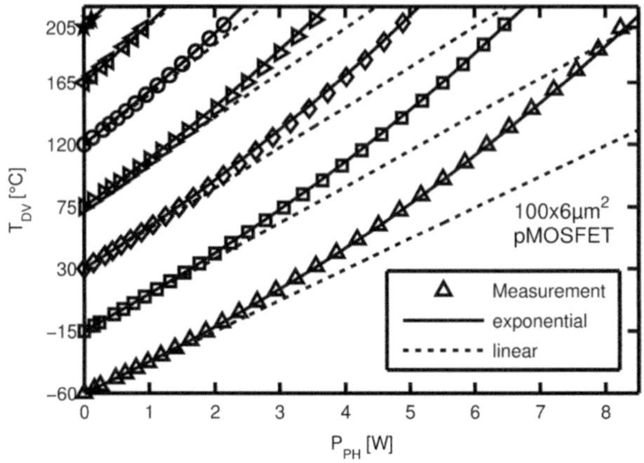

Fig. 2.3 Dependence of the device temperature on the power supplied to the poly-heater within the maximum accessible temperature range. The different *marker symbols* indicate different chuck temperatures. The increase of the device temperature is not linear

$$R^{\mathrm{th}}(T) = \frac{\mathrm{d}T}{\mathrm{d}\dot{Q}} \equiv \frac{\mathrm{d}T}{\mathrm{d}P} = T'(P). \tag{2.1}$$

This formula is a differential equation for the temperature as a function of power, which is in our case $T_{\mathrm{DV}}(P_{\mathrm{PH}})$, and opens a way to calculate the temperature directly when the temperature-dependent thermal resistance $R^{\mathrm{th}}_{\mathrm{sub}}(T)$ of the substrate is known. In particular, for silicon the thermal resistance depends on temperature as $R^{\mathrm{th}}(T) \propto T^{1.324}$ which leads to a rather complicated solution to (2.1) (not shown). However, a linear approximation of this power law introduces an error which is below the resolution limit for the measurement of the thermal resistance. Consequently, the thermal resistance can be approximated as

$$R^{\mathrm{th}}_{\mathrm{sub}}(T) = R^{\mathrm{th}}_{\mathrm{sub},0}\,(1 + \alpha(T - T_0)) \tag{2.2}$$

with three constants $R^{\mathrm{th}}_{\mathrm{sub},0}$, α, and T_0. By combining (2.1) and (2.2), with the requirement that the device temperature equals the chuck temperature T_{chuck} when there is no heater power supplied, we obtain

$$T(P) = T_0 - \frac{1}{\alpha} + \left(\frac{1}{\alpha} + T_{\mathrm{chuck}} - T_0\right) \exp\left(\alpha R^{\mathrm{th}}_{\mathrm{sub},0} P\right). \tag{2.3}$$

This means the functional dependence of the device temperature on the power supplied to the heater is an exponential function.

With (2.3) it is now possible to calculate the device temperature directly from the power supplied to the heater. However, the coefficients for the thermal resistance must be known. Using literature values can make the extrapolation erroneous because the thermal resistance can vary largely with the doping level [8] and only limited information exists about the thermal resistance of the interface between the

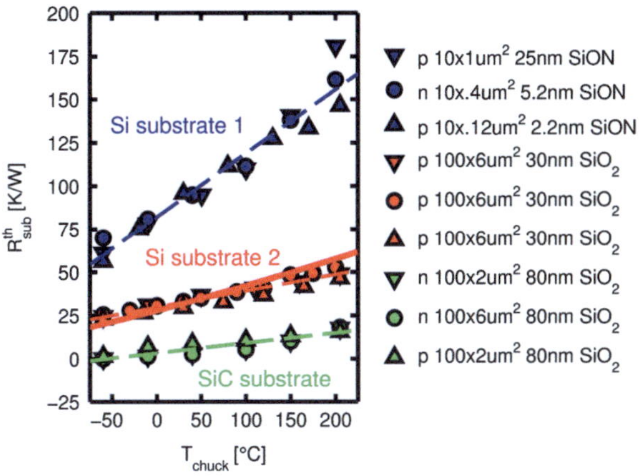

Fig. 2.4 Experimentally determined thermal resistances of several different devices from different technologies. In the legend, n and p refer to the MOSFET channel type, respectively, the micrometer size is the width times the length of the device, and the nanometer size is the thickness of the gate oxide. The *thin dashed lines* are linear fits of the data points of a group, while the *thick solid line* displays the theoretical dependence of $R^{\mathrm{th}} \propto T^{1.324}$

wafer and the chuck [9, 10]. As such, it appears beneficial to *measure* the thermal resistance for the given wafer/chuck system. This can be achieved by applying only little power to the poly-heater and measuring the change of the device temperature for different chuck temperatures

$$R^{\mathrm{th}}_{\mathrm{sub}}(T_{\mathrm{chuck}}) = \frac{T_{\mathrm{DV}}(P_{\mathrm{PH}}) - T_{\mathrm{chuck}}}{P_{\mathrm{PH}}} \tag{2.4}$$

to acquire $R^{\mathrm{th}}_{\mathrm{sub},0}$, α, and T_0. We measured several different devices of different technology and different substrate types as depicted in Fig. 2.4. The type and thickness of the substrate have the largest impact on the thermal resistance. For all investigated technologies the data can be reasonably well approximated by a linear relationship, further supporting the previously stated assumptions. The application of the extrapolation method (2.3) can now be compared within the temperature range of the thermo chuck to measurement data, as depicted in Fig. 2.5. The exponential equation (2.3) estimates the device temperature with only a few percent relative error.

Since previous investigations have shown that the thermal resistance of common semiconductors keep their functional dependence of the change of the thermal resistance with temperature until the melting or sublimation point [11–13], it is a safe assumption that (2.3) will hold also for the high temperature regime. As such it can be assumed that the small relative error of the method will also apply for temperature ranges much above the highest temperature of the chuck system.

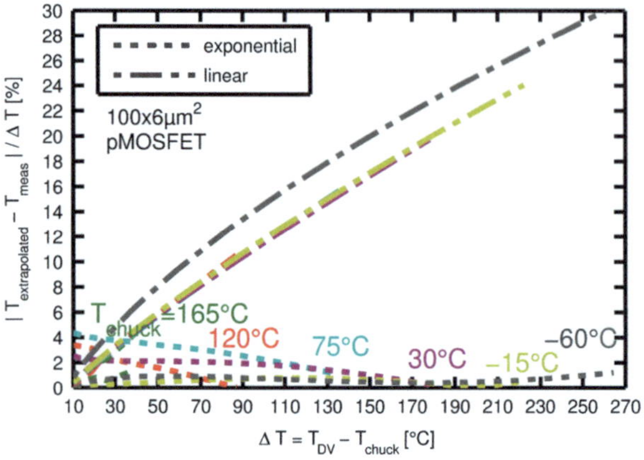

Fig. 2.5 Relative error of a linear extrapolation and the exponential extrapolation (2.3) from the first 10 °C at the respective chuck temperature. The underlying data is that of Fig. 2.3. The relative error for the linear extrapolation increases with increasing power supplied to the poly-heater. In contrast, for the method (2.3) the relative error stays well below a few percent

2.1.4 Heating and Cooling Dynamics

In this subsection we elaborate on the time-dependent heating and cooling dynamics of the device as a heating voltage/power is applied to or removed from the heater, respectively. The chuck (ambient) temperature was −60 °C during the following experiments. By applying a certain heating voltage to the poly-heater wires, the device temperature quickly elevates and stabilizes after a couple of seconds. On the other hand, when removing the heating voltage, the device cools down immediately. In order to determine the exact heating voltage necessary to reach a certain device temperature, the heater–device system was calibrated in the way discussed in the previous paragraphs. The output of this initial calibration were eight different heating voltages appropriate to heat the device from −60 to −40, −20, 0, 25, 50, 75, 100, and 125 °C. In the experiments illustrated in Fig. 2.6, different heating voltages were applied and later removed abruptly while recording in parallel the heater current for 100 s. From the heating voltage and current characteristics, the time-dependent power dissipation of the heater was calculated.

As can be seen in Fig. 2.6a, when turning the heater on, it takes up to 1 ms until the maximum power dissipation is reached. This delay time is mainly limited by the finite speed of the voltage source. Using our particular heater design, poly-heater supply voltages up to 34 V are required in order to overcome a temperature range of 185 K (−60 °C → 125 °C). After the voltage source has stabilized (>1 ms), the heater power tends to decrease slightly for a couple of seconds. This is because some time is needed to restore thermal equilibrium between heater, wafer, and thermo chuck. The decrease in power within the very first moments after turning on the heater is the greater the larger the temperature difference between heater and chuck. When turning the heater off, the heater power vanishes within approximately 1 ms. Again, this 1 ms is originated in the finite speed of the voltage source.

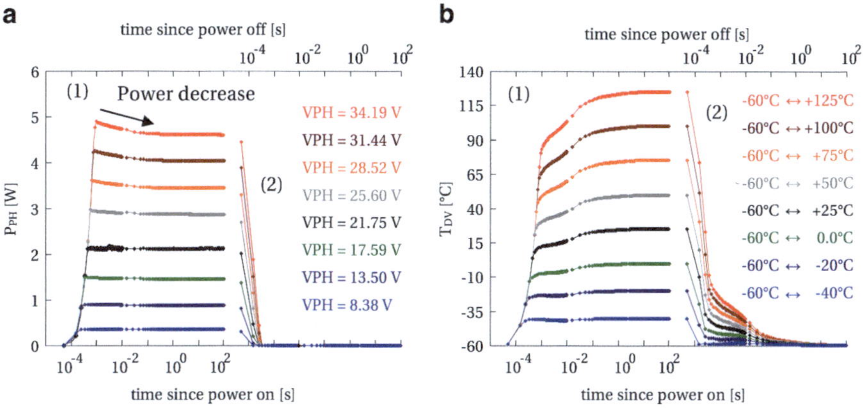

Fig. 2.6 (**a**) The heating power when turning the heater voltage abruptly on (1) and off (2). The chuck temperature was −60 °C. During the heater calibration specific voltages were determined to reach certain device temperatures. At the moment the heating voltage is turned on (1) or off (2), we record the heating current in parallel and calculate P_{PH}. (**b**) The heating and cooling characteristics of the device when turning the heater power abruptly on (1) or off (2). The heater/device/chuck system needs a couple of seconds to restore thermal equilibrium causing a shoulder in the device temperature at the very beginning of the heating and cooling procedure

In Fig. 2.6b the same experimental sequence as in Fig. 2.6a was performed but this time the drain current of the device was recorded as a representative for the Si/SiO_2 interface temperature. By using the results of Fig. 2.2a one can calculate the evolution of the device temperature T_{DV} from I_D.

As can be seen in Fig. 2.6b, when turning the heater on abruptly, it takes up to 10 s until the device has stabilized at its calibrated target temperature. The larger the temperature difference, the longer it takes to reach the target temperature. The larger time delay is due to the finite time interval necessary to restore thermal equilibrium a distance away from the actual heating source. The shoulder visible in the evolution of T_{DV} during heating is due to the power decrease illustrated in Fig. 2.6a. We remark that although heater power and heater temperature reach a maximum 1 ms after turning the heating voltage on, the device temperature does never exceed its target value due to the delayed thermal coupling between the poly-heater and the device. This is an important aspect since we do not want to subject the device to an elevated pre-stress at the moment the heater is turned on.

When turning the heater off, the situation is similar as during turn on. It takes a couple of seconds until the excess heat generated by the poly-heater can be removed by the thermo chuck. From Fig. 2.6b, we conclude that at an ambient temperature of −60 °C any temperature switch up to ±200 K can be executed with high accuracy within a time interval of maximal 10 s. In fact it takes approximately 0.1 s to reach the target temperature by 3%, 1 s to obtain a 1% accuracy and after 10 s the target temperature is adjusted by 0.1% which corresponds to the maximum resolution of the measurement.

2.1.5 *Summary of the Poly-heater Features*

In the previous subsections the features and performance of the in situ poly-heater measurement technique were discussed. The temperature calibration procedure for determining the poly temperature and device temperature was elaborated in detail for different ambient temperatures and power supplies. The thermal resistances of the poly-heater and the device were found to depend on the ambient temperature which is consistent with the nonlinear thermal conductivity reported for silicon. It was shown that a temperature range of more than 200 K can be bridged by additional power supply provided by the poly-heater. In particular, the ability of reaching device temperatures far beyond conventional thermo chuck ranges was pointed out. A thorough study on the heating and cooling dynamics of the device has revealed that a maximum time of 10 s is needed to switch the temperature within an interval of $\pm$200 K with a maximum precision of 0.1 K. Thereby, the heating and cooling procedure was found to be nearly independent of the difference between ambient temperature (chuck temperature) and target temperature. Equipped with these features the poly-heater tool exhibits a remarkable and unique tool for device reliability testing and characterization purposes which will be applied in the following sections for NBTI investigations.

2.2 The Principle of Degradation Quenching and Its Application for BTI Investigations

As demonstrated in the previous section, poly-heaters can be used to perform fast and reliable in situ heating on a single device on wafer level. The following section explains how such a feature can be used to perform NBTI stress at a certain stress temperature, which generates a certain degradation level, while the recovery itself can be studied at arbitrary recovery temperatures. By turning the heater on during stress and switching it off during recovery, the tool enables us (a) on the one hand to bring identically processed devices to the same degradation level and (b) on the other hand to fix a different temperature or vary the temperature in a defined way during recovery. By using this technique, our understanding of the recovery physics can be probed in a novel manner. Until now degradation and recovery mechanisms as well as the knowledge about relative contributions to the total threshold voltage shift and recovery are still controversial points in literature which require an unambiguous clarification in order to probe and formulate reliable degradation and recovery models [14, 15].

2.2.1 *Why Temperature Quenching?*

A trivial problem encountered in the observation of temperature effects in NBTI *stress* is the need to dispose of a set of (i) comparable devices that are brought to (ii) different degradation levels by different stress temperatures, but which are then (iii) characterized directly post stress at the same unique characterization temperature. In contrast, when temperature effects in NBTI *recovery* are studied, the following problems are encountered: one has to dispose a set of (i) comparable devices that are brought to (ii) the same degradation level by a unique stress temperature, but which are then (iii) characterized directly post stress at different characterization/recovery temperatures.

The first condition (comparable devices) may be solved by careful sample selection, involving thorough characterization before stress. Because stress and recovery dynamics are strongly temperature dependent, the second condition cannot be solved by classically available stress and characterization methods, if the third condition has to be maintained. Classical methods are either based on performing stress and characterization at the same temperature, or strictly separate stress and recovery by long, scarcely observable and basically undefined transition periods. While the first approach can provide the observation of recovery at very good time resolution [16], it obviously always violates condition (ii) if condition (iii) is fulfilled and vice versa.

To fulfill all three conditions at once, it is necessary to conserve the degradation level during cooling. Therefore, the temperature switch has to be fast, well controlled, and practically independent of the difference between stress temperature and recovery/characterization temperature. This demand cannot be fulfilled by a conventional thermo chuck system since the cooling duration of such systems is typically very long (>30 min) and strongly dependent on the target temperature. As a consequence, one has to deal either with *additional degradation* when maintaining the stress bias applied during cooling or with *uncontrolled recovery* when leaving the device floating during cooling. Also, contact difficulties arise due to thermal expansion of probe needles and metal pads which make it hard to maintain device biases during the temperature switch. In short, using the thermo chuck for temperature switches suffers from several systematic errors and drawbacks.

Our approach to harmonize all conditions and to get rid of the above-described technical difficulties is to make use of the poly-heater technique, cf. Fig. 2.7a. During stress a certain stress bias (V_{GS}) is applied to the gate and the (previously calibrated) heater generates an elevated device temperature (T_S) for a defined stress time t_S. Before initiating the recovery/characterization cycle, the heater is switched off, the device reaching ambient (chuck) temperature within a couple of seconds (t_D). In order to prevent any relaxation during the temperature switch, the stress bias remains applied within the delay time t_D. During t_D stress continues in an undefined way, but this additional degradation is negligible compared to the degradation occurring within the main stress period typically performed at a much higher stress temperature ($T_S \gg T_R$). This was verified in Fig. 2.7b where we have stressed different PMOS devices for 1,000 s at $T_S = 125\,°C$ and $E_{OX} = 6.0$ MV/cm and afterwards let them recover at $T_R = -60\,°C$ and V_{TH}. The cooling delay time t_D between turning off the heater and switching the gate bias from V_{GS} to V_{TH} was varied between 0 and 1,000 s. When using a delay time between 0 and 1 s, the device has not reached the target temperature T_R at the moment the stress bias is removed, cf. Fig. 2.6. Consequently, due to the larger temperature the measured

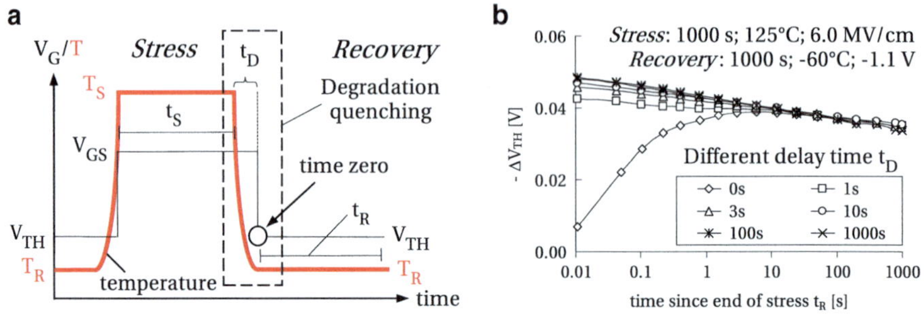

Fig. 2.7 (**a**) The principle of degradation quenching. Subsequently to the initial characterization phase at the analyzing temperature T_R and the gate bias V_{TH}, the heater is turned on, elevating the device temperature quickly toward the stress temperature T_S. Once at T_S, the stress phase is initiated by switching the gate bias from V_{TH} to V_{GS}. After the stress time t_S has elapsed, the heater is turned off. The stress bias remains applied for a delay time t_D until the device is at T_R (degradation quenching). The recovery cycle (t_R) is then initiated by switching the gate bias from V_{GS} to V_{TH}. (**b**) Threshold voltage shift as a function of the delay time t_D. Different PMOS devices were stressed for 1,000 s at $T_S = 125\,°C$ and $E_{OX} = 6.0\,MV/cm$. After degradation quenching, the V_{TH} shift was monitored at $T_R = -60\,°C$ and V_{TH} using different delay times t_D

V_{TH} shift is afflicted with an error affecting predominantly the first couple of seconds after removal of the stress bias. However, when using a delay time ≥ 3 s, we obtain similar recovery characteristics for arbitrary delay times suggesting (i) that 3 s is sufficiently enough to reach the target temperature and (ii) that we can safely neglect additional degradation or recovery during t_D provided the stress temperature exceeds the recovery temperature by far. Since the cooling time of the poly-heater–device system is nearly independent of the temperature difference ($T_S - T_R$), statement (ii) and (iii) are fulfilled simultaneously independent of the stress or recovery/characterization temperature provided T_R is significantly lower than T_S.

In the following the conservation of the degradation level during the temperature switch will be called “degradation quenching.” Degradation quenching after Negative bias temperature stress (NBTS) allows to switch the device temperature *first* from its stress level to its recovery level and *then* triggers V_{TH} recovery by switching the gate bias from the stress level to the threshold voltage of the device. Having demonstrated that degradation quenching can be achieved by using the poly-heater technique, we use the method to investigate the role of temperature in NBTI recovery.

2.2.2 *The Temperature Dependence of BTI Recovery*

To investigate the temperature dependence of NBTI recovery, the following experimental procedure was performed on different PMOS devices by making use of the previously described degradation quenching method. During stress, the heater generates a defined interface temperature of 125 °C and a certain stress bias is

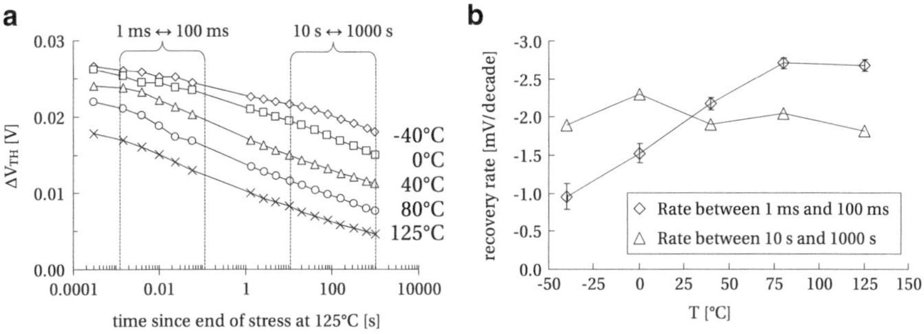

Fig. 2.8 (**a**) Recovery of the threshold voltage shift recorded at different temperatures after stressing all samples at $E_{OX} = 5.5\,\text{MV/cm}$ and $T_S = 125\,°\text{C}$. Recovery conditions: $V_{GR} = -1.1\,\text{V}$; $V_{DR} = -2.5\,\text{V}$; $T_R = -40$, 0, 40, 80, and 125 °C. (**b**) Temperature-dependent recovery rate between 1 and 100 ms (*diamonds*) and between 10 and 1,000 s (*triangles*)

applied to the gate, subjecting different devices to an oxide field of approximately $E_{OX} = 5.5\,\text{MV/cm}$. During stress, source and drain were at 0 V. Stress field, time, and temperature were identical for all samples, creating a unique degradation level of each device at the end of the stress time t_S which was 1,000 s.

After the 1,000 s had elapsed, the heating current was taken away and the device cooled down rapidly toward the individual ambient temperature which was defined by the temperature of the underlying thermo chuck. When intending to study recovery at −40 °C, the chuck has to be at that temperature already before stress. Naturally, the lower the base temperature of the thermo chuck, the greater the required power supply for the poly-heater to reach the unique stress temperature of 125 °C, the lower the base temperature of the thermo chuck. One second after the heater was turned off, the gate bias was switched to the threshold voltage ($V_{GR} = -1.1\,\text{V}$) of the device. While the switch of the device temperature terminates the stress, the switch of the gate bias initiates the recovery cycle (t_R). In parallel to the gate bias switch, the drain bias was set to its read-out value ($V_{DR} = -2.5\,\text{V}$) in order to measure the recovery of the saturation drain current. The transition from stress to read-out bias conditions required approximately 200 µs and was limited solely by the speed of the voltage unit. An additional 100 µs was needed for the measurement. Thus, the first current value at the individual recovery temperature was recorded about 300 µs after removal of the stress voltage. The time-dependent evolution of the saturation drain current was later converted into a stress/recovery induced threshold voltage shift, cf. [17].

The result of this temperature quenched recovery measurement is illustrated in Fig. 2.8a. The unique stress temperature, supplied by the poly-heater, was 125 °C. The individual recovery temperatures were −40, 0, 40, 80, and 125 °C. Stress and recovery durations were 1,000 s respectively. As can be seen in Fig. 2.8a, the recovery curves look quite similar except for a temperature-dependent offset which was already present at the first measurement point recorded 300 µs after removal of the stress field. Although the recovery temperature varies by more than 160 K with respect to the individual analyzing temperatures, there is no significant long-term temperature dependence visible in the recovery slopes. The recovery traces are nearly parallel.

In Fig. 2.8b the recovery rate per decade was evaluated more precisely for the first 100 ms and for the last two decades of the recovery traces. On closer inspection, there is a temperature dependence visible within the first 100 ms right after stress. We observe an increasing slope of the recovery curves with increasing temperatures. While at $-40\,^{\circ}$C the amount of recovery is only 1 mV per decade, it is about three times larger for temperatures above $80\,^{\circ}$C. A reason for the initial temperature dependence might be the fact that the device was probably not exactly at the target temperature due to a slightly too short cooling delay time (t_D) of only 1 s, cf. Fig. 2.7b. A few seconds after the termination of stress all traces become nearly parallel independent of T_R. After 10 s the decrease of the threshold voltage shift has leveled off to about 2 mV/dec for all samples. This holds at least for two or three decades in time.

The offset can be explained qualitatively by assuming an inelastic tunneling process and a homogeneous distribution of trap energy levels which are responsible for the observed log-like recovery traces. In such a model, lowering the analyzing temperature would increase all recovery time constants simultaneously, thereby shifting the entire recovery curve to larger times [18].

It has to be mentioned that a possibly remaining small offset could be also explained by the temperature-dependent position of the Fermi level at the read out gate bias $V_{GR} = -1.1$ V. At low temperatures, the Fermi level is pinned close to the valence band edge. When increasing the temperature (at constant gate bias), the Fermi level moves a little bit closer toward midgap. Considering creation of interface states within the silicon bandgap as a result of NBTS, their charge state (occupation probability) would be governed by the position of the Fermi level. Consequently, at lower temperatures more of them tend to be positively charged causing a slightly larger threshold voltage shift. This temperature-dependent variation of the Fermi level is a systematic error which is, however, believed to be much too small to explain the full offset.

In summary, the obtained recovery characteristics are quite surprising. A crucial point here is the temperature independence of the recovery rate (slope of the recovery curves in a semi-logarithmic plot) which challenges recovery models based on hydrogen diffusion. For instance, in dispersive diffusion models hydrogen is believed to be stored in traps within the oxide, at the interface or somewhere else close to the interface. In order to re-passivate stress-induced damage during recovery, the H atoms or H_2 molecules have to be released from there and overcome a thermodynamic barrier. At lower temperatures the probability of release as well as the diffusion rate of hydrogen is much lower. If such a mechanism would be the controlling process, one would expect freezing of recovery at $-40\,^{\circ}$C. However, time-dependent recovery is still observed at similar rates even at temperatures as low as $-40\,^{\circ}$C. The same argument also holds for the switching of hydrogen from a bonding to an antibonding Si–H configuration, as proposed by Tuttle [19]. Since the transfer from a bonding to an antibonding configuration is a thermodynamical process, it should be highly temperature activated.

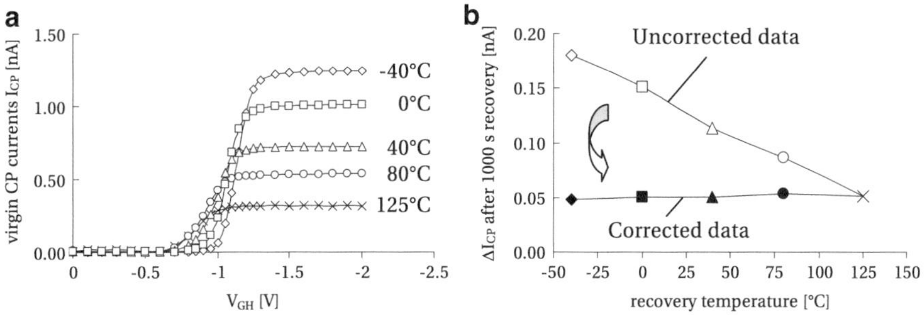

Fig. 2.9 (**a**) Virgin CP current characteristics recorded at the individual recovery temperatures. The lower the temperature, the higher the CP signal due to a larger profiled active energy range (ΔE_{CP}). (**b**) The remaining CP current degradation at the end of the constant bias recovery phases performed at different temperatures. CP setup: $V_{GB} = 1.0\,V$; $V_{GH} = -2.0\,V$; $f = 500\,kHz$; $t_r = t_f = 375\,ns$. The remaining CP signal is the larger the lower the temperature, however, when accounting for the temperature-dependent active energy interval (ΔE_{CP}), the obtained differences in the uncorrected ΔI_{CP} data (*open symbols*) is removed. The corrected data (*full symbols*) reveals a similar remaining degradation level of the interface after 1,000 s constant bias recovery at different temperatures

In addition to ΔV_{TH} shifts, we have investigated the role of interface states in the recovery process. Therefore, we have performed CP measurements at the end of the 1,000 s lasting constant bias recovery phases. Considering that the obtained ΔV_{TH} shifts are considerably different at the end of each recovery phase, one would expect a similar difference in the remaining CP current if interface state re-passivation would be the dominating recovery mechanism. Such a difference is actually obtained at the end of the recovery, cf. uncorrected data in Fig. 2.9b. The $-40\,°C$ data shows a considerably larger ΔI_{CP} than for example the $125\,°C$ data. However, when taking the temperature dependence of CP into account, the differences in the maximum CP currents at the end of the recovery vanish, cf. corrected data in Fig. 2.9b. We have corrected the original data for different analyzing temperatures by referencing to the initial offsets in the CP currents of the unstressed devices, cf. Fig. 2.9a. We remark that all samples showed a similar CP current before stress, when recorded at the same temperature. This result indicates that either no interface state recovery takes place at all, or interface state recovery is independent of temperature.

2.2.3 Identically Stressed Devices Subjected to Abrupt Temperature Switches

In Fig. 2.10a the V_{TH} recovery is illustrated for four different PMOS devices. All devices were stressed at an oxide field of $5.5\,MV/cm$ and at a temperature of $125\,°C$ for 1,000 s. Three reference devices recovered at a constant temperature of -40, 40, and $125\,°C$, respectively. The fourth device was subjected to two abrupt

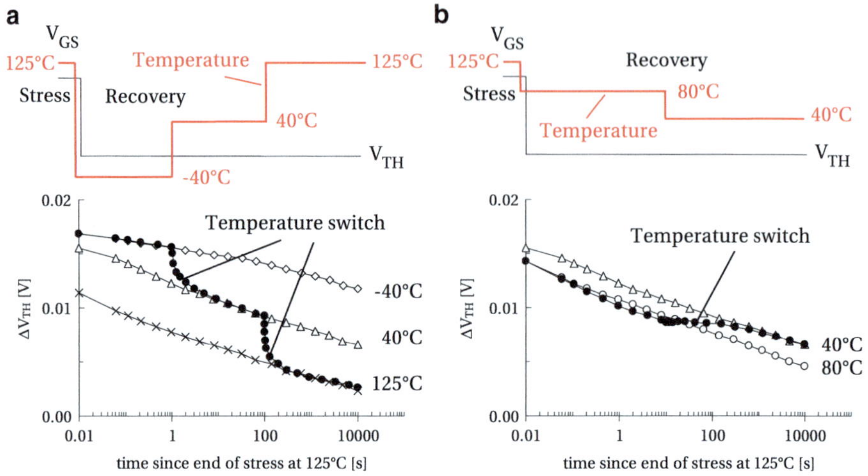

Fig. 2.10 (**a**) Two step heating performed during constant bias recovery. Reference measurements at −40, 40, and 125 °C are illustrated by *open diamonds/triangles/crosses*. Recovery can be accelerated twice as we heat the device abruptly from −40 to 40 °C (after 3 s) and from 40 to 125 °C (after 100 s), cf. *full symbols*. (**b**) Cooling performed during constant bias recovery. Reference measurements at 40/80 °C are illustrated by *open triangles/circles*. Lowering the recovery temperature (80 °C → 40 °C) leads to frozen recovery until the cooled (*full symbols*) reaches the reference curve at 40 °C

temperature switches by making use of the poly-heater technique. Right after stress it recovered for 1 s at −40 °C, then for 100 s at 40 °C, and finally for another 10,000 s at 125 °C providing two decades of inspection at each temperature. As can be seen in Fig. 2.10a, when elevating the device temperature abruptly during recovery, ΔV_{TH} relaxation is immediately accelerated approaching the reference curves after a couple of seconds. Such temperature accelerated recovery at constant gate bias conditions can definitely not be ascribed to elastic tunneling. In Fig. 2.10b we have performed the complementary experiment to Fig. 2.10a: at first, the device recovered at 80 °C for 10 s. Afterwards, the heater power was lowered so that the device cooled down to 40 °C. While heating accelerates recovery, it can be seen that cooling leads to frozen recovery for a certain interval of time. Indeed, recovery does not proceed before the cooled measurement curve reaches the 40 °C reference curve. Again, frozen recovery at constant gate bias conditions cannot be ascribed by an elastic hole trapping model.

2.2.4 *Conclusions*

Based on the upper key experiments performed on identically stressed PMOS devices, one may draw the following conclusions on the temperature dependence of ΔV_{TH} and CP current recovery:

1. Threshold voltage recovery is accelerated considerably by elevating the temperature. On the other hand, when decreasing the temperature during recovery, the degradation level remains frozen for a certain interval of time.
2. The mechanism causing ΔV_{TH} recovery at elevated temperature is a true chemical relaxation process which is not reversible by subsequent device cooling.
3. CP current recovery is only marginally influenced by either heating or cooling, suggesting interface state repassivation to play only a minor role in the recovery as long as the gate bias is maintained constant around the V_{TH}.
4. The recovery rates of the V_{TH} shift and the CP current (recorded under continuous gate pulsing conditions) are widely independent of temperature. This holds at least for long-term recovery measurements recorded between 1 and 1,000 s after the termination of stress.

Since NBTI shows both bias and temperature dependence, but our measurements support neither elastic tunneling nor interface state repassivation, a different mechanism has to be responsible for the observed recovery characteristics. Because bias dependence is totally incompatible with a diffusion process of neutral hydrogen species, we take the dependence of the V_{TH} recovery on the read-out voltage as an indication for a trapping/detrapping phenomenon and attempt to expand the idea of elastic carrier exchange to a temperature sensitive model. As opposed to elastic tunneling, inelastic phonon-assisted tunneling is temperature dependent [20]. Oxide defects and valence band electrons having different energetic positions cannot exchange carriers elastically. However, if they gain energy from lattice vibration (through phonons), they may pass the thermodynamical tunneling barrier ΔE_B with a certain temperature-dependent probability [21].

The lifetime of a single trap can be expressed by an Arrhenius law:

$$\tau(\Delta E_B, T_R) = \tau_0 \exp\left(\frac{\Delta E_B}{k_B T_R}\right), \tag{2.5}$$

where $\tau(\Delta E_B, T_R)$ is the inelastic tunneling lifetime of a single trap, τ_0 is the pseudo-elastic tunneling exchange time between a trap and a substrate carrier at a barrier height zero, ΔE_B is the thermodynamical tunneling barrier, k_B is the Boltzmann constant, and T_R is the analyzing temperature. At a constant temperature the time constants of different traps are solely determined by their individual barrier heights ΔE_B.

When assuming NBTI recovery to be mainly determined by the neutralization of positive oxide defects via electron capture from the silicon substrate (respectively hole emission into the silicon substrate), the observed threshold voltage recovery can be interpreted as a continuous decay of traps with different barrier heights ΔE_B.

Based on this idea, we can schematically illustrate the ΔV_{TH} recovery curve for three different traps having different thermodynamical barrier heights, cf. Fig. 2.11: According to their individual barrier heights, each trap has a certain characteristic

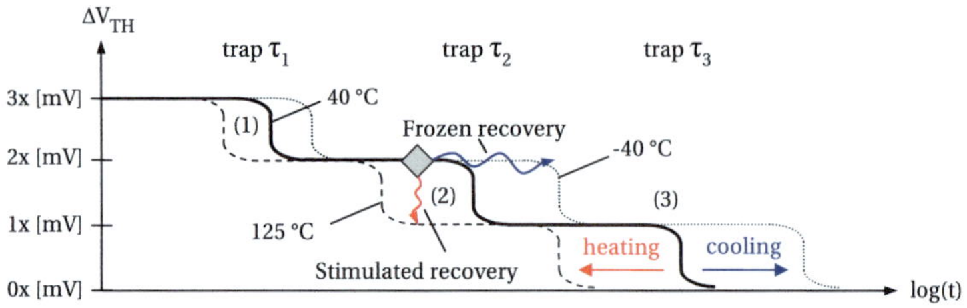

Fig. 2.11 First order model of temperature-dependent recovery effects using a simple picture of three different traps having three different time constants, barrier heights, respectively. Each trap is assigned to an arbitrary threshold voltage shift of $1\times$ mV. Heating or cooling shifts the recovery curve to the left or the right (shorter or longer time constants) resulting in stimulated or frozen recovery. The *diamond* indicates a hypothetic temperature switching event

time constant at which it recovers with maximum probability. A variation of temperature ($T_{R1} \longrightarrow T_{R2}$) impacts all time constants in parallel thereby shifting the plateaus along the time axis in log scale.

The respective shift in time for a trap with barrier height ΔE_B can be calculated as

$$\log\left(\tau(\Delta E_B, T_{R1})\right) - \log\left(\tau(\Delta E_B, T_{R2})\right) = \frac{\Delta E_B}{k_B T_{R1}} - \frac{\Delta E_B}{k_B T_{R2}}. \tag{2.6}$$

For $T_{R2} > T_{R1}$, the plateaus will shift to the left since the time constants of all traps will decrease, for $T_{R2} < T_{R1}$ the time constants increase leading to a shift to the right. Note that the widths of the plateaus are proportional to ΔE_B. The trap level which recovers first (τ_1) has the lowest barrier (ΔE_{B1}) and is therefore least temperature dependent. On the other hand, trap levels with higher barriers (ΔE_{B2} and ΔE_{B3}) depend stronger on temperature which results in a more significant temperature impact on the plateau broadness.

In this first order model, heating or cooling the device during recovery leads to stimulated recovery (at the diamond, stepping from the solid to the dashed line in Fig. 2.11, cf. Fig. 2.10a) or frozen recovery (at the diamond, stepping from the solid to the dotted line in Fig. 2.11, cf. Fig. 2.10b) compatible with our measurement results. When further assuming that the barrier ΔE_B itself can be lowered by a bias change, the model covers also bias change experiments and includes "mathematically" elastic tunneling in the limit $\Delta E_B = 0$. Furthermore, homogeneously distributed thermodynamical barriers would lead to a large variety of time constants resulting in a large number of small steps like the ones described in Fig. 2.11. In a realistic experiment (large device), one would therefore expect large amounts of small steps to be smeared out as a straight line in a $\log(t)$ diagram consistent with our recovery experiments.

Acknowledgements This work was jointly funded by the Austrian Research Promotion Agency (FFG, Project No. 831163) and the Carinthian Economic Promotion Fund (KWF, contract KWF-1521|22741|34186).

References

1. W. Muth, W. Walter, in *Proc.ESSDERC* (2007), pp. 1251–1262
2. C. Schluender, R.P. Vollertsen, W. Gustin, H. Reisinger, in *Proc.ESSDERC* (2007), pp. 131–134
3. T.K. Kang, C.S. Wang, K.C. Su, Jpn.J.Appl.Phys. **46**, 7639 (2007)
4. C.S. Wang, W.C. Chang, W.S. Ke, C.T. Chiang, C.F. Lee, K.C. Su, in *Proc.SSDM* (2005), pp. 580–581
5. C.S. Wang, W.C. Chang, W.S. Ke, K.C. Su, in *Proc.IIRW* (2006), pp. 136–138
6. H. Köck, V. Košel, C. Djelassi, M. Glavanovics, D. Pogany, Microelectron.Reliab. **49**, 1132 (2009)
7. A. Kelleha, W. Lane, IEEE Trans.Nucl.Sci. **43**, 997 (1996)
8. W. Liu, M. Asheghi, J. Appl. Phys. **98**, 123523 (2005)
9. A. Cardoso, A.K. Srivastava, J. Vac. Sci. Tech. B **19**, 397 (2001)
10. H. Ibele, K. Reitinger, in *IEEE Semiconductor Wafer Test Workshop* (2005)
11. P. Leturcq, J.M. Dorkel, A. Napieralski, E. Lachiver, Trans. Elec. Dev. **34**, 1147 (1987)
12. C.J. Glassbrenner, G.A. Slack, Phys. Rev. **134**, A1058 (1964)
13. G.A. Slack, J. Appl. Phys. **35**, 3460 (1964)
14. T. Aichinger, M. Nelhiebel, T. Grasser, in *Proc.ESREF* (2008), pp. 1178–1184
15. T. Aichinger, M. Nelhiebel, T. Grasser, in *Proc.IRPS* (2009), pp. 2–7
16. H. Reisinger, O. Blank, W. Heinrigs, A. Mühlhoff, W. Gustin, C. Schlünder, in *Proc.IRPS* (2006), pp. 448–453
17. B. Kaczer, V. Arkhipov, R. Degraeve, N. Collaert, G. Groeseneken, M. Goodwin, in *Proc.IRPS* (2005), pp. 381–387
18. G. Pobegen, T. Aichinger, M. Nelhiebel, T. Grasser, in *IEDM Tech. Dig.* (2011), pp. 27.3.1–27.3.4
19. B. Tuttle, Phys.Rev.B **59**, 12884 (1999)
20. S. Ganichev, E. Ziemann, W. Prettl, I. Yassievich, A. Istratov, E. Weber, Phys.Rev.B **61**, 61 (2000)
21. W. Gös, M. Karner, S. Tyaginov, P. Hehenberger, T. Grasser, in *Proc.SISPAD* (2008), pp. 69–72

Chapter 3
Statistical Characterization of BTI-Induced High-k Dielectric Traps in Nanoscale Transistors

Tahui Wang, Jung-Piao Chiu, and Yu-Heng Liu

Abstract Statistical behavior of BTI-induced high-k dielectric traps in nanometer MOSFETs is characterized. We measure individual trapped charge emission times and single-trapped charge-induced V_t shifts in BTI recovery. Statistical distributions of BTI trap characteristics such as trap spatial and energy distributions and trapped charge activation energy in emission are extracted. We compare the amplitudes of BTI and RTN single-charge-induced ΔV_t. BTI-induced ΔV_t exhibits a larger amplitude distribution tail. An explanation will be given by use of 3D atomistic numerical simulation. In addition, we find that V_t degradation in BTI stress exhibits two stages. The first stage has logarithmic stress time dependence and is believed due to the charging of preexisting high-k dielectric traps. The second stage follows power-law time dependence, which is attributed to dielectric trap creation.

3.1 Introduction

The aggressive CMOS scaling has been reaching the physical limit of conventional SiO_2 MOSFETs as a result of significant direct tunneling current through ultrathin oxides. High-permittivity (high-k) gate dielectrics have emerged as a post-SiO_2 solution. Bias temperature instability (BTI) has been recognized as one of the most important reliability issues in ultrathin gate oxide CMOS devices because of its large impact on performance and reliability in digital and analog circuits. The use of high-k gate dielectrics even expedites BTI degradation [1, 2]. Unlike most reliability effects, BTI-induced V_t degradation recovers partly after the removal of stress. In most of earlier works, BTI-induced degradation/recovery was characterized in large-area devices, and studies were based on average and continuous V_t evolutions [3–5].

T. Wang (✉) • J.-P. Chiu • Y.-H. Liu
Department of Electronics Engineering, National Chiao Tung University, Hsinchu, Taiwan
e-mail: twang@cc.nctu.edu.tw; clouders.ee96g@nctu.edu.tw; henryliu0306.eecs96@nctu.edu.tw

T. Grasser (ed.), *Bias Temperature Instability for Devices and Circuits*,
DOI 10.1007/978-1-4614-7909-3_3, © Springer Science+Business Media New York 2014

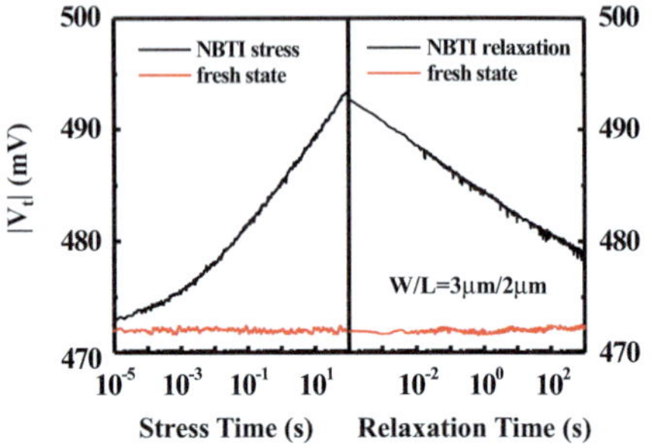

Fig. 3.1 Continuous V_t evolutions in NBTI stress and relaxation in a large-area (W/L = 3 μm/2 μm) high-k pMOSFET. V_g is −1.8 V in stress. T = 25 °C

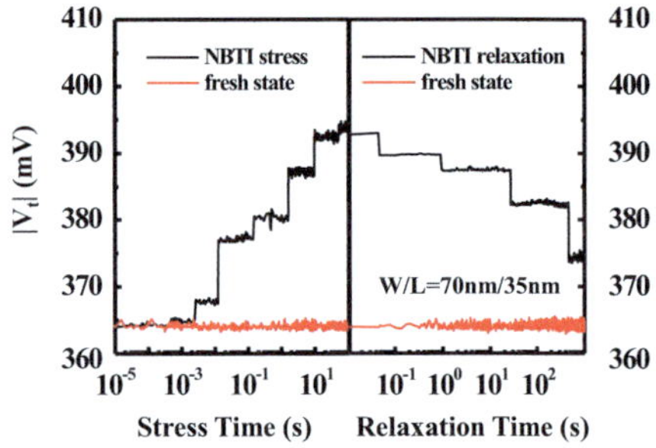

Fig. 3.2 Stepwise V_t evolutions in NBTI stress and relaxation in a small-area (W/L = 70 nm/35 nm) high-k pMOSFET. V_g is −1.8 V in stress. T = 25 °C. The abrupt V_t shifts represent single-charge trapping and detrapping

Typical V_t evolutions in a large-area high-k gate dielectric pMOSFET in BTI stress and relaxation are shown in Fig. 3.1. The V_t degradation and recovery start from a microsecond range.

Note that conventional measurement (e.g., by Agilent 4156), which usually takes a few seconds between stress and recovery transitions, is unable to catch an initial transient in a μs to ms range and may significantly underestimate the magnitude of a transient effect. Owing to recent improvements in measurement techniques [6–8], a measurement delay can be reduced to μs (e.g., by Agilent B1500) to avoid information missing during a switching transient.

In contrast to large-area devices, we found that BTI-induced V_t degradation and recovery in nanometer transistors proceed in discrete steps [8–12] due to augmentation of single-charge effects in scaled devices. Example of V_t evolutions in a small-area device (W/L = 70 nm/35 nm) is shown in Fig. 3.2. In the figure, each abrupt V_t change in stress/recovery V_t traces is caused by single-charge creation/detrapping in gate dielectrics. Due to the discrete nature of V_t evolutions, we are able to measure individual charge creation/detrapping times and the magnitudes of single-charge-induced V_t shifts. Statistical characterization of BTI traps in nanoscale devices helps gain insight into mechanisms in BTI stress and recovery as well as trap characteristics such as trap density, trap energy, and spatial distributions and activation energy in trapped charge emission. Furthermore, ΔV_t evolution traces in BTI recovery are reproducible by repeated trap refill and trapped charge emission. Figure 3.3 shows measured emission time distribution by charging and discharging the same trap 105 times. An exponential distribution $\exp(-t/\tau_e)$

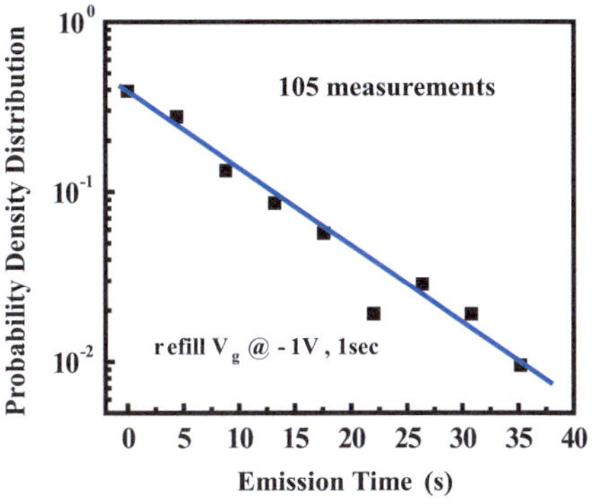

Fig. 3.3 Trapped hole emission time probability distribution in an NBTI stressed sample. The distribution is obtained by repeatedly charging and discharging the same trap 105 times. The trap refill condition is V_g at -1.0 V for 1 s

is shown, where t is an actual charge emission time. A characteristic emission time (τ_e) of the trap is then obtained by taking an average of all measured emission times. By taking advantage of the reproducible feature in BTI recovery, we are able to characterize the temperature and the electric field dependence of individual trapped charge emissions.

Sections 3.2–3.4 are focused on the statistical characterization of trapped charge emissions in BTI recovery. BTI and RTN amplitudes are compared in Sect. 3.5. A discussion on two-stage V_t degradation in BTI stress is given in Sect. 3.6.

3.2 Individual Trapped Charge Emissions in BTI Relaxation

We characterize BTI recovery in high-k (HfSiON) gate dielectric and metal gate MOSFETs. The devices have a gate length of 35 nm, a gate width of 70 nm, and an effective oxide thickness of $\sim$1.0 nm. Schematic diagram for BTI recovery transient measurement is shown in Fig. 3.4a. In NBTI, pMOSFETs are stressed at $V_{g,\text{stress}} = -1.8$ V for 100 s. The recovery characterization scheme is similar to [13], i.e., in a relaxation–measurement–relaxation sequence, as shown in Fig. 3.4b. Both stress and recovery are performed at room temperature. In measurement phase, the drain voltage $V_{d,\text{meas}}$ is -0.05 V and the gate voltage $V_{g,\text{meas}}$ is chosen such that a pre-stress drain current is $\sim$500 nA. Drain current variations (ΔI_d) are recorded using Agilent B1500 with a switch delay time less than 1 μs. A corresponding ΔV_t is obtained from a measured ΔI_d divided by a transconductance (g_m).

To check on Si surface trap creation in BTI stress, we monitor transconductance and subthreshold swing (S) degradations during stress. Pre-stress and post-stress subthreshold $I_d - V_g$ are shown in Fig. 3.5. An almost parallel shift is noted, suggesting that V_t degradation is mainly caused by trapped charge creation in the bulk of gate dielectrics rather than surface traps. Both S and g_m degradations are less than 5% after the stress. We also compare S and g_m before and after 1,000 s recovery and they are almost identical. For simplicity, a constant g_m is used when converting a ΔI_d into a ΔV_t. Figure 3.6 shows example of ΔI_d and V_t traces in BTI

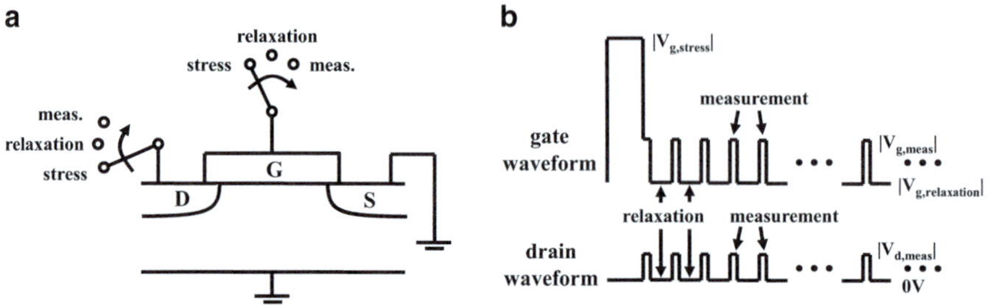

Fig. 3.4 (**a**) Schematic diagram for BTI recovery transient characterization. (**b**) The waveforms applied to the gate and the drain in stress and in recovery phases

Fig. 3.5 Log(I_d) versus V_g plots before and after 100 s NBTI stress at $V_g = -1.8$ V in a pMOSFET

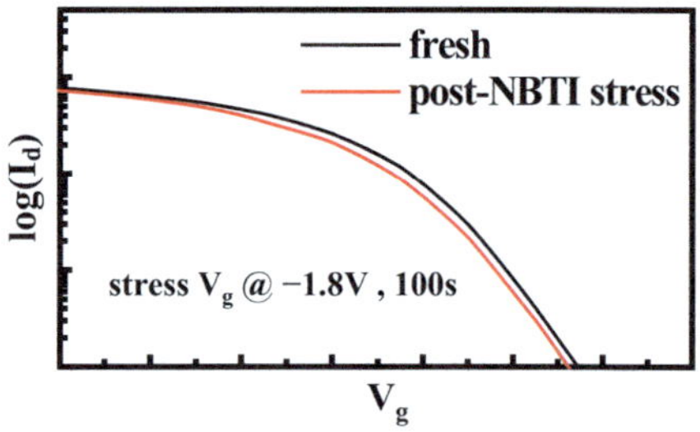

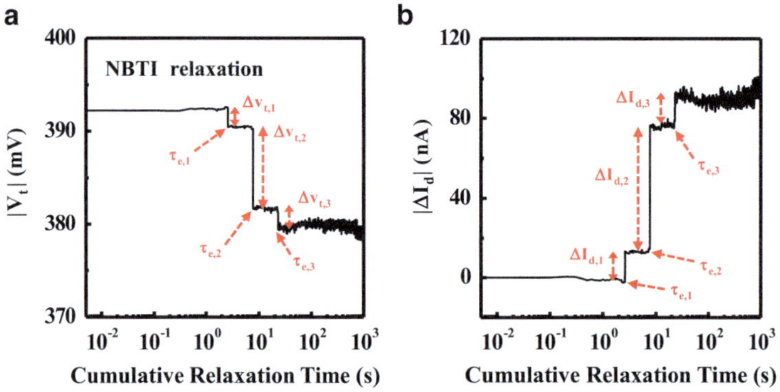

Fig. 3.6 (**a**) Example V_t trace in NBTI relaxation. $\tau_{e,1}$, $\tau_{e,2}$, and $\tau_{e,3}$ are the first, the second, and the third trapped hole emission times, respectively. $\Delta v_{t,i}$ ($i = 1, 2, 3$) represents a single-emitted charge-induced threshold voltage shift. (**b**) Corresponding ΔI_d trace in NBTI relaxation

recovery. A small letter ($\Delta v_{t,i}$) denotes a single-emitted charge-induced V_t shift, where *i* denotes an emission sequence number. A capital letter (ΔV_t) is a total V_t shift after relaxation. In nanoscale MOSFETs, nonuniform 3D electrostatics and the discreteness and the randomness of substrate dopants determine current percolation paths in a channel. Thus, each trapped charge has specific Δv_t amplitude depending on its position in a channel. In repeated recovery measurements, one can use voltage step heights to discern individual trapped charges.

3.3 Statistical Characteristics of BTI Trapped Charges

3.3.1 Single-Charge-Induced Δv_t Amplitudes

We measure and record the magnitudes of single-charge-induced Δv_t in BTI stress/recovery V_t traces in 170 pMOSFETs. RTN signals are not counted. The magnitude distributions of the Δv_t are plotted in Fig. 3.7. The measurement resolution is about 1 mV. Voltage steps with Δv_t less than 1 mV are not recorded. The collected Δv_t from stress traces and from recovery traces have a similar distribution, characterized by an exponential function $f(|\Delta v_t|) = \exp(-|\Delta v_t|/\sigma_{amp})/\sigma_{amp}$ with a σ_{amp} of 3.3 mV. A straight line with a slope of 3.3 mV is drawn to serve as a reference. The exponential function is an empirical formula. The origin and the dispersion of the Δv_t have been studied thoroughly. In such small devices, single-charge-induced Δv_t cannot be estimated from its distance to a gate electrode by using a 1D capacitance equation $C = \varepsilon/d$ because of a strong random dopant-induced current percolation effect. The exponential distribution is realized due to the percolation effect [14–17]. A 3D atomistic numerical device simulation shows a similar Δv_t probability function [15].

3.3.2 Trapped Charge Emission Time Distribution

Individual trapped charge emission times are clearly defined in recovery V_t traces, for example, $\tau_{e,1}$, $\tau_{e,2}$, and $\tau_{e,3}$ in Fig. 3.6. We collect the first three emitted charge characteristic times ($\tau_{e,i}$, $i = 1, 2, 3$) from about 170 devices. The emission times scatter over several decades of time. The probability density functions (PDFs) of the $\log(\tau_{e,i})$, $i = 1, 2, 3$, are shown in Fig. 3.8 [11]. Figure 3.8a, b refers to NBTI and PBTI, respectively. The mean ($\langle\log(\tau_{e,i})\rangle$) and the standard deviation of the distributions are indicated in the figure. The dots are measurement data and the curves are calculated from a Monte Carlo simulation. The Monte Carlo simulation

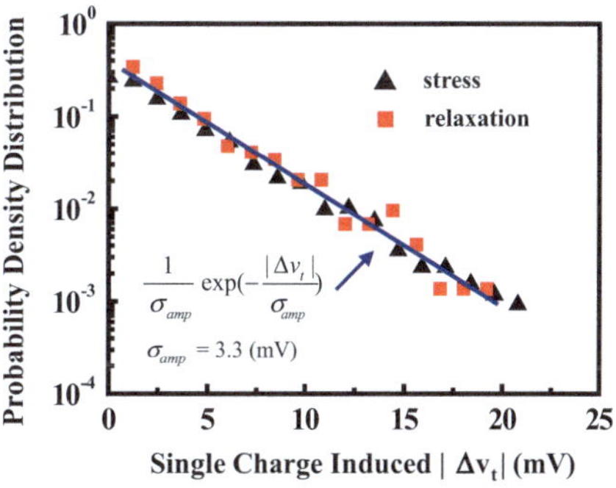

Fig. 3.7 The magnitude distributions of single-charge-induced Δv_t collected from NBTI stress and recovery V_t traces in 170 pMOSFETs. The *solid line* is drawn as a reference

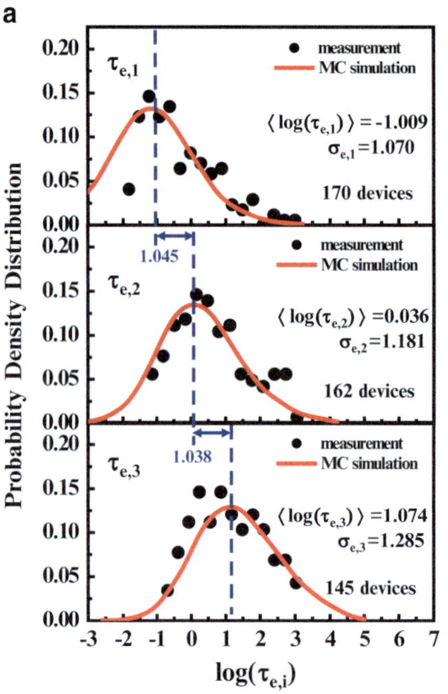

Fig. 3.8 The probability density distributions of the first three trapped charge emission times in BTI relaxation. $\tau_{e,1}$, $\tau_{e,2}$ and $\tau_{e,3}$ are the first, the second and the third trapped charge emission times, respectively. The mean ($\langle\log(\tau_{e,i})\rangle$) and the standard deviation ($\sigma_{e,i}$) of the distributions are indicated in the figure. The *dots* represent measurement results and the *solid lines* are from Monte Carlo simulation (Sect. 3.3.5). (**a**) Trapped hole emissions in NBTI relaxation in ~170 pMOSFETs. (**b**) Trapped electron emissions in PBTI relaxation in ~77 nMOSFETs

will be described in Sect. 3.3.5. The relationship between the $\langle\log(\tau_{e,i})\rangle$ ($i = 1$, 2, 3) is identified. The mean increases with a sequence number i approximately by the same amount, for example, $\langle\log(\tau_{e,i+1})\rangle - \langle\log(\tau_{e,i})\rangle \approx 1.04$ in NBTI-stressed pMOSFETs.

The histogram of the measured charge emission times is plotted in a $\log(\tau_e)$ scale in Fig. 3.9 [11]. A rather uniform distribution from 10^{-2} to 10^3 s is obtained, implying a log(t) dependence of a recovery ΔV_t in a large-area device. The result is consistent with Fig. 3.1. Note that a recovery ΔV_t saturates at a longer time (Sect. 3.4). This means that the distribution in Fig. 3.9 should fall off at a longer relaxation time which is out of our measurement period.

3.3.3 BTI Recovery and Trapped Charge Emission Model

Regarding BTI recovery mechanisms, various models were proposed in the past. Hydrogen back diffusion in the framework of the reaction–diffusion (RD) model was first proposed to explain an NBTI recovery phenomenon [18]. A ΔV_t relaxation

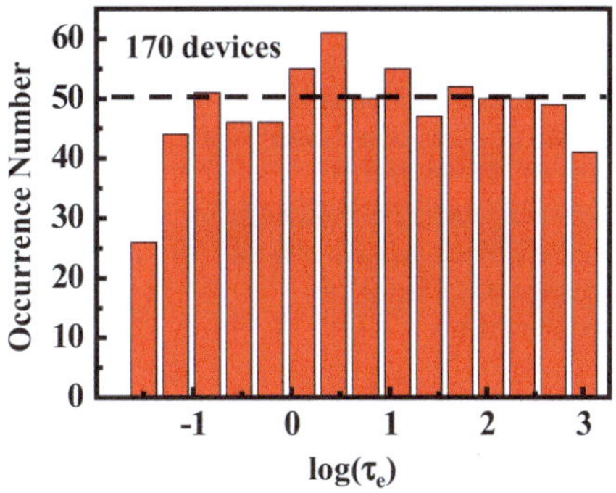

Fig. 3.9 Histogram of trapped hole emission times in a log(τ_e) scale collected from 170 pMOSFETs. The occurrence number distribution is rather uniform in a measurement period from 10^{-2} to 10^3 s

of the form of $\Delta V_t(t) = V_t(0)(1 - \sqrt{t/2t_0}/\sqrt{1+t/t_0})$ is anticipated from the back diffusion theory. We used a fast transient technique to characterize NBTI recovery and found a log(t) dependence of ΔV_t on relaxation time in a time span from 10^{-3} to 10^1 s [8–10]. Furthermore, we characterized individual trapped charge emissions in small-area devices and proposed a thermally assisted charge tunnel detrapping (ThAT) model for a recoverable component of the ΔV_t in relaxation. The log(t) dependence was also reported in [3] where the authors proposed a dispersive transport model within the RD framework to overcome the apparent deficiency of the RD model with respect to relaxation time dependence. Alam et al. ascribed a log(t) recovery transient to a fast charge detrapping process on top of slower N_{it} re-passivation as encapsulated by the RD model [4, 5]. More recently, a time-dependent defect spectroscopy method was used to investigate NBTI recovery, and the authors concluded that NBTI recovery is due to thermally assisted trapped hole emission and no diffusion process is involved [12].

In NBTI relaxation, there are three possible paths for trapped hole emission [8], namely, (a) Frenkel–Poole emission, (b) ThAT to the gate electrode, and (c) ThAT to the Si substrate. To identify a major path in an emission process, we characterize the dependence of a trapped hole emission time on recovery gate voltage and temperature. As pointed out earlier, a characteristic hole emission time can be measured by repeatedly charging and discharging a trap. An average of 15 repeated measurements at each V_g and temperature is taken. To avoid new trap creation in trap refilling, a small refill $|V_g|$ of 1 V is used and, the refill time is 1 s. Figure 3.10 shows the temperature dependence of $\langle\tau_e\rangle$ of two traps. The extracted activation energy (E_a) from the Arrhenius plot is about 0.52 and 0.56 eV. Figure 3.11 shows the dependence of $\langle\tau_e\rangle$ on a recovery $|V_g|$. The measured $\langle\tau_e\rangle$ is nearly constant for $|V_g| < |V_t|$ and increases with a recovery $|V_g|$ for $|V_g| > |V_t|$.

The detrapping path (a) is ruled out, because the E_a in Frenkel–Poole emission should be about the trap energy (>2 eV), and the measured E_a is only ~0.54 eV. Path (b) is also excluded, because a more negative recovery V_g would accelerate charge emission, giving a shorter charge emission time. The measured trend of $\langle\tau_c\rangle$ versus V_g is just opposite. As a result, (c) is identified as the path of trapped charge

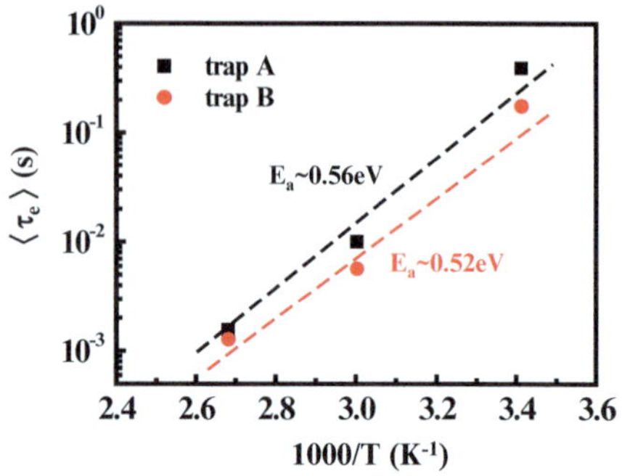

Fig. 3.10 Dependence of a trapped hole emission time $\langle \tau_e \rangle$ on recovery temperature in two devices (trap A and trap B). The extracted activation energy is 0.52 and 0.56 eV. Each data point is an average of 15 repeated measurements

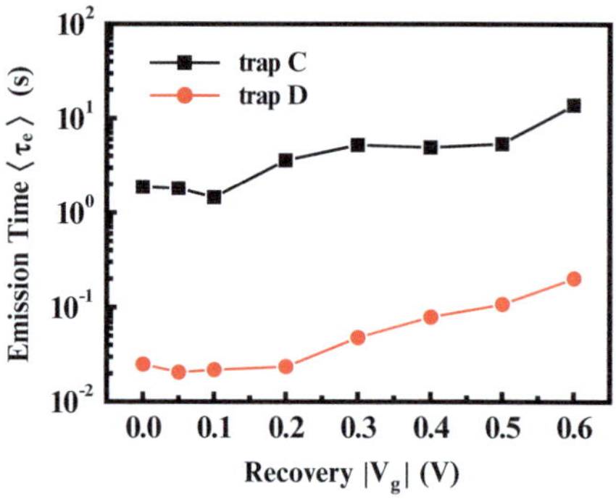

Fig. 3.11 Dependence of a trapped hole emission time $\langle \tau_e \rangle$ on recovery gate voltage in two devices (trap C and trap D)

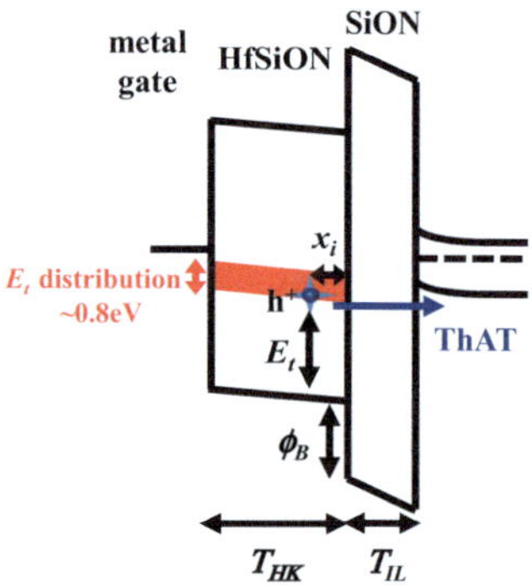

Fig. 3.12 Schematic representation of the band diagram of a high-k/metal gate pMOSFET in relaxation. x_i and E_t represent a trapped charge position and energy

emission in relaxation. Moreover, the temperature dependence suggests the role of thermal process in trapped charge tunnel emission possibly due to multi-phonon absorption. An analytical trapped charge emission time model based on the ThAT is therefore developed. In the following, trapped charge emission characteristics are analyzed based on the ThAT model. Trapped charge energy and spatial distributions and its activation energy distribution in the ThAT model are extracted. An energy band diagram of a high-k pMOSFET in NBTI relaxation is illustrated in Fig. 3.12 [11]. According to the WKB approximation, a trapped hole emission time is formulated as [8]

$$\tau_{e,i} = \tau_0 \exp(\frac{E_a}{kT}) \exp(\alpha_{IL} T_{IL}) \exp(\alpha_k x_i) \tag{3.1}$$

and

$$\tau_0^{-1} = N_v(1 - f_v) v_{th} \sigma_0 \tag{3.1a}$$

$$\alpha_{IL} = \frac{2\sqrt{2m_{IL}{}^* q(E_t + \phi_B)}}{\hbar} \tag{3.1b}$$

$$\alpha_k = \frac{2\sqrt{2m_{HK}{}^* q E_t}}{\hbar} \tag{3.1c}$$

where the pre-factor τ_0 is expressed in Eq. (3.1a). N_v is the effective density of state in Si valence band, f_v is a valence-band hole occupation probability (Fermi–Dirac distribution) in Si substrate at an energy aligned to the trapped charge, $N_v(1-f_v)$ is the amount of available states in Si substrate for out-tunneling holes from high-k traps. σ_0 and E_a are the trap cross section and activation energy. T_{IL} is an interfacial layer thickness, x_i denotes a trapped charge distance to the HK/IL interface, and E_t is a trapped charge energy. Other variables have their usual definitions. The hole tunneling mass used in this work is $m_{IL}{}^* = 0.41 m_0$ [19] and $m_{HK}{}^* = 0.18\, m_0$ [20]. The hole occupation probability (f_v) is a function of V_g in recovery. A larger recovery $|V_g|$ gives a higher channel hole concentration and thus a larger f_v. In addition, an effective tunneling barrier for trapped hole emission increases with a larger $|V_g|$. These two factors result in an increased hole emission time at a larger recovery $|V_g|$. As the recovery V_g reduces below the threshold voltage, f_v approaches zero, and the hole emission time becomes almost independent of V_g, as shown in Fig. 3.11. Due to the dispersion of E_a, E_t, and x_i, trapped holes hop out of the gate dielectric via quantum tunneling sequentially and in a dispersive manner.

3.3.4 Trapped Charge Spatial and Energy Distributions

According to Eq. (3.1), a tunneling front moves in a speed of $d = 2.3/\alpha_k$ per decade of time. The density of removable trapped charges (N_t) in relaxation therefore can be extracted from the emission occurrence number versus log(τ_e) in Fig. 3.9 as follows:

$$\begin{aligned} N_t &= \frac{\text{no. of emitted charges/device/decade}}{W \times L \times d} \\ &= \frac{\alpha_k \times (\text{no. of emitted charges/device/decade})}{2.3 \times W \times L} \end{aligned} \tag{3.2}$$

Since the number of emitted charges exhibits a uniform distribution in each decade of time approximately in Fig. 3.9, we obtain a constant removable trapped charge density N_t in space. With respect to a trapped charge energy distribution, we calculated a voltage drop across an interfacial oxide in NBTI stress by a two-dimensional numerical device simulation [21]. The purpose of the simulation is to estimate an upper bound of NBTI trapped hole energy distribution in the high-k layer. The calculated voltage drop across the IL is about 0.8 V at a stress V_g of −1.8 V. Thus, we assume that removable trapped holes are uniformly distributed in an energy range of 0–0.8 eV above the Si valence-band edge, corresponding to an E_t value of 2.7–3.5 eV with respect to the valence-band edge of the HfSiON. This assumption is supported partly by a charge pumping measurement result [22]. The calculated value of α_k is from 7.2 to 8.1 nm^{-1} in the range of E_t. For simplification, we used an average $\bar{\alpha}_k$ (=7.65 nm^{-1}) in Eq. (3.2) and obtained an N_t of $1.3 \times 10^{18}\ \text{cm}^{-3}$. An average distance ($\Delta x$) between two adjacent trapped charges in the gate-to-substrate direction is about $\Delta x = 1/WLN_t \sim 0.32$ nm [11].

In addition, the ratio of the emission times of two consecutive emitted trapped holes is

$$\begin{aligned}\langle \log(\tau_{e,i+1})\rangle - \langle \log(\tau_{e,i})\rangle &= \frac{1}{2.3} \times [\bar{\alpha}_k \cdot (\langle x_{i+1}\rangle - \langle x_i \rangle)] \\ &\equiv \frac{1}{2.3} \times (\bar{\alpha}_k \cdot \Delta x) \end{aligned} \tag{3.3}$$

Equation (3.3) shows that the mean of the $\log(\tau_{e,i})$ increases with i by an amount of $\alpha_k \Delta x/2.3 = 1.06$, without regard to activation energy. Equation (3.3) is consistent with the measurement result in Fig. 3.8a.

3.3.5 Activation Energy Distribution

First, we need to clarify the role of an electric field in trapped charge emission in BTI relaxation. Our emission time model [Eq. (3.1)] does not have explicit electric field dependence. In contrary, an RTN emission time model or an NBTI model in [12] shows exponential electric field dependence, i.e., exp ($-xqF/kT$). The major difference is in that removable trapped holes in our model are assumed to have energy in a range above the valence-band edge in relaxation (Fig. 3.12), while trapped charges in an RTN model or in [12] are within the silicon band gap. Our assumption is reasonable because of a large voltage drop across the IL in BTI stress. The measurement results of the electric field dependence in Fig. 3.11 and in literature [8, 12] also do not support exponential electric field dependence in NBTI recovery. Since an electric field is a secondary effect in NBTI recovery, a nonuniform electric field effect due to 3D electrostatics and discrete dopant charges is unimportant. The wide spread of the trapped charge emission times is therefore believed mainly due to (1) the dispersion of activation energy in emission resulting

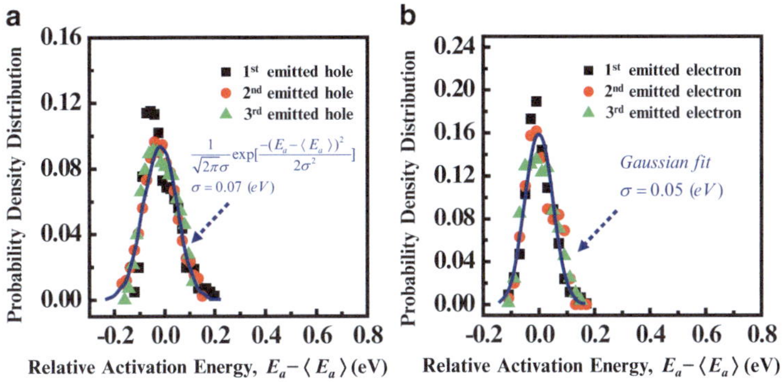

Fig. 3.13 Relative activation energy ($E_a - \langle E_a \rangle$) distributions extracted from the $\tau_{e,1}$, $\tau_{e,2}$, and $\tau_{e,3}$, respectively, in (**a**) NBTI in pMOSFETs and (**b**) PBTI in nMOSFETs. The *solid lines* represent a Gaussian-distribution fit

from different bond states and a local structural strain and (2) the spread of trapped charge energy and a trap depth. From Eq. (3.1), E_a can be expressed as

$$E_a = kT[2.3\log(\tau_{e,i}) - 2.3\log(\tau_0) - \alpha_{IL}T_{IL} - \alpha_k x_i]. \tag{3.4}$$

Activation energy, trap energy, and trap depth are three independent variables. For given distributions of τ_e, E_t, and x, it is a mathematical problem to derive an E_a distribution. For simplification, average values of α_k and x_i are used in Eq. (3.4) to extract an E_a distribution from the measured $\tau_{e,i}$. The average distance between two consecutive emitted charges is 0.32 nm. Relative E_a distributions extracted from $\tau_{e,1}$, $\tau_{e,2}$, and $\tau_{e,3}$ are shown in Fig. 3.13 for NBTI (a) and PBTI (b), respectively. A reasonably good match between them is obtained. The good match ascertains Eq. (3.1) and implies that activation energy is the cause of the spread of the τ_e. It should be remarked that the distortion of the E_a distribution of $i = 1$ is understood because our recovery measurement starts with a time delay of 5 ms. Some emitted charges with very short τ_e (<5 ms) are not counted. In Fig. 3.13, the extracted E_a distributions can be approximated by a Gaussian distribution [11]. An appropriate τ_0 is chosen such that the mean of an E_a distribution in pMOSFETs is about 0.54 eV to be consistent with Fig. 3.10. The solid lines in Fig. 3.13 represent a Gaussian distribution fit with a standard deviation of 0.07 eV in pMOSFETs and 0.05 eV in nMOSFETs.

To examine the validity of the E_a extraction, we recalculate the τ_e's distributions based on an extracted E_a distribution by a Monte Carlo method. In the Monte Carlo procedure, the number of removable trapped charges in each device is selected according to a Poisson distribution [14] with an average number of $N_t WLT_{HK}$, where T_{HK} is the thickness of a high-k dielectric. The use of a Poisson distribution here is an approximation, and its validity has been discussed in [23]. Then, removable trapped charges are randomly placed in the high-k layer. Trapped charges in the IL are not considered since our measurement period (5 ms to 10^3 s) does not match the range of IL trap time constants. For each trapped charge, an E_t is randomly selected in a range from 2.7 to 3.5 eV and an E_a is selected according to the distribution in Fig. 3.13. With a trapped charge location, energy, and activation energy, an emission time is calculated according to Eq. (3.1). An emission sequence number is then assigned to each trapped charge according to its calculated emission time. A trapped charge with the shortest τ_e has $i = 1$, the second shortest one has $i = 2$, and so on. The Monte Carlo simulated $\tau_{e,i}$ ($i = 1, 2, 3$) distributions (solid lines) are plotted in Fig. 3.8. A reasonably good agreement between simulation and measurement is obtained. The broadening of the $\log(\tau_{e,i})$ with a sequence number i in Fig. 3.8 can be partly explained as follows. We rearrange the terms in Eq. (3.4) and obtain the following equation:

$$\log(\tau_{e,i}) = \frac{1}{2.3}\left[\frac{E_a}{kT} + 2.3\log(\tau_0) + \alpha_{IL}T_{IL} + \alpha_k x_i\right]. \tag{3.5}$$

As i increases, the tunneling distance x_i is larger and the variance of the term ($\alpha_k x_i$) in the right-hand side increases and so does the variance of $\log(\tau_{e,i})$.

Although our model can reproduce the measured emission time distributions in Fig. 3.8 well, we do not exclude the possibility of other combinations of the trapped charge distributions which may still come to the data in Fig. 3.8, for example, a fixed trap depth at the IL/high-k interface and a broader activation energy distribution. But considering that pure oxide and high-k pMOSFETs both have a similar log(t) recovery characteristic, we believe that NBTI created traps are more likely to distribute uniformly (or very broadly) in a gate dielectric. Finally, we would like to remark that the above conclusions about BTI trap characteristics such as uniform trap spatial and energy distributions and a Gaussian-like activation energy distribution are reached based on a measurement dataset of $\Delta v_t > 1$ mV. Our conclusions should be applied to BTI traps with $\Delta v_t < 1$ mV as well. The reason is that the magnitude of Δv_t is dependent on a percolation path in the channel and is nothing to do with trap behavior.

3.3.6 BTI Relaxation in nMOSFETs

Generally speaking, PBTI in nMOSFETs exhibits similar features as NBTI in pMOSFETs. Single-trapped electron-induced Δv_t follows an exponential distribution, too. Similarly to pMOSFETs, the PDFs of the first three trapped-electron emission times in about 77 nMOSFETs are shown in Fig. 3.8b. Note that the average spacing between two consecutive electron emission times, i.e., $\langle\log(\tau_{e,i+1})\rangle - \langle\log(\tau_{e,i})\rangle$, is about 0.6 in nMOSFETs. The smaller spacing in nMOSFETs indicates a higher electron trap density after PBTI stress. Following the same extraction procedure, we obtain relative activation energy distribution in trapped-electron emission in Fig. 3.13b. The mean and the standard deviation of the activation energy distribution are smaller than those in NBTI recovery.

3.4 Recovery ΔV_t Distribution and Its Temporal Evolutions

BTI recovery has been exploited in several circuit techniques to alleviate BTI severity in memory and logic circuits [24]. To enlarge a design window, the integration of BTI statistical characteristics into a circuit simulation is needed in modern CMOS circuit design. In this section, we characterize an overall BTI recovery-induced ΔV_t distribution in a large number of small-area devices. A statistical ΔV_t model based on the ThAT combined with single-charge-induced Δv_t distribution is used to calculate an entire ΔV_t distribution and its temporal evolutions in BTI recovery [11]. We measure threshold voltage shifts at different recovery times in a number of BTI stressed pMOSFETs. The number of emitted holes and a total threshold voltage shift (ΔV_t) in each device are recorded. Figure 3.14 shows the measurement results at a recovery time of 0.1 s, 10 s, and 1,000 s, respectively. The y-axis is a total ΔV_t in recovery and the x-axis is the number of emitted holes. Each data point represents a device. A straight line with a slope of 3.3 mV, i.e., an average single-charge-induced V_t shift, is drawn in the figure as a reference. The measurement data scatter along the lines. The ΔV_t and the number distributions broaden with recovery time. An average of recovery V_t traces in 170 devices is plotted in Fig. 3.15 showing a log(t) dependence of ΔV_t. Figure 3.16 shows measured ΔV_t evolutions in large-area devices at two different stress V_g. The ΔV_t obeys a log(t) dependence in an initial period of relaxation and then gradually saturates. For a larger amount of stress (a stress $|V_g|$ of 1.7 V), the recovery $\Delta V_t(t)$ possesses a larger slope, and the log(t) dependence persists in a longer period of relaxation time.

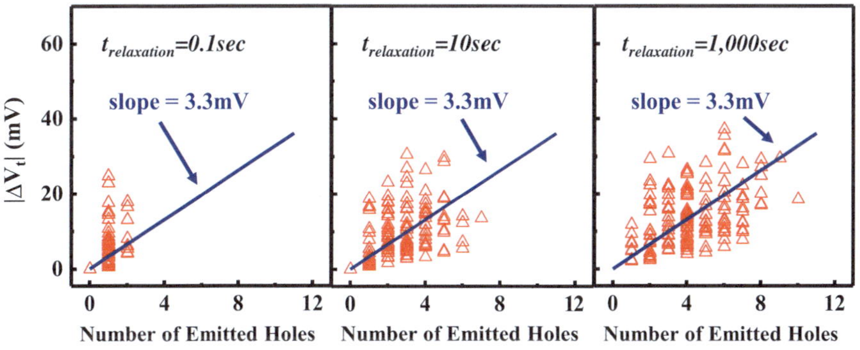

Fig. 3.14 A total V_t shift (ΔV_t) versus number of emitted trapped holes in a device at a relaxation time of 0.1, 10, and 1,000 s. Each data point represents a device. A *straight line* with a slope of 3.3 mV is drawn as a reference

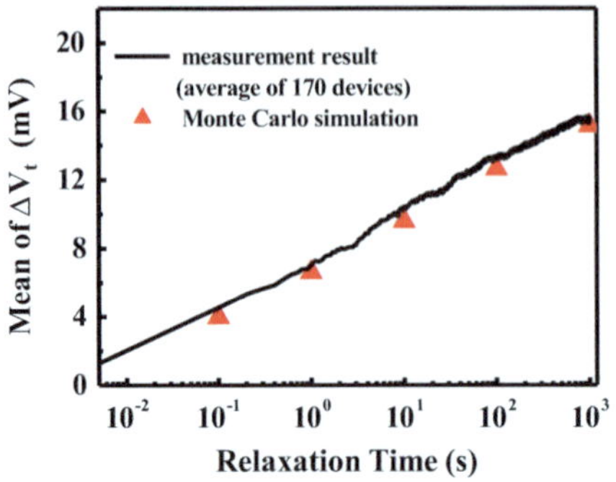

Fig. 3.15 The evolution of the ΔV_t with NBTI relaxation time. The *solid line* represents an average of measured recovery ΔV_t traces in 170 small-area (W/L = 70 nm/35 nm) pMOSFETs. The recovery measurement in small area devices has a 5 ms delay time. The symbols are the mean of Monte Carlo simulated ΔV_t distributions. A logarithmic time dependence of the ΔV_t is obtained in the measurement period

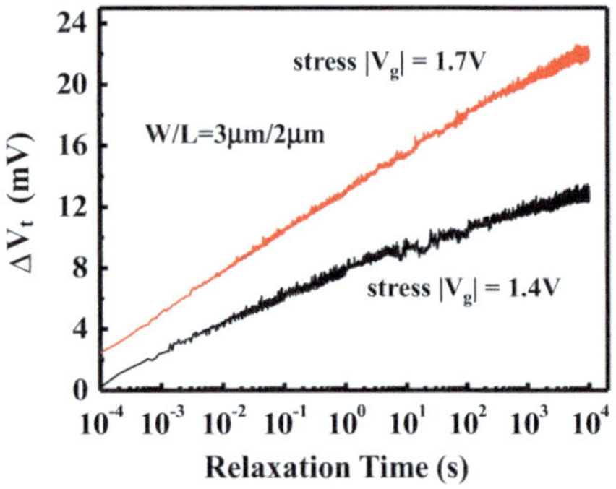

Fig. 3.16 ΔV_t evolutions with relaxation time in large-area (W/L = 3 μm/2 μm) pMOSFETs with two different stress V_g (−1.4 and −1.7 V). The ΔV_t gradually saturates at a relaxation time of 10^4 s at a stress V_g of −1.4 V

A Monte Carlo ΔV_t model based on the ThAT and the extracted trapped charge spatial, energetic, and activation energy distributions was developed. The simulation flowchart is shown in Fig. 3.17 [11]. At a recovery time t_r, the number of emitted charges (N) is computed by counting all the charges with $\tau_{e,i}$ less than t_r. For each emitted charge, a Δv_t is randomly selected based on the distribution $f(|\Delta v_t|) = \exp(-|\Delta v_t|/\sigma_{amp})/\sigma_{amp}$ with $\sigma_{amp} = 3.3$ mV. A total ΔV_t is then calculated as $\Delta V_t = \sum_{i=1}^{N} \Delta v_{t,\,i}$. In total, 5×10^5 devices are simulated. The simulated and measured ΔV_t distributions are shown in Fig. 3.18 at a recovery time of $t_r = 0.1$, 10, and 1,000 s. Our model is in good agreement with measurement. The mean and the variance of the modeled and the measured ΔV_t distributions versus relaxation time are shown in Figs. 3.15 and 3.19, respectively [11]. Our model can reproduce the log(t) dependence in large-area devices as well as an overall ΔV_t distribution and its temporal evolutions in small-area devices.

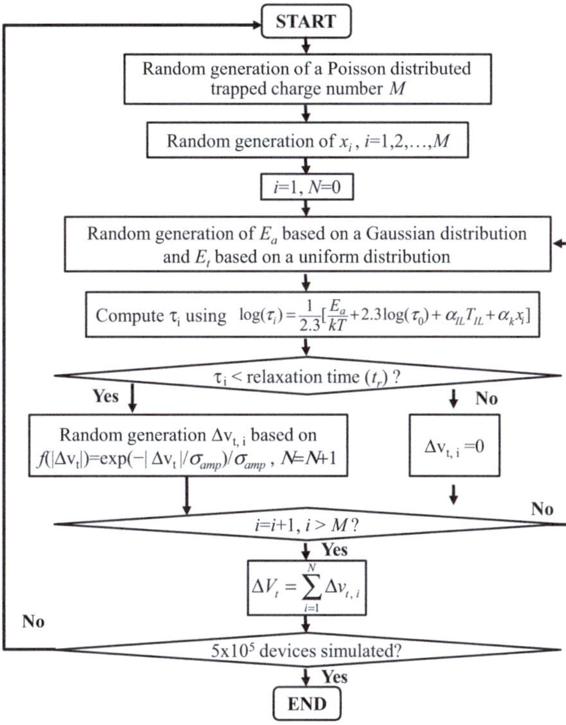

Fig. 3.17 Simulation flowchart of a Monte Carlo based ΔV_t distribution model for NBTI relaxation

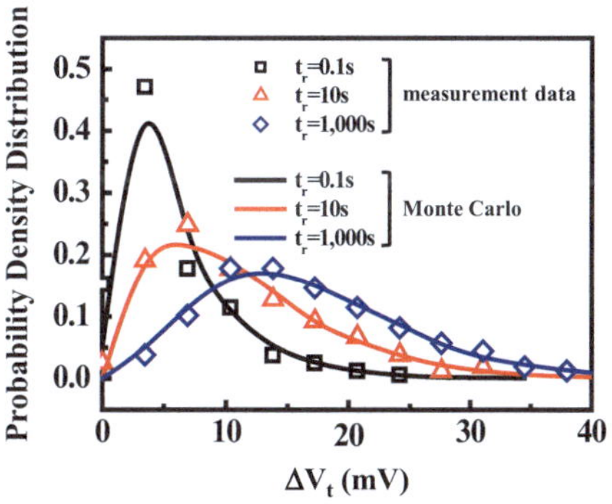

Fig. 3.18 The probability distributions of an NBTI recovery induced ΔV_t in 70 nm × 35 nm pMOSFETs from measurement (*symbols*) and from a Monte Carlo simulation (*lines*). ~170 devices were measured. The recovery time is 0.1, 10, and 1,000 s

In short, the log(t) behavior in BTI relaxation in large-area devices is a result of a uniform spatial distribution of trapped charges created by BTI stress while E_a and E_t distributions affect a ΔV_t distribution and its temporal evolutions in small-area devices.

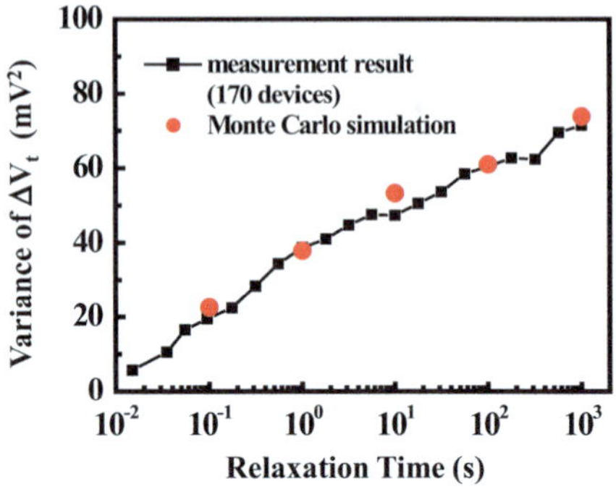

Fig. 3.19 Time evolutions of the variance of an NBTI recovery induced ΔV_t distribution in 70 nm × 35 nm pMOSFETs

3.5 Comparison of BTI and RTN Amplitudes

Single-charge trapping/detrapping-induced threshold voltage fluctuations in RTN and BTI have been widely explored. Similar phenomenology and connections have been found in both of them. In this section, we compare the amplitudes of RTN and BTI single-charge-induced Δv_t. This study gives clues about RTN and BTI trap generation. The complementary cumulative probability distributions of measured BTI and RTN amplitudes are plotted in Fig. 3.20a [16]. The two-level RTN amplitudes are measured in fresh devices. BTI apparently has a broader amplitude distribution ($\sigma_{amp} = 3.34$ mV) than RTN ($\sigma_{amp} = 1.12$ mV). Our findings suggest that BTI has a larger impact on CMOS reliability than RTN. Moreover, we compare RTN amplitudes in pre-stress and post-stress devices. The result is shown in Fig. 3.20b. The post-stress one also has a significantly larger Δv_t tail [16].

To explore the physics that BTI stress-created charges have larger amplitudes, we performed a 3D atomistic Monte Carlo simulation for both RTN and NBTI. In RTN simulation, substrate dopants are randomly and discretely placed in a simulated device, and an RTN trap position is randomly selected in the channel. The random placement of an RTN trap is based on an assumption that RTN traps in fresh devices (e.g., process-induced traps) have a uniform distribution in the channel. This assumption is actually verified by measurement. We extract an RTN trap lateral position in 124 devices by using a method similar to [25, 26] and plot their distribution along the channel in Fig. 3.21 [16]. The trap position distribution is rather uniform. In NBTI simulation, NBTI trapped charge creation is not uniform in space because of nonuniform 3D electrostatics and random and discrete placement of substrate dopants. In order to select a trapped charge position, we need to calculate a relative trap creation probability at each grid point in the surface of the channel during NBTI stress. According to the RD model [27] and assuming that a reaction phase dominates the process, NBTI trap generation rate, in the initial stage of stress (i.e., the trap density N_t is small), can be expressed by

$$\frac{dN_t}{dt} = k_F N_0 \tag{3.6}$$

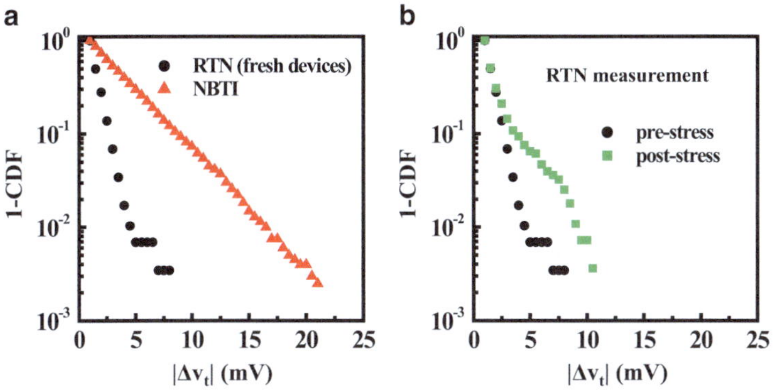

Fig. 3.20 (**a**) Complementary cumulative probability distributions of single-charge NBTI and RTN amplitudes. RTN is measured in fresh devices. (**b**) Complementary cumulative probability distributions of RTN amplitudes in fresh devices and in post-NBTI stress devices. The measured devices have a gate width of 80 nm and a gate length of 30 nm

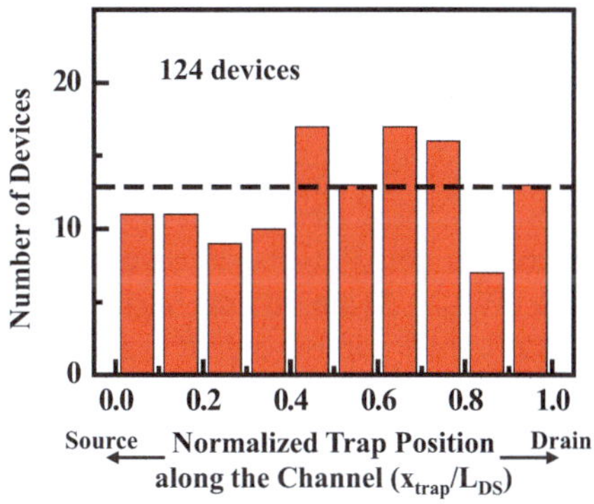

Fig. 3.21 RTN trap position distribution along the channel extracted from 124 devices. x_{trap} is a distance of a trap from the source and L_{DS} denotes a channel length

where N_0 is the total number of Si–H bonds. k_F is the Si–H dissociation rate constant, which is formulated as follows [27]:

$$k_F \propto p \cdot \exp(\frac{F}{F_0}) \quad (3.7)$$

where p is a channel surface hole concentration and F is a local electric field. The p and the F are obtained from a 3D atomistic device simulation. Thus, the relative trap creation probability at each point of the channel can be calculated from the product of p and $\exp(F/F_0)$. Our simulation result shows that the trap creation probability increases with a channel hole concentration [16]. In other words, a trap tends to be created in a high hole density region (i.e., a critical current path) in NBTI stress. In our NBTI simulation, we select a trapped hole position according to the calculated probability distribution, instead of a uniform probability distribution for RTN. The simulated RTN and NBTI amplitude distributions in the same device

Fig. 3.22 Simulated complementary cumulative probability distributions of RTN and NBTI amplitudes in the same dopant arrangement. The simulated device has a gate length of 30 nm and a gate width of 30 nm

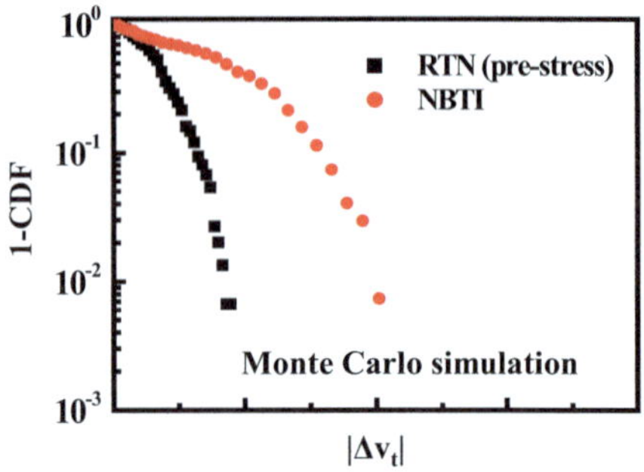

are compared in Fig. 3.22. The y-axis is a complementary cumulative probability distribution. Our simulation indeed confirms that NBTI possesses a larger Δv_t tail. The reason is that an NBTI charge tends to be created in a critical path and thus has a larger influence on a channel current and threshold voltage. A similar argument can be applied to post-stress RTN in Fig. 3.20b. In post-stressed devices, there exist two groups of RTN traps, process-induced traps (initial traps) and stress-created traps. The initial traps have a tight Δv_t distribution, while the stress-created traps have a broader one. The overall distribution in post-stress devices therefore has a larger tail.

It should be mentioned that it is not our intention to directly compare simulation (Fig. 3.22) and measurement results (Fig. 3.20a) because our simulator is not calibrated yet. Besides, to reduce 3D simulation time, we used a smaller device size (W/L = 30 nm/30 nm) in simulation. However, the trend that BTI has a larger Δv_t tail holds without regard to a device size. Our simulation result is different from the result in [15]. The difference is probably in a way to place a BTI trapped charge in simulation.

3.6 Two-Stage V_t Degradation in BTI Stress

Since post-stress high-k CMOS exhibits a large recovery effect from a μs to ms range, a switching delay in a conventional method might lead to significant underestimate of an initial BTI degradation. Throughout this section, a stress–measurement–stress technique is employed to measure BTI-induced degradation by using Agilent B1500.

It has been reported that BTI degradation is strongly influenced by gate dielectric processes [5]. The devices we used in this section have HfSiON/SiON gate dielectrics. The EOT is about 0.9 nm. Evolutions of NBTI-induced V_t degradation at different stress V_g are shown in Fig. 3.23. The V_t degradation initially evolves linearly in a log(t) scale. After a certain stress time, denoted by τ_{corner} in Fig. 3.23, accelerated degradation is observed. The two-stage BTI degradation was reported in [13, 28]. The accelerated degradation was also noticed in literature, whereas the authors attributed the imperfect log-time dependence to a nonuniform trap spatial

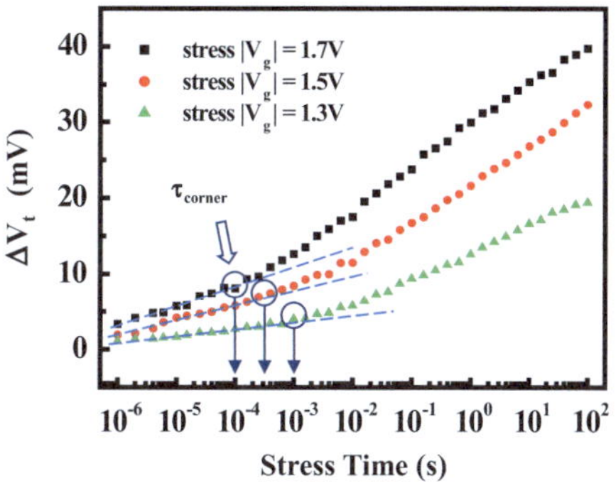

Fig. 3.23 Temporal evolutions of V_t degradation at different NBTI stress $|V_g|$ (1.3, 1.5 and 1.7 V), T = 25 °C. Two-stage NBTI degradation is observed. The transitional time is denoted by τ_{corner}. The degradation before τ_{corner} is referred to as the first stage degradation

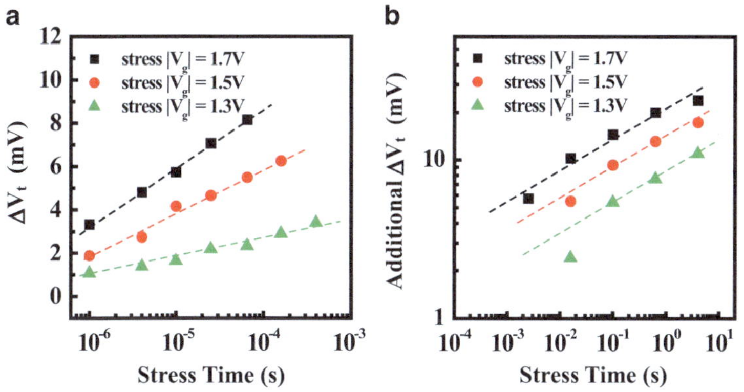

Fig. 3.24 Stress time dependence of NBTI degradation in the first stage (**a**) and in the second stage (**b**). The second-stage degradation is obtained by subtracting extrapolated first-stage degradation from measured ΔV_t

distribution [29]. The V_t degradation versus stress time before τ_{corner} (referred to as "the first stage" hereafter) and after τ_{corner} ("the second stage") are replotted in Fig. 3.24a, b, respectively. The second-stage degradation is obtained by subtracting the extrapolation of the first-stage degradation from the measured ΔV_t. Notably, the first-stage degradation has a log(t) dependence, while the second-stage degradation exhibits power-law time dependence with a power factor of ~0.19 without regard to a stress V_g. Stress temperature effect is also examined in Figs. 3.25 and 3.26. Three points should be mentioned. (1) At a higher stress temperature, the V_t degradation enters the second stage earlier or a smaller τ_{corner} (Fig. 3.26). (2) The first-stage degradation has negative stress temperature dependence. Nevertheless, the second stage shows an opposite trend, a positive temperature effect. A crossover of the V_t degradations at T = 25 °C and 100 °C is noticed in Fig. 3.25, and a larger stress $|V_g|$ results in an earlier crossover [13]. The crossover was also shown in [5]. The opposite temperature dependence implies that the dominant degradation mechanisms in the first and the second stages are different. (3) The degradation is driven into the second stage earlier at a higher stress $|V_g|$. For example, the corner time is around 10^{-3} s for $|V_g| = 1.3$ V and 2×10^{-4} s for $|V_g| = 1.7$ V in Fig. 3.26.

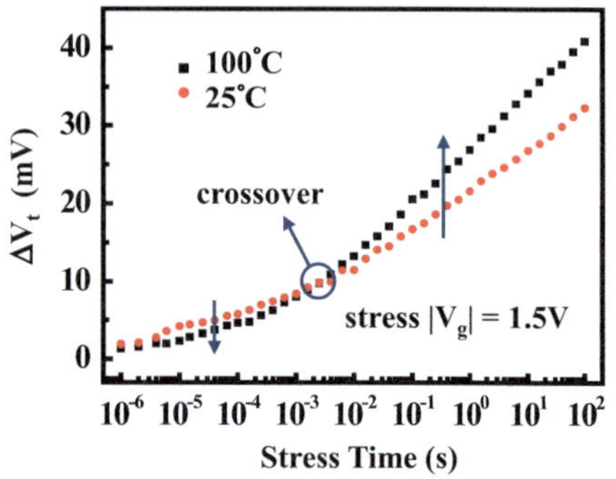

Fig. 3.25 Evolutions of NBTI degradation with stress time at two different stress temperatures, T = 25 and 100 °C. Stress $|V_g| = 1.5$ V. The first-stage degradation has negative temperature dependence, while the second-stage has positive temperature dependence

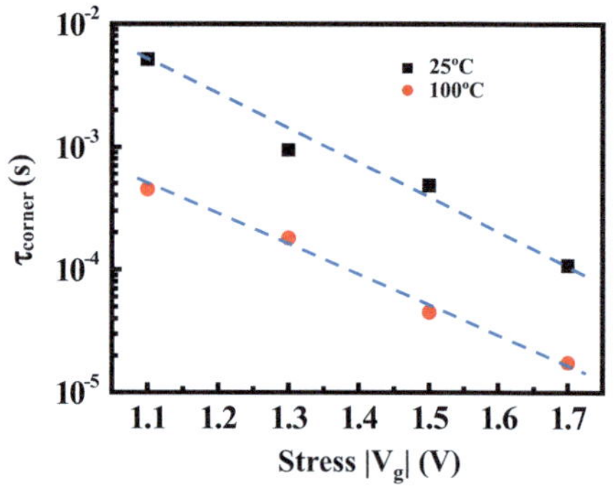

Fig. 3.26 Measured corner time τ_{corner} versus stress $|V_g|$ at T = 25 and 100 °C. The NBTI degradation is driven into the second stage earlier (a smaller τ_{corner}) at a higher stress $|V_g|$ and temperature

The log(t) degradation in the first stage suggests that charging [29] and concomitant discharging [9] of preexisting high-k traps dominate the first-stage V_t degradation. A higher stress $|V_g|$ leads to a larger hole tunneling probability, thus causing more severe V_t degradation. The cause of the negative temperature dependence is speculated as follows. Since high-k charge detrapping rate increases with temperature [9], a higher temperature results in a smaller net charge trapping rate and thus smaller V_t degradation. On the other side, new high-k traps are created during stress. At a certain stress time (the aforementioned "corner time"), newly created high-k trap density reaches a level comparable to or even more than preexisting ones. Charging and discharging of the preexisting traps are then no longer a dominant process. Thus, the V_t degradation is dictated by trap generation which has power-law stress time dependence [27, 30]. Furthermore, larger stress V_g and higher temperatures lead to faster high-k trap generation in the second stage (Figs. 3.23, 3.24, and 3.25) because of larger carrier fluency and energy [30, 31] and thus an accelerated thermochemical reaction for trap creation [27, 32]. As a result, the device V_t degradation is driven into the second stage earlier at higher stress V_g and/or temperature. In Fig. 3.23, the first-stage degradation amounts to about 20% of total degradation in a stress period of 100 s. As compared to our earlier work [13], the ratio of the first-stage degradation decreases possibly because of the reduction of initial high-k traps in this work.

3.7 Summary

A discrete feature in BTI recovery V_t evolutions due to individual trapped charge emissions in small-area high-k MOSFETs is reported. Single-charge emission times and induced V_t shifts are clearly defined. The recovery V_t evolution characteristics are reproducible by repeated trap refill and trapped charge emission. This single-charge characterization approach allows us to gain insight into BTI trap properties and a BTI recovery mechanism. We characterize the electric field and the temperature dependence of trapped charge emissions in BTI relaxation. A thermally assisted tunnel detrapping model based on the measured electric field and temperature dependence was proposed for BTI recovery. Statistical characterization of individual trapped charge emissions in BTI recovery in a large number of nanoscale devices is performed. Trapped charge emission time distributions are measured. BTI trap characteristics such as single-trapped charge-induced V_t shifts, trap density, trap spatial and energy distributions, and trapped charge activation energy in emission are extracted. Based on the extracted BTI trap distributions and the ThAT model, a Monte Carlo model is developed to calculate a statistical distribution of a BTI recovery induced ΔV_t. Our model can reproduce the measurement result of an overall recovery ΔV_t distribution and its time evolutions well.

As compared to RTN, BTI single-charge-induced ΔV_t possesses a larger amplitude distribution tail. The reason is that BTI trapped charges are created more likely in channel current percolation paths according to the RD model and thus have a larger influence on threshold voltage. With respect to BTI stress in high-k MOSFETs, two-stage V_t degradation is noticed. V_t degradation in the two stages has different temperature and stress time dependence. We believe that the first stage is caused by the charging of preexisting high-k traps and the second stage is attributed to new high-k trap creation.

References

1. S. Zafar, Y. H. Kim, V. Narayanan, C. Cabral Jr., V. Paruchuri, B. Doris, J. Stathis, A. Callegari and M. Chudzik, Tech. Dig. VLSI Symp. 2006, p. 23.
2. S. Zafar, A. Kumar, E. Gusev and E. Cartier, IEEE Trans. Device Mater. Reliab. **5**, 45 (2005).
3. B. Kaczer, V. Arkhipov, R. Degraeve, N. Collaert, G. Groeseneken and M. Goodwin, IEEE Int. Reliab. Phys. Symp. Proc. 2005, p. 381.
4. A. E. Islam, H. Kufluoglu, D. Varghese, S. Mahapatra and M. A. Alam, IEEE Trans. Electron Devices **54**, 2143 (2007).
5. S. Mahapatra, A. E. Islam, S. Deora, V. D. Maheta, K. Joshi and M. A. Alam, IEEE Int. Reliab. Phys. Symp. Proc. 2011, p. 614.
6. H. Reisinger, O. Blank, W. Heinrigs, A. Muhlhoff, W. Gustin and C. Schlunder, IEEE Int. Reliab. Phys. Symp. Proc. 2006, p. 448.
7. C. Shen, M.-F. Li, X. P. Wang, Y.-C. Yeo, and D.-L. Kwong, IEEE Electron Device Lett. **27**, 55 (2006).
8. T. Wang, C.T. Chan, C.J. Tang, C.W. Tsai, H.C.-H. Wang, M.H. Chi and D.D. Tang, IEEE Trans. Electron Devices **53**, 1073 (2006).

9. C. T. Chan, C. J. Tang, C. H. Kuo, H. C. Ma, C. W. Tsai, H. C. H. Wang, M. H. Chi, and Tahui Wang, IEEE Int. Reliab. Phys. Symp. Proc. 2005, p. 41.
10. C. T. Chan, H. C. Ma, C. J. Tang and T. Wang, Tech. Dig. VLSI Symp. 2005,p. 90.
11. J. P. Chiu, Y. H. Liu, H. D. Hsieh, C. W. Li, M. C. Chen and Tahui Wang, IEEE Trans. Electron Devices **60**, 978 (2013).
12. T. Grasser, H. Reisinger, P.-J. Wagner, F. Schanovsky, W. Goes and B. Kaczer, IEEE Int. Reliab. Phys. Symp. Proc. 2010, p.16.
13. C.T. Chan, C.J. Tang, T. Wang, H.C.-H. Wang and D.D. Tang, IEEE Trans. Electron Devices **53**, 1340 (2006).
14. B. Kaczer, T. Grasser, Ph. J. Roussel, J. Franco, R. Degraeve, L.-A. Ragnarsson, E. Simoen, G. Groeseneken and H. Reisinger, IEEE Int. Reliab. Phys. Symp. Proc. 2010, p. 26.
15. S. M. Amoroso, L. Gerrer, S. Markov, F. Adamu-Lema, and A. Asenov, in Proceedings of the ESSDERC. 2012, p. 109.
16. J. P. Chiu, Y. T. Chung, T. Wang, M. C. Chen, C. Y. Lu and K. F. Yu, IEEE Electron Device Lett. **33**, 176 (2012).
17. B. Kaczer, Ph. J. Roussel, T. Grasser and G. Groeseneken, IEEE Electron Device Lett. **31**, 411 (2010).
18. M.A. Alam, Tech. Dig. - Int. Electron Devices Meet. 2003, p. 345.
19. H. Y. Yu, Y. T. Hou, M. F. Li and D.-L. Kwong, IEEE Electron Device Lett. **23**, 285 (2002).
20. Y. T. Hou, M. F. Li, H. Y. Yu, Y. Jin and D.-L. Kwong, Tech. Dig. - Int. Electron Devices Meet. 2002, p. 731.
21. Technology Computer Aided Design (TCAD), Integrated Systems Engineering AG, Zurich.
22. Y. Y. Liao, S. F. Horng, Y. W. Chang, T. C. Lu, K. C. Chen, T. Wang and C. Y. Lu, IEEE Electron Device Lett. **28**, 828 (2007).
23. J.P. Chiu, C. W. Li and Tahui Wang, Appl. Phys. Lett. **101**, 082906 (2012).
24. T. Siddiqua and S. Gurumurthi, IEEE Trans. Very Large Scale Integration (VLSI) Syst. **20**, 616 (2012).
25. S. Lee, H.-J. Cho, Y. Son, D. S. Lee, and H. Shin, Tech. Dig. - Int. Electron Devices Meet. 2009, p. 763.
26. H.-C. Ma, Y.-L. Chou, J.-P. Chiu, Y.-T. Chung, T.-Y. Lin, T. Wang, Y.-P. Chao, K.-C. Chen, and C.-Y. Lu, IEEE Trans. Electron Devices **58**, 623 (2011).
27. M. A. Alam and S. Mahapatra, Microelectron. Reliab. **45**, 71 (2005).
28. C. T. Chan, C. J. Tang, Tahui Wang, H. C.-H. Wang, and D.D. Tang, Tech. Dig. - Int. Electron Devices Meet. 2005, p. 571.
29. A. Shanware, M. R. Visokay, J. J. Chambers, A. L. P. Rotondaro, H. Bu, M. J. Bevan, R. Khamankar, S. Aur, P. E. Nicollian, J. McPherson, and L. Colombo, IEEE Int. Reliab. Phys. Symp. Proc. 2003, p. 208.
30. R. Degraeve, A. Kerber, P. Roussell, E. Cartier, T. Kauerauf, L. Pantisano, G. Groeseneken, Tech. Dig. - Int. Electron Devices Meet. 2003, p. 935.
31. R. Degraeve, F. Crupi, D. H. Kwak, and G. Groeseneken, Tech. Dig. VLSI Symp. 2004, p. 140.
32. T. Yamaguchi, I. Hirano, R. Iijima, K. Sekine, M. Takayanagi, K. Eguchi, Y. Mitani, and N. Fukushima, IEEE Int. Reliab. Phys. Symp. Proc. 2005, p. 67.

Chapter 4
The Time-Dependent Defect Spectroscopy

Hans Reisinger

Abstract The time-dependent defect spectroscopy (TDDS) is an advancement of the technique to analyze random telegraph signals (RTS). RTS in the drain current of small-area MOSFETs has been used since the 1980s to study capture and emission times of charge carriers in individual traps in the gate insulator. These capture- and emission-time constants are the only electrically determined parameters of individual traps which provide information having the potential to identify the physical nature of these traps. The two main advantages of TDDS compared to RTS are that capture and emission times can be determined over a wide regime in gate bias, ranging from strong inversion to strong accumulation, and that TDDS signals from multiple traps can be analyzed more easily because they are less complex. This chapter is focused on explaining all the experimental aspects of TDDS, on preconditions with respect to samples, proper choice of stress and measuring parameters, data analysis, and limits due to instrumentation.

4.1 Introduction

The time-dependent defect spectroscopy (TDDS) is an experimental method to characterize the charge-capture and charge-emission times of single, individual defects in the insulator and in the semiconductor–insulator interface of MOSFETs. The TDDS technique has similarities to the technique to characterize random telegraph signals (or noise, RTS or RTN; see also Chap. 5) [1]: like RTS also TDDS analyzes steps in the drain current due to an *event* of capture or emission of charge in traps and—since the events are stochastic and thus require averaging—it also analyzes a given event *repeatedly* for many times. While RTS is done in

H. Reisinger (✉)
Infineon Technologies AG, Neubiberg, Germany
e-mail: Hans.Reisinger@infineon.com

T. Grasser (ed.), *Bias Temperature Instability for Devices and Circuits*,
DOI 10.1007/978-1-4614-7909-3_4, © Springer Science+Business Media New York 2014

quasi-thermal equilibrium at a constant gate voltage, TDDS switches the MOSFET space charge layer between inversion and depletion or accumulation. Thus TDDS in principle is an extension of the deep-level transient spectroscopy (DLTS) technique [2], adapted to small-area FETs. The new name TDDS has been introduced [3] because of the different analysis method of the acquired data: a frequently employed assumption in the DLTS is that all defects are charged after a filling pulse of a certain duration. This assumption is incorrect as—as we will see—the capture-time constant shows a very wide distribution, a fact observed as a nonsaturating behavior in the DLTS spectra.

The first TDDS measurement (though not named TDDS yet) was reported by Karwath et al. [4] in 1988. Karwath et al. observed steps in the drain current of an n-channel MOSFET due to charge emission from traps. The TDDS technique is actually quite simple and is no more demanding with respect to instrumentation, samples, or overall effort than RTS. Nevertheless its enhanced potential with respect to determine the field dependence of capture- and emission-time constants has not been recognized in 1988, and the technique has been forgotten. Only recently—as a consequence of the increasing interest in the negative bias temperature instability (NBTI) [5, 6] and in HiK gate stacks [7, 8] (also see Chaps. 3, 21–23)—some workgroups [9–14] have rediscovered TDDS. "Large-scale" TDDS measurements, that is, an analysis of the field and temperature dependencies of capture and emission, based on measurements of a large total number (1,000–100,000) of capture/emission events into the *same defect*, have been started and reported from 2009 on [3, 15–31]. The standard "DC-TDDS" measurements have recently been extended to stress under AC gate voltage [31]. AC stress experiments are able to give information about defects having properties which go beyond the simple two-level defect model which will be discussed in Sect. 4.3.

We will continue in Sect. 4.2 discussing the requirements and preconditions for doing RTS or TDDS measurements. Section 4.3 introduces a simplified defect model, which only partly describes the properties of real defects but will serve as a vehicle to understand the TDDS method(s) and provide a basic understanding of the phenomena. In Sect. 4.4 we will explain how the measurements intended to extract capture- and emission-time constants are done, and Sect. 4.5 will show some examples for "raw" TDDS data and for results extracted from repeated measurements. Since the focus of this chapter is primarily on the TDDS method and not on the results, only a few results for properties of selected defects under the influence of the electric field and temperature will be shown. Section 4.6 will give a brief comparison of the features of the related methods RTS measurement and TDDS. Considerations about data reduction will be done in Sect. 4.7. The subject of Sect. 4.8 will be error estimation and minimization. Section 4.9 will briefly discuss experimental setups, performance, problems, and limitations. Finally Sect. 4.10 will summarize in brief the new findings up to date about the physics of defects and NBTI.

4.2 Preconditions and Basics for TDDS

In a MOSFET, small enough to fulfill the condition that there are only few mobile carriers in the channel, the effect of one carrier from the channel being trapped may be easily detectable as a change of the drain current or of the threshold voltage. Figure 4.1 schematically illustrates the process and its effect on the drain current.

Trapping of a charge q sitting just *at* the semiconductor–insulator interface of a MOSFET, smeared out in lateral direction over the whole gate area, will cause a threshold shift ΔV_{th} of

$$\Delta V_{th} = q/C_{ox} \quad with \quad C_{ox} = \varepsilon \times A/t_{ox} \tag{4.1}$$

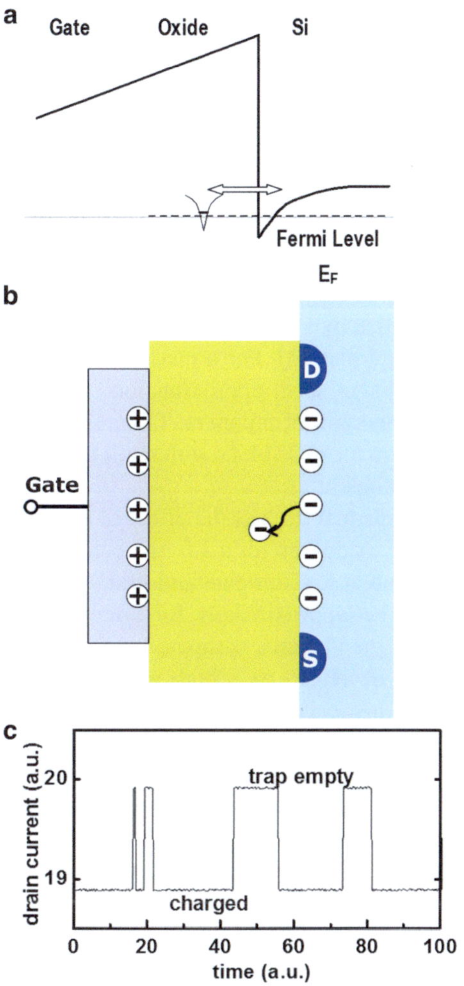

Fig. 4.1 Illustration of a negative charge moving at random from the channel into a trap in the vicinity of the Fermi level (capture) and reverse from the trap back into the channel (emission). (**a**) Conduction-band diagram illustrating trapping and de-trapping of a carrier from the channel into a trap at an energy near the Fermi level. (**b**) n-channel MOSFET with source/drain/gate terminals. Prior to capture of a negative charge in the oxide the number of carriers in the channel equals the number of charges in the gate (depletion charge neglected); after capture one (mobile) carrier in the channel is missing. Figure (**c**) shows the effect on the drain current at a const drain and gate voltage. The ratio of time at high level/time at low level is called mark/space ratio and is equal to τ_e/τ_c

Table 4.1 Useful numbers for some selected technology nodes: operation voltage *VDD*, minimal dimensions, SiO_2-equivalent insulator thickness EOT, specific and absolute gate capacitances, and numbers of free carriers and defects

Technology node	1 μm	10 nm	40 nm	16 nm
VDD (V)	3.3	1.2	1	0.8
Width = length in (μm)	1	0.1	0.04	0.16
EOT/nm	10	2.2	1	1
Specific capacitance (C/nF/cm^2)	345	1,568	3,450	3,450
Oxide capacitance C_{ox} (F)	3.45E − 15	1.57E − 16	5.52E − 17	8.83E − 18
Eox at VDD (MV/cm)	3.3	5.5	10.0	8.0
Number of carriers in channel at Eox = 5MV/cm	7.1E + 04	1.2E + 03	345	44
Number of active defects	1,000	10	1.6	0.3
ΔVth for single carrier (mV)	0.05	1.0	2.9	18.1

ΔV_{th} after Eq. (4.1). Assumption: defect density = 10^{11}/cm^2

where C_{ox} is the oxide (or insulator) capacitance, determined by its dielectric constant ε and the gate area A and the oxide thickness t_{ox}. In general q will be a positive (for p-channels) or negative (for n-channels) elementary charge. Equation (4.1) neglects the fact that the trapped charge is *not* two-dimensionally smeared out. In reality the charge is localized and the interaction with randomly distributed dopant atoms in the substrate brings about that real ΔV_{th}'s may be significantly smaller or larger than the value from Eq. (4.1). "Giant" defects may cause ΔV_{th}'s multiplied by a factor of 10 or even higher compared to Eq. (4.1) ([19, 32], and see Chaps. 7 and 13). For a qualitative discussion of the phenomena to be discussed, the "charge sheet approximation" Eq. (4.1) will be sufficient, however. To give an impression of numbers, Table 4.1 shows selected ΔV_{th} values and other useful numbers for MOSFETs with minimum dimensions for some past and actual CMOS technologies.

Let us first discuss the effect of the capture of just one carrier from the channel into an oxide trap for a 1.0 μm technology, as shown in Table 4.1: For a MOSFET operated at a given gate and drain voltage this trapping process means that there would be (approximately, for $Cox \ll$ capacitance of space charge layer) one carrier less in the channel, compared to the state before trapping. Thus the drain current would decrease. At a high gate voltage, corresponding to strong inversion, this would just cause a very small relative change in the drain current of about 1 in 10^4 (see Table 4.1). At a gate voltage around threshold this relative change would be higher, around 0.1%. The corresponding threshold shift would be less than 1 mV. Decreasing the gate voltage into the subthreshold regime would be a means to further increase the relative change. Even so, for any pre-1.0 μm technology at room temperature, the effect of trapping a single charge carrier is clearly below a reasonable experimental detection limit. Moreover, even at low gate voltages, there would be at least 100 defects being active (that is capturing and emitting) at the same time which would make the analysis of the V_{th} steps from any selected defect impossible. This is why the first RTS experiments could be done no earlier than 1984 by Ralls et al. [33], when the first small-area MOSFETs were available, together with digital data acquisition systems. Ralls et al. were using nMOSFETs with $\approx 0.1\ \mu m^2$ gate area at cryogenic temperatures. When we look at 100, 40, and 16 nm technologies in Table 4.1, we see that the relative change in the drain current and the ΔV_{th} values (due to capture/emission of one carrier) increase. For the 100 nm technology the RTS effect comes into the conveniently measurable 1% regime, and for the 16 nm technology "giants" may easily exceed 10% change in drain current or cause ΔV_{th}'s of more than 100 mV. On the other hand, with the gate area decreasing, it becomes more and more unlikely to detect a defect at all on a small gate area and thus might require a tedious prescreening of devices. Thus, nearly independent of the technology, the most promising device geometry to study RTS or TDDS is in the order of $W \times L = 50 \times 50\ nm^2$. For such an area one can expect to have about a handful of active traps available on a MOSFET.

4.3 Theoretical Background

The physics behind capture and emission of charge into defects and the properties of these defects is covered in Chaps. 9 and 10. In this chapter, focused on describing the technique and its capabilities, we just want to provide a very rough understanding of capture- and emission-time constants and probabilities and to discuss the equations [Eqs. (4.2), (4.3), and (4.4)] we need for the analysis of the experimental data.

Often elastic tunneling of electrons or holes or the Schottky–Read–Hall (SRH) recombination has been tried to explain capture and emission. Tunneling—for thin oxides and with realistic parameters—is far from being able to explain the long capture- and emission-time constants above 100 s which were experimentally observed [1, 34] and cannot explain the strong thermal activation with activation energies in the order of 1 eV either. The SRH theory, treating minority carriers with thermal velocity in bulk semiconductors, was never intended to describe processes involving space charge layers and therefore is not applicable. Most of the phenomena seen in TDDS can be explained by a defect model consisting of *two states*, a charged state and an uncharged one. With the defect sitting in the vicinity of the border between substrate and oxide [35], the energy level of the charged state can be shifted up and down with the applied gate voltage. The transition of the defect from one state to the other involves a structural change (structural relaxation; see [1, 36]) of the surrounding oxide matrix. This effect can explain that the two states are separated by a fairly large potential barrier in the order of 1 eV. During capture and emission this barrier has to be surmounted by thermal excitation, as illustrated in Fig. 4.2. This makes the transition from one state to the other a stochastic, thermally driven process like a chemical reaction. The transition rate is a function of the applied gate voltage, thus making capture/emission an electrochemical reaction. Actually, as will be shown in Chaps. 16 and 17, some defects may assume more than two states. The real nature and physics of the defects are more complicated than given by the two-state approximation. On the other hand,

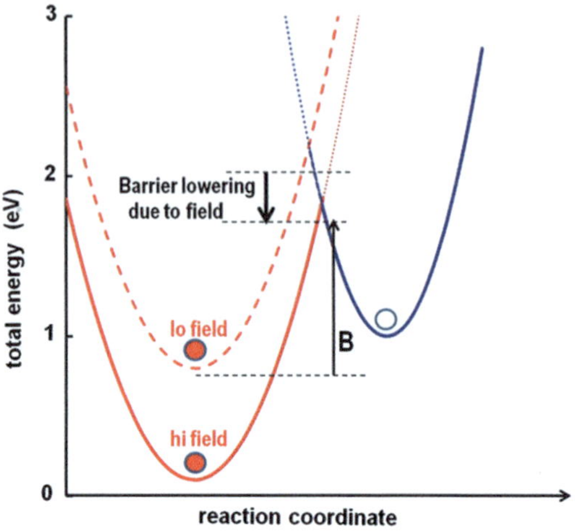

Fig. 4.2 Configuration coordinate diagram (schematic) of a defect in the uncharged state (*right*) and in the charged state for two different electric fields. A low electric field pulls the energy of the charged state down below the uncharged state and thus enables capture. A high field pulls the energy down further and also lowers the barrier B for capture

the two-state model is a good approximation for the phenomenological behavior of the standard defect, which is—for a given field and temperature—described by only one simple equation [Eq. (4.2)] and only two parameters, i.e., the capture- and the emission-time constant:

When the defect is in one state (this may be either the charged or the uncharged state) the transition probability P per time unit to switch to the other state is *time independent* and given by

$$P = 1/\bar{\tau} \propto \exp(-B/kT) \tag{4.2}$$

where $\bar{\tau}$ is the time constant corresponding to the probability P, and B is the potential barrier (=the activation energy). From the barrier energy in Eq. (4.2) it becomes plausible that the experimentally observed range of capture and emission time constants—best seen in capture- and emission-time maps of wide FETs [17, 37]—is very wide, from sub-μs to $>10^5$ s. Due to the amorphous nature of the oxide the barriers B vary, and a variation only from 1 to 1.4 eV causes a change in the Boltzmann factor by 5 orders of magnitude. It is obvious, but worth mentioning it, that the barriers for capture and emission in general will be different; thus capture and emission time constants will be different, dependent on gate bias.

As we will see, during TDDS, the gate voltage will preferably always be kept in the special case where capture and emission can be separated, that is, either in a pure "charge capture" condition or in a pure "charge emission" condition. In a pure stress condition $\bar{\tau}_C \ll \bar{\tau}_E$ is valid; that is, the equilibrium state is the "captured" state, and after the capture process has occurred any emission is very unlikely to happen. During the recovery condition the equilibrium state is the "emitted" state and the reversed condition $\bar{\tau}_C \gg \bar{\tau}_E$ is valid.

For these idealized cases we only have to deal with either only the capture process or only the emission process. For the time dependence during stress and recovery we then obtain simple exponential solutions for the occupancy *O(t) of the charged state* from the integration of Eq. (4.2).

For the stress condition we get an occupancy increasing with stress time:

$$O(t_s) = [1 - O(t_s = 0)] \times [1 - \exp(-t_s/\bar{\tau}_C)] \tag{4.3}$$

and for the recovery condition the occupancy will exponentially decay after

$$O(t_r) = O(t_r = 0) \times \exp(-t_r/\bar{\tau}_E) \tag{4.4}$$

In Eqs. (4.3) and (4.4), t_s and t_r are the stress and recovery times, respectively. The zeroes of t_s and t_r correspond to the time when switching from recovery voltage to stress and vice versa, respectively. $\bar{\tau}_C$ and $\bar{\tau}_E$ are the capture- and emission-time constants. $O(t=0)$ are the occupancies prior to starting stress or recovery. In general, the desired starting conditions prior to applying a stress or recovery pulse are $O=0$ or $O=1$, respectively. Obviously it does not make sense to start stress when the occupancy is already 90%, for instance. Thus the desired starting conditions have to be ensured by an appropriate electrical "preconditioning" of the device under test, that is, applying either zero or a high gate bias for a sufficiently long time to ensure an occupancy of zero or 100%, respectively. It should be noted that Eqs. (4.3) and (4.4) describe charging and discharging of an asymmetric RC element which is discussed in Fig. 4.15.

Prior to proceeding to the statistical analysis of capture and emission events of single defects we want to repeat and summarize the most important facts and assumptions:

1. Capture and emission are stochastic processes which are random and can happen at any time. Like for radioactive decay only probabilities for capture and emission can be given. Capture and emission (under a given condition, i.e., local electric field and temperature) each are characterized by a single parameter, $\bar{\tau}_C$ and $\bar{\tau}_E$, respectively. This is due to the fact that the transition probability p [Eq. (4.2)] is constant, i.e., time independent. As a consequence a resulting distribution of transition times is an *exponential distribution*. Note that differences from exponential distributions [38] may occur in principle if the probability p is *not constant*, for example, due to the fact that a defect can appear in more than two states.
2. Equations (4.3) and (4.4) are exact only for a two-state system under the assumption that the equilibrium occupancy under the stress or recovery condition will be unity or zero, respectively. The Fermi distribution function is neglected and assumed to be 1 during stress and 0 during recovery. This is a good approximation as long as the trap energy E_T is not within an energetic distance of <100 mV to the Fermi level E_F (<5% occupancy at T = 400 K). It should be noted that the validity of this approximation simplifies the data analysis but is not a necessary condition. Defects with E_T close to E_F cause an RTS signal (see Fig. 4.8) but are not likely to be observed at V_g near V_{th}.

3. Any defect has two states, and at any time is either occupied or unoccupied. Equations (4.3) and (4.4) give *average* occupation levels which one would approach by averaging over many of stress or recovery sequences done under the same conditions or averaging over many equivalent defects.
4. Any experiment averaging over a *finite* number of capture and emission events will yield time constants τ_C or τ_E. Due to the stochastic nature of the processes they in general will not be exactly equal to the *expectation values* $\bar{\tau}_C$ and $\bar{\tau}_E$ in Eqs. (4.3) and (4.4) (compare Sect. 4.8 about confidence limits). One would obtain these expectation values only by averaging over an infinite number of experiments.

4.4 The TDDS Technique

The purpose of TDDS is to extract the properties of single, individual defects in MOSFETs. The primary properties of a defect are its capture- and emission-time constant, both dependent on the local gate electric field and the temperature. In order to learn as much as possible about the physics of defects, and in order to enable a modeling of the reliability and degradation of the devices (under all possible stress and recovery conditions), it is desirable to extend the investigated gate voltage and temperature over a range as wide as possible. The number of investigated, different individual defects and types of defects for a given technology should be very large. This is desirable in order to model degradation for large, "analog" FETs and to model the ΔV_{th} variability of small FETs. Considering the wide distribution in capture- and emission-time constants, certainly the complete characterization of the properties of 100–1,000 defects would be required and desirable for this purpose. Complete characterization in this context means capture and emission to be measured over the full gate and drain voltage range from 0 to *VDD* and also at different temperatures. Since the characterization of only a handful of defects already is a matter of months, such a task would require an automated measurement of many small FETs in parallel, and at present is still an ambitious goal for the future.

For the following considerations the type of MOSFET (n- or p-channel) is irrelevant, and for the sake of simplicity we deal with an nMOSFET. All of the measurement examples are for pMOSFETs, which just would mean to flip the band diagrams (e.g., Fig. 4.16) upside down. Let us start with examining the behavior of a defect at three different gate voltages. Figure 4.3 gives a schematic illustration of the potential energy of a given defect in its two states, of the barriers, and of the capture and emission events. The right-hand column of Fig. 4.3 shows the potential energy of the defect vs. the reaction coordinate for three different gate voltages. When the defect state is in the left-hand potential minimum, the defect is charged (=captured state). This state is correlated with the low value of the drain current (see center column). In the right-hand potential minimum the defect is uncharged and the drain

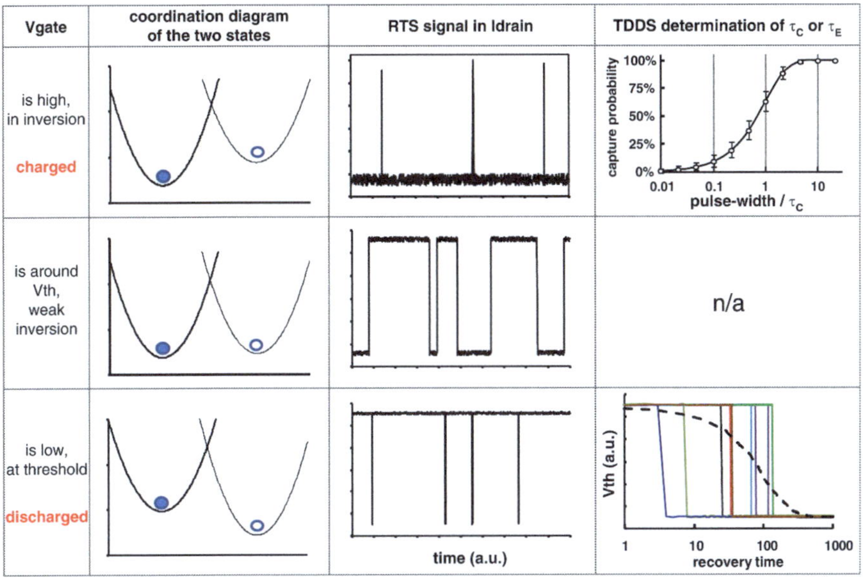

Fig. 4.3 From *left* to *right column*: C1: Configuration coordinate diagram of a defect, with the state of the defect in either the *left* (charged) or the *right* (uncharged) potential minimum. C2: corresponding random telegraph signal (RTS). C3: TDDS measurement. From *top* to *bottom row*: R1: Moderate stress voltage, where the defect charging probability would be 99%; ⇒ RTS–mark–space ratio = 1% and charging probability 99% after stress pulse time $\gg \bar{\tau}_C$; TDDS does a determination of $\bar{\tau}_C$ by applying stress pulses and testing the success rate for charging the defect. R2: a gate voltage in weak inversion, charging probability 50% and mark-space ratio 1:1; no TDDS determination of $\bar{\tau}_C$ or $\bar{\tau}_E$ in this regime. R3: Low gate voltage, defect discharged with probability 99%; ⇒ RTS–mark–space ratio = 99% and the defect is most likely (99%) emitting when ending stress; TDDS directly determines $\bar{\tau}_E$ by doing recovery traces after stress (see also Fig. 4.8)

current assumes its high value (=emitted state). Let us first look at the middle row of Fig. 4.3: The defect is assumed to have the same potential energy in both states, and this energy is assumed to be at the Fermi level.

Thus both states—ruled by Fermi–Dirac statistics, and neglecting any degeneracy—are occupied with the same probability; the defect spends equal amounts of time in both states, and the averaged mark/space ratio τ_e/τ_c of the RTS signal (middle column in Fig. 4.3) is unity. The defect captures and emits spontaneously and stochastically, driven by thermal excitation. That is, it switches continuously from one state to the other. Capture and emission times are ruled by a constant transition probability per time unit [see Eq. (4.2)].

All RTS measurements exclusively deal with states where the defect energy is close to the Fermi level and the defect switches spontaneously. For the special case assumed for the middle row of Fig. 4.3 the defect is just in equilibrium between capture and emission. By changing the gate voltage either more towards

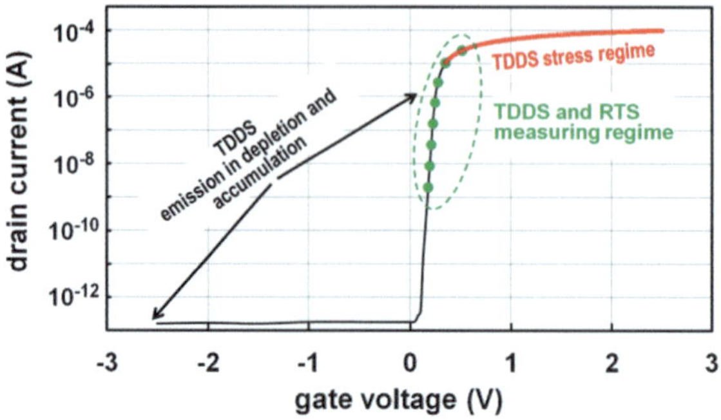

Fig. 4.4 I–V curve of a MOSFET meant to illustrate the different regimes employed in RTS and TDDS. Only the *green* (*dotted*) regime has a suitable drain current to measure emission or RTS. In the *red* (*thick*, inversion and strong inversion) capture of charge is done, and the regime below 0.5 V is used in TDDS for emission in dynamic TDDS (see Fig. 4.6c)

strong inversion (Fig. 4.3 upper row) or towards depletion (Fig. 4.3 bottom row) this equilibrium can be shifted towards the captured or the emitted direction, respectively. In general, a small change of the gate voltage ($\approx$200 mV for a 2 nm gate oxide) already will change the mark/space ratio by a factor in the order of 1,000 which makes measurement and analysis of RTS data over a wider bias range very difficult and finally impossible.

This is where the *TDDS* technique comes in. In contrast to the RTS measurement, where the MOSFET stays in quasi-equilibrium, during a TDDS measurement the MOSFET is actively forced to switch between two states or phases always: During the stress phase (phase 1) the gate voltage is high, thus switching the equilibrium state to the charged state. For TDDS in general the value of the gate voltage will be chosen to have this state occupied with a high probability or occupancy level, say >99% (compare Fig. 4.3 upper row). During the recovery phase (phase 2) the gate voltage is low, thus switching the equilibrium state to the uncharged state with a probability >99% (compare Fig. 4.3 lower row). There also must be a readout phase, where the actual measurement is done, which for the easiest case is identical to phase 2. Important differences between RTS and TDDS measurements will be treated in Sect. 4.6.

In order to illustrate the various regimes of the TDDS method, i.e., differing in the method to determine the capture- and emission-time constants, the IV curve of a MOSFET is drawn in Fig. 4.4. We have divided the IV curve into three different regimes:

1. The *strong-inversion* regime is the regime where the device is under BTI stress. The high gate voltage in this regime pulls the defects energetically below the quasi-Fermi level so that they can be charged. The density and the number of carriers in the channel drain are high ($\approx 10^{13}$/cm^2 or $\approx$1,000, respectively,

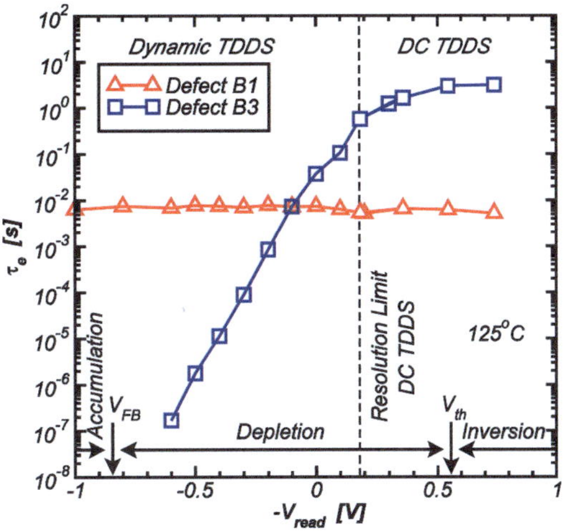

Fig. 4.5 An example for the wide range of the determination of the emission-time constants for two defects. By the "dynamic" TDDS technique, the range is extended to gate voltages far below threshold where no drain current can be measured. From [31]

compare Table 4.1). Noise and limited resolution in this regime normally prevent capture and emission processes to be directly measured as steps in the drain current.

2. In the *weak-inversion* and near-threshold regime only a negligible number of defects are in an energetic position to be charged. Most of the defects being charged in the strong-inversion regime will discharge in this regime (if previously been charged). This is the regime in which the actual emission processes during TDDS are *measured* (and where RTS measurements are done).
3. In the *depletion and accumulation* regime—like in the regime near threshold—defects previously charged are emitting their charges. Unlike in the regime near threshold these emission processes *cannot* be directly detected by measuring in the drain current because the drain current in this regime is close to zero.

For the sake of completeness we want to mention that in the accumulation regime processes like charging of defects from the gate side (pMOS PBTI [39, 40]) or charging from the substrate with opposite charge polarity may occur. But these effects will not be treated further here.

To summarize the description of measurement and extraction methods, the process *directly observable* during TDDS is only the *emission* of charges in *regime 2*. The other processes, which are capture of charges in regime 1 and discharge in regime 3, have to be studied indirectly.

Figure 4.5 shows a practical example illustrating the above regimes 1–3, and for the extraction of the emission times of two defects over a very wide regime in gate voltage, from accumulation over depletion into inversion. To clarify the procedures, the timing of the three different measuring sequences has been plotted in Fig. 4.6a–c.

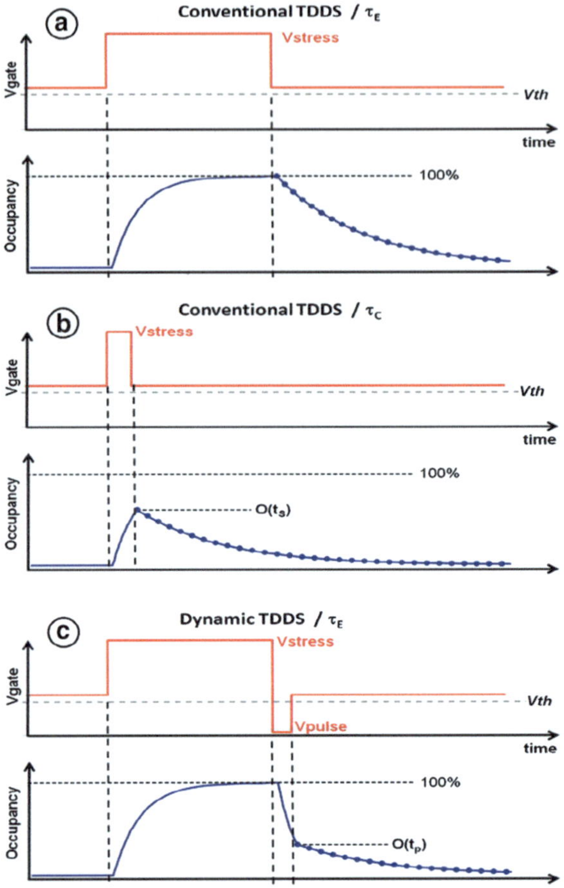

Fig. 4.6 Illustration of the three different measuring schemes used by TDDS (comp. also [41]). (**a**) *direct determination of* τ_E: the defect is charged to an occupancy close to 1 by switching the FET into inversion for a sufficiently long time (no measurement done during charging), and then the defect is discharged by switching the FET into weak inversion (threshold); the steps in Id caused by discharging are directly monitored during the measuring phase (see example Fig. 4.8). (**b**) *determination of* τ_C: same as in (**a**), but the charging pulse is shorter in order to produce an occupancy of about 10–.95%. Then the success rate of charging, i.e., the occupancy, is determined in a measuring phase. τ_C is calculated from Eq. (4.3). (**c**) the defect is charged like in a, and then a discharging pulse (accumulation) of a length t_P is applied. Then the "success rate" of discharging is tested by determining of $O(t_P)$ like in (**b**). The actual measuring of emission events is done in the dotted regime only

The determination of the capture- and emission-time constants $\bar{\tau}_C$ and $\bar{\tau}_E$ from the experimental data is based on Eq. (4.2) and straightforward. The full equations can be found in Sect. 4.8 about confidence limits.

4.5 Examples for Defect Analysis by TDDS

After having explained all necessary model conceptions, equations, and timing sequences in Sects. 4.3 and 4.4, we will continue with practical examples for the statistical analysis of defects.

Figure 4.7 shows a couple of examples for TDDS measurements. Only single recovery traces, measured after a stress pulse, are shown. Different individual traps can be distinguished by the defect-specific different step heights.

Figure 4.8 shows a further example for a TDDS measurement, using a pMOSFET with SiON gate oxide. In this example the stress pulses applied prior to measuring the recovery traces had two different lengths, i.e., 10 ms and 10 s, and about 7 (equivalent) recovery traces are shown after each stress condition. A couple of important things can be seen in Fig. 4.8: Four defects simultaneously can be analyzed from the data in Fig. 4.8. Each defect is characterized by its individual step height in V_{th}, which is like a *fingerprint* of the corresponding defect. This is of

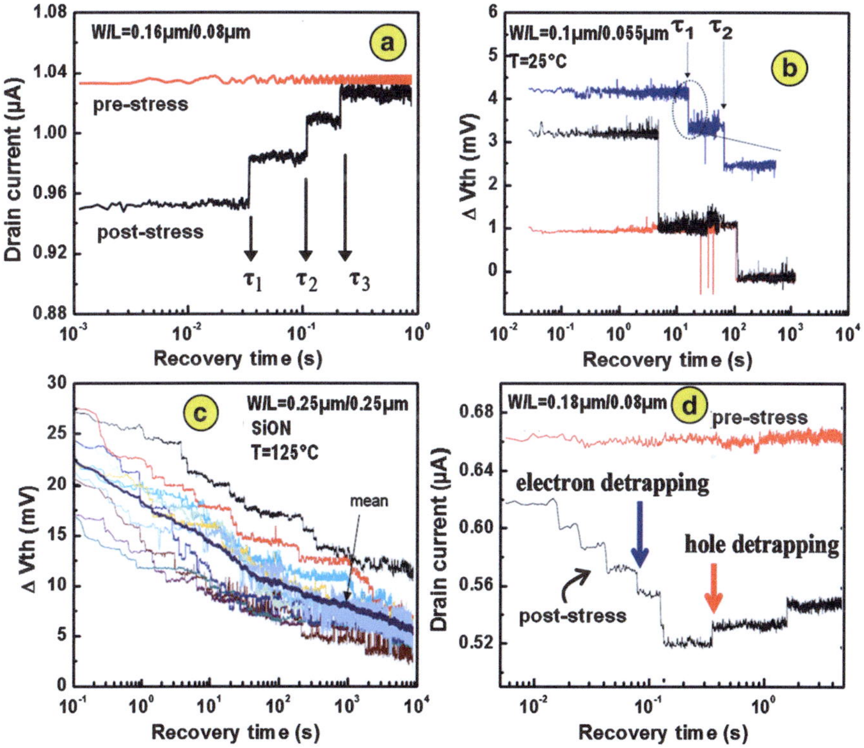

Fig. 4.7 Examples for TDDS from various authors and samples. (**a**) Pre- and post-stress current in a high-K MOSFET (data from [9]). (**b**) Hole de-trapping in small pMOSFETs (data from [11]). (**c**) SiON 1.4 nm, different pMOSFETs (data from [13]). (**d**) Electron and hole de-trapping after NBTI in pMOSFETs with a poly-silicon gate and a HfSiON–SiO_2 gate stack (data from [12])

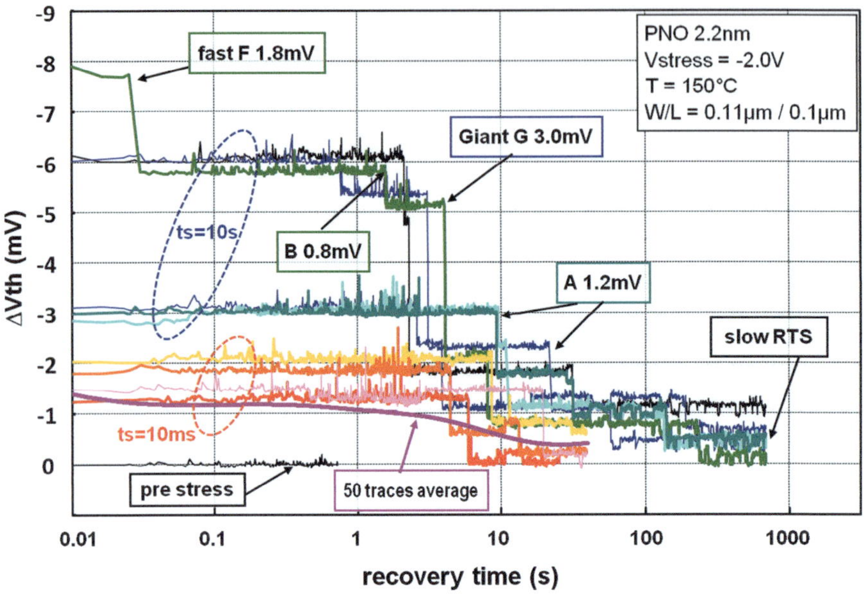

Fig. 4.8 Recovery traces recorded after repeated gate stress pulses with $t_S = 10$ ms and 10 s pulse length. Four different defects (named A, B, F, G) with different capture- and emission-time constants and step heights are charged. For the given gate area, ΔV_{th} after the charge sheet approximation [Eq. (4.1)] is 1 mV, the resolution is about 0.2 mV. Also shown is an average over 50 recovery traces after $t_S = 10$ ms, showing the exponential distribution Eq. (4.4)

uttermost importance for the analysis of a given defect. It should be noted that two or more defects having the *same* step height still could be distinguished and analyzed with acceptable certainty as long as their emission-time constants are differing by more than a factor of ≈100. (For a given defect the width of a measured distribution of emission times, containing 95% of the emission events, is roughly 2 decades; see Sect. 4.8.)

We first examine the recovery traces with stress time $t_s = 10$ ms. As seen, only a single defect named "A" with an emission time around 10 s happens to be active. The capture-time constant of this defect is <10 ms; thus its occupation probability is nearly unity after stress. The recovery of defect "A" is shown also as an average over 50 recovery traces. All the steps are smoothed out this way, and the average recovery is exponential following Eq. (4.4).

The "giant" defect in Fig. 4.8 sometimes is charged after the 10 s stress pulse, and sometimes is not, according to Eq. (4.3). Apparently it has a capture-time constant around 10 s. In the long-term part of the traces a slow RTS can be seen, with RTS time constants of roughly 100 s.

An example for the statistical determination of *emission*-time constants according to Eq. (4.4), for a selected defect, is shown in Fig. 4.9. Each emission time of this defect, extracted from a recovery trace like in Fig. 4.8, is plotted as a dot.

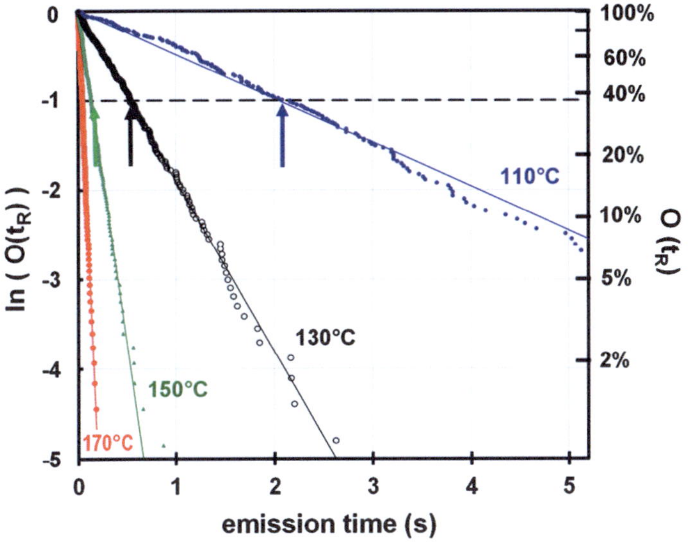

Fig. 4.9 Measured emission times, plotted as exponential distributions [Eq. (4.4)] for a single, selected defect. Same sample type as in Fig. 4.8. Each *dot* in the graph (256 dots per curve) corresponds to an emission event observed in a recovery trace. *Arrows* mark the average emission-time or emission-time constant. The emission-time constants decrease with increasing temperature

Fig. 4.10 Arrhenius plot for one defect from capture and emission (trap A from Fig. 4.8) and for another defect from emission. The *straight lines* and *small error bars* make clear that activation energies E_A can be determined with great precision and that thermal activation exactly follows an Arrhenius law

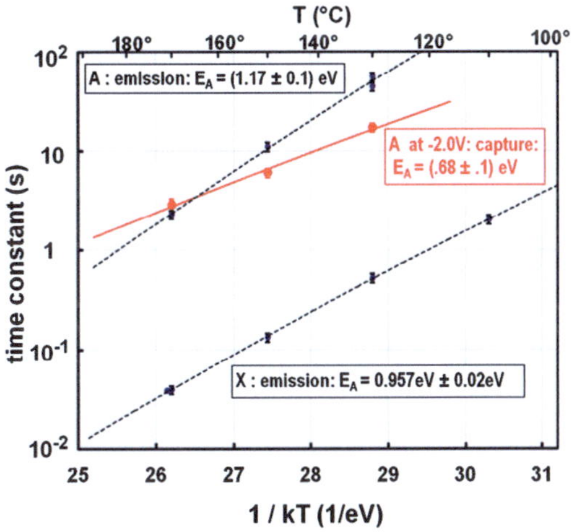

The fit to a straight line allows a reliable extraction of τ_e and also is a proof that the defects behave according to the expected exponential distribution and that *only one* defect is contained in the distribution. Mixing another similar defect into the same distribution would lead to a deviation from the straight line. Figure 4.10 shows the results from Fig. 4.9 as an Arrhenius plot. It should be noted that the thermal activation of the single defects *exactly* follows an Arrhenius behavior, in

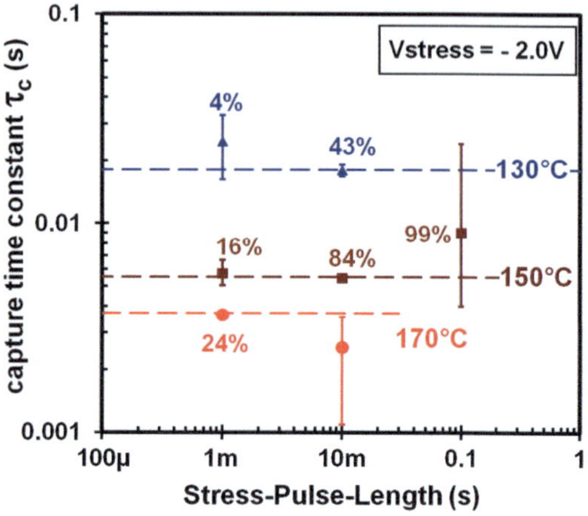

Fig. 4.11 Capture times for two different temperatures, determined from the measurement of the charging "success rate" (see labels) and Eq. (4.3), as illustrated in Fig. 4.6b. Same sample type as in Figs. 4.8 and 4.9

contrast to a conventional determination of thermal activation done on wide FETs [42]. The statistical TDDS analysis allows a very precise determination of activation energies E_A. We want to note at this point that the conventional determination of E_A delivers different and wrong values. This is due to the fact that a conventional NBTI measurement on a large FET measures a whole *ensemble* of defects and that—expressed in a much simplified way—speeding up degradation by increasing T also speeds up recovery, thus eliminating a part of the net effect. This is discussed in more detail in [15, 21, 23]. The E_A's found here are in agreement with the findings from ultrafast temperature changes, which avoid fast recovery [43, 44].

An example for the determination of *capture*-time constants τ_C according to Eq. (4.3) for a selected defect is shown in Fig. 4.11. As explained above, capture times cannot be measured directly but have to be determined indirectly by varying the stress pulse length and the determination of the capture rate. τ_C is given by Eq. (4.3), and the best regime of t_s and the corresponding confidence limits are discussed in Sect. 4.8.

The analysis shown in the figures above only analyzes the behavior of a single defect and does not contain information about the step height. In order to simultaneously analyze the discrete steps caused by the discharging of a dozen of defects, "spectral maps" have been introduced in [18]. Spectral maps provide a complete visualization simultaneously of all emission events from all traps, for a given charging or stress time, for instance, of hundreds of charging pulses and recovery traces like the "10-ms" traces in Fig. 4.8. Figure 4.12 explains with an example (for two recovery traces only) how *all* the emission events from a given experiment are mapped into a two-dimensional spectral map.

Figures 4.13 and 4.14 show examples demonstrating the potential of the spectral map representation of data. A spectral map contains both the emission times and the step heights from all recovery traces from a given stress condition (i.e., T, t_S, V_{stress}, V_d) and thus contains the *full information* from an experiment with many (hundreds or more) recovery traces.

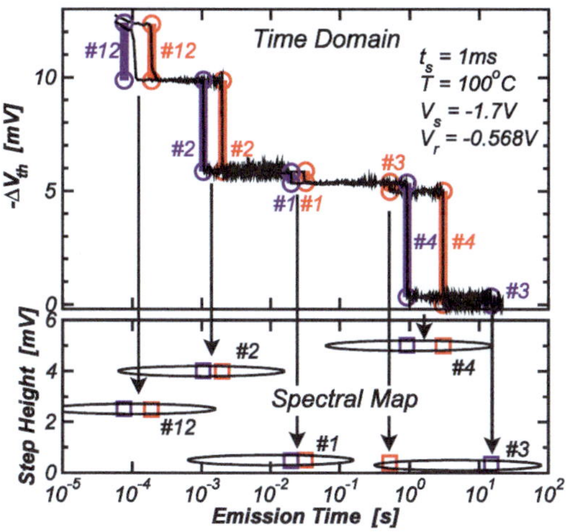

Fig. 4.12 An example showing how emission events are mapped into a spectral map. The *top* (time domain) shows two recovery traces with emission events from four different defects, with fastest to slowest emission-time constant ≈0.1 ms to ≈5 s, and step heights from 0.2 to 5mV. Each emission time and step heights for each emission event is transformed (see *arrows*) to a dot in the two-dimensional spectral map (*bottom*, spectral map). A real spectral map contains collected data from hundreds of recovery traces and delivers meaningful average emission-time constants

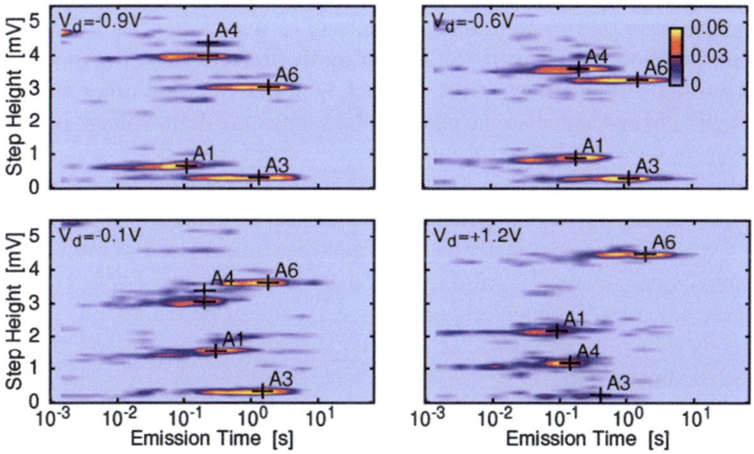

Fig. 4.13 Spectral maps from pMOSFETs with 2.2 nm PNO and W/L = 150 nm/100 nm. This series shows the drain bias sensitivity of the defect parameters at T = 125 °C, V_{stress} = −1.5 V, and t_s = 100 ms. When the drain bias is changed from −0.9 V to −0.6 V, −0.1 V, and eventually "+1.2 V" (symbolic for source and drain reversed), the step height of #1 increases from 0.8 to 2.1 mV, #4 decreases from 4 to 1 mV, and #6 increases from 3 to 4.3 mV. Note that the positions of the crosses marking τ_E also shift significantly and also τ_C changes as seen from the filling level of the clusters. Note that even though defects #1, #4, and #6 dramatically change their relative positions, reliable extraction of the defect parameters remains possible. For more results, see [3]

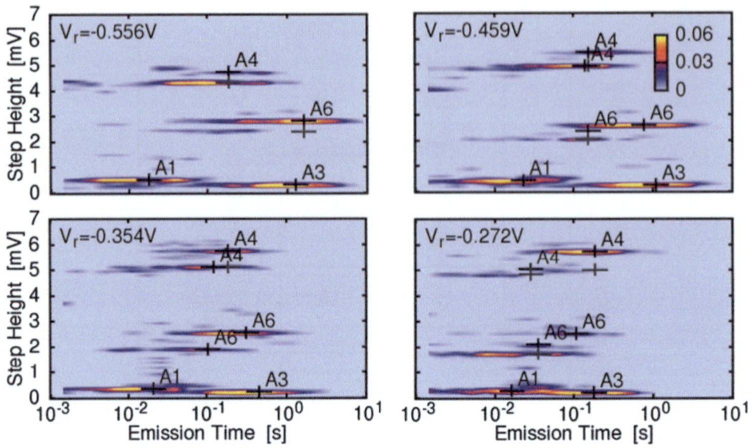

Fig. 4.14 Same device and T as in Fig. 4.13. This series shows the recovery gate bias dependence of the defect parameters. With decreasing gate bias, the emission times become shorter, visible by a shift to the left of the clusters in the spectral maps. Particularly #3 and #6 show a very strong bias dependence. Also note the strong splitting in #4 and #6 (see text). The sub-peaks corresponding to the same defect may also have different emission times, visible in the above example for #4 and #6. More results in [3]

Figure 4.13 shows a series of spectral maps taken at different drain biases. Varying the drain bias changes the local density of free carriers in the channel and the corresponding local electric field seen by the defect. Thus, depending on the lateral position of the defect in the channel, both the emission times and the step heights will change, and also the capture times when the drain voltage is on during stress.

Figure 4.14 shows, for the same device as in Fig. 4.13 and the same numbering of individual defects, a series of spectral maps taken at different recovery gate voltages. As seen, recovery becomes faster when the gate bias is decreased, or even driven into depletion. Quite interesting is also the fact that the step height of defect #4 changes to another value (#4′), depending on the actual state of defect #6, charged or not. Apparently defect #4 by chance is located close enough to defect #6 to cause an electrostatic interaction.

The focus of this chapter on TDDS is on the experimental technique and not on the results of the TDDS technique. Thus we will restrict ourselves to just giving the examples for results shown above. Other results of TDDS are given in Chaps. 3 and 23.

As already mentioned, TDDS allows to explore the capture- and emission-time constants of defects over the full range of gate voltage, from strong accumulation to strong inversion (see Fig. 4.5), to determine the lateral position of a defect by varying the local field (with varying the drain voltage or reversing drain and source), to determine the depth of defects in the oxide [35], to study the thermal activation of individual defects, and to study electrostatic interaction of defects [3]. We again

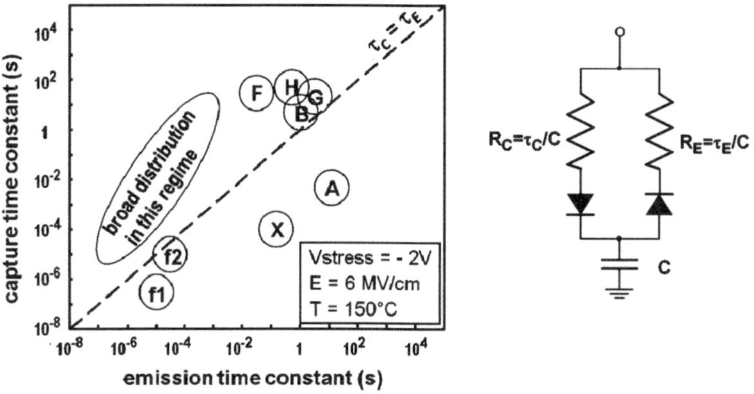

Fig. 4.15 *Right-hand side*: A discrete capture emission time map (data from same type of samples as in previous figures), showing the correlation and the wide distribution of capture and emission times for a couple of defects. *Left-hand side*: A circuit which is the exact equivalent of a single trap as described by Eqs. (4.3) and (4.4). The value of C represents the amount of charge in the trap (corresponding to the ΔV_{th} produced by the trap); R_C and R_E determine charging and discharging time constant of the capacitor

want to stress that all of the defects studied in SiON show anomalies, i.e., deviations from the simplified two-state model which can be explained only by introducing metastable defect states. Deviations from the two-state model are also seen in the results of AC-BTI of wide FETs [41].

Still the two-state model has proven itself as an approximation being reasonably useful for the practical simulation of degradation and recovery of large FETs containing many, i.e., a quasi-continuous distribution of τ_E's and τ_C's. Figure 4.15 shows a map with capture and emission times (CET map) of a couple of discrete defects, for a given stress and recovery condition. Though there are only very few defects in Fig. 4.14 the map already gives an impression of the wide distribution of τ's, with defects having τ_E's $\ll$ τ_C's and vice versa τ_E's $\gg$ τ_C's. As shown in Chap. 17 about CET maps in this book, and as justified by the TDDS results, a CET map with a continuous distribution of τ_E's and τ_C's can be also extracted by a conventional measurement from wide (say $W = 100\ \mu m$) FETs with thousands of defects. Such a continuous CET map has proven to provide a good tool to simulate degradation and recovery due to a stress signal with stress and recovery sequences of arbitrary lengths [17].

4.6 TDDS Vs. RTS

TDDS and RTS both investigate the same physical effect. Though some of the differences in the features of TDDS and RTS have been already mentioned in the previous sections, it makes sense to summarize the differences, advantages, and disadvantages again as a list of bullet points. It is important to note that RTS is a method automatically "included" by TDDS, that is, RTS measurements can be done with the identical setup/instruments and measurement software.

- General differences of the methods

  TDDS: does an active switching of the defect(s) (by applying a gate pulse when the defect is in quasi-thermal equilibrium) into an excited state and waits for the defect to relax into the—gate voltage dependent—ground state. This causes capture and emission processes to be *synchronized* with the gate pulses. Thus all capture and emission times are referenced to a relative zero time. This has advantages regarding the data analysis; it allows a de-convolution and separation of defects having by chance the same step height. It also allows data compression, as pointed out below. Practical: after the short emission times have occurred, the sampling rate can be decreased w/o loss of information. In the absence of RTS there is only *one emission event* from each defect in each trace.

  RTS: always keeps the MOSFET in a quasi-thermal equilibrium state. Short or long "mark" or "space" periods from a given defect can happen at any time and more than once. There is no "time zero"; thus the signals cannot be de-convoluted and data cannot be compressed. That is, the same (high) sampling rate has to be kept throughout the measuring trace.

- Regime of gate voltage

  TDDS: emission of charges from the channel from strong accumulation to depletion; capture of charge in weak and strong inversion. Note that the gate bias parameter space is two-dimensional for TDDS while it is one-dimensional for RTS.

  RTS: weak inversion only; accumulation not possible, strong inversion too noisy due to high number of carriers in channel.

- Type/size of MOSFET

  TDDS and RTS: both restricted to MOSFETs (n- and p-type, SiON or HiK) with a handful of active traps (see Table 4.1). Larger FETs with, say >10, active traps produce a signal too complex to be analyzed.

- Energetic position of defect

  TDDS: as shown in Fig. 4.16, any energetic position will be made available by sweeping the gate voltage from accumulation to strong inversion. Note: The energy with respect to the Fermi level *cannot* be directly determined, since TDDS always measures in a nonequilibrium state.

  RTS: as shown in Fig. 4.16, only defects with energetic positions in the vicinity of the Fermi level generate an RTS signal, when the MOSFET is in weak inversion. Note a major advantage of RTS: The energy with respect to the Fermi level *is directly determined* from the mark–space ratio [1] (Fig. 4.17).

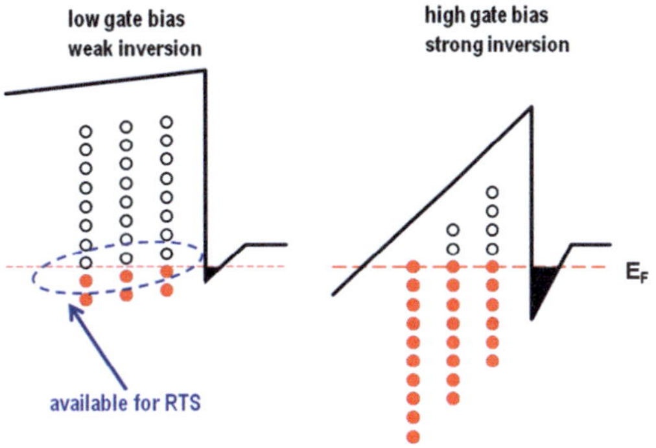

Fig. 4.16 Conduction band diagram (schematic) of an nMOSFET in weak and strong inversion. *Open* and *filled dots* represent neutral and charged traps, respectively. Only traps in the vicinity of E_F can be analyzed by RTS (*dashed ellipse*), while for TDDS all the *filled dots* in the *right-hand side* are available

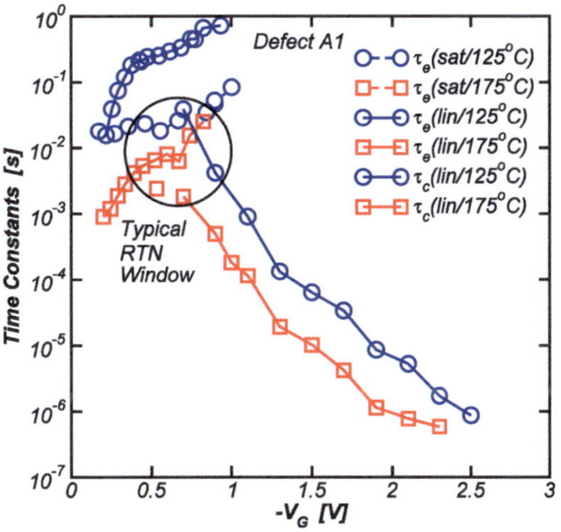

Fig. 4.17 An example for simultaneous determination of capture- and emission-time constants (at two temperatures, and emission in linear and saturation regime) of a defect in an extremely wide regime of time constants and gate bias. For comparison the RTS regime is encircled. RTS is limited by the condition $0.01 < \tau_E/\tau_C < 100$, approximately

- Data reduction

  Considerations regarding data reduction will be treated in the next section, and in principle data reduction is the same for TDDS and RTS, with the following important difference:

  TDDS: as pointed out above all recovery events are synchronized with the end of the stress pulse. A recovery event from a defect with an emission time constant of 1 ms, for example, will be discharged with a probability >99.9% after 7 ms. For all practical cases a relative accuracy in the determination of the emission time of 1% will be more than sufficient. A progressive data compression therefore can compress the acquired data. For the assumed 1% resolution in time (i.e., data points are equidistant on a log-time scale, and the time of each data point is the previous time multiplied by 1.01), this means a compression to 230 data points per decade or a total number of data points of about 2,000, assuming a recovery trace ranging from 1 μs to 1,000 s. Such a compression means a great alleviation in data analysis as well as a vast improvement in the amount of data to be stored.

Assuming an experiment running over a year, and producing recovery traces with a length of 1,000 s each, the amount of data to be stored would be only about 120 MB (with 2 bytes per data word).

RTS: in contrast to TDDS capture and emission are not synchronized with any time zero. Thus a fast event, say a 1 μs RTS pulse, can happen any time. No data compression—at least not prior to data reduction—is possible. This obviously complicates data reduction as well as data storage. Given the above example with an experiment running for 1 year, with a desired time resolution of 1 μs, the amount of data to be stored is 60,000 GB (about 60 hard disks; incredible in the 1980s, still a lot even for modern computers).

4.7 Data Reduction

We define "data reduction" as the process of extracting emission events by individual traps from the "raw" recovery traces. The first TDDS data by Karwath [4] and most of the RTS data of the 1980s certainly have been analyzed "manually," just by visual inspection of the stored signal. A typical, extended TDDS experiment may run over several months, and it typically produces a hundred spectral maps, each map consisting of hundreds of recovery traces. Thus up to 10,000 recovery traces have to be analyzed, which clearly rules out any manual data analysis. Setting up an automated data reduction scheme for TDDS does not bring about any problem, in principle. If the recovery traces are free of any RTS contributions, then the traces are monotonically decreasing with time (compare Fig. 4.8). Let us assume that ten defects are electrically active. This certainly is above the reasonable practically manageable upper limit for the number of defects discharging in the same recovery trace. Otherwise the signal gets too complex and noisy. With ten defects active, the corresponding drain current *Id* or V_{th} in the recovery trace could assume a number of $2^{10} = 1{,}024$ different discrete levels. In the ideal case, that is, without RTS, and the defects are discharging one by one, the number of discrete levels is reduced to 10. In reality the clusters (see Figs. 4.13 and 4.14) will have an overlap, so the defects will not discharge in the same order always. Thus the number of discrete levels is increased by a factor 2 for each overlap. Also note that the transitions from level to level can produce some extra, irregular points, due to the fact that transitions are assumed to be indefinitely fast, while the response of the FET and the data acquisition is not. These irregular points will be neglected here. As a consequence, all the Id values occurring in a recovery trace can be binned into a number of discrete bins, with the bins having the width of the noise amplitude. Then, in the ideal case, for the above example with ten defects, only ten bins would be populated per recovery trace, and the recovery trace would jump down from bin to bin. This way the data reduction would be very easy and straightforward. The considerations about data compression prior to the analysis have been pointed out in the last section.

Unfortunately recovery traces happen to be nonideal in many cases. That is, they are spoilt by RTS signals and are non-monotonous, which largely complicates the data analysis. Moreover the RTS signals easily might cause steps in the recovery signal to remain undetected or cause it to be assigned to the "wrong" defect. Especially when the number of events from a given defects is low, such errors may have dramatic effects on the extracted time constants (see Sect. 4.8). A sophisticated way to eliminate the errors introduced by the unwanted RTS signals is described in [45].

4.8 Confidence Limits and Minimization of Errors

Capture and emission in TDDS are stochastic processes. So whenever time constants are extracted, it obviously is mandatory to know and consider errors and upper and lower confidence limits. Moreover the TDDS experiments are time-consuming and thus an optimization of the parameters with respect to data output will help not to waste measuring time. The topic of error estimation is hardly addressed in any RTS or TDDS publication. Thus the purpose of this section is to treat the TDDS confidence limits and to show how to minimize them. In this section it is assumed that the experiments and the corresponding analysis of data—apart from their stochastic nature—are ideal. That is, any erroneous analysis, for example, due to complex, overlapping signals or noise, RTS, etc. is not considered and has to be treated separately. As mentioned above, TDDS uses two different methods for the extraction of time constants: (1) τ_E's in general are *directly* determined by measuring *emission times* in a recovery trace. (2) τ_C's are measured by *counting* capture events happening during a given stress pulse time. So this chapter is divided into two Sects. 4.8.1 and 4.8.2. The method to determine τ_E's in the accumulation regime is identical to method (2) and therefore is not treated separately.

4.8.1 Recovery

Let us assume that we want to determine the emission time constant τ_E for a given trap, from an experiment with a number of n repeatedly done stress pulses, each followed by measuring a recovery trace. In the ideal case, for a given trap, with an expectation value of the emission-time constant $\bar{\tau}_E$, when the occupancy O [compare Eqs. (4.3) and (4.4)] has been brought to 100%, an emission event can be detected for *each* of the n recovery traces. Then the measured distribution of emission times will be an exponential distribution (like shown in Fig. 4.9), following Eq. (4.4). The measured time constant τ_E will be just the average of the emission times. In reality the number of measured emission events m will be less than n for three reasons:

(i) The stress pulse might be too short to produce the desired average occupancy close to 100%. Thus for a fraction of the recovery pulses the given trap has not been occupied. Except for the fact that such a trace—for the given trap—is useless, it has no effect on the result, because Eqs. (4.5), (4.6), (4.7), and (4.8) do contain only m and not n.

(ii) The instrument has an initial "dead time," i.e., a delay t_D following the end of the stress pulse during which a proper measurement is not possible. For very short recovery times, shorter than t_D, the corresponding emission event will remain undetected.
(iii) The measuring time per recovery trace t_{MEAS} is not infinitely long. So, especially for long emission times, a finite number of emission events may not happen within t_{MEAS} and will also remain undetected.

For the case that t_{MEAS} is chosen long enough to have a ratio $t_{MEAS}/t_D > \approx 100$, the conditions (ii) and (iii) will not occur at the same time and thus can be treated separately.

For the ideal case, with $\bar{\tau}_E \gg t_D$ and $\bar{\tau}_E \ll t_{MEAS}$, the unknown emission-time constant τ_E is just equal to the measured average τ_M of all observed emission events:

$$\tau_E = \tau_M = \sum_m t_{E,i}/m \tag{4.5}$$

For the case discussed in (ii), with τ_E being comparable to t_D and $t_{MEAS} \gg \bar{\tau}_E$, that means that no emission events are missing due to a too short measuring time, we get the solution:

$$\tau_E = \tau_M - t_D \tag{4.6}$$

For the case discussed in (iii), for a measuring time being too short, there is no analytic solution to calculate τ_E from the measured value t_M. An approximate solution is given by

$$\tau_E = \tau_M \left\{ 1 + \exp\left[.658 \left(.5 - \frac{\tau_M}{t_{MEAS}} \right)^{-\frac{1}{3}} + 12 \left(\frac{\tau_M}{t_{MEAS}} \right) - 6.182 \right] \right\} \tag{4.7}$$

The approximation has a relative accuracy better than 5% as long as more than 10% of the emission events occur within the measuring time t_{MEAS}. If t_{MEAS} is longer than $5 * \bar{\tau}_E$, then the exponential correction term in Eq. (4.7) is less than .05 and no correction is required.

To avoid the limitations inflicted by Eqs. (4.5) and (4.7), it is good practice to extract τ_E by fitting the measured emission times to an exponential distribution as done in Fig. 4.9.

Let us assume that we do the above experiment, analyzing a number of emission times, many times in the same way. We will get a different value for τ_E each time we do the experiment, due to the stochastic nature of the process. For the case that the number of observations m is large ($m > 3$, see below), then the measured τ_E's will have a Gaussian distribution around the expectation value $\bar{\tau}_E$. For small values of m the confidence limits will be given by a chi-square distribution (see Table 4.2). For the evaluation of time constants it is sufficient to know the confidence

Table 4.2 Upper and lower *one-sigma* confidence limits for measured τ_E's

	m=	1	2	3	4	5	10	20	100
Chi-square	Upper	1.84	1.64	1.55	1.48	1.43	1.31	1.22	1.1
	Lower	0.105	0.35	0.45	0.52	0.57	0.69	0.78	0.9
Gaussian	Upper	2.00	1.71	1.58	1.50	1.45	1.32	1.22	1.10
	Lower	0.00	0.29	0.42	0.50	0.55	0.68	0.78	0.90

Example: for $m = 100$ 68% of the measured τ's lie within $0.90^* \bar{\tau}_E$ and $1.10^* \bar{\tau}_E$. So m = 100 achieves a reasonable ±10% relative error. For m = 1 (only one measurement done) the measured τ's lie within $0.1^* \bar{\tau}_E$ and $1.84^* \bar{\tau}_E$. So the real value of $\bar{\tau}_E$ may be a factor 10 higher than the measured value or roughly a factor 1.84 below the measured value (in 68% of all cases)

limits corresponding to $1 - \sigma$. 68% of the measured τ's will be within the $1 - \sigma$ confidence limits. In the Gaussian approximation the upper and lower measured $1 - \sigma$ limits are given by

$$\tau_{up,lo} = \bar{\tau} \times (1 \pm 1/\sqrt{m}). \tag{4.8}$$

Table 4.2 lists the upper and lower confidence factors [i.e., the values in bracket in Eq. (4.8)]. Table 4.2 shows that the Gaussian approximation is reasonably good except for values of $m < 3$.

To summarize Sect. 4.8.1, for just a rough estimation of time constants τ_E with a ±30% accuracy, an average over 10 measurements will be sufficient. The measurement of 100 emission times will ensure a statistical error of ±10%. Of course higher accuracies are feasible but hardly required for any purpose.

4.8.2 *Capture*

For the reasons explained above; the determination of capture-time constants τ_C must be done indirectly just by counting the "success rate" in charging a given trap by a test stress pulse of a given length. Let us assume that we do an experiment with a number of n stress pulses, then for a fraction m/n the charging of the trap would have been successful. For a correct result it has to be made sure that the trap was *unoccupied* prior to the stress pulse. The decision if the trap is occupied or not is done by observing the corresponding emission in a recovery trace following the stress pulse. For the sake of simplicity we will assume that each of the m emissions will be detected [see points (i)–(iii) in Sect. 4.8.1]. Due to the stochastic nature of capture, when doing the above experiment more than once, then the number of successful captures m will have a distribution around the expectation value $\bar{m}$. As an example, for a number of stress pulses $n = 200$ these distributions are shown in Fig. 4.18 for different stress pulse widths.

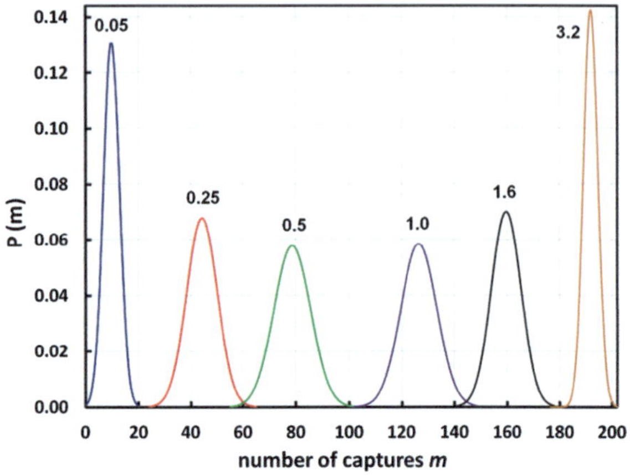

Fig. 4.18 Shown is the (discrete) probability to get m captures for different stress pulse widths (labels) t_s as parameter. t_s is given in units of the capture-time constant. The number of stress pulses is $n = 200$. $P(m)$ is approximated by a Gaussian distribution

For very small numbers m or $(n - m)$ the distributions in Fig. 4.18 are governed by Poisson distributions. For numbers m and $(n - m) \geq 5$ the distributions can be well approximated by Gaussian distributions, given by the expression:

$$P(m) = 1/(\sigma\sqrt{2\pi}) \times \exp[-(m - \bar{m})^2/(2 \times \sigma^2)] \tag{4.9a}$$

where the expectation value for the number of captures is given by

$$\bar{m} = n \times [1 - \exp(-t_s/\bar{\tau}_c)] \tag{4.9b}$$

and the standard deviation σ is given by a sum of two contributions:

$$\sigma = \sqrt{\left(\sqrt{\bar{m}} \times \exp(-t_s/\bar{\tau}_c)\right)^2 + \left(\sqrt{n - \bar{m}} \times [1 - \exp(-t_s/\bar{\tau}_c)]\right)^2} \tag{4.9c}$$

The final result for the upper and lower confidence limits of τ_C, determined from the measured number m, is

$$\tau_{C,up,lo} = -t_s/\ln[1 - (m \pm \sigma)/n] \tag{4.10}$$

The widths (at $1 - \sigma$) of the distributions from Fig. 4.18 are plotted again as the blue dash-dotted line in Fig. 4.19. As seen the width is zero for $t_s = 0$ and $t_s/\tau_c \gg 1$. As seen in Eq. (4.10), however, the precision of measured τ_C is determined by the *relative* width σ/m rather than sigma and m has to be transformed into τ by Eq. (4.10). This is why the error in τ diverges for $t_s/\tau_C \ll 1$ and $t_s/\tau_C \gg 1$, as seen in Fig. 4.19. In a real experiment t_s will be varied on a log-time scale, e.g., 4 t_s values per decade will be measured, each for 200 times, as indicated by the large open dots in Fig. 4.19.

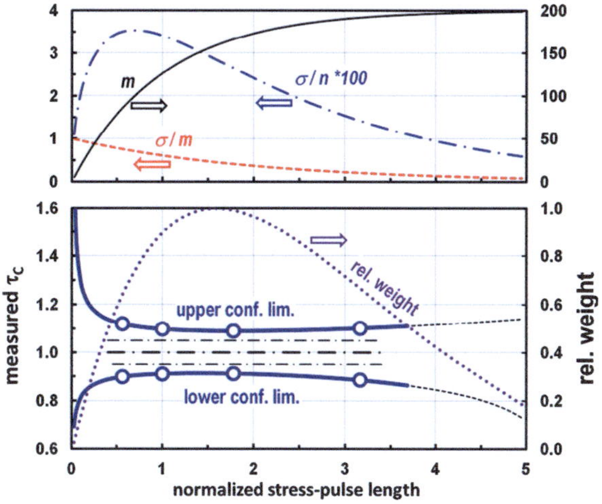

Fig. 4.19 Shown are, for $n = 200$, stress pulses as a function of the normalized stress pulse length t_S/τ_C: *Top*: number of captures m, variance σ/n, and relative variance σ/m. *Bottom*: upper and lower confidence limits; *thick*: regime where $5 < m < 195$ (Gaussian appr. valid). *Circles*: Example for measurements with spacing 4 per decade, $n = 200$ each measurements, resulting in the *final result* (*dash-dot black*) ±5% confidence limits [after Eq. (8.7)]; *magenta dotted*: relative weight factor $1/\Delta\tau^2$

Measurements with small values m or $n - m$ are prone to error and should be dropped when doing evaluations of τ's from multiple t_S values.

In practice, a determination of τ_C will not be done at a single value of t_S. A continuous series of t_S values will be tried and the ones with the best confidence limits will be selected, like the four values in the example in Fig. 4.19. For the final result of a value of τ_c from a series of j measured t_S's, we get

$$\tau_C = \sum_j (W_j * \tau_{C,j}) / \sum_j W_j \tag{4.10a}$$

The corresponding final error $\Delta\tau_C$ is summed up from the contributing error components, which are the differences between upper and lower confidence limits $\Delta\tau_C = (\tau_{C,up} - \tau_{C,lo})/2$

$$\Delta\tau_C = \sqrt{\sum_j (W_j{}^2 * \Delta\tau_{C,j}{}^2)} / \sum_j W_j \tag{4.10b}$$

The weight factors W are given by $W_j = 1/\Delta\tau_{C,j}{}^2$.

To summarize Sect. 4.8.2, there is a rather wide minimum in the statistical error of the determined τ_C around $\tau_C < t_S < 2^* \tau_C$. Several 100 stress pulses in total will be required to achieve a ±10% relative accuracy of the result.

4.9 Experimental Considerations

A vast number of experimental studies on NBTI, RTS, TDDS, and variability have been published over the last 5 years (2008–2012). Common to most of the electrical measurements on these subjects is that they require a precise, noise-free measurement of Id or V_{th}. For NBTI and TDDS the measurement delay t_D—the phase between end of stress and measuring Id—has to be kept as short as possible and the measurements should not be corrupted by instrument noise. In nearly all publications the source of noise, the measuring speed—why isn't it faster? what limits the resolution?—is hardly ever mentioned nor attempted to be discussed.

This is why in this section—though far from being comprehensive—we will try to give some hints on sources of errors and the limits of performance. Although some features of commercial instruments will be compared, the purpose is not to give a guideline for a TDDS setup or a selection of instruments. We just intend to draw the reader's attention to the critical points and try to provide a kind of benchmark. This section should help to answer the questions: How much better could be a dedicated and optimized setup, compared to an existing setup, for instance, built from standard equipment? What are the limits of performance one could possibly reach?

Figure 4.20 shows a measuring setup which—in its principle of operation—resembles the setups used by most experimenters. It consists of a MOSFET under test, a signal source for the gate voltage and an I-to-V converter to measure the source or drain current, followed by a digital data acquisition system. It should

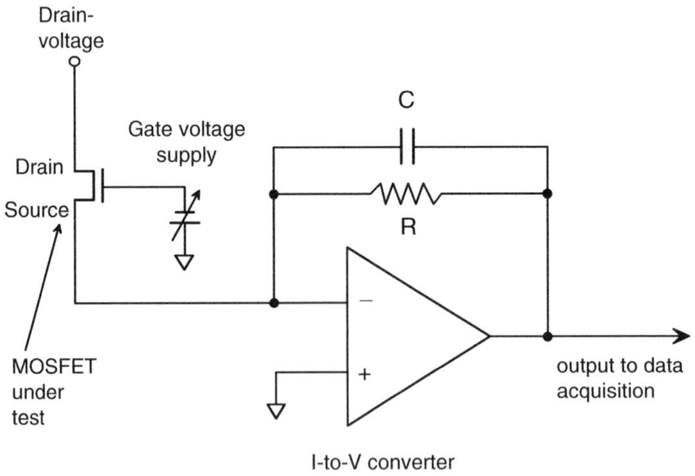

Fig. 4.20 A test circuit representing the principle of a commonly used experimental setup. This setup is also used to generate the data shown in the next figures and serves as reference performance data for experimental TDDS setups. Main component of the setup is an operational amplifier working as an I-to-V converter with amplification R and a feedback capacitor C for frequency compensation

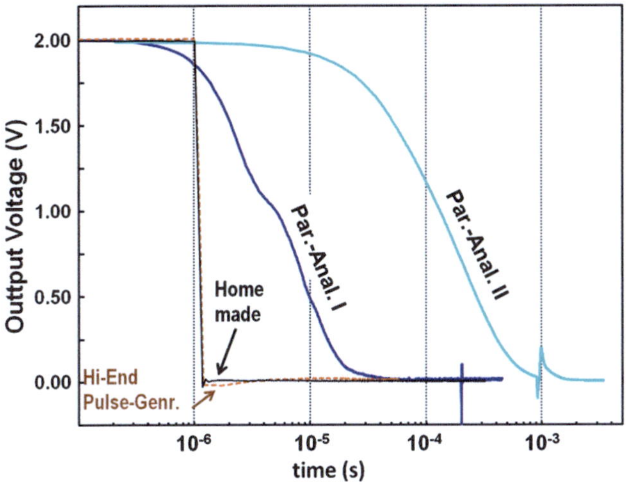

Fig. 4.21 Behavior of various signal sources (same as in Fig. 4.22) when switching from the stress voltage to the measuring voltage. As an example a transition from 2V to 0 was chosen. In order to be able to compare fast and slow signal sources in one plot a log-time scale has been chosen and the start of the switching event has been set to 10^{-6} s. The traces consist of single points taken with a sampling rate of 1/100 ns and the corresponding integration time of 100 ns per point

be noted that the setup in Fig. 4.20, despite its simplicity, is kind of optimized with regard to response time to gate voltage transients: Any unavoidable parasitic capacitances between the terminals gate/source/drain/substrate/ground do not deteriorate the speed since source and drain voltages are constant with respect to ground. We now will analyze the requirements and properties of the components shown in Fig. 4.20 separately:

(a) The gate signal source
The task of the gate signal source, for NBTI and TDDS measurements, is to switch from stress voltage to measuring voltage as fast as possible. It is clear that the measuring phase can only be started as soon as the signal generator has settled at the desired measuring voltage. During the measuring phase the signal should be highly stable and free of any fluctuations. The purpose of an NBTI or TDDS measurement is to determine V_{th} with high precision, and it is clear that any deviation of Vg from the desired voltage will end up directly (one to one) as an erroneous ΔV_{th} in the measurement. Figures 4.21 and 4.22 show a comparison of the switching behavior and the noise produced by different kinds of signal sources. As seen, pulse generators in general are designed to do a perfect switching speed faster than 100 ns, but also are doing undesired small voltage excursions in the 10 ms regime and for times up to 0.1 ms (see Fig. 4.21 and 4.22). As a consequence, for the given example of a pulse generator, TDDS V_{th} steps will not be reliably detected for recovery times below 0.1 ms. As shown in Fig. 4.21 the voltage sources in parameter analyzers may have pretty long settling times as long as 1 ms and in addition the signal is spoilt

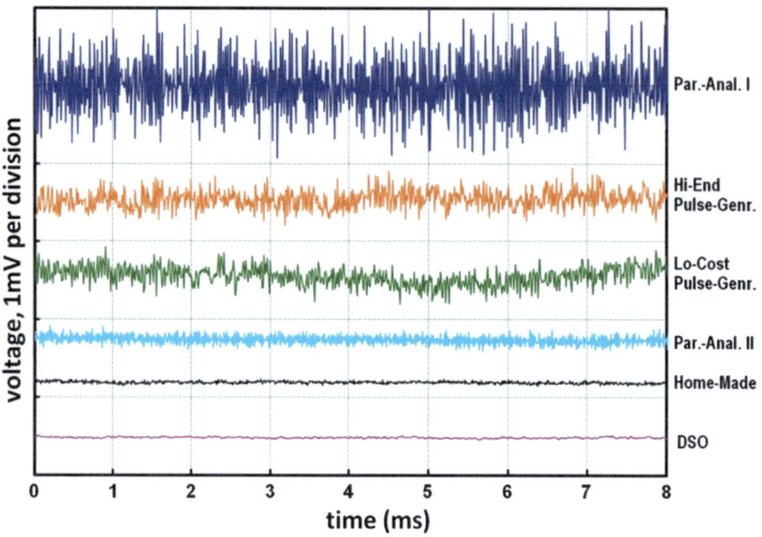

Fig. 4.22 Output noise, taken with a digital storage oscilloscope (DSO), from two different semiconductor analyzers, from two different pulse generators, from a homemade source, and from the DSO with grounded input which serves as the data acquisition system. The sampling rate for all the traces is 1/10 μs (100 points per division) and the corresponding integration time per point is 10 μs. Note that the noise amplitude is a function of the sampling frequency f (or integration time 1/f) and decreases with about $f^{-1/2}$

by short ms spikes in the 0.1 V regime. Shown as a reference signal source in Figs. 4.21 and 4.22 there is also a homemade signal source, consisting of two highly stable voltage sources. A solid state switch switches the output between the two voltage sources. Even for the short 100 ns integration time shown in Fig. 4.21 the maximum deviation from the set point is less than ±2 mV, and below ±0.1 mV for integration times >1 ms.

Figure 4.22 gives an illustration of the output noise from various signal sources. The same noise will be on a V_{th} determined by using these signal sources, in addition to the noise generated by the FET under test itself.

It should be mentioned that there is the chance to improve the properties of pulse generators by dividing the output voltage by employing passive attenuators, when this is possible.

(b) The current measurement

In the setup shown in Fig. 4.20 the source current is measured by a simple I-to-V converter based on a standard video OpAmp (GBW = 130 MHz, noise = $10nV/\sqrt{Hz}$). The input is ground referenced and there are no range switches. The design is simple and the performance with respect to the trade-off between speed and resolution can be regarded as a benchmark for any other current measuring instruments. The tests shown below have been done without a DUT, just using a fast switching current source. Thus all noise is generated by the I-to-V converter itself.

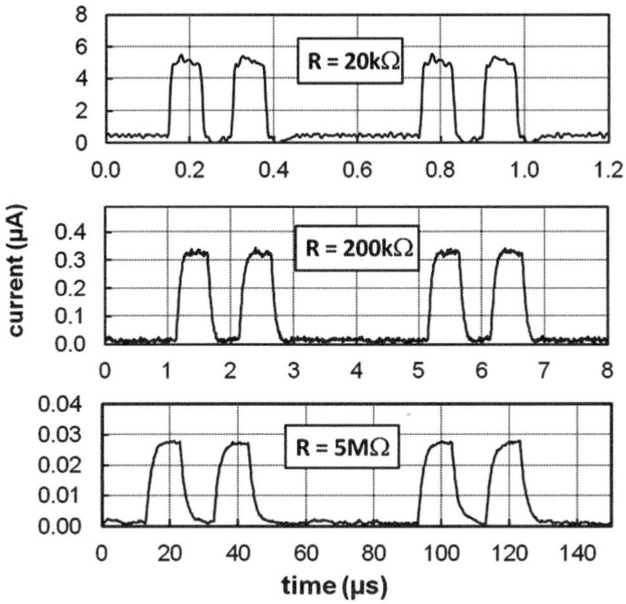

Fig. 4.23 Examples for the performance (i.e., time and current resolution) of I-to-V converters with some selected amplification factors *R*. Shown are two consecutive current pulses generated by artificial, perfectly rectangular shaped "RTS" pulses applied with a pulse generator with pulse width *W* and pulse spacing 2*W*. The vertical pulse amplitude and scale in each case have been chosen in a way that it is *20 times* the current resolution. In order to have the two pulses clearly distinguishable, *W* has been chosen to be *4 times* the settling time $\tau = 1/\omega_{\text{cutoff}}$ of the amplifier. The sample rate in each case has been set to sample a number of ≈20 data points per settling time τ

Figure 4.23 shows a few examples illustrating the time and current resolution of our I-to-V converter. Both resolutions are a function of the amplification factor *R* of the converter. We define the current resolution as the minimum detectable separation of two current levels, when the integration time is equal to 1/10 times the settling time τ.

Figure 4.24 shows the correlation between settling time, maximum frequency, and the current resolution. It should be noted that the current resolution for each range corresponds to an about constant (i.e., independent of "range" *R*) output noise V_{noise} of the converter of about 4 mV peak to peak. The current resolution then is given by V_{noise}/R. We think that the data in Fig. 4.24 could be improved by maybe another factor of 10 at maximum, if all components are optimized for the given frequency range. Nevertheless, data in Fig. 4.24 can be considered as being close to an upper performance limit and a benchmark for TDDS setups. Therefore, if any existing setup shows a considerably weaker performance than the one shown in Figs. 4.23 and 4.24 then there is room for improvement.

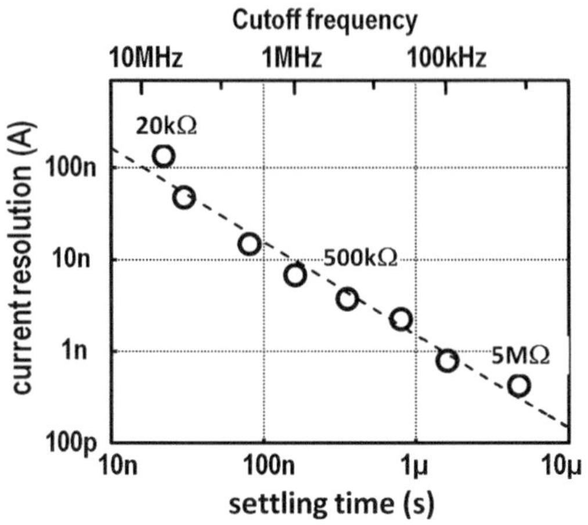

Fig. 4.24 Correlation between settling time and current resolution with different amplification factors R (i.e., current ranges) of the I-to-V converter as parameter (see labels). Data are taken from measurements like the ones shown in Fig. 4.23. The fastest settling time of ≈20 ns corresponds to 100 nA resolution and is limited by the GBW of the employed OpAmp of 130 MHz. Note that full scale of the I-to-V converter is about 10,000 times the current resolution, independent of the amplification factor R

(c) The data acquisition system

In general the digital data acquisition and data storage system will not be the component limiting the performance of a TDDS setup. The benefits of TDDS compared to RTS, regarding the requirements of the sampling rate, have been discussed in Sects. 4.6 and 4.7 already. An intelligent data compressing scheme will be able to minimize the amount of data to store. Commercial semiconductor parameter analyzers contain ADCs with 16- to 20-bit resolution which is much better than what is required for the “noisy” small FETs used for TDDS. On the other hand, the sampling rate of parameter analyzers might be too low for fast TDDS measurements. For most TDDS requirements a modern digital storage scope (DSO) will do a good job as a data acquisition system. Such a DSO has a reasonably noise-free pre-amp, a sufficient 12-bit voltage resolution, a sampling rate which can be set to any desired value, and a practically infinite storage depth (limited by the hard disk only) when programmed in an appropriate way.

There is one fact that is occasionally overlooked: Some data acquisition systems, in parameter analyzers (when doing log-time sampling), in PC acquisition cards, and also in DSOs, are doing the sampling in a way that the integration time is shorter than the sample interval. It is clear that making the integration time as long as possible, without any idle time, will improve the signal-to-noise ratio. Thus the experimenter should make sure that during the complete measuring phase of TDDS the data acquisition is always running and never idle.

4.10 Conclusion: The New Findings So Far from TDDS

The new TDDS technique has proven to be a powerful technique for the characterization of border traps in MOSFETs. Besides electron spin resonance (described in [46] and Chap. 9) electrical techniques probing the properties of single defects like RTS or TDDS are the only methods having the potential to reveal direct information on the physical nature of defects in gate oxide and interface. We conclude this chapter with a list of new findings that have been found by TDDS to date (end of 2012). Most of the points in the list are covered in detail by other chapters of this book; thus we give a very brief summary of facts only. We want to stress that most of the findings in the list could be only done by TDDS.

1. NBTI degradation and recovery have been shown to be due to charging and discharging of individual defects with a wide distribution of timescales. Both capture- and emission-time constants are temperature activated with E_A in the order of 1 eV, consistent with nonradiative multiphonon theory. Furthermore, the time constants are uncorrelated and can be very long ($>10^5$ s) also in thin oxides. No signs of a temperature-independent elastic tunneling process could be found. The defects responsible for the recoverable component of NBTI are identical to those causing RTS.
2. The capture-time constants show a very strong field dependence. Similarly, the bias dependence of the emission-time constant around V_{th} may be either weak or strong, depending on the configuration of the defect.
3. The total number of defect precursors is preexisting and hardly any signs of newly created defects could be found (actually one new defect appeared during months of stressing the same device) in short- to medium-term stress experiments at standard NBTI stress fields.
4. The existence of metastable states becomes obvious due to disappearing defects and transient RTS, and in anomalous AC behavior. Thus real defects are more complicated than simple "two-state defects."
5. TDDS results show that even for long-term stress times >100 s the recovery time for any given trap is independent of the preceding stress time, thus excluding degradation or recovery governed by diffusion.
6. TDDS yields a correct determination of activation energies E_A, in contrast to standard experiment with wide FETs.
7. The analysis of TDDS step heights helps to get a deeper understanding of variability of BTI degradation, which is important for SRAM failure and analog circuits.
8. The concept of universal recovery of NBTI which has been shown to be approximately in agreement with experimental data [47] is not supported by findings from TDDS.
9. The power law exponent n, describing NBTI degradation as $\Delta V_{th} \propto t_s{}^n$, with n around 0.15, has its origin in the distribution of capture-time constants rather than being controlled by diffusion of a hydrogen species in the gate stack.

Acknowledgment A part of this work has been supported by the European Commission under the 7th Framework Programme (Collaborative project MORDRED, Contract No. 261868). The author would like to acknowledge the cooperation with all the "MORDRED" partners, the close collaboration with IMEC and Prof. Tibor Grasser's group, and the help given by the Infineon reliability team, especially Karina Rott and Gunnar Rott.

References

1. M. J. Kirton and M. J. Uren, Adv. Phys., vol. **38**, pp. 367–468 (1989).
2. D. V. Lang, J. Appl. Phys., vol. **45**, 3023 (1974).
3. T. Grasser, H. Reisinger, P.-J. Wagner and B. Kaczer, Phys. Rev. B, vol. **82**, no. 24, 245318 (2010).
4. A. Karwath and M. Schulz, Appl. Phys. Lett. **52**, p. 634 (1988).
5. D. K. Schroder and J. A. Babcock, J. Appl. Phys **94**, pp. 1–18 (2003).
6. J. H. Stathis and S. Zafar, Mat. Res. **46**, no.2-4, p. 270 (2006).
7. G. D. Wilk, R. M. Wallace, and J. M. Anthony, J. Appl. Phys. **89**, 5243 (2001).
8. A. Kerber and E. A. Cartier, *IEEE Trans. Dev. Mat. Rel., vol.* 9, no. 2, p. 147.162 (2009).
9. C. T. Chan, H. C. Ma, C. J. Tang and T. Wang, VLSI Digest of tech. papers, p. 90 (2005).
10. C. T. Chan, C. J. Tang, C. H. Kuo, H. C. Ma, C. W. Tsai, H. C.-H. Wang, M. H. Chi, and T. Wang, Proc. Intl. Rel. Phys. Symp., p. 41 (2005).
11. V. Huard, C.R. Parthasarathy, and M. Denais, Intl. Integrated Reliability Workshop Final Report, p. 5 (2005).
12. H.C. Ma, J.P. Chiu, C.J. Tang, T. Wang and C.S. Chang, Proc. Intl. Rel. Phys. Symp., p. 51 (2009).
13. B. Kaczer, T. Grasser, J. Martin-Martinez, E. Simoen, M. Aoulaiche, Ph. J. Roussel, G. Groeseneken, Proc. Intl. Rel. Phys. Symp., p. 55 (2009).
14. T. Wang, C.-T. Chan, C.-J. Tang, C.-W. Tsai, H. C.-H. Wang, M.-H. Chi, and D. D. Tang, Transactions on Electron Devices, vol. 53, Issue 5, p. 1073 (2006).
15. H. Reisinger, T. Grasser and C. Schlünder, Intl. Integrated Reliability Workshop Final Report, p. 30 (2009).
16. T. Grasser, H. Reisinger, W. Goes, Th. Aichinger, Ph. Hehenberger, P.-J. Wagner, M. Nelhiebel, J. Franco, and B. Kaczer, IEDM Tech. Digest, p. 729 (2009).
17. H. Reisinger, T. Grasser, W. Gustin and C. Schlünder, proc. Intl. Rel. Phys. Symp., p. 7 (2010).
18. T. Grasser, H. Reisinger, P.-J. Wagner, F. Schanovsky, W. Goes and B. Kaczer, proc. Intl. Rel. Phys. Symp., p. 16 (2010).
19. B. Kaczer, T. Grasser, Ph. J. Roussel, J. Franco, R. Degraeve, L.-A. Ragnarsson, E. Simoen, G. Groeseneken and H. Reisinger, proc. Intl. Rel. Phys. Symp., p. 26 (2010).
20. V. Huard, proc. Intl. Rel. Phys. Symp., p. 33 (2010).
21. T. Grasser, B. Kaczer, W. Goes, H. Reisinger, Th. Aichinger, Ph. Hehenberger, P.-J. Wagner, F. Schanovsky, J. Franco, Ph. Roussel, and M. Nelhiebel, IEDM Tech. Digest, p. 82 (2010).
22. J. Martin-Martinez, B. Kaczer, M. Toledano-Luque, R. Rodriguez, M. Nafria, X. Aymerich, G. Groeseneken, proc Intl. Rel. Phys. Symp., p. XT4.1 (2011).
23. T. Grasser, B. Kaczer, W. Goes, H. Reisinger, P.J. Wagner, F. Schanovsky, J. Franco, M. T. Luque, M. Nelhiebel, Transactions on Electron Devices, vol. 58, Issue 11, p. 3652 (2011).
24. T. Grasser, Microelectronics Reliability, vol. 52, Issue 1, p.39 (2012).
25. C. Liu, R. Wang, J. Zou, R. Huang, C. Fan, L. Zhang, J. Fan, Y. Ai, Y. Wang, IEDM Tech. Digest, p. 23.6.1 (2011).
26. M. Toledano-Luque, B. Kaczer, Ph. J. Roussel, T. Grasser, G.I. Wirth, J. Franco, C. Vrancken, N. Horiguchi, G. Groeseneken, proc. Intl. Rel. Phys. Symp., p. 4A.2.1 (2011.)
27. T. Grasser, P.-J. Wagner, H. Reisinger, Th. Aichinger, G. Pobegen, M. Nelhiebel, and B. Kaczer, IEDM Tech. Digest, p. 618 (2011).

28. B. Kaczer, J. Franco, M. Toledano-Luque, Ph. J. Roussel, M. F. Bukhori, A. Asenov, B. Schwarz, M. Bina, T. Grasser, G. Groeseneken, proc. Intl. Rel. Phys. Symp., p. 5A.2.1 (2012).
29. T. Grasser, B. Kaczer, H. Reisinger, P.-J. Wagner, and M. Toledano-Luque, proc. Intl. Rel. Phys. Symp., p. XT.8.1 (2012).
30. J. Franco, B. Kaczer, M. Toledano-Luque, Ph. J. Roussel, J. Mitard, L.-Å. Ragnarsson, L. Witters, T. Chiarella, M. Togo, N. Horiguchi, G. Groeseneken, proc. Intl. Rel. Phys. Symp., p. 5A.4.1, (2012).
31. T. Grasser, H. Reisinger, K. Rott, M. Toledano-Luque, and B. Kaczer, IEDM Tech. Digest, p. 470, (2012).
32. M. F. Bukhori, T. Grasser, B. Kaczer, H .Reisinger, A. Asenov, IEEE Integrated Reliability Workshop Final Report, p. 76 (2010).
33. K. S. Ralls , W. J. Skocpol, L. D. Jackel, R. E. Howard, L. A. Fetter, R. W. Epworth, and D. M. Tennant, Phys. Rev. Lett., vol. **52**, p. 228, (1984).
34. J. P. Campbell, J. Qin, K.P. Cheung, L.C. Yu, J.S. Suehle, A. Oates, K. Sheng, proc. Intl. Rel. Phys. Symp., p. 382 (2009).
35. M. Toledano-Luque, B. Kaczer, Ph. J. Roussel, J. Franco, L. Å. Ragnarsson, T. Grasser, and G. Groeseneken, Appl. Phys. Lett. **98**, 183506 (2011).
36. A.M. Stoneham, Rep. Prog. Phys., vol. 44, p. 1251, (1981).
37. T. Grasser, P.-J. Wagner, H. Reisinger, Th. Aichinger, G. Pobegen, M. Nelhiebel, and B. Kaczer, IEDM Tech. Digest, p 618 (2011).
38. J. Zou, C. Liu, R. Wang, X. Xu, J. Liu, H. Wu, Y. Wang, R. Huang, IEEE Silicon Nanoelectronics Workshop (SNW), p. 1 (2012).
39. Ph. Hehenberger, H. Reisinger, T. Grasser, Intl. Integrated Reliability Workshop Final Report, p. 8 (2010).
40. K. Rott, H. Reisinger, S. Aresu, C. Schlünder, K. Kölpin, W. Gustin and T. Grasser, Microelectronics Reliability, vol. 52, Issues 9–10, p. 1891 (2012).
41. T. Grasser, K. Rott, H. Reisinger, P.-J. Wagner, W. Goes, F. Schanovsky, M. Waltl, M. Toledano-Luque, and B. Kaczer, proc. Intl. Rel. Phys. Symp., p. 2D.2.1, (2013).
42. B. Kaczer, V. Arkhipov, R. Degraeve, N. Collaert, G. Groeseneken, M. Goodwin, proc. Intl. Rel. Phys. Symp., p. 381 (2005).
43. Th. Aichinger, M. Nelhiebel and T. Grasser, proc. Intl. Rel. Phys. Symp., p. 55 (2009).
44. G. Pobegen, Th. Aichinger, M. Nelhiebel and T. Grasser, IEDM Tech. Digest, p. 614, (2011).
45. M. Waltl, P.-J. Wagner, H. Reisinger, K. Rott and T. Grasser, IEEE Integrated Reliability Workshop Final Report, p. 74 (2012).
46. J. P. Campbell, P. M. Lenahan, P.M.; A. T. Krishnan, S. Krishnan, IEEE Device and Materials Reliability, IEEE *Trans. Dev. Mat. Rel., vol. 6*, Issue: 2, p. 117 (2006).
47. S. Rangan, N. Mielke, E. C.C. Yeh, IEDM Tech. Digest, p. 14.3.1 (2003).

Chapter 5
Analysis of Oxide Traps in Nanoscale MOSFETs using Random Telegraph Noise

David J. Frank and Hiroshi Miki

Abstract This chapter describes the use of random telegraph noise (RTN) to obtain information about traps in highly scaled MOSFETs. A robust hidden Markov model (HMM) algorithm is presented to enable the accurate extraction of trap parameters from both single and multiple-trap signals. The results of a large number of measurements show that even in the absence of bias stress, RTN-generating traps can cause serious variation for high-k/metal gate (HKMG) FETs and that undoped channels do not reduce the problem. Trap time constants are shown to have wide ranging dependence on bias and temperature, leading to hysteretic behavior with time constants much longer than the circuit timescale. The impact of RTN on the stability of memory cells is also presented, along with experimental observations of these effects in SRAM arrays.

5.1 Introduction

When MOSFETs are fabricated, there are always some residual charge traps in the gate oxide. Consequently, even when the device has not been voltage stressed, traps are present. For large devices, these traps give rise to 1/f noise, but when the FETs are small enough, the noise can be resolved into discrete switching events. As first described by Ralls in 1984 [1], these discrete jumps cause the channel resistance to switch back and forth between two values, resembling telegraph signals with random timing, and are due to the discrete changes in the charge state of some particular trap. This phenomenon has come to be known as random telegraph noise

D.J. Frank (✉)
IBM T J Watson Research Center, Yorktown Heights, NY, USA
e-mail: djf@us.ibm.com

H. Miki
Central Research Laboratory, Hitachi, Kokubunji, Tokyo, Japan
e-mail: hiroshi.miki.tq@hitachi.com

T. Grasser (ed.), *Bias Temperature Instability for Devices and Circuits*,
DOI 10.1007/978-1-4614-7909-3_5, © Springer Science+Business Media New York 2014

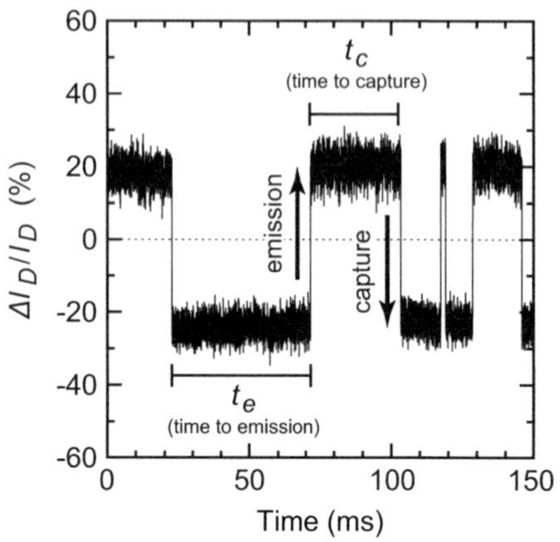

Fig. 5.1 Random telegraph noise signal, measured on an nFET, defining capture and emission times. I_D is the average drain current

(RTN) and is illustrated in Fig. 5.1 for an nFET with a simple 2-state trap. When the trap emits an electron, becoming neutral, the effective threshold voltage (V_T) decreases and the channel becomes more conductive, increasing the current. After a time t_c, the trap captures an electron, the effective V_T increases, and the current decreases. It stays in this new state for a time t_e until it emits an electron, and the cycle repeats.[1]

RTN is well described as a Markov process because the probability of a trap changing state (e.g., capturing an electron) depends only on its present state. It has no memory of past states. This is the basic definition of Markov processes. Furthermore, in keeping with this property, the holding times in each state are given by simple exponential distributions, with the characteristic time for t_e being τ_e and the characteristic time for t_c being τ_c.

In recent years RTN has come to be recognized as a significant source of variability in highly scaled MOSFETs [2–4]. Since it has been shown to cause effective V_T shifts that increase inversely with shrinking channel area and that can reach >100 mV in very small FETs, it is seen as especially problematic for future memory cells, because they make use of the smallest possible transistors.

More recently, it has been shown that most bias stress effects can be accounted for as due to the same sort of traps that cause RTN. The only difference is that bias stress measurements activate traps that require higher voltages and have longer time constants [5]. Thus, by carefully analyzing the RTN at low voltages in small FETs, we can learn a great deal about traps that can be applied to the higher voltage bias stress regime [6].

In this chapter we first briefly discuss techniques for measuring RTN and then describe analysis techniques for extracting information from the RTN time series,

[1]To be precise, this description corresponds to an acceptor-type trap, but we prefer not to focus on the trap types because some traps defy easy categorization. Instead, we simply use the preceding definitions of "capture" and "emission" for all traps when plotting $|I_D|$ versus time.

including particularly an HMM approach. Then we discuss the results of RTN measurements, including the amplitude distributions of RTN and the dependence of the characteristic times on the gate voltage, which leads to hysteretic effects and history-dependent logic switching times. Finally, we discuss RTN measurements in SRAM arrays and then conclude.

5.2 Measurement Techniques

Statistical analysis is essential in RTN measurement since both amplitude and time constants show remarkable variation from device to device. For example, amplitude is known to follow a long-tailed non-Gaussian distribution [2] while characteristic times vary over 10 or more orders of magnitude. Figure 5.2a shows a test array structure designed for easy measurement of large numbers of devices (27,000/die in [7]). As shown in Fig. 5.2b, the drain current (I_D) waveform is first converted to a

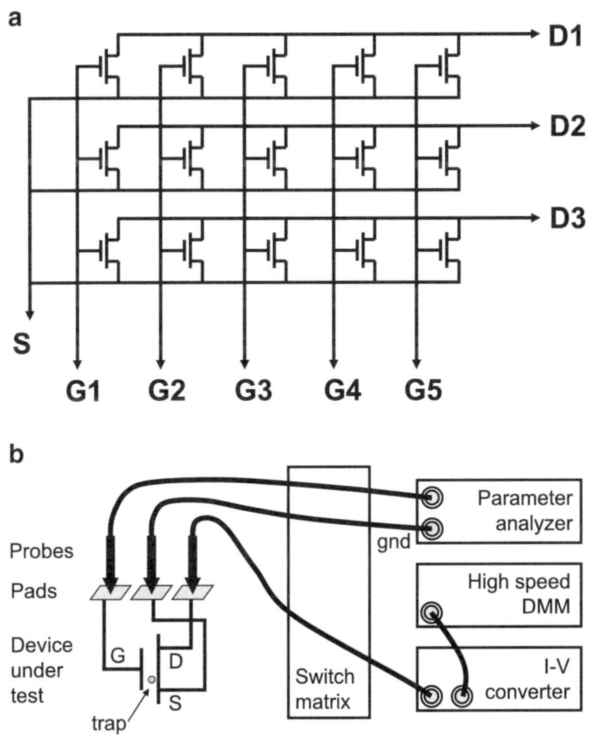

Fig. 5.2 (**a**) Schematic diagram of a small array of FETs, illustrating how one can achieve efficient use of probe pads (15 FETs per 9 probe pads, here) and (**b**) system for high-speed measurement of RTN time series. Adapted from [7]

voltage signal using a low-noise I–V converter, and the voltage signal is sampled by a high-speed digital multimeter. Alternatively, we have also used a fast measurement unit (Agilent 1530A), which enables a wide bandwidth of up to 1 MHz, to measure the RTN signals with flexible biasing conditions. In the work discussed in this chapter, we have used between 10^5 and several $\times 10^6$ sampling points for each time series. The time series were measured at a drain voltage of 50 mV with a gate voltage adjusted for the target I_D (e.g., 1 μA) to minimize effects of device variation other than RTN. All of the devices discussed here are gate-first high-k/metal gate SOI MOSFETs.

5.3 Analysis Techniques

5.3.1 Conventional Methods

The most widely used method to analyze an I_D time series is the histogram method [8]. As shown in Fig. 5.3a, if one makes a histogram of the number of times that a given value of I_D is observed, the counts show two peaks blurred by underlying Gaussian noise. The positions of the peaks correspond to the high and low levels of the RTN. If the noise is small and the peaks are well separated, one can readily assign the trap state for each sampling time and then determine the average time constants by averaging the holding durations for each state. For moderate noise levels (as in Fig. 5.3a), a more sophisticated method to determine the average time constant is given in [8]. The other frequently used graphical analysis technique is the so-called time-lag plot shown in Fig. 5.3b, where one plots the measured value of I_D at each time step against the measured value at the immediately preceding time step [9]. This results in two clusters of points, corresponding to the high and low RTN levels.

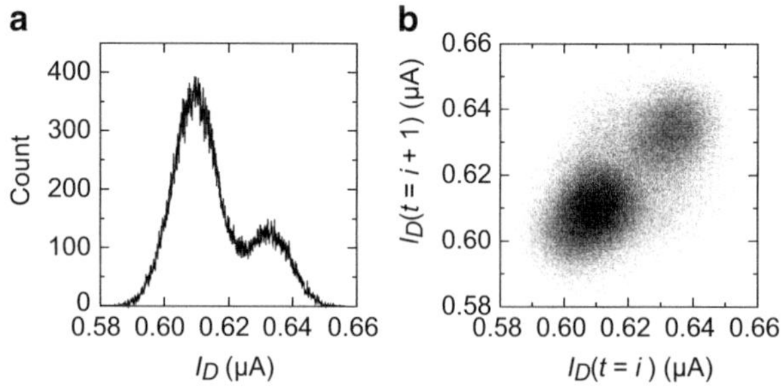

Fig. 5.3 Conventional RTN data analysis methods. (**a**) Histogram method, (**b**) lag plot (© 2012 IEEE. Reprinted, with permission, from [4])

Unfortunately, these methods tend to drop smaller traps because they are only effective when the trap amplitude is larger than the RMS noise level. In Fig. 5.3 only two peaks are readily seen, corresponding to a single trap, but in reality, the time series used to generate both of these plots is composed of responses from *two* traps, with $\Delta I = 22$ nA and 7 nA, thus the analyses should have resulted in four peaks or four clusters. Even though one might recognize the other smaller trap from the slightly asymmetric shape of the peaks (histogram) or the elongated clusters (lag plot), it is almost impossible to quantitatively extract the smaller amplitude trap in either conventional method.

5.3.2 *Hidden Markov Model*

A more reliable extraction method for RTN time constants and amplitudes is based on modeling the time series using a hidden Markov model [6, 10, 11]. Baum–Welch [12] and Viterbi [13] algorithms are used to find the optimal parameters and to obtain the most likely trap states at each time step, respectively. The algorithm can be implemented for any number of traps, but for the sake of illustration, we consider here RTN caused by two traps and one noise source.

We define the HMM as follows. Let $\vec{x} = (x_1, x_2, \ldots, x_n)$ be the values measured in a time series, a set of n independent uniformly spaced observations of a 2-trap system with an underlying Gaussian distribution of noise, and let $\vec{z} = (z_1, z_2, \ldots, z_n)$ and $\vec{Z} = (Z_1, Z_2, \ldots, Z_n)$ be the latent variables that specify which of the trap states has generated each observation, z's for the first trap and Z's for the second trap. Two-state traps are assumed here,[2] so each of the z's and Z's is either 1 or 2.

Define the transition probabilities as $P(z_i = j | z_{i-1} = k) = \gamma_{1kj}$ and $P(Z_i = j | Z_{i-1} = k) = \gamma_{2kj}$, for $j,k \in \{1,2\}$, and let the initial state probabilities be $P(z_1 = j) = \gamma_{1\_j}$ and $P(Z_1 = j) = \gamma_{2\_j}$, where $\gamma_{ij1} + \gamma_{ij2} = 1$, for $i \in \{1,2\}$ and $j \in \{\_, 1, 2\}$. In addition, define the additive noise as normally distributed:

$$P(x_i | z_i = j, Z_i = k) = N\left(x_i; \mu_0 + \delta_{2j}\mu_1 + \delta_{2k}\mu_2, \sigma\right),$$

where $j,k \in \{1,2\}$, and $N(x,\mu,\sigma) = \frac{1}{\sqrt{2\pi}\sigma}\exp\left(-\frac{1}{2}(x-\mu)^2/\sigma^2\right)$ is the normal distribution function, μ is the mean, σ is the standard deviation, and δ is the Kronecker delta.

The Baum–Welch algorithm is an expectation-maximization algorithm aimed at finding the best values for the set of parameters required in the preceding definitions:

$$\theta = \{\gamma_{111}, \gamma_{122}, \gamma_{1\_1}, \gamma_{211}, \gamma_{222}, \gamma_{2\_1}, \mu_0, \mu_1, \mu_2, \sigma\}.$$

[2] Some trap models involve three or four internal states but only two observable charge states. Those models can readily be incorporated into this framework, but are not explicitly considered here.

This set of parameters defines the characteristics of the traps and the noise. The algorithm considers the likelihood that the observed measurements $\vec{x}$ resulted from the parameters θ and from the state sequences $\vec{z}$ and $\vec{Z}$ and seeks to maximize the expectation value of this likelihood with respect to the set of parameters. This likelihood function is given by

$$\begin{aligned} L\left(\theta;\vec{x},\vec{z},\vec{Z}\right) &= P(\vec{z};\theta)\,P\left(\vec{Z};\theta\right)P\left(\vec{x}\middle|\vec{z},\vec{Z};\theta\right) \\ &= \prod_{i=1}^{n} \gamma_{1z_{i-1}z_i}\gamma_{2Z_{i-1}Z_i}N\left(x_i;\mu_0+\delta_{2z_i}\mu_1+\delta_{2Z_i}\mu_2,\sigma\right), \end{aligned}$$

where $\gamma_{iz_0z_1}$ is taken to mean $\gamma_{i\_z_1}$. To enable tractable calculations, the log of the likelihood is used:

$$\ln L\left(\theta;\vec{x},\vec{z}\right)=\sum_{i=1}^{n}\left[\begin{array}{l}\ln\gamma_{1z_{i-1}z_i}+\ln\gamma_{2Z_{i-1}Z_i}-\frac{1}{2}\left(\dfrac{x_i-\mu_0-\delta_{2z_i}\mu_1-\delta_{2Z_i}\mu_2}{\sigma}\right)^2 \\ -\ln\sigma-\frac{1}{2}\ln 2\pi\end{array}\right]\equiv\sum_{i=1}^{n}\ln(L_i).$$

The expectation value of $\ln(L)$ is the weighted average of $\ln(L)$ over all possible state sequences, where the weight for each sequence is the probability that that sequence could actually occur, given $\vec{x}$ and θ:

$$\begin{aligned} Q\left(\theta;\theta^{(t)}\right) &= E\left(\ln L\left(\theta;\vec{x},\vec{z},\vec{Z}\right)\right) \\ &= \sum_{z_1,Z_1}\sum_{z_2,Z_2}\cdots\sum_{z_n,Z_n} P\left(\vec{z},\vec{Z}\middle|\vec{x};\theta^{(t)}\right)\ln L\left(\theta;\vec{x},\vec{z},\vec{Z}\right) \end{aligned} \tag{5.1}$$

where the superscript t is the iteration index. To make the maximization tractable, the algorithm uses an iterative approach in which the parameters in the weights are assigned the values from the previous iteration (since the weights are too complicated to be usefully differentiated), and only the parameters in $\ln(L)$ are maximized (since they are readily differentiated). Then, one iterates

$$\theta^{(t+1)} \Leftarrow \underset{wrt\ \theta}{Max}\left(Q\left(\theta;\theta^{(t)}\right)\right).$$

According to Bayes' theorem, the weights can be rewritten as

$$P\left(\vec{z},\vec{Z}\middle|\vec{x};\theta^{(t)}\right)=\frac{P\left(\vec{x}\middle|\vec{z},\vec{Z};\theta^{(t)}\right)P\left(\vec{z},\vec{Z};\theta^{(t)}\right)}{P\left(\vec{x};\theta^{(t)}\right)}. \tag{5.2}$$

Expanding Eqs. (5.1) and (5.2), we obtain

$$Q\left(\theta;\theta^{(t)}\right)=E\left(\ln L\left(\theta;\vec{x},\vec{z},\vec{Z}\right)\right)$$

$$=\frac{\sum\limits_{i=1}^{n}\sum\limits_{z_1,Z_1}\sum\limits_{z_2,Z_2}\cdots\sum\limits_{z_n,Z_n}\prod\limits_{j=1}^{n}\left[\gamma_{1z_{j-1}z_j}^{(t)}\gamma_{2Z_{j-1}Z_j}^{(t)}N\left(x_j;\mu_0^{(t)}+\delta_{2j}\mu_1^{(t)}+\delta_{2k}\mu_2^{(t)},\sigma^{(t)}\right)\ln(L_i)\right]}{\sum\limits_{all\,\vec{z},\vec{Z}}\prod\limits_{j=1}^{n}\gamma_{1z_{j-1}z_j}^{(t)}\gamma_{2Z_{j-1}Z_j}^{(t)}N\left(x_j;\mu_0^{(t)}+\delta_{2j}\mu_1^{(t)}+\delta_{2k}\mu_2^{(t)},\sigma^{(t)}\right)}.$$

To evaluate this, it is useful to first break it down into smaller pieces. The large number of sums and products can be expressed as matrix operations such that the expectation value of functions that depend only on the states at a single point in time, z_i and Z_i, can be written in simplified form as

$$\overline{f(z_i,Z_i)}\equiv E(f(z_i,Z_i)\,|\vec{x},\theta)=\sum_{j,k\in\{1,2\}}f(j,k)\widehat{S}_{ijjkk},$$

and the expectation value of functions depending on consecutive time points, z_i, z_{i-1}, Z_i, and Z_{i-1}, can be written as

$$\overline{\overline{f(z_i,z_{i-1},Z_i,Z_{i-1})}}\equiv E(f(z_i,z_{i-1},Z_i,Z_{i-1})\,|\vec{x},\theta)=\sum_{\substack{j,k,l,m\\ \in\{1,2\}}}f(j,k,l,m)\widehat{S}_{ijklm}.$$

Here, the single and double overbars distinguish the two different types of expectation values, and $\widehat{S}_{ijklm}=\frac{\vec{A}_i^{\mathrm{T}}\cdot\hat{I}_{kjml}\cdot\vec{B}_i}{\vec{A}_1^{\mathrm{T}}\cdot\vec{v}_0}=\overline{\overline{\delta_{z_ij}\delta_{z_{i-1}k}\delta_{Z_il}\delta_{Z_{i-1}m}}}$ is a product of vectors and matrices, where $\hat{I}_{kjml}$ is the 4×4 matrix with 1 for the element that corresponds to the $k\to j$ transition for the first trap and the $m\to l$ transition for the second trap and has zeros for all of the other elements. The A's and B's are matrix product chains:

$$\vec{A}_i^{\mathrm{T}}=\begin{bmatrix}1&1&1&1\end{bmatrix}\cdot\hat{N}_{x_n}\cdot\left(\hat{T}_0\cdot\hat{N}_{x_{n-1}}\right)\cdots\left(\hat{T}_0\cdot\hat{N}_{x_i}\right)$$
$$\vec{B}_i=\left(\hat{T}_0\cdot\hat{N}_{x_{i-1}}\right)\cdots\left(\hat{T}_0\cdot\hat{N}_{x_1}\right)\cdot\vec{v}_0.$$

(Note that $\vec{A}_n^{\mathrm{T}}=\begin{bmatrix}1&1&1&1\end{bmatrix}\cdot\hat{N}_{x_n}$ and $\vec{B}_1=\vec{v}_0$ and that $\vec{A}_i^{\mathrm{T}}\cdot\vec{B}_i=\vec{A}_1^{\mathrm{T}}\cdot\vec{v}_0$ for all i.) The basic elements of these chains are the transition probability matrix, the Gaussian probability matrix, and the initial state probability vector:

$$\hat{T}_0=\begin{bmatrix}\gamma_{111}^{(t)}\gamma_{211}^{(t)}&\gamma_{121}^{(t)}\gamma_{211}^{(t)}&\gamma_{111}^{(t)}\gamma_{221}^{(t)}&\gamma_{121}^{(t)}\gamma_{221}^{(t)}\\ \gamma_{112}^{(t)}\gamma_{211}^{(t)}&\gamma_{122}^{(t)}\gamma_{211}^{(t)}&\gamma_{112}^{(t)}\gamma_{221}^{(t)}&\gamma_{122}^{(t)}\gamma_{221}^{(t)}\\ \gamma_{111}^{(t)}\gamma_{212}^{(t)}&\gamma_{121}^{(t)}\gamma_{212}^{(t)}&\gamma_{111}^{(t)}\gamma_{222}^{(t)}&\gamma_{121}^{(t)}\gamma_{222}^{(t)}\\ \gamma_{112}^{(t)}\gamma_{212}^{(t)}&\gamma_{122}^{(t)}\gamma_{212}^{(t)}&\gamma_{112}^{(t)}\gamma_{222}^{(t)}&\gamma_{122}^{(t)}\gamma_{222}^{(t)}\end{bmatrix}$$

$$\hat{N}_x = \begin{bmatrix} N\left(x;\mu_0^{(t)},\sigma^{(t)}\right) & 0 & 0 & 0 \\ 0 & N\left(x;\mu_0^{(t)}+\mu_1^{(t)},\sigma^{(t)}\right) & 0 & 0 \\ 0 & 0 & N\left(x;\mu_0^{(t)}+\mu_2^{(t)},\sigma^{(t)}\right) & 0 \\ 0 & 0 & 0 & N\left(x;\mu_0^{(t)}+\mu_1^{(t)}+\mu_2^{(t)},\sigma^{(t)}\right) \end{bmatrix}$$

$$\vec{\upsilon}_0 = \begin{bmatrix} \gamma_{1\_1}^{(t)}\gamma_{2\_1}^{(t)} \\ \gamma_{1\_2}^{(t)}\gamma_{2\_1}^{(t)} \\ \gamma_{1\_1}^{(t)}\gamma_{2\_2}^{(t)} \\ \gamma_{1\_2}^{(t)}\gamma_{2\_2}^{(t)} \end{bmatrix}.$$

These are for two 2-state traps. For m k-state traps, the vectors would have k^m elements, and the matrices would be $k^m \times k^m$.

Using the preceding equations, $E(\ln L)$ can be evaluated as

$$Q\left(\theta\,\middle|\,\theta^{(t)}\right) = E\left(\ln L\left(\theta;\vec{x},\vec{z},\vec{Z}\right)\right) = \sum_{i=2}^{n}\sum_{\substack{j,k,l,m\\ \in\{1,2\}}} \overset{\frown}{S}_{ijklm}\ln\left(\gamma_{1jk}\gamma_{2lm}\right) + \sum_{\substack{j,k\\ \in\{1,2\}}} \overset{\frown}{S}_{1jjkk}\ln\left(\gamma_{1\_j}\gamma_{2\_k}\right)$$
$$-\frac{1}{2\sigma^2}\sum_{i=1}^{n}\sum_{\substack{j,k\\ \in\{1,2\}}} \overset{\frown}{S}_{ijjkk}\left(x_i-\mu_0-\delta_{2j}\mu_1-\delta_{2k}\mu_2\right)^2 - n\ln\sigma - \tfrac{n}{2}\ln(2\pi)$$

The next step is to maximize Q with respect to the 10 parameters in θ. For the transition probabilities, this gives (using $\overset{\frown}{S}{}^{(t)}_{11111}+\overset{\frown}{S}{}^{(t)}_{12211}+\overset{\frown}{S}{}^{(t)}_{11122}+\overset{\frown}{S}{}^{(t)}_{12222}=1$)

$$\gamma_{1\_1}^{(t+1)} = \sum_{k\in\{1,2\}} \overset{\frown}{S}{}^{(t)}_{111kk} = 1-\gamma_{1\_2}^{(t+1)} \quad \text{and} \quad \gamma_{2\_1}^{(t+1)} = \sum_{k\in\{1,2\}} \overset{\frown}{S}{}^{(t)}_{1kk11} = 1-\gamma_{2\_2}^{(t+1)},$$

and

$$\gamma_{1jk}^{(t+1)} = \frac{\left\langle \overline{\overline{\delta_{jz_{i-1}}\,\delta_{kz_i}}} \right\rangle_2}{\left\langle \overline{\overline{\delta_{jz_{i-1}}}} \right\rangle_2} \quad \text{and} \quad \gamma_{2jk}^{(t+1)} = \frac{\left\langle \overline{\overline{\delta_{jZ_{i-1}}\,\delta_{kZ_i}}} \right\rangle_2}{\left\langle \overline{\overline{\delta_{jZ_{i-1}}}} \right\rangle_2},$$

where the average

$$\langle u_i\rangle_\alpha \equiv \frac{1}{n+1-\alpha}\sum_{i=\alpha}^{n} u_i$$

is summed starting at $i=2$ for transition probabilities. The unspecified dependencies in the expectation values must be summed over. For example, $\overline{\overline{\delta_{jz_{i-1}}\,\delta_{kz_i}}} = \sum_{l,m\in\{1,2\}} \overset{\frown}{S}_{ijklm}$ and $\overline{\overline{\delta_{jz_{i-1}}}} = \sum_{k,l,m\in\{1,2\}} \overset{\frown}{S}_{ijklm}$.

For the amplitudes (μ_0,μ_1,μ_2), the maximum can be found by solving the matrix equation

$$\begin{bmatrix} \langle x_i \rangle_1 \\ \left\langle x_i \overline{\delta_{2z_i}} \right\rangle_1 \\ \left\langle x_i \overline{\delta_{2Z_i}} \right\rangle_1 \end{bmatrix} = \begin{bmatrix} 1 & \left\langle \overline{\delta_{2z_i}} \right\rangle_1 & \left\langle \overline{\delta_{2Z_i}} \right\rangle_1 \\ \left\langle \overline{\delta_{2z_i}} \right\rangle_1 & \left\langle \overline{\delta_{2z_i}} \right\rangle_1 & \left\langle \overline{\delta_{2z_i} \delta_{2Z_i}} \right\rangle_1 \\ \left\langle \overline{\delta_{2Z_i}} \right\rangle_1 & \left\langle \overline{\delta_{2z_i} \delta_{2Z_i}} \right\rangle_1 & \left\langle \overline{\delta_{2Z_i}} \right\rangle_1 \end{bmatrix} \cdot \begin{bmatrix} \mu_0^{(t+1)} \\ \mu_1^{(t+1)} \\ \mu_2^{(t+1)} \end{bmatrix},$$

where here the averages are summed from $i = 1$. Using these amplitudes, the value of σ at the maximum can be found:

$$\sigma^{(t+1)} = \sqrt{\frac{1}{n} \sum_{i=1}^{n} \sum_{\substack{j,k \\ \in\{1,2\}}} \widehat{S}_{ijjkk} \left(x_i - \mu_0^{(t+1)} - \delta_{2j} \mu_1^{(t+1)} - \delta_{2k} \mu_2^{(t+1)} \right)^2}.$$

Having obtained these parameter values, one uses them to compute improved estimates for the matrices, continuing this loop until the values are sufficiently converged.

After iterating to convergence, the expectation values of the trap states are

$$\overline{\delta_{2z_i}} = 1 - \overline{\delta_{1z_i}} = \sum_{k \in\{1,2\}} \widehat{S}_{i22kk} \quad \text{and} \quad \overline{\delta_{2Z_i}} = 1 - \overline{\delta_{1Z_i}} = \sum_{k \in\{1,2\}} \widehat{S}_{ikk22}.$$

The algorithm also yields the transition probabilities, from which the characteristic capture and emission times can readily be obtained, as

$$\tau_c^{(1)} = \frac{\Delta t}{\gamma_{121}} \quad \text{and} \quad \tau_e^{(1)} = \frac{\Delta t}{\gamma_{112}},$$

for the first trap, and similarly for the second, where Δt is the time between consecutive points in the time series, and state 1 corresponds to the state in which the trap contains a captured electron.

Once all 10 parameters in θ are optimized, one could in principle determine the probability of each sequence of trap states and thus identify the one sequence that is most likely. The number of candidate sequences is, however, so huge (4^n) that one-by-one calculation is intractable. The Viterbi algorithm [13] is used to manage this problem utilizing the memoryless characteristic of RTN. Starting from the most likely final states (z_n, Z_n) determined by their expectation values, the algorithm traces the most likely path back to the beginning (z_1, Z_1). At each step the calculation only involves the present point and the immediately preceding point, because the transition probability between neighboring points is independent of trap states other than the two points under consideration. This reduces the computational complexity to of order n.

Usually the Viterbi algorithm would proceed from the first point in the time series, but strictly speaking, the path obtained from this algorithm is not always the most likely near its starting point, since the state of the starting point of the trace

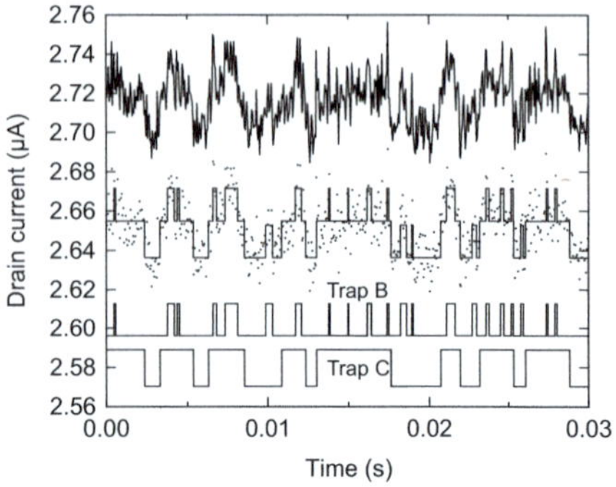

Fig. 5.4 Waveform (*top*) and extracted RTN states (*bottom* two traces). Center trace is the combined trap states overlaying the data points. This data is for the $V_g = 0.725$ V point in Fig. 5.5 (Adapted from [4])

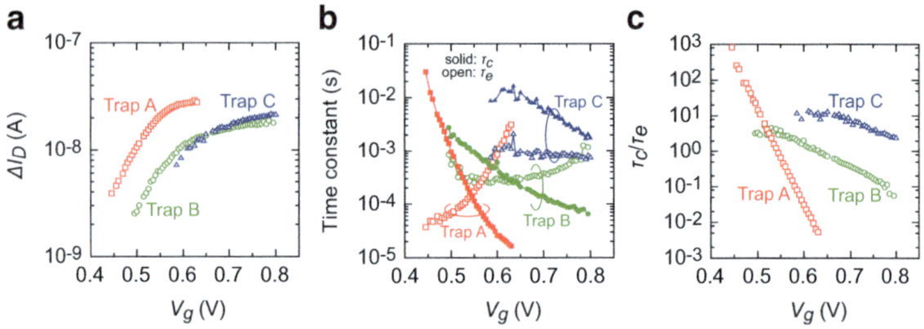

Fig. 5.5 Example of gate-voltage dependence of extracted RTN parameters. (**a**) ΔI_D, (**b**) τ_e and τ_c, and (**c**) τ_e/τ_c. Sampling time $\Delta t = 10$ μs and total time = 0.1 s. Adapted from [4]

is not absolutely defined but estimated from expectation values. To accommodate some of the experiments discussed in Sect. 5.5, which focus on the trap states at the beginning of the time series, we choose to trace the path in the backward direction, namely, from (z_n, Z_n) to (z_1, Z_1), so that it will be more accurate at the beginning.

Figure 5.4 illustrates the application of this HMM algorithm to a particular piece of an RTN time series. The top trace shows the measured time series, while the bottom traces show the trap states extracted using this algorithm. One obtains not only the step heights corresponding to the traps but also their timing. It is difficult to tell by eye what trap state(s) to assign to this time series, but the algorithm correctly finds that it is two distinct traps with very similar amplitudes. This is confirmed by considering Fig. 5.5, which shows a whole set of extracted parameters for this same FET, where a separate time series has been measured for every 5 mV increment in the gate voltage, and then the HMM extraction has been applied to each. The smooth variation of the amplitude and time constants with changing gate voltage supports the validity of the extraction, since each time series is different and each extraction is independent of the others. The data in Fig. 5.4 occurs at $V_g = 0.725$ V in Fig. 5.5c, a bias regime in which only traps B and C are active.

As can be seen, Fig. 5.5 is an example of an FET with three RTN-generating traps, which are successfully sorted out by this algorithm. The gate-voltage dependence of the amplitude and time constants for each trap is like a fingerprint, uniquely identifying the trap. As will be discussed later, a wide range of behaviors

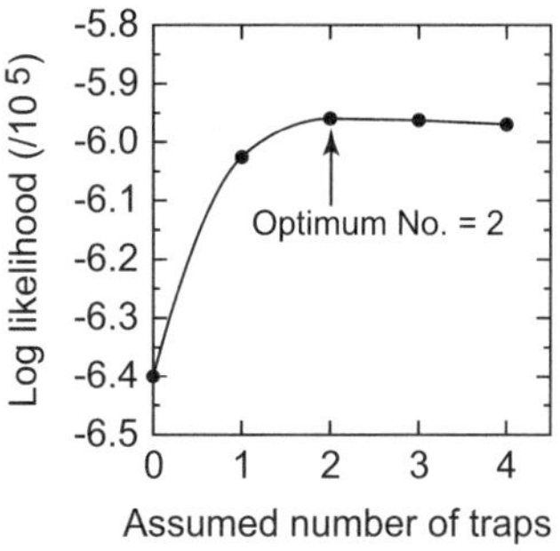

Fig. 5.6 Change of log likelihood as a function of assumed trap number (© 2012 IEEE. Reprinted, with permission, from [4])

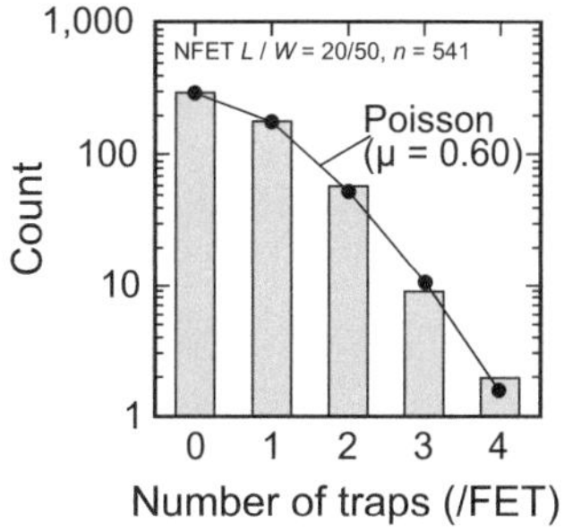

Fig. 5.7 Distribution of number of traps in a device (© 2012 IEEE. Reprinted, with permission, from [4])

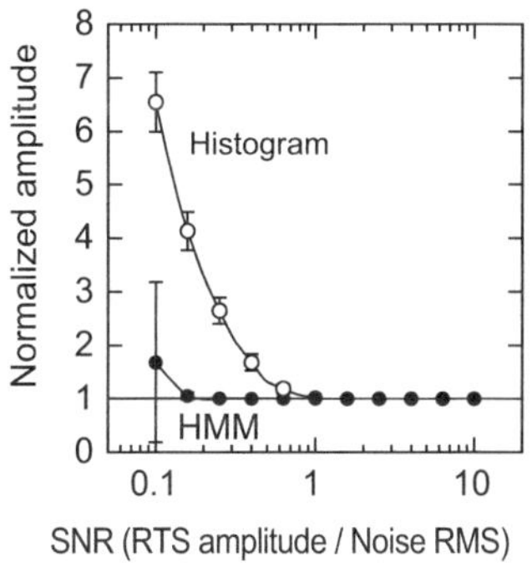

Fig. 5.8 Comparison of interference of white noise in amplitude extraction. For these synthetic time series the correct normalized amplitude is 1 (© 2012 IEEE. Reprinted, with permission, from [4])

can be found. The algorithm can successfully extract fits to a large number of traps, the primary limitation being computing time, which goes as $n \times 2^{2m}$, where n is the number of points in the time series and m is the number of 2-state traps. One can determine the optimal number of traps to fit a given device by looking for the maximum log likelihood, as illustrated in Fig. 5.6. Collecting data from the analysis of a large number of FETs, we have found that the number of traps in each device follows a Poisson distribution, as shown in Fig. 5.7 for 20 × 50 nm nFETs.

The HMM extraction method is a significant improvement over the histogram and lag methods of analyzing RTN. As shown in Fig. 5.8, for white noise it can accurately extract RTN amplitudes at a 5× lower signal-to-noise ratio (SNR) than the histogram method. We have also shown that it can accurately extract the trap time constants down to SNR = 0.2, while the histogram approach fails below

3.0 [11]. In the time domain, HMM can accurately resolve RTN parameters for time constants between ~5× the sampling period and ~1/10 the total measurement time, for SNR = 1. The lower bound occurs because the sampling does not capture all of the transitions. The upper bound occurs because the number of observed transitions is insufficient. These effects can be seen in Fig. 5.5: trap A isn't seen above 0.64 V because its τ_e becomes too short, trap C isn't seen below 0.6 V because its τ_c is too long, and trap B isn't seen below 0.5 V because its amplitude is too low.

The HMM approach is somewhat vulnerable to correlated signals, such as 1/f noise and sinusoidal pickup. This is not a significant problem, though, since these can be distinguished in the extracted parameters because they always have $\tau_c \approx \tau_e$, independent of gate bias [11].

5.4 Amplitude in HKMG and Undoped Channel

The amplitude of RTN is the most important parameter in assessing its impact on circuit operation. In the simplest electrostatic view, the ΔV_T caused by an RTN trap could be estimated as the voltage required to change the device charge by one electron: $q/A_{ch}C_{ox}$, where q and C_{ox} are elementary charge and gate-oxide capacitance per unit area, respectively, and A_{ch} is the channel area. Although ΔV_T can be modified by the distance of a trap from the oxide/silicon interface, the depth effect should always result in ΔV_T smaller than $q/A_{ch}C_{ox}$. Thus, one would guess that ΔV_T would follow a normal distribution with a tail on the low side. However, as shown in Fig. 5.9a, the observed distribution is entirely different than the simple estimation; it follows a wide distribution with a very long tail to *higher* values.

This wide distribution occurs both in per device data and in per trap data. In our work [2] and that of Realov [10], the distribution is found to be log-normal, as in Fig. 5.9a, while other workers [14–16] using different types of measurements have

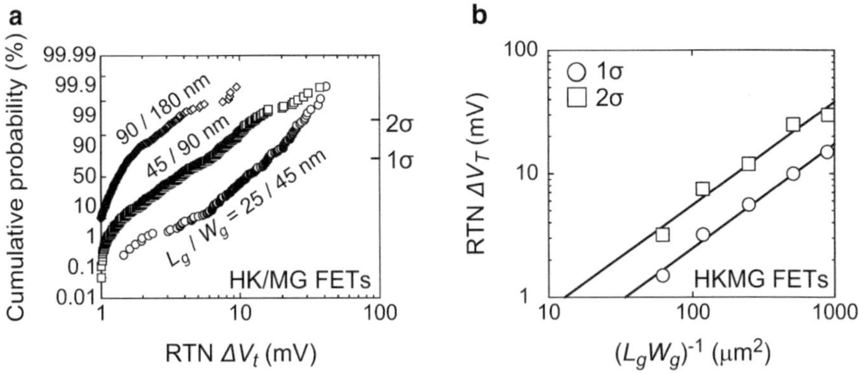

Fig. 5.9 Dimensional dependence of the distribution of per device RTN amplitude in HKMG nFETs. (**a**) Log-normal cumulative distribution for several sizes. (**b**) Dependence of the 1 and 2 sigma points on 1/area (© 2012 IEEE. Reprinted, with permission, from [4])

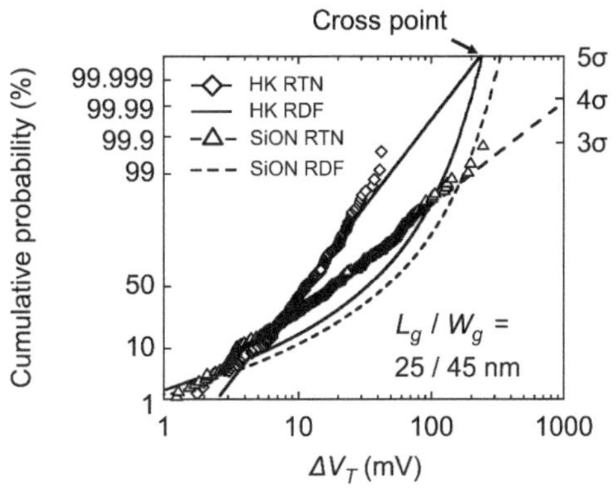

Fig. 5.10 Comparison between RTN and RDF threshold voltage shifts for PDSOI nFETs (© 2012 IEEE. Reprinted, with permission, from [4])

found trap amplitude distributions that are better characterized as being exponential. There is not presently any clear explanation for the different results.

This wide distribution is one of the most notable features of RTN and is thought to be caused at least in part by percolation and multiple-trap effects [17, 18]. Percolation occurs because the discrete randomly located dopants in highly scaled FETs cause local variations of the electrostatic potential in the channel, resulting in randomly located channels through which the drain current must flow. If a trap is strategically located near a bottleneck in one of these percolation paths, its changing charge state will modulate the drain current more than expected from the simple electrostatic view, forming the long tail to higher values. It should be noted that one consequence of this model is that it is very difficult to deduce the position of a trap in a FET using the amplitude. In addition, when the average number of traps in a FET is small (e.g., <1), there are still some FETs with extra traps, due to Poisson statistics. These multiple-trap amplitudes add, further stretching the tail of the distribution on a per device basis.

The amplitude is found to grow rapidly as the dimensions of FETs scale down. In particular, this is true of the high-amplitude tails, as shown in Fig. 5.9b, where 1σ and 2σ values are plotted as a function of reciprocal active area. This strong dependence on the dimension is thought to be attributable to scaling down of gate capacitance with no change of trapped charge ($=q$). This leads to a serious issue in future scaled devices, since conventional sources of variability such as random dopant fluctuation (RDF) only scale in proportion to $1/(\text{area})^{0.5}$. In other words, RTN grows faster than RDF. In 22 nm technology, however, device variability is still dominated by RDF due to increased C_{ox} and reduced trap density in optimized HKMG technology (Fig. 5.10).

If we look to the future, beyond the 22 nm node, another enabling technology is needed to reduce RTN. One hope is that RTN may decrease in future undoped channel devices due to the absence of drain current percolation paths in the absence of doping. Figure 5.11 compares tail distributions for doped partially depleted (PD) SOI and undoped extremely thin (ET) SOI near V_T and in strong inversion. The slopes at the tail are plotted as a function of gate overdrive in Fig. 5.12, showing that

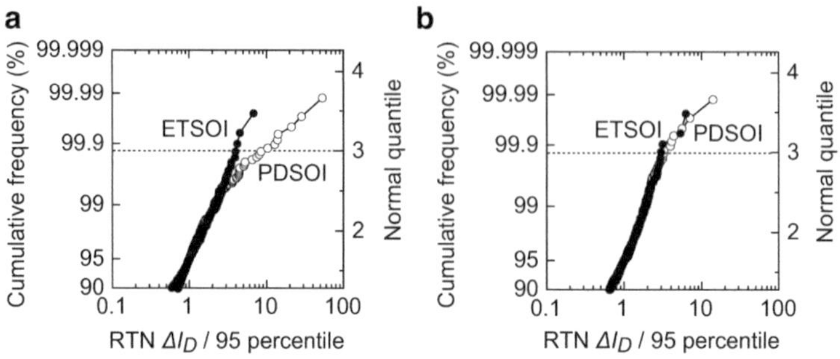

Fig. 5.11 Tail distribution of ΔI_D in ET-SOI and PD-SOI nFETs ($L/W = 20/50$ nm). (**a**) Near threshold, $V_g \sim V_T$, and (**b**) above threshold, with $V_g - V_T \sim 0.3$ V (© 2012 IEEE. Reprinted, with permission, from [4])

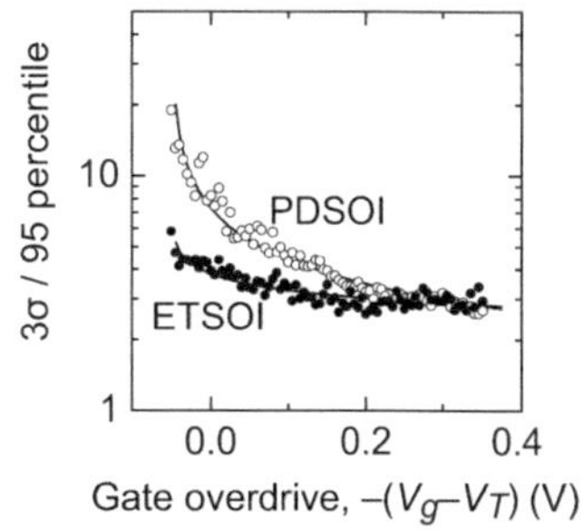

Fig. 5.12 Ratio of 3σ RTN amplitude to 1.64σ (95% point) RTN amplitude in nFETs ($L/W = 20/50$ nm) as a function of gate overdrive (© 2012 IEEE. Reprinted, with permission, from [4])

improvement for undoped channels is only found near V_T, but not at higher voltage. This indicates that the discrete dopant percolation path effect is only operative in subthreshold. At strong inversion it disappears due to screening, but we believe that in its place RTN amplitude is enhanced by gate-stack-induced current percolation paths. Consequently, RTN needs to be watched in future generations even when the channel is not intentionally doped.

For simplicity, we have discussed here the scaling and doping effects solely using nFETs, but these comments also apply to pFETs, since we have measured both and have not seen any clear difference in RTN between n- and pFETs as long as the fundamental structures of their HKMG stack and channel are shared. Hence, we do not explicitly discuss the difference regarding the channel polarity in the following sections.

5.5 Gate-Voltage Dependence and Hysteretic Effect

Average time-to-capture (τ_c) and time-to-emission (τ_e) are reciprocals of transition probability per unit time from empty state to occupied and occupied to empty, respectively. Hence, V_g exponentially modulates the two time constants because it

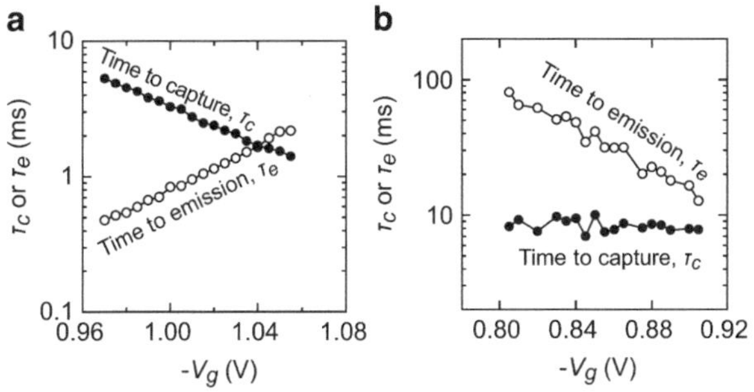

Fig. 5.13 Dependence of time constants on V_g for (**a**) positively and (**b**) negatively coupled RTN traps in pFETs (© 2012 IEEE. Reprinted, with permission, from [4])

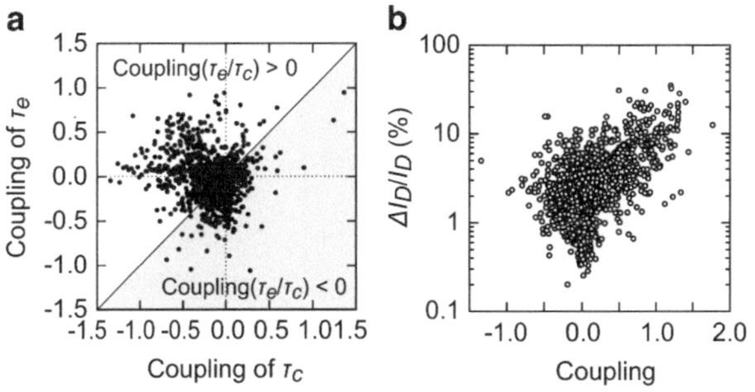

Fig. 5.14 (**a**) Scatter plot of couplings of τ_e and τ_c. (**b**) Correlation between ΔI_D and coupling of τ_e/τ_c to V_g. Samples are pFETs (© 2012 IEEE. Reprinted, with permission, from [4])

linearly changes the potential barrier between the channel and a trap. Figure 5.13a shows the most frequently observed dependence, wherein τ_c decreases with $|V_g|$ while τ_e increases. This coupling to V_g is reasonable since a carrier should be likely to stay at a trap when the potential of the trap is decreased. However, reversely coupled traps [19] are also found (Fig. 5.13b), which reveals the remarkable variety of traps in HKMG. We define the coupling factor of a time constant to V_g as the slope of $\ln(\tau)$ versus $|V_g|$ normalized by kT/q. Figure 5.14a shows a scatter plot of couplings of τ_c and τ_e, from which it can be seen that the couplings of the ratio of the two time constants (τ_e/τ_c) are usually positive (i.e., above the diagonal line), but a number of negatively coupled traps are also found. Among the variety of the couplings, strongly positive coupling is important as it often accompanies large amplitude, as shown in the correlation plot in Fig. 5.14b.

The temperature dependence of individual traps has been studied by measuring time series on the same set of FETs at different temperatures. The data shows that the V_g coupling is generally unchanged with temperature and that both τ_e and τ_c undergo similar temperature acceleration [20], as seen in the example in Fig. 5.15. This seems reasonable when one considers that RTN is only observable when its time-constant ratio is within a few orders of magnitude of unity (see

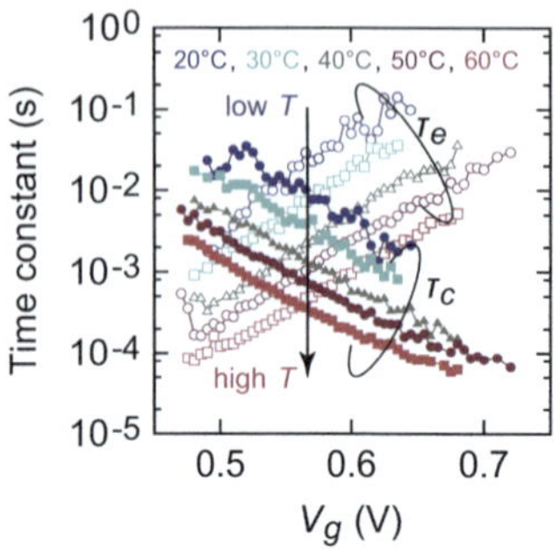

Fig. 5.15 Time constant as a function of V_g at temperatures from 20 to 60 °C. The sample is an undoped nFET with $L/W = 20/50$ nm. Adapted from [4]

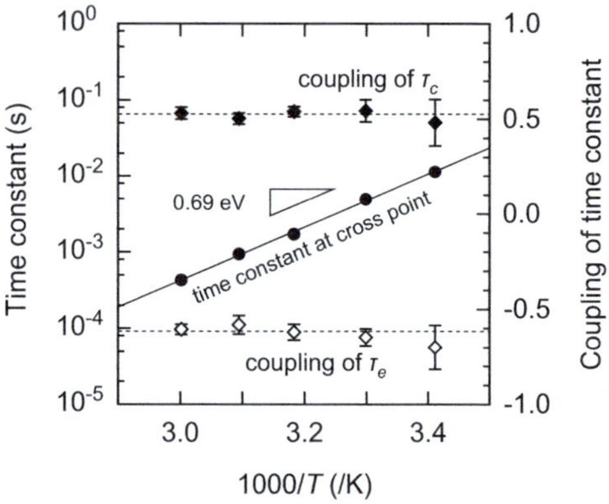

Fig. 5.16 Temperature dependence of time constants at intersection points (where $\tau_c = \tau_e$) and couplings of time constants to V_g. The sample is an undoped nFET, $L/W = 20/50$ nm (© 2012 IEEE. Reprinted, with permission, from [4])

Fig. 5.5c). Hence, the energy difference between captured and empty states is small (<~0.1 eV) in the narrow window for which RTN is visible. This leads to similar barrier height for capture and emission, resulting in the almost identical temperature acceleration and the unchanged coupling. The activation energy of the time constant at the cross point is 0.69 eV, as shown in Fig. 5.16, and presumably corresponds to the thermal barrier between the capture and emission states. The temperature dependence of threshold voltage may account for the 20 mV movement of the cross point ($\tau_c = \tau_e$) in Fig. 5.15.

The significance of the V_g dependence is found in the transient characteristics at turn-on or turnoff of FETs. If a given trap has a V_g dependence with a positive coupling as in Fig. 5.13a, the trap tends to be empty when $|V_g|$ is low since $\tau_e \ll \tau_c$. On the other hand the trap tends to be filled when $|V_g|$ is high since $\tau_e \gg \tau_c$. When a device is switched from low to high $|V_g|$, the I_D immediately after switching depends on the previous state (hysteretic effect). If the trap is empty at the previous low state, it tries to capture a carrier. The carrier capture is not immediate but takes τ_c at the high $|V_g|$. Using cycled waveforms (Fig. 5.17), we investigated the hysteretic transient (Fig. 5.18a). In this example, I_D always starts at the higher current RTN state, which reveals that the trap occupancy at the onset is defined in the previous off state. Time-sliced histograms (Fig. 5.18b) show that the distribution formed during the off state transforms into the on-state distribution.

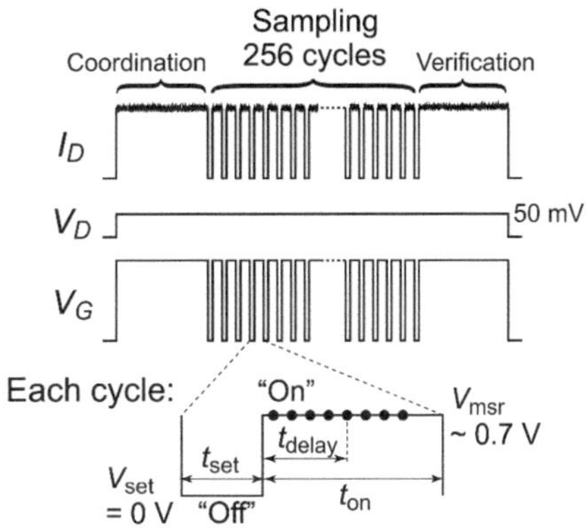

Fig. 5.17 Pulse sequence to evaluate hysteretic behavior (© 2012 IEEE. Reprinted, with permission, from [4])

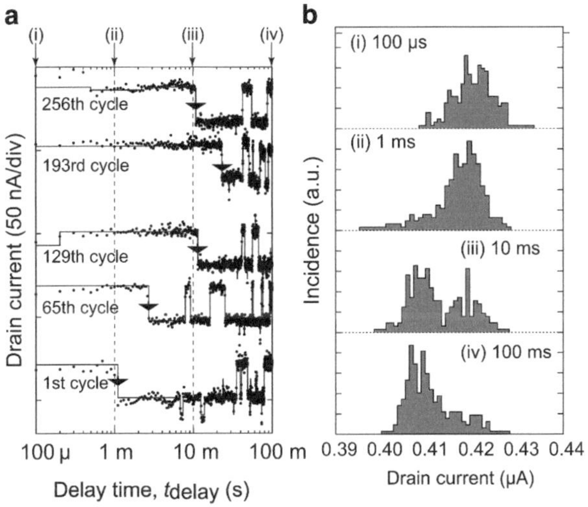

Fig. 5.18 Typical waveforms of turn-on transient of one nFET ($L/W = 25/25$ nm), and its time-sliced I_D histograms for 256 waveforms (© 2012 IEEE. Reprinted, with permission, from [4])

The duration needed to achieve the new thermal steady state after switching is found to be τ_s defined by $1/(1/\tau_e + 1/\tau_c)$ [21]. This regime is explored in Fig. 5.19. For the data in this figure, V_g is first set to 0.7 V (=high), falls down to V_{set} ranging from -0.2 V to 0.6 V (=low), then goes back to 0.7 V for I_D measurement. The probability of finding the high RTN state at the beginning of the measurement time, based on 256 transient measurements, is shown in Fig. 5.19a as a function of duration at the low state. As can be seen, for V_{set} in the range from 0.2 V to 0.6 V, the time required to reach the new steady state is essentially constant, whereas the probability level associated with that new steady state depends on V_{set} since the level is given by $1/(1 + \tau_e/\tau_c)$ at the V_{set} [21]. The duration, τ_s, and the probability level are confirmed to show excellent agreement with the

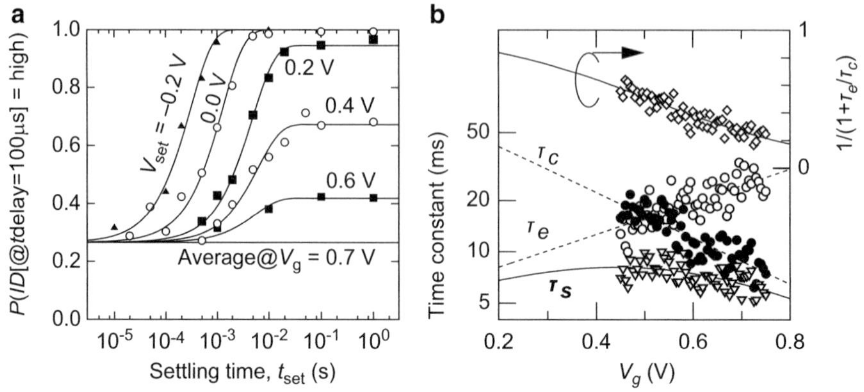

Fig. 5.19 (**a**) Probability of high I_D as a function of t_{set} after settling at various V_{set}. (© 2012 IEEE. Reprinted, with permission, from [4].) (**b**) Duration needed to achieve new steady state after switching (τ_s) and probability level at the new state $[1/(1+\tau_e/\tau_c)]$, calculated using steady-state RTN parameters (τ_e and τ_c). The sample is an nFET with $L/W = 25/25$ nm

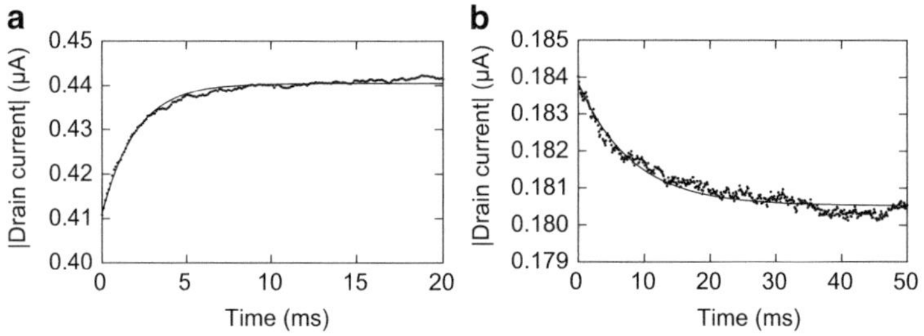

Fig. 5.20 Ensemble average of 256 transient waveforms after releasing stress voltage, for a device with an RTN trap that is (**a**) positively and (**b**) negatively coupled to V_g. The sample is a pFET with $L/W = 25/25$ nm (© 2012 IEEE. Reprinted, with permission, from [4])

calculated values using conventional steady-state RTN measurement within this range (Fig. 5.19b). However, if V_{set} is at or below threshold voltage (i.e., lower than 0.2 V), the emission process is significantly accelerated, and the duration to steady state becomes much shorter. This measurement is useful for measuring τ_e in the subthreshold regime, where steady-state RTN measurement is impossible because ΔI_D is too small compared to the noise floor at the necessary sampling rate.

A hysteretic effect similar to Fig. 5.18 can also be observed in transients from higher V_g to lower, as in conventional BTI assessments. The variety of V_g couplings shown in Fig. 5.14a leads to diverse responses. In particular, degradation after stress release is observed for FETs with a negatively coupled trap, in addition to the usual recovery for positively coupled traps (Fig. 5.20).

Both positive and negative hysteretic effects appear or disappear depending on whether the holding time at the previous state is shorter or longer than τ_s, leading

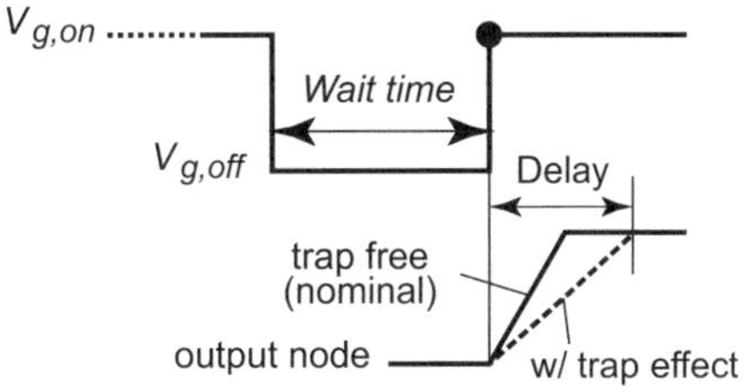

Fig. 5.21 V_g waveform used for MC simulation of delay at turn-on (© 2012 IEEE. Reprinted, with permission, from [4])

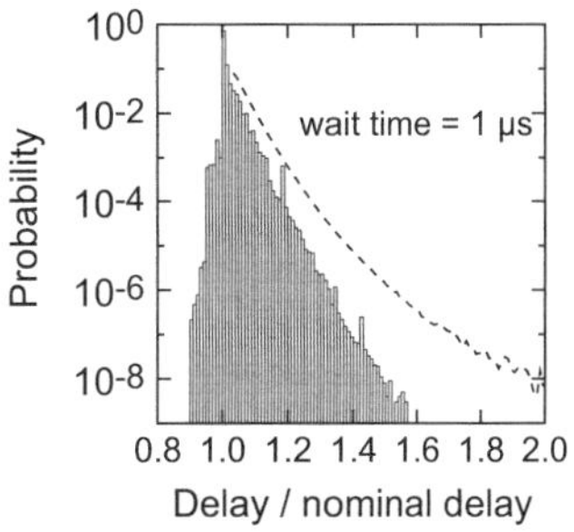

Fig. 5.22 Simulated distribution of delay at 27 °C for $L/W = 14/35$. *Bars*: traps are sampled from experimental dataset; *broken line*: using extrapolated ΔV_T (© 2012 IEEE. Reprinted, with permission, from [4])

to uncertainty in I_D. This may be a significant problem for circuits, since it leads to uncertainty in switching delay, due to these RTN traps, as indicated schematically in Fig. 5.21. We have estimated this uncertainty using the experimentally extracted time, voltage, and amplitude parameters for 600 traps from 1,200 measured nFETs. Monte Carlo simulation reveals that the timing uncertainty amounts to 50% of nominal value, for 14 nm minimum width devices operated at low V_{DD} (Fig. 5.22) [20].

5.6 Assessment of RTN in SRAM Arrays

Since SRAM cells contain the smallest FETs that are usually fabricated, they are expected to be the circuits most impacted by RTN effects. The inverse dependence of RTN on channel area dictates that these very small FETs will have increasingly large V_T shifts due to RTN as technology is scaled down. In addition, since the number of SRAM cells on a chip may well be in the millions, the distribution of RTN is very wide, increasing the likelihood of very large V_T shifts. For these reasons, many people have investigated RTN effects in SRAMs [22–25].

The way in which RTN affects SRAM cells is illustrated in Fig. 5.23. In (b) several N curves for the cell are plotted. N curves are obtained by attaching a voltage source, v_N, to the internal node of the cell, as indicated in (a), sweeping its voltage, and measuring the current that is supplied to the node. When the word line is off (low), the cell has a wide operating margin and two stable operating points ($i_N = 0$), at $v_N = 0$ and at $v_N = V_{CS}$, corresponding to the top N curve, where V_{CS} is the cell supply voltage. (Note that the center point is unstable because the slope at zero

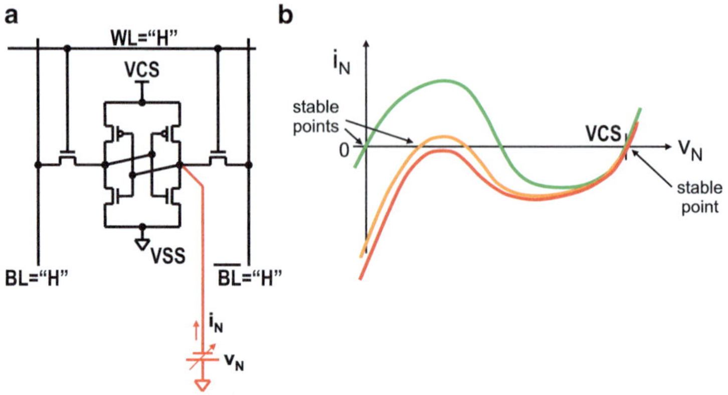

Fig. 5.23 (**a**) Schematic diagram of an SRAM cell, showing how an IV probe is connected to obtain the "N" curve. (**b**) "N" curves—current versus voltage for the circuit in (**a**) for three conditions: *green* is for WL low (cell not being accessed), *orange* is for WL high (cell being accessed) for a low margin cell, and *red* is for WL high for a cell with no margin—the only stable point is at V_{CS} (Adapted from [11])

crossing is negative.) When the word line is ON, however, the margin is reduced because the two nFETs on the right-hand side are competing with each other. Due to random process variations, some cells will have quite low margin, as indicated by the middle curve in (b), which still has two stable points, but just barely. If such a cell, which already has a low margin, also contains an RTN trap, then it may be that when the trap changes state it will shift the N curve to below the axis (to the lowest curve), causing the low stable state to disappear and the cell to switch to the high-voltage state. Thus RTN, though small, can cause SRAM read failures. The situation for write failures is similar, but different FETs are involved.

To evaluate the situation quantitatively, we performed circuit-level simulation of SRAM using our measured distribution of RTN V_T shifts and found a characteristic feature of RTN, which is illustrated in Fig. 5.24. In the absence of RTN or other noise, the probability of read/write failure as a function of V_{CS} is shown in (a). As V_{CS} decreases, the cell always works until it reaches its minimum operating voltage V_{min} and then it always fails for voltages below that. If RTN is present in one of the FETs, the situation becomes like (b). For V_{CS} below the empty-trap V_{min}, the cell always fails, as in (a), and for V_{CS} above the filled-trap V_{min}, the cell always works. For voltages in between, the cell works some of the time and fails some of the time, randomly, according to the RTN statistics. Thus, we expect there to be a plateau in the cell failure probability versus voltage. In the real case, there is additional variation, due to shot noise and other effects, so something like (c) is expected, with slopes rather than vertical lines.

We successfully observed this feature by repeatedly setting and measuring all of the cells in a 4-Mb macro [25]. This was done up to 1,000 times for each cell at each V_{CS} value, and V_{CS} was stepped in 1 or 2 mV increments. For 10 selected cells,

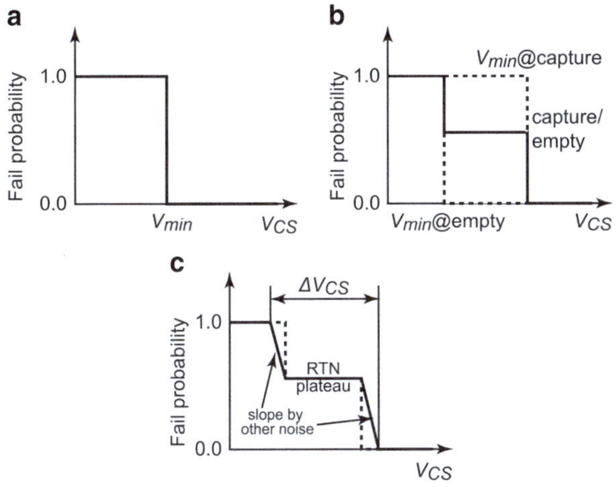

Fig. 5.24 Failure probability of consecutive read/write of SRAM cells (**a**) is in the absence of noise, (**b**) is in the presence of a single RTN trap, and (**c**) is in the presence of both an RTN trap and other noise (© 2012 IEEE. Reprinted, with permission, from [4])

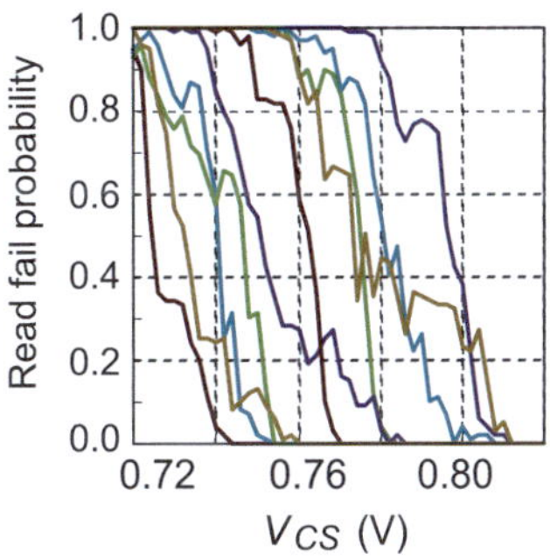

Fig. 5.25 Examples of the read failure probability as a function of V_{CS} for 10 specific cells that show RTN plateaus (© 2012 IEEE. Reprinted, with permission, from [4])

Fig. 5.25 shows the fraction of times that the cell failed versus V_{CS}. As can be seen, these cells exhibit noisy plateaus in their probability functions, just as predicted in Fig. 5.24c. These results are analyzed statistically in Fig. 5.26 by plotting the cumulative distribution function of the total width of the probability transition, ΔV_{CS} (see Fig. 5.24c). It follows a log-normal distribution at high percentile, and its areal dependence is close to 1/area, both of which are characteristic of RTN, leading us to conclude that this is indeed an RTN effect. Based on these results, a guardband voltage was calculated, which is the extra voltage that one should add to the measured V_{min} to obtain a safe reliable operating voltage. As shown in Fig. 5.27, the guardband is expected to increase greatly in future scaled-down designs because of the increasing RTN amplitudes. Since RTN is a major component of BTI effects in general, it is to be expected that the necessary guardband for all BTI effects taken together will be even larger than this.

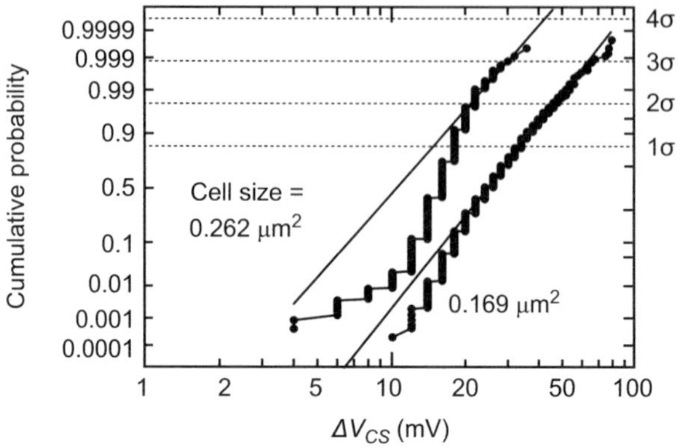

Fig. 5.26 Cumulative probability of ΔV_{CS} for two different cell areas. Smaller cell is composed of FETs with ~0.5 area (© 2012 IEEE. Reprinted, with permission, from [4])

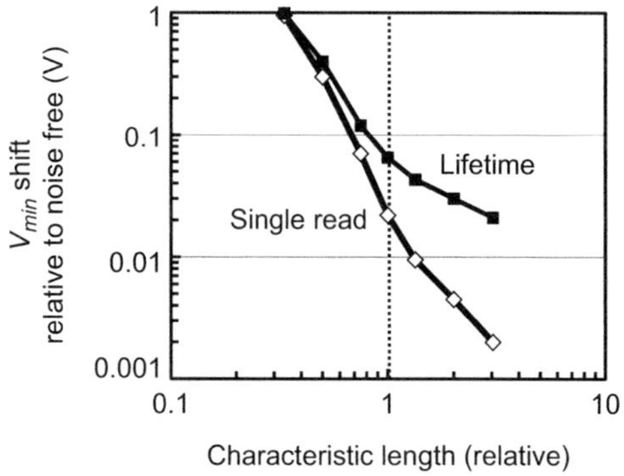

Fig. 5.27 Calculated required excess guardband due to noise (© 2012 IEEE. Reprinted, with permission, from [4])

5.7 Summary

As-fabricated traps and defects in HKMG MOSFETs give rise to random telegraph noise effects in highly scaled devices, even in the absence of gate-voltage stress. The effective V_T shifts associated with this RTN can have significant impact on CMOS circuits, as we have shown. They can degrade the operating margins of SRAM cells and can potentially cause large history-dependent delay variations in the switching time of logic gates. Since RTN amplitudes increase with scaling as the inverse of the area, RTN may be a serious reliability issue for future FET technologies. Another aspect of this reliability concern is that the fluctuating nature of RTN makes it difficult to screen for RTN-affected devices, since a defective device may pass functional test one time and fail it another. This increases the importance of understanding the temporal behavior of RTN traps.

We have described in detail a Baum–Welch-based algorithm for solving the hidden Markov model for the trap states in an RTN time series and have shown that it can reliably extract not only the trap amplitudes but also their characteristic time constants. The results of analyzing a large number of traps indicate that there is a very wide distribution of trap parameters, including both amplitudes and voltage dependencies. Measurements across temperature show that the characteristic times shrink with increasing temperature but that the voltage coupling coefficients are nearly independent of temperature. It is also observed that the trap amplitude distributions are nearly independent of whether the FET channel is doped or not, indicating that dopant-induced percolation does not play a significant role.

This information is very useful not only for understanding RTN but also for understanding bias stress effects in general, since many, if not most, BTI effects are thought to arise from the behaviors of RTN-like traps. To the extent that RTN is a subset of BTI, it should be the case that BTI margins will be sufficient to cover RTN effects in circuits, although this may be mostly a matter of making the BTI margins large enough. On the other hand, the wide distribution of RTN amplitudes observed in highly scaled FETs suggests that BTI can also be expected to show a similarly wide range of magnitudes when it is measured on very small FETs.

Acknowledgments We gratefully acknowledge the help and support of the many colleagues who contributed to the work discussed here, including N. Tega, M. Yamaoka, A. Bansal, M. Kobayashi, K. Cheng, C. P. D'Emic, Z. Ren, S. Wu, J.-B. Yau, Y. Zhu, M. A. Guillorn, D.-G. Park, W. Haensch, E. Leobandung, and K. Torii, and also the efforts of the staff of the IBM MRL and the Albany Nanotechnology facilities, where the samples were fabricated.

References

1. K. S. Ralls, W. J. Skocpol, L. D. Jackel, R. E. Howard, L. A. Fetter, R. W. Epworth, and D. M. Tennant, Phys. Rev. Lett. **52**, 228 (1984).
2. N. Tega, H. Miki, Z. Ren, C. P. D'Emic, Y. Zhu, D. J. Frank, J. Cai, M. A. Guillorn, D.-G. Park, W. Haensch, and K. Torii, in *2009 IEDM Tech. Dig.*, (IEEE), p. 771.
3. K. Takeuchi, IEICE Trans. Electron. **E95-C**, 414 (2012).
4. H. Miki, N. Tega, M. Yamaoka, D. J. Frank, A. Bansal, M. Kobayashi, K. Cheng, C. P. D'Emic, Z. Ren, S. Wu, J-B. Yau, Y. Zhu, M. A. Guillorn, D.-G. Park, W. Haensch, E. Leobandung, and K. Torii, in *2012 IEDM Tech. Dig.*, (IEEE), p. 450.
5. T. Grasser, H. Reisinger, W. Goes, T. Aichinger, P. Hehenberger, P.-J. Wagner, M. Nelhiebel, J. Franco, and B. Kaczer, in *2009 IEDM Tech. Dig.*, (IEEE), p. 729.
6. H. Miki, M. Yamaoka, N. Tega, Z. Ren, M. Kobayashi, C. P. D'Emic, Y. Zhu, D. J. Frank, M. A. Guillorn, D.-G. Park, W. Haensch, and K. Torii, in *2011 Symp. VLSI Tech.*, (IEEE and JSAP), p. 148.
7. N. Tega, H. Miki, F. Pagette, D. J. Frank, A. Ray, M. J. Rooks, W. Haensch, and K. Torii, in *2009 Symp. VLSI Tech.*, (IEEE and JSAP), p. 50.
8. Y. Yuzhelevski, M. Yuzhelevski, and G. Jung, Rev. Sci. Instrum. **71**, 1681 (2000).
9. T. Nagumo, K. Takeuchi, S. Yokogawa, K. Imai, and Y. Hayashi, in *2009 IEDM Tech. Dig.* (IEEE), p. 759.
10. S. Realov and K. L. Shepard, in *2010 IEDM Tech. Dig.*, (IEEE), p. 624.
11. D. J. Frank, in 2012 IRPS, Reliability Physics Tutorial Notes, Anaheim, CA (unpublished).
12. L. Rabiner, Proc. IEEE **77**, 257 (1989).
13. G.D. Forney, Jr., Proc. IEEE **61**, 268 (1973).

14. K. Takeuchi, T. Nagumo, S. Yokogawa, K. Imai, and Y. Hayashi, in *2009 Symp. VLSI Tech.*, (IEEE and JSAP), p. 54.
15. A. Ghetti, C. M. Compagnoni, A. S. Spinelli, and A. Visconti, IEEE Trans. Electron Devices **56**, 1746 (2009).
16. B. Kaczer, Ph.J. Roussel, T. Grasser, and G. Groeseneken, IEEE Electron Dev. Lett. **31**, 411 (2010).
17. A. Asenov, R. Balasubramaniam, A. R. Brown, and J. H. Davies, IEEE Trans. Electron Devices **50**, 839 (2003).
18. N. Tega, H. Miki, T. Osabe, A. Kotabe, K. Otsuga, H. Kurata, S. Kamohara, K. Tokami, Y. Ikeda, and R. Yamada, in *2006 IEDM Tech. Dig.*, (IEEE), p. 491.
19. T. Nagumo, K. Takeuchi, T. Hase, and Y. Hayashi, in *2010 IEDM Tech. Dig.*, (IEEE), p. 628.
20. H. Miki, M. Yamaoka, D. J. Frank, K. Cheng, D.-G. Park, E. Leobandung, and K. Torii, in *2012 Symp. VLSI Tech.*, (IEEE and JSAP), p. 137.
21. H. Miki, N. Tega, Z. Ren, C. P. D'Emic, Y. Zhu, D. J. Frank, M. A. Guillorn, D.-G. Park, W. Haensch, and K. Torii, in *2010 IEDM Tech. Dig.*, (IEEE), p. 620.
22. M. Agostinelli, J. Hicks, J. Xu, B. Woolery, K. Mistry, K. Zhang, S. Jacobs, J. Jopling, W. Yang, B. Lee, T. Raz, M. Mehalel, P. Kolar, Y. Wang, J. Sandford, D. Pivin, C. Peterson, M. DiBattista, S. Pae, M. Jones, S. Johnson, and G. Subramanian, in *2005 IEDM Tech. Dig.*, (IEEE), p. 655.
23. S. O. Toh, Y. Tsukamoto, Z. Guo, L. Jones, T.-J. King Liu, and B. Nikolic, in *2009 IEDM Tech. Dig.*, (IEEE), p. 767.
24. K. Takeuchi, T. Nagumo, K. Takeda, S. Asayama, S. Yokogawa, K. Imai, and Y. Hayashi, in *2010 Symp. VLSI Tech.*, (IEEE and JSAP), p. 189.
25. M. Yamaoka, H. Miki, A. Bansal, S. Wu, D. J. Frank, E. Leobandung, and K. Torii, in *2011 IEDM Tech. Dig.*, (IEEE), p. 745.

Chapter 6
BTI-Induced Statistical Variations

Stewart E. Rauch III

Abstract In this section, we discuss the statistics of BTI shift. It is now well known that the BTI mechanism will alter both mean and variance of the threshold voltage V_T (as well as that of other device parameters) of a group of MOSFETs under stress. There are two parts—extrinsic and intrinsic—to the induced variations, just as there are to the variations of the unstressed device characteristics. The extrinsic part is key to understanding the variations of BTI-induced shifts of performance (such as F_{MAX} [maximum product clock frequency]) among a population of chips. With deep scaling, the intrinsic contribution to chip-to-chip performance shift variations is increasing. Intrinsic variations will induce device mismatch shift, a potential concern for analog circuits. In addition, it turns out that this random fluctuation component has become extremely important to the problem of random SRAM bit reliability failures due to cell stability degradation.

6.1 Device Variability in Integrated Circuits

Due to the imperfection of semiconductor manufacturing processing, individual device characteristics deviate from the desired nominal values. These deviations or variations can be broadly divided into two classes based on the correlations between devices [1]. We will refer to these as "extrinsic" and "intrinsic" process variation. Extrinsic process variation (EPV) arises from the lack of perfect control of fabrication conditions. For example, the tools used to make the gate oxide or gate stack are unable to produce a perfectly uniform layer across a wafer. Additionally, the tool process may slowly change with time. If multiple tools are used in a particular manufacturing line for a certain process, then there will be tool to tool

S.E. Rauch III (✉)
IBM Microelectronics, Hopewell Jct., NY, USA
e-mail: s.rauch@ieee.org

T. Grasser (ed.), *Bias Temperature Instability for Devices and Circuits*,
DOI 10.1007/978-1-4614-7909-3_6, © Springer Science+Business Media New York 2014

differences. Thus there will be intra-die, die to die across the wafer, wafer to wafer in a lot, and lot-to-lot components to the overall variation. By definition, the correlation between the characteristics of devices with identical layouts on a die will become perfect as the spatial distance approaches zero between the two devices. This will happen regardless of the size of the devices.

On the other hand, intrinsic process variation (IPV) or *random local fluctuation* (RLF) between devices does not disappear at small separation; in fact, it effectively has no spatial dependence. This local variation arises from the inherent atomistic nature of matter in small dimensions. Due to the implacable scaling down of semiconductor devices, considerations such as lattice spacing, and the discrete charge of the electron are increasingly important. The most important example is called *random dopant fluctuation* (RDF) or *random discrete doping* (RDD). In the continuous limit, the V_T of a MOSFET is controlled by the spatial doping density in the well (along with the gate insulator thickness and the gate work function). But this doping is not really continuous; it actually is formed by discrete atoms at discrete locations. Doping by ion implantation (nowadays the most common technique), for example, basically consists of repeatedly firing energetic dopant ions at the silicon surface through openings. The number of ions that enter any particular opening must be an integer. Invariably, the distribution of this number for any set opening size will be Poisson. In addition, the ions will not magically become uniformly spaced; clumps and gaps will form randomly. Any subsequent drive-in anneal redistributes the dopants by a random diffusion mechanism. The net effect is to cause an intrinsic, local device V_T distribution, which scales inversely with device active area ($L \times W$)—smaller size results in more variation. This sort of scaling effect is a common characteristic of IPV.

There are many cases in analog circuits (for instance, differential amplifiers and current mirrors) in which it is desirable that the characteristics of two or more devices be closely matched. The normal design strategy to maximize the matching is to place the pair or group of identically laid-out devices in direct proximity. This will minimize differences due to EPV. Also, because of IPV, the size of the devices must be made large enough to achieve the required degree of matching.

In random logic circuits, individual device matching is usually much less important, although, with scaling, it will gain significance. Circuit performance commonly depends on the delays of logic paths (chains of gates) which are the sums of single gate delays. Thus there is some statistical averaging of multiple device characteristics. However, EPV is extremely important, as it controls the overall chip performance distribution. There is one very considerable exception to this, and that is the SRAM cell. Every cell (of millions on each chip) must operate on its own, so there is no averaging tendency. And since for density purposes the devices in the SRAM cell are made as small as possible, the intrinsic variations are large and tend to dominate over the extrinsic variations.

6.2 BTI Extrinsic Process Variation

Just as for unstressed devices, this variation arises from the variation of the fabrication process conditions themselves. BTI shift is very sensitive to compositional details of the gate oxide, or gate stack, and will vary across wafers and between wafers and lots. This is largely independent from the initial device characteristics, but not all of the variation is totally independent or unpredictable. For instance, if the BTI-induced device V_T shift can be expressed as

$$\Delta V_T = A\left(\frac{V_{GS}}{T_{INV}}\right)^p \exp\left(-\frac{E_A}{kT}\right) t^n, \tag{6.1}$$

then it can be seen that variation of T_{INV} (EOT) will cause variation in the shift. In general, the BTI shift will be dependent on observable (and hopefully monitored) process variables, such as T_{INV}, W, L, and initial V_T. These are (at least in principle) predictable based on the process parameters for a particular die. However, there are hidden, or unpredictable, variations which are not due to observable process parameters. An example of this would be properties of the trapping sites in the dielectric, such as their areal density. These effects would appear as variations of the pre-factor A. Thus, A itself is a distribution with a mean and sigma.

ΔV_T is often assumed to have a log-normal distribution. Even though A itself may be normally distributed, ΔV_T will appear to be log-normal due to the multiplicative nature of the BTI mechanism:

$$\ln(\Delta V_T) = \ln(A) + f(Stress\ Conditions) + n \times \ln t \tag{6.2}$$

Assume

$$A \sim \mathrm{N}\left(\mu, \sigma^2\right). \tag{6.3}$$

If the device sample size at each stress condition is relatively modest (let's say <50–100) and the distribution of A is relatively tight (for instance, $\sigma/\mu = 0.1$–0.2), then for each stress cell, we cannot distinguish between normal and log-normal behavior. That is,

$$\ln(A)\ is\ approx \sim \mathrm{N}\left(\ln\mu, \sigma^2/\mu^2\right). \tag{6.4}$$

$$\text{Therefore, } \ln(\Delta V_T)\ is\ approx \sim \mathrm{N}\left(\ln(\mu + f + n\ln t), \sigma^2/\mu^2\right). \tag{6.5}$$

For an example of experimental extrinsic ΔV_T distributions, see Fig. 6.1. These distributions are from a stress sample of about 50 relatively large area PFETs ($L \times W = 40$ nm $\times$ 10 μm). The data for each readout time is adequately fit by a log-normal distribution with the common log-normal $\sigma = 0.15$.

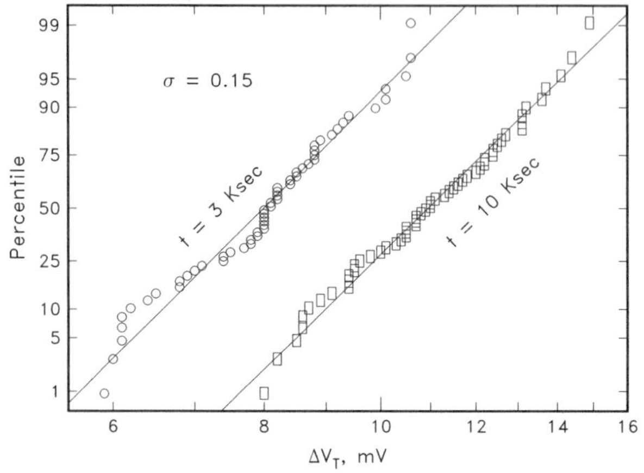

Fig. 6.1 Measured ΔV_T at two readout times of large area PFETs fit to log-normal distributions with a common sigma

Thus, *if the device area is large enough to ignore intrinsic variations*, a reasonable statistical regression approach is to fit ΔV_T in the log domain (i.e., use $\ln(\Delta V_T)$ as the independent variable). It should be remembered that time points (readouts) for each device stressed are correlated and cannot be treated as independent measurements. Doing so could lead the estimated confidence bounds on the fit parameters to be much lower than they should be [2].

6.3 BTI Random Local Fluctuations

It is commonly accepted that NBTI degradation is due to interface state generation and/or accumulation of bulk oxide charge [3, 4]. Because NBTI is caused by discrete charges, there should be intrinsic fluctuations induced. This is very analogous to the case of random dopants. For dopants, this process is called *random dopant fluctuation* (RDF), and for NBTI it is sometimes referred to as *random charge fluctuation* (RCF).

6.3.1 Atomistic Simulation

The classical view of MOS electrostatics is a simple one. In an MOS capacitor, for example, the gate and interface charges can be approximated by uniform charge sheets and the depletion region as a uniform spatial charge density. This allows an analytic one-dimensional solution to Poisson's equation in terms of the surface potential, ψ_S. To simulate a MOSFET structure, this can be built up to two dimensions by numerically solving Poisson's equation and the continuity and current density equations. Unfortunately, to realistically simulate RDF and RCF effects, it is necessary to introduce dopants (and trapped charges) as point charges with random locations and to perform a fully three-dimensional solution. This is more difficult computationally, but recently the so-called *Atomistic Simulators* have been extensively used to study random fluctuation effects on small devices [5–7]. In general, these simulations have shown how irregular the surface potential is in a sufficiently scaled device (sub-0.1 μm in length and width). The drain current flows more in a percolation manner than in the classical picture of uniform current density. This has important ramifications to the statistics of BTI. For more detail, see [42].

6.3.2 BTI-Induced Analog Mismatch Shift

A common practice in analog design is to use matched devices in balanced circuits (e.g., current mirrors, differential pairs, etc.). These circuits tend to be relatively insensitive to a matched V_T shift of the pair, but are much more sensitive to increases in the V_T mismatch between the pair. The *mismatch shift* is the change in mismatch or difference between electrical characteristics of two devices, A and B, over time. For example, the V_T mismatch between A and B is $V_{TA} - V_{TB}$. Typical BTI stress can be used to predict the mean V_T shift for the matched pair and any mismatch induced by differential stress conditions [8]. But even for identical use conditions and completely identical devices, BTI-induced mismatch shift may be enough to cause circuit failure. And for FETs normally operating at a low gate bias (for which the BTI shift would be small), there may be other circuit operation modes (such as power-down mode) which expose the devices to a relatively high V_{GS} [9]. The NBTI-induced mismatch shift concern for analog circuits has been reported as early as 2001 [8], but early authors were unable to measure any significant mismatch shift due to balanced stress, because of the relatively large gate area (W/L = 20/0.32 μm in [8]) and/or a low analog stress condition [10]. The first experimental evidence and simple model for RCF effects due to NBTI did not appear in the literature until 2002 [11]. This model is summarized here. (For convenience, the convention in this chapter is to treat the PFET V_T as normally positive.) By common practice, δ denotes the mismatch between the paired devices. The definition of V_T mismatch is, of course, $\delta V_T \equiv V_{TA} - V_{TB}$, for the A and B devices in the pair. As usual, Δ denotes the difference between time t and time 0 (shifts in time due to stress). V_T was measured under saturation conditions, since saturation mode is typical under both analog operation and digital switching conditions.

The variance of the induced V_T mismatch shift, $\mathrm{Var}(\Delta\delta V_T)$, due to the random fluctuations in the number of induced charges is calculated as follows.

Since a purely symmetric stress ($V_{DS} = 0$) is applied, there is no source/drain asymmetry or pinch-off. Let us assume that the NBTI damage mechanism is purely channel area related, not a perimeter effect, and that the induced charge is at the SiO_2–Si channel interface. (Note that this is not true for PBTI in high-K metal gate (high-K metal gate) stacks containing HfO_2 or another high-K dielectric since the electron trapping appears to be in the high-K layer.) Then, the total number of charges induced by stress, ΔN, is

$$\Delta N = \frac{1}{q}\frac{\Delta V_T}{C_{GATE}} = \frac{\varepsilon_{OX} A_G \Delta V_T}{q T_{OX,EFF}}, \tag{6.6}$$

where A_G (= $L_{EFF}W_{EFF}$) is the effective gate area (more properly, the *channel* area) and $T_{OX,EFF}$ is the equivalent oxide thickness corresponding to C_{GATE} in strong inversion and includes poly-Si gate depletion effects. For PBTI in high-K metal gate, $T_{OX,EFF}$ should be replaced by the equivalent oxide thickness from the gate to the centroid of the trapped electrons. Note: We will see that this simple charge sheet model is challenged by later authors.

ΔN is assumed to follow a Poisson distribution.

$$\text{Therefore, } Var(\Delta N) = Mean(\Delta N) = \frac{\varepsilon_{OX} A_G Mean(\Delta V_T)}{q T_{OX,EFF}}, \tag{6.7}$$

$$\text{and } Var(\Delta\delta V_T) = 2Var(\Delta N)\left(\frac{q T_{OX,EFF}}{\varepsilon_{OX} A_G}\right)^2 = \frac{K_0 T_{OX,EFF} Mean(\Delta V_T)}{A_G}, \tag{6.8}$$

$$\text{where } K_0 \equiv \frac{2q}{\varepsilon_{OX}} = 9.3\ mV\mu m. \tag{6.9}$$

The factor of 2 in K_0 comes from the fact that the δ is comparing two devices.

Fluctuations in the spatial distribution of induced charges also introduce random V_T variations [8, 9]. This was shown by Asenov [5] through atomistic simulation of random discrete doping. For his conditions, he estimated the V_T variance due to dopant spatial distribution to be about equal to that due to number fluctuation. To account for this additional V_T variation, an additional empirical constant K_1 was introduced:

$$Var(\Delta\delta V_T) = \frac{K_1 K_0 T_{OX,EFF} Mean(\Delta V_T)}{A_G}. \tag{6.10}$$

K_1 was experimentally determined, but to be consistent with Asenov, the constant K_1 was expected to be about 2. The measured value in reference [11] was $K_1 = 2.7$, a bit higher but not unreasonable.

This model was experimentally supported by three different device types and gate areas from 0.009 to 10 μm^2, as shown graphically in Fig. 6.2.

It has been further quantitatively supported by Agostinelli et al. [12] and La Rosa et al. [13]. It should be recognized that these data have generally been based on relatively slow measurement techniques, so there may have been substantial recovery.

Rauch [11] further derives a model for β mismatch shift. The parameter β is a measure of the device current drive for a given gate overdrive. A simplistic drain current model in saturation is $I_D = \beta(-V_{GS} - V_T)^m$, where m ~ 1.5–2.

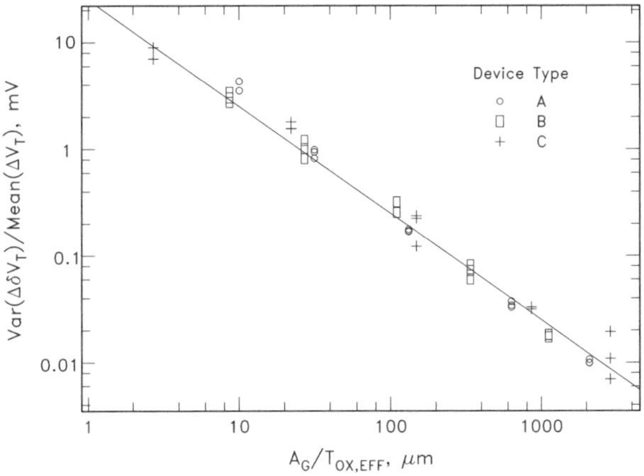

Fig. 6.2 Ratio of the variance of the V_T mismatch shift to the mean V_T shift versus $A_G/T_{OX,EFF}$. The *dashed line* is the model prediction for $K_1 = 2.7$ (After [11])

Since β is a multiplicative factor, β mismatch is defined as

$$\delta\beta \equiv 1 - \frac{\beta_A}{\beta_B}. \tag{6.11}$$

Starting with an assumed relation between mean β after shift to ΔN,

$$\beta \approx \frac{\beta_0}{1 + \alpha\Delta N/A_G}, \tag{6.12}$$

where the parameter α is an empirical constant. Then,

$$\frac{\Delta\beta}{\beta_0} \approx \alpha\Delta N/A_G. \tag{6.13}$$

Thus, for $\Delta\beta/\beta_0 << 1$,

$$Var(\Delta\delta\beta) \approx \left(\frac{2\alpha^2}{A_G{}^2}\right) Var(\Delta N) = \left(\frac{2\alpha^2}{A_G{}^2}\right) \frac{\varepsilon_{OX} A_G Mean(\Delta V_T)}{qT_{OX,EFF}}. \tag{6.14}$$

Again, the factor of 2 comes from the pair comparison. Simplifying and introducing the constant K_0,

$$Var(\Delta\delta\beta) \approx \left(\frac{4\alpha^2}{A_G T_{OX,EFF}}\right) \left(\frac{\varepsilon_{OX}}{2q}\right) Mean(\Delta V_T) = \frac{4\alpha^2 Mean(\Delta V_T)}{K_0 A_G T_{OX,EFF}}. \tag{6.15}$$

We introduce the empirical constant K_2, which leads to

$$Var(\Delta\delta\beta) = \frac{4K_2\alpha^2 Mean(\Delta V_T)}{K_0 A_G T_{OX,EFF}}. \tag{6.16}$$

The empirical result reported in reference [11] for K_2 is 12, a rather large value that perhaps needs more theoretical and experimental support. There has been little attention paid in the literature to this β statistical model. To illustrate the effect of β variation on drain current variation, let us consider the V_{GS} dependence of drain current mismatch, defined similarly to that of β as

$$\delta I_D \equiv 1 - \frac{I_{DA}}{I_{DB}}. \tag{6.17}$$

If I_D can be approximated by β times a function of overdrive ($-V_{GS} - V_T$), then for small variations,

$$\delta I_D \approx -\frac{\partial I_D}{\partial V_{GS}} \frac{\delta V_T}{I_D} + \delta\beta = -\frac{g_m}{I_D}\delta V_T + \delta\beta. \tag{6.18}$$

(g_m is also taken as a positive number.)
Therefore,

$$Var(\delta I_D) \approx \left(\frac{g_m}{I_D}\right)^2 Var(\delta V_T) + Var(\delta\beta) \tag{6.19}$$

(if V_T mismatch and β mismatch are independent).

And for the I_D mismatch shift,

$$Var(\Delta\delta I_D) \approx \left(\frac{g_m}{I_D}\right)^2 Var(\Delta\delta V_T) + Var(\Delta\delta\beta) \tag{6.20}$$

(if V_T mismatch shift and β mismatch shift are independent).

An example is given here for a group of PFET pairs sampled from an RF CMOS technology (Weff = 0.33 μm, Leff = 0.14 μm, sample size = 80 pairs) (see Fig. 6.3). This graph compares the V_{GS} dependencies of the variance of the I_D mismatch shift after 100 s on NBTI stress with the variance of the T0 I_D mismatch. The dashed lines are the variances due to the V_T mismatch, and the solid lines are the totals given by Eqs. (6.19) and (6.20). It can be observed that the impact of β mismatch shift to I_D mismatch shift is lesser than the impact of the β mismatch to I_D mismatch despite the large "K_2" factor. This is because the β shift due to NBTI, especially in the saturation regime, is generally small and is often neglected entirely.

This β variance would at most have second-order effects on gain mismatch in a differential pair, current matching in a current mirror, or the q-point of a differential pair. The consensus in the literature is that these NBTI-induced mismatch shift effects are only a concern for analog circuits under limited circumstances—two examples are exposure to high gate bias during power-down modes [9] and especially sensitive circuits such as the LSB of digital to analog (D/A) converters [12].

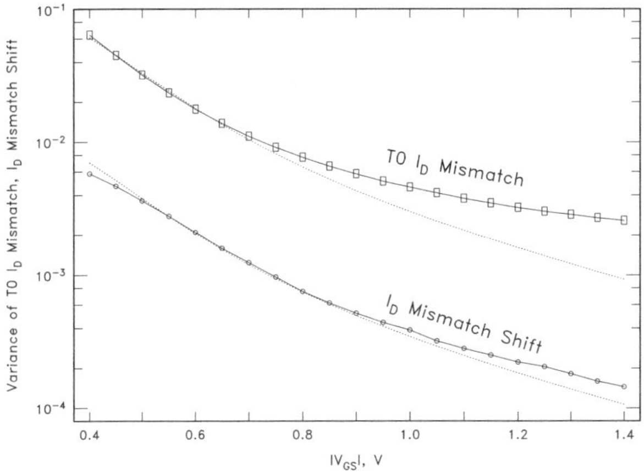

Fig. 6.3 Variances of NBTI-induced I_D mismatch shift and T0 I_D mismatch

6.3.3 SRAM Stability Concerns

Once NBTI-induced random charge fluctuation statistical effects were known, it was soon realized that these statistics (and not only the mean shift) are important to an accurate understanding of NBTI-induced read disturb failures in typical CMOS SRAMs [13, 14]. The MOS devices typically used in advanced SRAM cell designs are generally near minimum geometry in both length and width; hence, gate areas are extremely small, and NBTI-induced mismatch shift is large. NBTI degradation introduces additional variations of the Vth and β of the pMOSFET pull-up transistors which worsen the cell stability. Note that the shift in total device variability is much smaller than the mismatch shift. SRAM stability shift due to NBTI is discussed further in [43].

6.3.4 BTI-Induced ΔV_T Distributions

The first discussion in the literature of actual NBTI-induced ΔV_T distributions was by Rauch [15] from which the following discussion is derived.

If the NBTI-induced V_T shift process is examined more rigorously from a statistical point of view, the following general conclusions can be inferred:

1. It is well known that NBTI degradation is due to interface state generation and accumulation of bulk oxide charge [3, 4]. Thus the underlying stochastic process *is* discrete.
2. It is now recognized that there are concurrent interface state re-passivation (a prominent feature of the *Reaction–Diffusion Model* [16]) and oxide trap

detrapping processes occurring (as shown by fast recovery effects [17]). This means that the threshold voltage shift should be observed to occasionally reverse direction (increase and decrease with time) and even change sign (net negative charge delta) under stress. This has been observed by several authors [18, 19]. The result is that the process controlling the shift is not strictly a single Poisson one, but is actually the difference between two Poisson processes.

3. When a charge is created or destroyed, the magnitude of threshold voltage shift is not fixed, but is dependent on its position with respect to other NBTI-induced charges and random dopant positions [5, 20]. There is a discrete V_T step when a charge is created or destroyed, but the magnitude of the step is itself randomly distributed. Therefore, the creation and destruction processes are each modeled as a *compound Poisson process* (also called a *generalized Poisson process*). This type of stochastic process is defined as follows [21]:

$$\Delta V_T(t) = \sum_{i=1}^{N(t)} S_i \tag{6.21}$$

where N(t) is a Poisson process and S_i are independent random variables, also independent of N(t). The model is then

$$\Delta V_T(t) = \sum_{i=1}^{N_C(t)} S_i - \sum_{i=N_C+1}^{N_C(t)+N_D(t)} S_i. \tag{6.22}$$

By the way, in the literature this is sometimes asserted to be a Skellam process [22] (the same as defined by Eq. (6.21), where instead N(t) ~ Skellam). This is *not* rigorously true, even though the net number of charges does follow a Skellam distribution. However, in practice, there may be little difference, except for near zero and negative values of shift.

If any given charge state is both created and destroyed, it cancels out, and so only unique creations and destructions need to be considered.

The mean shift is

$$\mu\left(\Delta V_T\right) = \left(\mu_C - \mu_D\right)\mu\left(S\right). \tag{6.23}$$

But

$$\mu\left(S\right) = \frac{qT_{OX,EFF}}{\varepsilon_{OX} A_G}, \tag{6.24}$$

so that, as required assuming a charge sheet model,

$$\mu\left(\Delta V_T\right) = \frac{qT_{OX,EFF}}{\varepsilon_{OX} A_G}\mu\left(\Delta N\right) \equiv K_Q \mu\left(\Delta N\right). \tag{6.25}$$

If C and D are uncorrelated, the variance of V_T shift is

$$Var(\Delta V_T) = (\mu_C + \mu_D)\left(\mu^2(S) + Var(S)\right). \tag{6.26}$$

If we define a normalized random variable $U = S/K_Q$, then U has a mean of 1. Thus

$$\mu(\Delta V_T) = K_Q(\mu_C - \mu_D), \tag{6.27}$$

and

$$Var(\Delta V_T) = K_Q{}^2(\mu_C + \mu_D)(1 + Var(U)). \tag{6.28}$$

We now define the dispersion factor ϕ as

$$\phi \equiv \frac{Var(\Delta V_T)}{K_Q \mu(\Delta V_T)} = \frac{(\mu_C + \mu_D)}{(\mu_C - \mu_D)}(1 + Var(U)). \tag{6.29}$$

ϕ represents the "over-dispersion" compared to what would result from a simple Poisson process. Comparing with prior equations, $K_1 = \phi$. Asenov presents a plot (Fig. 12, [5]) from which we can estimate $Var(U) \sim 0.8$–1.1. (There is no D process for random dopants.) Then, to match the experimental value of $\phi \sim 2.7$, $\mu_D/\mu_C \sim 0.12$–0.20. This ratio of detrapping is assumed to derive from positive correlation between the full C and D processes contrary to assumption. If we split C and D into correlated (α) and uncorrelated parts (C' and D')

$$\begin{aligned} C &= \alpha + C' \\ D &= \alpha + D', \end{aligned} \tag{6.30}$$

then the mean and variance of the ΔV_T distribution will be

$$\mu(\Delta V_T) = K_Q(\mu_{C'} - \mu_{D'}), \tag{6.31}$$

$$\text{and } Var(\Delta V_T) = K_Q{}^2(\mu_{C'} + \mu_{D'})(1 + Var(U)). \tag{6.32}$$

Thus, only the uncorrelated parts should be considered. Later we will see that the μ_D/μ_C ratio is probably even smaller, and most of this extra variance is actually due to a higher effective value of K_Q.

To investigate the ΔV_T distribution for very small devices, we shall use NBTI data from advanced SRAM-sized PFETs with a gate area about 1/3 of the minimum in [11]. For these devices, $W_{eff} \times L_{eff} \sim 90$ nm $\times$ 37 nm, and $T_{OX.EFF} \sim 2.0$ nm ($K_Q \sim 2.8$ mV), and sample size = 147. The NBTI stress condition is as follows: T = 140C, $V_{GS} = -2.0$ V. Measurements are slow V_{GS} sweeps, with various readout times from 10 s to 10 ks. For this very small device, the variation in ΔV_T is

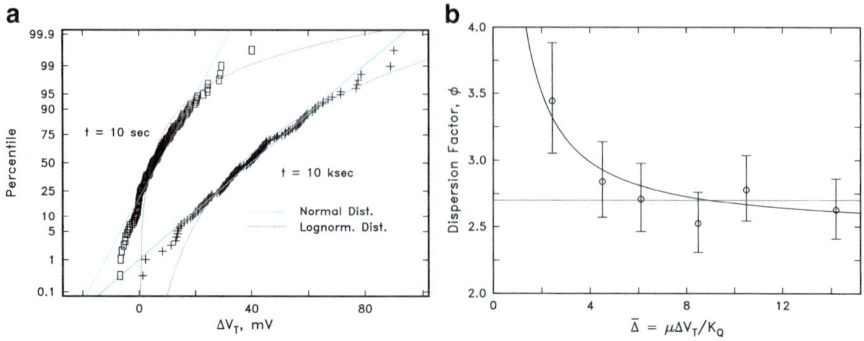

Fig. 6.4 (**a**) Measured ΔV_T distributions (*points*) compared to normal (*solid lines*) and log-normal (*dashed lines*) distributions. After [15]. (**b**) Dispersion factor versus normalized mean threshold voltage shift. *Dashed* and *solid lines* are two different models for dispersion explained in the text

dominated by the random fluctuation effect, and the effect of process variation is small. This can be checked by comparing the variances of V_T mismatch shift and of raw V_T shift. There is a ratio of two between these if the process variation has a negligible contribution. Figure 6.4a shows the measured ΔV_T distributions for readout times of 10 and 10,000 s. Comparison to normal and log-normal distributions with the same mean and variance demonstrates that the normal underestimates and the log-normal overestimates the high ΔV_T tails of the actual distributions.

Figure 6.4b shows the calculated dispersion factor, ϕ, versus the normalized average shift at each readout. The error bars represent the (statistical) standard errors of the ϕ estimates. The constant value of 2.7 (dashed line) is the prediction of [11]. This value is a reasonable fit for all except the first readout data. The solid line is the simple linear model, $\mu_D = a + b\mu_C$, with a = 0.45 and b = 0.13. It was felt that the reason for the dispersion factor to decrease with time is the increasing correlation of the D process with the C process. However, as we shall see, the reliance on the charge sheet value of K_Q is misleading, and much of this extra dispersion is probably not due to the D process after all.

6.3.4.1 Asymptotic Approximation

Even without knowledge of the exact U (normalized step size) distribution, an asymptotic approximation to the compound distribution for large N can be made. Since the points of high shift are the most important from a practical point of view, we first approximate the difference of two compound Poisson processes with a single compound Poisson process with the same dispersion factor. That is,

$$\Delta V_T(t) \approx K_Q \sum_{i=1}^{\Delta N(t)} U_i' \tag{6.33}$$

where U' has a mean of 1 and a variance $= \phi - 1$. Now for large ΔN by the Central Limit Theorem, the distribution of the sum $\sum_{i=1}^{\Delta N} U_i'$ approaches a normal distribution with $\mu = \Delta N$ and $\sigma = \sqrt{\Delta N(\phi - 1)}$. Since the joint probability of ΔV_T and ΔN is given by

$$p_{\Delta V_T,\Delta N}(x,m) = p_{\Delta V_T|\Delta N}(x|m)\,P_{\Delta N}(m), \tag{6.34}$$

then,

$$p_{\Delta V_T}(x) = \sum_{m=0}^{\infty} p_{\Delta V_T,\Delta N}(x,m) = \sum_{m=0}^{\infty} p_{\Delta V_T|\Delta N}(x|m)\,P_{\Delta N}(m). \tag{6.35}$$

Therefore, the PDF is approximately

$$f(\Delta V_T) \approx e^{-\bar{\Delta}} \sum_{m=0}^{\infty} \frac{\varphi\left(\frac{\Delta - m}{\sqrt{m(\phi-1)}}\right)}{\sqrt{m(\phi-1)}} \frac{\bar{\Delta}^m}{m!}, \tag{6.36}$$

where $\Delta \equiv \Delta V_T / K_Q$, $\bar{\Delta} \equiv \mu(\Delta)$, and φ is the Standard Normal PDF. For $m = 0$, φ is taken to be the delta function $\delta(x)$. The corresponding CDF is

$$F(\Delta V_T) \approx e^{-\bar{\Delta}} \sum_{m=0}^{\infty} \Phi\left(\frac{\Delta - m}{\sqrt{m(\phi-1)}}\right) \frac{\bar{\Delta}^m}{m!}, \tag{6.37}$$

where Φ is the Standard Normal CDF. For $m = 0$, Φ is taken to be the unit step function $H(x)$.

We will refer to this distribution as *Cpn*.

This CDF is compared to measured ΔV_T distributions in Fig. 6.5. The value of ϕ used is from the model of Fig. 6.4b (solid line). The predicted asymptotic CDF agrees with the measured data remarkably well even for relatively low shift levels (low ΔN). The agreement for $\Delta V_T < 0$ must be considered fortuitous, since the nonzero value of F in this region arises from $U' < 0$, which is physically unrealistic. (These physically stem from the D process.)

6.3.4.2 Exponential Step Size Distribution

There now exists more information regarding the step size distribution. Recently the connection has been realized between BTI and random telegraph noise (RTN) or random telegraph signal (RTS) [23–25]. Both exhibit discrete ΔV_T steps when a charge is captured or emitted from a trap. If the device size (channel area) is small enough, these steps can be directly measured. Both experimental and simulation results suggest an exponential distribution, at least for a large range of ΔV_T [26–28].

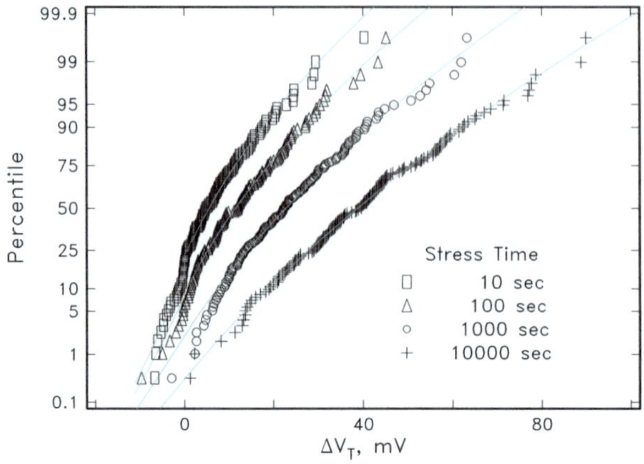

Fig. 6.5 Comparisons of the predicted approximate CDF (*lines*) with measured ΔV_T distributions (*points*) for the asymptotic approximation (Cpn)

The derivation of $F(\Delta V_T)$ with an exponential step size distribution follows [29]: If the step size $S \sim \text{Exp}(\eta)$,

$$f(S;\eta) = \left(\frac{1}{\eta}\right)\exp - \left(\frac{S}{\eta}\right). \tag{6.38}$$

Use of an exponential distribution puts a simplifying restriction on the BTI statistics. Since it has only one parameter, η, its mean and variance cannot be independently varied. In fact,

$$\mu(S) = \eta, \text{ and } Var(S) = \eta^2 = [\mu(S)]^2. \tag{6.39}$$

Thus the normalized S variable $U = S/\mu(S) = S/\eta$ has a mean of 1 and a variance of 1. This neatly fits within our requirement for Var(U) given in Sect. 6.3.4. f(U) is just $\exp(-U)$. The parameter η replaces K_Q.

Then

$$Var(\Delta V_T) = \eta^2 (\mu_C + \mu_D)(1 + Var(U)) = 2\eta^2 (\mu_C + \mu_D). \tag{6.40}$$

And

$$\frac{Var(\Delta V_T)}{\mu(\Delta V_T)} = 2\eta \frac{(\mu_C + \mu_D)}{(\mu_C - \mu_D)}, \tag{6.41}$$

$$\phi \equiv \frac{Var(\Delta V_T)}{\eta \mu(\Delta V_T)} = 2\frac{(\mu_C + \mu_D)}{(\mu_C - \mu_D)}. \tag{6.42}$$

The distribution of the sum of m-independent exponentially distributed variables is called an "Erlang" distribution (a special case of the gamma distribution).

$$\left(\sum_{i=1}^{m} U_i\right) \sim Erlang\,(m,1)\,. \tag{6.43}$$

The PDF and CDF are given by

$$f\,(x;m) = \frac{e^{-x}x^{m-1}}{(m-1)!}, F\,(x;m) = P\,(m,x)\,, \tag{6.44}$$

where P is the regularized gamma function given by

$$P\,(m,x) \equiv \frac{\int_0^x y^{m-1}e^{-y}dy}{(m-1)!}. \tag{6.45}$$

The PDF for the creation process is then

$$f(\Delta V_T)_C = e^{-\bar{\Delta}}\left(\delta(x) + \sum_{m=1}^{\infty} \frac{e^{-\Delta}\Delta^{m-1}\bar{\Delta}^m}{(m-1)!m!}\right), \tag{6.46}$$

and the CDF

$$F(\Delta V_T)_C = e^{-\bar{\Delta}}\left(H\,(\Delta) + \sum_{m=1}^{\infty} P\,(m,\Delta)\,\frac{\bar{\Delta}^m}{m!}\right), \tag{6.47}$$

for $\Delta \equiv \Delta V_T\,/\,\eta$ and $\bar{\Delta} \equiv \mu\,(\Delta) = \mu_C$.

This is the so-called *Cpe* distribution [30].

In many cases, $\mu_C >> \mu_D$, and the D process can be ignored, as we shall see.

Value of η and the Validity of the Charge Sheet Approximation. Many authors have found that the experimental value for the mean ΔV_T step size η is much larger than the charge sheet value, η_0 (or K_Q) [7, 28, 31]. Most experimental values in the literature for η/η_0 vary from ~1.5 to >3, although some results are consistent with $\eta/\eta_0 \sim 1$ [32]. If we apply the Cpe distribution to the data of Sect. 6.3.4.1 above, we must adjust η to 3.4 mV (~1.2 η_0, more on this later). Simulations also suggest $\eta/\eta_0 \sim$ 1–2 [28, 33]. Certainly, the effect of a single point charge is hardly the same as for that amount of charge smeared out over the entire channel area. As we know, atomistic device simulations show that the dopant position fluctuations induce large local variations in the surface potential. These local variations, of course, contribute to the step size distribution. Because of nonlinearity, the mean step size may increase, as well. Simulations by several authors support a well doping dependence; more dopants lead to more variation of step height. There are several other possible effects that can increase the effective value of η: (1) Since direct step size measurements have a lower resolution limit, generally of 1 mV or so, there may be a "hidden" set of ΔV_T steps near zero. In this case, these near-zero steps can be neglected, and the effective result is essentially the same as a lower number of steps with a higher η. (2) It is also conceivable that the trapping sites are correlated to the initial surface potential fluctuations. If the trapping probability is higher at those sites that have a larger inversion charge density—the very spots that are most sensitive to a trapped charge—the effective η will be enhanced.

The effective value of η can be deduced from measurements in two ways. The S distribution can be measured directly by using the *Time-Dependent Defect Spectroscopy* (TDDS) technique [34]. For more information on TDDS, see [44].

Also, if the D process is neglected, and assuming an exponential step height distribution,

$$\eta = \frac{1}{2}\left(\frac{Var(\Delta V_T)}{\mu(\Delta V_T)}\right). \tag{6.48}$$

This should be applied to stresses with fairly large shifts (not early readouts).

When statistical parameters are needed for a technology under development before measurements are possible, it would be recommended that a value of $\eta/\eta_0 = 1.5$–2 be used for NBTI. When calculating η_0, T_{OX} should include any gate depletion effects, and L approximate the effective channel length (not include S/D overlaps). For PBTI in high-K metal gate stacks, the T_{OX} value will depend on the position of the trapped charge centroid. η_0 will likely be significantly lower than for NBTI. Toledano-Luque et al. [31] reported a bimodal (double exponential) step height distribution in high-K NFETs—one η comparable to the PFET value and one about ¼ of this. This may indicate trapping at the SiO_2-Si channel interface as well as in the high-K layer. There is another potential complication for PBTI. If the trapped charge is widely distributed throughout the high-K layer, this represents an additional source of variation. This point will be discussed later.

6.3.4.3 Refitting Data with Cpe Distributions and $\eta > K_Q$

To illustrate the Cpe distributions with both the C and D processes in play, it is fit to the data of Sect. 6.3.4.1. As opposed to the Cpn assumptions (namely, K_Q or η_0, the charge sheet value), a value of μ_D constant (or even decreasing) with time is found to be consistent with the data. For $\mu_D = 0.45$, $\eta = 3.4$ mV ($\eta/\eta_0 \sim 1.2$), the predicted dispersion factors are a reasonable match to data as shown in Fig. 6.6a, based on

$$\phi \equiv \frac{Var(\Delta V_T)}{\eta\mu(\Delta V_T)} = 2\left(1 + \frac{2\mu_D}{\bar{\Delta}}\right). \tag{6.49}$$

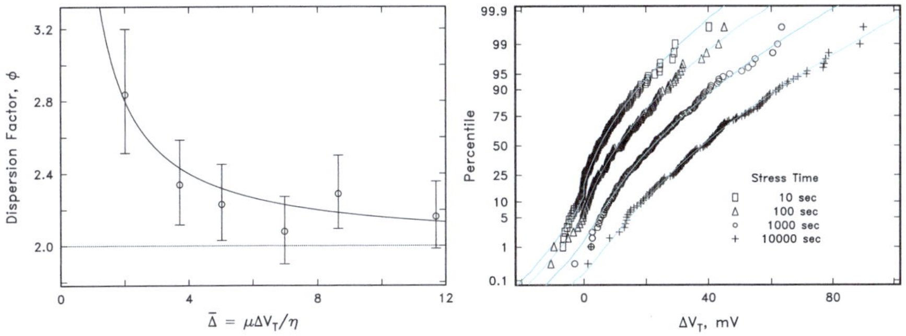

Fig. 6.6 (**a**) Dispersion factor versus normalized mean shift for $\eta = 3.4$ mV. The *solid line* is the expected value for a constant $\mu_D = 0.45$ and *dotted line* for $\mu_D = 0$. (**b**) Comparisons of the predicted Cpe CDF (*lines*) with measured ΔV_T distributions (*points*)

Also with these values, the negative parts of the predicted ΔV_T distributions at the shorter times were also comparable with data, as seen in Fig. 6.6b. This demonstrates that the Cpe distribution is consistent with V_T shift data and leads to additional simplification. A constant μ_D can be interpreted simply as a random telegraph signal. ($\mu_C = \mu_D =$ constant describes random telegraph noise; thus, the net BTI trapping process is then $\mu_{BTI} = \mu_C - \mu_D$, which is just the mean increase in the number of charged states.)

6.3.4.4 Comparison of Cpn and Cpe Distributions

Both the Cpn and Cpe distributions are sufficient to fit experimental data up to the 99 percentile regime. But we will show that the two diverge at larger percentiles. Even though the first and second moments (mean and variance) are the same, the third moments (skewness) are different. A positive skewness denotes a non-normality that will add probability to large ΔV_T events. This may be important for SRAM considerations, since the distribution of a very large number of devices is involved.

Skewness is defined as

$$Skew(Z) \equiv \frac{E\left((Z-\mu(Z))^3\right)}{(Var(Z))^{\frac{3}{2}}}. \tag{6.50}$$

Since the skewness is the normalized third moment, $Skew(\Delta V_T) = Skew(\Delta)$.

$$\text{If } \Delta \approx \sum_{i=1}^{N} U_i, \text{ then } E\left((\Delta - \mu_\Delta)^3\right) = \mu_N E\left(U^3\right). \tag{6.51}$$

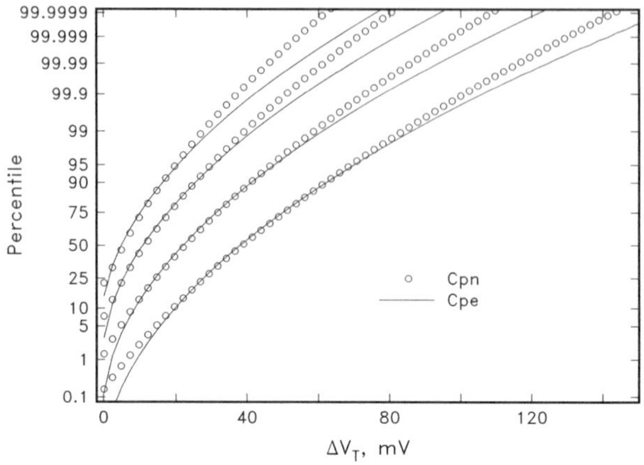

Fig. 6.7 Comparison of example Cpe and Cpn distributions, including high tails

Since μ(U) = Var(U) = 1,

$$E\left(U^3\right) = E\left(\left(U-1\right)^3\right) + 3Var(U) + 1$$

$$= (Var(U))^{\frac{3}{2}} Skew(U) + 3Var(U) + 1 = Skew(U) + 4 \tag{6.52}$$

$$= Skew(U) + 4 \tag{6.53}$$

$$\Rightarrow Skew(\Delta) = \frac{E\left(U^3\right)}{2^{\frac{3}{2}} \mu_N{}^{\frac{1}{2}}} = \frac{[Skew(U)+4]}{2^{\frac{3}{2}} \mu_N{}^{\frac{1}{2}}}. \tag{6.54}$$

The skewness of a normal distribution is 0. The skewness of the exponential distribution is 2. Therefore, for Cpn,

$$Skew(\Delta) = \sqrt{\frac{2}{\mu_N}}, \tag{6.55}$$

and for Cpe,

$$Skew(\Delta) = \frac{3}{2}\sqrt{\frac{2}{\mu_N}}. \tag{6.56}$$

Thus the skewness of the Cpe distribution is 1.5× that of the Cpn, and the upper tail will extend to higher values of ΔV_T. This is illustrated in Fig. 6.7. Considerable deviation between the two distributions can be observed above the 99 percentile points.

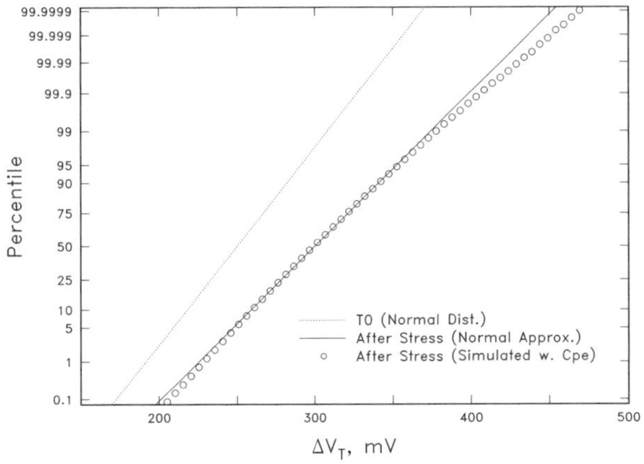

Fig. 6.8 Example of a predicted V_T distribution change due to NBTI showing non-normal effects

Fisher et al. [35] measured a BTI-induced ΔV_T distribution for >8,000 samples and compared it with a Cpn prediction. Deviations at the higher percentiles were evident. It is clear that a Cpe distribution would fit their data better.

6.3.4.5 Implications of Non-normality

The product level impact of this non-normality will depend on the situation. For analog matching or uncensored SRAM, it is the entire V_T distribution that is important. The V_T distribution after BTI is the convolution of the virgin device V_T distribution with the ΔV_T distribution. This convolution will mitigate the non-normal effects due to BTI. Here is an example which is a severe, but not unrealistic, case. We consider a minimum, or near minimum, size device ($\eta = 4$ mV) having a T0 V_T distribution (assumed to be normal) with a mean of 250 mV and a sigma of 25 mV. The NBTI-induced ΔV_T distribution has a mean of 50 mV. The initial and final V_T distributions and the normal approximation to the final distribution are shown in Fig. 6.8.

In the upper tail, the exact final V_T distribution can be observed to deviate from its normal approximation starting at about the 1,000 ppm level. At the 1 ppm level, there is ~15 mV error, and at the 0.1 ppm level, about 20 mV. This amount of deviation is unlikely to be important for analog matching applications because:

1. The mismatch (δV_T) distribution will be closer to normal than that of the V_T itself, since it is the difference of two independent random variables.
2. A minimum ground-rule device is atypical for matching applications.
3. The number of instances per chip is typically moderate.

For an uncensored SRAM, there may be some significant, but probably not severe, non-normal effects due to competing trends:

1. The initial I_{CRIT} distribution is typically quite wide, which dilutes the non-normality of ΔV_T.
2. SRAM devices tend to be at or near ground-rule minimum.
3. Typically, there is an extremely large number (millions) of instances per chip.

However, in the case of a censored (a V_{min} margin is applied at test; failing SRAMs are discarded) SRAM, it was demonstrated in [13] that the low I_{CRIT} tail after NBTI is due almost completely to the ΔV_T distribution itself (unconvoluted with any initial distribution.) Therefore, under certain conditions, non-normal effects may be much more significant (perhaps as much as 10× or more in cell failure rate due to stability.)

6.4 Other Sources of BTI Intrinsic Process Variation

6.4.1 Work Function Fluctuations in High-K Metal Gate

Random dopant fluctuation has been the dominant cause of device variations for typical poly-Si gate nitrided-oxide FETs. However, there are other sources of variation besides random discrete doping, such as *line edge roughness* (LER), and intrinsic T_{OX} variation. And as CMOS technology progresses, there are additional possible effects to consider, including *fin height variation* and *work function variation* (WFV) [36]. Since finfets are generally un-doped, or lightly doped, random discrete doping will be sharply mitigated or eliminated [37, 38]. WFV may emerge as a dominant variability mechanism in the future. It also is expected to exacerbate BTI variability.

If the metal gate in a high-K metal gate stack is polycrystalline with random small grains, then the local work function will depend on the crystal orientation of each grain. This will lead to work function variation, also referred to as *work function fluctuation* (WFF) or *metal gate granularity* (MGG). Simulated random telegraph signal step heights show an interaction with work function variation which is similar to that with random discrete doping and also result in an approximately exponential distribution [39]. Since metal gate work function is known to affect BTI, there is also the potential for local correlation, driving further BTI variability [40].

6.4.2 Variability of Charge Depth: PBTI in High-K

As mentioned above, PBTI in a high-K metal gate stack entails electron trapping in the high-K bulk, rather than at the SiO_2–Si channel interface. This trapping may be distributed through the depth of the high-K layer more or less uniformly [41]. Since the step height from a trapped electron depends on the depth, this represents an additional source of variability. The extra variance due to this contribution can easily

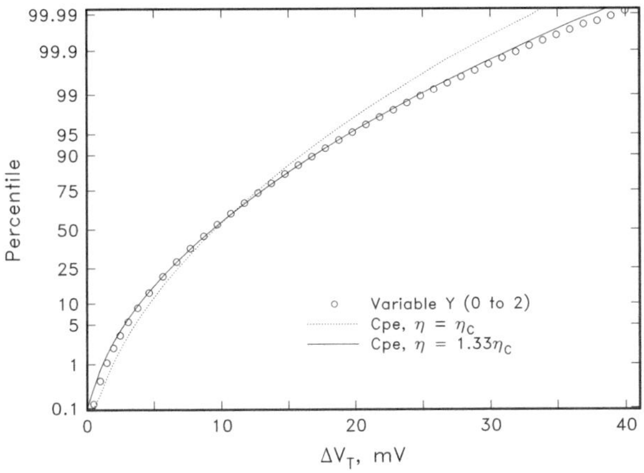

Fig. 6.9 Comparison of ΔV_T distribution with variable depth with two Cpe distributions: one with the same mean step height and one with 4/3 this value

be estimated. Let us suppose that the charge trapping depth is a uniform distribution from 0 (the gate) to the total high-K thickness (high-K–SiO_2 interface). The mean depth, or charge centroid, is half of this maximum thickness. Therefore, this effect can be included by replacing the normalized step height variable U by the product YU, where Y ~ Uniform (0, 2). Now,

$$\mu(\Delta V_T) = \mu_N \eta_C, \text{ and } Var(\Delta V_T) = \mu_N {\eta_C}^2 (1 + Var(YU)), \tag{6.57}$$

where η_c is the mean step height for Y = 1 (the centroid).

$$Var(YU) = {\mu_Y}^2 Var(U) + {\mu_U}^2 Var(Y) + Var(Y)Var(U). \tag{6.58}$$

Using $\mu_Y = 1$, $\mu_U = 1$, Var(Y) = 1/3, and Var(U) = 1,

$$Var(YU) = \frac{5}{3}, \text{ and } Var(\Delta V_T) = \frac{8}{3}\mu_N {\eta_C}^2. \tag{6.59}$$

The resultant ΔV_T distribution is not Cpe because the step height distribution is no longer exponential. However, we can approximate the ΔV_T distribution by a standard Cpe with the same mean and variance with the substitutions

$$\mu_N' = \frac{3}{4}\mu_N, \quad \eta' = \frac{4}{3}\eta_C. \tag{6.60}$$

An example is plotted in Fig. 6.9 for $\mu_N = 10$ and $\eta_c = 1$ mV. The subsequent ΔV_T distribution (circles) is compared to Cpe's with $\mu_N = 10$, $\eta_c = 1$ mV and $\mu_N = 7.50$, $\eta_c = 1.33$ mV. At least for μ_N values not too small, the approximate Cpe is a reasonable fit.

6.5 Statistical Interactions of BTI-Induced Intrinsic and Extrinsic Variations

Let us examine the variance of the total shift distribution if it has intrinsic and extrinsic variation components. The total variation in the parameter shift, a random variable X, can be separated into the two parts:

$$X = R + P, \tag{6.61}$$

where R, defined as the difference between the shift and the local mean shift, is the random variable due to random local fluctuations and P is the random variable due to extrinsic process variation. Thus the mean of R is 0 by definition. The mean of P is just the overall mean shift. Note that R is not independent of P, since the variance of R depends on P. We divide the samples into k subsamples or cells, such as die, over which we can assume that P is constant. Let m_i be the mean shift and n_i be the number of identical devices in cell i, and N = grand total number of devices. For the jth device in the ith cell,

$$x_{i,j} \equiv r_{i,j} + m_i. \tag{6.62}$$

(Let i and j be very large.)

The grand variance is then given by

$$V \equiv \frac{\sum_{i=1}^{k}\sum_{j=1}^{n_i}(x_{i,j}-M)^2}{N} = \frac{\sum_{i=1}^{k}\sum_{j=1}^{n_i}(r_{i,j}+m_i-M)^2}{N} \tag{6.63}$$

$$= \frac{\sum_{i=1}^{k}\sum_{j=1}^{n_i}\left[r_{i,j}{}^2 + 2r_{i,j}(m_i-M) + (m_i-M)^2\right]}{N}. \tag{6.64}$$

The variance of $r_{i,j}$ in cell i is $Var(R_i) = \frac{\sum_{j=1}^{n_i} r_{i,j}{}^2}{n_i}$. But we also know that $Var(R_i) = am_i$, because this is a property of the R distribution. Thus $\sum_{j=1}^{n_i} r_{i,j}{}^2 = n_i a m_i$. Of course, $\sum_{j=1}^{n_i} r_{i,j} = 0$.

$$V = \frac{a\sum_{i=1}^{k} n_i m_i + \sum_{i=1}^{k} n_i(m_i-M)^2}{N}. \tag{6.65}$$

Using the definition of the overall or grand mean M,

$$M = \frac{\sum_{i=1}^{k} n_i m_i}{N}, \tag{6.66}$$

and the variance of P

$$Var(P) = \frac{\sum_{i=1}^{k} n_i(m_i - M)^2}{N}, \tag{6.67}$$

$$\text{yielding } V = aM + Var(P) = Var(R) + Var(P). \tag{6.68}$$

Another way of showing this in a probabilistic (as opposed to statistical) way is to start with the law of total variance:

$$Var(X) = E\left(Var(X\,|P)\right) + Var\left(E(X\,|P)\right). \tag{6.69}$$

Since

$$E\left(Var(X\,|P)\right) = E\left(aP\right) = aM \text{ and } Var\left(E(X\,|P)\right) = Var(P). \tag{6.70}$$

we arrive at the same equation:

$$V = aM + Var(P) = Var(R) + Var(P). \tag{6.71}$$

Even though R and P are not independent, the variance of their sum is still the sum of their variances. This is due to the Poisson-like character of the R distribution, variance is proportional to mean.

We will now show that even though the second moment, variance, behaves as if the two components were independent, this is not true of the third (normalized) moment, skewness (previously defined). The skewness of the sum of two independent random variables is

$$Skew\left(Z_1 + Z_2\right) = \frac{[Var(Z_1)]^{3/2} Skew(Z_1) + [Var(Z_2)]^{3/2} Skew(Z_2)}{[Var(Z_1 + Z_2)]^{3/2}}. \tag{6.72}$$

The grand skewness S of X is

$$S \equiv \frac{\sum_{i=1}^{k} \sum_{j=1}^{n_i} \left(x_{i,j} - M\right)^3}{NV^{3/2}} = \frac{\sum_{i=1}^{k} \sum_{j=1}^{n_i} \left(r_{i,j} + m_i - M\right)^3}{NV^{3/2}} \tag{6.73}$$

$$= \frac{\sum_{i=1}^{k} \sum_{j=1}^{n_i} \left[r_{i,j}{}^3 + 3r_{i,j}{}^2\left(m_i - M\right) + 3r_{i,j}(m_i - M)^2 + \left(m_i - M\right)^3\right]}{NV^{3/2}}. \tag{6.74}$$

$$\text{Now we use } Skew(R_i) \equiv \frac{\sum_{j=1}^{n_i} r_{i,j}{}^3}{n_i[am_i]^{3/2}} \Rightarrow \sum_{j=1}^{n_i} r_{i,j}{}^3 = Skew(R_i) n_i (am_i)^{3/2}. \tag{6.75}$$

$$\text{and } Skew(R) \equiv \frac{\sum_{j=1}^{n_i} Skew(R_i) n_i (am_i)^{3/2}}{N[Var(R)]^{3/2}}, \tag{6.76}$$

yielding

$$S = \frac{\sum_{i=1}^{k} \left[S_i n_i [Var(R_i)]^{3/2} + 3an_i m_i (m_i - M) + (m_i - M)^3 \right]}{NV^{3/2}} \tag{6.77}$$

$$= \frac{\left[[Var(R)]^{3/2} Skew(R) + 3aVar(P) + [Var(P)]^{3/2} Skew(P) \right]}{V^{3/2}}. \tag{6.78}$$

Thus the skewness will be increased over that of two independent variables, due to the correlation between the cell mean and cell variance.

Assuming a normal distribution for P, Skew(P) = 0. If we further assume a Cpe R distribution,

$$a = 2\eta, \quad Var(R) = 2\eta\mu_P, \quad Skew(R) = \frac{3}{2}\left(\frac{2\eta}{\mu_P}\right)^{\frac{1}{2}}, \tag{6.79}$$

and

$$S = Skew(R) \frac{\left(1 + 2\frac{Var(P)}{Var(R)}\right)}{\left(1 + \frac{Var(P)}{Var(R)}\right)^{3/2}} = \frac{3}{2}\sqrt{\frac{2}{\mu_N}} \frac{\left(1 + \frac{Var(P)}{\eta\mu_P}\right)}{\left(1 + \frac{Var(P)}{2\eta\mu_P}\right)^{3/2}}. \tag{6.80}$$

If R and P were independent, the grand skewness would be

$$S = Skew(R)\left(1 + \frac{Var(P)}{Var(R)}\right)^{-3/2}. \tag{6.81}$$

The maximum increase of S from Skew(R) is when Var(P) = ½ Var(R). At this point S/Skew(R) = $(32/27)^{1/2} \sim 1.09$. For independent R and P, S/Skew(R) would be only $(2/3)^{3/2} = 0.544$, or half of the correlated value.

This demonstrates that the increase in skewness, while only a second-order effect, may influence the ΔV_T distributions at the 99.9th percentile level or higher.

6.6 Summary

1. There are two parts to BTI-induced device variation—extrinsic and intrinsic.
2. The extrinsic part contributes to across chip, chip-to-chip, wafer-to-wafer, and lot-to-lot variation. All device sizes are affected equally. The major reliability impact is to introduce an additional variation to the chip performance distribution after time.

3. The intrinsic part causes variation between any two devices regardless of proximity. Variations scale inversely with device area. Although this may impact chip performance shift with future scaling, the two main reliability implications to CMOS circuits have been identified to date:
   - An FET device mismatch drift which affects balanced analog circuits.
   - A broadening of the V_T distribution of the devices in SRAM cells, which contributes to SRAM stability failures.
4. The induced variances in V_T and β shift are proportional to the mean V_T shift and T_{OX}/A_G.
5. ΔV_T distributions are not normal (or log-normal).
6. The asymptotic analytic ΔV_T distribution function for any arbitrary step height distribution is compound Poisson—normal (Cpn).
7. The V_T step height distribution is approximately exponential, as shown both experimentally and by simulation. This distribution has only one parameter, η.
   - The resulting analytic distribution function is compound Poisson—exponential (Cpe).
   - The value of η is increased by random discrete doping and other effects.
8. The impact of the non-normality of the ΔV_T distribution is predicted to be relatively minor for analog mismatch, but can significantly increase SRAM cell stability failure rate.
9. The positive correlation of the extrinsic and intrinsic process variations of BTI does not affect the variance of the total ΔV_T variation; however, it does increase the skewness of the total distribution. Inclusion of this effect in SRAM stability simulations may improve its accuracy.

References

1. K. Bernstein et al., IBM J. Res. & Dev., v. 50, p. 433 (2006).
2. D. Cochrane and G. Orcutt, J. Amer. Stat. Assoc., v.44, p.32 (1949).
3. C. Blat et al., JAP, v.69, p.1712 (1991).
4. S. Ogawa et al., JAP, v.77, p.1137 (1995).
5. A. Asenov, IEEE TED, v. 45, p.2505 (1998).
6. A. Asenov et al., 2011 Eur. Des. Automation and Test Conf.
7. B. Kaczer et al., IEEE EDL, v.31, p.411 (2010).
8. Y. Chen et al., 2001 IRW Fin. Rep., p.41.
9. R. Thewes et al., IEEE TED, v.45, p 2505 (1998).
10. C. Schlünder et al., Microelec. Rel., v.45, p.39 (2005).
11. S. Rauch, IEEE TDMR, v.2, p. 89 (2002).
12. M. Agostinelli et al., Microelec. Rel., v.46, p.63 (2005).
13. G. La Rosa et al., Proc. 2006 IRPS, p.274.
14. A. Haggag et al, Proc. 2007 IRPS, p.452.
15. S. Rauch, IEEE TDMR, v.7, p.524 (2007).
16. M. Alam and S. Mahapatra, Microelec. Rel., v.45, p.71 (2005).

17. H. Reisinger et al., Proc. 2006 IRPS, p.448.
18. M. Agostinelli et al., Proc. 2005 IRPS, p.529.
19. B. Bindu et al., 2009 IIRW Final Report, p.94.
20. K. Takeuchi, 1998 Symp. VLSI Tech. Dig., p.72.
21. D. Snyder, *Random Point Processes*, Wiley (1975).
22. J. Skellam, J. Royal Stat. Soc., Series A, v.109, p. 296 (1946).
23. T. Grasser et al., IEEE TED, v. 58, p.3652 (2011).
24. T. Grasser et al., 2009 IEDM, p.729.
25. V. Huard et al., 2005 Proc. IIRW, p.5.
26. A. Ghetti et al., Proc. 2008 IRPS, p.610.
27. A. Ghetti et al., IEEE TED, v.56, p.1746 (2009).
28. J. Franco, 2012 IRPS, p.5A.4.1.
29. B. Kaczer et al., Proc. 2010 IRPS, p.26.
30. H. Alexandersson, J. Climate and Appl. Meteor., v.24, p.1285 (1985).
31. M. Toledano-Luque et al., 2011 VLSI Symp. Dig., pp 152.
32. V. Huard et al., Proc. 2008 IRPS, p.289.
33. M. Bukhori et al., 2010 IIRW Fin Rep., p76.
34. T. Grasser et al., Proc. 2010 IRPS, p.16.
35. T. Fischer et al., 2008 Eur. Sol.-St. Dev. Res. Conf., p.51.
36. K. Ohmori et al., 2008 IEDM Dig., p.409.
37. H. Dadgour et al., IEEE TED, v.57, p.2504 (2010).
38. S. Markov et al., 2011 IEEE Int. SOI Conf., p. 3.2.
39. X. Wang et al., 2012 IEEE SISPAD, p.256.
40. S. Rasouli et al., 2010 IEEE/ACM ICCAD, p.714.
41. C. Young et al., IEEE TED, v.56, p.1322 (2009).
42. S.M. Amoroso, L. Gerrer, F. Adamu-Lema, S. Markov, A. Asenov, in *Statistical Study of Bias Temperature Instabilities by Means of 3D 'Atomistic' Simulation*, ed. by T. Grasser. Bias Temperature Instability for Devices and Circuits (Springer, New York, 2013)
43. J. Martin-Martinez, R. Rodriguez, M. Nafria, *Simulation of BTI Related Time-Dependent Variability in CMOS Circuits*, ed. by T. Grasser. Bias Temperature Instability for Devices and Circuits. (Springer, New York, 2013)
44. H. Reisinger, *The Time Dependent Defect Spectroscopy*, ed. by T. Grasser. Bias Temperature Instability for Devices and Circuits (Springer, New York, 2013)

Chapter 7
Statistical Distribution of Defect Parameters

B. Kaczer, M. Toledano-Luque, J. Franco, and P. Weckx

Abstract The statistics of bias temperature instability (BTI) is derived within the "defect-centric" paradigm of device degradation. This paradigm is first briefly reviewed, drawing on similarities between BTI and random telegraph noise (RTN). The impact of a single trap on FET threshold voltage V_{th} is then shown to follow an exponential distribution with the expectation value η. The properties of η, such as its area and gate oxide thickness dependences, are discussed. The statistics of *multiple* defects is then developed, assuming (1) the single-trap exponential distribution and (2) a Poisson distribution of the number of traps in each device. The properties of the resulting time-dependent total ΔV_{th} statistics and its moments are then treated. Finally, the combined time-dependent and time-zero statistics of the total threshold voltage V_{th} is discussed, together with its properties and a brief example of its implications for circuit performance metrics.

7.1 Introduction

The large, micrometer-sized FET devices of the past CMOS technologies were considered identical in terms of electrical performance. Similarly, the application of a given stress resulted in an identical parameter shift in all devices. With the gradual downscaling of the FET devices, the oxide dielectric was the first to reach nanometer dimensions, thus introducing the first stochastically distributed reliability mechanism—the time-dependent dielectric breakdown [1]. With the shrinking of *lateral* device dimensions to atomic levels, variation between devices appeared due to effects such as random dopant fluctuation and line edge roughness [2]. This phenomenon, now routinely considered in circuit design, is referred to as initial, as-fabricated, or *time-zero variability* [3, 4].

B. Kaczer (✉) • M. Toledano-Luque • J. Franco • P. Weckx
Imec, Kapeldreef 75, 3001 Leuven, Belgium
e-mail: kaczer@imec.be; toleda@imec.be; francoj@imec.be; weckx@imec.be

T. Grasser (ed.), *Bias Temperature Instability for Devices and Circuits*,
DOI 10.1007/978-1-4614-7909-3_7,

Analogously to time-zero variability, application of a fixed stress in deeply scaled devices results in additional statistical *distributions* of the parameter *shifts* [5, 6]. This is referred to as *time-dependent* variability. The overall variability of deeply scaled devices is therefore caused by a combination of time-dependent variability effects and the time-zero variability. Correctly describing the time-dependent and the overall statistical distributions is therefore crucial for correctly predicting the reliability of future deeply downscaled technologies.

In this chapter, we attempt to convey the basic principles necessary for understanding time-dependent variability and its link with time-zero variability. We limit ourselves to discussing only a single but crucial FET parameter, the threshold voltage and its behavior (instability) during device operation—the so-called bias temperature instability (BTI) [7, 8]. We have argued that this instability can be understood as an extended (non-steady state) case of device (channel current) noise, which takes the form of random telegraph noise (RTN) in deeply scaled devices [9–11]. We discuss only device-to-device variability effects caused by random or "local" variations, and we note the links between random time-zero and time-dependent variations [12, 13]. Variations due to processing ("systematic" variations), especially processing of the device gate stack, will cause additional (and often linked) time-zero and time-dependent variations in device parameters. This aspect is, however, beyond the scope of this chapter. Finally, in order not to overcomplicate the explanation, only constant stress is discussed here. The more general case of an arbitrary workload, mandatory in clocked digital designs, is discussed, e.g., in [9, 14].

7.1.1 Defect-Centric Paradigm

The reduction of FET device dimensions to nanometer scales implies that literally, only a handful of defects are present in each device, while each defect has a substantial impact on the device operation. This constitutes a "paradigm shift" from the days of micrometer-sized devices in which charge was only considered in the form of *continuous* densities. Here, we argue that in deeply scaled devices many degradation mechanisms, including BTI and RTN, are best understood in terms of the impact of a small *ensemble of individual* (charged) traps. This "bottom-up" approach to device reliability is already being advocated by several groups [2, 11, 15–23]. (We like to compare this shift to the evolution of statistical mechanics from thermodynamics (the laws of averages) in the nineteenth century.) We show that when the properties of individual defects and their impact on the device are understood, this "defect-centric" view *naturally* yields the correct description of time-dependent variability [10, 24].

In general, each device can be characterized by the number of defects n in its gate oxide. Only the occupied (charged) defects are assumed to influence the channel current. The occupation of each defect can then be determined from the voltage and temperature-dependent (1) capture and (2) emission times, *particular to each*

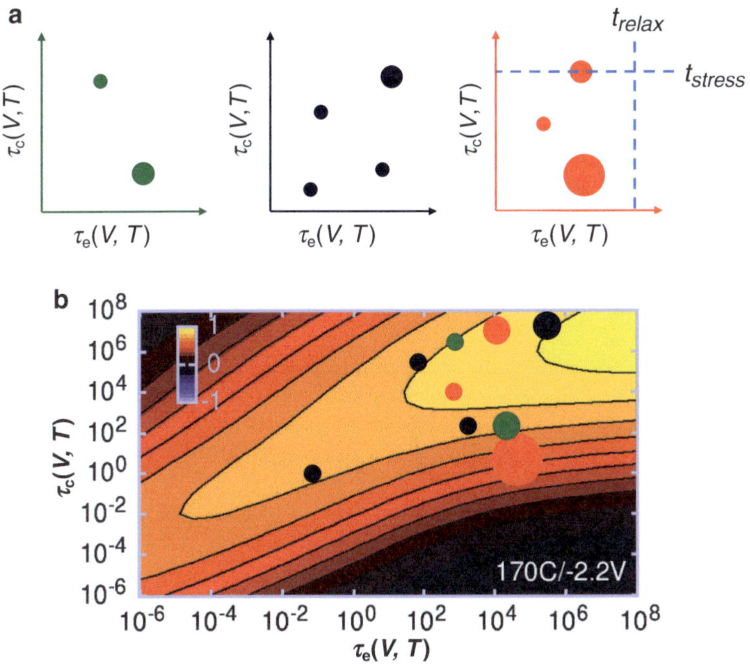

Fig. 7.1 (**a**) BTI degradation in deeply scaled devices can be described in terms of the total number of traps in each device (*circles*), their (voltage and temperature dependent) capture and emission times τ_c and τ_e, and their impact on the device (demarked by the size of the circle). Three FET instances are schematically illustrated. Traps are likely to be charged (occupied) when stress bias is applied and $t_{stress} > \tau_c$ and discharged (unoccupied) when the stress bias is removed and $t_{relax} > \tau_e$. (**b**) An example of a capture and emission time (CET) map [25] representing the probability density of finding a trap with a particular combination of τ_c and τ_e in a large device with a large number of traps. Defects will appear in individual deeply scaled devices with this probability

defect. Each defect is also known to (3) impact the channel current differently. The three properties are schematically illustrated for three devices in Fig. 7.1a [10]. The picture is easily extensible with additional defect parameters to cover, e.g., generation of defects during device operation. A link with large devices is shown in Fig. 7.1b. The continuous capture and emission time map, discussed in [26], describes BTI in terms of probability (density) of finding a defect with certain capture and emission times [25–27].

From Fig. 7.1a, it is already evident that each deeply scaled device will respond uniquely to workload dependence. This is the basis of time-dependent variability in the defect-centric paradigm. This view is further schematically summarized in Fig. 7.2. Figure 7.2a illustrates the degradation in the large devices of the past. Subjected to the same stress, all devices behave identically, and their lifetime (time to the failure criterion) can be characterized by a single, "average" number [10].

Each deeply scaled device, on the other hand, will respond differently depending on its particular trap configuration. At a constant bias condition, i.e., a steady state,

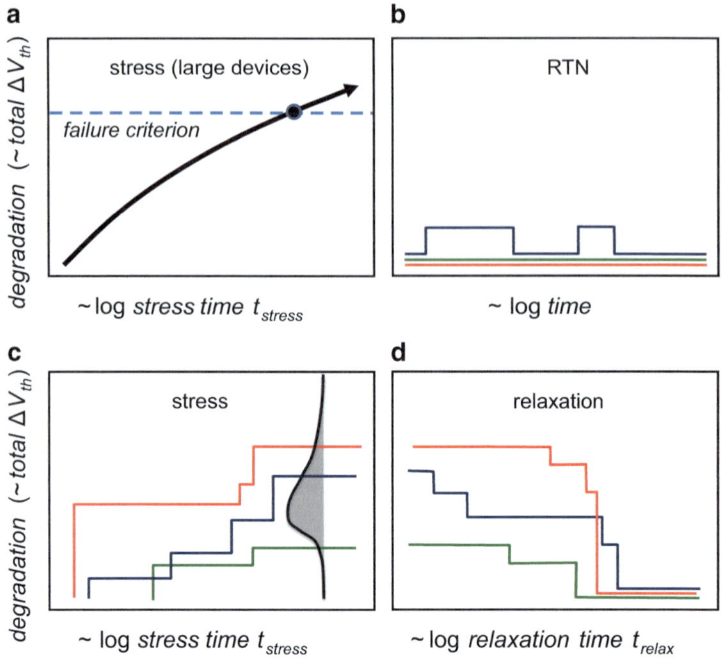

Fig. 7.2 (**a**) All large devices behave identically during stress and are assumed to fail when reaching the projected "hard" degradation criterion. (**b**) At constant bias, only a small subset of defects in deeply scaled devices will be active at constant bias conditions, resulting in RTN signal. (**c**) Progress of degradation during constant stress in the three devices in Fig. 7.1a (only capture events are shown). The origin of the BTI variability and the distribution of degradation at a given time are apparent. (**d**) The corresponding relaxation in the three devices when the stress bias is removed

some oxide traps with suitable time constants will interact with carriers in the channel. Consequently, some of the devices will manifest RTN (Fig. 7.2b) [28]. When stressed, the defects will be preferentially charged according to their capture times (Fig. 7.2c). When the perturbation is removed, the defects will one by one emit and the devices will return back to the steady state (Fig. 7.2d) [10]. This is the essence of the extended measure-stress-measure (eMSM) [29] and the time-dependent defect spectroscopy (TDDS) techniques (see [30]). Because the capture and emission times are widely distributed over many decades in time, so will be the degradation and relaxation processes shown in Fig. 7.2c, d.

7.1.2 Chapter Organization

The defect temporal properties (i.e., the capture and emission times) constitute a complex physical system and are discussed elsewhere [31]. In this chapter, we develop the statistics of degradation at a particular moment in time, as illustrated in Fig. 7.2c. It will be shown that this can be done by removing the defect kinetics from the picture, with the degradation being described simply by its mean value.

The chapter is structured as follows. In Sect. 7.2, we show that the impact of a *single* trap on FET threshold voltage V_{th} can be described by an exponential distribution with the expectation value η. The properties of η, such as its area and gate oxide thickness dependences, are then discussed. The statistics of *multiple* defects is then developed in Sect. 7.3, assuming (1) the aforementioned exponential distribution and (2) a Poisson distribution of the number of traps in each device. The properties of the resulting time-dependent total ΔV_{th} statistics and its moments are then treated. In Sect. 7.4, the combined time-dependent and time-zero statistics of the total threshold voltage V_{th} is discussed, together with its properties and a brief example of its implications for circuit performance metrics.

7.2 Individual-Trap ΔV_{th} Distribution

Figure 7.3a shows a typical result of the extended MSM measurement [29] in multiple pFETs [10]. As already reported previously [6, 7, 9, 27, 32, 33], clear steps caused by single discharge events are visible in the negative BTI (NBTI) relaxation transients. The individual down-steps Δv_{th}, together with the corresponding emission times, represent an *individual signature* (a "fingerprint") of each defect, which, e.g., allows tracing its properties under various stress conditions [27, 28].

A cumulative plot of emission step heights such as those in Fig. 7.3a from $N_{devices} = 72$ pFET devices is shown in Fig. 7.3b. The plot demonstrates that the distribution of the "BTI" relaxation steps Δv_{th} is exponential, with their probability distribution function (PDF) being

$$f_\eta(\Delta v_{th}) = \frac{1}{\eta}\exp\left(-\frac{\Delta v_{th}}{\eta}\right), \tag{7.1}$$

where the scaling factor η is the mean Δv_{th} value for a *single* charge. The cumulative distribution function (CDF) corresponding to Eq. (7.1) is then

$$F_\eta(\Delta v_{th}) = 1 - \exp\left(-\frac{\Delta v_{th}}{\eta}\right). \tag{7.2}$$

Furthermore, when plotting

$$\frac{N_{total}}{N_{devices}}\left[1 - F_\eta(\Delta v_{th})\right] = N_T \exp\left(-\frac{\Delta v_{th}}{\eta}\right), \tag{7.3}$$

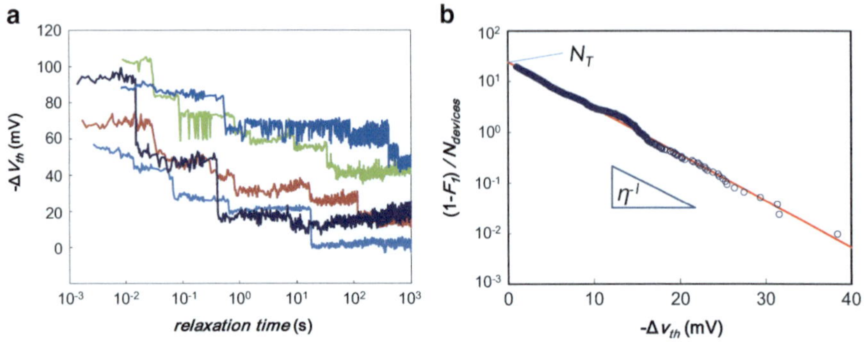

Fig. 7.3 (**a**) A typical experimental result of the eMSM sequence obtained on multiple 90 × 35 nm^2 high-k/metal gate devices (cf. Fig. 7.2d) [10]. Each device behaves differently, resulting in large time-dependent variability. Steps of varying heights due to single discharge events are clearly visible. Also note some defects producing RTN signal. (**b**) A complementary CDF plot of all transient step heights detected in 72 devices (*symbols*) shows a clear exponential distribution (line, maximum likelihood fit). When normalized by the number of devices, the intercept with the *y*-axis gives the average number of defects per device N_T that emitted in the measured relaxation interval

as in Fig. 7.3b, where N_{total} is the total number of steps detected in all devices in the measured relaxation interval, the average number of active defects per device N_T can be readily extracted [34].

We note that the exponential distribution has been repeatedly reported for RTN amplitudes [35–37]. This similarity further strengthens the link between RTN and BTI [10, 28]. The exponential distribution of single-charge Δv_{th} can be understood if nonuniformities in the pFET channel due to random dopant fluctuations (RDF) and other variability sources are considered [2]. The threshold voltage of such a device corresponds to the formation of a conduction (percolation) path in the random potential between the device source and drain (Fig. 7.4a) [37]. To the zeroth order, depending on the position of the oxide trap occupied after the BTI stress, the conduction path can be either unaffected or obstructed by the newly charged defect. The drop in the current has to be compensated by an increase of the gate voltage, resulting in the observed Δv_{th}. The more likely former case results in the large number of small Δv_{th} steps (bulk of the distribution in Fig. 7.3b) while the unlikely latter case yields a small number of "killer traps" (tail of distribution in Fig. 7.3b).

As a side note, we remark that many fundamental properties of the channel percolation can be qualitatively studied with a simple model depicted in Fig. 7.4b [10]. In this model, a mesh of "elementary" FETs with random V_{th}'s, representing variations in the local potential, is set up in SPICE to represent the channel of the deeply scaled FET. The simple percolation model correctly reproduces, e.g., the normal distribution of initial threshold voltages V_{th0} and the variance of V_{th0}

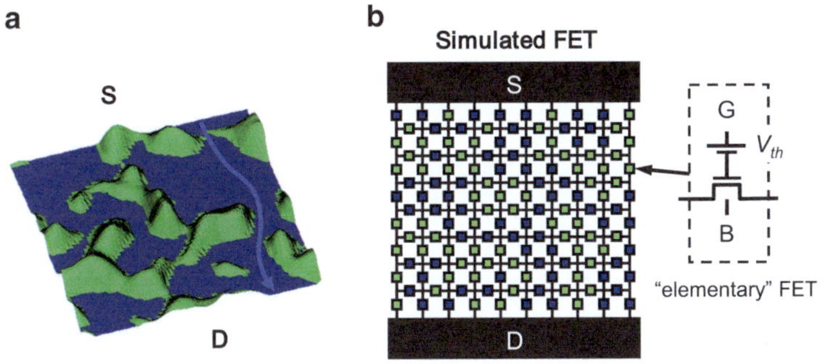

Fig. 7.4 (**a**) An illustration of a percolation path in a random potential (from [38]) such as that between FET source and drain. (**b**) A mesh of "elementary" FETs with random V_{th}s (voltage source in series with gate) representing (**a**) can be readily solved with SPICE

scaling reciprocally with the channel area A (Pelgrom's rule) [4]. Moreover, it approximately reproduces the exponential distribution of single-charge impact and also scaling of η with device area, discussed next.

7.2.1 Properties of η

As will be shown below, the average impact of a single defect η on the threshold voltage is a fundamental parameter determining the variability of deeply scaled technologies. It has been shown to scale as

$$\eta \cong \frac{t_{inv} N_A^{\alpha}}{A}, \tag{7.4}$$

where t_{inv} is the oxide thickness corresponding to capacitance in inversion, N_A the channel doping, and A the area of the device channel. The exponent α has been observed to be around 0.5 in simulations [37]. Its relation with the impact of a single charge in the charge sheet approximation,

$$\eta_0 = \frac{q}{C_{ox}}, \tag{7.5}$$

has been discussed, e.g., in [7, 39, 40]. In Eq. (7.5), q is the elementary charge and C_{ox} the gate oxide capacitance (in Farads) in inversion. The value of η can vary between multiples to a fraction of η_0, as also noted in [41]. The value of η will be reduced for oxide defects closer to the gate [40, 42].

Figure 7.5a demonstrates η scales reciprocally with device gate area, in agreement with Eq. (7.4) [7]. Over the whole measured range, η is observed ~2× higher than the expected single-charge η_0. Since time-dependent variance will be shown

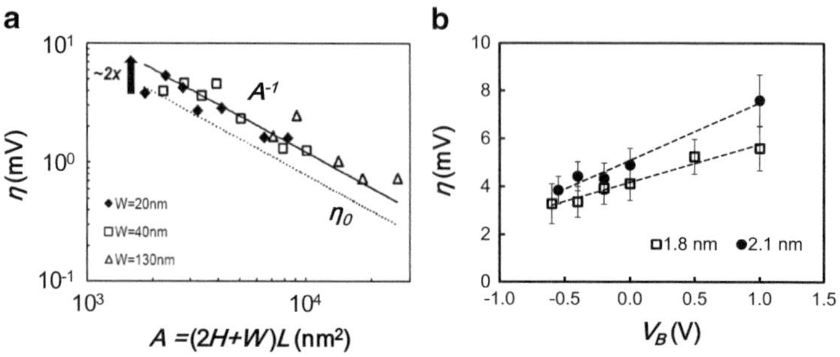

Fig. 7.5 (**a**) The average step height η scales inversely with A ($\eta \propto A-1$) on Si pFinFETs (high-k/MG, $t_{inv} \approx 1.7$ nm) with varying fin width W and gate length L (fin height H is fixed). Each point is extracted from a set of multiple devices with identical dimensions as in Fig. 7.3b. (**b**) The average step height η values extracted from distributions measured for varying back-bias V_B on two wafers with identical Si/SiON/Poly-Si planar pFETs and identical doping levels but slightly different oxide thicknesses ($t_{inv} = \sim 1.8$ and ~ 2.1 nm). Thicker oxide increases η while forward (reverse) back bias reduces (increases) the depletion width and thus reduces (increases) η, as per Eq. (7.4)

below to depend on η, using Eq. (7.5) as a substitute for this parameter can lead to underestimating the time-dependent variability component.

Figure 7.5b then demonstrates that η decreases with oxide thickness and can be changed by back bias, which effectively modulates the number of charged dopants in the channel. Both of those observations are in line with Eq. (7.4) [13].

7.3 Total ΔV_{th} Distribution: Time-Dependent BTI Variability

If the lateral locations of n successively trapped charges are assumed to be uncorrelated, the overall threshold voltage shift will be

$$\Delta V_{th} = \sum_{i=1}^{n} \Delta v_{th,i}. \tag{7.6}$$

The distribution of ΔV_{th} can be expressed as a convolution of individual exponential distributions [Eq. (7.1)], with the PDF and the CDF respectively described by

$$f_{\eta,n}(\Delta V_{th}) = \frac{n}{n!}\frac{\Delta V_{th}^{n-1}}{\eta^n}\exp\left(-\frac{\Delta V_{th}}{\eta}\right) \tag{7.7}$$

and

$$F_{\eta,n}(\Delta V_{th}) = 1 - \frac{n}{n!}\Gamma\left(n, \frac{\Delta V_{th}}{\eta}\right). \tag{7.8}$$

Here, Γ is the incomplete gamma function.

An actual population of stressed devices will consist of devices with a *different* number n of visible oxide defects in each device. That number will be Poisson distributed [9, 10, 32]. (Here we disregard the small fraction of device population manifesting RTN around V_{th0}—that assumption will lead to more complex statistics [5, 6, 43].) The *total* ΔV_{th} distribution can be therefore obtained by summing distributions F_n weighted by the Poisson probability

$$p_{N_T,n} = \frac{e^{-N_T} N_T^n}{n!}. \tag{7.9}$$

In Eq. (7.9), N_T is the mean number of defects in the FET gate oxide and is related to the oxide trap (surface) density N_{ot} as $N_T = W\,L\,N_{ot}$ (i.e., N_T is not an integer).

This line of reasoning then results in the *total* ΔV_{th} CDF given by

$$H_{\eta,N_T}(\Delta V_{th}) = \sum_{n=0}^{\infty} p_{N,n} F_{\eta,N}(\Delta v_{th}) = \sum_{n=0}^{\infty} \frac{e^{-N} N^n}{n!} \left[1 - \frac{n}{n!}\Gamma\left(n, \frac{\Delta V_{th}}{\eta}\right)\right]. \tag{7.10}$$

The corresponding PDF is

$$h_{\eta,N_T}(\Delta V_{th}) = e^{-N}\left[\delta(\Delta V_{th}) + \frac{N}{\eta}\exp\left(-\frac{\Delta V_{th}}{\eta}\right) {}_0F_1\left(2; \frac{N}{\eta}\Delta V_{th}\right)\right], \tag{7.11}$$

where the hypergeometric function ${}_0F_1(2;x)$ can be also written in terms of the modified Bessel function I_l as ${}_0F_1(2;x) = x^{-1/2} I_1(2x^{1/2})$. The Dirac $\delta(\Delta V_{th})$ term represents the fraction of devices with 0 V shift [36], which decreases with increasing N_T. The CDF of Eq. (7.10) is plotted in Fig. 7.6a for several values of N_T. For comparison, measured total ΔV_{th} distributions from [6] are excellently fitted by Eq. (7.10), supporting the presented approach. Application of Eq. (7.10) to high-κ devices is given in [44].

7.3.1 Properties of the Total ΔV_{th} Distribution

The advantages of describing the total ΔV_{th} distribution in terms of Eqs. (7.10) and (7.11) are their relative simplicity and tangibility of the variables, while the analytical description allows further statistical treatment [10, 24]. The mean of the above-derived distribution is

$$\langle \Delta V_{th}(t) \rangle = \eta N_T(t), \tag{7.12}$$

i.e., it should be independent of FET gate area A provided N_T and η are respectively directly and inversely proportional to A [12]. The variance of the distribution is then

$$\sigma^2_{\Delta V_{th}}(t) = 2\eta^2 N_T(t), \tag{7.13}$$

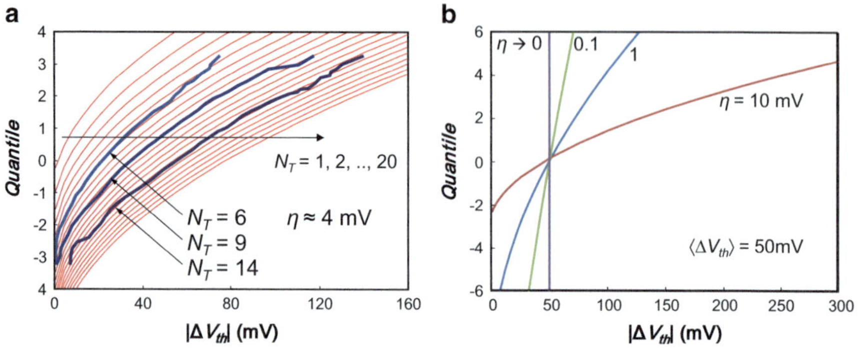

Fig. 7.6 (**a**) Total ΔV_{th} distribution [Eq. (7.10)] for the average number of defects N_T from 1 to 20 (*thin lines*) rescaled to fit experimental distributions from Fig. 10 of [6], with the corresponding values of N_T and η readily extracted. (**b**) Illustration of how the shape of the distribution in Eq. (7.10) changes with η for a fixed value of $\langle V_{th}\rangle = 50$ mV [i.e., N_T is obtained from Eq. (7.12)]. Large η is expected in small-area devices, devices with thick gate oxide, and devices with large channel potential variations [cf. Eq. (7.4)]

i.e., it increases with decreasing gate area. The *relative* deviation $\sigma_{\Delta Vth}/\langle V_{th}\rangle = (2/N_T)^{1/2}$ is decreasing with increasing N_T. Note that the factor of 2 [6] is rigorously derived when the second moment is extracted by symbolically integrating Eq. (7.11) multiplied by $(\Delta V_{th} - \langle V_{th}\rangle)^2$. For completeness, the higher moments of this distribution are the skewness

$$6\eta^3 N_T(t) \tag{7.14}$$

and kurtosis

$$12\eta^4 \left[N_T^2(t) + 2N_T(t)\right]. \tag{7.15}$$

Equations (7.12) and (7.13) allow us expressing both N_T and η in terms of more "circuit designer-friendly" parameters $\langle V_{th}\rangle$ and $\sigma^2_{\Delta Vth}$ as

$$N_T(t) = 2\frac{\langle \Delta V_{th}(t)\rangle^2}{\sigma^2_{\Delta Vth}} \tag{7.16}$$

and

$$\eta = \frac{\sigma^2_{\Delta Vth}}{2\langle \Delta V_{th}(t)\rangle}. \tag{7.17}$$

This implies that both N_T and η can be extracted from the first two moments of a measured total BTI V_{th} distribution [10], without having to characterize individual step heights as done in Fig. 7.3b.

A convenient way to discuss time-dependent variance ${\sigma_{\Delta Vth}}^2$ is to express it, with the help of Eqs. (7.12) and (7.13), in terms of one technology parameter η and one design parameter $\langle V_{th}\rangle$ as

$$\sigma^2_{\Delta V_{th}}(t) = 2\eta \left\langle \Delta V_{th}(t) \right\rangle . \tag{7.18}$$

Equation (7.18) allows removing the (complex) degradation kinetics (time dependence) from the consideration and expressing the degradation only in terms of the average degradation $\langle \Delta V_{th}\rangle$ [6, 12, 45]. The technology-dependent parameter η can be then obtained independently for each specific technology, either from single-emission measurements (Fig. 7.3b), measurements of total ΔV_{th} (BTI) distributions [Eq. (7.17)], or from considerations based on links with time-zero variance, given in the next subsection.

Figure 7.6b illustrates how the shape of the total ΔV_{th} distribution in Eq. (7.10) will change with η for a fixed value of $\langle \Delta V_{th}\rangle = 50$ mV. The distribution is wide and strongly non-normal for large values of η. As η decreases, corresponding to device area increase or gate oxide thickness decrease [cf. Eq. (7.4)], the variance of the distribution decreases (cf. Eq. (7.18) and [44]). The limiting case of $\eta \to 0$ then represents the "classical" interpretation of reliability in which all identical large devices can be described by a single, average value of degradation [32].

7.4 Total V_{th} Distribution: Time-Zero and Time-Dependent Variabilities

Design of modern ULSI circuits requires factoring in the time-zero variability. The time-zero threshold voltage V_{th0} is typically assumed to be normally distributed with σ_{Vth0}. The latter quantity is technology scale as [3, 4]

$$\sigma^2_{V_{th0}} \propto \frac{t_{ox}\sqrt{N_A}}{A}. \tag{7.19}$$

The threshold voltage in each device in the designed circuit is thus

$$V_{th}(t) = V_{th0} + \Delta V_{th}(t), \tag{7.20}$$

where the second term on the right-hand side represents the time-dependent degradation during circuit operation discussed in the previous section. Measured distributions of $V_{th}(t)$ at time-zero and after degradation inducing $\langle \Delta V_{th}\rangle = {\sim}20$ and ${\sim}50$ mV are illustrated in Fig. 7.7a. As expected, the variance of the V_{th} distribution increases with time, as given simply by

$$\sigma^2_{V_{th}}(t) \cong \sigma^2_{V_{th0}} + \sigma^2_{\Delta V_{th0}}(t). \tag{7.21}$$

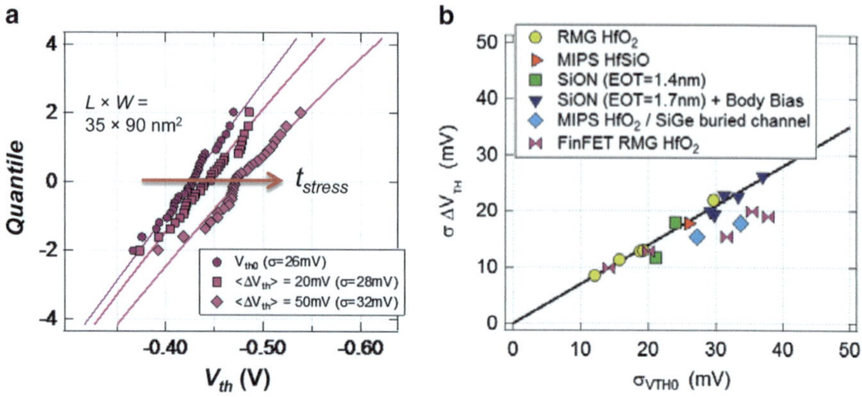

Fig. 7.7 (**a**) Measured initial and post-stress distributions of V_{th} for 32 pFETs (*symbols*) are well fitted by the total V_{th} distribution in Eq. (7.22) (*lines*). The variance is seen to increase with time [Eq. (7.21)]. (**b**) The spread of the time-dependent BTI ΔV_{th} distribution $\sigma_{\Delta Vth}$ is correlated with time-zero variability σ_{Vth0} of fresh pFET devices. For $\langle \Delta V_{th} \rangle = 50$ mV used in the plot, the ratio is ~0.65 for pFET [12]

The total $V_{th}(t)$ distribution (CDF) itself is then a convolution of time-zero and time-dependent components:

$$K_{\langle V_{th0}\rangle,\sigma_{Vth0},\langle \Delta V_{th}\rangle,\sigma_{\Delta Vth0}}(V_{th}) = \int_0^\infty H_{\eta,N_T}(V) g_{\langle V_{th0}\rangle,\sigma_{th0}}(V_{th} - V)dV. \tag{7.22}$$

Here, g is the normal (Gaussian) PDF with mean $\langle V_{th0}\rangle$ and standard deviation σ_{Vth0}. Note that η and N_T in H [Eq. (7.10)] can be calculated from $\langle \Delta V_{th}\rangle$ and $\sigma_{\Delta Vth}$ as per Eqs. (7.16) and (7.17). The resulting distribution is shown in Fig. 7.7a.

7.4.1 Correlation of Time-Zero and Time-Dependent Variances

We have recently observed [12] that, independently of technology, the time-zero and time-dependent variances σ_{Vth0} and $\sigma_{\Delta Vth}$ are correlated, as documented in Fig. 7.7b. The strong correlation between these two quantities constituting the right-hand side of Eq. (7.21) suggests that identical sources are responsible for time-zero and time-dependent variability—as also reflected in the algebraic similarity between σ_{Vth0} [Eq. (7.19)] and $\sigma_{\Delta Vth}$ [Eqs. (7.18) and (7.4)].

From the dependence in Fig. 7.7b, we can derive a simple empirical rule for pFET devices

$$\sigma^2_{\Delta V_{th}}(t) \cong \frac{\langle \Delta V_{th}(t)\rangle}{100mV}\sigma^2_{V_{th0}}. \tag{7.23}$$

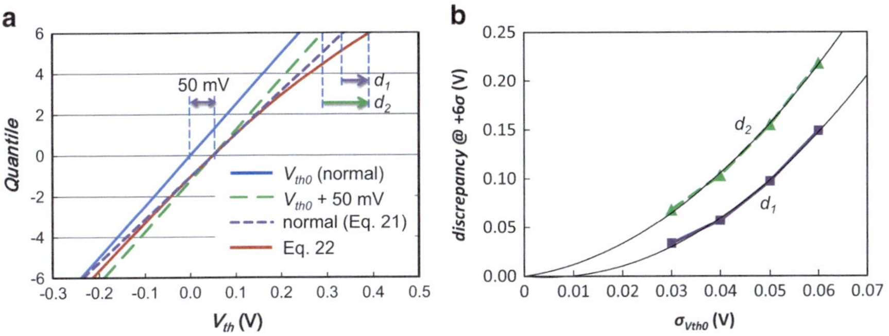

Fig. 7.8 (**a**) Three ways of adding time-dependent variability to time-zero distribution (*solid straight line*): (1) a simple shift of $\langle \Delta V_{th}(t) \rangle$ in all devices (*dashed line*), (2) a normal (Gaussian) with σ_{Vth} given by Eq. (7.21) (*dotted line*), and (3) the "correct" statistics given by Eq. (7.22) (*solid curve*). Equation (7.23) and $\langle \Delta V_{th}(t) \rangle = 50$ mV were used when constructing this plot. Discrepancies with respect to the "correct" statistics [Eq. (7.22)] at 6σ are demarcated. (b) The discrepancies from (a) plotted vs. varying time-zero deviations. *Lines*: parabolic fits

Equation (7.23) shows that time-dependent variability increases as degradation, expressed through $\langle \Delta V_{th}(t) \rangle$, progresses during circuit operation. Equation (7.23) also allows quickly estimating $\sigma_{\Delta Vth}$ if σ_{Vth0} is known for existing technology (or even estimated for upcoming technologies).

We also note for completeness that Eq. (7.23) can be combined with Eq. (7.18) to get the empirical dependence (for pFETs)

$$\eta \cong \frac{\sigma^2_{V_{th0}}}{200mV}. \tag{7.24}$$

Equation (7.24) allows obtaining η from time-zero variability and vice-versa.

7.4.2 Implications of the Combined Variabilities

The total V_{th} distribution described by Eq. (7.22) is compared in Fig. 7.8a with more "naive" ways of introducing time-dependent variability. Normal (Gaussian) distribution with σ_{Vth} given by Eq. (7.21) is often used to describe the total V_{th} distribution. From Fig. 7.8a, it is evident that this is at best a nonphysical approximation, as the total V_{th} normal approximation is bound to cross the time-zero normal distribution at low percentiles—not possible if only $\Delta V_{th} > 0$V (positive BTI, i.e., PBTI) or $\Delta V_{th} < 0$V (NBTI) is assumed. At higher percentiles, the normal total V_{th} distribution underestimates the expected variation. This discrepancy will increase as the distributions become wider (i.e., the deviations increasing), documented in Fig. 7.8b. The same discrepancy is also plotted for comparison for

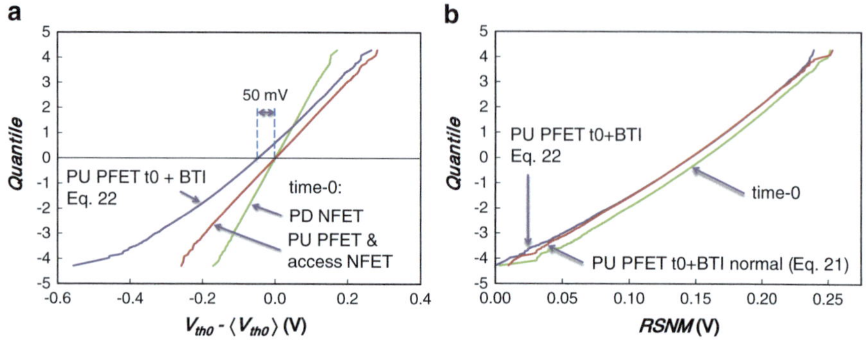

Fig. 7.9 (**a**) Time-zero V_{th} distributions of one pull-down 3-fin nFinFET and one pull-up 1-fin pFinFET (identical to that of the access 1-fin access nFinFETs) and combined time-dependent and time-zero V_{th} distributions for the pull-up 1-fin pFinFET [Eq. (7.22)]. (**b**) Read signal-noise-margin distribution calculated at time-zero and after degradation assuming normal statistics and statistics given by Eq. (7.22). The impact of correct statistics is visible at -4σ

the most naive case of all devices undergoing exactly the same $\langle \Delta V_{th}(t) \rangle$ shift. For example, for $\sigma_{Vth0} = \sim 40$ mV (corresponding to the smallest pFET in 28 nm technology), the underestimation of the maximum variation of V_{th} at 6 σ is $\sim$60 mV.

The use of the statistics developed above is finally illustrated on the simple example of a 6T SRAM cell. Time-zero variability is assumed in all six transistors, while time-dependent BTI variability is introduced only into the two pull-up pFETs, again with $\langle \Delta V_{th}(t) \rangle = 50$ mV (Fig. 7.9a). Figure 7.9b then shows the read signal-noise-margin (RSNM) calculated for these cells at time-zero and after the $\langle \Delta V_{th}(t) \rangle = 50$ mV degradation. From Fig. 7.9b, it is apparent that compared to the normal distribution with variance given by Eq. (7.21), the use of the correct statistics [Eq. (7.22)] results in a reduced figure-of-merit already at -4σ, implying higher expected failure during operation. Even higher impact is expected at -6σ. Further discussion of circuit-related issues can be found in [21–23].

7.5 Conclusions

The statistics of bias temperature instability (BTI) has been rigorously derived within the "defect-centric" paradigm of device degradation. The analytical description should prove useful for both reliability data analysis and simulations of deeply scaled CMOS circuitry.

Acknowledgments The authors gratefully acknowledge the invaluable input of Philippe Roussel for statistics; Prof. Tibor Grasser for general discussions and trap kinetics in particular; Prof. Guido Groeseneken for discussion and support; and Prof. Francky Catthoor, Dr. Praveen Raghavan, and

Halil Kukner for a circuit designer's perspective. This work was performed as part of imec's Core Partner Program. It has been in part supported by the European Commission under the 7th Framework Programme (collaborative project MORDRED, contract No. 261868).

References

1. R. Degraeve, G. Groeseneken, R. Bellens, M. Depas, and H. E. Maes, *Int. Electron Devices Meeting Tech. Dig.*, 863–866 (1995).
2. A. Asenov, S. Roy, R. A. Brown, G. Roy, C. Alexander, C. Riddet, C. Millar, B. Cheng, A. Martinez, N. Seoane, D. Reid, M. F. Bukhori, X. Wang, and U. Kovac, "Advanced simulation of statistical variability and reliability in nano CMOS transistors", *Int. Electron Devices Meeting Tech. Dig.* (2008).
3. K. J. Kuhn, M. D. Giles, D. Becher, P. Kolar, A. Kornfeld, R. Kotlyar, S. T. Ma, A. Maheshwari, and S. Mudanai, IEEE T. Electron Dev. **58**, 2197 (2011).
4. M. J. M. Pelgrom, A. C. J. Duinmaijer, and A, P. G. Welbers, *IEEE J. Solid-State Circ.***24**, 1433–1440 (1989).
5. S. E. Rauch, *IEEE T. Dev. Mat. Rel.***7**, 524 (2007).
6. V. Huard C. Parthasarathy, C. Guerin, T. Valentin, E. Pion, M. Mammasse, N. Planes, L. Camus, *Proc.* IEEE *Int. Rel. Phys. Symp.*, 289 (2008).
7. J. Franco, B. Kaczer, M. Toledano-Luque, Ph.J. Roussel, T. Grasser, J. Mitard, L. Å. Ragnarsson, L. Witters, T. Chiarella, M. Togo, N. Horiguchi, M.F. Bukhori, A. Asenov, and G. Groeseneken, *Proc. IEEE Int. Reliab. Phys. Symp.*, (2012).
8. B. Kaczer, J. Franco, M. Toledano-Luque, Ph. J. Roussel, M. F. Bukhori, A. Asenov, B. Schwarz, M. Bina, T. Grasser, G. Groeseneken, *Proc. IEEE Int. Reliab. Phys. Symp.* (2012).
9. B. Kaczer, T. Grasser, J. Martin-Martinez, E. Simoen, M. Aoulaiche, Ph.J. Roussel and G. Groeseneken, *Proc. IEEE Int. Reliab. Phys. Symp. (IRPS) Proc.*, 55 (2009).
10. B. Kaczer, T. Grasser, Ph. J. Roussel, J. Franco, R. Degraeve, L.-A. Ragnarsson, E. Simoen, G. Groeseneken and H. Reisinger, *Proc. IEEE Int. Reliab. Phys. Symp. (IRPS) Proc.*, 26 (2010).
11. T. Grasser, B. Kaczer, W. Goes, H. Reisinger, T. Aichinger, P. Hehenberger, P.-J. Wagner,; F. Schanovsky, J. Franco, P. J. Roussel, and M. Nelhiebel, "Recent advances in understanding the bias temperature instability", *Int. Electron Devices Meeting Tech. Dig.*, 4.4.1–4.4.4 (2010).
12. M. Toledano-Luque, B. Kaczer, J. Franco, Ph.J. Roussel, M. Bina, T. Grasser, M. Cho, P. Weckx, and G. Groeseneken, accepted to *Proc. VLSI Symp.*, 2013.
13. J. Franco, B. Kaczer, M. Toledano-Luque, Ph. J. Roussel, G. Groeseneken, B. Schwarz, M. Bina, M. Waltl, P.-J. Wagner, T. Grasser, *IEEE Int. Reliab. Phys. Symp. (IRPS) Proc.* (2013).
14. P. Weckx, B. Kaczer, M. Toledano-Luque, T. Grasser, Ph. J. Roussel, H. Kukner, P. Raghavan, F. Catthoor, and G. Groeseneken, *Proc. IEEE Int. Reliab. Phys. Symp. (IRPS) Proc.*, (2013).
15. B. Kaczer, T. Grasser, J. Franco, M. Toledano-Luque, Ph. J. Roussel, M. Cho, E. Simoen, G. Groeseneken, J. Vac. Sci. Technol. B **29**, 01AB01 (2011).
16. B. Kaczer, S. Mahato, V. Valduga de Almeida Camargo, M. Toledano-Luque, Ph. J. Roussel, T. Grasser, F. Catthoor, P. Dobrovolny, P. Zuber, G. Wirth, and G. Groeseneken, , *Proc. IEEE Int. Reliab. Phys. Symp.*, XT.3.1–XT.3.5 (2011).
17. V. Huard, N. Ruiz, F. Cacho, and E. Pion, *Microel. Reliab.***51**, 1425–1439 (2011).
18. A. E. Islam and M. A. Alam, *J.Comput.Electron.***10**, pp. 341–351 (2011).
19. M. Nafria, R. Rodriguez, M. Porti, J. Martin-Martinez, M. Lanza, and X. Aymerich, *Int. Electron Devices Meeting Tech. Dig.*, 6.3.1–6.3.4 (2011).
20. M. Toledano-Luque, B. Kaczer, J. Franco, Ph. J. Roussel, T. Grasser, and G. Groeseneken, Microel. Reliab. **52**, 1883–1890 (2012).
21. K.B. Sutaria, J.B. Velamala, V. Ravi, Y. Cao (2013) Multi-level reliability simulation for IC design. In: T. Grasser (ed.) Bias temperature instability for devices and circuits. Springer, New York

22. G. Wirth, Y. Cao, J.B. Velamala, K.B. Sutaria, T. Sato (2013) Charge trapping in MOSFETS: BTI and RTN modeling for circuits. In: T. Grasser (ed.) Bias temperature instability for devices and circuits. Springer, New York
23. J. Martin-Martinez, R. Rodriguez, M. Nafria (2013) Simulation of BTI related time-dependent variability in CMOS circuits. In: T. Grasser (ed.) Bias temperature instability for devices and circuits. Springer, New York
24. B. Kaczer, Ph. J. Roussel, T. Grasser and G. Groeseneken, *IEEE Electron Device Lett.***31**, 411 (2010).
25. T. Grasser, P.-J. Wagner, H. Reisinger, Th. Aichinger, G. Pobegen, M. Nelhiebel, and B. Kaczer, *Int. Electron Devices Meeting Tech. Dig.*, 27.4.1–27.4.4 (2011).
26. T. Grasser (2013) The capture/emission time map approach to the bias temperature instability. In: T. Grasser (ed.) Bias temperature instability for devices and circuits. Springer, New York
27. H. Reisinger, T. Grasser, W. Gustin, and C. Schlünder, *Proc. IEEE Int. Reliab. Phys. Symp.*, 7 (2010).
28. T. Grasser, H. Reisinger, W. Goes, Th. Aichinger, Ph. Hehenberger, P.-J. Wagner, M. Nelhiebel, J. Franco, and B. Kaczer, *Int. Electron Devices Meeting Tech. Dig.* (2009).
29. B. Kaczer, T. Grasser, P.J. Roussel, J. Martin-Martinez, R. O'Connor, B.J. O'Sullivan and G. Groeseneken, *Proc. IEEE Int. Reliab. Phys. Symp. (IRPS) Proc.* , 20 (2008).
30. H. Reisinger (2013) The time dependent defect spectroscopy. In: T. Grasser (ed.) Bias temperature instability for devices and circuits. Springer, New York
31. T. Grasser, Microel. Reliab. **52**, 39–70 (2012).
32. M. Toledano-Luque, B. Kaczer, Ph.J. Roussel, J. Franco, T. Grasser, C. Vrancken, N. Horiguchi, and G. Groeseneken, *Proc. IEEE Int. Reliab. Phys. Symp.*, 4A.2.1–4A.2.8 (2011).
33. T. Grasser, H. Reisinger, P.-J. Wagner, F. Schanovsky, W. Goes, and B. Kaczer, *Proc. IEEE Int. Reliab. Phys. Symp.*, 16 (2010).
34. M. Toledano-Luque, B. Kaczer, J. Franco, P. J. Roussel, T. Grasser, T. Y. Hoffmann, and G. Groeseneken, *Proc. VLSI Symp.*, 152–153 (2011).
35. A. Asenov, R. Balasubramaniam, A. R. Brown, and J. H. Davies, *IEEE T. Electron Dev.***50**, 839 (2003).
36. K. Takeuchi, T. Nagumo, S. Yokogawa, K. Imai, and Y. Hayashi, *Proc. VLSI Symp. Tech.*, 54 (2009).
37. A. Ghetti, C. M. Compagnoni, A. S. Spinelli, and A. Visconti, *IEEE T. Electron Dev.***56**, 1746–1752 (2009).
38. http://www.ibiblio.org/e-notes/Perc/contour.htm
39. K. Sonoda, M. Tanizawa, K. Ishikawa, and Y. Inoue, *Int. Conf. Simulation of Semiconductor Processes and Devices (SISPAD)*, 19–22 (2011).
40. J. Franco, B. Kaczer, M. Toledano-Luque, Ph. J. Roussel, T. Kauerauf, J. Mitard, L. Witters, T. Grasser, and G. Groeseneken, IEEE T. Electron. Dev. 60, 405 (2013).
41. J. Franco, B. Kaczer (2013) NBTI in (Si)Ge channel devices. In: T. Grasser (ed.) Bias temperature instability for devices and circuits. Springer, New York
42. M. Toledano-Luque, B. Kaczer, Ph. J. Roussel, J. Franco, L. Å. Ragnarsson, T. Grasser, and G. Groeseneken, *Appl. Phys. Lett.***98**, 183506 (2011).
43. S.E. Rauch III (2013) BTI induced statistical variations. In: T. Grasser (ed.) Bias temperature instability for devices and circuits. Springer, New York
44. M. Toledano-Luque, B. Kaczer (2013) Characterization of individual traps in high-κ oxides. In: T. Grasser (ed.) Bias temperature instability for devices and circuits. Springer, New York
45. A. Kerber and T. Nigam, IEEE Int. Reliab. Phys. Symp. (IRPS) Proc. (2013).

Chapter 8
Atomic-Scale Defects Associated with the Negative Bias Temperature Instability

Jason P. Campbell and Patrick M. Lenahan

Abstract We utilize magnetic resonance measurements to identify the fundamental atomic-scale defect structures involved in the negative bias temperature instability. In gate stacks composed of pure silicon dioxide, we find a degradation mechanism directly involving P_{b0} and P_{b1} defect centers (silicon dangling bond defects in which the silicon is back-bonded to three other silicon atoms precisely at the silicon/silicon dioxide interface). We observe that, in pure SiO_2-based devices, the generation of these interface defects is catalyzed by the generation of E' center bulk dielectric defects (silicon dangling bond defects in which the silicon is back-bonded to oxygen atoms). These observations are the first to indicate a prominent role for E' centers in the negative bias temperature instability for pure silicon dioxide-based devices. In gate stacks composed of plasma-nitrided oxides, we identify a degradation mechanism which is dominated by the generation of a new defect center which we identify as a K_N center. K_N centers are silicon dangling bond defects in which the silicon is back-bonded to three nitrogen atoms with second-nearest neighbor atoms likely including oxygen. K_N centers are located within the amorphous silicon oxy-nitride and electrically behave as both interface states as well as bulk dielectric defects (serve as both recombination and tunneling sites). In these plasma-nitrided gate stacks, the negative bias temperature instability does not involve the generation of P_{b0}, P_{b1}, or E' centers. These collective observations provide a useful fundamental understanding with which to critically examine the current and future negative bias temperature instability framework.

J.P. Campbell (✉)
National Institute of Standards and Technology, 100 Bureau Drive MS 8120, Gaithersburg, MD 20899, USA
e-mail: jason.campbell@nist.gov

P.M. Lenahan
The Pennsylvania State University, 212 Earth and Engineering Sciences Building, University Park, PA 16802, USA
e-mail: pmlesm@engr.psu.edu

T. Grasser (ed.), *Bias Temperature Instability for Devices and Circuits*,
DOI 10.1007/978-1-4614-7909-3_8,

8.1 Introduction

The negative bias temperature instability (NBTI) is a troubling reliability issue that has puzzled researchers for quite some time [1]. An enormous effort has been devoted to careful observations of NBTI-induced device parametric shifts. These clever studies have established a substantial set of device-level observables which limit the framework used to understand this elusive phenomenon. This evolving understanding has forced NBTI researchers to (often begrudgingly) redefine their understanding. The relatively new observation that NBTI-induced degradation recovers rapidly upon the cessation of stress [2] has spawned revolutions in both measurement methodologies and modeling. While these efforts have pushed the NBTI community towards a more complete understanding of NBTI, there is still a surprising lack of direct analytical evidence needed to properly assess the current NBTI understanding.

This book is largely filled with expert analysis on various aspects NBTI-induced device-level shifts and their implications for various NBTI mechanisms. Somewhat disparately, the purpose of this chapter is to instead provide direct experimental evidence with regard to the atomic-scale defects and mechanisms involved in NBTI. This evidence comes from magnetic resonance measurements. It is hoped that this information will provide an additional set of "boundary conditions" in the still evolving understanding of NBTI.

This chapter is organized as follows. Section 8.2 provides a brief overview of the NBTI phenomenon for completeness. The other contributors of this book will certainly provide more detailed parametric descriptions of NBTI as well as detailed historical perspectives. We refer the reader to these other chapters for a more detailed description. Section 8.3 provides a brief background on the experimental methods employed in this work, primarily magnetic resonance. Magnetic resonance is not widely utilized in reliability physic studies. Consequently, we provide some additional background regarding the underlying principles of these magnetic resonance measurements so that our observations can be better understood. Section 8.4 discusses our observations of NBTI in pure SiO_2-based devices. In these devices, we observe that NBTI generates both P_{b0} and P_{b1} centers (prototypical interface states). We also examine their densities of states and find that they are consistent with the observed device-level shifts associated with NBTI. We also note the observation of NBTI-induced E' center bulk dielectric defects, after somewhat harsher stress conditions. This observation serves as the impetus for Sect. 8.5. In this section, we discuss on-the-fly ESR measurements which provide strongly support a catalyzing role for E' centers in pure SiO_2 devices. Section 8.6 discusses our observations of NBTI in plasma-nitrided oxide (PNO)-based devices. Surprisingly, we do not observe the NBTI-induced generation of P_{b0}, P_{b1}, or E' centers. Instead, we observe the generation of a new defect which we unambiguously identify as a K_N center (silicon dangling bond defect in which the silicon is back-bonded to three nitrogen atoms). We conclusively identify the structure of this defect as well as estimate its density of states. We also conclusively show that this defect structure participates as both an interface state and a bulk dielectric defect. Section 8.7 details our, somewhat crude, observations regarding generation and recovery of these defects in both pure SiO_2 and PNO devices. Section 8.8 discusses our observations on the atomic-scale implications of the addition of fluorine as a method to NBTI-harden devices. Section 8.9 discusses our observations of NBTI in high-k-based devices. Finally, Sect. 8.10 provides a discussion on how these observations mesh with the current understanding of NBTI as well as some concluding remarks.

8.2 What Is NBTI?

NBTI manifests itself as a negative threshold voltage (V_{th}) shift and degradation in drive current in pMOSFET devices subject to negative gate bias at elevated temperatures [3]. Aggressive gate oxide scaling, less aggressive operating voltage scaling, and the addition of nitrogen to the gate dielectric have exacerbated NBTI [4]. Since NBTI-induced parametric shifts eventually lead to circuit failure, a detailed understanding of the NBTI mechanism is required. Despite nearly forty years of research [5], a fundamental understanding of NBTI is far from complete.

An examination of the simplest transistor characteristic equations illustrates the origins of the NBTI V_{th} shift. The threshold voltage of a pMOSFET is given by [3]:

$$V_{th} = \phi_{MS} - \frac{qN_{ot}}{C_{OX}} - \frac{qN_{it}}{C_{OX}} - 2\phi_F - \frac{\left|\sqrt{4\varepsilon_s \phi_F q N_D}\right|}{C_{OX}} \tag{8.1}$$

Assuming that the substrate doping (N_D) and the oxide capacitance (C_{OX}) are both constant during a given negative bias temperature stress (NBTS) condition, the observed shift in threshold voltage (ΔV_{th}) must arise from a change in the numbers of oxide and/or the interface trapped charges (ΔN_{ot} and ΔN_{it}) [3]. A very large body of NBTI research has shown that oxide (N_{ot}) and interfacial defects (N_{it}) are responsible for the NBTI-induced degradation. However, the role of each type of defect is still actively debated [6].

Defining the roles of NBTI-induced interface states and/or bulk traps has proven difficult, in part, because NBTI damage recovers upon stress cessation [2]. Thus, all measurements of the phenomenon are in some way influenced by recovery [7]. This makes simple quantification of NBTI degradation a never-ending struggle [8]. This rapid recovery is difficult to capture using conventional electrical characterization techniques and often obscures the relative balance of interface states and/or bulk traps [9, 10].

8.3 Experimental Methods

In this section, we review the combination of electrical and magnetic resonance measurements which we have used to study the atomic-scale defects involved in NBTI. We employ DC gate-controlled diode recombination current (DC-IV)

[11, 12] measurements to monitor NBTI-induced changes in device interface state density (D_{it}). These electrical measurements are combined with very sensitive electrically detected magnetic resonance (EDMR) measurements to allow for identification of specific NBTI-induced atomic-scale defects. The combination of these two types of measurements (DC-IV and EDMR) provides a direct correlation between the NBTI-induced electrical damage and the generation of specific atomic-scale defects [13].

Magnetic resonance is the only technique with the analytical power and sensitivity to probe the physical and chemical nature of the defects involved in NBTI [10, 14, 15]. One of these magnetic resonance techniques, conventional electron spin resonance (ESR), was first utilized by Fujieda et al. to examine the atomic-scale defects involved in NBTI [16]. However, their pioneering experiments involved large area ($\sim$1 cm^2) blanket dielectric capacitor structures on p-type substrates (used for nMOSFETs not pMOSFETs) [16]. We have circumvented [17–21] the sample requirements of conventional ESR (large area and simple structures) by utilizing two very sensitive (EDMR) techniques called spin-dependent recombination (SDR) and spin-dependent tunneling (SDT). These measurements rely on ESR-induced changes in device currents. Thus, the atomic-scale defects identified in these measurements are directly linked to NBTI-induced degradation in fully processed devices.

Since these techniques are not widely utilized in the reliability physics community, we begin our discussion by briefly reviewing them.

8.3.1 The DC-IV Technique

The DC-IV technique is a quasi-DC current versus voltage measurement performed on a MOSFET configured as a gate-controlled diode (source and drain contacts are shorted) [11, 12]. Shorting the source and drain forms a diode between the source/drain and substrate. With this diode slightly forward biased, a measurement of the device substrate current as a function of gate voltage yields a peak in the substrate current [11, 12]. This peak in substrate current is dominated by recombination current through interface states [11, 12]; varying the gate voltage changes the surface potential and, consequently, the relative populations of charge carriers present at the interface. The peak in substrate current occurs at the gate voltages which correspond to equal populations of electrons and holes at the interface (depletion) [11, 12]. This is so because recombination is most effective when equal numbers of electrons and holes exist at the location of the deep level recombination centers. The substrate current is less at gate voltages corresponding to unequal populations of electrons and holes at the interface (accumulation and inversion) [11, 12]. This concept is schematically illustrated in Fig. 8.1.

A quantitative derivation of the current versus voltage behavior was first presented in a series of papers by Fitzgerald and Grove [11, 22]. In this derivation, the peak in the gate-controlled diode measurement was related to the interface state

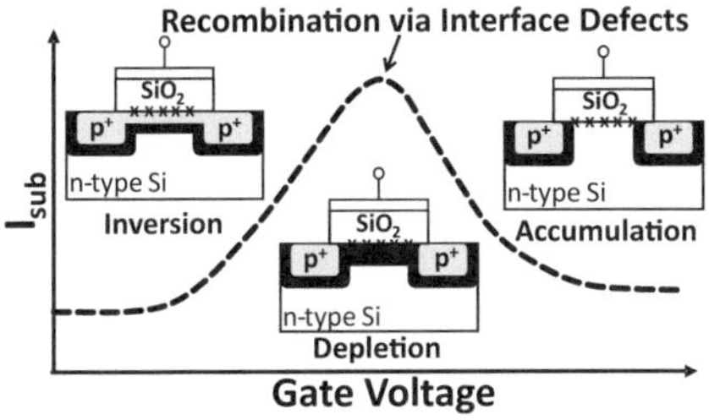

Fig. 8.1 Schematic of the DC-IV measurement. For a slightly forward-biased source/drain to substrate diode, sweeping the gate voltage results in a peaked substrate current due to recombination through interface states. This peak occurs near depletion

density (D_{it}) in a very straightforward manner. Fitzgerald and Grove's derivation begins by assuming that the recombination or generation events, in silicon, occur at deep level defects within the silicon band gap and that the process can be described by the Shockley–Read–Hall (SRH) model [23, 24] for recombination-generation. When the SRH model is applied to a gate-controlled diode structure, Fitzgerald and Grove [11] showed that the substrate recombination current (I_{SUB}) is given by

$$I_{SUB} = Aq\sigma_s v_{th} \left[\int_{E_V}^{E_C} \frac{D_{it}(E)dE}{p_S + n_S + 2n_i \cosh\left(\frac{E-E_i}{kT}\right)} \right] \left[p_S n_S - n_i^2 \right] \tag{8.2}$$

where A is the effective gate area; q is the electronic charge; σ_s is the geometric mean of the electron and hole capture cross sections; v_{th} is the thermal velocity of electrons; E_C and E_V denote the energies of the conduction and valence band edges, respectively; $D_{it}(E)$ is the interface state density as a function of energy between E_C and E_V; p_S and n_S are the hole and electron concentrations at the surface; n_i is the intrinsic number of carriers; E is the energy level of the surface recombination centers; E_i is the intrinsic Fermi level; k is Boltzmann's constant; and T is temperature. p_S and n_S for a pMOS device are given by

$$p_S \cong \frac{n_i^2}{N_D} \exp\left(\frac{-q\phi_S}{kT}\right) \exp\left(\frac{q\,|V_F|}{kT}\right) \tag{8.3}$$

$$n_S \cong N_D \exp\left(\frac{q\phi_S}{kT}\right) \tag{8.4}$$

where N_D is the substrate doping, ϕ_S is the surface band bending, and V_F is the forward bias applied to the source/drain to substrate junction. Note that increasing V_F increases the number of minority carrier holes (p_S) available for recombination (in a pMOS device).

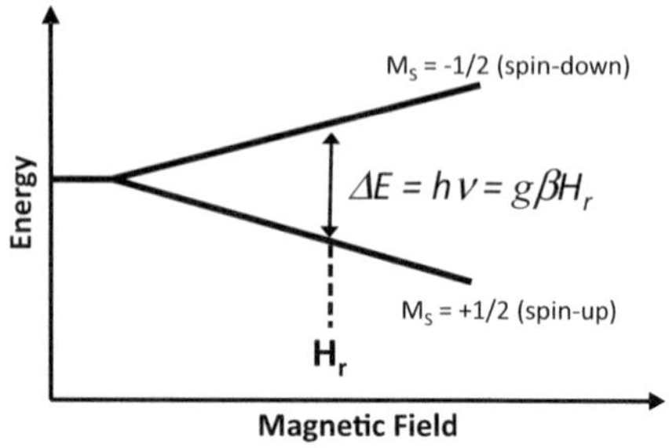

Fig. 8.2 Schematic representation of Zeeman splitting for the simplest case of unpaired electrons

With the assumptions of an energy-independent capture cross section and energy-independent D_{it} near mid-gap, Fitzgerald and Grove [11] showed that Eq. (8.2) can be approximated by

$$I_{SUB} = \left(\frac{1}{2}\right) qn_i\sigma_S v_{th} D_{it} Aq\left|V_F\right| \exp\left(\frac{q\left|V_F\right|}{2kT}\right) \text{ for } \left|V_F\right| > kT/q \qquad (8.5)$$

This equation allows for a straightforward extraction of interface state densities based on the peak in the substrate current. The energy window in which recombination occurs is controlled by V_F and scales as approximately $q|V_F|$. The implications of changing V_F and the recombination energy window to the DC-IV measurement are discussed later in Sect. 8.6.

8.3.2 *Electron Spin Resonance*

Since both of the EDMR techniques (SDR and SDT) used in this study are based on ESR theory, it is useful to first examine the basic principles of the technique. ESR measurements are sensitive to defects with unpaired electrons [15, 25, 26]. Since an electron is a charged particle with intrinsic angular momentum, it can qualitatively (and only qualitatively) be thought of as a negatively charged particle which is spinning on an axis [15, 25, 26]. In this qualitative picture, the spinning charged particle produces a magnetic field and an associated magnetic moment. In the absence of any external magnetic fields, these magnetic moments are randomly oriented [25, 26]. However, the application of a large external magnetic field tends to align any unpaired electrons such that the electron's magnetic moments align either parallel (spin-up, $M_S = +½$) or antiparallel (spin-down, $M_S = -½$) to the applied field [25, 26]. This polarization of the electrons splits the energy of the spin system into two different levels [15, 25, 26]. This is known as the Zeeman effect [25, 26]. At thermodynamic equilibrium, the lower energy level is more populated and corresponds to the spin-up state [15, 25, 26]. The higher energy level corresponds to the spin-down state [15, 25, 26]. The Zeeman energy splitting is illustrated in Fig. 8.2.

In addition to the external (polarizing) magnetic field, a second high-frequency (microwave) electromagnetic field is also applied to the system. If the product of Planck's constant and the frequency of the oscillating field ($h\nu$) equals their Zeeman splitting, resonance can occur [15, 25, 26]. At resonance, the electrons can absorb energy and "flip" spin orientation. For the simplest case, the resonance condition is described as [15, 25, 26]:

$$h\nu = g_e \beta H \tag{8.6}$$

where h is Planck's constant, ν is the frequency of microwaves added to the spin system, g_e is the free-electron g-value ($g_e = 2.002319$), β is the Bohr magneton, and H is the large applied magnetic field. The resonance condition described in Eq. (8.6) is for an isolated unpaired electron. In reality, the local environment of the unpaired electron changes the resonance condition. For material systems relevant to NBTI, the deviations from the resonance condition are almost entirely due to spin-orbit coupling and electron-nuclear hyperfine interactions [15, 25, 26].

8.3.2.1 Spin–Orbit Coupling

Spin–orbit coupling alters the resonance condition of Eq. (8.6) by the addition of an effective local magnetic field due to the electron's orbital angular momentum about the nucleus [15, 25, 26]. This can be *qualitatively* described using the Bohr atomic model in which electrons orbit the nucleus in a circular path [15, 25, 26]. Even though the electron is "orbiting the nucleus," from the perspective of the electron, the positively charged nucleus appears to orbit the electron. The circular orbit of the positively charged nucleus about the electron generates an additional local magnetic field. This alters the resonance condition for any electron "orbiting" a nucleus (e.g., an electron trapped in the dangling bond of a point defect). The spin-orbit coupling effect is included in the resonance condition by replacing the free-electron (g_e) with the g-matrix, g_{ij} [15, 25, 26]. The g_{ij} values of an electron trapped in a dangling bond of a point defect will deviate from g_e as excited states are mixed with the ground state [15, 25, 26]. The g-matrix is essentially a second rank tensor which is dependent on the orientation of the defect and the external magnetic field. In practice, (singular) g-values are typically reported for *specific* magnetic field orientations with respect to crystallographic orientations. These deviations help to identify the structure of these point defects.

8.3.2.2 Electron–Nuclear Hyperfine Interactions

Another important source of deviation from the resonance condition is due to electron-nuclear hyperfine interactions [15, 25, 26]. This occurs when an unpaired electron is located close to a nearby *magnetic* nucleus. The important magnetic nuclei involved in this work include silicon and nitrogen. Silicon naturally occurs

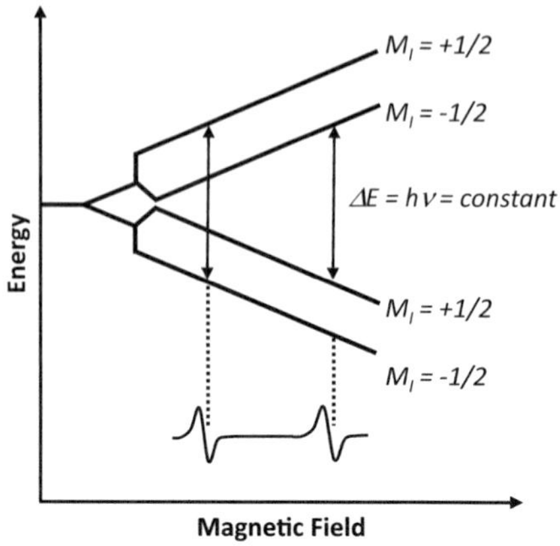

Fig. 8.3 Schematic illustration of the Zeeman splitting which is altered by the presence of a nearby magnetic nucleus with spin (I = ½). The diagram also schematically illustrates the expected spectrum due to this nuclear hyperfine splitting

with a 95.3% abundance of nonmagnetic nuclei and a 4.7% spin ½ nucleus corresponding to the ^{29}Si isotope [25, 26]. Nitrogen has a magnetic isotope with near 100% natural abundance; this ^{14}N isotope has a nuclear spin of 1 [25, 26]. A spin ½ magnetic nucleus has two possible orientations in the applied magnetic field while a spin 1 magnetic nucleus has three possible orientations in the applied magnetic field [15, 25, 26].

An unpaired electron nearby a nucleus with a net magnetic moment feels an additional local magnetic field due to that nuclear magnetic moment [15, 25, 26]. This local field splits the Zeeman levels into $(2I+1)$ additional levels, where I is the nuclear spin [25, 26]. This results in $(2I+1)$ additional resonant lines at $(2I+1)$ magnetic field values centered about the original resonant field [25, 26]. The Zeeman splitting and resonance spectra for a nucleus with spin ½ are illustrated in Fig. 8.3. Note that all of the resonance transition arrows in Fig. 8.3 all correspond to the same energy, corresponding to one value of $h\nu$. Thus, when a system is subject to microwave irradiation of a frequency, ν, the additional electron-nuclear hyperfine spectra occur at different applied fields.

As discussed above, the electron–nuclear hyperfine interaction alters the resonance condition by introducing an additional local magnetic field. If a single magnetic nucleus is involved with the unpaired electron, this altered resonance condition is given by [15, 25, 26]:

$$H = \frac{h\nu}{g\beta} + m_I A \tag{8.7}$$

where m_I is the nuclear spin quantum number and A is the electron-nuclear hyperfine matrix. It is important to note that not all nuclei have a magnetic moment. If no magnetic nuclei are present, the $m_I A$ term is zero [15, 25, 26]. The electron-nuclear hyperfine interaction (A) can be expressed in terms of an isotropic (A_{iso}) and an anisotropic (A_{aniso}) component. The isotropic component is a measure of the s-character of the unpaired electron's wave function [15, 25, 26]. In these studies,

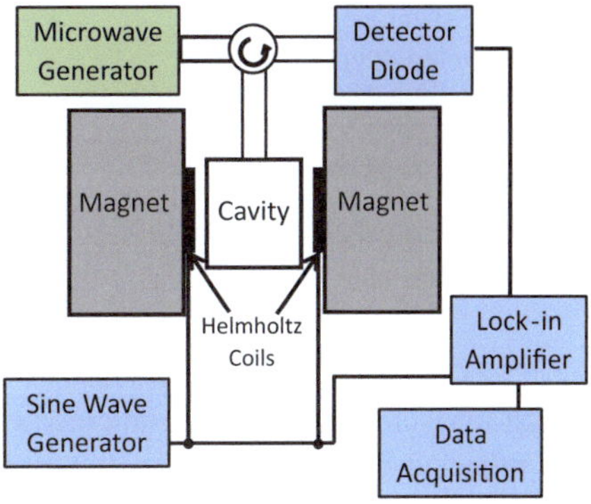

Fig. 8.4 Simplified schematic diagram of an ESR spectrometer

the anisotropic component is a measure of the p-character of the unpaired electron's wave function [15, 25, 26]. The geometry of a p-orbital dictates that the magnetic field interaction with an unpaired electron in a p-type orbital should be anisotropic. That is, the interactions are different if the magnetic field is aligned parallel or perpendicular to the symmetry axis of the p-orbital.

We assume that the wave function for an unpaired electron trapped in a dangling bond orbital consists of a linear combination of atomic orbitals (3s and 3p for the defects relevant to this study) [15]. A_{iso} and A_{aniso} can be theoretically calculated for 100% 3s and 100% 3p wave functions, respectively [15, 25, 26]. A comparison of the measured A_{iso} and A_{aniso} values with these calculated values gives a rough estimate of the s- and p-character of a defect's dangling bond wave function [15, 25, 26]. Additionally, the summation of the defect's s- and p-characters is a measure of the unpaired electrons localization to that dangling bond orbital.

Fortunately, the A_{iso} and A_{aniso} components of the hyperfine interaction can be related to the *measurable* interactions with the field parallel ($A_{||}$) and perpendicular to ($A_{\perp}$) the unpaired electron's orbital symmetry axis. The electron-nuclear hyperfine modification to the resonance condition provides an extremely useful tool to determine information about a point defect's physical and chemical nature.

8.3.2.3 The Electron Spin Resonance Spectrometer

A schematic diagram of the ESR spectrometer is shown in Fig. 8.4. The spectrometer consists of three main components: (1) an electromagnet, which is used to apply the polarizing magnetic field; (2) a microwave generator, which provides the necessary energy ($h\nu$) to allow spin flipping; and (3) a microwave cavity or resonator, which allows for efficient coupling of the microwave irradiation to the sample under study [26]. Typical ESR measurements involve X-band microwave irradiation (8–10 GHz) coupled to either TE_{102} or TE_{104} resonators at a magnetic field of approximately 3,500 G. The resonance detection scheme involves a microwave detector diode and a lock-in amplifier [26]. The microwave detector

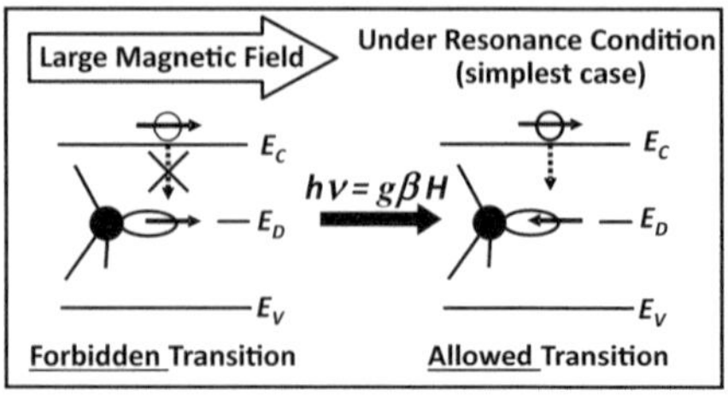

Fig. 8.5 Schematic illustration of Lepine's model for SDR. In this model, device substrate current is modified due to (1) an ESR-induced spin-flipping event and (2) the Pauli exclusion principle

diode measures the reflected microwave power exiting the resonator. Changes in the reflected power indicate microwave absorption or resonance. This absorption is monitored using a lock-in amplifier. The lock-in sample modulation is provided by a pair of Helmholtz coils on the walls of the resonator. A sample is placed in the resonator and the microwave irradiation is critically coupled to the sample at a constant frequency corresponding to the resonant frequency of the cavity. The reflected power is then monitored as a function of swept magnetic field. When this technique is applied to MOS systems, the sensitivity (10^{10} spins/G of line width) requires fairly large area ($\sim$1 cm^2) sample sizes. The requirement that the sample under study be critically coupled to the microwave adds additional restrictions on sample area and conductivity.

8.3.3 Spin-Dependent Recombination

Spin-dependent recombination (SDR) is an electrically detected ESR technique first demonstrated by Lepine in 1972 [27]. In SDR, the samples under study include fully processed devices. The ESR-induced spin flipping acts to modify device recombination current. The ESR-induced change to this recombination current is measured as a function of magnetic field. This results in an ESR-like spectrum which is due to the deep level defects participating in recombination. This technique provides an extremely large increase in sensitivity over ESR and allows for measurements in fully processed devices [28].

A brief, only qualitatively correct, explanation of SDR provided by Lepine [27] is useful in understanding our measurements. The Lepine model for SDR is schematically illustrated in Fig. 8.5. Lepine's model [27] combines both the Shockley–Read–Hall (SRH) model [23, 24] for recombination and the Pauli exclusion principle. A SRH recombination event occurs when a conduction electron is captured by a deep level defect and then a hole is captured at the same defect site. (The recombination sequence could, of course, also be reversed.) In SDR, a transistor is biased so that the source/drain to substrate current is dominated by recombination through interface defects (this corresponds to the peak in the DC-IV [11, 12]). The device, thus configured, is placed in a large slowly varying DC magnetic field which partially polarizes the spins of the conduction electrons, holes, and deep level defects. If a deep level defect and a charge carrier have the same spin orientation, the Pauli exclusion principle forbids charge capture by the deep level defect because the electrons must have different spin quantum numbers to occupy the same orbital. When a paramagnetic deep level's electron spin resonance condition is satisfied, the defect's electron spins are "flipped." Flipping the spins increases the probability of opposite spin orientations between deep level defects and charge carriers, thus increasing the recombination current. This increase in recombination current, which is spin dependent, is what is measured in SDR.

Lepine's model provides simple insight into the physics governing SDR but predicts a spin-dependent change in the recombination current of approximately 1 part in 10^6 at the fields and temperatures utilized in our study [27]. SDR measurements often involve much larger current changes, sometimes 1 part in 10^4 or larger [13, 28]. A more physically accurate (but more complex) description of SDR has been proposed by Kaplan, Solomon, and Mott (KSM) [29]. KSM extends Lepine's model to consider the coupling of two spins prior to the actual recombination events [29]. The two spins in our work correspond to a spin at the deep level site and the spin of a charge carrier. Once coupled, the two spins can either participate in charge capture or dissociate [29]. The KSM model assumes that capture mostly involves singlet pairs (pairs in which the electrons have opposite spin orientation without the help of magnetic resonance) [29]. The triplet (pairs in which the electrons have the same spin orientation) recombination rate is assumed to be negligible [29]. If resonance occurs while the two electrons are coupled, the triplet becomes a singlet and capture occurs. The size of SDR effect predicted by KSM can be relatively large [29] and is more consistent with experimental data on most devices. In the KSM model, the size of the SDR effect depends upon the relationship between the coupling time of the pair and the spin–lattice relation time, or T_1 [29]. The KSM analysis leads to the prediction of an SDR effect that is (to first order) magnetic field independent [29]. In other words, the magnetic field polarization of unpaired electrons has little effect on the recombination. It should be noted that the original KSM model assumes a coupling between conduction band and valence band electrons and holes [29]. In reality, and as we have discussed, the coupling is generally between conduction or valence band charge carriers and a deep level defect [13, 30]. The details of these SDR models and their relevance to MOS devices are discussed elsewhere [13, 30].

8.3.4 Spin-Dependent Tunneling

SDT is very similar to SDR. The difference being that in SDT, one measures a spin-dependent *tunneling* current. In the SDT measurement, a device is biased so that the substrate current is dominated by a trap-assisted tunneling current through the gate dielectric. The device, thus configured, is placed in a large slowly varying DC magnetic field which partially polarizes the spins of the electrons, holes, as well as the spins of paramagnetic deep level defects participating in trap-assisted tunneling in the dielectric. If, for example, an unpaired electron in the tunneling defect and a gate valence electron both have the same spin orientation, the Pauli exclusion principle forbids a tunneling event through the center. However, satisfaction of the resonance condition can "flip" the spin orientation of the unpaired electron in the tunneling center and increases the probability of opposite spin orientations between tunneling center and gate valence electrons, thereby increasing the tunneling current. The increase in trap-assisted tunneling current, which is spin-dependent, is what is measured in SDT [31–33]. The SDT observations presented in this work [17, 18] are consistent with spin-dependent trap-assisted tunneling through "near-interface" defects. This mechanism, as it pertains to this work, is further discussed in Sect. 8.6. Also note that a more detailed treatise on possible SDT mechanisms is found elsewhere [32].

8.3.5 The SDR/SDT Spectrometer

The major differences between an ESR and SDR/SDT spectrometer include sample type and detection scheme. In SDR and SDT, a fully processed transistor is coupled to the microwaves (instead of a simply processed sample suitable for ESR). The spin-dependent modification to the device recombination (tunneling) current is what is detected in SDR (SDT). This current is detected with the help of a trans-impedance preamplifier and the lock-in detection scheme. This measurement apparatus allows for very high sensitivity measurements. The SDR sensitivity can, under some circumstances, be better than 10^3 paramagnetic defects [28].

8.4 NBTI in Pure SiO_2-Based pMOSFETs

This section illustrates how SDR measurements have been utilized to directly observe the atomic-scale defects associated with NBTI in fully processed *pure* SiO_2 devices [19–21, 34, 35]. (Here, "pure" indicates that the dielectrics are SiO_2 with only hydrogen present as a significant impurity.) We also correlate the SDR defect observations with DC-IV measurements [11, 12] of NBTI-induced interface state density. The combination of DC-IV and SDR measurements links specific atomic-scale defects to the device electronic properties. Our measurements in this chapter detail the correlation between NBTI-induced interface state generation and the generation of P_{b0} and P_{b1} interfacial silicon dangling bonds [19–21, 34, 35]. (P_{b0} and P_{b1} are both silicon dangling bond defects in which the central silicon atom is back-bonded to three other silicon atoms precisely at the Si/SiO_2 interface.) Our results clearly show that NBTI (observed within our measurement window) is dominated by the generation of these P_{b0} and P_{b1} interface defects [19–21, 34, 35]. After a harsher NBTS, we also noted the observation of E' center oxide defects [19–21, 35]. Although our initial E' observations did not allow for a definitive assignment of E' center's role in NBTI, we suggested, on the basis of these results, that the E' defects are responsible for the inversion layer hole capture process [19–21, 35]. We also suggested that the presence of the charged E' centers then caused the transfer of hydrogen atoms from previously passivated P_b-H sites to E' sites. Straightforward statistical mechanics arguments indicated that this process is essentially inevitable because, upon hole capture, the E' center opens up with a neutral silicon dangling bond on one side of the positively charged E' vacancy and a positively charged silicon vacancy on the other side. The presence of the unpassivated E' silicon dangling bond in the oxide and the fully passivated silicon dangling bond at the Si/dielectric boundary provides a thermodynamically unstable situation [19–21, 35]. These ideas are further explored in Sect. 8.5.

8.4.1 Experimental Details

Our examination of NBTI in SiO_2 devices includes two different types of pure SiO_2-based pMOSFETs. The first type includes large area ($\approx$41,000 μm^2) 7.5 nm SiO_2 devices. The devices are fabricated in a gated-diode configuration where the source and drain are shorted together. The second type involves very large area ($\approx 1 \times 10^6$ μm^2) 48 nm SiO_2 power pMOSFETs. These devices involve a large gate overhang that extends over a lightly doped region. This device layout slightly alters the DC-IV characteristic curves but still allows for SDR observations of NBTI-induced atomic-scale defects. Both types of devices were subject to various NBTS conditions. Following stress, all of the devices were subject to a temperature quench in which the stressing temperature is reduced to room temperature over approximately 4 min. while the gate bias stress is maintained. We have found this method to be fairly effective at "locking-in" the NBTI-induced damage, rendering it more readily observable in the SDR/DC-IV measurements [19, 20, 36]. Interface state densities were monitored using the DC-IV measurement [11, 12]. NBTI-induced changes in the DC-IV-derived interface state density were correlated to SDR measurements detailing the generation of specific defect structures. SDR measurements were made at room temperature with a custom-built SDR/SDT spectrometer. SDR measurements were calibrated using a strong pitch spin standard. All DC-IV and SDR measurements in this section were made with +0.33 V applied to the source/drain to substrate diode.

8.4.2 NBTI in 7.5 nm SiO_2 pMOSFETs

7.5 nm SiO_2 devices were subject to an NBTS of −5.7 V at 140°C for 250,000 s. Pre- and post-NBTS gate-controlled diode DC-IV measurements are shown in Fig. 8.6. NBTS induces a large increase in the peak substrate current (D_{it}).

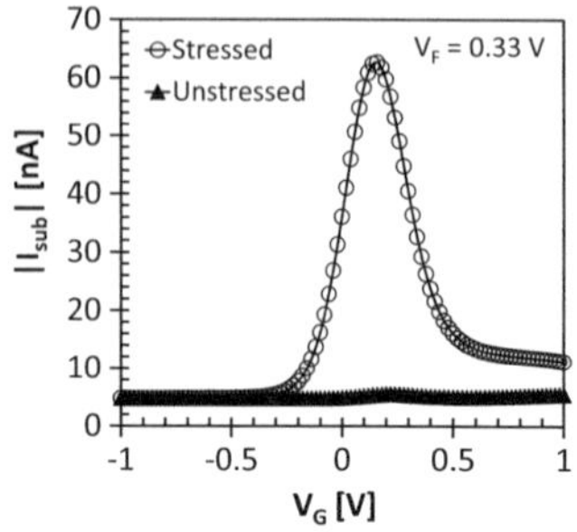

Fig. 8.6 Gate-controlled diode DC-IV measurements on a 7.5 nm pMOSFET before and after the application of NBTS (−5.7 V at 140°C for 250,000 s) (Reproduced from [19])

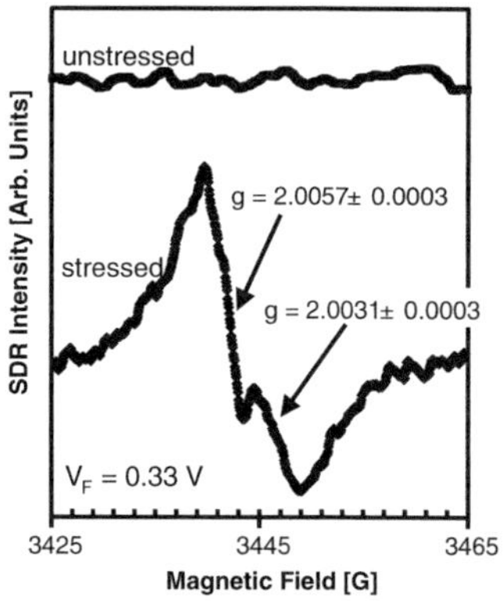

Fig. 8.7 SDR traces of 7.5 nm pMOSFETs with the magnetic field vector perpendicular to the (100) surface both before and after the application of NBTS (−5.7 V at 140°C for 250,000 s) (Reproduced from [19])

Following the analysis of Fitzgerald and Grove [11], D_{it} values were extracted for pre-NBTS (7×10^9 cm^{-2} eV^{-1}) and post-NBTS (5×10^{11} cm^{-2} eV^{-1}) using a mean capture cross section of $\sigma_s = 2 \times 10^{-16}$ cm^2. Figure 8.7 illustrates the corresponding pre- and post-NBTS SDR traces with the magnetic field vector perpendicular to the (100) surface. The interface state density in the unstressed device is below the SDR detection limit. After NBTS, we observe the generation of two strong signals at g = 2.0057 ± 0.0003 and g = 2.0031 ± 0.0003. The observed g-values at this magnetic field orientation allow us to attribute the g = 2.0057 signal to P_{b0} centers and the g = 2.0031 signal to P_{b1} centers. We note that the size of the SDR effect ($\Delta I/I \approx 5 \times 10^5$) is consistent with the KSM model [29].

P_{b0} and P_{b1} defects are both silicon dangling bond defects in which the central silicon atom is back-bonded to three other silicon atoms precisely at the Si/SiO_2 boundary [15, 28, 37–41]. Figure 8.8 illustrates a schematic drawing of P_{b0} and P_{b1} centers. The main differences between the two defects are in the dangling bond axes of symmetry [15, 28, 37–39, 41] and electronic densities of states [40, 43–46]. The P_{b0} dangling bond orbital points along the $\langle 111 \rangle$ directions [15, 28, 37] while the P_{b1} dangling bond orbital points approximately along the $\langle 211 \rangle$ directions [38, 39, 41]. The observation of NBTI-induced P_{b0}/P_{b1} defects is consistent with the defects' densities of states [15, 43, 44, 46, 47]. Both the (111) Si P_b center and (100)

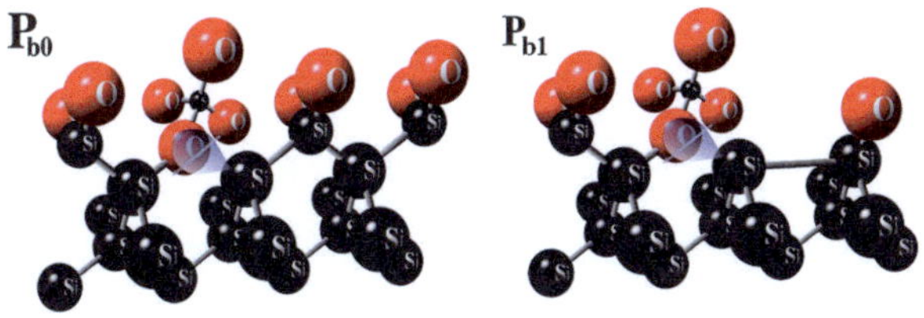

Fig. 8.8 Schematic drawing of the P_{b0} and P_{b1} Si/SiO_2 interface defects (Reproduced from [42])

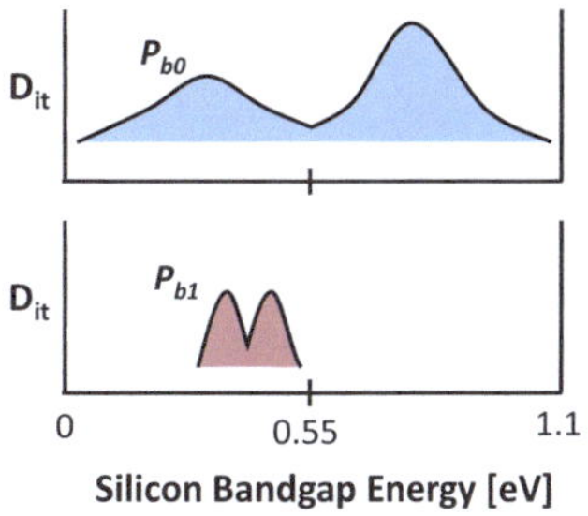

Fig. 8.9 Schematic representation of the density of states of the P_{b0} and P_{b1} defects as a function of band-gap energy. The sketches provide only a crude semiquantitative representation (Reproduced from [19])

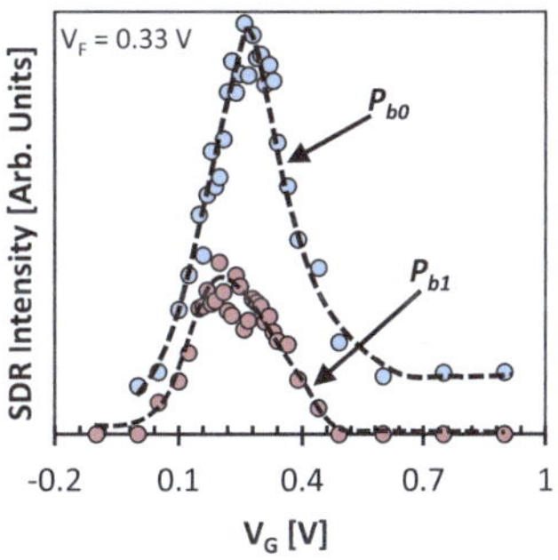

Fig. 8.10 SDR derived P_{b0} and P_{b1} signal amplitudes as a function of applied gate bias for the 7.5 nm pMOSFET subject to NBTS (−5.7 V at 140°C for 250,000 s). The *dashed lines* are included as merely a guide for the eye (Reproduced from [19])

Si analog, the P_{b0} center, have a broadly peaked density of states centered about mid-gap with the "+/0" and "0/−" transitions separated by about 0.7 eV [40, 44, 46]. Utilizing conventional ESR measurements, it has been shown [43] that the P_{b1} has a considerably different density of states with the "+/0" and "0/−" transitions separated by only a few tenths of an eV and shifted towards the lower part of the gap. A schematic illustration of the P_{b0} and P_{b1} densities of states is illustrated in Fig. 8.9.

Figure 8.10 illustrates the NBTI-induced P_{b0} and P_{b1} SDR signal amplitudes as a function of gate bias. The dashed lines are only a guide for the eye. Figure 8.10 clearly illustrates that both P_{b0} and P_{b1} centers have significant densities of states near mid-gap. The correspondence in gate voltage between the DC-IV peak (Fig. 8.6) and the SDR amplitude peaks (Fig. 8.10) provides further support to the concept that the dominating NBTI-induced defects in these SiO_2-based devices are P_{b0} and P_{b1} centers.

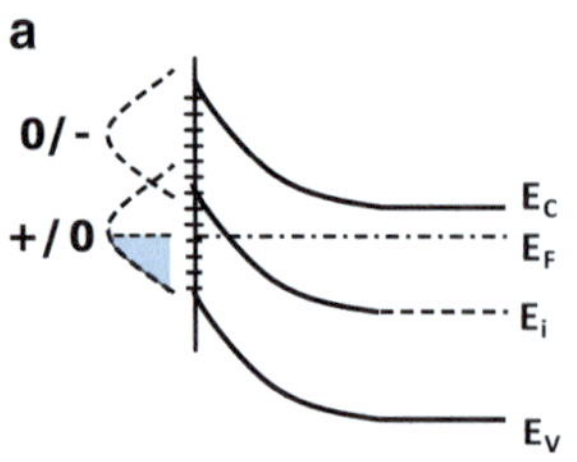

Fig. 8.11 Schematic representations of the (**a**) P_{b0} and (**b**) P_{b1} density of states for p-type MOS capacitor biased in inversion (Reproduced from [19])

Fig. 8.12 Gate-controlled diode DC-IV measurements on a 7.5 nm pMOSFET before and after the application of NBTS (−5.7 V at 200°C for 20,000 s) (Reproduced from [19])

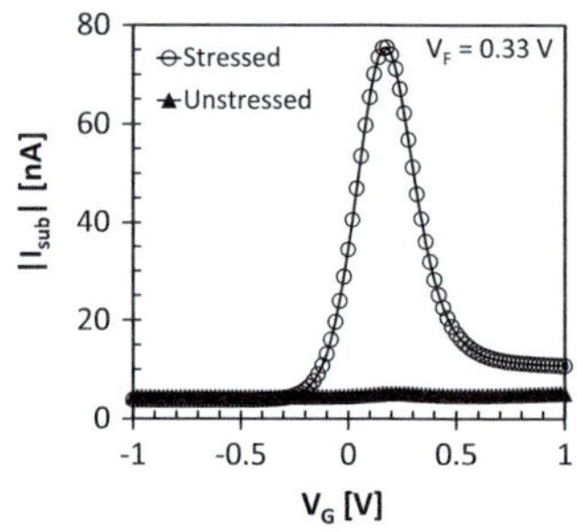

The assignment of P_{b0} and P_{b1} defects as the NBTI-induced defects in these SiO_2 devices is consistent with the observed NBTI-induced negative shift in threshold voltage. Figure 8.11 schematically illustrates, for a pMOS device biased in inversion, that both the P_{b0} and P_{b1} centers would be positively charged. The presence of the P_{b0}/P_{b1} defects would result in the expected NBTI-induced shift in threshold voltage. The fairly high densities of interface states seen in this study could account for quite significant threshold voltage shifts.

In an effort to further investigate the defects associated with NBTI, another 7.5 nm device was subject to a higher temperature NBTS condition (−5.7 V at 200°C for 20,000 s). The pre- and post-NBTS gate-controlled diode DC-IV measurements are shown in Fig. 8.12. Again, it is clear that the NBTS induces a large increase in the peak substrate current and interface state density. The Fitzgerald/Grove analysis [11] yields a pre-NBTS D_{it} of 7×10^9 cm^{-2} eV^{-1} and a post-NBTS D_{it} of 7×10^{11} cm^{-2} eV^{-1} assuming $\sigma_s = 2 \times 10^{-16}$ cm^2.

Figure 8.13 illustrates the corresponding pre- and post-NBTS (−5.7 V at 200°C for 20,000 s) SDR spectra with the magnetic field perpendicular to the (100) surface. Similar to the 140°C stressing case, we observe both P_{b0} and P_{b1} interface defect signals ($g = 2.0060 \pm 0.0003$ and $g = 2.0033 \pm 0.0003$, respectively). Within experimental error, the observed P_{b0} and P_{b1} g-values are the same as in the 140°C NBTS. In addition, we almost certainly observe a third signal (close to our sensitivity limit) at $g = 2.0007 \pm 0.0003$. We attribute this signal to an E' (gamma)

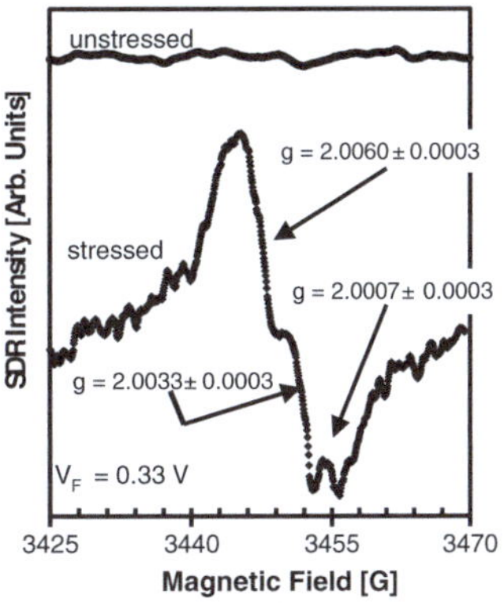

Fig. 8.13 SDR traces of 7.5 nm pMOSFETs with the magnetic field vector perpendicular to the (100) surface both before and after the application of NBTS (−5.7 V at 200°C for 20,000 s) (Reproduced from [19])

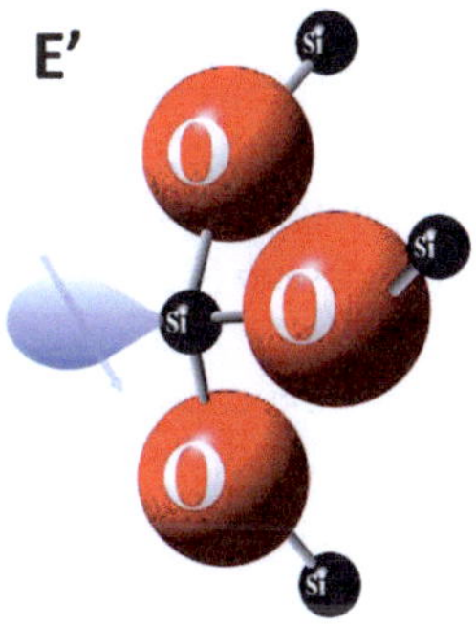

Fig. 8.14 Schematic representation of a neutral E' center oxide defect (Reproduced from [42])

center oxide defect. E' defects are oxygen vacancy centers [15] in which the unpaired electron is localized on a silicon atom which is back-bonded to oxygen atoms within the amorphous oxide region. Figure 8.14 illustrates a schematic drawing of a neutral E' center. E' centers (or any bulk dielectric defects) observed in an SDR measurement must be located very near the Si/SiO_2 interface because only "near-interface" oxide defects would be able to interact with the interface traps (via tunneling) to alter the recombination current [13, 30]. Since we are only observing the "near-interface" E' centers, the fact that the E' signal is much weaker than the P_{b0} and P_{b1} signals does not necessarily indicate that the E' density is significantly lower than the P_{b0}/P_{b1} defects. There may be more E' centers distributed further into the oxide.

Unfortunately, these measurements, by themselves, do not allow for a conclusive determination of the role of E' defects in NBTI. Since we are only able to detect the E' signal in the devices subjected to the harsher stressing condition (−5.7 V, 200°C for 20,000 s), we cannot rule out the possibility that, under these circumstances, the E' centers have been generated via hot hole oxide injection [48]. Nevertheless, our E' observations suggest that they may play an important role in NBTI. It is worth noting that E' centers have levels near the middle of the SiO_2 band gap at energies

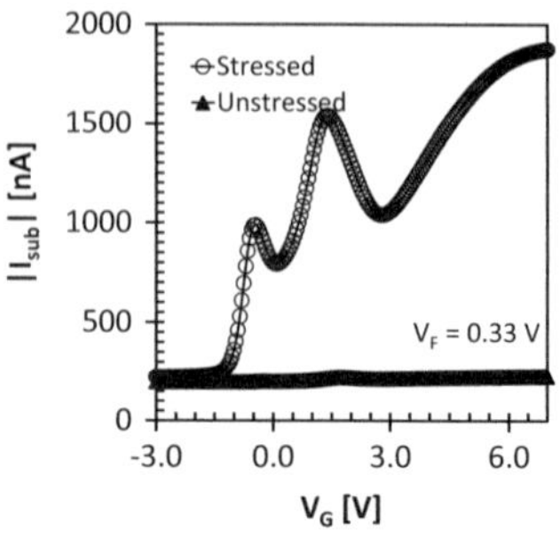

Fig. 8.15 Gate-controlled diode DC-IV measurements on a 48 nm pMOSFET before and after the application of NBTS (−25 V at 175°C for 100,000 s)

appropriate for hole capture from the silicon inversion layer [49]. There is extensive experimental evidence demonstrating that E' centers can capture holes [15, 50]. There is also clear experimental evidence that, after hole capture, E' centers can "crack" molecular hydrogen; a process which could lead to Si–H bond dissociation at the Si/dielectric boundary [50]. As mentioned above, most NBTI theories involve the capture of an inversion layer hole which then leads to Si/SiO_2 interface Si–H dissociation [4, 51–53]. Although these initial E' center results could reasonably be viewed with some skepticism, they do suggest a potential path in which an inversion layer hole capture event can lead to eventual Si/SiO_2 Si–H bond breaking. This concept of an NBTI process controlled by the generation of E' centers is revisited in more detail in Sect. 8.5.

8.4.3 NBTI in 48 nm SiO_2 Devices

We have also examined the NBTI response in much thicker pure SiO_2 power transistors. These devices also exhibit a similar NBTI response as observed in the 7.5 nm SiO_2 devices. These devices were subject to an NBTS condition of −25 V at 175°C for 100,000 s. Figure 8.15 shows both the pre- and post-stress DC-IV characteristic curves. The geometry of these devices (gate extension over lightly doped regions) results in two DC-IV peaks [54]. The DC-IV double peak has been reported in the literature for devices with gate oxides extending over lightly doped regions [54]. The DC-IV peak at $V_G = -0.5$ V is associated with interface states located in the channel while the DC-IV peak at $V_G = 1.4$ V is associated with interface states in the drain overlap region [54]. After NBTS, we observe an increase in both of these substrate current peaks. The Fitzgerald and Grove analysis [11] leads to pre-NBTS $D_{it} = 6 \times 10^9$ cm^{-2} eV^{-1} and $D_{it} = 8 \times 10^9$ cm^{-2} eV^{-1} for the $V_G = -0.5$ V and $V_G = 1.4$ V peaks, respectively. Post-NBTS, we observe an increase in substrate current peaks which correspond to $D_{it} = 2 \times 10^{11}$ cm^{-2} eV^{-1} and $D_{it} = 4 \times 10^{11}$ cm^{-2} eV^{-1} for the $V_G = -0.5$ V and $V_G = 1.4$ V peaks, respectively. All the extracted D_{it} values assume $\sigma_s = 2 \times 10^{-16}$ cm^2.

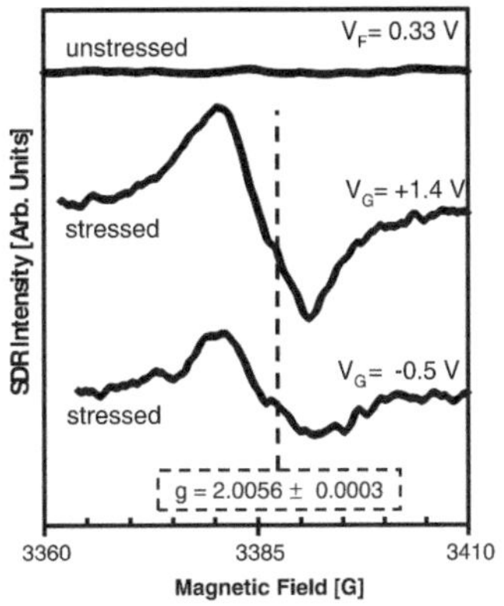

Fig. 8.16 SDR traces of 48 nm pMOSFETs with the magnetic field aligned parallel to the ⟨100⟩ direction both before and after the application of NBTS (−25 V at 175°C for 100,000 s)

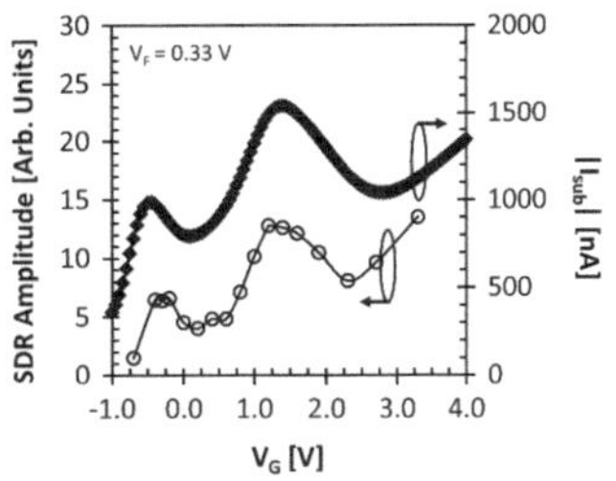

Fig. 8.17 SDR signal amplitude (*lower curve*) and DC-IV-derived substrate current (*upper curve*) as a function of gate bias for the 48 nm pMOSFET subject to an NBTS of −25 V at 175°C for 100,000 s

Figure 8.16 illustrates the corresponding SDR measurements for $V_G = -0.5$ V and $V_G = +1.4$ V with the magnetic field aligned parallel to the ⟨100⟩ surface normal. The interface state density in the unstressed device is below our SDR detection limit. After NBTS, we observe the generation of a strong signal at g = 2.0056 ± 0.0003 for $V_G = 1.4$ V and $V_G = -0.5$ V. The observed g-value at this magnetic field orientation indicates that this signal is due to P_{b0} centers. We note that the same defect signature is observed with $V_G = 1.4$ V and $V_G = -0.5$ V. This is an indication that NBTI generates the same defects in the channel and lightly doped interface regions of the device.

Figure 8.17 illustrates the P_{b0} SDR signal amplitudes as a function of gate bias as well as the DC-IV characteristic curve for the post-NBTS pMOSFET of Figs. 8.15 and 8.16. The close correspondence in gate voltage between the peaks in DC-IV and SDR measurements is another indication that the P_{b0} defects dominate the electronic NBTI-induced interface state density response.

From these observations, we conclude that the NBTI response in the 48 nm pMOSFETs is dominated by P_{b0} interface state generation. It is somewhat interesting that we observe both P_{b0} and P_{b1} defects in the 7.5 nm devices but only

P_{b0} defects in the 48 nm devices. The presence of P_{b0} and P_{b1} defects at the Si/SiO_2 interface is thought to be a result of strain relief. However, the reason for preferential generation of one of these defects over the others is still not understood.

8.4.4 Summary

In both the 7.5 nm and 48 nm SiO_2 pMOSFETs, DC-IV measurements indicate that the NBTI response is dominated by the generation of relatively large densities of interface states. The corresponding SDR measurements identify these NBTI-induced interface states as P_{b0} and P_{b1} defects in the 7.5 nm devices and P_{b0} defects in the 48 nm devices. Our observation of NBTI-induced E' centers suggests that they may play some additional role in NBTI. However, since relatively harsh NBTS was required to observe the E' centers, it is not possible to conclude with certainty what role they play under technologically relevant conditions.

8.5 The Role of E' Centers in NBTI in Pure SiO_2-Based Devices

In Sect. 8.4, we noted that our observations suggested that E' centers could play a very important (catalyzing) role in the NBTI process, at least in pure SiO_2-based devices. We proposed [19, 42, 55] that the NBTI process might be triggered by the capture of silicon inversion layer holes, which simple statistical mechanics arguments [55] indicate would lead to subsequent P_b center generation via loss of hydrogen at P_b-H precursor sites. The hole capture process can create a positively charged E' site in which one side is a neutral singly occupied silicon dangling bond and the other a positively charged diamagnetic silicon. These arguments linking E' hole capture and P_b generation have been detailed elsewhere in the context of NBTI [56] and earlier in the context of radiation damage [57–59]. Recently, Grasser et al. [60] developed a comprehensive quantitative two-stage model for NBTI. In this model, NBTI is also triggered by inversion layer hole capture at an E' center precursor site (a neutral oxygen vacancy). The presence of the oxide silicon dangling bond created in this process (the neutral side of the E' center) then triggers the creation of poorly recoverable defects (P_b centers) via an E'/P_b center hydrogen exchange. The comprehensive quantitative model of Grasser et al. [60] expands upon the earlier qualitative arguments [19, 42, 55] and explains NBTI degradation over a wide range of bias and stress temperature, the observed asymmetry between stress and recovery, and the strong sensitivity to bias and temperature during recovery. Central to the model is the prediction that paramagnetic E' centers will be present during stress and will, for the most part, very quickly recover upon removal of stress (similar to [19, 42]).

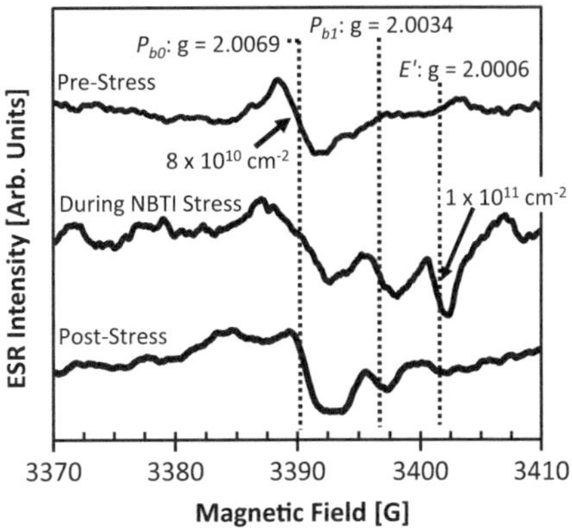

Fig. 8.18 Three ESR traces taken before (*top*), during (*middle*), and after NBTI stress (*bottom*). Note the clear generation of an E' signal during NBTI stress (*middle*) (Reproduced from [62])

The data we provided in support of our initial suggestion for a catalyzing role for E' centers in NBTI were somewhat tenuous [19, 21, 42] for two reasons. First, SDR does not permit observations at significant negative gate bias. To obtain reasonable SDR sensitivity, the stress biasing conditions must be shifted towards depletion, so that the electron and hole quasi-Fermi levels are split more or less symmetrically about the intrinsic Fermi level at the Si/SiO_2 interface [15, 30]. If the qualitative proposals of Campbell et al. and the model of Grasser et al. are correct, this would invariably lead to significant recovery of the E' centers [61]. In the earlier SDR measurements [19, 21, 42], most of the E' centers would be electrically neutralized and thus no longer paramagnetic. Second, even under optimized biasing conditions, SDR is only marginally adequate for E' center detection because E' centers are less effective recombination centers than Si/SiO_2 P_b centers; only those E' centers close to the interface contribute to SDR [30]. However, conventional ESR permits E' center detection at any gate bias, provided the E' center is paramagnetic [15]. The E' center oxygen vacancy would be paramagnetic when positively charged. To overcome these obstacles inherent to SDR measurements of E' centers, an on-the-fly approach has been developed in which conventional ESR measurements are performed during negative bias stressing of MOS structures at elevated temperature [62]. Note that, unlike prior ESR studies, this on-the-fly approach allows for the observation of NBTI defects void of any recovery contamination.

The samples used in this study [62] are large area Si/SiO_2 blanket capacitor structures with 49.5 nm thermally grown SiO_2 oxides which were treated with a post-oxidation forming gas anneal. ESR measurements were performed before, during, and after the sample was subjected to a modest NBTI stress of -25 V (oxide field < 5 MV/cm) at 100°C. Negative bias was applied to the sample utilizing corona ions to provide a virtual gate [46]. The gate bias was monitored before and after stress with a Kelvin probe. The thick (49.5 nm) oxides were chosen to ensure a uniform gate bias over the measurement time (several hours). A quartz dewar apparatus was utilized to heat the sample inside the microwave resonant cavity.

Figure 8.18 illustrates three ESR traces taken on the sample before, during, and after NBTI stress [62]. Each trace was signal averaged for several hours. Although

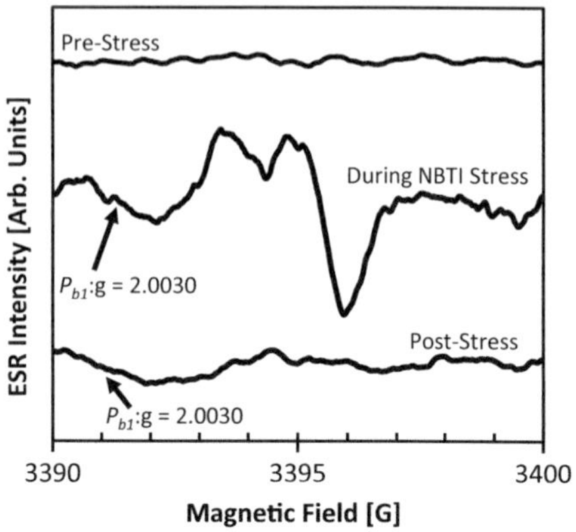

Fig. 8.19 Three ESR traces taken before (*top*), during (*middle*), and after NBTI stress (*bottom*). In these traces, the spectrometer settings are optimized to observe E' centers. Note the clear generation of an E' spectrum during stress (*middle*) and its subsequent recovery post-stress (*bottom*) (Reproduced from [62])

these measurements are slow, recovery is nonexistent since the stress conditions remain constant throughout the measurement. The spectrometer settings used were chosen to permit the observation of both Si/SiO_2 P_b centers and SiO_2 E' centers and are not optimized for either defect; the E' center density is underrepresented in these traces (a significant difference in E' and P_b spin–lattice relaxation times leads to this under-representation) [46]. In the pre-stress case (top), we observe a weak single line spectrum with $g = 2.0069 \pm 0.0003$ which is likely due to P_{b0} Si/SiO_2 interface states. During NBTI stress (middle), we observe the clear generation of Si/SiO_2 P_{b1} centers ($g = 2.0034 \pm 0.0003$) and SiO_2 E' centers ($g = 2.0006 \pm 0.0003$). Upon removal of the stress, the E' center signal completely recovers while some of the P_{b1} centers remain. This result *clearly* indicates that positively charged oxygen vacancy sites (E' centers) are generated during stress and quickly disappears once the stress is removed [62]. Although the model predicts very fast E' center recovery, the time resolution of our measurements is slow. Thus, we cannot determine how quickly the recovery occurs.

As mentioned previously, the spectrometer settings used in Fig. 8.18 were chosen to permit the observation of both Si/SiO_2 P_b centers and SiO_2 E' centers and are not optimized for either defect. To further demonstrate that E' centers are present during NBTI stressing, Fig. 8.19 shows three ESR traces taken on the sample before, during, and after NBTI stressing [62]. In this figure, the spectrometer settings are optimized for the observation of E' centers. During NBTI stress (middle), we observe a powder pattern signal consistent with an E' center. Upon removal of the NBTI stress (bottom), the E' signal completely disappears. Figure 8.20 provides additional evidence linking this to an E' center via comparison with a commercially available E' spin standard [62]. Note the very close correspondence between the two spectra.

These observations are consistent with and most strongly support the earlier proposal [19, 21, 42] that NBTI is triggered by the tunneling of electrons from a neutral E' center precursor to unoccupied valence band states. These results are also fully consistent with the NBTI model of Grasser et al. [60]. Both the earlier

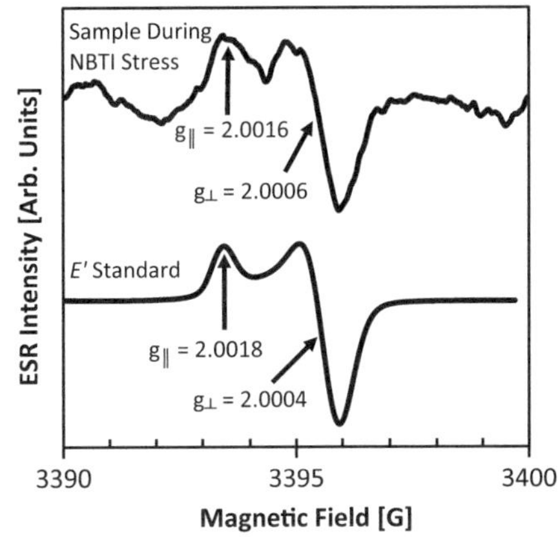

Fig. 8.20 Comparison of the "during NBTI stress" spectra (*top*) and a commercially available E' standard. The standard sample signal to noise ratio is much higher because the standard has orders of magnitude more E' centers. Note the close correspondence between the g-values and line shapes (Reproduced from [62])

qualitative arguments and the more recent quantitative work utilize the fact that the presence of unpassivated E' silicon dangling bonds in the presence of large numbers of passivated P_b center silicon dangling bonds is thermodynamically unstable. The Gibb's free energy of the P_b-H/E' dangling bond system would be lowered by the exchange of hydrogen from P_b-H to E' dangling bond states, generating interface traps.

The Gibb's free energy, G, is expressed as follows: $G = H - T\,S$, where H is enthalpy, T is absolute temperature, and S is entropy. The enthalpy of a silicon–hydrogen bond at the silicon/dielectric interface is very nearly the same as that for a silicon–hydrogen bond in the oxide. Thus, the transfer of a hydrogen from a passivated P_b site at the interface to an E' silicon dangling bond in the oxide will do little to the enthalpy term in Gibb's free energy [19, 21, 42, 55–59]. However, consider the entropy, $S = k\,\ln(\Omega)$, where k is Boltzmann's constant and Ω is the number of microscopic states yielding the macroscopic state. The transfer of a hydrogen atom from the essentially perfectly passivated Si/SiO_2 interface P_b centers to the essentially perfectly unpassivated oxide E' centers inevitably results in a very large increase in the system's configurational entropy. This process is discussed at considerable length in multiple publications [19, 21, 42, 55–59], but the underlying principle can be addressed quite simply. Consider the perfectly passivated silicon dangling bonds at the interface ($\sim 10^{12}/cm^2$). If the system is perfectly passivated, then $\Omega = 1$. Suppose that, at random, one site becomes depassivated. Therefore, in a square centimeter there are 10^{12} possibilities. Consider a second depassivation: that would be $10^{12} \times [(10^{12} - 1)/2]$ possibilities. A similar argument follows for an equal number of perfectly unpassivated E' oxide silicon dangling bonds. The Gibb's free energy is inevitably lowered by the quite substantial transfer of hydrogen atoms away from the P_b-H sites to the E' sites, because the increase in configurational entropy is inevitably very large. Thus, at least in pure SiO_2 devices there is overwhelming evidence which indicates that the earlier suggestions [19, 21, 42] and later modeling [60] of a catalyzing role for E' centers in NBTI are quite valid.

8.6 NBTI in Plasma-Nitrided Oxide pMOSFETs

This section discusses SDR and SDT magnetic resonance measurements utilized to directly observe the atomic-scale defects of NBTI in fully processed 2.3 nm PNO devices [17, 19, 42, 63]. We compare the SDR and SDT measurements with DC-IV observations of NBTI-induced changes in interface state densities as well as trap-assisted tunneling current measurements. We also make meaningful comparisons between these PNO observations and the observations on the 7.5 nm pure SiO_2 devices discussed earlier. Our results definitively identify the dominating atomic-scale defect in the 2.3 nm PNO devices as a K_N center (not a P_{b0} or P_{b1} center) [17, 42, 63]. K_N centers are silicon dangling bond defects in which the central silicon is back-bonded to three other nitrogen atoms [17, 42, 63]. The K_N centers are in the "near-interface" region of the dielectric and have a fairly narrow peaked effective density of states near the middle of the band gap [42, 63]. We find that NBTI-induced K_N centers participate in both spin-dependent recombination and tunneling processes [42, 63]. This is an indication that the same defect can act like an interface state and a bulk dielectric tunneling center [42, 63]. These collective observations may help explain why the addition of nitrogen enhances the NBTI phenomenon. They may also explain why NBTI experts can report conflicting views on the roles of NBTI-induced interface states and bulk traps.

8.6.1 Experimental Details

This study involves both 2.3 nm equivalent oxide thickness (EOT) PNO large area pMOSFETs ($\sim$40,000 μm^2) and the fairly well-understood 7.5 nm SiO_2 devices [19–21] discussed earlier. The SiO_2 devices are used as a control to compare and contrast the thinner PNO devices. The SiO_2 devices were subject to an NBTS of -5.7 V at 140°C for 210,000 s. The PNO devices were subject to an NBTS of -2.6 V at 140°C for 180,000 s. Interface state densities (D_{it}) were monitored using DC-IV measurements [11, 12]. As will be shown in the following sections, variations in the SDR and DC-IV recombination energy window provide deeper understanding of the defects involved. SDR and SDT measurements were made at room temperature with a custom-built SDR/SDT spectrometer. SDR/SDT measurements were calibrated using a strong pitch spin standard. Immediately following the stress, each device was subject to a post-NBTS temperature quench step which involves cooling the device to room temperature (over the span of several minutes) while the negative stressing bias is maintained. As discussed earlier, this step has been found to be fairly effective at "locking-in" the NBTI-induced damage, rendering it observable in SDR/SDT [19, 20, 36]. Also, to ensure that further recovery was minimal during subsequent measurements, all DC-IV, SDR, and SDT measurements were taken at least 4 h post-NBTS/temperature quench.

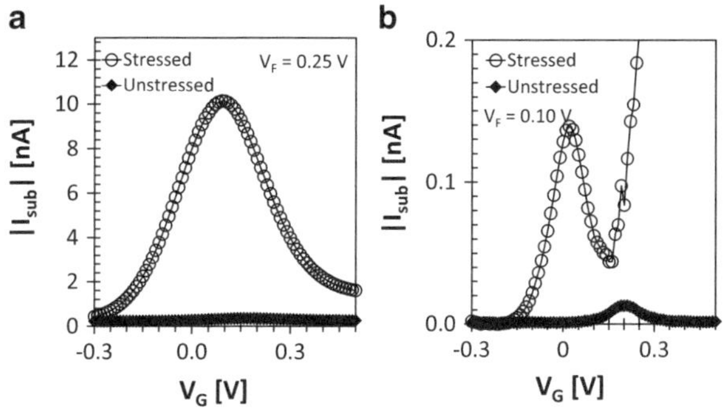

Fig. 8.21 Pre- and post-NBTS DC-IV measurements on (**a**) 7.5 nm SiO_2 pMOSFET (−5.7 V at 140°C for 210,000 s) and (**b**) 2.3 nm PNO pMOSFET (−2.6 V at 140°C for 180,000 s) (Reproduced from [42])

8.6.2 *Comparisons of NBTI in 7.5 nm SiO_2 and 2.3 nm PNO pMOSFETs: Different Defects*

Figure 8.21 illustrates the pre- and post-NBTS DC-IV characteristic curves for a 7.5 nm SiO_2 and a 2.3 nm EOT PNO-based pMOSFET. In the 7.5 nm SiO_2 device (Fig. 8.21a), NBTS clearly generates an increase in the peak substrate current corresponding to an increase in D_{it}. The Fitzgerald and Grove analysis [11] leads to a pre-stress $D_{it} \approx 6 \times 10^9$ cm^{-2} eV^{-1} and a post-stress $D_{it} \approx 4 \times 10^{11}$ cm^{-2} eV^{-1} assuming $\sigma_s = 2 \times 10^{-16}$ cm^2. In the 2.3 nm PNO device (Fig. 8.21b), NBTS also generates an increase in peak substrate current and D_{it}. Fitzgerald and Grove derived D_{it} values were not extracted for these devices because of the uncertainty in the mean capture cross section. However, we note an approximately one order of magnitude increase in the peak substrate current which should correspond to approximately one order of magnitude increase in D_{it}. It is clear that NBTS generates a very large increase in D_{it} in both the SiO_2 and PNO devices. We note that the shapes of the DC-IV characteristic curves for the SiO_2 and PNO devices are somewhat different. The SiO_2 DC-IV curve is consistent with recombination while the PNO DC-IV is due to both recombination and tunneling; tunneling dominates at the most positive gate voltages. As one would anticipate, the PNO tunneling component is most dominant when the source/drain to substrate diode forward bias (V_F) is reduced to approximately zero. (In this case, the recombination is turned off.)

Figure 8.22 illustrates post-NBTS magnetic resonance results for the 7.5 nm SiO_2 and 2.3 nm PNO-based devices with the magnetic field parallel to the ⟨100⟩ surface normal. The spectrometer settings in each case are optimized to yield the true line shape of the spectra. In this figure, SDR spectra are plotted as a function of g-value to illustrate differences in NBTI-induced defects in the SiO_2

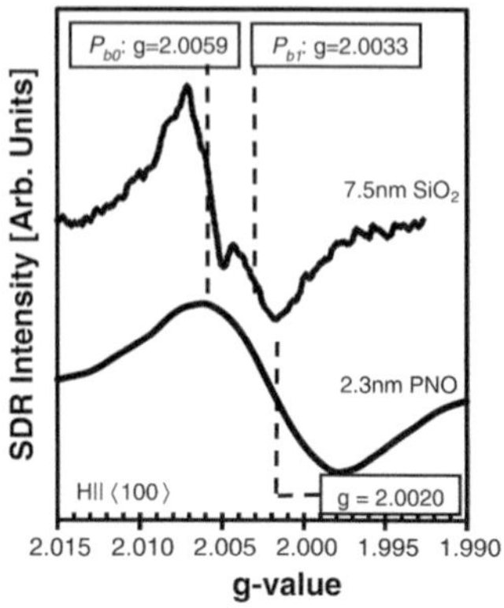

Fig. 8.22 Optimized SDR traces of NBTI-induced defects in 7.5 nm SiO_2 and 2.3 nm PNO-based devices as a function of g-value. NBTI generates different defects in SiO_2 and PNO devices (Reproduced from [42])

and PNO devices. NBTS in 7.5 nm SiO_2-based devices generates two SDR signals at $g = 2.0059 \pm 0.0003$ and $g = 2.0033 \pm 0.0003$ corresponding to P_{b0} and P_{b1} Si/SiO_2 interface defects respectively [21]. The P_{b0} and P_{b1} defects have been shown to be the dominating interface defects in Si/SiO_2 systems under many circumstances [15] and have also been shown earlier to play a large role in NBTI in thicker SiO_2-based devices [16, 19–21]. In 2.3 nm PNO-based devices, NBTS generates a much broader (≈14 G wide) SDR signal at $g = 2.0020 \pm 0.0003$. Measurements (not shown) with spectrometer settings optimized to observe narrower signals within the broad $g = 2.0020$ signal revealed no other components. If P_{b0}/P_{b1} centers were present in significant quantities in the PNO-based devices, the overlapping P_{b0}/P_{b1} signal would be observed to the left (higher g-value) of the apparently symmetric $g = 2.0020$ signal. It is clear that this is not the case. Thus, NBTS in thinner PNO-based devices generates this new defect at $g = 2.0020$ to such an extent that any P_{b0}/P_{b1} generation is completely obscured. From these DC-IV/SDR measurements, it is clear that NBTI generates different defects in the PNO devices. The following subsections of this manuscript examine the electronic properties and physical identity of this NBTI-induced center in the PNO devices.

8.6.3 PNO Defect: Physical Location

As discussed in Sect. 8.3, the relationship between an experimentally observed g-value and the orientation of the applied magnetic field is an indication of the defect's local environment. Figure 8.23 illustrates the g-value versus device orientation with respect to the applied magnetic field for the P_{b0} center as well as for the new PNO defect. In this figure, θ is defined as the angle between the magnetic field vector and the $\langle 100 \rangle$ surface normal. Figure 8.23a illustrates the P_{b0} g-value orientation dependence adapted from earlier ESR work [64]. The P_{b0} signal splits into three lines (each corresponding to a $\langle 111 \rangle$ direction) that are clearly a function

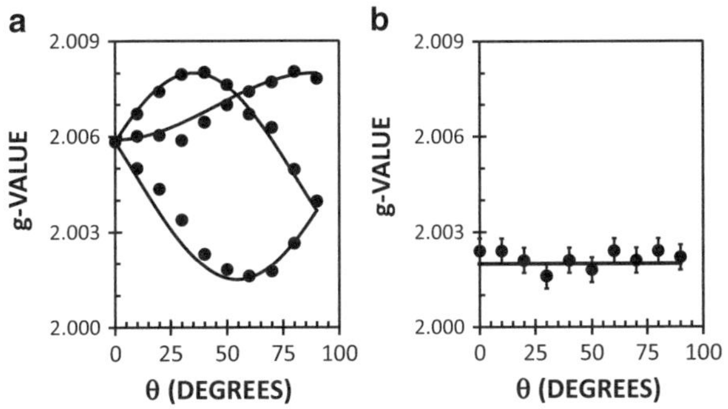

Fig. 8.23 Adapted ESR g-value angular dependence for the Si/SiO$_2$$P_{b0}$ interfacial defect [64] and (**b**) SDR g-value angular dependence for the NBTI-induced 2.3 nm PNO defect (Reproduced from [42])

of orientation (anisotropic) [64]. If a defect was located precisely at the interface, one would expect it to have an orientation dependent g-value because of the long range order present precisely at the interface [15, 25]. Rotation of the device in the magnetic field changes the alignment of the local magnetic field experienced by the defect relative to the applied field. This alters the defect's resonance condition resulting in an anisotropic g-value [15, 25]. Figure 8.23b illustrates the g-value orientation dependence of the NBTI-induced g = 2.0020 signal in the PNO devices. The g-value is independent of the device orientation (isotropic). If a defect was located in an amorphous dielectric, one would expect the defect to have an equal distribution in all orientations [15, 25]. In this case, rotation of the device in the magnetic field alters the local magnetic fields of all the defects, but the total effect would be (on average) zero, resulting in an unchanging g-value [15, 25]. Therefore, our observation of an isotropic g = 2.0020 (Fig. 8.23b) is a strong indication that the dominant NBTI-induced defect in the PNO devices is located within the amorphous dielectric. However, the NBTI-induced increase in both the DC-IV peak and the SDR spectrum indicates that this defect is located close enough to the interface to participate in recombination. Therefore, we conclude that the defect is located within the dielectric but close (probably $\leq$1 nm) to the interface (the "near-interface" region). However, we also note that in these 2.3 nm PNO devices, the "near-interface" region may account for a large percentage of the gate dielectric.

8.6.4 PNO Defect: Participation in SDR and SDT

Figure 8.24 illustrates the post-NBTS magnetic resonance spectra for a 2.3 nm pMOSFET taken with $V_F = -0.2$ V and $V_F = +0.2$ V. In these measurements, the magnetic field is parallel to the $\langle 100 \rangle$ surface normal. Both spectra clearly show the generation of the same PNO defect ($\approx$14 G wide signal at g = 2.0020 $\pm$ 0.0003).

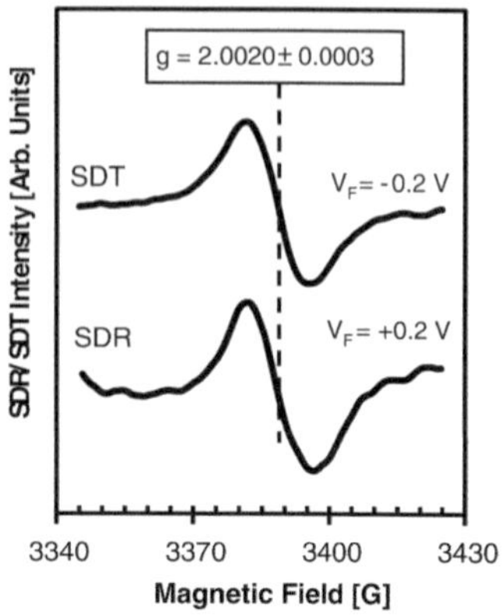

Fig. 8.24 SDT ($V_F = -0.2$ V) and SDR ($V_F = +0.2$ V) spectra of PNO defect. Both measurements clearly show the generation of a single dominating signal at g = 2.0020. The spectrometer gain is 100 times greater for the $V_F = -0.2$ V spectrum (Reproduced from [42])

The spectrometer gain for the $V_F = -0.2$ V spectrum is 100 times larger than the gain of the $V_F = +0.2$ V spectrum. Magnetic resonance measurements utilize microwave irradiation to satisfy the ESR resonance condition. In these particular measurements, the microwaves induce a (+0.2 V) bias between the source/drain to substrate diode. When a source/drain to substrate bias (V_F) of −0.2 V is applied to compensate for the microwave-induced biasing, the diode is completely turned off, and the device substrate current is dominated by tunneling current. Thus, the spectrum observed with $V_F = -0.2$ V is dominated by a spin-dependent tunneling current through the gate dielectric. However, when $V_F = +0.2$ V, the source/drain to substrate diode is turned on (sufficient numbers of interfacial electrons and holes), and the device substrate current is dominated by recombination. Thus, the spectrum observed with $V_F = +0.2$ V corresponds to spin-dependent recombination current through interface traps. The observation of the same defect signature at $V_F = -0.2$ V and at $V_F = +0.2$ V indicates that *the same defect participates in both spin-dependent tunneling and spin-dependent recombination.*

To further examine the role of the PNO defect in SDT and SDR phenomena, we have made extensive SDT/SDR measurements as a function of V_F. These measurements are shown in Fig. 8.25a as the normalized SDR and SDT amplitudes as a function of V_F. The normalization is achieved by dividing the spin-dependent modification to the tunneling and recombination currents (ΔI) by the non-spin-dependent DC current (I). We observe that the ΔI/I response steadily increases as V_F is decreased. The largest ΔI/I effect is seen for $V_F \leq -0.2$ V. As discussed above, for these values of V_F the source/drain to substrate diode is turned off, and the observed spectrum is dominated by SDT. For larger positive values of V_F, the source/drain to substrate diode is turned on, and the observed spectrum is dominated by SDR. In an attempt to delineate between SDR and SDT, Fig. 8.25b plots the modification to the spin-dependent currents (ΔI) as a function of V_F. The ΔI response is relatively

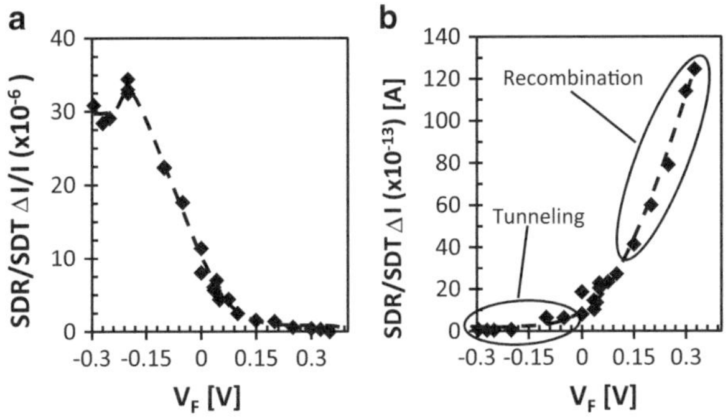

Fig. 8.25 (**a**) Normalized size of the SDR and SDT effect for the PNO defect as a function of V_F. The normalization is plotted as the modification to the SDR/SDT current (ΔI) divided by the DC source/drain to substrate diode current (I). (**b**) The modification to the SDR/SDT current (ΔI) as a function of V_F. These figures illustrate that the PNO defect clearly participates in both SDT and SDR phenomena (Reproduced from [42])

Fig. 8.26 Normalized size (ΔI/I) of the SDR and SDT effects as a function of V_G for $V_F = +0.2$ V and $V_F = -0.2$ V (Reproduced from [42])

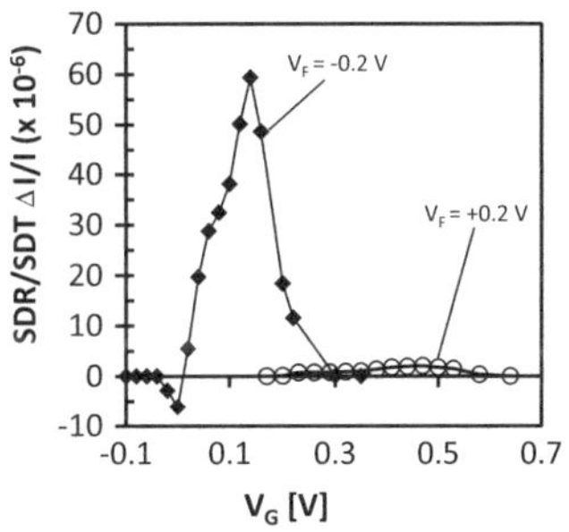

constant for negative values of V_F and increases sharply for positive values of V_F. Unlike ΔI/I, the ΔI of the SDT current should be nearly independent of V_F (it should only depend upon the gate voltage). Therefore, the increase in ΔI for positive values of V_F is due to an increase in SDR current. Thus, the spin-dependent current is dominated by tunneling for negative values of V_F and by recombination for larger positive values of V_F. It is difficult to determine the exact transition between SDT and SDR. However, since the ΔI value increases dramatically as V_F is increased (≈25× larger at $V_F = +0.1$ V, ≈60× larger at $V_F = +0.2$ V, and ≈120× larger at $V_F = +0.3$ V), we somewhat arbitrarily take SDT as the dominant mechanism for $V_F < +0.1$ V and SDR as the dominant mechanism for $V_F > +0.1$ V. Although our approach to separating the relative contributions of SDT and SDR is imprecise, we believe the argument is conclusive.

We have also examined the effect of V_G on the SDR and SDT phenomena. Figure 8.26 illustrates the SDR ($V_F = +0.2$ V) and SDT ($V_F = -0.2$ V) ΔI/I responses as a function of gate voltage (V_G). Since interface recombination can only

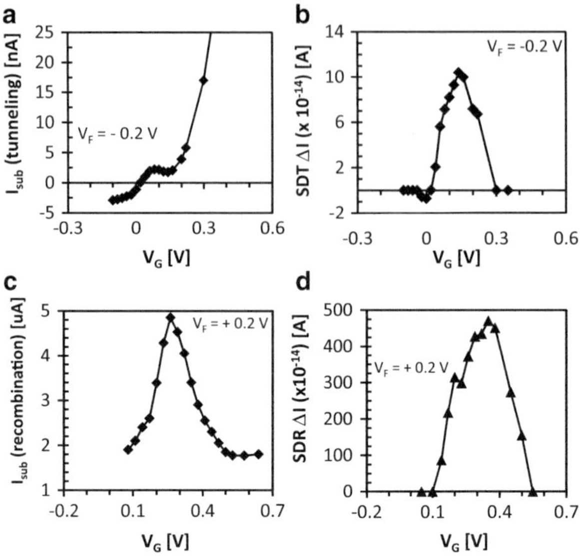

Fig. 8.27 DC substrate currents and modifications to the spin-dependent currents (ΔI) for SDT with $V_F = -0.2$ V (**a**) and (**b**) and for SDR with $V_F = +0.2$ V (**c**) and (**d**). The close correspondence between the peaks in the substrate currents and ΔI values is an indication that the PNO defect dominates both the device tunneling centers (**a**) and (**b**) and interfaces states (**c**) and (**d**) (Reproduced from [42])

take place in depletion and is maximized when there is equal numbers of electrons and holes at the interface [11, 22], the SDR ΔI/I will be largest for the gate voltages corresponding to a depleted channel. The SDR ΔI/I should be much less when V_G corresponds to an accumulated or inverted channel (unequal numbers of electrons and holes at the interface). This is exactly what we observe. The SDR ΔI/I is broadly peaked near depletion ($V_G \approx +0.45$ V) and is consistent with the expected SDR behavior. However, the SDT ΔI/I response is different; it is a narrower function of V_G and exhibits negative ΔI/I values. Since SDT should only occur when V_G bends the bands to an energetically favorable alignment of tunneling trap levels, one would expect the ΔI/I response to be narrower for SDT than for SDR. The negative ΔI/I values correspond to a reversal in the measured SDT signal polarity and sign of the DC current. We interpret this reversal in SDT signal polarity to a band alignment favorable for a reversal in direction of the SDT current.

Figure 8.27 illustrates the device tunneling ($V_F = -0.2$ V) and recombination ($V_F = +0.2$ V) currents along with the SDT ($V_F = -0.2$ V) and SDR ($V_F = -0.2$ V) ΔI responses of the PNO defect as a function of V_G. We observe that the peak in the device tunneling current closely corresponds to the peak SDT ΔI response (Fig. 8.27a, b) and that the peak in the device recombination current closely corresponds to the peak in the SDR ΔI response (Fig. 8.27c, d). This correspondence is another strong indication that the PNO defect acts as both the dominant tunneling center and interface state in these devices.

It is important to note that the maximum size of the SDT effect ($\Delta I/I \cong 6 \times 10^{-5}$) is much larger than the SDR effect ($\Delta I/I \cong 2 \times 10^{-6}$) observed in these PNO samples. The increased size of the SDT effect provides a considerable improvement in measurement sensitivity over SDR. Also, the ΔI/I SDT value we measure is smaller than what we would observe if more microwave power were available in our system because the signal amplitude is still increasing linearly at the maximum available microwave power (150 mW) of the spectrometer. Large SDR effects are typically explained in terms of a pairing of two spins (KSM model) [29]. In this case, the spins would be the charge carriers and the deep level defects spins. The size of the observed SDT effect is more consistent with that predicted by a KSM-like model in which there is a pairing between tunneling charge carriers and charge carriers trapped in PNO defect tunneling centers. However, the size of the observed SDR effect is smaller than the size predicted by the KSM model and smaller than most SDR measurements reported in the literature [13]. The size of the SDR effect in the PNO devices is more consistent with that predicted by the Lepine [27] model for SDR in which recombination occurs without a pairing of spins prior to capture. This implies that when the PNO defect is participating in interfacial recombination, there is little coupling between conduction or valence band charge carriers and deep level interfacial defect spins prior to the actual recombination event. This correspondence between the PNO defect's recombination response and the Lepine model may be consistent with the fact that the PNO defects are not located at the interface.

Considering that the PNO devices' gate dielectrics are relatively thin (2.3 nm), it is possible that the SDT mechanism involves something similar to an interface trap to interface trap tunneling mechanism similar to that proposed by Nicollian [65] to describe low-voltage stress-induced leakage currents. We have already shown that the PNO defects are physically located not at the interface but in the "near-interface" region of the dielectric. Therefore, we speculate that our SDT measurements involve a spin-dependent trap-assisted variation of the interface state to interface state tunneling mechanism which involves tunneling through the gate dielectric through NBTI-induced PNO "near-interface" defects. Very recently, there has been a considerable effort to further explore the mechanism of SDT. The SDT mechanism was originally thought to proceed through a series of deep level defects in the bulk of the dielectric. In this scenario, manipulation of the deep level bulk defects was responsible to the spin-dependent modification to the tunneling current. However, an SDT process in which the tunneling current can be modified by the spin dependence of interface or "near-interface" defects is likely to dominate in thinner dielectrics. Recent efforts confirm the somewhat qualitative analysis presented here [32].

8.6.5 PNO Defect: Density of States

We have shown that large variations in V_F can change the nature of the observed spin-dependent current from tunneling to recombination. However, restricting V_F to values in which SDR is the dominant mechanism ($V_F \geq +0.1$ V) also allows for a

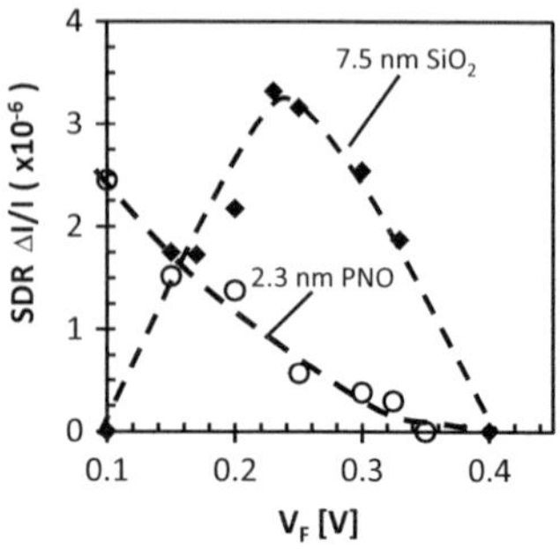

Fig. 8.28 Normalized size (ΔI/I) of the SDR effect as a function of V_F for the SiO_2 and PNO devices. The different responses indicate different densities of states (Reproduced from [42])

very useful comparison between the thicker SiO_2 and thinner PNO devices. Since SDR is a spin-dependent modification to the DC-IV measurement and V_F controls the DC-IV recombination energy window, or the energy range over which interface traps may contribute to the effect, differences in the size of the SDR effect as a function of V_F reflect differences in the densities of states of the different defects in these two types of devices.

Figure 8.28 illustrates the size of the post-NBTS SDR effect as a function of source/drain to substrate forward bias (V_F) for the SiO_2 and PNO devices. For the 7.5 nm SiO_2 devices, the SDR derived ΔI/I is broadly peaked at $V_F \approx +0.25$ V. This relatively weak dependence on V_F is consistent with a *relatively* flat density of states near the middle of the band gap. This is so because increasing V_F increases the energy window in which recombination occurs [11]. If the density of states near the middle of the band gap is *relatively* flat (like that expected in an interface dominated by P_{b0} and P_{b1} defects), increasing V_F allows more interface states to participate in the recombination process. Thus, the size of the SDR effect (seen via ΔI/I) increases. This behavior continues until V_F is large enough that the source/drain to substrate forward bias diffusion current overwhelms the recombination current and the denominator in ΔI/I will dominate. Thus, for an interface that is dominated by defects with a relatively flat density of states (P_{b0}/P_{b1}), the SDR ΔI/I should be broadly peaked as a function of V_F. (Note that our conclusions here are qualitative to semiquantitative in detail; the combined P_{b0} and P_{b1} density of states is not really flat at all but *relatively* flat in the sense that it is not sharply peaked.) However, the SDR ΔI/I response is quite different for the PNO devices. In the PNO case, ΔI/I steadily decreases as V_F is increased. This is consistent with a narrower density of states fairly near the middle of the band gap. Following the arguments above, if the density of states near the middle of the band gap is sharply peaked, increasing V_F beyond a point, approximately corresponding to the peak width, does not increase the number of defects participating in recombination (the number of defects in essentially constant). Instead, increasing V_F only increases the forward bias diffusion current and the denominator of the ΔI/I term dominates. Thus, for an interface with a sharply peaked density of states near the middle of the band gap, the SDR ΔI/I should steadily decreases as V_F is increased.

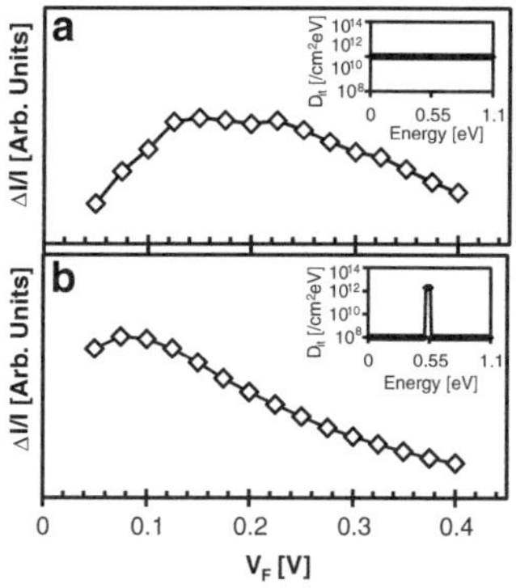

Fig. 8.29 Simulated ΔI/I DC-IV recombination current as a function of V_F for (**a**) 7.5 nm SiO_2 device with a flat density of states through the gap and (**b**) 2.3 nm PNO device with a very narrow density of states centered about the middle of the gap. The figure *insets* show the simulated densities of states in each case (Reproduced from [42])

Since our SDR measurements utilize a spin-dependent modification to the DC-IV measurement, a brief review of DC-IV theory may be useful in understanding this result [11, 12]. A careful review of Eqs. (8.2), (8.3), and (8.4) provides the necessary tools to look at this problem more quantitatively. (Note that increasing V_F between the source/drain to substrate p/n junction increases the number of minority carrier holes (p_S) available for recombination (in a pMOS device).) An examination of the integral in Eq. (8.2) helps explain the V_F-derived active recombination window. For $|E\text{-}E_i|$ values small enough to ensure that the $p_S + n_S$ term is greater than the $2n_i \cosh[(E\text{-}E_i)/kT]$ term, the denominator of the integral is nearly constant and relatively small. However, as $|E\text{-}E_i|$ increases, the $\cosh[(E\text{-}E_i)/kT]$ term blows up and overwhelms the V_F-derived $p_S + n_S$ term. This can be thought of as effectively closing the recombination window. Careful analysis of the integrand of Eq. (8.2) using Eqs. (8.3) and (8.4) leads to an effective DC-IV and SDR recombination energy window of $\sim q|V_F|$, centered about the middle of the band gap.

Simulations of the DC-IV recombination current using Eqs. (8.2), (8.3), and (8.4) were carried out for various values of V_F and various densities of states. By dividing the simulated DC-IV peak current by the baseline substrate current, an analogous ΔI/I behavior is observed. Figure 8.29a illustrates the simulated ΔI/I behavior for a 7.5 nm gate oxide with a flat density of states through the gap. Figure 8.29b illustrates the simulated ΔI/I behavior for a 2.3 nm gate oxide with a very narrowly peaked density of states centered about the middle of the gap. The insets of both figures show the simulated densities of states. The similarities between simulated ΔI/I (Fig. 8.29) and measured SDR ΔI/I (Fig. 8.28) suggest a common origin. The DC-IV-derived D_{it} values as a function of V_F for the SiO_2 and PNO devices are also consistent with this trend. D_{it} values, averaged over the energy window, for the SiO_2 device (Fig. 8.30a) are approximately constant as V_F is increased while the D_{it} values, averaged over the energy window, for the PNO devices (Fig. 8.30b) decreases as V_F is increased. This trend is what one would anticipate from our above analysis.

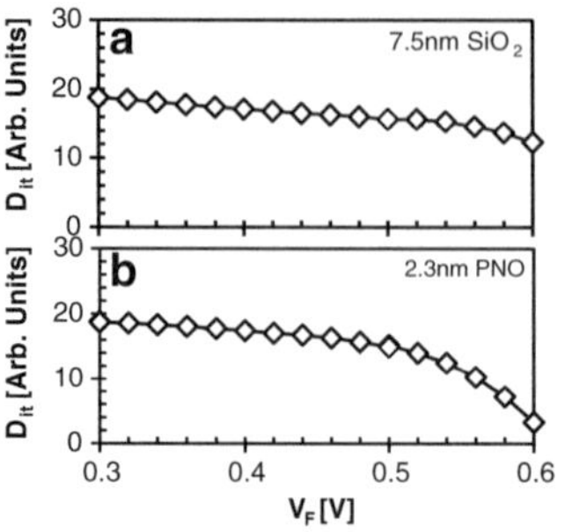

Fig. 8.30 DC-IV-derived average D_{it} values over the energy window for the (**a**) SiO_2 and (**b**) PNO devices as a function of source/drain to substrate diode forward bias (V_F). The D_{it} values were obtained from DC-IV measurements and Eq. (8.2). As V_F, and the recombination energy window, increases, D_{it} for the SiO_2 device is essentially constant while D_{it} for the PNO device decreases. This is consistent with the PNO defect having a narrowly peaked density of states (Reproduced from [42])

Krishnan et al. [10] and Stathis et al. [66] have reported results which suggest different NBTI-induced density of states in SiO_2 and PNO-based devices. Stathis et al. [66] speculate that the addition of nitrogen changes the dangling bond structure of the NBTI-induced defects. Our observations are consistent with this conclusion. (However, in detail, our conclusions differ with regard to density of states.) Our observations directly demonstrate that there are different dangling bond defects in PNO- and SiO_2-based devices and, as discussed in the next subsection, the PNO defects involve silicon atoms coupled to nitrogen atoms. Our observations also suggest a narrower density of states for the PNO defect than the P_{b0}/P_{b1} defects in SiO_2 devices. Figure 8.31 illustrates a schematic representation of the density of states of NBTI-induced defects in thicker SiO_2 and thinner PNO-based devices. ESR measurements indicate that the P_{b0}/P_{b1} composed density of states in thicker SiO_2 systems approximately corresponds to the schematic sketches of Fig. 8.31a [38, 40, 41, 43–47]. The P_{b0}/P_{b1} effective density of states would be the summation of the P_{b0} and P_{b1} defect levels yielding, in comparison, a *relatively* flat density of states near the middle of the band gap as schematically shown in Fig. 8.31b. Densities of states qualitatively similar to this pattern are typically reported in measurements of Si/SiO_2 interface traps [67]. Our results strongly suggest that the PNO defect has a narrower density of states near the middle of the band gap (Fig. 8.31c). (The schematic sketch does not show the PNO defect trap levels extending into the dielectric for simplicity of presentation.) These measurements cannot distinguish between donor and amphoteric nature of the defects. However, the observation of a narrow density of states and the known shift in threshold voltage are both consistent with, but do not prove, that NBTI-induced defects in thin nitrided devices have a narrow donor level, as has been suggested recently [10].

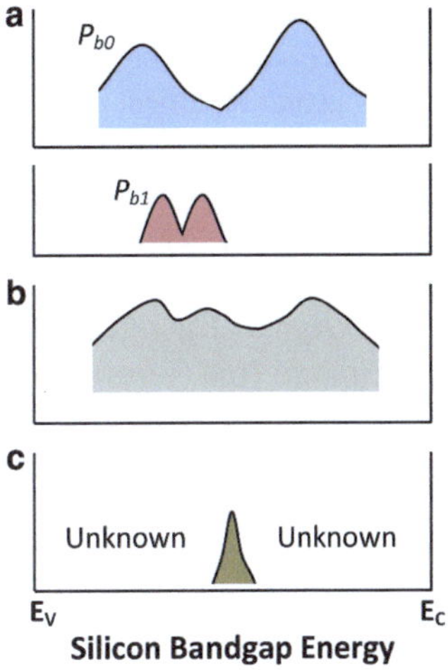

Fig. 8.31 Schematic representation of the (**a**) Si/SiO_2 P_{b0} and P_{b1} interface defect densities of states and (**b**) the effective P_{b0}/P_{b1} density of states illustrating the relatively flat distribution near the middle of the band gap. (**c**) Illustrates a schematic representation of the PNO defect's density of states near the middle of the band gap (Reproduced from [42])

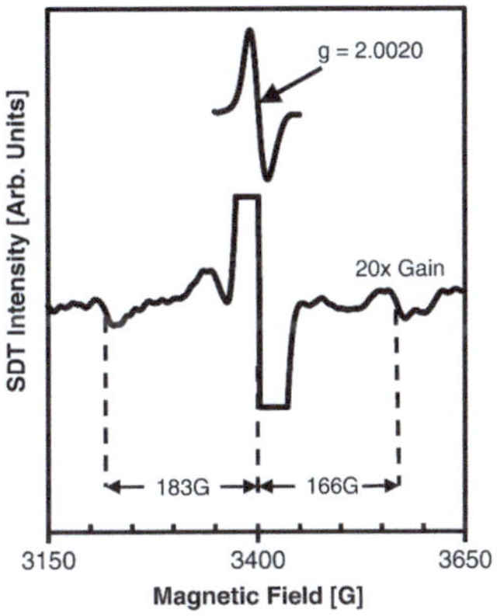

Fig. 8.32 Wide scan SDT spectra of the PNO defect. Increased spectrometer gain (×20) and extensive signal averaging reveal two ^{29}Si hyperfine side peaks (Reproduced from [17])

8.6.6 PNO Defect: ^{29}Si Hyperfine Identification (K_N Centers)

The high sensitivity observed in the SDT measurement provides the means to examine the local environment of the PNO defect via hyperfine interactions with nearby magnetic nuclei. Figure 8.32 illustrates the wide scan post-stress SDT spectra ($V_F = 0.0$ V) of the PNO defect with the magnetic field perpendicular to the (100) surface of the device. Extensive signal averaging plus the enhanced sensitivity of SDT has allowed for the observation of two weak satellite lines as well as a third small signal shifted ≈45 G below the central line. While the identity of the third smaller signal is not yet known, the observation of the two satellite side peaks and the $g = 2.0020$ central signal provides the necessary information to identify the physical and chemical nature of the PNO defect.

Both of the weak satellite lines have a peak-to-peak width of approximately 25 G and are virtually identical in amplitude. The low field line is 183 G below the central line, and the high field line is 166 G above the central line (349 G splitting). Each of the side peaks has an integrated intensity of ≈2% of the dominant central signal. The relative sizes of the central signal and side peaks *unequivocally identify the NBTI-stressed defect as a silicon dangling bond.* Most silicon nuclei (95.3%) are not magnetic; however, a small fraction (4.7% ^{29}Si) are magnetic and have a nuclear spin of ½ [25]. Thus, nearly all (95.3%) of the silicon dangling bond spectrum would appear as a single dominant signal while a small fraction (4.7%) of the silicon dangling bond spectrum would appear as two satellite side peaks. Each of these satellite side peaks would appear as ≈2.3% of the integrated intensity of the central signal. Within experimental error, this is exactly what is observed for the PNO defect. No other element could provide this 2.3%/95.3% pattern.

The small asymmetry in the hyperfine side peak splitting occurs when the electron-nuclear interaction is large [25]. The asymmetry can be calculated by the Breit-Rabi correction: $\delta \approx (\Delta H)^2/4H_0$, where δ is the asymmetry, ΔH is the hyperfine splitting, and H_0 is the resonant field without the nuclear interaction [25]. The observed hyperfine asymmetry (8.5 G) is consistent with a calculated Breit-Rabi correction ($\Delta H = 349$ G at $H_0 = 3400$ G) of $\delta \approx 9$ G.

The hyperfine interaction is generally expressed in a three by three matrix, A, often called the hyperfine "tensor" [25]. For a defect with axial symmetry, the parallel ($A_{||}$) and perpendicular ($A_{\perp}$) components of this "tensor" can be related to the defect's electronic wave function in terms of an isotropic (A_{iso}) and anisotropic (A_{aniso}) component. A_{iso} and A_{aniso} are measures of the s- and p-character of the defect's wave function, respectively [25]. Our observations allow for a moderately accurate measurement of A_{iso} and a very crude estimate of A_{aniso} [25, 68]. For the ideal case, A_{iso} and A_{aniso} are given [68] by

$$A_{iso} = \Delta H - \frac{1}{2}A_{aniso} \tag{8.8}$$

$$A_{aniso} = \frac{2}{3}\,(\text{hyperfine line width}) \tag{8.9}$$

The exact contributions of the anisotropic hyperfine coupling to the width of the hyperfine lines are difficult to determine because of additional broadening due to super-hyperfine interactions with second-nearest neighbor nitrogen atoms which each have a nuclear magnetic moment and a nuclear spin of 1 [25, 68]. However, a rough upper limit on the hyperfine line width is given by the measured peak-to-peak line width (25 G) while a rough lower limit is given by difference between the measured hyperfine line width and the central signal line width (25 G – 14 G = 11 G). This analysis yields a rather crude $A_{aniso} \approx (2/3)(11 \text{ to } 25\text{ G}) \approx 12 \pm 5$ G, and a more precise $A_{iso} = 349$ G - (12 G/2) = 343 G.

Table 8.1 Measured magnetic resonance parameters of important silicon dangling bond defects (Note that the *K* center and K_N center parameters are nearly the same) (Reproduced from [17])

	Centerline width (G)	g-value[a]	Hyperfine splitting (G)
P_{b0} [28]	$\cong 3$	2.0059	105
E'_{γ} [69]	$\cong 2$	2.0005	424
K center [68]	$\cong 14$	2.0028	358
NBTI defect (K_N center)	$\cong 14$	2.0020	349

[a]For the E', *K*, and K_N centers, this is the simple zero-crossing g-value, but for P_{b0}, it represents the *g* at a specific magnetic field orientation. In these measurements, H || ⟨100⟩ surface normal

The s-character and p-character of the observed defect can be roughly estimated by comparing the measured A_{iso} and A_{aniso} values with the A_{iso} and A_{aniso} values calculated for 100% 3s and 100% 3p silicon wave functions [25, 68]. The calculated A_{iso} for a 100% 3s silicon wave function and A_{aniso} for 100% 3p silicon wave function are 1639.3 G and 40.75 G, respectively [25]. The measured $A_{iso} = 343$ G indicates that the s-character of the NBTI-stressed defect is about 343 G/1639.3 G $\cong$ 21%, while the measured $A_{aniso} = 12 \pm 5$ G indicates that the p-character of the NBTI-stressed defect is of order 12 ± 5 G/40.75 G $\approx 29 \pm 12\%$. The sum of the s- and p-contributions to the defect's electronic wave function, or localization, is $50 \pm 12\%$. Note that this is a *very rough* estimate.

Table 8.1 compares the measured center line width, g-value, and hyperfine line splitting of the PNO defect with other silicon dangling bond defects (P_{b0}, E', and *K* centers) which might conceivably be present in these devices. As discussed above, P_{b0} defects are interfacial silicon dangling bond defects in which the central silicon is back-bonded to three other silicon atoms [15, 28, 37, 38, 40]. They, along with the closely related P_{b1} defects, account for most of the Si/SiO_2 interface states in SiO_2 MOSFETs [15, 28, 37, 38, 40]. E' centers are oxide silicon dangling bond defects in which the central silicon is back-bonded to oxygen atoms [69]. They dominate the electronic properties of irradiated and high-field stressed SiO_2 [15, 50, 69]. K centers are silicon dangling bond defects found in silicon nitride in which the central silicon is back-bonded to nitrogen atoms [68, 70]. They dominate charge trapping in Si_3N_4 films [68, 70, 71]. An examination of Table 8.1 reveals that the NBTI-stressed defect spectrum and the *K* center spectrum are very similar. Thus, we conclude that the PNO defect is a silicon dangling bond in which the silicon is back-bonded to nitrogen atoms that we refer to as K_N for NBTI.

The differences between the K_N and *K* center spectra are small but clearly larger than experimental error. These small deviations are likely due to slightly different bonding environments and probably the larger band gap in the PNO dielectric. In Si_3N_4, the *K* center's nitrogen atoms are bonded almost exclusively to silicon atoms [68, 70], while the SiON K_N center's nitrogen atoms may be bonded to one or more oxygen atoms. This small change in second-nearest neighbor bonding and possibly the larger band gap (the gap in Si_3N_4 is about 5 eV) are very likely responsible for the small deviations we observe between the *K* and K_N centers.

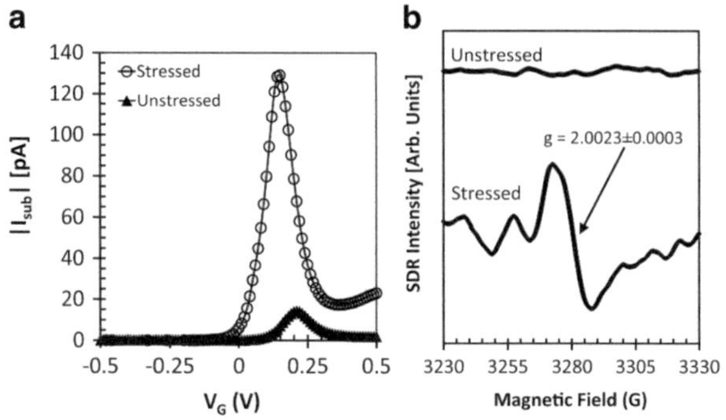

Fig. 8.33 Pre- and post-NBTS (**a**) DC-IV and (**b**) SDT measurement for a 2.3 nm PNO device subject to a quite moderate NBTS of -1.7 V at 140°C for 10^6 s with H || $\langle 100 \rangle$

8.6.7 K_N Center: Long-Term Stress

As is true with the majority of reported NBTI observations, the acceleration of the NBTI stress is a concern. Quite simply, one has to be sure that the acceleration of the NBTI stress to values which are compatible with near-term observations does not introduce different physical mechanisms which are absent at operation conditions. In an attempt to check the universality of and NBTI mechanism dominated by K_N center (at least for these PNO devices), we subject a 2.3 nm PNO device to a much less accelerated NBTS of -1.7 V at 140°C for 10^6 s. Note that this stress voltage corresponds to an dielectric field of only ≈6 MV/cm. Figure 8.33 illustrates the pre- and post-stress DC-IV (a) and pre- and post-stress SDT (b) measurements for a PNO device subject to this much more moderate stress. As can be clearly seen in Fig. 8.33, this moderate stress generates the same K_N center defect ($g = 2.0023 \pm 0.0003$) and that this defect is generated in lower densities, which is reflected in the lower signal to noise ratio. Thus, we conclude that the observation of K_N centers is consistent with an intrinsic NBTI mechanism and is not a product of accelerated stress (at least for these devices) [72].

8.6.8 Nitrogen-Enhanced NBTI Physics

The presence of the K_N center in PNO devices indicates that fundamental differences in the dielectric chemistry strongly influence the electronic properties of the interface and near-interface regions. The bonding mismatch between Si and SiO_2 is almost certainly the main reason that P_{b0}/P_{b1} defects exist. At a typical Si/SiO_2 interface, P_{b0}/P_{b1} defects are passivated with hydrogen atoms. The relatively easy dissociation of the hydrogen atoms from P_b precursors is an important factor in interface defect generation in NBTI [4, 10, 52, 53]. Our observations demonstrate the role of this mechanism in SiO_2-based devices [19–21]. However, the mechanism in the thinner PNO-based devices is different. The P_{b0}/P_{b1} defects were not observed post-stress. Our results indicate that the plasma-nitridation process must preferentially create large numbers of K_N center precursors. (It is possible that weaker P_{b0}/P_{b1} signals may be buried under the K_N signal.) The presence of large numbers of K_N precursors is almost certainly the main reason why nitridation so strongly enhances the NBTI response. Since it is known that K centers can be hydrogen passivated [73], one might speculate that NBTS liberates hydrogen from the K_N centers in much the same way that it is liberated from Si/SiO_2 P_{b0}/P_{b1} defects. However, our results suggest somewhat different physical mechanisms are involved.

8.6.9 NBTI-Induced Interface States Versus Bulk Traps

The observation of K_N centers in both SDR (participation in interfacial recombination) and SDT (participation in tunneling) demonstrates that K_N centers can act as both interface states and "near-interface" dielectric tunneling centers. Variations in the plasma-nitridation processing are thought to control the nitrogen profile within the gate dielectric [74]. We speculate that variations in the nitrogen profile (and the consequent proximity of the K_N centers to the Si/SiON interface) are a major source of conflicting conclusions regarding NBTI-induced interface states and fixed oxide charge generation. If the PNO processing creates $\mathrm{K_N}$ center precursors very near the interface, a post-NBTI electrical analysis would indicate interface state generation. But, if the PNO processing creates K_N center precursors in the "near-interface" or bulk dielectric regions, a post-NBTI electrical analysis would likely indicate some combination of interface state and bulk dielectric defect generation. So, it is possible for the same defect to be responsible for seemingly different aspects of the NBTI degradation. (In a 2.3 nm gate dielectric, it may be difficult to distinguish between "near-interface" and "bulk" dielectric defects.)

8.6.10 Model for NBTI in PNO Oxides

The SDR/SDT results obtained on PNO dielectrics are difficult or impossible to reconcile with reaction–diffusion models. One can envision a model somewhat like outlined by Campbell et al. for pure SiO_2 devices in which the NBTI process is triggered by the tunneling of a valence band hole to a dielectric defect, in this case a K center precursor instead of an E' center precursor [75, 76]. Such a model would make sense if the K center precursor involves a silicon-silicon bond in which one of the silicon atoms is back-bonded to nitrogen atoms. The K center precursor almost certainly has an energy level slightly below the silicon valence band edge because the recent EDMR results of Campbell et al. [42, 63] and Ryan et al. [32] directly demonstrate that the K centers have defects within the lower part of the silicon band gap. As discussed earlier, a significant negative bias results in the presence of holes in the silicon valence band; the holes allow tunneling events between near-interface K centers and the silicon to yield positively charged K center sites. Drawing an analogy with the well-understood E' defects (observed in pure SiO_2 devices), the K center site opens up with hole occupation [75, 76]. One side of the defect center is, after the hole capture event, a neutral silicon dangling bond; the other side of the defect center is a positively charged silicon. As in the E' case, a substantial structural relaxation occurs as a result of this process. It can readily be shown that the tunneling process leads to a power lawlike behavior which is, at least, plausibly consistent with the widely reported time dependence in NBTI [75, 76].

8.6.11 Summary

We have employed purely electrical DC-IV measurements as well as SDR and SDT magnetic resonance measurements on 7.5 nm SiO_2 and 2.3 nm PNO pMOSFETs. Our results indicate that the dominating NBTI-induced defects in the SiO_2 and PNO devices are different. NBTI in SiO_2 devices is dominated by P_{b0} and P_{b1} Si/SiO_2 interfacial silicon dangling bond defects. NBTI in the PNO devices is dominated by K_N centers. Our measurements show that the K_N centers are located in the "near-interface" region and they participate in both SDR and SDT phenomena. (The "near-interface" region could be a large fraction of a 2.3 nm oxide.) An examination of the K_N defects SDR versus V_F response strongly suggests that the K_N defects have a narrowly peaked density of states around the middle of the silicon band gap. The realization that K_N defects (not P_b centers) dominate NBTI in the PNO devices may help explain NBTI's enhancement in nitrided devices, as well as conflicting reports of NBTI-induced interface states and/or bulk traps.

8.7 Recovery

In this section, we utilize SDR and DC-IV measurements as a function of time to examine the role of these atomic-scale defects in NBTI recovery [2, 7–9, 20, 77]. Our SDR and DC-IV measurement approach is not fast enough to observe the entire NBTI recovery phenomenon which begins immediately after cessation of stress. However, the fact that recovery continues for at least 10^5 s after stress removal [9, 77] allows for a qualitative SDR/DC-IV examination of the atomic-scale defects involved in NBTI recovery. We present DC-IV and SDR measurements in pure SiO_2 and PNO devices which show that NBTI recovery results in a partial reduction in interface state density and a corresponding reduction in the density of the atomic-scale defects observed in SDR. In pure SiO_2 devices, we observe that the D_{it} recovery is accompanied by a partial reduction in P_{b0} defect density. In PNO devices, we observe that the D_{it} recovery is accompanied by a partial reduction in K_N center density.

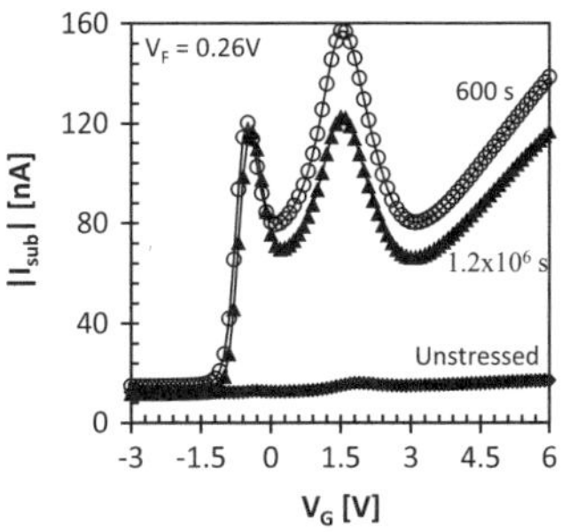

Fig. 8.34 DC-IV measurements taken pre- and post-NBTS (−25 V, 150°C, for 100,000 s). The 600 s curve denotes the DC-IV measurement taken 600 s after the temperature quench step. The 1.2×10^6 s curve denotes the DC-IV measurement taken 1.2×10^6 s after the temperature quench step

8.7.1 Experimental Details

Our NBTI recovery measurements involve pure SiO_2 and PNO devices. The SiO_2 devices are very large area ($\approx 1 \times 10^6$ μm^2) 48 nm SiO_2 power pMOSFETs. The PNO devices are large area ($\approx$41,000 μm^2) 2.3 nm PNO pMOSFETs. These SiO_2 and PNO devices are the same as those discussed in Sects. 8.4 and 8.6, respectively. In these recovery experiments, the devices were subject to a NBTS condition followed by a temperature quench step. The temperature quench reduces the temperature of the device to room temperature over the span of approximately 4 min while the gate bias is maintained. This step is thought to "lock in" the NBTI damage [77] rendering it observable in the DC-IV and SDR measurements. DC-IV and SDR measurements were taken pre-NBTS as well as a function of time post-NBTS/temperature quench. Recovery is observed as a reduction in the DC-IV-derived D_{it} and SDR signal amplitude (which is proportional to defect density). This approach allows for a crude investigation of the specific atomic-scale defects involved in NBTI recovery.

8.7.2 NBTI Recovery in 48 nm SiO_2 pMOSFETs

48 nm SiO_2 devices were subject to an ex situ NBTS of −25 V, 150°C for 100,000 s. Following the NBTS and temperature quench step, both SDR and DC-IV measurements were taken over a period of approximately 2 weeks. In all measurements, the source/drain to substrate junction is biased at +0.26 V. Figure 8.34 illustrates the DC-IV measurements for the unstressed device as well as measurements taken 600 s and 1.2×10^6 s post-NBTS/temperature quench. Since the NBTS/temperature quench was applied ex situ, 600 s elapsed before the device was loaded into the spectrometer, and the first DC-IV measurement was complete. Consequently, the

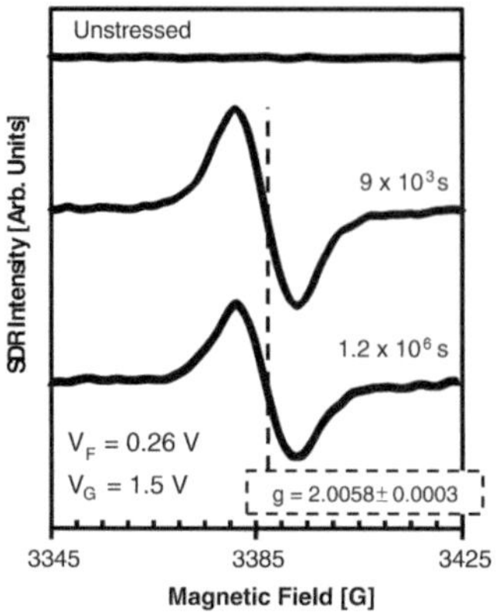

Fig. 8.35 SDR spectra taken both pre- and post-NBTS (−25 V, 150°C, for 100,000 s). After stress we note the generation of a strong $g = 2.0058 \pm 0.0003$ signal corresponding to P_{b0} defects. The 9×10^3 s curve denotes the SDR measurement completed after 9×10^3 s post-temperature quench. The 1.2×10^6 s curve denotes the SDR measurement completed 1.2×10^6 s after the temperature quench step. These spectra correspond to the $V_G = 1.5$ V DC-IV peak

600 s measurement was the first measurement taken post-NBTS/temperature quench (the 1.2×10^6 s measurement was the last measurement taken). As discussed in Sect. 8.4, the geometry of these devices (gate extension over lightly doped regions) results in two DC-IV peaks. The DC-IV peak at $V_G = -0.5$ V is associated with interface states located near the center of the channel while the DC-IV peak at $V_G = -1.5$ V is associated with interface states in the drift regions adjacent to the source and drain. For both gate voltages, it is clear that NBTS generates an increase in the peak substrate current which corresponds to an increase in D_{it}. The Fitzgerald and Grove analysis [11] leads to pre-NBTS $D_{it} = 6 \times 10^9\ \text{cm}^{-2}\ \text{eV}^{-1}$ and $D_{it} = 8 \times 10^9\ \text{cm}^{-2}\ \text{eV}^{-1}$ for the $V_G = -0.5$ V and $V_G = 1.5$ V peaks, respectively. Post-NBTS (600 s measurement), we observe an increase in peak substrate current which correspond to $D_{it} = 2 \times 10^{11}\ \text{cm}^{-2}\ \text{eV}^{-1}$ and $D_{it} = 3 \times 10^{11}\ \text{cm}^{-2}\ \text{eV}^{-1}$ for the $V_G = -0.5$ V and $V_G = 1.5$ V peaks, respectively. All the extracted D_{it} values assume $\sigma_s = 2 \times 10^{-16}\ \text{cm}^2$. After 1.2×10^6 s of recovery (approximately 2 weeks), the D_{it} corresponding to the $V_G = -0.5$ V peak exhibits little to no recovery, while the D_{it} corresponding to the $V_G = +1.5$ V peak recovers approximately 20%.

Figure 8.35 illustrates the corresponding SDR measurements for the unstressed device: the *first measurement completed* post-NBTS/temperature quench (9×10^3 s) and the last measurement (1.2×10^6 s). These SDR measurements were taken with the magnetic field aligned parallel to the $\langle 100 \rangle$ surface normal. The SDR measurements were taken with $V_G = 1.5$ V (not $V_G = -0.5$ V) because of the improvement in SDR sensitivity corresponding to this voltage (see Sect. 8.4). As expected, we observe that NBTS clearly generates a dominating signal at $g = 2.0058 \pm 0.0003$ which we attribute to P_{b0} defects. The P_{b0} signal amplitude (proportional to the number of defects) decreases by approximately 20% between the first and last measurements.

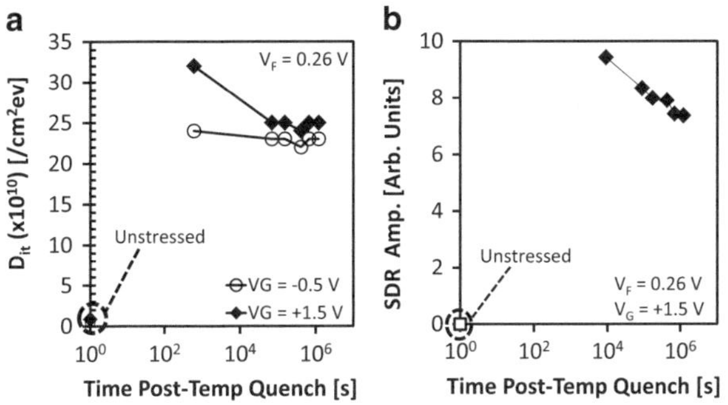

Fig. 8.36 (**a**) DC-IV-derived D_{it} values on the stressed devices as a function of time post-temperature quench. (**b**) SDR signal amplitude (scales with P_{b0} defect density) on the stressed devices as a function of time post-temperature quench

Figure 8.36 illustrates the (a) DC-IV-derived D_{it} and (b) SDR signal amplitudes as a function of time post-NBTS/temperature quench. We observe that the $V_G = -0.5$ V peak displays little or no D_{it} recovery, an observation which is still puzzling. We speculate that the geometry of the device is such that recovery occurs quicker near the center of the channel (perhaps faster than is captured by the first DC-IV measurement). We observe that the $V_G = +1.5$ V peak recovers approximately 20%. This reduction in interface state density is accompanied by a 20% recovery in P_{b0} defect density. This correlation between D_{it} and P_{b0} recovery strongly suggests a common origin.

8.7.3 *NBTI Recovery in 2.3 nm PNO pMOSFETs*

The relatively long stress/measurement delay indicative of the recovery measurements on the pure SiO_2 devices can be further reduced by performing the NBTS and temperature quench in situ (hot air is piped into the spectrometer's microwave cavity via an evacuated quartz dewar to provide the elevated temperature). This in situ stressing scheme is utilized to further investigate recovery in the PNO devices over a somewhat shorter time scale.

Figure 8.37 illustrates the DC-IV characteristic curves taken pre- and post-NBTS (−2.8 V 140°C for 10,000 s) as well as during the recovery process ($V_F = +0.33$ V). The NBTS generates a large increase in D_{it} (approximately an order of magnitude). A subsequent DC-IV measurement after ≈5 h of SDR measurement shows a reduction in D_{it} associated with recovery. In an attempt to induce further recovery, the device was then subjected to a 24-h bake (150°C with 0 V on the gate electrode). A DC-IV measurement post-bake shows only a modest amount of recovery. It is possible that the lack of significant post-bake recovery is due to the temperature

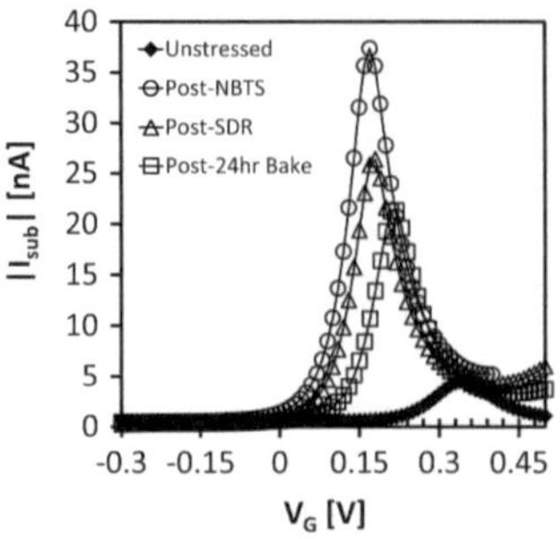

Fig. 8.37 DC-IV measurements of 2.3 nm PNO pMOSFET pre- and post-NBTS (−2.8 V at 140°C for 10,000 s) as well as post-SDR measurement and post-24 h bake (150°C with 0 V on the gate electrode). This indicates that D_{it} exhibits a recovery process post-NBTS ($V_F = +\ 0.33$ V) (Reproduced from [20])

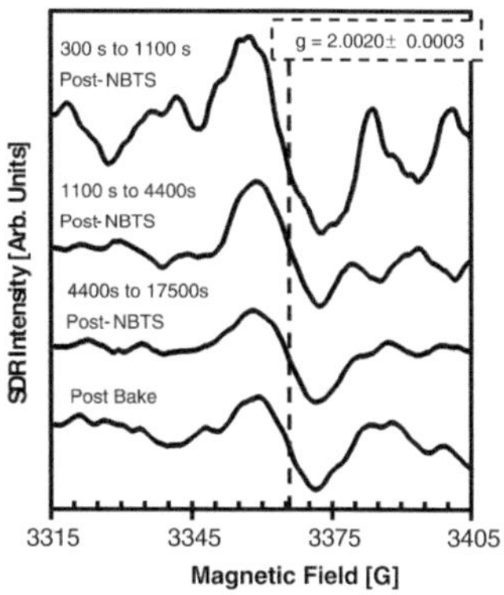

Fig. 8.38 SDR traces of 2.3 nm PNO pMOSFET post-NBTS (−2.8 V at 140°C for 10,000 s) at various time steps which shows that the K_N center density exhibits NBTI recovery (signal size scales with the density of defects) (Reproduced from [20])

quench post-NBTS. It is also possible that the lack of post-bake recovery is a product of the modest bake [78, 79]. Regardless, the NBTI-induced damage in this case is, to some degree, "locked-in" during the temperature quench and is relatively unaffected by additional high temperature treatments [77]. Figure 8.38 illustrates the corresponding SDR measurements ($V_F = 0.33$ V) taken with the magnetic field aligned parallel to the $\langle 100 \rangle$ direction at various post-NBTS time steps. The NBTS generates a signal at $g = 2.0020 \pm 0.0003$ which corresponds to a K_N center [17]. The amplitude of the K_N center signal (which is proportional to the number of defects) decreases with time post-NBTS. Figure 8.39 summarizes the D_{it} and SDR signal intensity recovery. Since the observed K_N center signal shows a recovery behavior that qualitatively scales with the observed D_{it} degradation, these results are a strong indication that this defect is involved in the NBTI recovery process in these devices.

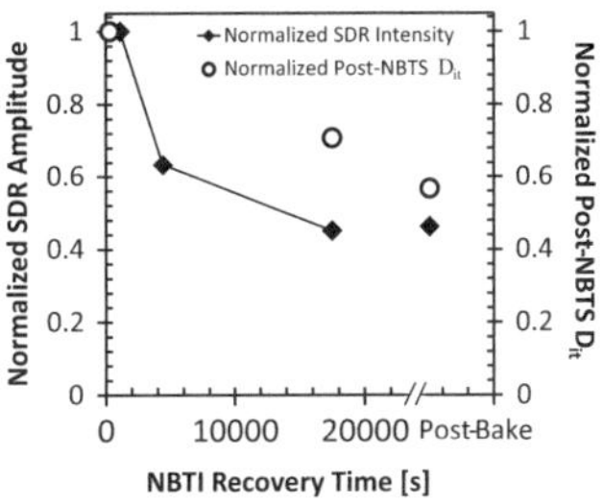

Fig. 8.39 Normalized SDR signal amplitude and normalized post-NBTS D_{it} as a function of NBTI recovery time illustrating the NBTI recovery process qualitatively tracks with the density of the K_N centers (Reproduced from [20])

8.7.4 Summary

We report DC-IV and SDR measurements taken as a function of time during NBTI recovery that allow for a qualitative observation of the atomic-scale defects involved in NBTI recovery in SiO_2 and PNO devices. In 48 nm SiO_2 devices, the DC-IV recovery in D_{it} is accompanied by a recovery in P_{b0} defect density. In PNO devices, the DC-IV recovery in D_{it} is accompanied by a recovery in K_N center defect density. These observations suggest an NBTI recovery mechanism which involves a partial P_{b0} and K_N repassivation in SiO_2 and PNO devices, respectively.

8.8 Fluorine's Impact on NBTI

This section discusses the atomic-scale defect implications to NBTI in devices which have been subject to fluorine incorporation [80, 81]. In thicker SiO_2 devices, fluorine incorporation is thought to "toughen up" the interface by replacing the weaker hydrogen passivation of P_b centers (Si–H) with a stronger fluorine passivation (Si–F) [82]. The stronger passivation is thought to be the reason that fluorine incorporation reduces NBTI in thicker SiO_2 devices [83, 84]. In this section, we investigate a series of three thicker (7.5 nm) SiO_2 pMOSFETs ($\approx$40,000 μm^2) which have undergone a range of gate dielectric fluorine treatments. We compare these results to that of the pure SiO_2 devices which have not been subject to fluorine exposure. SDR and DC-IV measurements were made before and after identical NBTS sequences ($V_g = -5.7$ V at 140°C for 250,000 s). Following NBTS, all devices were subjected to the same temperature quench step discussed earlier.

Figure 8.40 illustrates the pre- and post-NBTS DC-IV characteristics for the pure SiO_2 device case (a) as well as a representative measurement set from the fluorinated device case (b). Note that NBTS in the fluorinated devices all resulted in similar DC-IV characteristics. Following the Fitzgerald and Grove analysis [11] (with $\sigma_s = 2 \times 10^{-16}$ cm^2), the pre- and post-NBTS D_{it} for the pure SiO_2 device (a) were 7×10^9 cm^{-2} eV^{-1} and 5×10^{11} cm^{-2} eV^{-1}, respectively. For the fluorinated devices (b), the pre- and post-NBTS D_{it} were $\approx 1 \times 10^{10}$ cm^{-2} eV^{-1} and 1×10^{11} cm^{-2} eV^{-1}, respectively. The reduction in post-NBTS D_{it} (compared to the nearly identical pure SiO_2 device) was observed in all three sets of fluorinated

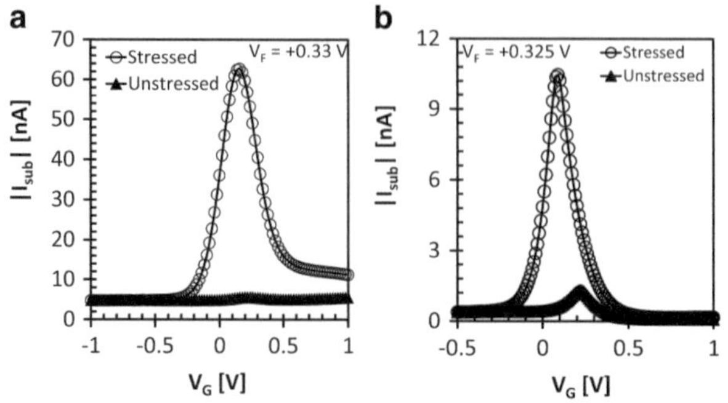

Fig. 8.40 Comparison of the pre- and post-stress (5.7 V at 140°C for 250,000 s) DC-IV measurements for the pure SiO_2 pMOSFETs (**a**) and a representative DC-IV measurement set for pMOSFETs which were subject to a fluorination treatment (**b**) (Reproduced from [81])

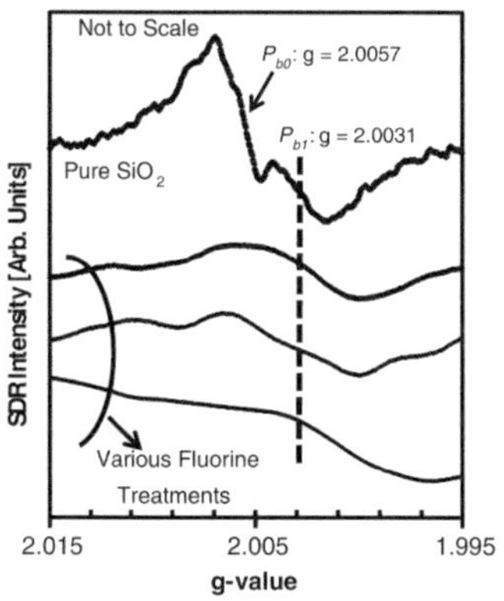

Fig. 8.41 Post-NBTS (−5.7 V at 140°C for 250,000 s) SDR measurements on 7.5 nm pMOSFETS. The top trace is from the pure SiO_2 device while the bottom three traces are from devices which have all received various fluorine treatments. The fluorinated traces are scaled and offset to highlight the common P_{b1} signal. Their respective amplitudes are not to scale (Reproduced from [81])

SiO_2 devices. Again, this is consistent with other reports indicating a positive NBTI effect in gate stack which are subject to fluorination [83, 84]. Figure 8.41 represents a comparison of the post-NBTS SDR measurements taken on the pure SiO_2 device of Fig. 8.6 as well as the three fluorinated devices. In all these traces, the magnetic field is parallel to the [100] Si/SiO_2 interface normal. Pre-NBTS defect spectra were all below the detection limit. As discussed earlier in Sect. 8.4, NBTS in the pure SiO_2 device results in overlapping P_{b0} and P_{b1} interface state signals ($g = 2.0057 \pm 0.0003$ and $g = 2.0031 \pm 0.0003$, respectively). However, in all the fluorinated devices, we observe the NBTI-induced generation of a weaker single broad line with g-value ranging from 2.0026 ± 0.0003 to 2.0033 ± 0.0003. Note that the absorption axis is not to scale for this figure since the traces have been offset and scaled to allow for a better comparison of the P_{b1} signals. In general, the g-values for the fluorinated devices are roughly consistent with a P_{b1} center. Note also the somewhat broader than expected width of the signal may indicate the presence of nearby fluorine nuclei [25].

Although the fluorinated devices of Fig. 8.41 are very similar structurally and were stressed identically to the pure SiO_2 device, their NBTI response is very different. In the fluorinated devices, there is no indication of P_{b0} center generation. This observation suggests that the incorporation of fluorine can selectively passivate P_{b0} precursors. This observation might help to explain the diminished interface state generation observed (electrically) in other recent reports of NBTI-stressed fluorinated devices [83, 84]. A review of Figs. 8.9 and 8.31 indicates that the P_{b0} and P_{b1} defects have different densities of states, and the preferential generation of P_{b1} defects will likely result in a larger threshold voltage shift in proportion to the total number of P_{b1} states. That is, a higher percentage of P_{b1} centers will likely be positively charged when the pMOSFET transistor is on. This larger effect per defect is, of course, more than compensated by the smaller total number of P_{b1} centers. Also, as has been noted in some of the earlier fluorine literature, these results make sense in terms of the relative strengths of Si–H and Si–F bond energies [85]. Considering these results (Figs. 8.40 and 8.41), it is clear that, in pure SiO_2 devices, NBTI hardening can be achieved by striving to make devices with a dominant P_{b0} interface defect distribution and then fluorination. However, as was mentioned earlier in Sect. 8.4, the preferential generation of either P_{b0} or P_{b1} defects is a delicate strain relief response which is currently very poorly understood.

Also note that our K_N defect observations in PNO devices may also help explain somewhat puzzling observations regarding fluorine's impact on NBTI. Fluorination of nitride devices has been observed to have little or no effect on NBTI [86]. The realization that NBTI in PNO devices creates K_N defects (not primarily interfacial P_b centers) may help explain this. The introduction of fluorine in nitrided devices probably does replace some interfacial Si–H bonds with Si-F bonds. However, interfacial Si–H depassivation does not dominate NBTI in these devices. If the fluorine incorporation in nitrided devices does not passivate the dominating NBTI-induced defects, it will be ineffective at suppressing NBTI.

8.9 NBTI in HfO_2-Based pMOSFETs

This section discusses SDR measurements utilized to directly observe the atomic-scale defects of NBTI in fully processed 1.2 nm EOT HfO_2-based pMOSFETs [87]. We compare the SDR measurements with DC-IV observations of NBTI-induced changes in interface state densities. The devices were subject to short-term room temperature NBTS as well as a longer term elevated temperature NBTS. Post-stress, we observe an increase in interface state density and the corresponding generation of different atomic-scale defect structures in the two differently stressed devices.

Figure 8.42 illustrates the DC-IV measurements for the pre-stress ($D_{it} \approx 4 \times 10^9$ cm^{-2} eV^{-1}), post-short-term stress (−2.0 V, 25°C, for 5 s) ($D_{it} \approx$

Fig. 8.42 A comparison of DC-IV on HfO_2-based pMOSFETs for an unstressed device, a device subjected to a (−2.0 V/25°C/5 s) NBTS, and a device subjected to a (−1.8 V/140°C/10,000 s) NBTS (Reproduced from [87])

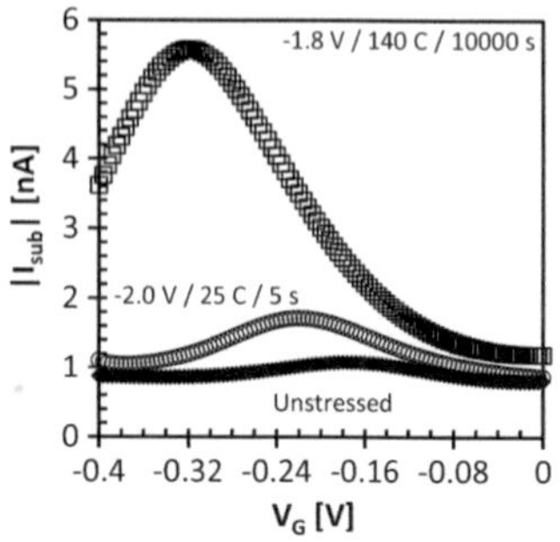

Fig. 8.43 SDR spectra before and after the (−2.0 V/25°C/5 s) NBTS (Reproduced from [87])

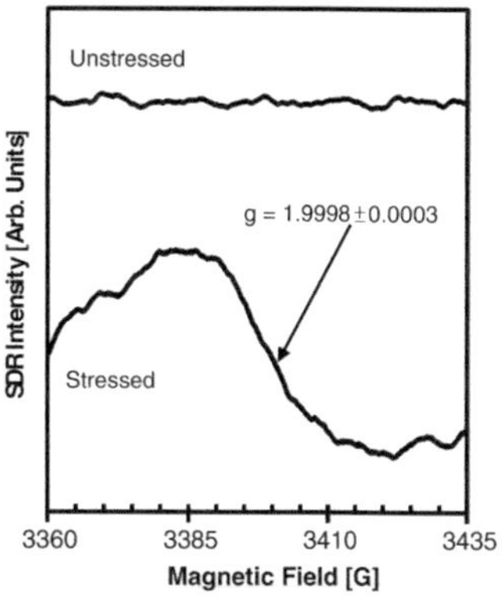

1.6×10^{10} cm^{-2} eV^{-1}), and post-long-term stress (−1.8 V, 140°C, 10,000 s) ($D_{it} \approx 8.2 \times 10^{10}$ cm^{-2} eV^{-1}) cases. We were forced to stress at a slightly lower voltage of −1.8 V because, at an elevated temperature, stressing for more than a few seconds tended to induce dielectric breakdown. The voltage shift in the peaks is likely due to buildup of charge in the dielectric as a result of stressing.

Figure 8.43 illustrates SDR traces before and after the short-term room temperature stress. The short stress creates a very broad SDR line (peak-to-peak width about 35 G) with g = 1.9998 ± 0.0003. The fairly large shift in g from the free-electron value ($g_e = 2.00232$) in the short-term stress case (g = 1.9998) as well as the large line width suggests that this brief stress quickly activates a type of defect specific to the Hf-based dielectric stack (It is unlike any signal reported in conventional SiO_2 devices [15].) Figure 8.44 illustrates the SDR traces before and after the longer term elevated temperature stress. The longer term SDR result is a significantly narrower line of about 23 G with a somewhat higher g-value (g = 2.0026 ± 0.0003). This is clearly different than the SDR spectrum observed in the short-term stress and is still of unknown origin. The line widths and g-values indicate that both the short- and long-term stress-induced "interface" trapping defects are almost certainly *not* located precisely at the Si/dielectric boundary. If this were to be the case, the spectra would presumably resemble those of P_b or K_N centers [15] The width of the

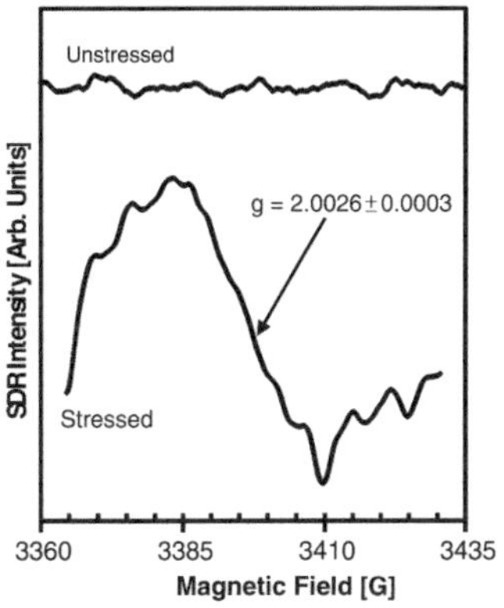

Fig. 8.44 SDR spectra before and after the (−1.8 V/140°C/10,000 s) NBTS. Note that the long-term high temperature stressing creates a significantly different SDR response than the brief room temperature stressing (Reproduced from [87])

lines, in both cases, probably indicates that the defects involve some interaction with hafnium atoms. Because Hf atoms have outer shell d-orbital electrons and relatively large nuclei, large spin–orbit coupling effects would tend to broaden the SDR spectrum [88]. If this is the case, the result indicates that there is at least limited diffusion of some hafnium atoms across the interfacial layer close to the Si/dielectric boundary.

8.10 Perspectives and Concluding Remarks

At this point in the chapter, we are hopeful that the reader has a new found appreciation for magnetic resonance measurements and their power to help understand the fundamentals of device reliability. However, it is entirely reasonable that the reader is left wondering how all these spectra (graphs of wiggly lines) mesh with the arguments typically used in an NBTI discussion (i.e., power law time exponent of V_{th} degradation, temperature acceleration, field acceleration, etc.). However, armed with this atomic-level NBTI defect understanding, we are hopeful that the reader is better equipped to draw his or her own conclusions about NBTI models.

For example, the NBTI literature is filled with works either advocating or opposing an NBTI description which involves the reaction–diffusion model [4, 52, 53]. The reaction–diffusion model framework does lead to a description of NBTI degradation which is similar to experimental observations. Also, the central idea of a Si–H depassivation which forms silicon dangling bond defects at the Si/SiO_2 interface is quite consistent with our P_{b0}/P_{b1} observations in the pure SiO_2 devices. However, it is not clear how our observations of NBTI-induced E' centers could fit into this paradigm. More importantly, it is perhaps less surprising to the reader why a model which does not involve E' centers struggles to describe recovery. (Reversing the Si–H depassivation process simply does not capture the necessary physics to align with observations [8].) Thus, at least for NBTI in pure SiO_2-based devices, a model similar to the earlier suggestions of Campbell and Lenahan [19, 21, 42,

55, 56] and later model development of Grasser et al. [60] should be seriously considered.

In this chapter, we reviewed quite clear and unambiguous evidence demonstrating that NBTI in PNO-based devices involves the generation of defects different from those utilized in the vast majority of NBTI literature models (reaction–diffusion variants or even an E' catalyzing two-stage model). As discussed in great length in Sect. 8.6, the necessary K_N center details required to build a physically correct NBTI model for nitrided devices have all been firmly explored.

Recently, a semiquantitative model which incorporates the K_N center physics derived from the magnetic resonance measurements discussed herein was proposed for NBTI in nitrided devices [75, 76]. While the universality of this model is still open for evaluation, it is important to note that this model (built using physics completely different from reaction–diffusion) predicts ΔV_{th} degradation trends quite similar to those from the reaction–diffusion framework. The similarities between these two predictions illustrate that multiple models based upon significantly different physical phenomena can yield results generally consistent with ΔV_{th} measurements over some time frame. It is the hope of the authors that their results contained herein will deepen the reliability of community's understanding of the underlying physical mechanisms responsible for NBTI and contribute to the construction of predictive models more consistent with underlying physical mechanisms.

Acknowledgements We gratefully acknowledge the contributions to this work by Dr. Jason T. Ryan, Dr. Corey J. Cochrane, Dr. Anand T. Krishnan, and Dr. Srikanth Krishnan.

References

1. D.K. Schroder, Microelectron. Rel. **47**(6) 841 (2007).
2. M. Ershov, S. Saxena, H. Karbasi, S. Winters, S. Minehane, J. Babcock, R. Lindley, P. Clifton, M. Redford, and A. Shibkov, Appl. Phys. Lett. **83**(8) 1647 (2003).
3. D.K. Schroder and J.A. Babcock, J. Appl. Phys. **94**(1) 1 (2003).
4. M.A. Alam and S. Mahapatra, Microelectron. Rel. **45**(1) 71 (2005).
5. B.E. Deal, M. Sklar, A.S. Grove, and E.H. Snow, J. Electrochem. Soc. **114**(3) 266 (1967).
6. S. Mahapatra, K. Ahmed, D. Varghese, A.E. Islam, G. Gupta, L. Madhav, D. Saha, and M.A. Alam IEEE Int. Reliability Phys. Symp., 1 (2007).
7. T. Grasser, B. Kaczer, P. Hehenberger, W. Goes, R. O'Connor, H. Reisinger, W. Gustin, and C. Schlunder IEEE Int. Electron Devices Meet., 801 (2007).
8. H. Reisinger, U. Brunner, W. Heinrigs, W. Gustin, and C. Schlunder, IEEE Trans. Device Mater. Reliab. **7**(4) 531 (2007).
9. T. Grasser, W. Gos, V. Sverdlov, and B. Kaczer IEEE Int. Reliability Phys. Symp., 268 (2007).
10. A.T. Krishnan, C. Chancellor, S. Chakravarthi, P.E. Nicollian, V. Reddy, A. Varghese, R.B. Khamankar, and S. Krishnan IEEE Int. Electron Devices Meet., 705 (2005).
11. D.J. Fitzgerald and A.S. Grove, Surf. Sci. **9** 347 (1968).
12. A. Neugroschel, C.T. Sah, K.M. Han, M.S. Carroll, T. Nishida, J.T. Kavalieros, and Y. Lu, IEEE Trans. Electron Dev. **42**(9) 1657 (1995).
13. M.A. Jupina and P.M. Lenahan, IEEE Trans. Nuc. Sci. **36**(6) 1800 (1989).
14. A.E. Islam, G. Gupta, S. Mahapatra, A.T. Krishnan, K. Ahmed, F. Nouri, A. Oates, and M.A. Alam IEEE Int. Electron Devices Meet., 329 (2006).

15. P.M. Lenahan and J.F. Conley, J. Vac. Sci. Technol. B. **16**(4) 2134 (1998).
16. S. Fujieda, Y. Miura, M. Saitoh, E. Hasegawa, S. Koyama, and K. Ando, Appl. Phys. Lett. **82**(21) 3677 (2003).
17. J.P. Campbell, P.M. Lenahan, A.T. Krishnan, and S. Krishnan, Appl. Phys. Lett. **91** 133507 (2007).
18. J.P. Campbell, P.M. Lenahan, A.T. Krishnan, and S. Krishnan IEEE Int. Reliability Phys. Symp., 503 (2007).
19. J.P. Campbell, P.M. Lenahan, A.T. Krishnan, and S. Krishnan, IEEE Trans. Dev. and Mat. Rel. **6**(2) 117 (2006).
20. J.P. Campbell, P.M. Lenahan, A.T. Krishnan, and S. Krishnan IEEE Int. Reliability Phys. Symp., 442 (2006).
21. J.P. Campbell, P.M. Lenahan, A.T. Krishnan, and S. Krishnan, Appl. Phys. Lett. **87**(20) 204106 (2005).
22. A.S. Grove and D.J. Fitzgerald, Solid State Electron. **9** 783 (1966).
23. R.N. Hall, Phys. Rev. **87**(2) 387 (1952).
24. W. Shockley and W.T. Read, Phys. Rev. **87**(5) 835 (1952).
25. J.A. Weil, J.R. Bolton, and J.E. Wertz, *Electron Paramagnetic Resonance: Elementary Theory and Practical Applications* (John Wiley & Sons, New York, NY, 1994).
26. J.E. Wertz and J.R. Bolton, *Electron Spin Resonance: Elementary Theory and Practical Applications* (McGraw-Hill, New York, 1972).
27. D.J. Lepine, Phys. Rev. B. **6**(2) 436 (1972).
28. J.W. Gabrys, P.M. Lenahan, and W. Weber, Microelectron. Eng. **22**(1–4) 273 (1993).
29. D. Kaplan, I. Solomon, and N.F. Mott, Journal De Physique Lettres **39**(4) L51 (1978).
30. P.M. Lenahan and M.A. Jupina, Colloids and Surfaces **45** 191 (1990).
31. Y. Miura and S. Fujieda, Jpn. J. Appl. Phys., Part 1 **40**(4B) 2840 (2001).
32. J.T. Ryan, P.M. Lenahan, A.T. Krishnan, and S. Krishnan, J. Appl. Phys. **108**(6) 2010).
33. J.H. Stathis, Appl. Phys. Lett. **68**(12) 1669 (1996).
34. J.P. Campbell, P.M. Lenahan, A.T. Krishnan, and S. Krishnan IEEE Int. Integrated Reliability Workshop, 118 (2004).
35. J.P. Campbell, P.M. Lenahan, A.T. Krishnan, and S. Krishnan IEEE Int. Integrated Reliability Workshop, 1 (2005).
36. T. Aichinger, M. Nelhiebel, and T. Grasser, Microelectronics Reliability **48** 1178 (2008).
37. K.L. Brower, Zeitschrift Fur Physikalische Chemie Neue Folge **151** 177 (1987).
38. T.D. Mishima and P.M. Lenahan, IEEE Trans. Nuc. Sci. **47**(6) 2249 (2000).
39. T.D. Mishima, P.M. Lenahan, and W. Weber, Appl. Phys. Lett. **76**(25) 3771 (2000).
40. E.H. Poindexter, G.J. Gerardi, M.E. Rueckel, P.J. Caplan, N.M. Johnson, and D.K. Biegelsen, J. Appl. Phys. **56**(10) 2844 (1984).
41. A. Stesmans, B. Nouwen, and V.V. Afanas'ev, Phys. Rev. B. **58**(23) 15801 (1998).
42. J.P. Campbell, P.M. Lenahan, C.J. Cochrane, A.T. Krishnan, and S. Krishnan, IEEE Trans. Dev. and Mat. Rel. **7**(4) 540 (2007).
43. J.P. Campbell and P.M. Lenahan, Appl. Phys. Lett. **80**(11) 1945 (2002).
44. G.J. Gerardi, E.H. Poindexter, P.J. Caplan, and N.M. Johnson, Appl. Phys. Lett. **49**(6) 348 (1986).
45. P.M. Lenahan and P.V. Dressendorfer, J. Appl. Phys. **54**(3) 1457 (1983).
46. P.M. Lenahan and P.V. Dressendorfer, J. Appl. Phys. **55**(10) 3495 (1984).
47. P.M. Lenahan and P.V. Dressendorfer, Appl. Phys. Lett. **41**(6) 542 (1982).
48. D. Varghese, S. Mahapatra, and M.A. Alam, IEEE Electron Dev. Lett. **26**(8) 572 (2005).
49. E.P. O'Reilly and J. Robertson, Phys. Rev. B. **27**(6) 3780 (1983).
50. J.F. Conley, P.M. Lenahan, H.L. Evans, R.K. Lowry, and T.J. Morthorst, J. Appl. Phys. **76**(5) 2872 (1994).
51. S. Chakravarthi, A.T. Krishnan, V. Reddy, C.F. Machala, and S. Krishnan IEEE Int. Reliability Phys. Symp., 273 (2004).
52. K.O. Jeppson and C.M. Svensson, J. Appl. Phys. **48**(5) 2004 (1977).
53. S. Ogawa and N. Shiono, Phys. Rev. B. **51**(7) 4218 (1995).

54. B.B. Jie, M.F. Li, C.L. Lou, W.K. Chim, D.S.H. Chan, and K.F. Lo, IEEE Electron Dev. Lett. **18**(12) 583 (1997).
55. P.M. Lenahan, Microelectron. Eng. **69**(2–4) 173 (2003).
56. P.M. Lenahan, Microelectron. Rel. **47**(6) 890 (2007).
57. J.F. Conley, P.M. Lenahan, B.D. Wallace, and P. Cole, IEEE Trans. Nuc. Sci. **44**(6) 1804 (1997).
58. P.M. Lenahan and J.F. Conley, Appl. Phys. Lett. **71**(21) 3126 (1997).
59. P.M. Lenahan and J.F. Conley, IEEE Trans. Nuc. Sci. **45**(6) 2413 (1998).
60. T. Grasser, B. Kaczer, W. Goes, T. Aichinger, P. Hehenberger, and M. Nelhiebel IEEE Int. Reliability Phys. Symp., 33 (2009).
61. T. Grasser, B. Kaczer, W. Goes, H. Reisinger, T. Aichinger, P. Hehenberger, P.-J. Wagner, F. Schanovsky, J. Franco, M. Toledano-Luque, and M. Nelhiebel, IEEE Trans. Electron. Dev. **58**(11) 3652 (2011).
62. J.T. Ryan, P.M. Lenahan, T. Grasser, and H. Enichlmair, Appl. Phys. Lett. **96**(22) 2010).
63. J.P. Campbell, P.M. Lenahan, A.T. Krishnan, and S. Krishnan, J. Appl. Phys. **103**(4) 2008).
64. Y.Y. Kim and P.M. Lenahan, J. Appl. Phys. **64**(7) 3551 (1988).
65. P.E. Nicollian, M. Rodder, D.T. Grider, P. Chen, R.M. Wallace, and S.V. Hattangady IEEE Int. Reliability Phys. Symp., 400 (1999).
66. J.H. Stathis, G. LaRosa, and A. Chou IEEE Int. Reliability Phys. Symp., 1 (2004).
67. E.H. Nicollian and J.R. Brews, *MOS (Metal Oxide Semiconductor) Physics and Technology* (John Wiley & Sons, New York, 1982).
68. P.M. Lenahan and S.E. Curry, Appl. Phys. Lett. **56**(2) 157 (1990).
69. D.L. Griscom, E.J. Friebele, and G.H. Sigel, Solid State Commun. **15**(3) 479 (1974).
70. W.L. Warren and P.M. Lenahan, Phys. Rev. B. **42**(3) 1773 (1990).
71. D.T. Krick, P.M. Lenahan, and J. Kanicki, J. Appl. Phys. **64**(7) 3558 (1988).
72. J.P. Campbell, P.M. Lenahan, A.T. Krishnan, and S. Krishnan IEEE Int. Integrated Reliability Workshop, 12 (2007).
73. V.J. Kapoor, R.S. Bailey, and H.J. Stein, J. Vac. Sci. Technol. A. **1**(2) 600 (1983).
74. P.A. Kraus, K. Ahmed, T.C. Chua, M. Ershov, H. Karbasi, C.S. Olsen, F. Nouri, J. Holland, R. Zhao, G. Miner, and A. Lepert Symp. on VLSI Technol., 143 (2003).
75. P.M. Lenahan IEEE Int. Integrated Reliability Workshop, 90 (2009).
76. P.M. Lenahan, J.P. Campbell, A.T. Krishnan, and S. Krishnan, IEEE Trans. Device Mater. Reliab. **11**(2) 219 (2011).
77. S. Rangan, N. Mielke, and E.C.C. Yeh IEEE Int. Electron Devices Meet., 341 (2003).
78. C. Benard and J.-L. Ogier, IEEE Int. Integrated Reliability Workshop, 7 (2008).
79. A.A. Katsetos, Microelectronics Reliability **48** 1655 (2008)
80. J.T. Ryan, P.M. Lenahan, A.T. Krishnan, S. Krishnan, and J.P. Campbell IEEE Int. Integrated Reliability Workshop, 137 (2008).
81. J.T. Ryan, P.M. Lenahan, A.T. Krishnan, S. Krishnan, and J.P. Campbell IEEE Int. Reliability Phys. Symp., 988 (2009).
82. P.J. Wright and K.C. Saraswat, IEEE Trans. Electron Dev. **36**(5) 879 (1989).
83. T.B. Hook, E. Adler, F. Guarin, J. Lukaitis, N. Rovedo, and K. Schruefer, IEEE Trans. Electron Dev. **48**(7) 1346 (2001).
84. C.H. Liu, M.T. Lee, C.-Y. Lin, J. Chen, K. Schruefer, J. Brighten, N. Rovedo, T.B. Hook, M.V. Khare, S.-F. Huang, C. Wann, T.-C. Chen, and T.H. Ning IEEE Int. Electron Devices Meet., 861 (2001).
85. L. Pauling, *Appendix VIII of General Chemistry* (Dover, New York, 1998).
86. Y. Mitani, M. Nagamine, H. Satake, and A. Toriumi IEEE Int. Electron Devices Meet., 509 (2002).
87. C.J. Cochrane, P.M. Lenahan, J.P. Campbell, G. Bersuker, and A. Neugroschel, Appl. Phys. Lett. **90**(12) 2007).
88. G. Bersuker, C.S. Park, J. Barnett, P.S. Lysaght, P.D. Kirsch, C.D. Young, R. Choi, B.H. Lee, B. Foran, K. van Benthem, S.J. Pennycook, P.M. Lenahan, and J.T. Ryan, J. Appl. Phys. **100**(9) 2006).

Chapter 9
Charge Properties of Paramagnetic Defects in Semiconductor/Oxide Structures

V.V. Afanas'ev, M. Houssa, and A. Stesmans

Abstract This chapter overviews charge properties of oxide and interface paramagnetic defects in semiconductor/oxide entities, primarily Si/SiO_2, as revealed by correlative analysis of electron spin resonance data and electrical analysis. The role of dangling bond defects in electrical degradation phenomena induced by bias-temperature stress or irradiation is discussed. In particular, the importance of hydrogen/proton-mediated interactions is underlined on the basis of the available experimental evidence.

9.1 Introduction

Stress-induced degradation of metal-oxide-semiconductor (MOS) devices is in many cases associated with generation of additional charges at the semiconductor/insulator interface or inside the insulating oxide itself. Usually the charges observed in MOS structures are classified as being of either ionic or electronic origin with the latter being further separated into the interface-trapped charge and the oxide-trapped charge [1] to account for contributions of specific interface-related or oxide imperfections, respectively. These imperfections, e.g., intrinsic defects or impurity centers, may become charged under a broad variety of stress conditions which include not only the negative bias-temperature (NBT) stress but also hot-carrier injection or Fowler–Nordheim tunneling stress, irradiation with high-energy photons, and ion bombardment. As a result, *identification* of charge trapped entities has remained in the focus of research for decades as it represents the basic element in *understanding* the physics of the stress- or irradiation-induced device degradation.

V.V. Afanas'ev (✉) • M. Houssa • A. Stesmans
Laboratory of Semiconductor Physics, Department of Physics and Astronomy, University of Leuven, Celestijnenlaan 200 D, 3001 Leuven, Belgium
e-mail: Valeri.Afanasiev@fys.kuleuven.be; Michel.Houssa@fys.kuleuven.be; Andre.Stesmans@fys.kuleuven.be

T. Grasser (ed.), *Bias Temperature Instability for Devices and Circuits*, DOI 10.1007/978-1-4614-7909-3_9, © Springer Science+Business Media New York 2014

Furthermore, besides identification, proper quantification of the traps is required to provide technology with a meaningful feedback, which would allow for steering trap removal aimed at better device reliability.

The goal of the present Chapter is to overview charge trapping properties of defects commonly encountered in semiconductor/oxide structures, primarily in Si and SiO_2-based systems, as revealed by quantitative correlations between trap densities inferred from electrical measurements and densities of paramagnetic defects detected in electron spin resonance (ESR) experiments.

9.2 Methodology of Correlative Analysis

Thanks to its high sensitivity, the ESR spectroscopy represents a unique instrument for atomic identification of paramagnetic defects as well as their quantification in absolute terms which enables comparison to the density of charges or charge traps encountered in MOS structures after electrical stress or irradiation. Two major approaches to establish a relationship of particular paramagnetic defects to electrical behavior of MOS structures have been developed so far. First, the classical derivative-absorption ESR spectroscopy applied to semiconductor/interface structures can be taken to the extreme sensitivity in the range 10^{11}–10^{12} centers (spin $S = ½$) per cm^2 of interface area, reaching the charge density range typically encountered after degradation of a MOS structure (10^{11}–10^{12} elemental charges/cm^2). This sensitivity level is achieved by performing ESR measurements at low temperature (close to the liquid He temperature range) in combination with bundling sample slices in a stack of 10–20 pieces to enhance the total interface area per sample in the ESR probing cell (cavity). Furthermore, by referencing the g-values and intensities of investigated ESR signals to those of a co-mounted marker sample, e.g., Si:P ($g = 1.99869 \pm 0.00002$ at 4.3 K), one solves the problem of absolute calibration which is a prerequisite for any correlative analysis. Typical absolute accuracy obtained in the determination of paramagnetic defect densities is about 20%, while the relative accuracy of 10% is in reach. The major drawback of the conventional ESR approach consists in the laborious and time-consuming sample preparation and complicated measurement procedures with extensive signal averaging. Also, in its conventional realization, the ESR spectroscopy can hardly be applied to finished MOS devices because of the general presence of highly doped and/or metal-covered areas which unacceptably impairs the microwave resonator quality factor, thus aggravating sensitivity problems associated with the small active area of a device.

As more device-oriented alternative, the electrically detected magnetic resonance (EDMR) methods, such as spin-dependent recombination (SDR) or spin-dependent current (SDC) spectroscopy, have been developed over the last two decades [2–6]. These techniques make use of spin-dependent interactions between mobile charge carriers and paramagnetic defects and observe the ESR through modulation of the device current, e.g., the recombination current or the photocurrent, by an external magnetic field while sending an appropriate microwave radiation to the

MOS device. The possibility to reach high sensitivity (orders of magnitude higher than conventional ESR) on a device-sized sample is routinely indicated as the major advantage of the EDMR approach. However, since the measurements of spin-dependent current concern mostly defects located in semiconductor region(s), only limited information regarding defects responsible for charge trapping can be obtained in this way. Furthermore, the amplitude of EDMR signal reflects variations in the rate of spin-dependent electron transitions contributing to the measured current that is a function not only of the density of the corresponding defects but also of the local concentration of mobile charge carriers and an a priori unknown cross section of the carrier-defect interaction. As a result, quantification of defect densities within the EDMR methodology becomes problematic with no reliable defect densities reported for MOS systems so far. This density determination factor leaves us with the conventional ESR analysis as the most viable option if considering the problem of atomic identification of traps and charged defects in semiconductor/insulator structures.

In contrast to ESR, electrical measurements can provide information regarding charge densities in MOS structures and devices in several ways. First, the net density of fixed or trapped oxide charges (Q_{ot}) can be calculated from the shift of flatband voltage (V_{fb}) on the capacitance-voltage (CV) curve of an MOS structure or from the threshold voltage (V_{th}) shift of a MOS transistor [7]. The correction for the centroid of the charge in-depth distribution can be done on the basis of oxide etch-back experiments [7–9], by performing measurements on samples series with different thicknesses of the insulator [10] or using charge profiling based on internal photoemission measurements [8, 11].

Quantification of the interface trap density per unit area (N_{it}) is more challenging since not only the trap density energy distribution $D_{it}(E)$ needs to be found across the entire semiconductor bandgap width, but also the semiconductor surface potential has to be determined as a function of gate bias in a separate measurement [7]. The latter becomes problematic when the MOS structure exhibits a leakage as this hampers measurements of the quasi-static CV curve. This difficulty can partially be solved by performing the ac capacitance measurements upon forming an inversion layer at the periphery of the metal gate which enables to attain the low-frequency response [12]. However, taken all together, these measurements result in a considerable error in the N_{it} determination (30–50%).

To improve the absolute accuracy of interface trap density measurements to the level attained by ESR analysis ($\leq$20%), one may wish to exclude the need of the surface potential determination. This can be done by combining high-frequency CV measurements at low temperature, e.g., 77 K, on n- and p-type MOS capacitors with identical oxide thickness (*d*) [13–15]: The difference between V_{fb} values of the n- and p-type MOS capacitors contains the contributions corresponding to the difference in the Fermi level, E_F, in n- and p-type semiconductor and that of the interface traps [16]:

$$V_{fb}(n-type) - V_{fb}(p-type) = \frac{1}{q}E_F(n-type) - \frac{1}{q}E_F(p-type) + \frac{qN_{it}}{\varepsilon_0 \kappa}d. \quad (9.1)$$

In Eq. (9.1) q is the elemental charge, N_{it} is the interface trap density integrated over the energy interval between E_F(n-type) and E_F(p-type), ε_0 is the vacuum permittivity, and κ is the relative permittivity (dielectric constant) of the insulating layer. In the case of measurements at 77 K, the Fermi level difference term on the right-hand side of Eq. (9.1) is close to the semiconductor bandgap width where one should take into account the Si gap widening from 1.12 eV at 300 K to 1.16 eV at 77 K: The observed N_{it} corresponds to *the total interface trap charge density* integrated over the entire Si bandgap (cf. bottom panel in Fig. 5 in [15]). Then, by comparing 77 K CV traces of n- and p-type Si samples, one can determine the total interface trap density. To the best of our knowledge and experience, this method provides the most accurate N_{it} values with the typical error being below 10%.

9.3 Interface Traps from ESR Analysis

In this section we will address the ESR identification of interface charge trapping centers in silicon-based MOS structures as affected by sample processing, in particular, by hydrogen passivation. It is useful starting the discussion from silicon dangling bond defects (P_b-type centers) because they not only represent the model case of correlative analysis but were also shown to be the key players in the NBT instabilities at Si/SiO_2 interfaces [17–19]. Further, we will shortly discuss recent results of ESR and electrical characterization of germanium dangling bond centers at $Si_{1-x}Ge_x/SiO_2$ interfaces and their interaction with hydrogen. This topic becomes more and more important as the high-mobility $Si_{1-x}Ge_x$ channels become more frequently used in practical transistor structures.

9.3.1 $P_{b(0)}$ Centers

The P_b-type centers [specifically P_{b0} at the (100)Si/SiO_2 interface] have been identified by ESR [20, 21] as trivalent interfacial silicon centers, denoted as interfacial $Si_3{\equiv}Si\bullet$ entities where the dot symbolizes an unpaired electron in a $sp^3\langle 111\rangle$-like orbital, generally referred to as the dangling bond (DB). The proposed atomic configurations of these centers are schematically shown in Fig. 9.1. These defects are commonly encountered at interfaces of oxidized silicon as well as in a broad variety of other interfaces including Si/Si_3N_4, Si/high-κ metal oxides, and Si/a–Si:H. Such a universal appearance has placed the $P_{b(0)}$ defects in the focus of MOS reliability research for nearly three decades. As result of extensive experimental efforts, this center represents the best characterized interface defect up to the date.

With the ESR spectroscopic properties of $P_{b(0)}$ defects extensively reviewed in the literature [19, 21], we will focus on the correlation between the density of these paramagnetic centers and the density of charge traps encountered at interfaces

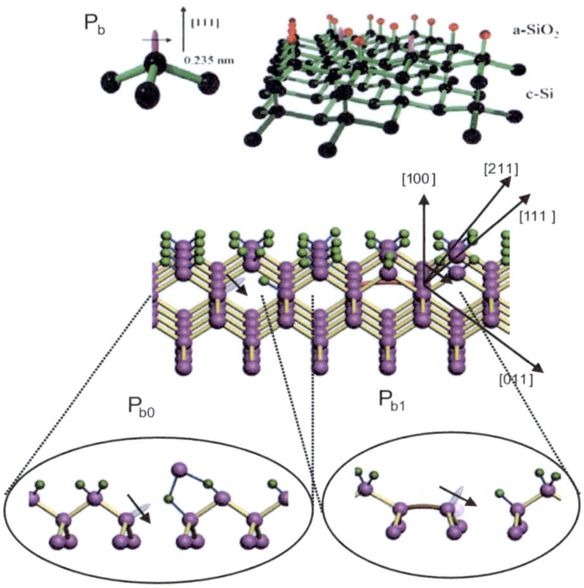

Fig. 9.1 Atomic configurations of silicon dangling bond defects (P_b-type centers) at (111)Si/SiO_2 (*top*) and (100)Si/SiO_2 (*bottom*) interfaces. *Arrows* indicate crystallographic directions in the Si substrate

of silicon. The key analytic power for such an approach consists in the possibility to vary the density of $P_{b(0)}$ defects (provided by ESR) by processing means while monitoring the charge trap density electrically. In this sense, the $P_{b(0)}$-type centers provide an excellent possibility for investigation because their density can be changed in a controllable way by a variety of technological means.

First of all, the $P_{b(0)}$ centers may be (chemically) passivated by attaching a hydrogen atom to the Si DB. This passivation leads to a diamagnetic (ESR-inactive) defect state accompanied with attendant elimination of the trap levels from the silicon bandgap. Thus, the number of passivated DB states may be correlated with the removal of a specific fraction from the interface trap energy distribution $D_{it}(E)$ or, else, with the variation in the total decrease in N_{it}. Importantly, hydrogen passivation of $P_{b(0)}$ centers is completely reversible, i.e., the passivating H atom can be removed by thermal annealing [22] or irradiation [23, 24]. Partial depassivation of $P_{b(0)}$ centers is also observed upon exposure to atomic H [25], hot-carrier injection [19], NBT stress [17, 19], etc., and represents the major mechanism of Si/SiO_2 interface degradation in terms of charge trap generation.

Next, the density of inherently occurring $P_{b(0)}$ centers is strongly sensitive to the crystallographic orientation of the silicon face, being minimal for the (100) plane, followed by the more defective (111) and (110) faces [26]. This allows one to correlate changes in the crystallographically sensitive $D_{it}(E)$ distributions, or in the N_{it}, with the variation of $P_{b(0)}$ density determined by ESR. As an example, in Fig. 9.2 are shown the $D_{it}(E)$ distributions observed at interfaces of differently

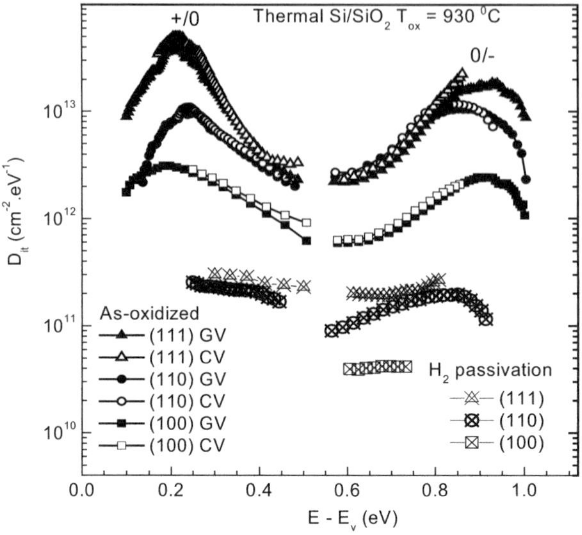

Fig. 9.2 $D_{it}(E)$ profiles of Si/SiO_2 interfaces derived from CV (*solid symbols*) and ac conductance-voltage (GV, *open symbols*) methods in Si/SiO_2 samples fabricated on (100), (110), and (111) faces of Si. Results for the as-oxidized samples (no H-passivation) and those subjected to H_2 passivation (30 min anneal in 1.1 atm H_2 at 400 °C) are shown for comparison [15]

oriented silicon crystals with thermally grown oxide on top [15]. The $D_{it}(E)$ profiles exhibit the well-known double-peak shape corresponding to the charge transition of the DBs from positive to neutral (peak below the silicon midgap) and from neutral to negative (peak above the silicon midgap) [19, 27, 28]; it refers to the amphoteric electrical nature of the DB traps ($P_{b(0)}$ centers). Besides revealing the qualitative correlation between the densities of interface traps and the earlier mentioned changes in the density of $P_{b(0)}$ centers, comparison between the H-free (as-oxidized) and H-passivated samples indicates that most of the initially observed interface traps become inactivated upon annealing in H_2. Therefore, as far as the interface between silicon and its thermal oxide is concerned, the $P_{b(0)}$ centers represent the most significant source of interface traps.

The same conclusion can be reached when comparing the total density of interface traps, N_{it}, as measured in the as-oxidized and H-passivated samples using the low-temperature CV technique. As can be seen from Fig. 9.3, the density of eliminated interface traps goes hand in hand with passivation of $P_{b(0)}$ defects which are passivated to a density below the detection limit of ESR ($\approx 10^{11}$ cm^{-2}). Furthermore, in agreement with the suggested amphoteric electrical behavior of the Si DB defects, the trap density at interfaces of both (100)Si and (111)Si integrated across the gap of silicon, i.e., including both (+/0) and (0/−) transition peaks shown in Fig. 9.2, is nearly twice the $P_{b(0)}$ density. However, at (110)Si/SiO_2 interfaces, N_{it} appears to be lower than that ideally expected from the amphoteric nature of the $P_{b(0)}$ center. In particular, when the temperature of oxidation is lowered, N_{it} in

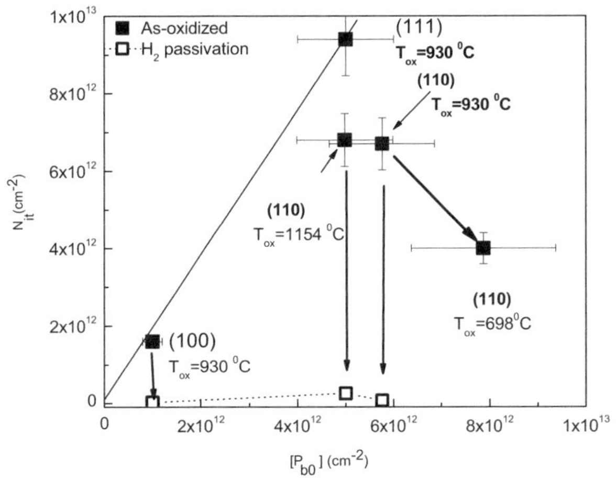

Fig. 9.3 Plot of the electrically active defect density N_{it} at Si/SiO_2 interfaces versus the density $[P_{b(0)}]$ obtained from ESR. The data for the $(110)Si/SiO_2$ interfaces grown at T_{ox}=698 °C and 1,154 °C are also shown for comparison. The *solid line* denotes the ratio $N_{it}/[P_{b(0)}] \approx 2$, expected from isolated amphoteric centers. The *bold arrow* indicates the trend found in $(110)Si/SiO_2$ when the oxidation temperature is decreased. *Open symbols* show N_{it} values observed in the hydrogen-passivated (30 min anneal in 1.1 atm H_2 at 400 °C) samples [15]

$(110)Si/SiO_2$ shows a trend to decrease despite increasing density of $P_{b(0)}$ centers, suggesting partial electrical inactivation of these defects. Though the exact reason for this behavior remains unknown, the fact that this effect is observed both in n- and p-type (110)Si MOS capacitors has led to the suggestion that Si DBs at this silicon face are formed in pairs (or clusters). Charging one of the traps by capturing an electron or a hole will then prevent the subsequent filling of the neighboring defect by a charge carrier of the same charge sign because of Coulomb repulsion [15].

Among other technological factors affecting the density of $P_{b(0)}$ centers and, accordingly, the associated interface traps, one may indicate, besides the silicon surface orientation and treatment in H_2, the oxidation temperature [29, 30], post-oxidation annealing treatments [13, 31], and strain in the surface Si layer [32]. In all these cases, it appears possible to establish a correlation between the density of $P_{b(0)}$ defects and the interface trap-related properties of the Si/SiO_2 interfaces.

In returning to the discussion on the NBT instability mechanisms, it is worth to consider the interaction of $P_{b(0)}$ centers not only with molecular but also with atomic hydrogen (H^0). This interaction includes two reactions [25], essentially as given below, in which unpassivated DBs are spontaneously saturated by H^0, (first reaction equation) while the passivated ones may become depassivated due to the energetically favorable formation of molecular H_2:

$$Si_3 \equiv Si\bullet + H^0 \rightarrow Si_3 \equiv Si - H, \tag{9.2}$$

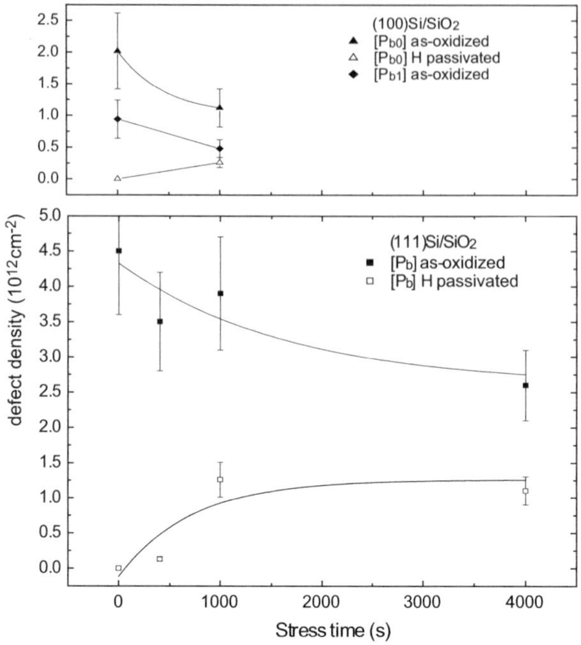

Fig. 9.4 Densities of ESR-active P_b-type centers as a function of NBT stress time measured on hydrogen-passivated and as-oxidized (depassivated) (100)Si/SiO_2 (*top panel*) and (111)Si/SiO_2 (*bottom panel*) interfaces

$$Si_3 \equiv Si - H + H^0 \rightarrow Si_3 \equiv Si \bullet + H_2. \tag{9.3}$$

As the result of balance between these competing chemical reactions, under permanent supply of atomic hydrogen at 300 K, one reaches the steady state with $\sim$50% of Si DBs being passivated. (Both reactions are exothermic with a small or no activation energy involved.) This kind of behavior has been reported upon exposure of Si/SiO_2 entities to H^0 generated by a remote plasma source [25] or, else, exposure to $H^+(H_2O)_n$ ions generated by corona discharge in room air [33]. In the latter case H^0 is generated at the oxide surface when the hydronium-like ions ($Si_2{=}OH^+$) are neutralized by electrons tunneling from the silicon substrate.

These observations of the $P_{b(0)}$ interactions with atomic hydrogen appear to be relevant to the Si/oxide interface degradation upon NBT stressing since the pattern of the $P_{b(0)}$ density variation closely resembles that caused by exposure to H^0: As one can see from the variation in the $P_{b(0)}$ defect density upon NBT stress time shown in Fig. 9.4 for both (100)Si/SiO_2 and (111)Si/SiO_2 interfaces [34], in the H-free (as-oxidized) samples, the stressing results in a lower density of $P_{b(0)}$ defects. At the same time, if starting from the H-passivated Si/SiO_2 interfaces, one observes partial activation of the DBs both by ESR and electrically [34]. Interestingly, upon long NBT stress, the densities of $P_{b(0)}$-s approach the earlier mentioned $\approx$50%

passivation level, albeit not quite. Apparently, there is only a limited H^0 in the MOS capacitors, which prevents both the passivated and depassivated (H-free) samples from reaching the same steady-state $P_{b(0)}$ level observed in the gate-free Si/SiO_2 structures after a "massive" exposure to atomic hydrogen. The importance of this observation is that it questions the relevance of the standard NBT instability picture based on H release from the interfacial $P_{b(0)}$s.

Taken together, these experimental results point to the presence in the Si MOS structures of a hydrogen reservoir which under NBT stressing releases atomic hydrogen into the oxide. With the possibility that the dopants (B, P, As) present in the silicon substrate serve as the H-storage sites firmly excluded by the experimentally observed boron deactivation effect (cf. Fig. 3 in [34]), other options must be considered. Among the suggestions made in the literature, we find the interaction with H-containing ambient molecules, e.g., H_2O [19], hydrogeneous species dissolved in the metal electrode of the MOS capacitor [35–37], or hydrogeneous species present at the metal/oxide interface [35, 38]. We will return to the discussion regarding possible hydrogen sources in the silicon MOS system later in this chapter.

9.3.2 P_{b1} Centers at the (100)Si/SiO_2 Interface

In addition to the above-discussed $P_{b(0)}$ defects, thermally grown (100)Si/SiO_2 exhibits another interfacial paramagnetic center termed P_{b1}. This center has been extensively characterized by ESR [19, 21, 39] which reveals a Si DB as the kernel of this imperfection. However, from the observed deviation of the apex sp^3-like hybrid direction from the normal $\langle 111 \rangle$ Si-Si bond directions of Si crystal, as it is shown in Fig. 9.1, it has been concluded that the P_{b1} center is probably associated with the DB of an interfacial Si atom with at least one backbond under strain (the strained Si–Si dimer model $\equiv$Si—Si$^{\bullet}$=Si$_2$).

The interaction of the P_{b1} center with hydrogen resembles that of the $P_{b(0)}$ defects, the passivation occurring with similar activation energy value. There are, however, two features in the behavior of the P_{b1} center which make it distinctively different from the $P_{b(0)}$. First, the P_{b1} center ESR appearance requires that the silicon oxidation or annealing has been performed at a higher temperature ($\geq$400 °C); substantial P_{b1} generation needs the (100)Si/SiO_2 system to have "seen" a thermal step at sufficiently high temperature, suggesting that the P_{b1} formation would require a minimum level of oxide (interface) relaxation. In the (100)Si samples with low-temperature or deposited oxides, including important cases of high-κ insulators, generally no measurable density of these defects is found.

Second, perhaps more pertinently to the discussion about NBT instability, the available experimental results suggest that the P_{b1} center does not behave as a charge trap at the (100)Si/SiO_2 interface. There are two major experimental observations. First, despite the significantly different ESR appearance between P_{b0} and P_{b1}

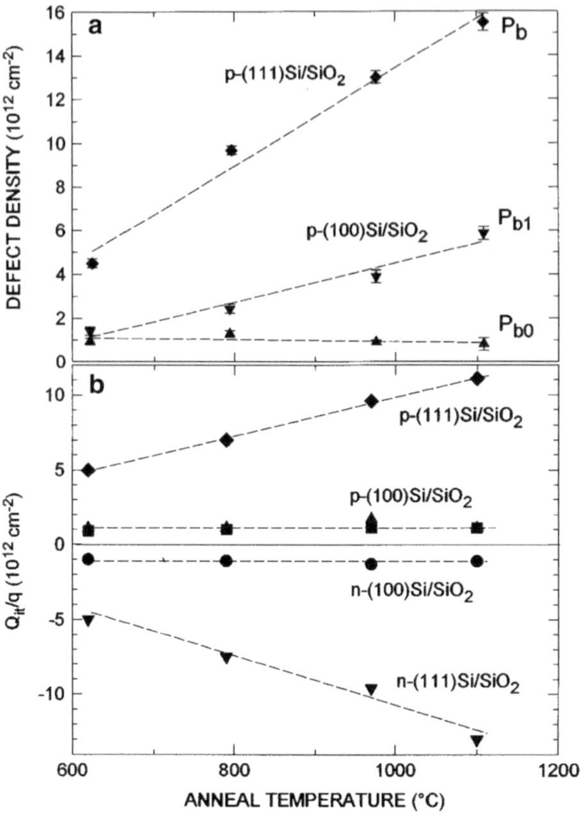

Fig. 9.5 (**a**) Densities of P_b, P_{b0}, and P_{b1} interface defects in thermal p-type (100)Si/SiO_2 and (111)Si/SiO_2 as a function of the post-oxidation vacuum anneal temperature as found from ESR experiments; (**b**) Areal density of interface state charge measured using the low-temperature CV technique on various sets of co-processed p- and n-type (100)Si/SiO_2 and (111)Si/SiO_2 structures

centers indicative of a difference in wave function of the unpaired electrons, no component of the $D_{it}(E)$ profile specific to P_{b1} (as compared to P_{b0}) can be found [40, 41]. This result excludes the later proposal placing the P_{b1} energy level close to Si midgap [42, 43]. Furthermore, the annealing-induced variations in the density of paramagnetic P_{b1} centers are not reflected in the N_{it} evolution as can be seen from the comparison of ESR and electrical results shown in Fig. 9.5 [13]. By contrast, the annealing-induced changes in the $P_{b(0)}$ density are scrupulously followed by the corresponding N_{it} change. These experiments suggest that the P_{b1} center by itself does not provide a significant contribution to the traps observed at the (100)Si/SiO_2 interface after stressing.

9.3.3 D-Centers in Amorphous or Disordered Silicon

As close relatives of $P_{b(0)}$ defects, D-centers represent threefold coordinated silicon centers ($Si_3{\equiv}Si\bullet$) in an amorphous or disordered surrounding; in one interpretation, they are described as P_b-type centers (cf. Fig. 9.1) in a disordered Si environment. These defects are usually absent at the device-grade silicon/oxide interfaces processed at high temperature. However, if growth of silicon at low temperature is involved, e.g., in a-Si:H layers of solar cell structures, D-centers may contribute to the charge trap density [16]. In particular, the frequently used approach to realize passivation of high-mobility semiconductor channels, e.g., Ge[44] or GaAs[45], by making use of a thin silicon interlayer may lead to the formation of disordered silicon regions containing D-centers. It is possible then that, similarly to $P_{b(0)}$ centers, the D-centers will also contribute to the density of interface traps observed after NBTI stressing as in happens in the SiGe channel devices [46].

9.3.4 Ge P_{b1} Center

Among recent developments, the ESR detection and identification of a GeP_{b1} center at the (100)$Si_{1-x}Ge_x/SiO_2$ interfaces of the SiGe-on-insulator structures prepared by the condensation growth method [47] represents a major advancement as it concerns non-silicon DB defects. Though detection of hyperfine interactions appears to be problematic, still the defect could be identified, to a large degree of certainty, on the basis of g-values and g-tensor symmetry, in combination with theoretical insight: As key part, it has been assigned to an unpaired electron residing in an sp^3-like hybrid on an interfacial Ge atom in a configuration resembling the earlier discussed P_{b1} center at the (100)Si/SiO_2 interface. Similarly to its silicon counterpart, the GeP_{b1} centers, at least a major part, can be reversibly passivated and depassivated by annealing in H_2 and in vacuum, respectively.

Electronic properties of the GeP_{b1} center appear, however, dramatically different from those of the SiP_{b1} center which cannot be correlated with charge trapping. From the quantitatively matching densities of paramagnetic centers and negative charges in (100)$Si/SiO_2/Si_{1-x}Ge_x/SiO_2$ entities with 16–40 nm thin $Si_{1-x}Ge_x$ layers (cf. Fig. 9.6 for $0.28 \le x \le 0.93$), it has been found that the unpassivated GeP_{b1} centers electrically behave as acceptors [48]. The energy level of this acceptor state lies close to the SiGe valence band top edge. Upon passivation of these DBs by hydrogen, the negative charge is reduced following the corresponding variations in the density of ESR-active GeP_{b1} centers. Conversely, by using annealing in vacuum, the densities of both negatively charged and paramagnetic centers can be restored to the same original value. Because of its charge, the GeP_{b1} center potentially represents the source of charge instability in devices with high-mobility $Si_{1-x}Ge_x$ channels.

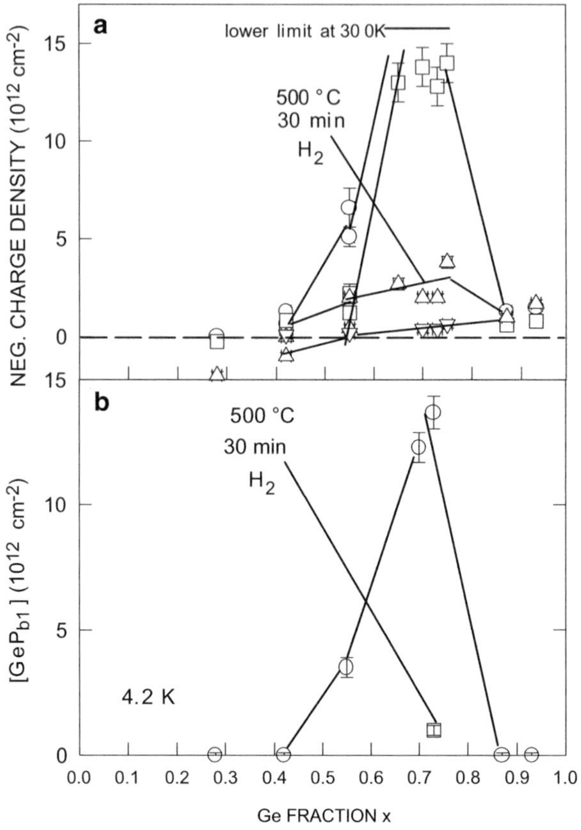

Fig. 9.6 (**a**) Charge density determined in as-prepared condensation-grown (100)Si/SiO_2/$Si_{1-x}Ge_x$/SiO_2 samples from the CV curve shift at 300 (*open circle*) and 77 K (*open square*) as compared to the samples passivated in H_2 at 500 °C for 30 min with CV measurements conducted at 300 (*open triangle*) and 77 K (*open inverted triangle*), respectively; (**b**) areal density of Ge DBs (Ge P_{b1} centers) determined from ESR measurements at 4.2 K

It also needs to be added that these results pertaining to the GeP_{b1} centers are obtained at interfaces with insulating SiO_2. A considerable number of attempts to detect Ge DB centers at the interfaces of SiGe or Ge with other insulators using *conventional* ESR spectroscopy failed to find any ESR-active centers with density exceeding the detection limit of ($\approx 5 \times 10^{11}$ cm^{-2} for the linewidth corresponding to the GeP_{b1} center and a sample of amenable size). At the same time, the density of electrically active interface traps in these samples is found to be in the range 10^{12}–10^{13} cm^{-2}. The discrepancy between the density of ESR- and electrically active centers at these interfaces suggests that the vast majority of the observed interface traps are associated with defects of different nature [49, 50]. It is also possible that the low density of Ge DB centers at interfaces of Ge or SiGe with oxides on top results from relaxation of the strain at the Ge/oxide interface through viscous flow of GeO_2 formed at the initial stages of oxide growth (or deposition) [51]. In the condensation growth technology, the GeO_2-assisted strain relaxation mechanism is excluded because the processing temperature is intentionally set above the limit of the thermal stability of Ge oxides; no Ge–oxide remains. As a result, the $Si_{1-x}Ge_x$ layer appears to be under considerable compressive strain induced by a robust medium (SiO_2). Due to the influence of this strain, a high density of DB defects is generated as can be seen from the density values indicated in Fig. 9.6.

9.4 Oxide Charges

This section deals with the origin of positive oxide charges commonly encountered in insulating SiO_2 upon NBT stressing or irradiation. The major focus will be on the role of paramagnetic oxide defects in generation of this charge and, as it becomes clear from the discussion concerning interface traps, on the hydrogen-related charging phenomena. Based on the analysis of available experiments results, hydrogen-related (protonic) species will be identified as the major source of the oxide-trapped positive charge.

9.4.1 ESR-Active Oxide Defects

Over several decades of research, ESR spectroscopy has succeeded in identification of several paramagnetic centers in thermally grown oxide on silicon. The most frequently encountered ESR-active defect is the E_γ'-center (with zero-crossing g-value $g_c = 2.00055$ for the powder pattern observed in amorphous SiO_2) associated with an unpaired electron localized on a threefold coordinated silicon atom in the oxide (the $O_3{\equiv}Si\bullet$ entity, considered as generic entity for the whole family of E'-type centers). The electrical appearance of this defect will be discussed later in more detail. Another representative of the E' family, usually denoted as E_δ' center of $g_c \simeq 2.0020$ [52, 53], is typically found is oxygen-deficient oxides [54] and associated with positive charges [55]. Recent experiments on quantification of the hyperfine structure of this center suggest its relationship to an electron delocalized over a cluster of 4 or 5 silicon atoms [56, 57]. However, the E_δ' center is rarely found in device-grade gate oxides and is an unlikely candidate to provide substantial contribution to the NBT instabilities.

Another intrinsic paramagnetic center, termed the EX center, is characterized by $g = 2.0025$ and accompanied by a hyperfine-induced 14 G split doublet [58]. This defect has been tentatively assigned to an unpaired spin delocalized over several oxygen atoms formally at the site of a Si-vacancy in SiO_2. In another view, it has been pictured as an agglomerate of four oxygen-related hole centers (OHCs). Accordingly, in its electrical appearance, the EX center represents a fixed positive charge [59]. However, as generation of EX centers is usually observed in thin oxides grown in a specific temperature range around 700–800 °C, it is rarely found in the gate SiO_2 usually processed at a higher temperature. It also needs to be mentioned here that the densities of both E_δ' and EX defects are at best in the low 10^{11} cm^{-2} range. Therefore, these imperfections are unlikely to provide the major contribution to the NBT- or injection-induced instabilities in Si MOS devices.

9.4.2 E_γ'-Centers in Thermal SiO_2

The E_γ'-centers have been long known to appear upon irradiation of thermally grown SiO_2 layers under a broad variety of conditions (see, e.g., [53]). However, much controversy is still found in the literature concerning the net charge state of the total of the defect site. Initial results have shown that the increase in the E_γ'-center density upon irradiation or Fowler–Nordheim stressing corresponds to the density of positive charge found in the degraded oxide [60–62]. However, stress-induced positive charge was later reported to appear without any detectable E' density at all [63]. In turn, a high density of E'-centers ($\approx 10^{14}$ cm^{-2}) has been found upon irradiation of Si MOS capacitors with Al top electrodes without a comparable density of positive charge [35]. The direct correspondence of the densities of E'-centers and positive charge has been questioned [64] as a universal feature of these defects on the basis of the observation that differently grown layers of thermal SiO_2 show different density ratios of E'-centers and charged centers, leading the authors of [64] to the suggestion that the E'-centers may be present in thermal SiO_2 both as positively charged and neutral entities. This hypothesis echoed the earlier suggestion of Helms and Poindexter [19] about the copresence of the neutral E'-centers (resembling $O_3{\equiv}Si{\bullet}$ entities at the surface of SiO_2) and their positively charged counterparts corresponding to oxygen vacancy ($O_3{\equiv}Si{\bullet}\ {}^{+}Si{\equiv}\ O_3$ defects) in c-SiO_2.

In order to check on the validity of the proposed ambivalent electrical appearance of E'-centers in thermal SiO_2, an analysis has been carried out of the correlation between the density of these centers and that of positive charge generated upon irradiation by 10-eV photons [65]. This was done using the well-known dependence of the trapped positive charge on the strength of the electric field present in the oxide during irradiation: While the optical absorption in SiO_2 is virtually insensitive to the field, the saturation level of the trapped positive charge is determined by the balance between hole trapping and their annihilation by simultaneously generated electrons. In its turn, the annihilation cross section exhibits a strong dependence on

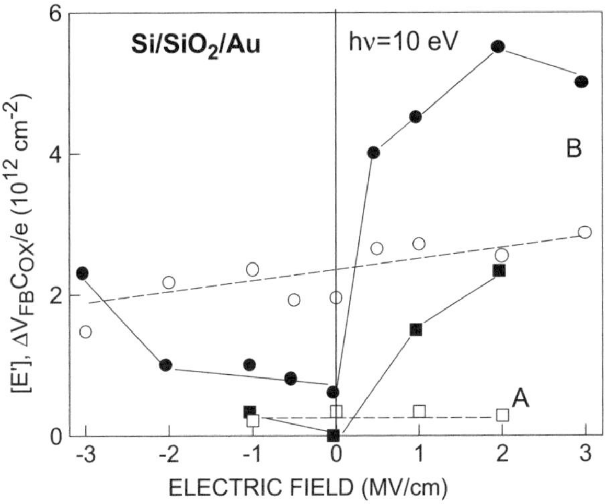

Fig. 9.7 Densities of E'-centers (*open circle, open square*) and the positive charge (*filled circle, filled square*) observed after injection of 10^{15} electron/hole pairs cm^{-2} (10-eV photon irradiation) as a function of the strength of the applied electric field over the SiO_2 oxide for MOS structures with the oxide in the as grown state (A: *open square, filled square*) and after degradation by a high-temperature annealing (B: *open circle, filled circle*) [65]. *Lines* are guides to the eye

the electric field (cf. data compiled in Fig. 11.2.1 in [11]). This leads to a significant variation of the positive charge density with changing orientation and strength of the electric field applied during irradiation, as illustrated in Fig. 9.7 (solid symbols). However, the density of E'-centers detected by ESR in physically the same samples appears to be not only different in its absolute values but also to follow an entirely different trend as shown by open symbols in Fig. 9.7. These results indicate that the transition of the E'-centers to the diamagnetic state is not related to the annihilation of positive charge by electrons, i.e., the E'-defect bears no positive charge by itself. Taken into account that this lack of correlation is found for differently processed SiO_2 layers with largely different densities of E'-centers and positive charges, the neutral electrical appearance of E'-defects emerges as the universal property of the entire E'-bath [65].

To explain the experimental observations, we proposed a model in which holes are trapped on hydrogen-passivated E'-precursors ($O_3{\equiv}Si{-}H$ centers) [66, 67]: While the E'-center appears to remain a neutral $O_3{\equiv}Si{\bullet}$ entity upon hole trapping, the positive charge must then be carried away by a proton. Clearly, this hypothesis brings back the question about the role hydrogenic species play in charging phenomena within the Si/SiO_2 system, which will be addressed later in this section. In the advanced model, the E'-centers still represent the key players in the generation of radiation-induced positive charge in SiO_2, but now they are seen as the source of protons rather than positively charged entities in itself.

Experiments on proton injection in SiO_2 from a low-energy ion beam reveal that, independent of the SiO_2 growth and annealing conditions, proton trapping occurs with high probability, approaching 100% [9, 67]. This observation explains the earlier indicated correlation between the density of E'-centers and the density of trapped positive charge since the rate of the proton generation upon hole trapping increases proportionally to the density of E'-precursors. However, when considering the observations over a wider range of hole-injection doses and different electric fields, the absence of quantitative correlation becomes evident. In some cases, under influence of different metallization schemes, even an anticorrelation between the E'-density and that of positive charge can be observed (cf. Fig. 2 in [66]).

What remains yet unclear is the role the E'-centers play in the phenomena induced by NBT stressing. On the one hand, there are reports indicating generation of paramagnetic E'-states upon electrical or NBT stress [68]. On the other hand, NBT instabilities have been observed with no detectable density of E' appearing (ESR detection limit $\approx(0.5-1)\times10^{11}$ cm^{-2}) [34]. Furthermore, no E'-centers or their precursors can be detected in ultrathin ($d < 4$ nm) thermal oxides on Si or in the Si/SiO_x/high-κ oxide stacks not subjected to supplemental oxidation, while upon NBT stressing, generation of traps and charges is clearly observed. Two possible explanations of this discrepancy can be proposed. First, more pertinent to thick oxides studied in the past, the application of stressing voltages exceeding the bandgap of SiO_2 (8.9 eV) may cause impact ionization of the oxide. Moreover, hot electrons arriving during NBT stress at the Si/SiO_2 interface may cause hole injection from silicon. Then, the generated or injected holes will interact with $O_3{\equiv}Si{-}H$ precursors activating them to the paramagnetic $O_3{\equiv}Si\bullet$ entity. Second, the generation of ESR-active E'-centers may occur as the result of a process accompanying the major NBT instability reactions but not determining them. For instance, an interface ionization of hydrogen in the vicinity of the silicon surface may generate mobile protons [69]. These particles, as suggested by the proton injection experiments [9], may not only ESR activate the existing E' precursors but also facilitate rupture of strained bonds in the amorphous oxide network.

9.4.3 Protonic Charges in SiO_2

From the overviewed results, it becomes clear that in addition to the intrinsic defects in the Si/SiO_2 system, hydrogen-related species play a pivotal role in the electrical degradation phenomena. One can find much more experimental evidence in the literature indicating that not only NBT instabilities but also electron trapping [70, 71], anomalous positive charge induced by electron injection [70, 72], and interface trap generation [73–76] are associated with the presence of H-containing defects and mobile species. For instance, the presence of positive (protonic) charges, as under discussion, were hypothesized already by Revesz [77], leading to the model of McLean describing metastable radiation-induced charges in SiO_2 [78]. However, unlike the ESR-active defects, the hydrogen-related charges often escape atomic identification and can be only assessed by electrical means. Furthermore, if considering instability mechanisms, one needs to trace the sources of hydrogen supply rather than the total amount of hydrogen present in the sample. This problem remains largely unsolved up to the date.

In order to investigate the correlation of charge instabilities with the presence of hydrogenic species, a number of hydrogen-detection approaches have been used. First, one may study the correlation of charging phenomena with the density of H-containing bonds (Si–H, O–H) detected by infrared absorption spectroscopy [71], However, the latter technique requires a considerable volume of material to be probed. Next, the incorporation of hydrogen, and accordingly its impact, may be varied by changing the sample growth or annealing processing, though the result of such "selective doping" approach is not obvious in the case of hydrogen. For instance, the standard post-metallization anneal of Si/SiO_2 structures in forming gas (10% H_2 + 90% N_2) may actually reduce the amount of hydrogen in the MOS structure [36] rather than increase it. The density of trapped positive charge (protons) decreases accordingly (cf. Fig. 2 in [66]).

Next, various isotope marker techniques ranging from the tritium radioactive tracer [79] method to the deuterium isotope mass effect [80–86] may be used to reveal the involvement of hydrogen motion in the charging process. Though these methods provide the most convincing indication of hydrogen involvement in the rate-limiting step of the charging reaction, it provides no insight on the atomic nature of the involved interaction sites of hydrogen atoms or protons.

One detection technique selectively sensitive to the presence of atomic hydrogen has been proven to be particularly useful in establishing the pattern of hydrogen-related instabilities in Si MOS structures. It has been first discovered experimentally [87, 88] and later supported by theoretical results [89] that atomic hydrogen may deactivate boron acceptors in silicon. The corresponding variation in the concentration of active B acceptors can easily be monitored using the inversion capacitance (C_{inv}) value measured on the same high-frequency CV curve [84] than that is used to monitor the trapped charge density. This provides a simple way to correlate the oxide charging and the presence of H^0 in the Si/oxide structures.

To illustrate the method, we address its application to two types of oxide layers, with different density of traps, which were prepared in the same way as the samples studied in [34] (cf. Fig. 9.7) and denoted in the same way as samples A and B [66]. The key results are presented in Fig. 9.8a showing the variations in the effective trapped charge density found from the V_{fb} shift as a function of the density of holes injected into SiO_2 layers followed by electron injection used to annihilate the trapped positive charge. In the same panel (a) are also plotted variations in the density of E'-centers as measured by ESR, while panel (b) shows the corresponding relative decrease in the relative capacitance (δC_{inv}) caused by deactivation of boron dopants in the p-type silicon substrate. These data show that H^0 release accompanies hole injection and qualitatively correlates with the positive charge buildup. Moreover, what is even more interesting is the observation of H^0 liberation upon annihilation of the trapped positive charge by electron injection as indicated by even more enhanced B deactivation [additional decrease in C_{inv} shown in Fig. 9.8b]. This observation leaves little doubt that the positive charge caused by

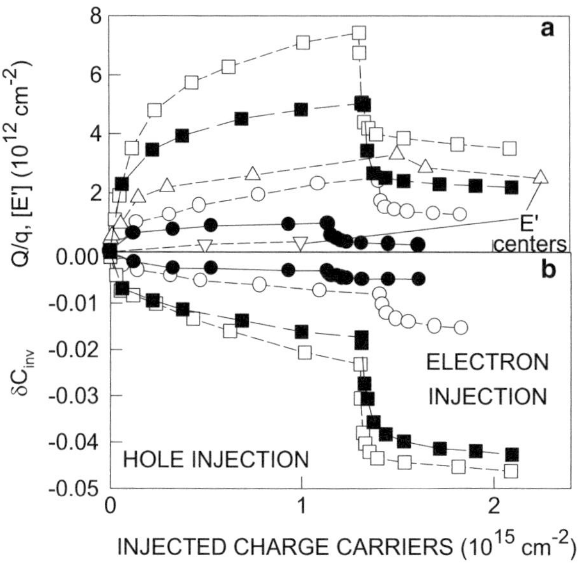

Fig. 9.8 Density of positively charged centers (**a**) and variation of inversion capacitance (**b**) as a function of injected electron/hole density in MOS structures of type A (*open circle*) and B (*open square*). The data for the samples subjected to post-metallization annealing in H_2 at 400 °C (30 min) are shown by *filled symbols*. The observed E'-center density observed by ESR is shown for samples A (*open inverted triangle*) and B (*open triangle*) in panel (**a**)

"hole trapping" in reality represents the trapped proton entity. This result explains the diamagnetic nature of this charge as well as the possibility to trap a density of positive charge that is a multiple of the E'-center density: The E'-centers are known to interact with H^0, and even H_2, at room temperature [90, 91] and, therefore, can be "recycled" during charge injection each producing multiple protons. For more details regarding the positive charge trapping and its correlation with atomic hydrogen behavior, the reader is referred to the original publications [9, 66].

The stable trapping of protons in SiO_2 raises an important issue about the proton generation mechanisms since these allow the available sources of hydrogen to enhance the positive charge generation. The above-discussed hole trapping mechanism

$$O_3 \equiv Si - H + h^+ \rightarrow O_3 \equiv Si \bullet + H^+, \tag{9.4}$$

represents the process with the highest hole capture cross section of (3–4) $\times$ 10^{-14} cm^2 and plays the major role in the irradiation-induced effects [35]. However, it is also possible to form H^+ by an alternative reaction once H^0 arrives to the oxide layers close to the Si/SiO_2 interface: There is a vast experimental evidence indicating formation of near-interfacial donor states under conditions when holes cannot reach the Si/SiO_2 interface, e.g., under purely electron injection [70, 72, 92], atomic H exposure [76], or electron–hole pair generation at the oxide surface [35].

The formed positive charge, referred to as "the anomalous positive charge" [70], is resistant to neutralization by electrons and can be traced down to slow donor interface states with energy levels in the upper half of the silicon bandgap [92]. Unlike P_b-related interface traps, these states slowly anneal (relax) already at room temperature [93]. Two essential experimental observations point toward the relationship of these donor states to trapped protons [73, 75]:

- The annealing of these defects was found to correlate with the release of H^0 as detected using the boron deactivation method.
- The annealing rate of these donors is controlled by their charge state. They anneal at 300 K if their energy level is shifted below the Fermi level of silicon, i.e., when the donors are neutral. By contrast, they appear to be stable over extended time (10^6 s at least) if the Fermi level in Si lies below the energy level of the donors, i.e., when they are kept positively charged. This behavior closely resembles the stability pattern of a hydronium ion in water, $(H_2O)H^+$.

These observations suggest a similar proton bonding configuration in SiO_2: A $Si_2{=}OH^+$ center with the proton bonded to the lone-pair electron cloud of a bridging oxygen atom. This bonding scheme suggests an ESR-inactive bonded state, still to be confirmed experimentally. However, there are a number of theoretical treatments indicating the feasibility of the hydronium-like proton bonding in SiO_2. Moreover, since the amorphous oxide network is expected to contain Si–O–Si bridges with a wide distribution in bridge angles, a wide range of proton bonding energies is expected as well. The latter is, indeed, observed in annealing experiments directed to observe removal of protons trapped in the oxide at different temperatures: The highest binding energy, of 2.3–2.4 eV, is found for proton-induced charges generated by annealing ("the oxide fixed charge") [94, 95] and may be associated with the first layer of O atoms bonded to the Si crystal [96, 97]. The charge observed upon proton injection is found to anneal with an activation energy of $\approx$1.7 eV [9], while the annealing of the "trapped holes" requires 1.4 eV activation energy [98]. Finally, the transport of mobile protons in the oxide bulk [99] and annealing of charges induced by low-temperature hole injection [100] require only a 0.6–0.7 eV activation.

The analysis of the in-depth distribution of the trapped positive charge [9, 11, 101] reveals that most of the stable positive charge (protons) is located in the oxide layer close to the silicon substrate, pointing toward the influence of interface strain. Using a carbon gate electrode to block electron photo-injection [101], the detailed charge density profiles can be obtained as shown in Fig. 9.9 for 80-nm thick thermal oxides grown on silicon using two different oxidation methods. The same values of the trapped charge density (of about 3×10^{18} cm^{-3}) close to the Si/SiO_2 interface in these two samples reflects the Coulomb limit of the trapped charge density, while the real density of the oxide network sites suitable for the proton trapping is by far higher. Therefore, one may expect a variety of SiO_2 network sites with different ability to immobilize a proton or a hydrogen atom to be present, even in deep downscaled MOS devices. Combined with a wide distribution

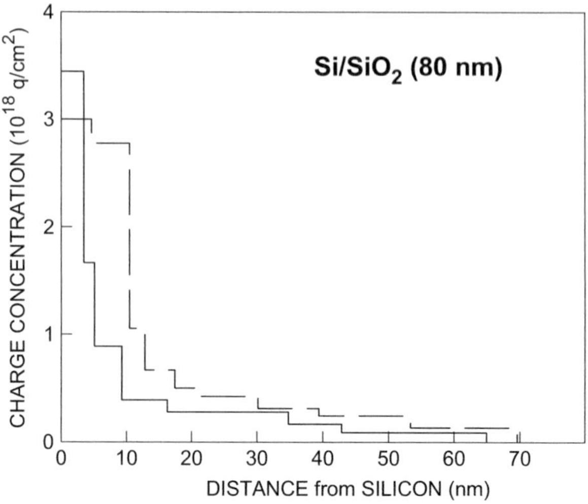

Fig. 9.9 Concentration profiles of positive charges observed after photogeneration of holes in a 80-nm thick oxides thermally grown on (100)Si in dry O_2 at 1,000 °C (*solid line*) or in $O_2 + 1\%$ HCl at 1,150 °C (*dashed line*). The partial charge neutralization measurements were performed using IPE of electrons from the silicon substrate and by applying a 10-nm thick amorphous carbon injection-blocking interlayer between the SiO_2 layer and the semitransparent Au field electrode

of trap-to-silicon distances shown in Fig. 9.9, the hydrogen species immobilized on these network sites may account for the wide range of the trapping/emission time constants and the different influence of the individual traps on the channel current found in experiments [102]. Due to the already discussed metastability of the traps related to the hydrogen bonded in SiO_2, some of these traps may be annealed and disappear after certain time. However, since the corresponding SiO_2 network sites remain intact, traps with identical parameters may "reappear" upon repetitive BT stressing provided the hydrogen supply is not exhausted.

Since the trapped proton states are formed even without holes being injected into the valence band of SiO_2, it is reasonable to suggest that these species are generated by ionization of atomic hydrogen near the Si/SiO_2 interface, either by emitting an electron to Si or, else, by trapping a hole from the semiconductor valence band. Note that during the NBT stress, the orientation of the externally applied electric field not only favors both indicated processes of interface ionization but also shifts the energy levels of the proton-related donor states above the Fermi level of silicon, ensuring their maximal stability [73]. The only "ingredient" needed to facilitate the formation of these charges as well as depassivation of the P_b-H centers is a source of atomic hydrogen in the Si/oxide/gate stack.

9.5 Conclusions

On the basis of the above described experimental observations, we can identify three sources of NBT stress-induced traps and charges in silicon-based MOS systems:

- Amphoteric interface traps associated with interfacial dangling bonds (P_b and P_{b0} centers). The appearance of these defects is determined by dissociation of the bond of Si with the passivating hydrogen atom in the corresponding $Si_3{\equiv}Si$–H precursor states, which can occur as the result of interaction with atomic hydrogen or hot carriers.
- Donor-like interface traps with large spread in recharging time constants related to hydrogen atoms bonded in the oxide at different distances from the silicon surface. The suggested bonding configuration $Si_2{=}OH^+$ resembles the hydronium ion and, if kept neutral (upon electron trapping), dissociates by releasing H^0, thus leading to metastability of the observed defect pattern. Generation of these traps is controlled by the supply of atomic hydrogen and, possibly, protons to the Si/SiO_2 interface.
- Trapped "fixed" positive charge is mainly associated with protons. The major sources of these are the E'-center precursors in the oxide ($O_3{\equiv}Si$–H entities) interacting with injected holes and the H-atoms ionized at the interface with silicon.

This identification of the traps and charges suggests that the description and modeling of the stress-induced degradation must involve interactions of at least three mobile particles (electrons, holes, hydrogen atoms), two defect precursors associated with Si DBs, one at the surface of Si and the other in the oxide, and probably high, but yet unknown, a density of proton trapping sites suggested to be strained Si–O–Si bridges. Even if the contribution of E'-centers can be excluded as suggested by the absence of any measurable density of these defects in ultrathin oxides, the overall picture remains very complex. The major challenge in clarifying the degradation behavior lies with experimental identification of the "reservoirs" of hydrogen and the mechanisms which can supply H^0 or H^+ under particular stressing conditions.

This work was partially supported by the EU FP7 project MORDRED (grant No. 261868).

References

1. B. E. Deal, J. Electrochem. Soc. **127**, 979 (1980).
2. R. L. Vranch, B. Henderson, and M. Pepper, Appl. Phys. Lett. **52**, 1161 (1988).
3. J. T. Krick, P. M. Lenahan, and G. J. Dunn, Appl. Phys. Lett. **59**, 3437 (1991).
4. M. S. Brandt and M. Stutzmann, Appl. Phys. Lett. **61**, 2569 (1992).
5. J. H. Stathis, Appl. Phys. Lett. **68**, 1669 (1996).

6. S. Baldovino, S. Nokhring, G. Scarel, M. Fanciulli, and M. S. Brandt, J. Non-Cryst. Solids **322**, 168 (2003)
7. E. H. Nicollian and J. R. Brews, *MOS Physics and Technology* (Wiley, New York, 1982).
8. V. V. Afanas'ev and V. K. Adamchuk, Progr. Surf. Sci. **47**, 301 (1994).
9. V. V. Afanas'ev, F. Ciobanu, G. Pensl, and A. Stesmans, Solid-State Electron. **46**, 1816 (2002).
10. V. V. Afanas'ev and A. Stesmans, J. Appl. Phys. **95**, 2518 (2004).
11. V. V. Afanas'ev, *Internal Photoemission Spectroscopy,* (Elsevier, Oxford, 2008).
12. Y. G. Fedorenko, L. Truong, V. V. Afanas'ev, and A. Stesmans, Appl. Phys. Lett. **84**, 4771 (2004).
13. A. Stesmans and V. V. Afanas'ev, Phys. Rev. B **57**, 10030 (1998).
14. Y. G. Fedorenko, L. Truong, V. V. Afanas'ev, A. Stesmans, Z. Zhang, and S. A. Campbell, J. Appl. Phys. **98**, 123703 (2005).
15. N. H. Thoan, K. Keunen, V. V. Afanas'ev, and A. Stesmans, J. Appl. Phys. **109**, 013710 (2011).
16. N. H. Thoan, M. Jivanescu, B. O'Sullivan, L. Pantisano, I. Gordon, V. V. Afanas'ev, and A. Stesmans, Appl. Phys. Lett. **100**, 142101 (2012).
17. G. J. Gerardi, E. H. Poindexter, M. Harmatz, W. L. Warren, E. H. Nicollian, and A. H. Edwards, J. Electrochem. Soc. **138**, 3765 (1991).
18. C. E. Blat, E. H. Nicollian, and E. H. Poindexter, J. Appl. Phys. **69**, 17121 (1991).
19. C. R. Helms and E. H. Poindexter, Rep. Prog. Phys. **57**, 791 (1994).
20. P. J. Caplan, E. H. Poindexter, B. E. Deal, and R. R. Razouk, J. Appl. Phys. **50**, 5847 (1979).
21. E. H. Poindexter, Semicond. Sci. Technol. **4**, 961 (1989).
22. A. L. Stesmans, Phys. Rev. B **61**, 8393 (2000).
23. A. Pusel, U. Wetterauer, and P. Hess, Phys. Rev. Lett. **81**, 645 (1998).
24. T. Vondrak and X.-Y. Zhu, Phys. Rev. Lett. **82**, 1967 (1999).
25. J. H. Stathis and E. Cartier, Phys. Rev. Lett. **72**, 2745 (1994).
26. G. J. Gerardi, E. H. Poindexter, P. J. Caplan, and N. M. Johnson, Appl. Phys. Lett. **49**, 348 (1986).
27. J. A. Miller, C. Blat, and E. H. Nicollian, J. Appl. Phys. **66**, 716 (1989).
28. P. K. Hurley, B. J. O'Sullivan, V. V. Afanas'ev, and A. Stesmans, Electrochem. Solid State Lett. **8**, G44 (2005).
29. A. Stesmans, Phys. Rev. B **48**, 2418 (1993).
30. A. Stesmans and V. V. Afanas'ev, Appl. Phys. Lett. **77**, 1469 (2000).
31. A. Stesmans and V. V. Afanas'ev, Phys. Rev. B **54**, 11129 (1996); J. Vac. Sci. Technol. B 16, 3108 (1998).
32. A. Stesmans, D. Pierreux, R. J. Jaccodine, M. T. Lin, and T. J. Delph, Appl. Phys. Lett. **82**, 3038 (2003).
33. A. Stesmans and V. V. Afanas'ev, Appl. Phys. Lett. **82**, 2835 (2003).
34. M. Houssa, A. Stesmans, V. V. Afanas'ev, M. Aoulaiche, G. Groeseneken, and M. M. Heyns, Appl. Phys. Lett. **90**, 043505 (2007).
35. V. V. Afanas'ev, J. M. M. de Nijs, P. Balk, and A. Stesmans, J. Appl. Phys. **78**, 6481 (1995).
36. J. Krauser, F. Wulf, M. A. Briere, J. Steiger, and D. Braunig, Microelectron Eng. **22**, 65 (1993).
37. K. Maser, U. Mohr, R. Leihkauf, K. Ecker, U. Beck, D. Grambole, R. Grotzschel, F. Herrmann, J. Krauser, and A. Weidinger, Microelectron Eng. **48**, 139 (1999).
38. V. V. Afanas'ev, J. M. M. de Nijs, and P. Balk, J. Non-Cryst. Solids **187**, 248 (1995).
39. A. Stesmans and V. V. Afanas'ev, Phys. Rev. B **58**, 15801 (1998).
40. M. J. Uren, K. M. Brunson, J. H. Stathis, and E. Cartier, Microelectron Eng. **36**, 219 (1997).
41. M. J. Uren, V. Nayar, K. Bruinson, C. J. Anthony, J. H. Stathis, and E. Cartier, J. Electrochem. Soc. **145**, 683 (1998).
42. T. D. Mishima, P. M. Lenahan, and W. Weber, Appl. Phys. Lett **76**, 3771 (2000).
43. J. P. Campbell and P. M. Lenahan, Appl. Phys. Lett. **80**, 1945 (2002).
44. G. Pourtois, M. Houssa, B. De Jager, B. Kaczer, F. Leys, M. Meuris, M. Caymax, G. Groeseneken, and M. M. Heyns, Appl. Pjus. Lett. **91**, 023506 (2007).

45. S. Koveshnikov, W. Tsai, I. Ok, J. C. Lee, V. Torkanov, M. Yakimov, and S. Oktyabrsky, Appl. Phys. Lett. **88**, 022106 (2006).
46. J. Franco et al., Int. Electron. Device Meeting 2010. p. 4.1. (2010).
47. A. Stesmans, P. Somers, and V. V. Afanas'ev, Phys. Rev. B **79**, 195301 (2009).
48. V. V. Afanas'ev, M. Houssa, A. Stesmans, R. Souriau, R. Loo, and M. Meuris, Appl. Phys. Lett. **95**, 222106 (2009).
49. V. V. Afanas'ev, Y. Fedorenko, and A. Stesmans, Appl. Phys. Lett. **87**, 032107 (2005).
50. V. V. Afanas'ev and A. Stesmans, Mater. Sci. Semicond. Proc. **9**, 764 (2006).
51. M. Houssa, G. Poutois, M. Caymax, M. Meuris, M. M. Heyns, V. V. Afanas'ev and A. Stesmans, Appl. Phys. Lett. **93**, 161909 (2008).
52. D. L. Griscom and E. J. Friebele, Phys. Rev. B **34**, 7524 (1986).
53. D. L. Griscom, In: *Glass:Science and Technology.* Edited by D. K. Uhlmann and N. J. Kreidl (Academic, New York, 1990), p. 199.
54. K. Vanheusden and A. Stesmans, Appl. Phys. Lett. **62**, 2406 (1993).
55. W. L. Warren, D. M. Fleetwood, M. R. Shaneyfelt, J. R. Schwank, P. S. Winokur, and R. A. B. Devine, Appl. Phys. Lett. **62**, 3330 (1993).
56. G. Buscarino, S. Angello, and F. M. Geraldi, Phys. Rev. Lett. **94**, 125501 (2005).
57. M. Jivanescu, A. Stesmans, and V. V. Afanas'ev, Phys. Rev. B **83**, 094118 (2011).
58. A. Stesmans and F. Scheerlinck, Phys. Rev. B **51**, 4987 (1995).
59. A. Stesmans, F. Scheerlinck, and V. V. Afanas'ev, Appl. Phys. Lett. **64**, 2282 (1994).
60. P. M. Lenahan and P. V. Dressendorfer, J. Appl. Phys. **55**, 3495 (1984).
61. H. S. Witham and P. M. Lenahan, IEEE Trans. Nucl. Sci. **34**, 1147 (1987).
62. Y. Y. Kim and P. M. Lenahan, J. Appl. Phys. **64**, 3551 (1988).
63. L. P. Trombetta, G. J. Gerardi, D. J. DiMaria, and E. Tierney, J. Appl. Phys. **64**, 2434 (1988).
64. J. F. Conley Jr., P. M. Lenahan, H. L. Evans, R. K. Lowry, and T. J. Morthorst, J. Appl. Phys. **76**, 2872 (1994).
65. V. V. Afanas'ev and A. Stesmans, J. Phys.: Condens. Matter 12, 2285 (2000).
66. V. V. Afanas'ev and A. Stesmans, Europhys. Lett. **53**, 233 (2001).
67. V. V. Afanas'ev, G. J. Adriaenssens, and A. Stesmans, Microelectron. Eng. **59**, 85 (2001).
68. J. T. Ryan, P. M. Lenahan, T. Grasser, H. Eichlmair, Appl. Phys. Lett. **96**, 223509 (2010).
69. V. V. Afanas'ev and A. Stesmans, Mater. Sci. Eng. B **58**, 56 (1999).
70. F. J. Feigl, D. R. Young, D. J. DiMaria, S. Lai, and J. Calise, J. Appl. Phys. **52**, 5665 (1981).
71. A. Hartstein and D. R. Young, Appl. Phys. Lett. **38**, 631 (1981).
72. R. A. Gdula, J. Electrochem. Soc. **123**, 42 (1976).
73. J. M. M. de Nijs, K. G. Druif, V. V. Afanas'ev, E. van der Drift, and P. Balk, Appl. Phys. Lett. **65**, 2428 (1994).
74. E. Cartier and J. H. Stathis, Microelectron. Eng. **28**, 3 (1995).
75. K. G. Druijf, J. M. M. deNijs, E. van der Drift, V. V. Afanas'ev, E. H. A. Granneman, and P. Balk, J. Non-Cryst. Solids **187**, 206 (1995).
76. R. E. Stahlbush, In: *Physics and Chemistry of SiO_2 and Si-SiO_2 Interface –III.* Edited by H. Z. Massoud, E. H. Poindexter, and C. R. Helms. ECS Series Vol. 96 (1), 525 (1996).
77. A. G. Revesz, IEEE Trans. Nucl. Sci. **24**, 2102 (1977); J. Electrochem. Soc. 126, 122 (1979).
78. F. B. McLean, IEEE Trans. Nucl. Sci. **27**, 1651 (1980).
79. E. H. Nicollian, C. N. Berglund, P. F. Schmidt, and J. M. Andrews, J. Appl. Phys. **42**, 5654 (1971).
80. R. Gale, H. Chew, F. J. Feigl, and C. W. Magee, In: *The Physics and Chemistry of SiO_2 and the Si-SiO_2 Interface.* Edited by C. R. Helms and B. E. Deal (Plenum, New-York, 1988), p.177.
81. N. S. Saks and R. W. Rendell, IEEE Trans. Nucl. Sci. **39**, 2220 (1992).
82. N. S. Saks and R. W. Rendell, Appl. Phys. Lett. **61**, 3014 (1992).
83. V. V. Afanas'ev, J. M. M. de Nijs, and P. Balk, Appl. Phys. Lett. **66**, 1738 (1995).
84. V. V. Afanas'ev, A. G. Revesz, G. A. Brown, and H. L. Hughes, J. Electrochem. Soc. **142**, 1983 (1995).
85. J. W. Lyding, K. Hess, and I. C. Kizilyalli, Appl. Phys. Lett. **68**, 2526 (1996).

86. M. Houssa, M Aoulaiche, S. DeGendt, G. Groeseneken, M. M. Heyns, and A. Stesmans, Electrochem. Solid State Lett. **9**, G10 (2006).
87. C. T. Sah, J. Y. Sun, and J. J. Tzou, Appl. Phys. Lett. **43**, 204 (1983).
88. J. I Pankove, D. E. Carlson, J. E. Berkeyheiser, and R. O. Wance, Phys. Rev. Lett. **51**, 2224 (1983).
89. G. G. Deleo and W. B. Fowler, Phys. Rev. B **31**, 6861 (1985).
90. B. J. Mrstik and R. W. Rendell, Appl. Phys. Lett. **59**, 3012 (1991).
91. R. E. Stahlbush, A. H. Edwards, D. L. Griscom, and B. J. Mrstik, J. Appl. Phys. **73**, 658 (1993).
92. V. V. Afanas'ev, J. M. M. de Nijs, and P. Balk, J. Appl. Phys. **76**, 7990 (1994).
93. R. E. Stahbush, E. Cartier, and D. A. Buchanan, Microelectron. Eng. **28**, 15 (1995).
94. V. V. Afanas'ev and A. Stesmans, Appl. Phys. Lett. **72** , 79 (1998).
95. V. V. Afanas'ev and A. Stesmans, Phys. Rev. Lett. **80**, 5176 (1998).
96. V. V. Afanas'ev and A. Stesmans, Phys. Rev. B. **60**, 5506 (1999).
97. V. V. Afanas'ev and A. Stesmans, Mat. Sci. Eng. B **58**, 56 (1999).
98. Y. Li and C. T. Sah, J. Appl. Phys. **78**, 3156 (1995).
99. R. E. Stahbush, R. K. Lawrence, and H. L. Hughes, IEEE Trans. Nucl. Sci. **45**, 2398 (1998).
100. S. Fujieda, J. Appl. Phys. **89**, 3337 (2001).
101. V. K. Adamchuk and V. V. Afanas'ev, Progr. Surf. Sci. **41**, 111 (1992).
102. T. Grasser, H. Reisinger, P.-J. Wagner, and B. Kaczer, Phys. Rev. B **82**, 245318 (2010).

Chapter 10
Oxide Defects

Jian F. Zhang

Abstract This work reviews the recent progress in understanding defects in gate oxides, including acceptor-like electron traps, donor-like hole traps, and process-induced positive charges. Traps can be either as-grown or generated by electrical stresses and their differences will be pointed out. The physical mechanism responsible for trap creation will be examined and the two damaging species are identified: hydrogenous species and free holes in oxides. The key properties of traps will be reported, including trapping kinetics, capture cross sections, effective densities, energy levels, and physical locations. The impact of different types of traps on device performance will be discussed. The dielectrics covered by this work include SiO_2, SiON, HfO_2/SiON, and HfSiON/SiON and attentions will be paid to the similarity and differences between SiON and Hf-dielectric/SiON stack.

10.1 Introduction

The defects play a key role in the performance and reliability of MOS devices, and they have been investigated ever since the first generation of CMOS technology was developed in the early 1960s [1, 2]. As the technology progresses and the transistor dimensions are downscaled, the main reliability issues also change. In 1970s, the top reliability issue was the contamination, such as mobile ions and induced instability [3], which was overcome by using clean-room technology. In 1980s, the operation voltage was maintained at 5 V as downscaling continued, resulting in higher electrical field in the device. Hot carriers were limiting the lifetime of nMOSFETs [4, 5] and became the main reliability issue. When the gate oxide became thinner than 3 nm, gate leakage became considerable under operation bias. The defects built

J.F. Zhang (✉)
School of Engineering, Liverpool John Moores University, Byrom Street,
Liverpool L3 3AF, UK
e-mail: J.F.Zhang@Ljmu.ac.uk

T. Grasser (ed.), *Bias Temperature Instability for Devices and Circuits*,
DOI 10.1007/978-1-4614-7909-3_10, © Springer Science+Business Media New York 2014

up in gate oxides and the time-dependent dielectric breakdown (TDDB) became the main reliability concern in 1990s [6]. As the nitrogen density in gate oxides rises, the bias temperature instability (BTI) can lead to shorter pMOSFETs' lifetime than that of nMOSFETs and has attracted many attentions in 2000s [7, 8], the topic of this book. In the future, the degradation-induced time-dependent device variability will be a major issue [9].

On the mechanism responsible for BTI, it has been proposed that the PBTI of nMOSFETs with high-k dielectric stack is dominated by electron trapping [8, 10]. Agreement has not been reached on the mechanism of NBTI [11–19] and more details are given in [84] and [85]. The early NBTI test was performed on relatively thick oxide (>4 nm) under moderate electrical field (<7 MV/cm) with negligible carrier injection into the oxide [2, 11, 12] and hydrogen diffusion was proposed to limit the degradation rate [12]. This reaction–diffusion model also was used for thin oxides where substantial carrier injection occurred without [13] or with [14] taking hole trapping into account. More recently, the proposed models include a capture/emission time map model [16, 17], as described in details in [86], and a two-stage model, where interface state creation follows hole trapping [18]. An as-grown-generation (AG) model was proposed, where the effect of filling as-grown defects is combined with that of generating new defects [19].

Although defects can exist in both gate oxides and at the oxide/silicon interface, this work will focus on the defects in the oxides only. In oxides, there are acceptor-like electron traps and donor-like hole traps. Both of them can be either as-grown or created by electrical stresses. The understanding of these defects is still incomplete and this work will summarize their reported properties, including trapping kinetics, capture cross section, effective densities, energy levels, and spatial locations. The generation mechanism will be explored and the damaging species are identified. The impact of different types of defects on device performance will be discussed.

In addition to hole trapping and stress-induced positive charges, processing can also form positive charges in the oxides. It will be shown that they originate from hydrogenous species and can be either mobile or fixed. Their properties and relations with hole traps will be briefly reviewed.

This work mainly covers the experimental results measured by electrical techniques, which do not give direct information on the atomistic structure of the defects. The atomistic structure of the defects is addressed in [87]. The intention is to let the results speak for themselves and speculations will be limited. It will focus on the results reported by the author's research group, in anticipation that the results from other groups will be reviewed by their authors in this book.

10.2 Electron Traps

Electron traps are acceptor-like defects in oxides: negative when charged by stresses and neutral in a fresh device. For device-grade SiO_2 or SiON with poly-Si gate, there is little as-grown, in another term, preexisting, electron traps. Electrical stresses,

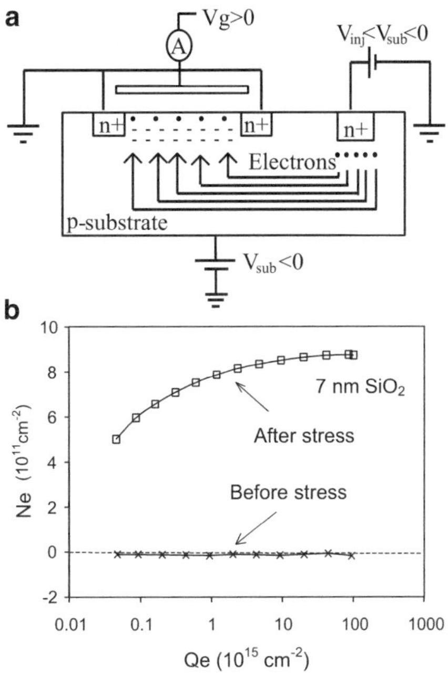

Fig. 10.1 Electron trap generation by electrical stresses (**a**) shows the substrate hot electron injection (SHE) technique and (**b**) compares the trapping before and after electrical stresses. Qe is evaluated from Eq. (10.1). Ne is the effective density of trapped electrons [20]

however, can generate electron traps. In this section, the generation mechanism and damaging species will be explored first, followed by reporting the properties of generated traps. Attentions will then be paid to as-grown electron traps, which can be significant for high-k stack. Finally, the impact of different types of electron traps on device performance will be briefly discussed.

10.2.1 Mechanism of Electron Trap Generation

The concept of trap generation is well known, but not often clearly defined. If a trap is as-grown, the trap is already there in a fresh device and the time for filling it the first time is the same as that for refilling it after neutralization. A trap is defined as "generated" if the stress time or carrier fluency required for filling it for the first time is clearly longer or larger than that for refilling. It takes longer the first time, since the trap is not there initially and has to be generated before filling [20, 21].

One example is given in Fig. 10.1 [20, 21]. To separate the trap filling from the trap generation, the generation during the filling step should be minimized. This was achieved by using the substrate hot electron injection (SHE) technique, which allows electrons being injected under relatively low oxide field. As illustrated in Fig. 10.1a, a pn junction next to an nMOSFET was forwardly biased with $Vinj < Vsub < 0$. The electrons from this pn junction were attracted towards the oxide/substrate interface,

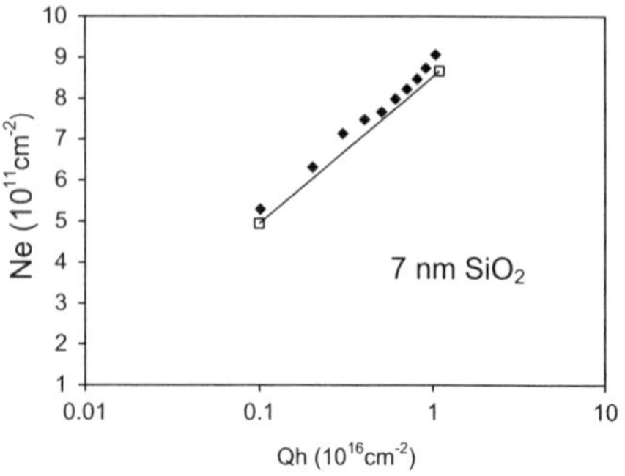

Fig. 10.2 Electron trap generation in two devices: electron–hole recombination for symbol *open square* is one order of magnitude less. Qh is the number of injected holes during stress [21]

accelerated in the space charge region, and some of them were injected into the oxide. The electron fluency, Qe, is evaluated from [8, 20, 21],

$$Qe = \int_0^{tinj} Jgdt/q + \Delta VthCox/q,$$

where tinj is the injection time, Jg the gate current per unit area, ΔVth the trapping-induced threshold voltage shift, Cox the oxide capacitance per unit area, and q one electron charge.

Figure 10.1b shows that when SHE was performed on a fresh device, there is little electron trapping, confirming that as-grown electron traps are negligible in a SiO_2 layer with poly-Si gate and there is little trap generation during the filling step. After electrical stress, however, electron trapping becomes substantial, because of the trap generation during the stress.

At least four models have been proposed for electron trap creation: electron–hole recombination [22], high oxide field [23], hydrogen-oxide interaction [24], and free hole-oxide interaction [25]. They will be examined one by one.

Electron–hole Recombination Model: It is proposed that traps were created by the energy released when a trapped hole captures an electron in oxides [22]. By using the substrate hole injection technique, the number of electron–hole recombination can be controlled. Figure 10.2 shows that an increase of electron–hole recombination by one order of magnitude has little effects on the number of generated electron traps.

High-Field Model: It is proposed that traps were created by the high oxide field-induced energy directly [23]. This model predicts that an increase of oxide field always leads to a higher number of generated electron traps. Figure 10.3, however, clearly shows that this is not the case.

Free Hole-Oxide Interaction Model: It is proposed that the interaction of free holes with the oxide causes trap creation [25, 26] and the electrical field over the oxide only plays the role of facilitating hole injection. The mobility of holes in the oxides is six orders of magnitude smaller than that of electrons [27] and this slow

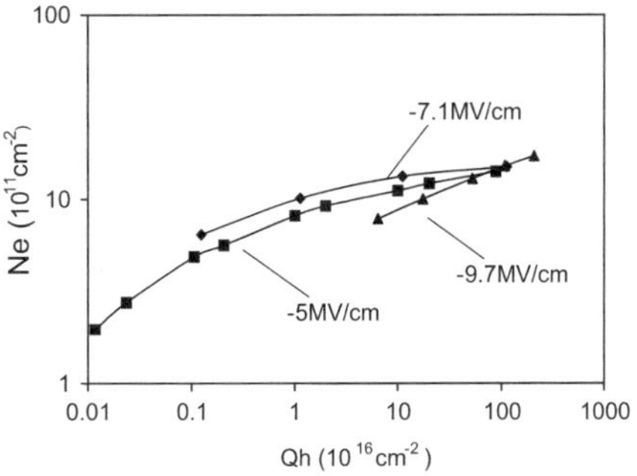

Fig. 10.3 Insensitivity of electron trap generation to oxide field during stress by substrate hole injection [21]

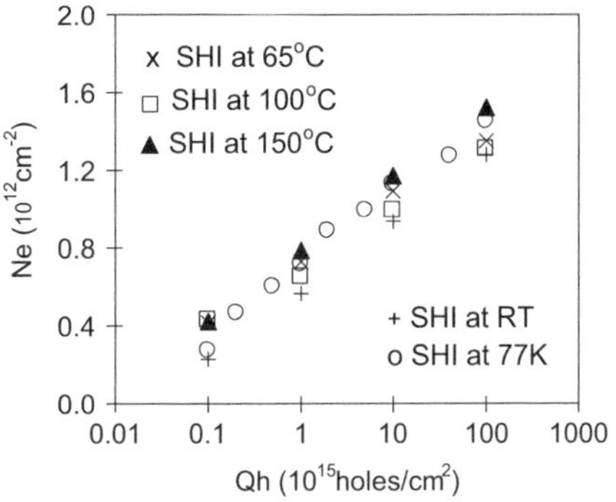

Fig. 10.4 Insensitivity of electron trap generation to temperature during stress by substrate hole injection [26]

movement causes "fatigue" of oxides. The details of the generation process remain unknown. Figure 10.4 shows that injection of holes alone can create electron traps and this generation process is insensitive to temperature [26]. It should be pointed out that at liquid nitrogen temperature of 77 K, hydrogen species are frozen and cannot contribute to degradation [26, 28].

Hydrogen-Oxide Interaction Model: Electrical field over oxides again only plays a role of facilitating the release of hydrogenous species [21, 24]. When electrons arrive at the anode, i.e., the positive terminal of gate oxide capacitor, they drop into the conduction band of the anode and the released energy can be used to free hydrogenous species in the device. When the hydrogenous species travel through oxides, they create electron traps. The details of these hydrogenous species and their reaction with the oxide remain to be resolved. However, there is little doubt that hydrogen can create electron traps. Figure 10.5 shows that for the same amount of injected holes, addition of electron injection substantially increases trap generation through hydrogen release [21, 24]. Unlike hole-induced electron trap generation, the hydrogen-induced generation is a thermally activated process [29].

In summary, electron traps can be generated through the interaction of oxide with either holes or hydrogenous species. Their relative importance will depend on their concentration and stress conditions.

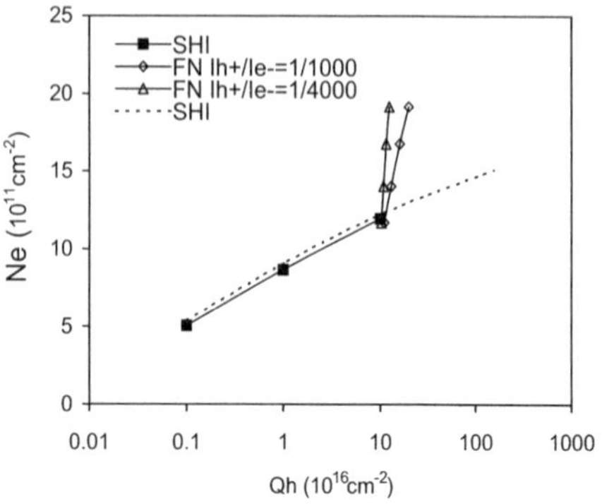

Fig. 10.5 Addition of electron injection at a hole/electron ratio of 1/1,000 and 1/4,000 enhances trap generation [21]

10.2.2 Properties of Generated Electron Traps

Capture Cross Sections: One key property for electron traps is the capture cross section, which is their effective physical size in oxides. To extract it, a trapping model is needed and the first-order model is well known [25, 30]:

$$Ne = \sum_{i=1}^{m} Ni[1 - \exp(-\sigma i Qe)], \tag{10.1}$$

where Ne is the effective density of trapped electrons by assuming all traps are at the oxide/Si interface [25, 30]. Ni is the saturation level of traps with a capture cross section of σi. Qe is the number of electrons injected into the oxide that can fill traps. "m" is the number of discrete capture cross sections existing in oxides.

It must be pointed out that the above model is only applicable when trap generation and detrapping are negligible in comparison with trapping during filling, since the model does not include either the generation or detrapping processes. Test conditions should be tuned to clearly separate the trap filling phase from the generation and detrapping. Under many stress conditions, such as high-field tunneling, trapping, detrapping, and trap generation occur simultaneously and this model must not be used to extract capture cross sections in such cases.

Another challenge for applying this model is how to select the number of capture cross sections, i.e., the value of "m." Some researchers [31] doubted the correctness of this model, since it could be used to fit any data by increasing "m" and the number of fitting parameters. Such "curve-fitting exercise" gave wrong values for the capture cross sections and erodes confidence in the validity of the model.

After suppressing generation and detrapping during trap filling, Fig. 10.6a shows that two capture cross sections are needed to fit the generated electron traps [30]. To avoid abusing this model, one must justify an increase of "m" above m = 1. In Fig. 10.6b, the trapping after different stress levels are compared. The first impression is that the trapping at Qe = 4×10^{14} cm^{-2}, i.e., Ne1, increases substantially, while further trapping beyond Qe = 4×10^{14} cm^{-2} only increases modestly. Figure 10.6c compares the two saturation levels extracted by using Eq. (10.1)

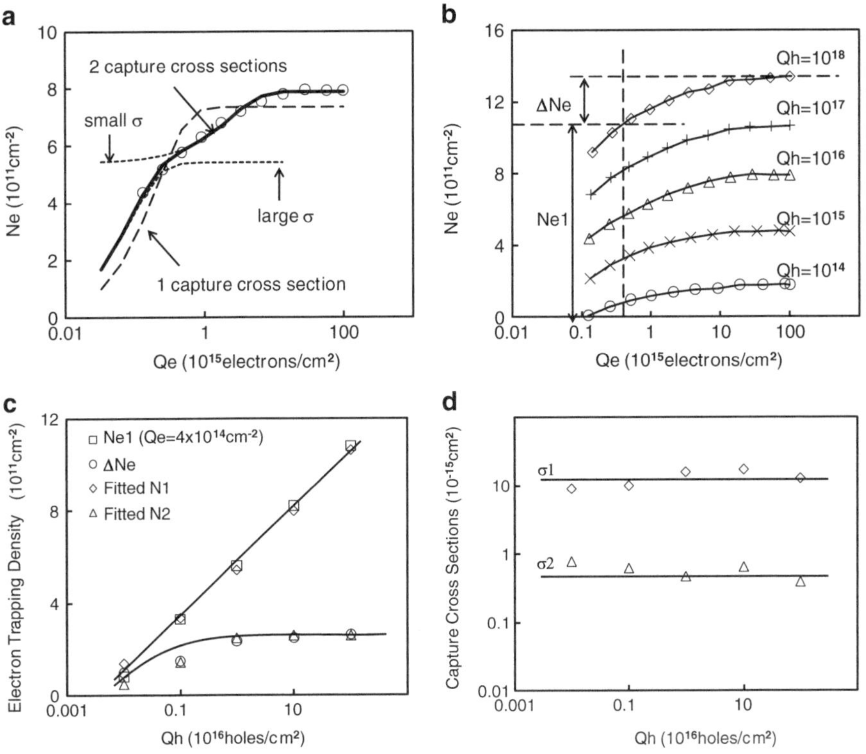

Fig. 10.6 (**a**) Fitting the generated electron trap with two capture cross sections. (**b**) Shows trapping after different stress levels. Ne1 is the trapping level at Qe = 4×10^{14} cm^{-2} and ΔNe is the further trapping for Qe > 4×10^{14} cm^{-2}. (**c**) Compares the extracted N1 and N2 with Ne1 and ΔNe. (**d**) Shows that the extracted capture cross sections are insensitive to stress levels [30]

with the directly observed values and the good agreement gives confidence to the existence of two capture cross sections. Figure 10.6c also shows that the N1 for σ1 can vary independently from the N2 for σ2 [29, 30]: The N1 increases continuously with stress, while N2 does not. The two extracted capture cross sections are well separated, in the order of 10^{-14} cm^2 and 10^{-15}–10^{-16} cm^2, and Fig. 10.6d shows that they are insensitive to stress levels. These support the existence of two capture cross sections for the generated electron traps and the validity of the first-order model.

Figure 10.7a compares the trapping kinetics for electron traps generated by different types of stresses [30]. Holes are the dominant damaging species during the substrate hole injection (SHI), while hydrogen was released by the electrons under Fowler-Nordheim tunneling (FNI). There is little difference in the trapping kinetics, indicating that the same types of electron traps were created under different stress conditions [25, 30]. Figure 10.7b compares the trapping in oxides of different thicknesses. The two data sets can be fitted within the same capture cross sections [30].

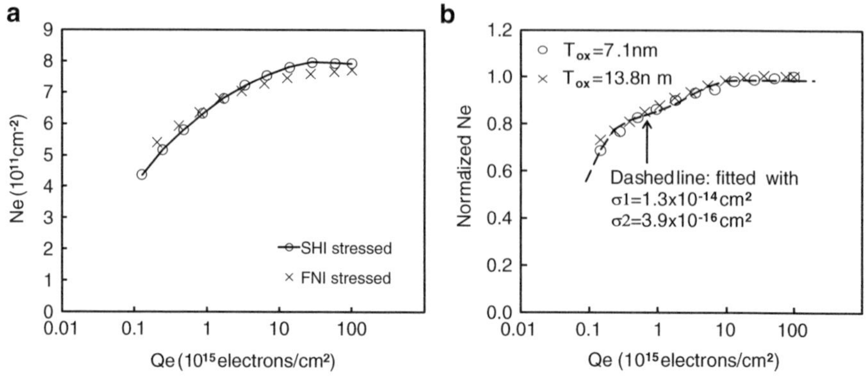

Fig. 10.7 The insensitivity of trapping kinetics for electron traps generated under different stress conditions (**a**) and in oxides of different thicknesses (**b**) [30]

Fig. 10.8 Volume density of created electron traps [29]

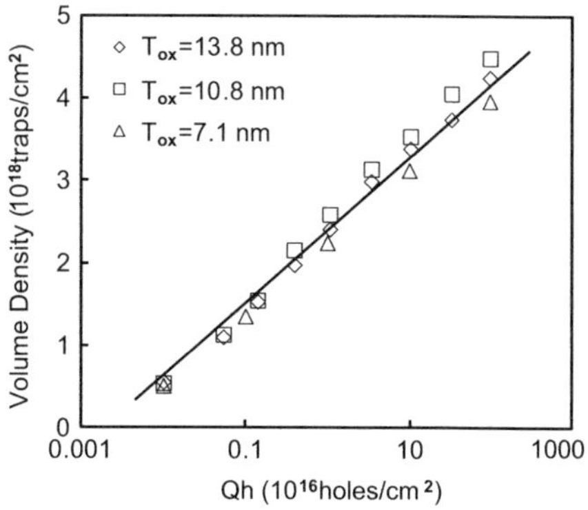

Spatial Location: Figure 10.8 shows that the volume density of generated electron traps is insensitive to the oxide thickness, supporting a uniform spatial distribution on a macroscopic scale. For a device-grade oxide, it has been suggested that the electron trap generation is intrinsic and randomly distributed in oxides [6]. A random spatial distribution will give a constant volume density on a macroscopic scale, in agreement with Fig. 10.8.

Energy Levels: Figure 10.9a gives the trapping level for the same amount of injected electrons under different oxide field strength, i.e., Eox. An increase of Eox allows shallow traps to detrap through tunneling. When Eox reaches 8 MV/cm, however, the detrapping stops and some deep traps can keep their electrons even under Eox $>$ 10 MV/cm. Figure 10.9b shows that the generation of shallow traps is insensitive to temperature, while the creation of deep traps is thermally accelerated, supporting that shallow and deep traps are of different types.

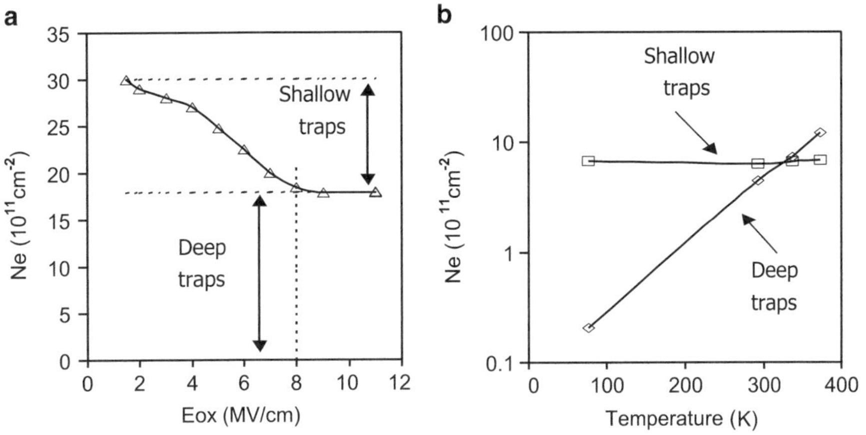

Fig. 10.9 (**a**) Filling the generated traps by injecting 10^{17} cm^{-2} electrons into the oxide under different oxide field. The injection under low Eox was achieved by accelerating electron in the substrate (see Fig. 10.1a). (**b**) The dependence of shallow and deep trap generation on temperature [29]

10.2.3 As-Grown Electron Traps

SiO_2 and SiON: As-grown electron traps in SiO_2 are extrinsic and can originate from a range of contaminants, such as metal ions, dopants, water, and other hydrogenous species. The ions can be positively charged and it was reported that they are coulombic attractive electron traps with a capture cross section as large as 10^{-12} cm^2 at low oxide field [3, 32]. One feature of these coulombic electron traps is that their capture cross section reduces for higher electrical field [3]. Although these huge traps have been eliminated in a device-grade SiO_2 or SiON, they could appear when new materials and dielectrics are used during process development and one example is given in Fig. 10.10.

Arsenium, a donor for silicon, was reported to be an electron trap in oxides and a good correlation between the doping density in oxides and electron traps were observed [34]. An exposure of SiO_2 to water introduces electron traps with a well-defined capture cross section around 10^{-17} cm^2 [35]. When aluminum was used as the gate, hydrogen could be released from its interface with SiO_2 during post-metallization anneal, resulting in electron traps with capture cross sections in the order of 10^{-17}–10^{-18} cm^2 [25, 36, 37]. The SiO_2 prepared under low temperature contains more hydrogenous species and, in turn, more as-grown electron traps and annealing at high temperature, e.g., 960 °C, greatly reduce these traps [38]. The use of poly-Si gate, clean-room technology, and process control and optimization has effectively suppressed these extrinsic as-grown electron traps in SiO_2 and SiON for modern CMOS technologies.

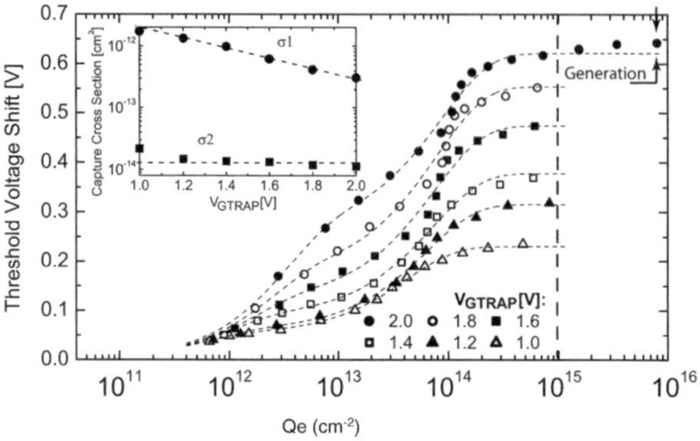

Fig. 10.10 Electron trapping-induced threshold voltage shift for an HfO_2/SiO_2/Si-cap/Ge/Si-substrate stack. The *inset* shows that the large capture cross section reduces for a higher V_{GTRAP} used to fill the traps [33]

Fig. 10.11 Electron trapping during the first and second filling for an HfO_2/SiON stack [42]

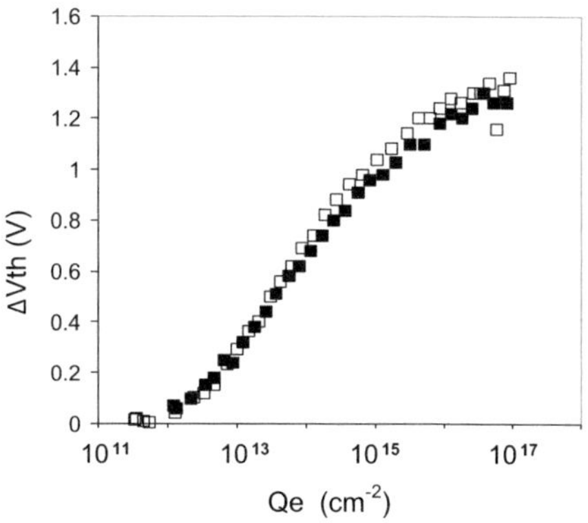

High-k/SiON Stack: The as-grown electron trapping can be substantial in a high-k/SiON stack even if the fabrication was carried out under ultra clean environment. They were a major instability issue for nMOSFETs during the early stage of high-k/SiON development [39–41]. Figure 10.11 shows that electron trapping can induce a gate voltage shift over 1 V [42]. Their first filling is as fast as the second, so that they are as-grown, rather than generated. In the following, their measurement issues will be addressed and their properties will be reported.

Measurement Issues: Three issues are measurement speed, impact of trapping on electron injection, and dependence on conduction mechanism.

Electron trapping in thin dielectric layers is highly dynamic and sensitive to measurement delay. For conventional quasi-DC parameter analyzers, it takes milliseconds for measuring one point and Fig. 10.12 shows that trapping can be substantially underestimated due to detrapping during the measurement. More details of the fast measurement technique are given in [88]. The presence of

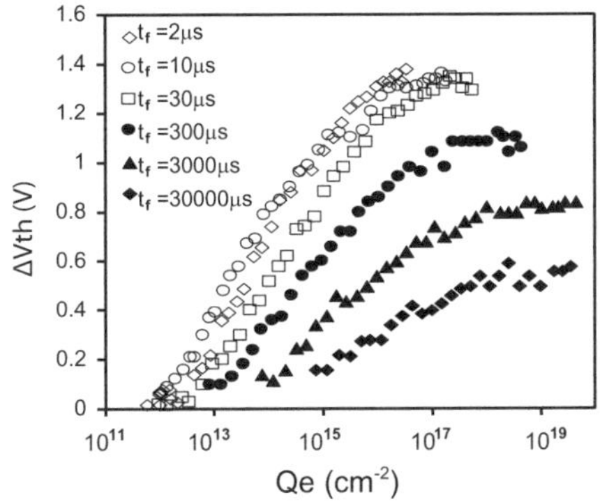

Fig. 10.12 Dependence of electron trapping on measurement time for an HfO_2/SiON stack [42]

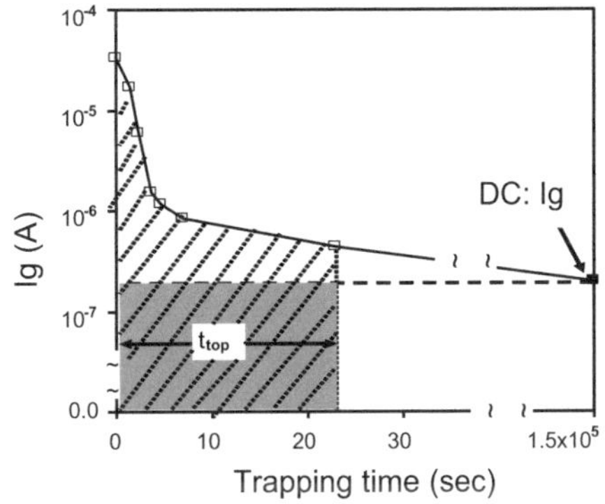

Fig. 10.13 Electron trapping reduces Ig substantially. The conventionally measured DC Ig is marked out [40]

detrapping also makes the recorded data unsuitable for extracting capture cross sections. Figure 10.12 shows that a measurement speed of microseconds is needed to suppress detrapping [40, 42].

To use the first-order model (1) for extracting the capture cross section, one must know the number of electrons available for trapping in the dielectric. As trapping continues, negative space charges buildup in the oxide, which reduces the field near cathode and in turn the gate current Ig. Figure 10.13 shows that Ig can reduce by two orders of magnitude in tens of microseconds [33, 40]. As illustrated in Fig. 10.13, the conventionally measured quasi-DC Ig does not take this large Ig transience into account, so that it underestimates Ig during the trapping period. If this quasi-DC Ig is used to calculate Qe, Qe will be underestimated by one order of magnitude approximately. This will lead to an overestimation of capture cross section by a factor of 10 and a correction like Fig. 10.13 is essential [40].

Another issue is that not every electron passing through the stack can fill traps. Figure 10.14a shows that electron trapping at 25 °C is higher than that under 110 °C for the same level of electron fluency. This is, however, an artifact, since the thermally enhanced conduction shown in the inset of Fig. 10.14b actually does not contribute to trapping. When the trapping at 110 °C is plotted against the electron fluency at 25 °C, Fig. 10.14b shows that trapping is insensitive to temperature [39].

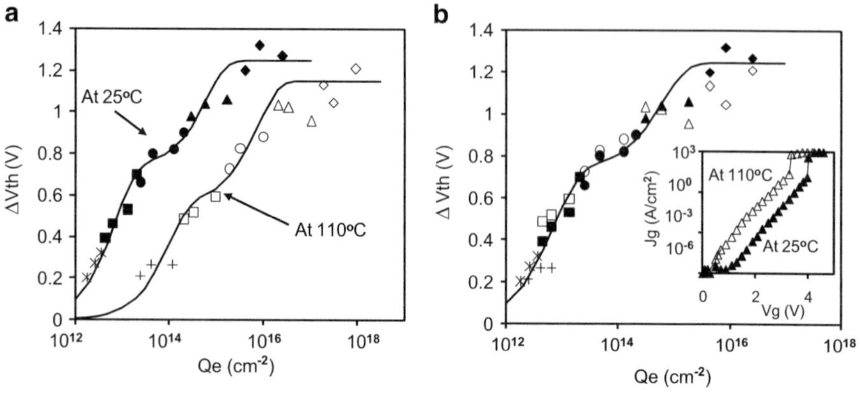

Fig. 10.14 Electron trapping at different temperatures. (**a**) Qe was calculated from the Ig in the *inset* of (**b**). (**b**) Qe at 25 °C was used for both sets of data [39]

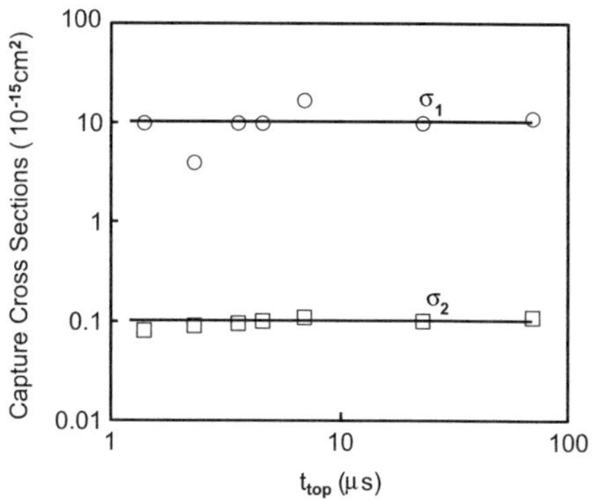

Fig. 10.15 The capture cross sections of as-grown electron traps in HfO_2/SiON stack [40]

Properties: After taking the impact of trapping on Ig into account, the extracted capture cross sections are given in Fig. 10.15 [40]. Two well-separated capture cross sections are in the order of $\sigma_1 = 10^{-14}$ cm^2 and $\sigma_2 = 10^{-16}$ cm^2, respectively. The σ is not sensitive to the stress and fabrication conditions, although the trap density is [40]. These two capture cross sections are similar to those for the electron traps generated in SiO_2 by electrical stress [29, 30].

On the spatial location, since there are little as-grown electron traps in SiON [20, 21, 30], two assumptions for their locations in the Hf-dielectric/SiON stack are in the Hf-dielectric layer uniformly or at the Hf-dielectric/SiON interface. To test them, the thickness of Hf-dielectric is varied without changing the interfacial layer (IL). Figure 10.16 shows that both propositions do not agree with the test data [8, 42]. Good agreement is obtained by assuming traps being in the central region of Hf-dielectric layer. As the thickness of Hf-dielectric is downscaled, electron trapping reduces sharply for a properly processed stack [8, 42]. One may speculate that thicker Hf-dielectric layer is easier to crystallize and traps may be formed along grain boundaries.

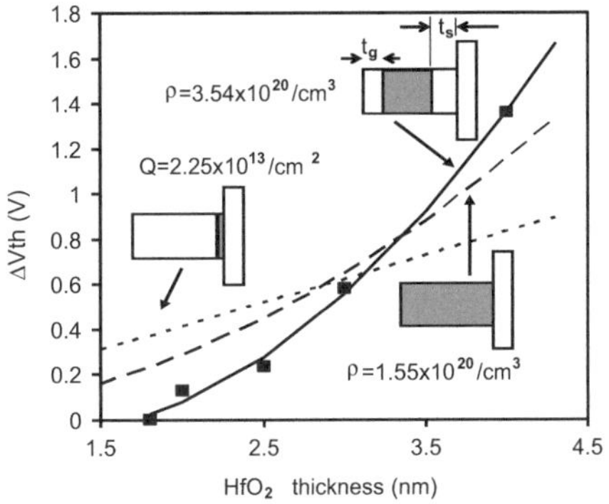

Fig. 10.16 Dependence of traps on HfO_2 thickness. The data supports that traps are in the bulk of HfO_2 layer [8]

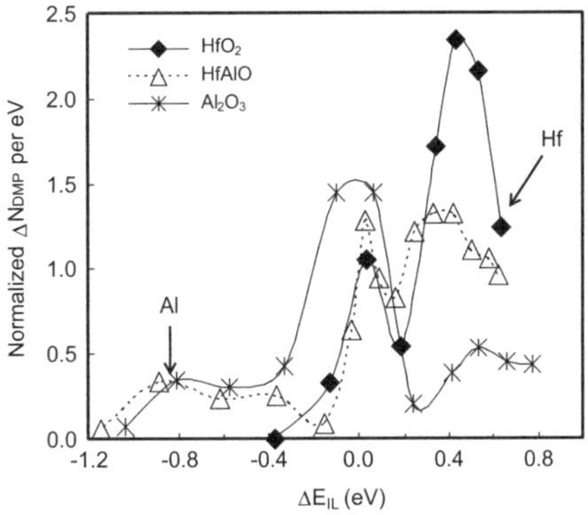

Fig. 10.17 Energy distribution of as-grown electron traps in high-k-dielectric layer. ΔE_{IL} is the energy drop over the interfacial layer, with reference to Si Ec at the Si/IL interface [43]

The energy distribution is given in Fig. 10.17. Most traps in HfO_2 are above the conduction band edge of silicon, making them readily dischargeable under a negative gate bias [43]. The energy level is relatively deep in Al_2O_3.

10.2.4 Impacts on Device Performance

SiO_2 and SiON: The gate SiON used in modern CMOS technologies is generally less than 2 nm and the generated electron traps cannot hold their electrons steadily due to effective detrapping by tunneling. As a result, the electron trapping-induced PBTI is insignificant for SiON when compared with NBTI. The generated electron traps, however, can lead to a stress-induced leakage current (SILC) and a time-dependent dielectric breakdown (TDDB) by forming a conduction path between two electrodes [6]. The defects responsible for TDDB must possess the following signatures: the thermally accelerated generation, non-saturation with stress time, and a capture cross section in agreement with the defect size obtained by fitting the TDDB data. Figure 10.9b shows that the creation of shallow traps is not thermally accelerated, but the generation of deep traps is. Figure 10.6c shows that the trap of a capture cross section of 10^{-15}–10^{-16} cm^2 saturates, but the trap of 10^{-14} cm^2

Fig. 10.18 Dependence of as-grown electron trapping-induced threshold voltage shift on processes for 2 nm HfO_2/SiON [40]

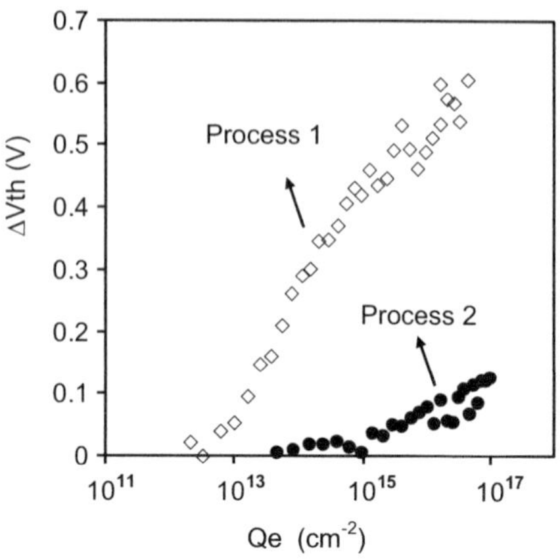

does not. The size of the latter also agrees with the defect size extracted from the TDDB data [6]. As a result, the deep trap of a capture cross section of 10^{-14} cm^2 has all the signatures required for TDDB.

Hf-Dielectric/SiON Stack: Although the trapping density can be reduced substantially by using sub-2 nm Hf-dielectric layer and Hf-silicates, as-grown electron trapping-induced PBTI of nMOSFETs is a reliability concern. The as-grown electron trap density is sensitive to detailed processing conditions [40, 44] and process optimization is essential. For example, Fig. 10.18 shows that two 2 nm HfO_2 fabricated by different processes can have large differences in their as-grown electron trap density [40].

10.3 Hole Traps

Although as-grown electron traps are negligible in SiO_2 or SiON, there are substantial amounts of as-grown hole traps in them, e.g., over 5×10^{12} cm^{-2} [45, 46]. This means that neither electron traps nor hole traps are amphoteric [25]. Electrical stresses can also create new hole traps and their complex behavior has caused a lot of confusions. Clear evidences will be presented to show that different types of hole traps are created and a framework will be proposed for positive charges in oxides, which is applicable to SiO_2, SiON, and high-k/SiON stack.

10.3.1 As-Grown Hole Traps

As-grown hole traps (AHT) will not be charged under a relatively low oxide field, e.g., $\leq$5 MV/cm, so that they cannot be responsible for the NBTI reported under similar field by the early works [2, 12]. To fill them at $Eox = -5$ MV/cm, substrate

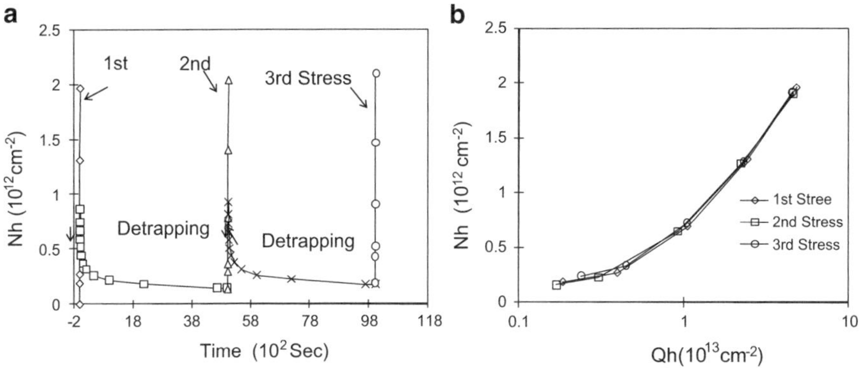

Fig. 10.19 The reversible trapping and detrapping of as-grown hole traps in SiO_2. (**a**) Shows the test sequence. (**b**) Compares the first and the subsequent trapping [47]

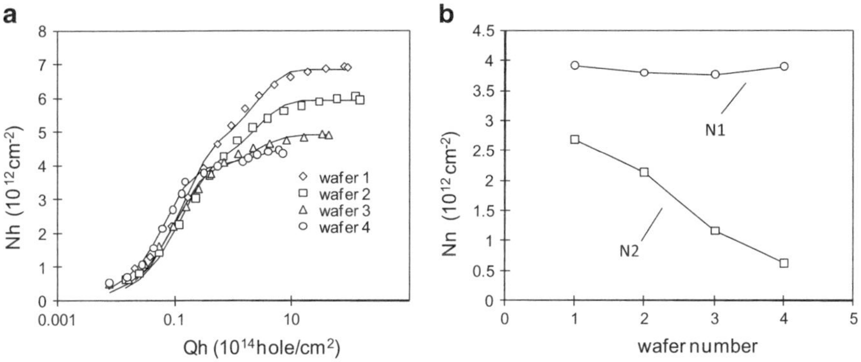

Fig. 10.20 (**a**) Filling as-grown hole traps in SiON processed under four different conditions. The *solid lines* are fitted with the first-order model. (**b**) The extracted effective density for the two capture cross sections. N1 is for the larger trap [48]

hot holes must be used [45, 46], indicating that they have an energy level below the top edge of Si valence band, i.e., Ev. Figure 10.19 shows that their first and subsequent fillings are the same and confirm that they are as-grown.

Similar to electron traps, Fig. 10.20a shows that hole trapping also follows the first-order model given in Eq. (10.1) and the two extracted capture cross sections are in the order of 10^{-13}–10^{-14} cm^2 and 10^{-15} cm^2, respectively [47]. To justify the presence of two discrete capture cross sections, Fig. 10.20b shows that the effective density of the large trap, i.e., N1, remains stable when processing conditions vary, while that of the small trap, i.e., N2, changes substantially [48]. An observation of Fig. 10.20a shows that the trapping is insensitive to process conditions at low hole fluency but varies considerably at high hole fluency, supporting the presence of two discrete capture cross sections. It is proposed that oxygen vacancies are hole traps [48–52] and the smaller trap is hydrogen related [50].

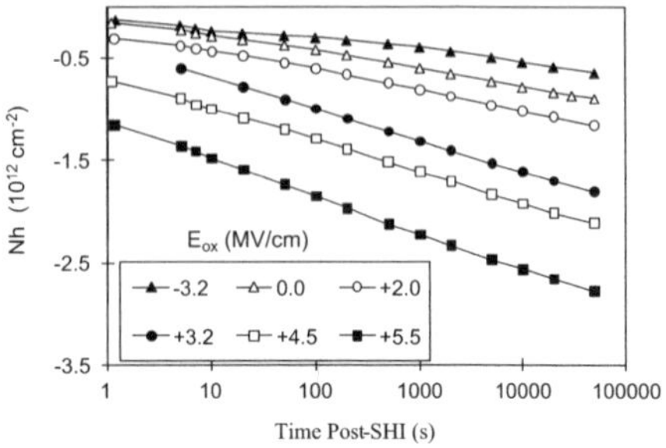

Fig. 10.21 Detrapping of AHT under different gate bias for a 13.8 nm SiO_2 [54]

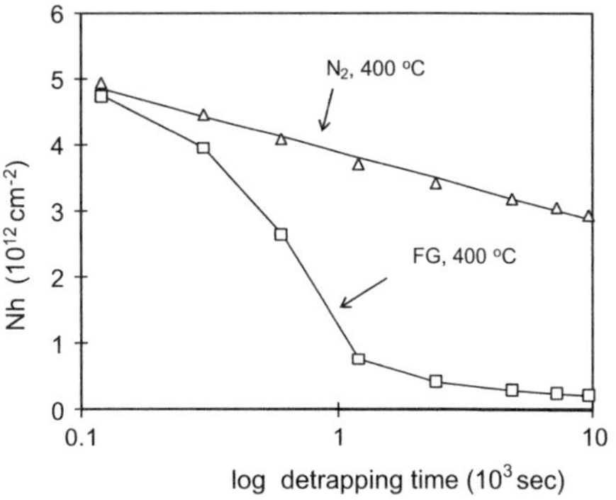

Fig. 10.22 Detrapping of AHT at 400 °C in different ambient gases for a 7 nm SiO_2 [50]

Unlike the random spatial distribution of generated electron traps, the location of AHT is biased towards the oxide/substrate interface [47, 53, 54]. This can be seen from the asymmetric detrapping against gate bias polarity. Figure 10.21 shows that detrapping under $Vg > 0$ is more efficient than that under $Vg < 0$, since trapped holes are closer to the substrate [54]. In fact, the detrapping under $Eox = -3.2$ MV/cm is slower than under $Vg = 0$, indicating that detrapping is negligible from gate.

The detailed spatial distribution of AHT is still missing. For thin SiON (e.g., <2 nm), the detrapping of AHT completes in seconds [8, 55], unlike the generated hole traps whose detrapping is more difficult, as detailed in Sect. 10.3.3. For thick SiO_2 (e.g., >7 cm), however, the detrapping of AHT lasts at least for days and its completion was not observed within a practical test time [50, 54]. Figure 10.22 shows that over half of the trapped holes can survive after an exposure to 400 °C in N_2 for 10^4 s in a 7 nm SiO_2 [50, 56], indicating that hole traps can be nanometers away from the interface. The linear detrapping with logarithmic time is a signature that detrapping is through carrier tunneling and the tunneling time increases exponentially with distance from the interface. Apart from tunneling, detrapping can also be achieved by interacting with H_2 and the resultant H^+ carried away the positive charge. Figure 10.22 confirms that the reaction with H_2 has a single characteristic time [50, 56].

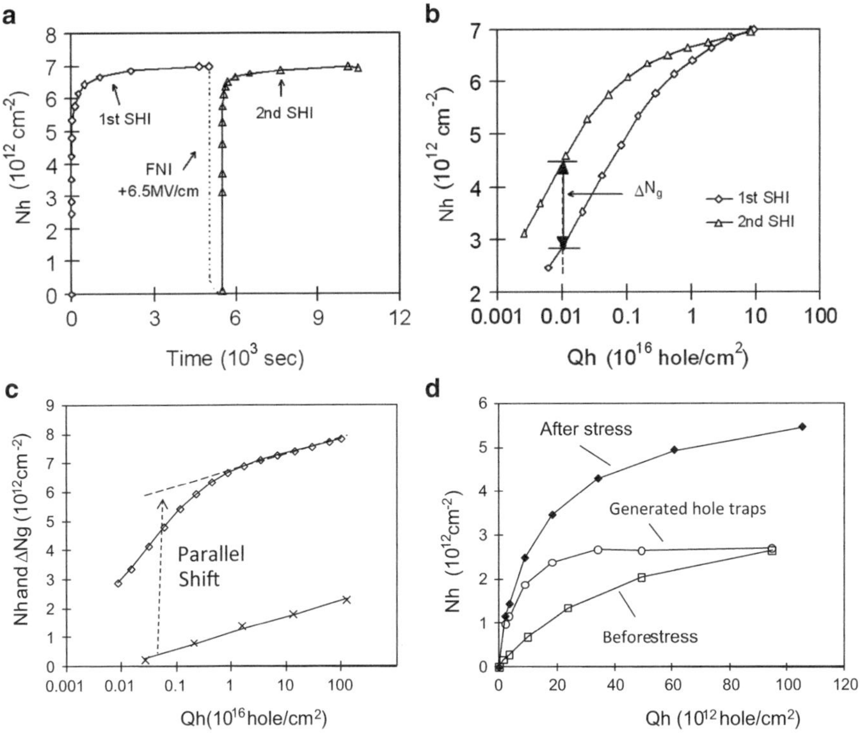

Fig. 10.23 Generation of hole traps (**a**) shows the test procedure, (**b**) compares hole trapping during the first and secnd hole injection, (**c**) shows that the generated traps increases with stress and controls trapping at high stress level, and (**d**) shows that generated traps have a single capture cross section [47]

10.3.2 Hole Trap Generation

Hole trap generation was not as well known as electron trap generation and early work [57] suggested that hole traps were fixed by fabrication conditions, probably because the high level of AHT masked out the generation. Figure 10.23, however, shows that hole traps can also be created. Following a stress and neutralization in Fig. 10.23a, the refilling in Fig. 10.23b is higher by ΔNg due to the new traps created by the preceding stress [47]. Figure 10.23c shows the expected increase of ΔNg with stress level. Importantly, this increase in ΔNg can fully explain the non-saturation of hole trapping as stress continues. The trapping at high stress level is controlled by the generation process, rather than filling. If the data set "◇" were used to fit the first-order model (1), it would result in small capture cross sections, which is an artifact. Figure 10.23d shows that the real σ of generated hole traps is around 10^{-13} cm^2.

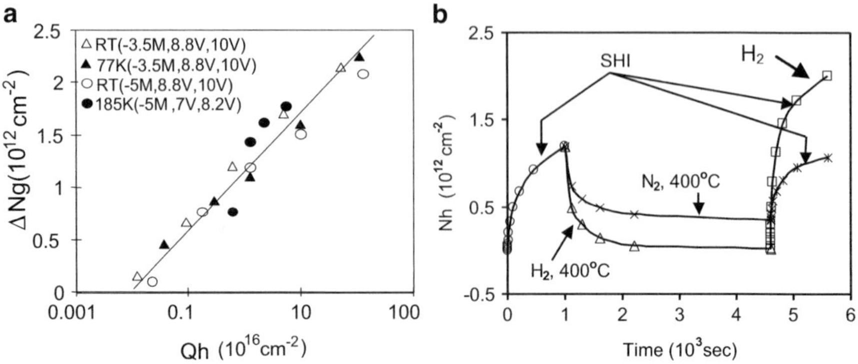

Fig. 10.24 Species for hole trap generation. (**a**) Shows hole injection can create hole traps [48]. (**b**) Shows that an exposure to H^+/H can generate hole traps [58]

Fig. 10.25 Dependence of positive charging on Vg polarity for devices with metal and poly-Si gates [60]

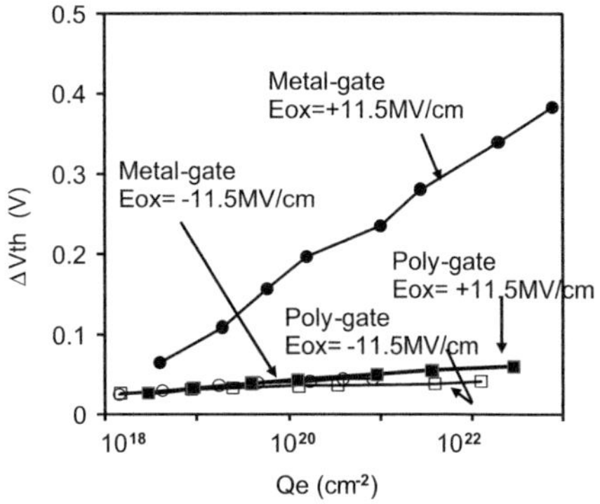

Generation Mechanism: Similar to electron trap generation, hole trap can be generated through the interaction of oxides with either free holes or hydrogen [48, 58]. When hole injection dominates stress, Figure 10.24a shows that the generation is insensitive to temperature down to 77 K, where hydrogen transportation is frozen [28, 48]. In Fig. 10.24b, a device was exposed to H_2 at 400 °C after hole trapping. The trapped holes can crack H_2 into H^+/H [56, 58, 59], which is reactive and leads to the higher subsequent trapping by creating new hole traps. When N_2 replaced H_2, Fig. 10.24b shows that there is no hole trap generation, confirming that the damaging species is hydrogen related [58].

Figure 10.25 shows that positive charging is substantially higher when metal gate is used as the anode ($Vg > 0$), when compared with the stress under $Vg < 0$ [60]. It appears that hydrogen species at the metal gate/dielectric interface can be released by electrons [60]. With poly-Si gate, positive charging becomes insensitive to Vg polarity. Figure 10.26 shows the generated hole traps for five wafers with different levels of nitridation. The generation is insensitive to nitridation.

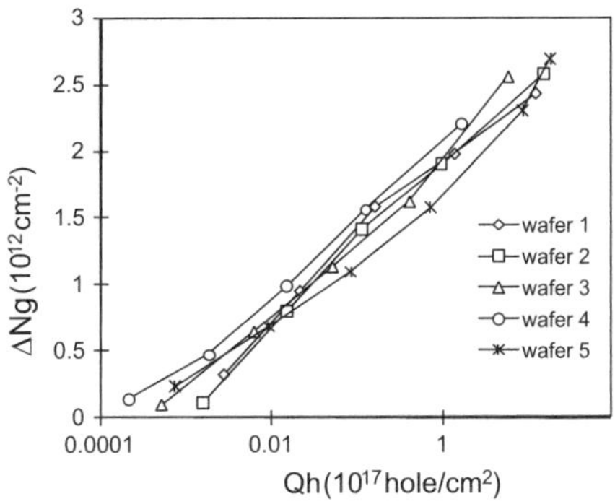

Fig. 10.26 Hole trap generation in five wafers of different nitridation levels [48]

10.3.3 A Framework for Positive Charges in Oxides

Positive charges (PCs) in oxides can behave in a complex way and different names have been used by different researchers, typically to capture only one of their many features. It is often for a researcher to call different types of PCs by the same name and this causes confusions. In the following, clear evidences will be given to show that different types of PCs exist in oxides and a framework will be proposed to clarify the confusions. Although this framework was initially proposed when devices were stressed by the substrate hole injection (SHI) [61], it is also applicable to the PCs formed under other stress conditions, including NBTI, and it will be shown that the same types of PCs were created under different stress conditions [8, 62–67]. Based on this framework, the inadequacies of the names used by early works will be pointed out.

The Framework: PCs in oxides consist of three different types: as-grown hole traps (AHT), cyclic positive charges (CPC), and antineutralization positive charges (ANPC) [8, 61, 62]. In Fig. 10.27a, substrate hole injection (SHI) was performed first under $Eox = -5$ MV/cm and the trapped holes built up. All trapped holes were then neutralized by an electron injection under $Eox = +6.5$ MV/cm. This was followed by applying $Vg > 0$ and $Vg < 0$ alternatively with $Eox = \pm 5$ MV/cm. Some positive charges can be repeatedly neutralized under $Vg > 0$ and recharged under $Vg < 0$, so that they are called as cyclic positive charge (CPC). CPC has an energy level around Ec (Fig. 10.27b, c). Part of PCs has energy level above Ec, so that their neutralization is more difficult than charging and they are referred to as antineutralization positive charge (ANPC).

As mentioned in the section on as-grown hole traps, AHT has energy level below Ev, allowing them being neutralized easily [8, 61, 62]. AHT cannot be charged under $Eox = -5$ MV/cm without switching on substrate hot hole injection (SHI) and is responsible for the difference in PCs under $Eox = -5$ MV/cm with and without SHI in Fig. 10.27a.

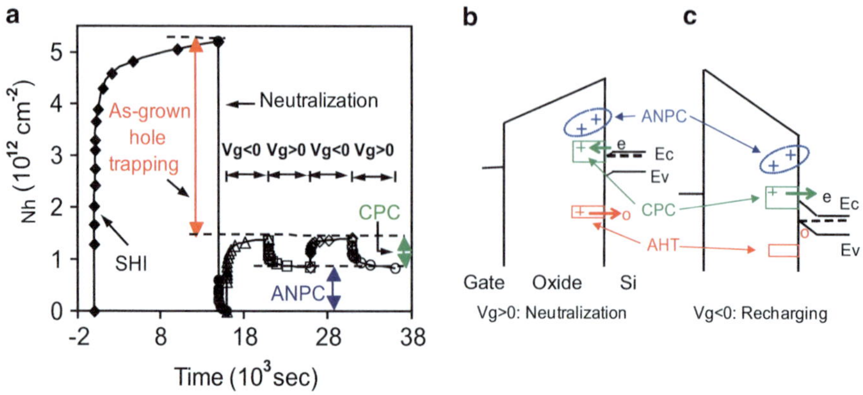

Fig. 10.27 A framework for PCs in oxides. (**a**) Shows the test sequence. SHI was carried out under Eox = −5 MV/cm and the neutralization was under Eox = +6.5 MV/cm. Vg < 0 was also under Eox = −5 MV/cm without switching on SHI and Vg > 0 is under Eox = +5 MV/cm. (**b**) and (**c**) show the three types of PCs during neutralization and recharging, respectively [61]

Fig. 10.28 Dependence of CPC and ANPC on stress time [63]

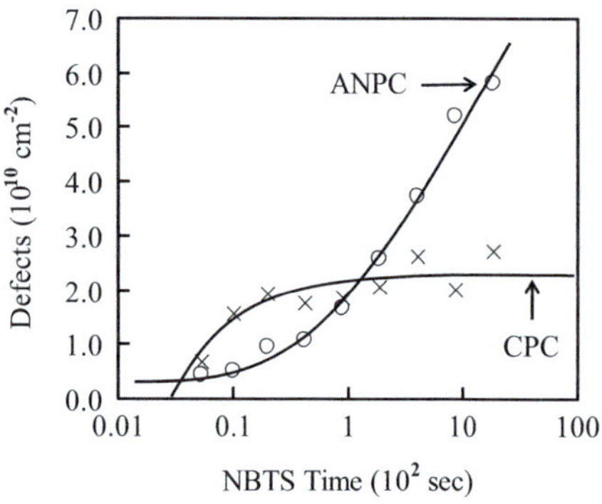

Justification of the Framework: The differences in the charging and discharging properties for the three types of PCs have been presented in Fig. 10.27, but further evidences are needed to confirm that they are different types of PCs. Figure 10.28 shows that the generation of CPC clearly saturates, but ANPC does not [61, 63]. Since ANPC is above Ec, its neutralization should be thermally accelerated as illustrated in Fig. 10.29. CPC is neutralized by tunneling, which should be insensitive to temperature. This prediction agrees with the results in Fig. 10.29a, b [8, 64, 65].

Figure 10.30a shows that when AHT was filled in considerable numbers, CPC and ANPC remains negligible and they only appear after a heavy stress. Both CPC and ANPC are the generated hole traps, therefore [61]. Although Fig. 10.23d shows that they have a capture cross section of $\sim 10^{-13}$ cm^2, their energy levels make them chargeable without substrate hole injection, as shown in Fig. 10.27, unlike the AHT. In Fig. 10.30b, CPCs and ANPCs were charged first and the SHI was then switched

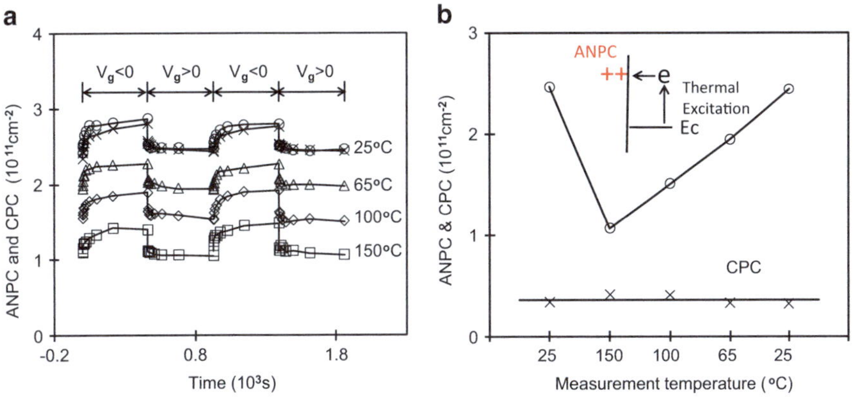

Fig. 10.29 Dependence of ANPC and CPC on measurement temperatures. After stress, ANPC and CPC were measured in a temperature sequence of 25, 150, 100, 65, and 25 °C [64]

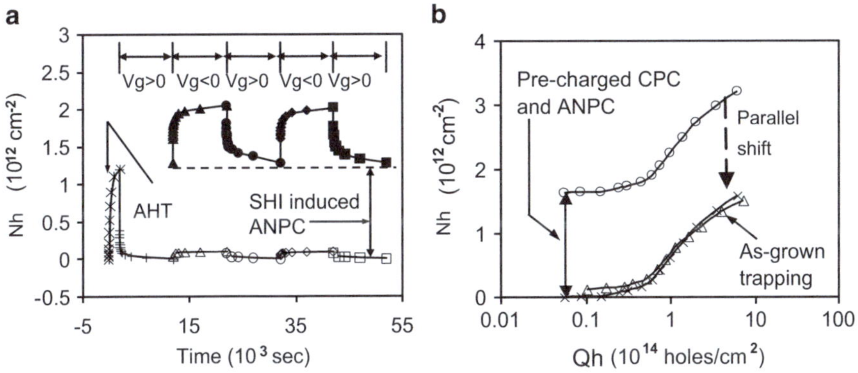

Fig. 10.30 (**a**) AHT can be charged with negligible CPC and ANPC. The *filled symbols* were the CPC and ANPC after a heavy stress. (**b**) The same amount of AHT were charged with/without generating and then charging CPC and ANPC [61]

on to fill the AHT. When compared with the filling of a fresh device, the results show that the same number of AHT were filled after generating and charging CPC and ANPC [61]. This independence of AHT from CPC and ANPC supports that they originate from different types of defects.

This framework of PCs is applicable to PCs formed under different stress conditions and in samples prepared under different process conditions. Figure 10.27 was obtained after stressing a 5.5 nm SiO_2 under substrate hot hole injection [61]. Figure 10.31a were obtained from a 2.7 nm SiON under negative bias temperature stress (NBTS) [64], while Fig. 10.31b were obtained from a HfO_2/SiON with an equivalent oxide thickness of 1.13 nm after NBTI stress [66].

The similarity of PCs in SiON and HfO_2/SiON stack can be understood, since the PCs in the stack are dominated by the interfacial SiON layers [8, 67]. Figure 10.32

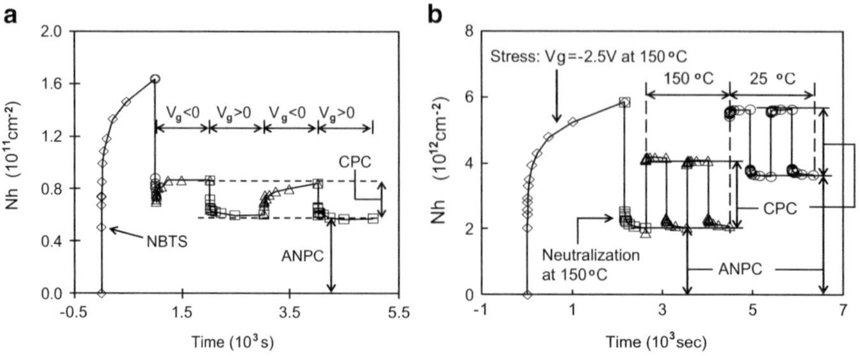

Fig. 10.31 Three different types of PCs generated by negative bias temperature stress in a 2.7 nm SiON (**a**) [64] and an HfO_2/SiON stack of EOT = 1.13 nm (**b**) [66]

Fig. 10.32 PCs in HfSiON/SiON stack is dominated by the SiON layer [8]

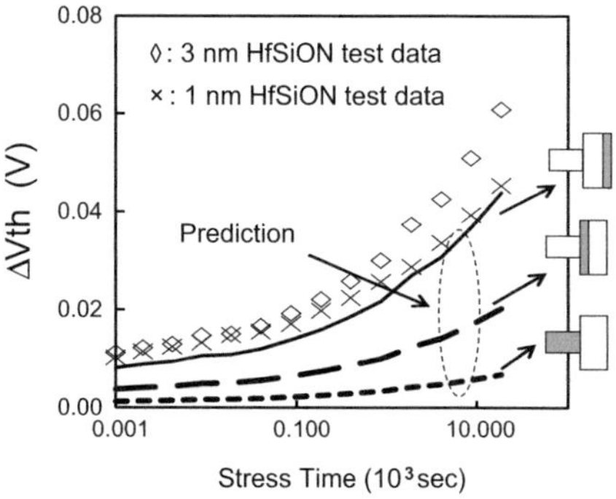

gives the result obtained from HfSiON/SiON stacks, where the HfSiON thickness varies for the same SiON thickness. It shows that test data does not agree with the propositions that PC is dominated by Hf-dielectric or piles up at the Hf-dielectric/SiON interface. Good agreement is achieved by assuming that PC is near to the substrate interface [8, 67]. As a result, in contrast with electron trapping in Fig. 10.16, a reduction of Hf-dielectric thickness will not reduce PC and the different spatial locations of PC and electron traps also rule out that they originate from the same defect.

Naming the Positive Charges: After clarifying that there are three different types of PCs in the oxide, we look at the shortcomings of the names used in early works. One of them is "anomalous positive charges" [68]. It was noted that during electron injection, "anomalous PCs" were formed, which could not be neutralized, compensated electron trapping, and results in a "turnaround" of the net trapping density [25, 68]. These anomalous PCs could be neutralized at higher temperature, a signature of the ANPC in the framework [25, 61, 64]. The term "anomalous" does not describe the charging/discharging behavior of PC directly, while "ANPC" tells readers that these PCs are difficult to neutralize and is therefore preferred.

Another term used is the “border traps” [69], indicating PCs are located somewhere between interface and oxide bulk. This term is too general and does not separate the three types of PCs, since all of them are located somewhere between interface and oxide bulk. The name also does not specify that the defects are donor-like.

“Slow states” is another name used to indicate that the charging and discharging of PCs are slower than the interface states [70]. It is again too general, since different types of PCs can be slow in different ways. For instance, CPC can be slow in both charging and discharging through tunneling. AHTs will be faster than CPC to neutralize, but slower than CPC to charge. For thin oxides, ANPC can remain charged. The term “slow states” will not inform readers these differences.

Other names used include “switching oxide traps” [71] and “switching hole traps” [72]. These names only tell readers that hole traps can communicate with substrate electrically.

To describe the three different types of PCs in oxides, it is necessary to use three names. AHT is a name conventionally used for preexisting hole traps and is kept in this framework. For the first time, this framework separates the generated hole traps into two types and new names should be used to capture their features. The direct and the most important impacts of defects on devices are their charging and discharging properties and the name should reflect these properties. CPC is used since these PCs are cyclic with similar charging and discharging speed. ANPC is used to directly refer to the difficulty in neutralizing them.

10.3.4 Impacts on the NBTI of pMOSFETs

For thin oxides, positive charges (PC) are located within tunneling distance from electrodes. The charging is highly dynamic and substantial recovery can occur during measurements [55, 73]. AHT, with its energy level below the valence band edge, is the least stable defects and will contribute to the recovery. A higher stress bias will fill the AHT further below Ev and increase the recovery. CPC also contributes to recovery under $Vg \geq 0$. Some ANPC, however, has the energy level beyond the reach of free electrons in Si and can survive the recovery, contributing to the so-called permanent component. The lower the temperature, the more ANPC is charged. This induces a temperature-dependent instability issue even for the same number of defects [61, 64, 65] and can cause circuit failure when temperature changes, for example, during the switching on of equipment. Moreover, ANPC is the only type of PCs that does not saturate as stress increases and will play an increasingly important role for longer time. This explains the increase of permanent/recoverable component ratio with stress time.

To suppress the recovery, the on-the-fly (OTF) [64, 73, 74] and ultra-fast pulse techniques (UFP) [73, 75, 76] were developed. The OTF technique measures the degradation at stress gate bias, while the UFP probes the degradation at threshold voltage level. Figure 10.33 shows that the threshold voltage shift, ΔVth, obtained by

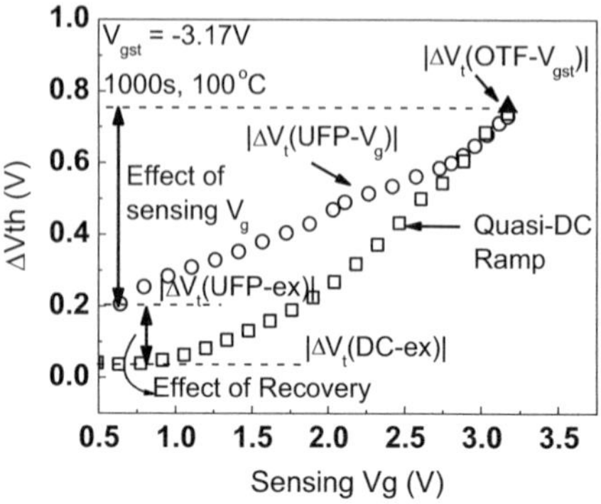

Fig. 10.33 A comparison of ΔVth measured by different techniques and their gap can be bridged by the Vg sensing effect [73]

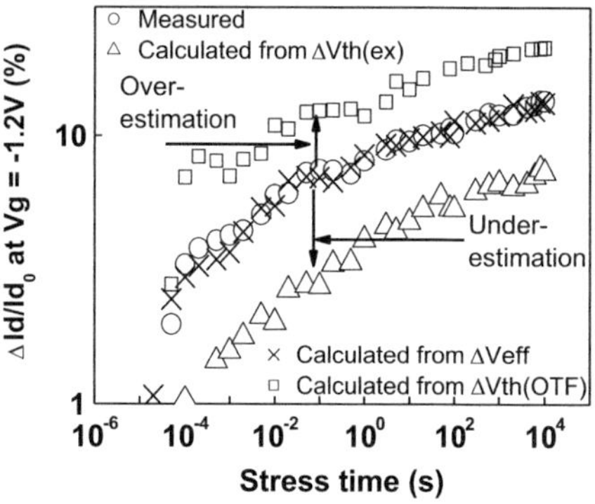

Fig. 10.34 A comparison of measured ΔId/Id with that predicted from ΔVth(UFP-ex), and ΔVth(OTF), and ΔVth(UFP at Vg = Op. Bias) [77]

these two techniques is different, because of the different sensing Vg used. The gap between them is fully covered by ΔVth measured at different Vg. A higher sensing |Vg| allows AHTs further below Ev being charged, leading to a larger |ΔVth|. Under a given sensing Vg, an increase of measurement time leads to further recovery, as PCs further in oxides were neutralized through tunneling. Figure 10.33 shows that the ΔVth measured by the conventional quasi-DC technique, ΔVt(DC-ex), can substantially underestimate the NBTI and care must be exercised when using it for device lifetime definition.

Since the positive charging increases for higher |Vg|, the ΔVth measured by either OTF and UFP techniques cannot be used to predict the current degradation, i.e., ΔId/Id, under operation bias. The stress bias used in a typical test is higher than the operation bias and Fig. 10.34 shows that the ΔVth measured by the OTF at the stress bias overestimates the ΔId/Id. In a digital circuit, the operation bias is higher than threshold voltage and Fig. 10.34 shows that the ΔVth measured at Vth by extrapolation underestimates ΔId/Id [77]. To correctly calculate ΔId/Id, the ΔVth should be measured with the operation bias as the sensing Vg.

It is well known that NBTI follows a power law kinetics, when the ΔVth was measured from quasi-DC transfer characteristics with a sensing Vg near to Vth [47, 73]. Once the recovery is suppressed, however, NBTI kinetics will not follow a

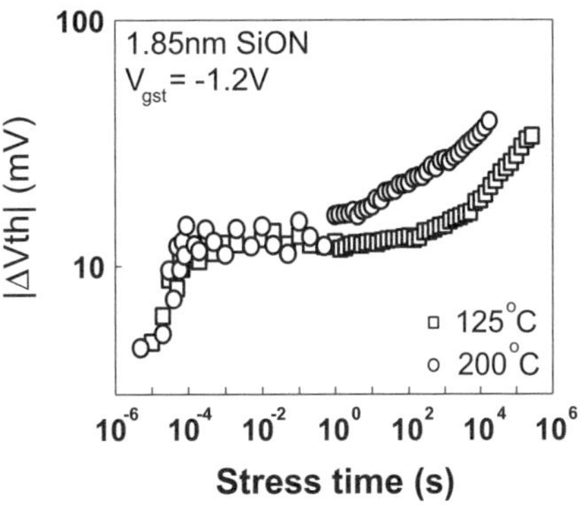

Fig. 10.35 ΔVth is measured under a stress bias of Vg = −1.2 V on the fly for different stress temperatures [19]

power law [19, 76, 78, 79]. For example, Fig. 10.35 shows the NBTI kinetics cannot be described by a power law, when devices were stressed and the degradation was measured at an operation bias of Vg = −1.2 V [19]. After a rapid rise initially, a plateau appears, which is caused by the saturation of filling as-grown hole traps. To support this statement, stresses were carried out at different temperatures and Fig. 10.35 shows that the plateau height is insensitive to temperature. This agrees with the saturation of filling AHT, which has a fixed number of hole traps, independent of stress temperature. After the plateau, ΔVth starts rising again by creating new defects and the generation is thermally activated. As the generation through free hole-oxide interaction is insensitive to temperature, the results show that hydrogen must be released during NBTI stress, which generates defects through thermal acceleration.

Since the NBTI kinetics no longer follows a simple power law, a new kinetic model is needed. This new kinetics should contain at least two terms: one for AHT and one for defect generation. As filling AHT follows the first-order model and new defect generation follows a power law, it is proposed [19] that

$$\Delta V_{th} = At^n + c(1 - e^{-t/t*}) \tag{10.2}$$

where A, n, c, and t* are constants under a given stress condition and their values extracted for six different processes are given in Table 10.1 [19].

Figure 10.36 shows that the test data follows the new model well over many orders of stress time. Fitting test data is a relatively easy task, but using it to predict the future where test data are not available is more challenging. For a model to be of value, it should be able to make predictions. The device lifetime is typically in years, while the test time is practically limited to weeks. A model is required to predict two orders of magnitude in time ahead, therefore. To test the prediction ability of this model, the test data in the last two orders of magnitude in time ("□" in Fig. 10.36) were not used to fit the model. The fitted parameters were then used to predict these data. Figure 10.36 shows that the prediction accuracy is good [19].

A reliable NBTI kinetics under operation bias allows device lifetime being estimated from a single test. As shown in Fig. 10.36, for a permitted ΔVth, the

Table 10.1 The wafers and the fitted parameters at 125 °C [19]

Processes	Gate Dielectrics	A (mV/s)	n	c (mV)	t* (μs)
A	1.85 nm 12 s Plasma SiON	0.23	0.36	11.91	30
B	1.4 nm Plasma SiON	7.70	0.13	18.39	4,390
C	2.7 nm Thermal SiON	26.27	0.07	22.07	80
D	2.0 nm 45 s Plasma SiON	4.10	0.12	7.49	740
E	2.0 nm 20 s Plasma SiON	3.91	0.12	1.89	110
F	TiN, ALCVD 2.0 nm/1 nm HfSiON/SiON	12.21	0.14	40.86	30

t* is the characteristic time for trap-filling

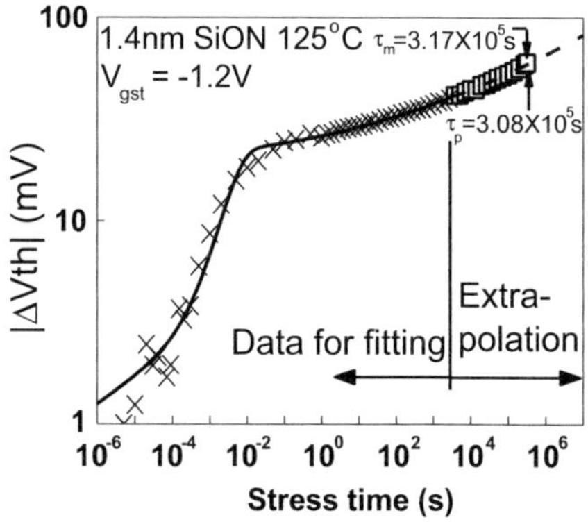

Fig. 10.36 Fitting test data with model (2) for the symbol *multiplication sign* and then using the fitted model to predict the symbol *open square* [19]

lifetime can be estimated from the fitted kinetics. It should be pointed out that this lifetime is for the worst case with zero recovery [19].

It should be pointed out the plateau in some test samples is not as obvious as that in Fig. 10.35. The as-grown filling and the generation phases may not be clearly separated in time and in many cases an inflection is observed rather than a plateau and one example is given in Fig. 10.37 [19]. The underlying physical processes, however, are the same: filling as-grown trap gives the initial fast degradation and the generation of new defects are responsible for the gradual degradation at longer time. Figure 10.37 also shows that the same kinetics can be applied for lifetime prediction.

Finally, to confirm that similar kinetic behavior can be observed from the results published by other groups, Fig. 10.38a gives one example where an inflection appears [78], while Fig. 10.38b gives one example where the plateau can be observed [79].

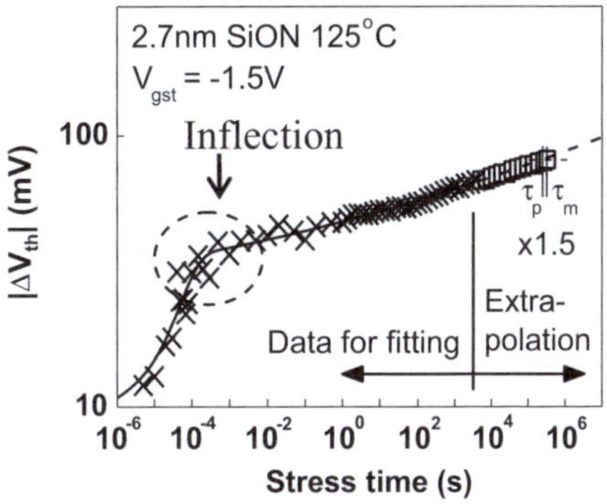

Fig. 10.37 An inflection is formed when filling as-grown trap is not clearly separated from trap generation [19]

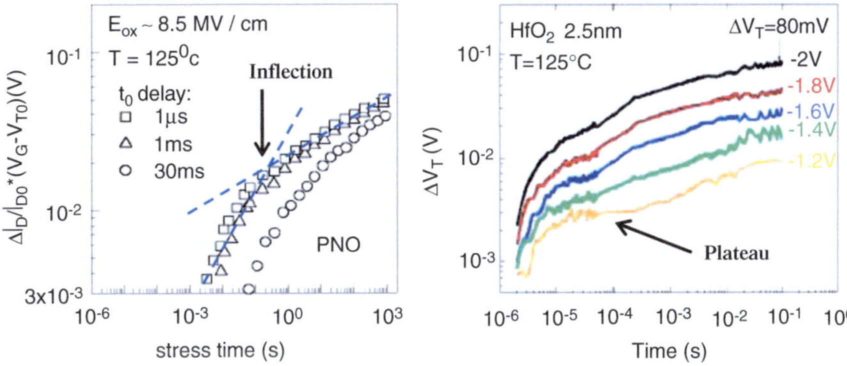

Fig. 10.38 Observation of "inflection" (**a**) [78] and "plateau" (**b**) [79] from results reported by other groups

10.4 Process-Induced Positive Charges

Apart from electrical stress-induced positive charges, PCs can also be formed in oxides during processing [80–82]. One example of the process-induced positive charges (PIPC) is given in Fig. 10.39 [80]. Both mobile and fixed PIPC are formed by annealing a device in H_2 ambient at 450 °C. It should be noted that the direction of PIPC swing in Fig. 10.39 is opposite to that of CPC in Fig. 10.31: the effective density of PIPC increases, but CPC decreases, under $Vg > 0$. This increase is caused by driving the mobile PIPC toward the substrate, rather than an increase in the number of PIPC [80].

To explore the origin of PIPC, Fig. 10.40 shows that there is little PIPC after exposing a device to N_2. When the gas was switched to H_2 at the same temperature, PIPC became substantial, supporting that PIPC originates from hydrogen-related species [80].

Unlike the hole traps that can be neutralized under either high field or temperature, PIPC cannot be neutralized. In Fig. 10.41, an electron injection of $\sim 10^{17}$ cm^{-2}

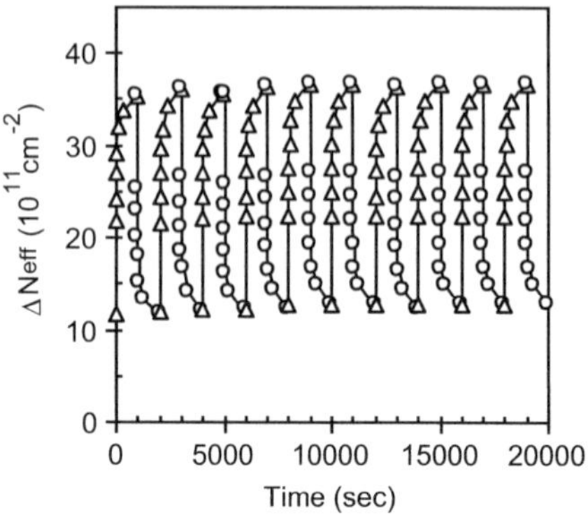

Fig. 10.39 Formation of process-induced positive charges (PIPC). The symbols *open triangle* and *open circle* are under Eox = +3 and −3 MV/cm, respectively [80]

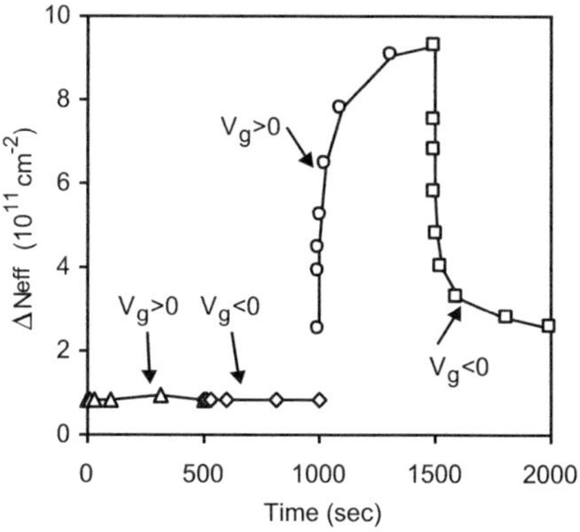

Fig. 10.40 Impact of ambient gas on PIPC. *Open triangle* and *open diamond* was recorded after N_2 exposure and *open circle* and *open square* after H_2 exposure at 450 °C [80]

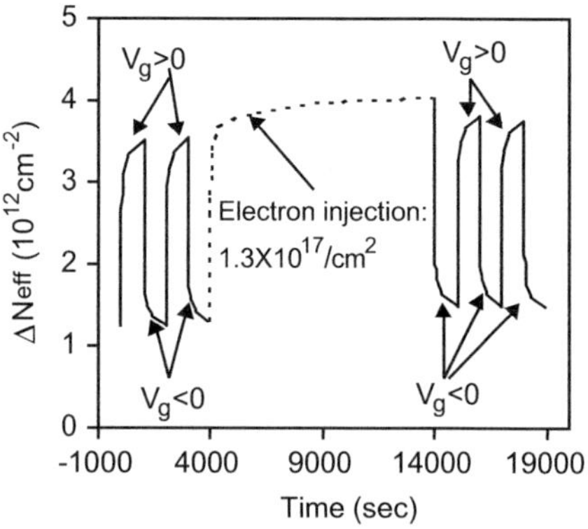

Fig. 10.41 Electron injection under Vg > 0 does not neutralize the PIPC [80]

was carried out under Vg > 0, which is adequate to neutralize all hole traps in this device. It, however, has not neutralized PIPC. In fact, the effective density of PIPC increases during the electron injection, as mobile PIPC moves towards substrate interface under Vg > 0.

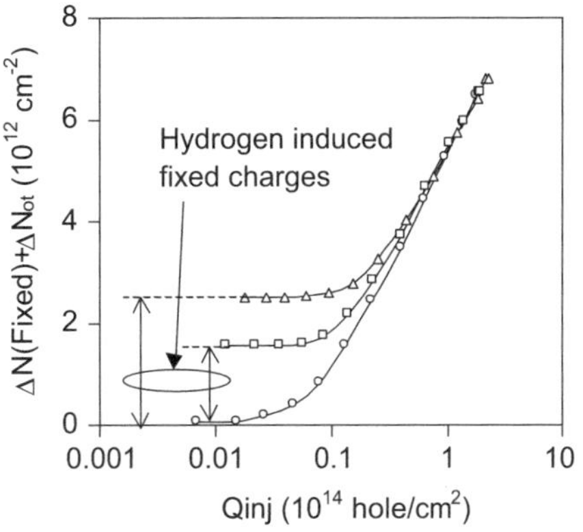

Fig. 10.42 PIPC does not increase hole trapping. PIPC occupies hole traps and form fixed charges [50]

The details of the hydrogenous species responsible for PIPC remain unknown. Hydrogen plays a complex role in the devices and can be either reactive or inactive [50, 80]. As reported in Sect. 10.3.2 (see Fig. 10.24), H_2 can be cracked into H^+/H by reacting with a trapped hole. H^+ and H are highly reactive: They generate both electron and hole traps. Moreover, they anneal interface states at 400 °C but create interface states at room temperature [45, 46, 56]. The PIPC-related hydrogenous species, however, have no effects on interface states [80], so that they are not H^+. On the relation between PIPC and hole traps, Fig. 10.42 shows that an increase of PIPC does not increase the total PCs. As a result, PIPC does not create hole traps. The PIPC-related hydrogenous species can occupy a hole trap and form the fixed charges [50]. These fixed charges, however, are different from the stress-induced ANPC [58].

PIPC has been observed in gate SiO_2, the SOI buried SiO_2, and Hf-dielectric/SiON stack [80–82]. It is sensitive to process conditions and can vary considerably even for devices on the same wafer [80, 81]. An increase of oxidation temperature to 1,100 °C has been reported to enhance PIPC [83] and Boron can play a catalyzing role [80]. The out-diffusion of some species from the gate edges increases PIPC for Hf-dielectric/SiON stack [82]. Through process optimization, PIPC can become negligible.

10.5 Conclusions

This work focuses on reviewing our recent progress in understanding oxide defects, including electron traps, hole traps, and process-induced positive charges. For SiO_2 and SiON with a poly-Si gate, there is little as-grown electron traps. Electrical stresses, however, can create electron traps. The electron-trapped hole recombination and high oxide field do not create the trap directly. The damaging species have been found to be free holes and hydrogen. The trapping follows the first-order model with two well-separated capture cross sections in the order of

10^{-14} and 10^{-15}–10^{-16} cm^2, respectively. They have a wide distribution in energy and are randomly distributed in space. For thin SiON used in modern CMOS technologies, these generated electron traps do not form steady charging due to efficient detrapping through tunneling. The buildup of the trap of 10^{-14} cm^2 leads to the oxide breakdown.

In contrast, there can be substantial amount of as-grown electron traps in Hf-dielectric. Their trapping is highly dynamic and measurement time has to be reduced to microseconds to suppress detrapping during measurement. They have capture cross sections similar to that of generated electron traps in SiO_2 and are located in the central region of Hf-dielectric layer. A reduction of Hf-dielectric layer thickness can dramatically reduce their density. These as-grown electron traps are sensitive to processing conditions and can cause a positive bias temperature instability (PBTI) for nMOSFETs.

There are substantial amounts of as-grown hole traps (AHT) in SiO_2 and SiON. Their filling also follows the first-order model with two capture cross sections in the order of 10^{-13}–10^{-14} and 10^{-15} cm^2, respectively. The effective density of AHT often reaches $>5 \times 10^{12}$ cm^{-2}, even in thin SiON. Their energy level is below Si Ev and their spatial location is biased towards the oxide/substrate interface.

The interaction between the oxide and free holes or hydrogenous species generates two types of hole traps: cyclic positive charges (CPC) and antineutralization positive charges (ANPC). CPC has energy level around Si Ec and can be charged and discharged through electron tunneling, which is insensitive to temperature. In contrast, ANPC has energy level above Si Ec and its neutralization increases for higher temperature. For Hf-dielectric/SiON stacks, positive charging is dominated by the interfacial SiON layer and the same framework can be applied.

Positive charges in oxides play a major role in NBTI of pMOSFETs. For thin oxides used in modern CMOS technologies, on one hand, the charging of AHT and CPC is highly dynamic, sensitive to the sensing Vg, and contributing to recovery. On the other hand, ANPC can survive neutralization and is the only type of PCs that do not saturate as stress increases. To predict the current degradation under an operation bias, the degradation should be measured at the same bias to avoid overestimation and to suppress recovery. After suppressing recovery, NBTI does not follow a power law. It is dominated by AHT initially, before defect generation becomes important. An as-grown-generation (AG) model is proposed for NBTI, which includes both filling AHT and defect generation. This model can be used not only to fit test data but also to predict ahead of time where test data are not available, allowing estimating device lifetime.

Finally, process can also induce positive charges (PIPC) in oxides, without electrical stress. The PIPC can be both mobile and fixed and originate from hydrogenous species. Unlike the hydrogen released by electrical stresses, these PIPC-related hydrogen species neither interact with the oxide/Si interface nor create traps in oxides. They are stable and cannot be neutralized by electron injection or anneal. They can occupy hole traps to form fixed charges. PIPC has been reported in gate SiO_2, buried SiO_2, and Hf-dielectric/SiON stacks. They are sensitive to process conditions and can be suppressed through process optimization.

Acknowledgments The author acknowledges the contribution of I. S. Al-Kofahi, H. K. Sii, W. Zhang, C. Z. Zhao, M. H. Chang, Z. Ji, X. F. Zheng, M. B. Zahid, L. Lin, M. Duan, B. Tang, and B. Benbakhti of Liverpool John Moores University and G. Groeseneken, B. Kaczer, R. Degraeve, L. Pantisano, S. De Gendt, and M. Heyns of IMEC. Test samples were supplied by IMEC. This work was funded by the EPSRC of UK under the grant numbers of EP/C003071/1, EP/C003098/1, EP/C003101/1, and EP/I012966/1.

References

1. L. M. Terman, Solid-State Electron. 5, 285 (1962).
2. B. E. Deal, M. Sklar, A. S. Grove, and E. H. Snow, J. Electrochem. Soc. 114, 266 (1967).
3. D. J. DiMaria, in The Physics of SiO_2 and its Interface, S. T. Pantelides, Ed. New York: Pergamon, 160 (1978).
4. W. Weber and R. Thewes, Semicond. Sci. Technol. 10, 1432 (1995).
5. J. F. Zhang and W. Eccleston, IEEE Trans. Elec. Dev. 42, 1269 (1995).
6. R. Degraeve, G. Groeseneken, R. Bellens, J. L. Ogier, M. Depas, P. J. Roussel, and H. E. Maes, IEEE Trans. Elec. Dev. 45, 904 (1998).
7. T. Grasser, B. Kaczer, W. Goes, H. Reisinger, T. Aichinger, P. Hehenberger, P. J. Wagner, F. Schanovsky, J. Franco, M. T. Luque, M. Nelhiebel, IEEE Trans. Elec. Dev. 58, 3652 (2011).
8. J. F. Zhang, Microelectron Eng. 86, 1883 (2009).
9. B. J. Cheng, A. R. Brown, and A. Asenov, IEEE Elec. Dev. Lett. 32, 740 (2011).
10. G. Bersuker, J. H. Sim, C. S. Park, C. D. Young, S. Nadkarni, R. Choi, and B. H. Lee, in Proc. IRPS, 179 (2006).
11. K. O. Jeppson and C. M. Svensson, J. Appl. Phys. 48, 2004 (1977).
12. S. Ogawa, M. Shimaya, and N. Shiono, J. Appl. Phys. 77, 1137 (1995).
13. M. A. Alam, in Proc. IEDM Tech. Dig. 345 (2003).
14. S. Mahapatra, V. D. Maheta, A. E. Islam and M. A. Alam, IEEE Trans. Elec. Dev. 56, 236 (2009).
15. Z. Q. Teo, D. S. Ang, and C. M. Ng, IEEE Electron Dev. Lett. 31, 269 (2010).
16. T. Grasser, P. J. Wagner, H. Reisinger, Th. Aichinger, G. Pobegen, M. Nelhiebel and B. Kaczer, in Proc. IEDM, 618 (2011).
17. H. Reisinger, T. Grasser, K. Ermisch, H. Nielen, W. Gustin and C. Schlunder, in Proc. IRPS, 597 (2011).
18. T. Grasser, B. Kaczer, W. Goes, Th. Aichinger, Ph. Hehenberger and M. Nelhiebel, in Proc. IRPS, 33 (2009).
19. Z. Ji, L. Lin, J. F. Zhang, B. Kaczer, and G. Groeseneken, IEEE Trans. Elec. Dev. 57, 228 (2010).
20. W. D. Zhang, J. F. Zhang, M. J. Lalor, D. R. Burton, G. Groeseneken and R. Degraeve, Semicond. Sci. Technol. 18, 174 (2003).
21. W. D. Zhang, J. F. Zhang, M. Lalor, D. Burton, G. Groeseneken, and R. Degraeve, Microelectronic Eng. 59, 89 (2001).
22. I.C. Chen, S. Holland, C. Hu, J. Appl. Phys. 61, 4544 (1987).
23. D. J. Dumin, J. R. Maddux, R. S. Scott, R. Subramoniam, IEEE Trans. Elec. Dev. ED-41, 1570 (1994).
24. D. J. DiMaria, J. H. Stathis, J. Appl. Phys. 89, 5015 (2001).
25. J. F. Zhang, S. Taylor, W. Eccleston, J. Appl. Phys. 71, 725 (1992).
26. M. H. Chang and J. F. Zhang, Semicond. Sci. and Technol. 19, 1333 (2004).
27. R. C. Hughes, Solid-State Electron. 21, 251 (1978).
28. N. S. Saks, R. B. Klein, and D. L. Griscom, IEEE Trans. Nuclear Sci. 35, 1234 (1988).
29. W. D. Zhang, J. F. Zhang, C. Z. Zhao, M. H. Chang, G. Groeseneken and R. Degraeve, IEEE Elec. Dev. Lett. 27, 393 (2006).

30. M. H. Chang, J. F. Zhang, and W. D. Zhang, IEEE Trans. Elec. Dev. 53, 1347 (2006).
31. D. R. Walters and J. J. van der Schoot, J. Appl. Phys. 58, 831(1985).
32. T. H. Ning, J. Appl. Phys. 47, 3203 (1976).
33. B. Benbakhti, J. F. Zhang, Z. Ji, W. Zhang, J. Mitard, B. Kaczer, G. Groeseneken, S. Hall, J. Robertson, and P. Chalker, IEEE Elec. Dev. Lett. 33, 1681 (2012).
34. R. F. DekKeersmaecker and D. J. DiMaria, J. Appl. Phys. 51, 1085 (1980).
35. E. H. Nicollian, C N. Berglund, P. F. Schmidt, and J. M. Andrews, J. Appl. Phys. 42, 5654 (1971).
36. P. Balk, Paper 111, The Electrochem. Soc. Meeting, Buffalo, NY, Oct. 10–14 (1965).
37. J. F. Zhang, S. Taylor, and W. Eccleston, J. Appl. Phys. 71, 5989 (1992).
38. J. F. Zhang, S. Taylor, and W. Eccleston, J. Appl. Phys. 72, 1429 (1992).
39. C. Z. Zhao, M. B. Zahid, J. F. Zhang, G. Groeseneken, R. Degraeve, and S. De Gendt, Microelectronic Eng. 80, 366 (2005).
40. C. Z. Zhao, J. F. Zhang, M. B. Zahid, B. Govoreanu, G. Groeseneken, and S. De Gendt, J. Appl. Phys. 100, Art. no.093716 (2006).
41. Z. Ji, J. F. Zhang, W. Zhang, G. Groeseneken L. Pantisano, S. De Gendt, M. M. Heyns, Appl. Phys. Lett. 95, Art. No. 263502 (2009).
42. J. F. Zhang, C. Z. Zhao, M. B. Zahid, G. Groeseneken, R. Degraeve, and S. De Gendt, IEEE Elec. Dev. Lett. 27, 817 (2006).
43. X. F. Zheng, W. D. Zhang, B. Govoreanu, J. F. Zhang, and J. Van Houdt, IEEE Trans. Elect. Dev. 57, 2484 (2010).
44. M. B. Zahid, R. Degraeve, J. F. Zhang, G. Groeseneken, Microelectronic Eng. 84, 1951 (2007).
45. J. F. Zhang, H. K. Sii, G. Groeseneken, and R. Degraeve, IEEE Trans. Elec. Dev. 47, 378 (2000).
46. J. F. Zhang, I. S. Al-kofahi, and G. Groeseneken, J. Appl. Phys. 83, 843 (1998).
47. J. F. Zhang, H. K. Sii, G. Groeseneken, and R. Degraeve, IEEE Trans. Elec. Dev. 48, 1127 (2001).
48. J. F. Zhang, H. K. Sii, A. H. Chen, C. Z. Zhao, M. J. Uren, G. Groeseneken and R. Degraeve, Semicond. Sci. and Technol. 19, L1 (2004).
49. T. Grasser, B. Kaczer, W. Goes, Th. Aichinger, Ph. Hehenberger, M. Nelhiebel, Microelectronic Eng. 86, 1876 (2009).
50. J. F. Zhang, C. Z. Zhao, G. Groeseneken, R. Degraeve, J. N. Ellis, and C. D. Beech, Solid-State Electronics 46, 1839 (2002).
51. H. S. Witham and P. M. Lenahan, Appl. Phys. Lett. 51, 1007 (1987).
52. J. F. Zhang, C. Z. Zhao, G. Groeseneken, and R. Degraeve J. Appl. Phys. 93, 6107 (2003).
53. D. J. DiMaria, Z. A. Weinberg, and J. M. Aitken, J. Appl. Phys. 48, 898 (1977).
54. I. S. Al-kofahi, J. F. Zhang and G. Groeseneken, J. Appl. Phys. 81, 2686 (1997).
55. J. F. Zhang, Z. Ji, M. H. Chang, B. Kaczer, and G. Groeseneken, in Proc. IEDM Tech. Dig. 817 (2007).
56. J. F. Zhang, H. K. Sii, R. Degraeve, and G. Groeseneken, J. Appl. Phys. 87, 2967 (2000).
57. M. M. Heyns and R. F. De Keersmaecker, Mater. Res. Soc. Symp. Proc. 105, 205 (1988).
58. C. Z. Zhao and J. F. Zhang, J. Appl. Phys. 97, Art. no. 073703 (2005).
59. C. Z. Zhao, J. F. Zhang, G. Groeseneken, R. Degraeve, J. N. Ellis, and C. D. Beech, J. Appl. Phys. 90, 328 (2001).
60. C. Z. Zhao, J. F. Zhang, M. B. Zahid, G. Groeseneken, R. Degraeve, and S. De Gendt, Appl. Phys. Lett. 89, Art.No. 023507 (2006).
61. J. F. Zhang, C. Z. Zhao, A. H. Chen, G. Groeseneken and R. Degraeve, IEEE Trans. Elec. Dev. 51, 1267 (2004).
62. C. Z. Zhao, J. F. Zhang, G. Groeseneken and R. Degraeve, IEEE Trans. Elec. Dev. 51, 1274 (2004).
63. M. H. Chang and J. F. Zhang, in Proc. ECS Symp. Silicon nitride, Silicon Dioxide Thin Insulating Films, and Other Emerging Dielectrics VIII, PV 2005–01, 293 (2005).
64. M. H. Chang and J. F. Zhang, J. Appl. Phys. 101, Art. no. 024516 (2007).
65. J. F. Zhang, M. H. Chang, and G. Groeseneken, IEEE Elec. Dev. Lett. 28, 298 (2007).

66. C. Z. Zhao, J. F. Zhang, M. H. Chang, A. R. Peaker, S. Hall, G. Groeseneken, L. Pantisano, S. De Gendt, and M. Heyns, IEEE Trans. Elec. Dev. 55, 1647 (2008).
67. J. F. Zhang, M. H. Chang, Z. Ji, L. Lin, I. Ferain, G. Groeseneken, L. Pantisano, S. De Gendt, and M. M. Heyns, IEEE Elec. Dev. Lett. 29, 1360 (2008).
68. D. R. Young, E. A. Irene, D. J. DiMaria, R. F. De Keersmaecker, and H. Z. Massoud, J. Appl. Phys. 50, 6366 (1979).
69. D. M. Fleetwood, Microelectron. Reliab. 42, 523 (2002).
70. S. K. Lai and D. R. Young, J. Appl. Phys. 52, 6231 (1981).
71. A. J. Lelis and T. R. Oldham, IEEE Trans. Nucl. Sci. 41, 1835 (1994).
72. Y. Gao, D. S. Ang, C. D. Young, and G. Bersuker, Proc. IRPS 5A.5.1 (2012).
73. Z. Ji, J. F. Zhang, M. H. Chang, B. Kaczer, and G. Groeseneken, IEEE Trans. Elec. Dev. 56, 1086 (2009).
74. M. Denais, A. Bravaix, V. Huard, C. Parthasarathy, G. Ribes, F. Perrier,Y. Rey-Tauriac, and N. Revil, in IEDM Tech. Dig., 109 (2004).
75. T. Yang, M. F. Li, C. Shen, C. H. Ang, C. Zhu, Y.-C. Yeo, G. Samudra, S. C. Rustagi, M. B. Yu, and D. L. Kwong, in VLSI Symp. Tech. Dig., 92 (2005).
76. A. E. Islam, E. N. Kumar, H. Das, S. Purawat, V. Maheta, H. Aono, E. Murakami, S. Mahapatra, and M. A. Alam, in IEDM Tech. Dig., 805 (2007).
77. J. F. Zhang, Z. Ji, L. Lin, and W. Zhang, Proc. of IEEE 10th Int. Conf. on Solid-State and Integrated-Circuit Technol., 1600 (2010).
78. E. N. Kumar, V. D. Maheta, S. Purawat, A. E. Islam, C. Olsen, K. Ahmed, M. A. Alam, and S. Mahapatra, in Proc. IEDM Tech. Dig., 809 (2007).
79. M. Rafix, X. Garros, G. Ribes, G. Ghibaudo, C. hobbs, A. Zauner, M. Muller, V. Huard, C. Ouvrard, in Proc. IEDM Tech. Dig., 825 (2007).
80. J. F. Zhang, C. Z. Zhao, G. Groeseneken, R. Degraeve, J. N. Ellis, and C. D. Beech, J. Appl. Phys. 90, 1911 (2001).
81. C. Z. Zhao, J. F. Zhang, M. H. Chang, A. R. Peaker, S. Hall, G. Groeseneken, L. Pantisano, S. De Gendt, and M. Heyns, J. Appl. Phys. 103, Art. No. 014507 (2008).
82. M. H. Chang, C. Z. Zhao, Z. Ji, J. F. Zhang, G. Groeseneken, L. Pantisano, S. De Gendt, M. M. Heyns, J. Appl. Phys. 105, Art. no. 054505 (2009).
83. K. Vanheusden, W. L. Warren, R. A. B. Devine, D. M. Fleetwood, J. R. Schwank, M. R. Shaneyfelt, P. S. Winokur, and Z. J. Lemnios, Nature 386, 587 (1997).
84. D. S. Ang, Understanding negative-bias temperature instability from dynamic stress experiments, in *Bias Temperature Instability for Devices and Circuits*, ed. by T. Grasser (Springer, Heidelberg, 2013).
85. S. Mahapatra, A comprehensive modeling framework for DC and AC NBTI, in *Bias Temperature Instability for Devices and Circuits*, ed. by T. Grasser (Springer, Heidelberg, 2013).
86. T. Grasser, The capture/emission time map approach to the bias temperature instability, in *Bias Temperature Instability for Devices and Circuits*, ed. by T. Grasser (Springer, Heidelberg, 2013).
87. J. P. Campbell, P. M. Lenahan, Atomic scale defects associated with the negative bias temperature instability, in *Bias Temperature Instability for Devices and Circuits*, ed. by T. Grasser (Springer, Heidelberg, 2013).
88. A. Kerber, E. Cartier, Bias temperature instability characterization methods, in *Bias Temperature Instability for Devices and Circuits*, ed. by T. Grasser (Springer, Heidelberg, 2013).

Chapter 11
Understanding Negative-Bias Temperature Instability from Dynamic Stress Experiments

Diing Shenp Ang

Abstract The purpose of this chapter is to summarize key experimental evidences on the important role of hole trapping on negative-bias temperature instability (NBTI). For a long time, the focus of this research topic had been on interface degradation driven by hydrogen transport and hole trapping was regarded as a side effect arising out of fast measurement techniques proposed to mitigate the effect of recovery on measurement data. In recent studies, we showed that the threshold voltage (V_t) fluctuations one typically observed under dynamic NBTI were mainly the result of hole trapping and not hydrogen-transport-driven interface-state generation/passivation proposed earlier. In particular, the cyclical V_t shifts and constant V_t recovery are inconsistent with the basic principle of the hydrogen transport model. Such behaviors are better described in terms of hole trapping/detrapping at preexisting oxide defects. We have also shown that interface degradation during NBTI stressing has no apparent impact on bulk (oxide) trap generation, i.e., interface trap generation does not lead to bulk trap generation. This result raises further questions on the validity of the hydrogen transport mechanism and the long-standing hypothesis on hydrogen-induced bulk trap generation and gate oxide breakdown. Finally, it is shown that the transient hole trapping responsible for the V_t shift fluctuations could be transformed into more permanent trapped holes under NBTI stressing. The extent of transformation is accelerated by a high oxide field and temperature. An excellent correlation with stress-induced leakage current indicates that such transformation underlies the generation of bulk traps reported by earlier studies.

D.S. Ang (✉)
School of Electrical and Electronic Engineering, Nanyang Technological University, Singapore
e-mail: EDSAng@ntu.edu.sg

T. Grasser (ed.), *Bias Temperature Instability for Devices and Circuits*, 287
DOI 10.1007/978-1-4614-7909-3_11, © Springer Science+Business Media New York 2014

11.1 Introduction

Negative-bias temperature instability (NBTI) refers to the progressive shift of transistor parameters (such as the threshold voltage (V_t) and transconductance) under the combined effect of a negative gate voltage and elevated temperature. From a relatively unknown problem in the past, NBTI has become one of the most crucial front-end reliability issues of advanced complementary metal–oxide–semiconductor (MOS) technology. The reasons for its importance are fourfold: (1) Scaling of the gate oxide thickness has led to a substantial increase of the oxide field E_{ox}. NBTI is found to exhibit a power-law dependence on the oxide field ($\sim {E_{ox}}^{3.3}$) [1], which makes this issue especially critical for the ultrathin gate p-channel MOS field-effect transistor (p-MOSFET). (2) The change from a buried-channel to a surface-channel structure for the control of short-channel effects has placed inversion holes in the p-MOSFET directly at the SiO_2/Si interface. Although the basic role of holes remains unclear, they are no doubt one of the main "ingredients" of the NBTI problem as such an instability problem did not exist in the SiO_2 gate n-MOSFET. (3) The introduction of nitrogen into the gate oxide to suppress boron penetration (from the p^+ polysilicon gate) and gate tunneling current. NBTI of the nitrided SiO_2 or oxynitride gate p-MOSFET is shown to be significantly worse than that of the SiO_2 gate p-MOSFET [2] and this is believed to be due to a higher density of nitrogen-related hole traps in the oxynitride gate dielectric [3]. (4) The adoption of high dielectric-constant or high-k gate oxides and alternative channel materials (e.g., Ge, InGaAs, etc.) to sustain Moore's law scaling in future technology nodes. The interface quality in these gate stacks is inherently poorer than that of the SiO_2/Si, thus making NBTI an even more critical problem for these novel gate stack technologies. Apart from NBTI, high-k gate stacks are also found to be susceptible to positive-bias temperature instability or PBTI caused by electron trapping in the high-k gate dielectric. This chapter is focused on the NBTI problem only.

In spite of the long history of NBTI (known since the inception of the MOS technology in the 1960s), a complete understanding of the physical mechanism(s) remains elusive, even for the conventional oxynitride gate p-MOSFET. One of the main reasons for the lack of understanding is the substantial "blackout" period (~1977–1999) during which research activities on NBTI were practically nonexistent. Before the dual-gate technology in the early 2000s, the NBTI problem was largely mitigated by the buried p-channel, in which the inversion holes were situated away from the oxide/Si interface. Moreover, the apparent success of the reaction–diffusion (R-D) model [4, 5] gave the impression that the basic physics of NBTI was already understood. Indeed, NBTI virtually became synonymous with the R-D model and research effort following the revival of interest in the early 2000s was mostly focused on addressing inconsistencies that arose from the introduction of fast measurement techniques. For instance, the small time exponent (typically less than 0.1) of as-measured drift curve was ascribed to the parasitic hole trapping effect. It was argued that consistency with the R-D model could be maintained by excluding the parametric shift contributed by hole trapping [6, 7].

The R-D model [4, 5] or its dispersive transport counterpart had remained the focus of NBTI modeling for many years due to their perceived ability to be able to describe not only the stress but also the recovery phenomena in a single framework. Only until very recently did studies convincingly reveal that it was hole trapping and not hydrogen transport that dominated the NBTI recovery observed under pulsed gate condition [8–22]. The purpose of this chapter is to present a brief review of recent results from dynamic stress experiments showing "non-R-D" characteristics of NBTI [8, 10]. Specifically, we discuss evidence of a cyclical behavior of threshold voltage (V_t) shift, which implies that it is hole trapping and detrapping by a given group of preexisting oxide defects, and not the transport of hydrogen to and fro the oxide/Si interface, that determine dynamic NBTI. We also summarize evidence which shows no apparent relationship between interface and bulk trap generation under NBTI stressing [19], contrary to the hypothesis that ascribes bulk trap generation to hydrogen species released from the interface degradation or Si–H bond dissociation process [23, 24]. This will be followed by a summary of results showing the transformation of transient hole trapping into more permanent trapped holes and its role on bulk trap generation [14, 16, 18, 20–22].

11.2 "Non-R-D" Characteristics of NBTI Recovery

Some authors argued that NBTI was consistent with the R-D model after excluding the hole trapping effect (assumed to dominate the initial degradation and saturate after $\sim$1 s of stressing) [6, 7]. This was based primarily on the ability to reproduce a power-law exponent of 1/6 for the "corrected" drift curve by subtracting the contribution of hole trapping estimated at a stress time of $\sim$1 s. While the R-D model had no difficulty describing the stress data, major issues had existed for the recovery data. As pointed out by Reisinger et al. [25], non-negligible recovery already happened several microseconds after the stress was stopped, and the as-measured NBTI recovery curve spanned many orders of time ($\sim10^{-6}$–10^5 s) with recovery apparently continuing after 10^5 s. On the other hand, simulation based on the R-D model showed that recovery, accounting for $\sim$90% of the overall, began several milliseconds after the termination of stress and lasted only for 2–3 orders of time interval. This behavior is in sharp contrast to the experimentally observed ultrafast recovery which begins almost instantaneously after stress and exhibits a logarithmic time dependence lasting more than ten orders of time interval [Fig. 3, 25]. Various authors suggested that the discrepancy between experiment and theory may be reconciled using a dispersive proton transport model [26–28]. A dispersive transport model was also able to describe the temperature-dependent time exponent observed by some authors [27, 28]. As it was possible to reconcile the discrepancy between experiment and theory, most studies before 2009 supported the hydrogen transport mechanism.

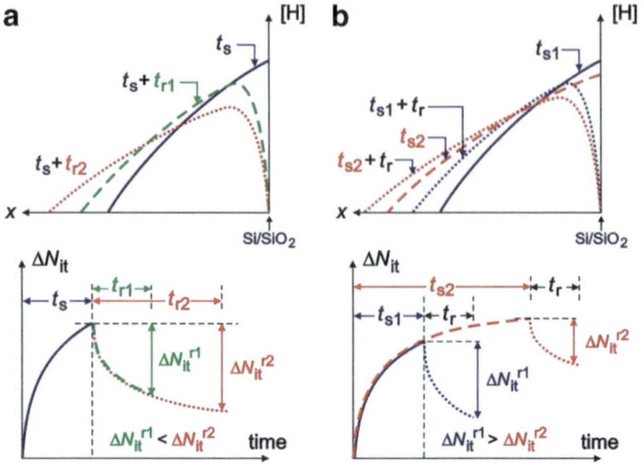

Fig. 11.1 Schematic illustration of the evolution of hydrogen concentration profile in the gate and the associated Si/SiO_2 interface-state density N_{it} variation in a dynamic NBTI cycle. (**a**) The stress time t_s is fixed while the relaxation time t_r is varied ($t_{r2} > t_{r1}$); (**b**) the relaxation time t_r is fixed while the stress time t_s is varied ($t_{s2} > t_{s1}$). In either case, the hydrogen profile continues to move away from the Si/SiO_2 interface during relaxation [13]

It should, however, be pointed out that nearly all studies on the R-D model or its variants[1] had focused either only on the stress regime or a limited number of stress/relaxation cycles. Almost no attention was paid to the self-limiting recovery inherent in the hydrogen transport model, although this feature was already manifested in simulation results presented in earlier studies. For instance, it was shown by Krishnan et al. [29] that the concentration of hydrogen near the interface decreased steadily with the number of stress/relaxation cycles. This implies that recovery via the repassivation of dangling Si bonds would decrease correspondingly. The physics of self-limiting recovery is illustrated schematically in Fig. 11.1 [13]. Following the initial recovery involving hydrogen present exactly at the interface, longer term recovery is driven by a second diffusion front created by the depletion of near-interface hydrogen during the initial stage of recovery. However, the original diffusion front which drives hydrogen away from the interface, so that interface traps may be successfully generated during stressing, still exists. Thus, even during recovery, some hydrogen is being driven away from the interface. When the transistor is stressed and relaxed numerous times, the recovery per relaxation cycle would decrease steadily (Fig. 11.2a; filled circle) [8, 10, 13] since the amount of hydrogen that could move back to the interface within a given relaxation interval is gradually reducing.

On the other hand, the V_t shift recovery per cycle measured experimentally exhibits a totally different trend (Fig. 11.2a; open circle). No decrease is observed

[1]For simplicity, subsequent usage of the term "R-D model" is assumed to encompass the dispersive transport model.

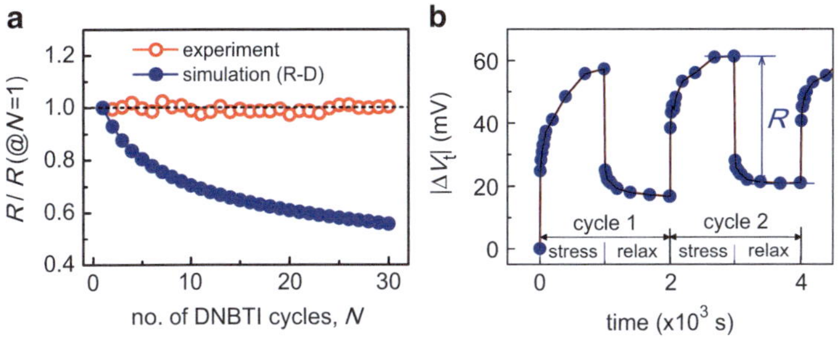

Fig. 11.2 (**a**) Comparison of the threshold voltage V_t shift recovery per cycle or R (see *left panel*) measured experimentally with that obtained from R-D model simulation [8, 10, 13]. (**b**) Variation in the magnitude of ΔV_t of a 1.7 nm EOT oxynitride gate p-MOSFET subjected to repeated stressing/relaxation

after as many as 30 stress/relaxation cycles, corresponding to a cumulative time of 6×10^4 s. The difference between the V_t recovery of the last cycle and that of the first cycle is comparable to the resolution limit of V_t measurement ($\sim$1 mV). The constant V_t recovery is in sharp contrast to the progressive decrease predicted by the R-D model (Fig. 11.2a). Since self-limited recovery is a fundamental property of a transport model, reconciliation with the constant recovery observed experimentally is deemed not possible. The discrepancy raises a big question on the validity of the R-D model for NBTI.

It is important to emphasize that the self-limiting recovery would not show up clearly if only one or a few relaxation cycles are simulated. Despite a relatively long cumulative time of 4×10^3 s (i.e., at the end of the second dynamic cycle), the simulated recovery of the second cycle is only marginally reduced ($\sim$7%) as compared to the first cycle (Fig. 11.2a). This implies that self-limiting recovery is also not obvious for any simulation study that involves numerous but short stress and relaxation intervals. It should further be mentioned that the extent of self-limiting recovery may be suppressed (i.e., the gradual decrease of recovery diminished) by the use of a much higher hydrogen diffusivity during relaxation (than that during stress) such that most of the hydrogen could return to the interface. But such an attempt is not physical.

It is interesting to note the highly similar recovery curves from the first to the last relaxation cycles (Fig. 11.3a) [12, 13]. The fluctuation of V_t shift can also be observed to become cyclical after the initial few stress/relaxation cycles, i.e., the stress-induced V_t shift is almost equal to the V_t shift recovery in each cycle (Fig. 11.3b) [10, 12, 13]. In conjunction with the constant recovery per cycle (Fig. 11.2a), these results imply that it is the capture and emission of holes by the same group of preexisting oxide defects (and not hydrogen transport) that give rise to the repetitive V_t fluctuations observed under dynamic NBTI. Studies on small area p-MOSFETs have also revealed steplike V_t recovery characterized by consistent step heights and time constants, believed to be due to the emission of holes from individual oxide traps charged by a prior stress [9, 11].

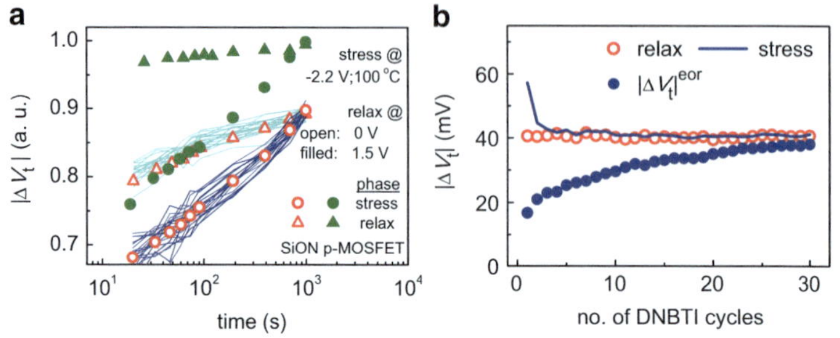

Fig. 11.3 (**a**) Line: Evolution of V_t shift during individual stress and relaxation phases. For the former, data for the 10th to 30th cycles are shown. As for the latter, data for the 1st to 30th cycles are shown. The symbols denote the average value. Although the time-dependent drifts are different during stress and relaxation, the end points coincide, indicating that the total V_t shifts are nearly always equal. The highly similar drift curves of individual stress and relaxation phases are observed regardless of the gate relaxation voltage (only the average value is shown for the case of a positive gate relaxation voltage). (**b**) Comparison of the total V_t shift for the individual stress and relaxation phases. The initial difference (<5 cycles) may be ascribed to non-negligible interface-state generation, which is relatively permanent. But this degradation mechanism slows down considerably between successive cycles in the later stage (see evolution of V_t shift which did not recover at the end of each relaxation phase, i.e., $|\Delta V_t|^{eor}$; *circle*), giving a clear manifestation of the cyclical V_t shift behavior [10, 12, 13]

It should be mentioned that some studies [30, 31] have modeled the spread in the time of steplike V_t recovery in the small area p-MOSFET in terms of stochastic transport of hydrogen back to the oxide/Si interface. The approach is similar to the dispersive transport model proposed earlier and recovery should likewise be self-limited. But repeating the stress/relaxation sequence hundreds of times on a given p-MOSFET has yielded consistent recovery step heights and time constants for individual defects [9, 11], which fail to support the inherent self-limited recovery of a transport model. Moreover, Grasser et al. [9] has shown that while stochastic hydrogen transport could lead to discrete steplike recovery, the statistics are in disagreement with experimental data. The latter is always exponentially distributed, whereas the R-D model predicts a much wider distribution which moves with time.

11.3 Uncorrelated Interface and Bulk Trap Generation

The similarity between NBTI and TDDB (time-dependent dielectric breakdown) stresses (except for the magnitude of the oxide field) and the apparent success of the R-D model prior to 2009 had prompted some authors [23, 24] to ascribe the bulk trap generation observed under NBTI stressing to the interaction between the oxide network and hydrogen released from the interface-state generation (Si–H bond dissociation) process. The proposed relationship between bulk and interface trap generation appeared consistent with the long-standing hypothesis on oxide trap generation under TDDB stressing—the anode hydrogen release (AHR)

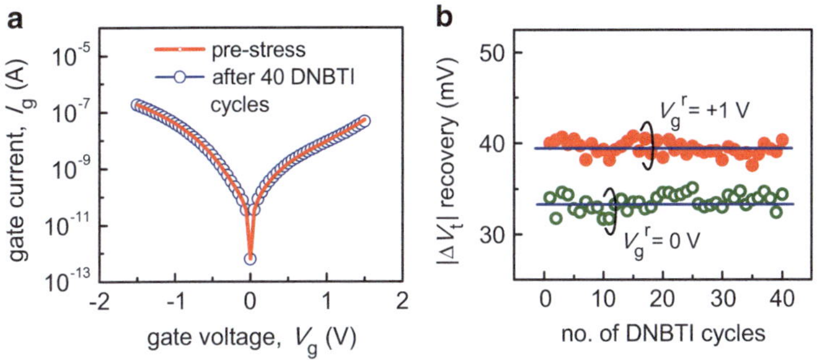

Fig. 11.4 (**a**) Comparison of gate leakage current before and after 40 DNBTI cycles; (**b**) V_t shift recovery per cycle as a function of the number of DNBTI cycles. The larger recovery under a positive gate relaxation voltage $V_g{}^r$ is a result of the detrapping of deep-level trapped holes [19]

mechanism.[2] In this section, we summarize recent experimental results which show, on the contrary, no apparent relationship between interface and bulk trap generation under NBTI stressing [19]. Specifically, evidence showing almost no bulk trap generation in spite of significant interface degradation under NBTI stressing will be presented. A possible explanation for the discrepancy with earlier works is given.

In Fig. 11.4a, the gate tunneling current I_g of an ultrathin oxynitride gate p-MOSFET (equivalent oxide thickness, EOT = 1.7 nm) before and after 40 stress/relaxation cycles are compared. No increase of I_g can be observed, which implies no generation of bulk traps despite a cumulative stress time of 4×10^4 s.[3] This result is in agreement with the constant V_t recovery (Fig. 11.4b), confirming that there is no observable increase of oxide trap density and the V_t fluctuations under gate pulsing arise from hole trapping/detrapping at preexisting oxide traps, as concluded previously.

On the other hand, a steady rise of the remnant V_t shift (i.e., the relatively permanent part of V_t shift which did not recover) at the end of each relaxation phase is evident (Fig. 11.5), reaching ~30 mV at the end of the 40th stress/relaxation cycle. It should be mentioned that part of the remnant V_t shift obtained under 0 V gate recovery voltage is due to holes trapped at preexisting deep-level oxide traps [32, 33]. The results in Fig. 11.5 are obtained under a +1 V gate recovery voltage, under which the contribution from deep-level trapped holes has been excluded (cf. Fig. 11.4b which shows a larger V_t shift recovery, due to increased hole detrapping, under +1 V as compared to 0 V gate recovery voltage). Since no

[2]Under the AHR model, the dissociation of the Si–H bonds is believed to be caused by hot electrons which arise from the gate tunneling current.

[3]A separate study involving p-MOSFETs with a thicker SiO_2 gate oxide (2.8 nm) and prestress I_g on the order of 10^{-12} A at 1 V also showed no increase of I_g after 5×10^4 s of NBTI stressing at comparable oxide field [Fig. 7, 19]. This rules out the possibility that any increase of I_g in Fig. 11.5 is masked out by the relatively high prestress I_g of the thinner oxynitride gate oxide.

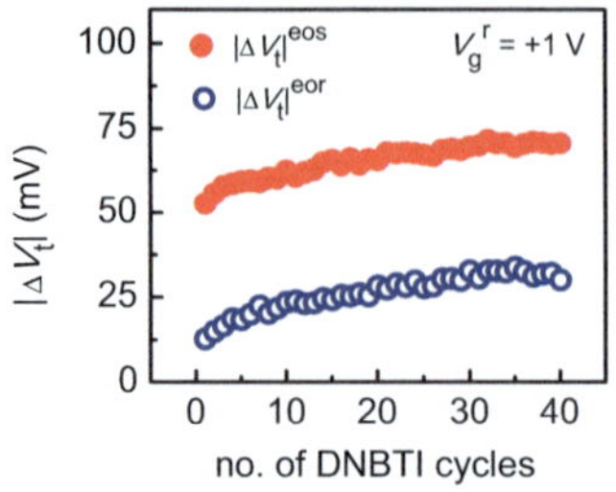

Fig. 11.5 Evolution of $|\Delta V_t|^{eos}$ and $|\Delta V_t|^{eor}$, the respective V_t shift at the end of stress (eos) and end of relaxation (eor) phase, of an ultrathin oxynitride (EOT = 1.7 nm) gate p-MOSFET subjected to repetitive stress/relaxation cycling. $|\Delta V_t|^{eos}$ measures the total positive trapped charge in the gate oxide prior to the onset of relaxation while $|\Delta V_t|^{eor}$ gives the part of the V_t shift which did not recover at the end of each relaxation phase [19]

bulk trap generation is observed, the non-negligible remnant V_t shift must be due to defects located at or very near the oxide/Si interface. The remnant V_t shift of 30 mV corresponds to a sizeable interface-state density of 4×10^{11} cm^{-2}, indicating that significant interface-state generation has occurred under the stress condition applied. The results show clearly that the substantial interface degradation induced by NBTI stressing did not result in any apparent bulk trap generation.

It is therefore essential to address the apparent discrepancy between this set of results, which show no bulk trap generation in spite of significant interface degradation, and the earlier observation of bulk trap generation under NBTI stressing [23, 24]. It should be noted that there has been no explicit evidence confirming that the bulk trap generation observed in earlier studies stemmed from hydrogen released from the interface degradation process. The inference was drawn based on the correlated increase of stress-induced leakage current (SILC) and charge pumping current (I_{cp}) [23] or subthreshold swing (SS) [24], on the *presumption* that the interface defects sensed by the I_{cp} or SS are dangling Si bonds or P_b centers that result from Si–H bond dissociation. While the P_b center is a major source of defects at the SiO_2/Si interface, it may not be the case for the more advanced gate oxide which incorporates a non-negligible amount of nitrogen and fluorine (from BF_2 implantation). The latter has been shown to lessen the NBTI effect [3], and it is believed that this occurs through the formation of stronger Si–F bonds in place of the Si–H bonds. Electron-spin resonance (ESR) study [34] of a post-stressed plasma-nitrided gate p-MOSFET has also not found any signal related to the P_b center but a totally different and dominant signal related to a near-interface oxygen vacancy defect (the k_N center). The evidence implies that the NBTI of modern p-MOSFETs may be dominated by defects other than the P_b center.

It should be mentioned that a near-interface oxygen vacancy defect could effectively function as an interface trap during electrical measurement and be sensed by techniques such as the charge pumping and subthreshold swing. At the same time, these defects could function as trap-assisted tunneling centers, i.e., oxide traps that result in a higher gate leakage current. As will be shown in the next section, it is possible for oxide traps to be generated under certain NBTI stress condition[4] that gives rise to an apparent relationship between SILC and interface-state generation, especially in the ultrathin gate p-MOSFET [23, 24].

11.4 Transient-to-Permanent Trapped-Hole Transformation

As discussed in the second section, the recovery of V_t shift per cycle is constantly independent of the number of times the device is stressed and relaxed. This evidence strongly suggests that the observed V_t shift fluctuation is a result of the capture and emission of holes by a similar group of preexisting oxide defects that always respond under a given experimental condition. As will be shown in this section, a gradual decrease of the V_t shift recovery per cycle can be observed, especially under high oxide stress field and temperature [14, 16, 18, 20–22]. We present evidence which shows that the decrease is a result of a part of the transient (recoverable) trapped holes being transformed into more permanent ones. The linear correlation between the decrease of V_t recovery and SILC lends support to our hypothesis that the transformation is a key mechanism for the permanent bulk trap generation observed by some early studies [23, 24].

In Fig. 11.6a [18], the V_t shifts at the end of stress (eos) and at the end of relaxation (eor) phases, $|\Delta V_t|^{eos}$ and $|\Delta V_t|^{eor}$, respectively, of two similar devices subjected to numerous dynamic gate cycles are plotted. In one case, the usual dynamic NBTI test was repetitively applied, such as in the case of Fig. 11.2. In the second case, an intermediate constant voltage stressing (CVS) phase was inserted in-between two successive dynamic NBTI cycles. For instance, upon the completion of the first dynamic NBTI cycle, the device was subjected to CVS, at a much higher oxide stress field than that applied during the NBTI stress. The dynamic NBTI test was resumed after the CVS and the DNBTI-CVS sequence was repeated numerous times. As can be seen from Fig. 11.6b [18], the V_t recovery per cycle decreases with the number of times the DNBTI-CVS sequence was repeated, in contrast to the case where the CVS phase was omitted.

An explanation for the decrease of V_t recovery per cycle under repeated DNBTI-CVS testing may be found in Fig. 11.6a, in which the evolution of $|\Delta V_t|^{eos}$ and $|\Delta V_t|^{eor}$ is compared to those of the case where the CVS phase was excluded. It should be mentioned that $|\Delta V_t|^{eos}$ measures the total positive trapped charge (trapped holes and interface trapped charge) in the gate oxide whereas $|\Delta V_t|^{eor}$ probes the fraction of positive trapped charge remaining after each relaxation phase, i.e., the part which is relatively permanent and did not recover. Apart from the initially larger $|\Delta V_t|^{eos}$ for the DNBTI-CVS test, it can be seen that the subsequent evolution of $|\Delta V_t|^{eos}$ is comparable to that of the DNBTI-only (i.e., CVS

[4] The extent of oxide defect generation depends on the processing conditions and may vary substantially from one case to another for a given stress condition.

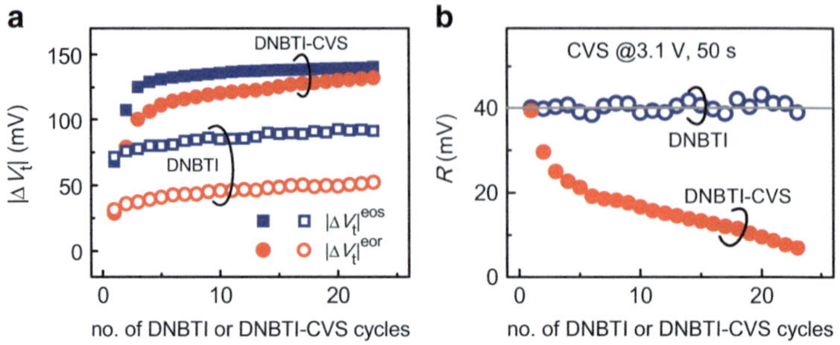

Fig. 11.6 (**a**) Evolution of $|\Delta V_t|^{eos}$ and $|\Delta V_t|^{eor}$ of two similar p-MOSFETs but subjected to different test patterns. In one case (DNBTI), only repeated NBTI stress/relaxation cycling was applied. The gate stress voltage was -1.8 V and the gate relaxation voltage was 0 V. The stress and relaxation phase each lasted 1×10^3 s. In the other case (DNBTI-CVS), a constant voltage stressing phase follows the DNBTI test. The CVS was carried out at a much higher gate stress voltage of 3 V but only for a short interval of 50 s. The DNBTI test was repeated after the CVS phase and the sequence repeated. (**b**) Comparison of the V_t shift recovery per cycle under the DNBTI and DNBTI-CVS test sequences [18]

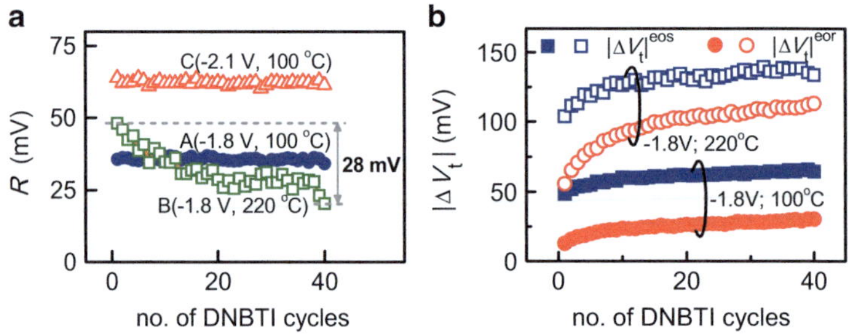

Fig. 11.7 (**a**) Evolution of V_t shift recovery per cycle under different temperatures. The gate stress voltages are as indicated. (**b**) Evolution of $|\Delta V_t|^{eos}$ and $|\Delta V_t|^{eor}$ at 100 and 220 °C, with the gate stress voltage fixed at -1.8 V [22]

excluded) test. This implies that the incremental total positive trapped charge in both cases are rising at similar rates, i.e., there is no substantial additional positive charge generation except in the initial stage of the DNBTI-CVS test. On the other hand, the $|\Delta V_t|^{eor}$ of the DNBTI-CVS test increases more rapidly as compared to that of the DNBTI-only test. This implies that the portion of relatively permanent positive trapped charge is increasing at a faster rate in the former. Since the total positive trapped charge is increasing at comparable rates, the results therefore indicate that in the case of the DNBTI-CVS test, the CVS phase has gradually transformed a part of the recoverable trapped holes into a more permanent form.

A similar observation is obtained when the temperature of the DNBTI test is increased, as shown in Fig. 11.7 [22]. In this case, no intermediate CVS was performed and the DNBTI conditions for the two cases were the same except for the temperature. One experiment was performed at 220 °C and the other at 100 °C.

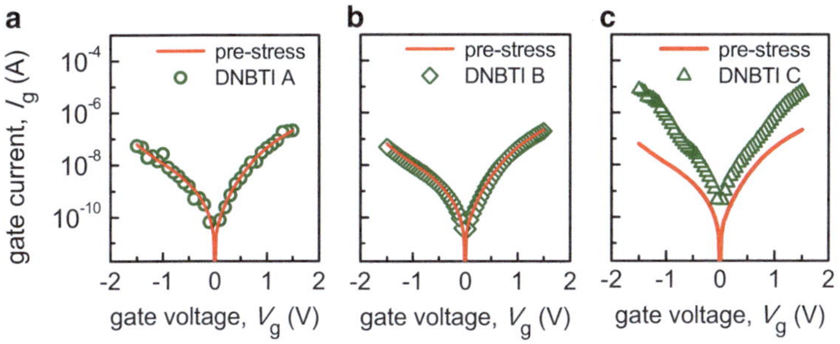

Fig. 11.8 Gate current I_g versus gate voltage V_g for different DNBTI conditions (**a**, **b**, and **c** correspond to the conditions stated in Fig. 11.7) [22]

For the former, the V_t recovery per cycle gradually decreases while that of the latter remains constant (as in Fig. 11.2). Figure 11.7b shows that the reduction in the V_t recovery per cycle stems from the transformation of a portion of the recoverable (or transient) hole trapping into a more permanent form.

It should be mentioned that Grasser et al. [35] also observed a decrease in the V_t shift recovery or R of the non-nitrided SiO_2 gate p-MOSFET at higher temperatures (~200 °C). Time-dependent defect spectroscopy study on an oxynitride gate p-MOSFET at elevated temperatures similarly revealed a halt in the hole emission event at a given defect for an extended period [36].

In another set of experiments, gate current measurement was also made at the end of the relaxation interval to examine the impact of such transformation on the gate oxide integrity and the results are shown in Fig. 11.8 [22]. Almost no SILC can be observed when the V_t recovery per cycle remains constant (as in Fig. 11.4). However, whenever a decrease of the V_t recovery per cycle is observed, SILC generation is also observed—the two are found to exhibit a very good linear correlation as shown in Fig. 11.9. Similar observations apply to the case of DNBTI-CVS testing depicted in Fig. 11.6. As the increase of SILC implies the generation of bulk traps, the linear correlation leads us to conclude that the bulk traps originate from the transient-to-permanent transformation of hole trapping. We believe such transformation (and not Si–H bond dissociation) underlies the observation of bulk trap generation reported in earlier studies [23, 24]. It should be emphasized that similar results are obtained for high-k gate p-MOSFETs (Figs. 11.10 and 11.11) [14, 16, 20, 21] implying the observed phenomenon is generic across different gate oxide materials.

Before concluding, we would like to provide a plausible explanation for the observed hole trapping transformation, based upon the current understanding on the nature of hole traps in the traditional SiO_2 and oxynitride gate dielectrics as well as the high-k gate dielectrics. Oxygen vacancies V_O's are a major source of hole traps in these gate dielectrics. Upon the capture of a hole, rearrangement of the atoms around the defect occurs to accommodate the loss of an electronic charge. We expect such structural relaxation to entail the overcoming of a certain energy

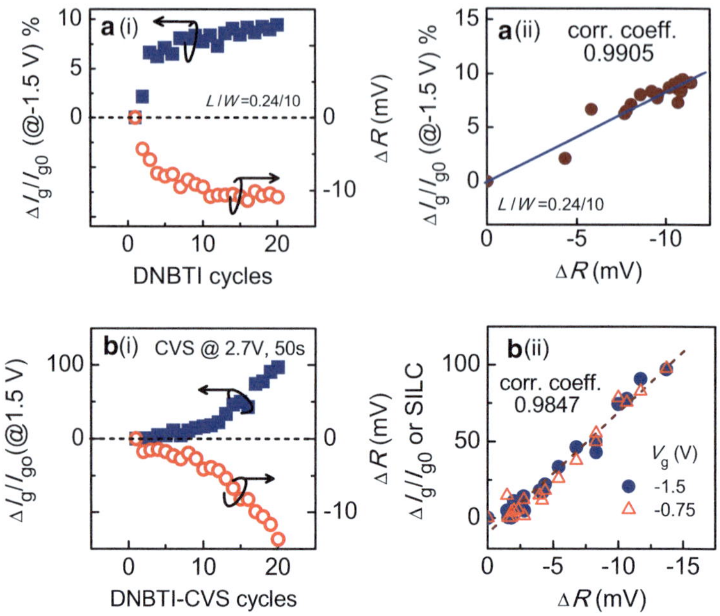

Fig. 11.9 (**a**-i) At a high temperature (220 °C), the decrease in the V_t shift recovery per cycle is seen to track the increase of stress-induced leakage current very closely. (**a**-ii) An excellent linear correlation can be observed between the two. (**b**-i) and (**b**-ii) Similar results hold for the DNBTI-CVS testing described earlier [18, 22]

barrier determined by the local oxide chemistry. If such a barrier is not overcome, permanent structural change could not occur and the hole trapping is temporary, i.e., the trapped hole is readily reemitted when the electrical stressing is removed. For instance, ab initio simulation has revealed a category of Vo's, called the $E_\delta{}'$ center, in the SiO_2 [37, 38]. Upon the capture of a hole, structural relaxation of the Si−Si dimer of the $E_\delta{}'$ center is marginal (in fact the remaining electron is shared between the two Si atoms). The positively charged, singly bond ed Si−Si dimer reverts to the neutral, doubly bonded Si=Si dimer immediately upon the reintroduction of an electronic charge to the defect site. On the other hand, there also exists a group of V_O's called the $E_\gamma{}'$ centers for which a substantially greater structural relaxation occurs upon the capture of a hole. The energy barrier for structural relaxation to occur in the $E_\gamma{}'$ center is presumably smaller than that for the $E_\delta{}'$. Indeed, examination of the local bonding network has revealed that the $E\gamma'$ center is usually found in places where a void exists to allow one of the Si atoms to move and become back-bonded to a nearby oxygen atom to reach an energetically favorable state. One expects the energy barrier to be more easily overcome under a high oxide field and temperature, which introduce greater distortion to the oxide network.

Similar ab initio simulation study on the oxynitride [39] and HfO_2 has been made by our group and results show that a larger percentage of the V_Os in these gate dielectrics exhibit substantial structural relaxation following the capture of

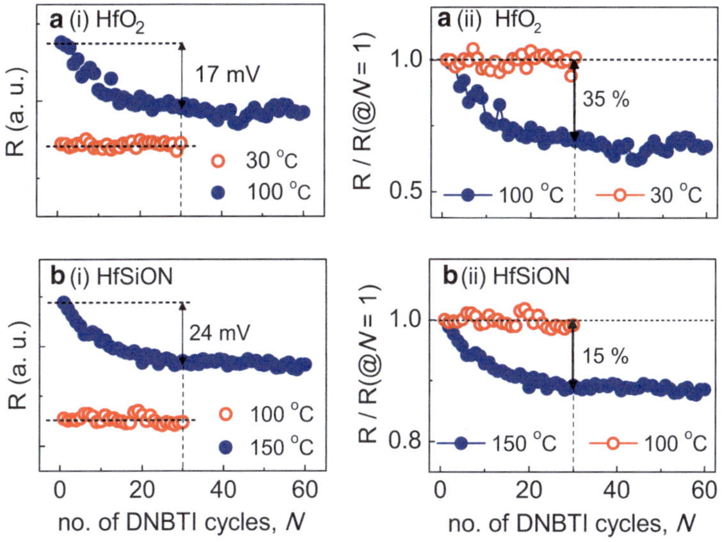

Fig. 11.10 Evolution of V_t shift recovery per relaxation phase or R as a function of the number of DNBTI cycles for the (**a**) HfO_2 and (**b**) HfSiON p-MOSFETs. The R remains constant at 30 °C for HfO_2 and 100 °C for HfSiON. For both dielectrics, the R is observed to decrease at higher temperatures. Diagrams on the *right* show R normalized with respect to that of the first DNBTI cycle. R decreases by ~35% for the HfO_2 p-MOSFET and ~15% for the HfSiON p-MOSFET after 30 DNBTI cycles [20]

holes as compared to the SiO_2. For the oxynitride, it was found that a neighboring nitrogen atom could facilitate the structural relaxation of V_O through N–Si bond formation involving the former's lone pair (a mechanism we believe to be similar to the back bonding of Si to a nearby O atom in the case of the $E_\gamma{}'$ center). As for the HfO_2, we attribute the result to its greater ionicity and the stronger electron–lattice interaction. These simulation results may explain (1) why V_Os in the plasma-nitrided gate oxide following NBTI stressing could be observed using the con-ventional ESR method [34] whereas the same method applied on the SiO_2 yielded a negative result. Recently, V_Os were successfully observed in the SiO_2 with ESR measurement carried out on the fly (i.e., with the stress continuously applied) [40]. (2) The oxide field and temperature thresholds for SILC or bulk trap generation and for the decrease of the V_t recovery per cycle in the HfO_2 gate p-MOSFET are much lower than those for the SiO_2 gate devices [14, 16, 20, 21].

11.5 Summary

An overview of recent key experimental evidences highlighting thc importance of hole trapping in NBTI is presented. The role of hole trapping was unappreciated (and in fact regarded as a parasitic effect) when the focus was largely on the

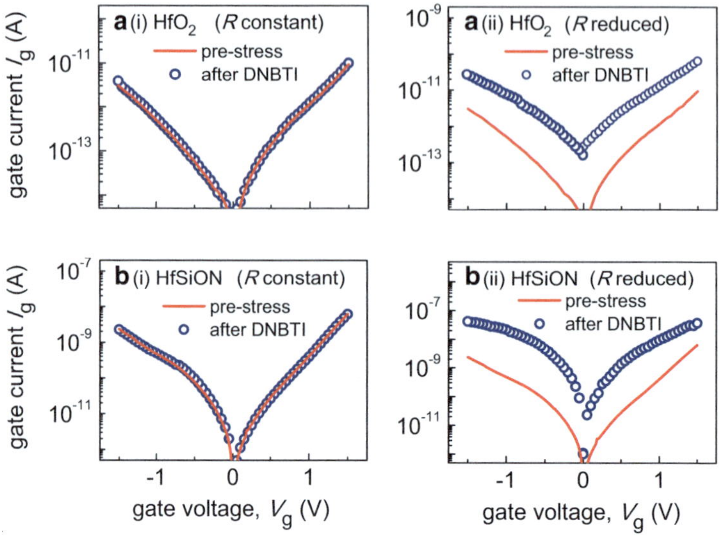

Fig. 11.11 The impact of 30 DNBTI cycles on SILC. No apparent SILC is observed for cases when R is constant, in spite of a substantial increase of $|\Delta V_t|^{eor}$. This implies that the $|\Delta V_t|^{eor}$ is mainly a result of interface defect generation. However, for cases when R is decreased, a very significant SILC is observed, indicating that bulk traps have been generated. This means that the decrease of R is correlated to bulk trap generation in high-k gate dielectrics, as in the SiO_2 and oxynitride gate dielectrics [20]

R-D model in the past. But as revealed in recent studies on DNBTI, the transport model is unable to explain the cyclical fluctuation and constant recovery of the V_t shift. These observations are better described in terms of hole trapping/detrapping at a group of preexisting oxide traps which consistently respond under a given set of experimental conditions. Moreover, experimental evidence which challenges a previous hypothesis that oxide trap generation is driven by hydrogen released from the dissociation of Si–H bonds (at the interface) is presented. In particular, no apparent bulk trap generation is observed in spite of significant interface degradation. Instead, bulk trap generation is observed to correspond to the reduction of V_t recovery or transient hole trapping, implying that the former stems from the transformation of transient-to-permanent hole trapping.

References

1. H. Reisinger, R. P. Vollertsen, P. J. Wagner, T. Huttner, A. Martin, S. Aresu, W. Gustin, T. Grasser, and C. Schlünder, in *Intl. Integrated Reliab. Workshop Final Report*, p. 1 (2008).
2. N. Kimizuka, K. Yamaguchi, K. Imai, T. Iizuka, C. T. Liu, R. C. Keller, and T. Horiuchi, in *Proc. Symp. VLSI Tech.*, p. 92 (2000).
3. Y. Mitani, M. Nagamine, H. Satake, and A. Toriumi, in *IEDM Tech. Dig.*, p. 509 (2002).

4. K. O. Jeppson and C. M. Svensson, *J. Appl. Phys.*, **48**, 2004 (1977).
5. S. Ogawa and N. Shiono, *Phys. Rev. B*, **51**, 4218 (1995).
6. J. H. Lee, W. H. Wu, A. E. Islam, M. A. Alam, and A. S. Oates, in *Proc. Intl. Reliab. Phys. Symp.*, p. 745 (2008).
7. S. Mahapatra, V. D. Maheta, A. E. Islam, and M. A. Alam, *IEEE Trans. Electron Dev.*, **56**, 236 (2009).
8. Z. Q. Teo, D. S. Ang, and K. S. See, in *IEDM Tech. Dig.*, p. 737 (2009).
9. T. Grasser, H. Reisinger, W. Goes, T. Aichinger, P. Hehenberger, P. J. Wagner, M. Nelhiebel, J. Franco, and B. Kaczer, in *IEDM Tech. Dig.*, p. 681 (2009).
10. Z. Q. Teo, D. S. Ang, and C. M. Ng, *IEEE Electron Dev. Lett.*, **31**, p. 269 (2010).
11. H. Reisinger, T. Grasser, W. Gustin, and C. Schlünder, in *Proc. Intl. Reliab. Phys. Symp.*, p. 7 (2010).
12. Z. Q. Teo, D. S. Ang, and C. M. Ng, *IEEE Electron Dev. Lett.*, **31**, p. 656 (2010).
13. D. S. Ang, Z. Q. Teo, T. J. J. Ho, and C. M. Ng, *IEEE Trans. Dev. & Mat. Reliab.*, **11**, p. 19 (2011).
14. Y. Gao, D. S. Ang, A. A. Boo, and Z. Q. Teo, *Microelectron. Eng.*, **88**, p. 1392 (2011).
15. Z. Q. Teo, A. A. Boo, D. S. Ang, and K. C. Leong, in *Proc. Intl. Reliab. Phys. Symp.*, p. 935 (2011).
16. Y. Gao, A. A. Boo, Z. Q. Teo, and D. S. Ang, in *Proc. Intl. Reliab. Phys. Symp.*, p. 943 (2011).
17. T. Grasser, P.-J. Wagner, H. Reisinger, Th. Aichinger, G. Pobegen, M. Nelhiebel, and B. Kaczer, in *IEDM Tech. Dig.*, p. 618 (2011).
18. A. A. Boo, D. S. Ang, Z. Q. Teo, and K. C. Leong, *IEEE Electron Dev. Lett.*, **33**, p. 486 (2012).
19. T. J. J. Ho, D. S. Ang, A. A. Boo, Z. Q. Teo, and K. C. Leong, *IEEE Trans. Electron Dev.*, **59**, p. 1013 (2012).
20. Y. Gao, D. S. Ang, C. D. Young, and G. Bersuker, in *Proc. Intl. Reliab. Phys. Symp.*, p. 5A.5.1 (2012).
21. Y. Gao and D. S. Ang, in *Proc. Intl. Symp. Phys. & Failure Analysis of I.C.*, p. 4.4.1 (2012).
22. A. A. Boo and D. S. Ang, *IEEE Trans. Electron Dev.*, **59**, p. 3133 (2012).
23. V. Huard, F. Monsieur, G. Ribes, and S. Bruyere, in *Proc. Intl. Reliab. Phys. Symp.*, p. 178 (2003).
24. S. Tsujikawa and J. Yugami, *IEEE Trans. Electron Dev.*, **53**, p. 51 (2006).
25. H. Reisinger, O. Blank, W. Heinrigs, A. Mühlhoff, W. Gustin, and C. Schlünder, in *Proc. Intl. Reliab. Phys. Symp.*, p. 448 (2006).
26. T. Grasser, W. Gös, V. Sverdlov, and B. Kaczer, in *Proc. Intl. Reliab. Phys. Symp.*, p. 268 (2007).
27. B. Kaczer, V. Arkhipov, M. Jurczak, and G. Groeseneken, *Microelectron Eng.*, **80**, p. 122 (2005).
28. M. Houssa, M. Aoulaiche, S. De Gendt, G. Groeseneken, M. M. Heyns, and A. Stesmans, *App. Phys. Lett.*, **86**, p. 093506 (2005).
29. A. T. Krishnan, C. Chancellor, S. Chakravarthi, P. E. Nicollian, V. Reddy, A. Varghese, R. B. Khamankar, and S. Krishnan, in *IEDM Tech. Dig.*, p. 688 (2005).
30. S. Choi et al., in Proc. Intl. Workshop on Compact Modeling, p. 33 (2012).
31. S. Choi, Y. J. Park, C.-K. Baek, and S. Park, in *Proc. Intl. Conf. Simulation of Semicond. Process & Dev.*, p. 185 (2012).
32. D. S. Ang and S. Wang, *IEEE Electron Dev. Lett.*, **27**, p. 914 (2006).
33. D. S. Ang, S. Wang, G. A. Du, and Y. Z. Hu, *IEEE Trans. Dev. & Mat. Reliab.*, 8, p. 22 (2008).
34. J. P. Campbell, P. M. Lenahan, A. T. Krishnan, and S. Krishnan, *J. Appl. Phys.*, **103**, p. 044505 (2008).
35. T. Grasser, H. Reisinger, P.-J. Wagner, B. Kaczer, *Phys. Rev. B*, **82**, p. 245318 (2010).
36. T. Grasser, P.-J. Wagner, H. Reisinger, Th. Aichinger, G. Pobegen, M. Nelhiebel, and B. Kaczer, in *IEDM Tech. Dig.*, 2011, p. 618 (2011).

37. Z.-Y. Lu, C. J. Nicklaw, D. M. Fleetwood, R. D. Schrimpf, S. T. Pantelides, *Phys. Rev. Lett.*, **89**, p. 285505 (2002).
38. T. Uchino and T. Yoko, *Phys. Rev. B*, **74**, p. 125203 (2006).
39. C. J. Gu, D. S. Ang, and Z. Q. Teo, *ECS Trans.*, **35**, p. 125 (2011).
40. J. T. Ryan, P. M. Lenahan, T. Grasser, and H. Enichlmair, in *Proc. Intl. Reliab. Phys. Symp.*, p. 43 (2010).

Part II

Chapter 12
Atomistic Modeling of Defects Implicated in the Bias Temperature Instability

Al-Moatasem El-Sayed and Alexander L. Shluger

Abstract Capture and emission of carriers by point defects in gate dielectrics, such as SiO_2 and HfO_2, and at their interfaces with the substrate are thought to be responsible for performance and reliability issues in MOS devices, particularly dielectric degradation and the bias temperature instability (BTI). Ultra-thin silicon dioxide films are present at the interface between Si and high-κ oxides; thus it is hoped that understanding the defects in silica which contribute to BTI will also aid the reliability of devices containing high-κ oxides. This chapter reviews the state of the art of modeling oxygen deficiency defects implicated in both electron and hole trapping in amorphous silica (a-SiO_2).

12.1 Introduction

Point defects in gate oxides have been implicated in many reliability problems in electronic devices, including random telegraph noise (RTN), bias temperature instability (BTI), stress-induced leakage current as well as in degrading the device performance by reducing the channel mobility and increasing the static power dissipation by enhancing the gate current [1]. In particular, electron and hole trapping/de-trapping at point defects in amorphous silica (a-SiO_2) gate oxide layers has been observed under a variety of conditions. Although the electronics industry is implementing more complex high-κ oxides for use as the gate dielectric, an

A.-M. El-Sayed (✉)
Department of Physics and Astronomy, University College London, Gower Street, London WC1E 6BT, UK
e-mail: al-moatasem.el-sayed.10@ucl.ac.uk

A.L. Shluger
Department of Physics and Astronomy and London Centre for Nanotechnology, 17-19 Gordon Street, London WC1H 0AH, UK
e-mail: a.shluger@ucl.ac.uk

T. Grasser (ed.), *Bias Temperature Instability for Devices and Circuits*,
DOI 10.1007/978-1-4614-7909-3_12, © Springer Science+Business Media New York 2014

ultra-thin SiO_2 layer persists in stacks containing these oxides too. Therefore it is imperative to understand how point defects in a-SiO_2 may contribute to BTI and device degradation. However, experimental identification of defects is a significant challenge, even more so as the oxide layers become thinner. The development of computational modeling tools allows one to create and characterize atomistic models of defects in materials, aiding their experimental identification and characterization.

Amorphous silica is a versatile material used not only in micro- and nano-electronics but in optical fibers and other optical devices, where defects play a crucial role in determining the properties of materials. Historically, the microelectronics community has borrowed many of the models of radiation-induced point defects in amorphous silica, which have been investigated by the glass community for over 56 years, producing electron spin resonance (ESR) data [2–4] and optical studies of pure and doped silica glass in bulk and fiber-optic forms. Many of the radiation-induced defects intrinsic to pure and B-, Al-, Ge-, and P-doped silicas have been reviewed in [5, 6]. Spectroscopic properties of hydrogen-related radiation defects in SiO_2-based glasses have been recently studied in a series of papers [7, 8]. The nanometer thickness of silica films in novel MOS devices and the presence of interfaces makes the spectroscopic studies of defects in these films extremely challenging, and a lot of our understanding is derived from electrical measurements and theoretical models [1].

For example, ESR studies show that the positive charge generated in relatively thick irradiated a-SiO_2 films correlates with the E' center creation [9–11]. The RTN [12] and BTI effects have also been associated with electron and hole trapping at oxygen vacancies in a-SiO_2 [13, 14]. However, a definite connection between the defects either identified or presumed to exist in silica films and the charge trapping, stress-induced leakage current and BTI in real MOS devices is still missing. Extracting atomistic information about electrically active defects is important for optimizing the stack manufacturing process and device reliability.

In this chapter we present a brief overview of atomistic modeling techniques that have been used over the past decade in SiO_2 defect research and models of oxygen deficient centers in a-SiO_2 implicated in reliability problems of MOS devices.

12.2 Atomistic Modeling Techniques

Computational modeling of structural and functional materials across all scales is a rapidly growing research area underpinning many applications and strongly benefiting from the growth in computer power [15]. Until relatively recently modeling a transistor mainly involved solving the Poisson and current continuity equations; no knowledge of the atomic structure was needed. Due to the aggressive scaling of transistors, the number of atoms in each device and hence the number of defects in the gate oxide is being reduced dramatically and measuring single-carrier discharges involving only a handful of defects is now a common practice [16]. This has resulted in significant time-dependent variability of devices manifesting in BTI and reduced reliability of the device. The convergence of the ever reducing size of devices with growing computational power means that it should become possible to carry out atomistic modeling of real devices in the near future using quantum mechanical methods. Below we discuss some basic features of such calculations.

A popular method for the calculation of the electronic structures of defects is Density Functional Theory (DFT) along with periodic boundary conditions or embedded cluster models of a solid material [17, 18]. These two models differ by the boundary conditions imposed on a system's wavefunction. Initially, the periodic model was developed for calculations of the electronic structure and properties of ideal periodic crystals, whereas the cluster model was introduced from molecular calculations. Both models gradually developed to provide accurate description of defect structures and processes.

The translational invariance of atoms in an ideal infinite crystal has been exploited in calculations of their band structure with great success [19]. Using large periodic cells one can also simulate a solid with periodically translated point defects. Calculations in the periodic model allow one to determine the geometric structures of defects, such as oxygen vacancies, in different charge states as well as their energy levels in the gap of an oxide [20]. They have obvious limitations as the size of periodic cell imposes constraints on the extent and character of lattice deformation around a defect and the complexity of interfaces which can be modeled. These limitations are gradually reduced as the size of periodic cells continues to increase.

Modeling defects in disordered solids and at interfaces between crystallites in polycrystalline systems often requires more flexibility than periodic models can provide. Also, optical absorption and ESR properties of defects are more readily calculated within molecular type models. For example, in many calculations of defects in a-SiO_2 the amorphous network has been represented by small molecular fragments, including the local environment of a defect and terminated by hydrogen atoms, e.g., [21, 22]. This molecular model is, however, too rough. It can be improved by taking into account the interaction of the cluster including a defect and neighboring ions with the rest of the host lattice as well as the perturbation of the lattice by the defect, and the reciprocal effect of the lattice polarization on the defect itself. This is achieved by constructing an external potential into which the cluster is then embedded and by surrounding the cluster by an infinite polarisable lattice treated using the classical force field method. Such a potential is called an embedding potential, and the model—the embedded cluster model [23, 24]. Some examples of the implementation of this approach and applications to defects in amorphous and polycrystalline oxides are discussed in [25–29].

Understanding the origins of different defect configurations in amorphous and polycrystalline materials and predicting their properties requires correlating the local structural characteristics of oxygen sites in the material, such as bond lengths, ring size, dihedral angles, with structural models and properties of defects created at these sites. Using theoretical tools one can build a model of an amorphous structure and then produce defects in a selection of sites and build a distribution of defect parameters. Theory usually deals with some idealized models of amorphous structure, such as that of continuum random network [30]. These can be generated by simulating a melt-quench procedure using classical or quantum mechanical molecular dynamics (MD) techniques [31–33] and the periodic model with large periodic cells. The DFT calculations are still computationally expensive so that one can routinely model systems of up to 1,000 atoms and perform MD calculations for a few picoseconds, dependent on available computational power. Many of the existing DFT simulations of defects in amorphous silica use 72 or 192 atoms in a periodic cell [34–36]. Using classical inter-atomic potentials one can model systems containing millions of atoms and simulations can be run for up to a microsecond, but no electronic structure information is then provided. The number of atoms in a periodic cell is important for modeling rare defects or defect configurations in amorphous structure. The length of MD simulation determines the rate of quenching of silica melt and hence the quality of the amorphous structure produced. Using very high rates one can create unrealistic structures and defects [32].

This approach has been used to study the distribution of properties of peroxy linkage defects [37], oxygen diffusion [38] in a-SiO_2, and more recently to analyze the structures and characteristics of different types of E' centers in silica [34, 35, 39–41]. This has proved very illuminating but only allows one to build good statistics of defect properties in thermal equilibrium and if a large number of sites in an amorphous structure is used in the statistical sampling. In those cases where defects are produced by irradiation and/or samples are only subjected to a partial anneal, only representative defect configurations can be reliably predicted in a selection of sites in the amorphous structure employed.

12.3 Oxygen Vacancy Defects in a-SiO_2

As mentioned above, oxygen vacancies in a-SiO_2 have been implicated in electron trapping/de-trapping processes involved in RTN and BTI in MOS devices. Below we discuss models of different charge states of oxygen vacancies in a-SiO_2. We note that the notation E' center in this chapter is reserved for the paramagnetic centers created by trapping a hole on a neutral oxygen vacancy. We use the notation V^q for other charge states of the oxygen vacancy in a-SiO_2. The models of four charge states of oxygen vacancies in a-SiO_2 are shown schematically in Fig 12.1.

12.3.1 Neutral Oxygen Vacancy

The neutral oxygen vacancy in SiO_2 is implicated as the fundamental positive charge trap [1]. It is formed when an oxygen atom is missing between two Si atoms in a regular SiO_2 network. Calculations of the oxygen vacancy in SiO_2 demonstrate that the two Si atoms displace toward the vacant oxygen site with the formation of a Si–Si bond. This center is often referred to as the oxygen deficient center I (ODC I) and is diamagnetic [6].

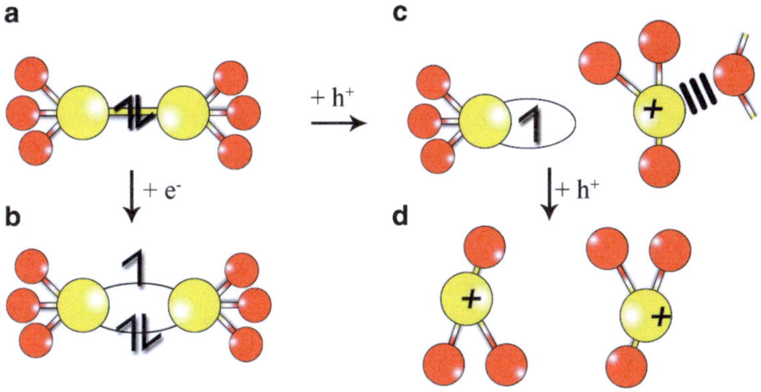

Fig. 12.1 Schematic representation of different charge states of the oxygen vacancy in a-SiO_2. Si atoms represented by *yellow spheres* and O atoms represented by *red spheres*. A neutral vacancy (**a**) can trap an electron and transform into a negatively charged vacancy (**b**) or trap a hole and transform into a positively charged vacancy (**c**). Double ionization of a neutral oxygen vacancy or trapping an extra hole by a positively charged vacancy induces strong distortion of the flexible amorphous network and creation of a V^{2+} center (**d**). The singly positively charged vacancy shown in (**c**) has several configurations discussed below. Trapping of an electron by the doubly positively charged vacancy (**d**) can lead to formation of a back-projected configuration of E′ center, discussed in Sect. 12.3.3

Calculations of the neutral oxygen vacancy using ab initio and semi-empirical methods within molecular cluster, embedded cluster, and periodic models predict the Si–Si bond length to be between 2.3 and 2.7 Å. This bond length is close to the Si–Si distance in elemental silicon and strongly reduced with respect to the initial Si–Si distance of ≈3.2 Å, reflecting the flexibility of the SiO_2 network and the strength of the Si–Si bond. The SiO_2 network relaxation associated with the formation of a neutral oxygen vacancy is known to be long-ranged. Embedded cluster calculations in α-quartz by Sulimov et al. [18] and in a-SiO_2 by Mukhopadhyay et al. [40] show that the displacements of ions from their equilibrium positions are still very significant up to 5 Å away from the defect and propagate as far as 13 Å from the vacancy site both in quartz and in a-SiO_2 [18, 40]. The long-range network relaxation has been shown to be asymmetric due to the absence of inversion symmetry in quartz and strongly depends on the local environment of vacancies in a-SiO_2 [18], indicating that the local environment strongly affects the position of the double occupied level of the vacancy, which is located at ≈1.0 eV above the top of the valence band of SiO_2.

12.3.2 Singly Positively Charged Oxygen Vacancies

The ab initio calculations used to explain the origins of 1/*f* noise and thermally stimulated current in MOS devices [1, 33, 39] suggested that it is caused by the thermally activated capture and emission of carriers at O vacancy centers near the

Si/SiO_2 interface. A similar mechanism has been proposed for explaining NBTI; however, the nature of electron/hole traps involved has not been fully understood [9, 16, 42].

Both phenomena involve hole capture from the Si substrate by a neutral vacancy in the oxide. The results of [1,33,39] identify these defects with at least two kinds of the positive oxygen vacancy configurations in un-irradiated and irradiated devices. The first is a "dimer" vacancy configuration (see Fig. 12.1a), often called the E'_δ center [36, 43]. The second is a neutral or positively charged center, which has two configurations. One is referred to in the literature as the E'_γ center (in which one of the Si atoms relaxes through the plane of its oxygen neighbors) and forms a dipole after the electron capture (see Fig. 12.1c). The other is shown to be a different kind of configuration, which does not form a dipole after the electron capture (see Fig. 12.1b).

The prevailing model of the E' center in SiO_2 first introduced by Rudra and Fowler in α-quartz [44] is similar to the puckered configuration (PC) shown in Fig. 12.2c. In this model one of the Si atoms accommodates an unpaired electron on a dangling bond whereas the second Si traps a hole and relaxes through the plane of its three neighboring oxygen atoms. Importantly this configuration is stabilized in α-quartz because the puckering Si forms a bond with a so-called back-oxygen ion (shown schematically in Fig. 12.1b). Allan and Teter were the first to demonstrate that along with the PC there is the dimer configuration (DC) [see Fig. 12.2a], wherein the two silicon atoms that face the vacancy relax very little away from their ideal positions in the perfect quartz lattice [45]. Most calculations predict the PC to be lower in energy than the DC and there is no clear experimental evidence that the DC is significantly populated in α-quartz [39, 40, 46].

Contrary to α-quartz, several atomistic models have been proposed for the E' center in a-SiO_2 [39, 40, 47]. E' center is now understood in a more general sense as a Si dangling bond with an unpaired electron, which can exist on its own as a neutral defect or as a part of a positively charged oxygen vacancy (as in Fig. 12.2 b–d), or as an even more complex defect associated with a proton or delocalized by four or five Si atoms [48]. These defects are labeled by subscripts α, β, γ, δ to distinguish species characterized by different EPR parameters (see, for example, [49]). However, their structural models are still controversial. The barrier for the transformation between PC and DC configurations of the E' center in α-quartz are shown to be small (0.1–0.5 eV) [33, 39]. However, in a-SiO_2 the very existence of such barrier depends on a local configuration of the defect. Moreover, the so-called back-oxygen atom which helps to stabilize the PC in α-quartz is often located much further away in a-SiO_2. This means that there is no back oxygen to facilitate puckering of the Si ion through the plane of its neighboring three O ions in many locations in a-SiO_2. Ab initio calculations reveal that the PC and a range of intermediate configurations shown in Fig. 12.2b are the most probable configurations of the E' center in a-SiO_2 [40]. The flexibility of the a-SiO_2 network gives rise to a range of such configurations characterized by different short- and long-range network relaxation and wide distribution of positions of defect levels in the gap. It also gives rise to a very different configuration, termed the back-projected

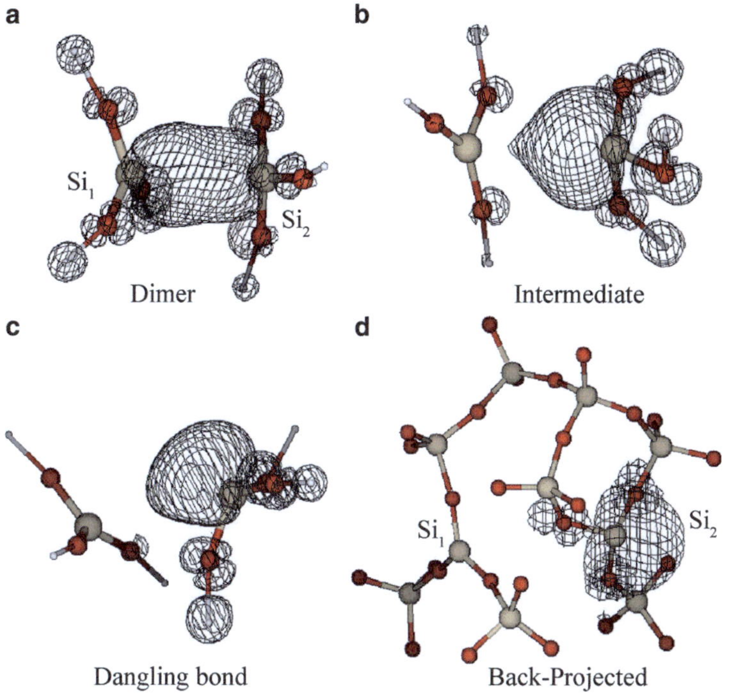

Fig. 12.2 Different configurations of the singly positively charged oxygen vacancy, the E′ center, showing the spin densities represented on wire-frames. The Si atoms are represented by the *bigger white spheres* while the O atoms are the *smaller dark spheres*. (**a**) Dimer configuration where two Si atoms form a weak bond. (**b**) Intermediate configuration, where an unpaired electron is mostly localized on one Si atom. (**c**) Puckered configuration, where the unpaired electron is localized on a three-coordinated Si while the other Si of the vacancy, trapping the hole, is displaced through the plane of its three oxygen neighbors. (**d**) Back-projected configuration, where the spin density of the unpaired electron localized on three-coordinated Si atom is directed away from the vacancy

configuration, comprising an unpaired electron in a positively charged oxygen vacancy in a-SiO_2 shown in Fig. 12.2d. Understanding the possible mechanism of formation of this configuration requires first describing a doubly positively charged vacancy.

12.3.3 Doubly Positively Charged Oxygen Vacancies

In addition to the singly charged defects, one may expect that further ionization of (or hole trapping by) the positively charged E′ centers could produce transient V^{2+} centers in SiO_2. A doubly positively charged oxygen vacancy, V^{2+}, has been considered theoretically in [44, 50, 51]. The doubly ionized vacancy is characterized by a significant local structural relaxation resulting in a 4 Å Si–Si distance where

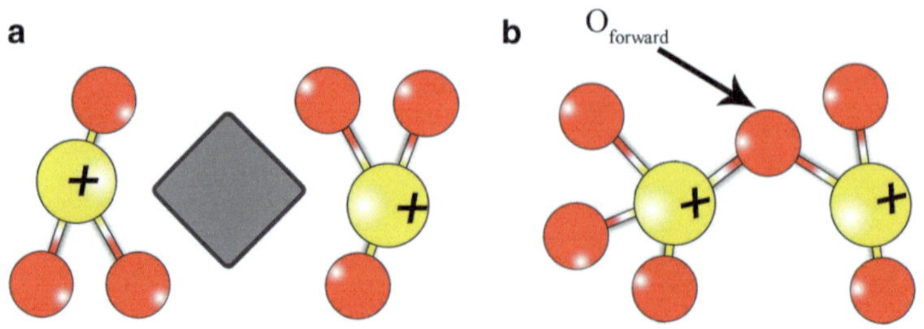

Fig. 12.3 Schematic of the doubly positively charged vacancies in SiO_2. The vacancy is represented by the *gray diamonds*. (**a**) shows the configuration in which both Si atoms relax through the planes of their oxygen neighbors. (**b**) shows the second configuration in which one Si atom bonds with the $O_{forward}$ atom which is the nearest neighbor of the other Si atom

the oxygen atom has been removed [44]. Chadi showed that the structures of the V^+ and V^{2+} centers in α-quartz are similar and that the V^{2+} center has a puckered configuration, where one of the three-coordinated Si atoms moves through the plane of its three basal oxygens and forms a fourth bond with a distant oxygen atom from the back-ring [50]. These calculations have demonstrated that the double ionization of the oxygen vacancy in quartz induces a significant rearrangement of nuclei extending into the lattice.

The final structure of the relaxed defect in a-SiO_2 strongly depends on the local environment as well as the initial Si–Si distance and positions of the, so called, back oxygens. The most prominent feature of this relaxation corresponds to the puckering of both Si atoms around the vacancy backwards through the plane of their basal oxygens and creation of bonds with distant oxygen atoms from the back rings (see Fig. 12.3a). The second type corresponds to the relaxation of only one of the Si atoms back through the plane of its basal oxygen atoms, where it produces a bond with the distant O atom (see Fig. 12.3b).

The V^{2+} centers create an unoccupied electronic level in the band gap. As mentioned above, the actual position of this level strongly depends on the distance between the two Si ions and the details of relaxation of the surrounding amorphous network. For the 20 defect configurations considered in [52], the energies of the one-electron defect levels with respect to the top of the valence band are situated between 4.7 and 6.4 eV.

The diamagnetic nature of V^{2+} centers makes their experimental detection difficult, although they can participate in radiation and stress-induced transformations of the E′ centers and other defects in SiO_2. Calculations of their optical absorption energies for different configurations predict the optical excitations from the states in the valence band into the unoccupied defect state in the band gap [52]. Due to the delocalized nature of the valence band states involved in the strongest optical excitations, the transitions have relatively weak oscillator strengths, between 0.003 and 0.05. The calculated energies for the optical transitions are distributed in a wide range between 4.5 and 6.5 eV.

12.3.4 Back-Projected Configurations of the E′ Center

The doubly positively charged vacancies described above can take part in further transformations by trapping electrons and converting into E′ centers. However, not all of them will result in the familiar configurations shown in Fig. 12.2a–c. Injecting an extra electron into the V^{2+} configurations produced defect configurations similar to those of the E'_γ center only in about half of the configurations studied [52]. These configurations are characterized by the relaxation of one of the Si atoms bearing the hole toward the vacancy while the other one is located in the plane of its basal oxygen atoms (Fig. 12.1c). The electronic structure of these centers corresponds to full localization of an unpaired spin on a Si ion projected towards the vacant oxygen site. They are characterized by the typical value of the isotropic hyperfine constant of 42 mT and a slightly orthorhombic g-tensor.

Notably, in half of the configurations the Si atom becomes three-coordinated while keeping its position inside the back ring. At the same time, the other Si atom remains four-coordinated bonded with a distant oxygen atom from the back-ring. The electronic structure of this defect corresponds to the localization of an unpaired electron on a three-coordinated silicon with the spin density directed backwards from the oxygen vacancy, hence it is termed a back-projected configuration (see Figs. 12.2d and 12.4). The calculated hyperfine coupling constant for this defect is equal to 42 mT, which is similar to that for the E'_γ center.

We note that the total energy of the back-projected E′ center is higher than that of the corresponding E'_γ center [40]. Therefore, the system is in the local minimum and can relax into the global minimum overcoming a barrier. The calculated barrier for converting the back-projected E′ center into the E'_γ-center configuration is around

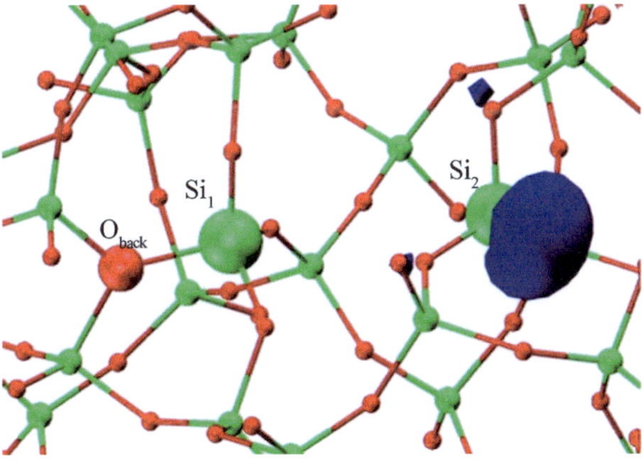

Fig. 12.4 The back-projected configuration of the E′ center. Si atoms represented by *green spheres* and O atoms represented by *red spheres*. Si_1 and Si_2 are the Si atoms of the vacancy. O_{back} is the back oxygen. The spin density of the localized electron is shown in *blue*

1.2 eV [40]. This relatively high barrier hampers thermally activated conversion and the center can be stable at room temperature.

These results demonstrate that different treatment of a-SiO_2 samples and MOS devices can lead to formation of different types of centers. In particular, V^{2+} centers can be easily formed under irradiation and then filled by electrons. This occurs either as a result of recombination of radiation-induced defects and/or carriers, or by tunneling from Si. Experimental verification of these finding is difficult. We note that the ESR properties of back-projected centers are consistent with those of the recently observed E'_{α} center in bulk a-SiO_2 [53]. The existence of back-projected configurations of E' center has been suggested first by Griscom et al. on the basis of ESR measurements on silica glass samples [54].

12.3.5 Singly Negatively Charged Oxygen Vacancies

There have been suggestions that the neutral oxygen vacancy can serve as a precursor to both positively and negatively charged vacancy states (see Fig. 12.1b, d) in SiO_2 (see, for example, [9,55]). The experimental observations suggest, however, that the electron capture cross section of the neutral electron traps is 2–3 orders of magnitude smaller than that of the hole [9]. Small capture cross-section values may be one of the reasons why V^- centers proved to be difficult to observe and identify experimentally.

On the theoretical front, earlier molecular cluster calculations suggested that the great flexibility of the SiO_2 lattice could promote trapping of an extra electron by a neutral vacancy [56, 57], which has recently been confirmed by more detailed calculations for V^- centers in α-quartz [43]. These calculations demonstrated that the neutral vacancy forms a very shallow precursor site for an extra electron; before relaxation this electron is predominantly localized in the vicinity of the vacancy. This suggests a small cross-section for the electron trapping from the conduction band. However, owing to the flexibility of the silica structure, the fully relaxed structure is lower in energy by about 1.7 eV. The most prominent feature of the lattice relaxation is the increase of the Si–Si distance in the vacancy by approximately 0.2 Å. An unpaired electron is almost equally localized on the Si ions neighboring the vacancy (see Fig. 12.5). The V^- center has the isotropic hyperfine coupling constant of 32 mT and an orthorhombic symmetry of the g-tensor.

Further calculations for four V^- configurations in a-SiO_2 confirmed these conclusions [52]. Kimmel et al. have found that the values of the vertical electron affinities of V^0 vary from 0.5 to 1.6 eV, indicating that some of the vacancies may have relatively large cross sections for electron trapping. The geometric relaxation of the defect strongly depends on the local environment of the vacancy and leads to a lowering of the total energy of the system by 0.8–2.1 eV.

The relaxed electron affinities (these are equal to thermal ionization energies into the conduction band) vary between 2.0 and 3.3 eV, indicating that the V^- center is a deep electron trap. The unpaired electron of the V^- center is delocalized

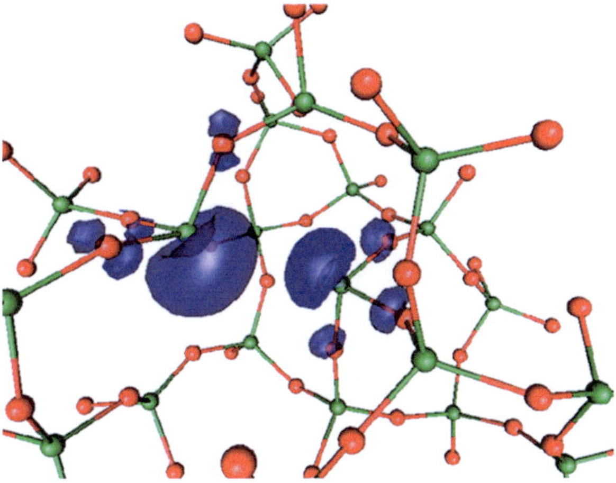

Fig. 12.5 Spin density distribution in one of the negatively charged O vacancies in a-SiO_2

in approximately equal manner over the two Si atoms neighboring the vacancy (see Fig. 12.5). The negatively charged vacancy is characterized by the isotropic hyperfine coupling which varies from 32.8 to 37.4 mT depending on the local geometry of the defect and an orthorhombic g-tensor. It has been noted that there are quite strong variations of the components of the g-tensor between different configurations.

12.4 Role of Hydrogen

The role of hydrogen in the BTI-related processes is not fully understood with several models having been proposed [58, 59]. It is well established that the hydrogen atom and molecule interact only very weakly with the silica network and can diffuse with very small barriers [60, 61]. Dependent on the chemical potential, a hydrogen atom may adopt a positively or negatively charged state, forming strong bonds with a bridging oxygen ion or a fully coordinated Si ion in the network, respectively. The formation of these bonds is accompanied by a strong relaxation of the surrounding network. These defects, as well as the interaction of a hydrogen atom with the neutral oxygen vacancy, have been modeled in detail by Blöchl in the attempt to understand the mechanism of trap-assisted leakage current [36]. These calculations have demonstrated that the hydrogen atom can incorporate into a neutral oxygen vacancy in α-quartz in three charge states: positive, neutral, and negative. It has been suggested that this, so-called hydrogen bridge defect (see Fig. 12.6), can scrve as a good candidate for relaying an electron from a metal electrode into a silicon substrate and thus can be responsible for hot-electron degradation in thin MOS oxides [62].

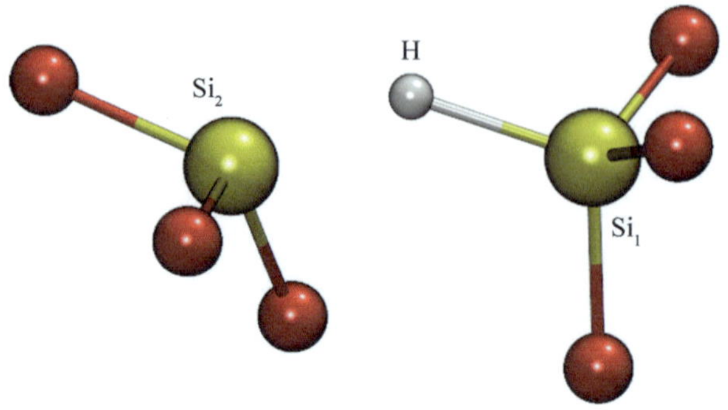

Fig. 12.6 Model of the hydrogen bridging E′ center, identified as the E'_4 center in α-quartz. The Si atoms are shown as *yellow spheres*, O atoms as *red spheres* and H atoms as *white spheres*

In Si/SiO_2/Si systems with a buried oxide, it is known that the incorporation of H_2 molecules results in the formation of protons [60]. Calculations by Lopez et al. show that E′ centers in SiO_2 can act as active sites for the cracking of hydrogen molecules and releasing protons into the oxide [63]. These calculations demonstrate that a hydrogen molecule can break into two separate hydrogen atoms as a result of interaction with an E′ center, causing the formation of two Si–H bonds. The formation of Si–H bonds causes the E′ center to unpucker and both Si–H bonds to face each other. A proton can then come off one of the Si atoms (the Si atom that had originally puckered through the plane of its neighbors) and bind to the nearest oxygen ion. This reduces the total energy of the system by 0.2 eV, resulting in a more stable configuration. The proton can then diffuse away, the barrier to which is measured experimentally as 0.81 eV [64].

Amorphous silica samples can contain three-coordinated Si atoms with a dangling bond passivated by hydrogen. Afanas'ev et al. argued [65] that trapping of a hole by the Si–H bond of $O_3 \equiv$Si–H can lead to the formation of a dangling bond on $O_3 \equiv$Si and release a proton. If the proton diffuses away, the $O_3 \equiv$Si· entity with the dangling bond is equivalent to a neutral E′ center.

Ling et al. carried out periodic DFT calculations of amorphous SiO_2 to investigate how Si–H bonds could serve as precursors to E′ centers [47]. They studied 20 models of bulk amorphous SiO_2, each containing 216 atoms, and showed that the energy required for breaking an Si–H bond with the H atom moving to an interstitial position is ≈4.2 eV. However, if an hole is added to the system, the Si–H bond could be broken if the hole becomes localized on the bond. This results in the formation of a three-coordinated Si atom with an unpaired electron and a proton which is bound initially to the nearest oxygen atom (see Fig. 12.7). The calculations in 20 a-SiO_2 models showed that the barrier for dissociating the proton from the three-coordinated Si and attaching it to one of the nearest oxygen ions (see Fig. 12.7) ranges from 0 to 0.5 eV, dependent on the initial distance of the proton to the oxygen to which it binds.

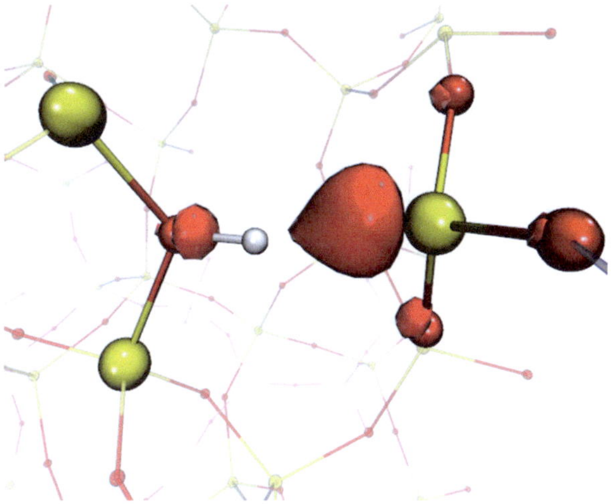

Fig. 12.7 Neutral E′ center showing the unpaired electron localized on a three-coordinated Si atom and a proton bound to an oxygen atom in a hydronium-like configuration. Si atoms represented as *yellow spheres* and O atoms as *red spheres*

The calculated values of the isotropic hyperfine interaction between the ^{29}Si nucleus and the unpaired electron localized on the Si atom range from 40.0 to 47.8 mT. These values are close to those usually associated with the E′ center in a-SiO_2 [66]. The localization of the unpaired electron on a three-coordinated Si atom renders the properties of this defect similar to those of the E′ center, although the presence of a proton makes some difference on the defect's properties. The isotropic component of the hyperfine interaction between the proton bound in an hydronium-like configuration (see Fig. 12.7) and the unpaired electron localized on the Si atom was calculated to be 1.6 mT, similar to the 1.05 mT satellite signals that are seen in the experiment [7]. The specific position of the proton affects the electron states of the defect: at the shortest separation they move down by about 0.5 eV with respect to the long separation exceeding 0.7 nm.

12.5 Conclusions

We outlined the wide range of geometric configurations associated with the oxygen vacancy in a-SiO_2 corresponding to four possible charge states. The positions of the occupied defect levels in the band gap of a-SiO_2 corresponding to different charge states of oxygen vacancy as well as to the hydrogen bridge defect in a-SiO_2 are summarized in a schematic in Fig. 12.8. The levels are broadened to emphasize the fact that the disorder in the a-SiO_2 network strongly affects the geometric and electronic structures of the defects. However, the real extent of this effect is not well understood and the broadening in Fig. 12.8 should not be treated as

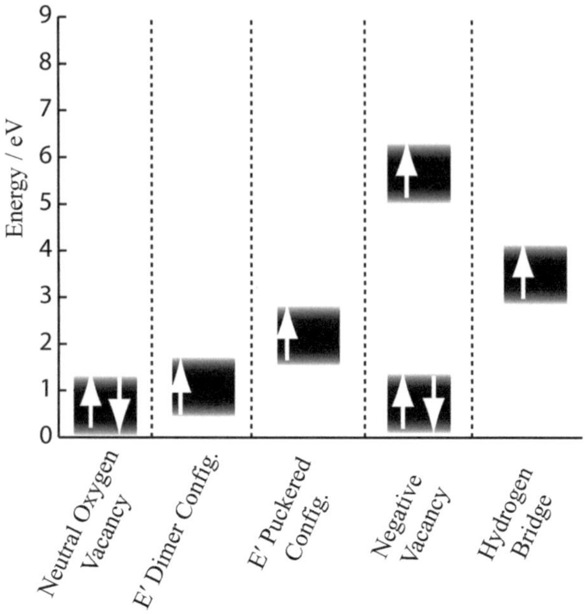

Fig. 12.8 Schematic showing the defect levels of vacancy related defects in a-SiO_2. Each defect level is broadened by an arbitrarily chosen value of 1.0 eV. This broadening is a schematic representation of the effect of the variable long-range order of the amorphous network

a true representation of how the energies of the electronic states are distributed. For each particular defect this distribution is affected not only by the disorder of the amorphous network, but also by the position of the defect with respect to the interfaces with semiconductor and metal electrodes, confinement of the silica layer, interaction with other defects and by the bias. In relation to BTI, these factors also affect the defect diffusion barriers, optical absorption energies and ESR parameters [41, 47, 52, 67].

The effect of disorder on defect levels has been studied by several groups [40, 46, 52]; however, due to high computational costs, systematic studies started to appear only recently. In particular, Anderson et al. recently studied a variety of defects in a-SiO_2, including the neutral oxygen vacancy, five-coordinated Si, three-coordinated Si, three-coordinated O, and two-coordinated Si [34]. To understand the effects of the amorphous network on the variability of defect properties, 120 uncorrelated models of a-SiO_2 were generated containing either 192 atoms or 191 atoms (to study oxygen vacancies). The structures were obtained using classical MD with all structures subsequently optimized at the DFT level. Statistical distributions of defect energies and their switching levels have been calculated for a range of defects demonstrating the width of distributions of about 1 eV as reflected in Fig. 12.8.

Even as the industry moves toward high-κ dielectrics, the question of the microscopic origins of BTI in SiO_2 devices persists. A thin SiO_2 layer (on the order of 1 nm) in high-κ devices has been shown to interact with high-κ dielectric, inducing a rather high density of oxygen vacancy-related defects in this SiO_2 layer. Although models of oxygen vacancies in these ultra-thin silica layers are expected to remain the same as discussed above, the structure and electrical levels of these defects will strongly depend on their position in the layer. Identifying these defects and studying their properties remain a significant challenge for both experiment and theory.

Acknowledgements We would like to acknowledge the financial support from EPSRC and the EU FP7 project MORDRED (FP7-NMP-2010-SMALL-4). The authors are grateful to V. Afanas'ev, A. Asenov, G. Bersuker, M.B. Watkins, S. Ling, S.R. Bradley, and F. Schanovsky for useful discussions.

References

1. Fleetwood, D.M., Pantelides, S.T., Schrimpf, R.D.: Defects in microelectronic materials and devices. CRC Press (2009)
2. Weeks, R.A.: J. Appl. Phys. **27**, 1376 (1956)
3. Weeks, R.A., Nelson, C.M.: Trapped electrons in irradiated quartz and silica. 2. Electron spin resonance. J. Am. Ceram. Soc. **43**, 399 (1960)
4. Nelson, C.M., Weeks, R.A.: Trapped electron centers in irradiated quartz and silica. 1. Optical absorption. J. Am. Ceram. Soc. **43**, 396–399 (1960)
5. Griscom, D.L.: Trapped electron centres in silica. J. Non-Cryst. Solids **357**, 1945–1962 (2011)
6. Skuja, L.: Optically active oxygen-deficiency-related centers in amorphous silicon dioxide. J. Non-Cryst. Solids **239**, 16–48 (1998)
7. Skuja, L., Kajihara, K., Hirano, M., Hosono, H.: Hydrogen-related radiation defects in SiO_2-based glasses. Nucl. Instrum. Methods Phys. Res. B **266**, 2971–2975 (2008)
8. Messina, F., Cannas, M.: Photochemical generation of E' centres from Si-H in amorphous SiO_2 under pulsed ultraviolet laser radiation. J. Phys.: Condens. Matter **18**(43), 9967–9973
9. Walters, M., Reisman, A.: Radiation-induced neutral electron trap generation in electrically biased insulated gate field effect transistor gate insulators. J. Electrochem. Soc. **138**, 2756–2762 (1991)
10. Lelis, A.J., Oldham, T.R.: Time dependence of switching oxide traps. IEEE Trans. Nucl. Sci. **41**, 1835–1839 (1994)
11. Conley, J.F., Lenahan, P.M., Lelis, A.J., Oldham, T.R.: Electron Spin Resonance evidence E' centers can behave as switching oxide traps. IEEE Trans. Nucl. Sci. **42**, 1744–1749 (1995)
12. Wagner, P., Aichinger, T., Grasser, T., Nelhiebel, M., Vandamme, L.: in *Proceedings of the International conference on Noise and Fluctuations*. pp. 621–624 (2009)
13. Grasser, T., Kaczer, B., Göes, W., Aichinger, T., Hehengerber, P., Nelhiebel, M.: Understanding Negative Bias Temperature Instability in the Context of Hole Trapping. Microelectron. Engineering **86**, 1876–1882 (2009)
14. Schroder, D.K.: Microelectron. Reliab. **47**, 41–852 (2006)
15. Leach, A.R.: Molecular modelling: Principles and applications, pp. 359–362. Pearson Prentice Hall (2001)
16. Grasser, T., Kaczer, B., Goes, W., Reisinger, H., Aichinger, T., Hehenberger, P., Wagner, P.J., Schanovsky, F., Franco, J., Luque, M.T., Nelhiebel, M.: The Paradigm Shift in Understanding the Bias Temperature Instability: From Reaction-Diffusion to Switching Oxide Traps. IEEE Trans. Electr. Dev. **58**(11), 3652–3666 (2011)
17. Martin, R.: Electronic structure: Basic theory and practical methods, p. 173. Cambridge University Press, Cambridge (2004)

18. Sulimov, V.B., Sushko, P.V., Edwards, A.H., Shluger, A.L., Stoneham, A.M.: Phys. Rev. B **66**, 024108 (2002)
19. Kittel, C.: Introduction to Solid-State Physics. Wiley, New York (1976)
20. Van de Walle, C., Neugebauer, J.: First-principles calculations for defects and impurities: Applications to III-Nitrides. J. Appl. Phys. **95**(8), 3851–3879 (2004)
21. Pacchioni, G., Ieranò, G.: Phys. Rev. B **57**, 818–832 (1998)
22. Uchino, T., Yoko, T.: Phys. Rev. B **68**, 041201(R)–1–4 (2003)
23. Sushko, P.V., Mukhopadhyay, S., Mysovsky, A.S., Sulimov, V.B., Taga, A., Shluger, A.L.: Structure and properties of defects in amorphous silica: new insight from embedded cluster calculations. J. Phys.: Condens. Matter **17**, S2115–S2140 (2005)
24. Muñoz Ramo, D., Gavartin, J.L., Shluger, A.L., Bersuker, G.: Phys. Rev. B **75**, 205336–1–12 (2007)
25. Pandey, R., Vail, J.M.: F-type centres and hydrogen anions in MgO: Hartree-Fock ground states. Journal of Physics: Condensed Matter **1**(17), 2801 (1989)
26. Vail, J.M.: Theory of electronic defects - Applications to MgO and alkali-halides. J. Phys. Chem. Solids **51**(7), 589–607 (1990)
27. Sousa, C., Illas, F.: On the accurate prediction of the optical absorption energy of F-centers in MgO from explicitly correlated ab initio cluster model calculations. J. Chem. Phys. **115**(3), 1435–1439 (2001)
28. Donnerberg, H., Birkholz, A.: Ab initio study of oxygen vacancies in $BaTiO_3$. J Phys.: Condens. Matter **12**(38), 8239 (2000)
29. Sushko, P.V., Shluger, A.L., Catlow, C.R.A.: Relative energies of surface and defect states: ab initio calculations for the MgO (001) surface. Surf. Sci. **450**(3), 153–170 (2000)
30. Wright, A. C.: Defects in SiO_2 and related dielectrics: Science and Technology, p. 1. Kluwer Academic Publisher (2000)
31. Pramod Vedula, R., Anderson, N.L., Strachan, A.: Phys. Rev. B **85**, 205209–1–11 (2012)
32. Vollmayr, K., Kob, W., Binder, K.: Phys. Rev. B **54**, 15808–15827 (1996)
33. Nicklaw, C.J., Lu, Z.Y., Fleetwood, D.M., Schrimpf, R.D., Pantelides, S.T.: The structure, properties and dynamics of oxygen vacancies in amorphous SiO_2. IEEE Trans. Nucl. Sci. **49**, 2667–2673 (2002)
34. Anderson, N.L., Pramod Vedula, R., Schultz, P.A., Van Ginhoven, R.M., Strachan, A.: First-Principles Investigation of low energy E' center precursors in amorphous silica. Phys. Rev. Lett. **106**, 206402–1–4 (2011)
35. Anderson, N.L., Pramod Vedula, R., Schultz, P.A., Van Ginhoven, R.M., Strachan, A.: Defect level distributions and atomic relaxations induced by charge trapping in amorphous silica. Appl. Phys. Lett. **100**, 172908–1–3 (2012)
36. Blöchl, P.: Phys. Rev. B **62**, 6158 (2000)
37. Szymanski, M.A., Shluger, A.L., Stoneham, A.M.: Phys. Rev. B **63**, 224207–1–9 (2001)
38. Bongiorno, A., Pasquarello, A.: Oxygen diffusion through the disordered oxide network during silicon oxidation. Phys. Rev. Lett. **88**, 125901–1–4 (2002)
39. Lu, Z.Y., Nicklaw, C.J., Fleetwood, D.M., Schrimpf, R.D., Pantelides, S.T.: Structure, properties, and dynamics of oxygen vacancies in amorphous SiO_2. Phys. Rev. Lett. **89**, 285505–1–4 (2002)
40. Mukhopadhyay, S., Sushko, P.V., Stoneham, A.M., Shluger, A.L.: Phys. Rev. B **70**, 195203–1–10 (2004)
41. Mukhopadhyay, S., Sushko, P.V., Stoneham, A.M., Shluger, A.L.: Phys. Rev. B **71**, 235204–1–9 (2005)
42. Donaldio, D., Bernasconi, M., Boero, M.: Ab initio simulations of photoinduced interconversions of oxygen deficient centers in amorphous silica. Phys. Rev. Lett. **87**, 195504–1–4 (2001)
43. Sushko, P.V., Mukhopadhyay, S., Stoneham, A.M., Shluger, A.L.: Oxygen vacancies in amorphous silica: Structure and distribution of properties. Microelectron. Eng. **80**, 292–295 (2005)
44. Rudra, J.K., Fowler, W.B.: Phys. Rev. B **35**, 8223–8230 (1987)

45. Allan, D.C., Teter, M.P.: Local density approximation total energy calculations for silica and titania structure and defects. J. Amer. Ceram. Soc. **73**(11), 3247–3250 (1990)
46. Pantelides, S.T., Lu, Z.Y., Nicklaw, C., Bakos, T., Rashkeev, S.N., Fleetwood, D.M., Schrimpf, R.D.: The E′ center and oxygen vacancies in SiO_2. J. Non-Cryst. Solids **354**, 217–223 (2008)
47. Ling, S., El-Sayed, A.M., Lopez-Gejo, F., Watkins, M.B., Afanas'ev, V., Shluger, A.L.: A computational study of Si-H bonds as precursors for neutral E′ centres in amorphous silica and at the Si/SiO_2 interface. Microelectron. Engineering **109**, 310–313 (2013)
48. Jivanescu, M., Stesmans, A., Afanas'ev, V.V.: Phys. Rev. B **83**(9) (2011)
49. Skuja, L.: Defects in SiO_2 and related dielectrics: Science and Technology, p. 73. Kluwer Academic Publisher (2000)
50. Chadi, D.J.: Negative-U property of oxygen vacancy defect in SiO_2 and its implication for the E′ center in α-quartz. Appl. Phys. Lett. **83**, 437–439 (2003)
51. Roma, G., Lymoge, Y.: Phys. Rev. B **70**, 174101–1–8 (2004)
52. Kimmel, A.V., Sushko, P.V., Shluger, A.L., Bersuker, G.: Positive and negative oxygen vacancies in amorphous SiO_2. ECS Trans. **19**, 3–18 (2009)
53. Buscarino, G., Agnello, S., Gelardi, F.M.: ^{29}Si hyperfine structure of E′ center in amorphous silicon dioxide. Phys. Rev. Lett. **97**, 135502–1–4 (2006)
54. Griscom, D., Cook, M.: ^{29}Si super-hyperfine interactions of the E′ center: a potential probe of range-II order in silica glass. J. Non-Cryst. Solids **182**, 119–134 (1995)
55. Robertson, J.: High dielectric constant gate oxides for metal oxide Si transistors. Rep. Prog. Phys. **69**, 327–396 (2006)
56. Yip, K.L., Fowler, W.B.: Phys. Rev. B **11**, 2327–2338 (1975)
57. Courtot-Descharles, A., Paillet, P., Leray, L.J.: Theoretical study using density functional theory of defects in amorphous silicon dioxide. J. Non-Cryst. Solids **245**, 154–160 (1999)
58. Tsetseris, L., Fleetwood, D.M., Schrimpf, R.D., Zhou, X.J., Batyrev, I.G., Pantelides, S.T.: Hydrogen effects in MOS devices. Microelectron. Engineering **84**, 2344–2349 (2007)
59. Krishnan, A.T., Chakravarthi, S., Nicollian, P., Reddy, V., Krishnan, S.: Negative bias temperature instability mechanism: The role of molecular hydrogen. Appl. Phys. Lett. **88**(15), 153518 (2006)
60. Vanheusden, K., Warren, W.L., Devine, R.A.B., Fleetwood, D.M., Schwank, J.R., Shaneyfelt, M.R., Winokur, P.S., Lemnios, Z.J.: Non-volatile memory device based on mobile protons in sio_2 thin films. Nature **386**(6625), 587–589 (1997)
61. Godet, J., Pasquarello, A.: Proton diffusion mechanism in amorphous SiO_2. Phys. Rev. Lett. **97**, 155901 (2006)
62. Schanovsky, F., Gös, W., Grasser, T.: Multiphonon hole trapping from first principles. J. Vac. Sci. Technol. B **29**,
63. Vitiello, M., Lopez, N., Illas, F., Pacchioni, G.: H_2 cracking at SiO_2 defect centers. J. Phys. Cem. A **104**(20), 4674–4684 (2000)
64. Vanheusden, K., Warren, W., Devine, R.: H^+ and D^+ associated charge buildup during annealing of Si/SiO_2/Si structures. J. Non-Cryst. Solids **216**(0), 116–123 (1997)
65. Afanas'ev, V.V., Stesmans, A.: J. Phys.: Condens. Matter **12**(10), 2285 (2000)
66. Jani, M.G., Bossoli, R.B., Halliburton, L.E.: Phys. Rev. B **27**, 2285–2293 (1983)
67. Capron, N., Broqvist, P., Pasquarello, A.: Migration of oxygen vacancy in HfO_2 and across the HfO_2/SiO_2 interface: A first principles investigation. Appl. Phys. Lett. **91**, 192905–1–6 (2007)

Chapter 13
Statistical Study of Bias Temperature Instabilities by Means of 3D "Atomistic" Simulation

Salvatore Maria Amoroso, Louis Gerrer, Fikru Adamu-Lema, Stanislav Markov, and Asen Asenov

Abstract This chapter presents a comprehensive simulation study of the reliability performance in contemporary bulk MOSFET devices. With the CMOS technology entering in the nanoscale era, the statistical variability due to random dopant fluctuations plays a critical role in determining the transistor reliability performance. As a consequence, in contemporary devices, reliability and variability cannot be considered anymore as separate concepts. The reliability has to be reinterpreted as a time-dependent form of variability. In the first part of this chapter we introduce computational models and methods for modelling the reliability phenomena in presence of statistical variability. In particular we present both a frozen-time and a dynamical approach, showing details of their implementation and verification. In the second part of the chapter we report a broad set of simulation results highlighting the importance of variability in reliability evaluation of nanoscale devices. In particular we analyse the impact of variability on the single transistor and on many different transistors in presence of a single trapped charge. Then we show the effects related to multiple trapped charges. Finally the statistical results obtained using the frozen-time and the dynamical methods are compared in terms of accuracy in predicting the statistical dispersion in threshold voltage shifts.

S.M. Amoroso (✉) • L. Gerrer • F. Adamu-Lema • S. Markov
University of Glasgow, Oakfield Avenue, G12 8LT Glasgow, UK
e-mail: salvatore.amoroso@glasgow.ac.uk; louis.gerrer@glasgow.ac.uk;
fikru.adamu-lema@glasgow.ac.uk; figaro@hku.hk

A. Asenov
Gold Standard Simulations Ltd, G12 8QQ Glasgow, UK
e-mail: asen.asenov@glasgow.ac.uk

T. Grasser (ed.), *Bias Temperature Instability for Devices and Circuits*,
DOI 10.1007/978-1-4614-7909-3_13,

13.1 Introduction

One of the biggest challenges that reliability-aware design methodology has to face today, in order to keep pace with the aggressive transistor scaling, is the statistical variability associated with the discreteness of charge and granularity of matter in nanometre scale CMOS transistors [1–4]. Several works in the last few years have assessed the intrinsic connection between random fluctuations and reliability performance, showing that reliability-related parameters, e.g. the device lifetime, have to be reinterpreted as stochastic variables [5–14]. These aspects are addressed in detail in [15, 16]. Corroborated by a surge of new experimental evidences, an important paradigm shift has recently identified the oxide traps as the uniquely responsible entity leading to phenomena like random telegraph noise (RTN) and bias temperature instabilities (BTI) [1, 17–22]. In contrast, BTI dynamics had previously been attributed to reaction–diffusion phenomena in the oxide [23–30]. This aspect is discussed in detail in [31].

The intrinsic interplay between reliability and variability can be understood considering the percolative nature of the source-to-drain conduction in nanoscale MOSFETs arising from the potential variations associated with the random dopant fluctuations (RDF) in the channel [32, 33] and other statistical variability sources, like line edge roughness (LER) [34] and metal grain granularity (MGG) [35–37]. The device reliability is related to the trap formation and the subsequent charge trapping phenomena in the gate oxide: it is clear that a charge trapped over a percolative conduction path will have a large impact on the device threshold voltage shift, while the impact will be much less if the charge is trapped over a region which already has low current density. This is clearly demonstrated in Fig. 13.1, where the 3D simulated carrier density is shown highlighting the percolative nature of conduction between source and drain and the impact of a trap located over the preferential conduction path.

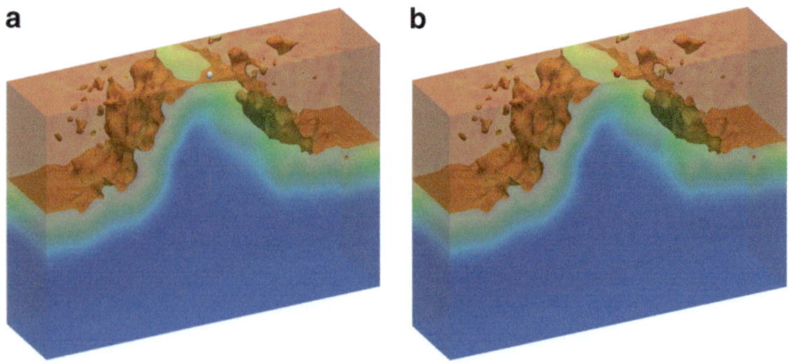

Fig. 13.1 (**a**) Example of percolative conduction path between source and drain in an atomistically doped 25MOSFET device. (**b**) The effect of a trapped charge closing off the preferential conduction path during the threshold voltage reading operation

Therefore, the importance of developing three-dimensional (3D) physics-based simulation tools to fully understand the phenomenology of reliability in presence of statistical variability becomes evident. The overall drift-diffusion simulation framework that we have developed for the reliability evaluation of nanoscale MOSFETs in presence of statistical variability is reported in Fig. 13.2 and is presented in greater detail in the rest of this chapter. Note that the oxide breakdown will not be taken into account in the following, although it could be easily included in our simulation framework once a physics-based model is provided. In the next sections we will discuss the modifications of the 3D-atomistic drift-diffusion simulator GARAND [38], in order to deal with the charge trapping in the gate oxide. Next, we will introduce the Kinetic Monte Carlo (KMC) engine developed to reproduce the stochastic nature of charge injection/emission from the channel to the gate oxide during the BTI stress/relaxation. Finally we will show results obtained using both frozen-time and dynamical simulation approaches.

13.2 Charge Trapping in a Drift-Diffusion Framework

In this section we report the drift-diffusion-based simulation framework developed to study the reliability in presence of RDF-induced variability in decananometer MOSFET devices. We first explain how the discrete traps have been introduced in the existing University of Glasgow/GSS quantum-corrected drift-diffusion simulator GARAND [38]: this allows us to perform frozen-time simulations, as reported in Sect. 13.3. Then, we introduce a dynamical simulation approach based on a KMC engine: this allows us to perform time-dependent reliability evaluations, as reported in Sect. 13.4.

13.2.1 Introduction of Discrete Oxide Traps in GARAND

In our approach, a discrete trap is described by means of three spatial coordinates (x_T,y_T,z_T), one energy level (E_T), a capture cross section (σ_T) and an occupancy status (Occ_T), as reported in Table 13.1. At this stage, we are considering only one kind of traps featuring an amphoteric (tri-state) behaviour with neutral status when unoccupied. Moreover the cross section is approximated as a constant and independent from the trap occupancy. In the rest of the chapter $\sigma_T = 10^{-14}\,\text{cm}^2$ is adopted [39,40]. This description can be refined and generalized once experimental data and ab initio simulations are available for the conception of more elaborate and accurate mesoscopic models.

In order to model the electrostatic effect of the trapped charge when a trap is occupied, the charge is assigned to the surrounding nodes of the device discretisation grid using the Cloud-In-Cell (CIC) technique, in the way that ionized impurities in

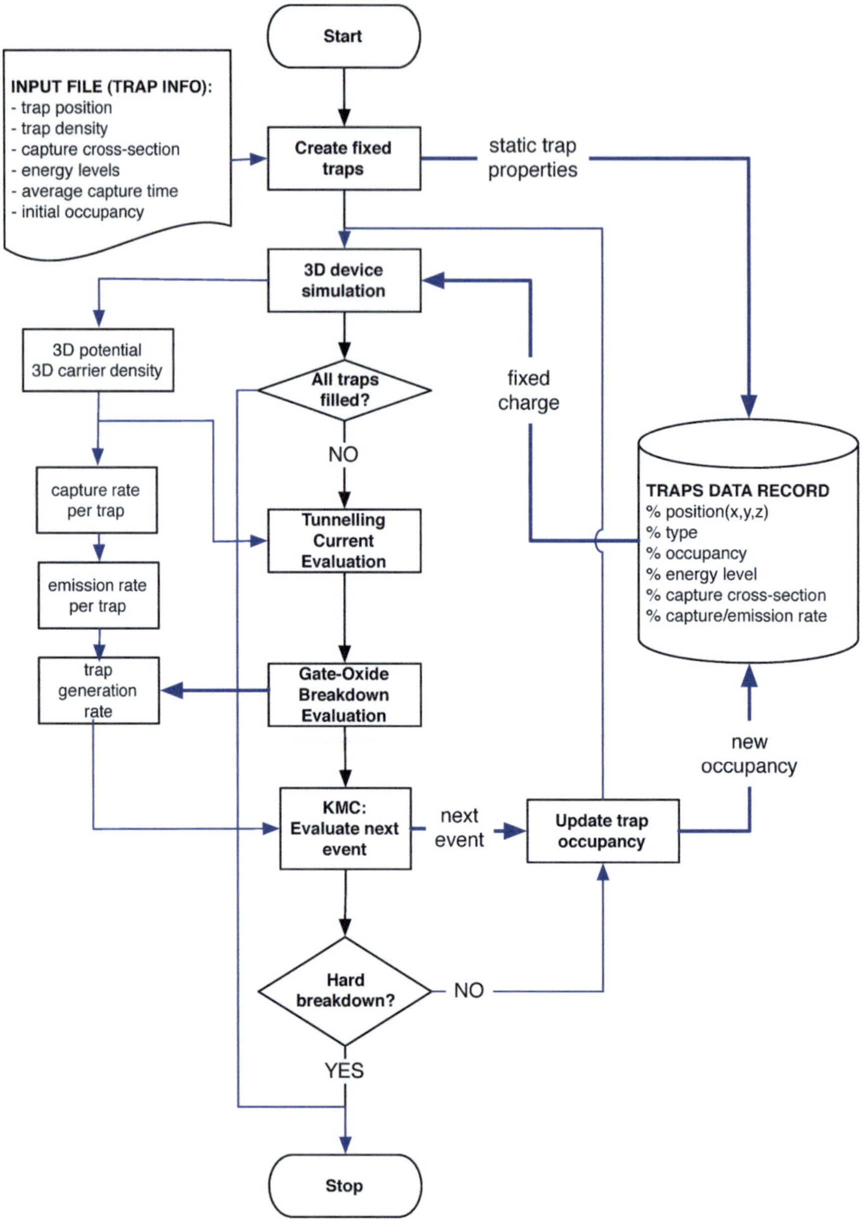

Fig. 13.2 General drift-diffusion computational framework for the reliability evaluation of nanoscale MOSFETs

Table 13.1 Discrete oxide trap definition in GARAND simulator

Trap parameter	Description
x_T, y_T, z_T (nm)	Positional coordinates
Kind	Amphoteric
E_T (eV)	Energy level
σ_T (cm^2)	Capture cross section
Occ_T (−1,0,+1)	Trap occupancy
τ_c (cm^2)	Capture time constant
τ_e (cm^2)	Emission time constant

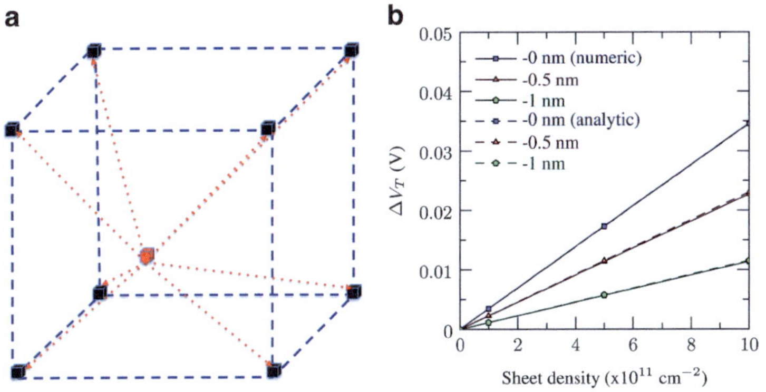

Fig. 13.3 (**a**) Schematic representation of Cloud-In-Cell charge assignment; (**b**) threshold shift ΔV_T versus charge density, with the position of the sheet charge (relative to the Si-oxide interface) as a parameter. The results from the analytical expression for uniform sheet charge (*line*) are overlapped by the numerical solution for fractional charges distributed uniformly on the device grid (*dash*)

Si are treated [33]. In this approach, the charge associated with a single trap is spread over the eight surrounding mesh points using a weight function inversely proportional to the distance between the trap and the mesh point positions, as:

$$w_{xm} = \begin{cases} 1 - |x_i - x_m|, & \text{if } |x_i - x_m| < 1. \\ 0, & \text{otherwise} \end{cases} \tag{13.1}$$

where x_m is the coordinate of the mesh node and xi the trap position. Figure 13.3a schematically illustrates this method of charge assignment.

Figure 13.3b verifies the implementation of the scheme, by showing that a collection of fractional point-charges placed in a plane above the Si-oxide interface induce the same threshold shift VT as predicted by the analytical expression $\Delta V_T = -Q_S/C_{ox}$, for the equivalent sheet density of charge Q_S, considering the ideal oxide sheet-capacitance C_{ox}.

Two other important parameters associated with a single trap are its capture time constant (τ_c) and emission time constant (τ_e): these variables depend on the trap

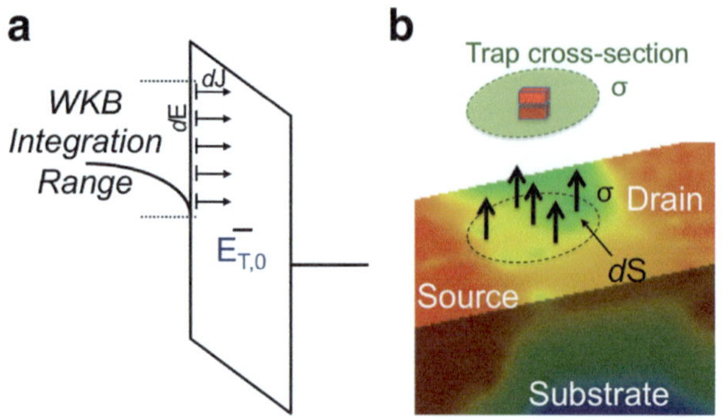

Fig. 13.4 Schematic representation of capture time constants calculation using WKB tunnelling approximation

spatial position, on the trap energy level, on the trap capture cross section and on the applied bias conditions according to the physics-based model presented in [11,41–44]:

$$\tau_c = exp\left(\frac{E_A}{kT}\right)\frac{q}{\int J(x,y)dxdy} \tag{13.2}$$

$$\tau_e = \tau_c exp\left(-\frac{E_T - E_F}{kT}\right) \tag{13.3}$$

In Eq. (13.2) the capture time constant is computed by evaluating, under the Wentzel–Kramer–Brillouin (WKB) approximation, the tunnel current density $J(x,y)$ that reaches the trap from the channel, and integrating it over an area equal to the trap cross section, as sketched in Fig. 13.4. Please note that the integration in energy starts from the maximum between the trap level and the silicon conduction band bottom. The exponential pre-factor in Eq. (13.2) models a multi-phonon assisted capture, as done in [45]: in this context E_A represents the activation energy of the capture process. This correction is necessary to take into account the experimentally observed temperature dependence of the capture time [45–50]. However, the dependence of E_A from the electric field is neglected in our model. This dependence is necessary to carefully model the time constants behaviour, especially at high gate voltages [51]. A quadratic field correction to this term can be straightforwardly implemented as in [48], or a more refined correction can be introduced following [51]. The emission time constant is obtained from the capture time constant through Eq. (13.3) in order to respect a detailed balance [52, 53]: to this aim the local difference between the trap energy level in the oxide and the quasi-Fermi level in the channel has to be calculated. Because of the non-negligible trap energy level shift following a trapping event in nanoscale devices, the SRH statistics may not be fully appropriate for modelling the capture/emission balance and corrective terms due to Coulomb blockade [46] or lattice relaxation

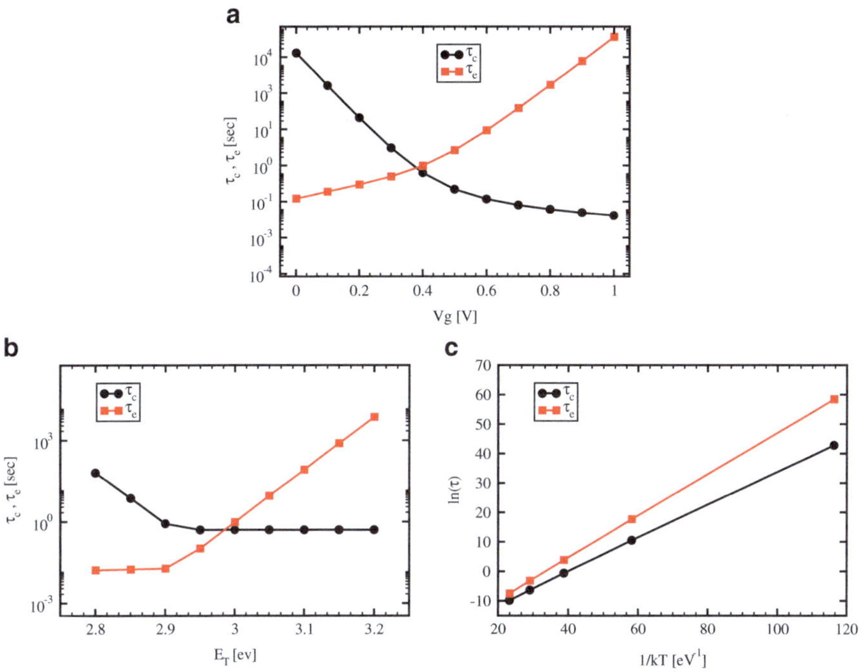

Fig. 13.5 (**a**) Capture and emission time constants as a function of applied gate voltage (E_T = 3.0 eV below SiO_2 conduction band, E_A = 0.5 eV, T = 300 K); (**b**) capture and emission time constants as a function of trap energy depth below the SiO_2 conduction band (V_G = 0.4 V, E_A = 0.5 eV, T = 300 K); (**c**) capture and emission time constants as a function of temperature (V_G = 0.4, E_T = 3.2 eV, E_A = 0.5)

effects [48, 51] have been proposed in literature. However, the aim of this work is to stress the importance of variability in reliability simulation of decananometer MOSFETs and not to present an ultimate model for the capture and emission time constants ruling the BTI behaviour. Figure 13.5a shows the dependence of capture and emission time constants on the applied gate voltage. Both capture and emission time constants strongly depend on the applied gate bias, with capture (emission) time constant decreasing (increasing) when the gate bias increases for the case of the analysed n-channel MOSFET (this dependence is even stronger, above threshold voltage, when the field dependence of E_A is introduced [48]). Further, Fig. 13.5b shows that the emission strongly depends on the traps energy position E_T, while the capture time remains practically constant if the trap level in the oxide is below the quasi-Fermi in the channel; if the trap level is above the Fermi level, then the capture time constant start to increase because the energy integration used in WKB tunnelling starts from the energy level and not from the silicon conduction band bottom (see Fig. 13.4 and relative comments). As a consequence of Eq. (13.3), when the capture time increases, the emission time departs from its pure exponential behaviour with respect to energy level. Finally, Fig. 13.5c shows that both capture time and emission time decrease with the increase of temperature. Note that while the capture time has an activation energy $E_A = 0.5\,\text{eV}$, assigned as a model parameter (see Eq. (13.2)), the emission time exhibits a larger activation energy (0.7 eV) because of the additional contribution given by the difference between the trap energy level and the Fermi level in the channel (see Eq. (13.3)). This computational framework allows the simulation of frozen-time reliability in presence of variability, as reported in Sect. 13.3.

13.2.2 Dynamic Charge Trapping and Traps Generation: KMC Approach

In the previous paragraph we have introduced a computational framework that allows the frozen-time simulation of reliability: this means that for a given time we can assume that our devices have an average number of filled traps with a Poissonian statistical dispersion and study the device performance for that given level of degradation. The previous approach neglects two important aspects related to the temporal evolution of the device degradation:

1. The existing traps are not instantaneously filled at a given time, but the charge trapping/detrapping is ruled by stochastic processes having time constants that depend on the electrostatic conditions (and in turn on the occupancy status of the available traps at a given time).
2. The number of traps in the gate oxide does not remain constant but increases with time according to dynamics that are accelerated by the stress bias conditions, or decreases due to annealing phenomena.

We start considering the first point, assuming a fixed number of traps (constant in time) is present in the device. Figure 13.6 shows the simulation procedure developed for the statistical dynamical analysis of device reliability. An outer Monte Carlo loop is used to gather results on thousands of microscopically different devices: after defining the stochastic configuration of atomistic dopants and oxide traps, the 3D electrostatics and the drift-diffusion (DD) equations are solved to obtain the threshold voltage (V_T) of the fresh device. An inner KMC loop is then used to simulate the stochastic charge-injection process from substrate to oxide traps: once the cell electrostatics is solved for typical stress bias conditions, the tunnelling current density $J(x,y)$ reaching each trap is calculated over the channel area in the WKB approximation and the average capture time constant $\langle\tau_c\rangle$ is computed for each trap through Eq. (13.2). The average emission time constant $\langle\tau_e\rangle$ is derived for each trap by Eq. (13.3). Then, for each trap, the stochastic capture times τ_c and τ_e are drawn from two exponential distributions with average value $\langle\tau_c\rangle$ and $\langle\tau_e\rangle$, respectively. Based on these random times, the KMC engine chooses the lucky trap and the corresponding trapping or detrapping event. After each event the device V_T is calculated again and, if the failure limit (e.g. $\Delta V_T = 30\,\text{mV}$) is not reached, the internal Monte Carlo loop is repeated.

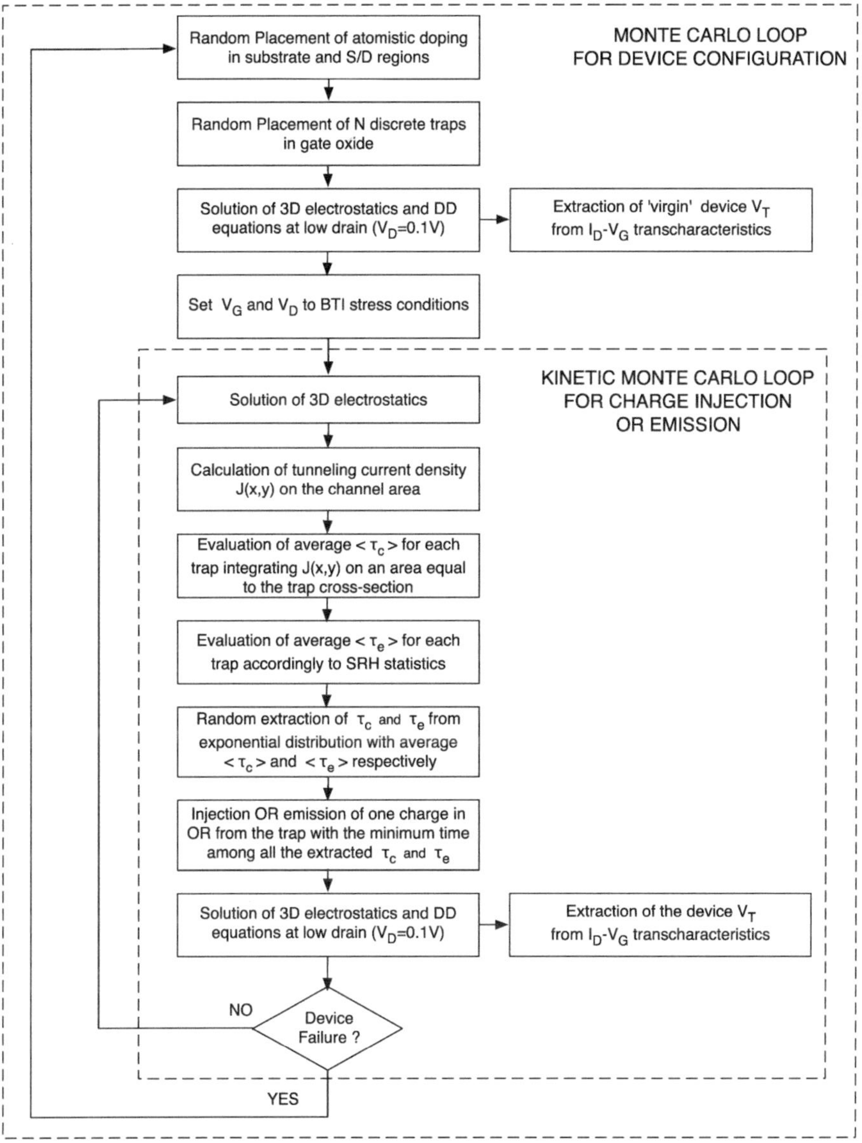

Fig. 13.6 Simulation procedure for dynamical reliability simulation including the Kinetic Monte Carlo loop

The KMC engine is implemented in a way that a general number of traps and a general number of events can be handled. With this flexibility in mind, we can introduce in the list of possible events also the trap formation event in addition to the trapping and detrapping events. The algorithm consists of the following steps:

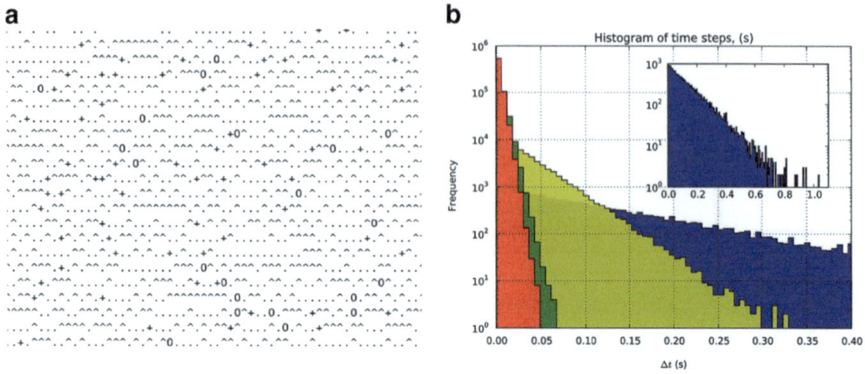

Fig. 13.7 (**a**) Random sequence of four events as obtained from the KMC algorithm. Each event is represented by a symbol (1 = "^", 2 = ".", 3 = "*", 4 = "0"); (**b**) frequency distribution of the inter-event time steps

1. Set the time to zero
2. Calculate $\langle \tau_c \rangle$ and $\langle \tau_e \rangle$ for each possible trap (event) from Eqs. (13.2) to (13.3)
3. Calculate the cumulative distribution function (total rate):

$$R_i = \sum_{j=1}^{i} \left(\frac{1}{\tau_j} \right) \qquad i = 1, 2, 3, \ldots N$$

4. Generate random number r and decide which trap (event) to fill by

$$R_{i-1} < rR_N \leq R_i$$

5. Fill or empty trap i
6. Update the time: get another random number r and compute capture time

$$\tau(i) = \tau(i-1) + \Delta\tau$$

   where

$$\Delta\tau = \frac{1}{R_N} ln \left(1/r' \right)$$

7. Is the process complete?

$$\text{NO} \Longrightarrow \text{go to step 2} \qquad \text{YES} \Longrightarrow \text{Finish}$$

Figure 13.7a shows a graphical example of a random sequence of four types of events (that we can imagine to be, for example, capture, emission, trap formation and trap annealing), with assigned average rates of 200 (symbol ^), 270 (symbol .), 30 (symbol *) and 8 (symbol 0) Hz. The picture reports a fragment of the output of the simulation illustrating the random order in which the different events appear.

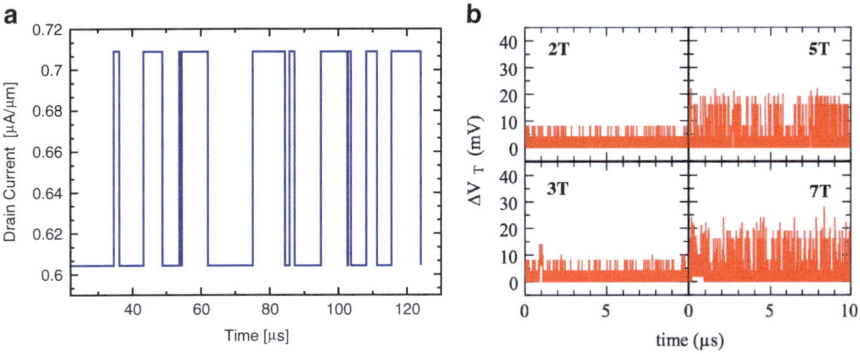

Fig. 13.8 (**a**) RTN trace for the analysed 25 nm MOSFET device for $V_G = 0.15$ V; (**b**) RTN traces for the case a multiple traps in the device

This order is associated with the random numbers used in the KMC engine. Simulated time was 33 ms. The total counts of events of each type (505,104, 680,379, 75,292, 20,131) reflects the proportions between the scattering rates in the KMC algorithm.

In order to verify the accuracy of the method used for implementing the KMC algorithm, we report in Fig. 13.7b the simulated inter-event time step obtained in the computational experiment of Fig. 13.7a. The distribution of inter-event time steps is confirmed to be exponential, as expected from the step 6 of the algorithm which drawn each event time from exponential distributions. Note that the stochasticity of the time steps is governed by the second random number (r) used in the algorithm, which is carefully derived in a manner that ensures it is uncorrelated with the random number governing the order of events (r).

This computational framework allows the time-dependent simulation of reliability in presence of variability. Please note that a similar approach has been already used in [54], albeit neglecting the impact of 3D electrostatic and atomistic variability. In the remaining of this section we report just some example of dynamical simulation results to show the capability of this tool, while a detailed statistical analysis will be presented in Sect. 13.4. The simplest situation illustrates a process related to a single trap that can capture or emit a charge: this is the typical situation describing the RTN phenomenon affecting the drain current of MOSFET over time. Figure 13.8a shows a simulated RTN trace on our 25 nm MOSFET template described in Sect. 13.3: the drain current continues to oscillate between a high state (empty trap) and a low state (filled trap) with random time intervals obtained by our KMC procedure. In the same way, this methodology can be used to simulate the dynamic behaviour of more than one trap, as shown in Fig. 13.8b where we report the threshold voltage shifts in time for the same device in presence of 3, 5 and 7 traps in the gate oxide, as indicated.

Furthermore, the same procedure reported in Fig. 13.6 can be used to simulate BTI. In fact, if we consider the case with more than one trap and apply stress bias

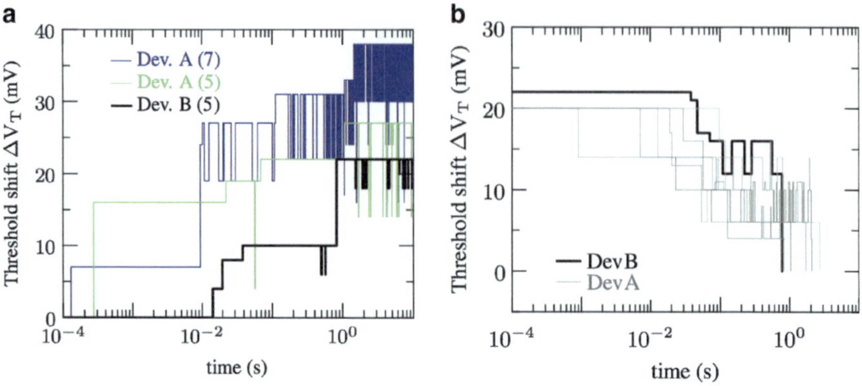

Fig. 13.9 (**a**) BTI stress simulation: charging traces are stochastic based on RDF and number of traps; (**b**) BTI relaxation: discharging traces are stochastic based on RDF and number of traps

conditions (e.g. $V_G = V_{DD}$) instead of V_T-reading conditions, we obtain the BTI stress traces reported in Fig. 13.9a. Note that even maintaining fixed the number of traps (i.e. ignoring trap creation/annealing) a large variation in the time to failure (e.g. reaching a predetermined ΔV_T) is obtained due to the RDF-induced statistical variability and to the fluctuation in the number of traps from device to device. In the same way, after the BTI charging, we can recover part of the degradation by applying relaxation bias conditions (e.g. $V_G = V_T$). The result is shown in Fig. 13.9b, where once again the role of statistical variability in reliability evaluation of nanoscale MOSFET devices clearly appears.

The generality of the KMC procedure allows us to deal with the second problem mentioned at the beginning of this paragraph: the number of traps in the gate oxide tends to increase during stress. The trap formation can be straightforwardly introduced in the KMC routine, treating it as a new event, the rate of which may depend on the applied gate voltage, current flux, temperature, etc. [55–58]: the KMC engine then will choose the sequence of events taking into account charge capture, charge emission, and new trap formation. In similar way, the trap annealing event can be introduced in the class of possible events [59]. However, in the rest of the chapter we will focus our attention on the variability affecting the trapping/detrapping events, neglecting, in first approximation, the trap formation and annealing.

13.3 Frozen-Time Simulation of Reliability in Presence of Variability

In this section we present frozen-time simulation results of reliability using a 25 nm bulk n-channel MOSFET as a test structure. After a brief description of the template transistor, we will show the frozen-time statistical results analysing the impact

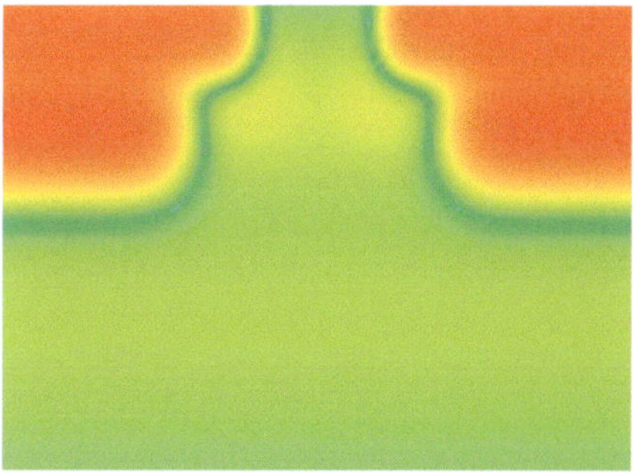

Fig. 13.10 Net doping profile for the template n-channel 25 nm bulk MOSFET

of variability in a single transistor and on many transistors, highlighting also the effects related to many trapped charges in the gate oxide. In the following we will consider RDF [4, 32, 33, 60–62] as main source of variability, neglecting the impact of LER [34, 63] and MGG [35, 37].

It is also important to mention that the impact of trapped charge on mobility is not taken into account in this analysis, as a drift-diffusion approximation is adopted to solve the continuity equation. A refined evaluation of the impact of charge trapping on carriers mobility may require a Monte Carlo approach to the charge transport [64], that is beyond the scope of this work. As a final remark, note that the density-gradient quantum corrections adopted in GARAND allow to obtain simulation results that show very low sensitivity to the mesh-size [33, 65].

Our simulation study has been carried on using an n-channel bulk MOSFET with a physical gate length of 25 nm, representative for the 22/20 nm CMOS technology generation. This device has been created using the Gold Standard Simulations process simulator, ION [38]. This process simulator has been designed in order to allow the generation of realistic doping profiles that closely match those obtained from full process simulation. ION uses published data for the stopping distances of ions in matter, and associated projected range and straggle parameters to create complex doping profiles. It also simulates the doping activation/annealing process. The 25 nm bulk MOSFET has been designed following the prescriptions of the ITRS-2010 update and subject to realistic physical constraints. The device features a high-dielectric gate stack with 0.85 nm EOT and has metal gate. The device structures and doping profile are shown in Fig. 13.10, while the main geometric and electrical parameters are summarized in Table 13.2.

13.3.1 Single Transistor Variability

In order to study the reliability in presence of RDF-induced statistical variability, we first analyse how the capture/emission time constants and the threshold voltage shift associated with a single trap are stochastically dispersed over the channel area of a single atomistically doped device (chosen to be representative of the average behaviour). Figure 13.11a shows that the threshold voltage shift (ΔV_T) is largely

Table 13.2 Geometrical and doping parameters for the template n-channel 25 nm bulk MOSFET

Parameter	Value	Description
L_g (nm)	25	Physical gate length
EOT (nm)	0.85	Equivalent oxide thickness
x_j (nm)	15	S/D extensions
N_A ($\times 10^{18}$ cm^{-3})	4.5	Channel doping
V_{dd} (V)	1	Supply voltage
I_{off} (nA)	100	Off current
I_{on} (μA)	1,351	Drive current
Spacer (nm)	24	Spacer length

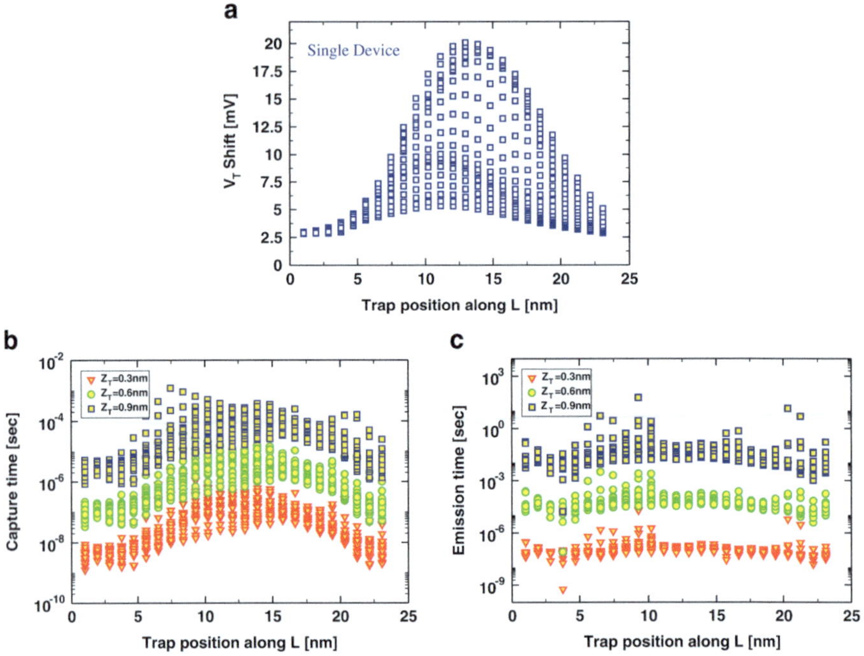

Fig. 13.11 (**a**) V_T shift as a function of the trap position along the channel length; (**b**) capture time constant along the channel length for three different z_T and $V_g = V_T$; (**c**) emission time constant along the channel length for three different z_T and $V_g = V_T$

dispersed over the channel area (z_T fixed at 0.3 nm from channel interface). The dispersion reaches its maximum at the centre of the channel and then decreases towards the source and drain regions, where the electrostatic screening coming from the high charge density in the S/D extension overlap-regions lowers the fluctuation amplitudes due to random dopants. Similar considerations hold for the capture time constants (Fig. 13.11b), while the emission time constants are less affected by variability for a fixed vertical position z_T (Fig. 13.11c). However, it should be noted that still a large dispersion of both capture and emission time constants is present if we take into account also the z_T variations.

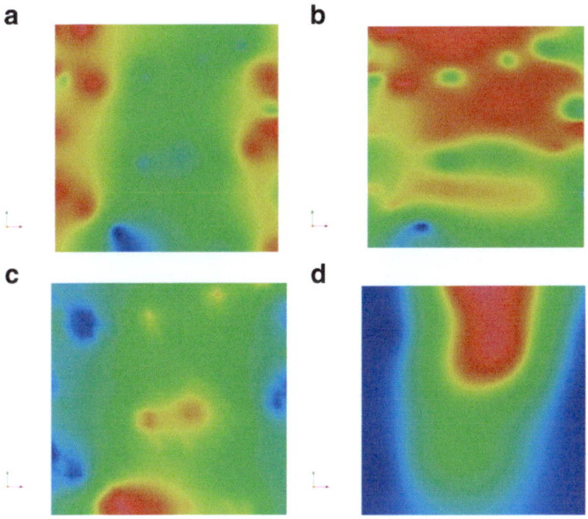

Fig. 13.12 (**a**) Carrier concentration in the channel area at threshold of a given device; (**b**) corresponding current density in the channel area, at threshold; (**c**) capture time constant over the channel area; (**d**) threshold voltage shift over the channel area

It is also interesting to analyse if any correlation exists between the capture time constants and the ΔV_T. For the discussion so far, one could expect some kind of correlation between the two variables, given the envelope-shapes of the dispersions of the two variables with the lateral position. Indeed we stated that the most effective traps in terms of ΔV_T are that having position over a percolative conduction path, i.e. over regions of the channel with high carrier concentration. Because of Eq. (13.2), the same regions with high carrier concentration give rise also to fast capture events, as shown in [8, 12] (and also visible in the 2D maps of Fig. 13.12a, c). However, it is worth noting that high carrier concentration is only a necessary condition (and not sufficient) to have a continuous percolative path between source and drain: this is clearly demonstrated comparing Fig. 13.12a with b, where we report the simulated carried density and current density over the channel area at threshold bias conditions. This clearly means that ΔV_T is correlated with the presence of a percolative path (compare 2D maps in Fig. 13.12b and d), but completely uncorrelated with the carrier concentration in the channel, as confirmed by the results reported in Fig. 13.13a. As a consequence, ΔV_T and the capture time constants are uncorrelated variables, as shown in Fig. 13.13b.

13.3.2 Many Transistors Variability

In this paragraph we will analyse the variability in the reliability figures of merit from transistor to transistor. To this aim, we have carried out simulations of an ensemble of 1,000 microscopically different devices, considering in each of them

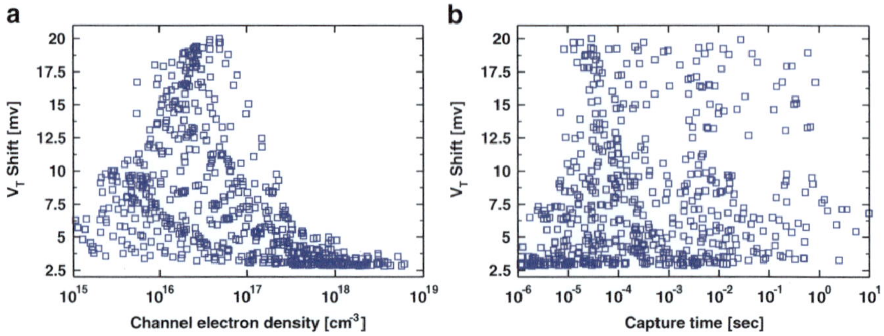

Fig. 13.13 (**a**) Threshold voltage shift as a function of the carrier concentration in the channel; (**b**) threshold voltage shift as a function of the trap capture time constant

Fig. 13.14 Threshold voltage shift along the channel length for 1,000 different atomistic devices

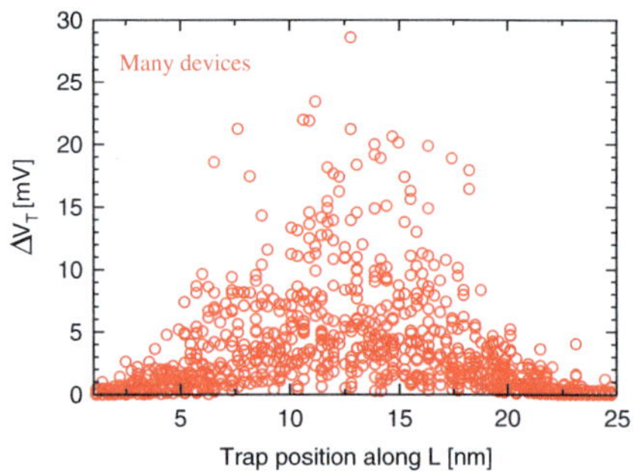

the presence of a single trap. In this case, both the trap position (x_T,y_T, z_T) and its energy level E_T are chosen from uniform random distribution, in such a way to evaluate the impact on the variability of ΔV_T and capture/emission time constants. The E_T range is chosen to be $3.15 \pm 0.15\,\text{eV}$ below the oxide conduction band. Figure 13.14 shows that the single-trap-induced ΔV_T along the channel length obtained on many devices exhibits a similar behaviour of that obtained within a single device. This also highlights that the variability of the (x_T,y_T) position is dominant over the variability in the vertical z_T position. It is worth reminding that the variability on the energy position has no impact on ΔV_T.

In Fig. 13.15a we report the cumulative distribution of the ΔV_T (solid line). The exponential behaviour of the distribution is evident and is a clear fingerprint of the percolative conduction regime in the channel (Fig. 13.15b). It is worth mentioning that whenever the conduction loses its percolative behaviour (i.e. at high gate bias, when the conduction is quite uniform because of screening effects on the discrete dopants) the cumulative distribution is not anymore a pure exponential, as already experimentally shown in [7] and as evident from our simulation results obtained on a uniformly doped device (Fig. 13.15a, dashed lines).

Figure 13.16a shows that the capture/emission time constants are also statistically dispersed from device to device. The shape of the distribution is similar to that

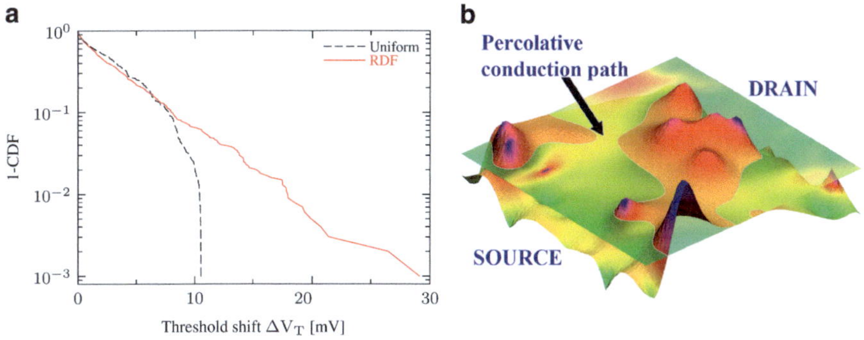

Fig. 13.15 (**a**) Cumulative probability distribution of threshold voltage shifts for 1,000 atomistic devices (*solid*) and for 1,000 continuously doped devices (*dashed*); (**b**) simulated potential fluctuation in the channel: the plane helps to visualize the presence of percolation paths

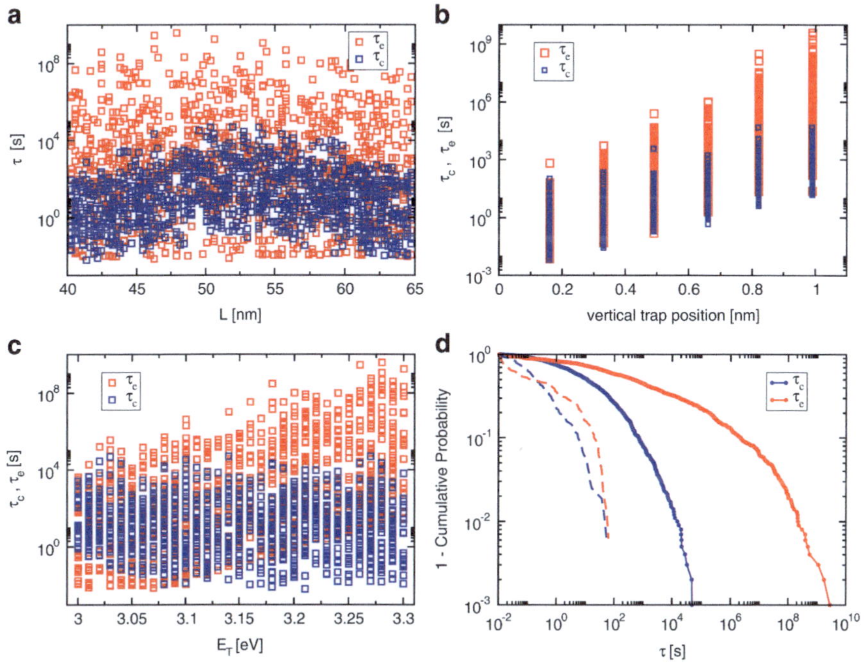

Fig. 13.16 (**a**) Capture and emission constants as function of the trap position along the channel length for 1,000 atomistic devices; (**b**) capture and emission constants as a function of the trap vertical position; (**c**) capture and emission time constants as a function of the trap energy depth; (**d**) cumulative distribution of capture and emission time constants for 1,000 atomistic devices (*solid lines*). Cumulative distribution of the sub-population having traps only at the interface (*dashed lines*)

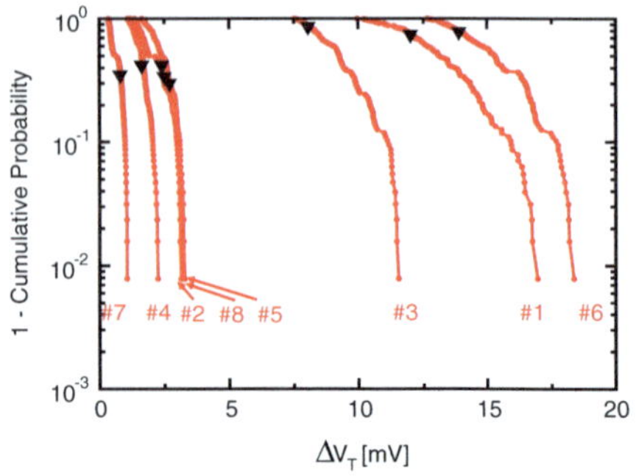

Fig. 13.17 Cumulative distribution of ΔV_T due to each single trap in presence of other seven traps in the oxide. The *black symbols* indicate the ΔV_T value given by each trap when not affected by the influence of other traps

obtained with the analysis on single device reported in Fig. 13.11b. However in this case the variability is much larger because both the z_T and E_T variation are taken into account. In fact Fig. 13.16b shows that both capture and emission time constants increase with increasing z_T: the former is due to the increase in the tunnelling barrier width, the latter is due to the proportionality of τ_e to τ_c imposed by Eq. (13.3) in our model, and because the increase in z_T also leads to an increase in the average difference between E_T and E_F. Additionally, Fig. 13.16c shows that the capture time is not correlated with E_T, while the emission time constants are positively correlated with E_T.

Finally, in Fig. 13.16d we show the cumulative distribution of capture and emission time constants for the 1,000 analysed atomistic devices. It is evident that considering the fluctuations of z_T and E_T, in addition to that of x_T and y_T, has a dramatic impact on the variability of time constants. The impact is especially strong on the emission times that span a range of values of 11 orders of magnitude. It is obvious that if the traps involved in the BTI phenomenon are only interfacial traps ($z_T = 0$), then the variability due to z_T is suppressed and the overall variability in the time constants is reduced, as shown in Fig. 13.16d. Please note that the extreme reduction in emission time variability is due to the fact that fixing z_T we are reducing also the variability in the absolute energy of the traps.

13.3.3 Many Traps Effects

We now consider the case in which more than one charge is trapped in the gate oxide. The aim of this paragraph is to analyse how the properties of the single trap change under the influence of neighbouring occupied traps. This point is of utmost importance for the BTI dynamic behaviour analysis: the same identical trap can give a completely different response depending on the influence of the other traps. Figure 13.17 shows the cumulative distribution of ΔV_T due to one single trap in presence of seven other traps. Note that a single atomistic device is employed in this simulation experiment, and that there are 2^{N-1} possible configurations of the remaining traps (leading to 128 data-points per trap). It is clear that the ΔV_T

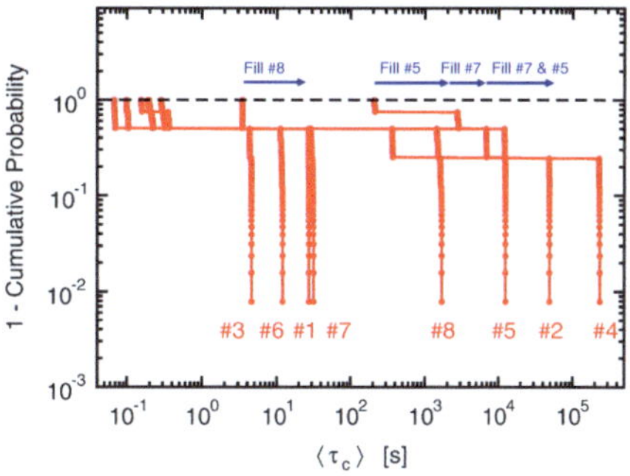

Fig. 13.18 Cumulative distribution of capture time constant of each single trap in presence of other seven traps in the oxide

value given by each trap is influenced by the occupancy status of the other traps and that this influence largely depends on the position of the given trap. Please note the influence of neighbour traps can give either a positive or a negative contribution to the ΔV_T expressed by the single trap in absence of perturbation. Additionally, Fig. 13.18 shows that the average capture time constant of each trap is even more strongly influenced by the occupancy status of the neighbouring traps. The difference in the ΔV_T and capture time behaviours is due to the fact that the capture time is calculated during the stress operation (that is at fixed gate bias, so that a filled trap has an effective impact on channel electrostatics), while ΔV_T is calculated during the reading operation (that is at fixed drain current, so that the conduction profile is reorganized at each trapping event in order to sustain the reading current).

13.4 Dynamical Simulation of Reliability in Presence of Variability

The results presented in the previous section are obtained with a frozen-time technique, i.e. neglecting the dynamics of charge trapping detrapping and assuming that at a certain time a certain number of charges (Poissonianly distributed) are trapped in the oxide. In order to study the charge trapping/detrapping dynamics, the full KMC procedure presented in Sect. 13.2.2 has to be used.

Figure 13.19 shows the dynamical simulation results for 200 atomistic devices (featuring 24 Poissonianly distributed traps) under BTI stress conditions ($V_G = 1$ V). It clearly demonstrates how reliability performance is strongly affected by statistical variability. The *time-to-device failure* (defined as the time necessary to reach a limit ΔV_T) becomes a stochastic variable for nanoscale devices. In other words, *reliability has to be reinterpreted as a time-dependent form of variability in contemporary CMOS technology*. In this picture it is highlighted how devices with the same number of traps can show completely different reliability performance, depending on the relative location of traps with respect to the substrate dopants. From the analysis of the data of Fig. 13.19 we can better understand the stochastic behaviour

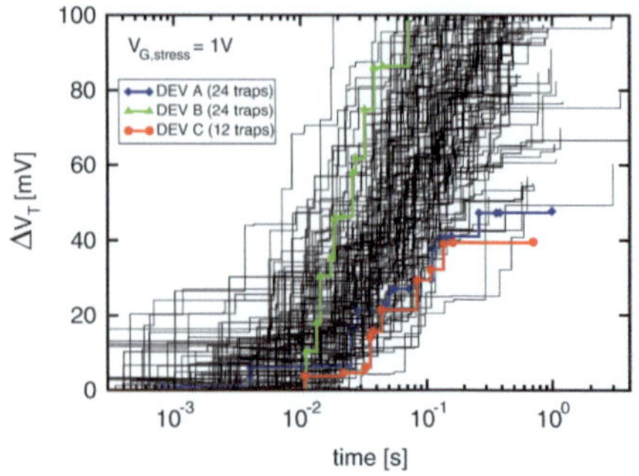

Fig. 13.19 Threshold voltage shift under BTI stress conditions (V_G =1 V) for the analysed Bulk MOSFET. An average of 24 traps is present on each device, with the actual stochastic number being Poissonianly distributed

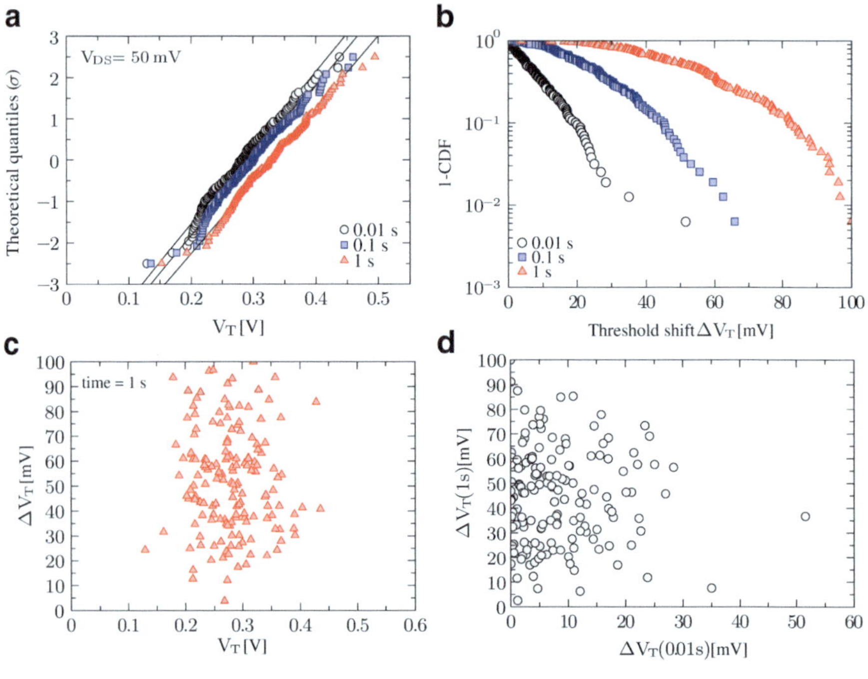

Fig. 13.20 (**a**) Normal probability plot of the threshold voltage at three different BTI stressing times (V_D = 50 mV, V_G = 1.0 V; (**b**) cumulative distribution of the threshold voltage shift at three different BTI stressing times; (**c**) scatter plot of ΔV_T (after 10 s stress) as a function of the initial device V_T; (**d**) scatter plot of ΔV_T (after 0.01 s stress) as a function of ΔV_T (after 10 s stress)

of these devices under BTI stress conditions. Figure 13.20a shows the threshold voltage distribution at three different stress times: the average of the distribution increases as a consequence of the charge trapping, but the standard deviation is barely modified. To further investigate the effects of the charge trapping we report in Fig. 13.20b the cumulative distribution of the threshold voltage shifts obtained at the same stressing times of Fig. 13.20a. It appears that the dispersion of threshold voltage shifts increases with the stressing time, with extreme transistors able to reach

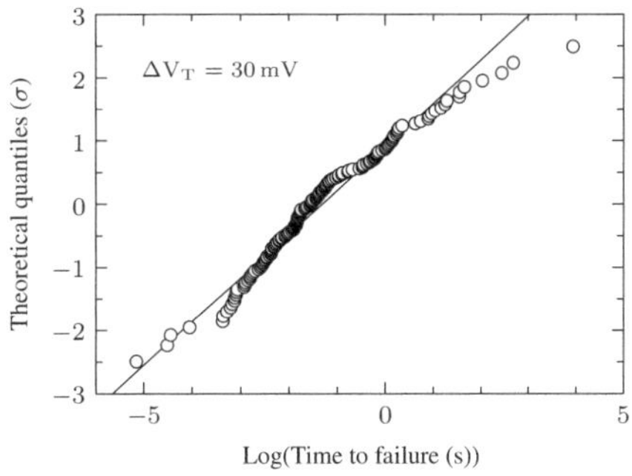

Fig. 13.21 QQ plot of the time-to-device-failure (failing condition $\Delta V_T = 30\,\mathrm{mV}$)

nearly 100 mV of shift after 1 s of stress. It is also interesting to show that the ΔV_T reached at a certain stress time is completely uncorrelated with the initial device threshold voltage (Fig. 13.20c). Moreover, Fig. 13.20d shows that the ΔV_T reached after a stress time t_1 and the ΔV_T reached after a subsequent stress time t_2 are also uncorrelated, demonstrating that the reliability performance evolves in a stochastic way for the same device: a device that starts with bad reliability performance can show a better behaviour later and, vice versa, a device that starts with good reliability performance can show a worse behaviour later. This also suggests that RTN and BTI performance are not correlated, as already demonstrated in [20, 21].

Another important information we can extract from the dynamical simulations in Fig. 13.19 is the *time-to-device-failure* defined as the time necessary for a given device to reach the maximum tolerable threshold voltage shift (e.g. 30 mV): Fig. 13.21 shows that this reliability figure of merit is statistically dispersed over 5 order of magnitudes, demonstrating that any reliability projection based on the average device is completely useless for the sake of a reliability-robust design. Further, Fig. 13.21 shows that the time-to-device-failure does not follow a Gaussian distribution. This is a caveat to take into account when compact analytical model is used to estimate the reliability figure of merit distributions of nanoscale devices.

The same KMC algorithm can easily be used to simulate the BTI relaxation after stress. The relaxation phase has received, in the last few years, a great experimental and modelling attention [1, 48, 66–68], mainly because the very long relaxation tails—of almost logarithmic nature [69–71]—cannot be successfully described by the reaction–diffusion model [71] and could hold the key to unravel the underlying NBTI mechanism.

Figure 13.22 shows an example of stochastic BTI relaxation obtained from 1,000 simulations of one randomly selected device. Starting from this curves it is possible to build the so-called *Time-Dependent Defect Spectroscopy* (TDDS) maps, introduced for the first time in [72, 73] and successfully used to characterize the properties of the discrete traps underlying the BTI [72, 73]. Figure 13.23 shows an example of TDDS map obtained from the relaxation curves of Fig. 13.22. This map allows to identify and characterize the defects involved in BTI over a very large relaxation time range. In this way both fast and slow traps can be studied in detail. Each trap appears with an emission time constant distributed over several orders of

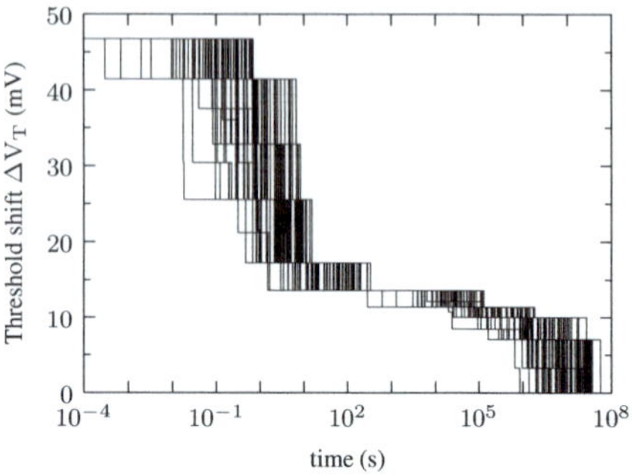

Fig. 13.22 BTI relaxation curves for one randomly selected device featuring eight traps. The stochastic nature of discrete charge emission makes the BTI relaxation curve never equal to itself when repeated several times

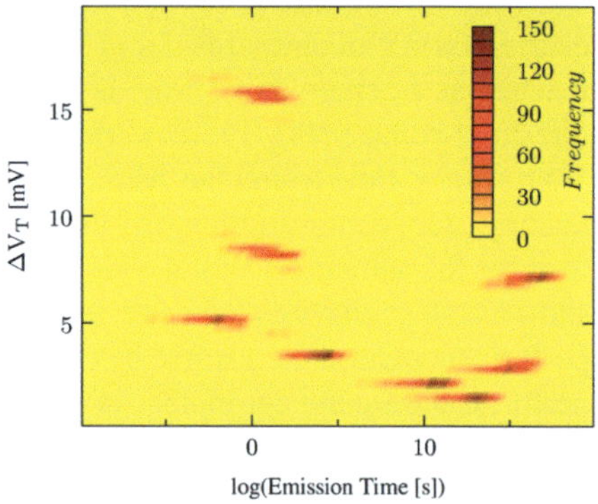

Fig. 13.23 Time-Dependent Defect Spectroscopy (TDDS) map obtained, for the device in Fig. 13.22, following the procedure reported in [72]

magnitude, as a result of the stochastic nature of discrete charge emission. Further, also the ΔV_T associated with each trap may show slightly different values depending on the interaction with other traps (Fig. 13.17). A detailed review of TDDS theory and application is given in [74].

In order to conclude this chapter, it may be very useful to compare the frozen-time and the dynamical simulation results. It is obvious that the frozen-time approach presented in Sect. 13.2 cannot provide information regarding the charge trapping dynamics as, for example, the time-to-device-failure value and its dispersion. However, we can at least compare the frozen-time and the KMC dynamical methods in terms of accuracy in providing results of the ΔV_T amplitude dispersion. It has been shown in [42, 75] that the charge trapping dynamics can lead to sub-Poissonian dispersion of the ΔV_T amplitude. If this is the case, the frozen-time approach will lead to an over-estimation of the ΔV_T dispersion, because in this method the charges are always distributed according to the Poissonian statistics governing the distribution of number of traps in the ensemble of devices, hence neglecting any deviation in the number of trapped charge due to charge injection dynamics. However, Fig. 13.24 shows that, in the range of interest of BTI phenomenon (i.e. $\Delta V_T < 30\,\text{mV}$), the ΔV_T dispersion predicted by the KMC approach follows a Poissonian behaviour. This means that the electrostatic feedback provided by the trapped charges is not strong enough to give rise to a reduction

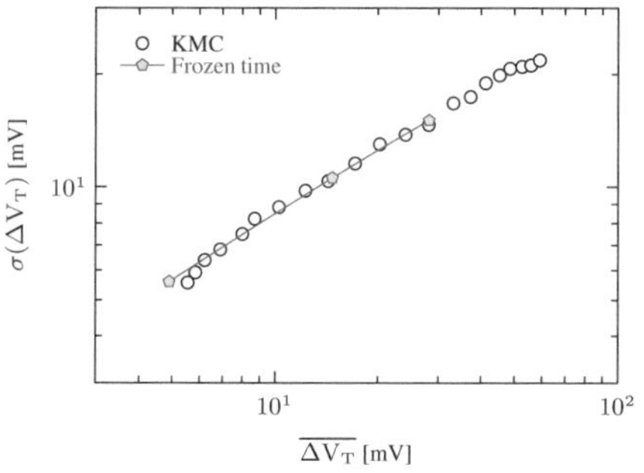

Fig. 13.24 ΔV_T dispersion as a function of average ΔV_T, comparing the dynamical KMC and the frozen-time simulation results. KMC results are obtained slicing the BTI traces (for 200 devices with 24 traps, Poissonianly distributed) at several stressing times. Frozen-time results are obtained placing 1, 3 and 6 trapped charges in the oxide (for 200 devices with charges Poissonianly distributed)

of the dispersion of the injected carriers and, in turn, a reduction of the ΔV_T dispersion. This result is fundamental for the development of analytical and practical models that account for the time-dependent statistical variability in devices and circuits [10, 76, 77].

13.5 Conclusions

We have presented a thorough simulation study of the reliability performance in contemporary bulk MOSFET devices. Because gate length has reached the decananometer region, the statistical variability due to RDF plays a role of utmost importance in determining the device reliability performance. Indeed, in nanoscale devices, reliability and variability cannot be seen anymore as separate concepts, but reliability has to be reinterpreted as a time-dependent form of variability. In the first part of the chapter we have shown the computational models and methods developed and implemented at the University of Glasgow for modelling the reliability phenomena in presence of statistical variability. In particular we have presented a frozen-time and dynamical KMC approach, showing details of their implementation and verification. In the second part of the chapter we have shown a broad set of simulation results highlighting the importance of variability in reliability evaluation of nanoscale devices. In particular we have analysed the impact of the variability on the single transistor and on many different transistors in presence of a single trapped charge. Then we have shown the effects related to multiple trapped charges. Finally the statistical results offered by the frozen-time and the KMC methods have been compared in terms of accuracy in predicting the statistical dispersion in threshold voltage shifts. The drift-diffusion-based computational tools and methodologies presented in this chapter represent a valuable instrument to shine light on the intricate phenomenology of reliability in contemporary devices.

Acknowledgements The results presented here have been obtained through extensive research activity funded by the European Community through the FP7 project "MORDRED" no. 261868.

References

1. B. Kaczer, T. Grasser, J. Martin-Martinez, E. Simoen, M. Aoulaiche, P. J. Roussel and G. Groeseneken. Proc. IRPS 2009, 152–153 (2009).
2. B. Kaczer, T. Grasser, P. J. Roussel, J. Franco, R. Degraeve, L. Ragnarsson, E. Simoen, G. Groesenekenand and H. Reisinger. Proc. IRPS 2010, 26–32 (2010).
3. M. Toledano-Luque, B. Kaczer, J. Franco, P. Roussel, T. Grasser, T. Hoffmann and G. Groeseneken. IEEE VLSI Technology 2011, 152–153 (2011).
4. A. Asenov, S. Roy, R. A. Brown, G. Roy, C. Alexander, C. Riddet, C. Millar, B. Cheng, A. Martinez, N. Seoane, D. Reid, M. F. Bukhori, X. Wang and U. Kovac. IEDM Tech. Dig. 2008 (2008).
5. M. F. Bukhori, S. Roy and A. Asenov. Microelectron. Reliab. 48, 1549–1552 (2008).
6. J. Franco, B. Kaczer, M. Toledano-Luque, P. J. Roussel, J. Mitard, L.-Ã. Ragnarsson, L. Witters, T. Chiarella, M. Togo, N. Horiguchi and G. Groeseneken. Proc. IRPS 2012, 1–6 (2012).
7. B. Kaczer, J. Franco, M. Toledano-Luque, P. J. Roussel, M. F. Bukhori, A. Asenov, B. Schwarz, M. Bina, T. Grasser and G. Groeseneken. Proc. IRPS 2012, 152–153 (2012).
8. A. Mauri, N. Castellani, C. M. Compagnoni, A. Ghetti, P. Cappelletti, A. S. Spinelli and A. L. Lacaita. IEDM Tech. Dig. 2011, 405–408 (2011).
9. G. D. Panagopoulos and K. Roy. IEEE Trans. Electron Devices 58, 2337–2345 (2011).
10. A. R. Brown, V. Huard and A. Asenov. IEEE Trans. Electron Devices 57, 2320–2323 (2010).
11. S. M. Amoroso, F. Adamu-Lema, S. Markov, L. Gerrer and A. Asenov. Proc IWCE 2012, 1–4 (2012).
12. F. Adamu-Lema, S. Amoroso, S. Markov, L. Gerrer and A. Asenov. Proc.of SISPAD. 2012, 101–104 (2012).
13. S. Markov, L. Gerrer, F. Adamu-Lema, S. Amoroso and A. Asenov. Proc. SISPAD 2012, 157–160 (2012).
14. F. Adamu-Lema, C. M. Compagnoni, S. M. Amoroso, N. Castellani, L. Gerrer, S. Markov, A. S. Spinelli, A. L. Lacaita and A. Asenov. IEEE Trans. Elec. Dev. 60, 833–839 (2013).
15. B. Kaczer, M. Toledano-Luque, J. Franco, P. Weckx (2013) Statistical distribution of defect parameters. In: T. Grasser (eds) Bias temperature instability for devices and circuits. Springer, New York
16. M. Toledano-Luque, B. Kaczer (2013) Characterization of individual traps in high-κ oxides. In: T. Grasser (ed.) Bias temperature instability for devices and circuits. Springer, New York
17. A. Asenov, R. Balasubramaniam, A. Brown and J. Davies. IEEE Trans.Electron Devices 50, 334–336 (2003).
18. T. Grasser, B. Kaczer, W. Goes, H. Reisinger, T. Aichinger, P. Hehen-berger, P. Wagner, F. Schanovsky, J. Franco, M. T. Luque, and M. Nelhiebel. IEEE Trans. Electron Devices 58, 3652–3666 (2011).
19. M. Toledano-Luque, B. Kaczer, E. Simoen, R. Degraeve, J. Franco, P. J. Roussel, T. Grasser and G. Groeseneken. Proc. IRPS 2012, 1–6 (2012).
20. S. M. Amoroso, L. Gerrer, S. Markov, F. Adamu-Lema and A. Asenov. Proc. of ESSDERC 2012, 109–112 (2012).
21. S. M. Amoroso, L. Gerrer, S. Markov, F. Adamu-Lema and A. Asenov. Solid-State Electronics - Accepted for publication (2013).
22. L. Gerrer, S. Markov, S. Amoroso, F. Adamu-Lema and A. Asenov. Microelectron. Reliab. 52, 1918–1923 (2012).
23. S. Ogawa and N. Shiono. Phys. Rev. B 51, 4218–4230 (1995).

24. M. A. Alam. IEDM Tech. Dig. 2003, 345–348 (2003).
25. D. K. Schroder and J. A. Babcock. J. Appl. Phys. 94, 1–18 (2003).
26. M.A.Alam and S.Mohapatra. Journal of Microelectronics Reliability 71–81 (2005).
27. J. H. Stathis and S. Zafar. Microelectron. Reliab. 46, 270–286 (2006).
28. T. Yamamoto, K. Uwasawa and T. Mogami. IEEE Trans. Elec. Dev. 46, 921–926 (1999).
29. D. K. Schroder and J. A. Babcock. Microelectron. Reliab. 47, 841–852 (2007).
30. S. Mahapatra, M. A. Alam, P. B. Kumar, T. R. Dalei, D. Varghese and D. Saha. Microelectron. Eng. 80, 114–121 (2005).
31. T. Grasser (2013). The capture/emission time map approach to the bias temperature instability. In: T. Grasser (eds) Bias temperature instability for devices and circuits. Springer, New York
32. A. Asenov, A. R. Brown, J. H. Davies, S. Kaya and G. Slavcheva. IEEE Trans. Electron Devices 50, 1837–1852 (2003).
33. G. Roy, A. R. Brown, F. Adamu-Lema, S. Roy and A. Asenov. IEEE Trans. Elec. Dev. 53, 3063–3069 (2006).
34. D. Reid, C. Millar, S. Roy and A. Asenov. IEEE Trans. Electron Devices 57, 2801–2807 (2010).
35. A. R. Brown, J. Watling and A. Asenov. J. Comput.Electron 5, 333–336 (2006).
36. H. Dadgour, K. Endo and K.Banerjee. IEDM, Tech. Dig. 2008, 705–708 (2008).
37. A. R. Brown, N. Idris, J. Watling and A. Asenov. IEEE Electron Device Lett 31, 1199–1201 (2010).
38. www.goldstandardsimulations.com.
39. D. A. Buchanan, M. V. Fischetti and D. J. DiMaria. Phys. Rev. B 43, 1471–1486 (1991).
40. M. Lax. Phys. Rev. 119, 1502–1523 (1960).
41. S. M. Amoroso, A. Maconi, A. Mauri, C. M. Compagnoni, E. Greco, E. Camozzi, S. Vigano', P. Tessariol, A. Ghetti, A. S. Spinelli and A. L. Lacaita. IEDM Tech. Dig. 2010, 540–543 (2010).
42. S. M. Amoroso, A. Maconi, A. Mauri, C. M. Compagnoni, A. S. Spinelli, and A. L. Lacaita. IEEE Trans. Elec. Dev. 1864–1871 (2011).
43. N. Castellani, C. M. Compagnoni, A. Mauri, A. S. Spinelli and A. L. Lacaita. IEEE Trans. Electron Devices 59, 2488–2494 (2012).
44. C. M. Compagnoni, N. Castellani, A. Mauri, A. S. Spinelli and A. L. Lacaita. IEEE Trans. Elec. Dev. 59, 2495–2500 (2012).
45. M. Kirton and M. Uren. Advances in Physics 38, 367–468 (1989).
46. A. Palma, A. Godoy, J. A. Jimenez-Tejada, J. E. Carceller and J. A. Lopez-Villanueva. Phys. Rev. B 56, 9565–9574 (1997).
47. J. P. Campbell, J. Qin, K. P. Cheung, L. C. Yu, J. S. Suehle, A. Oates and K. Sheng. Proc. IRPS 2009, 382–388 (2009).
48. T. Grasser, H. Reisinger, W. Goes, T. Aichinger, P. Hehenberger, P.-J. Wagner, M. Nelhiebel, J. Franco and B. Kaczer. IEDM Tech. Dig. 2009, 729–732 (2009).
49. T. Nagumo, K. Takeuchi, T. Hase and Y. Hayashi. IEDM Tech. Dig. 2010, 628–631 (2010).
50. M. Toledano-Luque, B. Kaczer, P. Roussel, M. J. Cho, T. Grasser and G. Groeseneken. J. Vac. Sci. Technol. B 01AA04 (2011).
51. T. Grasser, B. Kaczer, W. Goes, T. Aichinger, P. Hehenberger and M. Nelhiebel. Proc. IRPS 2009, 33–44 (2009).
52. K. S. Ralls, W. J. Skocpol, L. D. Jackel, R. E. Howard, L. A. Fetter, R. W. Epworth and D. M. Tennant. Phys. Rev. Lett. 52, 228–231 (1984).
53. K. K. Hung, P. K. Ko, C. Hu and Y. C. Cheng. IEEE Electron Device Lett. 11, 90–92 (1990).
54. T. Grasser, B. Kaczer, W. Goes, H. Reisinger, T. Aichinger, P. Hehenberger, P.-J. Wagner, F. Schanovsky, J. Franco, P. Roussel, and M. Nelhiebel. IEDM Tech. Dig. 2010, 82–85 (2010).
55. I. Chen, S. E. Holland and C. Hu. IEEE Trans. Electron Devices 32, 413–422 (1985).
56. D. J. DiMaria. J. Appl. Phys. 8707–8715 (2000).
57. D. Ielmini, A. S. Spinelli, A. L. Lacaita and M. van Duuren. IEEE Trans. Electron Devices 51, 1288–1295 (2004).
58. D. Ielmini, A. S. Spinelli, A. L. Lacaita and M. van Duuren. IEEE Trans. Electron Devices 51, 1281–121 (2004).

59. T. Grasser, T. Aichinger, G. Pobegen, H. Reisinger, P.-J. Wagner, J. Franco, M. Nelhiebel and B. Kaczer. Proc. IRPS 2011, 605–613 (2011).
60. A. Asenov. IEDM Tech.Dig. 1999 (1999).
61. A. Asenov, A. R. Brown and J. Watling. Solid-State Electronics 47, 1141–1145 (2003).
62. G. D. Panagopoulos and K. Roy. IEEE Trans. Electron Devices 58, 391–403 (2011).
63. A.Asenov, S.Kaya and A.R.Brown. IEEE Trans.Electron Devices 50, 1254–1260 (2003).
64. C. Alexander, A. R. Brown, J. R. Watling and A. Asenov. Solid-State Electron 733–739 (2005).
65. A. Asenov, G. Roy, C. Alexander, A. R. Brown, J. Watling and S. Roy. IEEE Conference on Nanotechnology 2004, 334–336 (2004).
66. T. Grasser, W. Goes, V. Sverdlov and B. Kaczer. Proc. IRPS 2007, 268–280 (2007).
67. H. Reisinger, T. Grasser and C. Schlunder. Proc. IIRW 2009, 30–35 (2009).
68. H. Reisinger, T. Grasser, W. Gustin and C. Schlünder. Proc. IRPS 2010, 7–15 (2010).
69. B. Kaczer, V. Arkhipov, R. Degraeve, N. Collaert, G. Groeseneken and M. Goodwin. Proc. IRPS 2005, 381–387 (2005).
70. M. Denais, A. Bravaix, V. Huard, C. Parthasarathy, G. Ribes, F. Perrier, Y. Rey-Tauriac, and N. Revil. IEDM, Tech. Dig. 2004, 109–112 (2004).
71. H. Reisinger, O. Blank, W. Heinrigs, A. Mühlhoff, W. Gustin and C. Schlünder. Proc. IRPS 2006, 448–453 (2006).
72. T. Grasser, H. Reisinger, P.-J. Wagner, W. Goes, F. Schanovsky and B. Kaczer. Proc. IRPS 2010, 16–25 (2010).
73. M. Toledano-Luque, B. Kaczer, E. Simoen, P. J. Roussel, A. Veloso, T. Grasser and G. Groeseneken. Microelectronic Eng. 1243–1246 (2011).
74. H. Reisinger (2013) The time dependent defect spectroscopy. In: T. Grasser (eds) Bias temperature instability for devices and circuits. Springer, New York
75. C. M. Compagnoni, A. S. Spinelli, R. Gusmeroli, S. Beltrami, A. Ghetti and A. Visconti. IEEE Trans. Elec. Dev. 55, 2695–2702 (2008).
76. B. Kaczer, P. J. Roussel, T. Grasser and G. Groeseneken. IEEE Elec. Dev. Lett. 411–413 (2010).
77. B. Cheng, A. R. Brown and A. Asenov. IEEE Elec. Dev. Lett. 32, 740–742 (2011).

Chapter 14
A Comprehensive Modeling Framework for DC and AC NBTI

Souvik Mahapatra

Abstract This chapter presents a comprehensive modeling framework to explain DC and AC NBTI experiments. The framework consists of uncorrelated contribution from interface trap generation, along with hole trapping in process-related preexisting and stress-induced generated bulk insulator traps. A wide range of experimental data, such as time evolution of degradation during and after DC stress, very long-time stress experiments, AC degradation as a function of pulse duty cycle and frequency, and measurement speed dependence of DC and AC NBTI, can be successfully explained for devices having wide range of gate insulator processes. Model equations and parameters have been listed.

14.1 Introduction

In this chapter, DC and AC Negative Bias Temperature Instability (NBTI) measurements are explained by using a comprehensive modeling framework of uncorrelated contributions from interface trap generation (ΔN_{IT}), together with hole trapping in preexisting (ΔN_{HT}) and generated (ΔN_{OT}) bulk insulator traps [1]. Generation and recovery of interface traps are modeled using the H/H_2 Reaction–Diffusion (RD) model [2, 3]. Hole trapping and detrapping are modeled using the 2-energy level model [4, 5]. Generation of bulk insulator traps is modeled using an empirical formula, although its contribution has been found to be negligible in thin gate insulator devices under use condition [1, 6]. It will be shown that the proposed comprehensive modeling framework can explain the following NBTI features with consistent set of model parameters: (1) time evolution of threshold voltage shift (ΔV_T) during DC stress and (2) recovery of ΔV_T after DC stress, (3) time evolution

S. Mahapatra (✉)
Department of Electrical Engineering, Indian Institute of Technology Bombay,
Mumbai 400076, India
e-mail: souvik@ee.iitb.ac.in

T. Grasser (ed.), *Bias Temperature Instability for Devices and Circuits*, 349
DOI 10.1007/978-1-4614-7909-3_14,

of ΔV_T at long stress time (t-stress) obtained from High Temperature Operating Life (HTOL) measurements, dependence of ΔV_T during AC stress on (4) pulse frequency and (5) pulse duty cycle, (6) impact of measurement speed on DC and AC measurements, as well as (7) gate insulator process dependence of DC and AC experiments.

It is important to note that several alternative models have been proposed to explain NBTI experiments and are reviewed in [1]. Some reports propose only ΔN_{HT} to be responsible for NBTI [7–9] and are inconsistent with direct experimental evidence of ΔN_{IT} obtained using Charge Pumping (CP) and DCIV measurements[1] [1, 5, 6, 11–16]. On the other hand, only ΔN_{IT}-based models [3, 12, 17] can explain data from relatively slower measurements but are inconsistent with ultrafast measurements [18, 19] and gate insulator process dependence of NBTI [1, 5, 6, 13, 15, 16, 19, 20]. Therefore, models that consider both ΔN_{IT} and ΔN_{HT} have gained prominence [1, 5, 6, 13, 15, 16, 21, 22].

Of these models, the 2-stage or 4-energy well model involving switching hole traps [21] relies on strong coupling between ΔN_{IT} and ΔN_{HT} and has been recently found to be incapable of simultaneously predicting DC stress and recovery, process dependence, as well as AC experiments [1, 5, 15, 23]. The model relying on uncorrelated ΔN_{IT} and ΔN_{HT} but suggests no recovery of ΔN_{IT} as proposed in [22] is inconsistent with direct experimental evidence of ΔN_{IT} recovery [1, 5, 6]. As mentioned, the proposed comprehensive model framework [1] assumes generation and recovery of both ΔN_{IT} and ΔN_{HT} during and after NBTI stress, and these underlying components are assumed to be completely uncorrelated. Furthermore, uncorrelated contribution due to ΔN_{OT} is added to take care of Time-Dependent Dielectric Breakdown (TDDB) like situations [24] for larger gate stress bias ($V_{G,STR}$) in devices having thicker gate insulators [1, 15].

The comprehensive NBTI modeling framework will be discussed first. Prediction of ΔV_T time evolution during DC stress and recovery on p-MOSFETs having different gate insulator processes will be shown next. This will be followed by the description of a simple NBTI model for prediction of long-time DC stress data for different gate insulator processes. The impact of gate insulator process on long-time NBTI parameters will be modeled. The validity of underlying ΔN_{IT} and ΔN_{HT} components and their gate insulator process dependence will be verified by using

[1]During CP measurement [10], the gate of the MOSFET is repetitively pulsed from inversion to accumulation, source and drain are grounded, and the DC current due to electron–hole recombination in traps at or near the Si/SiO_2 interface is measured at the substrate. CP measurements are done before and after NBTI stress in the conventional measure-stress-measure (MSM) mode and increase in CP current after stress indicates trap generation. In DCIV measurement [11], the source and drain junctions of the MOSFET are forward biased, the gate is swept from accumulation to inversion, and the DC current due to electron–hole recombination in traps at or near the Si/SiO_2 interface is measured at the substrate. Increase in DCIV current after NBTI stress indicates trap generation. Like CP, DCIV measurement is also done in the MSM mode. Both measurements suffer from recovery issues [12] and scan traps at a particular portion of the energy bandgap, and corrections due to measurement delay and bandgap differences are needed before obtained ΔN_{IT} can be compared to ΔV_T obtained from fast or ultrafast IV measurements [1, 5, 13].

independent experiments. This will be followed by a discussion on AC frequency and duty cycle dependence of NBTI. Finally, the chapter will be concluded by summarizing key results.

14.2 Description of the Comprehensive Modeling Framework

In this section, the H/H_2 RD model for ΔN_{IT} will be discussed first, which will be followed by a description of the 2-energy well model for ΔN_{HT}. The empirical model for ΔN_{OT} will be shown next. Model parameters used for prediction of different DC and AC experimental results will be listed. The total ΔV_T is calculated by summing individual subcomponents as $\Delta V_T = \Delta V_{IT} + \Delta V_{HT} + \Delta V_{OT}$, where $\Delta V_{IT} = q\ \Delta N_{IT}/C_{OX}$, $\Delta V_{HT} = q\ \Delta N_{HT}/C_{OX}$, and $\Delta V_{OT} = q\ \Delta N_{OT}/C_{OX}$; q is the electronic charge and C_{OX} is the gate capacitance.

14.2.1 H/H_2 Reaction–Diffusion (RD) Model for Interface Traps

Figure 14.1 illustrates the H/H_2 RD model [2, 3] for N_{IT} generation and recovery during and after NBTI stress. During stress at a gate bias of $V_{G,STR}$, inversion layer hole tunnels into and is captured by interfacial Si–H bond; the weakened bond then gets broken by thermal excitation [25, 26]. The released H diffuses out and reacts with another available H to form molecular H_2, which eventually diffuses out. The broken Si- bond forms N_{IT} at the Si/SiO_2 interface. Two versions of RD model have been proposed in the literature. In the conventional H/H_2 RD model [2, 26], the "other" H is commonly ascribed to that coming from another broken Si–H bond at the Si/SiO_2 interface, as shown in Fig. 14.1, although it can as well come from a broken bond at the SiO_2/poly-Si interface or from poly-Si grain boundary. In poly H/H_2 RD model [3], released H from Si/SiO_2 interface reacts with another Si–H bond at SiO_2/poly-Si interface to form H_2, as shown in Fig. 14.1. During recovery at a reduced gate bias of $V_{G,REC}$, H_2 diffuses back towards the Si/SiO_2 interface, monomerizes into H, and reacts with the broken Si- bond to passivate N_{IT}. Alternatively, the returning H_2 can react with broken Si- at the SiO_2/poly-Si interface, and the resulting H can passivate the broken Si- at the Si/SiO_2 interface.

As illustrated in Table 14.1 [1, 6, 26], a set of coupled partial differential equations can be solved to model the breaking and passivation of Si–H bonds at the Si/SiO_2 interface (1); H to H_2 dimerization and monomer formation for the conventional model (2), or creation and dissociation of H_2 at the SiO_2/poly-Si interface for the poly model (3); and diffusion of H and H_2 (4). The fixed and adjustable model

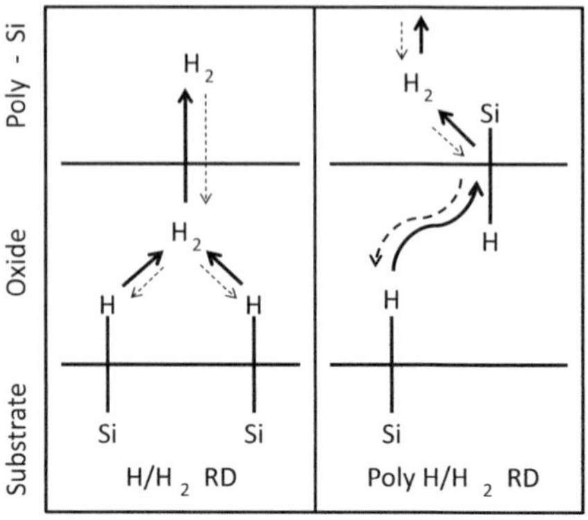

Fig. 14.1 Schematic of conventional H/H$_2$ RD and poly H/H$_2$ RD models

parameters along with their voltage dependence[2] and Arrhenius temperature (T) activation are also described in Table 14.1. Only two device-dependent adjustable parameters ($k_{F0(S)}$ and Γ_{IT}) are needed to predict ΔN_{IT} contribution during DC and AC experiments across different gate insulator processes, and these are listed in Table 14.2. Figure 14.2 shows the time evolution of N_{IT} during stress and recovery simulated using the conventional and poly H/H$_2$ models [1, 27]. For the model parameters listed in Table 14.1, both models produce similar time evolution of ΔN_{IT} during stress and recovery[3] as shown in Fig. 14.2. A detailed comparison of the conventional and poly H/H$_2$ RD models including the impact of model parameters have been discussed in [27].

Note that a full solution of the time evolution of N_{IT} generation and recovery requires solving the RD model equations, as done in Fig. 14.2, and is used in the next section to predict experimental data. However, meaningful insight about the predictive capability of the H/H$_2$ RD model can be obtained by noting the long-time analytic solution during stress that yields [26]

$$\Delta N_{IT} = (k_F N_0 / k_R)^{2/3} (k_H / k_{H2})^{1/3} (D_{H2} t - \text{stress})^{1/6} Z$$

and suggests power law time evolution of N_{IT} with time exponent $n = 1/6$, where t-stress is stress time and other model parameters are listed in Table 14.1. To verify trap generation during NBTI, Fig. 14.3 shows time evolution of ΔN_{IT} measured directly using the CP method [13]. A power law time evolution has been observed, however with much larger n (than 1/6), which increases with

[2]It is well known that NBTI degradation depends on gate oxide field (E_{OX}) and not stress gate bias ($V_{G,STR}$) [14]. For simplicity and without loss of generality, power law gate voltage (V_G) dependence has been used [6].

[3]Both conventional and poly H/H$_2$ RD models show similar results when implemented in 1D. However, differences appear when simulations are performed at higher (>1D) dimensions. Reasons for these differences have been discussed in detail elsewhere [27]. All simulated ΔN_{IT} results shown in this chapter are calculated by using the poly H/H$_2$ RD model.

Table 14.1 Equations and device-independent model parameters for conventional H/H_2 RD and poly H/H_2 RD models

(1) $\frac{dN_{IT(s)}}{dt} = k_{F(s)}\left(N_{0(s)} - N_{IT(s)}\right) - k_{R(s)}N_{IT(s)}N_H^{(s)}$

(2a) $\frac{\delta}{2}\frac{dN_H^{(s)}}{dt} = D_H\frac{dN_H^{(s)}}{dx} + \frac{dN_{IT(s)}}{dt} - \delta k_H\left[N_H^{(s)}\right]^2 + \delta k_{H2}N_{H2}^{(s)}$

(2b) $\frac{\delta}{2}\frac{dN_{H2}^{(s)}}{dt} = D_{H2}\frac{dN_{H2}^{(s)}}{dx} + \frac{\delta}{2}k_H\left[N_H^{(s)}\right]^2 - \frac{\delta}{2}k_{H2}N_{H2}^{(s)}$

(3) $\frac{dN_{IT(p)}}{dt} = k_{F(p)}\left(N_{0(p)} - N_{IT(p)}\right)N_H^{(p)} - k_{R(p)}N_{IT(p)}N_{H2}^{(p)}$

(4a) $\frac{dN_H}{dt} = D_H\frac{d^2N_H}{dx^2} - k_HN_H^2 + k_{H2}N_{H2}$

(4b) $\frac{dN_{H2}}{dt} = D_{H2}\frac{d^2N_{H2}}{dx^2} + \frac{1}{2}k_HN_H^2 - \frac{1}{2}k_{H2}N_{H2}$

$k_{F(s)} = k_{F0(s)}(V_G - V_{T0})^{\frac{3}{2}\Gamma_{IT}}e^{-\frac{E_{Akf}}{kT}}$ $\quad k_{R(s)} = k_{R0(s)}e^{-\frac{E_{Akr}}{kT}}$

$k_H = k_{H0}e^{-\frac{E_{AkH}}{kT}}$ $\quad k_{H2} = k_{H20}e^{-\frac{E_{AkH2}}{kT}}$

$D_H = D_{H0}e^{-\frac{E_{ADH}}{kT}}$ $\quad D_{H2} = D_{H20}e^{-\frac{E_{ADH2}}{kT}}$

$k_{F(p)} = k_{F0(p)}(V_G - V_{T0})^{\frac{3}{2}}\Gamma_{IT}e^{-\frac{E_{Akf}}{kT}}$ $\quad k_{R(p)} = k_{R0(p)}e^{-\frac{E_{Akr}}{kT}}$

Subscript (s) and (p) denotes Si/SiO_2 and SiO_2/p-Si interface respectively.
Equations (1), (2) and (4) are for conventional H/H_2 RD model.
While (1), (2), (3) and (4) represents poly H/H_2 RD model.
$k_{F(s)}$, $k_{F(p)}$ is Si–H bond breaking reaction rate constants;
$k_{R(s)}$, $k_{R(p)}$ is Si–H bond annealing reaction rate constants;
$N_{IT(s)}$, $N_{IT(p)}$ is interface trap density; $N_{0(s)}$, $N_{0(p)}$ is initial Si–H bond density;
$N_H^{(s)}$, $N_H^{(p)}$ is atomic hydrogen density near the interface;
$N_{H2}^{(s)}$, $N_{H2}^{(p)}$ is molecular hydrogen density;
N_H and N_{H2} are concentration of atomic and molecular hydrogen respectively;
D_H and D_{H2} are diffusivities of atomic and molecular hydrogen respectively;
k_H and k_{H2} are generation and dissociation rates of H_2 respectively;
δ is interfacial thickness (~1.5 Å)

Parameters

$E_{A\,kf} = 0.175$ eV; $E_{A\,kr} = 0.2$ eV; $k_{H0} = 8.56\ cm^3/s$; $E_{A\,kH} = 0.3$ eV; $k_{H20} = 507e5s^{-1}$;
$E_{A\,kH2} = 0.3$ eV; $E_{A\,DH} = 0.2$ eV; $Ea_{DH2} = 0.58$ eV; $k_{F0(p)} \approx 3/5{*}k_{F0(s)}$.
$k_{F0(s)}$ and Γ_{IT} are device dependent parameters (see Table 14.2).

H/H_2RD model

$k_{R0(s)} = 9.9e{-}7$; $D_{H0} = 9.56e{-}8\ cm^2/s$ and $D_{H20} = 3.5e{-}5\ cm^2/s$.

Poly H/H_2RD model

$k_{R0(s)} = 9.9e{-}5$; $k_{R0(p)} = 8e{-}4$; $D_{H0} = 1.5e{-}5\ cm^2/s$ and $D_{H20} = 9.5e{-}5\ cm^2/s$.

Table 14.2 SiON p-MOSFET device details and device-dependent RD model parameters used for model prediction of experimental results

Device	D1	D2	D3
N%	17	23	43
EOT (Å)	23.5	14	14.5
Variable parameters			
$k_{F0(S)}$	1.1	13	320
Γ_{NIT}	4.3	4.3	3.1

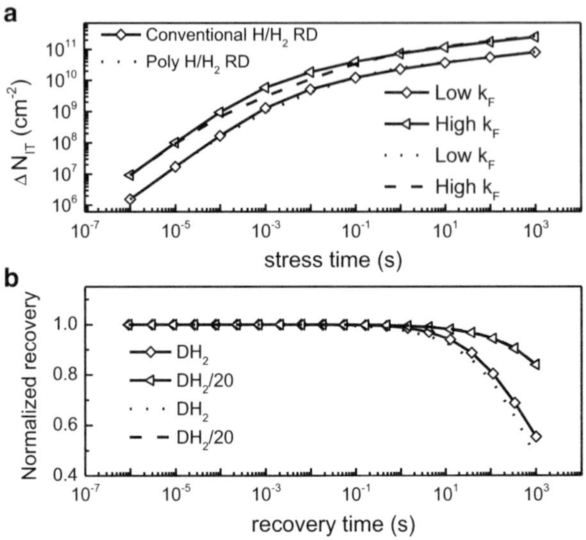

Fig. 14.2 Simulated time evolution of N_{IT} during stress and recovery using conventional H/H_2 RD and poly H/H_2 RD models

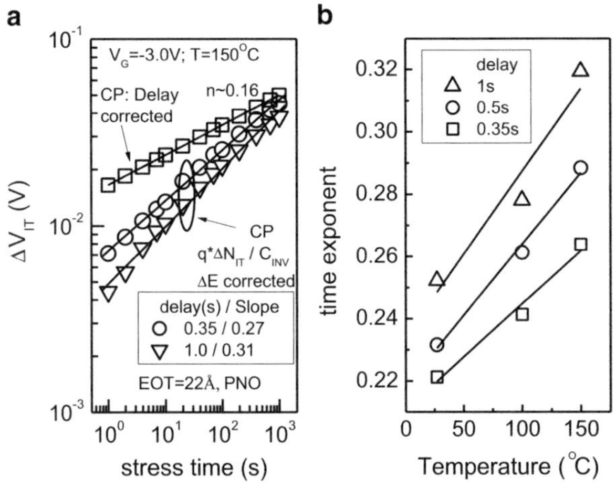

Fig. 14.3 Time evolution of ΔN_{IT} during stress measured using CP method for different measurement delay (*left panel*). Data after delay correction are also shown. Data from [13]. Measured time exponent *n* as a function of stress T from CP measurements for different measurement delay. Data from [12]

increase in measurement delay. Figure 14.3 also plots measured *n* as a function of stress T for different measurement delay, which shows higher *n* for higher T and larger measurement delay [12]. It is now well known that CP method suffers from measurement delay artifacts (higher N_{IT} recovery at higher T and for larger

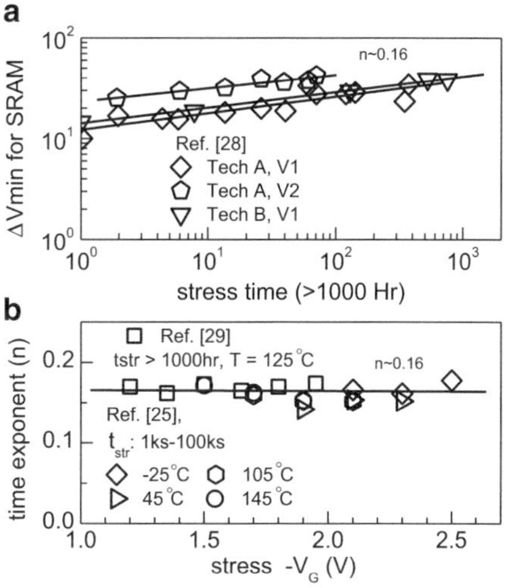

Fig. 14.4 Time evolution of NBTI degradation during long-time HTOL test (*top*) and measured power law time exponent (*n*) at long stress time, obtained from different production quality devices. Data from [25, 28, 29]

measurement delay increases *n*) as it is implemented in the MSM mode, which needs to be corrected [12]. Once corrected, time evolution of ΔN_{IT} shows $n \sim 1/6$ power law dependence, as shown in Fig. 14.3, consistent with RD model prediction[4] [13].

As a further validation, Fig. 14.4 shows (a) time evolution of NBTI during HTOL test and (b) extracted exponent *n* for very long stress time, obtained from production quality devices [25, 28, 29]. The use of production quality devices ensures well-optimized gate stacks, with negligible ΔN_{HT} contribution arising out of trapping in preexisting defects [13, 30], and the use of relatively lower $V_{G,STR}$ ensures ΔN_{OT} is negligible [14, 15]. Therefore, data shown in Fig. 14.4 are dominated by N_{IT} generation. Indeed, consistent with the prediction of the RD model, $n \sim 1/6$ time exponent has been observed in broad range of experiments. It is important to realize that the prediction of $n \sim 1/6$ time exponent is an intrinsic feature of the H/H_2 RD model and is obtained with no (*zero*) adjustable parameter. The ability to predict a key feature of NBTI experiment, relevant for lifetime prediction at long stress time and technology qualification, clearly establishes the validity and robustness of the H/H_2 RD model.

The long-time analytical solution of RD model also suggests that t-stress to obtain a particular ΔN_{IT} at different stress T is inversely proportional to D_{H2}, provided T activation of k_F equals k_R and that of k_H equals k_{H2}, which can be justified by invoking detailed balance. Figure 14.5 shows measured ΔN_{IT} using DCIV method as a function of t-stress for different stress T. Measured ΔN_{IT} data are plotted after delay correction and show identical time exponent ($n \sim 1/6$) at

[4]As CP method scans traps only at the central portion of the energy bandgap, further corrections are needed before ΔV_{IT} (obtained from ΔN_{IT}) can be compared to ΔV_T from fast IV measurements [13].

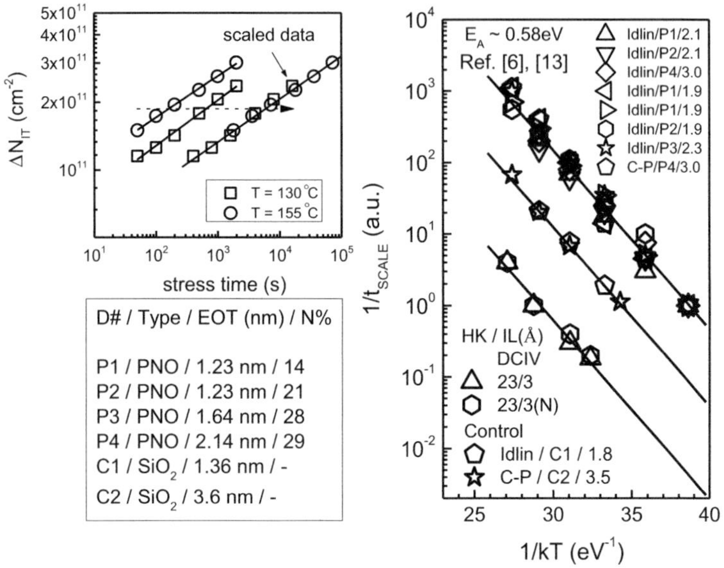

Fig. 14.5 Time evolution of ΔN_{IT} for different stress T, measured using DCIV method and plotted after delay correction (*left panel*). Data scaling (by a factor t_{SCALE}) along X-axis to a universal relation is also shown. T activation of the scaling factor t_{SCALE}, obtained from different measurements and from different devices (*right panel*). Data from [13, 27]

different T. The invariance of n across stress T suggests Arrhenius T activation of the N_{IT} generation process [12]. Measured data is scaled along t-stress (X-axis) to universal relation, as shown, and the scaling factor (t_{SCALE}) is obtained for each stress T. Figure 14.5 also plots the T activation of $1/t_{SCALE}$ obtained by performing the above exercise in different devices [13, 27]. Time evolution of ΔN_{IT} at different stress T has been obtained directly from DCIV and CP measurements and also indirectly from ΔV_T measurements using suitable devices with low preexisting defects and suitable stress condition with low $V_{G,STR}$ that, respectively, ensures negligible contribution from ΔN_{HT} and ΔN_{OT} [13–15, 30]. Note that such diverse set of devices shows universal T activation energy of $E_A \sim 0.6$ eV and suggests molecular H_2 diffusion [31], which is consistent with long-time power law time exponent of $n = 1/6$ shown in Fig. 14.4. Therefore, measured time and T dependence of ΔN_{IT} at long stress time is consistent with H/H_2 RD model prediction.

Figure 14.6 demonstrates the impact of gate insulator processes on N_{IT} generation measured using CP and DCIV methods [15]. ΔN_{IT} is measured in plasma and thermal nitrided SiON p-MOSFETs, stressed for identical time but under different stress E_{OX}. Note that thermal SiON shows higher ΔN_{IT} than the plasma SiON device, and both methods show similar results. Therefore, Figs. 14.3, 14.4, 14.5, and 14.6 unequivocally establish the presence of N_{IT} generation during NBTI stress, which gets impacted by gate insulator processes, and, once recovery artifacts are corrected, shows power law time evolution as per H/H_2 RD model solution. As will

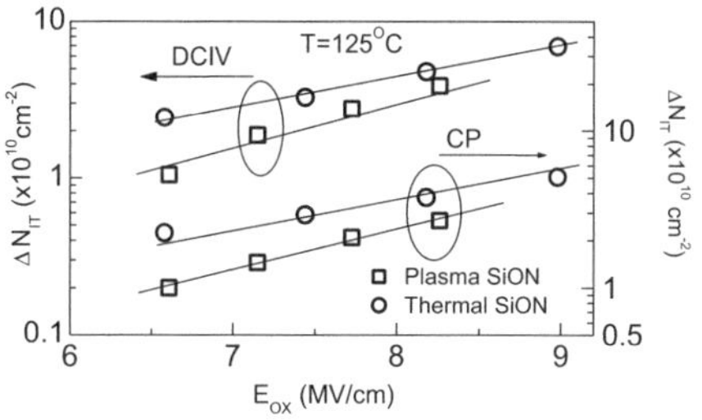

Fig. 14.6 Measured ΔN_{IT} at fixed t-stress as a function of E_{OX}, obtained using CP and DCIV methods in plasma and thermal SiON p-MOSFETs. Data from [15]

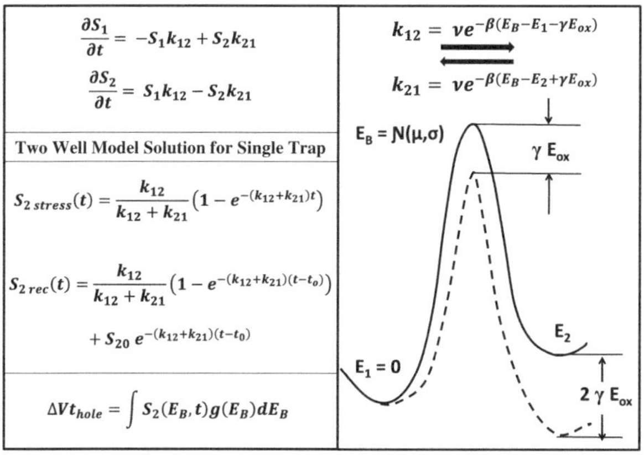

Fig. 14.7 Schematic of 2-energy well model and model equations

be discussed in the next section, ΔN_{IT} contribution must be taken into account for modeling ΔV_T during NBTI stress. However, a full solution of ΔV_T time evolution under different devices and experimental conditions needs additional contribution from ΔN_{HT} and ΔN_{OT} and is discussed below.

14.2.2 Two-Energy Well Model for Trapped Holes

Figure 14.7 shows the schematic of the 2-energy well model [1, 4, 5]. The energy levels of the trap before and after hole trapping are, respectively, represented by two energy wells E_1 and E_2, where the energy of the ground state E_1 is chosen as a reference at $E_1 = 0$. An energy barrier of height E_B is seen by the holes between the two energy states. The present implementation neglects tunneling between the wells and only considers thermionic emission over the barrier E_B. The barrier height E_B and the energy of the trapped state E_2 change with respect to the reference E_1 when a negative gate bias (V_G) is applied. E_B is reduced by "γ E_{OX}" and E_2 by "2 γ E_{OX}," where E_{OX} is the oxide field and γ is the difference in "q x" between the two wells, x being the generalized coordinate [5].

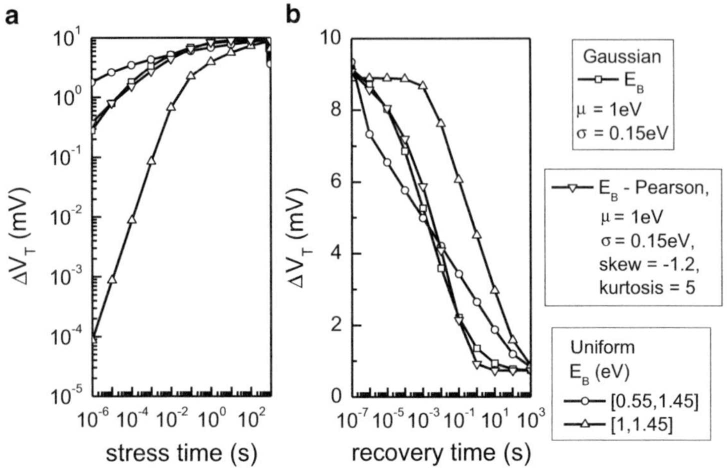

Fig. 14.8 Time evolution of ΔN_{HT} during stress and recovery, simulated using different distributions of the energy barrier (E_B). Data from [5]

The rate constants for transition from well 1 to 2 (k_{12}) and back (k_{21}) and corresponding rate equations are also shown in Fig. 14.7 [5], where $\beta = 1/(k_BT)$, k_B is the Boltzmann constant, and ν is the attempt to escape frequency having a typical value of the order of 10^{12}–10^{13} s^{-1} [4]. The rate equations are solved to get the occupancies of the wells S_1 and S_2 during stress and recovery, S_2 being the charged state that contributes to ΔV_{HT}. Note that Fig. 14.7 shows the well 2 occupancy S_2 for a single hole trap having a fixed barrier height E_B during stress and recovery. In reality, the gate insulator of a device has multiple traps distributed in position and energy, and a 2-well system has to be solved for each of these traps. The trap distribution can be modeled using a distribution $g(E_B)$ of the barrier height E_B. Hence, ΔV_{HT} can be calculated by integrating $[q\,N_0/C_{OX}]*[S_2\,g(E_B)]$ over the entire range of E_B, where C_{OX} is the gate capacitance [4, 5].

Figure 14.8 shows the time evolution of ΔV_{HT} during stress and recovery, calculated using the 2-well model with Gaussian, Pearson, and uniform distributions of the barrier E_B [5]. The time evolution of ΔV_{HT} has been found to be a strong function of the range and shape of the E_B barrier distribution for both stress and recovery. Note that the trapping and detrapping time constants of the traps, proportional to the inverse of k_{12} and k_{21}, respectively, are dependent on the E_B value. Slow traps having large time constants are associated with large E_B, while those associated with small E_B have small time constants and can be defined as fast traps. Therefore, a chosen E_B distribution skewed towards high values of E_B would show slow stress and recovery, whereas the trends would be opposite for a distribution having greater number of traps with low E_B values. Both the Gaussian and Pearson distributions with the range of E_B as given in Fig. 14.8 are able to provide ΔV_{HT} time evolution needed for predicting measured time evolution of ΔV_T as discussed in the next section. The uniform distribution has been used to

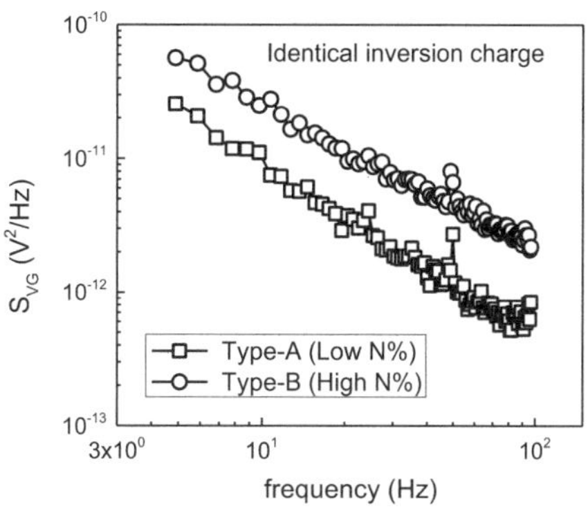

Fig. 14.9 Measured input refereed noise (S_{VG}) as a function of frequency, measured using flicker noise method in plasma (Type-A) and thermal (Type-B) SiON p-MOSFETs. Measurements are done in prestress, and gate bias is chosen to achieve identical inversion charges during measurement. Data from [15]

illustrate the effect of the range of E_B [minima, maxima] on time evolution of ΔV_{HT} during stress and recovery. Uniform distribution with the range [0.55 eV, 1.45 eV] shows faster saturation and recovery compared to the distribution with range [1 eV, 1.45 eV], as the former consists of more fast traps with small E_B values compared to the latter.

Gate insulator processes have a strong influence on preexisting hole traps in the gate insulator bulk, which can be independently estimated by using flicker noise measurement in prestress. Figure 14.9 shows input-referred noise (S_{VG}) as a function of frequency, measured in plasma and thermal SiON p-MOSFETs [15]. The gate voltage during measurements has been carefully chosen to achieve identical inversion layer charges. Thermal SiON shows higher S_{VG}, suggesting higher density of preexisting bulk hole traps compared to the plasma SiON device. As discussed in detail in [32] and also later in this section, thermal SiON and other Type-B SiON devices, having larger preexisting hole traps in the gate insulator, show fast hole trapping-dominated NBTI when compared to plasma SiON and other Type-A SiON devices. Fast hole trapping results in substantial degradation in the sub 1 ms time scale that shows negligible T dependence for Type-B devices and also higher long-time degradation that shows lower time exponent n and T activation E_A for Type-B compared to Type-A devices.

Table 14.3 lists the 2-well model parameters used to calculate the ΔV_{HT} component of ΔV_T for prediction of stress and recovery experiments across various gate insulator devices as shown in the next section [5]. A Gaussian E_B distribution has been used. The parameters N_0, γ, ν, and E_2, although different for different gate stacks, are fixed for a given device as stress V_G or T is varied. Only the E_B distribution parameters (μ and σ) are varied with T, and their T activation energy values are also listed in Table 14.3. Moreover, the value of N_0 is consistent with the atomic N_2 content (N%) in the oxide as N_0 increases with increase in N%, which is consistent with independent flicker noise measurements [30]. The model with calibrated parameters suggests fast hole trapping and detrapping as shown in the next section, with trapping being much faster than detrapping, and largest time constant of ~10 s has been observed in the case of detrapping during NBTI recovery. Note that predicted time constants are consistent with independent verification from Random Telegraph Noise (RTN) measurements [33].

Table 14.3 Parameters for 2-energy well model, used for prediction of experimental results

Model parameters	D1	D2	D3
γ (C m)	7.58×10^{-11}	6.21×10^{-11}	3.17×10^{-11}
ν (s^{-1})	10^{13}	10^{13}	5×10^{13}
N_0 (cm^{-2})	5.24×10^{11}	5.62×10^{12}	2.86×10^{12}
E_2 (eV)	0.216	0.209	0.122
E_B Distribution parameters: Mean: $\mu = \mu_0 e^{-\frac{E_{A1}}{kT}}$; Sigma: $\sigma = \sigma_0 e^{-\frac{E_{A2}}{kT}}$			
μ_0 (eV)	1.82	1.92	1.93
E_{A1} (eV)	0.0206	0.0224	0.0221
σ_0 (eV)	0.521	0.644	0.432
E_{A2} (eV)	0.0427	0.0497	0.0248

Table 14.4 Empirical equations and model parameters for bulk trap generation during stress and hole detrapping from generated bulk traps during recovery

For stress
$\Delta V_{OT} = \frac{q}{COX} C \left(1 - e^{\left(-\left(\frac{t}{n}\right)^{\beta_{OT}}\right)} \right); n = \eta (V_G - V_{T0} - \Delta V_T)^{\frac{\Gamma_{OT}}{\beta_{OT}}} e^{\left(\frac{E_{AOT}}{kT\beta_{OT}}\right)}$
Where
$\eta = 5 \times 10^{12}$, $\beta_{OT} = 0.36eV$, $\Gamma_{OT} = 9$, $E_{AOT} = 0.15eV$
C is the device dependent parameter
For recovery
$\Delta V_{OT} = \frac{q}{C_{OX}} B^* \left(e^{\left(-\left(\frac{t}{\tau_r}\right)\beta_r\right)} \right)$
Where $\beta_r = 0.13$ & $\tau_r = 1 \times 10^{-3}$s
Value of B* is adjusted to match the start of recovery with end of stress

14.2.3 Empirical Model for Bulk Oxide Traps

Table 14.4 lists the equations and parameters governing ΔN_{OT} during NBTI stress and recovery. Although negligible for thin gate insulators and at use condition, contribution due to ΔN_{OT} needs to be accounted for completeness, especially to model TDDB-like situations [24] involving NBTI stress using high $V_{G,STR}$ in relatively thicker gate insulators [1, 6, 15]. Note that ΔN_{OT} during NBTI stress can be independently accessed by measuring stress-induced leakage current [34], and it is desirable to suitably choose $V_{G,STR}$ during stress to minimize its contribution to prevent power law time exponent contamination at long stress time [14]. Nevertheless, for accurate modeling, contribution due to ΔN_{OT} must be included, when needed, for complete prediction of ΔV_T time evolution during and after NBTI stress, as shown in the next section.

14.3 Prediction of DC Stress and Recovery Experiments

In this section, the comprehensive modeling framework of the previous section will be used to predict the time evolution of ΔV_T during and after NBTI stress. Experiments have been performed in silicon oxynitride (SiON) p-MOSFETs having different gate insulators [20], shown in Table 14.2 of the previous section, by using the ultrafast on-the-fly (UF-OTF) I_{DLIN} method [19, 20]. The devices have different equivalent oxide thickness (EOT) and N_2-related preexisting bulk trap density in the gate insulator. The density of preexisting bulk traps have been independently verified by flicker noise measurements [30]. Compared to device D1, device D2 has thicker EOT and slightly lower N% while device D3 has much larger N% and similar EOT. D1 and D2 are defined as Type-A devices, while D3 as Type-B device. The impact of gate insulator nitridation on NBTI has been discussed in [32]. It has been shown that Type-A SiON devices have lower N_2 density at the Si/SiON interface compared to Type-B devices. The use of an ultrafast method ensures measured degradation is free from recovery-related artifacts [12]. Although described in detail in [32], the UF-OTF I_{DLIN} technique [19] is briefly described hereinafter to help understand measured results.

During stress, I_{DLIN} is recorded "on the fly" as the gate of the p-MOSFET remains at $V_{G,STR}$, and the first I_{DLIN} measurement, defined as $I_{DLIN}(t_0)$, is usually done within $t_0 = 1$ ms (for conventional method [35]) or $t_0 = 1$ μs (for ultrafast method [19]) of the application of $V_{G,STR}$, where t_0 is defined as the time-zero delay. I_{DLIN} degradation ($\Delta I_{DLIN}(t) = I_{DLIN}(t_0) - I_{DLIN}(t)$) is converted to ΔV_T by mobility correction as explained in [36]. The following points are to be noted for OTF I_{DLIN} measurements as used in stress. First, as $V_{G,STR}$ is not removed during measurement, the difference between ultrafast ($t_0 = 1$ μs) and conventional ($t_0 = 1$ ms) OTF data is due to the capture (for $t_0 = 1$ μs) or non-capture (for $t_0 = 1$ ms) of degradation in the sub 1 ms time scale, and not due to any recovery issues. Furthermore, as the degradation is calculated by assuming t_0 data as being unstressed, resultant ΔV_T at shorter t-stress is found to be lower than actual due to this t_0 subtraction artifact, the effect being more for Type-B devices having larger degradation in the sub 1 ms time scale. During recovery, the gate is dropped to $V_{G,REC}$ after stress for a certain duration at $V_{G,STR}$, and I_{DLIN} is sampled for the required duration. ΔV_T is calculated as discussed above, where $I_{DLIN}(t_0)$ is obtained from prestress measurements at $V_G = V_{G,REC}$. Time evolution of ΔV_T during stress will be modeled first and will be followed by modeling of ΔV_T recovery after NBTI stress.

14.3.1 Prediction of Stress

Figure 14.10 shows measured ΔV_T in different devices as a function of t-stress, obtained using OTF method with t_0 delay of 1 μs and 1 ms [20]. All devices were stressed at identical T and E_{OX}, obtained by adjusting $V_{G,STR}$. Figure 14.11 shows

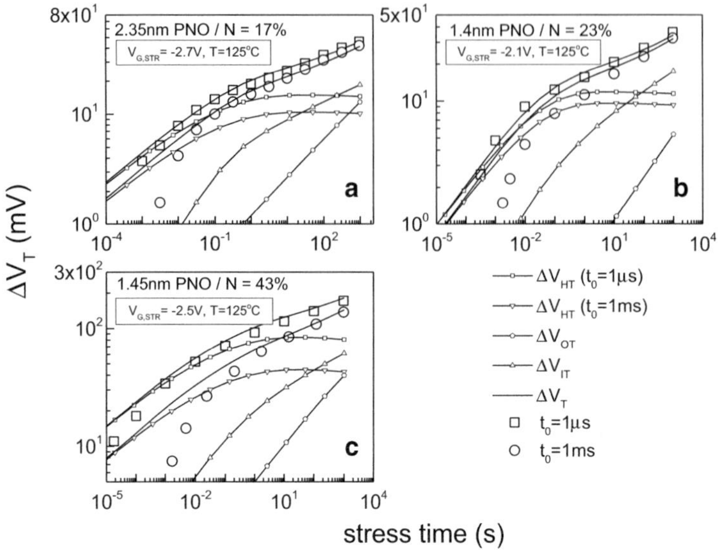

Fig. 14.10 Time evolution of measured ΔV_T during stress for different SiON devices listed in Table 14.2. Overall model prediction and contributions due to ΔN_{IT}, ΔN_{HT}, and ΔN_{OT} are also shown. Data from [5]

ΔV_T time evolution measured by $t_0 = 1$ μs OTF in these devices, stressed at different $V_{G,STR}$ (resulting in different E_{OX}) and T [20]. Although discussed in [32], the following observations are made regarding the impact of gate oxide nitridation on NBTI. For a particular choice of t_0 delay, as well as stress E_{OX} and T, Type-B D3 device shows higher ΔV_T than Type-A D1 and D2 devices. For all devices, larger ΔV_T is captured by $t_0 = 1$ μs OTF, and the difference between 1 μs and 1 ms t_0 delay data is more prominent for Type-B device especially at short, sub 1 ms t-stress. For $t_0 = 1$ μs OTF, Type-B device shows large ΔV_T at sub 1 ms t-stress, which has negligible T dependence. Finally when compared to Type-A devices, Type-B device shows lower power law time exponent (*n*), T activation (E_A), and E_{OX} acceleration factor (Γ_E) at long stress time.

Calculated ΔV_{IT}, ΔV_{HT}, and ΔV_{OT} components and overall ΔV_T for different devices are shown in Fig. 14.10. As mentioned before, ΔN_{IT} is calculated using the H/H_2 RD model, ΔN_{HT} by the 2-energy well model, and ΔN_{OT} using an empirical expression [1]. The prediction of $V_{G,STR}$ and T-dependent data for different devices is shown in Fig. 14.11. The following points can be noted about model prediction. Compared to device D1, device D2 has higher contribution from ΔN_{OT} due to thicker EOT and larger $V_{G,STR}$, while device D3 has higher contribution from ΔN_{HT} due to larger density of preexisting hole traps, which has been independently verified by flicker noise [30]. For all devices, measured 1 μs and 1 ms data can be predicted by *identical* ΔN_{IT} and ΔN_{OT} but different ΔN_{HT} that saturates at long time. The only difference in ΔN_{HT} between $t_0 = 1$ μs and 1 ms measurements confirms that short

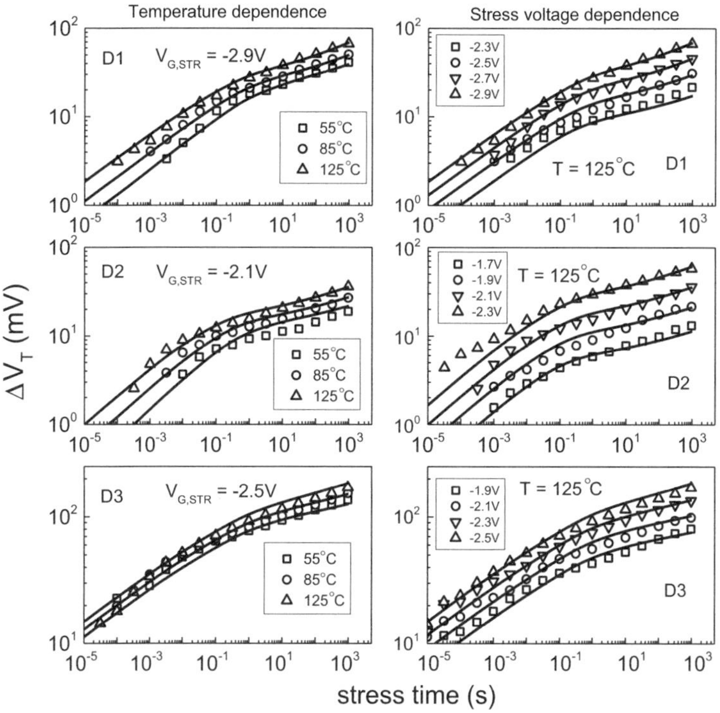

Fig. 14.11 Time evolution of measured ΔV_T during stress at different $V_{G,STR}$ and T for different SiON devices listed in Table 14.2, and prediction by the comprehensive model. Data from [5]

time NBTI is dominated by fast hole trapping in preexisting bulk insulator traps. Larger ΔN_{HT} for Type-B D3 device explains larger difference between 1 μs and 1 ms t_0 delay data at sub 1 ms t-stress. Note that the difference between model prediction and measurement at shorter t-stress, especially for $t_0 = 1$ ms data for Type-B device, can be attributed to lower measured ΔV_T due to t_0 subtraction artifact as mentioned before. As ΔN_{HT} saturates at longer t-stress ($n \sim 0$ in a log–log plot), higher ΔN_{HT} captured for $t_0 = 1$ μs OTF results in lower time exponent n when compared to $t_0 = 1$ ms OTF for all devices. Similarly, larger ΔN_{HT} for Type-B D3 device results in lower n at longer t-stress compared to Type-A D1 and D2 devices. Since ΔN_{HT} is a fast process having weak T activation, Type-B device shows large but weak T activated ΔV_T at sub 1 ms t-stress when ΔV_T is dominated by ΔN_{HT}. At longer t-stress, ΔN_{HT} with lower T activation gets added to ΔN_{IT} having higher T activation and reduces T activation of overall ΔV_T. Since Type-B D3 has larger ΔN_{HT} contribution, it results in lower T activation for ΔV_T at longer t-stress compared to Type-A D1 and D2 devices. Therefore, the proposed framework can explain NBTI degradation for different stress V_G and T, different measurement speed, and on devices having different gate insulator processes.

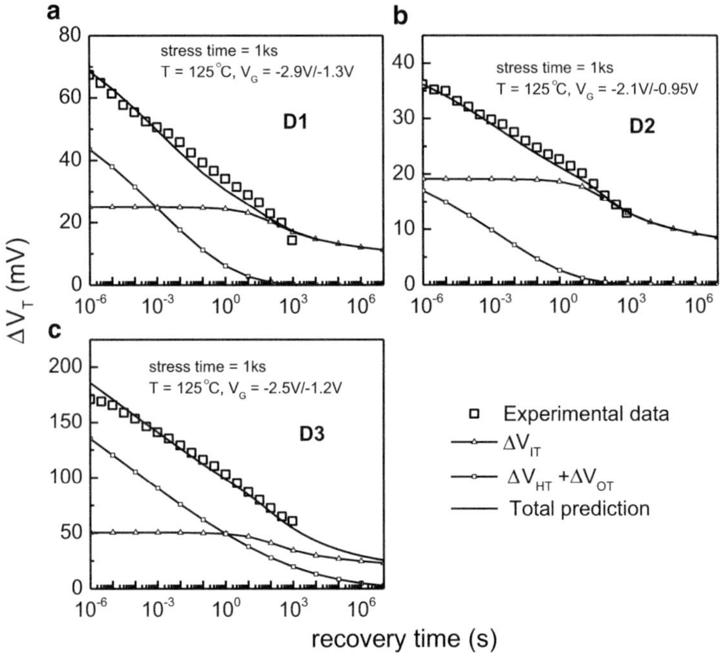

Fig. 14.12 Time evolution of measured ΔV_T during recovery following stress for different SiON devices listed in Table 14.2. Overall model prediction and contributions due to ΔN_{IT}, ΔN_{HT}, and ΔN_{OT} are also shown. Data from [5]

14.3.2 Prediction of Recovery

Figure 14.12 shows the time evolution of ΔV_T in Type-A D1, D2, and Type-B D3 SiON devices, listed in Table 14.2, measured using the $t_0 = 1$ μs OTF method during recovery at $V_{G,REC}$, following NBTI stress at $V_{G,STR}$ for duration t-stress. Calculated ΔV_T recovery is also shown together with ΔN_{HT} and ΔN_{OT} recovery contributions due to fast hole detrapping from preexisting and generated traps, obtained, respectively, by 2-energy well model and empirical expression. Recovery of ΔN_{IT} calculated by the RD model solution is also shown. The magnitudes of ΔN_{IT}, ΔN_{HT}, and ΔN_{OT} at the beginning of recovery are equal to those calculated at the end of stress. For all devices, short time recovery has been accurately predicted by ΔN_{HT} and ΔN_{OT} contributions. The long-term part of recovery is accurately predicted by RD model solution for ΔN_{IT}, with suitable modification in H_2 diffusivity to take into account geometrical effects [15] as discussed hereinafter.

During stress, H_2 diffuses out from Si/SiON interface towards the gate insulator and beyond with a particular diffusivity, and during recovery, it diffuses back towards Si/SiON interface with the same diffusivity. However, while returning, H_2 must "find" a broken Si- bond to passivate and hence has to spend some

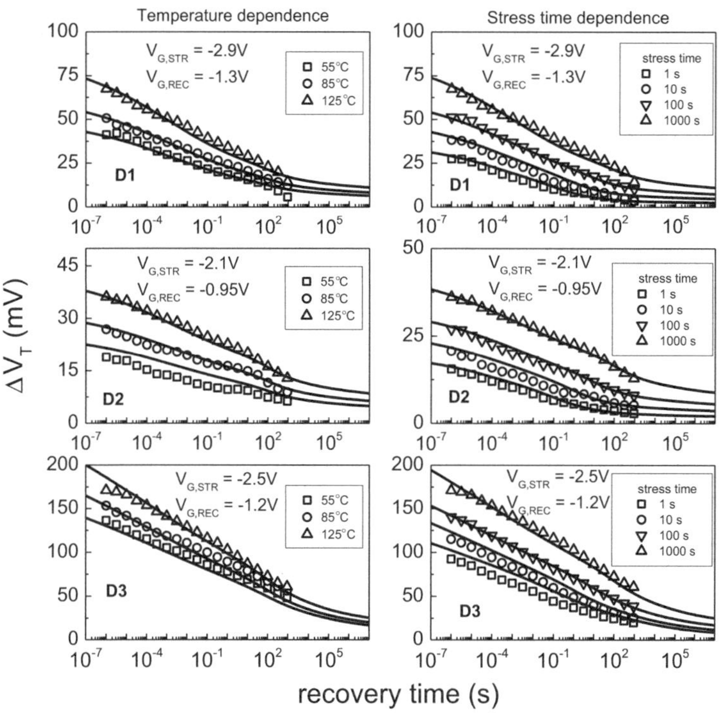

Fig. 14.13 Time evolution of measured ΔV_T during recovery following stress at different stress time and T, for different SiON devices listed in Table 14.2, and prediction by the comprehensive model. Data from [5]

time hopping near the Si/SiON interface. This hopping-induced delay would be larger at longer recovery time when only fewer broken Si- bonds remain available for passivation. Accurate prediction of this phenomenon requires 3D stochastic simulation of large area device, which has a high computational overhead. However, in an approximate 1D implementation, this hopping-induced delay can be *effectively captured* by simulating ΔN_{IT} recovery with a reduction in H_2 diffusivity only during recovery[5] as shown in Fig. 14.12 [1, 5, 15].

To demonstrate the universality of this scheme, Fig. 14.13 shows ΔV_T recovery measured using $t_0 = 1$ μs OTF for different Type-A and Type-B SiON devices,

[5] In [15], ΔN_{IT} recovery is simulated by using conventional H/H_2 RD model with different reduced diffusivity of H_2, and the net solution is arrived at by taking a weighted average of these solutions. A somewhat different approach is used in [5], where poly H/H_2 RD model is used, and the H_2 diffusivity is reduced by the expression $D_{H2} = D_{H20}/(1 + 10\ (t/t\text{-stress}))$ during recovery, where D_{H20} is the diffusivity value used during stress, t-stress is total stress time, and t is recovery time. The later scheme has been used in this chapter to simulate ΔN_{IT} recovery for devices having different gate insulator processes.

described in Table 14.2 of the previous section, stressed at different t-stress and T, and model prediction by the comprehensive framework. It can be clearly seen that the framework can predict ΔV_T recovery for different experimental conditions and on devices having different gate insulator processes.

14.4 Model for Long-Time DC Stress

At longer t-stress, ΔN_{IT} shows power law time dependence with time exponent $n = 1/6$, while ΔN_{HT} saturates ($n \sim 0$ in a log–log plot). Therefore, time evolution of ΔV_T at longer t-stress can be modeled using simpler analytical equations listed in Table 14.5 [6]. Once again, ΔV_T is modeled by using uncorrelated contribution from ΔN_{IT}, ΔN_{HT}, and, when applicable, ΔN_{OT}. The framework will be used to predict NBTI data in both SiON and high-k metal gate (HKMG) p-MOSFETs. The relative contribution of ΔN_{IT} and ΔN_{HT} used to model ΔV_T for different devices will be verified using independent measurements, and their impact on long-time NBTI parameters (n and E_A) will also be discussed.

14.4.1 SiON Device Results

Figure 14.14 shows measured ΔV_T using $t_0 = 1$ μs and 1 ms OTF in different devices stressed at different $V_{G,STR}$ and T. Device details, fixed parameters, and device-dependent adjustable parameters are listed in Table 14.6 [27]. Compared to Type-A device D1, Type-A device D2 has thicker EOT and larger ΔN_{OT} contribution due to the higher $V_{G,STR}$, while Type-B device D3 has higher N% and hence larger ΔN_{HT} contribution. Once again, for a given device and stress condition, $t_0 = 1$ μs and 1 ms OTF data can be modeled using identical ΔN_{IT} and ΔN_{OT} but different ΔN_{HT}, with

Table 14.5 Empirical equations and model parameters for prediction of long-time NBTI degradation during stress

Simplified model for stress			
$\Delta V_{IT} = \frac{q}{C_{OX}} A (V_G - V_{T0})^{\Gamma_{IT}} e^{-\frac{E_{AIT}}{kT}} t^{\frac{1}{6}}$			
Where: $E_{AIT} = \left(\frac{2}{3}\left(E_{A\,kf} - E_{A\,kr}\right) + \frac{E_{ADH2}}{6}\right)$			
$\Delta V_{HT} = \frac{q}{C_{OX}} B (V_G - V_{T0})^{\Gamma_{HT}} e^{-\frac{E_{AHT}}{kT}}$			
$\Delta V_{OT} = \frac{q}{C_{OX}} C \left(1 - e^{\left(-\left(\frac{t}{n}\right)\beta_{OT}\right)}\right)$			
Where: $n = \eta (V_G - V_{T0})^{\frac{\Gamma_{OT}}{\beta_{OT}}} e^{-\frac{E_{AOT}}{kT}}$			
A, B, C andΓ_{IT} $(= \Gamma_{HT})$*are device dependent parameters.*			
$E_{A\,kf} = 0.175$ eV	$E_{A\,kr} = 0.2$ eV	$E_{A\,DH2} = 0.58$ eV	$E_{A\,HT} = 0.03$ eV
$\Gamma_{OT} = 9$	$E_{A\,OT} = 0.15$ eV	$\beta_{OT} = 0.36$	$\eta = 5 \times 10^{12}$

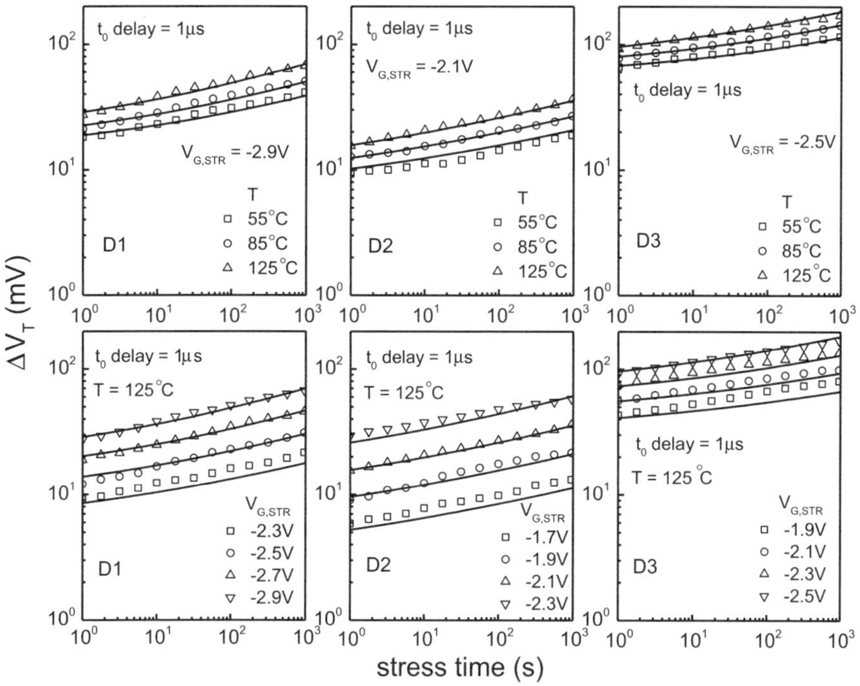

Fig. 14.14 Time evolution of measured ΔV_T during long-time stress at different T and $V_{G,STR}$ for different SiON devices listed in Table 14.6, and prediction by the simplified model described in Table 14.5. Data from [27]

Table 14.6 SiON p-MOSFET device details, universal and device-dependent model parameters used for model prediction of long-time stress data

Device	D1	D2	D3
N%	17	23	43
EOT(Å)	23.5	14	14.5
Variable parameters (device dependent parameters)			
A ($\times 10^{10}$)	2.5	12	110
B ($\times 10^{10}$) ($t_0 = 1$ μs)	0.8	2.6	60
B ($\times 10^{10}$) ($t_0 = 1$ ms)	0.4	1.5	25
C($\times 10^{13}$)	2.4	15	130
$\Gamma_{IT} = \Gamma_{HT}$	4.3	4.3	2.2

higher ΔN_{HT} obtained for $t_0 = 1$ μs OTF measurements. For a given measurement speed, ΔV_T obtained in wide variety of devices can be modeled using only four device-dependent adjustable parameters as shown in Table 14.6 [27].

Figure 14.15 shows (a) time exponent *n* and (b) T activation E_A measured using $t_0 = 1$ μs OTF at longer t-stress in SiON devices having different atomic N% in the gate stack [20]. Note that *n* and E_A show very similar N% dependence; their values remain constant for N ~ 30% and reduce for higher N%. Figure 14.15c shows flicker noise measured in prestress as a function of N% [30]. The magnitude of

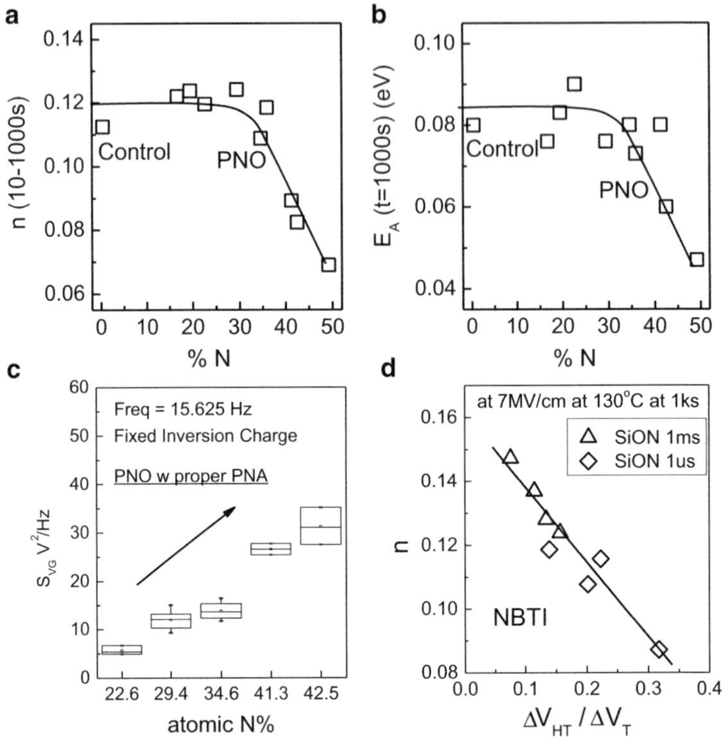

Fig. 14.15 (**a**) Measured time exponent (n) and (**b**) T activation energy (E_A) as a function of atomic N% in the gate insulator for different plasma-nitrided SiON p-MOSFETs. Data from [20]. (**c**) Measured flicker noise data in prestress as a function of N% for these devices [30]. (**d**) Correlation of measured n versus calculated fractional hole trap contribution. Data from [5]. *Lines* are guide to eye

input-referred noise increases with increase in N% and suggests larger density of process-related hole traps. As ΔN_{HT} saturates at longer t-stress and the saturated ΔN_{HT} has smaller E_A ($\sim$0.03 eV, refer to Table 14.6), larger relative contribution of ΔN_{HT}, when added to ΔN_{IT} (which has long-time $n \sim 1/6$ and $E_A \sim 0.1$ V, refer to Table 14.6), reduces the n and E_A of overall ΔV_T as shown in Fig. 14.15a, b. Figure 14.15d shows long-time n measured using $t_0 = 1$ μs and 1 ms OTF as a function of saturated ΔV_{HT} for different devices. Note that ΔV_{HT} (normalized to ΔV_T) has been obtained from modeling long-time ΔV_T data for different devices, similar to that shown in Fig. 14.14. A universal correlation has been obtained for all devices and across different measurement speed. This is consistent with hole trapping being a fast process, and faster OTF captures larger ΔN_{HT} and therefore reduces long-time n, more so for devices having higher N% and when measured using a faster OTF method which captures early part of the degradation. Note that the reduction of n and E_A with higher N% can only be modeled using uncorrelated

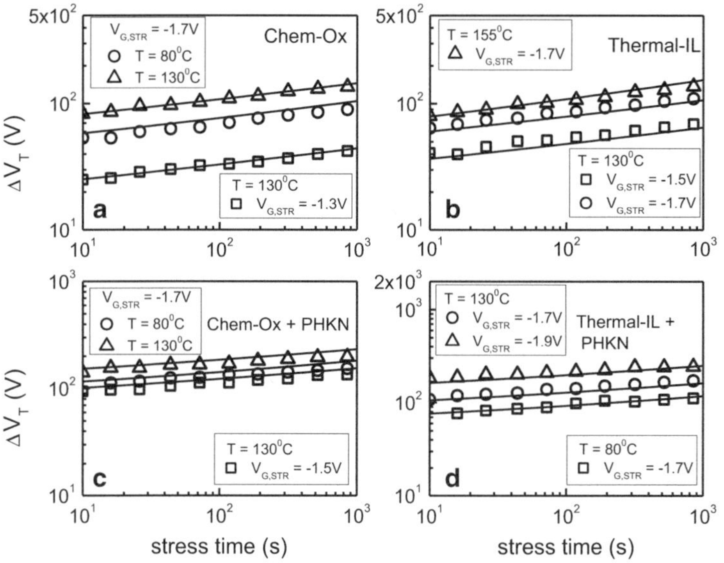

Fig. 14.16 Time evolution of measured ΔV_T during long-time stress at different T and $V_{G,STR}$ for different HKMG devices listed in Table 14.7, and prediction by the simplified model described in Table 14.5. Data from [16]

ΔN_{IT} and ΔN_{HT}, with larger relative increase in ΔN_{HT} with increase in N%. Note that the 2-stage model assumes strongly correlated ΔN_{IT} and ΔN_{HT} [21] and therefore cannot explain such process dependence of NBTI parameters.

14.4.2 HKMG Device Results

The framework of uncorrelated ΔN_{IT} and ΔN_{HT} can also predict measured ΔV_T data in HKMG devices. Figure 14.16 shows time evolution of ΔV_T at longer t-stress for different devices and different stress V_G and T, measured using single-point drop-down method with 1 ms delay [16]. The model described in Table 14.5 has been used to predict measured data as also shown in Fig. 14.16. Table 14.7 describes the device details and fixed and device-dependent model parameters. The HKMG gate stacks have (A) chemical oxide (Chem-Ox) [37] and (B) Rapid Thermal Process (RTP)-based Thermal Oxide (TO) [38] as the interlayer (IL), Atomic Layer Deposition (ALD)-based HfO_2 HK and TiN MG. The stacks were subjected to post HK nitridation (PHKN) [39] that results in (C) nitrided Chem-Ox IL and (D) nitrided TO IL stacks. Note that NBTI in HKMG devices is governed by the SiO(N) interlayer (IL) and the HK layer simply acts as a voltage divider [40]. Due to low-voltage drop across IL as $V_{G,STR}$ gets divided between IL and HK layers, and due to large voltage de-acceleration factor for ΔN_{OT} [15], the contribution due to ΔN_{OT}

Table 14.7 HKMG p-MOSFET device details, universal and device-dependent model parameters used for model prediction of long-time stress data

	Chem-OX	Thermal-IL	Chem-Ox + PHKN	Thermal-IL + PHKN
EOT (Å)	10.3	9.1	8.6	8
A	3e13	4.5e13	4.5e13	9.5e13
$\Gamma_{IT} = \Gamma_{HT}$	2.6	3	1.9	2.5
B	5e12	7e12	2e13	3.8e13

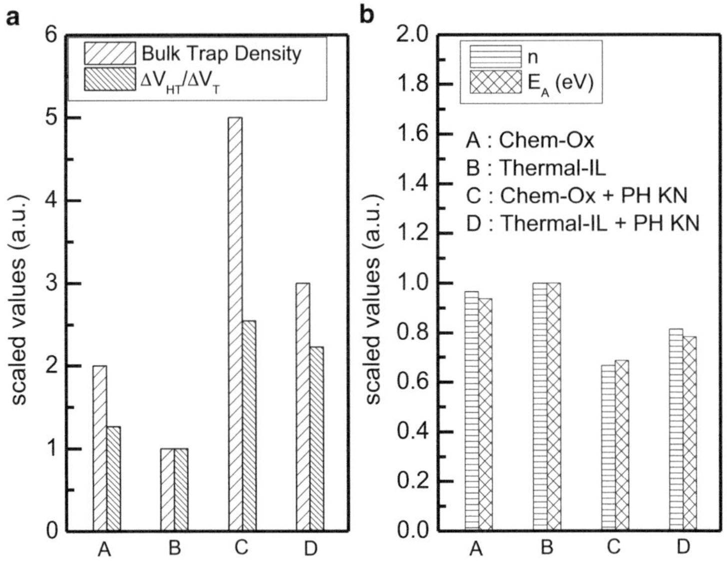

Fig. 14.17 (**a**) Measured prestress hole trap density using flicker noise and calculated ΔN_{HT} contribution, as well as (**b**) measured time exponent *n* and T activation E_A, for different HKMG devices listed in Table 14.7. Data from [16]

has been found to be negligible in these HKMG stacks. Therefore, long-time ΔV_T for different stress V_G and T can be modeled using only three device-dependent parameters as shown in Table 14.7.

Figure 14.17a shows process-induced IL hole trap density measured using flicker noise in prestress in stacks A through D [16]. Calculated saturated ΔN_{HT} contribution for these stacks, using the model described in Table 14.5, and parameters in Table 14.7 are also shown, extracted at a fixed E_{OX} of 7 MV/cm for all devices. Note that higher trap density is observed for Chem-Ox IL stack A compared to TO IL stack B and is consistent with higher ΔN_{HT} contribution calculated for these stacks. Larger trap density is measured and larger ΔN_{HT} is calculated for nitrided stacks C and D compared to their non-nitrided counterparts A and B, respectively, and is consistent with SiON results [30]. However, nitrided Chem-Ox IL stack C shows higher trap density compared to TO IL stack D, which is also consistent with calculated relative ΔN_{HT} contribution in these stacks.

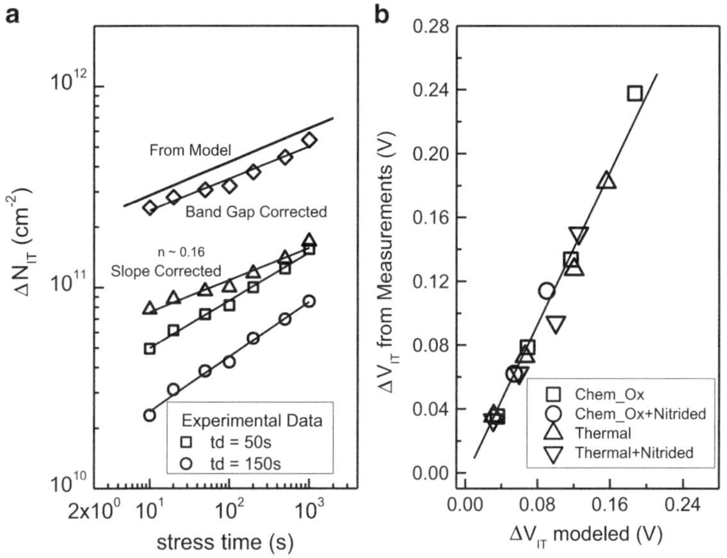

Fig. 14.18 (**a**) Time evolution of measured ΔV_{IT} (from ΔN_{IT}) during stress, measured using DCIV method with different delay. Corrections due to delay and bandgap differences are shown (see text). ΔV_{IT} contribution calculated using the model is also shown. (**b**) Comparison of measured and calculated ΔV_{IT} for different HKMG devices listed in Table 14.7. Data from [5]

Figure 14.17b shows measured n and E_A at long t-stress for these stacks. As mentioned before, ΔN_{HT} saturates at long t-stress and has weak T activation. Therefore, devices having larger ΔN_{HT} contribution would show lower n and E_A of overall ΔV_T measured at long t-stress. Note that Chem-Ox stacks A and C show lower n and E_A compared to TO stacks B and D, respectively, while nitrided stacks C and D show lower n and E_A compared to their non-nitrided counterparts A and B, and these relative trends are consistent with relative magnitude of ΔN_{HT} shown in Fig. 14.17a. Therefore, impact of gate insulator processes on NBTI parameters, especially the role of ΔN_{HT} contribution, is independently verified.

Figure 14.18a shows time evolution of ΔV_{IT}, obtained from direct measurement of ΔN_{IT} using the DCIV method [5]. DCIV measurements probe ΔN_{IT} only since ΔN_{OT} is negligible in HKMG stacks as discussed above. Note that DCIV measurement [11] is done in the conventional MSM mode and hence suffers from recovery artifacts due to measurement delay [12]. As shown in Fig. 14.18a, ΔV_{IT} recovery results in lower magnitude and higher time exponent n when measured using DCIV with larger delay. Therefore, measured ΔV_{IT} should be corrected for delay, which has been done by using the universality of ΔN_{IT} recovery [41]. Delay correction results in increase in ΔV_{IT} magnitude and reduction in the time exponent to $n \sim 1/6$, the latter being consistent with RD model prediction. Moreover, DCIV scans interface traps at a part of the bandgap centered around the midgap [11], which also needs to be corrected and is shown in Fig. 14.18a. The final corrected ΔV_{IT}

from DCIV measurements is quite close to the ΔV_{IT} contribution of overall ΔV_T obtained by the model of Table 14.5 and parameters listed in Table 14.7, as also shown in Fig. 14.18a. Similar exercise has been done on different HKMG devices A through D, described Table 14.7. Figure 14.18b correlates ΔV_{IT} from DCIV measurements (after correction) to ΔV_{IT} used in the ΔV_T model for different $V_{G,STR}$. Close correlation of predicted ΔV_{IT} to measured values justifies the correctness of the ΔV_{IT} model and model parameters. Therefore, underlying ΔN_{IT} and ΔN_{HT} contributions of overall ΔV_T have been verified by other independent measurements and justify the validity of the proposed framework.

14.5 Prediction of AC Degradation

In the previous section, DC NBTI has been modeled by using uncorrelated contribution from trap generation and trapping. It has been shown that degradation during NBTI stress and recovery of degradation after the stress is removed are due to fast trapping and detrapping of holes and relatively slower generation and passivation of interface traps. As degradation during the on phase of the AC pulse would recover during the off phase, both factors influence AC NBTI degradation as discussed below.

Figure 14.19 shows ΔN_{IT} measured using the DCIV method and plotted as a function of AC pulse duty cycle (PDC) [1, 5]. Both DC and AC data are measured for identical t-stress, which implies that the effective t-stress for AC stress is lower than DC and depends on PDC of the gate pulse. Note that measured data from

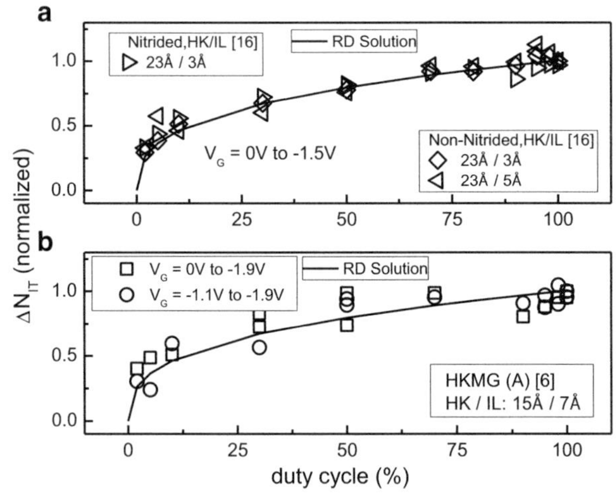

Fig. 14.19 AC pulse duty cycle dependence of ΔN_{IT}, measured using DCIV method in different HKMG devices, and also for different pulse low-level conditions. RD model solution is also shown. Data from [5]

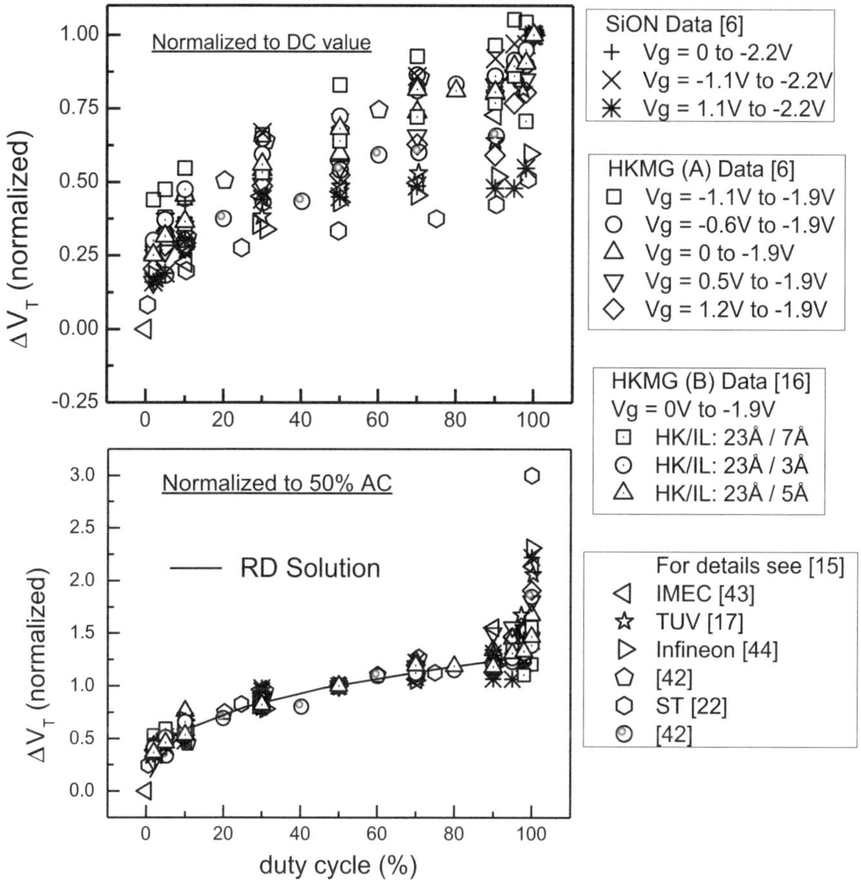

Fig. 14.20 AC pulse duty cycle dependence of ΔV_T, normalized to (***top***) DC and (***bottom***) 50% AC data for each dataset, obtained from different sources (see text)

different HKMG devices and pulse low bias conditions, when normalized to their respective DC values, demonstrate a universal PDC dependence as shown. The solid line represents the prediction of PDC dependence of ΔN_{IT} as obtained using RD model simulation, with equations and parameters shown in Table 14.1. Note that RD model can accurately predict the PDC dependence of interface trap generation during AC stress.

Figure 14.20a shows measured ΔV_T for AC stress versus PDC and normalized to DC, obtained from different devices, for different pulse low conditions, and also from different sources [1, 5, 6, 15, 17, 22, 42–44]. Unlike the universality observed in the PDC dependence of ΔN_{IT} as shown in Fig. 14.16, PDC-dependent ΔV_T data from various sources demonstrate a large scatter. Interestingly, as shown in Fig. 14.20b, same data when re-normalized to the 50% PDC value for each device [15] demonstrate universality up to ∼80% PDC and large scatter for higher PDC

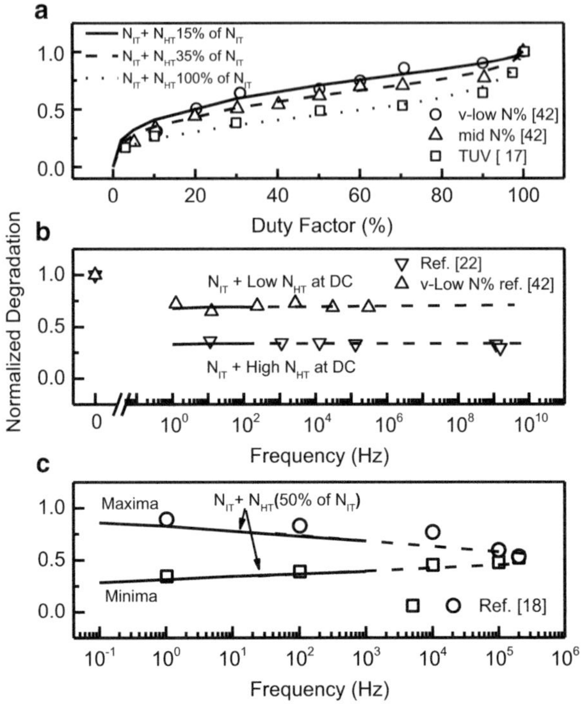

Fig. 14.21 Measured (**a**) AC pulse duty cycle dependence of ΔV_T for different devices obtained at pulse minima (end of last full cycle) and AC pulse frequency dependence of ΔV_T for (**b**) slow measurements on different devices obtained at pulse minima and (**c**) fast measurements obtained at pulse maxima (end of last half cycle) and minima. Model predictions are shown using *lines*

close to DC. The universality of the re-normalized data up to ~80% PDC can be predicted by RD model solution as shown by the solid line. Since ΔN_{IT} shows universality for all values of PDC and ΔV_T shows universality up to ~80% PDC, and since this universal behavior can be predicted by RD model solution for ΔN_{IT}, it can be concluded that the spread in ΔV_T for higher PDC close to DC, as shown in Fig. 14.20b, or the large scatter in the entire PDC-dependent data as shown in Fig. 14.20a, is due to difference in ΔN_{HT} between different devices. As discussed in detail in [1, 15], the spread in data from different sources can be ascribed to differences in device quality, mostly preexisting bulk trap density, and differences in measurement speed between DC and AC experiments, which result in different relative contributions of ΔN_{IT} and ΔN_{HT}. Since hole trapping and detrapping are fast processes, trapped holes during the pulse on phase get detrapped during the pulse off phase, unless PDC is very large and hole detrapping gets suppressed due to insufficient time. Therefore, the impact of ΔN_{HT} becomes appreciable only for large PDC close to DC.

Figure 14.21a shows AC PDC dependence of ΔV_T measured in devices having different N% in the gate stack and normalized to the DC value [17, 42]. ΔV_T shows a typical "S"-shaped dependence with PDC, although with different AC/DC ratio for different devices. Also shown in Fig. 14.21a is the overall model prediction consisting of RD model solution for ΔN_{IT} and 2-energy well model solution for ΔN_{HT} recorded at the minima of the AC gate pulse (end of last full cycle). Model parameters are identical to those used to predict DC stress and recovery data, and ΔN_{OT} is assumed to be negligible due to the use of relatively lower $V_{G,STR}$ as discussed before. The model can predict experimental data with relatively different ΔN_{IT} and ΔN_{HT} for different devices, where the relative magnitudes of ΔN_{IT} and ΔN_{HT} have been adjusted *only at DC*, and verifies the validity of the proposed framework of uncorrelated ΔN_{IT} and ΔN_{HT}.

Figure 14.21b, c shows ΔV_T as a function of AC pulse frequency (f) for different devices and measurement conditions. Data in [22] and [42] are measured using slower measurement method and show f independence but with different AC/DC ratio, as shown in Fig. 14.21b. Data in [18] are measured using faster method and show f dependence for data recorded at maximum of gate pulse (end of last half cycle), while f dependence is negligible for data recorded at the pulse minima (end of last full cycle), as shown in Fig. 14.21c. As hole trapping and detrapping are fast processes as discussed above, ΔV_T at 50% PDC obtained at pulse minima is dominated by ΔN_{IT}. Therefore, the difference in AC/DC ratio between [22] and [42] is due to difference in ΔN_{HT} measured at DC stress. Also note that AC f independent degradation is a natural prediction of the RD model for interface traps, as discussed in [45]. Indeed, as shown in Fig. 14.21b, the f independence can be captured by AC RD model solution for ΔN_{IT}, with relatively different ΔN_{HT} added *only to DC* data to account for process-dependent preexisting hole traps in the gate insulator bulk [1, 5].

However, as shown in Fig. 14.21c, some amount of ΔN_{HT} would be captured during AC stress, especially at lower f and when measured at the pulse maxima using a faster method [18]. ΔV_T measured at the maximum and minimum of the AC pulse, respectively, show strong and weak f dependence, which can be predicted by RD model solution for ΔN_{IT} combined with 2-energy well model solution for ΔN_{HT}, now applied for both DC and AC data, with identical model parameters used for prediction of DC stress and recovery [1, 5]. Therefore in addition to PDC dependence, the framework of uncorrelated trapping and trap generation can explain f dependence of AC NBTI for different measurement conditions and on devices having different gate insulator processes.

Note that the proposed framework assumes completely uncorrelated ΔN_{IT} and ΔN_{HT} to explain AC NBTI experiments, with model equations and parameters that are fully consistent with DC NBTI. Although not explicitly discussed in this chapter, it has been described in detail in [1, 5] that the 2-stage model [21], which assumes strongly coupled ΔN_{IT} and ΔN_{HT}, cannot explain the AC pulse duty cycle and frequency-dependent NBTI data with consistent set of model parameters. As mentioned earlier in this chapter and explained elsewhere [1, 5], the 2-stage model also cannot explain DC stress and recovery with consistent set of model parameters. Finally, the recently proposed multistate-model framework [46, 47] exclusively relies on hole trapping-detrapping to explain AC NBTI, which is not consistent with strong and direct experimental evidence of generation and recovery of interface traps during DC as well as AC NBTI. Therefore, the multistate-model framework of [46, 47] is not physically justifiable and needs modification to become consistent with NBTI experimental features.

14.6 Summary

NBTI degradation and recovery during DC and AC experiments are modeled using uncorrelated contribution from interface trap generation and hole trapping in preexisting and generated bulk insulator traps. Interface trap generation and recovery are relatively slower processes and have been modeled using the poly H/H_2 Reaction–Diffusion (RD) model. Hole trapping and detrapping are relatively faster processes and have been modeled using the 2-energy well model. Generation of bulk traps has been found to be negligible for thinner gate insulators and lower stress gate bias and is modeled by empirical equations. Model equations are described in detail and calibrated model parameters have been listed. The framework can predict NBTI degradation and recovery during and after DC stress, NBTI degradation for very long stress time, dependence of NBTI on AC pulse duty cycle and frequency, impact of measurement speed on DC and AC degradation, as well as the impact of gate insulator processes on DC and AC NBTI with consistent set of model parameters. It is important to note that the RD model can predict interface trap generation and recovery under such wide range of experimental conditions and devices with only two adjustable parameters. The proposed framework has been successfully demonstrated to explain NBTI in both SiON and HKMG devices, and the validity of underlying trap generation and trapping components have been verified by independent experiments.

The author would like to acknowledge contribution by Nilesh Goel, Kaustubh Joshi, and Subhadeep Mukhopadhyay towards preparation of this chapter. Special thanks to M. A. Alam, Sujay Desai, and A. E. Islam for useful discussions.

References

1. S. Mahapatra, N. Goel, S. Desai, S. Gupta, B. Jose, S. Mukhopadhyay, K. Joshi, A. Jain, A. E. Islam, and M. A. Alam, IEEE Trans. Electron Devices, 60, 901 (2013)
2. S. Chakravarthi, A. Krishnan, V. Reddy, C. F. Machala, and S. Krishnan, Proc. Int. Reliab. Phys. Symp., 273 (2004)
3. A. T. Krishnan, C. Chancellor, S. Chakravarthi, P. E. Nicollian, V. Reddy, A. Varghese, R. B. Khamankar, S. Krishnan, and L. Levitov, Proc. Int. Electron Dev. Meet., 688 (2005).
4. J. R. Jameson, W. Harrison, P. B. Griffin, J. D. Plummer, and Y. Nishi, J. Appl. Phys., 100, 124104–1 (2006)
5. S. Desai, S. Mukhopadhyay, N. Goel, N. Nanaware, B. Jose, K. Joshi and S. Mahapatra, Proc. Int. Reliab. Phys. Symp., XT.2.1-XT.2.11 (2013)

6. K. Joshi, S. Mukhopadhyay, N. Goel and S. Mahapatra, Proc. Int. Reliab. Phys. Symp., 5A.3.1 (2012)
7. D. S. Ang, and S. Wang, IEEE Electron Dev. Lett., vol. 27, 914–916 (2006)
8. D. Ielmini, M. Manigrasso, F. Gattel, and G. Valentini, Proc. Int. Reliab. Phys. Symp., 26–32, (2009)
9. H. Reisinger, T. Grasser, K. Ermisch, H. Nielen, W. Gustin, C. Schlunder, Proc. Int. Reliab. Phys. Symp., 6A.1.1-6A.1.8 (2011)
10. G. Groeseneken, H. E. Maes, N. Beltran,R. F. De Keersmaecker, , IEEE Trans. Electron Devices, 31, 42–53 (1984)
11. A. Neugroschel, G. Bersuker, R. Choi, C. Cochrane, P. Lenahan, D. Heh, C. Young, C. Y. Kang, B. H. Lee, R. Jammy, Proc. Int. Electron Dev. Meet., 1–4, (2006)
12. D. Varghese, D. Saha, S. Mahapatra, K. Ahmed, F. Nouri, and M. A. Alam, Proc. Int. Electron Dev. Meet., 684–687 (2005)
13. S. Mahapatra, K. Ahmed, D. Varghese, A. E. Islam, G. Gupta, L. Madhav, D. Saha, M. A. Alam, Proc. Int. Rel. Phys. Symp., 1–9 (2007)
14. S. Mahapatra, P. Kumar, and M. A. Alam, IEEE Trans. Electron Devices, vol. 51, 1371–1379, (2004)
15. S. Mahapatra, A. Islam, S. Deora, V. Maheta, K. Joshi, A. Jain, and M. Alam, Proc. Int. Rel. Phys. Symp., 6A.3.1 –6A.3.10 (2011)
16. K. Joshi, S. Hung, S. Mukhopadhyay, V. Chaudhary, N. Nanaware, B. Rajmohnan, T. Sato, M. Bevan, A. Wei, A. Noori, B. Mc.Dougal, C. Ni, G. Saheli, C. Lazik, P. Liu, D. Chu, L. Date, S. Datta, A. Brand, J Swenberg, and S. Mahapatra, Proc. Int. Reliab. Phys. Symp., (2013)
17. T. Grasser, B. Kaczer, and W. Goes, Proc. Int. Reliab. Phys. Symp., 28 (2008)
18. C. Shen, M. F. Li, C. E. Foo, T. Yang, D. M. Huang,A. Yap, G. S. Samudra, and Y. C. Yeo, Proc. Int. Electron Dev. Meet., 12.5.1, (2006)
19. E. N. Kumar, V. D. Maheta, S. Purawat, A. E. Islam, C. Olsen, K. Ahmed, M. Alam and S. Mahapatra, Proc. Int. Electron Dev. Meet., 809 (2007)
20. V.D. Maheta, C. Olsen, K. Ahmed, and Souvik Mahapatra, IEEE Trans. Electron Devices, vol. 55, 1630–1638, (2008)
21. T. Grasser, B. Kaczer, W. Goes, T. Aichinger, P. Hehenberger and M. Nelhiebel, Proc. Int. Reliab. Phys. Symp. 33, (2009)
22. V. Huard, Proc. Int. Reliab. Phys. Symp., 33–42, (2010)
23. S. Gupta, IRPS 2012 S. Gupta, B. Jose, K. Joshi, A. Jain, M. A. Alam, and S. Mahapatra, Proc. Int. Reliab. Phys. Symp., XT.3.1-XT.3.6., (2012)
24. M. A. Alam, Jeff Bude, and A. Ghetti, Proc Int. Reliab. Phy. Symp., 21–26, (2000)
25. A. Islam, G. Gupta, S. Mahapatra, A. T. Krishnan, K. Ahmed, F. Nouri, A. Oates, and M. A. Alam, Proc. Int. Electron Dev. Meet., 1 –4, (2006)
26. A. E. Islam, H. Kufluoglu, D. Varghese, S. Mahapatra, and M. A. Alam, IEEE Trans. Electron Devices, 54, 2143(2007)
27. T. Naphade, N. Goel, P. R. Nair and S. Mahapatra, Proc. Int. Reliab. Phys. Symp., XT.5.1–XT.5.11 (2013)
28. A. Haggag, G. Anderson, S. Parohar, D. Burnett, G. Abeln, J. Higman and M. Moosa, Proc. Int. Reliab. Phys. Symp., 452–456,(2007)
29. C. L. Chen, Y. M. Lin, C. J. Wang, and K. Wu, Proc. Int. Reliab. Phys. Symp., 704–705, (2005)
30. G. Kapila, N. Goyal, V. D. Maheta, C. Olsen, K. Ahmed, and S. Mahapatra, Proc. Int. Electron Dev. Meet., 1–4, (2008)
31. M. L. Reed and J. D. Plummer, J. Appl. Phys., vol. 63, 5776–5793, (1988)
32. S. Mahapatra, FEOL and BEOL process dependence of NBTI, in *Bias Temperature Instability for Devices and Circuits*, ed. by T. Grasser (Springer, Heidelberg, 2013).
33. H. Miki, N. Tega, Zhibin Ren, C.P. D'Emic, Yu Zhu, D.J. Frank, M. A. Guillorn, Dae-Gyu Parks, W. Haensch, K. Torii, Proc. Int. Electron Dev. Meet., 28.1.1-28.1.4, (2010)
34. S. Takagi, N. Yasuda, A. Toriumi, IEEE Trans. Electron Devices, 46, 335 (1999)
35. S. Rangan, N. Mielke, and E. C. C. Yeh, Proc. Int. Electron Dev. Meet., 341–344 (2003)

36. A. E. Islam, V. D. Maheta, H. Das, S. Mahapatra and M. A. Alam, Proc. Int. Reliab. Phys. Symp., 87, (2008)
37. K. Choi, H. Jagannathan, C. Choi, L. Edge, T. Ando, M. Frank, P. Jamison, M. Wang, E. Cartier, S. Zafar, J. Bruley, A. Kerber, B. Linder, A. Callegari, Q. Yang, S. Brown, J. Stathis, J. Iacoponi, V. Paruchuri and V. Narayanan, Proc. Symp. on VLSI Technology, 138–139, (2009)
38. M. J. Bevan, R. Curtis, T. Guarini, W. Liu, S.C.H. Hung, H. Graoui, Proc. of Advanced Thermal Processing of Semiconductors (RTP), 154–156, (2010)
39. C. Olsen, US Patent 017 596 1A1, (2004)
40. Cartier, IEDM 2011 E. Cartier, A. Kerber, T. Ando, M. M. Frank, K. Choi, S. Krishnan, B. Linder, K. Zhao, F. Monsieur, J. Stathis, and V. Narayanan, Proc. Int. Electron Dev. Meet., 18.4.1-18.4.4, (2011)
41. T. Grasser, W. Gos, V. Sverdlov, B. Kaczer, Proc. Int. Reliab. Phys. Symp., 268–280, (2007)
42. A. E. Islam, S. Mahapatra, S. Deora, V. D. Maheta, and M. A. Alam, Proc. Int. Electron Dev. Meet., 1–4. (2009)
43. R. Fernández, B. Kaczer, A. Nackaerts, S. Demuynck,R. Rodríguez, M. Nafría, and G. Groeseneken, Proc. Int. Electron Dev. Meet., 12.6.1, (2006)
44. H. Reisinger, T. Grasser, W. Gustin and C. Schlünder, Proc. Int. Reliab. Phys. Symp., 7–15, (2010).
45. M. A. Alam, Proc. Int. Electron Dev. Meet., 14.4.1–14.4.4, (2003)
46. T. Grasser, B. Kaczer, H. Reisinger, P.-J. Wagner, M. Toledano-Luque, Proc. Int. Reliab. Phys. Symp., 7–15, XT.8.1–XT.8.7, (2012)
47. T. Grasser, H. Reisinger, K. Rott, M. Toledano-Luque, B. Kaczer, Proc. Int. Electron Dev. Meet., 470–473, (2012)

Chapter 15
On the Microscopic Limit of the RD Model

Franz Schanovsky and Tibor Grasser

Abstract The popular reaction–diffusion model for the negative bias temperature instability is discussed from the viewpoint of stochastic chemical kinetics. We present a microscopic formulation of the reaction–diffusion model based on the reaction–diffusion master equation and solve it using the stochastic simulation algorithm. The calculations are compared to the macroscopic version as well as established experimental data. The degradation predicted by the microscopic reaction–diffusion model strongly deviates from the macroscopic version and the experimentally observed behavior. Those deviations are explained as necessary consequences of the physical processes involved. The presented results show the impact of the unphysical assumptions in the reaction–diffusion model. Further, we generally question the suitability of the mathematical framework of reaction rate equations for a reactive-diffusive system at the given particle densities.

15.1 Introduction

The first model for NBTI was put forward by Jeppson and Svensson in 1977 [1]. This model was based on the following ideas, which are illustrated in Fig. 15.1. Due to the lattice mismatch between silicon and silicon dioxide, some of the silicon atoms do not have an oxygen neighbor. A silicon atom in this situation has one unpaired valence electron, which is called a dangling bond. This dangling bond is visible in electronic measurements as it gives rise to states within the band-gap [2]. During the manufacturing process the wafer is exposed to a hydrogen-rich atmosphere so that hydrogen atoms can penetrate through the oxide and passivate the silicon dangling bonds, leading to a removal of the band-gap states.

F. Schanovsky (✉) • T. Grasser
Institute for Microelectronics, Gußhausstraße 27-29/E360, 1040 Vienna, Austria
e-mail: schanovsky@iue.tuwien.ac.at; grasser@iue.tuwien.ac.at

T. Grasser (ed.), *Bias Temperature Instability for Devices and Circuits*,
DOI 10.1007/978-1-4614-7909-3_15, © Springer Science+Business Media New York 2014

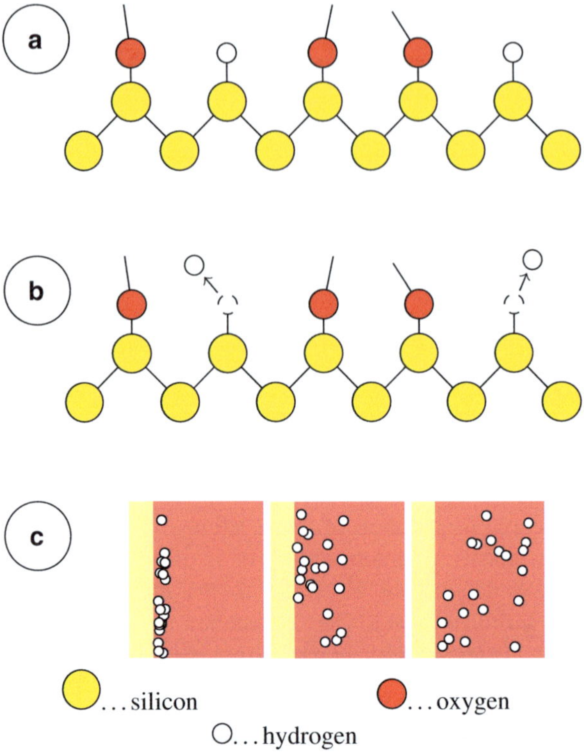

Fig. 15.1 The basic concept behind the reaction–diffusion model for NBTI. (**a**) Silicon dangling bonds at the Si–SiO_2 interface are initially passivated by hydrogen atoms. (**b**) During stress, hydrogen atoms are liberated leaving behind the unpassivated silicon dangling bonds which degrade the device properties. (**c**) The time evolution is determined by the depopulation of the interface due to the flux of hydrogen into the oxide

During stress, the presence of holes at the interface and the increased temperature leads to a liberation of the hydrogen atoms. The remaining silicon dangling bonds become electrically active carrier traps. According to the model, the depassivation and repassivation of dangling bonds at the interface reach an equilibrium in a very short time [3, 4], and it is the constant flux of hydrogen atoms (or some hydrogenic species) away from the interface that determines the temporal evolution of the degradation. Because of the two proposed stages—the electrochemical reaction at the interface and the subsequent diffusion of the hydrogenic species—this model bears the name *reaction–diffusion* (RD) model.

The mathematical framework of the model is based on a macroscopic description using a rate equation for the interface reaction and a Fickian diffusion equation for the motion of the hydrogen in the oxide. Central actors are the density of depassivated silicon dangling bonds at the interface $N_{\mathrm{it}} = [\mathrm{Si}^*]$, and the concentration of hydrogen in the oxide $H = [\mathrm{H}](x,t)$ and at the interface $H_{\mathrm{it}} = [\mathrm{H}](0,t)$.

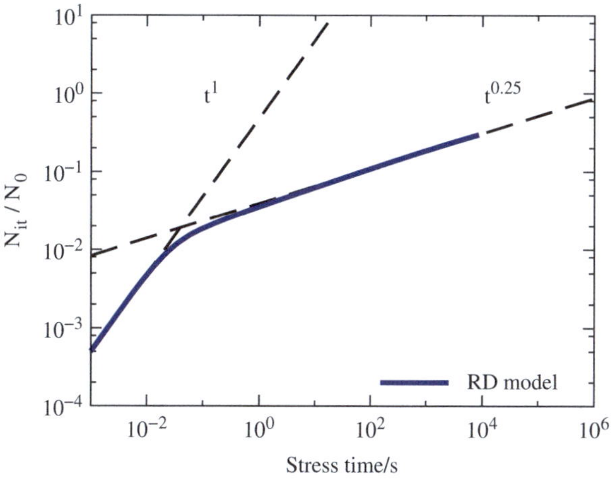

Fig. 15.2 Basic features of the degradation predicted by the RD model for NBTI. In the initial phase, the depassivation reaction with rate k_f dominates, giving rise to a degradation that increases linearly with time. After the depassivation and repassivation reactions have reached an equilibrium, the degradation is determined by the flux of hydrogen, which gives rise to a power-law with an exponent of $1/4$

During degradation, a fraction N_{it} of the initially passivated silicon dangling bonds $N_0 = [\mathrm{SiH}]_0$ is depassivated according to

$$\frac{\partial N_{it}}{\partial t} = k_f(N_0 - N_{it}) - k_r N_{it} H_{it}, \tag{15.1}$$

with the depassivation (forward) rate k_f and the repassivation (reverse) rate k_r. The hydrogen liberated at the interface then diffuses into the oxide as

$$\frac{\partial H}{\partial t} = -D\frac{\partial^2 H}{\partial x^2} \tag{15.2}$$

with the diffusion coefficient D. The RD model became popular among reliability engineers as it features a simple mathematical description and a small set of parameters which have a sound physical interpretation. Most importantly, as shown in Fig. 15.2, this model predicts a constant-stress degradation that initially grows linearly with time and then follows a power-law of the form [3, 4]

$$N_{it}(t) = \sqrt{\frac{k_f N_0}{2k_r}}(Dt)^{1/4}. \tag{15.3}$$

This power-law degradation corresponded well with experimental results of the 1970s.

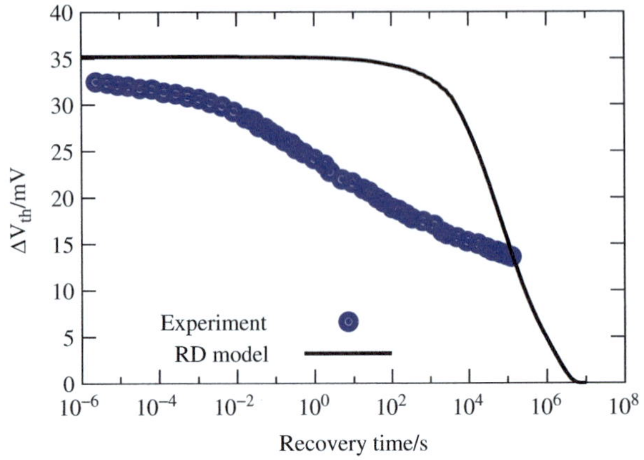

Fig. 15.3 Typical recovery trace as predicted by the RD model for NBTI using (15.4)–(15.6), which is similar for all variants of the RD mechanism. The comparison with experimental data [7] shows that the RD predicted recovery occurs much too late and proceeds much too fast

In later experiments, power-law exponents were found that differed from the $1/4$ prediction of the model. These findings led to a modification of the original RD model to account for different diffusing species such as H_2 [5]. For almost four decades, the reaction–diffusion idea was the unquestioned standard interpretation for NBTI until around 2005 NBT recovery moved into the focus of the scientific attention. The experiments showed that NBTI recovery starts immediately (even before a microsecond) after the removal of stress and extended over several decades, continuing even after more than 10^5 s [6, 7]. This behavior stands in strong contrast to the predictions of the RD model, which predicts a recovery that proceeds within four decades, centering around the duration of the preceding stress phase [8, 9]. A comparison of a typical experimental NBT recovery trace and the corresponding prediction of the RD model is shown in Fig. 15.3. Several extensions to the RD model have been put forward, such as dispersive transport of the hydrogenic species [4, 6], but none could give the observed experimental behavior. The current state-of-the-art RD-based modeling supplements the RD theory with empirical hole-trapping expressions. It is assumed that short-time (1 s) degradation and recovery is dominated by hole trapping into oxide and interface defects, while the long-term degradation and recovery are determined by the RD mechanism [10–13]. The RD theory employed in these modeling efforts is the *modified* RD model [14–16] that has been developed as an extension of the classical RD models and explicitly considers diffusion of H and H_2 and their interconversion reactions. Classical models assume an instantaneous transition between the liberated interfacial hydrogen and the diffusing species, usually H_2 [5]. The reactions present in the modified RD model are the interface reaction $SiH \rightleftharpoons Si^* + H$, the dimerization reaction $2H \rightleftharpoons H_2$, and the diffusion of both species. The mathematical framework is an extension of (15.1) and (15.2) [15, 16],

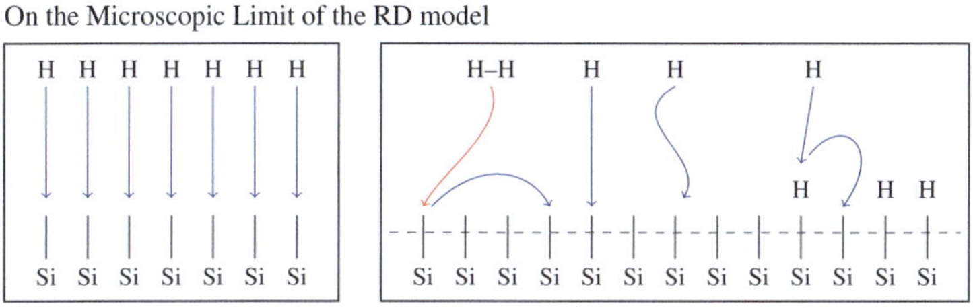

Fig. 15.4 (*Left*) According to Mahapatra et al. [11], the inability of the macroscopic reaction–diffusion model (15.4)–(15.6) to predict the experimentally observed NBTI recovery is due to the one-dimensional description of the diffusive motion which makes it too easy for the hydrogen to find its dangling bond. (*Right*) A correct description of the three-dimensional atomic motion, so the argument, leads to much richer repassivation kinetics and thus to a distribution of repassivation times

$$\frac{\partial N_{\mathrm{it}}}{\partial t} = k_{\mathrm{f}}(N_0 - N_{\mathrm{it}}) - k_{\mathrm{r}} N_{\mathrm{it}} H_{\mathrm{it}}, \tag{15.4}$$

$$\frac{\partial H}{\partial t} = -D\frac{\partial^2 H}{\partial x^2} - k_H H^2 + k_{H_2} H_2, \tag{15.5}$$

$$\frac{\partial H_2}{\partial t} = -D_2\frac{\partial^2 H_2}{\partial x^2} + \frac{k_H}{2} H^2 - \frac{k_{H_2}}{2} H_2, \tag{15.6}$$

with the additional parameters k_H and k_{H_2} which are the reaction rates for dimerization and atomization, respectively. Again the motion of H and H_2 is described by a simple diffusion law with the corresponding diffusion coefficients D and D_2 [17]. The combination of this modified reaction–diffusion model with empirical hole-trapping somewhat improves the match with experimental DC and AC stress data. The failure of the RD model to properly describe NBT recovery is shifted out of the time window of some experiments, but essentially remains.

Quite recently it was claimed that the misprediction of recovery is due to the one-dimensional description of the diffusing species in the macroscopic model (15.5) and (15.6) [11]. As illustrated in Fig. 15.4, it was suggested that this formulation makes it too easy for the hydrogen atom to find a dangling bond to passivate because the one-dimensional diffusion considers only two options of motion: forward and backward jumping. In a higher-dimensional description the diffusion and reaction kinetics are much richer:

1. The atoms can move in all three dimensions equally likely, leading to a distribution of arrival times at the interface during recovery.
2. H_2-molecules dissociate at a dangling bond, creating a passivated dangling bond and a free hydrogen atom that does not immediately find another dangling bond to passivate.
3. Hydrogen atoms arriving later have to hover along the interface to find an unoccupied dangling bond.

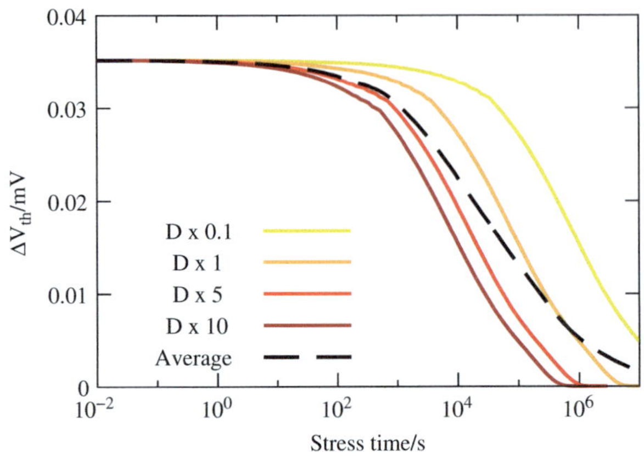

Fig. 15.5 Numerically calculated recovery traces for different diffusion coefficients during recovery and the average of these traces. In accord with [11], this average trace shows a recovery that proceeds over more time-scales than the individual traces

A simple estimate of the recovery in this hypothetical three-dimensional model is given in [11]. This estimate tries to mimic the different repassivation kinetics arising in the atomic description within the framework of the usual macroscopic RD model. To account for the longer "effective" recovery paths, the diffusion coefficients in the macroscopic model are reduced by different factors during recovery and the resulting recovery traces are averaged. Although this approach gives a recovery that proceeds over more time scales, as shown in Fig. 15.5, no derivation for the quasi-three-dimensional description is given and its physical validity is at least questionable. One of our targets is to test the claims of [11] within a firm theoretical framework.

We have derived and implemented a microscopic formulation of the RD model [18, 19], in order to study the behavior of the RD mechanism on the atomic scale. This effort was made not only to test the claims of [11], but also to investigate general issues of the rate-equation-based description in the context of MOS reliability. As a literature study reveals, reaction–diffusion systems have been studied in numerous scientific communities from both the theoretical and the experimental side for more than a century [22–28]. Although the mathematical framework of the RD model (15.4)–(15.6) seems physically sound and the description using densities and rate equations is commonly considered adequate, it is a well-known and experimentally confirmed result of theoretical chemistry that the partial differential equation-based description of chemical kinetics breaks down for low concentrations [22]. Additionally, in reaction–diffusion systems bimolecular reactions, such as the passivation and the dimerization reaction, require a certain proximity of the reactant species, termed *reaction radius* [23, 28]. Usually the elementary bimolecular reactions happen almost instantaneously and it is the required collision, i.e., the reduction of the distance between two reactants below the reaction radius, which is the rate-limiting step [22]. In chemical kinetics, these reactions are called *diffusion-limited* or *diffusion-controlled* reactions [24].

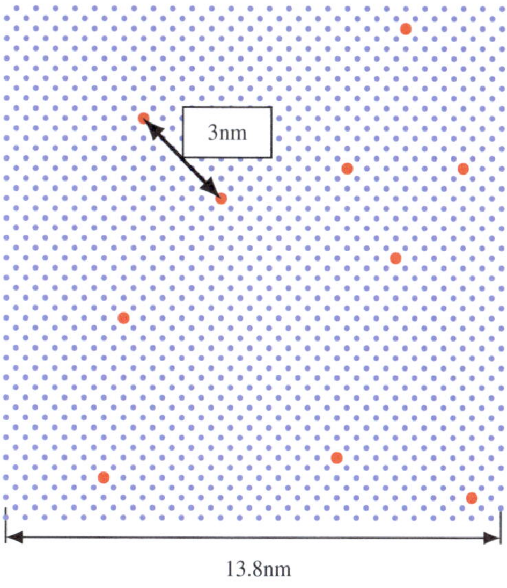

Fig. 15.6 A random distribution of ten dangling bonds on a silicon (100) surface corresponding to a dangling bond density of $5 \times 10^{12}\,\text{cm}^{-2}$, which is a common assumption for the number of bonding defects at the Si–SiO_2 interface [15, 20, 21]. The surface silicon atoms are shown in *blue*, the dangling bonds in *red*. At this density the average distance between two dangling bonds is ∼4.5 nm

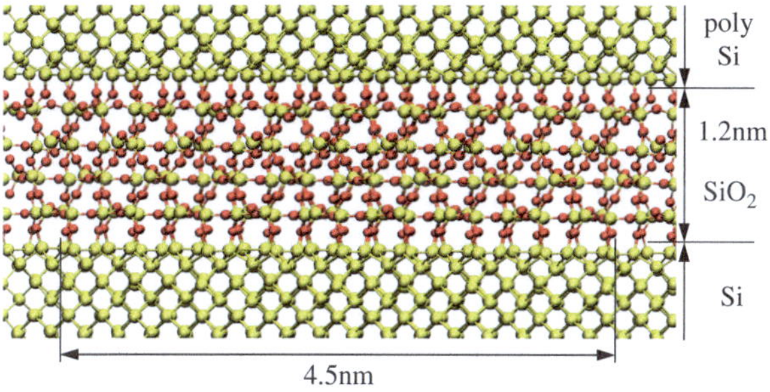

Fig. 15.7 An idealized atomic model of an MOS structure and the average dangling bond distance of 4.5 nm, which spans several interstitial positions. It is intuitively clear that an elementary reaction between particles separated by this distance is strongly influenced by diffusion

It is easy to show that diffusion must play a dominant role in the bimolecular reactions in the RD model for NBTI. The density of bonding defects on oxidized silicon (100) surfaces is about $1 \times 10^{12}\,\text{cm}^{-2}$ [20]. Figure 15.6 schematically shows a uniform random distribution of dangling bonds on a silicon (100) surface that corresponds to a density of $5 \times 10^{12}\,\text{cm}^{-2}$, which is a usual assumption for N_0 [15, 21] in the RD model (15.1) or (15.4). The average distance between two nearest neighbors at this density is $d = N_0{}^{-1/2} \approx 4.5\,\text{nm}$. An atomic model of the Si–SiO_2 interface as in Fig. 15.7 shows that two points separated by this distance have a large number of atoms in between. The assumption of an elementary reaction over this distance is clearly inappropriate, so any reaction between particles of this separation must involve a diffusive step.

Once established by the atomic viewpoint above, the diffusive influence on the bimolecular reactions leads to contradictions in the RD model and its physical interpretation. The predicted degradation of the RD model that is compatible with experimental data is only obtained if the hydrogen atoms that are liberated during stress compete for the available dangling bonds and dimerize at a certain rate. Both requirements involve diffusion over distances much larger than the nearest neighbor distance, which takes about 2 s at a commonly assumed diffusion coefficient of $D = 10^{-13}\,\mathrm{cm}^2/\mathrm{s}$ [15, 21]. The reaction radius ρ_{H} of the dimerization reaction can be estimated from the Smoluchowski theory for irreversible bimolecular reactions [23, 25, 26]

$$\rho_{\mathrm{H}} = \frac{k_H}{4\pi D}. \tag{15.7}$$

While a reasonable reaction radius is in the regime of the average radius of the oxide interstitials, which is about 4 Å [29], the application of (15.7) to published dimerization rates gives values ranging from 70 μm for the parametrization of [21] to thousands of kilometers for other parametrizations [15]. Although both values for ρ_{H} seem quite unreasonable, they only indicate a limited physical validity of the selected parametrization. An evaluation of the physical validity of the reaction–diffusion model itself requires a more detailed study using a computational model that properly treats the stochastic chemical kinetics involved. For the present study of the microscopic properties of the RD mechanism we have developed an atomistic reaction–diffusion simulator, which is described in the following.

15.2 Stochastic Description of Reaction–Diffusion Systems

Our microscopic RD model attempts to mimic the proposed mechanisms of the reaction–diffusion model at a microscopic level. The basic actors are H atoms, H_2 molecules, and the silicon dangling bonds at the interface. The investigations are carried out at the stochastic chemistry level. Several approaches have been used in the chemical literature for the stochastic simulation of reaction–diffusion systems [25, 26]. These approaches can roughly be categorized as grid-based methods or grid-less methods [26], owing to the description of the diffusion of the reactants, see Fig. 15.8. Grid-less methods propagate the coordinates of the diffusing species through Newton's equations of motion, quite similarly to molecular dynamics methods. Instead of explicitly treating all atoms of the solvent and their effect on the trajectory of the diffusors, the motion of the diffusing particles is perturbed by an empirical random force to generate a Brownian motion. Bimolecular reactions happen at a certain rate as soon as two reaction partners approach closer than a given radius. Although this technique suffers from its sensitivity to the time-step and the specific choice of the random force, it is a popular choice for the simulation of reaction–diffusion processes in liquid solutions where real molecular-dynamics

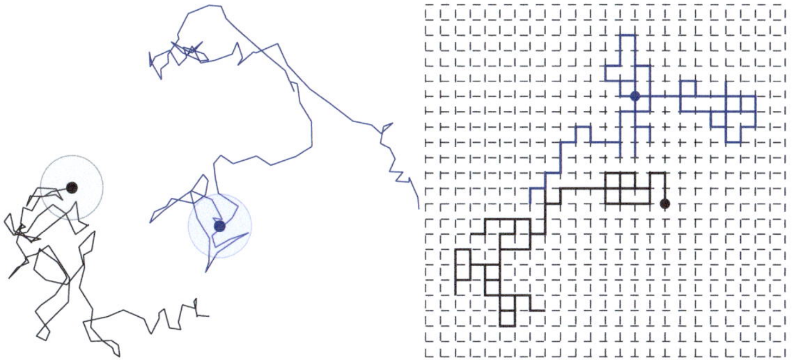

Fig. 15.8 Schematic illustration of the stochastic modeling approaches to reaction–diffusion systems. *(Left)* In grid-less methods, a molecular trajectory is generated from the transient solution of Newton's equations of motion as in molecular dynamics simulation. The interaction with the solvent is modeled by a random force that acts on the diffusors. Bimolecular reactions occur when two particles approach closer than the reaction radius (indicated as *circles* around the particles) *(Right)* In grid-based methods diffusion proceeds as jumps between the sub-domains defined by the grid. Bimolecular reactions occur when two particles occupy the same grid-point

simulations are not feasible [25, 26]. In grid-based methods the simulated volume is divided into small domains and each diffusing particle is assigned to a specific domain. The motion of the diffusors proceeds as hopping between the grid-points. In these models the bimolecular reactions happen at a certain rate as soon as two reactants occupy the same sub-volume. The advantage of this approach is that it can be formulated on top of the chemical master equation. This equation can be solved without artificial time-stepping, as explained in the following section. A problem of the grid-based method that is repeatedly discussed in chemical literature is the choice of the spatial grid as it induces a more or less unphysical motion in liquid solutions. Additionally, the probability to find two particles on the same grid point and in consequence the rate of bimolecular reactions are quite sensitive to the volume of the sub-domains [26].

In the reaction–diffusion model for NBTI, the diffusion of the particles proceeds inside a solid-state solvent. Contrary to diffusion in gases or liquids, the motion of an impurity in a solid-state host material proceeds via jumps between metastable states as illustrated in Fig. 15.9. This hopping diffusion is understood as a hopping process over energetic barriers. In the case of H or H_2, which do not react with the host atoms, these barriers arise from the repelling Coulomb interaction between the electron clouds of the host lattice and the diffusor. The minima of the potential energy surface are thus the interstitial positions of the host lattice [30, 31]. In between the jumps, the motion of the atom is randomly vibrational rather than diffusive. This discreteness of motion not only strongly suggests the use of a grid-based method, where the grid points are interstitial positions of the host lattice, but also induces a natural discretization into the reaction–diffusion equations. As a consequence, the description based on macroscopic diffusion equations in the RD

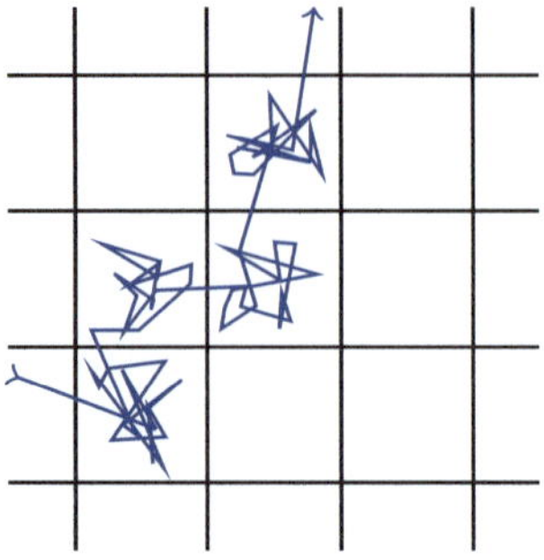

Fig. 15.9 Schematic trajectory of an inert interstitial atom diffusing in a solid-state host material. The potential energy barriers of the host material are indicated as black grid. The diffusion itself proceeds via jumps between the interstitial positions. In between the jumps, the atom vibrates randomly around an energetic minimum

model (15.5) and (15.6) are only valid at distances that are much larger than the interstitial radius and it has to be assumed that at very short distances a description using hopping diffusion is more accurate.

15.3 The Chemical Master Equation

From the considerations of the previous section we conclude that the most appropriate description of the physics considered in the present work is obtained from the reaction–diffusion master equation approach [25–28]. Within the natural lattice of interstitial positions the actors of our RD system exist in well-defined and discrete states. Once the chemical states and reactions that comprise the chemical system under consideration are defined, their dynamics can be described as a random process that switches between the states [32, 33]. Mathematically, the state of the chemical system is described as a vector $\vec{x}$. In addition, a set of reaction channels is established, which cause the transitions between the discrete states of this vector. Due to the unpredictable nature of the dynamics of the microstates, the time at which a reaction takes place is not a deterministic quantity. Instead, if the chemical system is in a given state $\vec{x}_\alpha$ at time t, for every reaction channel γ a reaction rate c_γ can be defined, so that $c_\gamma \mathrm{d}t$ is the probability of the reaction taking place between t and $t + \mathrm{d}t$ [33]. Different chemical states have different reaction rate constants for their reaction channels. These reaction rate constants depend only on the current state of the chemical system irrespective of the previous states of the system. In this case a function can be defined for every reaction channel that assigns a specific rate to a specific state $c_\gamma = a_\gamma(\vec{x}_\alpha)$. These functions are called the *propensity functions* [33]. The change induced by the reaction channel γ is described using the state change vector $\vec{v}_\gamma$. The thus formulated model describes a memory-less random process with discrete states, which is usually called a Markov process [34]. The removal

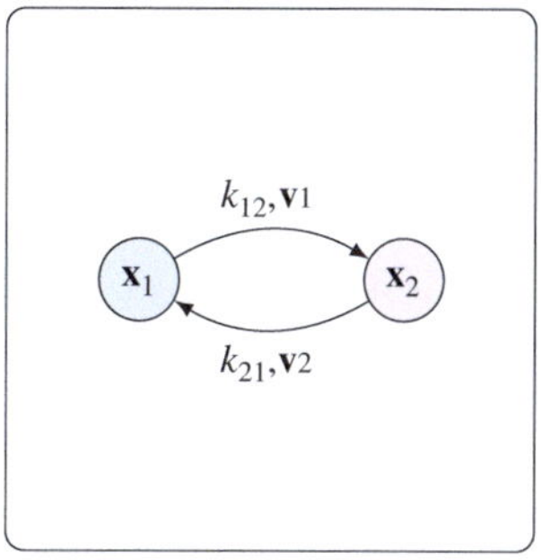

Fig. 15.10 Illustration of the example system discussed in the text. The system exists in one of the two states $\vec{x}_1$ and $\vec{x}_2$. Transitions between these states are caused by two reaction channels with the state-change vectors $\vec{v}_1$ and $\vec{v}_2$, and the associated rates k_{12} and k_{21}

of memory from the system occurs through the thermal equilibration, which is assumed to happen much faster than the chemical reactions. According to the theory of stochastic chemical kinetics [32, 33], the evolution of this system over time can then be described by a chemical master equation

$$\frac{\partial P(\vec{x},t)}{\partial t} = \sum_{\gamma=1}^{\Gamma} [a_\gamma(\vec{x}-\vec{v}_\gamma)P(\vec{x}-\vec{v}_\gamma,t) - a_\gamma(\vec{x})P(\vec{x},t)], \quad (15.8)$$

where $P(\vec{x},t) = P(\vec{X} = \vec{x},t|\vec{x}_0,t_0)$ is the probability that the stochastic process $\vec{X}(t)$ equals $\vec{x}$ at time t, given that $\vec{X}(t_0) = \vec{x}_0$.

The master equation approach can be illustrated using the simple example of a system with two states $\vec{x}_1$ and $\vec{x}_2$ [34], see Fig. 15.10. The system has two reaction channels 1 and 2, which connect the two states through the state change vectors $\vec{v}_1$ and $\vec{v}_2$ as

$$\vec{x}_1 + \vec{v}_1 = \vec{x}_2 \quad \text{and} \quad \vec{x}_2 + \vec{v}_2 = \vec{x}_1 \quad (15.9)$$

The propensity functions a_1 and a_2 assume the form

$$a_1(\vec{x}_1) = k_{12}, \qquad a_1(\vec{x}_2) = 0, \quad (15.10)$$

$$a_2(\vec{x}_1) = 0, \text{and} \quad a_2(\vec{x}_2) = k_{21}. \quad (15.11)$$

The master equation for this system consequently reads

$$\frac{\partial P(\vec{x}_1,t)}{\partial t} = k_{21}P(\vec{x}_2,t) - k_{12}P(\vec{x}_1,t) \quad (15.12)$$

$$\frac{\partial P(\vec{x}_2,t)}{\partial t} = k_{12}P(\vec{x}_1,t) - k_{21}P(\vec{x}_2,t) \quad (15.13)$$

As the system can only exist in one of the two states at a time, it follows that

$$P(\vec{x}_1,t) = 1 - P(\vec{x}_2,t) = p(t), \quad (15.14)$$

which reduces the master equation of the two-state system to

$$\frac{\partial p(t)}{\partial t} = k_{21}(1 - p(t)) - k_{12}p(t). \tag{15.15}$$

This is the rate-equation of the two-state system, which is equivalent to the master equation for this simple example.

Within the theoretical framework of the chemical master equation, all the microphysical details elaborated in the previous section are now contained in the propensity functions a_γ and the state-change vectors $\vec{v}_\gamma$ for the Γ reaction channels.

15.4 States and Reactions in the Microscopic RD Model

The main actors of the microscopic RD model are the H atoms and the H_2 molecules. The state vector $\vec{x}$ of the system consequently contains the interstitial positions and bonding states of all actors. The reactions employed in our simulations are the hopping transport between interstitial sites, the passivation/depassivation reaction, and the dimerization/atomization reaction. These reactions are treated as elementary reactions and are formalized in the reaction channels given in Fig. 15.11. The stochastic chemical model is solved using the stochastic simulation algorithm (SSA) explained in Sect. 15.5.

In the microscopic RD model employed in this work the interstitial sites form a regular and orthogonal three-dimensional grid and the hopping rates for the diffusors are assumed to be constant in accord with the isotropic and non-dispersive diffusion underlying the conventional macroscopic RD model [4]. In a real SiO_2 of a MOS transistor the amorphous structure will of course lead to a random network of interstitial sites [29] with a variety of hopping rates and a more complex topology. However, as the power-law degradation predicted by the macroscopic RD model requires a constant diffusion coefficient, these variations must be assumed unimportant [17] in order to obtain agreement with the established model. As illustrated in Fig. 15.12, the simulation region in our calculations is a rectangular box which extends to infinity normal to the Si–SiO_2 interface and has closed lateral boundaries. The Si–SiO_2 interface itself is represented by a special region at the bottom of the simulation box where selected interface sites have the ability to bond or release a diffusing hydrogen atom, see Figs. 15.11 and 15.12. The positions of the dangling bond sites in the interface region are picked randomly, similar to Fig. 15.6.

As mentioned above, the choice of the grid size requires special attention as it determines the probability of the bimolecular reactions. The interstitial size of amorphous silica has been calculated for molecular dynamics generated atomic structures and is about 4 Å [29]. We take this value as the physically most reasonable grid size.

Once the microscopic model is defined, the relation to the macroscopic RD model (15.4)–(15.6) has to be established. Using the number of dangling bonds in the simulation box n_{DB}, the number of hydrogen atoms passivating a dangling bond

	Reaction	Macroscopic	Microscopic	Illustration
a.	$Si^* + H \rightarrow SiH$	$k_r N_{it} H_{it}$	$\frac{k_r}{h^3} n_{DBi} n_{Hi}$	
b.	$SiH \rightarrow Si^* + H$	$k_f(N_0 - N_{it})$	$k_f n_{p,i}$	
c.	$H: I_1 \rightarrow I_2$	$-D \frac{\partial^2 H}{\partial x^2}$	$\frac{D}{h^2} n_{Hi}$	
d.	$H_2: I_1 \rightarrow I_2$	$-D_2 \frac{\partial^2 H_2}{\partial x^2}$	$\frac{D_2}{h^2} n_{H_2 i}$	
e.	$2H \rightarrow H_2$	$k_H H^2$	$2\frac{k_H}{h^3} n_{Hi}(n_{Hi} - 1)$	
f.	$H_2 \rightarrow 2H$	$k_{H_2} H_2$	$k_{H_2} n_{H_2 i}$	

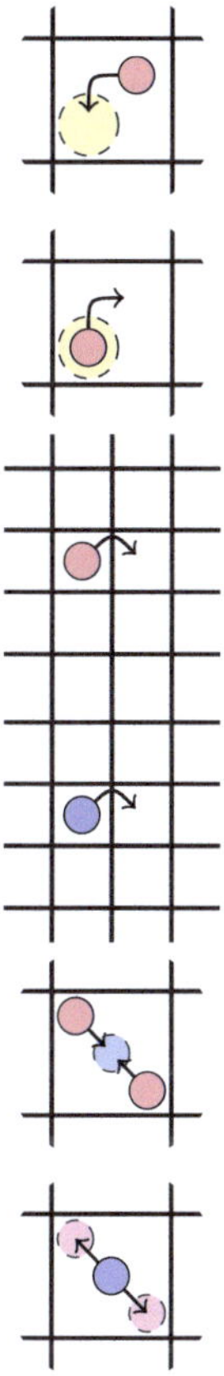

...Dangling bond ...H ...H_2

Fig. 15.11 Reaction channels and propensities in the microscopic RD model along with their macroscopic counterpart. (**a**) The dangling bonds are represented by special sites at the bottom of the simulation box. Empty dangling bond sites can be passivated by a free hydrogen atom. (**b**) Occupied dangling bond sites do not offer a bonding reaction channel, they can only emit their hydrogen atom. (**c, d**) Within the bulk SiO_2, the atoms or molecules are allowed to jump from an interstitial I_1 to any neighboring site I_2. (**e**) When two hydrogen atoms occupy the same interstitial position, they can undergo a dimerization at rate k_H and form H_2. (**f**) Each hydrogen molecule dissociates at a rate k_{H_2} back into two hydrogen atoms. For interstitial site i, n_{DBi} is the number of (depassivated) dangling bonds, $n_{p,i}$ is the number of passivating hydrogen atoms, n_{Hi} is the number of free hydrogen atoms, and $n_{H_2 i}$ is the number of hydrogen molecules. h denotes the step size of the spatial grid

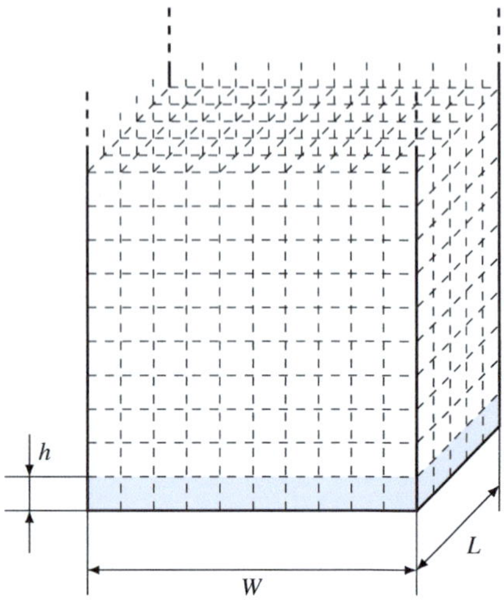

Fig. 15.12 The simulation structure for the microscopic RD model employed in this work is a bounded region of width W and length L and infinite extension in z-direction, i.e., normal to the interface. The silicon dangling bonds are connected to special interstitial positions in the $Si–SiO_2$ interface region at the bottom of the simulation box (*blue/gray*). The interstitial positions are assumed to form an orthogonal lattice with constant jump-width and constant diffusion coefficients

n_p and the numbers n_{Hi} of H and $n_{H_2 i}$ of H_2 at interstitial i, this relation is obtained from the discretization induced by the grid [21] as

$$N_0 = \frac{n_{DB}}{WL}, \tag{15.16}$$

$$N_{it} = \frac{n_{DB} - n_p}{WL}, \tag{15.17}$$

$$H(x_i) = \frac{n_{Hi}}{V_i}, \tag{15.18}$$

$$H_2(x_i) = \frac{n_{H_2 i}}{V_i}, \tag{15.19}$$

where W, L and h are illustrated in Fig. 15.12 and V_i is the volume of interstitial i which is $V_i = h^3$ in this work. The relation between the rates of the macroscopic model and the microscopic propensity functions are given in Fig. 15.11. Initially, all hydrogen atoms are passivating silicon dangling bonds

$$n_p(t=0) = n_{DB}, \tag{15.20}$$

in accordance with the assumptions of the macroscopic RD model.

15.5 Solution of the Master Equation

Now that the chemical states and reactions are defined we can calculate the time evolution of the chemical system from the chemical master equation (15.8). As explained above, this equation is a stochastic differential equation which assigns a

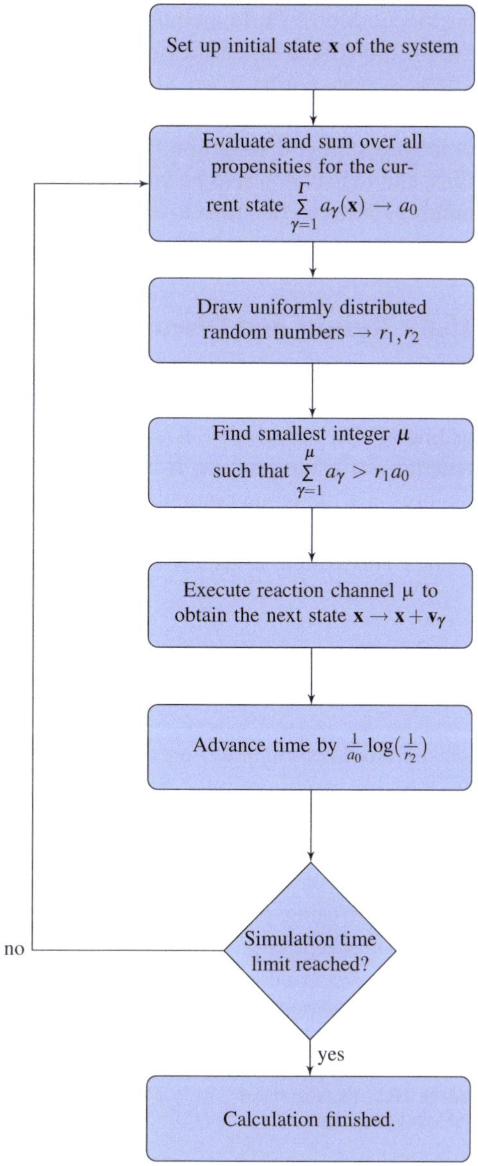

Fig. 15.13 Sketch of the stochastic simulation algorithm (SSA) [32]. The algorithm generates a realization of the stochastic process described by the chemical master equation (15.8)

probability at time t to any state vector $\vec{x}$, given that $\vec{X} = \vec{x}_0$ at $t = t_0$. As in the simple example above, for a system with a small set of states $\{\vec{x}_1, \ldots, \vec{x}_\Omega\}$, a direct solution can be attempted, which results in a coupled system of Ω differential equations [34]. However, in many situations the number of states will be large or even infinite, rendering a direct solution of the master equation unfeasible or even impossible. A feasible alternative is the SSA [32] explained in Fig. 15.13, which is also known as the kinetic Monte Carlo method. Instead of solving the differential equation (15.8), a realization of the stochastic process $\vec{X}$ is generated using pseudo-random numbers. The SSA does not have any algorithmic parameters and is a mathematically exact description of the system defined by the states and reaction channels [32]. Averages of the probability distribution $P(\vec{x},t)$ are trivially calculated over several simulation runs, although care has to be taken to ensure the randomness of the pseudo-random numbers between two runs to avoid correlation effects.

15.6 Results and Discussion

Two different systems have been studied in detail: a model system and a "real-world-example." The model system is used to study the general features of the microscopic reaction–diffusion process. It is parametrized in order to clearly show all relevant features at a moderate computational effort. The parametrization of the real-world system is based on a published parametrization of the modified reaction–diffusion model. This system is used to relate our microscopic model to published data.

15.6.1 General Behavior of the Microscopic RD Model

The parametrization that is used to study the general behavior is given in Table 15.1. As the time evolution in the SSA proceeds reaction by reaction, the channels with the fastest rates determine the execution time. The computational effort scales linearly with the number of particles in the system, which is determined by the lateral extent of the simulation box. As the dangling bonds and in consequence also the diffusing particles are uniformly distributed along the interface at any time in our calculations, the reflecting boundary conditions in our calculations are equivalent to periodic boundary conditions. Thus, our calculations correspond to an infinitely extended Si–SiO_2 interface and the lateral box size only determines the

Table 15.1 Parameters of the model system

Reaction	Propensity (s^{-1})
Depassivation	0.5
Passivation	4×10^4
Dimerization	2×10^5
Atomization	5
H-hopping	100
H_2-hopping	100

The parameters have been selected to enable a study of the different regimes of the microscopic RD model at moderate computational expense. The rates are given in terms of the microscopic model as in Fig. 15.11

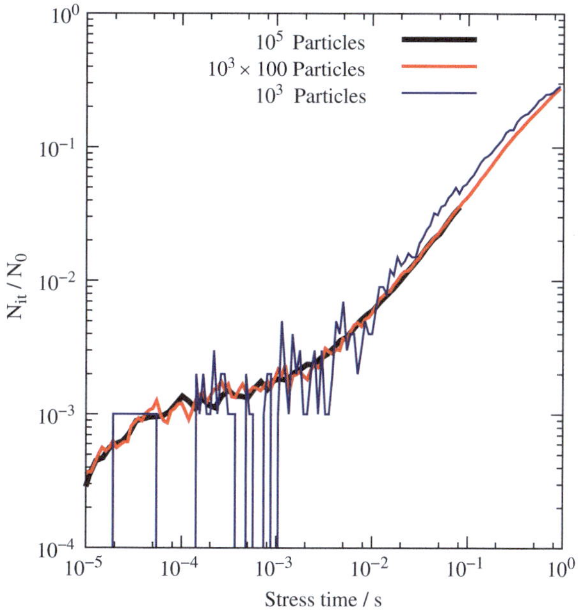

Fig. 15.14 As the interaction between the diffusing particles is small at early degradation times, the computation can be parallelized by averaging over several simulation runs. The figure shows that a calculation with 10^5 particles is equivalent to the average of 100 calculation runs with 10^3 particles. The result of a single 10^3 particle run is shown for comparison. k_f was increased by a factor of 100 for this calculation, in order to obtain smooth curves from the 10^5 particle run

resolution, i.e., the noise level, of the degradation curves. The computational effort scales roughly linearly with the simulated time, which means exponential scaling for the logarithmic abscissa that is used for BTI degradation curves. The choice of the lateral extent of the simulation box is thus based on a trade-off between accuracy and computational speed and has to be adapted for the study of the different degradation regimes.

At early degradation times the low degradation level requires a high resolution, i.e., a large number of particles is required to obtain smooth results. Fortunately, as reactions between the hydrogen atoms or between hydrogen atoms and neighboring dangling bonds do not happen in this regime, a good parallelization can be obtained by averaging over separate simulation runs, see Fig. 15.14.

The earliest degradation times are dominated by the depassivation of the silicon dangling bonds leading to a linear increase of the degradation, which is equivalent to the initial "reaction limited" degradation of the macroscopic RD model [3]. However, the degradation predicted by the microscopic RD model quickly saturates as an equilibrium forms between depassivation and repassivation *for each dangling bond separately*. In the absence of any diffusion the time evolution of the number of hydrogen atoms passivating a silicon dangling bond is given by

$$\frac{\partial n_p}{\partial t}(t) = -k_f n_p(t) + \frac{k_r}{h^3}(n_{DB} - n_p(t)) \quad (15.21)$$

$$n_p(t = 0) = n_{DB} \quad (15.22)$$

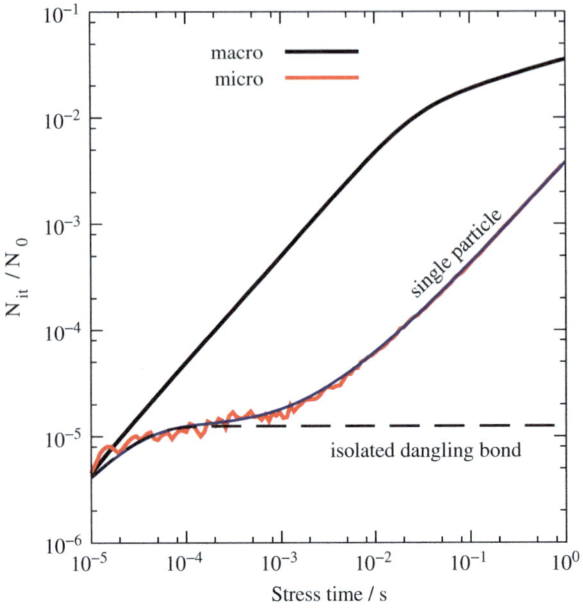

Fig. 15.15 Comparison of the microscopic RD model with 25,000 particles averaged over 50,000 runs to its macroscopic counterpart, the single-particle expressions (15.24) and (15.25) and the isolated dangling-bond equilibration (15.23). The earliest degradation times are dominated by the equilibration between the depassivation and passivation reaction at every dangling bond. Around 1 ms, the departure of the hydrogen atoms from the dangling bond site begins but the interaction between the diffusors is still negligible

with the solution

$$n_{\mathrm{p}}(t) = n_{\mathrm{DB}} - \frac{n_{\mathrm{DB}} k_{\mathrm{f}}}{k_{\mathrm{f}} + \frac{k_{\mathrm{r}}}{h^3}} \left(1 - e^{-\left(k_{\mathrm{f}} + \frac{k_{\mathrm{r}}}{h^3}\right)t}\right). \tag{15.23}$$

A comparison of the microscopic RD model and (15.23) is shown in Fig. 15.15. The initial behavior of the microscopic RD model stands in stark contrast to the degradation in the macroscopic model where the linear regime continues until a global equilibrium has formed at the interface.

As this initial behavior takes a central position in our further discussion, it requires a deeper analysis. The microscopic single-particle regime can be accurately described using rate equations as it does not contain any second-order reactions. The required equations are basically those of the RD model, but as every hydrogen atom can be assumed to act independently, the expressions for the hydrogen bonding as well as the competition for dangling bonds are neglected. As the kinetic behavior in this regime is strongly determined by the first diffusive steps of the hydrogen atoms, the diffusion part of this approximation must have the same interstitial topology as the microscopic model. As all hydrogen atoms act independently, only one atom and

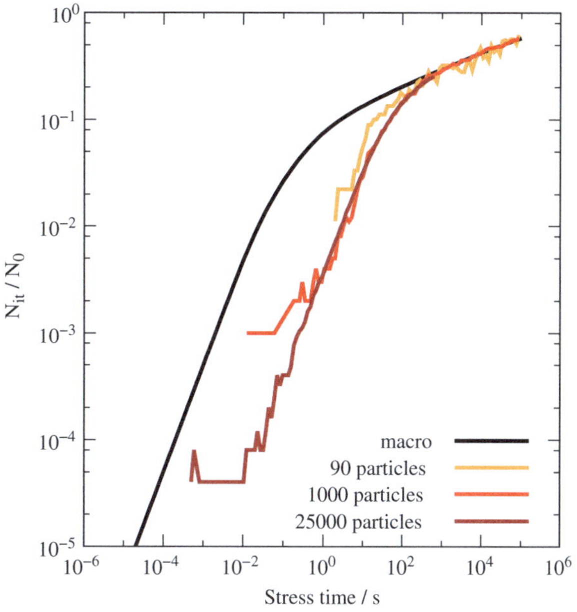

Fig. 15.16 Due to the larger fraction of depassivated dangling bonds, the number of diffusing particles can be reduced for long-term simulations. Three microscopic calculations are compared to the macroscopic result. The 25,000 particle simulation clearly shows the transition between the single-particle and the macroscopic diffusion-limited regime. The 1,000 particle calculation captures the transition region but is too noisy for $t < 100\,\mathrm{ms}$. The 90 particle simulation captures the macroscopic regime with reasonable accuracy

one dangling bond need to be considered. The interface reaction and the diffusion of the hydrogen atom is thus described as

$$\frac{\partial n_{\mathrm{P}}}{\partial t} = -k_{\mathrm{f}} n_{\mathrm{P}} + \frac{k_{\mathrm{r}}}{h^3} n_{\mathrm{DB}} n_{\mathrm{H0}} \text{ and} \tag{15.24}$$

$$\frac{\partial n_{\mathrm{H}i}}{\partial t} = \sum_{j \in \mathscr{N}(i)} \frac{D}{h^2} (n_{\mathrm{H}j} - n_{\mathrm{H}i}), \tag{15.25}$$

respectively, where $\mathscr{N}$ denotes the set of neighboring interstitials to i. Figure 15.15 compares the microscopic RD calculation with the approximations for the different regimes at early degradation times, which shows that the single-particle approximation perfectly matches the behavior of the full model in the initial phase.

After the atoms have traveled sufficiently long distances, the interaction between the particles becomes relevant and the single-particle approximation becomes invalid. In Fig. 15.16 this is visible as a transition away from the single-particle behavior toward the macroscopic solution between 1 s and 1 ks. As the fraction of depassivated dangling bonds in this regime is much higher than during early degradation times, the results are not as strongly influenced by the noise of the SSA calculation. Consequently the number of particles can be reduced for longer simulation times, which makes the prediction of long-term degradation possible.

Finally, Fig. 15.17 compares the microscopic RD model to the macroscopic version over the course of one complete stress cycle, where the microscopic curve was obtained by combining calculations of different accuracy, as explained above.

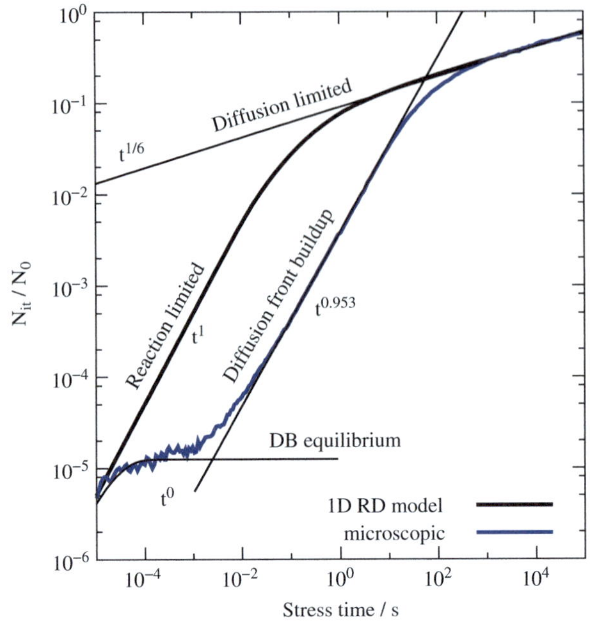

Fig. 15.17 Comparison of all regimes of the microscopic RD model to the degradation predicted by the macroscopic RD model. Obviously there is a large discrepancy between the two descriptions and the behavior of the physically more reasonable microscopic model is not experimentally observed

Instead of the three regions which arise from the macroscopic RD model—reaction-limited, equilibration, and diffusion-limited—the H–H_2 microscopic description has four to five regimes depending on the particular parametrization:

- The earliest degradation times ($t < 20\,\mu\text{s}$ in this case) are dominated by the depassivation of dangling bonds. In this regime, the microscopic and the macroscopic model give identical degradation behavior.
- After the passivation and depassivation have reached an equilibrium between k_f and k_r separately for each Si–H bond, the fraction of depassivated dangling bonds remains constant until the diffusion of the hydrogen atoms becomes dominant. This regime only shows when the individual hydrogen atoms are considered and consequently is not obtained from any model based on rate equations.
- As more and more hydrogen atoms leave their initial position, the degradation is determined by the buildup of a diffusion front along the Si–SiO_2 interface and the equilibration between the dangling bonds. This regime has a very large power-law exponent of almost one[1] that is not experimentally observed. The stress time

[1]In our earlier studies on two-dimensional systems this exponent was around 0.8 [19], owing to the topology dependence of this regime.

range in which this regime is observed depends on the average distance between two dangling bonds, the diffusion coefficient, and the interstitial size.

- As the bimolecular reactions become relevant, the macroscopic diffusion-limited regime begins to emerge. For some parametrizations we have observed a time window in which the initial diffusion-limited regime has the typical $t^{1/4}$-form that arises from the classical RD model without H_2 [19]. In this case the dimerization rate is reduced by the diffusive step and the H diffusion dominates the degradation until a sufficient amount of H_2 has formed.

The initial single-particle phase of the degradation is a remarkable feature of the microscopic model. As it is incompatible with experimental data and very sensitive to the parametrization, its relevance for real-world reliability projections has to be investigated. For this purpose we have run calculations based on a published parametrization of the reaction–diffusion model for NBTI, see Sect. 15.6.4.

15.6.2 Recovery

In agreement with our investigations on two-dimensional systems [18, 19], the three-dimensional stochastic motion of the hydrogen atoms *does not influence* the recovery behavior of the system after long-term stress, which contradicts the suggestions of [11]. As shown in Fig. 15.18, a longer relaxation transient is only obtained if the preceding stress phase does not show a power-law regime. As the system comes closer to the macroscopic degradation behavior, the recovery in the microscopic model also approaches the macroscopic version, which is incompatible with experimental data [7, 8, 35]. This behavior is to be expected as the $t^{1/6}$ degradation regime requires an equilibration and thus a quasi-one-dimensional behavior. The recovery proceeds on a timescale that is at least two orders of magnitude longer than the stress time. The lateral search of hydrogen atoms for unoccupied dangling bonds was suggested to dominate at the end of the recovery. However, due to the logarithmic time scale on which recovery is monitored, the equilibration along the interface has negligible impact at the end of the recovery trace if this equilibration proceeds about two orders of magnitude faster. Thus, the hovering of hydrogen atoms along the interface does not influence the shape of the recovery transient.

15.6.3 Approximations in the Macroscopic Model

After the microscopic RD theory Fig. 15.11 has been established and its general behavior has been investigated, one can use this framework to analyze the assumptions that are implicit to the macroscopic RD model (15.4)–(15.6), which is still widely considered to be an adequate approximation.

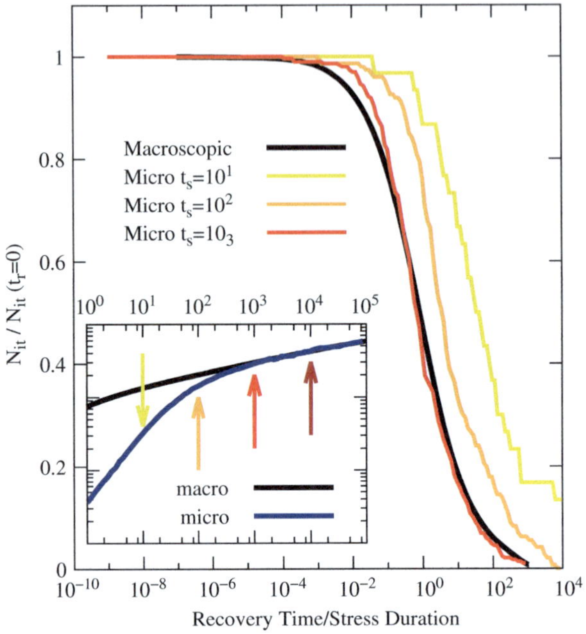

Fig. 15.18 Recovery transients for different stress times. As the degradation transient approaches the macroscopic diffusion limited regime (see inset), the recovery comes closer to the macroscopic recovery, leading to a perfect match as soon as the degradation assumes the experimentally relevant $t^{1/6}$ form

The most obvious approximation in the macroscopic RD model is the one-dimensional description of diffusion. While this may seem to be appropriate as boundary effects in the diffusion of both H and H_2 are negligible, it tacitly introduces the assumption of lateral homogeneity along the interface. This homogeneity includes the following assumptions:

- *All* the liberated hydrogen atoms at the interface (H_{it} in (15.1) and (15.4)) compete instantaneously with *all* the other free interfacial hydrogen atoms for *all* the available dangling bonds.
- *All* the pairs of hydrogen atoms at a certain distance from the Si–SiO_2 interface are equally likely to undergo dimerization and form H_2, independently of their spatial separation.

As was shown above, a hydrogen atom liberated during stress initially stays in the vicinity of its original dangling bond and thus the lateral homogeneity has to be considered a long-term approximation. It is accurate when the diffusion of hydrogen has led to enough intermixing so that there is no significant variability in the concentration of free hydrogen along the interface. Following [33], this condition can be called "lateral well-stirredness" of the system.

The second and more delicate approximation in the macroscopic RD model is the mathematical description using rate- and diffusion-equations. In the microscopic

Table 15.2 The parameters employed in the real-world simulations

k_f	$3\,\mathrm{s}^{-1}$
k_r	$6 \times 10^{-13}\,\mathrm{cm}^3\,\mathrm{s}^{-1}$
k_H	$5.6 \times 10^{-11}\,\mathrm{cm}^3\,\mathrm{s}^{-1}$
k_{H_2}	$95.4\,\mathrm{s}^{-1}$
D	$10^{-13}\,\mathrm{cm}^2\,\mathrm{s}^{-1}$
D_2	$1.8126 \times 10^{-14}\,\mathrm{cm}^2\,\mathrm{s}^{-1}$
N_0	$5 \times 10^{12}\,\mathrm{cm}^{-2}$

The parameter set is based on the values published in [15] but was slightly modified to give the same degradation behavior with physically more reasonable k_r and k_H

RD model, the rate at which an atom at the interface passivates a dangling bond depends not only on the rate k_r but also on the probability of finding this atom at the position of the dangling bond. In the macroscopic model the precondition of having an unoccupied dangling bond at the interface is described multiplicatively as $k_r N_{it} H_{it}$. At early times during degradation, when each hydrogen atom still resides near its dangling bond, this term introduces an unphysical self-interaction where each hydrogen atom competes with *itself* for its dangling bond. As the root of this problem lies in the assumptions implicit to a formulation based on rate-equations, the error is also present in a macroscopic model with three-dimensional diffusion. As explained in [19], this means that a rate-equation-based RD model will not accurately describe the degradation at early times even if higher-dimensional diffusion and discrete dangling bonds are considered.

Similar to the passivation rate, the rate at which H_2 is formed in the microscopic RD model depends on both the dimerization rate k_H and the probability of finding two hydrogen atoms which occupy the same interstitial position. In the macroscopic RD model, this dimerization reaction is modeled as $k_H \mathrm{H}^2$. As thoroughly explained in [22], this approximation is only valid for large numbers of particles, as the number of pairs of hydrogen atoms in an interstitial goes as $N(N-1)$ which can only be approximated as N^2 if N is sufficiently large.

All in all, the macroscopic RD model can only be considered a valid approximation of the microscopic RD model for very long stress times and a sufficient amount of liberated hydrogen atoms. The time it takes for the macroscopic approximation to become valid, however, may exceed the time range in which it is usually applied, depending on the parametrization.

15.6.4 A Real-World Example

To study the behavior of the atomistic model for a real-world example, we compare to the measurements of Reisinger et al. [7] using the parametrization of Islam et al. [15] in a modified form, see Table 15.2. Figure 15.19 shows the results of our

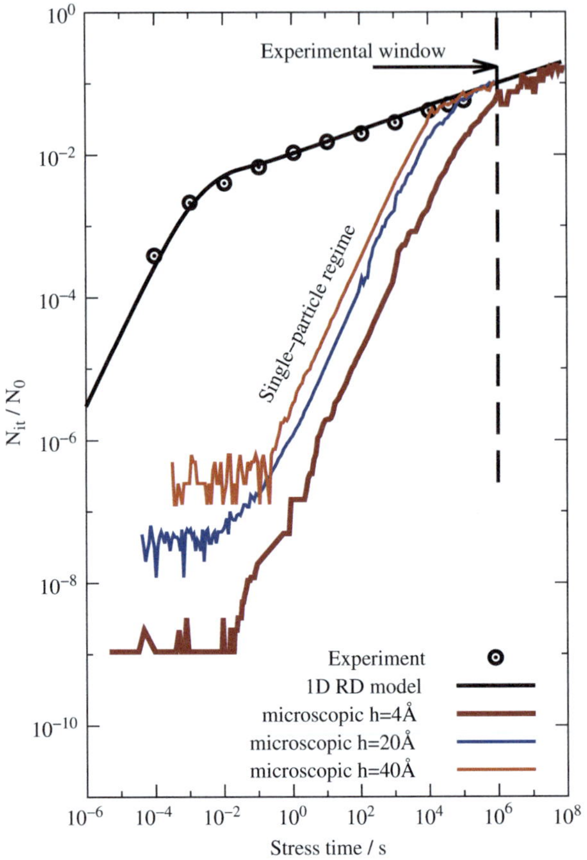

Fig. 15.19 The degradation transient predicted by the microscopic RD model for four interstitial sizes compared to the macroscopic one-dimensional model and experimental data. Using the parameters in Table 15.2, the prediction of the microscopic RD model is completely incompatible with the experimental data as the onset of the $t^{1/6}$ regime is delayed beyond 10^8 s (about 3 years) for a reasonable interstitial size of 4 Å. Increasing the interstitial size reduces the effect as it increases the effective reaction radius for the bimolecular reactions. However, even for unphysically large interstitial sizes, the onset of the $t^{1/6}$ regime is delayed to 10^4 s ($h = 40$ Å) or 10^5 s ($h = 20$ Å)

calculations for several interstitial sizes. While the macroscopic one-dimensional RD model fits the data very well, the kinetic Monte Carlo data shows a completely different behavior. Again, the single-particle regime is clearly present. However, due to the low density of dangling bonds at the interface, the single-particle regime dominates the degradation for a large part of the stress time. For a realistic interstitial size of 4 Å [29, 36], the onset of the $t^{1/6}$ regime lies far beyond the experimental window of 10^5 s. When the interstitial size is increased, the onset of the $t^{1/6}$ regime moves to earlier times, which is due to the increase of the reaction radius for the bimolecular reactions as explained above. For the given parameter set, an interstitial size of $h = 2\,\text{nm}$, which is the total thickness of the oxide of the device under consideration [7], is required to at least have the $t^{1/6}$ regime touch the experimental window.

A shift of the onset of the experimentally observed regime to earlier times at a realistic interstitial size requires a dramatic increase of either the hydrogen diffusion coefficient or the availability of free hydrogen near the interface. An increase of the hydrogen diffusion coefficient, however, breaks the dominance of H_2 flux over the flux of atomic hydrogen and changes the predicted degradation away from the experimentally observed $t^{1/6}$ towards $t^{1/4}$. Increasing the availability of hydrogen at the interface by adjusting the ratio k_f/k_r causes similar problems, as the H_2 diffusion coefficient has to be lowered in order to give the same overall degradation.

This indicates that in the given microscopic model it is impossible to obtain the experimentally observed $t^{1/6}$ degradation within the experimental window at a reasonable interstitial size.

15.6.5 Increased Interface Diffusion

The behavior predicted by the microscopic model is completely incompatible with any experimental data, while the description is much closer to the physical reality than the macroscopic RD model. Only two interpretations are possible to resolve this dilemma. Either the ability of the macroscopic RD model to fit degradation measurements has to be regarded as a mathematical artifact without physical meaning, or the structure of the Si/SiO_2 interface somehow accelerates the lateral equilibration considerably. We investigated the second option more closely by considering first-principles calculations that have shown a lowering of diffusion barriers for hydrogen (molecules) along the Si/SiO_2-interface as compared to the bulk SiO_2 [37]. These findings indicate that the motion of hydrogen might proceed at a much higher rate along the interface. As a higher diffusivity at the interface aids the lateral equilibration, it might be the sought process that makes the one-dimensional RD model physically meaningful. To account for it in our microscopic model, we applied different diffusion coefficients D_I and D_B in the interface region and in the bulk, respectively.

As can be seen in Fig. 15.20, the increase of the interface diffusion coefficient accelerates the degradation during the initial phase as it increases the transport of hydrogen atoms away from their dangling bonds. Interestingly, even if the interface diffusion coefficient is increased by four orders of magnitude there is no $t^{1/6}$ behavior visible, but instead the degradation takes on the typical $t^{1/4}$ behavior of a hydrogen-only reaction–diffusion model. While the competition for dangling bonds sets in earlier for increased interface diffusion coefficients, the formation of H_2 is not accelerated in the same way. Inspection of the atomic diffusion shows that the acceleration of the dimerization is much less pronounced as the liberated hydrogen atoms constantly leave the interface region into the bulk where the diffusion proceeds slower and the collision rate is reduced. Only in the limit

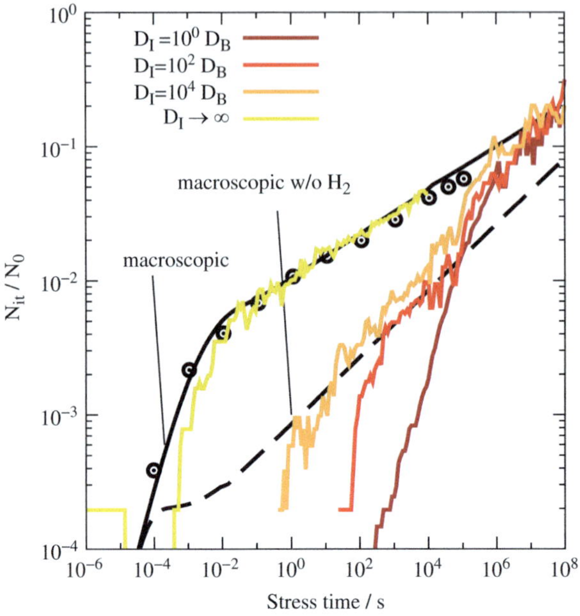

Fig. 15.20 For increasing D_{I}, the departure of hydrogen atoms from their dangling bond sites starts earlier, leading to an increased degradation at earlier times. Comparison to the classical RD model without H_2 formation shows that competition for dangling bonds sets in after about 100 s, leading to a $t^{1/4}$ degradation. The formation of H_2 is slowly accelerated by the increased D_{I} and only for $D_{\text{I}} \to \infty$, the macroscopic behavior is obtained

of $D_{\text{I}} \to \infty$ will the microscopic RD model match the experimentally observed behavior. Although these extremely high interface diffusion coefficients lack any physical justification, this is still closer to the physical reality than the assumption of immediate equilibration along the Si–SiO_2-interface at any depth that is inherent to the usually employed one-dimensional macroscopic RD model.

As a side note we remark that in a real wafer, a nearly infinite diffusion coefficient along the Si/SiO_2-interface would make the hydrogen spread out through the waver during stress. This would again alter the degradation slope and give rise to cross-talk between neighboring devices that would be measurable, but has never been reported.

15.7 Related Work

Four other scientific groups have put forward microscopic RD models recently [21, 38, 39] and interestingly those investigations find a reasonable agreement between their microscopic description and the macroscopic RD model. In the work of Islam et al. [21] the atomic description is basically equivalent to the work presented here but is built upon a one-dimensional foundation which carries the same implicit assumptions as the macroscopic model. Clearly this model cannot capture the effects discussed in this chapter as those are solely due to higher-dimensional effects. From a physical point of view, however, the one-dimensional approximation lacks justification considering the results presented above.

The work of Choi et al. [39] considers the three-dimensional diffusion of the particles based on a grid-less stochastic formulation. Although the degradation in that work seems to match the macroscopic RD model quite well at first sight, also strong discrepancies arise between the two for longer stress times. Interestingly, for situations where the approach presented above predicts a degradation far below the prediction of the macroscopic model, the degradation predicted by Choi et al. overshoots the macroscopic model considerably. Only for an enormous density of dangling bonds or a very large reaction radius the macroscopic behavior is obtained, in accord with our results. The degradation behavior in [39] initially follows $N_{\mathrm{it}}(t) = k_{\mathrm{f}}t$, which suggests that the depassivated hydrogen atoms instantly leave the reaction radius of their respective dangling bond. The following excessively high power-law exponent suggests that the repassivation of the silicon dangling bonds is somehow inhibited in this formulation. The most likely explanation for this behavior is a too low resolution of the time-stepping, in combination with the physically unjustifiable description of the diffusive motion.

The work of Panagopoulos and Roy [38] uses a grid-based stochastic RD model that seems to be compatible with our description. The surprisingly good agreement between their results and the macroscopic RD model may be an artifact of the employed method which is based on an adaptive time-stepping. Also, the paper states that the passivation reaction occurs if a hydrogen atom is "close" to a dangling bond. This indicates an artificial capture radius, but this is not explicitly stated. Also, the grid spacing is not given in the paper and its physical relevance is not discussed. However, as shown by our calculations, an unphysically large grid spacing strongly promotes bimolecular reactions and thus induces a degradation behavior that is (falsely) compatible with the macroscopic RD model.

Finally, Naphade et al. [40] recently presented a stochastic version of the poly H/H_2 RD model. In this approach it is assumed that the diffusion of H is restricted to the oxide, while the diffusion of H_2 happens in the gate contact and beyond. The large H diffusion coefficient in these calculations in combination with the large grid size of 1 nm reduces the effect of the diffusion limitation on the bimolecular reactions. Although this model is formulated on a stochastic description, and is in better agreement with the experimental data, its physical validity is again questionable due to the assumed H diffusion coefficient of $10^{-5}\,\mathrm{cm}^2/\mathrm{s}$. At this diffusion coefficient, the hydrogen diffusion front would extend to 1 cm after 100 ks of stress. Even if the assumption that hydrogen is unable to penetrate into the gate contact holds, the isotropic nature of the diffusion process would lead to a lateral diffusion whose front quickly exceeded the dimensions of the MOS device. This would lead to a considerable out-diffusion of hydrogen from the gate area, resulting in a sharp increase of N_{it} and again a destruction of the $t^{1/6}$ power-law degradation. In the calculations of Naphade et al., however, this effect is not present due to the reflecting lateral boundary conditions which in this case are not justifiable anymore. Apart from these issues, the poly H/H_2 RD model shares the shortcomings of all RD-based models with respect to the prediction of NBTI recovery. Based on our calculations with increased interfacial diffusion coefficients, which corresponds to the increased oxide diffusion coefficient in the model of Naphade et al., we expect this problem to be also present in the stochastic version.

15.8 Conclusion

Our work shows that the reaction–diffusion model for the negative bias temperature instability, which has been used for nearly 40 years to interpret experimental data, has a number of inherent assumptions on the underlying physics that lack any physical justification. Those are:

1. *Continuous diffusion in the sub-nm regime.* Diffusion of neutral hydrogen atoms and H_2 proceeds via jumps between the interstitial sites of the host material. Positional changes that are smaller than about 4 Å are atomic vibrations around an equilibrium position and thus not diffusive in nature. This is especially relevant as in the macroscopic modified H–H_2 RD model, the onset of the power-law regime is quite discretization dependent.
2. *Instantaneous well-stirredness along the interface.* The one-dimensional macroscopic RD model, which gives the experimentally relevant $t^{1/6}$ behavior, inherently assumes that all hydrogen atoms that are liberated during stress instantaneously compete with all other hydrogen atoms at the interface for available dangling bonds or dimerize with each other. However, at typically assumed dangling bond densities of $5 \times 10^{12}\,\mathrm{cm}^{-2}$, the distance between two dangling bonds will be about 4.5 nm. At a depassivation level of 1% this means that the average initial distance between two hydrogen atoms is even in the range of 45 nm. The reduction of this distance to the typical H_2 bonding distance of 0.7 Å [30] needs to be overcome by a diffusion step, which takes about 200 s at a diffusion coefficient of $10^{-13}\,\mathrm{cm}^2/\mathrm{s}$.
3. *Rate-equation-based description.* It is well established in chemical literature that bimolecular reactions are not sufficiently described by reaction rate equations if the particle numbers are small. In a reaction rate equation system it is for instance possible for 0.5 H atoms to form 0.25 H_2, which is physically meaningless. An accurate description in the limit of small particle numbers is only obtained from an atomistic description.

We have implemented a stochastic three-dimensional modified reaction–diffusion model for NBTI to study the degree to which a more realistic description changes the predicted behavior. The model is theoretically well founded on the theory of stochastic chemical kinetics and is understood as a consequent realization of the physical picture behind the reaction–diffusion theory.

The degradation predicted by the microscopic model features a unique new initial regime in which the motion of each hydrogen atom is completely independent from the others. This regime features a strongly increased power-law exponent that is not observed experimentally, yet it is a necessary consequence of the liberation of hydrogen during stress. Application of the atomic RD model to a real-world example shows that for a realistic jump width it is impossible to obtain the experimentally observed behavior due to the apparent diffusion limitation of the dimerization and passivation rates. The match of the microscopic model with the macroscopic version and experimental data can be improved by using an increased diffusion coefficient at the interface. However, the required diffusion coefficients are many orders of magnitude above $10^{-9}\,\mathrm{cm}^2/\mathrm{s}$, which corresponds to a diffusion length of $100\,\mu\mathrm{m}$ after 100 ks. The lateral diffusion of the hydrogen in this case would reach way beyond the dimensions of individual microelectronic devices, leading to cross talk and a dramatically increased degradation due to the loss of hydrogen.

The recovery predicted by the microscopic model matches the macroscopic counterpart as soon as the previous degradation has entered the classical diffusion-limited regime. This behavior is due to the prerequisite that the system has to be equilibrated along the interface before the $t^{1/6}$ regime can emerge. As the recovery happens on much larger time-scales than the stress duration, lateral equilibration effects are invisible in recovery traces. A distribution of arrival times as predicted by the simple estimate using different diffusion coefficients during recovery as in [11] could not be found.

In summary, our study of the microscopic limit reveals a number of serious problems in the traditional mathematical formulation of the reaction diffusion model for NBTI, rendering all variants that are based on partial differential equations physically meaningless. In a physically meaningful microscopic version of the model, no experimental feature remains that can be accurately predicted. The apparent match of the RD models with experimental data must therefore be considered a mathematical artifact without any physical background.

Acknowledgements This work has received funding from the EC's FP7 grant agreement NMP.2010.2.5-1 (MORDRED).

References

1. K. Jeppson, C. Svensson, J.Appl.Phys. **48**(5), 2004 (1977)
2. D.K. Schroder, Microelectronics Reliability **47**, 841 (2007)
3. H. Kufluoglu, M. Alam, IEEE Trans.Electron Devices **53**(5), 1120 (2006). DOI 10.1109/TED.2006.872098
4. T. Grasser, W. Goes, B. Kaczer, IEEE Trans.Device and Materials Reliability **8**(1), 79 (2008). DOI 10.1109/TDMR.2007.912779
5. S. Ogawa, N. Shiono, Physical Review B **51**(7), 4218 (1995)
6. B. Kaczer, V. Arkhipov, R. Degraeve, N. Collaert, G. Groeseneken, M. Goodwin, In *Proc. Intl.Rel.Phys.Symp.* (2005), pp. 381–387

7. H. Reisinger, O. Blank, W. Heinrigs, A. Mühlhoff, W. Gustin, C. Schlünder, In *Proc. Intl.Rel.Phys.Symp.* (2006), pp. 448–453
8. T. Grasser, W. Goes, V. Sverdlov, B. Kaczer, In *Proc. Intl.Rel.Phys.Symp.* (2007), pp. 268–280. DOI 10.1109/RELPHY.2007.369904
9. T. Grasser, B. Kaczer, W. Gös, H. Reisinger, T. Aichinger, P. Hehenberger, P.J. Wagner, F. Schanovsky, J. Franco, P.J. Roussel, M. Nelhiebel, In *Proc. Intl.Electron Devices Meeting* (2010), pp. 82–85
10. S. Mahapatra, V.D. Maheta, A.E. Islam, M.A. Alam, IEEE Trans.Electron Devices **56**(2), 236 (2009)
11. S. Mahapatra, A. Islam, S. Deora, V. Maheta, K. Joshi, A. Jain, M. Alam, In *Proc. Intl.Rel.Phys.Symp.* (2011), pp. 6A.3.1 –6A.3.10. DOI 10.1109/IRPS.2011.5784544
12. S. Mahapatra, A. Islam, S. Deora, V. Maheta, K. Joshi, M. Alam, In *Proc. Intl.Symp. on Physical and Failure Analysis of Integrated Circuits* (2011), pp. 1–7. DOI 10.1109/IPFA.2011.5992794
13. K. Joshi, S. Mukhopadhyay, N. Goel, S. Mahapatra, In *Proc. Intl.Rel.Phys.Symp* (2012), pp. 5A.3.1–10
14. A. Islam, H. Kufluoglu, D. Varghese, S. Mahapatra, M. Alam, IEEE Trans.Electron Devices **54**(9), 2143 (2007). DOI 10.1109/TED.2007.902883
15. A.E. Islam, H. Kufluoglu, D. Varghese, M.A. Alam, Appl.Phys.Lett. **90**(8), 083505 (2007). DOI: 10.1063/1.2695998. URL http://dx.doi.org/doi/10.1063/1.2695998
16. H. Kufluoglu, M. Alam, IEEE Trans.Electron Devices **54**(5), 1101 (2007)
17. A. Islam, H. Kufluoglu, D. Varghese, M. Alam, Appl.Phys.Lett. **90**(1), 083505 (2007)
18. F. Schanovsky, T. Grasser, In *Proc. Intl.Integrated Reliability Workshop* (2011), pp. 17–21
19. F. Schanovsky, T. Grasser, In *Proc. Intl.Rel.Phys.Symp* (2012), pp. XT.10.1–6
20. A. Stesmans, B. Nouwen, V.V. Afanas'ev, Phys. Rev. B **58**, 15801 (1998). DOI 10.1103/PhysRevB.58.15801. URL http://link.aps.org/doi/10.1103/PhysRevB.58.15801
21. A. Islam, M. Alam, J.Comp.Elect. pp. 1–11 (2011). URL http://dx.doi.org/10.1007/s10825-011-0369-4. 10.1007/s10825-011-0369-4
22. D.A. McQuarrie, J.Appl.Prob **4**(3), 413 (1967)
23. P. Hänggi, P. Talkner, M. Borkovec, Rev.Mod.Phys **62**(2), 251 (1990)
24. S. Torquato, C.L.Y. Yeong, J.Chem.Phys. **106**, 8814 (1997)
25. S.S. Andrews, D. Bray, Phys.Biol. **1**, 137 (2004)
26. R. Erban, S.J. Chapman, Phys.Biol. **6**, 046001 (2009)
27. S.A. Isaacson, D. Isaacson, Physical Review E **80**, 066106 (2009)
28. D. Fange, O.G. Berg, P. Sjöberg, J. Elf, Proc.Nat.Acad.Sci. **107**(46), 19820 (2010)
29. G. Malavasi, M.C. Menziani, A. Pedone, U. Segre, Journal of Non-Crystalline Solids **352**(3), 285 (2006). DOI 10.1016/j.jnoncrysol.2005.11.022. URL http://www.sciencedirect.com/science/article/pii/S0022309305007994
30. P.E. Blöchl, Physical Review B **62**(10), 6158 (2000)
31. A. Bongiorno, L. Colombo, F. Cargnoni, Chem.Phys.Lett. **264**, 435 (1997)
32. D. Gillespie, J.Comp.Phys. **22**, 403 (1976)
33. D.T. Gillespie, in *Proc. Int. Conf. Form. Meth. Sys. Bio.* (Springer-Verlag, Berlin, Heidelberg, 2008), SFM'08, pp. 125–167. URL http://dl.acm.org/citation.cfm?id=1786698.1786704
34. T. Grasser, Microelectronics Reliability **52**(1), 39 (2012). DOI 10.1016/j.microrel.2011.09.002
35. V. Huard, M. Denais, C. Parthasarathy, Microelectronics Reliability **46**(1), 1 (2006)
36. B. Tuttle, **61**(7), 4417 (2000)
37. S.T. Pantelides, L. Tsetseris, S. Rashkeev, X. Zhou, D. Fleetwood, R. Schrimpf, Microelectronics Reliability **47**(6), 903 (2007). DOI DOI: 10.1016/j.microrel.2006.10.011. URL http://www.sciencedirect.com/science/article/pii/S0026271406003817
38. G. Panagopoulos, K. Roy, IEEE Trans.Electron Devices **58**(8), 2337 (2011). DOI 10.1109/TED.2011.2148720
39. S. Choi, Y. Park, C.K. Baek, S. Park, in *Proc. Simu.Semicond.Proc.Dev.* (2012), pp. 185–188
40. T. Naphade, N. Goel, P.R. Nair, S. Mahapatra, in *Proc. Intl.Rel.Phys.Symp.* (2013)

Chapter 16
Advanced Modeling of Oxide Defects

Wolfgang Goes, Franz Schanovsky, and Tibor Grasser

Abstract During the last couple of years, there is growing experimental evidence which confirms charge trapping as the recoverable component of BTI. The trapping process is believed to be a non-radiative multiphonon (NMP) process, which is also encountered in numerous physically related problems. Therefore, the underlying NMP theory is frequently found as an important ingredient in the youngest BTI reliability models. While several different descriptions of the NMP transitions are available in literature, most of them are not suitable for the application to device simulation. In this chapter, we will present a rigorous derivation that starts out from the microscopic Franck–Condon theory and yields generalized trapping rates accounting for all possible NMP transitions with the conduction and the valence band in the substrate as well as in the poly-gate. Most importantly, this derivation considers the more general quadratic electron–phonon coupling contrary to several previous charge trapping models. However, the pure NMP transitions do not suffice to describe the charge trapping behavior seen in time-dependent defect spectroscopy (TDDS). Inspired by these measurements, we introduced metastable states, which have a strong impact on the trapping dynamics of the investigated defect. It is found that these states provide an explanation for plenty of experimental features observed in TDDS measurements. In particular, they can explain the behavior of fixed as well as switching oxide hole traps, both regularly observed in TDDS measurements.

16.1 Introduction

For a long time, the research in bias temperature instability (BTI) was dominated by variants of the reaction–diffusion (RD) model [1–8], discussed in [9]. In the course of the last decade it was realized that the concept of the RD model cannot

W. Goes (✉) • F. Schanovsky • T. Grasser
Institute for Microelectronics, TU Vienna, Gusshausstrasse 27-19, 1040 Vienna, Austria
e-mail: goes@iue.tuwien.ac.at; schanovsky@iue.tuwien.ac.at; grasser@iue.tuwien.ac.at

T. Grasser (ed.), *Bias Temperature Instability for Devices and Circuits*,
DOI 10.1007/978-1-4614-7909-3_16,

explain BTI [8, 10, 11]. At the same time, a new measurement technique called time-dependent defect spectroscopy (TDDS) emerged, which indicated that some sort of charge trapping is involved in BTI. This method is capable of detecting single charge emission events from individual defects [12–18] in recovery traces that last up to a few hundred seconds. Thus TDDS allows for the analysis of the recoverable component of BTI and opened the doors toward in-depth investigations of the physical trapping mechanism underlying BTI. For a detailed description of this measurement method see [19].

First variants of charge trapping models relied on elastic hole tunneling of holes between the substrate and oxide defects [20–24]. However, these models show a negligible temperature dependence, which is in contrast to what has been observed experimentally. Other variants were based on the famous Shockley–Read–Hall (SRH) model [25] and modified to account for the tunneling effect [26] and the thermal activation of BTI [27, 28]. For the latter, transition barriers were phenomenologically introduced to reproduce the observed temperature dependence. They were reasoned by non-radiative multiphonon (NMP) transitions but were not rigorously derived from a microscopic theory [29–39]. The underlying theory provides a rigorous framework for the description of the charge transfer process between the substrate and the oxide defects in BTI. Hence, this theory forms the basis of our multi-state model and will be discussed in detail at first. Subsequently they will be simplified to make them applicable for analytical calculations.

Furthermore, TDDS studies demonstrated that the trapping dynamics must involve metastable states as well as thermal but field-insensitive transitions. This observation suggested a bistable BTI defect, which features quite complicated trapping dynamics including two-step capture and emission processes. This kind of defect also allows for different transition paths, which can explain the dual trap behavior seen in TDDS. For validation of this new model, we will evaluate the simulation results to the experimental data obtained from TDDS.

16.2 Benchmarks for a BTI Model

As a result of the continuous downscaling of the device geometries, single charge detrapping events have become visible as discrete steps in the BTI recovery curves. These steps came into the focus of scientific interest so that measurement techniques, such as TDDS, have become frequently employed. The TDDS relates these steps to several single charging or discharging events [12, 13, 18] of defects and therefore allows for the analysis of individual oxide defects and their trapping behavior. As such, the findings from TDDS [14–17] are used as criteria for the development of an atomistic BTI model and are listed in the following:

(i) The plot in Fig. 16.1 reveals that the defects exhibit a strong, nearly exponential stress voltage dependence of τ_c. Empirically, this dependence can be described by $\exp(-c_1 F_{ox} + c_2 F_{ox}^2)$. However, it differs from defect to defect, implying that it is related to certain defect properties.

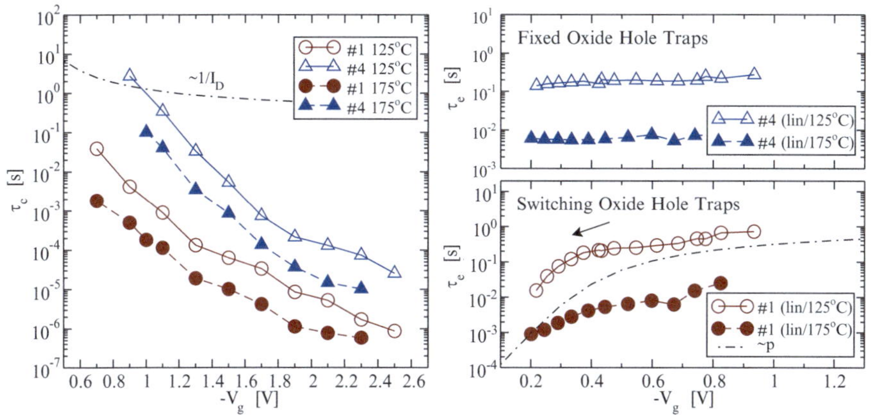

Fig. 16.1 *Left*: The capture time constants τ_c as a function of V_g for two defects at different temperatures extracted from a single device. *Open and closed symbols* mark measurements carried out at 125 °C and 175 °C, respectively. The τ_c curves show a strong field acceleration and temperature activation. However, the observed field acceleration does not follow the $1/I_d \approx 1/p$ dependence (*dot-dashed line*) as predicted by the conventional SRH model. *Right*: The emission time constants τ_e for single defects gathered from the TDDS for varying recovery gate voltages. The two distinct field dependences (*upper and lower panel*) suggest the existence of two types of defects present in the oxide. The defect #1 shows different field behaviors depending on whether the device is operated in the linear or the saturation regime during the measurement (not shown here). This suggests that the electrostatics within the device are responsible for the two distinct field dependences. It is noteworthy that the drop in τ_e goes hand in hand with the decrease in the interfacial hole concentration p (*dot-dashed line*)

(ii) The time constant plots show a marked temperature dependence, which becomes obvious by the downward shift of the τ_c curves at higher temperatures. The activation energies extracted from Arrhenius plots are about 0.6 eV.
(iii) One type of the oxide defects ("fixed oxide hole traps") has a τ_e that remains unaffected by changes in V_g [40, 41].
(iv) The other type ("switching oxide hole traps") shows a drop in τ_e toward lower V_g [40–42].
(v) The τ_e of both types shows a temperature activation with a large spread (0.6–1.4 eV).

Furthermore, it was found that several TDDS recovery traces display random telegraph noise (RTN) when studying a device at certain bias conditions [14]. After a while, this RTN signal vanishes and does not reoccur during the remaining measurement time. The termination of the noise signal is ascribed to hole traps which change to their neutral charge state and remain therein. This kind of noise is termed temporary RTN [14] (tRTN) since it occurs only for a limited amount of time. A similar phenomenon called anomalous RTN (aRTN) was discovered earlier by Kirton and Uren [27]. Therein, electron traps were observed, which repeatedly produce noise for random time intervals. During the interruptions of this RTN signal, the defects dwell in their negative charge state generating no RTN noise signal. The behavior of these traps was interpreted by the existence of a metastable defect state. Unfortunately, there exist only a quite limited amount of noise data so that no reliable statistics can be generated. Nevertheless, it is viewed as a stringent requirement that the sought BTI model can also capture these noise phenomena in principle.

16.3 Previous Modeling Attempts

Early BTI modeling attempts relied on the classical reaction–diffusion model [1, 2, 5–7] or variants thereof [3, 5, 8] accounting for dispersive diffusion [3, 8] and three-dimensional effects [10, 11]. Even though these models are still popular, it has been demonstrated that the underlying concept cannot describe the basic feature of BTI (see [9]). As an alternative explanation for BTI, charge trapping based on elastic electron tunneling was previously suggested. However, this process exhibits a far too weak temperature dependence as compared to measurements. The next evolution of trapping models rested upon SRH theory combined with elastic tunneling, thereby mimicking an inelastic and thus temperature-activated trapping process. To its disadvantage, the underlying trapping process is not specified within the general SRH framework and can therefore not be linked to simulations based on well-founded atomistic theories. A prototype version of this SRH model was proposed by McWhorter [26], who extended the SRH equations by the factor $\exp(x_\mathrm{t}/x_0)$ in order to account for the effect of electron tunneling. Since this model suffers from a weak temperature dependence of τ_c and small time constants, Kirton and Uren [27] incorporated a term with field-independent energy barriers ΔE_b. This "ad hoc" introduction of barriers has been motivated by the theory of non-radiative multi-phonon transitions (NMP) process [38]. However, Kirton and Uren did not provide a detailed theoretical derivation based on NMP theory. Nevertheless, their work must be regarded as a substantial improvement in the interpretation of charge trapping at semiconductor–oxide interfaces. In this variant, the capture and emission time constants read

$$\tau_\mathrm{c} = \tau_0 \exp\left(\frac{x_\mathrm{t}}{x_0}\right) \exp(\beta \Delta E_\mathrm{b}) \frac{N_\mathrm{v}}{p} \begin{cases} 1, & E_\mathrm{t} > E_\mathrm{v} \\ \exp(-\beta \Delta E_\mathrm{t}) \exp(\beta q_0 F_\mathrm{ox} x_\mathrm{t}), & E_\mathrm{t} < E_\mathrm{v} \end{cases} \quad (16.1)$$

$$\tau_\mathrm{e} = \tau_0 \exp\left(\frac{x_\mathrm{t}}{x_0}\right) \exp(\beta \Delta E_\mathrm{b}) \begin{cases} \exp(\beta \Delta E_\mathrm{t}) \exp(-\beta q_0 F_\mathrm{ox} x_\mathrm{t}), & E_\mathrm{t} > E_\mathrm{v} \\ 1, & E_\mathrm{t} < E_\mathrm{v} \end{cases} \quad (16.2)$$

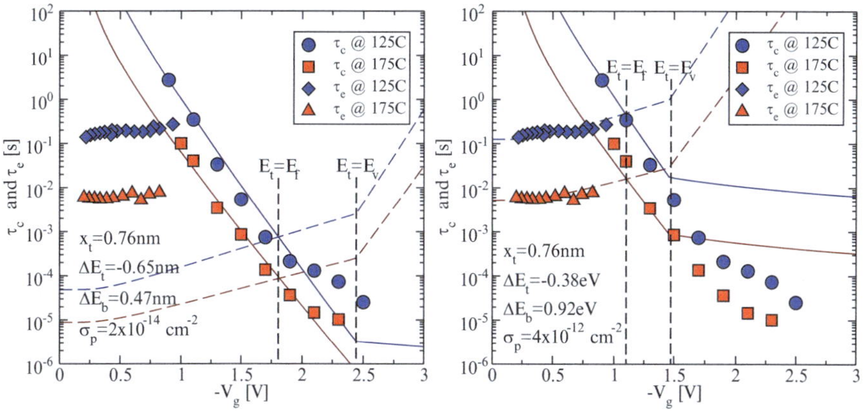

Fig. 16.2 Two fits of the Kirton model to the TDDS data. The symbols stand for the measurement data and the lines represent the simulated time constants. *Left*: When the Kirton model is optimized to the hole capture times τ_c, reasonable fits can be achieved but τ_e is predicted three orders of magnitudes too low. *Right*: Alternatively, a good agreement can be obtained for the hole emission times τ_e but with a strong mismatch of the capture times τ_c for $E_t > E_v$. From this it is concluded that the Kirton model is not capable of fitting τ_c and τ_e at the same time

where the trap level E_t is defined as

$$E_t(x_t) = E_v + \underbrace{E_{t,0} - E_{v,0}}_{=\Delta E_t} - q_0 x_t F_{ox} \tag{16.3}$$

with $E_{t,0}$ and $E_{v,0}$ denoting the trap level and the valence band edge in the absence of an electric field.

The behavior of the model with respect to the temperature and the oxide field is illustrated in the left plot of Fig. 16.2 (left). When the trap level lies below the valence band edge ($E_t < E_v$), τ_c shows an exponential field dependence. At low gate biases, the breakdown of the inversion layer gives rise to a drop in the hole concentration and in consequence to a strong increase in τ_c. Comparing the model to the experimental TDDS data, this exponential behavior allows for reasonably good, approximative fits of τ_c but is still incompatible with the observed curvature in τ_c (see the left fit in Fig. 16.2). τ_e is experimentally observed to be field insensitive, which goes hand in hand with (16.2) based on Boltzmann statistics. However, when accurate Fermi–Dirac statistics (as implemented in device simulators) are employed, the emission times exhibit a weak field dependence that agrees reasonably well with the behavior of fixed oxide hole traps (constant emission times) but is incompatible with the behavior of switching oxide hole traps (a drop at weak oxide fields). Alternatively, when τ_e is optimized in the Kirton model (see right fit in Fig. 16.2), a reasonable fit can be achieved but at the same time a strong mismatch arises for τ_c in the range $E_t > E_v$. Furthermore, Fig. 16.2 reveals that the introduction of ΔE_b yields the required temperature activation and larger time constants in agreement with the points (ii) and (v) of the TDDS findings. Even though the model can reproduce several features seen in the TDDS data separately—except for the curvature in τ_c—no reasonable agreement with the whole set of measurement data can be achieved.

16.4 NMP Transitions Between Single States

Contrary to the previously discussed charge trapping models, the non-radiative multiphonon (NMP) theory [37–39] relies on a solid physical foundation. Its understanding requires the knowledge of fundamental microscopic theories, which are briefly discussed in the following. In the Huang–Born approximation, a certain atomic configuration is split into a system of electrons and nuclei, which are described by two separated Schrödinger equations.

$$\left\{\hat{T}_e + \hat{V}_{ee}(\mathbf{r}) + \hat{V}_{en}(\mathbf{r};\mathbf{R}) + \hat{V}_{nn}(\mathbf{R})\right\} \varphi_i(\mathbf{r};\mathbf{R}) = V_i(\mathbf{R})\varphi_i(\mathbf{r};\mathbf{R}) \tag{16.4}$$

$$\left\{\hat{T}_n + V_i(\mathbf{R})\right\} \eta_{i\alpha}(\mathbf{R}) = E_{i\alpha}\eta_{i\alpha}(\mathbf{R}) \tag{16.5}$$

These equations contain Coulomb contributions from the electron–electron ($\hat{V}_{ee}$), electron–nucleus ($\hat{V}_{en}$), and nucleus–nucleus ($\hat{V}_{nn}$) interactions as well as the kinetic energies of the electrons ($\hat{T}_e$) and the nuclei ($\hat{T}_n$). The electronic Hamiltonian in (16.4) depends on the electronic ($\mathbf{r}$) and the nuclear ($\mathbf{R}$) degrees of freedom, where the latter only enter parametrically. The solution $V_i(\mathbf{R})$ of the electronic Schrödinger equation (16.4) corresponds to the energy of a certain atomic configuration and acts as a potential for the nuclei in the Schrödinger equation (16.5). Therefore, $V_i(\mathbf{R})$ is usually referred to as the adiabatic potential energy. In the Huang–Born approximation, the nuclei of the atoms are treated as a system of quantum mechanical particles with quantized states $\eta_{i\alpha}$ and discrete energies $E_{i\alpha}$. Also the wavefunction of the composite electron–nucleus system is split into an electronic $\varphi_i(\mathbf{r};\mathbf{R})$ and nuclear $\eta_{i\alpha}(\mathbf{R})$ part, denoted the electronic and the vibrational wavefunction, respectively.

In the case of charge trapping in BTI, one deals with a process that is frequently termed "charge transfer reaction" in the theoretical literature. Such a kind of process must be described by a system consisting of all atoms involved. Since the trapped charge carrier is exchanged between the defect and the substrate, the system includes the atoms surrounding the BTI defect as well as the atoms in the substrate. Altogether, these atoms span a $3N$-dimensional space with N being the number of considered atoms. The adiabatic potential energy surface in this configurational space is usually visualized in a configuration coordinate diagram (see Fig. 16.3). Therein, the atomic positions are reduced to a one-dimensional quantity called configuration coordinate, which allows to describe the correlated motion of atoms, such as lattice relaxation. In these plots, the adiabatic potential energy surfaces assume an almost parabolic shape for small atomic displacements and are thus usually approximated by harmonic quantum oscillators in solid state theory.

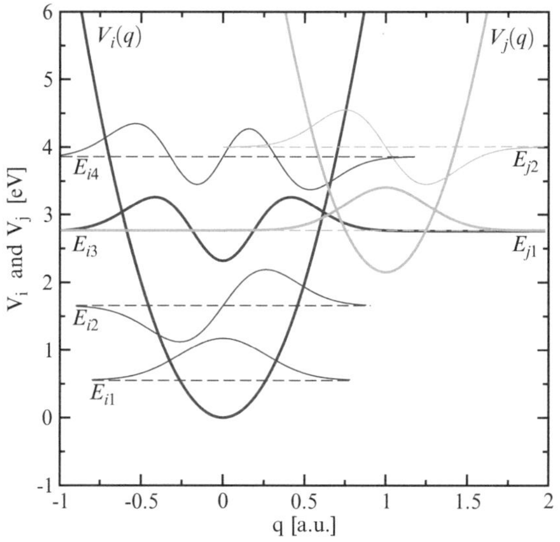

Fig. 16.3 Adiabatic potentials involved in a charge transfer reaction. Each of the two parabola corresponds to one charge state of the defect where the left (V_i) one represents the initial and the right (V_j) one the final charge state. Their corresponding wavefunctions and eigenenergies are depicted as *solid* and *dashed lines*, respectively. An NMP transition only occurs when the initial and the final energies coincide as it is the case for E_{i3} and E_{j1}. Then the overlap of their corresponding vibrational wavefunctions enters the calculation of the lineshape function f_{ij} and consequently determines the NMP transition probability

During a charge trapping process, the defect changes from the charge state i to j, where each of the charge states is represented by its own adiabatic potential in the configuration coordinate diagram (see Fig. 16.3). The NMP transition rate k_{ij} is then derived from first-order time-dependent perturbation theory using the Franck–Condon approximation [37,39,43,44].

$$k_{ij} = A_{ij} f_{ij} \tag{16.6}$$

$$A_{ij} = \frac{2\pi}{\hbar} |\langle \varphi_i | V' | \varphi_j \rangle|^2 \tag{16.7}$$

$$f_{ij} = \operatorname*{ave}_{\alpha} \sum_{\beta} |\langle \eta_{i\alpha} | \eta_{j\beta} \rangle|^2 \tag{16.8}$$

Here, "ave" stands for the thermal average over all initial states "α" and the sum runs over the final states "β". A_{ij} is the electronic matrix element with the adiabatic operator as a perturbation V' and is associated with a simple electronic transition. The Franck–Condon factor $|\langle \eta_{i\alpha} | \eta_{j\beta} \rangle|^2$ in (16.8) only gives a contribution when the initial and the final state have the same energy. If this is the case, this factor is calculated as the overlap integral of the two vibrational wavefunctions "$i\alpha$" and "$j\beta$" and corresponds to the respective transition probability (cf. Fig. 16.3). Calculating the thermal average over the initial states α and summing over the final

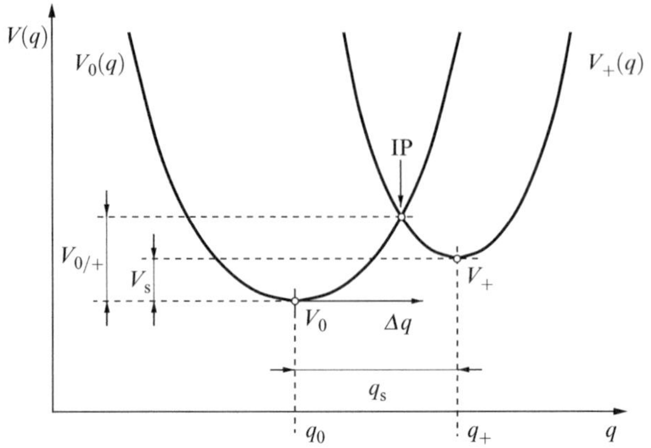

Fig. 16.4 The configuration coordinate diagram for an NMP transition. The adiabatic potentials for the initial and the final states are denoted as $V_0(q)$ and $V_+(q)$, respectively. They are defined by their corresponding minima V_0 and V_+ located at their equilibrium configurations q_0 and q_+, respectively. To simplify the mathematical calculations, the axis origin is shifted into the energy minimum V_0

states β yield the lineshape function f_{ij} that will be found to govern the gate bias and temperature dependence of the NMP transition rate. In solids the eigenspectrum $E_{i\alpha}$ is usually densely spaced so that there are numerous possible transitions from the initial charge i to the final charge state j. This lineshape function has its largest contributions from those energies that lie close to the intersection point (IP) of the adiabatic potentials. Around this point, the lineshape function is assumed to have a Dirac peak in the classical limit [45]. This assumption allows for simple analytical expressions that can be conveniently employed for device simulation.

In the following, the NMP transition rates will be derived for a defect which changes between its neutral (0) and its positive (+) charge state upon hole trapping or detrapping. The corresponding initial ($i = 0$) and final ($j = +$) potential energy surface can be expressed as

$$V_0(q) = c_0(q - q_0)^2 + V_0 = c_0 \Delta q^2 + V_0 \tag{16.9}$$

$$V_+(q) = c_+(q - q_+)^2 + V_+ = c_+(\Delta q - q_s)^2 + V_0 + V_s \tag{16.10}$$

using the quantities defined in Fig. 16.4 and the shorthands $\Delta q = q - q_0$, $q_s = q_+ - q_0$, and $V_s = V_+ - V_0$. c_0 and c_+ denote the curvature of the adiabatic potentials for the neutral and the positively charged defect, respectively. Without loss of generality, V_0 can be chosen to be zero and will thus be neglected from now on. Note that the two parabolas are characterized by different curvatures ($c_0 \neq c_+$), implying that there exist two intersection points given by

$$\Delta q_{1,2} = \frac{c_+ q_s \pm \sqrt{c_0 c_+ q_s{}^2 + V_s(c_0 - c_+)}}{c_+ - c_0}. \tag{16.11}$$

In the literature, this case is usually referred to as quadratic electron–phonon coupling. For equal curvatures ($c_0 = c_+ = c$), linear electron–phonon coupling is obtained, which yields only one intersection point located at

$$\Delta q_1 = \frac{V_s/c + q_s^2}{2q_s}. \tag{16.12}$$

The classical lineshape function for hole capture is obtained from

$$f_{0/+}(c_0, c_+, q_s, V_0, V_+) = Z^{-1} \int_q e^{-\beta V_0(q')} \delta\left(V_0(q') - V_+(q')\right) dq' \tag{16.13}$$

with the partition function

$$Z = \int_q e^{-\beta V_0(q')} dq'. \tag{16.14}$$

In accordance with the classical limit, the Dirac delta function in (16.13) ensures that the integral is only evaluated at the intersection point of the two parabolas. Using the integration rule for Dirac delta functions, this integral evaluates to

$$\int_{\Delta q} e^{-\beta V_0(\Delta q')} \delta\left(V_0(\Delta q') - V_+(\Delta q')\right) d(\Delta q')$$
$$= \frac{e^{-\beta c_0 \Delta q_1^2}}{|2c_0\Delta q_1 - 2c_+(\Delta q_1 - q_s)|} + \frac{e^{-\beta c_0 \Delta q_2^2}}{|2c_0\Delta q_2 - 2c_+(\Delta q_2 - q_s)|} \tag{16.15}$$

and the partition function simplifies to

$$\int_{-\infty}^{+\infty} e^{-\beta c_0 \Delta q'^2} d(\Delta q') = \sqrt{\frac{\pi}{c_0 \beta}}. \tag{16.16}$$

Inserting (16.15) and (16.16) into the definition of the lineshape function (16.13) leads to [45]

$$f_{0/+}(c_0, c_+, q_s, V_0, V_+) = f_{0/+}(c_0, c_+, q_s, V_s)$$
$$= \frac{1}{2}\sqrt{\frac{c_0\beta}{\pi}} \left(\frac{e^{-\beta c_0 \Delta q_1^2}}{|c_0\Delta q_1 - c_+(\Delta q_1 - q_s)|} + \frac{e^{-\beta c_0 \Delta q_2^2}}{|c_0\Delta q_2 - c_+(\Delta q_2 - q_s)|} \right). \tag{16.17}$$

Keep in mind that the lineshape function may also vanish ($f_{0/+} = 0$) when the two parabolas do not share a common intersection point. For linear electron–phonon coupling ($c = c_0 = c_+$), the above expression reduces to

$$f_{0/+}(c, q_s, V_0, V_+) = f_{0/+}(c, q_s, V_s) = \frac{1}{2}\sqrt{\frac{c\beta}{\pi}} \frac{e^{-\beta c \Delta q_1^2}}{c|q_s|} \tag{16.18}$$

with

$$\Delta q_1 = \frac{V_s/c + q_s{}^2}{2q_s}. \tag{16.19}$$

It is emphasized here that the lineshape function is most strongly affected by the exponential term, where the expression $c_0 \Delta q_{1,2}{}^2$ can be identified with the energy barrier from the minimum V_0 to the saddle point IP (cf. Fig. 16.4). This NMP transition barrier can be expressed as

$$\begin{aligned} V_{0/+} &= V_0(\Delta q_{1,2}) \\ &= \frac{c_0 q_s^2}{(\frac{c_0}{c_+} - 1)^2} \left(1 \pm \sqrt{\frac{c_0}{c_+} + \frac{V_s(\frac{c_0}{c_+} - 1)}{c_+ q_s^2}} \right)^2, \end{aligned} \tag{16.20}$$

or

$$V_{0/+} = \left(\frac{V_s + c q_s^2}{2\sqrt{c} q_s} \right)^2 \tag{16.21}$$

for linear electron–phonon coupling. For hole emission the roles of the initial and the final states are reversed. The corresponding lineshape function $f_{+/0}$ and the NMP barrier $V_{+/0}$ are of the same form as in (16.17) and (16.20), respectively, but have their subscripts "0" and "+" exchanged.

As will be demonstrated in Sect. 16.7, the NMP transition barrier varies strongly with the temperature and the gate bias and therefore governs the trapping behavior of BTI defects. In the following calculations, the above analytical expressions for the lineshape function are preferred to the Franck–Condon overlap factors since they can be easily implemented in simple device simulators at computational feasible costs.

Next, the NMP theory has to be specified for the situation of charge capture and emission in MOSFETs. Therefore, the energy minima V_0 and V_+ at the potential energy surfaces must be linked to the energy of the transferred electron in the band energy diagram before and after an NMP transition. In a simplified picture, it can be envisioned that only the energy of the transferred electron changes while the energy of the other electrons ($\tilde{V}_0$) remains unaffected. In the following, we discuss a hole capture[1] process, during which an electron is emitted from the energy level E_t of a trap into an energy level E in the substrate valance band state. Then the energy minima V_0 and V_+ can be expressed as

$$V_0 = \tilde{V}_0 + E_t \tag{16.22}$$

$$V_+ = \tilde{V}_0 + E \tag{16.23}$$

[1] It is stressed that the term "hole capture" refers to either a capture of hole from the valence band into a trap or an emission of an electron from the trap into the valence band. Keep in mind that both of these processes are equivalent from a physical point of view.

with $\tilde{V}_0$ being the energy of the system minus the energy of the transferred electron. The NMP transition rate is then written as

$$k_{0/+} = A_{0/+}(E) f_{0/+}(c_0, c_+, q_s, E - E_t). \tag{16.24}$$

The unknown auxiliary quantity $\tilde{V}_0$ cancels out in the lineshape function, which only depends on the energy difference

$$V_s = V_+ - V_0 = E - E_t \,. \tag{16.25}$$

The trap wavefunction in the electronic matrix element $A_{0/+}(E)$ is strongly localized around the defect so that the integrand in (16.7) has its largest contribution at the defect site and $A_{0/+}(E, x_t)$ can be approximated by

$$\begin{aligned} A_{0/+}(E, x_t) &= A_0 |\langle x_t | \varphi \rangle|^2 = A_0 |\varphi(x_t)|^2 \\ &= A_1 \lambda(E, x_t) \,. \end{aligned} \tag{16.26}$$

Here, A_0 is a not further specified prefactor and $\varphi(E)$ stands for the channel wavefunction with an energy E. The electronic matrix element is governed by the exponential decay of the channel wavefunction and can be approximated using a WKB factor $\lambda(E, x_t)$ for the implementation in simple device simulators.

16.5 NMP Transition with a Whole Band of States

So far, the theoretical foundation for NMP transitions between two certain states has been discussed. In BTI, however, the oxide defects interact with the whole conduction or valence band of the substrate so that the current formulation of the NMP processes must be extended to account for transitions with a multitude of band states at different energies E. For this reason, one has to introduce a summation over all possible valence band states n in (16.24). Since the valence band states form a continuous spectrum, this summation can also be transformed to an integral over a density of states [46].

$$\sum_n \rightarrow \Omega \int_{-\infty}^{E_v} D_p(E) \mathrm{d}E \tag{16.27}$$

Using the above transformation, the NMP hole capture rate can be expressed as

$$k_{0/+}^{pc} = \Omega \int_{-\infty}^{E_v} D_p(E) A_{0/+}(E, x_t) f_{0/+}(c_0, c_+, q_s, E - E_t) \mathrm{d}E \,. \tag{16.28}$$

The density of states $D_p(E)$ can be calculated using a simple expression based on the parabolic band approximation

$$D_p(E) = \sum_v \frac{g_v m_{pv}}{\hbar^3 \pi^2} \sqrt{2m_{pv}(E - E_c)}\,, \tag{16.29}$$

where g_v is the degeneracy of the vth valence band valley and m_p its corresponding effective hole mass. Alternatively, the density of states may originate from a more sophisticated Schrödinger–Poisson solver that allows for quantized states E_{vk} arising from the one-dimensional confinement of the charge carriers in the inversion layer.

$$D_p(E) = \sum_v \frac{g_v m_{pv}}{\hbar^2 \pi} \sum_k \Theta(E - E_{vk}) \tag{16.30}$$

Next, the hole occupancy of the band states (f_p for E) and electron occupancy of the trap state (f_t for E_t) have to be taken into account. Then, the resulting NMP transition rates read

$$k_{0/+}^{pc} = \Omega \int_{-\infty}^{E_v} D_p(E) f_p(E, E_f) A_{0/+}(E, x_t) f_{0/+}(c_0, c_+, q_s, E - E_t) f_t \mathrm{d}E\,. \tag{16.31}$$

For the case of electron emission,[2] the electron is emitted into the substrate conduction band and thus $D_p(E)$ must be replaced by $D_n(E)$.

$$k_{0/+}^{ne} = \Omega \int_{E_c}^{+\infty} D_n(E) f_p(E, E_f) A_{0/+}(E, x_t) f_{0/+}(c_0, c_+, q_s, E - E_t) f_t \mathrm{d}E \tag{16.32}$$

The configuration coordinate diagrams of both processes are combined in Fig. 16.5, which now covers all electron or hole transitions from the defect into the substrate. Interestingly, the final states span an energy spectrum V_+ that can be identified with band energy diagram including the conduction as well as the valence band. Each of these states is associated with a distinct position of its adiabatic potential $V_+(q)$ and thus has a different NMP barrier height along with a different transition probability according to the lineshape function in the transition rates (16.31) and (16.32) (cf. Fig. 16.6). For hole capture (case A), the defect has to undergo an NMP transition from the parabola $V_0(q)$ to the parabola $V_+(q)$. This transition occurs the fastest when $V_+(q)$ cuts the minimum of $V_0(q)$. Then the corresponding transition barrier $V_{0/+}$ is negligible and the lineshape function $f_{0/+}(E)$ reaches its maximum value. When hole emission is considered (case B), the roles of the initial and the final states

[2] Note that electron emission corresponds to hole capture into the substrate conduction band.

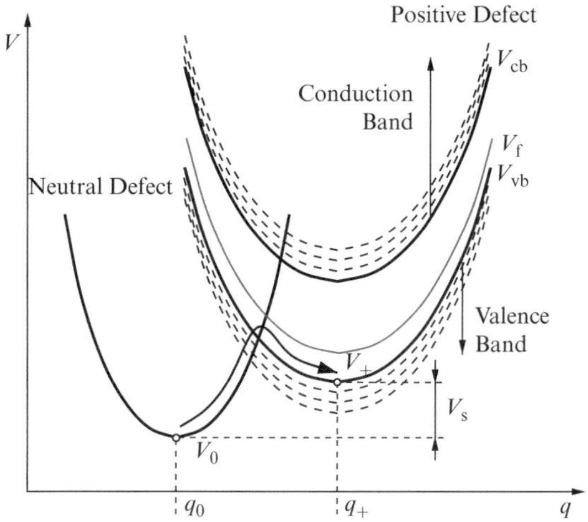

Fig. 16.5 A combined configuration coordinate diagram for hole capture and electron emission. According to the relation $V_+ = \tilde{V}_0 + E$, an electron located in an energetically higher band state E is represented by higher adiabatic potential $V_+(q)$. As a consequence, the upper and the lower family of curves constitute the set of adiabatic potentials V_+ associated with the conduction and valence band, respectively. It is noted that this configuration coordinate diagram remains unchanged for hole emission and electron capture and can therefore be used for both processes. As such, this diagram covers all possible NMP transitions of the considered defect with the substrate

are reversed so that the NMP transition proceeds from the adiabatic potential $V_+(q)$ to $V_0(q)$. Then the corresponding lineshape function $f_{+/0}(E)$ peaks when $V_+(q)$ is cut in its minimum. Note that the maximal transition rates for hole capture and emission are associated with different energy levels E, which are frequently referred to as the switching trap levels[3] in literature [47–54]. However, they should not be confused with the thermodynamic trap levels E_t that enter SRH-like formulations of the charge transfer process used here. The thermodynamic trap level (case C) is associated with the energy level E, at which the hole capture and emission are balanced and the two lineshape functions $f_{0/+}(E)$ and $f_{+/0}(E)$ assume the same value (cf. Fig. 16.6). In the configuration coordinate diagram, this is the case for the situation when the minima of adiabatic potentials $V_0(q)$ and $V_+(q)$ are at the same height. Note that special importance is attached to this energy level with respect to the equilibrium occupancy of the defect. If the Fermi level is located above the thermodynamic level, the dominating trapping process is hole emission and the defect becomes neutral. However, when the Fermi level falls below the thermodynamic level, the hole capture rate exceeds the hole emission rate and the defect becomes occupied by a hole.

[3]Note that the same term "switching trap level" is also used for the thermodynamic trap level for a switching oxide hole trap introduced in Fig. 16.1.

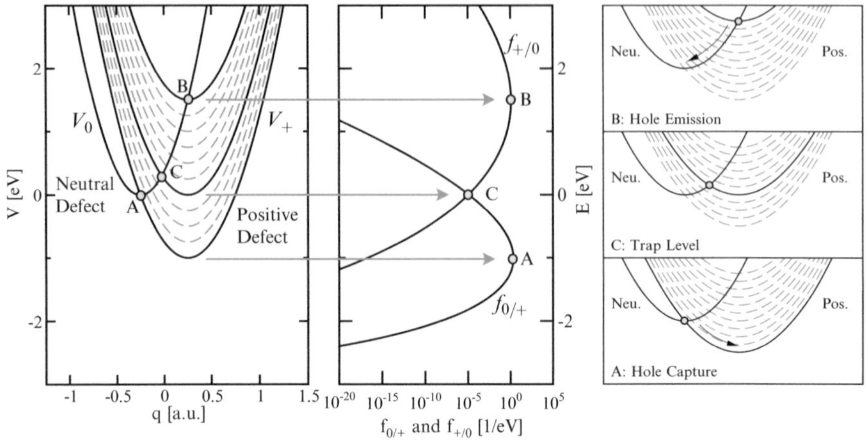

Fig. 16.6 Configuration coordinate diagram (*left*) for a continuum of adiabatic potentials $V_+(q)$, the corresponding lineshape functions (*middle*), and sketches of the cases A, B, and C (*right*). For hole capture, the lineshape function $f_{0/+}(E)$ reaches its maximal value when $V_+(q)$ intersects the minimum of $V_0(q)$ and thus the NMP transition has a vanishing barrier $V_{0/+}$ (case A). When changing from the configuration coordinate diagram (*left*) to the lineshape function (*middle*), the adiabatic potentials are converted to electron energies according to (16.23). For the hole emission, the analogous considerations apply as for hole capture. Now the intersection point must lie in the minimum of $V_+(q)$, giving rise to the peak of the lineshape function $f_{+/0}(E)$ (case B). If the minima of both parabolas coincide, the barriers for both directions have the same heights, which leads to equaling NMP transition rates (case C)

In semiconductor theory—especially when NBTI in pMOS transistors is considered—the trapping dynamics is preferentially described in the "hole picture." In this case the hole is emitted from a continuum of states where its energy in the initial state is undefined. By contrast, the hole energy is exactly specified by the trap level E_t in the final state (cf. Fig. 16.7). As a consequence, the trap level E_t and the band states E change their roles. Furthermore, the energy axis of the charge carriers is inverted so that the energy spectrum of V_+ in Fig. 16.5 is flipped in the hole picture in Fig. 16.7.

$$V_0 = \tilde{V}_0 - E \qquad (16.33)$$

$$V_+ = \tilde{V}_0 - E_t \qquad (16.34)$$

The energy difference of the adiabatic potentials is then given by

$$V_s = V_+ - V_0 = E - E_t\,, \qquad (16.35)$$

implying that the same activation energy is required as in the electron picture. Following the same derivation as for the electron picture, the NMP transition rate for hole capture reads

$$k_{0/+}^{\mathrm{pc}} = \Omega \int_{-\infty}^{E_v} D_p(E) f_p(E, E_f) A_{0/+}(E, x_t) f_{0/+}(c_0, c_+, q_s, E - E_t) f_t \mathrm{d}E\,. \qquad (16.36)$$

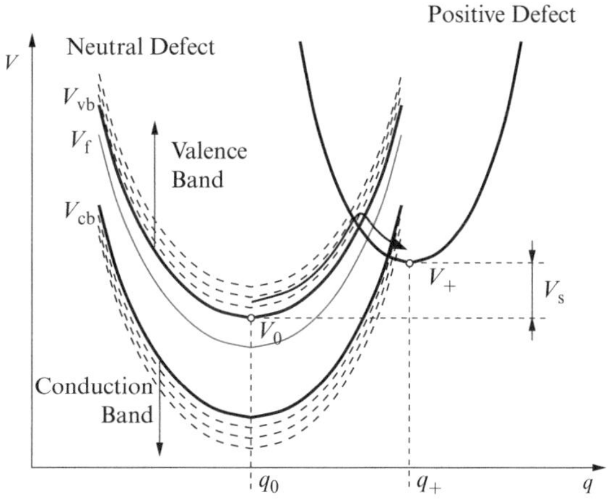

Fig. 16.7 The same configuration coordinate diagram as in Fig. 16.5 but in the "hole picture." Note that the energy scale of the charge carriers and thus the band diagram is inverted compared to the "electron picture." Furthermore, the energy of the transferred charge carrier is now undefined for the initial state since the hole is in one of the valence band states. By contrast, it can be specified by E_t when the hole is trapped

It is remarked that the electronic matrix elements $A_{0/+}(E,x_t)$ in the hole picture and in the electron picture equal since they are determined by the same channel wavefunction. Using the approximation (16.26), the whole set of NMP trapping rates can be written as

$$k^{nc} = k_0^n \int_{E_c}^{+\infty} D_n(E) f_n(E,E_f) \lambda(E,x_t) f_{+/0}(c_+,c_0,q_s,E_t - E) dE \tag{16.37}$$

$$k^{ne} = k_0^n \int_{E_c}^{+\infty} D_n(E) f_p(E,E_f) \lambda(E,x_t) f_{0/+}(c_0,c_+,q_s,E - E_t) dE \tag{16.38}$$

$$k^{pc} = k_0^p \int_{-\infty}^{E_v} D_p(E) f_p(E,E_f) \lambda(E,x_t) f_{0/+}(c_0,c_+,q_s,E - E_t) dE \tag{16.39}$$

$$k^{pe} = k_0^p \int_{-\infty}^{E_v} D_p(E) f_n(E,E_f) \lambda(E,x_t) f_{+/0}(c_+,c_0,q_s,E_t - E) dE\,, \tag{16.40}$$

where the quantities $k_0^{n/p}$ are used as shorthands for the product of the prefactors Ω and A_1. "n" and "p" refer to electrons or holes while "c" and "e" stand for capture and emission processes, respectively. It has to be noted that the integrands of the above rate equations are usually sharply peaked due to the strong exponential dependences of the occupancies $f_p(E)$ and $f_n(E)$ as well as the lineshape functions $f_{0/+}(E)$ and $f_{+/0}(E)$. Hence, these integrals are solved numerically using adaptive integration schemes in order to keep the computation costs low and to ensure a sufficient accuracy of the computed rates.

The above set of rate equations can also be modified to the case where the defect exchanges charge carriers with the poly-gate by replacing the band edges and the Fermi level with their respective values at the poly-gate. They can also be adapted for an electron trap, whose charge state switches between neutral and negative. As such, these rate equations form the basis for charge trapping involving the substrate as well as the gate and could consequently also cover trap-assisted tunneling occurring via NMP transitions.

16.6 Huang–Rhys Parameter

The employed NMP theory was initially derived for the fluorescence and absorption spectra of gases and solids, where the Huang–Rhys factor S was introduced to obtain compact analytical solutions [37]. This quantity corresponds to the number of absorbed or emitted phonons during an optical transition and thereby characterizes the shape of two adiabatic potentials $V_0(q)$ and $V_+(q)$. For quadratic electron–phonon coupling, the adiabatic potentials are represented by two parabolas that are shifted against each other and have different curvatures. To define them, we introduce the quantities S and R (see Fig. 16.8), which are defined as follows:

$$S\hbar\omega = c_0 q_s^2 \tag{16.41}$$

$$R^2 = \frac{c_0}{c_+}. \tag{16.42}$$

Using the above substitutions, the NMP transition barrier in (16.20) can be rewritten as

$$V_{0/+}(V_s) = \frac{S\hbar\omega}{(R^2-1)^2}\left(1 \pm R\sqrt{\frac{S\hbar\omega + V_s(R^2-1)}{S\hbar\omega}}\right)^2. \tag{16.43}$$

The prefactor $\xi_{0/+}(\Delta q)$ of the exponential term in (16.17) is of the form

$$\xi_{0/+}(\Delta q_{1,2}) = \sqrt{\frac{\beta c_0}{4\pi}}\frac{1}{|c_0\Delta q - c_+(\Delta q - q_s)|} \tag{16.44}$$

and can be expressed as

$$\xi_{0/+}(V_s) = \sqrt{\frac{\beta}{4\pi}}\frac{R}{\sqrt{S\hbar\omega + V_s(R^2-1)}}. \tag{16.45}$$

For linear electron–phonon coupling, one obtains the frequently applied result

$$V_{0/+}(V_s) = \frac{(V_s + S\hbar\omega)^2}{4S\hbar\omega} \tag{16.46}$$

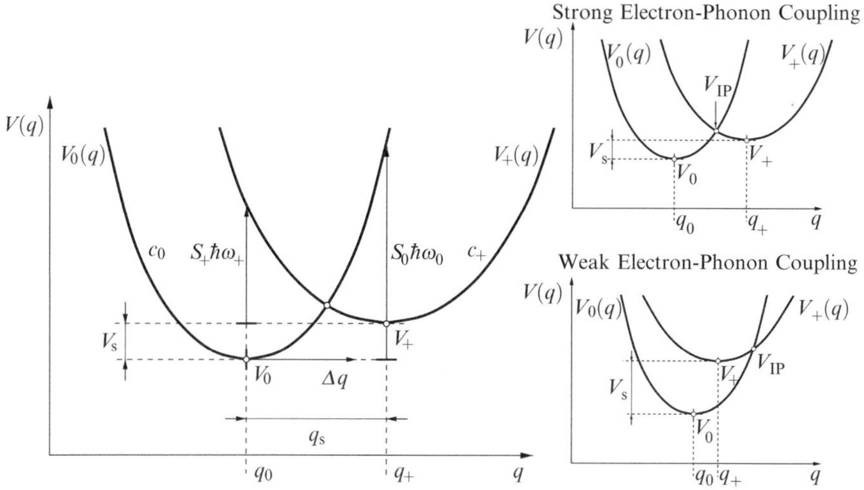

Fig. 16.8 *Left*: The configuration coordinate diagram including the Huang–Rhys factors S_0 and S_+. The adiabatic potentials are often defined as harmonic oscillators of the form $V_0(q) = 1/2 M\omega_0^2(q-q_0)^2 + V_0$ and $V_+(q) = 1/2 M\omega_+^2(q-q_+)^2 + V_+$, where ω_0 and ω_+ are their respective oscillator frequencies. For an optical transition, the energy delivered by the photon must equal the energy difference $V_0(q_+) - V_+$, which is indicated by the upwards arrow and can be expressed as an integral multiple S_0 of $\hbar\omega_0$. In analogy, $S_+\hbar\omega_+$ equals the energy difference $V_+(q_0) - V_0$. In the remainder of this chapter, $S_0\hbar\omega_0$ and $S_+\hbar\omega_+$ will be replaced by $S\hbar\omega$ and $R^2 S\hbar\omega$, respectively. *Right*: Strong (*top*) and weak (*bottom*) electron–phonon coupling. In the first case the parabolas are positioned such that the intersection point is situated in between their minima while in the second case one parabola lies inside the other and the intersection point is located beside the two minima

for the NMP transition barrier with the prefactor

$$\xi_{0/+}(\Delta q) = \sqrt{\frac{\beta c_0}{4\pi}} \frac{1}{|c_0 \Delta q - c_+(\Delta q - q_s)|} = \sqrt{\frac{\beta}{4\pi}} \frac{1}{\sqrt{S\hbar\omega}} . \tag{16.47}$$

16.6.1 Analytical Expressions for the NMP Rates

A second order expansion of the expression (16.43) delivers

$$V_{0/+}(V_s) \approx \frac{S\hbar\omega}{(1+R)^2} + \frac{R}{1+R} V_s + \frac{R}{4S\hbar\omega} V_s^2 . \tag{16.48}$$

If the curvatures c_0 and c_+ differ, the quantity R deviates from unity. Since R also enters the above expression for the barrier height, the ratio of the curvatures has a

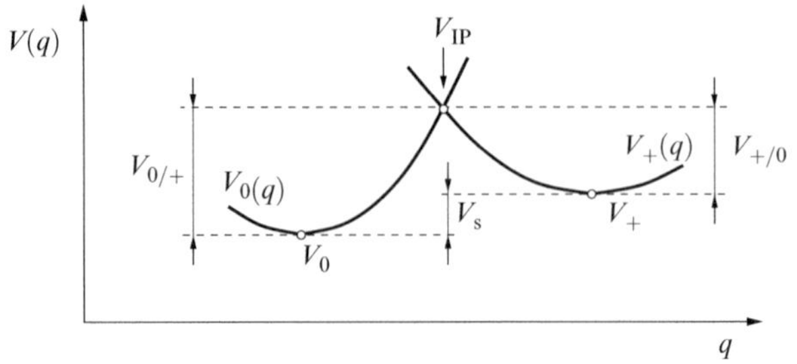

Fig. 16.9 The hole capture ($V_{0/+}$) and emission ($V_{+/0}$) barrier for an NMP transition. The barrier heights are calculated as the energy differences between the corresponding minimum and the intersection point in the hole picture, which yields $V_{0/+} = V_{\mathrm{IP}} - V_0 = V_{\mathrm{IP}} - \tilde{V}_0 + E$ and $\Delta V_{+/0} = V_{\mathrm{IP}} - V_+ = V_{\mathrm{IP}} - \tilde{V}_0 + E_{\mathrm{t}}$ for the capture and the emission barrier, respectively

strong impact on the NMP transition rates (cf. Fig. 16.9). As in the previous section, V_{s} can be expressed as

$$V_{\mathrm{s}} = V_+ - V_0 = E - E_{\mathrm{t}} = \underbrace{E - E_{\mathrm{v}}}_{=-\Delta E} + E_{\mathrm{v}} - E_{\mathrm{t}} \tag{16.49}$$

so that (16.48) can be rewritten as

$$V_{0/+}(\Delta E) \approx \frac{S\hbar\omega}{(1+R)^2} + \frac{R}{1+R}(E_{\mathrm{v}} - E_{\mathrm{t}} - \Delta E) + \frac{R}{4S\hbar\omega}(E_{\mathrm{v}} - E_{\mathrm{t}} - \Delta E)^2. \tag{16.50}$$

In the case of strong electron–phonon coupling (see Fig. 16.8) $S\hbar\omega \gg |E_{\mathrm{v}} - E_{\mathrm{t}} - \Delta E|$ holds and the third term of (16.50) can be neglected. In order to evaluate the integral in the hole capture rate (16.39), the following assumptions must be made:

- Assuming the parabolic-band approximation, the valence band density of states (16.29) is given by $D_{\mathrm{p}}(E) = D_{\mathrm{p},0}\sqrt{\Delta E}$ with $D_{\mathrm{p},0}$ being an energy-independent prefactor.
- The occupancy $f_{\mathrm{p}}(E, E_{\mathrm{f}})$ follows Boltzmann statistics.
- The WKB factor is approximated by the factor $\exp(-x_{\mathrm{t}}/x_0)$ with the tunneling length x_0.
- The lineshape function is dominated by the exponential barrier term so that the prefactor $\xi_{0/+}$ can be neglected to first order.

With the above simplifications, the hole capture rate (16.39) evaluates to

$$\begin{aligned} k^{\mathrm{pc}} &= k_0^{\mathrm{p}} \int_{-\infty}^{E_{\mathrm{v}}} D_{\mathrm{p}}(E) f_{\mathrm{p}}(E, E_{\mathrm{f}}) \lambda(E, x_{\mathrm{t}}) \exp(-\beta V_{0/+}(\Delta E)) \mathrm{d}E \\ &= k_0^{\mathrm{p}}(1+R)^{3/2} p \exp(-x_{\mathrm{t}}/x_0) \exp\left(-\beta\left(\frac{S\hbar\omega}{(1+R)^2} - \frac{R}{1+R}\Delta E_{\mathrm{t}}\right)\right), \end{aligned} \tag{16.51}$$

where the hole density p is given by the expression

$$p = D_{\mathrm{p},0} \exp\big(\beta(E_{\mathrm{v}} - E_{\mathrm{f}})\big)\, \beta^{-3/2}\, \Gamma(3/2) \tag{16.52}$$

with $\Gamma(x)$ being the Gamma function. Motivated by the similarity to the rate equations in the standard SRH theory, the prefactor k_0^{p} has been identified with the hole thermal velocity $v_{\mathrm{th,p}}$ times a hole capture cross-section σ_{p}. The lengthy expression in the exponent of the last term of (16.51) can be related to the hole capture barrier $\varepsilon^{\mathrm{pc}}$, which is evaluated for $E = E_{\mathrm{v}}$.

$$\frac{S\hbar\omega}{(1+R)^2} + \frac{R}{1+R} E_{\mathrm{v}} - E_{\mathrm{t}} = V_{0/+}\Big|_{\Delta E=0} = \varepsilon^{\mathrm{pc}} \tag{16.53}$$

This is actually surprising since the NMP transition barrier $V_{0/+}(\Delta E)$ is a function of the hole energy E per definition. However, for strong electron–phonon coupling, the rate integral (16.39) delivers its largest contribution close to the valence band edge ($\Delta E = 0$) so that the barrier $V_{0/+}(E)$ can be approximated by $V_{0/+}(E_{\mathrm{v}})$. As a consequence, the hole capture rate simplifies to

$$k^{\mathrm{pc}} = v_{\mathrm{th,p}} \sigma_{\mathrm{p}} (1+R)^{3/2} \exp(-x_{\mathrm{t}}/x_0)\, p \exp(-\beta \varepsilon^{\mathrm{pc}})\,. \tag{16.54}$$

The hole emission rate is derived from (16.40) using the two relations: First, the electron occupation function can be replaced by

$$f_{\mathrm{n}}(E, E_{\mathrm{f}}) = f_{\mathrm{p}}(E, E_{\mathrm{f}}) \exp\big(-\beta(E - E_{\mathrm{f}})\big)\,. \tag{16.55}$$

Second, the ratio of the exponential barrier terms (see Fig. 16.9) gives

$$\exp(-\beta V_{+/0})/\exp(-\beta V_{0/+}) = \exp(-\beta\,(E_{\mathrm{t}} - E)) \tag{16.56}$$

for each band state E. Inserting both relations in (16.40) and using the same assumptions as before yields the hole emission rate

$$\begin{aligned} k^{\mathrm{pe}} &= v_{\mathrm{th,p}} \sigma_{\mathrm{p}} \int_{+\infty}^{E_{\mathrm{v}}} D_{\mathrm{p}}(E) f_{\mathrm{n}}(E, E_{\mathrm{f}}) \lambda(E, x_{\mathrm{t}}) \exp(-\beta V_{0/+})\, \mathrm{d}E \\ &= v_{\mathrm{th,p}} \sigma_{\mathrm{p}} (1+R)^{3/2} \exp(-x_{\mathrm{t}}/x_0)\, p \exp(-\beta \varepsilon^{\mathrm{pc}}) \exp(-\beta\,(E_{\mathrm{t}} - E_{\mathrm{f}}))\,. \end{aligned} \tag{16.57}$$

Interestingly, (16.54) and (16.57) closely resemble the rates obtained from the standard SRH theory except from the exponential barrier terms and even have the same shape as those of Kirton and Uren. However, the NMP transition barriers derived above are calculated from the intersection point of two adiabatic potentials—in this case parabolas—and thus reflect their gate bias dependence governed by the energy separation between the trap level and the valence band edge according to (16.53). Even though they rely on a series of approximations, they contain the main physics involved in charge trapping. As such, they promote the understanding of the gate bias and temperature tendencies in charge trapping and allow compact analytical expressions for the assumption of strong electron–phonon coupling.

16.7 State Diagram of the Multi-State Model

The NMP transition rates derived in the previous sections describe charge transfer reactions, i.e., the pure charge trapping or detrapping processes. However, the TDDS studies revealed that some defects are found to disappear on the spectral maps. This observation can only be reasoned by the existence of metastable states, in which the oxide defects dwell for a certain amount of time. Furthermore, the TDDS also reveals gate bias-independent transitions that cannot be related to charge transfer reactions. These transitions are associated with an activation over thermal barriers, leaving the charge state of the defect unchanged. Both observations suggest a bistable defect, which has an additional metastable configuration (marked by primes) that appears in two charge states (cf. Fig. 16.10). This means that the defect features two neutral (1, $1'$) and two positive (2, $2'$) charge states (cf. Fig. 16.10), where thermal transitions allow for transitions between same charge states ($1 \leftrightarrow 1'$ and $2 \leftrightarrow 2'$) and NMP transitions between opposite charge states ($1 \leftrightarrow 2'$ and $2 \leftrightarrow 1'$). The bistable defect described above is the heart of the "multi-state model" and will be discussed in detail in the following.

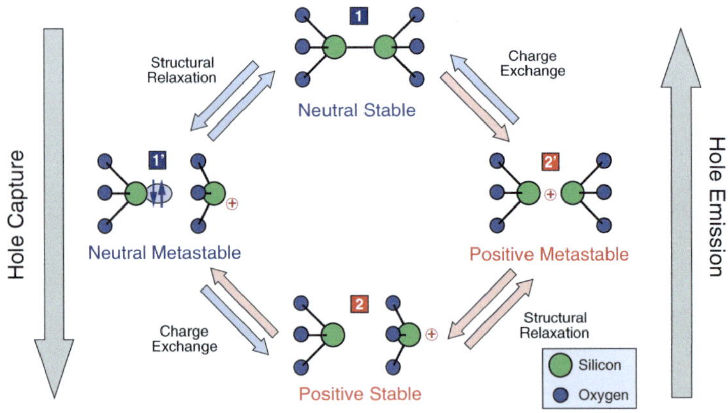

Fig. 16.10 State diagram of the multi-state model. The defect is present in a stable neutral (1) and a stable positive (2) charge state, where each of them has a second metastable state marked by a prime ($1'$, $2'$). The NMP transitions $1 \leftrightarrow 2'$ and $2 \leftrightarrow 1'$ occur between different charge states while the thermal transitions $1 \leftrightarrow 1'$ and $2 \leftrightarrow 2'$ between same charge states. Note that the transitions between the stable states are of main interest since they correspond to the experimentally measured capture and emission times in BTI. However, they involve intermediate states, which are metastable and important for the gate-bias and temperature dependence of the overall transition. The stick-and-ball models correspond to the configurations of a possible defect candidate, i.e., the oxygen vacancy, which is only shown for illustration purpose

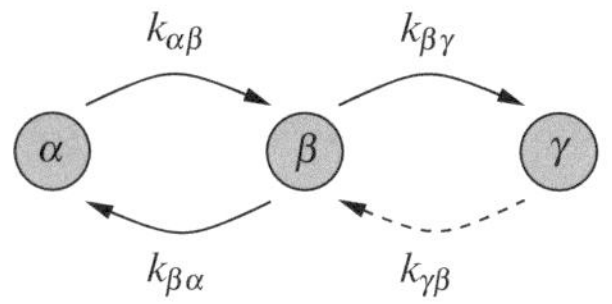

Fig. 16.11 The state diagram for a two-step process from the state α to γ. The first passage time of such a process is calculated by (16.59). Consider that the transition rate $k_{\gamma\beta}$, indicated by the *dashed arrow*, does not enter this equation

Such defects [55] show complex dynamics between those four states and must be correctly treated using homogeneous continuous-time Markov chain theory [56]. This theory rests upon the assumption that the future transitions between the states do not depend on the past of the investigated system. This assumption is justified as long as the defect relaxes after each transition by interacting with its environment, thereby losing the memory of its past. In fact, this is the case for both pure thermal and NMP transitions disregarding special theories, such as recombination-enhanced defect reaction. The time evolution of such a defect system is described by a first-order differential equation termed the Master equation.

$$\partial_t \pi_i(t) = \sum_{j \neq i} \pi_j(t) k_{ji} - \sum_{i \neq j} \pi_i(t) k_{ij} \tag{16.58}$$

Here, $\pi_i(t)$ is the time-dependent occupation probability that the defect is in state i and k_{ij} denotes the transition rate from state i to state j. When going from a single to a multitude of defects, the occupation probabilities must be averaged and become occupancies. The resulting rate equations, which are of the same form as the above Master equation, are usually solved in device simulators in order to predict the degradation for large area devices. Those kind of simulations can also account for the fact that the defect properties vary from trap to trap. The wide distributions of the defect properties arise from the amorphous defect environments but also come from the random dopant fluctuations, which have increasingly attracted scientific interest during the last several years [18, 57–62]. (For a detailed discussion of this topic, the interested reader is referred to [63].) For a comparison to the TDDS data, one is primarily interested in the transition times between stable states. The metastable states will only be occupied temporarily and are not observable in experiments. However, they gain their relevance for the overall gate bias and temperature dependence of two-step processes. The transitions between stable states are obtained from first-passage times. For a two-step process, the transition time from a state α to a state γ over a state β (cf. Fig. 16.11) reads

$$\tau_{\alpha\gamma} = \frac{k_{\alpha\beta} + k_{\beta\gamma} + k_{\beta\alpha}}{k_{\alpha\beta} k_{\beta\gamma}} = \frac{1}{k_{\alpha\beta}} + \frac{1}{k_{\beta\gamma}} + \frac{1}{k_{\beta\gamma}} \frac{k_{\beta\alpha}}{k_{\alpha\beta}}. \tag{16.59}$$

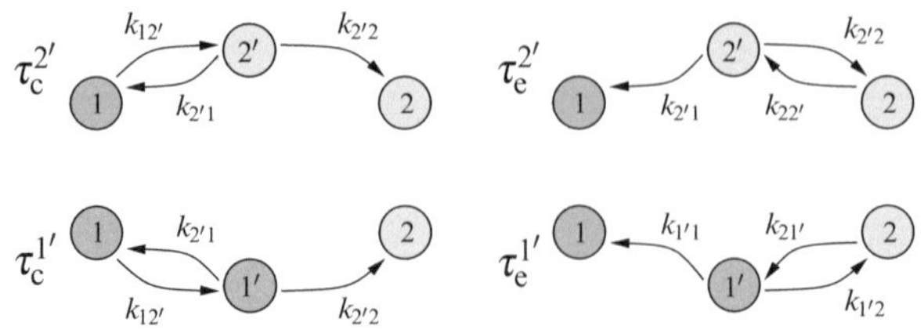

Fig. 16.12 Simplified state diagrams of hole capture and emission over the metastable states $1'$ and $2'$. The superscripts of τ denote the intermediate state, which has been passed through during a complete capture or emission event. Note that there exist two competing pathways for a hole capture event, namely one over the intermediate state $1'$ and one over $2'$. Of course, the same holds true for a hole emission event

The multi-state model with its four states allows for four distinct transition pathways (see Fig. 16.12), whose first-passage times are listed below:

$$\tau_c^{2'} = \frac{1}{k_{12'}} + \frac{1}{k_{2'2}} + \frac{1}{k_{2'2}} \frac{k_{2'1}}{k_{12'}} \tag{16.60}$$

$$\tau_c^{1'} = \frac{1}{k_{11'}} + \frac{1}{k_{1'2}} + \frac{1}{k_{1'2}} \frac{k_{1'1}}{k_{11'}} \tag{16.61}$$

$$\tau_e^{2'} = \frac{1}{k_{22'}} + \frac{1}{k_{2'1}} + \frac{1}{k_{2'1}} \frac{k_{2'2}}{k_{22'}} \tag{16.62}$$

$$\tau_e^{1'} = \frac{1}{k_{21'}} + \frac{1}{k_{1'1}} + \frac{1}{k_{1'1}} \frac{k_{1'2}}{k_{21'}} \tag{16.63}$$

The transition barriers for the partial rates can be extracted from the configuration coordinate diagram of the bistable defect (see Fig. 16.13). The bistability of the defect is reflected in the double-well shape of the adiabatic potentials. The transitions $T_{1\leftrightarrow 1'}$ and $T_{2\leftrightarrow 2'}$ are thermally activated and do not vary with the applied gate bias. According to transition state theory, they can be expressed as

$$k_{11'} = \nu_0 \exp(-\beta \varepsilon_{11'}) \tag{16.64}$$

$$k_{1'1} = \nu_0 \exp(-\beta \varepsilon_{1'1}) \tag{16.65}$$

$$k_{22'} = \nu_0 \exp(-\beta \varepsilon_{22'}) \tag{16.66}$$

$$k_{2'2} = \nu_0 \exp(-\beta \varepsilon_{2'2}) \tag{16.67}$$

where the barriers ε_{ij} are defined in Fig. 16.13 and ν_0 is the attempt frequency, which is typically of the order $10^{13}\,\mathrm{s}^{-1}$. The NMP transition rates are evaluated using (16.37)–(16.40), which contain lineshape functions and thus depend on V_s. The energy minima in the configuration coordinate diagram of Fig. 16.13 are given by

$$V_1 = \tilde{V}_0 - E \tag{16.68}$$

$$V_{2'} = \tilde{V}_0 + \varepsilon_{\mathrm{T}2'} - E_\mathrm{t} \tag{16.69}$$

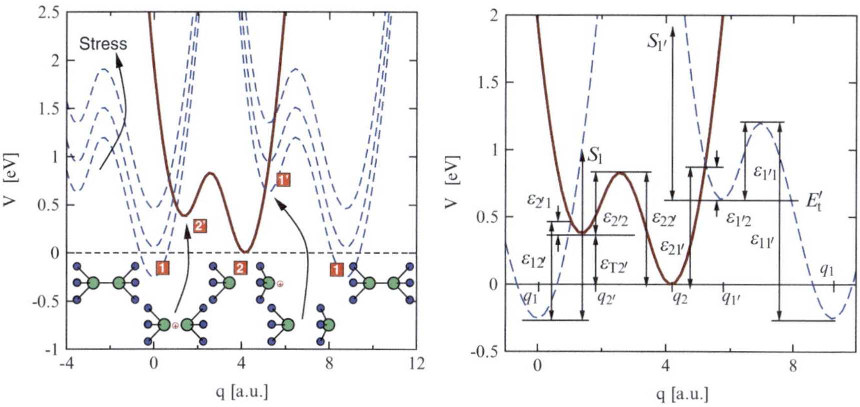

Fig. 16.13 *Left*: A schematic of the configuration coordinate diagram for a bistable defect. The *solid* and the *dashed lines* represent the adiabatic potentials for a defect in its positive and neutral charge state, respectively. The energy minima correspond to the stable or metastable defect configurations, labeled 1, 1′, 2, and 2′. The present configuration coordinate diagram describes the exchange of holes with the valence band and thus is associated with a hole capture or emission process. The stick-and-ball models display a defect in its various stable and metastable configurations for illustration purpose. *Right*: Definitions of the used energies and barriers in the multi-state model. Recall that two adiabatic potentials must be shown for one transition. It is assumed that an alternative transition pathway with an additional crossing point exists in the multidimensional atomic configuration space. In order to show both intersections (related to the transitions $1 \leftrightarrow 2'$ and $2 \leftrightarrow 1'$) in one configuration coordinate diagram, the "neutral" potential must be plotted twice. Obviously, $\varepsilon_{22'} = \varepsilon_{2'2} + \varepsilon_{T2'}$ holds

$$V_2 = \tilde{V}_0 - E'_t \tag{16.70}$$

$$V_{1'} = \tilde{V}_0 - E \tag{16.71}$$

in the hole picture. Here, the V_i stands for the adiabatic potentials with i being one of the states in Fig. 16.10. Furthermore, the hole is assumed to be energetically located at the valence band edge. It is emphasized that the energy $\varepsilon_{T2'}$ must be added to $\tilde{V}_0$ to obtain the correct energy minimum of state $2'$.

$$E_t \to E_t - \varepsilon_{T2'} \tag{16.72}$$

As a consequence, $\varepsilon_{T2'}$ modifies the energy differences V_s extracted from the configuration coordinate diagram

$$V_{12'} = V_{2'} - V_1 = E - E_t + \varepsilon_{T2'} \tag{16.73}$$

$$V_{1'2} = V_2 - V_{1'} = E - E'_t \tag{16.74}$$

and enters the NMP rates

$$k_{12'} = v_{\mathrm{th,n}}\sigma_{\mathrm{n}} \int_{E_{\mathrm{c}}}^{+\infty} D_{\mathrm{n}}(E) f_{\mathrm{p}}(E,E_{\mathrm{f}}) \lambda(E) f_{0/+}(c_0, c_+, q_{\mathrm{s}}, \underbrace{E - E_{\mathrm{t}} + \varepsilon_{\mathrm{T2'}}}_{=V_{12'}}) \mathrm{d}E$$

$$+ v_{\mathrm{th,p}}\sigma_{\mathrm{p}} \int_{-\infty}^{E_{\mathrm{v}}} D_{\mathrm{p}}(E) f_{\mathrm{p}}(E,E_{\mathrm{f}}) \lambda(E) f_{0/+}(c_0, c_+, q_{\mathrm{s}}, \underbrace{E - E_{\mathrm{t}} + \varepsilon_{\mathrm{T2'}}}_{=V_{12'}}) \mathrm{d}E \tag{16.75}$$

$$k_{2'1} = v_{\mathrm{th,n}}\sigma_{\mathrm{n}} \int_{E_{\mathrm{c}}}^{+\infty} D_{\mathrm{n}}(E) f_{\mathrm{n}}(E,E_{\mathrm{f}}) \lambda(E) f_{+/0}(c_+, c_0, q_{\mathrm{s}}, \underbrace{E_{\mathrm{t}} - \varepsilon_{\mathrm{T2'}} - E}_{=-V_{12'}}) \mathrm{d}E$$

$$+ v_{\mathrm{th,p}}\sigma_{\mathrm{p}} \int_{-\infty}^{E_{\mathrm{v}}} D_{\mathrm{p}}(E) f_{\mathrm{n}}(E,E_{\mathrm{f}}) \lambda(E) f_{+/0}(c_+, c_0, q_{\mathrm{s}}, \underbrace{E_{\mathrm{t}} - \varepsilon_{\mathrm{T2'}} - E}_{=-V_{12'}}) \mathrm{d}E \tag{16.76}$$

$$k_{1'2} = v_{\mathrm{th,n}}\sigma_{\mathrm{n}} \int_{E_{\mathrm{c}}}^{+\infty} D_{\mathrm{n}}(E) f_{\mathrm{p}}(E,E_{\mathrm{f}}) \lambda(E) f_{0/+}(c_0, c_+, q_{\mathrm{s}}, \underbrace{E - E'_{\mathrm{t}}}_{=V_{1'2}}) \mathrm{d}E$$

$$+ v_{\mathrm{th,p}}\sigma_{\mathrm{p}} \int_{-\infty}^{E_{\mathrm{v}}} D_{\mathrm{p}}(E) f_{\mathrm{p}}(E,E_{\mathrm{f}}) \lambda(E) f_{0/+}(c_0, c_+, q_{\mathrm{s}}, \underbrace{E - E'_{\mathrm{t}}}_{=V_{1'2}}) \mathrm{d}E \tag{16.77}$$

$$k_{21'} = v_{\mathrm{th,n}}\sigma_{\mathrm{n}} \int_{E_{\mathrm{c}}}^{+\infty} D_{\mathrm{n}}(E) f_{\mathrm{n}}(E,E_{\mathrm{f}}) \lambda(E) f_{+/0}(c_+, c_0, q_{\mathrm{s}}, \underbrace{E'_{\mathrm{t}} - E}_{=-V_{1'2}}) \mathrm{d}E$$

$$+ v_{\mathrm{th,p}}\sigma_{\mathrm{p}} \int_{-\infty}^{E_{\mathrm{v}}} D_{\mathrm{p}}(E) f_{\mathrm{n}}(E,E_{\mathrm{f}}) \lambda(E) f_{+/0}(c_+, c_0, q_{\mathrm{s}}, \underbrace{E'_{\mathrm{t}} - E}_{=-V_{1'2}}) \mathrm{d}E\,. \tag{16.78}$$

The above NMP transition rates along with the thermal transition rates (16.64)–(16.67) enter the expressions of the capture and emission times (16.60)–(16.63) that are comparable to time constants observed in the TDDS data. In the next section, they will be used to evaluate the multi-state model against the TDDS data and allow a verification of this model.

16.8 Model Evaluation

As outlined in Sect. 16.2, TDDS experiments measure the response of single defects to different gate voltage or temperature conditions. Based on these data, they give insight into the behavior of single defects and can thus reveal whether a BTI trapping

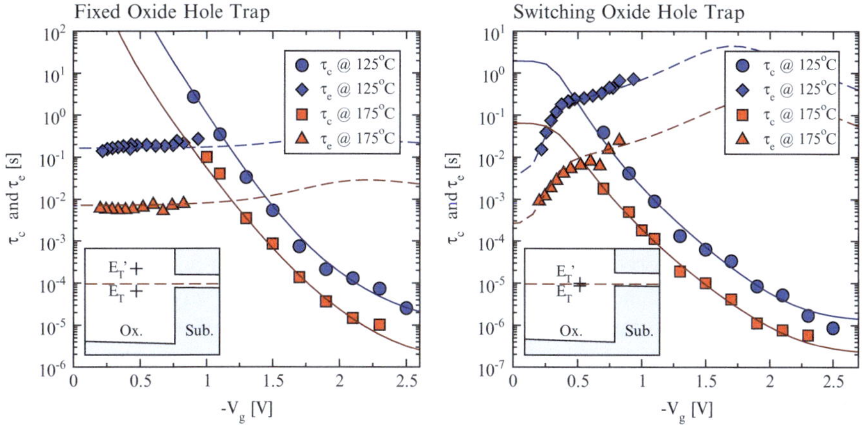

Fig. 16.14 *Left*: The capture (*solid lines*) and emission (*dashed lines*) times of a fixed oxide hole trap as a function of the gate bias. The symbols stand for the measurement data and the lines represent the simulation results of the multi-state model. The latter are shown to be in remarkable agreement with the experimental data. The inset (*bottom left*) depicts the band diagram of a MOSFET with the trap levels E_t and E_t' for the case when no bias is applied to the gate. Under these conditions the trap level E_t' is located far above the substrate Fermi level and the emission time remains unaffected by the gate bias. This fact eventually characterizes this defect as a fixed oxide hole trap *Right*: The same but for a switching oxide hole trap as presented in the Sect. 16.2. Compared to the fixed oxide hole trap, it shows a strong gate bias dependence of τ_e at small gate biases. In contrast to a fixed oxide hole trap, the Fermi level and the trap level E_t' coincide there, resulting in the strong sensitivity of τ_e to V_g

model reflects the physics of real defects. The time constant plots in Fig. 16.14 depict a fit of the multi-state model against the time constants extracted from the TDDS measurement data. The following calculations are carried out on a device simulator that delivers the band energy diagram for the devices used in the TDDS measurements. With these data, the thermal and the NMP transition rates were evaluated, which were subsequently used to calculate the capture and emission times. In these simulation, we accounted for the exchange of charge carriers with the substrate as well as the gate from the conduction and the valence band. An evaluation of the TDDS checklist is given below:

(i) The curvature in τ_c is reproduced by the multi-state model.
(ii) τ_c shows a marked temperature activation over the whole range of V_g, visible as a parallel upward shift.
(iii) In general, the multi-state model yields field-insensitive τ_e as displayed in Fig. 16.14 (left). It is important to note here that at larger oxide fields this model also predicts an exponential dependence, which has also been observed for some defects in RTN measurements [31].
(iv) The multi-state model also allows for a field-dependent τ_e provided that the substrate Fermi level and the trap level E_t' are separated by only a few hundredth of an electron Volt at small V_g (cf. Fig. 16.14, right).
(v) In both cases, τ_e is thermally activated.

The above checklist demonstrates that the multi-state model can reproduce the key features of the hole capture and emission process correctly, strongly indicating that the multi-state model can describe the physics of the defects seen in TDDS.

16.9 Discussion of the Multi-State Model

In Sect. 16.7, we derived a full set of rate equations that can accurately describe charge trapping within the multi-state model. However, they rely on complicated integrals which obscure the gate bias and temperature-dependent behavior of defects. For this reason, we also provide analytical expressions that promote understanding of the essential physical behind the mathematical framework.

Following the derivation in Sect. 16.6, the NMP transition rates can be written as

$$k_{12'} = v_{\mathrm{th,p}}\sigma_{\mathrm{p}}(1+R)^{3/2}\lambda(E_{\mathrm{v}})p\exp(-\beta\varepsilon_{12'}) \tag{16.79}$$

$$k_{2'1} = v_{\mathrm{th,p}}\sigma_{\mathrm{p}}(1+R)^{3/2}\lambda(E_{\mathrm{v}})p\exp(-\beta\varepsilon_{12'})\exp(-\beta(E_{\mathrm{t}}-\varepsilon_{\mathrm{T2'}}-E_{\mathrm{f}})) \tag{16.80}$$

$$k_{1'2} = v_{\mathrm{th,p}}\sigma_{\mathrm{p}}(1+R')^{3/2}\lambda(E_{\mathrm{v}})p\exp(-\beta\varepsilon_{1'2}) \tag{16.81}$$

$$k_{21'} = v_{\mathrm{th,p}}\sigma_{\mathrm{p}}(1+R')^{3/2}\lambda(E_{\mathrm{v}})p\exp(-\beta\varepsilon_{1'2})\exp(-\beta(E_{\mathrm{t}}-E_{\mathrm{f}})) \tag{16.82}$$

with

$$\varepsilon_{12'} = \frac{S_1\hbar\omega_1}{(1+R_1)^2} + \frac{R_1}{1+R_1}(E_{\mathrm{v}}-E_{\mathrm{t}}+\varepsilon_{\mathrm{T2'}}) \tag{16.83}$$

$$= \frac{S_1\hbar\omega_1}{(1+R_1)^2} - \frac{R_1}{1+R_1}(\Delta E_{\mathrm{t}}-\varepsilon_{\mathrm{T2'}}) + \frac{R_1}{1+R_1}q_0x_{\mathrm{t}}F_{\mathrm{ox}} \tag{16.84}$$

$$\varepsilon_{1'2} = \frac{S_{1'}\hbar\omega_{1'}}{(1+R_{1'})^2} + \frac{R_{1'}}{1+R_{1'}}(E_{\mathrm{v}}-E'_{\mathrm{t}}) \tag{16.85}$$

$$= \frac{S_{1'}\hbar\omega_{1'}}{(1+R_{1'})^2} - \frac{R_{1'}}{1+R_{1'}}\Delta E'_{\mathrm{t}} + \frac{R_{1'}}{1+R_{1'}}q_0x_{\mathrm{t}}F_{\mathrm{ox}} \tag{16.86}$$

using (16.3). In analogy to the derivation of the exact NMP transition rates (16.68)–(16.78), the trap level E_{t} must again be referenced to the minimum $2'$ according to (16.69). This reference of E_{t} is required in the calculation of the NMP barriers (16.84) and (16.86) as well as the last term of (16.80) following the concept outlined in Fig. 16.9. With the thermal transitions (16.64)–(16.67) and the above expression of the NMP rates (16.79)–(16.82), the capture and emission times (16.60)–(16.63) read

$$\tau_{\mathrm{c}}^{2'} = \tau_{\mathrm{c,min}}^{2'} + \tau_{\mathrm{p0}}\frac{N_2}{p}\exp\left(\beta\frac{R_1q_0x_{\mathrm{t}}F_{\mathrm{ox}}}{1+R_1}\right) + \tau_{\mathrm{c,min}}^{2'}\frac{N_1}{p}\exp(\beta q_0x_{\mathrm{t}}F_{\mathrm{ox}}) \tag{16.87}$$

$$\tau_{\mathrm{c}}^{1'} = \tau_{\mathrm{c,min}}^{1'} + \tau_{\mathrm{p0}} \frac{N_3}{p} \exp\left(\beta \frac{R_{1'} q_0 x_{\mathrm{t}} F_{\mathrm{ox}}}{1+R_{1'}}\right) \tag{16.88}$$

$$\tau_{\mathrm{e}}^{2'} = \tau_{\mathrm{e,min}}^{2'} + \tau_{2'} \exp\left(-\beta \frac{q_0 x_{\mathrm{t}} F_{\mathrm{ox}}}{1+R_1}\right) \tag{16.89}$$

$$\tau_{\mathrm{e}}^{1'} = \tau_{1'} \exp\left(-\beta \frac{q_0 x_{\mathrm{t}} F_{\mathrm{ox}}}{1+R_{1'}}\right) + \tau_{\mathrm{e,min}}^{1'} \left(1 + \exp\left(\beta (E_{\mathrm{t}}' - E_{\mathrm{f}})\right)\right) \tag{16.90}$$

using the definitions

$$N_1 = N_{\mathrm{v}} \exp\left(\beta (\varepsilon_{\mathrm{T2'}} - \Delta E_{\mathrm{t}})\right) \tag{16.91}$$

$$N_2 = \frac{N_{\mathrm{v}}}{(1+R_1)^{3/2}} \exp\left(\beta \frac{S_1 \hbar\omega_1}{(1+R_1)^2}\right) \exp\left(-\beta \frac{R_1 (\Delta E_{\mathrm{t}} - \varepsilon_{\mathrm{T2'}})}{1+R_1}\right) \tag{16.92}$$

$$N_3 = \frac{N_{\mathrm{v}}}{(1+R_{1'})^{3/2}} \exp\left(\beta \frac{S_{1'} \hbar\omega_{1'}}{(1+R_{1'})^2}\right) \exp\left(-\beta \frac{R_{1'}}{1+R_{1'}} \Delta E_{\mathrm{t}}'\right) \\ \times \left(1 + \exp\left(\beta (\Delta E_{\mathrm{t}}' - \Delta E_{\mathrm{t}})\right)\right) \tag{16.93}$$

$$\tau_{2'} = \frac{\tau_{\mathrm{p0}}}{(1+R_1)^{3/2}} \exp\left(\beta \frac{S_1 \hbar\omega_1}{(1+R_1)^2}\right) \exp\left(\beta \frac{\Delta E_{\mathrm{t}} - \varepsilon_{\mathrm{T2'}}}{1+R_1}\right) \\ \times \left(1 + \exp(\beta \varepsilon_{\mathrm{T2'}})\right) \tag{16.94}$$

$$\tau_{1'} = \frac{\tau_{\mathrm{p0}}}{(1+R_{1'})^{3/2}} \exp\left(\beta \frac{S_{1'} \hbar\omega_{1'}}{(1+R_{1'})^2}\right) \exp\left(\beta \frac{\Delta E_{\mathrm{t}}'}{1+R_{1'}}\right) \tag{16.95}$$

$$\tau_{\mathrm{c,min}}^{2'} = 1/k_{2'2} \tag{16.96}$$

$$\tau_{\mathrm{e,min}}^{2'} = 1/k_{22'} \tag{16.97}$$

$$\tau_{\mathrm{c,min}}^{1'} = 1/k_{11'} \tag{16.98}$$

$$\tau_{\mathrm{e,min}}^{1'} = 1/k_{1'1} \tag{16.99}$$

$$\tau_{\mathrm{p0}} = \frac{1}{\sigma_{\mathrm{p}} v_{\mathrm{th,p}} N_{\mathrm{v}} \lambda(E_{\mathrm{v}})}. \tag{16.100}$$

Recall that the hole capture process can proceed from state 1 over one of the metastable states $2'$ or $1'$ to the final state 2 according to the state diagram of Fig. 16.12. The corresponding capture time constants are denoted as $\tau_{\mathrm{c}}^{2'}$ and $\tau_{\mathrm{c}}^{1'}$, respectively, and will be discussed in the following. If the transition pathway $T_{1\to 2'\to 2}$ is preferred, the capture time constant has the same shape as (16.59).

$$\tau_{\mathrm{c}}^{2'} = \frac{k_{12'} + k_{2'1} + k_{2'2}}{k_{12'} k_{2'2}} \tag{16.101}$$

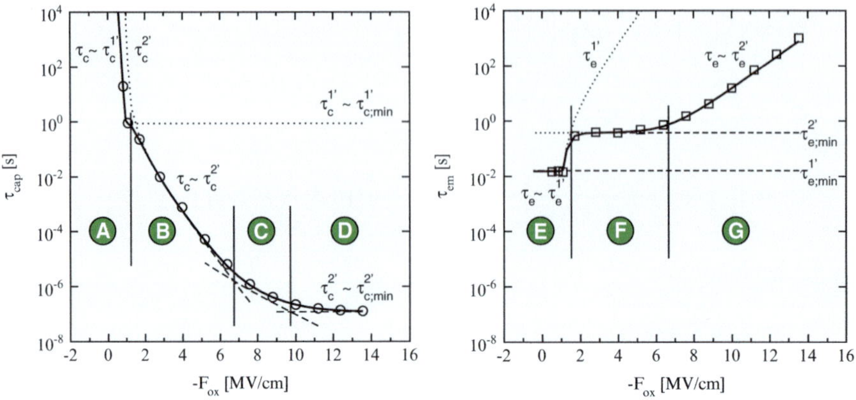

Fig. 16.15 *Left*: The calculated hole capture time constants as a function of the oxide field. The different regimes of τ_c (A, B, C, and D) are separated by the *thin vertical lines* and labeled by the *circles* with the capital letters. The *dotted curves* τ_c^i show the capture processes over a metastable state i. The field dependence of τ_c within a certain regime is shown by the *dashed curve*. *Right*: The same but for the hole emission time constants with the regimes (E, F, and G)

$$= \underbrace{\frac{1}{k_{12'}}}_{D} + \underbrace{\frac{1}{k_{2'2}}}_{C} + \underbrace{\frac{1}{k_{2'2}}\frac{k_{2'1}}{k_{12'}}}_{B} \tag{16.102}$$

Each summand in the nominator can be dominant, leading to (16.60), which is characterized by three distinct regimes, namely B, C, and D in Fig. 16.15. At extremely high negative oxide fields (regime D), $k_{12'}$ is the dominant rate meaning that the transition[4] $T_{1\to 2'}$ proceeds much faster than $T_{2'\to 2}$ (cf. Fig. 16.16). Thus complete capture process ($T_{1\to 2'\to 2}$) is controlled by the second transition $T_{2'\to 2}$, which is much slower and has a time constant of $\tau_{c,\min}^{2'}$. Since this second step is only thermally activated, $\tau_c^{2'}$ does not depend on the oxide field. This is consistent with (16.87), in which both exponential terms become negligible at extremely high negative oxide fields. At moderate negative oxide fields (regime C), the rate $k_{12'}$ approaches the order of $k_{2'1}$ and even falls below $k_{2'2}$. In this case the thermal transition $T_{2'\to 2}$ immediately follows the hole capture process from the state 1 to $2'$. As a result, the trapping kinetics are governed by the forward rate of the NMP process $T_{1\to 2'}$. Then $\tau_c^{2'}$ shows an exponential oxide field dependence, which is reflected in the second term of (16.87). At low negative oxide fields (regime B), $k_{12'}$ is already outbalanced by its reverse rate $k_{2'1}$ (see Fig. 16.16) and the ratio of both rates determines the oxide field dependence. This gives an increased exponential

[4] Keep in mind that the term "transition" does not refer to the duration of the physical process itself, such as the time it takes an electron to tunnel through an energy barrier. It rather denotes the mean time until the physical process takes place and the defect change its state.

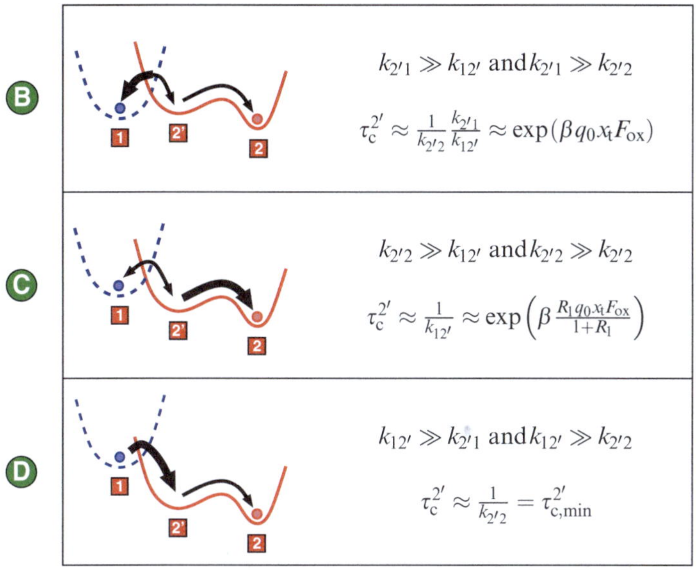

Fig. 16.16 A schematic representation of adiabatic potentials in the regimes B, C, and D. The *arrows* show the transitions involved in the hole capture process. Their thicknesses indicate the magnitude of their rates. This means that the *thinner arrows* are associated with larger transitions times and thus governs the oxide field and temperature dependence of the complete capture process $T_{1\to 2}$. With higher oxide fields (B $\to$ D) the potential of the neutral defect (*dashed line*) is raised relative to that of the positive defect (*solid line*). This is associated with an increase of $k_{12'}$ and a decrease of the reverse rate $k_{2'1}$. In contrast to the charge transfer reactions $T_{1\to 2'}$ and $T_{2\to' 1}$, the thermal transition $T_{2'\to 2}$ is not affected by the oxide field

slope originating from the third term of (16.87). *The transitions between these three regimes are smooth so that the capture time becomes curved in its time constant plots (cf. Fig. 16.16).* It emphasized here that the curvature in the capture times are one of the most obstinate feature for BTI modeling and has only been reproduced by the multi-state model so far.

However, if the transition over the metastable state $1'$ is favored (regime A), the capture time constant can be formulated using first-passage times:

$$\tau_c^{1'} = \frac{k_{11'} + k_{1'1} + k_{1'2}}{k_{11'} k_{1'2}} \qquad (16.103)$$

Since the metastable state $1'$ is situated above the state 1 by definition, $k_{1'1} \gg k_{11'}$ holds. Therefore, the expression (16.103) can be approximated by

$$\tau_c^{1'} \approx \underbrace{\frac{k_{1'1}}{k_{11'} k_{1'2}}}_{A''} + \underbrace{\frac{1}{k_{11'}}}_{A'}, \qquad (16.104)$$

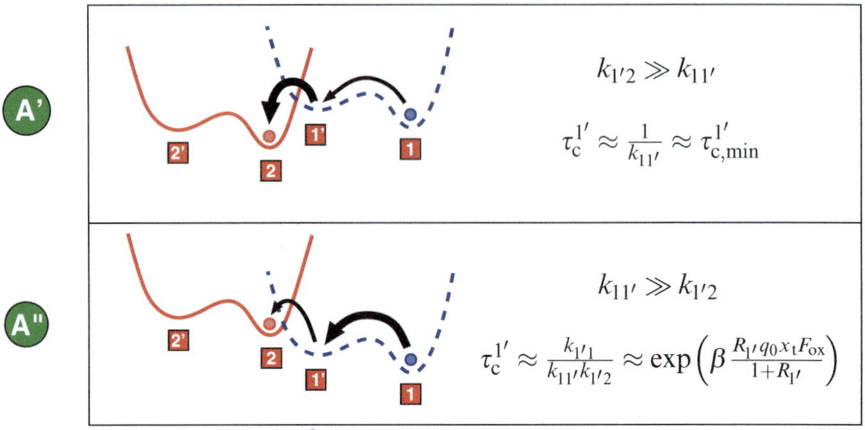

Fig. 16.17 The same as in Fig. 16.16 but for the regimes A' and A" of the oxide field dependence of $\tau_c^{1'}$

which is characterized by only two regimes (A' and A") now. At negative oxide fields (regime A'), the state $1'$ is located high (see Fig. 16.17) so that the transition rate $k_{1'2}$ is large compared to $k_{11'}$. Then the first term of expression (16.104) vanishes and the field-dependent transition $T_{1\to 1'}$ with a time constant of $\tau_{c,min}^{1'}$ dominates $\tau_c^{1'}$ in (16.104). When reducing the oxide field, the state $1'$ is shifted downwards in the configuration coordinate diagram, thereby decreasing the transition rate $k_{1'2}$. At a certain oxide field, $k_{1'2}$ falls below $k_{1'1}$ and the first term of the expression (16.104) becomes dominant (regime A"). As a consequence, $\tau_c^{1'}$ is governed by the field-dependent transition $T_{1'\to 2}$, which is reflected in the exponential term of the expression (16.88). The transition between A' and A" yields a kink, which is visible in $\tau_c^{1'}$ (cf. Fig. 16.15) but not in the overall hole capture τ_c time given by

$$\frac{1}{\tau_c} \approx \frac{1}{\tau_c^{1'}} + \frac{1}{\tau_c^{2'}} . \tag{16.105}$$

So far, this transition has not been observed in TDDS experiments, which is why the regimes A' and A" are not differentiated in Fig. 16.16.

Also the hole emission process has the possibility to proceed over either the state $1'$ or $2'$, with $\tau_e^{1'}$ and $\tau_e^{2'}$ being the corresponding emission time constants (see Fig. 16.18). For the transition pathway over $2'$, the emission time constant can be expressed as:

$$\tau_e^{2'} = \frac{k_{22'} + k_{2'2} + k_{2'1}}{k_{22'}k_{2'1}} \tag{16.106}$$

Since $k_{2'2} \gg k_{22'}$ applies, $\tau_e^{2'}$ has only two regimes, labeled with the capital letters F and G in Fig. 16.15.

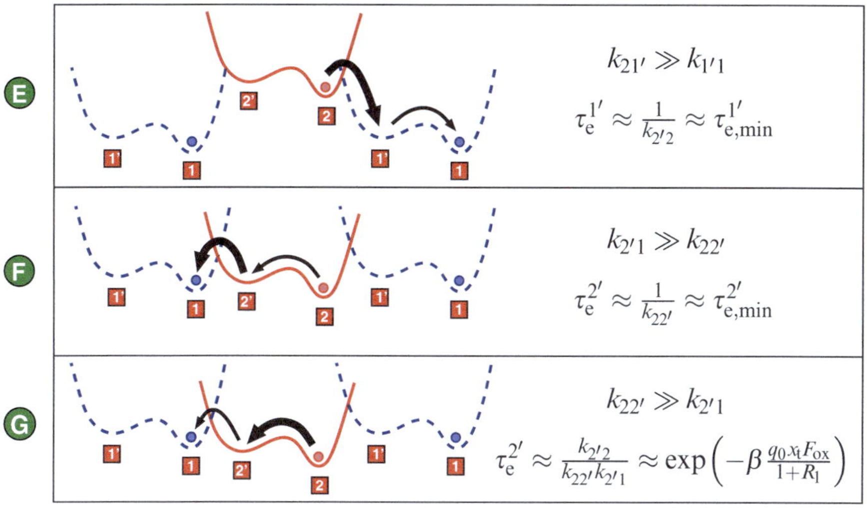

Fig. 16.18 The same as in Fig. 16.16 but for the regimes E, F, and G of the oxide field dependence of τ_e

$$\tau_e^{2'} \approx \underbrace{\frac{k_{2'2}}{k_{22'}k_{2'1}}}_{G} + \underbrace{\frac{1}{k_{22'}}}_{F} \tag{16.107}$$

At high negative oxide fields (regime G), the state 1 is shifted upwards so that $k_{22'}$ is the dominant rate and the field-dependent NMP transition $T_{2'\to 1}$ controls the transition $T_{1\to 2'\to 2}$. The oxide field dependence $T_{2'\to 1}$ is reflected in the second term of (16.89). At moderate negative oxide fields (regime F), the transition $T_{2'\to 1}$ proceeds much faster than $T_{2\to 2'}$. Thus, $\tau_e^{2'}$ is determined by the field-insensitive transition $T_{2\to 2'}$ with a time constant of $\tau_{e,min}^{2'}$. It is pointed out that *the regime F can give an explanation for the field-independent emission time constants observed for fixed oxide hole traps (cf. Fig. 16.14 left)*. This is a direct consequence of the assumed bistability of the defect in the multi-state model.

At a low oxide field (regime E), the state $1'$ is further shifted down, which speeds up the transition $T_{2\to 1'}$ and allows the pathway over the metastable state $1'$. The corresponding emission time constant $\tau_e^{1'}$ is then given by

$$\tau_e^{1'} = \frac{k_{21'} + k_{1'2} + r_{1'1}}{k_{21'}k_{1'1}} \ . \tag{16.108}$$

For a sufficiently large barrier $\varepsilon_{1'1}$, the rate $k_{1'1}$ becomes negligible compared to $k_{21'}$ and $k_{1'2}$ and the above equation simplifies to

$$\tau_e^{1'} = \frac{1}{k_{1'1}} + \frac{k_{1'2}}{k_{21'}k_{1'1}} \ . \tag{16.109}$$

In this case, the state diagram reduces to a subsystem that includes the states $1'$ and 2 and is marginally disturbed by the rate $k_{1'1}$. In this subsystem the states $1'$ and 2 can be assumed to be in quasi-equilibrium

$$f_{1'}k_{1'2} = f_2 k_{21'} \tag{16.110}$$

and the condition $f_{1'} + f_2 = 1$ is met. Then the trap occupancy $f_t' = f_{1'}$ is given by

$$f_{1'} = \frac{1}{1+\frac{k_{21'}}{k_{1'2}}} = \frac{1}{1+\exp(\beta(E_t' - E_f))} \,. \tag{16.111}$$

From this equation, it follows that the condition $k_{1'2} = k_{21'}$ is equivalent to $E_t' = E_f$. Furthermore, this equation can also be used to simplify (16.90) to

$$\tau_e^{1'} = \tau_{1'} \exp\left(-\beta \frac{q_0 x_t F_{ox}}{1+R_{1'}}\right) + \frac{\tau_{e,min}^{1'}}{f_{t'}} \,. \tag{16.112}$$

If E_t' falls below E_f at a certain relaxation voltage, the state $1'$ becomes occupied and the emission time $\tau_e^{1'}$ is determined by the field-independent transition $T_{1'\to 1}$ with the time constant $\tau_{e,min}^{1'}$. By contrast, if E_t' is raised above E_f, the state $1'$ is underpopulated thereby slowing down the hole emission process. This occupancy effect is reflected in the second term, which is sensitive to changes in E_f.

The overall hole emission time τ_e follows approximately from

$$\frac{1}{\tau_e} \approx \frac{1}{\tau_e^{1'}} + \frac{1}{\tau_e^{2'}} \tag{16.113}$$

and is depicted in Fig. 16.15. At a certain oxide field, when the state $1'$ is shifted below state 2, $\tau_e^{1'}$ reaches its minimum value and falls below $\tau_e^{2'}$. *The resulting drop in τ_e is observed as the field dependence characterizing fixed oxide hole traps at weak oxide fields (cf. Fig. 16.14 right).* The drop in τ_e occurs when the minimum of the state $1'$ passes that of state 2, and is thus related to the exact shape of the configuration coordinate diagram. It is emphasized here that in the multi-state model the bistability of the defect allows for fixed as well as switching oxide hole traps while there is no explanation for these two kinds for defects in other models.

In summary, several features observed in the TDDS data have been quantitatively reproduced as shown in Sect. 16.8 and qualitatively understood following the above discussion based on analytical expressions. As such, this model can be regarded as a suited model to describe hole trapping in BTI.

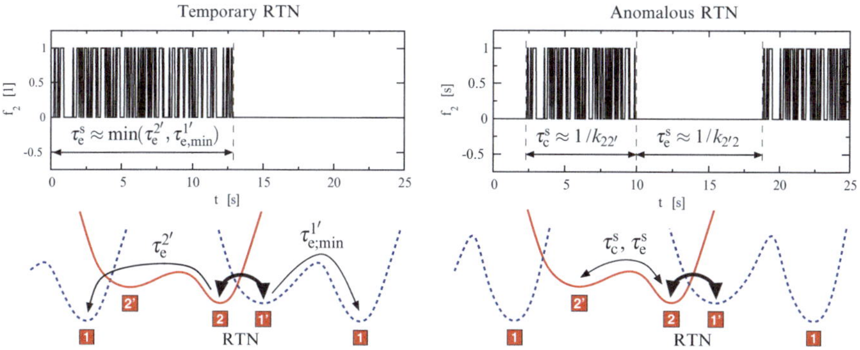

Fig. 16.19 *Top left*: The hole occupancy during tRTN. At $t = 0$ the stress voltage has been removed and the defect is in its positive state 2. After a time τ_e^s the defect ceases to produce noise. *Bottom left*: Configuration coordinate diagram for a tRTN defect. The *thick arrow* indicates the fast switches between the states 2 and $1'$ related to the occurrence of noise. The possibilities to escape from these states are shown by the *thin arrows*. *Top right*: Electron occupancy during aRTN. *Bottom right*: Configuration coordinate diagram for an aRTN defect. Since this defect is an electron trap, the *solid* and the *dashed line* correspond to the negative and the neutral charge state of the defect, respectively. The *double-sided thick arrow* is associated with aRTN while the *thin one* represents the transitions into and out of the metastable state $2'$

16.10 Noise

So far it has been shown that the multi-state model accounts for all features seen in the time constant plots for the fixed as well as the switching oxide hole traps. Beyond that, the model can also give an explanation for tRTN observed in TDDS (see Sect. 16.2). The generated noise stems from defects switching back and forth between states 2 and $1'$. The associated charge transfer reactions $T_{2\leftrightarrow 1'}$ do not involve any intermediate states and are therefore simple NMP processes. It is remarked here that the transitions $T_{2\leftrightarrow 1'}$ require the energy minima 2 and $1'$ in the configuration coordinate diagram to be on approximately the same level at the relaxation voltage. This is only the case for a group of defects whose energy minima 1 and $1'$ are energetically not far separated. In the TDDS measurements, the investigated devices are stressed at a high V_g so that the defects are forced from the state 1 into the state 2 or $1'$. During this step, the defects undergo the transition $T_{1\rightarrow 2'\rightarrow 2}$ into the state 2 or even further into $1'$. The other direct pathway $T_{1\rightarrow 1'}$ into the state $1'$ or 2 is assumed to go over a large barrier $\varepsilon_{11'}$. Therefore, the transition $T_{1\rightarrow 1'}$ proceeds on much larger timescales compared to $T_{1\rightarrow 2'\rightarrow 2}$ and can be neglected. After stressing, the recovery traces are monitored at low V_g or F_{ox}, respectively, at which the energy minima of the states 2 and $1'$ coincide and noise is produced. However, the state 1 is thermodynamically preferred due to its energetically lower position compared to the states 2 and $1'$. When the defect returns to its initial state 1, the RTN signal disappears with a time constant of τ_e^s. The corresponding transition could be either $T_{2\rightarrow 2'\rightarrow 1}$ or $T_{1'\rightarrow 1}$ with a time constant of $\tau_e^{2'}$

or $\tau_{e,min}^{1'}$, respectively (cf. Fig. 16.19). The termination of the noise signal after a time period of τ_e^s is determined by the minimum of these time constants. Consider that the NMP barriers $\varepsilon_{21'}$ and $\varepsilon_{1'2}$ must not be too large since otherwise trapping events will occur too fast and are therefore not detected using a conventional measurement equipment.

Interestingly, there also exists a type of defect which repeatedly produces noise for stochastically distributed time intervals (see Sect. 16.2). This kind of noise was observed for electron traps [27] in nMOSFETS and is referred to as aRTN. Just as in the case of tRTN, the noise signal is generated by charge transfer reactions between the states 2 and $1'$. The recurrent pauses of the noise signal (see Fig. 16.19) originate from transitions into the metastable state $2'$, which is electrically indistinguishable from the state 2. These interruptions correspond to the time during which the defect dwells in this state and no charge transfer reaction can take place. Thereby it has been presumed that the NMP transition $T_{2'\to 1}$ occurs on larger time scales than the return to the state 2 through the transition $T_{2'\to 2}$. The slow capture time constant τ_c^s in Fig. 16.19 defines the mean time interval during which noise is observed. Its value is given by the inverse of the transition rate $1/k_{22'}$. The slow emission time constant $\tau_e^s = 1/k_{2'2}$ corresponds to the mean time interval until the next noise period starts.

One should keep in mind that when adopting the concept of aRTN to hole traps in pMOSFET, it may also explain the tRTN behavior seen in TDDS measurements. During TDDS stress, this sort of defects are forced into one of the states 2 and $1'$ where they produce an RTN signal. As in aRTN, they undergo a transition to the metastable state $2'$ thereby stopping to produce a noise signal. However, this special sort of defects is characterized by a slow emission time constant τ_e^s, which is much larger than the typical measurement time of TDDS. As a consequence, the next transition back to the state 2 and the subsequent noise period are shifted out of the experimental time window of TDDS and will not be recorded during the measurement run. According to this explanation, tRTN can also be explained as a stimulated variant of aRTN.

In summary, the multi-state model can account for the features from the time constant plots and is consistent with the observation of tRTN as well as aRTN. This fact is presented here since it is regarded as an additional support for the validity of this model.

16.11 Conclusion

With the departure from the established reaction–diffusion model, charge trapping in BTI has recently attracted scientific interest. Therefore, the nature of charge trapping has remained vaguely understood for a long time and has been intensively studied within our group. In this chapter we presented a detailed derivation of our charge trapping model, termed multi-state model, in which the focus was on correctly modeling microscopic processes involved in BTI. In order to support understanding of the tendencies in this model, we have also given analytical expressions, which still capture the main physics underlying charge trapping in BTI.

For the verification of our model, we have chosen the TDDS technique since it allows to analyze the behavior of single defects. The evaluation of our multi-state model was based on five criteria including the curvature in the capture times, the gate bias and temperature dependences, and the fixed as well as the switching oxide hole trap behavior. So far, all these features have only been reproduced by the multi-state model, which strongly indicates that this model is based on correct assumptions. Interestingly, the model gives also an explanation for temporary and anomalous RTN, thereby further corroborating its validity.

Acknowledgements This work has received funding from the Austrian Science Fund (FWF) project n° 23390-N24 and the European Communities FP7 n° 261868 (MORDRED).

References

1. K.O. Jeppson and C.M. Svensson, "Negative Bias Stress of MOS Devices at High Electric Fields and Degradation of MNOS Devices," *J.Appl.Phys.*, vol. 48, no. 5, pp. 2004–2014, 1977.
2. S. Ogawa and N. Shiono, *Phys.Rev.B*, vol. 51, no. 7, pp. 4218–4230, 1995.
3. B. Kaczer, V. Arkhipov, R. Degraeve, N. Collaert, G. Groeseneken, and M. Goodwin, "Disorder-Controlled-Kinetics Model for Negative Bias Temperature Instability and its Experimental Verification," in *Proc.IRPS*, 2005, pp. 381–387.
4. S. Zafar, "Statistical Mechanics Based Model for Negative Bias Temperature Instability Induced Degradation," *J.Appl.Phys.*, vol. 97, no. 10, pp. 1–9, 2005.
5. M. Houssa, M. Aoulaiche, S. De Gendt, G. Groeseneken, M.M. Heyns, and A. Stesmans, "Reaction-Dispersive Proton Transport Model for Negative Bias Temperature Instabilities," *Appl.Phys.Lett.*, vol. 86, no. 9, pp. 1–3, 2005.
6. M.A. Alam, H. Kufluoglu, D. Varghese, and S. Mahapatra, "A Comprehensive Model for pMOS NBTI Degradation: Recent Progress," *Microelectron.Reliab.*, vol. 47, no. 6, pp. 853–862, 2007.
7. S. Chakravarthi, A.T. Krishnan, V. Reddy, C.F. Machala, and S. Krishnan, "A Comprehensive Framework for Predictive Modeling of Negative Bias Temperature Instability," in *Proc.IRPS*, 2004, pp. 273–282.
8. T. Grasser, W. Goes, and B. Kaczer, "Towards Engineering Modeling of Negative Bias Temperature Instability," in *Defects in Microelectronic Materials and Devices*, D. Fleetwood, R. Schrimpf, and S. Pantelides, Eds., pp. 1–30. Taylor and Francis/CRC Press, 2008.
9. F. Schanovsky and T. Grasser (2013) On the microscopic limit of the RD model. In: T. Grasser (eds) Bias temperature instability for devices and circuits. Springer, New York
10. F. Schanovsky and T. Grasser, "On the Microscopic Limit of the Reaction-Diffusion Model for Negative Bias Temperature Instability," in *Proc.IIRW*, 2011, pp. 17–21.
11. F. Schanovsky and T. Grasser, "On the Microscopic Limit of the Modified Reaction-Diffusion Model for Negative Bias Temperature Instability," in *Proc.IRPS*, 2012, pp. XT.10.1–6.
12. B. Kaczer, T. Grasser, J. Martin-Martinez, E. Simoen, M. Aoulaiche, Ph.J. Roussel, and G. Groeseneken, "NBTI from the Perspective of Defect States with Widely Distributed Time Scales," in *Proc.IRPS*, 2009, pp. 55–60.
13. H. Reisinger, T. Grasser, and C. Schlünder, "A Study of NBTI by the Statistical Analysis of the Properties of Individual Defects in pMOSFETs," in *Proc.IIRW*, 2009, pp. 30–35.

14. T. Grasser, H. Reisinger, P.-J. Wagner, and B. Kaczer, *Phys.Rev.B*, vol. 82, no. 24, pp. 245318, 2010.
15. P.-J. Wagner, T. Grasser, H. Reisinger, and B. Kaczer, "Oxide Traps in MOS Transistors: Semi-Automatic Extraction of Trap Parameters from Time Dependent Defect Spectroscopy," in *Proc.IPFA*, 2010, pp. 249–254.
16. T. Grasser, H. Reisinger, P.-J. Wagner, F. Schanovsky, W. Goes, and B. Kaczer, "The Time Dependent Defect Spectroscopy (TDDS) for the Characterization of the Bias Temperature Instability," in *Proc.IRPS*, 2010, pp. 16 –25.
17. H. Reisinger, T. Grasser, W. Gustin, and C. Schlünder, "The Statistical Analysis of Individual Defects Constituting NBTI and its Implications for Modeling DC- and AC-Stress," in *Proc.IRPS*, 2010, pp. 7–15.
18. B. Kaczer, T. Grasser, Ph.J. Roussel, J. Franco, R. Degraeve, L.A. Ragnarsson, E. Simoen, G. Groeseneken, and H. Reisinger, "Origin of NBTI Variability in Deeply Scaled PFETs," in *Proc.IRPS*, 2010, pp. 1095–1098.
19. H. Reisinger (2013) The time dependent defect spectroscopy. In: T. Grasser (eds) Bias temperature instability for devices and circuits. Springer, New York
20. V. Huard, C. Parthasarathy, N. Rallet, C. Guerin, M. Mammase, D. Barge, and C. Ouvrard, "New Characterization and Modeling Approach for NBTI Degradation from Transistor to Product Level," in *Proc.IEDM*, 2007, pp. 797–800.
21. T.L. Tewksbury, *Relaxation Effects in MOS Devices due to Tunnel Exchange with Near-Interface Oxide Traps*, Ph.D. Thesis, MIT, 1992.
22. I. Lundstrom and C. Svensson, "Tunneling to Traps in Insulators," *J.Appl.Phys.*, vol. 43, no. 12, pp. 5045–5047, 1972.
23. F.P. Heiman and G. Warfield, "The Effects of Oxide Traps on the MOS Capacitance," *IEEE Trans.Elect.Dev.*, vol. 12, no. 4, pp. 167–178, 1965.
24. S. Christensson, I. Lundström, and C. Svensson, "Low Frequency Noise in MOS Transistors — I Theory," *Sol.-St.Electr.*, vol. 11, pp. 797–812, 1968.
25. W. Shockley and W.T. Read, *Phys.Rev.*, vol. 87, no. 5, pp. 835–842, 1952.
26. A.L. McWhorter, "1/f Noise and Germanium Surface Properties," in *Sem.Surf.Phys.* RH Kingston (Univ Penn Press), 1957.
27. M.J. Kirton and M.J. Uren, "Noise in Solid-State Microstructures: A New Perspective on Individual Defects, Interface States, and Low-Frequency (1/f) Noise," *Adv.Phys.*, vol. 38, no. 4, pp. 367–486, 1989.
28. M. Masuduzzaman, A.E. Islam, and M.A. Alam, "Exploring the Capability of Multifrequency Charge Pumping in Resolving Location and Energy Levels of Traps Within Dielectric," *IEEE Elect.Dev.Let.*, vol. 55, no. 12, pp. 3421–3431, 2008.
29. A. Avellan, D. Schroeder, and W. Krautschneider, "Modeling Random Telegraph Signals in the Gate Current of Metal-Oxide-Semiconductor Field Effect Transistors after Oxide Breakdown," *J.Appl.Phys.*, vol. 94, no. 1, pp. 703–708, 2003.
30. M. Isler and D. Liebig, *Phys.Rev.B*, vol. 61, no. 11, pp. 7483–7488, 2000.
31. N. Zanolla, D. Siprak, P. Baumgartner, E. Sangiorgi, and C. Fiegna, "Measurement and Simulation of Gate Voltage Dependence of RTS Emission and Capture Time Constants in MOSFETs," in *Ultimate Integration of Silicon*, 2008, pp. 137–140.
32. R.R. Siergiej, M.H. White, and N.S. Saks, "Theory and Measurement of Quantization Effects on $Si - SiO_2$ Interface Trap Modeling," *Sol.-St.Electr.*, vol. 35, no. 6, pp. 843–854, 1992.
33. N.B. Lukyanchikova, M.V. Petrichuk, N.P. Garbar, E. Simoen, and C. Claeys, "Influence of the Substrate Voltage on the Random Telegraph Signal Parameters in Submicron *n*-Channel Metal-Oxide-Semiconductor Field-Effect Transistors under a Constant Inversion Charge Density," *Appl.Phys.A*, vol. 70, no. 3, pp. 345–353, 2000.
34. S. Makram-Ebeid and M. Lannoo, *Phys.Rev.B*, vol. 25, no. 10, pp. 6406–6424, 1982.
35. S.D. Ganichev, W. Prettl, and I.N. Yassievich, "Deep Impurity-Center Ionization by Far-Infrared Radiation," *Phys.Solid State*, vol. 39, no. 1, pp. 1703–1726, 1997.
36. S.D. Ganichev, I.N. Yassievich, V.I. Perel, H. Ketterl, and W. Prettl, *Phys.Rev.B*, vol. 65, pp. 085203, 2002.

37. K. Huang and A. Rhys, "Theory of Light Absorption and Non-Radiative Transitions in F-Centres," *Proceedings of the Royal Society of London. Series A*, vol. 204, pp. 406–423, 1950.
38. C.H. Henry and D.V. Lang, *Phys.Rev.B*, vol. 15, no. 2, pp. 989–1016, 1977.
39. K.V. Mikkelsen and M.A. Ratner, "Electron Tunneling in Solid-State Electron-Transfer Reactions," *Chemical Reviews*, vol. 87, no. 1, pp. 113–153, 1987.
40. Conley and Lenahan, "Electron Spin Resonance Evidence that E'_γ Centers Can Behave as Switching Traps," *IEEE Trans.Nucl.Sci.*, vol. 42, no. 6, pp. 1744–1749, 1995.
41. J.F. Conley Jr., P.M. Lenahan, A.J. Lelis, and T.R. Oldham, "Electron Spin Resonance Evidence for the Structure of a Switching Oxide Trap: Long Term Structural Change at Silicon Dangling Bond Sites in SiO_2," *Appl.Phys.Lett.*, vol. 67, no. 15, pp. 2179–2181, 1995.
42. A.J. Lelis and T.R. Oldham, "Time Dependence of Switching Oxide Traps," *IEEE Trans.Nucl.Sci.*, vol. 41, no. 6, pp. 1835–1843, 1994.
43. M. Lax, "The Franck-Condon Principle and Its Application to Crystals," *Journ.Chem.Phys.*, vol. 20, no. 11, pp. 1752–1760, 1952.
44. T.H. Keil, *Phys.Rev.*, vol. 140, no. 2A, pp. A601–A617, 1965.
45. F. Schanovsky, O. Baumgartner, V. Sverdlov, and T. Grasser, "A Multi Scale Modeling Approach to Non-Radiative Multi Phonon Transitions at Oxide Defects in MOS Structures," *Journ. of Computational Electronics*, vol. 11, no. 3, pp. 218–224, 2012.
46. S. Datta, *Quantum Transport — Atom to Transistor*, Cambridge University Press, 2005.
47. M.O. Andersson, Z. Xiao, S. Norrman, and O. Engström, "Model Based on Trap-Assisted Tunneling for Two-Level Current Fluctuations in Submicrometer Metal-Silicon-Dioxide Diodes," *Phys.Rev.B*, vol. 41, no. 14, pp. 9836–9842, 1990.
48. P.E. Blöchl and J.H. Stathis, "Hydrogen Electrochemistry and Stress-Induced Leakage Current in Silica," *Phys.Rev.Lett.*, vol. 83, no. 2, pp. 372–375, 1999.
49. P.E. Blöchl and J.H. Stathis, "Aspects of Defects in Silica Related to Dielectric Breakdown of Gate Oxides in MOSFETs," *Phys.B*, vol. 273-274, pp. 1022–1026, 1999.
50. W.B. Fowler, J.K. Rudra, M.E. Zvanut, and F.J. Feigl, *Phys.Rev.B*, vol. 41, no. 12, pp. 8313–8317, 1990.
51. W. Goes and T. Grasser, "First-Principles Investigation on Oxide Trapping," in *Proc.SISPAD*, 2007, pp. 157–160.
52. W. Goes and T. Grasser, "Charging and Discharging of Oxide Defects in Reliability Issues," in *Proc.IIRW*, 2007, pp. 27–32.
53. W. Goes, M. Karner, V. Sverdlov, and T. Grasser, "Charging and Discharging of Oxide Defects in Reliability Issues," *IEEE Trans.Dev.Mater.Rel.*, vol. 8, no. 3, pp. 491–500, 2008.
54. A. Alkauskas and A. Pasquarello, "Alignment of Hydrogen-Related Defect Levels at the $Si - SiO_2$ Interface," *Phys.B Condens.Matter*, vol. 401–402, pp. 546–549, 2007.
55. T. Grasser, "Stochastic Charge Trapping in Oxides: From Random Telegraph Noise to Bias Temperature Instabilities," *Microelectron.Reliab.*, vol. 52, no. 1, pp. 39–70, 2012.
56. O.C. Ibe, *Markov Processes for Stochastic Modeling*, Academic Press, 2009.
57. M. Bina, O. Triebl, B. Schwarz, M. Karner, B. Kaczer, and T. Grasser, "Simulation of Reliability on Nanoscale Devices," in *Proc.SISPAD*, 2012, pp. 109–112.
58. B. Kaczer, J. Franco, M. Toledano-Luque, Ph.J. Roussel, M.F. Bukhori, A. Asenov, B. Schwarz, M. Bina, T. Grasser, and G. Groeseneken, "The Relevance of Deeply-Scaled FET Threshold Voltage Shifts for Operation Lifetimes," in *Proc.IRPS*, 2012.
59. A. Asenov, "Random Dopant-Induced Threshold Voltage Lowering and Fluctuations in Sub-0.1 μm MOSFET's: A 3-D "Atomistic" Simulation Study," *IEEE Trans.Elect.Dev.*, vol. 45, no. 12, pp. 2505–2513, 1998.
60. A. Mauri, N. Castellani, C.M. Compagnoni, A. Ghetti, P. Cappelletti, A.S. Spinelli, and A.L. Lacaita, "Impact of Atomistic Doping and 3D Electrostatics on the Variability of RTN Time Constants in Flash Memories," in *Proc.IEDM*, 2011, pp. 17.1.1–17.1.4.
61. N. Sano, K. Matsuzawa, M. Mukai, and N. Nakayama, "On Discrete Random Dopant Modeling in Drift-Diffusion Simulations: Physical Meaning of "Atomistic" Dopants," *Microelectron.Reliab.*, vol. 42, no. 2, pp. 189–199, 2002.

62. A. Asenov, G. Slavcheva, A.R. Brown, J.H. Davies, and S. Saini, "Increase in the Random Dopant-Induced Threshold Fluctuations and Lowering in Sub-100 nm MOSFETs due to Quantum Effects: a 3-D Density-Gradient Simulation Study," *IEEE Trans.Elect.Dev.*, vol. 48, no. 4, pp. 722–729, 2001.
63. S.M. Amoroso (2013) Statistical study of bias temperature instabilities by means of 3D 'atomistic' simulation. In: T. Grasser (eds) Bias temperature instability for devices and circuits. Springer, New York

Chapter 17
The Capture/Emission Time Map Approach to the Bias Temperature Instability

Tibor Grasser

Abstract Recent results suggest that the bias temperature instability can in good approximation be understood as the collective response of an ensemble of independent defects. Although the kinetics of charge capture and defect creation clearly require the presence of charge carriers in the channel, they appear reaction rather than diffusion limited. While a number of peculiar features in these kinetics have been revealed recently, the most striking feature remains the wide distribution of reaction rates, or equivalently, time constants. By modeling the activation energies of the time constants via bivariate Gaussian distributions in what we call *capture/emission time maps*, a wide range of experimentally observed features can be explained in closed analytical form. Examples are the temperature- and bias-independent power-law time exponent during stress including saturation at longer times, the long logarithmic-like recovery traces, as well as differences and similarities between DC and AC stress.

17.1 Introduction

Numerous studies conducted over the last couple of decades have shown that at least two types of defects contribute to the bias temperature instability (BTI), namely oxide and interface defects [1–7]. Considerable evidence has piled up in recent years suggesting that oxide defects are mainly responsible for the recoverable component of BTI [8–11], while interface defects are mostly permanent in typical experimental windows [12–15].

Charge exchange between the channel and oxide defects has traditionally been modeled using a simple Shockley–Read–Hall (SRH) model [16]. The SRH model was originally developed for bulk defects, but later extended in an empirical manner

---

T. Grasser (✉)
TU Wien, Institute for Microelectronics, Gusshausstrasse 27-29, 1040 Wien, Austria
e-mail: grasser@iue.tuwien.ac.at

T. Grasser (ed.), *Bias Temperature Instability for Devices and Circuits*,
DOI 10.1007/978-1-4614-7909-3_17,

to describe oxide defects by the introduction of a WKB tunneling factor [17–20]. Detailed time-dependent defect spectroscopy (TDDS) studies have shown, however, that the oxide defects contributing to NBTI are of a more complicated nature [9, 10, 21, 22]. In particular, transitions between the different charge states are consistent with nonradiative multiphonon (NMP) processes, as has already been observed in random telegraph noise studies [23]. Furthermore, metastable states seem to be an essential aspect since they explain the switching trap behavior [10,21,24,25] as well as the frequency dependence of the capture time constant [26–30]. Nevertheless, the most intriguing feature appears to be the wide distribution of both the capture and the emission time constants [31]. These time constants may even be too short to be experimentally observable ($<1\,\mu$s) as well as extremely large (>1 ks). While the chemical nature of these oxide defects has not been unanimously identified [32–34], it is this distribution of time constants which essentially determines the typical recovery behavior of a device following bias temperature stress [31, 35–38].

Interface states, at least at SiO_2/Si interfaces, are most likely due to silicon dangling bonds at the silicon–insulator interface, known as P_b centers [33, 39, 40]. The creation dynamics are much harder to study experimentally since both capture and emission time constants are rather large. Also, in every BTI experiment the recoverable component R appears to overshadow the build-up of the permanent component P [15]. Thus, a number of attempts have been made at characterizing P:

- Application of measurement methods which (hopefully) dominantly measure interface states, such as charge-pumping techniques [13, 41–43].
- Attempts to remove R by for instance accelerating recovery by switching the gate voltage into accumulation.
- Attempts to guess from the dominant behavior of R on the underlying evolution of P (the *universal recovery* idea) [44, 45].

Unfortunately, all these methods introduce uncertainties:

- First, charge-pumping currents have to be converted to the typically measured threshold-voltage shifts used to characterize R to make the components comparable. However, it is not clear whether only interface states contribute to those recombination currents and how the density-of-states in the fraction of the bandgap visible to charge-pumping has to be extended to the remainder of the bandgap to allow for a meaningful comparison [15, 41, 46].
- Second, all attempts in removing R by the application of controlled discharge pulses appear to leave some unspecified remaining fraction of defects behind, since not all defects react to switches of the gate bias [24, 47]. Furthermore, the devices may show a tendency to go back to their pre-pulse rather than to their pre-stress state [15].
- Finally, while the universality appears to capture an interesting aspect of R, it is not clear what physical process is responsible for such a behavior and how accurate such an extraction scheme is.

As a consequence, we know much less about P than we know about R, making the available models more rudimentary. In particular, it remains controversial whether R and P are created in a coupled manner [48, 49] or not [11, 37, 50]. Nonetheless, similar to R, the creation/annealing time constants of P also show a wide distribution. This is also consistent with electron-spin-resonance data on creation and annealing of P_b centers [51].

Since the wide distribution of the time constants is responsible for both the build-up and the recovery of R and P, this distribution essentially determines the time-dependence of the degradation. As such, it appears natural to seek a description of BTI based on these distributions [31, 36–38]. A particularly useful observation in enabling a simple description is that despite their multi-state nature, charging and discharging of individual oxide traps responsible for R can be well described by an effective first-order process [20], at least for lower frequencies [29]. While not that much is known about P, available experimental data appear to indicate that the same is true for P [4, 28]. In the following we will summarize our recent attempts in developing such a model which describes the build-up of R and P as the collective action of a large number of individual defects, each described by a first-order process.

17.2 The Capture and Emission Times

In order to describe the defects, we first have to specify their capture and emission times as a function of bias and temperature. While the models used for oxide and interface traps are fundamentally different, they can still be approximately brought into the same mathematical form, yielding effective capture and emission times

$$\tau_c = \tau_0 e^{\beta \mathscr{E}_c} \quad \text{and} \quad \tau_e = \tau_0 e^{\beta \mathscr{E}_e} \tag{17.1}$$

with $\beta = 1/k_B T$, k_B the Boltzmann constant, and T the absolute temperature. In general, the effective time constant τ_0 will depend only weakly on bias and temperature, while the effective capture and emission barriers $\mathscr{E}_c$ and $\mathscr{E}_e$ can have a strong bias dependence. In the following, the assumptions required to bring available physical models for oxide and interface defects into the simple form (17.1) will be summarized.

17.2.1 Oxide Defects

We begin our discussion with oxide defects, which have been shown [10, 24, 29, 47] to have at least four states, 1, 1′, 2′, and 2, see Fig. 17.1. The unprimed states 1 and 2 are assumed to correspond to the stable equilibrium configuration in the neutral and positive charge states, while 1′ and 2′ are their metastable counterparts.

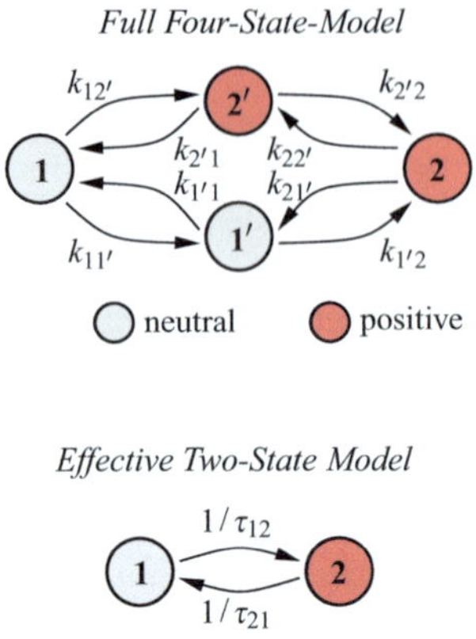

Fig. 17.1 *Top*: The four states of oxide defects extracted from DC TDDS experiments [10, 52]. Each defect has two stable states, 1 and 2, and possibly two metastable states $1'$ and $2'$. The metastable state $2'$ seems to be always present, while the existence of the metastable state $1'$ decides on whether the trap behaves like a fixed or a switching trap [24,53]. *Bottom*: An effective two-state approximation of the four-state defect using the first-passage times τ_{12} and τ_{21} [10, 20]

The transitions between the states are described by 8 rates, k_{ij}. In a first-order description, we neglect the switching state $1'$. This approximation is valid as long as the gate voltage remains above the threshold voltage but misses the rapid decrease of the emission time once the device is biased into depletion or accumulation [47]. Transitions between these states appear to be consistent with a Markov process, which in essence means that the defect forgets its past once it has arrived in a certain state. These transitions are stochastic processes, where the transition events for each individual transition are exponential distributed. The parameter of this distribution gives the mean transition time. Neglecting state $1'$, the first passage times [54, 55] for an overall transition from 1 to 2 define the effective capture and emission times as [20]

$$\tau_c = \frac{k_{12'} + k_{2'1} + k_{2'2}}{k_{12'}\, k_{2'2}} \quad \text{and} \quad \tau_e = \frac{k_{2'2} + k_{22'} + k_{2'1}}{k_{22'}\, k_{2'1}}, \tag{17.2}$$

which is not quite in the simple form (17.1) yet. While the first passage times exactly describe the mean of the overall distribution of the capture and emission times, replacing the four-state defect model (or three-state model in this case) by an effective two-state model approximates the distributions of the stochastic capture and emission events by exponential distributions [20]. Nonetheless, this appears to be an excellent approximation [10].

The physics behind the initial charge capture transition $1 \rightarrow 2'$ can be modeled at various levels of detail. In order to obtain the simplest results possible, we consider only the ground state of the neutral and the metastable positive configuration (E_1 and $E_{2'}$) and assume that all holes are located at the valence band edge E_V directly at

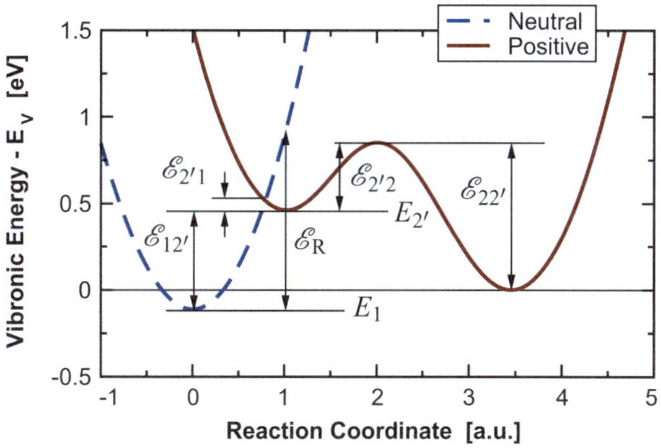

Fig. 17.2 Definition of the symbols required to describe the adiabatic defect potential of the simple model without the switching state $1'$. The vibronic energy, which is the sum of the electronic and vibrational energies, is shown relative to the substrate valence band in state 2, that is, $E_2 = 0$

the interface, see Fig. 17.2. The transitions are described using NMP theory [56–60] based on linear electron–phonon coupling [61], with the rates given in the classical (high-temperature) limit by

$$k_{12'} = p\, v_{\mathrm{th}}\, \sigma\, \mathrm{e}^{-x/x_0}\, \mathrm{e}^{-\beta \mathscr{E}_{12'}}, \tag{17.3}$$

$$k_{2'1} = p\, v_{\mathrm{th}}\, \sigma\, \mathrm{e}^{-x/x_0}\, \mathrm{e}^{-\beta \mathscr{E}_{12'}}\, \mathrm{e}^{-\beta E_{1\mathrm{F}}}, \tag{17.4}$$

where p is the surface hole concentration, v_{th} the thermal velocity, σ the capture cross section, x_0 the parameter in the simplified WKB tunneling expression, and $E_{1\mathrm{F}} = E_1 - E_{\mathrm{F}}$ the distance of the trap level from the Fermi-level. For linear electron–phonon coupling the NMP barrier is obtained as

$$\mathscr{E}_{12'} = \frac{(\mathscr{E}_{\mathrm{R}} + E_{2'1})^2}{4\mathscr{E}_{\mathrm{R}}} \tag{17.5}$$

where $E_{2'1} = E_{2'} - E_1$, $\mathscr{E}_{\mathrm{R}} = S\hbar\omega$ as the lattice relaxation energy, ω the oscillator frequency determined by the curvature of the parabolic adiabatic potential [62], and S the Huang–Rhys factor which gives the number of phonons required for the optical transition. For strong electron–phonon coupling ($\mathscr{E}_{\mathrm{R}} \gg E_{2'1}$), the quadratic dependence simplifies to $\mathscr{E}_{12'} = \mathscr{E}_{\mathrm{R}}/4 + E_{2'1}/2$, which we will use in the following for the derivation of the analytical results. It is convenient to express the flat-band defect energy levels E_{10} and $E_{2'0}$ relative to $E_{\mathrm{V}0}$, the flat-band valence band edge, by introducing $\mathscr{E}_1 = E_{10} - E_{\mathrm{V}0}$ and $\mathscr{E}_{2'} = E_{2'0} - E_{\mathrm{V}0}$. Assuming to first order that the charges trapped inside the oxide have only a small impact on the electric field, we have $E_{2'1} = E_{2'0} - E_{10} - \mathrm{q}xF = \mathscr{E}_{2'} - \mathscr{E}_1 - \mathrm{q}xF$, with x the distance of the trap into the oxide, F the oxide field, and E_{i0} the trap level for $F = 0$. The sign conventions are such that x is positive and F is positive for a negative bias at the gate (NBTI). Inserting the above into the rates delivers

$$k_{12'} = \quad p v_{\mathrm{th}}\,\sigma\,\mathrm{e}^{-x/x_0}\,\mathrm{e}^{-\beta(\mathscr{E}_{\mathrm{R}}+2\mathscr{E}_{2'})/4}\,\mathrm{e}^{+\beta(\mathscr{E}_1+\mathrm{q}xF)/2}, \tag{17.6}$$

$$k_{2'1} = N_{\mathrm{v}}\,v_{\mathrm{th}}\,\sigma\,\mathrm{e}^{-x/x_0}\,\mathrm{e}^{-\beta(\mathscr{E}_{\mathrm{R}}+2\mathscr{E}_{2'})/4}\,\mathrm{e}^{-\beta(\mathscr{E}_1+\mathrm{q}xF)/2}, \tag{17.7}$$

where Boltzmann statistics have been assumed for simplicity, $p = N_{\mathrm{v}}\exp(\beta E_{\mathrm{VF}})$. The barrier crossing rates for the transitions $2' \leftrightharpoons 2$ are expressed by a simple Arrhenius law with an attempt frequency $\nu = 10^{13}\,\mathrm{s}^{-1}$

$$k_{2'2} = \nu\mathrm{e}^{-\beta\mathscr{E}_{2'2}} \quad \text{and} \quad k_{22'} = \nu\mathrm{e}^{-\beta(\mathscr{E}_{2'2}+\mathscr{E}_{2'})}. \tag{17.8}$$

We proceed by rewriting the first passage times using the definitions $\tau_{ij} = 1/k_{ij}$ as

$$\tau_{\mathrm{c}} = \tau_{12'} + \tau_{2'2}\left(1 + \frac{\tau_{12'}}{\tau_{2'1}}\right) = \tau_{12'} + \tau_{2'2}\left(1 + \frac{N_{\mathrm{v}}}{p}\mathrm{e}^{-\beta(\mathscr{E}_1+\mathrm{q}xF)}\right), \tag{17.9}$$

$$\tau_{\mathrm{e}} = \tau_{22'} + \tau_{2'1}\left(1 + \frac{\tau_{22'}}{\tau_{2'2}}\right) = \tau_{22'} + \tau_{2'1}\left(1 + \mathrm{e}^{\beta\mathscr{E}_{2'}}\right). \tag{17.10}$$

This is an interesting result. (a) First, we see that τ_{c} at very high fields becomes bias independent and is only determined by the barrier between $2'$ and 2, $\tau_{\mathrm{c}} \approx \tau_{2'2}$. (b) Both time constants can potentially show a strong exponential bias dependence, via the dependence on $\tau_{12'}$ and $\tau_{2'1}$. (c) While both time constants $\tau_{12'}$ and $\tau_{2'1}$ depend on $\mathscr{E}_1$, this dependence is not normally relevant for τ_{e}, which is dominated by $\tau_{22'}$. (d) Finally, under typical NBTI conditions, recovery is measured at low F where $\tau_{2'1}$ is small, so $\tau_{\mathrm{e}} \approx \tau_{22'}$, that is, recovery is dominated by the barrier from 2 to $2'$. Only for biases lower than about the threshold voltage, the pathway $2 \leftrightharpoons 1' \rightarrow 1$ can be triggered when accessible (switching traps). As such hole emission even from below E_{V} will have a barrier since holes can no longer simply "bubble up" as in the SRH picture [20].

In principle, all parameters appearing in (17.9) and (17.10) are different for each defect, including the surface hole concentration p due to the random location of the current percolation paths [63, 64]. Unfortunately, not much is known at present about the nature of these distributions, so we have to invoke a few bold assumptions here: first, it has been demonstrated [65] that τ_{c} and x are uncorrelated for those defects contribution to RTN. Whether this also holds for NBTI is unknown at the moment, but we will nonetheless assume in the following $x = \bar{x}$, its average effective value. Lacking evidence to the contrary, all other parameters are assumed to follow a Gaussian distribution for simplicity. A particularly noteworthy issue is the following: since many parameters (e.g., $\mathscr{E}_{\mathrm{R}}$, $\mathscr{E}_1$, $\mathscr{E}_{2'}$) which control the defect behavior result from a certain defect constellation, it appears likely that some hidden correlations exist. Note also that for defects contributing to NBTI, $\mathscr{E}_1$ is typically smaller than zero, since the defect has to lie below the valence band to be initially neutral (to be more precise, the defect level has to lie below the Fermi-level at the read-out or recovery voltage).

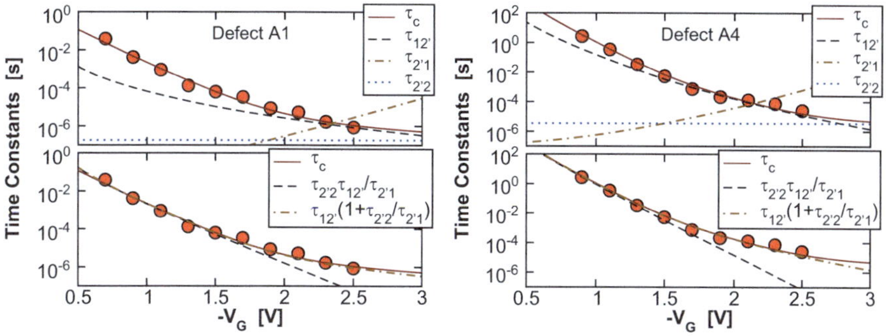

Fig. 17.3 The effective capture time τ_c is a function of all three partial rates. Shown are two defects from [10] together with a fit to the model

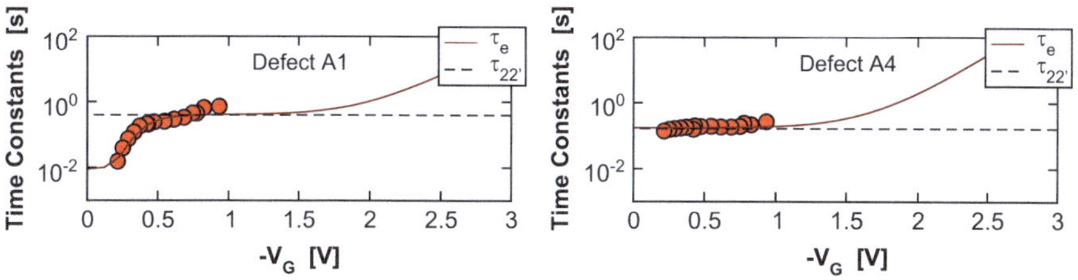

Fig. 17.4 The effective emission time τ_c is dominated by $\tau_{22'}$ for V_G above the threshold voltage. Shown are two defects from [10] together with a fit to the model. Defect A1 is a switching trap while the emission time of A4 appears independent of V_G. The switching behavior results from a backward transition via the pathway $2 \leftrightharpoons 1' \rightarrow 1$ but is ignored in the present discussion for simplicity. See [47] for an extended data set and modeling results toward lower V_G

17.2.1.1 Low Fields

For low fields during stress, Fig. 17.3 indicates that τ_c is dominated by the $\tau_{2'2}\tau_{12'}/\tau_{2'1}$ term, while Fig. 17.4 shows that τ_e is bias independent. Thus we have

$$\tau_c = \tau_{2'2}\frac{\tau_{12'}}{\tau_{2'1}} = \nu e^{-\beta(E_{VF}+\mathscr{E}_{2'2}+\mathscr{E}_1+q\bar{x}F)}, \tag{17.11}$$

$$\tau_e = \tau_{22'} \qquad = \nu e^{-\beta(\mathscr{E}_{2'2}+\mathscr{E}_{2'})} = \nu e^{-\beta\mathscr{E}_{22'}}. \tag{17.12}$$

As can be seen, τ_c depends exponentially on the electric field F. The above also implies that there is some explicit correlation between τ_c and τ_e due to the occurrence of $\mathscr{E}_{2'2}$ in both expressions in addition to the unknown hidden correlations in the parameters.

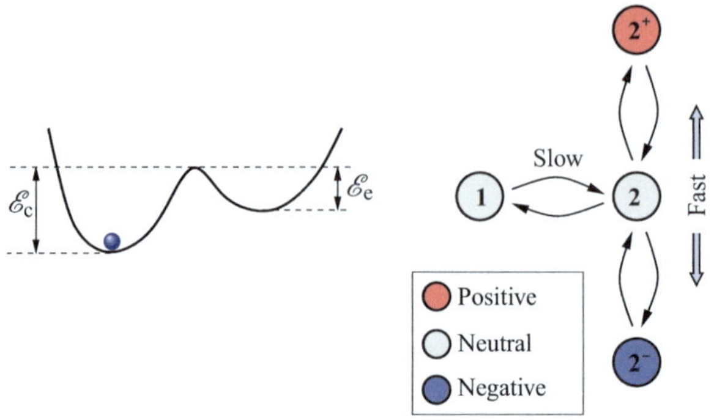

Fig. 17.5 *Left*: Simple double-well model used for the creation of interface states. *Right*: State diagram for interface states. The double-well model is used to describe the transitions $1 \leftrightharpoons 2$, while a change of the charge state can be obtained using a SRH model

17.2.1.2 Medium Fields

At medium stress fields, τ_c is basically given by $\tau_{12'}$ and we have $\tau_c = \tau_{12'}$ and $\tau_e = \tau_{22'}$. In this regime no obvious correlation exists and any experimentally observed correlation must be due to hidden correlations in the parameters. This issue is discussed in the next section using a simple model for the creation of the interface states.

17.2.1.3 Strong Fields

At strong fields, the capture time will be dominated by the bias-independent barrier $\mathcal{E}_{2'2}$ and thus become bias-independent. The emission time, on the other hand, will be dominated by the bias-dependent barrier $\mathcal{E}_{2'1}$ and increase significantly.

17.2.2 Interface Defects

Since inside typical measurement windows NBTI degradation is dominated by the activation and annealing of oxide defects, much less is known about interface states, the creation of which appears nonetheless universally acknowledged [11, 13, 37, 50, 66]. As the details of the creation dynamics are unclear at the moment, the creation of interface states from a precursor state 1 is typically modeled using a simple double-well model into the neutral state 2 [11, 37, 50, 67], see Fig. 17.5. Again, as with the hole trapping model, the barriers are assumed to be statistically distributed [4, 51, 67]. Alternatively, some groups advocate a reaction–diffusion mechanism [50, 68, 69], which we consider inadequate due to the lacking direct experimental evidence [70] and additional theoretical difficulties [71, 72]. The charge state of the amphoteric defect is then determined using an SRH model [73]. Since in this picture the charge state can change rapidly, we limit our attention to the creation process $1 \leftrightharpoons 2$. In the simple double-well model, we obtain the classical over-the-barrier rates

$$\tau_\mathrm{c} = \nu \mathrm{e}^{\beta \mathscr{E}_\mathrm{c}} \quad \text{and} \quad \tau_\mathrm{e} = \nu \mathrm{e}^{\beta \mathscr{E}_\mathrm{e}}. \tag{17.13}$$

As hinted at previously, we will study in the following the distributions of τ_c and τ_e for a large number of defects. Written in the form (17.13), no correlation between these two time constants would be obtained if $\mathscr{E}_\mathrm{c}$ and $\mathscr{E}_\mathrm{e}$ were independent random variables. However, given that the adiabatic potential describing the double-well is a result of the various forces acting on the atoms, such an independence is unlikely. Quite to the contrary, one can expect a hidden correlation between the parameters $\mathscr{E}_\mathrm{c}$ and $\mathscr{E}_\mathrm{e}$, since it is unlikely that changes in the defect configuration only impacts the barrier E_B without altering the levels E_1 and E_2. As such, if we choose to write $\mathscr{E}_\mathrm{c} = E_\mathrm{B} - E_1$ and $\mathscr{E}_\mathrm{e} = E_\mathrm{B} - E_2$, a distribution of E_B will affect both $\mathscr{E}_\mathrm{c}$ and $\mathscr{E}_\mathrm{e}$ since

$$\tau_\mathrm{c} = \nu \mathrm{e}^{\beta (E_\mathrm{B} - E_1)} \quad \text{and} \quad \tau_\mathrm{e} = \nu \mathrm{e}^{\beta (E_\mathrm{B} - E_2)} = \tau_\mathrm{c} \mathrm{e}^{\beta (E_1 - E_2)}. \tag{17.14}$$

Even in this case, the energies E_B, E_1, and E_2 cannot be expected to be independent. Nonetheless, to make the model even simpler, we assume that it is the quantities $E_\mathrm{B1} = E_\mathrm{B} - E_1$ and $E_{12} = E_1 - E_2$ that are independently distributed. The only justification we have at the moment is that this assumption appears to capture the essence of the experimental data, in particular the observed correlation between τ_c and τ_e.

17.3 The Capture/Emission Time Map

We now proceed from individual defects of either type to a large collection of both types. Assume we have a collection of independent defects with a distribution of capture and emission times. In the interval $[\tau_\mathrm{c}, \tau_\mathrm{c} + \mathrm{d}\tau_\mathrm{c}]$ and $[\tau_\mathrm{e}, \tau_\mathrm{e} + \mathrm{d}\tau_\mathrm{e}]$ the number of defects contributing to ΔV_th is $g(\tau_\mathrm{c}, \tau_\mathrm{e})\,\mathrm{d}\tau_\mathrm{c}\,\mathrm{d}\tau_\mathrm{e}$, where the capture/emission time distribution ("the map") g has dimension $\mathrm{V/s^2}$. Depending on the stressing history of the device, all defects with similar τ_c and τ_e values can be expected to have a similar occupancy. This occupancy, $h(\tau_\mathrm{c}, \tau_\mathrm{e})$, is 1 if all defects in that interval fully contribute to ΔV_th and 0 if they do not contribute at all. For the assumed first-order processes, h is simple to calculate as a consequence of arbitrarily switching gate voltages between a high and low level. Then, by multiplying h with g and integrating over the whole domain, ΔV_th can be calculated at any time, provided g remains constant. Apparently, this is roughly the case, although defect transformations have been occasionally observed [10,37,74–77], which will be neglected in the following.

Mathematically, the total ΔV_{th} is thus obtained by summing up the contributions of all defects with a particular combination of τ_{c} and τ_{e}, embodied by $g(\tau_{\text{c}},\tau_{\text{e}})\mathrm{d}\tau_{\text{c}}\mathrm{d}\tau_{\text{e}}$, weighted by the occupancy $h(\tau_{\text{c}},\tau_{\text{e}})$ as

$$\Delta V_{\text{th}}(t_{\text{s}},t_{\text{r}}) \approx \int_0^\infty \mathrm{d}\tau_{\text{c}} \int_0^\infty \mathrm{d}\tau_{\text{e}}\, g(\tau_{\text{c}},\tau_{\text{e}})\, h(\tau_{\text{c}},\tau_{\text{e}};t_{\text{s}},t_{\text{r}}). \tag{17.15}$$

As said before, the occupancy function h depends on the history of stress and recovery cycles the device has been exposed to and on the details of the physical process. A simple case is obtained for a collection of defects following first-order processes, which have been subjected to a DC stress phase of duration t_{s} and a recovery time t_{r},

$$h(\tau_{\text{c}},\tau_{\text{e}};t_{\text{s}},t_{\text{r}}) = \left(1-\mathrm{e}^{-t_{\text{s}}/\tau_{\text{c}}}\right)\mathrm{e}^{-t_{\text{r}}/\tau_{\text{e}}} \tag{17.16}$$

provided that the occupancy is 0 at the initial read-out voltage and 1 after a stress duration $t_{\text{s}} \gg \tau_{\text{c}}$. Note that τ_{c} is taken at the stress voltage, while τ_{e} is considered at the recovery voltage. To simplify the integration, we employ the approximation

$$h(\tau_{\text{c}},\tau_{\text{e}};t_{\text{s}},t_{\text{r}}) \approx H(t_{\text{s}}-\tau_{\text{c}})H(\tau_{\text{e}}-t_{\text{r}}). \tag{17.17}$$

where H is the unit step function. Although this approximation is somewhat crude, as the two transitions contained in h cover a decade in time, it gives us a very simple and intuitive connection between ΔV_{th} and g,

$$\Delta V_{\text{th}}(t_{\text{s}},t_{\text{r}}) \approx \int_0^{t_{\text{s}}} \mathrm{d}\tau_{\text{c}} \int_{t_{\text{r}}}^\infty \mathrm{d}\tau_{\text{e}}\, g(\tau_{\text{c}},\tau_{\text{e}}). \tag{17.18}$$

In words this means that ΔV_{th} is given by the sum of all defects charged until t_{s} but not yet discharged after t_{r}. Equation (17.18) can now be used to give a simple method for the extraction of g by simply taking the negative mixed partial derivative of a given ΔV_{th} stress/recovery data set [78],

$$g(\tau_{\text{c}},\tau_{\text{e}}) \approx -\frac{\partial^2 \Delta V_{\text{th}}(\tau_{\text{c}},\tau_{\text{e}})}{\partial\tau_{\text{c}}\,\partial\tau_{\text{e}}}. \tag{17.19}$$

Note that completely permanent defects with $\tau_{\text{e}} \to \infty$ do not show up in the CET map. Given the wide distribution of the defect time constants, it is advantageous to represent the CET map on logarithmic axes. Transformation of the variables gives [20]

$$\tilde{g}(\tau_{\text{c}},\tau_{\text{e}}) \approx -\frac{\partial^2 \Delta V_{\text{th}}(\tau_{\text{c}},\tau_{\text{e}})}{\partial\log(\tau_{\text{c}})\,\partial\log(\tau_{\text{e}})} = \tau_{\text{c}}\tau_{\text{e}}\, g(\tau_{\text{c}},\tau_{\text{e}}). \tag{17.20}$$

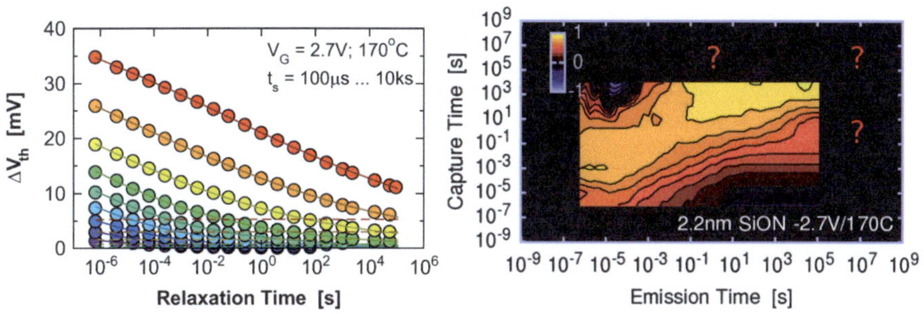

Fig. 17.6 *Left*: Experimental recovery data following increasing stress times (*symbols*). *Right*: The logarithmic CET map obtained by taking the numerical derivative of the data. Numerical integration results in the *solid lines* of the *left figure*. The *dashed lines* are obtained from a "permanent" component with $\tau_e > t_{relax,max}$, which is not directly available in the numerical CET map and has to be provided separately. The map is normalized and plotted using a signed log operator $\text{sign}(g)\log_{10}(1+\kappa|g/g_{max}|)/\log_{10}(1+\kappa)$ with $\kappa = 100$ to bring out all important details

While g gives the density of defects per unit time, for example information on how much ΔV_{th} is gained/lost in a second, $\tilde{g}$ gives the density on a logarithmic scale, for example on how much ΔV_{th} is gained/lost per decade.

An example of $\tilde{g}$ extracted from experimental $\Delta V_{th}(t_s, t_r)$ data is shown in Fig. 17.6. Note that while a correlation between τ_c and τ_e exists, it is weak and a significant density is obtained in the whole experimental window.

17.3.1 Occupancy Patterns

In (17.16), we have already given the occupancy of a defect after a certain stress and relaxation time under the assumption of an initial empty and finally fully occupied defect. This can be easily generalized as shown in the following.

17.3.1.1 DC Stress

Given that the defect has the occupancy $f(t_0)$ at time t_0, its occupancy after a stress time of duration t_s is

$$f(t_0 + t_s) = f_s + (f(t_0) - f_s)e^{-t_s k_s}, \tag{17.21}$$

while after an additional recovery time t_r one has

$$f(t_0 + t_s + t_r) = f_r + (f(t_0 + t_s) - f_r)e^{-t_r k_r}. \tag{17.22}$$

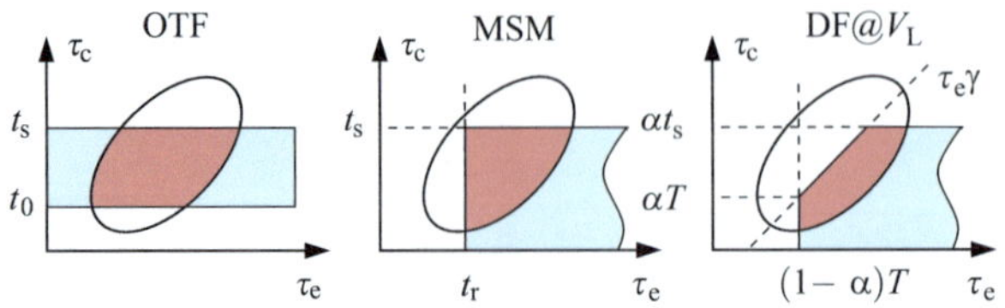

Fig. 17.7 Depending on the measurement scheme, a certain fraction of the CET map can contribute to $\Delta V_{\rm th}$. *Left*: An on-the-fly setup (OTF) misses $\tau_c < t_0$. *Middle*: A measure-stress-measure (MSM) setup misses $\tau_e < t_r$. *Right*: An AC stress results in a trapezoidal region

The occupancy after infinitely long stress would be f_s, while after an infinitely long recovery we have f_r. Both occupancies follow directly from the bias- and temperature-dependent rates

$$f_s = k_{12}/(k_{12}+k_{21})|_{\rm stress}, \qquad k_s = k_{12}+k_{21}|_{\rm stress}, \tag{17.23}$$

$$f_r = k_{12}/(k_{12}+k_{21})|_{\rm relax}, \qquad k_r = k_{12}+k_{21}|_{\rm relax}. \tag{17.24}$$

As initial condition we assume that all defects have their equilibrium occupancy at the off-voltage, $f(t_0=0)=f_r$ and measure only the deviation from f_r, which reads

$$\Delta f(t_s) = (f_s - f_r)(1-{\rm e}^{-t_s k_s}). \tag{17.25}$$

Upon termination of the stress, we have after a recovery time of t_r

$$\Delta f(t_s,t_r) = \Delta f(t_s){\rm e}^{-t_r k_r}, \tag{17.26}$$

which is of the same form as (17.16), except for the prefactor $(f_s - f_r)$. This prefactor cannot be extracted from macroscopic data and will be tacitly moved into the CET map g. Note that if the equilibrium occupancies differ from 0 or 1, this means that the defect produces RTN.

As before, if we assume now a collection of defects with distributed k_s and k_r, (17.25) and (17.26) can be used to calculate the occupancy of each defect after a stress time t_s and recovery time t_r: from (17.25) it follows that all defects with $k_s < 1/t_s$ will remain unoccupied, while (17.26) says that all defects which were occupied during stress will already be unoccupied again if $k_r > 1/t_r$. These two conditions describe a rectangular area in the CET map, shown in Fig. 17.7.

17.3.1.2 AC Stress

The above procedure can be easily generalized to digital on-off (AC) stress [36, 38, 79], with duty factor α and period T. After the first cycle we have the occupancies

$$s_1 = f_s + (f_r - f_s)s \tag{17.27}$$

$$r_1 = f_r + (s_1 - f_r)r \tag{17.28}$$

with $s = \exp(-\alpha T k_s)$ and $r = \exp(-(1-\alpha)T k_r)$. Continuing this scheme recursively for n cycles, we obtain a simple geometric series in $B = sr$, which eventually gives

$$\Delta s_n = (f_s - f_r)(1 - B^n)\frac{1-s}{1-B}, \tag{17.29}$$

$$\Delta r_n = \Delta s_n r. \tag{17.30}$$

After a certain stress time t_s, the cycle number is obtained via $n = \lceil t_s/T \rceil$. Since n will be a large integer number in practical cases, we consider it a continuous variable to simplify the notation, $n \approx t_s/T$. The dominant term in (17.29) is B^n, or

$$B^n = (sr)^n = \mathrm{e}^{-nT(\alpha k_s + (1-\alpha)k_r)} \doteq \mathrm{e}^{-n\alpha T k_{AC}} \tag{17.31}$$

with $k_{AC} \doteq k_s + k_r/\gamma$ and $\gamma = \alpha/(1-\alpha)$. Thus, we have

$$\Delta s_n = (f_s - f_r)(1 - \mathrm{e}^{-\alpha t_s k_{AC}})\frac{1 - \mathrm{e}^{-\alpha T k_s}}{1 - \mathrm{e}^{-\alpha T k_{AC}}}. \tag{17.32}$$

Equation (17.32) gives the occupancy of a certain defect with effective rates k_s and k_r after a stress time t_s. The first exponential factor gives a transition from 1 to 0 when $\alpha t_s \approx 1/k_{AC}$, thereby giving the upper bound of the trapezoidal region shown in Fig. 17.7. For small k_s and k_r, the last term can be approximated using $\exp(-x) \approx 1 - x$ as

$$\frac{1 - \mathrm{e}^{-\alpha T k_s}}{1 - \mathrm{e}^{-\alpha T k_{AC}}} \approx \frac{k_s}{k_s + k_r/\gamma}. \tag{17.33}$$

This term results in the diagonal of the trapezoidal region [36]. To see this, take a fixed k_r (or τ_e), for which this term becomes 0 for small k_s (large τ_c) and 1 for large k_s (small τ_c), with the transition occurring roughly at $k_s = k_r/\gamma$, or $\tau_c = \tau_e \gamma$.

17.3.2 The Capture Time Map

Occasionally, we are not that much interested in the details of recovery, for example when we want to determine the worst-case degradation under constant bias stress. We can then simplify the problem to a certain degree by collapsing the τ_e axis of the full distribution $g(\tau_c, \tau_e)$. For instance, a typical measure-stress-measure (MSM) setup will require a certain delay t_M with which the degradation can be determined. Thus, in order to calculate the degradation at a certain stress time t_s measured with a certain measurement delay t_M, we integrate over the τ_e axis starting from t_M until infinity, see Fig. 17.7. This includes the contribution of all defects $\tau_\mathrm{e} > t_\mathrm{M}$ because they have not yet emitted their charge. We therefore define the capture time map as

$$g_\mathrm{c}(\tau_\mathrm{c}, t_\mathrm{M}) = \int_{t_\mathrm{M}}^{\infty} g(\tau_\mathrm{c}, \tau_\mathrm{e}) \, \mathrm{d}\tau_\mathrm{e} \tag{17.34}$$

which completely determines ΔV_th as

$$\Delta V_\mathrm{th}(t_\mathrm{s}, t_\mathrm{M}) = \int_0^{t_\mathrm{s}} g_\mathrm{c}(\tau_\mathrm{c}, t_\mathrm{M}) \, \mathrm{d}\tau_\mathrm{c} = G_\mathrm{c}(t_\mathrm{s}, t_\mathrm{M}) \tag{17.35}$$

because $G_\mathrm{c}(0, t_\mathrm{M})$ must vanish. In fact, if we chose to normalize $g_\mathrm{c}(\tau_\mathrm{c}, t_\mathrm{M})$, it would be just like the probability density function of ΔV_th while $G(\tau_\mathrm{c})$ would correspond to the cumulative distribution function. However, as we shall see in the sequel, this analogy should not be taken too far, since g and g_c can have a negative sign if non-first-order processes are considered. This is for instance the case when the prediction of the reaction–diffusion model is cast into this formalism. Also, the loss of defects over time may result in negative entries in g. While these more subtle points will not be discussed in the following, they may prove crucial in the near future and are the reason why g is referred to as *map* rather than *distribution*.

In delay-free experiments, which have become known as on-the-fly (OTF) measurements [80, 81], the measurement delay is zero and the capture time map covers the whole τ_e axis. In practice, however, a delay-free experiment requires determination of a reference value for the calculation of ΔV_th. This reference value is determined with a certain delay t_M at the stress voltage, which corresponds to

$$\Delta V_\mathrm{th}^\mathrm{OTF}(t_\mathrm{s}, t_\mathrm{M}) = \int_{t_\mathrm{M}}^{t_\mathrm{s}} g_\mathrm{c}(\tau_\mathrm{c}, 0) \, \mathrm{d}\tau_\mathrm{c}. \tag{17.36}$$

As a result, even OTF measurements do not capture all defects as the lower part of the τ_c axis is missed. The opposite is true for MSM setups which cover the complete τ_c axis but only a part of the τ_e axis. The difference between the two setups is visualized in Fig. 17.7.

Equations (17.35) and (17.36) now provide a simple procedure for the extraction of $g_\mathrm{c}(\tau_\mathrm{c}, t_\mathrm{M})$ from a given $\Delta V_\mathrm{th}(t_\mathrm{s}, t_\mathrm{M})$,

$$g_\mathrm{c}(\tau_\mathrm{c}, t_\mathrm{M}) = \frac{\mathrm{d}\Delta V_\mathrm{th}(\tau_\mathrm{c}, t_\mathrm{M})}{\mathrm{d}\tau_\mathrm{c}} \quad \text{and} \quad g_\mathrm{c}(\tau_\mathrm{c}, 0) = \frac{\mathrm{d}\Delta V_\mathrm{th}^\mathrm{OTF}(\tau_\mathrm{c}, t_\mathrm{M})}{\mathrm{d}\tau_\mathrm{c}}, \tag{17.37}$$

the first including the delay of the MSM measurement while the second being valid for OFT data with $\tau_\mathrm{c} > t_\mathrm{M}$.

17.3.3 The Logarithmic Capture/Emission Time Map

So far we have defined the two-dimensional capture/emission time map as well as its reduced one-dimensional counterpart, the capture time maps. These maps give the density of defects having certain time constants on a *linear* axis. As we shall see in the sequel, it is useful to transform the density onto *logarithmic* axes. Such a transformation is inspired by the typically observed power-law degradation behavior, which corresponds to a straight line on a double logarithmic plot, as well as by the typically observed logarithmic recovery. In particular, the latter implies that about the same amount of charge is lost per decade in time, see Fig. 17.6.

We start by introducing

$$\theta_\mathrm{c} = \log(\tau_\mathrm{c}/\tau_0) \qquad \text{and} \qquad \theta_\mathrm{e} = \log(\tau_\mathrm{e}/\tau_0)\,, \tag{17.38}$$

with a suitably chosen τ_0. For the time being, τ_0 serves the purpose of a normalization constant, while θ is merely the logarithm of a normalized time constant. However, as physical models for the time constants can usually be cast into the form $\tau = \tau_0 \exp(\theta)$, this already implies the basic structure of the physical model, as hinted at in (17.1).

Instead of integrating over τ, we rewrite the integration of g as an integration over θ

$$\begin{aligned}\Delta V_\mathrm{th}(t_\mathrm{s},t_\mathrm{r}) &= \int_0^{t_\mathrm{s}} \mathrm{d}\tau_\mathrm{c} \int_{t_\mathrm{r}}^{\infty} \mathrm{d}\tau_\mathrm{e}\, g(\tau_\mathrm{c},\tau_\mathrm{e}) \\ &= \int_{-\infty}^{\log(t_\mathrm{s}/\tau_0)} \mathrm{d}\theta_\mathrm{c} \int_{\log(t_\mathrm{r}/\tau_0)}^{\infty} \mathrm{d}\theta_\mathrm{e}\, g(\tau_0 \mathrm{e}^{\theta_\mathrm{c}},\tau_0 \mathrm{e}^{\theta_\mathrm{e}})\, \tau_0 \mathrm{e}^{\theta_\mathrm{c}}\, \tau_0 \mathrm{e}^{\theta_\mathrm{e}}.\end{aligned} \tag{17.39}$$

With $\theta_\mathrm{s} = \log(t_\mathrm{s}/\tau_0)$, $\theta_\mathrm{r} = \log(t_\mathrm{r}/\tau_0)$, and

$$\tilde{g}(\theta_\mathrm{c},\theta_\mathrm{e}) = g(\tau_0 \mathrm{e}^{\theta_\mathrm{c}},\tau_0 \mathrm{e}^{\theta_\mathrm{e}})\, \tau_0 \mathrm{e}^{\theta_\mathrm{c}}\, \tau_0 \mathrm{e}^{\theta_\mathrm{e}} \tag{17.40}$$

we can finally write

$$\Delta V_\mathrm{th}(t_\mathrm{s},t_\mathrm{r}) = \int_{-\infty}^{\theta_\mathrm{s}} \mathrm{d}\theta_\mathrm{c} \int_{\theta_\mathrm{r}}^{\infty} \mathrm{d}\theta_\mathrm{e}\, \tilde{g}(\theta_\mathrm{c},\theta_\mathrm{e}) = \tilde{G}(\theta_\mathrm{s},\infty) - \tilde{G}(\theta_\mathrm{s},\theta_\mathrm{r}). \tag{17.41}$$

Equation (17.40) handles the transformation from the linear to the logarithmic scale, with $\tilde{g}(\theta_\mathrm{c},\theta_\mathrm{e})$ as the logarithmic CET map. Conversely, we have the inverse transformation

$$g(\tau_\mathrm{c},\tau_\mathrm{e}) = \frac{\tilde{g}(\log(\tau_\mathrm{c}/\tau_0),\log(\tau_\mathrm{e}/\tau_0))}{\tau_\mathrm{c}\tau_\mathrm{e}}. \tag{17.42}$$

Similarly, the transformation rules for the linear and logarithmic capture time maps are

$$\tilde{g}_{\mathrm{c}}(\theta_{\mathrm{c}}) = g_{\mathrm{c}}(\tau_0 \mathrm{e}^{\theta_{\mathrm{c}}})\, \tau_0 \mathrm{e}^{\theta_{\mathrm{c}}} \qquad \text{and} \qquad g_{\mathrm{c}}(\tau_{\mathrm{c}}) = \frac{\tilde{g}_{\mathrm{c}}(\log(\tau_{\mathrm{c}}/\tau_0))}{\tau_{\mathrm{c}}}. \tag{17.43}$$

With the logarithmic capture time map, ΔV_{th} can be obtained as

$$\Delta V_{\mathrm{th}}(t_{\mathrm{s}}) = \int_{-\infty}^{\theta_{\mathrm{s}}} \tilde{g}_{\mathrm{c}}(\theta_{\mathrm{c}})\, \mathrm{d}\theta_{\mathrm{c}} = \tilde{G}_{\mathrm{c}}(\theta_{\mathrm{s}}). \tag{17.44}$$

17.3.4 Properties of the Capture Time Map

So far we have derived theoretical relations which describe the connections between the various maps and the experimentally observed degradation. No assumptions on their functional forms have been made. Naturally, any experimentally observed degradation and recovery behavior will require a unique CET map.

For simplicity, we start with the capture time map, which can be calculated from a known $\Delta V_{\mathrm{th}}(t_{\mathrm{s}})$. Experimentally, two functional forms of $\Delta V_{\mathrm{th}}(t_{\mathrm{s}})$ are of importance, namely the logarithmic degradation, $\log(t_{\mathrm{s}}/t_0)$, particularly for short-time data [82, 83], and the power-law t_{s}^n [50, 68]. The capture time maps required to produce such a time behavior will be derived in the following.

17.3.4.1 Logarithmic Time Behavior

Assume that the experimentally observed degradation follows a logarithm in time,

$$\Delta V_{\mathrm{th}}(t_{\mathrm{s}}) = A \log(t_{\mathrm{s}}/t_0) \tag{17.45}$$

starting from a certain time $t_{\mathrm{s}} \geq t_0$ as sketched in Fig. 17.8. In NBTI data, the point t_0 is typically outside the measurement window [83]. According to (17.37) we obtain

$$g_{\mathrm{c}}(\tau_{\mathrm{c}}) = \frac{\mathrm{d}\Delta V_{\mathrm{th}}(\tau_{\mathrm{c}})}{\mathrm{d}\tau_{\mathrm{c}}} = \frac{A}{\tau_{\mathrm{c}}}, \tag{17.46}$$

which corresponds to the p.d.f. of a log-uniform distribution. This is easier to see when g_{c} is transformed on the logarithmic θ_{c} axis using (17.43)

$$\tilde{g}_{\mathrm{c}}(\theta_{\mathrm{c}}) = g_{\mathrm{c}}(\tau_0 \mathrm{e}^{\theta_{\mathrm{c}}})\, \tau_0 \mathrm{e}^{\theta_{\mathrm{c}}} = A \tag{17.47}$$

for $\theta_{\mathrm{c}} \geq \log(t_0/\tau_0)$ and shown in Fig. 17.8.

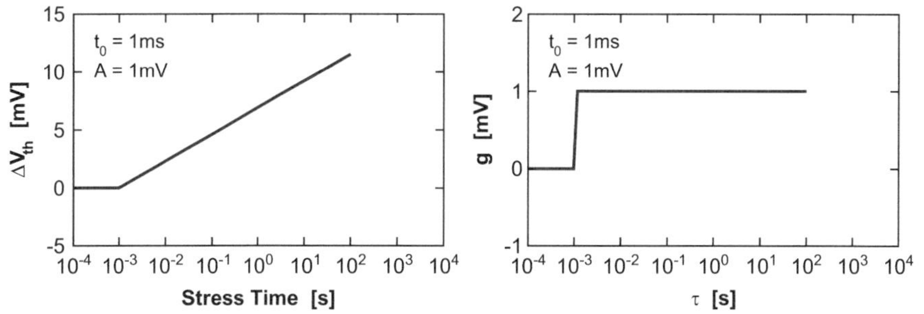

Fig. 17.8 *Left*: Logarithmic time evolution of ΔV_{th}. *Right*: The corresponding logarithmic capture time map

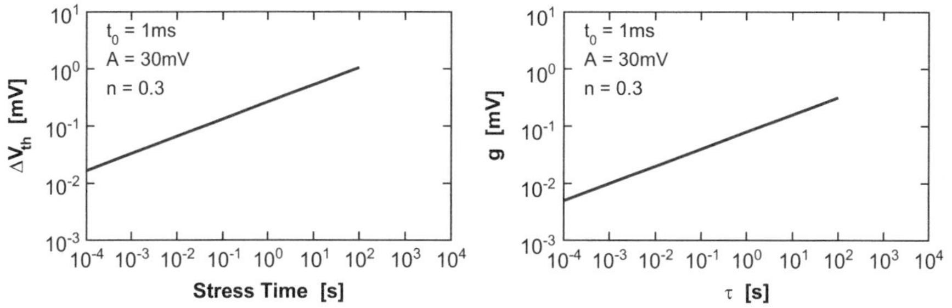

Fig. 17.9 *Left*: Power-law time evolution of ΔV_{th}. *Right*: The corresponding logarithmic capture time map

17.3.4.2 Power-Law Time Behavior

Assume now that the degradation follows a power-law in time,

$$\Delta V_{\mathrm{th}}(t_{\mathrm{s}}) = A\left(\frac{t_{\mathrm{s}}}{t_0}\right)^n \tag{17.48}$$

as shown in Fig. 17.9. By making use of (17.37) as before, we obtain

$$g(\tau_{\mathrm{c}}) = \frac{\mathrm{d}\Delta V_{\mathrm{th}}(\tau_{\mathrm{c}})}{\mathrm{d}\tau_{\mathrm{c}}} = A\frac{n}{t_0^n}\frac{1}{\tau_{\mathrm{c}}^{1-n}} \tag{17.49}$$

for $t_{\mathrm{s}} \geq t_0$. On the logarithmic axis we have

$$\tilde{g}(\theta_{\mathrm{c}}) = An\left(\frac{\tau_0}{t_0}\right)^n \mathrm{e}^{n\theta_{\mathrm{c}}}. \tag{17.50}$$

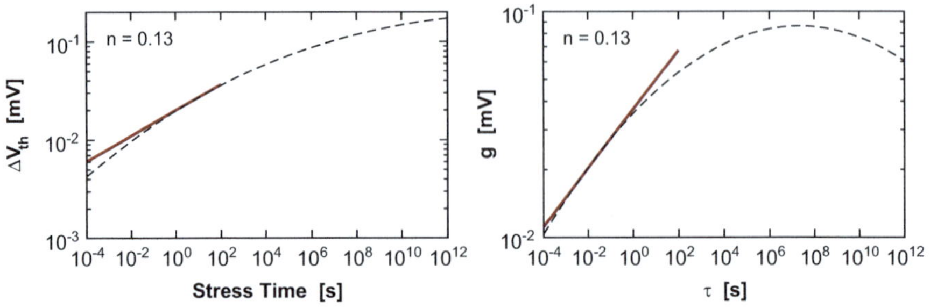

Fig. 17.10 *Left*: Power-law time evolution (*solid line*) as a short-time property of a Gaussian distribution (*dashed line*) ΔV_{th}. *Right*: The corresponding logarithmic capture time map

This can be written as a function of $\tau_c(\theta_c)$ as

$$\tilde{g}(\tau_c) = A\frac{n}{t_0^n}\tau_c^n. \tag{17.51}$$

In words, a power-law degradation in time requires a logarithmic density which increases following a power-law in τ_c with the same exponent n, see Fig. 17.9.

17.3.4.3 Discussion

In summary, the two cases of the logarithmic and power-law time-dependence will result from the following distributions

$$\begin{aligned} &\text{Logarithmic}: \quad g_c(\tau_c) \sim 1/\tau_c \qquad && \tilde{g}_c(\theta_c) \sim \text{const.} \\ &\text{Power}-\text{Law}: \quad g_c(\tau_c) \sim 1/\tau_c^{1-n} \qquad && \tilde{g}_c(\theta_c) \sim \mathrm{e}^{n\theta_c} = \tau_c^n \end{aligned}$$

The power-law exponent typically observed is rather small, say $n = 0.15$, which results in $g(\tau_c) \sim 1/\tau_c^{0.85}$. This is reminiscent to the problem of $1/f$ noise [84]: theoretically, a uniform distribution in θ results in $1/f$ noise. Experimentally, however, one often sees something more like $1/f^{\alpha}$, with exponents close to unity, which then would correspond to a "power-law" distribution.

The fundamental question that springs to mind is how these distributions will behave for larger τ_c. For example, a perfect power-law requires an indefinitely increasing $\tilde{g}_c$, which is clearly not a sensible option. It is thus important to realize that $\tilde{g}_c$ measured over a limited time window can only provide some local snapshot of a more general distribution, which eventually has to saturate and fall off. A natural example for such a distribution would be a Gaussian distribution on a logarithmic scale, see Fig. 17.10. As will be discussed in more detail below, a wide Gaussian distribution will produce a power-law in time over many decades, albeit with a slight curvature. Indeed, while such a deviation from the power-law can also be attributed to the influence of the measurement delay [68, 85, 86], a curvature can be clearly observed also in long-term OTF data [49, 87–89].

17.3.5 *Physical Origin of the Capture Time Map*

As discussed above, one cannot expect the power-law degradation to continue indefinitely in time. The most obvious explanation would be that the (partial) distribution $e^{n\theta}$ is the tail of a more realistic distribution that, after having reached its peak, eventually levels off with increasing θ.

17.3.5.1 Tail of a Gaussian Distribution

The natural choice for such a distribution would be the Gaussian distribution of θ as $\tilde{g}_c(\theta_c) = \Delta V_{th}^{max} f(\theta_c)$ with

$$f_g(\theta) = \frac{1}{\sqrt{2\pi}\sigma}\exp\Big(-\frac{(\theta-\mu)^2}{2\sigma^2}\Big) \tag{17.52}$$

Taking the Taylor expansion of $\log(f)$ at some $\theta_0 < \mu$ we have

$$f_g(\theta) \approx f_g(\theta_0)e^{-n\theta_0}e^{n\theta} = \frac{1}{\sqrt{2\pi}\sigma}\exp\Big(\frac{\theta_0^2-\mu^2}{2\sigma^2}\Big)e^{n\theta} \tag{17.53}$$

with the power-law exponent

$$n = \Big[\frac{1}{f_g(\theta)}\frac{df_g(\theta)}{d\theta}\Big]_{\theta=\theta_0} = \frac{\mu-\theta_0}{\sigma^2}. \tag{17.54}$$

Recall that $\theta_0 < \mu$ was assumed, so $n > 0$ as it should be.

We have already shown that for a certain region around θ_0 a Gaussian distribution results in a power-law in time. The full time evolution including the curvature and eventual saturation can be obtained from $\tilde{g}_c(\theta_c) = \Delta V_{th}^{max} f_g(\theta_c)$ as

$$\Delta V_{th}(t_s) = \frac{\Delta V_{th}^{max}}{2}\mathrm{erfc}\Big(\frac{\mu - \log(t_s/\tau_0)}{\sqrt{2}\sigma}\Big). \tag{17.55}$$

17.3.5.2 Tail of a Logistic Distribution

Rather than using a Gaussian distribution, which results in awkward error functions when integrated, the use of the logistic distribution

$$f_l(\theta) = \frac{1}{s}\frac{\exp\Big(\frac{\mu-\theta}{s}\Big)}{\Big(1+\exp\Big(\frac{\mu-\theta}{s}\Big)\Big)^2} \tag{17.56}$$

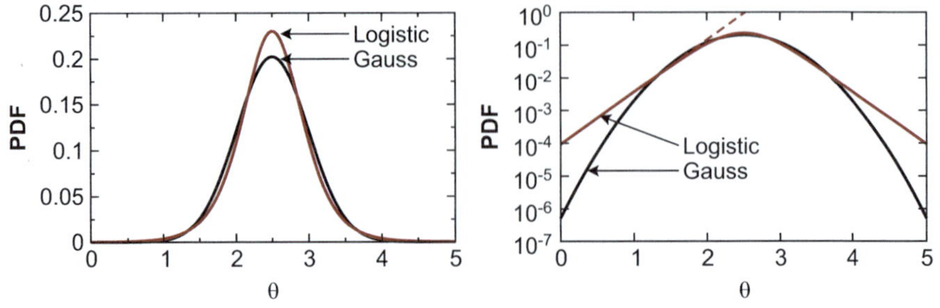

Fig. 17.11 *Left*: On a linear scale, the logistic distribution appears to be very similar to a Gaussian distribution. For this comparison the same mean and standard deviation ($\mu = 2.5$ and $\sigma = 0.5$) were chosen. A slight increase in the standard deviation of the logistic distribution to 0.56 would further increase the visual similarity (not shown). *Right*: On a logarithmic y-axis, one can see that the tails of the logistic distribution are linear in θ while for the Gaussian distribution they depend quadratically on θ. As a consequence, the Gaussian distribution always has a curvature on a log-lin plot

has been suggested [90], with mean μ and parameter s. The standard deviation of the logistic distribution is $\sigma = s\pi/\sqrt{3}$. When plotted using a linear y-axis, the logistic distribution appears like a Gaussian hump, see Fig. 17.11. In contrast to the Gaussian distribution, however, it can be easily integrated and results in the Fermi function

$$F_{\mathrm{l}}(\theta) = \frac{1}{1 + \exp\left(\dfrac{\mu - \theta}{s}\right)}. \tag{17.57}$$

Because of this property, the logistic distribution is sometimes called Fermi-derivative distribution [90]. The most tempting choice to explain the power-law is to assume $\theta \ll \mu$ [90], for which the logistic distribution can be approximated by

$$f_{\mathrm{l}}(\theta) \approx \frac{1}{s}\exp\left(\frac{\theta - \mu}{s}\right), \tag{17.58}$$

which would perfectly correspond to the density $\tilde{g}(\theta)$ required for a power-law in time. Then, the power-law exponent would be given by $n = 1/s = \pi/\sqrt{3}\sigma$. Albeit tempting, we will see later that this is an unfortunate choice, since one may be lead to the wrong conclusion that the logistic distribution is incompatible with experimental data. To obtain the "correct" results, we expand the distribution in a more general fashion as in the Gaussian case, which leads to

$$n = \frac{1}{s}\frac{\exp\left(\dfrac{\mu - \theta}{s}\right) + 1}{\exp\left(\dfrac{\mu - \theta}{s}\right) - 1}. \tag{17.59}$$

The above will only give the conventional $n \approx 1/s$ for θ far below the mean, that is, $\theta \ll \mu$. However, as will be shown in the following sections, it is only this more general form of the power-law exponent which is consistent with experimental data. In this general case, the time evolution of ΔV_{th} resulting from a logistic distribution is

$$\Delta V_{\mathrm{th}}(t_{\mathrm{s}}) = \frac{\Delta V_{\mathrm{th}}^{\max}}{1 + \mathrm{e}^{\mu/s}\left(\frac{\tau_0}{t_{\mathrm{s}}}\right)^{1/s}}. \tag{17.60}$$

17.3.6 Simple Thermal Activation Model

So far we have established that an exponential density $g(\theta_{\mathrm{c}}) = \mathrm{e}^{n\theta_{\mathrm{c}}}$ is required for a power-law in time. Also, this exponential density can be justified as the tail of either a Gaussian or a logistic distribution. What remains to be seen is the *physical meaning* of the relation $\tau_{\mathrm{c}} = \tau_0 \mathrm{e}^{\theta_{\mathrm{c}}}$. In other words, what physical model would give time constants of the form $\tau_{\mathrm{c}} = \tau_0 \mathrm{e}^{\theta_{\mathrm{c}}}$?

As already hinted at in the discussion on the physical models around (17.1), the most obvious choice that springs to mind is the Arrhenius law, $\tau_{\mathrm{c}} = \tau_0 \mathrm{e}^{\beta \mathscr{E}_{\mathrm{c}}}$. In that case, the physical meaning of θ would be given by the activation energy of the process,

$$\mathscr{E}_{\mathrm{c}} = \theta_{\mathrm{c}}/\beta. \tag{17.61}$$

Also, rather than assuming that θ_{c} is distributed according to a certain distribution, it appears more sensible to assume that it is the activation energy itself which is distributed. The difference between these two assumptions is fundamental, as in the latter case the distribution of θ_{c} will depend on temperature while in the former case it will not. Whichever option is correct can then be easily determined by verifying the "built-in" temperature dependence of the model with experimental data.

For the simple distributions discussed here, we only need to be concerned about the mean $\bar{\mathscr{E}}_{\mathrm{c}}$ and the standard deviation σ_{c} of the activation energy. It follows from basic statistical laws that the according moments of the transformed distribution θ are $\mu = \bar{\mathscr{E}}_{\mathrm{c}}/\mathrm{k_B}T$ and $\sigma = \sigma_{\mathrm{c}}/\mathrm{k_B}T$.

The fundamental question to answer here is whether these distributions are compatible with the experimentally observed temperature-independent power law exponents. At a first glance, this is anything but obvious and has led to claims that such distributions are incompatible with data. We start by writing the time evolution of a Gaussian distribution

$$\Delta V_{\mathrm{th}}(t_{\mathrm{s}}) = \frac{\Delta V_{\mathrm{th}}^{\max}}{2} \operatorname{erfc}\left(\frac{\bar{\mathscr{E}}_{\mathrm{c}} - \mathrm{k_B}T \log(t_{\mathrm{s}}/\tau_0)}{\sqrt{2}\sigma_{\mathrm{c}}}\right), \tag{17.62}$$

which can be approximated by a power law around a certain measurement window given by $\tau_0 \exp(\theta_0)$. The slope of this power law is obtained from (17.54) and (17.61) as

$$n = \frac{\bar{\mathscr{E}}_c/\mathrm{k_B}T - \theta_0}{(\sigma_c/\mathrm{k_B}T)^2} \tag{17.63}$$

and apparently depends on temperature. However, the important point to see here is that the measurement point θ_0 is determined by the experimental window and is therefore not temperature-dependent. If θ_0 were much smaller than $\bar{\mathscr{E}}_c/\mathrm{k_B}T$, n would be clearly temperature-dependent. If, however, θ_0 is say about half of $\bar{\mathscr{E}}_c/\mathrm{k_B}T$, the data will appear temperature-independent in a certain window around θ_0.

Let us now try to work out under what circumstances (17.63) can give a temperature-independent n and whether such a scenario makes physical sense. We start by assuming that we measure the degradation at two different temperatures, say T_1 and T_2. As we have to be consistent with the experimental observation that n is temperature-independent, we require n to have the same value at both temperatures,

$$\frac{\bar{\mathscr{E}}_c/\mathrm{k_B}T_1 - \theta_0}{(\sigma_c/\mathrm{k_B}T_1)^2} = \frac{\bar{\mathscr{E}}_c/\mathrm{k_B}T_2 - \theta_0}{(\sigma_c/\mathrm{k_B}T_2)^2}. \tag{17.64}$$

From this we see that a given measurement range around θ_0 determines the required mean activation energy $\bar{\mathscr{E}}_c$

$$\bar{\mathscr{E}}_c = \theta_0 \mathrm{k_B}(T_1 + T_2). \tag{17.65}$$

Then, in order to give a certain temperature-independent power law slope n, for instance $n = 1/6$, we can calculate the required σ_c from

$$n = \frac{\bar{\mathscr{E}}_c/\mathrm{k_B}T_1 - \theta_0}{(\sigma_c/\mathrm{k_B}T_1)^2} \tag{17.66}$$

as

$$\sigma_c^2 = \theta_0 \frac{\mathrm{k_B}^2}{n} T_1 T_2. \tag{17.67}$$

Long-term power-law exponents are usually determined in the range 100 s to 1 ks. This rather firmly sets the value of θ_0, for instance to $\theta_0 \approx \log(250\,\mathrm{s}/\tau_0)$. When we now assume $T_1 = 100\,^\circ\mathrm{C}$ and $T_2 = 200\,^\circ\mathrm{C}$, we obtain from (17.65) a mean activation energy of $\bar{\mathscr{E}}_c = 2.25\,\mathrm{eV}$. Finally, with $n = 1/6$ we obtain from (17.67) a standard deviation of $\sigma_c = 0.5\,\mathrm{eV}$. Both $\bar{\mathscr{E}}_c$ and σ_c appear sensible parameters of a distribution of activation energies in an amorphous oxide. The conditions on the parameters can be relaxed when we merely require a roughly temperature-independent n.

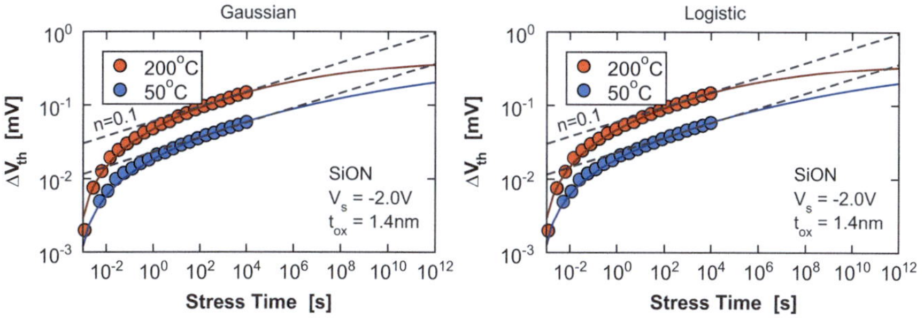

Fig. 17.12 OTF measurement on a 1.4 nm PNO device at two temperatures fitted by Gaussian (*left*) and a logistic (*right*) distribution of activation energies. Both fits are virtually indistinguishable in the measurement window, reproduce the temperature-independent power-law, but differ slightly in their long term prediction. The reference value of the measurement was obtained at $t_{\mathrm{M}} = 1\,\mathrm{ms}$, which is emulated in the fits by subtracting $\Delta V_{\mathrm{th}}(t_{\mathrm{M}})$ and visible as a rapid increase in ΔV_{th} for $t_{\mathrm{s}} > t_{\mathrm{M}}$

Similar conclusions can be drawn for the logistic distribution. However, in the approximation where the logistic distribution is expanded far away from the maximum, as is usually done in literature, the resulting power-law exponent n will be linearly temperature-dependent, $n = \mathrm{k_B}T/s$. As stated above, this is contrary to experimental observations. However, this must not be mistaken as a failure of the logistic distribution itself, but rather as a consequence of an unfavorable approximation. Suitable approximations can only be obtained when the distribution is expanded somewhere closer to the mean rather than in the tail, quite similar to the Gaussian case, see Fig. 17.12. While both fits have virtually the same quality inside the measurement window, they behave slightly different at longer times. Since the Gaussian distribution appears a more natural choice for the distribution of activation energies, it will be preferred in the following. Another reason is that the mathematical advantage of the logistic distribution cannot be exploited for the two-dimensional capture/emission time maps.

17.4 The Analytic CET Map

In the following we try to generalize our previous observations to derive an empirical analytical model for the CET map. The model is based on the following assumptions:

- The CET map will consist of two distributions, one describing the recoverable component R, the other the more permanent contribution P. Lacking firm evidence to the contrary, we take the simplest route and assume for the time being that these components are independent.

- We have also seen before that many experimental features like the power-law dependence can be captured by a Gaussian distribution for τ_c, which appears a natural choice.
- We assume that the effective activation energies are distributed, which results in a particular "built-in" temperature dependence of the model. The bias-dependence, on the other hand, must be added by making some parameters of the model bias-dependent [37].
- Visual inspection of the numerically extracted CET map in Fig. 17.6 shows that the emission times become larger with increasing capture times, implying a correlation between the two. The simplest way to express this mathematically is to write the activation energy of τ_e in the form $\mathscr{E}_e = \mathscr{E}_c + \Delta\mathscr{E}_e$, where $\Delta\mathscr{E}_e$ describes an uncorrelated part of $\mathscr{E}_e$. Again, we assume that $\Delta\mathscr{E}_e$ follows a Gaussian distribution.

In the following we assume that both oxide traps and interface traps can be written in the form (17.14)

$$\tau_c = \tau_0 e^{\beta \mathscr{E}_c} \quad \text{and} \quad \tau_e = \tau_c e^{\beta \Delta\mathscr{E}_e}. \tag{17.68}$$

Again, the only justification of this assumption will be the agreement with experimental data demonstrated later on. In order to proceed, we need to know the joint probability density function $g(\tau_c, \tau_e)$ which characterizes the distribution of both time constants. In general, all three quantities in the above, τ_0, $\mathscr{E}_c$, $\Delta\mathscr{E}_e$ will be distributed. RTN experiments [65] show no correlation between the depth of the defect into the oxide, which should essentially determine the distribution of τ_0 via the WKB factor, and τ_e and τ_c. Also, the time constants will depend much weaker on a distribution of τ_0 compared to a distribution of the energies. We therefore assume that the energy distribution to be the dominant contribution. As such, we need to find a model for the joint distribution $g(\mathscr{E}_c, \mathscr{E}_e)$, with $\mathscr{E}_e = \mathscr{E}_c + \Delta\mathscr{E}_e$. This distribution is easy to construct via the conditional "probability" $g(\mathscr{E}_e|\mathscr{E}_c)$ and noting that

$$g(\mathscr{E}_c, \mathscr{E}_e) = g(\mathscr{E}_e|\mathscr{E}_c)\, g(\mathscr{E}_c). \tag{17.69}$$

The conditional probability $g(\mathscr{E}_e|\mathscr{E}_c)$ is the probability of obtaining a certain value of $\mathscr{E}_e$ for a fixed $\mathscr{E}_c$. Since we assume $\Delta\mathscr{E}_e$ to be Gaussian distributed with standard deviation $\sigma_{\Delta e}$, we have

$$g(\mathscr{E}_e|\mathscr{E}_c) = \frac{1}{\sigma_{\Delta e}} \phi\Big(\frac{\mathscr{E}_e - (\mathscr{E}_c + \mu_{\Delta e})}{\sigma_{\Delta e}}\Big). \tag{17.70}$$

Thus, in total we obtain

$$g(\mathscr{E}_c, \mathscr{E}_e) = \frac{1}{\sigma_c \sigma_{\Delta e}} \phi\Big(\frac{\mathscr{E}_c - \mu_c}{\sigma_c}\Big) \phi\Big(\frac{\mathscr{E}_e - (\mathscr{E}_c + \mu_{\Delta e})}{\sigma_{\Delta e}}\Big) \tag{17.71}$$

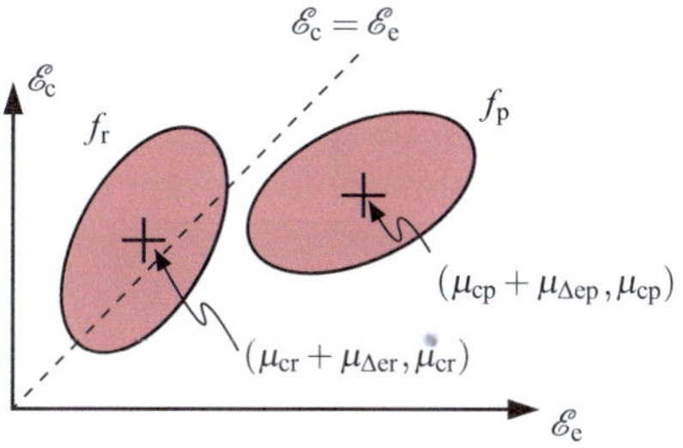

Fig. 17.13 The CET map is modeled in the activation-energy-space using two bivariate Gaussian distributions, one for the recoverable component, f_r, and one for the more permanent component, f_p

which is a bivariate Gaussian distribution. We will use such a bivariate Gaussian distribution to describe both the recoverable and the "permanent" part of the degradation as sketched in Fig. 17.13. The following properties of the above joint distribution are worth mentioning:

- The marginal distribution for $\mathscr{E}_e$, which is obtained by integrating $g(\mathscr{E}_c, \mathscr{E}_e)$ over $\mathscr{E}_c$, is a Gaussian with $\mu_e = \mu_c + \mu_{\Delta e}$ and $\sigma_e^2 = \sigma_c^2 + \sigma_{\Delta e}^2$.
- The correlation coefficient is $\rho = \sigma_c/\sigma_e$. Note that this correlation coefficient is a consequence of our Ansatz for $\mathscr{E}_e$ and thus not directly a parameter of the model.

By introducing the normalized variates

$$x(\mathscr{E}_e) = \frac{\mathscr{E}_e - (\mu_c + \mu_{\Delta e})}{\sigma_e} \quad \text{and} \quad y(\mathscr{E}_c) = \frac{\mathscr{E}_c - \mu_c}{\sigma_c}, \tag{17.72}$$

the bivariate Gaussian distribution (17.71) can be written in standard form

$$f(x,y,\rho) = \frac{\phi(y)}{\sqrt{1-\rho^2}}\,\phi\Big(\frac{x-\rho y}{\sqrt{1-\rho^2}}\Big). \tag{17.73}$$

In order to calculate the response to DC stress, we need the sum over all defects with $\tau_c < t_s$ and $\tau_e > t_r$, which corresponds to all defects being charged up to t_s and not yet discharged at t_r. By transforming t_s and t_r to their corresponding energies and then into our normalized (x,y) space as $a = x(\log(t_r/\tau_0)/\beta)$ and $b = y(\log(t_s/\tau_0)/\beta)$, the fraction of all defects contributing is given by the integral

$$\begin{aligned} F(a,b,\rho) &= \int_{-\infty}^{b} \mathrm{d}y \int_{a}^{\infty} \mathrm{d}x\, f(x,y,\rho) = \int_{-\infty}^{b} \mathrm{d}y\, \phi(y) \int_{a}^{\infty} \mathrm{d}x\, \phi\Big(\frac{x-\rho y}{\sqrt{1-\rho^2}}\Big) \\ &= \int_{-\infty}^{b} \phi(y) Q\Big(\frac{a-\rho y}{\sqrt{1-\rho^2}}\Big)\,\mathrm{d}y \end{aligned} \tag{17.74}$$

with the standard integral of the Gaussian distribution

$$Q(x) = \int_{x}^{\infty} \phi(x)\,\mathrm{d}t = \tfrac{1}{2}\Big(1 - \mathrm{erf}\Big(\frac{x}{\sqrt{2}}\Big)\Big) = \tfrac{1}{2}\,\mathrm{erfc}\Big(\frac{x}{\sqrt{2}}\Big). \tag{17.75}$$

Unfortunately, the integrand of $F(a,b,\rho)$ consists of a Gaussian function multiplied by an error function, which cannot be integrated in closed form. In fact, the calculation of bivariate normal integral poses a standard problem in statistics and numerous solutions to the problem have been proposed over the last decades [91–93]. However, most of these approximations are too crude for our purpose, since our expression needs to capture the integral over a wide range of times and temperatures. A slightly more involved yet simple approach has been suggested recently [94], which is based on approximating $\mathrm{erf}(x)$ in $Q(x)$ as

$$\mathrm{erf}(x) \approx 1 - \mathrm{e}^{-c_1 x - c_2 x^2} \tag{17.76}$$

for $x > 0$ with two fitting parameters c_1 and c_2. The values for $x < 0$ are obtained from $\mathrm{erf}(-x) = -\mathrm{erf}(x)$. A least squares fit in the interval $0 \leq x \leq 3$ gives $c_1 = 1.0950$ and $c_2 = 0.756508$ and a relative error smaller than 0.2% for $x > 0.34$ and smaller than 3% for $0 \leq x \leq 0.34$. The beauty of this approximation is a consequence of the fact that the Gaussian distribution when multiplied by an exponential of a second-order polynomial can be rearranged into a shifted and scaled distribution which can then be integrated and expressed as combinations of normal integrals

$$\Phi(x) = \int_{-\infty}^{x} \phi(x)\,\mathrm{d}t = \tfrac{1}{2}\left(1 + \mathrm{erf}\left(\frac{x}{\sqrt{2}}\right)\right) = \tfrac{1}{2}\,\mathrm{erfc}\left(-\frac{x}{\sqrt{2}}\right). \tag{17.77}$$

A slight price to pay comes from the piecewise integration for $x \leq 0$ and $x > 0$, which corresponds to $a \leq \rho b$ and $a > \rho b$. After some tedious but straightforward manipulations one obtains the slightly daunting but highly accurate expressions

$$\begin{aligned} F(a,b) = {} & \Phi(b) - \Phi\left(\frac{a}{\rho}\right) + \frac{1}{2r_2}\exp\left(\frac{r_1^2 - 2a^2C_2}{2r_2^2}\right) \\ & \times \left\{\exp\left(-a\frac{C_1}{r_2^2}\right)\Phi\left(\frac{a/\rho - r_1}{r_2}\right)\right. \\ & \left. + \exp\left(a\frac{C_1}{r_2^2}\right)\left[\Phi\left(\frac{a/\rho + r_1}{r_2}\right) - \Phi\left(\frac{b + r_1 - 2C_2\rho(a - b\rho)}{r_2}\right)\right]\right\} \end{aligned} \tag{17.78}$$

valid for shorter recovery times ($a \leq \rho b$), while for longer recovery times ($a > \rho b$) we have

$$F(a,b) = \frac{1}{2r_2}\exp\left(\frac{r_1^2 - 2a(C_1 + aC_2)}{2r_2^2}\right)\Phi\left(\frac{b - r_1 - 2C_2\rho(a - b\rho)}{r_2}\right)$$

with

$$C_1 = c_1\eta, \quad C_2 = c_2\eta^2, \quad \eta = 1/\sqrt{2(1-\rho^2)}, \tag{17.79}$$

$$r_1 = c_1\rho\eta, \quad r_2^2 = 1 + 2c_2(\rho\eta)^2. \tag{17.80}$$

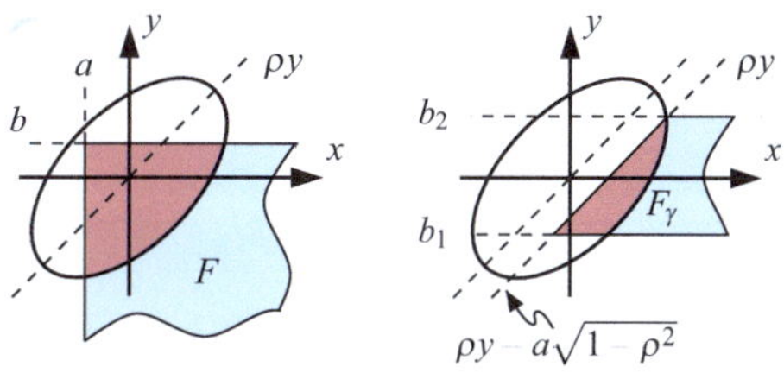

Fig. 17.14 The auxiliary integrals F and F_γ used to calculate the degradation after DC and AC stress

A slightly less accurate but more compact version is obtained by setting $c_1 = \sqrt{3}$ and $c_2 = 0$. By introducing $c = \sqrt{3}\eta$ we then obtain for $a \le \rho b$

$$F(a,b) = \Phi(b) - \Phi\Big(\frac{a}{\rho}\Big) + \tfrac{1}{2}\mathrm{e}^{c^2\rho^2/2} \times \Big\{\mathrm{e}^{-ca}\Phi\Big(\frac{a}{\rho} - c\rho\Big) + \mathrm{e}^{ca}\Big[\Phi\Big(\frac{a}{\rho} + c\rho\Big) - \Phi(b + c\rho)\Big]\Big\}, \tag{17.81}$$

while for $a > \rho b$ we have

$$F(a,b) = \frac{1}{2}\mathrm{e}^{c^2\rho^2/2}\mathrm{e}^{-ca}\Phi(b - c\rho). \tag{17.82}$$

With the above we can write ΔV_{th} after a DC stress of duration t_{s} and after a recovery time t_{r} as (see Fig. 17.7)

$$\Delta V_{\mathrm{th}}(t_{\mathrm{r}}, t_{\mathrm{s}}) = A_{\mathrm{r}} G_{\mathrm{r}}(t_{\mathrm{r}}, t_{\mathrm{s}}) + A_{\mathrm{p}} G_{\mathrm{p}}(t_{\mathrm{r}}, t_{\mathrm{s}}) \tag{17.83}$$

with the auxiliary functions describing the permanent and recoverable peaks

$$G_{\mathrm{r}}(t_{\mathrm{r}}, t_{\mathrm{s}}) = F\Big(\frac{\mathrm{k_B}T\log(t_{\mathrm{r}}/\tau_{0\mathrm{r}}) - \mu_{\Delta\mathrm{er}} - \mu_{\mathrm{cr}}}{\sigma_{\mathrm{er}}}, \frac{\mathrm{k_B}T\log(t_{\mathrm{s}}/\tau_{0\mathrm{r}}) - \mu_{\mathrm{cr}}}{\sigma_{\mathrm{cr}}}, \frac{\sigma_{\mathrm{cr}}}{\sigma_{\mathrm{er}}}\Big) \tag{17.84}$$

$$G_{\mathrm{p}}(t_{\mathrm{r}}, t_{\mathrm{s}}) = F\Big(\frac{\mathrm{k_B}T\log(t_{\mathrm{r}}/\tau_{0\mathrm{p}}) - \mu_{\Delta\mathrm{ep}} - \mu_{\mathrm{cp}}}{\sigma_{\mathrm{ep}}}, \frac{\mathrm{k_B}T\log(t_{\mathrm{s}}/\tau_{0\mathrm{p}}) - \mu_{\mathrm{cp}}}{\sigma_{\mathrm{cp}}}, \frac{\sigma_{\mathrm{cp}}}{\sigma_{\mathrm{ep}}}\Big), \tag{17.85}$$

where A_{r} and A_{p} give the maximum degradation obtainable from each peak. The limiting case of zero delay ($a \to -\infty$) is simply obtained as $F(a,b) = \Phi(b)$, in agreement with the discussion in Sect. 17.3.5.1.

If the experiment is carried out in an on-the-fly manner, we have zero delay ($a \to -\infty$). However, the degradation is measured relative to the value obtained after a certain measurement delay $t_{\mathrm{s}} = t_0$, see Fig. 17.7. Thus, we have

$$\Delta V_{\mathrm{th}}(0, t_{\mathrm{s}}) = A_{\mathrm{r}}\Big(G_{\mathrm{r}}(0, t_{\mathrm{s}}) - G_{\mathrm{r}}(0, t_0)\Big) + A_{\mathrm{p}}\Big(G_{\mathrm{p}}(0, t_{\mathrm{s}}) - G_{\mathrm{p}}(0, t_0)\Big). \tag{17.86}$$

Finally, for the calculation of AC stress with period T and a duty-factor γ we need the auxiliary integral (see Figs. 17.7 and 17.14)

$$
\begin{aligned}
F_{\gamma}(a,b_1,b_2,\rho) &= \int_{-b_1}^{b_2} \mathrm{d}y \int_{-a\sqrt{1-\rho^2}+\rho y}^{\infty} \mathrm{d}x\, f(x,y,\rho) \\
&= \int_{-b_1}^{b_2} \mathrm{d}y\, \phi(y)\, Q\Big(\frac{-a\sqrt{1-\rho^2}+\rho y-\rho y}{\sqrt{1-\rho^2}}\Big) \\
&= (\Phi(b_2)-\Phi(b_1))\,\Phi(a),
\end{aligned}
\tag{17.87}
$$

which is fortunately very simple to calculate. With the auxiliary functions F and F_{γ}, the total ΔV_{th} can be constructed at any time of an AC stress sequence. For instance, at the end of the V_{L} period, where the recovery time is $t_{\mathrm{r}} = (1-\alpha)T$, we would have

$$
\begin{aligned}
\Delta V_{\mathrm{th}}((1-\alpha)T,t_{\mathrm{s}}) = A_{\mathrm{r}}\Big(&G_{\mathrm{r}}((1-\alpha)T,\alpha T)+G_{\gamma\mathrm{r}}((1-\alpha)T,\alpha t_{\mathrm{s}})\Big)+ \\
&A_{\mathrm{p}}\Big(G_{\mathrm{p}}((1-\alpha)T,\alpha T)+G_{\gamma\mathrm{p}}((1-\alpha)T,\alpha t_{\mathrm{s}})\Big),
\end{aligned}
\tag{17.88}
$$

where $G_{\gamma\mathrm{r}}$ and $G_{\gamma\mathrm{p}}$ are defined analogously to G_{r} and G_{p}.

17.4.1 Bias Dependence

While the temperature dependence is inherently considered by the distribution of the activation energies, the bias dependence of the CET maps is modeled by assuming the amplitude of each component to follow $A = (V_{\mathrm{stress}}/V_{\mathrm{s0}})^m$, with the stress voltage V_{stress} and constants V_{s0} and m. Also, as previous studies on individual defects have shown [10], the mean values of the distribution are expected to approximately follow $\mu_{\mathrm{c}} = \mu_{\mathrm{c0}} + kV_{\mathrm{stress}}$ and $\Delta\mu_{\mathrm{e}} = \Delta\mu_{\mathrm{e0}} - kV_{\mathrm{stress}}$, with a constant k. The main effect of k is to shift the capture times toward shorter values without affecting the emission times. However, by fitting the model to experimental data, the effect of the bias on the mean values was found to be small and completely negligible for the permanent component, which is somewhat surprising given the strong exponential bias dependence of the individual defects.

17.4.2 Experimental Validation

Finally, the model is evaluated on a 2.2 nm SiON technology [95], where very detailed MSM data was acquired for the construction of the CET maps. Recording of each dataset required about 1–2 weeks. Figure 17.15 shows the analytic CET model, which contains all essential features visible in the numerical map shown in Fig. 17.6. In particular, the rightward slant of the distribution with increasing capture times, which previously necessitated the introduction of the higher-order polynomials for the mean and standard deviation of the single normal distribution [78], is

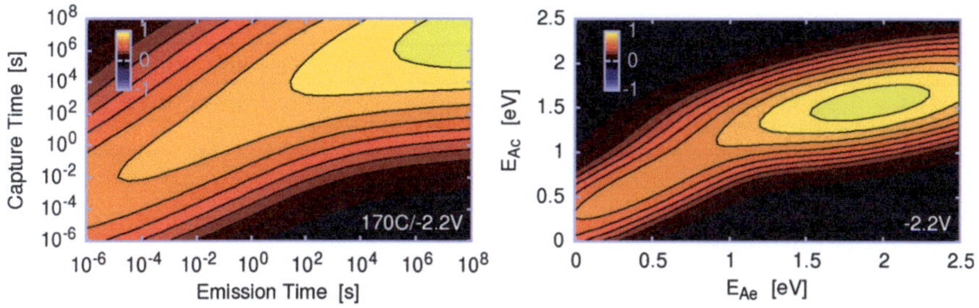

Fig. 17.15 *Left*: The analytic CET map extracted for the data shown in Fig. 17.16, which contains all essential features visible in Fig. 17.6. *Right*: The analytic activation energy map for the same data set

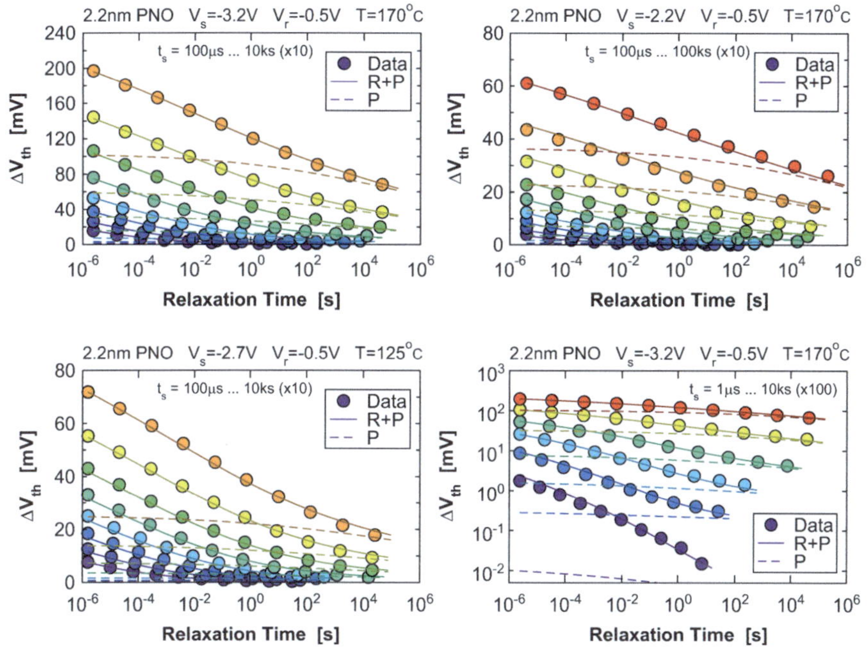

Fig. 17.16 Comparison of the analytic model using the activation energies of Fig. 17.15 to experimental data at different stress biases and temperatures. Excellent agreement is obtained for all stress and relaxation times in the extremely wide experimental window—also on a logarithmic scale (second column, *right-most figure*)

well captured by the superposition of two bivariate normal distributions. By simultaneously extracting the analytic distribution for a number of datasets recorded at different V_{stress} and T_{L}, a bias- and temperature-dependent analytic CET map is obtained. A convincing comparison of the analytic model to experimental data for a number of $V_{\mathrm{stress}}/T_{\mathrm{L}}$ combinations is given in Fig. 17.16 using the parameters of Table 17.1.

Table 17.1 Parameters used for the analytic CET maps of Figs. 17.15 and 17.16

	τ_0 (ns)	μ_c (eV)	σ_c (eV)	$\mu_{\Delta e}$ (eV)	$\sigma_{\Delta e}$ (mV)	V_{s0} (V)	m	k
R	98	0.55	0.43	−0.2	0.26	5.22	3.58	4.4×10^{-3} eV/V
P	0.59	1.6	0.31	0.32	0.48	3.04	3.74	0

17.5 Conclusions

Starting from a rigorous microscopic description of oxide defects and a somewhat less rigorous description of interface states, we have suggested a physics-based analytic model for BTI which covers DC, AC, and duty-factor dependent stress and the subsequent recovery as a function of stress voltage and temperature. Since the model is intuitively based on the occupancy of defects in the capture/emission time maps, it can be easily generalized to other more complicated stress/recovery patterns.

Acknowledgements This work would not have been possible without the invaluable support of B. Kaczer (imec) and H. Reisinger (Infineon Munich). Furthermore, inspiring discussions with W. Goes, P.-J. Wagner, and F. Schanovsky are gratefully acknowledged.

The research leading to these results has received funding from the FWF project n°23390-M24 and the European Community's FP7 project n°261868 (MORDRED).

References

1. M. Denais, V. Huard, C. Parthasarathy, G. Ribes, F. Perrier, N. Revil, and A. Bravaix, "Interface Trap Generation and Hole Trapping under NBTI and PBTI in Advanced CMOS Technology with a 2-nm Gate Oxide," *IEEE Trans.Dev.Mat.Rel.*, vol. 4, no. 4, pp. 715–722, 2004.
2. D.S. Ang, S. Wang, and C.H. Ling, "Evidence of Two Distinct Degradation Mechanisms from Temperature Dependence of Negative Bias Stressing of the Ultrathin Gate p-MOSFET," *IEEE Electron Device Lett.*, vol. 26, no. 12, pp. 906–908, 2005.
3. C.-T. Chan T. Wang AND, C.-J. Tang, C.-W. Tsai, H. Wang, M.-H. Chi, and D. Tang, "A Novel Transient Characterization Technique to Investigate Trap Properties in HfSiON Gate Dielectric MOSFETs—From Single Electron Emission to PBTI Recovery Transient," *IEEE Trans.Electron Devices*, vol. 53, no. 5, pp. 1073–1079, 2006.
4. V. Huard, M. Denais, and C. Parthasarathy, "NBTI Degradation: From Physical Mechanisms to Modelling," *Microelectronics Reliability*, vol. 46, no. 1, pp. 1–23, 2006.
5. S. Mahapatra, K. Ahmed, D. Varghese, A. E. Islam, G. Gupta, L. Madhav, D. Saha, and M. A. Alam, "On the Physical Mechanism of NBTI in Silicon Oxynitride p-MOSFETs: Can Differences in Insulator Processing Conditions Resolve the Interface Trap Generation versus Hole Trapping Controversy?," in *Proc. Intl.Rel.Phys.Symp. (IRPS)*, 2007, pp. 1–9.
6. J.P. Campbell, P.M. Lenahan, C.J. Cochrane, A.T. Krishnan, and S. Krishnan, "Atomic-Scale Defects Involved in the Negative-Bias Temperature Instability," *IEEE Trans.Dev.Mat.Rel.*, vol. 7, no. 4, pp. 540–557, 2007.
7. Th. Aichinger, M. Nelhiebel, S. Einspieler, and T. Grasser, "Observing Two Stage Recovery of Gate Oxide Damage Created under Negative Bias Temperature Stress," *J.Appl.Phys.*, vol. 107, pp. (024508–1)–(024508–8), 2010.

8. V. Huard, C. Parthasarathy, and M. Denais, "Single-Hole Detrapping Events in pMOSFETs NBTI Degradation," in *Proc. Intl.Integrated Reliability Workshop*, 2005, pp. 5–9.
9. H. Reisinger, T. Grasser, and C. Schlünder, "A Study of NBTI by the Statistical Analysis of the Properties of Individual Defects in pMOSFETs," in *Proc. Intl.Integrated Reliability Workshop*, 2009, pp. 30–35.
10. T. Grasser, H. Reisinger, P.-J. Wagner, W. Goes, F. Schanovsky, and B. Kaczer, "The Time Dependent Defect Spectroscopy (TDDS) for the Characterization of the Bias Temperature Instability," in *Proc. Intl.Rel.Phys.Symp. (IRPS)*, May 2010, pp. 16–25.
11. V. Huard, "Two Independent Components Modeling for Negative Bias Temperature Instability," in *Proc. Intl.Rel.Phys.Symp. (IRPS)*, May 2010, pp. 33–42.
12. M. Denais, V. Huard, C. Parthasarathy, G. Ribes, F. Perrier, D. Roy, and A. Bravaix, "Perspectives on NBTI in Advanced Technologies: Modelling & Characterization," in *Proc. ESSDERC*, 2005, pp. 399–402.
13. D.S. Ang, "Observation of Suppressed Interface State Relaxation under Positive Gate Biasing of the Ultrathin Oxynitride Gate p-MOSFET Subjected to Negative-Bias Temperature Stressing," *IEEE Electron Device Lett.*, vol. 27, no. 5, pp. 412–415, 2006.
14. Th. Aichinger, M. Nelhiebel, and T. Grasser, "Unambiguous Identification of the NBTI Recovery Mechanism using Ultra-Fast Temperature Changes," in *Proc. Intl.Rel.Phys.Symp. (IRPS)*, 2009, pp. 2–7.
15. T. Grasser, Th. Aichinger, G. Pobegen, H. Reisinger, P.-J. Wagner, J. Franco, M. Nelhiebel, and B. Kaczer, "The 'Permanent' Component of NBTI: Composition and Annealing," in *Proc. Intl.Rel.Phys.Symp. (IRPS)*, Apr. 2011, pp. 605–613.
16. W. Shockley and W.T. Read, *Physical Review*, vol. 87, no. 5, pp. 835–842, 1952.
17. A.L. McWhorter, "1/f Noise and Germanium Surface Properties," *Sem.Surf.Phys*, pp. 207–228, 1957.
18. F.P. Heiman and G. Warfield, "The Effects of Oxide Traps on the MOS Capacitance," *IEEE Trans.Electron Devices*, vol. 12, no. 4, pp. 167–178, 1965.
19. M. Masuduzzaman, A.E. Islam, and M.A. Alam, "Exploring the Capability of Multifrequency Charge Pumping in Resolving Location and Energy Levels of Traps Within Dielectric," *IEEE Trans.Electron Devices*, vol. 55, no. 12, pp. 3421–3431, 2008.
20. T. Grasser, "Stochastic Charge Trapping in Oxides: From Random Telegraph Noise to Bias Temperature Instabilities," in *Microelectronics Reliability*, 2012, vol. 52, pp. 39–70.
21. T. Grasser, H. Reisinger, W. Goes, Th. Aichinger, Ph. Hehenberger, P.J. Wagner, M. Nelhiebel, J. Franco, and B. Kaczer, "Switching Oxide Traps as the Missing Link between Negative Bias Temperature Instability and Random Telegraph Noise," in *Proc. Intl.Electron Devices Meeting (IEDM)*, 2009, pp. 729–732.
22. T. Grasser, H. Reisinger, P.-J. Wagner, and B. Kaczer, *Physical Review B*, vol. 82, no. 24, pp. 245318, 2010.
23. A. Palma, A. Godoy, J. A. Jimenez-Tejada, J. E. Carceller, and J. A. Lopez-Villanueva, *Physical Review B*, vol. 56, no. 15, pp. 9565–9574, 1997.
24. J.F. Conley Jr., P.M. Lenahan, A.J. Lelis, and T.R. Oldham, "Electron Spin Resonance Evidence that E'_γ Centers can Behave as Switching Oxide Traps," *IEEE Trans.Nucl.Sci.*, vol. 42, no. 6, pp. 1744–1749, 1995.
25. P.M. Lenahan and J.F. Conley Jr., "What Can Electron Paramagnetic Resonance Tell Us about the Si/SiO_2 System?," *J.Vac.Sci.Technol.B*, vol. 16, no. 4, pp. 2134–2153, 1998.
26. C. Shen, M.-F. Li, X.P. Wang, H.Y. Yu, Y.P. Feng, A.T.-L. Lim, Y.C. Yeo, D.S.H. Chan, and D.L. Kwong, "Negative *U* Traps in HfO_2 Gate Dielectrics and Frequency Dependence of Dynamic BTI in MOSFETs," in *Proc. Intl.Electron Devices Meeting (IEDM)*, 2004, pp. 733–736.
27. T. Yang, C. Shen, M.-F. Li, C.H. Ang, C.X. Zhu, Y.-C. Yeo, G. Samudra, S.C. Rustagi, M.B. Yu, and D.-L. Kwong, "Fast DNBTI Components in p-MOSFET with SiON Dielectric," *IEEE Electron Device Lett.*, vol. 26, no. 11, pp. 826–828, 2005.
28. T. Grasser, B. Kaczer, H. Reisinger, P.-J. Wagner, and M. Toledano-Luque, "On the Frequency Dependence of the Bias Temperature Instability," in *Proc. Intl.Rel.Phys.Symp. (IRPS)*, Apr. 2012, pp. XT.8.1–XT.8.7.

29. T. Grasser, H. Reisinger, K. Rott, M. Toledano-Luque, and B. Kaczer, "On the Microscopic Origin of the Frequency Dependence of Hole Trapping in pMOSFETs," in *Proc. Intl.Electron Devices Meeting (IEDM)*, Dec. 2012, pp. 19.6.1–19.6.4.
30. W. Goes, F. Schanovsky, and T. Grasser (2013) Advanced modeling of oxide defects. In: T. Grasser (eds) Bias temperature instability for devices and circuits. Springer, New York
31. B. Kaczer, T. Grasser, J. Martin-Martinez, E. Simoen, M. Aoulaiche, Ph.J. Roussel, and G. Groeseneken, "NBTI from the Perspective of Defect States with Widely Distributed Time Scales," in *Proc. Intl.Rel.Phys.Symp. (IRPS)*, 2009, pp. 55–60.
32. V.V. Afanas'ev and A. Stesmans, "Proton Nature of Radiation-Induced Positive Charge in SiO_2 Layers on Si," *Eur.Phys.Lett.*, vol. 53, no. 2, pp. 233–239, 2001.
33. J.P. Campbell, P.M. Lenahan, A.T. Krishnan, and S. Krishnan, "Identification of the Atomic-Scale Defects Involved in the Negative Bias Temperature Instability in Plasma-Nitrided p-Channel Metal-Oxide-Silicon Field-Effect Transistors," *J.Appl.Phys.*, vol. 103, no. 4, pp. 044505, 2008.
34. J.T. Ryan, P.M. Lenahan, T. Grasser, and H. Enichlmair, "Observations of Negative Bias Temperature Instability Defect Generation via On The Fly Electron Spin Resonance," *Appl.Phys.Lett.*, vol. 96, no. 22, pp. 223509–1–223509–3, 2010.
35. B. Kaczer, S. Mahato, V. Camargo, M. Toledano-Luque, Ph.J. Roussel, T. Grasser F. Catthoor, P. Dobrovolny, P. Zuber, G. Wirth, and G. Groeseneken, "Atomistic Approach to Variability of Bias-Temperature Instability in Circuit Simulations," in *Proc. Intl.Rel.Phys.Symp. (IRPS)*, 2011, pp. 915–919.
36. H. Reisinger, T. Grasser, K. Ermisch, H. Nielen, W. Gustin, and Ch. Schlünder, "Understanding and Modeling AC BTI," in *Proc. Intl.Rel.Phys.Symp. (IRPS)*, Apr. 2011, pp. 597–604.
37. T. Grasser, P.-J. Wagner, H. Reisinger, Th. Aichinger, G. Pobegen, M. Nelhiebel, and B. Kaczer, "Analytic Modeling of the Bias Temperature Instability Using Capture/Emission Time Maps," in *Proc. Intl.Electron Devices Meeting (IEDM)*, Dec. 2011, pp. 27.4.1–27.4.4.
38. K. Zhao, J.H. Stathis, B.P. Linder, E. Cartier, and A. Kerber, "PBTI Under Dynamic Stress: From a Single Defect Point of View," in *Proc. Intl.Rel.Phys.Symp. (IRPS)*, Apr. 2011, pp. 372–380.
39. J.P. Campbell and P.M. Lenahan(2013) Atomic scale defects associated with the negative bias temperature instability. In: T. Grasser (eds) Bias temperature instability for devices and circuits. Springer, New York
40. V.V. Afanas'ev, M. Houssa, A. Stesmans (2013) Charge properties of paramagnetic defects in semiconductor/oxide structures. In: T. Grasser (eds) Bias temperature instability for devices and circuits. Springer, New York
41. G. Groeseneken, H.E. Maes, N. Beltran, and R.F. de Keersmaecker, "A Reliable Approach to Charge-Pumping Measurements in MOS Transistors," *IEEE Trans.Electron Devices*, vol. 31, no. 1, pp. 42–53, 1984.
42. M.-F. Li, D. Huang, C. Shen, T. Yang, W.J., W.J. Liu, and Z. Liu, "Understand NBTI Mechanism by Developing Novel Measurement Techniques," *IEEE Trans.Dev.Mat.Rel.*, vol. 8, no. 1, pp. 62–71, March 2008.
43. Ph. Hehenberger, Th. Aichinger, T. Grasser, W. Goes, O. Triebl, B. Kaczer, and M. Nelhiebel, "Do NBTI-Induced Interface States Show Fast Recovery? A Study Using a Corrected On-The-Fly Charge-Pumping Measurement Technique," in *Proc. Intl.Rel.Phys.Symp. (IRPS)*, 2009, pp. 1033–1038.
44. M. Denais, A. Bravaix, V. Huard, C. Parthasarathy, C. Guerin, G. Ribes, F. Perrier, M. Mairy, and D. Roy, "Paradigm Shift for NBTI Characterization in Ultra-Scaled CMOS Technologies," in *Proc. Intl.Rel.Phys.Symp. (IRPS)*, 2006, pp. 735–736.
45. T. Grasser and B. Kaczer, "Negative Bias Temperature Instability: Recoverable versus Permanent Degradation," in *Proc. ESSDERC*, 2007, pp. 127–130.
46. Th. Aichinger, M. Nelhiebel, and T. Grasser, "A Combined Study of p- and n-Channel MOS Devices to Investigate the Energetic Distribution of Oxide Traps after NBTI," *IEEE Trans.Electron Devices*, vol. 56, no. 12, pp. 3018–3026, 2009.

47. T. Grasser, K. Rott, H. Reisinger, P.-J. Wagner, W. Goes, F. Schanovsky, M. Waltl, M. Toledano-Luque, and B. Kaczer, "Advanced Characterization of Oxide Traps: The Dynamic Time-Dependent Defect Spectroscopy," in *Proc. Intl.Rel.Phys.Symp. (IRPS)*, Apr. 2013.
48. T. Grasser, B. Kaczer, W. Goes, Th. Aichinger, Ph. Hehenberger, and M. Nelhiebel, "Understanding Negative Bias Temperature Instability in the Context of Hole Trapping," Microelectronic Engineering, 2009, 86, 7–9, pp. 1876–1882
49. T. Grasser and B. Kaczer, "Evidence that Two Tightly Coupled Mechanism are Responsible for Negative Bias Temperature Instability in Oxynitride MOSFETs," *IEEE Trans.Electron Devices*, vol. 56, no. 5, pp. 1056–1062, 2009.
50. S. Mahapatra, A.E. Islam, S. Deora, V.D. Maheta, K. Joshi1, A. Jain, and M.A. Alam, "A Critical Re-evaluation of the Usefulness of R-D Framework in Predicting NBTI Stress and Recovery," in *Proc. Intl.Rel.Phys.Symp. (IRPS)*, 2011, pp. 614–623.
51. A. Stesmans, *Physical Review B*, vol. 61, no. 12, pp. 8393–8403, 2000.
52. H. Reisinger (2013) The time dependent defect spectroscopy. In: T. Grasser (eds) Bias temperature instability for devices and circuits. Springer, New York
53. A.J. Lelis and T.R. Oldham, "Time Dependence of Switching Oxide Traps," *IEEE Trans.Nucl.Sci.*, vol. 41, no. 6, pp. 1835–1843, Dec 1994.
54. D.T. Gillespie, *Markov Processes: An Introduction for Physical Scientists*, Academic Press, 1992.
55. O.C. Ibe, *Markov Processes for Stochastic Modeling*, Academic Press, 2009.
56. K. Huang and A. Rhys, "Theory of Light Absorption and Non-Radiative Transitions in F-Centres," *Proc.R.Soc.A*, vol. 204, pp. 406–423, 1950.
57. C.H. Henry and D.V. Lang, *Physical Review B*, vol. 15, no. 2, pp. 989–1016, 1977.
58. A.M. Stoneham, "Non-radiative Transitions in Semiconductors," *Rep.Prog.Phys.*, vol. 44, pp. 1251–1295, 1981.
59. W.B. Fowler, J.K. Rudra, M.E. Zvanut, and F.J. Feigl, *Physical Review B*, vol. 41, no. 12, pp. 8313–8317, 1990.
60. B.K. Ridley, *Quantum Processes in Semiconductors*, Oxford University Press, third edition, 1993.
61. C.S. Kelley, *Physical Review B*, vol. 20, no. 12, pp. 5084–5089, 1979.
62. F. Schanovsky, W. Goes, and T. Grasser, "Multiphonon Hole Trapping from First Principles," *J.Vac.Sci.Technol.B*, vol. 29, no. 1, pp. 01A2011–01A2015, 2011.
63. A. Asenov, R. Balasubramaniam, A.R. Brown, and J.H. Davies, "RTS Amplitudes in Decananometer MOSFETs: 3-D Simulation Study," *IEEE Trans.Electron Devices*, vol. 50, no. 3, pp. 839–845, 2003.
64. S.M. Amoroso (2013) Statistical study of bias temperature instabilities by means of 3D atomistic simulation. In: T. Grasser (eds) Bias temperature instability for devices and circuits. Springer, New York
65. T. Nagumo, K. Takeuchi, T. Hase, and Y. Hayashi, "Statistical Characterization of Trap Position, Energy, Amplitude and Time Constants by RTN Measurement of Multiple Individual Traps," in *Proc. Intl.Electron Devices Meeting (IEDM)*, 2010, pp. 628–631.
66. T. Yang, C. Shen, M.-F. Li, C.H. Ang, C.X. Zhu, Y.-C. Yeo, G. Samudra, and D.-L. Kwong, "Interface Trap Passivation Effect in NBTI Measurement for p-MOSFET with SiON Gate Dielectric," *IEEE Electron Device Lett.*, vol. 26, no. 10, pp. 758–760, 2005.
67. T. Grasser, B. Kaczer, Th. Aichinger, W. Goes, and Michael Nelhiebel, "Defect Creation Stimulated by Thermally Activated Hole Trapping as the Driving Force Behind Negative Bias Temperature Instability in SiO_2, SiON, and High-k Gate Stacks," in *Proc. Intl.Integrated Reliability Workshop*, Apr. 2008, pp. 91–95.
68. M.A. Alam, H. Kufluoglu, D. Varghese, and S. Mahapatra, "A Comprehensive Model for pMOS NBTI Degradation: Recent Progress," *Microelectronics Reliability*, vol. 47, no. 6, pp. 853–862, 2007.
69. S. Mahapatra (2013) A comprehensive modeling framework for DC and AC NBTI. In: T. Grasser (eds) Bias temperature instability for devices and circuits. Springer, New York

70. T. Grasser, B. Kaczer, W. Goes, H. Reisinger, Th. Aichinger, Ph. Hehenberger, P.-J. Wagner, F. Schanovsky, J. Franco, M. Toledano-Luque, and M. Nelhiebel, "The Paradigm Shift in Understanding the Bias Temperature Instability: From Reaction-Diffusion to Switching Oxide Traps," *IEEE Trans.Electron Devices*, vol. 58, no. 11, pp. 3652–3666, 2011.
71. F. Schanovsky and T. Grasser, "On the Microscopic Limit of the Modified Reaction-Diffusion Model for the Negative Bias Temperature Instability," in *Proc. Intl.Rel.Phys.Symp. (IRPS)*, Apr. 2012, pp. XT.10.1–XT.10.6.
72. F. Schanovsky and T. Grasser (2013) On the microscopic limit of the RD model. In: T. Grasser (eds) Bias temperature instability for devices and circuits. Springer, New York
73. Y. Yang and M.H. White, "Charge Retention of Scaled SONOS Nonvolatile Memory Devices at Elevated Temperatures," *Solid-State Electron.*, vol. 44, pp. 949–958, 2000.
74. Y. Gao, A.A. Boo, Z.Q. Teo, and D.S. Ang, "On the Evolution of the Recoverable Component of the SiON, HfSiON and HfO_2 P-MOSFETs under Dynamic NBTI," in *Proc. Intl.Rel.Phys.Symp. (IRPS)*, Apr. 2011, pp. 935–940.
75. Z.Q. Teo, A.A. Boo, D.S. Ang, and K.C. Leong, "On the Cyclic Threshold Voltage Shift of Dynamic Negative-Bias Temperature Instability," in *Proc. Intl.Rel.Phys.Symp. (IRPS)*, 2011, pp. 943–947.
76. M. Duan, J.F. Zhang, Z. Ji, W.D. Zhang, B. Kaczer, S. De Gendt, and G. Groeseneken, "Defect Loss: A New Concept for Reliability of MOSFETs," *IEEE Electron Device Lett.*, vol. 33, no. 4, pp. 480–482, 2012.
77. D.S. Ang (2013) Understanding negative-bias temperature instability from dynamic stress experiments. In: T. Grasser (eds) Bias temperature instability for devices and circuits. Springer, New York
78. H. Reisinger, T. Grasser, W. Gustin, and C. Schlünder, "The Statistical Analysis of Individual Defects Constituting NBTI and its Implications for Modeling DC- and AC-Stress," in *Proc. Intl.Rel.Phys.Symp. (IRPS)*, May 2010, pp. 7–15.
79. M. Toledano-Luque, B. Kaczer, Ph.J. Roussel, T. Grasser, G.I. Wirth, J. Franco, C. Vrancken, N. Horiguchi, and G. Groeseneken, "Response of a Single Trap to AC Negative Bias Temperature Stress," in *Proc. Intl.Rel.Phys.Symp. (IRPS)*, 2011, pp. 364–371.
80. M. Denais, A. Bravaix, V. Huard, C. Parthasarathy, G. Ribes, F. Perrier, Y. Rey-Tauriac, and N. Revil, "On-the-fly Characterization of NBTI in Ultra-Thin Gate Oxide pMOSFET's," in *Proc. Intl.Electron Devices Meeting (IEDM)*, 2004, pp. 109–112.
81. A. Kerber and E. Cartier (2013) Bias temperature instability characterization methods. In: T. Grasser (eds) Bias temperature instability for devices and circuits. Springer, New York
82. H. Reisinger, O. Blank, W. Heinrigs, A. Mühlhoff, W. Gustin, and C. Schlünder, "Analysis of NBTI Degradation- and Recovery-Behavior Based on Ultra Fast V_{th}-Measurements," in *Proc. Intl.Rel.Phys.Symp. (IRPS)*, 2006, pp. 448–453.
83. Ph. Hehenberger, P.-J. Wagner, H. Reisinger, and T. Grasser, "On the Temperature and Voltage Dependence of Short-Term Negative Bias Temperature Stress," *Microelectronics Reliability*, vol. 49, pp. 1013–1017, 2009.
84. M.J. Kirton and M.J. Uren, "Noise in Solid-State Microstructures: A New Perspective on Individual Defects, Interface States and Low-Frequency (1/f) Noise," *Adv.Phys.*, vol. 38, no. 4, pp. 367–486, 1989.
85. C. Shen, M.-F. Li, C. E. Foo, T. Yang, D.M. Huang, A. Yap, G.S. Samudra, and Y.-C. Yeo, "Characterization and Physical Origin of Fast V_{th} Transient in NBTI of pMOSFETs with SiON Dielectric," in *Proc. Intl.Electron Devices Meeting (IEDM)*, 2006, pp. 333–336.
86. T. Grasser, W. Goes, V. Sverdlov, and B. Kaczer, "The Universality of NBTI Relaxation and its Implications for Modeling and Characterization," in *Proc. Intl.Rel.Phys.Symp. (IRPS)*, 2007, pp. 268–280.
87. S. Rangan, N. Mielke, and E.C.C. Yeh, "Universal Recovery Behavior of Negative Bias Temperature Instability," in *Proc. Intl.Electron Devices Meeting (IEDM)*, 2003, pp. 341–344.
88. T. Grasser, W. Goes, and B. Kaczer, "Towards Engineering Modeling of Negative Bias Temperature Instability," in *Defects in Microelectronic Materials and Devices*, D. Fleetwood, R. Schrimpf, and S. Pantelides, Eds., pp. 399–436. Taylor and Francis/CRC Press, 2008.

89. G. Pobegen, T. Aichinger, M. Nelhiebel, and T. Grasser, "Understanding Temperature Acceleration for NBTI," in *Proc. Intl.Electron Devices Meeting (IEDM)*, Dec. 2011, pp. 27.3.1–27.3.4.
90. A. Haggag, W. McMahon, K. Hess, K. Cheng, J. Lee, and J. Lyding, "High-Performance Chip Reliability from Short-Time-Tests," in *Proc. Intl.Rel.Phys.Symp. (IRPS)*, 2001, pp. 271–279.
91. J.H. Cadwell, "The Bivariate Normal Integral," *Biometrica Trust*, vol. 38, no. 3/4, pp. 475–479, 1951.
92. R.J. Henery, "An Approximation to Certain Multivariate Normal Probabilities," *Journal of the Royal Statistical Society B*, vol. 43, no. 1, pp. 81–85, 1981.
93. D.R. Cox and N. Wermuth, "A Simple Approximation for Bivariate and Trivariate Normal Integrals," *International Statistical Review*, vol. 59, no. 2, pp. 263–269, 1991.
94. W.-J. Tsay and P.-H. Ke, "A Simple Approximation for Bivariate Normal Integral Based on Error Function and its Application on Probit Model with Binary Endogenous Regressor," *IEAS Working Paper No. 09-A011*, 2009.
95. T. Grasser, B. Kaczer, Ph. Hehenberger, W. Goes, R. O'Connor, H. Reisinger, W. Gustin, and C. Schlünder, "Simultaneous Extraction of Recoverable and Permanent Components Contributing to Bias-Temperature Instability," in *Proc. Intl.Electron Devices Meeting (IEDM)*, 2007, pp. 801–804.

Part III

Chapter 18
Impact of Hydrogen on the Bias Temperature Instability

Gregor Pobegen, Thomas Aichinger, and Michael Nelhiebel

Abstract The ability of hydrogen to saturate lattice imperfections, which arise naturally at the silicon–oxide interface due to the structural mismatch of the two materials, has already early motivated to connect H with the bias temperature instability. Consistently, ESR measurements after NBTS observed P_b center defects, i.e. previously H passivated interfacial dangling bonds on silicon atoms at the interface, which supports the assumption that H is detached from defect precursors during NBTS. In contrast, theoretical and experimental investigations on the Si–H bond dissociation energy revealed a rather large value, inconsistent with the low-energy nature of conventional NBTI test. We summarize several explanations to this problem and compare these ideas with studies where the amount of H near the interfacial layer is varied through particular process adjustments.

18.1 Introduction

The pervasive nature of hydrogen (H) in semiconductor processing and its ability to saturate lattice imperfections makes this element also a candidate to be responsible for reliability issues in semiconductor devices. In the case of negative bias temperature instability (NBTI), hydrogen became a suspect to cause this degradation effect already in the first papers concerning this topic [1]. This was due to the fact that researchers recognized through electrical measurements that bias temperature stress (BTS) creates, among other traps located within the gate oxide, also interface

G. Pobegen (✉) • M. Nelhiebel
KAI Kompetenzzentrum für Automobil- und Industrielektronik,
Europastraße 8, Villach, Austria
e-mail: gregor.pobegen@k-ai.at; michael.nelhiebel@k-ai.at

T. Aichinger
Infineon Technologies Austria AG, Siemensstraße 2, Villach, Austria
e-mail: thomas.aichinger@infineon.com

T. Grasser (ed.), *Bias Temperature Instability for Devices and Circuits*,
DOI 10.1007/978-1-4614-7909-3_18, © Springer Science+Business Media New York 2014

traps [1]. It probably appeared unlikely that the relatively low electric fields and temperatures during BTS are strong enough to disrupt intact parts of the interface between silicon (Si) and silicon dioxide (SiO_2). As such, it was very advantageous that from previous work it was already known that the lattice mismatch between Si and SiO_2 causes a reproducible number of electrically active traps at the interface and that these traps are usually neutralized by H during semiconductor processing [2, 3]. It was therefore suggested that BTS causes a depassivation of these H saturated traps at the interface, rather than the creation of new, previously inexistent, traps.

When people started to investigate the possible microscopic transitions of H in the Si–SiO_2 system in more detail they observed characteristics which challenged the conclusions drawn before. In particular, several independent experimental and theoretical studies showed that the chemical reaction of removing a hydrogen atom from a passivated defect at the interface requires an energy in the range of 2.5–3.3 eV [4–10]—a value which is basically inaccessible under typical NBTS conditions, because only a 10^{-9} fraction of hydrogen atoms would dissociate from Si for a 2.5 eV bond and typical experimental conditions (200 °C, 10^4 s). The main question regarding the role of hydrogen in NBTI is therefore this disagreement between the high energetic dissociation of the defect precursor and the low energetic nature of BTS. The following section treats several different suggestions which try to resolve this issue and adds some more aspects of the interaction of H with NBTI.

We remark at this point that a prerequisite to correctly understand the atomic transitions involving H for NBTI was only given a few years ago, when it was found that H is involved solely in a quasi-permanent part of NBTI degradation [11–15]. The quasi-permanent part is defined by large emission time constants of the defects at the edge of the feasible experimental window of approximately 10^6 s [12, 13, 15] or by the annealing of positive traps by driving the pMOSFET into accumulation and thus offering electrons at the interface [11, 12, 14, 15]. This work allowed to understand the preceding literature about H and NBTI from a different perspective. For that reason, we can emphasize that the statements within this chapter apply only to the quasi-permanent component of NBTI, which might be independent of a faster recovering and especially accumulation susceptible part of the NBTI degradation.

While there has been no doubt yet about the link between the quasi-permanent component of NBTI and H, there is still a discussion of whether the quasi-permanent component and the recoverable component are somehow connected [12, 16, 17] or not [13, 15, 18]. We will touch upon this issue only marginally and will concentrate on the interaction of H with the quasi-permanent part of NBTI.

18.2 H Related Defects in the Si–SiO_2 System

In order to be able to understand the transitions of H during NBTS we give a brief review of some of the previously suggested reactions of H with defects in the Si–SiO_2 system which are important for later considerations.

Table 18.1 Dissociation and passivation activation energy values for the P_b center family of defects at the Si–SiO_2 interface [6, 7, 9, 25, 26]

		$P_b^{(111)}$	$P_{b0}^{(100)}$	$P_{b1}^{(100)}$
Passivation	Mean	1.51	1.51	1.57
	Variance	0.06	0.14	0.15
Dissociation	Mean	2.83	2.86	2.91
	Variance	0.08	0.07	0.07

All values are given in electron-volt

18.2.1 Interface Dangling Bond Passivation

The most prominent effect of H on the Si–SiO_2 MOS system is the passivation of dangling bonds at Si atoms at the interface (P_b centers) [19, 20]. The existence of unsaturated bonds is due to the natural mismatch of the lattice of Si and the amorphous layer of thermally grown SiO_2. A maximum number of interfacial dangling bonds at the surface of Si in the extreme case when some sort of stress would hypothetically separate the Si from the SiO_2 can be estimated by using the value of the lattice constants of Si, which is precisely known as 0.5431 nm [21]. From this follows an atomic density of silicon atoms at the interface of $6.78 \times 10^{14}\,\mathrm{cm}^{-2}$ for (100)Si and $11.76 \times 10^{14}\,\mathrm{cm}^{-2}$ for (111)Si [21,22]. The lattice mismatch of Si and SiO_2 induces a two decades smaller number of about $10^{13}\,\mathrm{cm}^{-2}$ dangling bonds at the interface for both (111) and (100)Si [23, 24]. Optimization of the annealing process in forming gas (H_2 and N_2) or molecular hydrogen gas has brought the remaining electrically active dangling bond density down to about $10^9\,\mathrm{cm}^{-2}$. This results in an interface where approximately every hundredth atom at the Si side is passivated by an H atom and about every millionth of these could not find an H atom for passivation or a nearest neighbor of the SiO_2 and are left unsaturated. The mean distance between two H passivated Si dangling bonds is thereby in the nanometer range and two unsaturated bonds are separated by some hundred nanometers, respectively. This points out that even devices with just a few tens of nanometer width and length will have a few interface traps.

Detailed investigations of the passivation and dissociation kinetics of interface traps through ESR measurements revealed that due to the amorphous nature of the oxide the reaction limiting energy values are normally distributed [6,7,9,25,26] and not single-valued [4, 27, 28]. The passivation/dissociation energies are summarized in Table 18.1. The values for the dissociation of a P_bH complex are >2.83 eV and are therefore usually not reached for a semiconductor in operation. Also theoretical calculations for Si–H bonds in silicon revealed rather large activation energies for H depassivation [29]. A challenging question for NBTI is therefore how the strong Si–H bond can be broken at typical NBTS temperatures of 100–150 °C and low oxide fields in the range of 3–10 MV/cm. Various possible explanations will be given in Sect. 18.3.

It is worth noting that theoretical [5] as well as experimental [30] work suggested that the passivation/dissociation kinetics of P_bH complexes might depend on the charge state of the P_b center.

18.2.2 E' Center Interaction with H

Various oxide defects involving hydrogen can exist in SiO_2 [8]. Among all possibilities, H interacting with the E' center, a dangling bond on an Si atom threefoldly bonded to oxygen atoms within the SiO_2 [31–33], has been studied widely by ESR measurements. It has been shown that the interaction of the E' center with H may either lead to paramagnetic variants of the E' center [34–39] or passivate the dangling bond [40, 41].

The E' center may turn to one paramagnetic variant, the 74 G doublet, when one of the three oxygen atoms bonded to the silicon atom is exchanged with a hydrogen atom. The 10.4 G doublet is formed when one of the three oxygen atoms bonds to a hydrogen atom. These two variants were shown to form even at room temperature when the sample is exposed to molecular hydrogen gas [38]. The exact passivation/dissociation energy for the formation of these defect variants has remained hidden up to now. In contrast, it was suggested that the diffusion of molecular H is the limiting factor for the transformation of E' centers into their doublet variants [38]. Furthermore, the transformation process is not accompanied by a charge state transfer since both defect types are positive. As such, it is very unlikely that such a transformation is happening during NBTS, where positive charge is created.

However, the 10.4 and 74 G doublet variants were suggested to play a role in interface trap generation. Thereby, the exposure of an E' center rich SiO_2 layer created as many interface traps as doublet variants, suggesting that H_2 cracked at E' sites lead to dissociation of Si–H bonds at the interface via atomic hydrogen [42, 43]. This process involves the passivation of the E' center with hydrogen [40, 42, 43]. For this reaction a lower limit for the activation energy of 0.3 eV was obtained [42]. Consequently, this reaction may also occur at room temperature, as later experimentally proved [43]. Afanas'ev et al. [44] stated that it is the proton which carries the positive charge when an O_3Si–H complex is depassivated. They could show that the efficiency of proton trapping in SiO_2 is fairly independent of the type of oxide. They further observed that the H passivated E' center can convert a hole into a proton, which then is trapped efficiently in the SiO_2. All these transitions were linked to the instability mechanisms of semiconductor devices like hot carrier degradation or radiation-induced trap generation [43] but have not been discussed in the context of NBTI yet.

Yet another aspect of the H passivated E' center is the experimentally obtained electron trap level 3.1 eV below the conduction band edge of the SiO_2 [40], which is only 0.05 eV above the Si conduction band. Therefore, the defect could be a candidate responsible for PBTI.

18.2.3 Other Oxide Defects Interacting with H

The hydrogen bridge is worth mentioning at this point because it is suggested to be responsible for stress-induced leakage current (SILC) [41, 45], another severe degradation mechanism in the Si–SiO_2 system. The hydrogen bridge is formed when one hypothetically replaces an oxygen atom in SiO_2 through a hydrogen atom. Calculations showed indeed that the hydrogen bridge could be a source of ESR-inactive positive charged trapped in the oxide [8], as it is frequently observed after NBTS [1, 13]. It was shown that SILC and NBTI could be linked, as SILC can increase through intense NBTS [17, 46] and SILC can affect the recoverable component of NBTI [47]. Despite this indication for a connection between the hydrogen bridge and NBTI, its direct responsibility has only seldom been suggested [48].

A further possibility how H could be involved in defect creation during NBTS is in its ionized state as a proton. Experimental studies [44, 49–51] as well as theoretical calculations [8, 41, 52] have revealed that a proton may reside within the SiO_2 and may give rise to a positive fixed charge. Experimental work using ESR provided evidence that the proton bonds to an intact Si–O–Si complex through rearrangement of the surrounding amorphous SiO_2 lattice [49]. Theoretical calculations using density functional theory and supercells suspect the proton to bond to the oxygen vacancy [52] instead of an intact Si–O–Si complex. For that reason, the H atom could become trapped in the close vicinity of the P_b center after P_bH dissociation. This idea has already been formulated in the context of NBTI by [53].

18.3 P_bH Dissociation

In the following we give an overview about the possible transitions of H during NBTS which result in electrically active defects. Since the P_bH complex is at least partly involved in the creation of defects, we deal with the mechanisms which may reduce the large energy of the dissociation of the P_bH complex

$$\text{Si–H} \longrightarrow \text{Si}\cdot + \text{H}. \tag{18.1}$$

Direct physical evidence from ESR measurements [54, 55] identifying defects from the P_b center family as well as electrical measurements showing an increase in interface trap density [1, 56] have unquestionably shown that interface traps are created through NBTS. Since thermochemical measurements for bond dissociation energies suggest that the bond strength of Si–O is much larger (8.3 eV [57]) than that for Si–H (3.3 eV [58]) it is very likely that the P_bH defect precursor will be broken during NBTS rather than an intact Si–O bond at the interface. Consequently, the involvement of H in the creation of interface traps during NBTS appears as a requirement.

As already stated in Sect. 18.2, the dissociation energy of the P_bH complex was experimentally [4, 6, 7, 9, 25–28] as well as theoretically [5, 29, 59, 60] determined to be around 2.5–3.3 eV. We can estimate the time for the dissociation of P_bH through temperature only with the values of [4]: dissociation energy $E_\mathrm{d} = 2.6\,\mathrm{eV}$ and forward rate constant $k_{\mathrm{d}0} = 1.2 \times 10^{12}\mathrm{s}^{-1}$ (no distribution in the dissociation energy considered). For a rather small increase in the interface trap density of only $10^9\,\mathrm{cm}^{-2}$ at a typical NBTS temperature of 125 °C we obtain already around $10^{16}\,$s which correspond to about 3×10^8 years. As a result, the dissociation of P_bH under typical NBTS conditions is not possible due to temperature alone and other mechanisms must be involved in the process. In the following we will review several different approaches which attempt to resolve this issue.

18.3.1 Hole Capture

The most prominent argument for a decrease in the dissociation energy of the Si–H bond is the capture of a hole prior to bond dissociation

$$\text{Si–H} + h^+ \longrightarrow \text{Si}\cdot + \text{H}^+. \tag{18.2}$$

This argument is used in a rather vague form frequently [56,61,62] as if it would not need any detailed clarification. In fact, a consistent and detailed explanation of how and why a captured hole reduces the bond-dissociation energy of Si–H is unclear. Support for the idea stems mostly from density functional theory calculations for different defect types. The amount of dissociation energy reduction is thereby unclear and varies largely. The largest reduction is suggested for the Si–O bond in the context of TDDB, where it is stated that a capture of a hole reduces the bond-dissociation energy by 2 eV [63]. In the context of NBTI, the calculated decrease in the dissociation energy of Si–H is only 0.3 eV [10]. Accordingly, the applied electric stress field might only reduce the dissociation energy by another 0.1 eV, a reduction which is still far too little to make the dissociation possible under NBTS conditions.

We remark that, despite the challenging issue concerning the too large dissociation energy, it was suggested that the diffusion of the released H species determines the dynamics of NBTI instead of the reaction. The reaction–diffusion theory [1,56,61,62,64,65], which evolved from this idea, could, in spite of several attempts [12,14,66], not be used to consistently explain all features of experimental data. Hence, the focus of this work lies in the reaction process of H dissociation which must precede any discussion on diffusion.

18.3.2 Atomic Hydrogen

An energetically more favorable reaction than the direct dissociation of Si–H with the capture of a hole is the dissociation with the help of an H atom

$$\text{Si–H} + \text{H} \longrightarrow \text{Si} + \text{H}_2 \tag{18.3}$$

which forms molecular hydrogen. This transition has a calculated barrier energy in the range of 0.95 eV [67]. This means that the existence of atomic hydrogen or a proton (also positively charged variants of reaction (18.3) are possible [68]) in the vicinity of an H passivated interface trap may lead to bond dissociation. Possible source for the atomic hydrogen might be the metal layers above the device [69], H passivated dopants within Si [67] or H from H_2 cracked at E' centers [42,43,55,70] or at trapped holes [71,72].

The occurrence of transition (18.3) during NBTS experiences support from nuclear reaction analysis. This is a technique where gamma rays are detected which are emitted by the nuclear reaction [73]

$$^{1}\text{H} + ^{15}\text{N} \longrightarrow ^{12}\text{C} + ^{4}\text{He} + \gamma, \tag{18.4}$$

which takes place when bombarding a hydrogen containing material with $^{15}N^{2+}$ ions. This technique has revealed that NBTS causes an accumulation of H atoms near the Si–SiO_2 interface [74]. The H atoms concentration thereby peaks at a position roughly 4 nm away from the interface with a spread of about 8 nm [73]. This peak is already preexisting before stress and increases through NBTS [74]. Since the overall number of $^{15}N^{2+}$ ions has to be kept low to avoid a measurement-induced redistribution of H atoms, the work of Liu et al. [74] documented NBTS-induced changes in the number of H atoms per square centimeter only for the estimated position of the interface. They could still show that the increase of H atoms at the interface with NBTS is larger than the increase of interface traps, which means that H is transported from other places within the semiconductor towards the interface.

Reference [75] gives another hint that atomic hydrogen could be involved in the microscopic NBTI degradation mechanism. It was found that a positive bias phase at room temperature after NBTS can lead to an increase in interfacial recombination centers even though the conditions for PBTS are not given. According to their explanation, atomic hydrogen, previously released through NBTS, returns to the interface and creates additional interface traps through transition (18.3). In particular, the gated diode leakage current (measured by the direct current–current voltage (DCIV) technique) showed two distinct peaks where only one of them increased through the positive bias treatment. This coincided with ESR measurements which showed that the P_{b0} variant of the P_b center family on (100)Si–SiO_2 reacts more readily with H than the other P_{b1} variant. Furthermore, the ESR data of [75] indicated hyperfine peaks symmetrically around the P_{b0} center, rather than around the E' center, which would have indicated the 10.4 or 74 G variants of the E' center described in Sect. 18.2.2. As such, the P_{b0} center itself has a hyperfine interaction due to an H atom in the close vicinity, as, e.g., in an anti-bonding configuration [29,76] or in a neighboring Si–Si bond [9].

18.3.3 Trapped Hole in Vicinity

Another explanation for the creation of interface traps during NBTS stems from thermodynamical considerations [77]. Previous work showed that numerous E'

centers are created in the SiO_2 layer during NBTS through hole trapping [16,55,78] and that E' centers may become passivated with H [42, 43, 79]. The activation of E' centers during NBTS creates now a completely different environment for the H atoms bonded to Si atoms at the interface. The H atoms receive an increasing number of possible states where they could reside. Now, from a thermodynamic perspective, the H atoms need to occupy the newly created free sites to minimize the Gibbs free energy of the whole system [77]. Many of the H atoms of the passivated P_bH centers will transfer to the E' centers in the SiO_2, leaving interface traps behind. I.e. the creation of P_b centers can occur, despite the large barrier for the dissociation of the Si–H bond at the interface, when E' centers are situated close to the interface.

Studies which tried to identify such a transition used naturally ESR measurements on suitable devices. One common disadvantage thereby is that the ESR measurement is conducted after the devices are being stressed. But after stress, most of the E' centers either neutralize through emitting their positive charge (the commonly observed recovery) or become passivated by an H atom. Consistently, researchers observed either no E' signal [54, 69] or just a very small one close to the sensitivity limit of the equipment [55, 80]. Only later performed studies [78] in an on-the-fly manner *during* NBTS showed coherently that the E' center signal can be large and is as such presumably connected to NBTI, but vanishes quickly after termination of the stress.

The idea of the transition of H from a P_b to an E' center was combined with the Harry Diamond Laboratories (HDL) model for switching oxide traps [81–83] (to explain the recoverable component) to form the two-stage model for NBTI [16]. This model states that the transition of H from P_bH to NBTS activated E' centers in SiO_2 [38, 43] leads to a lock-in of the E' center [77, 84] and is summarized in Fig. 18.1.

18.3.4 Large Variance in Dissociation Energy

ESR studies [23, 25, 26, 85] revealed that the oxidation process of the SiO_2 layer (in particular the oxidation temperature) can impact the variance of the distribution of the Si–H bond-dissociation energy. The reason for the variance is thereby the configurational distribution of the close atomic environment of the P_b center. The variance reduces with higher oxidation temperature because the temperature-induced relaxation decreases the spread in configurational compositions for the P_b defects [85]. The largest reported variance for P_bH association is 0.11 eV for (111)Si–SiO_2 [85] and 0.15 eV for (100)Si–SiO_2 [25, 26]. If one proposes such a variance for the dissociation energy of the P_bH complex of 2.83 eV, only a negligible part of the P_bH could be dissociated at typical activation energies of NBTS.

However, there are two possible explanations how the dissociation energy, including the inherent spread, is further reduced. The first idea stems from studies concerning TDDB [86], where it is proposed that the dissociation energy is reduced through a polarization effect. This means that the electrical field pulls the negative

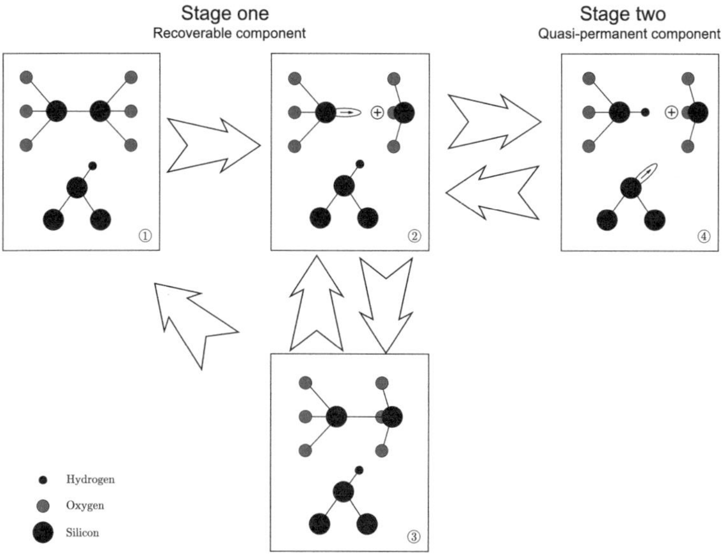

Fig. 18.1 The two-stage model for NBTI [16] combines the HDL model for switching oxide traps (stage one, recoverable component) with a transition of H from a P_bH defect to an E'_γ center to lock-in the positive charge (stage two, quasi-permanent component). The result of this transition is an equal amount of positive oxide charges and interface traps for the quasi-permanent component of NBTI

and positive charge centroids, and therefore the molecule itself, apart. This leads to stretching or compression of the atomic bond which reduces its dissociation energy. The dissociation energy reduces in a linear fashion by about 0.6eV for a large NBTI oxide field of $10\mathrm{MVcm}^{-1}$ [87]. The second possible reason is that for NBTI only one of the two vibrational modes of the Si–H bond is crucial [60,87]. This is because the bending mode is supposed to have a smaller dissociation energy (1.5eV) than the stretching mode (2.5eV) [29,59,60] of the Si–H bond. We remark that the proposed Fermi derivative distribution function [60, 88] is, except for the detailed shape of the tail, very similar to the normal distribution function and exhibits in particular equivalent mean and variance values. Both concepts together lead to a situation that there exists a distribution of activation energies around the mean value of about 0.9eV and a spread of up to 0.15eV. As a result, at least a low-energy fraction of the available Si–H bonds could be broken during NBTS [60, 87, 88].

If these assumptions are correct, one would need to observe the normal distribution of activation energies in NBTI stress and recovery data. Indeed, the normal distribution was directly measured through acceleration of the NBTS with high stress temperatures using the poly-heater [89]. Furthermore, a comprehensive model for NBTI which was tested for large ranges of temperature, bias, and time on several different technologies has the assumption of normally distributed activation energies as an indispensable requirement [90].

18.4 Impact of the H Passivation Degree on BTI

One elegant way to test the assumptions and models for the interaction of H with BTI is to vary the H content near the Si–SiO_2 interface to create varying numbers of H defect precursors [11,12,15,91–95]. Quite surprisingly, it appears that back end of line (BEOL) processes have a very large impact on the gate oxide reliability. Among all possibilities, three particular layers of BEOL processing have been shown to largely impact the NBTI susceptibility of devices.

A large impact is given by the thickness of the titanium (Ti) layer between the metal and the dielectric layers, i.e. by the amount of Ti within the BEOL stack. Ti was shown to gather H efficiently [92, 96, 97]. Therefore, the Ti suppresses the diffusion of H from H-rich layers located in a higher level of the stack towards the Si–SiO_2 interface. Time of flight secondary ion mass spectroscopy measurements depicted in Fig. 18.2 [12,98] as well as electrical measurements using charge pumping (CP) [12, 15] (c.f. Fig. 18.3) proved that the amount of electrically detectable defects is inversely proportional to the thickness of the Ti layers. Among those electrically active traps are not only interface traps but also positively charged oxide traps [15, 69, 91] which are passivated by H. Those positive oxide charges were identified to act as border traps with large emission and capture time constants in a frequency varying CP measurement as depicted in Fig. 18.4. This means the amount of Ti in the BEOL stack changes the H passivation degree near the Si–SiO_2 interface and therefore the number of electrically active interface and border traps.

Other layers of the BEOL process which impact the BTI reliability of devices are layers made from silicon nitride (SNIT) [91, 102]. SNIT is known to contain up to 30% hydrogen bound to Si and N atoms [91], also because it is usually

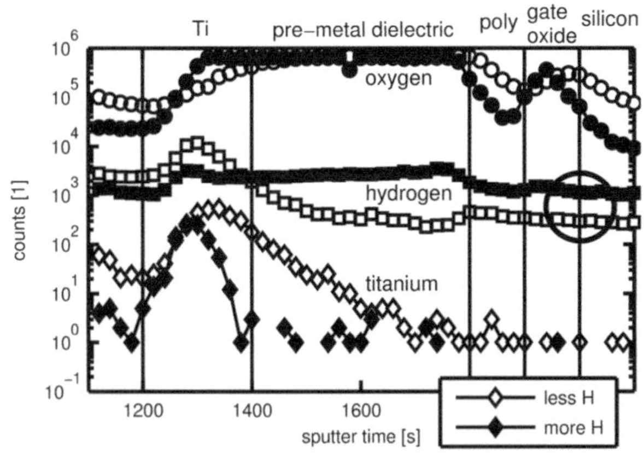

Fig. 18.2 Measured O, H, and Ti concentrations profile vertically through a device using TOF-SIMS [12]. The sputter time gives an estimate for the depth in the sample. The more H wafer has a thin Ti layer and therefore more H in the region of the gate oxide (*thick solid line circle*) compared to the less H wafer with a thick Ti liner

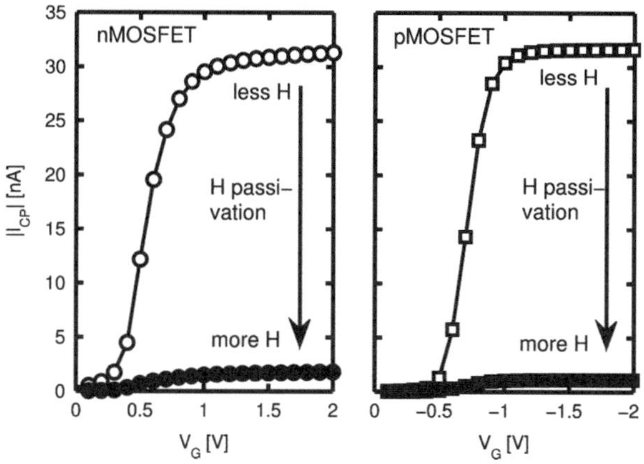

Fig. 18.3 Constant base level CP measurements [99] for n- and pMOSFET devices show that a thin Ti barrier (more H) leads to a more efficient passivation of CP detectable interface traps [11, 12, 15]

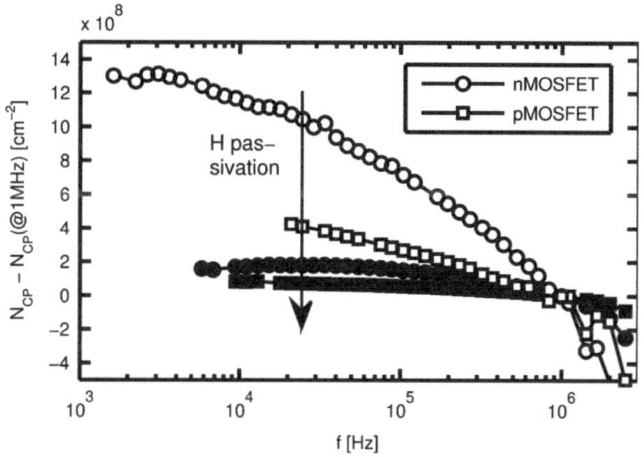

Fig. 18.4 An increase in the number of charges pumped per cycle $N_{CP} = I_{CP}/(qfA)$ with decreasing frequency hints for the existence of border traps with time constants approximately larger than $1/(2f)$ [15, 100, 101]. Both n- and pMOSFETs with more H near the Si–SiO_2 interface show less border traps than comparison devices [15]. The large difference in interface trap density N_{CP}(@1 MHz) was subtracted to emphasize the contribution of border traps to the CP signal

deposited on the wafer through chemical vapor deposition from SiH_4 and NH_3 gas. It is supposed that the SNIT layers lose some of the incorporated H during high temperature processing of the metalization. Consequently, if not hampered by diffusion barriers as the above described Ti, the H may diffuse towards the Si–SiO_2 interface where it either passivates or depassivates (depending on whether the H diffuses in its reactive atomic form or as molecular H_2) interface and border traps. Commonly, both processes take place simultaneously such that the device has an initially low density of interface traps but experience also larger degradation due to an increased precursor density [15, 91]. In general, it was shown that simultaneous presence of H_2 and trapped holes at elevated temperatures dramatically increases the P_bH density of the interface [71].

As observed only recently [94], also the type of metalization to contact the device can have an impact on the BTI performance. By comparing a process split for the same devices having either copper (Cu) or aluminum (Al) power metalization, it was found that devices with Al experience always larger drifts compared to device with Cu metalization. This was explained by the previously reported ability of Al to split H_2 in atomic hydrogen [103]. The reactive H moves, similar as if it was released from the SNIT layer, towards the interface where it passivates interface traps but creates also precursor defects for NBTI.

Another important topic is the influence of deuterium (D), the heavier natural isotope of hydrogen, as passivation through a deuterium forming gas anneal or deuterium implantation. Historically, it was first realized that hot carrier degradation is reduced when the device is annealed in a deuterium ambient [104–107]. This was motivated because D is harder to remove from the Si *surface* by using the tip of an scanning tunnel microscope compared to H [108]. From this idea, it was soon found that also NBTI is reduced through D anneal [13, 54, 109–113], even though occasionally contrary observations were made [114–116]. The argumentation for the decreased NBTI is, consistent with the ideas for hot carrier degradation, due to an apparently stronger Si–D bond compared to Si–H. In detail, the chemical bond of these two variants is not different because of the identical valence electron configuration. But the dissociation of Si–D is less probable because the vibrational bending mode of the Si–D bond is closer to the phonon frequencies of the Si lattice and because of this the excitation energy of the bond can be more easily conducted towards the semiconductor before the actual dissociation [117, 118]. The reduction of NBTI is always in the range of 0.6–0.8, which can be connected to the square root of the deuterium/hydrogen mass ratio of $1/\sqrt{2} \approx 0.71$, when diffusion is considered [13, 119]. A reason that a further reduction of NBTI is not obtained may lie in the relatively high activation energy of the replacement of Si–H through Si–D through D_2 [120] of 1.84 eV [121] compared to the 1.51 eV for the passivation with H_2.

To conclude, atomic H may not only stem from the environment during processing but may also be formed by cracking H_2 within the BEOL stack. The diffusion of H is influenced by Ti layers which has the largest impact on BTI reliability of devices. Consequently, we study the impact of Ti thickness thoroughly for positive as well as negative BTI.

18.4.1 Impact on Negative BTI

In order to understand the impact of the H content near the interface on negative BTI an elaborate electrical experiment was performed [12] which shows the bias and temperature-dependent recovery following logarithmically increasing stress times.

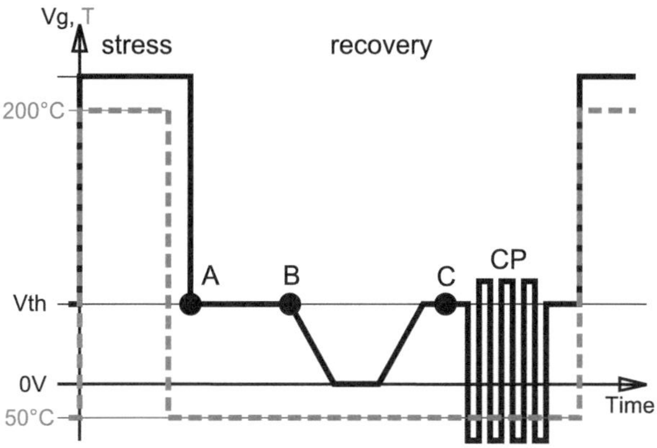

Fig. 18.5 Evolution of the bias at the gate and the device temperature. Point A corresponds to the ΔV_{TH} value right after stress, point B to the value after a constant bias recovery phase at the V_{TH} of -2 V, and point C to the value after bringing the device into accumulation by applying zero bias to the gate, respectively. After this a CP measurement was performed [12]

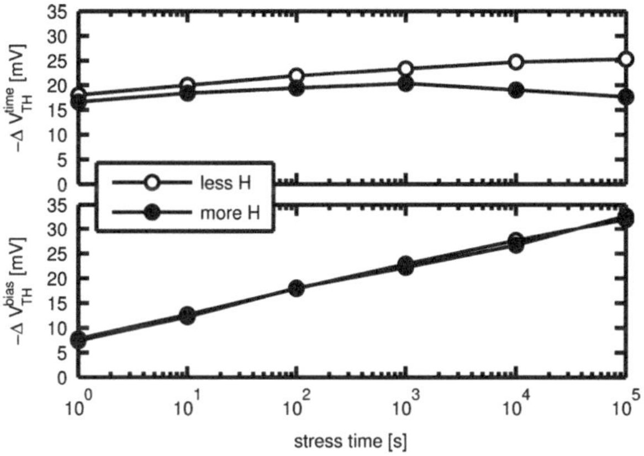

Fig. 18.6 *Upper plot*: Amount of recovery during the constant bias phase right after stress (difference between the ΔV_{TH} at point B (1 ks) and point A (50 ms) of Fig. 18.5) over stress time. *Lower plot*: Amount of recovery through the application of zero bias to the gate (difference between the ΔV_{TH} at point C and point B) over stress time. For both the bias and the time-dependent recovery the H content near the interface is rather insignificant [12]

The time diagram for the gate bias and the device temperature is given in Fig. 18.5. The constant bias recovery phase at the V_{TH} of the device right after stress allows to investigate whether the recovery is dependent on the H content near the interface or not. To identify this we depicted the amount of recovery during the constant bias recovery phase and the subsequent accumulation phase in Fig. 18.6. Through this experiment we are able to measure the accumulation-dependent and the time-dependent part of the recoverable component of NBTI independently. The result shows that both types are fairly independent of the H content near the interface.

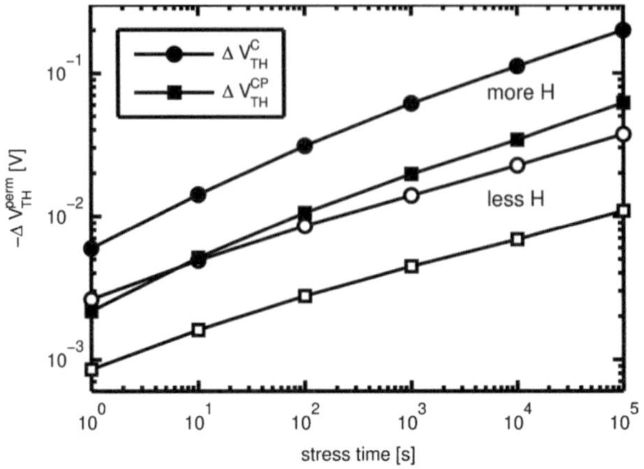

Fig. 18.7 The *circle symbols* correspond to the ΔV_{TH} which remains after applying zero bias to the gate (point C in Fig. 18.5) and the *square symbols* are the ΔV_{TH} contribution of interface traps calculated from the charge pumping current. By multiplying the CP data with a factor of three the differently measured ΔV_{TH} characteristics can be merged on one line for more H and one line for less H, respectively [12]

We can conclude that the recoverable component of NBTI, may it be defined either by the amount of charges which recover during a period of time or by the amount of charges which annihilate by driving the device into accumulation, is independent of the H content near the interface [11, 12, 15, 95].

After bringing the device into accumulation a CP measurement was performed. The as such obtained density of interface traps can be transformed into an according ΔV_{TH}^{IT} due to interface traps only [122] by assuming an amphoteric nature of interface traps [123] and a flat density of states energy profile [100]. This shift value can be compared to the remaining shift ΔV_{TH}^{perm} at point C of Fig. 18.5 after the application of zero bias at the gate as depicted in Fig. 18.7. It was found that a multiplication of ΔV_{TH}^{IT} with a factor of three lets the characteristic over stress time align with the one of ΔV_{TH}^{perm}. This shows that only about a third of the degradation of the quasi-permanent component visible in ΔV_{TH}^{perm} can be explained by interface traps and positively charged oxide traps are a definite contributor also to the quasi-permanent part of NBTI.

We found that the device with more H near the Si–SiO_2 interface drifts more compared to the device with less H at the interface [12, 15]. One could argue that the increase in drift is due to a larger number of defect precursors, namely Si–H bonds at the interface, which were passivated through the H treatment and are left unpassivated otherwise. Therefore, the sum of precursor defects and interface traps should be the same for both differently passivated devices and both devices should have a common maximum degradation level ΔV_{TH}^{max}. The direct measurement of ΔV_{TH}^{max} is presumably impossible because no reports of a complete saturation of the drift versus stress time characteristic have been published so far. The ΔV_{TH}^{max} can, however, be estimated when accelerating the NBTS with 400°C stress temperatures

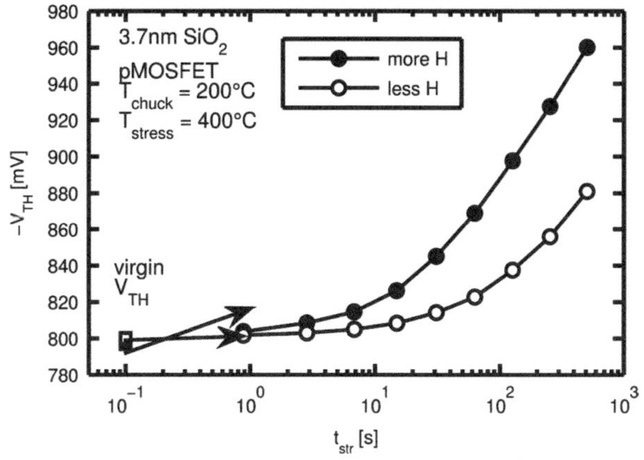

Fig. 18.8 Change of the threshold voltage of two devices with more or less H at the interface over the stress duration at a high stress temperature. The virgin difference in the V_{TH} of the two devices is inverted after already a second of intense NBTS at 400 °C and the more H device experiences much larger degradation. The temperature was switched during the experiment using the poly-heater [125]

[15, 89] as depicted in Fig. 18.8, by using the poly-heater (see Chap. 2 of this book [124]). The extreme acceleration shows that, when properly activated, the number of defects in the device with more H is larger than the number in the comparison device. The H passivation therefore increases the number of precursor defects for NBTI [12, 15, 71, 91, 92]. A possible explanation for this might be the creation of interface traps through atomic hydrogen, as described in more detail in Sect. 18.3.2, or the creation of different defect precursors within the SiO_2 by entering either an Si–Si or an Si–O bond (refer Sects. 18.2.2 and 18.2.3).

We remark that recent results [95] show that the H passivation degree changes only the number of available defect precursors and not any other parameters like the activation energy.

18.4.2 Impact on Positive BTI

For the sake of completeness and in order to test the assumptions done for *negative* BTI above, it is very interesting to study the influence of H on the *positive* BTI. We performed this investigation on nMOSFET devices with 30 nm thick gate oxides and n^{++} doped polycrystalline silicon gates in order to suppress possible tunneling of holes from the gate towards the Si–SiO_2 interface [126]. In this way we are convinced that only electrons are present at the Si–SiO_2 interface during stress and recovery. We observed [15] that the H content near the interface has the opposite impact on PBTI compared to NBTI, i.e. more H means less drift of the V_{TH}, as depicted in Fig. 18.9. Through detailed investigations on the virgin capacitance

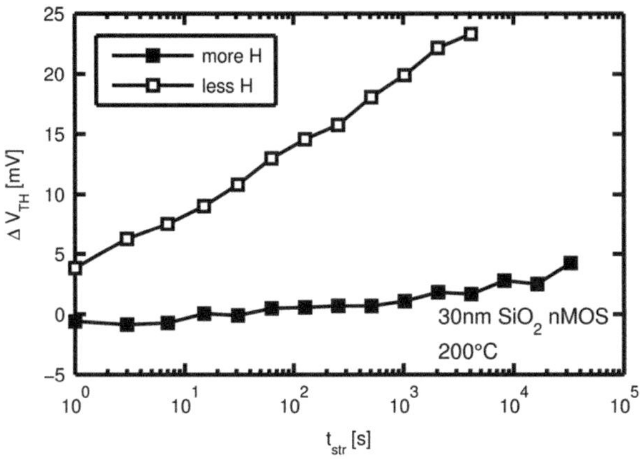

Fig. 18.9 The more H device experiences less PBTI drift compared to the less H device. The drift direction is positive, meaning that either negative charges are created or positive charges are lost through the stress [15]

voltage characteristics of the devices it was found that the apparent creation of negative charges is in fact the loss of positive charges through positive BTS [15]. Together with the observation that H passivates preexisting positive charges within the gate oxide, we concluded that a number of positive defects exist in the SiO_2 after fabrication and these defects may become annihilated either by an H atom during later steps of the processing or through electrons during PBTS [15]. The microscopic transitions which lead to this behavior are still nebulous up to now. Several possibilities can be deducted from the general discussion about defect transitions given in Sect. 18.2 but none of them can be unambiguously identified or declined because of the lack of appropriate experimental data.

18.5 Conclusions

The theoretically and experimentally suggested transitions of H at the Si–SiO_2 interface are manifold and might appear not even remotely comprehensible. However, in the context of NBTI, a central problem becomes evident when treating the topic in detail: How can the atomic bond of hydrogen to silicon be broken during the rather moderate conditions of a typical NBTI experiment, especially if several theoretical and experimental studies show that this bond has a rather large dissociation energy of around 2.8 eV? We presented four different assertions with completely different settings to explain how the dissociation energy is probably reduced.

A frequently used assumption in the context of NBTI is the reduction of the dissociation energy via hole trapping of the Si–H bond. This statement is only little supported by theoretical investigations and as such questionable.

More support can be found for the idea that atomic H dissociates the interfacial Si–H bond to create an H_2 molecule. The atomic H can thereby originate from various different sources, as H is stored and may be released during stress from layers within the upper metal stack, H passivated dopants or the SiO_2 itself. Atomic hydrogen may further create damage even after the end of stress, when it comes back to the interface after it has been released previously.

Various reports state that the Si–H bond dissociation occurs because of thermodynamical considerations, rather independent of the actual dissociation energy. The NBTS creates positively charged defects within the SiO_2 which are new possible sites for the H atom. Consequently, a transition of the H atom into the oxide happens because Gibbs free energy of the system must be minimal. This idea leads to models for NBTI which state that the permanent component of NBTI is due to locked-in positive oxide traps besides interface traps.

Increasing support from ESR measurements, and recently also from electrical measurements, is received for the argumentation that NBTI activates only a low-energy fraction of a distribution of activation energies for Si–H dissociation. The reason for the broad distribution around the large value for Si–H dissociation is thereby the amorphous nature of the thermally grown oxide and, inherently, the variable structural configuration of the defect.

The impact of H on the device performance can be tested by varying the amount of Ti in the BEOL stack, since Ti impedes the diffusion of H from H-rich layers situated above the device. An increased amount of H at the interface thereby decreases not only the interface trap density through the formation of Si–H but also the number of positively charged border traps with rather long carrier emission and capture time constants. But H does not only passivate existing defects, it creates also additional precursor defects. This is evident because the maximum drift level a device can reach increases with the H passivation degree. These additionally created defect precursors are activated during NBTS and form the quasi-permanent component of NBTI, visible as accumulation-independent defects with rather large annealing time constants. For positive BTI the impact of H is opposite, meaning that larger H passivation reduces PBTI. This is explained with the ability of H to passivate positive oxide border traps, which are therefore not available for neutralization with positive bias at the gate.

Acknowledgments This work was jointly funded by the Austrian Research Promotion Agency (FFG, Project No. 831163) and the Carinthian Economic Promotion Fund (KWF, contract KWF-1521|22741|34186).

References

1. K.O. Jeppson, C.M. Svensson, Journal of Applied Physics **48**, 2004 (1977)
2. G.L. Holmberg, A.B. Kuper, F.D. Miraldi, Journal of The Electrochemical Society **117**, 677 (1970)
3. K.H. Beckmann, N.J. Harrick, Journal of The Electrochemical Society **118**, 614 (1971)

4. K.L. Brower, Physical Review B **42**, 3444 (1990)
5. A.H. Edwards, Physical Review B **44**, 1832 (1991)
6. J.H. Stathis, Journal of Applied Physics **77**, 6205 (1995)
7. J.H. Stathis, Journal of Applied Physics **78**, 5215 (1995)
8. P. Bunson, M. Di Ventra, S. Pantelides, D. Fleetwood, R. Schrimpf, IEEE Transactions on Nuclear Science **47**, 2289 (2000)
9. A. Stesmans, Physical Review B **61**, 8393 (2000)
10. S.T. Pantelides, L. Tsetseris, S. Rashkeev, X. Zhou, D. Fleetwood, R. Schrimpf, Microelectronics Reliability **47**, 903 (2007)
11. T. Aichinger, M. Nelhiebel, S. Decker, T. Grasser, Applied Physics Letters **96**, 133511 (2010)
12. T. Aichinger, S. Puchner, M. Nelhiebel, T. Grasser, H. Hutter, in *IEEE International Reliability Physics Symposium* (2010), p. 1063
13. V. Huard, in *IEEE International Reliability Physics Symposium* (2010), pp. 33–42
14. T. Grasser, T. Aichinger, G. Pobegen, H. Reisinger, P.J. Wagner, J. Franco, M. Nelhiebel, C. Ortolland, B. Kaczer, in *IEEE International Reliability Physics Symposium* (2011), pp. 605–613
15. G. Pobegen, M. Nelhiebel, T. Grasser, in *IEEE International Integrated Reliability Workshop* (2012), pp. 54–58
16. T. Grasser, B. Kaczer, W. Goes, T. Aichinger, P. Hehenberger, M. Nelhiebel, in *IEEE International Reliability Physics Symposium* (2009), pp. 33–44
17. P.J. Wagner, B. Kaczer, A. Scholten, H. Reisinger, S. Bychikhin, D. Pogany, L.K.J. Vandamme, T. Grasser, in *IEEE International Integrated Reliability Workshop* (2012), pp. 60–64
18. T. Ho, D. Ang, A. Boo, Z. Teo, K. Leong, IEEE Transactions on Electron Devices **59**, 1013 (2012)
19. P.M. Lenahan, P.V. Dressendorfer, Applied Physics Letters **41**, 542 (1982)
20. P.M. Lenahan, P.V. Dressendorfer, Journal of Applied Physics **55**, 3495 (1984)
21. J.R. Ligenza, Journal of Physical Chemistry **65**, 2011 (1961)
22. K. Kushida-Abdelghafar, K. Watanabe, J. Ushio, E. Murakami, Applied Physics Letters **81**, 4362 (2002)
23. A. Stesmans, Physical Review B **48**, 2418 (1993)
24. A. Stesmans, B. Nouwen, V.V. Afanas'ev, Physical Review B **58**, 15801 (1998)
25. A. Stesmans, Applied Physics Letters **68**, 2723 (1996)
26. A. Stesmans, Applied Physics Letters **68**, 2076 (1996)
27. K.L. Brower, S.M. Myers, Applied Physics Letters **57**, 162 (1990)
28. K.L. Brower, Physical Review B **38**, 9657 (1988)
29. B. Tuttle, C.G. Van de Walle, Physical Review B **59**, 12884 (1999)
30. L.A. Ragnarsson, P. Lundgren, Journal of Applied Physics **88**, 938 (2000)
31. R.A. Weeks, Journal of Applied Physics **27**, 1376 (1956)
32. R.H. Silsbee, Journal of Applied Physics **32**, 1459 (1961)
33. F.J. Feigl, W. Fowler, K.L. Yip, Solid State Communications **14**, 225 (1974)
34. J. Vitko, Journal of Applied Physics **49**, 5530 (1978)
35. T. Takahashi, B.B. Triplett, K. Yokogawa, T. Sugano, Applied Physics Letters **51**, 1334 (1987)
36. B.B. Triplett, T. Takahashi, T. Sugano, Applied Physics Letters **50**, 1663 (1987)
37. T.E. Tsai, D.L. Griscom, Journal of Non-Crystalline Solids **91**, 170 (1987)
38. J. Conley, P. Lenahan, IEEE Transactions on Nuclear Science **39**, 2186 (1992)
39. P.M. Lenahan, J. J. F. Conley, Journal of Vacuum Science and Technology B **16**, 2134 (1998)
40. V.V. Afanas'ev, A. Stesmans, Applied Physics Letters **71**, 3844 (1997)
41. P.E. Bloechl, J.H. Stathis, Physical Review Letters **83**, 372 (1999)
42. Z. Li, S. Fonash, E. Poindexter, M. Harmatz, F. Rong, W. Buchwald, Journal of Non-Crystalline Solids **126**, 173 (1990)
43. J. Conley, J.F., P. Lenahan, IEEE Transactions on Nuclear Science **40**, 1335 (1993)
44. V. Afanas'ev, G. Adriaenssens, A. Stesmans, Microelectronic Engineering **59**, 85 (2001)
45. P.E. Bloechl, J.H. Stathis, Physica B: Condensed Matter **273–274**, 1022 (1999)

46. Y. Gao, D. Ang, C.D. Young, G. Bersuker, in *IEEE International Reliability Physics Symposium* (2012), pp. 5A.5.1–5A.5.5
47. A.A. Boo, D.S. Ang, Z.Q. Teo, K.C. Leong, IEEE Electron Device Letters **33**, 486 (2012)
48. F. Schanovsky, W. Goes, T. Grasser, Journal of Vacuum Science and Technology B **29**, 01A201 (2011)
49. V.V. Afanas'ev, A. Stesmans, Physical Review Letters **80**, 5176 (1998)
50. V.V. Afanas'ev, A. Stesmans, Physical Review B **60**, 5506 (1999)
51. V.V. Afanas'ev, A. Stesmans, Europhysics Letters **53**, 233 (2001)
52. S. Pantelides, S. Rashkeev, R. Buczko, D. Fleetwood, R. Schrimpf, IEEE Transactions on Nuclear Science **47**, 2262 (2000)
53. J.M. Soon, K.P. Loh, S.S. Tan, T.P. Chen, W.Y. Teo, L. Chan, Applied Physics Letters **83**, 3063 (2003)
54. S. Fujieda, Y. Miura, M. Saitoh, E. Hasegawa, S. Koyama, K. Ando, Applied Physics Letters **82**, 3677 (2003)
55. J. Campbell, P. Lenahan, A. Krishnan, S. Krishnan, IEEE Transactions on Device and Materials Reliability **6**, 117 (2006)
56. C.E. Blat, E.H. Nicollian, E.H. Poindexter, Journal of Applied Physics **69**, 1712 (1991)
57. S.W. Benson, Journal of Chemical Education **42**, 502 (1965)
58. A.E. Douglas, Canadian Journal of Physics **35**, 71 (1957)
59. C. Kaneta, T. Yamasaki, Y. Kosaka, Fujitsu Scientific and Technical Journal **39**, 106 (2003)
60. C. Guerin, V. Huard, A. Bravaix, Journal of Applied Physics **105**, 114513 (2009)
61. S. Ogawa, N. Shiono, Physical Review B **51**, 4218 (1995)
62. M.A. Alam, S. Mahapatra, Microelectronics Reliability **45**, 71 (2005)
63. J.W. McPherson, R.B. Khamankar, A. Shanware, Journal of Applied Physics **88**, 5351 (2000)
64. S. Mahapatra, P.B. Kumar, M.A. Alam, IEEE Transactions on Electron Devices **51**, 1371 (2004)
65. S. Mahapatra, A.E. Islam, S. Deora, V.D. Maheta, K. Joshi, A. Jain, M.A. Alam, in *IEEE International Reliability Physics Symposium* (2011), pp. 614–623
66. F. Schanovsky, T. Grasser, in *IEEE Integrated Reliability Workshop* (2011), pp. 17–21
67. L. Tsetseris, X.J. Zhou, D.M. Fleetwood, R.D. Schrimpf, S.T. Pantelides, Applied Physics Letters **86**, 142103 (2005)
68. S. Rashkeev, D. Fleetwood, R. Schrimpf, S. Pantelides, Physical Review Letters p. 165506 (2001)
69. M. Houssa, V.V. Afanas'ev, A. Stesmans, M. Aoulaiche, G. Groeseneken, M.M. Heyns, Applied Physics Letters **90**, 043505 (2007)
70. A.H. Edwards, Journal of Non-Crystalline Solids **187**, 232 (1995)
71. J.F. Zhang, H.K. Sii, R. Degraeve, G. Groeseneken, Journal of Applied Physics **87**, 2967 (2000)
72. R.E. Stahlbush, A.H. Edwards, D.L. Griscom, B.J. Mrstik, Journal of Applied Physics **73**, 658 (1993)
73. M. Wilde, M. Matsumoto, K. Fukutani, Z. Liu, K. Ando, Y. Kawashima, S. Fujieda, Journal of Applied Physics **92**, 4320 (2002)
74. Z. Liu, S. Fujieda, K. Terashima, M. Wilde, K. Fukutani, Applied Physics Letters **81**, 2397 (2002)
75. T. Aichinger, P. Lenahan, T. Grasser, G. Pobegen, M. Nelhiebel, in *IEEE International Reliability Physics Symposium* (2012), pp. XT.2.1 – XT.2.6
76. A. Alkauskas, A. Pasquarello, Physica B: Condensed Matter **401–402**, 546 (2007)
77. P. Lenahan, Microelectronic Engineering **69**, 173 (2003)
78. J. Ryan, P. Lenahan, T. Grasser, H. Enichlmair, in *IEEE International Reliability Physics Symposium* (2010), pp. 43–49
79. J.W. Lee, M. Tomozawa, R. MacCrone, Journal of Non-Crystalline Solids **354**, 3510 (2008)
80. J.P. Campbell, P.M. Lenahan, A.T. Krishnan, S. Krishnan, Applied Physics Letters **87**, 204106 (2005)

81. A. Lelis, J. Boesch, H.E., T. Oldham, F. McLean, IEEE Transactions on Nuclear Science **35**, 1186 (1988)
82. A. Lelis, T. Oldham, J. Boesch, H.E., F. McLean, IEEE Transactions on Nuclear Science **36**, 1808 (1989)
83. A. Lelis, T. Oldham, IEEE Transactions on Nuclear Science **41**, 1835 (1994)
84. P. Lenahan, Microelectronics Reliability **47**, 890 (2007)
85. A. Stesmans, Journal of Applied Physics **92**, 1317 (2002)
86. J.W. McPherson, V.K. Reddy, H.C. Mogul, Applied Physics Letters **71**, 1101 (1997)
87. V. Huard, C. Parthasarathy, N. Rallet, C. Guerin, M. Mammase, D. Barge, C. Ouvrard, in *IEEE International Electron Devices Meeting* (2007), pp. 797–800
88. V. Huard, M. Denais, C. Parthasarathy, Microelectronics Reliability **46**, 1 (2006)
89. G. Pobegen, T. Aichinger, M. Nelhiebel, T. Grasser, in *IEEE International Electron Device Meeting* (2011), pp. 27.3.1–27.3.4
90. T. Grasser, P. Wagner, H. Reisinger, T. Aichinger, G. Pobegen, M. Nelhiebel, B. Kaczer, in *IEEE International Electron Devices Meeting* (2011), pp. 27.4.1–27.4.4
91. M. Nelhiebel, J. Wissenwasser, T. Detzel, A. Timmerer, E. Bertagnolli, Microelectronics Reliability **45**, 1355 (2005)
92. T. Pompl, K.H. Allers, R. Schwab, K. Hofmann, M. Rochner, in *IEEE International Reliability Physics Symposium* (2005), pp. 388–397
93. T. Aichinger, On the role of hydrogen in silicon device degradation and metalization processing. Ph.D. thesis, TU Vienna (2010)
94. R. Stradiotto, G. Pobegen, M. Nelhiebel, in *IEEE International Integrated Reliability Workshop* (2012), pp. 65–69
95. G. Pobegen, M. Nelhiebel, T. Grasser, in *IEEE International Reliability Physics Symposium* (2013)
96. A.D. Marwick, J.C. Liu, K.P. Rodbell, Journal of Applied Physics **69**, 7921 (1991)
97. D.E. Woon, D.S. Marynick, S.K. Estreicher, Physical Review B **45**, 13383 (1992)
98. S. Puchner, Characterization of contaminations on semiconductor surfaces and thin layer systems with time of flight - secondary ion mass spectrometry. Ph.D. thesis, TU Vienna (2011)
99. T. Aichinger, M. Nelhiebel, IEEE Transactions on Device and Materials Reliability **8**, 509 (2008)
100. G. Groeseneken, H. Maes, N. Beltran, R. De Keersmaecker, IEEE Transactions on Electron Devices **31**, 42 (1984)
101. J.T. Ryan, L.C. Yu, J.H. Han, J.J. Kopanski, K.P. Cheung, F. Zhang, C. Wang, J.P. Campbell, J.S. Suehle, V. Tilak, J. Fronheiser, in *IEEE International Reliability Physics Symposium* (2011)
102. R. Sun, J. Clemens, J. Nelson, in *IEEE International Reliability Physics Symposium* (1980), pp. 244–251
103. G. Dunn, IEEE Electron Device Letters **10**, 333 (1989)
104. R.A.B. Devine, J.L. Autran, W.L. Warren, K.L. Vanheusdan, J.C. Rostaing, Applied Physics Letters **70**, 2999 (1997)
105. I. Kizilyalli, J. Lyding, K. Hess, IEEE Electron Device Letters **18**, 81 (1997)
106. K. Hess, L. Register, B. Tuttle, J. Lyding, I. Kizilyalli, Physica E: Low-dimensional Systems and Nanostructures **3**, 1 (1998)
107. E. Li, E. Rosenbaum, J. Tao, G.C.F. Yeap, M. Lin, P. Fang, in *IEEE International Reliability Physics Symposium* (1999), pp. 253–258
108. J.W. Lyding, K. Hess, I.C. Kizilyalli, Applied Physics Letters **68**, 2526 (1996)
109. N. Kimizuka, K. Yamaguchi, K. Imai, T. Iizuka, C.T. Liu, R. Keller, T. Horiuchi, in *Symposium on VLSI Technology* (2000), pp. 92–93
110. C. Liu, M. Lee, C.Y. Lin, J. Chen, K. Schruefer, J. Brighten, N. Rovedo, T. Hook, M. Khare, S.F. Huang, C. Wann, T. chiang Chen, T. Ning, in *International Electron Devices Meeting* (2001), pp. 39.2.1–39.2.4

111. V. Huard, F. Monsieur, G. Ribes, S. Bruyere, in *IEEE International Reliability Physics Symposium* (2003), pp. 178 –182
112. S. Fujieda, Y. Miura, M. Saitoh, Y. Teraoka, A. Yoshigoe, Microelectronics Reliability **45**, 57 (2005)
113. Y. Mitani, H. Satake, in *IEEE International Conference on Integrated Circuit Design and Technology* (2006), pp. 1–4
114. J. Wu, E. Rosenbaum, B. MacDonald, E. Li, J. Tao, B. Tracy, P. Fang, in *IEEE International Reliability Physics Symposium* (2000), pp. 27–32
115. T.B. Hook, R. Bolam, W. Clark, J. Burnham, N. Rovedo, L. Schutz, Microelectronics Reliability **45**, 47 (2005)
116. J.S. Lee, Transactions on Electrical and Electronic Materials **13**, 188 (2012)
117. C.G.V. de Walle, W.B. Jackson, Applied Physics Letters **69**, 2441 (1996)
118. R. Biswas, Y.P. Li, B.C. Pan, Applied Physics Letters **72**, 3500 (1998)
119. S. Zafar, B. Lee, J. Stathis, A. Callegari, T. Ning, in *Symposium on VLSI Technology* (2004), pp. 208–209
120. K. Cheng, K. Hess, J. Lyding, IEEE Electron Device Letters **22**, 441 (2001)
121. K. Cheng, K. Hess, J.W. Lyding, Journal of Applied Physics **90**, 6536 (2001)
122. T. Aichinger, M. Nelhiebel, S. Einspieler, T. Grasser, Journal of Applied Physics **107**, 024508 (2010)
123. P.V. Gray, D.M. Brown, Applied Physics Letters **8**, 31 (1966)
124. T. Aichinger, G. Pobegen, M. Nelhiebel, *Application of on-chip device heating for BTI investigations* (Springer Verlag, 2013), Chap. 2
125. T. Aichinger, M. Nelhiebel, S. Einspieler, T. Grasser, IEEE Transactions on Device and Materials Reliability **10**, 3 (2010)
126. G. Pobegen, T. Aichinger, T. Grasser, M. Nelhiebel, Microelectronics Reliability **51**, 1530 (2011)

Chapter 19
FEOL and BEOL Process Dependence of NBTI

Souvik Mahapatra

Abstract This chapter reviews different transistor processes that influence NBTI degradation. Effect of gate oxidation under dry and wet ambient, incorporation of nitrogen and fluorine, compressive stress, type of source-drain dopant atoms, hydrogen and deuterium post metallization anneal, use of different barrier metals, inter-metal dielectrics, and cap layers, as well as antenna charging are briefly reviewed. Due to its technological importance, a detailed review of nitrogen incorporation effect on NBTI, measured using ultrafast measurement method, is also done. The impact of nitrogen distribution profile in the gate insulator stack on NBTI degradation magnitude and parameters are studied in both SiON and HKMG devices. Key process dependent results are summarized.

19.1 Introduction

Negative Bias Temperature Instability (NBTI) has been found to be significantly influenced by different Front-End-Of-Line (FEOL) and Back-End-Of-Line (BEOL) process steps. The process impact of NBTI is of immense practical interest, as it allows the development of suitable process technologies for mitigation and control of device degradation during stress, which is a key factor in keeping circuit degradation under acceptable limits. A brief review of the impact of several FEOL and BEOL processes will be presented first. Due to its technological importance, a detailed study of the impact of gate oxide nitridation in Silicon Oxynitride (SiON) as well as High-k Metal Gate (HKMG)p-MOSFETs will be presented next. Finally, key results will be summarized.

S. Mahapatra (✉)
Department of Electrical Engineering, Indian Institute of Technology Bombay,
Mumbai 400076, India
e-mail: souvik@ee.iitb.ac.in

T. Grasser (ed.), *Bias Temperature Instability for Devices and Circuits*,
DOI 10.1007/978-1-4614-7909-3_19, © Springer Science+Business Media New York 2014

Although NBTI measurement methods and underlying physical mechanism are discussed in detail in earlier chapters of this book, a brief review is nevertheless presented hereinafter, which is necessary to understand the process dependence of NBTI. It is now believed that NBTI is due to uncorrelated contribution of the following underlying components [1]. Generation of donor-like interface traps (ΔN_{IT}) due to breaking of Si–H bonds at the Si/SiO_2 interface, hole trapping in process-related preexisting traps in the gate insulator bulk (ΔN_{HT}), and hole trapping in generated bulk insulator traps (ΔN_{OT}). Positive charges arising out of these traps cause negative threshold voltage shift (ΔV_T) during NBTI stress. Note that although alternative NBTI physical mechanisms have been proposed, as reviewed in [1], the above mechanism involving uncorrelated underlying components has been shown to explain a variety of NBTI DC and AC experimental data, including gate insulator process dependence, as discussed in [2]. Note that the presence of ΔN_{OT} is due to Time-Dependent Dielectric Breakdown (TDDB) like mechanism [3], and it is desirable to minimize its contribution by suitable choice of relatively lower stress gate bias ($V_{G,STR}$) [4, 5]. Therefore, when proper stress conditions are used, and especially for thin gate insulators used in logic devices, changes in ΔN_{IT} and/or ΔN_{HT} can explain the NBTI process impact.

It is now well known that NBTI degradation recovers substantially once $V_{G,STR}$ is removed [6, 7], and hence, the magnitude of measured ΔV_T depends on measurement speed [7–9]. NBTI recovery is due to fast detrapping of trapped holes and relatively slower passivation of generated interface traps [1]. Therefore, depending on the measurement speed, ΔV_T would be due to different fractions of underlying ΔN_{IT} and ΔN_{HT} components, provided ΔN_{OT} is kept low by choosing proper $V_{G,STR}$ [4]. Since hole detrapping is fast in thin gate insulators, as shown in [2], significantly lower ΔN_{HT} would be captured in conventional slow measurements,[1] and proper estimation of NBTI degradation would require the use of ultrafast measurement methods [7–9].

Finally, note that ΔN_{IT} is often directly characterized by Charge Pumping (CP)[2] [11] or similar methods and is compared to measured ΔV_T to estimate relative contribution of ΔN_{IT} and ΔN_{HT} to overall ΔV_T [12]. However, as discussed in [13], CP measurement scans interface traps at a part of the Si band gap centered

---

[1]In a typical NBTI experiment, transfer IV characteristics are measured before and after stress, and post-stress measurements are usually done after different durations of stress. Threshold voltage (V_T) is calculated before and after stress, and ΔV_T is obtained. This approach is known as the conventional Measure-Stress-Measure (MSM) method. As discussed later in this chapter, it is now well known that unless ultrafast IV measurements are employed [8], conventional MSM approach suffers from NBTI recovery-related issues [10].

[2]During CP measurement, gate of the MOSFET is repetitively pulsed from inversion to accumulation, source and drain terminals are grounded, and the DC current due to electron–hole recombination in traps at or near the Si/SiO_2 interface is measured at the substrate. CP measurements are done before and after NBTI stress in the conventional MSM mode, and increase in CP current after stress indicates trap generation. CP usually takes longer time than IV measurement and hence suffers from higher recovery-related issues [13].

around midgap and suffers from recovery issues as it is implemented in the MSM mode. Therefore, corrections due to measurement delay and band gap differences are required before ΔN_{IT} contribution obtained from CP is directly compared to ΔV_T from IV measurements, as shown in [13], and failure to do so would result in underestimation of ΔN_{IT} and overestimation of ΔN_{HT} [12] and incorrect estimation of NBTI process dependence.

19.2 Overview of Process Dependence

The FEOL and BEOL processes that impact NBTI degradation include gate oxide processing under different dry and wet ambient, incorporation of nitrogen (N_2) and fluorine (F_2) in the gate stack, effect of source-drain doping and compressive stress, type and nature of Inter-Metal Dielectric (IMD), barrier metal and cap layer, hydrogen (H_2) or deuterium (D_2) Post Metallization Anneal (PMA), and plasma charging effect in antenna devices. The impact of these processes on NBTI is briefly reviewed in this section.

19.2.1 Gate Oxidation

The most important process step that impacts NBTI is gate oxidation. Figure 19.1 shows measured ΔV_T due to NBTI stress in silicon dioxide (SiO_2) p-MOSFETs, with gate oxide grown using dry and wet oxidation processes [12, 14]. For identical stress condition, dry SiO_2 shows lower ΔV_T when compared to wet SiO_2, and this observation has been made by different groups. This suggests the detrimental effect of water-related hydroxyl (OH) or H_2 species, present in the wet gate oxide, on NBTI degradation. Note that since slow measurements have been used, observed

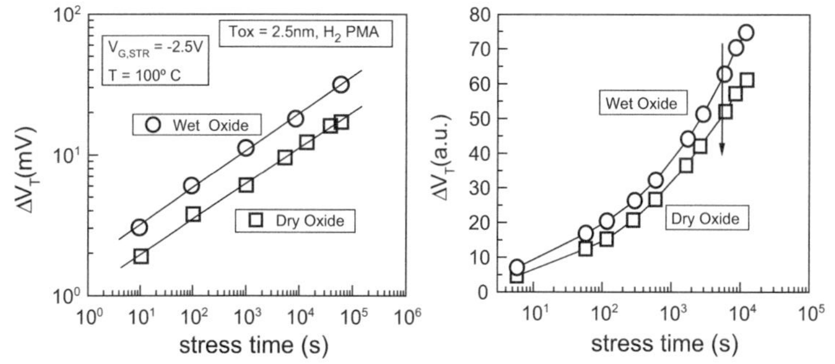

Fig. 19.1 Time evolution of ΔV_T during NBTI stress, measured in SiO_2 p-MOSFETs fabricated using dry and wet oxidation. *Left panel* data from [14], *right panel* from [12]

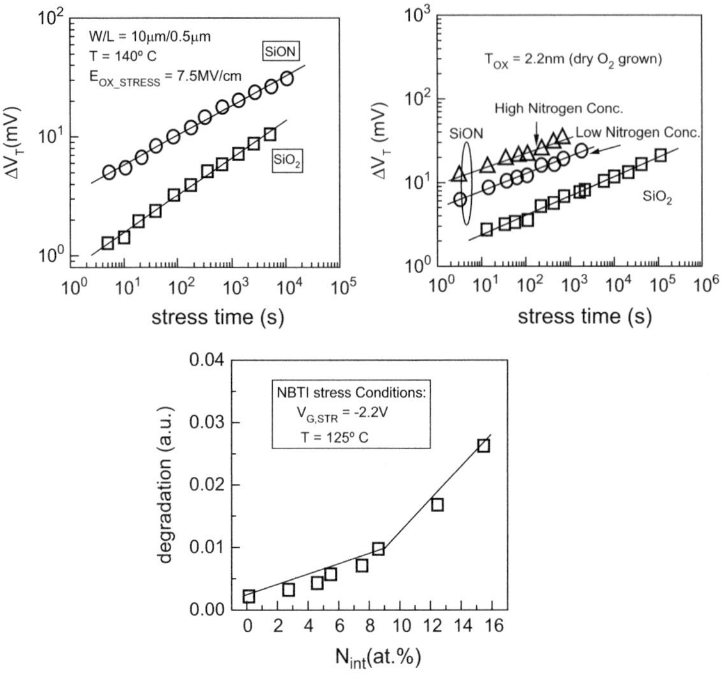

Fig. 19.2 Impact of nitridation on (**a**, **b**) time evolution of ΔV_T and on (**c**) degradation measured at a fixed time during NBTI stress. *Left panel* data from [15], *right panel* from [14], *bottom panel* from [16]

difference in ΔV_T is presumably due to difference in ΔN_{IT} as negligible ΔN_{HT} would be captured. Due to its superior NBTI immunity, majority of gate insulators have traditionally been grown in a dry ambient.

19.2.2 Gate Oxide Nitridation

Gate oxide process technology has evolved from pure SiO_2 to SiON insulators for Equivalent Oxide Thickness (EOT) scaling and has naturally motivated an extensive study on the effect of N_2 on NBTI. Figure 19.2 summarizes the effect of N_2 on NBTI observed by different groups [14–16]. SiON device shows higher ΔV_T when compared to SiO_2 device under similar NBTI stress [15], and the degradation increases with increase in N_2 content in the gate insulator [14, 16]. Once again since slow measurements have been used, the observed increase in ΔV_T should be due to increase in ΔN_{IT} as negligible ΔN_{HT} would be captured and is likely due to the N_2-induced enhancement of the Si–H bond breaking process as discussed in [17].

Figure 19.3 shows ΔV_T due to NBTI measured in p-MOSFETs with SiON gate stacks fabricated using different processes [18]. The NISS process where SiO_2

Fig. 19.3 Time evolution of ΔV_T during NBTI stress, measured in SiON p-MOSFETs fabricated using different nitridation techniques (see text). Data from [18]

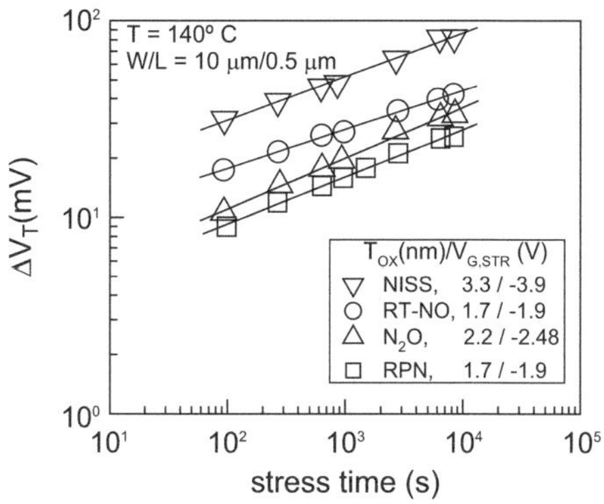

growth is done on N_2-implanted Si substrate shows highest degradation, followed by N_2O-based gate oxide formed using rapid thermal process and nitrogen oxide (N_2O)-based thermal gate oxide, while SiON made using the Remote Plasma Nitridation (RPN) process shows lowest degradation. Once again, note that differences in measured ΔV_T are likely due to differences in ΔN_{IT} as slow measurements have been used. In general for a particular process technology used for N_2 incorporation in the gate oxide, higher N_2 dose would result in higher NBTI and vice versa. However, the impact of different N_2 incorporation processes on measured degradation can be better understood by noting that NBTI is governed by N_2 density at or near the Si/SiON interface and is not influenced by the total integrated N_2 content in the gate insulator, as discussed in [19, 20].

To illustrate this very important feature, Fig. 19.4 shows N_2 density distribution profile in the gate insulator [19, 20], measured by X-ray Photoelectron Spectroscopy (XPS) [21] in different devices made using different gate insulator processes. NBTI degradation measured in these devices is also shown. In the top figure [19], the NO first process has higher peak and overall integrated N_2 content in the gate insulator but lower Si/SiON N_2 density compared to the NO last process. Lower NBTI-induced ΔV_T for the NO first process indicates that it is the N_2 density at the Si/SiON interface, and not the total N_2 content of the gate insulator, that governs NBTI. In the bottom figure [20], N_2 density at the Si/SiON interface increases, while the peak and total integrated N_2 content in the gate stack reduces, from process A through C. For a given N_2 content, NBTI degradation is lowest, which results in highest extrapolated lifetime for process A, while the degradation is highest and extrapolated lifetime is lowest for process C, and this is consistent with relative N_2 density at the Si/SiON interface for these devices.[3]

[3]NBTI lifetime is defined as the time needed to reach a particular degradation value and is usually determined by extrapolation of measured degradation. Higher measured degradation implies lower extrapolated lifetime and vice versa.

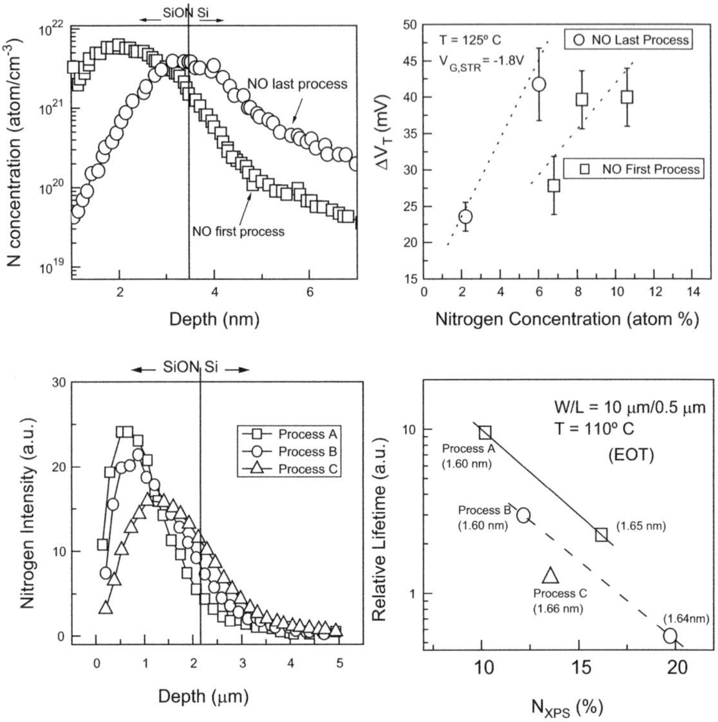

Fig. 19.4 Measured N_2 density distribution profiles in SiON gate insulators fabricated using different gate nitridation processes. Corresponding NBTI degradation and extrapolated lifetime are also shown. *Top panel* data from [19], *bottom panel* data from [20]

As NBTI is measured using slow method, once again, the differences in ΔV_T are largely due to difference in ΔN_{IT} caused by different amount of N_2 at or near the Si/SiON interface, presumably due to the N_2-related mechanism explained in [17].

The total integrated N_2 content in the gate stack helps in EOT scaling, while the N_2 density at the Si/SiON interface determines NBTI. Compared to thermal N_2O-based process, the RPN process results in lower Si/SiON interfacial N_2 density for identical integrated N_2 content in the gate insulator, and therefore is beneficial for both EOT scaling and NBTI, and has been universally adopted. To further illustrate the effect of N_2 distribution profile in the gate insulator on NBTI, Fig. 19.5 shows NBTI stress-induced ΔV_T and ΔN_{IT} for thermal SiON- and RPN-based plasma SiON devices, where the stress-induced increase in CP current (ΔI_{CP}) can be attributed to increase in N_{IT} [22]. Plasma SiON shows lower ΔV_T and ΔN_{IT} when compared to thermal SiON under similar stress condition. However, the difference in ΔV_T between the two processes is larger than the corresponding difference in ΔN_{IT}. In [22], this difference has been attributed to additional hole trapping in N_2-related traps in the gate insulator [23], which is indeed possible [13]. However, since

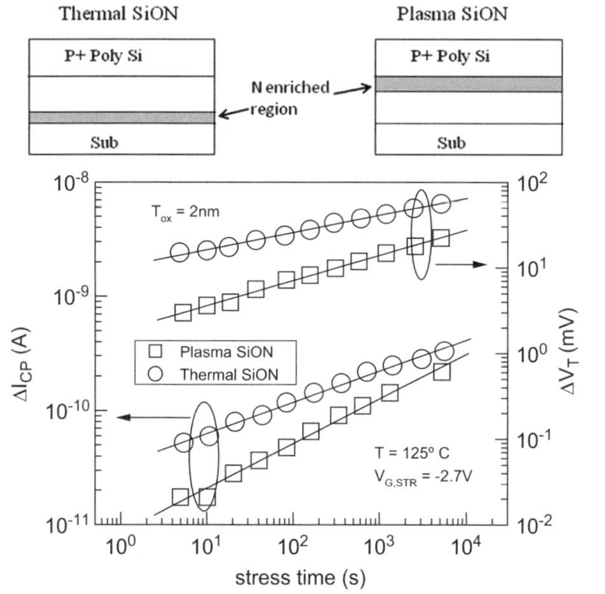

Fig. 19.5 Time evolution of ΔI_{CP} and ΔV_T during NBTI stress, measured in SiON p-MOSFETs with thermal and plasma nitrided gate insulators. Schematic of thermal and plasma SiON gate stacks showing peak N_2 position are also shown. Data from [22]

slower measurements have been used in [22], it is unlikely that significant ΔN_{HT} would be captured, and the difference between ΔV_T and ΔN_{IT} is probably due to larger recovery in CP measurements as discussed in [13].

For RPN process, NBTI magnitude depends not only on total N_2 dose but, more importantly, also on the effectiveness of Post Nitridation Anneal (PNA) [24]. Figure 19.6 shows measured ΔV_T due to NBTI in RPN-processed SiON devices having identical N_2 dose but subjected to different PNA [25]. It is evident that proper PNA is a crucial process step in significantly reducing NBTI, even though the RPN N_2 dose remained the same across all devices as shown in Fig. 19.6. The impact of gate oxide nitridation has been a very important research area and therefore will be discussed in more detail in the latter part of this chapter.

19.2.3 Fluorine Incorporation

It has been reported by several groups that NBTI degradation can be significantly reduced by F_2 incorporation in the gate stack [12, 20, 26]. Figure 19.7 shows NBTI stress induced ΔV_T and direct estimation of ΔN_{IT} as measured using the CP method for devices without and with F_2. In [12], the effect of F_2 incorporation is studied by using boron- and BF_2-implanted source-drain junctions, where F_2 diffuses from the

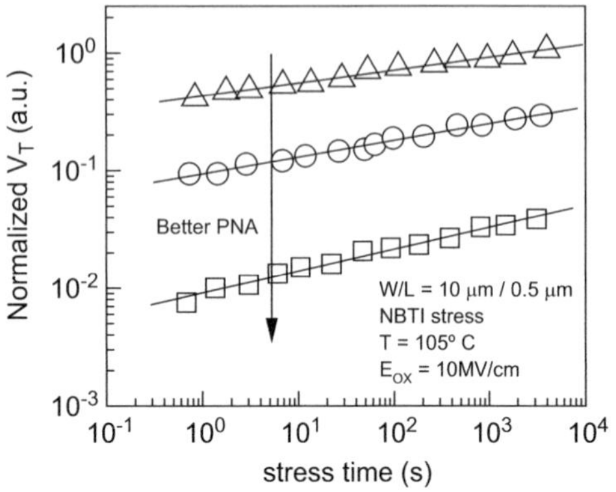

Fig. 19.6 Time evolution of ΔV_T during NBTI stress, measured in SiON p-MOSFETs with RPN gate insulators and different PNA. Data from [25]

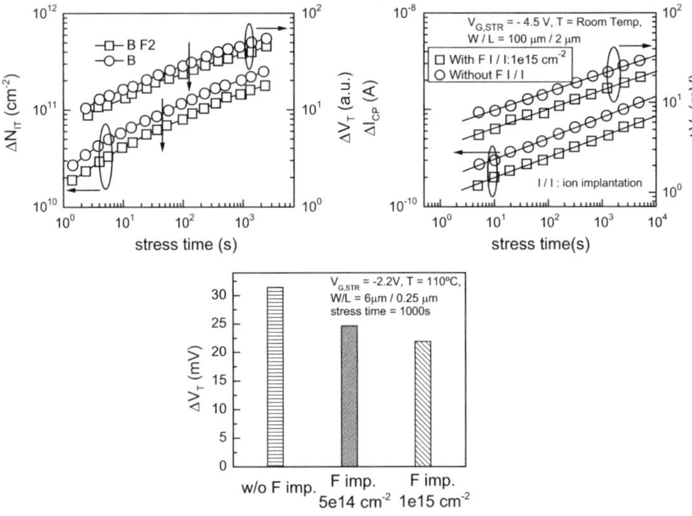

Fig. 19.7 Time evolution of ΔI_{CP} and ΔV_T during NBTI stress in p-MOSFETs fabricated without and with F_2 incorporated gate insulators. *Left panel* data from [12], *right panel* from [26]. *Bottom figure* shows ΔV_T for different F_2 dose, data from [20]

junctions to the gate oxide during subsequent junction activation anneal. In [26], F_2 is incorporated by using ion implantation after poly-Si deposition in the gate stack, while both methods have been used in [20]. Note that F_2 incorporation achieved by such diverse methods always reduces both ΔV_T and ΔN_{IT} as shown by different groups. Since slow measurements have been used, F_2-induced reduction in ΔV_T is largely due to reduction in ΔN_{IT}, and any possible impact of F_2 on ΔN_{HT} has not been captured. It has been suggested [26] that F_2 incorporation replaces some of the interfacial Si–H bonds with Si–F bonds that are harder to break during NBTI stress, and hence, the H_2 release is suppressed, which causes reduction in ΔN_{IT} and hence ΔV_T. The reduction in ΔV_T is enhanced by incorporation of higher F_2 species in the gate stack [20] and is also shown in Fig. 19.7.

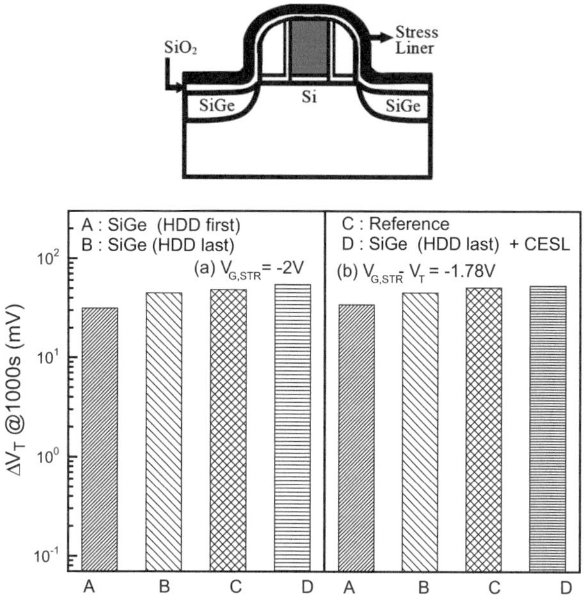

Fig. 19.8 Schematic showing different methods of applying compressive stress in p-MOSFETs. Measured ΔV_T during NBTI stress at constant gate bias and constant gate overdrive (see text), in SiON p-MOSFETs with different compressive stress. Data from [27]

19.2.4 Compressive Stress

Different compressive stress processes have been used to boost p-MOSFET channel hole mobility, and the impact of such process-induced stress on NBTI degradation has been studied [27, 28]. Figure 19.8 describes four different types of devices studied in [27]. The reference device has no stress. Compressive stress has been applied by Selective Epitaxial Growth (SEG) of SiGe source and drain regions, with SEG done either before or after the formation of Highly Doped Drain (HDD), respectively, resulting in HDD last and HDD first devices. For the HDD last device, a Compressive Etch Stop Liner (CESL) has also been added to create a mixed stress device. Figure 19.8 also shows NBTI stress-induced ΔV_T measured using conventional slow MSM method in devices having different process-induced stresses [27].

Note that measured ΔV_T is primarily due to ΔN_{IT} as slow measurements have been used. Compared to the reference device, HDD first device shows lower ΔV_T, HDD last device shows similar ΔV_T, and the mixed HDD last CESL device shows slightly higher ΔV_T, when different devices were subjected to NBTI stress at identical $V_{G,STR}$. However, when stressed at identical gate overdrive ($V_{G,STR} - V_{T0}$), as prestress threshold voltage (V_{T0}) has been found to be different among different devices, the mixed HDD last CESL device shows similar ΔV_T

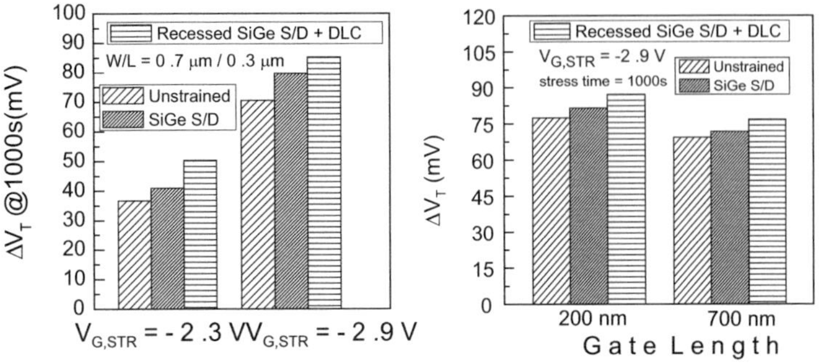

Fig. 19.9 NBTI stress-induced ΔV_T at constant stress gate bias, measured using ultrafast method in HKMG p-MOSFETs with different compressive stresses (*left panel*). Impact of channel length on ΔV_T during NBTI stress on devices without and with compressive stress (*right panel*). Data from [28]

compared to reference, while the HDD first device still shows slightly lower ΔV_T. It has been concluded that compressive stress up to a maximum of -1.5 GPa has no considerable effect on NBTI, and no effect of NBTI stress has been observed in measured channel length (L_{CH}) dependence or temperature (T) activation of degradation for both unstressed and compressively stressed devices [27].

In [28], the effect of compressive stress has been studied by using ultrafast measurements, which captures ΔV_T contributions due to both ΔN_{IT} and ΔN_{HT}. Devices were subjected to compressive stress by only SEG SiGe source-drain regions and mixed SEG and Diamond-Like Carbon (DLC) liner processes, the mixed process resulting in a compressive stress of >5 GPa. Figure 19.9 shows measured ΔV_T in different devices under identical $V_{G,STR}$.

Although not compared under identical gate overdrive as in [27], the SEG device shows almost similar ΔV_T, while the mixed SEG plus DLC device shows slightly higher ΔV_T compared to the reference device, once again suggesting negligible impact of compressive stress on NBTI. Both unstressed and stressed devices show increase in ΔV_T at smaller L_{CH} when NBTI stress is done at identical $V_{G,STR}$. This is presumably due to reduction in V_{T0} at smaller L_{CH} and hence increase in gate overdrive during NBTI stress. Note that negligible impact of compressive stress on NBTI degradation holds for both slow [27] and fast [28] measurements, which implies that process-induced compressive stress has no impact on either ΔN_{IT} or ΔN_{HT} components of NBTI degradation.

19.2.5 Post Metallization Anneal (PMA)

Figure 19.10 shows ΔV_T measured using slow MSM method in SiO_2 p-MOSFETs fabricated with PMA done in H_2 and D_2 ambient [12, 14]. Device with D_2 PMA shows lower degradation than the device having conventional H_2 PMA, and this

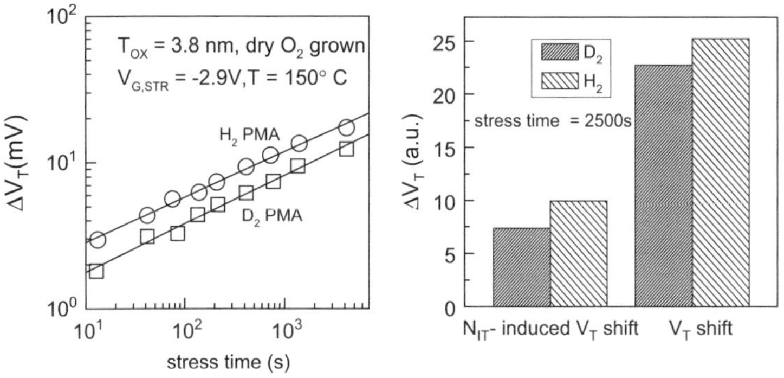

Fig. 19.10 Time evolution of ΔV_T during NBTI stress in SiO_2 p-MOSFETs with H_2 and D_2 PMA (*left panel*). Data from [14]. Measured ΔV_T and ΔN_{IT} at fixed stress time in SiO_2 p-MOSFETs with H_2 and D_2 PMA (*right panel*). Data from [12]

phenomenon is reported by different groups. Note that measured ΔV_T is due to ΔN_{IT} as slow measurements have been used. As discussed in [1], ΔN_{IT} is caused by de-passivation of Si–H (or Si–D, as the case may be) bonds at the Si/SiO_2 interface and subsequent diffusion of H_2 (or D_2) species. Hence, this process impact can be due to different strength of Si–H and Si–D bonds and/or different diffusivity of H_2 and D_2. The impact of H_2 and D_2 PMA on ΔN_{IT} measured directly using the CP method is also shown in Fig. 19.10 [12]. As expected, the device having D_2 PMA shows lower ΔN_{IT} compared to the conventional device with H_2 PMA and therefore explains the difference between H_2 versus D_2 PMA on measured NBTI degradation.

Note that the difference between H_2 and D_2 PMA devices as measured by ΔN_{IT} is slightly lower than that measured by ΔV_T, presumably due to larger recovery observed in CP measurements [13]. Although not widely adopted due to practical difficulty, D_2 PMA can indeed be used to reduce NBTI degradation.

19.2.6 Type of IMD, Barrier Metal, and Cap Layer

NBTI degradation is also influenced by different backend processes such as the type of IMD, barrier metals, and cap layers as described by using Fig. 19.11 [29]. All devices used in this study have identically grown SiON gate insulator. NBTI-induced degradation in saturated drain current (ΔI_{DSAT}) has been studied using conventional slow MSM method, and therefore measured NBTI is influenced by ΔN_{IT} alone.

Note that devices with Silicon Nitride (SiN) cap show higher degradation compared to devices with Silicon Carbide (SiC) cap, and this is irrespective of the type of IMD layer (SiO and SiLK). For SiN cap, SiO IMD shows higher NBTI

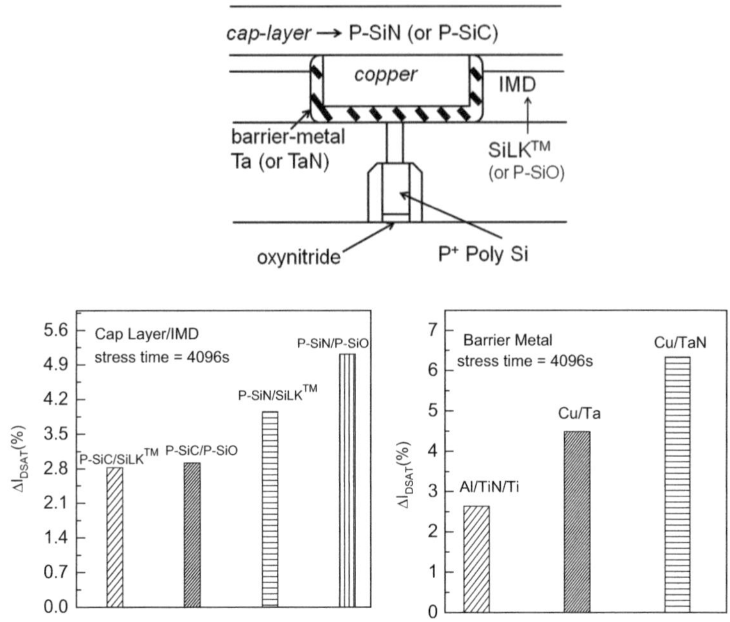

Fig. 19.11 Schematic of SiON p-MOSFET device cross section showing IMD, barrier metal, and cap layer. Measured NBTI degradation for different BEOL processes (see text). Data from [29]

compared to SiLK IMD, while the type of IMD has no effect for SiC-capped devices. Impact of these backend layers has been explained by measuring the water content in the gate oxide by Thermal Desorption Spectroscopy (TDS) method [30]. For SiN cap, higher water content in the gate stack has been found for SiO compared to SiLK IMD, while no water has been found for either type of IMD for the SiC-capped devices. As SiN cap layer prevents water to escape to upper layers during backend thermal anneal, water diffuses down to the gate oxide, and higher water content in SiO IMD results in higher water content in the gate oxide compared to SiLK IMD. Note that water is known to cause higher NBTI in wet compared to dry oxides [12, 14], and hence, devices with SiN cap and SiO IMD show highest NBTI. However, the SiC cap allows water to escape to upper layers during backend thermal anneal and therefore results in negligible water in the gate oxide, and therefore, the difference in water content of the IMD layer has no effect on NBTI as shown in Fig. 19.11.

Figure 19.11 also shows the effect of different barrier metals [29]. In this study, all devices have SiN cap and SiLK IMD and hence identical water content in the gate stack. Devices with Tantalum Nitride (TaN) barrier metal show highest degradation. As explained in [29], during SiN cap layer deposition, H_2 diffuses via the copper layer to the barrier metal, and the barrier metal acts as H_2 storage. During subsequent PMA, H_2 desorption takes place followed by H_2 diffusion into the gate

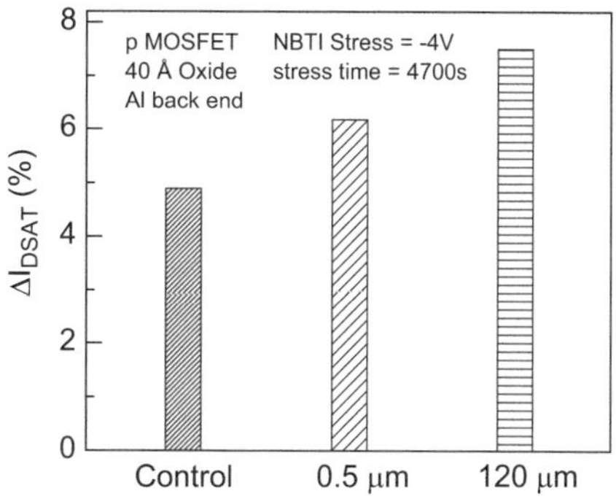

Fig. 19.12 Measured NBTI degradation in SiO_2 p-MOSFETs having different antenna perimeters. Data from [32]

dielectric, as negligible H_2 diffusion takes place via the SiN cap layer to upper layers due to lower diffusivity of H_2 in SiN. Secondary Ion Mass Spectroscopy (SIMS) measurements [31] have shown highest H_2 content for devices having TaN barrier, which corroborates with highest NBTI seen in these devices. It has also been shown [29] that devices that undergo PMA at higher T show larger NBTI due to higher H_2 desorption from the TaN barrier layer and enhanced diffusion into the gate oxide. Therefore, not only the type (H_2 or D_2) of PMA, the PMA process T also influences NBTI. As shown in Fig. 19.11, several backend processes influence NBTI and can be suitably modified to keep degradation under acceptable limits.

19.2.7 Antenna Charging Effect

NBTI degradation is also influenced by charging damage due to plasma-based processes and has been studied in both aluminum (Al) and copper (Cu) backend devices having different antenna perimeters [32]. Devices with larger antenna perimeter would undergo higher process-induced plasma charging damage and vice versa. Figure 19.12 shows measured ΔI_{DSAT} due to NBTI in reference as well as devices with small and large antenna structures. Higher degradation has been observed in the antenna devices, with devices having larger antenna showing higher degradation for both Al and Cu devices. Note that differences in ΔI_{DSAT} between different antenna structures is due to the differences in ΔN_{IT} as NBTI is measured in the conventional MSM mode by using slow measurements. It has been suggested that plasma charging damage results in higher density of Si-dangling bonds at the Si/SiO_2 interface, which gets passivated during H_2 PMA. Therefore compared to reference, antenna devices have higher density of Si–H bonds at the beginning of NBTI stress. As the reaction rate governing Si–H bond dissociation is proportional to density of Si–H bonds [1], antenna devices show higher ΔN_{IT} and hence higher degradation during NBTI stress.

19.3 Detailed Investigation on the Influence of Nitrogen

Although several front and backend processes impact NBTI degradation as summarized in the previous section, the incorporation of N_2 in the gate insulator plays a very important role and has naturally attracted much attention. The impact of N_2 on NBTI has also been reviewed in the previous section. However, as mentioned, all N_2 incorporation studies [12, 14–16] reviewed above have been done in the conventional MSM mode by using slower measurement methods to measure NBTI degradation. As also mentioned before, it is now well known that degradation occurred during NBTI stress substantially recovers after the removal of stress for measurement in conventional MSM mode [6, 7], which leads to several recovery-related artifacts[4] [10]. Moreover, slower measurement methods mostly capture the process influence on ΔN_{IT} and cannot capture that on ΔN_{HT}. To circumvent these issues, different fast [6] and ultrafast [7–9] measurement methods have been developed to capture recovery artifact-free NBTI. Therefore, it is important to revisit the impact of N_2 incorporation in the gate insulator on NBTI degradation using ultrafast measurements. Time evolution of degradation for different N_2-containing gate insulator processes will be discussed. This will be followed by a discussion on the impact of N_2 on NBTI parameters. Finally, gate insulator nitridation results from High-k Metal Gate (HKMG) devices will be discussed.

19.3.1 Time Evolution of Degradation

Figure 19.13 describes the Ultra-Fast On-The-Fly (UF-OTF) method [9] used to measure NBTI degradation from very short ($\sim\mu$s) to long stress time (t_{STR}) till the end of stress. The gate is connected to a Pulse Generator (PG), the source to an IV Converter (IVC) and Digital Storage Oscilloscope (DSO), and the drain to a switch, which in turn connects either to a DC Power Supply (DCPS) or to a Source-Measure Unit (SMU). A drain bias (V_D) of 100 mV is set in DCPS and SMU. The SMU and DSO are triggered first in the sampling mode respectively with 1 ms and 1 μs time intervals. The SMU then sends a hard trigger to the PG unit, which applies the gate pulse ($V_{G,STR}$) for the entire duration of stress. The drain remains connected to DCPS from the application of $V_{G,STR}$ to t-stress of 30 ms and is then connected

[4]NBTI recovery inherent in slower measurement techniques results in lower NBTI magnitude and higher power law time exponent (n) when degradation is plotted as a function of time in a log–log plot [8, 9]. As measured degradation is extrapolated to end of life to determine NBTI lifetime, slower measurements introduce significant error in estimated lifetime. The time exponent n obtained by slower measurements increases with temperature [33], which is also shown to be a recovery-related artifact [10], and results in erroneous conclusion regarding the physical mechanism responsible for NBTI degradation [33].

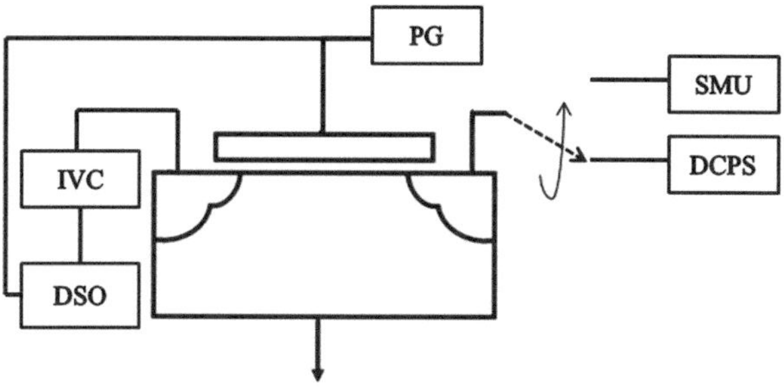

Fig. 19.13 Schematic of UF-OTF I_{DLIN} measurement setup used for characterization of NBTI degradation from 1 μs stress time. Refer to [9] for details

to SMU for the remaining duration of stress.[5] The source current is measured from the application of $V_{G,STR}$ to t-stress of 300 ms using the DSO, and the drain current from t-stress of 30 ms to the end of stress. The overlap t-stress period of 30–300 ms is used to calibrate the source and drain current.[6] Note that the gate always remains at $V_{G,STR}$ during the entire experimental duration, while the source and/or drain current are measured on the fly, and therefore, the method does not suffer from any recovery artifacts. Figure 19.14 shows the gate voltage (V_G) and the linear drain (=source) current (I_{DLIN}) captured at the initiation of stress. Note the rise in I_{DLIN} as $V_{G,STR}$ is applied, and the first I_{DLIN} data point after the initiation of $V_{G,STR}$ can be recorded within a minimum time-zero (t_0) delay of 1 μs. Figure 19.14 also shows I_{DLIN} degradation due to NBTI from 1 μs till the end of stress. Measurements are done on RPN-based Plasma Nitrided Oxide (PNO) and N_2O-based Rapid Thermal Nitrided Oxide (RTNO) SiON p-MOSFETs. RTNO device shows lower I_{DLIN} but higher degradation in I_{DLIN} due to NBTI stress when compared to PNO device.

NBTI degradation can be calculated using $\Delta V = (I_{DLIN}(t_0) - I_{DLIN}(t))/I_{DLIN}(t_0) * (V_{G,STR} - V_{T0})$, where $I_{DLIN}(t)$ is measured I_{DLIN} at different t-stress, $I_{DLIN}(t_0)$ is the first data point after the application of $V_{G,STR}$, and V_{T0} is prestress threshold voltage.[7] Figure 19.15 shows measured ΔV in PNO and RTNO devices as a function of t-stress for different t_0 delay, gate oxide field (E_{OX}), and stress T [9, 35].

[5]It has been observed [9] that the initial current transient measured by the DSO is affected due to RC time constant issues if SMU is connected to the drain. To avoid this issue, the DCPS is used at the drain for the early duration of stress, and the drain is later switched to SMU for long-time measurements.

[6]Caution should be applied while using UF-OTF method for ultrathin gate oxide devices with very high gate leakage, where source and drain currents can be significantly different, especially at higher $V_{G,STR}$.

[7]As described in [34], ΔV is directly proportional to ΔV_T, and correction due to mobility degradation is needed to convert ΔV to ΔV_T. While ΔV is a good indicator for a relative study of NBTI process dependence, it must be converted to ΔV_T before experimental results are compared to theory, as done in [1].

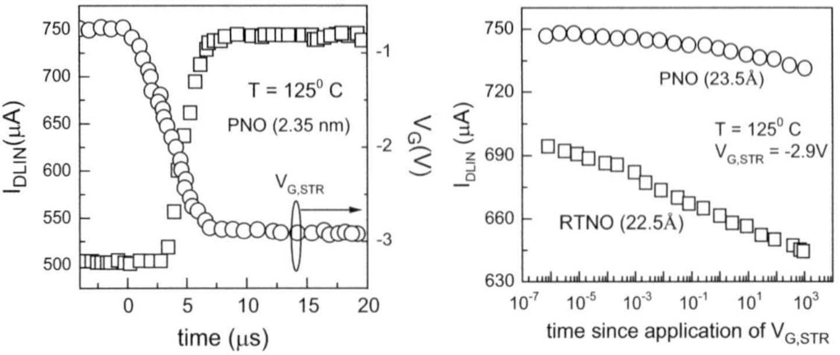

Fig. 19.14 Measured V_G and I_{DLIN} transients at the initiation of stress (*left panel*) and I_{DLIN} degradation from short to long stress time (*right panel*). Data taken from [9] for PNO and RTNO SiON p-MOSFETs

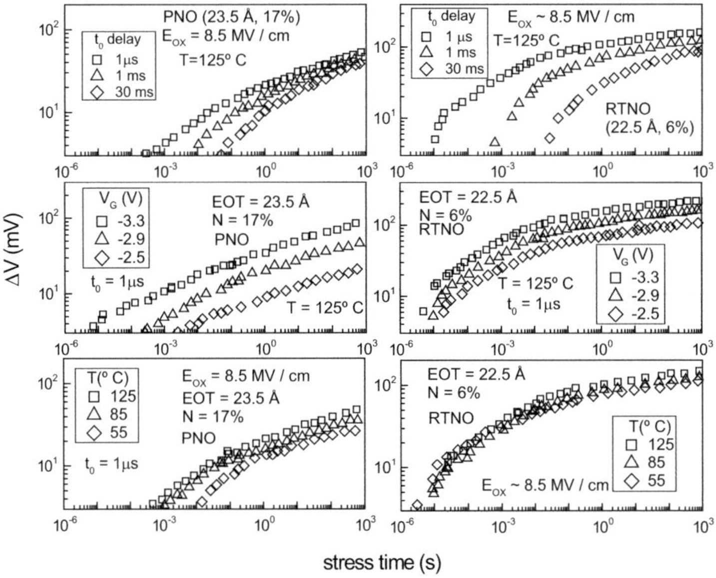

Fig. 19.15 Time evolution of NBTI degradation measured by using UF-OTF method with different t_0 delay (*top panel*) and for different stress V_G (*middle panel*) and T (*bottom panel*) measured using $t_0 = 1$ μs OTF. Data taken from [35] for PNO and RTNO SiON p-MOSFETs

Calculation of ΔV in Fig. 19.15a has been done using different t_0 delay values from Fig. 19.14 for the calculation of $I_{DLIN}(t_0)$, and $t_0 = 1$ μs has been used for ΔV calculation in Fig. 19.15b, c.

It can be seen that RTNO device shows much larger degradation compared to PNO device under identical t_0 delay, E_{OX}, and T. Significant degradation in the sub 1 ms time scale has been observed for the RTNO device, especially for $t_0 = 1$ μs, and the impact of t_0 delay is much larger for the RTNO device compared to PNO device. Compared to PNO device, RTNO device shows lower E_{OX} dependence (Γ_E) and

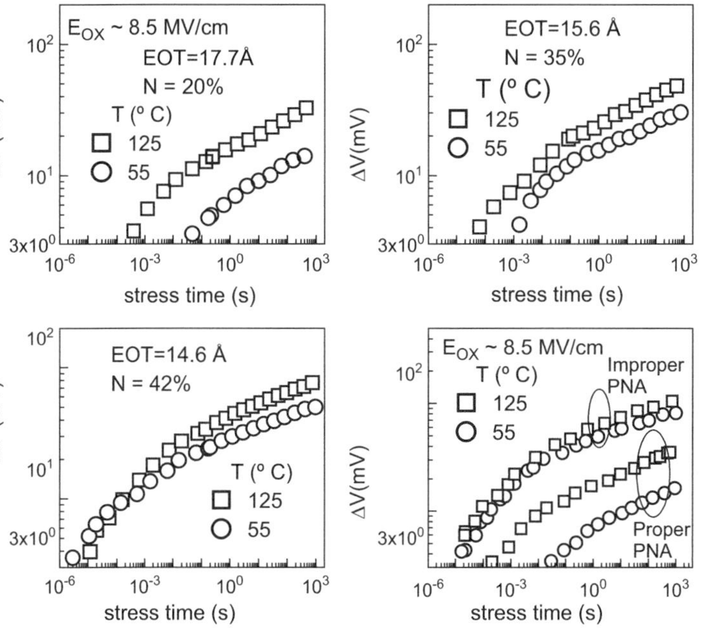

Fig. 19.16 Time evolution of NBTI degradation measured using $t_0 = 1$ μs OTF for different stress T. Data taken from [35, 36] for PNO SiON p-MOSFETs having different N_2 dose and proper PNA and also for a device having low N_2 dose but improper PNA

lower T activation energy (E_A) of degradation, and the sub 1 ms t-stress degradation for the RTNO device shows negligible T dependence. Higher NBTI degradation observed for RTNO compared to PNO device is consistent with other published results from conventional slow measurements [22] as discussed earlier in this chapter. It is also consistent with results obtained by using fast OTF method [13]. However, use of an ultrafast measurement method has uncovered important, but previously unappreciated NBTI features in the sub 1 ms time scale, and highlights several key differences in measured time evolution of NBTI degradation between RTNO and PNO devices.

Figure 19.16 shows time evolution of ΔV for different stress T, measured in PNO devices having different gate insulator N_2 dose and PNA [35, 36]. For PNO devices subjected to proper 2-step PNA [24], NBTI degradation increases and the T dependence reduces with increase in N_2 dose. Increase in NBTI degradation with increase in N_2 content of the gate stack is consistent with other published results discussed earlier in this chapter [12, 14–16]. The PNO device having very high N_2 dose shows significant NBTI degradation in the sub 1 ms time scale, and interestingly, this large sub 1 ms degradation also shows negligible T dependence, and these observations are exactly similar to RTNO device results shown previously in Fig. 19.15. In the absence of proper PNA, PNO device having lower N_2 dose shows higher NBTI than a PNO device with higher N_2 dose and proper PNA.

This is consistent with other published results [25] discussed earlier in this chapter, highlighting the importance of proper PNA in reducing NBTI degradation for PNO devices. Moreover, note that compared to the proper PNA device, the improper PNA device, in spite of having relatively lower N_2 dose, shows significant degradation in the sub 1 ms t-stress which has negligible T dependence, similar to RTNO, and very high N_2 dose PNO with proper PNA devices.

Figures 19.15 and 19.16 suggest that SiON devices can be broadly classified into two subcategories. PNO devices having relatively lower N_2 dose and proper PNA show negligible degradation in sub 1 ms time scale and lower overall degradation at longer t-stress, show higher T activation and E_{OX} acceleration for the entire stress duration, and are defined as Type-A devices. Note that Type-A devices have relatively lower N_2 density at the Si/SiON interface. On the other hand, RTNO, PNO with very high N_2 dose but proper PNA, and PNO of any N_2 dose without proper PNA are defined as Type-B devices that have higher N_2 density at the Si/SiON interface and show significant degradation in the sub 1 ms time scale and relatively larger overall degradation, relatively lower T activation and E_{OX} acceleration, and the high ΔV observed in sub 1 ms time scale always shows negligible T dependence. Although discussed in detail in [2], it is worth a mention that CP measurements show slightly higher ΔN_{IT} in Type-B compared to Type-A devices [5]. However, flicker noise measurements in prestress show much larger process-related SiON bulk traps for Type-B compared to Type-A devices [23]. Therefore, the difference in NBTI time dynamics between Type-A and Type-B devices is primarily due to fast hole trapping in process-related traps and results in large degradation in the sub 1 ms time scale which has negligible T dependence and also larger overall NBTI, shown in Figs. 19.15c and 19.16. As hole trapping is a fast process, Type-B devices show larger impact of measurement speed (t_0 delay) compared to Type-A devices, shown in Fig. 19.15a.

Figure 19.17 plots ΔV measured at a fixed t-stress as a function of E_{OX} for different PNO devices with proper PNA and different N_2 dose, RTNO, and RTNO + PNO devices [35]. The gate insulator for PNO devices show higher N_2 density close to the SiON/poly-Si interface and lower N_2 density at the Si/SiON interface, while RTNO device has high N_2 density at the Si/SiON interface. The starting base oxide thickness and the thermal and plasma N_2 dose of the RTNO + PNO device are adjusted to obtain N_2 density at the Si/SiON and SiON/poly-Si interface similar to that of RTNO and PNO-B device, respectively. For PNO devices, measured ΔV increases and Γ_E reduces with increase in N_2 dose from PNO-A through PNO-C. However, the RTNO + PNO device shows similar ΔV and Γ_E as the RTNO device, and much larger ΔV and lower Γ_E compared to the PNO-B device, once again suggesting NBTI being governed by N_2 density at the Si/SiON interface [19, 20]. Note, the PNO-C device has very high integrated N_2 content in the gate insulator, but still shows lower NBTI compared to RTNO and RTNO + PNO devices, which indicates that NBTI is not dependent on total N_2 content of the gate stack. Figure 19.17 suggests that NBTI can be reduced by reducing the N_2 density at the Si/SiON interface, irrespective of the total N_2 content in the gate stack. A proper 2-step PNA not only anneals any plasma damage caused during the RPN

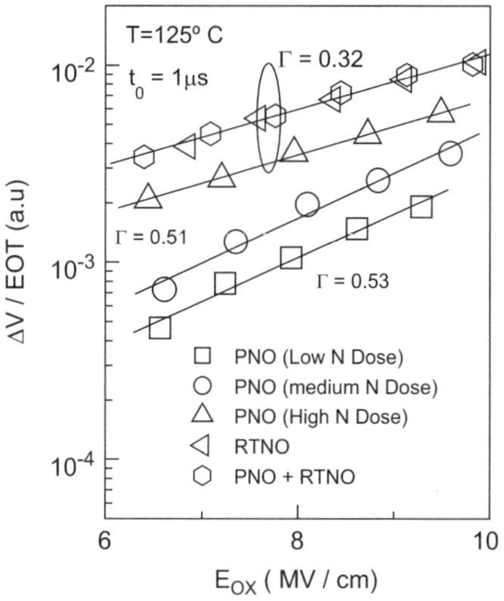

Fig. 19.17 NBTI degradation measured using $t_0 = 1$ μs OTF at fixed stress time, as a function of stress E_{OX}. Data taken from [35] for SiON p-MOSFETs having PNO, RTNO, and mixed PNO + RTNO gate insulators

step but also slightly re-oxidizes the Si/SiON interface and pushes N_2 away from the interface [24]. Therefore, in spite of having relatively higher N_2 dose, PNO with proper PNA device shows lower NBTI degradation compared to improper PNA device having lower N_2 dose, as shown in Fig. 19.16.

19.3.2 NBTI Parameters (*n*, E_A, Γ_E)

As shown in the previous section, time evolution of NBTI degradation is usually plotted in a log–log scale. Although definitely not true for the entire stress duration, measured degradation at longer stress duration (t-stress $\geq$ 10 s) can be fitted by using a power law time dependence having a time exponent *n*. The process dependence of the parameter *n* is of practical interest, as degradation plotted in a log–log scale is extrapolated to end of life for the determination of device lifetime.

Figure 19.18 shows extracted *n* from measured NBTI degradation in different PNO and RTNO devices, as a function of t_0 delay, stress E_{OX}, and T [35]. PNO devices have different EOT due to different starting base oxide thickness; have relatively lower, although different, N_2 content in the gate stack; and were subjected to proper PNA. For a particular stress (E_{OX} and T) and measurement (t_0 delay) condition, all PNO devices show similar values of *n*, which are much higher than that obtained for the RTNO device. Extracted n values reduce with reduction in t_0 delay; however, the variation of n with t_0 delay is within the error bar caused by error in $I_{DLIN}(t_0)$ measurement for $t_0 \leq 10$ μs. Therefore, *n* extracted from $t_0 = 1$ μs measurement can be used as a reliable parameter to compare process impact of NBTI across different devices.

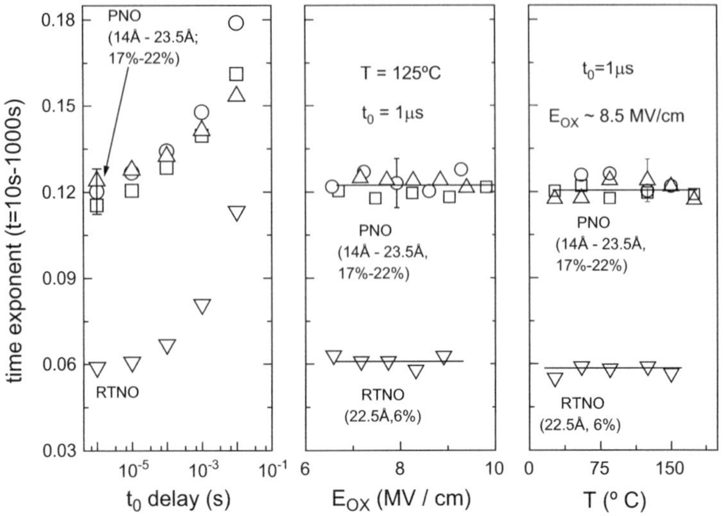

Fig. 19.18 Power law time exponent (n) extracted from long-time NBTI measured using $t_0 = 1$ μs OTF and plotted as function of t_0 delay, stress E_{OX}, and T. Data taken from [35] for PNO and RTNO SiON p-MOSFETs

For $t_0 = 1$ μs, extracted n values remain invariant across variations in T and E_{OX} for all PNO and RTNO devices. As mentioned earlier in this chapter, n increases with increase in T due to recovery-related artifacts when NBTI is measured in the MSM mode with slow measurement methods [10]. The invariance of n with T is therefore an indirect proof of the measurement technique being free of recovery-related issues. Moreover as also discussed before, since NBTI stress is also similar to TDDB stress [3], there remains a finite probability of the presence of ΔN_{OT}, which would result in an increase in n, especially at longer t-stress and at higher $V_{G,STR}$ [4, 37]. The invariance of n with E_{OX} (hence $V_{G,STR}$) therefore suggests negligible impact of ΔN_{OT} for the chosen stress conditions (E_{OX} or $V_{G,STR}$, T, and t-stress) shown in Fig. 19.18. Note that choice of proper stress conditions such that bulk trap generation can be minimized and a suitable measurement method that is free from recovery-related artifacts are the two most important criteria for reliable NBTI experiments and can be respectively verified by E_{OX} (or $V_{G,STR}$) and T independence of the extracted time exponent n.

Figure 19.19 shows extracted n as a function of stress E_{OX} from $t_0 = 1$ μs measurements for PNO with proper PNA devices having different N_2 dose, and PNO devices with different PNA conditions [35, 36]. Note that n remains invariant of stress E_{OX} for all devices and suggests negligible bulk trap generation. Extracted n values reduce with increase in N_2 dose for proper PNA devices, and lower n values are also observed for PNO devices with improper PNA.

Figures 19.18 and 19.19 suggest a reciprocal relationship between measured ΔV and extracted long-time n for different processes. Type-A devices show lower ΔV

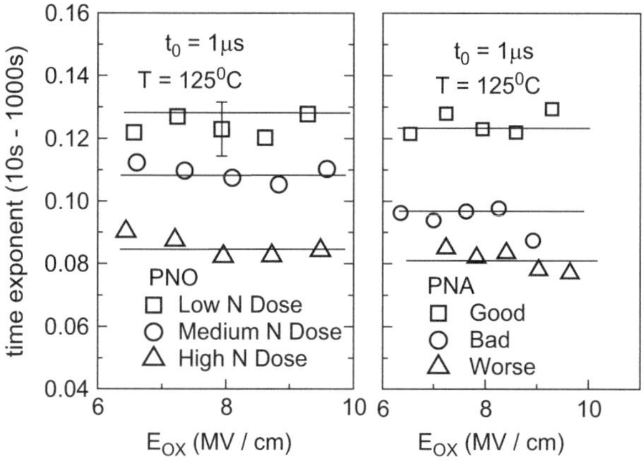

Fig. 19.19 Power law time exponent (n) extracted from long-time NBTI measured using $t_0 = 1$ μs OTF and plotted as function of stress E_{OX}. Data taken from [35] for PNO SiON p-MOSFETs having different N_2 dose and different PNA

and higher n, while Type-B devices show higher ΔV and lower n. As explained in [1, 5, 13] and discussed in [2], Type-B devices show larger ΔN_{HT} contribution that is fast and saturates at longer t-stress. As ΔV is due to both ΔN_{IT} and ΔN_{HT}, saturation of ΔN_{HT} at longer t-stress reduces n of overall ΔV. As shown in the previous section, there is also an inverse correlation of measured ΔV and T activation as gate insulator processes are varied. Type-B devices show higher ΔV and lower T activation of ΔV compared to Type-A devices. Note that this is consistent with the fact that saturated hole trapping shows negligible T dependence, and larger ΔN_{HT} in Type-B devices lowers the T activation of overall NBTI when measured at longer t-stress. Note that the T activation of NBTI can be obtained by measuring T dependence of ΔV at a fixed t-stress. As the time exponent n remains invariant of T when extracted at longer t-stress, the T dependence can be assumed to be Arrhenius activated [10], and the T activation energy E_A can be obtained. Finally, the E_{OX} acceleration factor Γ_E can be obtained by measuring E_{OX} dependence of ΔV at a fixed t-stress. As shown in Fig. 19.17, there is also an inverse correlation of ΔV and Γ_E, with devices having higher ΔV show lower Γ_E and vice versa. More work is needed to understand the physical mechanism responsible for the impact of Si/SiON N_2 density on Γ_E.

Figure 19.20 shows extracted NBTI parameters (n, E_A and Γ_E) from $t_0 = 1$ μs measurements versus atomic N content (N%) in the gate stack for a wide range of Type-A and Type-B SiON processes [35]. Note that all parameters show very similar N% dependence with variation in SiON processes. PNO with proper PNA devices show similar n, E_A, and Γ_E for N∼30% (Type-A), and their values reduce for higher N% (Type-B). The control SiO_2 device suffers from boron penetration and shows a slightly lower n and E_A (and higher ΔV, not shown) compared to Type-A devices. Type-B devices such as PNO with improper PNA, RTNO, and mixed RTNO + PNO show lower n, E_A, and Γ_E for a given N% when compared to the PNO with proper PNA device trend.

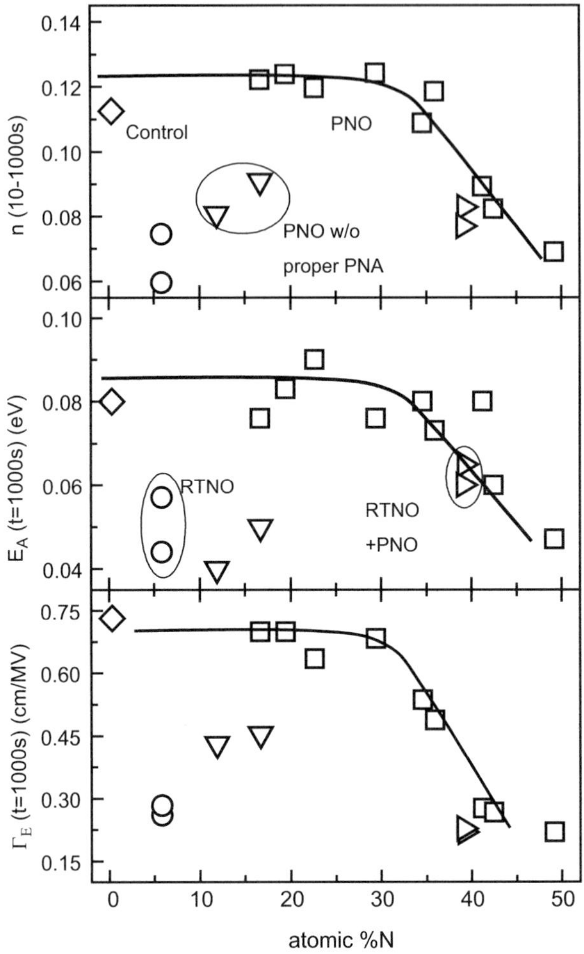

Fig. 19.20 Time exponent (n), T activation (E_A), and E_{OX} acceleration (Γ_E) extracted from long-time NBTI measured using $t_0 = 1$ μs OTF and plotted as function of atomic N% of the gate stack calculated using XPS. Data taken from [35] for SiO_2, PNO with and without proper PNA, RTNO, and mixed PNO + RTNO SiON p-MOSFETs

As discussed earlier in this chapter, NBTI is governed by the N_2 density at the Si/SiON interface and not by the overall N_2 content in the gate insulator. For PNO with proper PNA devices, although there is a direct correlation between the two, N_2 density at the Si/SiON interface remains low, due to proper PNA, up to a total N_2 content (N∼30% atomic) and increases beyond that, which results in increase in ΔV and reduction in n, E_A, and Γ_E. Irrespective of total N_2 content, all Type-B devices have higher N_2 density at the Si/SiON interface and hence show higher ΔV and lower n, E_A, and Γ_E compared to Type-A devices.

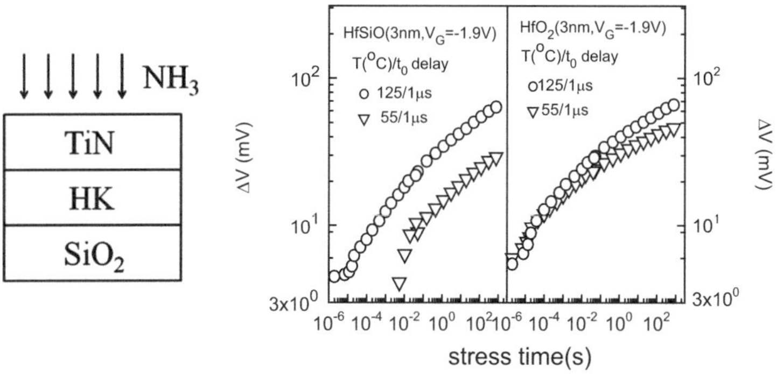

Fig. 19.21 Schematic description of post-HK nitridation in different HKMG stacks and time evolution of NBTI for different stress T, measured using $t_0 = 1$ μs OTF in HfSiO and HfO_2 stacks. Data from [38]

19.3.3 High-k Metal Gate (HKMG) Device Results

The impact of N_2 incorporation is studied in HKMG p-MOSFETs having SiO_2 interlayer (IL) and either hafnium silicate (HfSiO) or hafnium dioxide (HfO_2) High-k (HK) as dual-layer gate insulator stacks as shown in Fig. 19.21 [38]. The gate stacks for both types of HK materials have been fabricated with 1 nm thick IL and either 2 nm or 3 nm thick HK layers. These stacks have titanium nitride (TiN) metal gate (MG), which is followed by poly-Si deposition, and then subjected to Ammonia (NH_3) anneal, which introduces N_2 in the gate stack. Note that the presence of Si in the HfSiO HK layer helps the formation of Si–N bonds during NH_3 anneal, which in turn reduces N_2 diffusion into the IL [39, 40]. However, this is not the case for the HfO_2 HK, and NH_3 anneal results in large N_2 diffusion into the SiO_2 IL layer for these stacks.

Therefore, for identical post poly-Si deposition NH_3 anneal, N_2 content in the IL would be larger for HfO_2 stacks and these devices are expected to behave similar to Type-B SiON devices described in the previous section. On the other hand, lower N_2 penetration would occur in the IL for HfSiO stacks, and therefore, these devices would behave as Type-A SiON devices.

Figure 19.21 shows time evolution of ΔV obtained using $t_0 = 1$ μs measurements in 3 nm HfSiO and HfO_2 gate stacks for different stress T [38]. The HfSiO device shows clear T dependence of NBTI from short to long t-stress, while the HfO_2 device shows negligible T dependence in sub 1 ms t-stress and relatively weak T activation at longer t-stress.

Note that negligible T dependence of NBTI in sub 1 ms t-stress is consistent with higher N_2 in the IL of HfO_2-based stacks, and this feature has been also observed in RTNO SiON devices having higher N_2 density close to the Si/SiON interface. Therefore, N_2 incorporation in the gate stack impacts the time evolution of NBTI quite similarly for SiON and HKMG stacks.

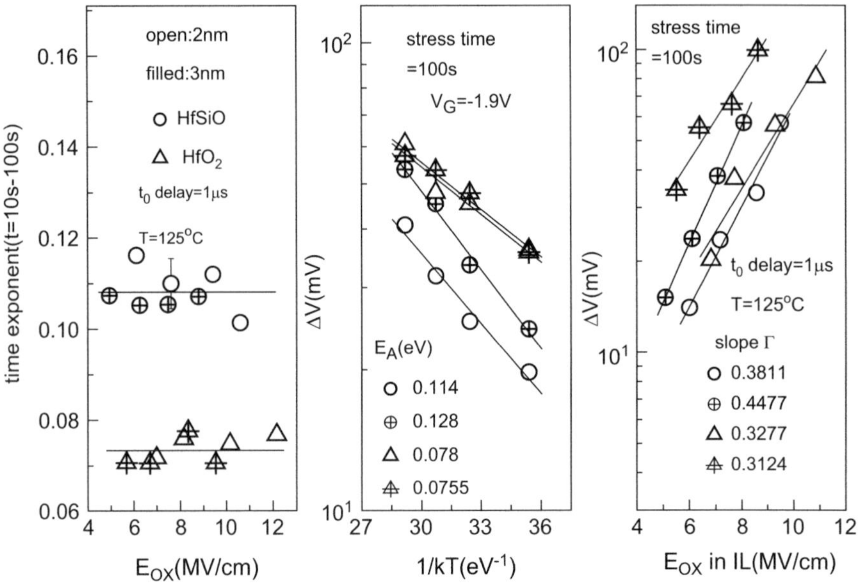

Fig. 19.22 Power law time exponent (n) as a function of E_{OX} and NBTI degradation versus stress T and E_{OX}, measured using $t_0 = 1$ μs OTF. Data from [38] for different HKMG p-MOSFETs having HfSiO- and HfO_2-based gate stacks

Figure 19.22 shows long-time power law time exponent n versus stress T and ΔV as a function of stress T and E_{OX}, obtained using $t_0 = 1$ μs measurements for different HfSiO and HfO_2 devices [38]. Note that HfSiO devices show lower ΔV and higher n, E_A, and Γ_E compared to HfO_2 devices, and such differences are similar to that observed between PNO and RTNO SiON devices. These results are once again consistent with higher N_2 density in the IL of HfO_2 compared to HfSiO stacks.

19.4 Summary

NBTI degradation is affected by several FEOL and BEOL process steps such as dry or wet oxidation, H_2 or D_2 PMA, gate oxide nitridation, incorporation of F_2, type of cap layer, IMD and barrier metals, compressive stress, type of source-drain dopant atoms, plasma charging, and gate antenna area. Due to its technological relevance owing to EOT scaling, gate oxide nitridation has attracted most attention. SiON devices show higher NBTI than SiO_2 devices, and in general, magnitude of NBTI increases with increase in N_2 content in the gate stack. However, it is now well established that rather than the total integrated N_2 content, N_2 density at or near the Si/SiON interface impacts NBTI. A comprehensive study by ultrafast measurements shows two broad types of SiON devices. Processes resulting in lower Si/SiON interfacial N_2 density result in lower magnitude of NBTI, but larger time exponent n, T activation $E_{A,}$ and E_{OX} acceleration Γ_E, and are classified as Type-A devices. Processes leading to higher Si/SiON interfacial N_2 density, for Type-B devices, result in higher NBTI, but lower n, $E_{A,}$ and Γ_E. These features are consistently observed in a wide range of SiON and HKMG devices. Interestingly, while not observed for Type-A devices, Type-B devices show substantial degradation in the sub 1 ms time scale when measured using ultrafast methods, and this additional degradation shows negligible T dependence. This and other process-dependent features are important signatures of the underlying NBTI physical mechanism and are discussed in [2].

The author would like to acknowledge contribution by Subhadeep Mukhopadhyay and Nilesh Goel towards the preparation of this chapter. Special thanks to E. Naresh Kumar, D. Varghese, V. D. Maheta, S. Deora, C. Olsen, and K. Ahmed for useful discussions and Applied Materials for supporting the gate insulator process dependent study.

References

1. S. Mahapatra, N. Goel, S. Desai, S. Gupta, B. Jose, S. Mukhopadhyay, K. Joshi, A. Jain, A. E. Islam and M. A. Alam, IEEE Trans. Electron Devices, **60**, 901 (2013)
2. S. Mahapatra, A comprehensive modeling framework for DC and AC NBTI, in *Bias Temperature Instability for Devices and Circuits*, ed. by T. Grasser (Springer, Heidelberg, 2013).
3. M. A. Alam, Jeff Bude, and A. Ghetti, Proc. Int. Rel. Phys. Symp., 21 (2000).
4. S. Mahapatra and M. A. Alam, Proc. Int. Electron Dev. Meet., 505 (2002)
5. S. Mahapatra, A. Islam, S. Deora, V. Maheta, K. Joshi, A. Jain, and M. Alam, Proc. Int. Rel. Phys. Symp., 6A.3.1 (2011)
6. S. Rangan, N. Mielke, and E. C. C. Yeh, Proc. Int. Electron Dev. Meet., 341 (2003)
7. H. .Reisinger, O. Blank, W. Heinrigs, A. Muhlhoff, W. Gustin, and C. Schlunder, Proc. Int. Rel. Phys. Symp.,448 (2006)
8. C. Shen, M. F. Li, C. E. Foo, T. Yang, D. M. Huang, A. Yap, G. S. Samudra, and Y. C. Yeo, Proc. Int. Electron Dev. Meet., 12.5.1(2006)
9. E. N. Kumar, V. D. Maheta, S. Purawat, A. E. Islam, C. Olsen, K. Ahmed, M. Alam and S. Mahapatra, Proc. Int. Electron Dev. Meet., 809 (2007)
10. D. Varghese, D. Saha, S. Mahapatra, K. Ahmed, F. Nouri, and M. A. Alam, Proc. Int. Electron Dev. Meet., 684 (2005)
11. G. Groeseneken, H. E. Maes, N. Beltran,R. F. De Keersmaecker,IEEE Trans. Electron Devices, **31**, 42(1984)
12. V. Huard, M. Denais, F. Perrier, N. Revil, C. Parthasarathy, A. Bravaix, and E. Vincent, Microelectron. Reliab., **45**, 83 (2005)
13. S. Mahapatra, K. Ahmed, D. Varghese, A. E. Islam, G. Gupta, L. Madhav, D. Saha, M. A. Alam, Proc. Int. Rel. Phys. Symp., 1(2007)
14. N. Kimizuka, K. Yamaguchi, K. Iniai, T. Iizuka, C. T. Liu, R. C. Keller, and T. Horiuchi, VLSI Tech. Symp., 92 (2000)
15. Y. Mitani, M. Nagamine, H. Satake and A. Toriumi, Proc. Int. Electron Dev. Meet., 509 (2002)
16. S. S. Tan, T. P. Chen, C. H. Ang, and L. Chan, IEEE Electron Dev. Lett., **25**, 504 (2004)
17. S. S. Tan, T. P. Chen, J. M. Soon, K. P. Loh, C. H. Ang, and L. Chan, Appl. Phys. Lett., **82**, 1881 (2003)

18. C. H. Liu, M. T. Lee, Chih-Yung Lin, J. Chen, K. Schruefer, J. Brighten, N. Rovedo, T. B. Hook, M. V. Khare, Shih-Fen Huang, C. Wann, Tze-Chiang Chen, T. H. Ning, Intl. Electron Dev. Meet., 39.2.1 (2001)
19. T. Sasaki, K. Kuwazawa, K. Tanaka, J. Kato, Dim-Lee Kwong, IEEE Electron Dev. Lett., **24**, 150 (2003)
20. M. Terai, K. Watanabe, and S. Fujieda, IEEE Trans. Electron Devices, **54**, 1658 (2007)
21. J. Hollander and W. L. Jolly, Acc. Chem. Res., **3**, 193 (1970)
22. Y. Mitani, H. Satake, A. Toriumi, IEEE Trans. Device Mater. Rel., **8**, 6 (2008)
23. G. Kapila, N. Goyal, V. D. Maheta, C. Olsen, K. Ahmed, and S. Mahapatra, Proc. Int. Electron Dev. Meet., 1 (2008)
24. C. Olsen, U.S. Patent 017 596 1A1 (2004).
25. K. Sakuma, D. Matsushita, K. Muraoka, and Y. Mitani, Proc. Int. Rel. Phys. Symp., 454 (2006)
26. Y. Mitani, T. Yamaguchi, H.Satake and A. Toriumi, Proc. Int. Rel..Phys. Symp., 226 (2007)
27. A. Shickova, B. Kaczer, P. Verheyen, G. Eneman, E. S. Andres, M. Jurczak, P. Absil, H. Maes, G. Groeseneken, IEEE Electron Dev. Lett., **28**, 242 (2007)
28. Bin Liu, Kian-Ming Tan, Ming-Chu Yang, and Yee-Chia Yeo, Proc. Int. Rel. Phys. Symp., 977 (2009)
29. A. Suzuki, K. Tabuchi, H. Kimura, T. Hasegawa, S. Kadomura, VLSI Tech. Symp., 216 (2002)
30. A. M. de Jong and J. W. Niemantsverdriet, Surface Sc., 233, **355** (1990)
31. A. Benninghoven, Surface Sc., **35**, 427 (1973)
32. A. T. Krishnan, V. Reddy, and S. Krishnan, Proc. Int. Electron Dev. Meet., 39.3.1 (2001)
33. B. Kaczer, V. Arkhipov, R. Degraeve, N. Collaert, G. Groeseneken and M.Goodwin, Proc. Int. Rel. Phys. Symp., 381 (2005)
34. A. E. Islam, V. D. Maheta, H. Das, S. Mahapatra and M. A. Alam, Proc. Int. Rel. Phys. Symp., 87 (2008)
35. V. Maheta, C. Olsen, K. Ahmed, and S. Mahapatra, IEEE Trans. Electron Devices, **55**, 1630 (2008)
36. V. D. Maheta, C. Olsen, K. Ahmed and S. Mahapatra, Proc. Int. Symp. On the Physical & Failure Analysis of Integrated Circuits, 1 (2008)
37. C. L. Chen, Y. M. Lin, C. J. Wang, and K. Wu, Proc. Int. Rel. Phys. Symp., 704 (2005)
38. S. Deora, V. D. Maheta, G. Bersuker, C. Olsen, K. Z. Ahmed, R. Jammy, S. Mahapatra, IEEE Electron Dev. Lett., **30**, 152 (2009)
39. N. Ikarashi, K. Watanabe, K. Masuzaki, T. Nakagawa, and M. Miyamura, J. Appl. Phys. **100**, 063507 (2006)
40. T. J. Park, J. H. Kim, J. H. Jang, K. D. Na, C. S. Hwang, and J. H. Yoo, J. Appl. Phys. **104**, 054101 (2008)

Chapter 20
Negative Bias Temperature Instability in Thick Gate Oxides for Power MOS Transistors

Ninoslav Stojadinović, Ivica Manić, Danijel Danković, Snežana Djorić-Veljković, Vojkan Davidović, Aneta Prijić, Snežana Golubović, and Zoran Prijić

Abstract Vast majority of recent extensive investigations of Negative Bias Temperature Instability (NBT) have been focused to the related phenomena in ultrathin gate dielectric layers of SiO2, SiON, and high-k materials. However, even though the gate oxides in nanometer scale technologies have been continuously thinned down, the interest in thick oxides has not ceased owing to widespread use of MOS technologies for the realization of power devices. Power MOSFETs are widely used as fast switching devices in home appliances and automotive, industrial, and military electronics. In a number of applications, these devices are routinely operated in the harsh environment and at high current and voltage levels, which lead to self-heating and/or increased fields, and thus favor NBTI. Accordingly, NBTI could be critical for reliable operation of power MOSFETs even though they have ultra-thick gate oxides. Our research over the past few years has been focused to degradation mechanisms in p-channel power Vertical Double-Diffused MOSFETs (VDMOSFETs) subjected to NBT stressing, including effects found during the post-stress annealing under the low gate bias and during the sequence of several NBT stress and low gate bias annealing steps. NBTI in n-channel power VDMOSFETs has been investigated as well. This chapter is aimed at revealing the main features of NBTI in thick gate oxides for power MOSFETs and reviews the work mentioned above with suitable reference to other published work. Peculiarities associated with

N. Stojadinović (✉) • I. Manić • D. Danković • V. Davidović • A. Prijić • S. Golubović • Z. Prijić
Faculty of Electronic Engineering, University of Niš, Aleksandra Medvedeva 14, 18000 Niš, Serbia
e-mail: Ninoslav.Stojadinovic@elfak.ni.ac.rs; ivica.manic@elfak.ni.ac.rs; danijel.dankovic@elfak.ni.ac.rs; vojkan.davidovic@elfak.ni.ac.rs; aneta.prijic@elfak.ni.ac.rs; snezana.golubovic@elfak.ni.ac.rs; zoran.prijic@elfak.ni.ac.rs

S. Djorić-Veljković
Faculty of Civil Engineering and Architecture, University of Niš, Aleksandra Medvedeva 14, 18000 Niš, Serbia
e-mail: snezana@gaf.ni.ac.rs

T. Grasser (ed.), *Bias Temperature Instability for Devices and Circuits*, 533
DOI 10.1007/978-1-4614-7909-3_20, © Springer Science+Business Media New York 2014

NBTI in thick oxides, such as the unusual post-stress generation of interface traps and rarely observed remarkable instability in n-channel devices are particularly addressed.

20.1 Introduction

The negative bias temperature instability (NBTI), which is commonly observed as threshold voltage shift (ΔV_T) in p-channel MOS transistors operated at elevated temperatures under increased gate oxide electric fields, has become one of the most critical degradation mechanisms in state-of-the-art CMOS technologies [1–6]. The phenomenon is related to the stress-induced generation of oxide-trapped charge and interface traps and has originally been noticed almost 50 years ago [7], but was not considered of great importance until recently because of the low electric fields used. However, the reliability issues associated with NBTI resurfaced in the past 15 years due to the convergence of several factors resulting from device scaling. These include the increase of operating temperature and gate oxide fields (gate oxides have been thinned below 2 nm without proportional scaling of supply voltages), the addition of nitrogen into the gate oxide to prevent boron penetration and reduce gate leakage, but at the expense of enhanced NBTI [8], and the potential replacement of the SiO_2 with high-k dielectrics, which allow for further reduction of equivalent oxide thickness, but are also susceptible to NBTI [9, 10].

Vast majority of recent extensive investigations of NBTI have, therefore, been focused to the related phenomena in ultrathin gate dielectric layers of SiO_2, SiON, and high-k materials [1–6, 8–10], and only few research groups seem to have addressed the NBTI in thick gate oxides [11–19]. However, though the gate oxides in nanometer scale technologies have been continuously thinned down, the interest in thick oxides has not ceased owing to widespread use of MOS technologies for the realization of power devices. Power MOSFETs are widely used as fast switching devices in home appliances and automotive, industrial, and military electronics. In a number of applications, these devices are routinely operated in the harsh environment and at high current and voltage levels, which lead to self-heating and/or increased fields, and thus favor NBTI. Accordingly, NBTI could be critical for reliable operation of power MOSFETs even though they have ultra-thick gate oxides. Our research over the past few years has been focused to degradation mechanisms and lifetime estimation in p-channel power Vertical Double-Diffused MOSFETs (VDMOSFET) [20, 21] subjected to NBT stressing, including effects found during the post-stress annealing under the low gate bias and during the sequence of several NBT stress and low gate bias annealing steps [22–25]. Also, threshold voltage instabilities under the pulsed NBT stress conditions have been analyzed and compared with static stress in terms of the effects on device lifetime [26, 27], and NBTI in n-channel power VDMOSFETs has been investigated as well [28]. This chapter, aimed at revealing the main features of NBTI in thick gate oxides for power MOSFETs, will review the work mentioned above with suitable reference to other published work. Peculiarities associated with NBTI in thick oxides, such as the unusual post-stress generation of interface traps and rarely observed remarkable instability in n-channel devices [25, 26, 28], will be particularly addressed.

The devices used in our studies were the commercial p-channel and n-channel power VDMOSFETs, designated as IRF9520 and IRF510, respectively, both built in standard Si-gate technology with assumed gate oxide thickness of 100 nm. The p- and n-channel devices had similar current/voltage ratings (6.8 A/100 V, 5.6 A/100 V, respectively) and approximately equal absolute values of threshold voltage before stressing (~3 V). The sets of p- and n-channel devices were stressed by specified negative gate voltage (−30, −35, −40, or −45 V) up to 2,000 h at 125, 150, and 175°C. Few devices were subjected to corresponding positive bias temperature (PBT) stress for comparison. The post-stress annealing under the low gate bias (−10 V, 0, +10 V) was performed at the same temperatures as during the stress. Electrical characterization of devices under test (DUT) during both stressing and annealing was performed by intermittent measurements of their transfer *I–V* characteristics as well as by charge pumping (CP) current measurements (triangular pulses, $f = 100$ kHz, $\Delta V_G = 2.6$ V, DTC = 50%). A traditional measure–stress–measure approach, where the stress (or bias anneal) voltage is removed from DUT to perform the measurement and is reapplied once the measurement has been done, was employed. The conventional equipment used was slow to avoid relaxation effects and could not capture most of the "fast" or "recoverable" component of NBTI, so the data to be shown here practically represent only the "slow" NBTI component, which is more or less permanent [15, 29–31]. The fast and on-the-fly NBTI measurement techniques [32–37], which have recently been developed for deep submicron MOS devices with ultrathin gate oxides where the magnitude of the stress voltage is comparable to the normal operating gate voltage, are not applicable in the case of VDMOSFETs. Namely, VDMOSFETs have much thicker gate oxide, which means the stress voltages required for accelerated NBTI investigations in these devices must be several times higher than their typical operating voltage, so the separate circuits for providing the stress voltage and performing the measurements are needed. We have recently developed a cost-effective stress and measurement setup suitable for NBTI investigations in power VDMOSFETs, which was shown to enable accurate interim measurements within the time window short enough to mitigate the relaxation effects and capture the faster part of the slow NBTI component [38], but have not yet managed to collect enough data with this setup and will therefore rely in this review on earlier results from previously used slow measurement approach. The above threshold region of the measured transfer *I–V* characteristics was used to extract device threshold voltage and calculate its changes during the stressing and annealing [39], while the subthreshold region of these characteristics was used to determine the corresponding changes in the densities of oxide-trapped charge and interface traps by means of subthreshold midgap (SMG) technique [40]. The measured CP characteristics were additionally used for independent calculations of stress-induced ΔN_{it} [41, 42].

20.2 NBT Stress Effects in p-Channel Devices

The threshold voltage shifts observed in p-channel VDMOSFETs during the NBT stress under various conditions are shown in Fig. 20.1. As it could be expected, more significant threshold voltage shifts were obtained at higher stress voltages and/or temperatures. Data analysis has shown that ΔV_T time dependencies follow the t^n power low, but with three different phases (as indicated by the dashed lines), which can be clearly distinguished depending on the value of parameter n. In the first (early) phase of stressing, n strongly depends on both bias and temperature, varying from 0.4 to 1.14. In the second phase, parameter n is almost independent on bias and temperature and equals approximately 0.25, as reported in other NBTI studies done with similar (slow) measurement approach on devices manufactured in various technologies [1, 2, 11]. Suitable mathematical analysis of the data shown in Fig. 20.1 leads to the following dependence of stress-induced ΔV_T on electric field, stress time, and temperature in the second stress phase [22]:

$$\Delta V_T = 3.04E^{2.05}t^{0.25}\exp(-0.24/kT). \tag{20.1}$$

Finally, at long stress times (third phase), parameter n becomes bias and temperature dependent again, gradually decreasing from 0.25 to 0.14, while ΔV_T tends to saturate. The relative values of ΔV_T in saturation after 2,000 h of NBT stressing were found to vary from 4.4% (125°C, −30 V) to near 20% (175°C, −45 V).

The underlying phenomenon leading to the above threshold voltage shifts in stressed devices is the stress-induced buildup of oxide-trapped charge (ΔN_{ot}) and interface traps (ΔN_{it}). Typical time dependencies of stress-induced ΔN_{ot} and ΔN_{it} for different stress voltages at the temperature of 150°C and for different stress

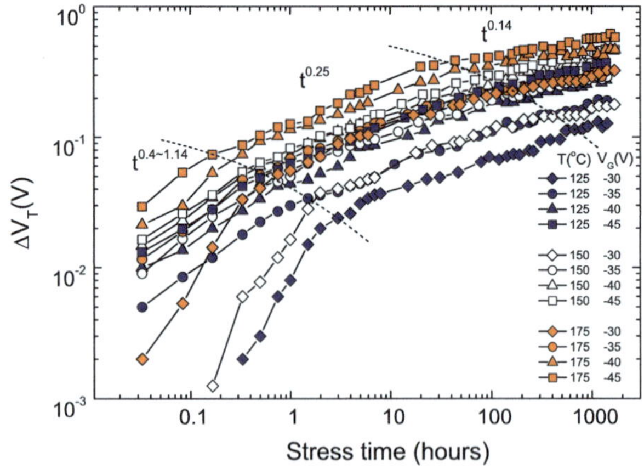

Fig. 20.1 Time dependencies of NBT stress-induced ΔV_T in p-channel power VDMOSFETs

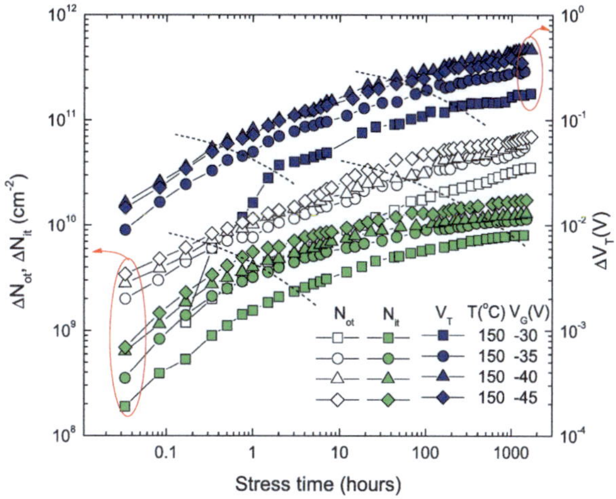

Fig. 20.2 Time dependencies of ΔN_{ot}, ΔN_{it}, and ΔV_T for different stress voltages at 150°C

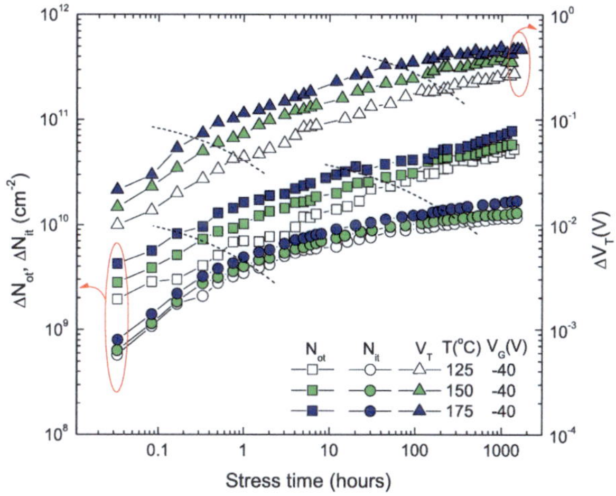

Fig. 20.3 Time dependencies of ΔN_{ot}, ΔN_{it}, and ΔV_T at various stress temperatures for $V_G = -40$ V

temperatures at stress voltage $V_G = -40$ V are shown in Figs. 20.2 and 20.3, respectively. Both figures include the corresponding ΔV_T data for comparison. Note that the stress phase transitions as indicated in Fig. 20.1 for ΔV_T are not clearly visible in the cases of ΔN_{ot} and ΔN_{it} time dependencies, but earlier established transitions are preserved in Figs. 20.2 and 20.3 (dashed lines) for the purpose of data analysis. It is important to note that, in the case of p-channel devices, SMG and CP techniques yielded similar values of ΔN_{it}, so the CP measurements practically were not necessary in this case (in n-channel devices, however, SMG and CP techniques yielded quite different values of ΔN_{it}, which will be addressed in details in Sect. 20.4). As can be seen in Figs. 20.2 and 20.3, the buildup of oxide charge is more significant than that of interface traps for every specific combination of stress voltage and temperature in all three stress phases. It can be noted that ΔN_{it} rapidly increases in the early phase, but slows down in the second phase and tends to saturate faster than ΔN_{ot}. The time dependencies of ΔN_{ot} in both figures look similarly shaped to those of ΔV_T, whereas such a strong correlation does not seem to exist between corresponding ΔV_T and ΔN_{it} dependencies, with disagreement becoming more pronounced as the NBT stressing advances into the second phase and especially further into the saturation. Therefore, ΔV_T time dependencies in power VDMOS devices seem to be mostly affected by NBT stress-induced buildup of oxide-trapped charge, which is not in agreement with early literature data emphasizing dominant role of stress-induced interface traps [1, 2, 11]. The procedure similar to one used to obtain ΔV_T dependence given by Eq. (20.1) was applied to ΔN_{ot} and ΔN_{it} data shown in Figs. 20.2 and 20.3, and the following dependencies of ΔN_{ot} and ΔN_{it} on NBT stress field, time, and temperature in the second stress phase were obtained [22]:

$$\Delta N_{ot} = 1.16 \cdot 10^{11} E^{2.44} t^{0.25} \exp(-0.21/kT), \tag{20.2}$$

$$\Delta N_{it} = 1.56 \cdot 10^{10} E^{2.11} t^{0.18} \exp(-0.15/kT). \tag{20.3}$$

Comparing Eqs. (20.1) and (20.2), it can be seen that ΔN_{ot} and ΔV_T follow the same $t^{0.25}$ time dependence, which confirms the above observation on dominant influence of oxide-trapped charge on V_T shift in NBT stressed p-channel power VDMOSFETs.

In addition to the above power law fitting functions, we have also fitted our data by the stretched exponential (SE) model [43, 44]. The SE model did not provide good fit to our data at stress times shorter than 1 h, probably because of the fact that our slow measurements failed to capture the fast NBTI component, which at short stress times likely dominates in overall degradation over the slow component [15]. In contrast, the SE fit was in good agreement with our data at longer stress times, over the entire second and third stress phases, and was thus useful for lifetime estimation [23]. It could be, however, even more plausible in the near future to try fitting with more recent analytical model, which is based on the occupancy of defects in the capture/emission time maps [45]. In this approach, the overall NBTI degradation is represented by the sum of two separate expressions for the fast and slow degradation components, so the latter could be suitable for fitting to our data which dominantly include slow component.

20.3 Post-Stress Annealing Effects

Following the above analysis of the behavior of threshold voltage and underlying changes in the densities of gate oxide-trapped charge and interface traps in p-channel VDMOSFETs subjected to NBT stress, we have tried to disclose the effects of post-stress and intermittent annealing on degradation associated with NBTI. More specifically, it was expected that low gate bias annealing after each of three stress phases observed could clarify the role of charged species in NBTI and provide an additional insight into the related phenomena. Accordingly, three sets of devices had been subjected to NBT stressing under typical conditions (−40 V, 150°C) for 1 h (end of first phase), 168 h (end of second phase), and about 2,000 h (deep third phase), respectively, for each set. After stressing, each set was divided into three subsets for 168 h of annealing at 150°C under the low gate bias of −10 V, 0, and + 10 V, respectively, for each subset.

The time dependencies of the V_T shift observed in three sets of devices subjected to above NBT stress and gate bias annealing schemes are shown in Fig. 20.4. As can be seen, annealing under the low negative gate bias did not cause any significant changes to V_T shifts induced by the preceding NBT stress. A small recovery, less than about 10%, seems to have been only achieved in devices stressed for 1 h, but one cannot be sure if the recovery in this case was real because of large scattering

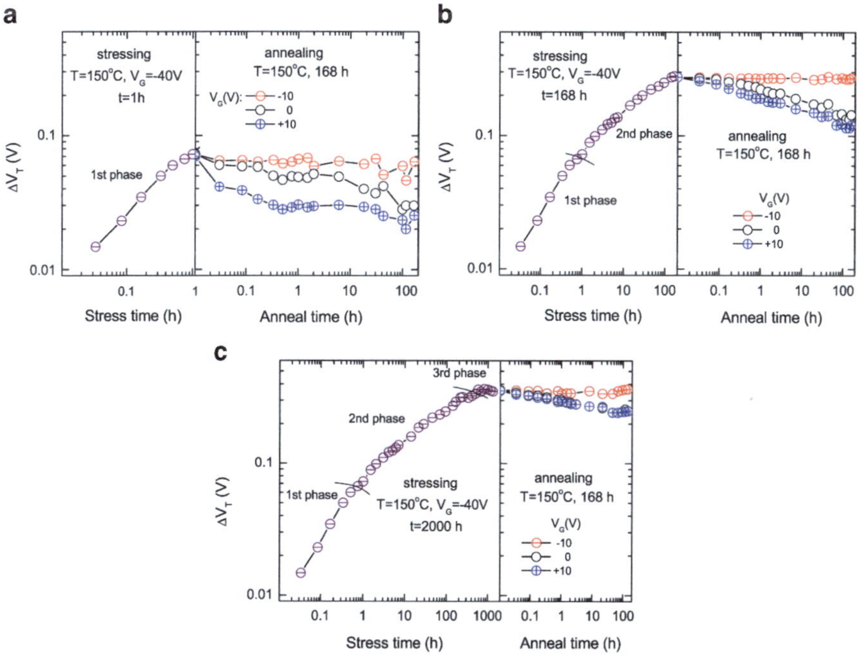

Fig. 20.4 V_T shifts during NBT stressing and low gate bias annealing in p-channel VDMOS devices stressed for (**a**) 1 h, (**b**) 168 h, and (**c**) 2,000 h

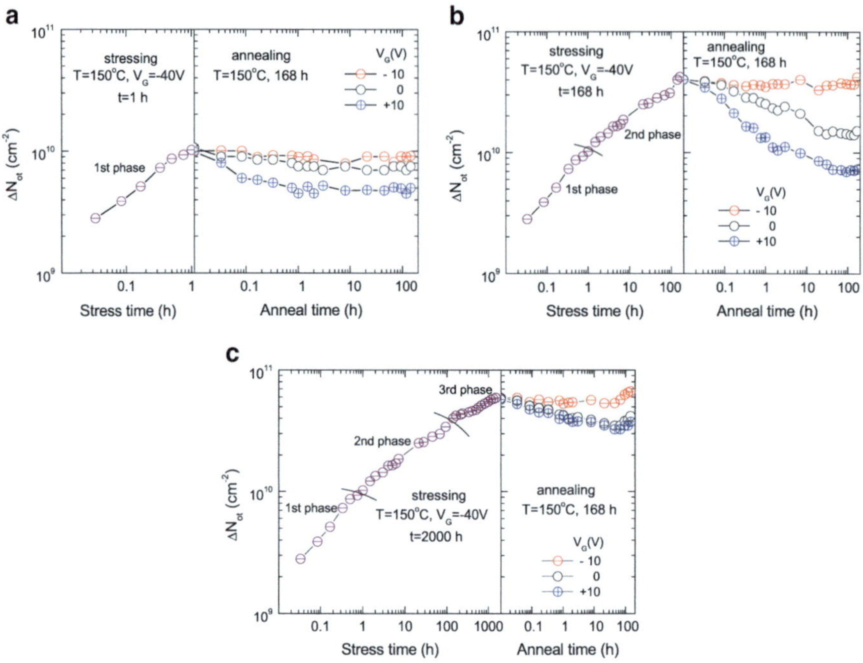

Fig. 20.5 ΔN_{ot} during NBT stressing and low gate bias annealing in p-channel VDMOS devices stressed for (**a**) 1 h, (**b**) 168 h, and (**c**) 2,000 h

of the annealing data (Fig. 20.4a). However, it must be noticed that the recovery achieved by annealing under the zero and especially positive gate bias was quite remarkable: the V_T after 168 h of annealing under the positive gate bias recovered almost 65% in devices stressed for 1 h, about 56% in devices stressed for 168 h, and about 30% in those stressed for 2,000 h.

The above observations generally apply to all three sets of devices, but it is possible to see two potentially important differences among the sets stressed for different times. Actually, the annealing plots shown in Fig. 20.4 strongly suggest that relative amounts of V_T recovery achieved by post-stress annealing decrease with extending the time of preceding NBT stress as well as that the differences in the amounts of post-stress recovery among the devices annealed under the zero and positive gate bias shrink with extending the stress time. These are clear indications that the effects of post-stress annealing depend not only on temperature and gate bias conditions but also on the status of the gate oxide and the SiO_2–Si interface found immediately after the stress, including the densities of stress-induced oxide-trapped charge and interface traps and their spatial and energy distributions, number of potential trapping sites and quantities of reacting species available after the stress, and quantity and distribution of new defects possibly created by preceding stress.

The underlying changes in the densities of gate oxide-trapped charge and interface traps are shown in Figs. 20.5 and 20.6, respectively. As can be seen, the

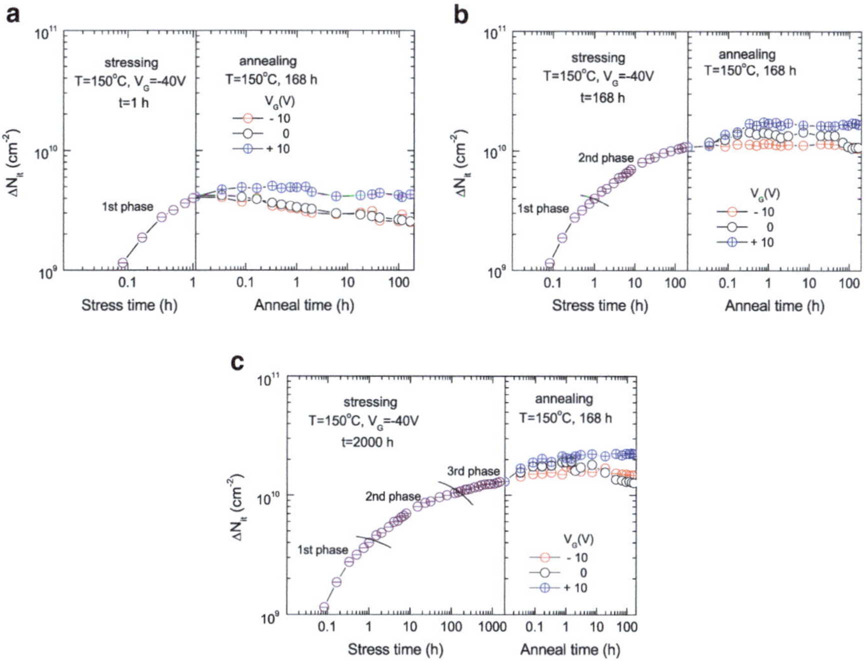

Fig. 20.6 ΔN_{it} during NBT stressing and low gate bias annealing in p-channel VDMOS devices stressed for (**a**) 1 h, (**b**) 168 h, and (**c**) 2,000 h

changes in the density of oxide-trapped charge generally are more significant than in that of interface traps, which is in line with our earlier observation on dominant role of oxide-trapped charge in shaping the ΔV_T time dependences in NBT-stressed VDMOSFETs (see Sect. 20.2). Moreover, it can be noticed that the shapes of ΔN_{ot} curves (Fig. 20.5) mostly follow those of ΔV_T (Fig. 20.4) for corresponding gate bias conditions during the subsequent annealing as well. However, in the case of interface traps (Fig. 20.4), this applies only to annealing under the negative bias, which does not seem to cause any changes to stress-induced ΔN_{it}. In contrast, ΔN_{it} in devices annealed under the zero and especially positive gate bias does not decrease like both ΔV_T and ΔN_{ot}, but continues its tendency from the stress period to increase. Therefore, it appears that annealing performed under the zero and positive gate bias removes the portion of NBT stress-induced oxide charge while creating new interface traps, in addition to those that have been created during the preceding NBT stress. This additional increase of ΔN_{it} begins shortly after replacing the high negative stress voltage with zero or $+10$ V gate bias and ends after just about 2 h of annealing, when ΔN_{it} in the case of positive bias saturates and remains nearly unchanged throughout the rest of 168 h post-stress annealing period, whereas in the case of zero bias, it gradually decreases down to the level found immediately after stressing. The post-stress growth of ΔN_{it} is least significant in devices stressed only for 1 h, where it is less than 10% in respect to ΔN_{it} value found at the end of the NBT stress, but it increases to remarkable 60% in devices stressed for 168 h and near 70% in 2,000 h-stressed ones. It is particularly interesting to note, for example, that 168 h of NBT stressing followed by just an hour of annealing (Fig. 20.6b) resulted into ΔN_{it} larger even than in the case of 2,000 h continuous NBT stress (Fig. 20.6c).

The post-stress generation of interface traps has frequently been observed in MOS devices exposed to various doses of irradiation and is well documented in the literature [46–51]. Degradation after termination of NBT stressing does not seem to have been observed so far in the case of thin oxides, where the density of NBT stress-induced interface traps was found either to remain unchanged after stress or to decrease with annealing time [1–5]. It was even shown that NBTI degradation can be completely annealed at somewhat higher temperature of 300°C [31, 52, 53]. However, the extension of degradation to the period after the end of NBT stress was observed in 30 nm thick SiO_2 films grown on hydrogen rich wafer [15]. Our results shown in Fig. 20.6, which were obtained by both SMG and CP techniques, clearly indicate that annealing under the zero or low positive gate bias after NBT stressing leads to formation of additional interface traps in p-channel VDMOSFETs with 100 nm thick gate oxide rather than to the recovery. These facts suggest that degradation after the NBT stressing might be only associated with thick gate oxides, which serve as reservoir of hydrogen-related reacting species required for both passivation and depassivation processes occurring at the SiO_2–Si interface during the stress and after the end of stress in similar manner as in the case of devices exposed to irradiation. Accordingly, some elements of the approach applied in standard models of irradiation damage [49–51, 54, 55] might be plausible in considering the NBTI in thick oxides, as will be discussed in Sect. 20.5.

Following the above results, it appeared interesting to examine what would happen if the annealed devices were stressed and annealed again. Accordingly, the set of IRF9520 devices, which in previous experiment had been stressed and annealed at 150°C, were restressed and re-annealed under the same conditions as in the previous experiment and finally were stressed once again for about 1,000 h. Devices were, therefore, subjected to a five-step sequence, which included three NBT stress steps interchanging with two intermediate bias annealing steps. The threshold voltage shifts and underlying changes in the densities of gate oxide-trapped charge and interface traps observed in devices subjected to the full sequence are shown in Figs. 20.7, 20.8, and 20.9, respectively, where the results for a first stress–anneal subsequence have been practically repeated from Figs. 20.4, 20.5, and 20.6 for comparison with those obtained during the next stress–anneal–stress subsequence.

There are few quite interesting features that can be observed in Figs. 20.7, 20.8, and 20.9. A general one is related to the role of oxide charge vs. that of interface traps. In addition to earlier finding on more significant buildup of oxide-trapped charge than that of interface traps, it can be seen in Figs. 20.7 and 20.8 that the shapes of ΔN_{ot} curves mostly follow those of ΔV_T for corresponding device subsets (as defined by the annealing bias) over the full sequence of NBT stress and bias annealing steps, suggesting that charge-trapping/detrapping processes occurring in thick oxide bulk could be of primary importance for NBTI in power

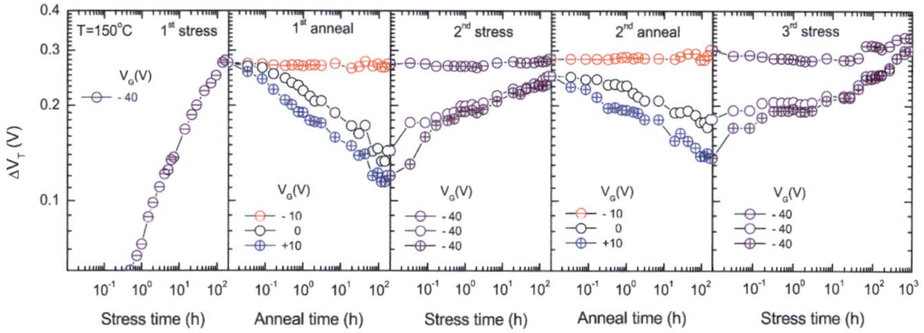

Fig. 20.7 V_T shifts in p-channel VDMOSFETs during the five-step sequence of NBT stressing and gate bias annealing

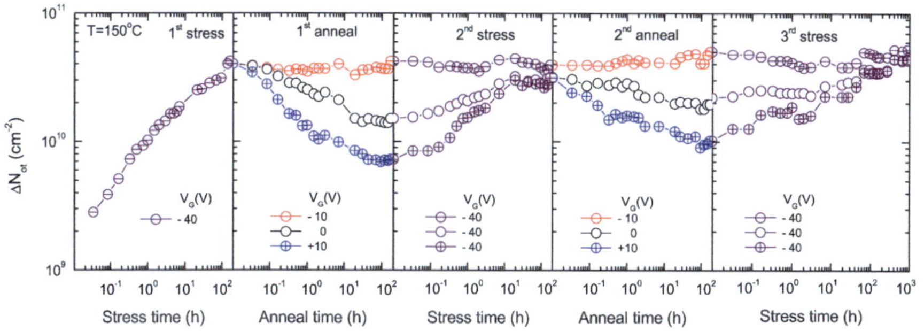

Fig. 20.8 ΔN_{ot} in p-channel VDMOSFETs during the five-step sequence of NBT stressing and gate bias annealing

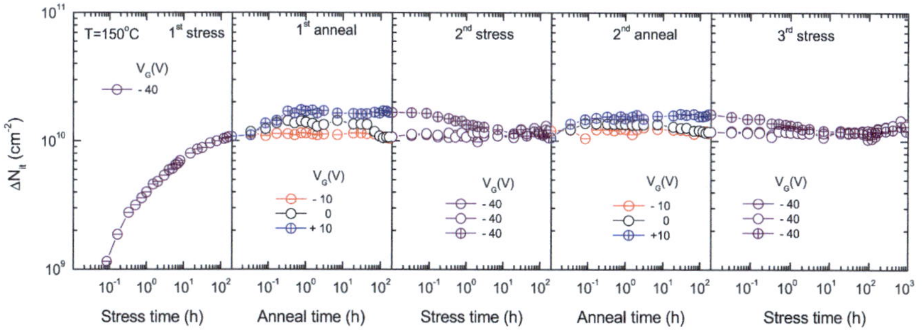

Fig. 20.9 ΔN_{it} in p-channel VDMOSFETs during the five-step sequence of NBT stressing and gate bias annealing

VDMOSFETs indeed. The other features are related to the specific gate bias applied during annealing. It can be seen that ΔV_T induced by the initial NBT stress, as well as the underlying ΔN_{ot} and ΔN_{it}, in the case of annealing under the low negative bias remain almost constant not only during the first annealing but also during the forthcoming subsequence of repeated stress and annealing steps. Further degradation in this case was only observed at the end of the final stress step, which was extended to 1,000 h, so it seems that negative bias annealing could conserve initial degradation for a while. On the other hand, initially created V_T shifts in devices annealed under the zero and positive gate biases suffered remarkable changes. As can be seen in Fig. 20.7, ΔV_T in both these subsets of devices decreased during each annealing, but increased again during the subsequent NBT stressing, with changes fading a little on repeating the stress–anneal steps. More significant decrease of ΔV_T was observed in devices annealed under the positive bias, but forthcoming increase during the next stressing was also higher in these, so the resulting V_T shifts found at the end of each stressing step in devices annealed under the zero and positive biases were equal. Similar behavior during the whole sequence is observed in the case of stress-induced ΔN_{ot} (Fig. 20.8): a remarkable decrease during annealing was followed by increase during the next stress, etc., with all changes being more significant in the case of devices annealed under the positive bias and fading a little on each repetition. The behavior of stress-induced ΔN_{it} in this case seems, however, more interesting. A remarkable post-stress increase of ΔN_{it} under the zero and especially positive gate bias has already been discussed (see above discussion of Fig. 20.6) and is confirmed in Fig. 20.9 (see first anneal data). The behavior of ΔN_{it} in devices annealed under the positive gate bias is most intriguing. In contrast to the expectation that stress repetition would lead to additional degradation, the second stress in this case actually leads to a decrease of ΔN_{it} down to approximately equal value as after the first stress, whereas second annealing had very similar effect as the first one and reproduced most of ΔN_{it} that were lost during the second stress step. In the final third stress step, ΔN_{it} decreased in a similar way as during the second stress and started slowly to increase only after prolonged stressing. As for devices annealed under the zero bias, during the first anneal step, ΔN_{it} initially increased as in the case of annealing under the positive bias, but then decreased down to the level found immediately after the initial stressing and remained almost unchanged during the second stress, which all repeated during the subsequent second anneal and third stress steps, respectively.

Summarizing the above considerations, it was shown that intermittent annealing under the low gate bias might have significant impact on overall NBT stress-induced degradation. Annealing under the negative bias maintains degradation at the level found after the initial NBT stress, whereas annealing under both zero and positive biases leads to apparent recovery of device threshold voltage. However, this recovery does not seem to be a true one because only ΔN_{ot} decreases while ΔN_{it} simultaneously increases. It was further shown that the changes in both N_{ot} and N_{it} observed during positive bias annealing were reversible, as the repetition of NBT stress after annealing restored most of the annealed oxide charge while removing the reversible component of interface traps. It is interesting to note, however, that all the changes shrink on each repetition of the stress–anneal subsequence, which is further illustrated in Fig. 20.10 showing in linear scale the evolution of ΔV_T during the full 5-step stress–anneal sequence in devices annealed under the positive bias. As can be seen, threshold voltage quickly recovers in the early stage of each annealing

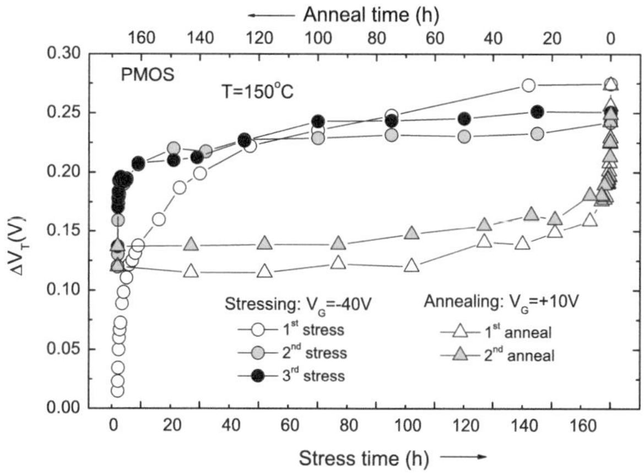

Fig. 20.10 Evolution of ΔV_T in p-channel VDMOSFETs during the full sequence of NBT stressing and positive bias annealing steps

step, but major portion of the shift induced by the initial NBT stressing is also quickly restored on repeating the stress. The changes in ΔV_T tend to decrease on each new repetition, indicating that nonreversible components of N_{ot} and N_{it} tend to increase. As a consequence, one may expect that stress-induced ΔV_T may remain within specific range (around 0.15 V in the case shown in Fig. 20.10) as long as the NBT stress and positive bias anneal conditions are frequently interchanged.

20.4 NBTI in n-Channel Devices

Most of the literature data accentuate that NBTI can be of importance only in p-channel MOSFETs [1–6], which seems to be in line with the nature of underlying stress-induced oxide-trapped charge and interface traps. The oxide-trapped charge is mostly positive in both p- and n-channel devices, whereas the net charge in interface traps depends on gate bias: it is positive in p-channel transistors, which are normally biased with negative gate voltage, but is negative in n-channel devices, which require positive gate bias to be turned on. Accordingly, threshold voltage shifts due to stress-induced oxide-trapped charge and interface traps in p- and n-channel MOSFETs can be expressed, respectively, as [54]

$$\Delta V_{Tp} = \frac{q\Delta N_{otp}}{C'_{ox}} + \frac{q\Delta N_{itp}}{C'_{ox}}, \tag{20.4}$$

$$\Delta V_{Tn} = -\frac{q\Delta N_{otn}}{C'_{ox}} + \frac{q\Delta N_{itn}}{C'_{ox}}, \tag{20.5}$$

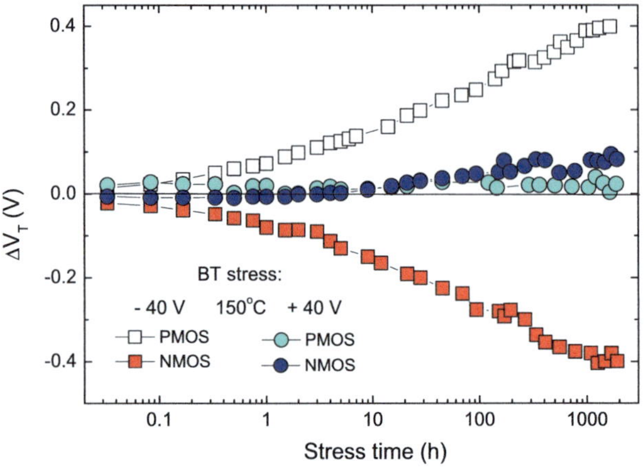

Fig. 20.11 Threshold voltage shifts during the typical NBT stress (150°C, −40 V) in p- and n-channel VDMOSFETs. Absolute value of threshold voltage in p-channel devices was found to increase, so the corresponding shift is shown as positive

where q is the elementary charge and C'_{ox} is the gate oxide capacitance per unit area. Assuming that NBT stress creates similar amounts of oxide-trapped charge and interface traps in both p- and n-channel devices, the net effect on threshold voltage, ΔV_T, must be greater in p-channel devices, as only in this case the positive oxide charge and positive interface charge are additive. Moreover, the n-channel devices are not operated under the negative gate bias, so the NBTI generally is not considered of importance in n-channel MOSFETs. However, arguing that high negative gate bias can be used in some automotive applications for faster turning the n-channel devices off, rather significant NBT stress-induced threshold voltage shifts have been found in n-channel trench DMOS transistors [14]. Our research on NBTI in power VDMOSFETs have led to a quite similar finding [28]. This is illustrated in Fig. 20.11, which clearly shows that NBT stress under typical conditions yields practically identical V_T shifts in p- and n-channel VDMOS devices. The corresponding PBT stress, however, does not seem to significantly affect threshold voltage in any of two device types.

The V_T shifts observed in n-channel VDMOSFETs during both NBT stressing and post-stress annealing under the low gate bias are shown in Fig. 20.12. For the purpose of comparison, n-channel devices were stressed for 168 h under the typical conditions (150°C, −40 V) and annealed in the same manner as earlier considered p-channel ones (see Sect. 20.3). As can be seen from comparing Figs. 20.4b and 20.12, NBT stress created V_T shifts of about 0.3 V in both p- and n-channel devices. In the case of p-channel devices (Fig. 20.4b), annealing under the negative gate bias did not cause any apparent change to the stress-induced ΔV_T, whereas annealing under the zero or positive gate bias led to the V_T recovery of over 50%. Positive bias annealing appeared most efficient to reduce the stress-induced shift, but the final recovery of threshold voltage was just a little higher than in the case of zero bias applied. On the

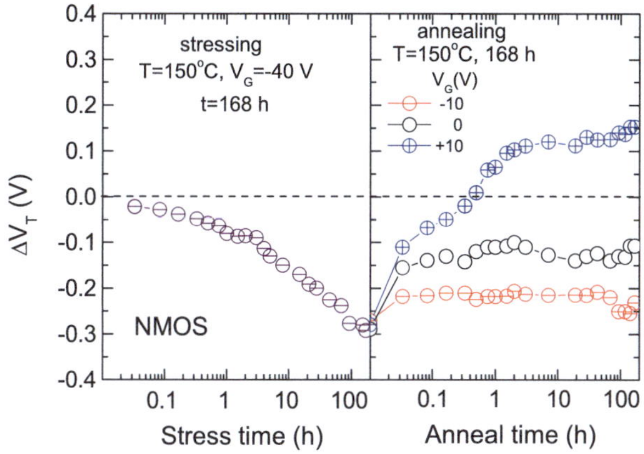

Fig. 20.12 V_T shifts during NBT stressing and gate bias annealing in n-channel VDMOSFETs

other hand, post-stress annealing in n-channel devices had clearly different effects on threshold voltage in each of three gate bias conditions applied (Fig. 20.12). The recovery was quite small under the negative and relatively significant under the zero bias, whereas positive bias annealing yielded full recovery of threshold voltage in just an hour, which was even followed by further increase (rebound) beyond the value that V_T had before the initial NBT stress. According to Fig. 20.4b, maximum variation of threshold voltage in p-channel devices, somewhat less than 0.3 V, was found at the end of the NBT stress, and the subsequent annealing under any bias did not increase this variation, but only reduced it. Threshold voltage in n-channel devices decreased during the stress also for 0.3 V, and annealing under the zero and negative gate bias reduced the stress-induced shift again. However, annealing under the positive gate bias (which is normal operation bias in n-channel devices) led not only to a full recovery of threshold voltage but also to its increase for about 0.15 V above the initial (prestress) value, so the total variation observed over the whole stress and anneal sequence was about 0.45 V. This clearly indicates that, if the n-channel devices were exposed to a negative gate bias and elevated temperature at any stage of their operation, the resulting instabilities of threshold voltage could be more serious than the corresponding instabilities found in p-channel devices.

The underlying changes in the densities of oxide-trapped charge and interface traps in n-channel devices, as determined by the SMG technique, are shown in Fig. 20.13 (the corresponding ΔN_{ot} and ΔN_{it} data for p-channel devices were shown in Figs. 20.5b and 20.6b, respectively). Regarding the ΔN_{ot}, it can be seen from Figs. 20.5b and 20.13a that NBT stress caused more significant increase in n-channel devices, whereas subsequent annealing in two types of devices had very similar effects: stress-induced ΔN_{ot} in both cases almost did not change during annealing under the negative gate bias, but decreased under the zero and, especially, under the positive bias. Similarly, comparison of ΔN_{it} data in Figs. 20.6b and 20.13b for p- and n-channel devices, respectively, shows that stress-induced increase of N_{it} also was higher in n-channel devices. The subsequent annealing under the

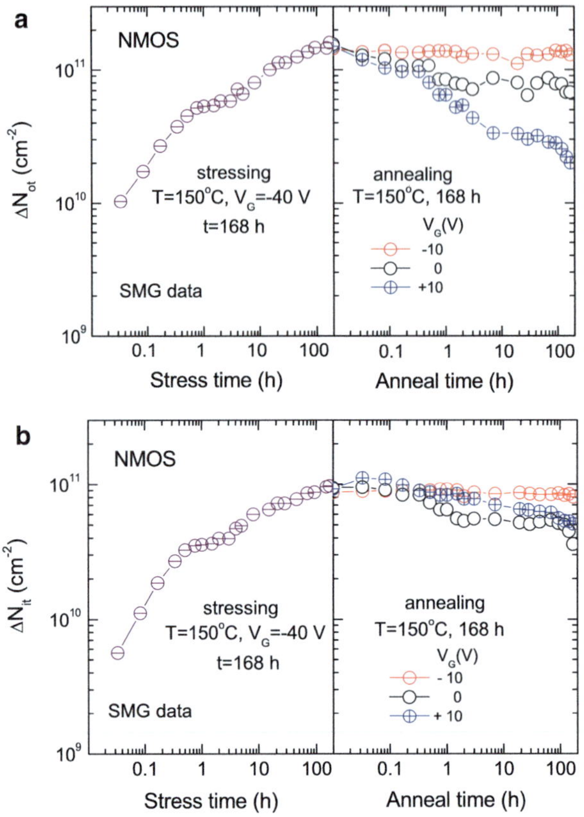

Fig. 20.13 SMG data for the changes in the densities of (**a**) oxide-trapped charge and (**b**) interface traps in n-channel VDMOSFETs during NBT stressing and low gate bias annealing

low negative gate bias did not cause apparent changes to stress-induced ΔN_{it} in both p- and n-channel devices. However, the effects of zero and positive bias annealing in n-channel VDMOSFETs differed from the effects in p-channel ones: rather remarkable post-stress increase in ΔN_{it} was observed in p-channel transistors (Fig. 20.6b), whereas ΔN_{it} in n-channel devices (Fig. 20.13b) made only a small increase in the early phase of annealing, which was followed by slow decrease. It is interesting to note in both types of devices that NBT stress created ΔN_{ot} higher than the corresponding ΔN_{it}, whereas positive bias annealing caused ΔN_{ot} to fall below the corresponding ΔN_{it}.

The above results have shown that NBT stress did not create similar amounts of oxide-trapped charge and interface traps in p- and n-channel VDMOSFETs. Instead, we have actually found that NBT stress created similar V_T shifts in two types of VDMOS devices (Fig. 20.11), so in our case it could have been expected that stress-induced ΔN_{ot} would be higher in n-channel devices, as confirmed by our data (Fig. 20.5b cf. Fig. 20.13a). However, the difference in stress-induced ΔN_{ot} between

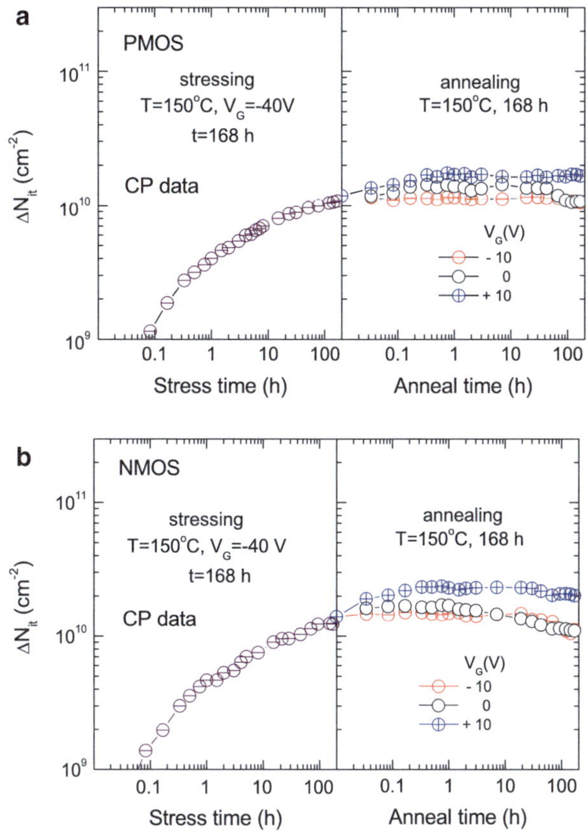

Fig. 20.14 CP data for the changes in the densities of interface traps in (**a**) p-channel and (**b**) n-channel VDMOSFETs during NBT stressing and low gate bias annealing

two device types seems too big, and it is even more surprising that stress-induced ΔN_{it} also was higher in n-channel devices (Fig. 20.6b cf. Fig. 20.13b). It should be noted, however, that ΔN_{ot} and ΔN_{it} shown in previous figures were obtained by SMG technique based on sweeping *I–V* measurements and may include contribution from border traps, also known as switching oxide traps [56–58]. These are oxide traps located near the SiO_2–Si interface, which at low measurement frequencies may easily exchange charge with the Si substrate and thus behave as interface traps, so the above analysis might not be quite appropriate. That is actually why we have additionally estimated ΔN_{it} in our devices by the CP technique, which is based on high-frequency measurements and is thus considered to sense only the interface traps and perhaps just few the fastest among switching oxide traps [41, 42]. Accordingly, it is expected that comparison of SMG and CP data for ΔN_{it} may provide information on the amount of stress-induced switching oxide traps. Our results on power VDMOSFETs obtained by CP technique are shown in Fig. 20.14, which reveals that NBT stress created nearly equal amounts of interface traps in p-channel (Fig. 20.14a) and n-channel devices (Fig. 20.14b), as well as that there was almost no difference in ΔN_{it} behavior between two types of devices during the post-stress annealing either. Comparing the SMG and CP data, the difference between the two techniques appears negligible in the case of p-channel devices (Fig. 20.6b cf. Fig. 20.14a), but is rather significant in the case of n-channel ones (Fig. 20.13b cf. Fig. 20.14b). It is particularly interesting to note that remarkable post-stress increase of ΔN_{it} under the low positive gate bias was observed by CP technique in both p- and n-channel devices (Fig. 20.14), whereas SMG technique found significant increase of ΔN_{it} after the end of stressing only in p-channel VDMOSFETs (Fig. 20.6), but not in n-channel ones (Fig. 20.13b). These observations suggest that NBT stress did not create any significant amount of switching oxide traps in p-channel devices, whereas the amount of stress-induced switching oxide traps in n-channel ones was rather high and could even dominate in ΔN_{it} values estimated by SMG technique.

It should be noted that the above conclusion derived from direct comparison of data obtained by the two techniques could be cast in doubt due to the fact that different measurement techniques scan different portions of the Si bandgap as well as due to the findings that CP method may enhance the recovery leading to significant relaxation of ΔN_{it} if the CP amplitude is chosen too large [15, 59–61]. Indeed, CP measurements scan the central portion of the bandgap, whose width ΔE may vary from 0.37 to 0.52 eV [61], whereas SMG technique probes the range from the midgap to the energy level corresponding to the threshold voltage. However, ΔE widths scanned by the CP and SMG techniques in the case of standard VDMOSFETs are similar (approximately 0.46 eV for CP and 0.42 eV for SMG) [62] and cannot be the cause for significant differences in ΔN_{it} values obtained by the two techniques in n-channel devices. We found good agreement in ΔN_{it} data between the SMG and CP techniques in p-channel devices, and our CP data for both types of devices have shown that ΔN_{it} did not decrease during the post-stress annealing but actually increased under the specific bias conditions, which all indicated that CP-induced relaxation was not significant in our case. It should be noted that CP technique has also been reported to give good agreement with subthreshold swing technique [61], which is (like the SMG technique used in our study) also based on transfer *I–V* characteristics measured in the subthreshold region. For all these reasons we believe that direct comparison of SMG and CP data in our case was not inappropriate.

The n-channel VDMOSFETs also were, in the same way as the p-channel devices, subjected to a five-step stress–anneal sequence, which included three NBT stress steps interchanging with two bias annealing steps. The full plots for time dependencies of ΔV_T, ΔN_{ot}, and ΔN_{it} in n-channel devices over the entire sequence (such as the plots in Figs. 20.7, 20.8, and 20.9 for p-channel devices) will not be shown here because they do not contain much novelty in comparison with previous findings. Briefly, the results for n-channel VDMOSFETs confirmed earlier results obtained on p-channel devices that annealing under the low negative bias could suppress further degradation during the subsequent stress and anneal steps,

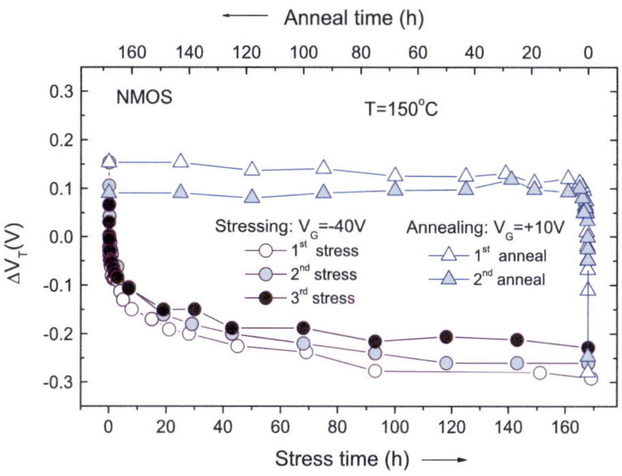

Fig. 20.15 Evolution of ΔV_T in n-channel VDMOSFETs during the full sequence of NBT stressing and positive bias annealing steps

practically preserving initial degradation at the level found after the first NBT stress. Also, the changes in stress-induced ΔV_T, ΔN_{ot}, and ΔN_{it} observed during annealing under the zero and positive gate biases were found to be mostly reversible like in the case of p-channel devices. For example, it is most interesting that each annealing under the positive gate bias led to the full recovery of stress-induced ΔV_T followed by rebound (like in Fig. 20.12), whereas each repetition of NBT stress restored most of the initial shift caused by the first NBT stress. This is further illustrated in Fig. 20.15, which shows evolution of ΔV_T during the full 5-step stress–anneal sequence in n-channel devices annealed under the positive bias. It can be seen that, in similar way as in the case of p-channel devices (Fig. 20.10), threshold voltage quickly recovers and even increases above its initial value in the early stage of each annealing step, but major portion of the negative shift induced by the initial NBT stress is also quickly restored on repeating the stress. The changes in ΔV_T show tendency to decrease on each new repetition like in p-channel devices, but comparison of the results shown in Figs. 20.10 and 20.15 confirms that overall variations of threshold voltage over the entire stress and anneal sequence can be greater in n-channel ones.

20.5 Mechanisms of Degradation

The microscopic origin behind the NBTI-related degradation is one of the most extensively discussed issues in publications on reliability research in modern MOS devices. The most common interpretations of NBTI include various forms of the hydrogen reaction–diffusion (RD) model, which was originally proposed by Jeppson and Svensson [63]. The model assumes that hydrogen species are released from previously passivated defects at SiO_2–Si interface and diffuse into the oxide, leaving behind interface traps [1, 2, 4, 63]. Arguing that trap-controlled hydrogen migration in the oxide results in dispersive transport behavior, a number of modified RD model versions to account for dispersive hydrogen motion were proposed [5, 6, 64–66]. Those modifications were aimed at improving the flexibility of the basic model to reconcile some experimental discrepancies, which were believed to originate from wide variations in the state-of-the-art gate dielectric technologies employed. There was, however, suggestion that interface trap creation could be reaction-controlled mechanism rather than diffusion-controlled one [3], and generation of positive charge in the oxide bulk due to hole trapping has been reported in addition to generation of interface traps [3, 5, 65, 67]. For some time there was a controversy on the role of trapped charge in NBTI [3, 68], but a number of studies strongly suggest that dominant contribution to degradation actually comes from the hole trapping [31, 69–73]. These findings have eventually led to the proposal of a new charge-trapping model, which links the NBTI degradation with the creation of switching oxide traps and is consistent with recovery data showing dispersion over the wide range of time [74, 75].

The results obtained on VDMOS devices, which were shown in Sects. 20.1–20.4, signify that major contribution to NBT stress-induced degradation in these devices also comes from the oxide-trapped charge. The other important feature of NBTI in ultra-thick gate oxides for power VDMOSFETs is additional generation of interface traps during the post-stress annealing under the positive gate bias, which was observed in both p- and n-channel devices. It is also important to note that NBT stress creates equal threshold voltage shifts in both device types, whereas subsequent annealing under the positive gate bias results in more significant overall instability in n-channel devices. Our results indicate strong bias dependence of the processes occurring over both stress and anneal periods, suggesting that one or more kinds of charged species have been involved. The holes induced and/or accumulated under the gate oxide must be among them, as only negative gate bias stress resulted into significant V_T shifts in both p- and n-channel VDMOSFETs. We believe that hydrogen, as the most common impurity in MOS devices, which is widely considered as the primary agent of instabilities associated with radiation damage [46–51], hot carrier injection, and high electric field stress [76–82], has to be considered in BTI as well. As mentioned above, various versions of the RD model have already been used to explain NBTI through stress-initiated electrochemical processes involving oxide and interface defects, holes, and hydrogen species [1, 2, 4–6, 64–66]. The impact of hydrogen on BTI is discussed in details in Chap. 18.

Qualitative similarity between the effects that we observed in p- and n-channel VDMOSFETs (which recently has been found in deeply scaled FETs with various gate dielectrics as well) [83, 84] suggests that similar or the same mechanisms could be responsible for NBTI in both device types. Let us begin with considering the processes occurring during the NBT stress. The buildup of oxide charge under the high negative oxide field at elevated temperatures can be attributed to hole trapping at oxygen vacancy defects near the SiO_2–Si interface [22]:

$$O_3 \equiv Si^{\bullet\bullet}Si \equiv O_3 + h^+ \rightarrow O_3 \equiv Si^{+\bullet}Si \equiv O_3. \tag{20.6}$$

Generation of interface traps also can be ascribed to high electric field, which at increased temperatures and in the presence of holes may dissociate the Si–H bonds at the SiO_2–Si interface [2, 60]:

$$Si_3 \equiv Si - H \leftrightarrow Si_3 \equiv Si^\bullet + H^\bullet. \tag{20.7}$$

The NBT stress-induced generation of interface traps through this reaction has been explained as a hole-assisted field-enhanced thermally activated process of Si–H bond breaking [60]. Hydrogen atoms released in reaction 20.7 are highly reactive and they also can dissociate the interfacial Si–H bonds, thus leading to additional creation of interface traps [1, 2, 51]:

$$Si_3 \equiv Si - H + H^\bullet \leftrightarrow Si_3 \equiv Si^\bullet + H_2. \tag{20.8}$$

Alternatively, the H atoms can react with holes from the substrate to form ions [51]:

$$H^\bullet + h^+ \rightarrow H^+, \tag{20.9}$$

which thereafter, drifting away from the interface under the negative oxide field, can dissociate the Si–H bonds in the gate oxide near the interface [2, 51]:

$$O_3 \equiv Si - H + H^+ \leftrightarrow O_3 \equiv Si^+ + H_2, \tag{20.10}$$

leading to creation of additional positively charged oxide defects. It should be noted that reactions 20.7, 20.8, and 20.10 may occur in both forward and reverse directions. As for the reverse reactions 20.7 and 20.8, they practically re-passivate interfacial Si–H bonds and are not expected in the early stage of stressing. Instead, $H^\bullet$ atoms released in reaction 20.7 are more likely to participate in reaction 20.9, while H_2 molecules released in 20.8 are likely to diffuse away from interface into the oxide, where they can be cracked on positively charged oxide defects through the reverse reaction 20.10. Note that multiple occurrences of reaction 20.10 in reverse and forward directions under the negative oxide field tend to move positive charge deeper into the oxide bulk with extending the stress time. Extended stressing gradually reduces the number of previously passivated interface defects available for dissociation, whereas the probability of re-passivation through the reverse reactions 20.7 and 20.8 gradually increases, so the stress-induced ΔN_{it} tend to saturate.

The basic assumption in the above consideration is that electric field applied during typical NBT stress is strong enough to dissociate interfacial Si–H bonds through the reaction 20.7, which also releases $H^\bullet$ atoms required for the reactions 20.8, 20.9, and 20.10. It has been argued, however, that removal of hydrogen from the Si–H bond requires an activation energy of about 2.4 eV [85], which is reduced to about 2.1 eV in the presence of holes, but is still much higher than activation energies associated with typical BT stress conditions [85]. Accordingly, it has been concluded that reaction 20.7 remains inactive under BTI conditions. Instead, it was proposed that the processes on the interface could be triggered by an alternative reaction, which involves H^+ ions originating from the semiconductor substrate and has much lower activation energy [85]. The other alternative proposal is that hydrogen- and water-related species trapped at the gate–SiO_2 interface after deposition of the metal (or polysilicon) gate could be cracked when exposed to a high electric field at high temperature, producing $H^\bullet$ atoms, which subsequently migrate towards the Si–SiO_2 interface to participate in reaction 20.8 [86]. Reaction 20.8 in this case assumes the main role in processes occurring at the Si–SiO_2 interface and may lead either to the creation of interface traps or their passivation, depending on whether it occurs in forward or reverse direction. However, there is still a possibility that hydrogen required for the above processes may originate from the Si–SiO_2 interface itself. Namely, it has been reported that the binding energies of the Si–H bonds exhibit Gaussian broadening [3, 87], which suggests that typical BTI conditions may suffice to break some of the weaker bonds and thus initiate degradation at the interface. In addition, it has also been reported that large background concentration of hydrogen may exist near the Si–SiO_2 interface [88] as well as that hydrogen in thick oxides may be released near the anode at gate oxide fields just above 1.5 MV/cm [76], which also suggests that the required hydrogen may originate from the Si–SiO_2 interface.

Regarding the processes occurring during the post-stress annealing under the low gate bias, positive charge that was found trapped near the SiO_2–Si interface immediately after ceasing the NBT stress may interact with interfacial Si–H bonds to be transformed into the interface traps [22, 24, 89]:

$$O_3 \equiv Si^{+\bullet}Si \equiv O_3 + Si_3 \equiv Si - H + e^- \rightarrow O_3 \equiv Si^{\bullet\bullet}Si \equiv O_3 + Si_3 \equiv Si^\bullet + H^\bullet. \tag{20.11}$$

In addition, the H^+ ions formed during the preceding NBT stress may dissociate interfacial Si–H bonds [2, 51]:

$$Si_3 \equiv Si - H + H^+ + e^- \leftrightarrow Si_3 \equiv Si^\bullet + H_2, \tag{20.12}$$

leading to an additional generation of interface traps during the post-stress annealing. These two reactions require electrons from the substrate and interface-oriented drift of H^+ ions, so they are not likely to occur under the negative gate bias. Accordingly, ΔN_{ot} and ΔN_{it} in the case of annealing at -10 V gate bias remain nearly constant in both p- and n-channel devices. However, both reactions are enhanced by positive oxide field, either external or local (due to oxide-trapped charge itself), leading to simultaneous decrease in ΔN_{ot} and increase in ΔN_{it} in devices annealed under the zero or $+10$ V gate bias. The results shown in Figs. 20.5 and 20.6 for p-channel devices, as well as those in Figs. 20.13a and 20.14b for n-channel ones, do not indicate the one-to-one correspondence between the decrease in ΔN_{ot} and simultaneous increase in ΔN_{it}, implying that some of the positively charged oxide defects could be simply neutralized by electrons from the substrate. Besides, the H_2 molecules released in reaction 20.12 may diffuse into the oxide to be cracked at positively charged traps through the reverse reaction 20.10 [51], neutralizing the oxide traps and creating additional H^+ ions for reaction 20.12. However, as the annealing progresses and ΔN_{it} exceeds ΔN_{ot}, the probabilities for the occurrence of reactions 20.11 and 20.12 are getting lower, so ΔN_{ot} tends to saturate and ΔN_{it} starts slowly to decrease, which suggests that $H^{\bullet}$ atoms and H_2 molecules released in these reactions begin to passivate interface traps through the reverse reactions 20.7 and 20.8. It must be noted here that, in addition to the presence of positive gate bias, increased temperature during annealing also seemed necessary to trigger reactions 20.11 and 20.12, as the room temperature annealing did not cause any apparent changes to NBT stress-induced ΔV_T, independently on whether the -10 V, zero, or $+10$ V gate bias was applied [90].

The above considerations can be extended to explain both qualitative and quantitative differences in threshold voltage behavior between p- and n-channel devices observed during the positive gate bias annealing (Fig. 20.4b cf. Fig. 20.12). Namely, looking into Eqs. (20.4) and (20.5) for the stress-induced threshold voltage shifts in p- and n-channel transistors, one can deduce that transformation of oxide charge trapped near the interface into interface traps through reaction 20.11 does not have any apparent impact on V_T shift in p-channel devices as both terms on the right-hand side in Eq. (20.4) are positive. Accordingly, the decrease of V_T shift in p-channel devices on annealing under the positive gate bias (Fig. 20.4b) cannot be attributed to reaction 20.11, but only to neutralization of the portion of the oxide traps with electrons from the substrate. In contrast, the occurrence of reaction 20.11 has double effect on stress-induced V_T shift in the case of n-channel devices. Namely, the decrease in ΔN_{otn} and simultaneous increase in ΔN_{itn} cumulatively contribute to a positive threshold voltage shift, which is further enhanced by oxide trap neutralization with electrons. As a result, the initial NBT stress-induced negative V_T shift in n-channel devices rapidly decreases during positive bias annealing down to zero and even turns into a positive shift once the ΔN_{itn} value exceeds that of ΔN_{otn} (Fig. 20.12).

Regarding the peculiarities associated with the repetition of the stress–anneal subsequence (Sect. 20.3), the processes identical or similar to those observed in the first stress–anneal subsequence are expected to occur during the repeated subsequence. However, the total number of potential trapping sites at the SiO_2–Si interface and in the oxide, and the amount of hydrogen species available for reactions as well, have been changed during the preceding stress and anneal steps, so all the changes are now less dramatic. Moreover, ΔN_{it} in devices annealed under the positive bias does not increase during the second stress but decreases, indicating that $H^{\bullet}$ atoms and H_2 molecules released during the preceding anneal step in reactions 20.11 and 20.12 now contribute to the passivation of interface traps through the reverse reactions 20.7 and 20.8 rather than to dissociation of the Si$-$H bonds. The processes during the second anneal and third stress steps are similar to those occurring in the first anneal and second stress steps, respectively, but all the changes shrink owing to nonreversible component of NBT stress-induced N_{ot} that cannot be annealed. The results in Fig. 20.9 show that ΔN_{it} tends to saturate at the end of each stressing step, which indicates that even high negative gate bias, pulling the holes and/or H^+ ions from the SiO_2–Si interface, may lead to nearly balanced processes at the interface (Si–H bond dissociation vs. interface trap passivation). This balance is disturbed on applying the positive gate bias, which redirects positive charge towards the interface causing N_{it} to increase, but is reestablished on applying the negative stress bias again. Seemingly, positive bias annealing removes the reversible component of N_{ot} and creates that of N_{it}, the latter being removed on redoing the NBT stress.

20.6 Conclusions

The main features of negative bias temperature instability in thick gate oxides found in power VDMOSFETs have been reviewed. The NBT stress was found to cause equal threshold voltage shifts in p- and n-channel devices. The underlying buildup of oxide-trapped charge in both device types was more significant than that of interface traps. Comparing the results for stress-induced interface traps obtained by the SMG and CP techniques, it was concluded that significant contribution to NBTI in n-channel VDMOSFETs could actually originate from the switching oxide traps. The effects of post-stress annealing performed at the same temperature as during the stress were strongly dependent on gate bias applied. Annealing under the low negative bias did not affect degradation found after initial NBT stress and appeared to suppress further changes during the subsequent repetitions of the stressing and/or annealing. However, annealing under the zero and especially positive gate bias removed significant portion of stress-generated oxide-trapped charge while creating comparably smaller (but also rather significant) amount of additional interface traps. These changes were reversible, as each repeated NBT stress regenerated most of the annealed oxide charge and removed interface traps created during the preceding anneal step, leveling their concentrations at values found after the initial NBT stress. Similar phenomena during annealing under the positive bias (removal of reversible component of stress-induced oxide charge and simultaneous generation of reversible component of interface traps) were found in both p- and n-channel devices, but the resulting effect on threshold voltage differed: only partial recovery was observed in p-channel devices, whereas threshold voltage in n-channel devices recovered completely and even increased beyond its prestress value. These results have shown that, if the n-channel devices were exposed to a negative gate bias and elevated temperature at any stage of their operation, the resulting instability could be even more serious than that in the case of p-channel devices.

The experimental results were discussed in terms of the mechanisms that include charge trapping–detrapping in the gate oxide and various stress-initiated electrochemical processes involving interface and oxide defects, holes, and hydrogen species as common impurity in MOS devices. The NBT stress-induced buildup of oxide-trapped charge was assumed mostly to be the consequence of hole trapping at near-interface oxide defects under the high negative field, while the generation of interface traps was ascribed to Si–H bond dissociation and/or passivation processes. The post-stress generation of interface traps under the positive oxide field was explained by the reversed drift direction of positively charged species, which initiated processes at the SiO_2–Si interface that were not likely to occur under the negative gate bias. The full recovery of threshold voltage and even rebound beyond the initial value, which was observed in n-channel devices during annealing under the positive gate bias, was ascribed to transformation of near-interface oxide-trapped charge into the interface traps.

Acknowledgments This work has been supported by the Ministry of Education, Science and Technological Development of the Republic of Serbia, under the grants No. OI171026 and No. TR32026.

References

1. S. Ogawa, M. Shimaya, and N. Shiono, J. Appl. Phys. **77**, 1137 (1995).
2. D. K. Schroder and J. A. Babcock, J. Appl. Phys. **94**, 1 (2003).
3. V. Huard, M. Denais, and C. Parthasarathy, Microelectron. Reliab. **46**, 1 (2006).
4. J. H. Stathis and S. Zafar, Microelectron. Reliab. **46**, 270 (2006).
5. A. E. Islam, H. Kufluoglu, D. Varghese, S. Mahapatra, and M. A. Alam, IEEE Trans. Electron Devices **54**, 2143 (2007).
6. T. Grasser, W. Goes, and B. Kaczer, IEEE Trans. Device Mater. Reliab. **8**, 79 (2008).
7. Y. Miura and Y. Matukura, Jpn. J. Appl. Phys. **5**, 180 (1966).
8. Y. Mitani, H. Satake, and A. Toriumi, IEEE Trans. Device Mater. Reliab. **8**, 6 (2008).
9. G. Ribes, J. Mitard, M. Denais, S. Bruyere, F. Monsieur, E. Vincent, and G. Ghibaudo, IEEE Trans. Device Mater. Reliab. **5**, 5 (2005).
10. A. Neugroschel, G. Bersuker, R. Choi, and B. H. Lee, IEEE Trans. Device Mater. Reliab. **8**, 47 (2008).
11. S. Gamerith and M. Pölzl, Microelectron. Reliab. **42**, 1439 (2002).
12. R. Entner, T. Grasser, O. Triebl, H. Enichlmair, and R. Minixhofer, Microelectron. Reliab. **47**, 697 (2007).
13. M. Alwan, B. Beydoun, K. Ketata, and M. Zoaeter, Microelectron. J. **38**, 727 (2007).
14. S. Aresu, W. Kanert, R. Pufall, and M. Goroll, Microelectron. Reliab. **47**, 1416 (2007).
15. T. Grasser, T. Aichinger, G. Pobegen, H. Reisinger, P.-J. Wagner, J. Franco, M. Nelhiebel, and B. Kaczer, in *Proc. Intl. Reliab. Phys. Symp. (IRPS)*, Monterey, CA, 2011, p. 605.
16. T. Aichinger, M. Nelhiebel, and T. Grasser, Microelectron. Reliab. **48**, 1178 (2008).
17. T. Aichinger, M. Nelhiebel, and T. Grasser, IEEE Trans. Electron Devices **56**, 3018 (2009).
18. T. Aichinger, M. Nelhiebel, S. Einspieler, and T. Grasser, J. Appl. Phys. **107**, 024508 (2010).
19. T. Aichinger, M. Nelhiebel, S. Decker, and T. Grasser, Appl. Phys. Lett. **96**, 133511 (2010).
20. B. Jayant Baliga, *Modern Power Devices* (John Wiley & Sons, New York, 1987).
21. V. Benda, J. Gowar, and D. A. Grant, *Power Semiconductor Devices* (John Wiley, New York, 1999).
22. N. Stojadinović, D. Danković, S. Djorić-Veljković, V. Davidović, I. Manić, and S. Golubović, Microelectron. Reliab. **45**, 1343 (2005).
23. D. Danković, I. Manić, S. Djorić-Veljković, V. Davidović, S. Golubović, and N. Stojadinović, Microelectron. Reliab. **46**, 1828 (2006).

24. D. Danković, I. Manić, S. Djorić-Veljković, V. Davidović, S. Golubović, and N. Stojadinović, Microelectron. Reliab. **47**, 1400 (2007).
25. I. Manić, D. Danković, S. Djorić-Veljković, V. Davidović, S. Golubović, and N. Stojadinović, Microelectron. Reliab. **49**, 1003 (2009).
26. N. Stojadinović, D. Danković, I. Manić, A. Prijić, V. Davidović, S. Djorić-Veljković, S. Golubović, and Z. Prijić, Microelectron. Reliab. **50**, 1278 (2010).
27. I. Manić, D. Danković, A. Prijić, V. Davidović, S. Djorić-Veljković, S. Golubović, Z. Prijić, and N. Stojadinović, Microelectron. Reliab. **51**, 1540 (2011).
28. D. Danković, I. Manić, V. Davidović, S. Djorić-Veljković, S. Golubović, and N. Stojadinović, Microelectron. Reliab. **48**, 1313 (2008).
29. T. Grasser, B. Kaczer, P. Hehenberger, W. Goes, R. O'Connor, H. Reisinger W. Gustin, and C. Schlünder, in *Proc. Intl. Electron Devices Meeting (IEDM)*, Washington, DC, 2007, p. 801.
30. T. Grasser and B. Kaczer, IEEE Trans. Electron Devices **56**, 1056 (2009).
31. V. Huard, in *Proc. Intl. Reliab. Phys. Symp. (IRPS)*, Anaheim, CA, 2010, p. 33.
32. M. Denais, V. Huard, C. Parthasarathy, G. Ribes, F. Perrier, N. Revil, and A. Bravaix, in *Proc. European Solid-State Device Research Conference (ESSDERC)*, Leuven, Belgium, 2004, p. 265
33. C. Shen, M. Li, X. Wang, Y. Yeo, and D. Kwong, IEEE Electron Device Lett. **27**, 55 (2006).
34. H. Reisinger, U. Brunner, W. Heinrigs, W. Gustin, and C. Schlünder, IEEE Trans. Device Mater. Reliab. **7**, 531 (2007).
35. M.-F. Li, D. Huang, C. Shen, T. Yang, W. J. Liu, and Z. Liu, IEEE Trans. Device Mater. Reliab. **8**, 62 (2008).
36. T. Grasser, P.-J. Wagner, P. Hehenberger, W. Goes, and B. Kaczer, IEEE Trans. Device Mater. Reliab. **8**, 526 (2008).
37. D. Brisbin and P. Chaparala, IEEE Trans. Device Mater. Reliab. **9**, 115 (2009).
38. A. Prijić, D. Danković, Lj. Vračar, I. Manić, Z. Prijić, and N. Stojadinović, Meas. Sci. Technol. **23**, 085003 (2012).
39. G. Ghibaudo, Electron. Lett. **24**, 543 (1988).
40. P. J. McWhorter and P.S. Winokur, Appl. Phys. Lett. **48**, 133 (1986).
41. G. Groeseneken, H.E. Maes, N. Beltran, and R. F. De Keersmaecker, IEEE Trans. Electron Devices **31**, 42 (1984).
42. P. Habaš, Z. Prijić, D. Pantić, and N. Stojadinović, IEEE Trans. Electron Devices **43**, 2197 (1996).
43. S. Zafar, B. H. Lee, and J. Stathis, IEEE Electron Device Lett. **25**, 153 (2004).
44. S. Zafar, A. Callegari, E. Gusev, and M. V. Fischetti, J. Appl. Phys. **93**, 9298 (2003).
45. T. Grasser, P.-J. Wagner, H. Reisinger. Th. Aichinger, G. Pobegen, M. Nelhiebel, and B. Kaczer, in *Proc. Intl. Electron Devices Meeting (IEDM)*, Washington DC, 2011, p.618.
46. D. L. Griscom, J. Appl. Phys. **58**, 2524 (1985).
47. D. B. Brown, IEEE Trans. Nucl. Sci. **32**, 3900 (1985).
48. F. B. McLean and H. E. Boesch Jr. IEEE Trans. Nucl. Sci. **36**, 1772 (1989).
49. N. S. Saks, C. M. Dozier, and D. B. Brown, IEEE Trans. Nucl. Sci. **35**, 3900 (1988).
50. R. E. Stahlbush, A. H. Edwards, D. L. Griscom, and B. J. Mrstik, J. Appl. Phys. **73**, 658 (1993).
51. D. M. Fleetwood, Microelectron. Reliab. **42**, 523 (2002).
52. A. A. Katsetos, Microelectron. Reliab. **48**, 1655 (2008).
53. C. Bernard, G. Math, P. Fornara, J. Ogier, and D. Goguenheim, Microelectron. Reliab. **49**, 1008 (2009).
54. T. P. Ma and P.V. Dressendorfer, *Ionizing Radiation Effects in MOS Devices and Circuits* (John Wiley & Sons, New York, 1989).
55. A. Stesmans, J. Appl. Phys. **88**, 489 (2000).
56. D. M. Fleetwood, M. R. Shaneyfelt, W. L. Waren, J. R. Schwank, T. L. Meisenheimer, and P. S. Winokur, Microelectron. Reliab. **35**, 403 (1995).
57. J. F. Conley Jr., P. M. Lenahan, A. J. Lelis, and T. R. Oldham, IEEE Trans. Nucl. Sci. **42**, 1744 (1995).
58. D. M. Fleetwood and N. S. Saks, J. Appl. Phys. **79**, 1583 (1996).

59. T. Yang, C. Shen, M. F. Li, C. H. Ang, C. X. Zhu, Y.-C. Yeo, G. Samudra, and D.-L. Kwong, IEEE Electron Device Lett. **26**, 758 (2005).
60. S. Mahapatra and M. A. Alam, IEEE Trans. Device Mater. Reliab. **8**, 35 (2008).
61. D. S. Ang, S. Wang, G. A. Du, and Y. Z. Hu, IEEE Trans. Device Mater. Reliab. **8**, 22 (2008).
62. G. S. Ristić, M. M. Pejović, and A. B. Jakšić, Appl. Surface Sci. **220**, 181 (2003).
63. K. O. Jeppson and C. M. Svensson, J. Appl. Phys. **48**, 2004 (1977).
64. M. Houssa, M. Aoulaiche, S. De Gendt, G. Groeseneken, M. M. Heyns, and A. Stesmans, Appl. Phys. Lett. **86**, 093506 (2005).
65. S. Zafar, J. Appl. Phys. **97**, 103709 (2005).
66. B. Kaczer, V. Arkhipov, R. Degraeve, N. Collaert, G. Groeseneken, and M. Goodwin, Appl. Phys. Lett. **86**, 143506 (2005).
67. T. Yang, C. Shen, M. F. Li, C. H. Ang, C. X. Zhu, Y.-C. Yeo, G. Samudra, S. C. Rustagi, M. B. Yu, and D.-L. Kwong, IEEE Electron Device Lett. **26**, 826 (2005).
68. D. Varghese, S. Mahapatra, and M. A. Alam, IEEE Electron Device Lett. **26**, 572 (2005).
69. D. S. Ang, IEEE Electron Device Lett. **27**, 412 (2006).
70. V. Huard, C. Parthasarathy, N. Rallet, C. Guerin, M. Mammase, D. Barge, and C. Ouvrard, in *Proc. Intl. Electron Devices Meeting (IEDM)*, Washington, DC, 2007, p. 797.
71. Z. Q. Teo, D. S. Ang, and K. S. See, in *Proc. Intl. Electron Devices Meeting (IEDM)*, Baltimore, MD, 2009, p. 737.
72. T. Grasser, H. Reisinger, P.-J. Wagner, W. Goes, F. Schanowsky, and B. Kaczer, in *Proc. Intl. Reliab. Phys. Symp. (IRPS)*, Anaheim, CA, 2010, p. 16.
73. D. S. Ang, Z. Q. Teo, T. J. J. Ho, and C. M. Ng, IEEE Trans. Device Mater. Reliab. **11**, 19 (2011).
74. T. Grasser, B. Kaczer, W. Goes, H. Reisinger, T. Aichinger, P. Hehenberger, P.-J. Wagner, F. Schanowsky, J. Franco, M. Toledano Luque, and M. Nelhiebel, IEEE Trans. Electron Devices **58**, 3652 (2011).
75. T. Grasser, Microelectron. Reliab. **52**, 39 (2012).
76. E. Cartier, Microelectron. Reliab. **48**, 201 (1998).
77. P. E. Blöchl and J. H. Stathis, Phys. Rev. Lett. **83**, 372 (1999).
78. R. Degraeve, B. Kaczer, and G. Groeseneken, Microelectron. Reliab. **39**, 1445 (1999).
79. P. E. Blöchl, Phys. Rev. B **62**, 6158 (2000).
80. D. J. DiMaria and J. H. Stathis, J. Appl. Phys. **89**, 5015 (2001).
81. E. Rosenbaum and J. Wu, Microelectron. Reliab. **41**, 625 (2001).
82. E. M. Vogel, M. D. Edelstein, and J. S. Suehle, J. Appl. Phys. **90**, 2338 (2001).
83. M. Toledano-Luque, B. Kaczer, E. Simoen, P. J. Roussel, A. Veloso, T. Grasser, and G. Groeseneken, Microelectron. Eng. **88**, 1243 (2011).
84. M. Toledano-Luque, B. Kaczer, J. Franco, P. J. Roussel, T. Grasser, T. Y. Hoffmann, and G. Groeseneken, in *Proc. VLSI Symposium*, Honolulu, HI, 2011, p. 152.
85. L. Tsetseris, X. J. Zhou, D. M. Fleetwood, R. D. Schrimpf, and S. T. Pantelides, Appl. Phys. Lett. **86**, 142103 (2005).
86. M. Houssa, V. V. Afanas'ev, A. Stesmans, M. Aoulaiche, G. Groeseneken, and M. M. Heyns, Appl. Phys. Lett. **90**, 043505 (2007).
87. A. Stesmans, Phys. Rev. B **61**, 8393 (2000).
88. N. H. Nickel, A. Yin, and S. J. Fonash, Appl. Phys. Lett. **65**, 3099 (1994).
89. N. Stojadinović, I. Manić, V. Davidović, D. Danković, S. Djorić-Veljković, S. Golubović, and S. Dimitrijev, Microelectron. Reliab. **45**, 115 (2005).
90. S. Djorić-Veljković, V. Davidović, and S. Golubović, in *Proc. 52nd Conf. Electronics, Telecommunications, Computers, Automation, and Nuclear Technique (ETRAN)*, Palić, Serbia, 2008, p. MO1.3.

Chapter 21
NBTI and PBTI in HKMG

Kai Zhao, Siddarth Krishnan, Barry Linder, and James H. Stathis

Abstract As the CMOS technology nodes progress aggressively into "nano" era, introduction of High-k Metal gate (HKMG) has became key to maintain the scaling trend. Much effort has been devoted to understand the reliability aspects of HKMG over the last decade. Especially in recent years since HfO2-based HKMG was first implemented in high-performance products, the understanding of device instability such as PBTI and NBTI associated with HKMG and gate stack integration has been advanced significantly. In this chapter, some of the latest learning of NBTI and PBTI in HKMG is reviewed. In the first part of the discussion, latest results of process interaction with BTI are reviewed and the key process knobs in HKMG, such as high-k thickness, interfacial layer thickness, nitrogen concentration at IL, and channel type, which modulate NBTI and PBTI are discussed. In the second part of the discussion, recent study of relaxation dynamics and AC behavior of PBTI/NBTI in HKMG is reviewed. The implications to accurate modeling of BTI V_t shift under realistic circuit operation conditions are discussed.

21.1 Introduction

As the CMOS technology nodes progress aggressively into "nano" era, introduction of High-k Metal gate (HKMG) has became key to maintain the scaling trend [1–4]. The unique advantage of HKMG over conventional SiON/poly-Si is to allow aggressive scaling of electrical inversion thickness (Tinv) for the need of device short channel effect (SCE) control and performance enhancement while

K. Zhao (✉) • S. Krishnan • B. Linder • J.H. Stathis
e-mail: kzhao@us.ibm.com

T. Grasser (ed.), *Bias Temperature Instability for Devices and Circuits*,
DOI 10.1007/978-1-4614-7909-3_21, © Springer Science+Business Media New York 2014

keeping the gate leakage constrained under relevant technology use conditions. Much effort has been devoted to understand the reliability aspects of HKMG over the last decade [5–25]. Especially in recent years since HfO2-based HKMG was first implemented in high-performance products, the understanding of device instability such as PBTI and NBTI associated with HKMG and gate stack integration has been advanced significantly [11–15]. While the exact underlying mechanism still remains controversial, NBTI has been known since SiON technology to be primarily driven by the interface state density and hole trapping in bulk defects; on the other hand, PBTI has been known as a unique reliability phenomenon associated with HKMG and is mostly driven by the electron trapping in preexisting oxygen vacancy states in the bulk high-k dielectric [16, 17]. In general, NBTI and PBTI are reliability hazards that result in circuit degradation through the product's lifetime. Because both NBTI and PBTI in HKMG exhibit complex dynamic behavior, their impact in circuits is strongly dependent on the operation environment [17–19]. To minimize the circuit degradation induced by NBTI and PBTI in HKMG, great effort has been given to understand the process dependence and to reduce BTI through process optimization during technology development. In addition, to accurately predict the circuit degradation due to BTI, a focal point has been to understand the underlying mechanism that governs the BTI dynamics in HKMG.

In this chapter, some latest learning of NBTI and PBTI in HKMG will be reviewed. The first part of discussion will be focused on the latest study of process interaction and to review some of the key process knobs in HKMG that modulates NBTI and PBTI. In the second part, some recent study of relaxation dynamics and AC behavior of PBTI/NBTI in HKMG will be discussed.

21.2 BTI in HKMG: Gate Stack Process Dependence

HKMG reliability is a well-documented area of research, with literature spanning over a decade [5–25]. The introduction of HKMG products into the field in servers and mobile chips, over the last few years [2–4], however, has required a substantial improvement in the understanding of the many different sources of transistor instability with use, particularly in PBTI and NBTI. Due to the large number of variables that contribute to device instabilities, process centering needs to be a lot quicker, feeding into both reliability and performance. This complexity has led to the need to innovate in the way these instabilities are now assessed. PBTI has been known to be driven by preexisting traps in the high-k dielectric, into which electrons tunneling through the gate oxide can be trapped, causing an increase in the threshold voltage. NBTI, alternately, is mostly affected by the interface state density as well as hole traps in the SiO(N) and is known to be largely influenced by nitrogen in the interface layer—the exact mechanisms that lead to NBTI are still fairly controversial and HKMG-specific NBTI has only added to the controversies. An accurate understanding and modeling of the device instabilities and their dependencies on process variations in the gate stack is generally essential

Table 21.1 Key parameters that influence PBTI and NBTI

PBTI	NBTI
High-k thickness	Channel type (Si vs. SiGe)
Interface thickness	Interface quality
Thermal budget	Nitrogen in interface
Gate leakage	Thermal budget

to accurate modeling of the device over its lifetime. In this part of the discussion, we look at the key process knobs that modulate NBTI and PBTI in HKMG. While much of this discussion deals with a "gate first"-like integration scheme, we also explore some basic BTI behavior of the replacement gate "gate-last" scheme. The high-k dielectric used here is HfO_2, for the most part, unless otherwise specified.

As reviewed already in the previous chapter [49], the ramp voltage BTI test can significantly reduce the BTI stress time and provide quicker feedback for process optimization [9, 21]. For this reason, ramp voltage BTI is the main technique that was used in this work to study the process dependence of BTI in HKMG. There are two key parameters—the voltage to a specified value of threshold voltage (Vt) shift (in this case, we use V_g to 50 mV V_t shift or V_{g50}) and the slope of the $V_t - V_g$ transfer curve [the slope equals the sum of the voltage acceleration factor (m) and the time exponent (n)]. The two attributes can be used to estimate lifetime at a reference bias (e.g., operating bias).

21.2.1 Key Process Parameters That Influence BTI

Table 21.1 summarizes the most important parameters that influence NBTI and PBTI in HKMG.

In the subsequent sections, we will look at the impact of some of these process parameters on PBTI and NBTI, respectively.

21.2.2 PBTI: Interface Thickness and High-k Thickness

Figure 21.1 illustrates how PBTI and the lifetime projection evolve as HK thickness and IL thickness change in HKMG devices.

As the interface thickness is changed from 10 to14Å, through a variety of methods (including changing the temperature of thermally grown oxides), PBTI V_{g50} increases along with an increase in the slope. The interface thickness, therefore, is the primary parameter to minimize PBTI. The slope, as previously mentioned, is a sum of the voltage acceleration factor and the time exponent of PBTI. With the time exponent (~0.17) forming a small fraction of the slope, the determining factor in the change of slope is the voltage acceleration factor. High-k thickness, when reduced from about 20 Å to about 14 Å, increases V_{g50}. The increase in V_{g50} is accompanied

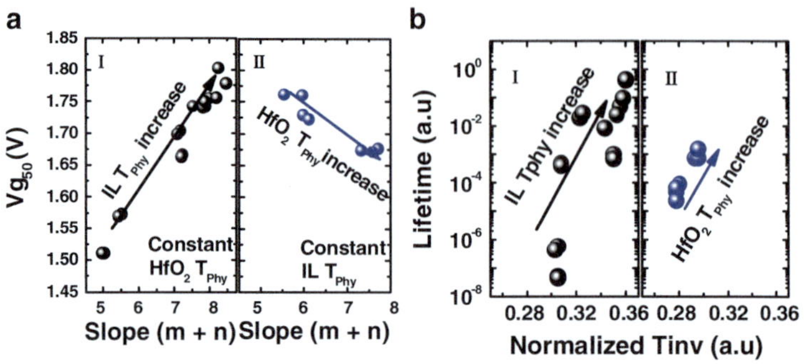

Fig. 21.1 (**a**) PBTI V_{g50} vs. slope as a function of interface layer thickness (I) and HfO_2 thickness (II)—the slope reduces rapidly with a reduction in interface layer and high-k thickness. (**b**) PBTI lifetime projection as a function of IL thickness at all operating biases (I); at an arbitrarily chosen operating bias, lifetime as a function of HfO_2 thickness (II)

Fig. 21.2 In Replacement Metal Gate HKMG, reducing high-k thickness by 4 Å improves PBTI lifetime by orders of magnitude at matched Tinv, while changing the interface has a gentler impact on lifetime

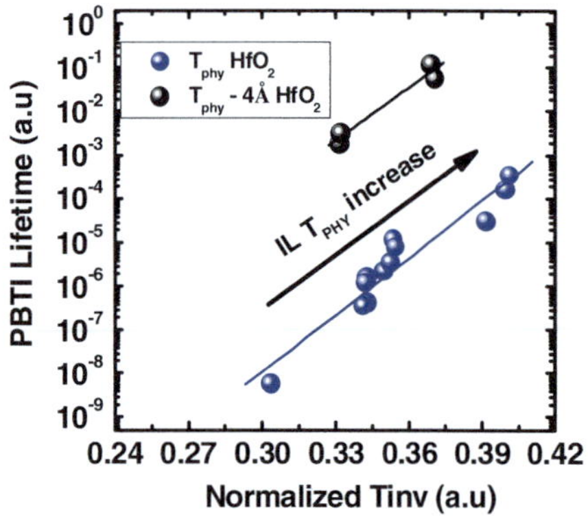

by a reduction in slope (or voltage acceleration factor). The lifetime at operating bias, therefore, can increase or decrease with increasing HfO_2 thickness based on which operating bias we extrapolate to.

For HKMG with Replacement Metal Gate integration, PBTI also exhibits a very strong dependence on interface layer and high-k thickness. The primary difference in gate-last HKMG is that reducing the high-k thickness has a much stronger impact in reducing the trap density and improves lifetime substantially (Fig. 21.2).

21.2.3 PBTI: Gate Leakage

While PBTI has been understood to be primarily due to electron trapping in preexisting traps in HK, relating PBTI to gate leakage is not a straightforward. This is mostly because the amount of trapped charge also depends on the number of trapping center in HK, which modulates with high-k thickness. However, for a given high-k thickness, PBTI can be correlated back to gate leakage as shown in Fig. 21.3.

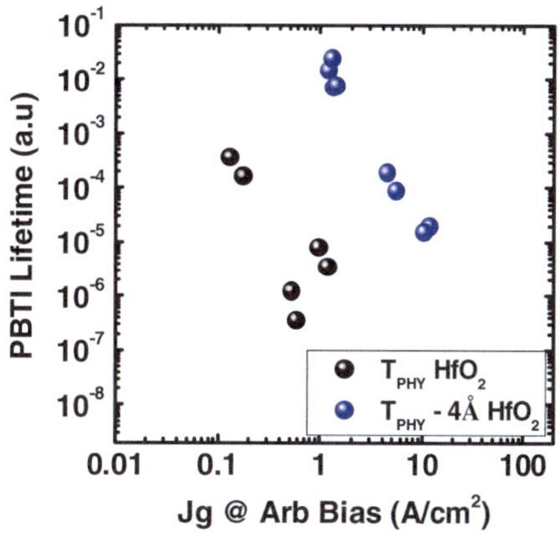

Fig. 21.3 PBTI dependence on gate leakage with given high-K thickness

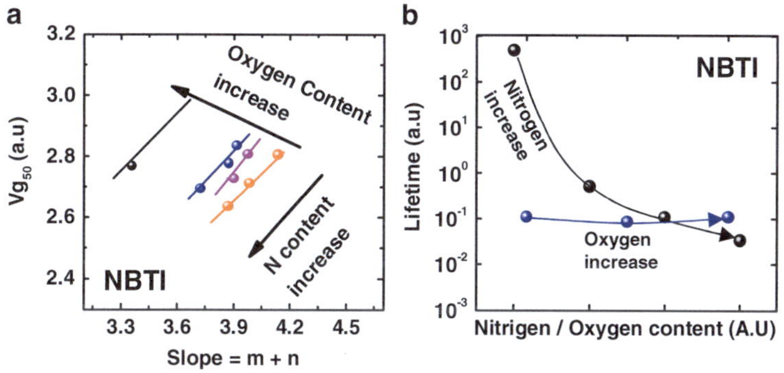

Fig. 21.4 (**a**) Increasing the nitrogen content in the interface layer reduces both V_{g50} and the slope, while increasing the oxygen content in the interface increases V_{g50}, but reduces the slope. (**b**) Nitrogen in the interface reduces lifetime, while the oxygen content in the interface does not appreciably affect the NBTI lifetime

21.2.4 *NBTI: Interface Quality, Thickness, and Nitrogen Quantity*

The sources for NBTI degradation in HKMG stacks are similar to the sources in previous silicon oxy-nitride-based technologies. The two key sources of NBTI are:

(a) Hydrogen atoms at the interface, which break under electric fields, causing positive charges to be created in the interface, leading to an increase in threshold voltage.

(b) Nitrogen in the interface layer, which acts as a trapping center for holes from the inversion layer in the channel. The impact of nitrogen and oxygen content is illustrated in Fig. 21.4 in gate-first-based HKMG devices. The nitrogen in the gate stack is increased using the temperature of thermal nitridation of the interface layer, while the oxygen content in the interface layer is increased using a thermal oxidation of the interface layer.

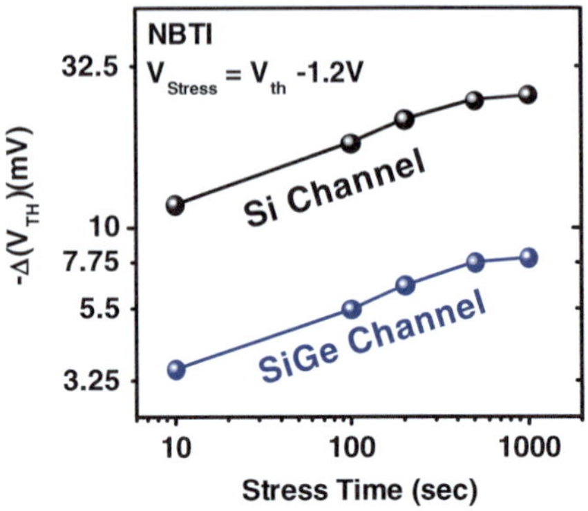

Fig. 21.5 Silicon germanium channels have lower NBTI than silicon channels do [22]

Increasing the nitrogen content in the interface layer reduces both V_{g50} and the slope, a reflection of the reduction in voltage acceleration factor. Increasing the thickness of the interface layer, however, has minor impacts on both the slope and V_{g50}. Nitrogen in the interface layer, therefore, reduces lifetime, while the oxygen content in the interface layer has no appreciable impact on NBTI extrapolations.

21.2.5 NBTI: Channel Type

While Silicon has dominated the device landscape till the 45 nm technology node, silicon germanium (SiGe) has been introduced as the channel material in the 32 nm technology node, due to improved threshold voltages in HKMG gate-first (and Replacement Metal Gate) technologies [22, 23]. Silicon germanium, additionally, improves NBTI substantially (Fig. 21.5). The NBTI improvement due to SiGe can be exploited by gate stack designers to add nitrogen into the gate stack for Tinv scaling.

21.2.6 NBTI: Thermal Budget

Thermal budget is a very significant modifier of NBTI behavior in HKMG stacks. Figure 21.6 illustrates the improvement in NBTI as we go from a low thermal budget gate stack to a high thermal budget gate stack.

NBTI and PBTI cause systematic degradation in nFETs and pFETs in HKMG technologies, causing circuit degradation due to threshold voltage shifts. PBTI is modulated primarily by the interface thickness and the high-k thickness. NBTI, on the other hand, depends strongly on the nitrogen quantity in the interface layer, the type of channel (Si vs. SiGe), and the thermal budget that the dielectric stack experiences. These process knobs can be used during technology development to reduce BTI effect. As technology continues to scale, new process innovations may be engaged to further reduce the BTI effect.

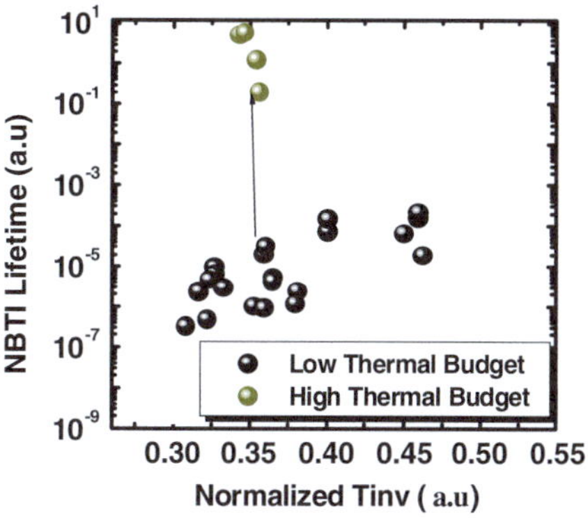

Fig. 21.6 NBTI lifetime projection of gate stacks with different thermal budgets. High thermal budget gate stacks have better NBTI than low thermal budget gate stacks

21.3 BTI Dynamics in HKMG: Relaxation Phenomena and AC Effects

While process optimization has been the driving force to reduce NBTI and PBTI effect in HKMG during technology development, modeling of the BTI effect accurately in realistic circuit operation condition is equally important. This requires good understanding of the underlying mechanism of NBTI and PBTI and their dynamics under DC and AC conditions. In this part of discussion, two important properties of BTI, relaxation phenomena and AC effects, will be discussed. Trapping/de-trapping of defects in HKMG is believed to be the main mechanism that governs the BTI dynamic characteristics.

21.3.1 Relaxation in BTI

Relaxation phenomenon in NBTI and PBTI has been long observed and studied in the past years [19, 25–32]. A transistor under stress experiences threshold voltage (*Vt*) increase. Immediately after releasing the stress voltage, part of the V_t shift during stress can be relaxed in a temporal fashion. Figure 21.7 shows a typical time trace of V_t shift during PBTI stress and relaxation. The typical time scale of the relaxation process spans over orders of magnitude in time from sub-usec to hours. The relaxation phenomena have profound implication to BTI dynamic behavior. As a result of the relaxation effect, a transistor under AC operation condition generally experiences less V_t shift than under DC condition over its lifetime. Detailed study of relaxation also can provide insight to the understanding of underlying mechanism that governs the BTI dynamics.

In this part of discussion, experimental data is collected from samples with hafnium-based high-k dielectric layer on a SiO_2 interlayer (IL) with TiN/poly-Si electrodes. Typical waveforms for DC stress and AC stress with 50% duty cycle are illustrated in Fig. 21.8.

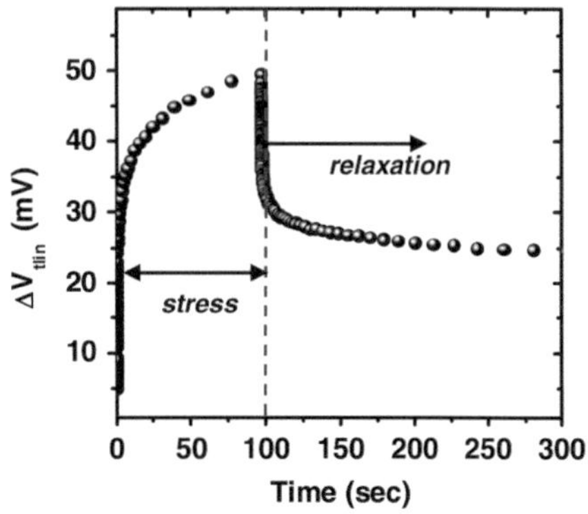

Fig. 21.7 Example of typical time trace of V_t shift during PBTI stress and relaxation

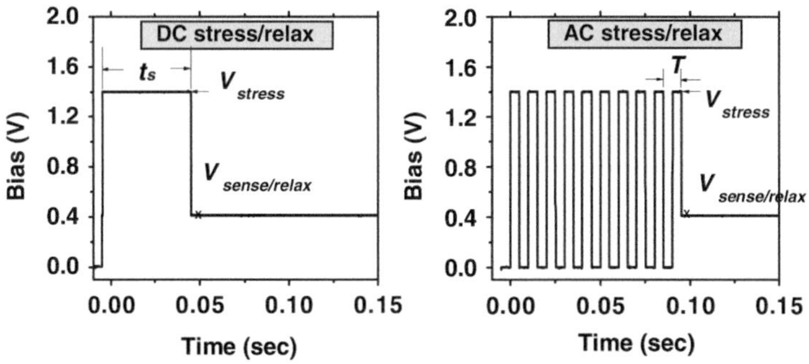

Fig. 21.8 Typical waveforms used to study PBTI relaxation after AC and DC stresses. Relaxation is measured ~300 μs after the stress is removed. For AC stress, the net stress time is defined as $t_s = N \times C \times T$

During AC stress, the gate bias is alternating between V_{stress} and zero. The net stress time is defined as $t_s = N \times C \times T$, where N is the total number of the stress cycles, C is duty cycle, and T is the period. In all cases, AC stress is compared to DC stress with the same net stress time t_s. To monitor the PBTI relaxation, a fresh device is stressed for t_s, and then the drain current I_d is measured at sense bias set approximately equal to the initial *Vt*. Finally, the V_t shift is calculated from the I_d degradation by comparing the I_d value with a reference I_d–V_g curve measured before the device was stressed. For both stress and relaxation measurements, the first I_d point is measured ~300 μs after the end of stress.

Figure 21.9 shows the comparison of PBTI and NBTI relaxation traces measured after different stress time t_s, under AC (100 Hz) and DC stresses. In this experiment, V_{stress} and V_{relax} were kept the same for all measurements. As can be seen, in both PBTI and NBTI, for the same stress time t_s, AC stress always results in less *Vt* shift. It also shows that the relaxation slope right after AC stress is much shallower than those after DC stress. At long relaxation time, the relaxation after DC and AC stresses gradually merges together. This indicates that after the same total stress time T_s, AC stress results in less trapped charges, thus less V_t shift than DC stress.

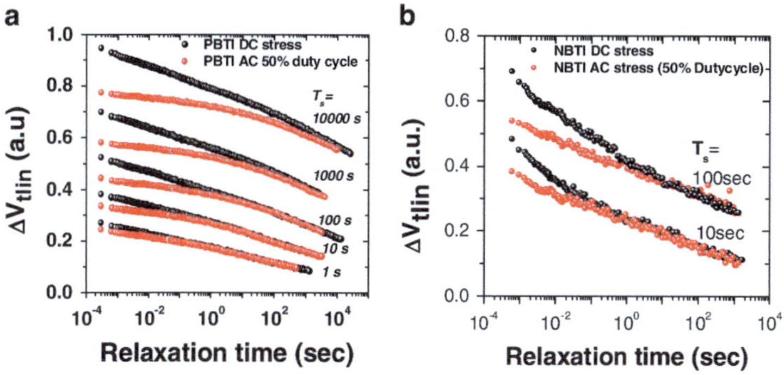

Fig. 21.9 (**a**) Relaxation traces measured after AC (100 Hz) and DC stresses with stress time changing from 1 to 10,000 s. (**b**) NBTI relaxation traces measured after AC (100 Hz) and DC stresses with stress time of 10 and 100 s

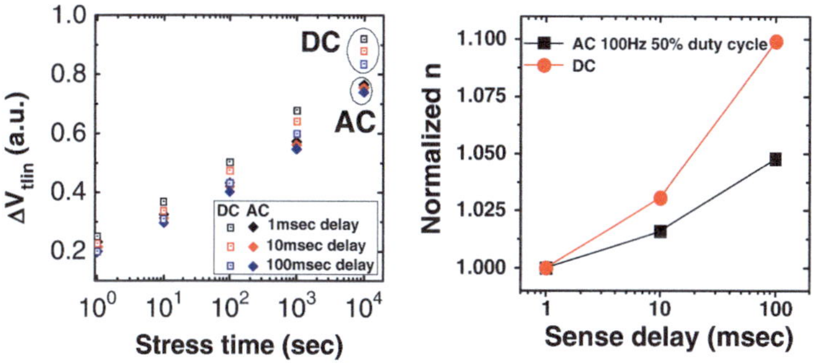

Fig. 21.10 PBTI V_t shift measured with different delay time under AC (100 Hz) and DC stresses, showing AC stress impact by measurement delay is much less than DC stress

As stress t_s increases, the difference between DC stress and AC stress also increases. The shallower slope at the beginning of AC relaxation suggests the traps responsible for V_t shift difference are mostly shallow traps with short emission time.

Because AC relaxation exhibits much shallower slope, the measurement sensitivity to delay time is reduced. As shown in Fig. 21.10, for the same measurement delay time, the relative difference in measured PBTI V_t shift is suppressed in the AC stress case.

The comparison of relaxation after AC and DC BTI stresses reveals that the stress and relaxation dynamics are strongly correlated. For both NBTI and PBTI in HKMG, the difference in AC and DC relaxations is driven by the different distribution of trap occupancy after AC and DC stresses. A simplified model is illustrated in Fig. 21.11. During DC and AC stresses, filling of trap states follows a distribution function $F(t_r, t_s)$ and $F'(t_r, t_s)$, where t_r is the emission time of traps and t_s is the total stress time. For the same amount of stress time, t_s, the difference between F and F' is caused by the additional de-trapping events occurring at the short relaxation intervals during AC stress. The relaxation interval of AC stress is

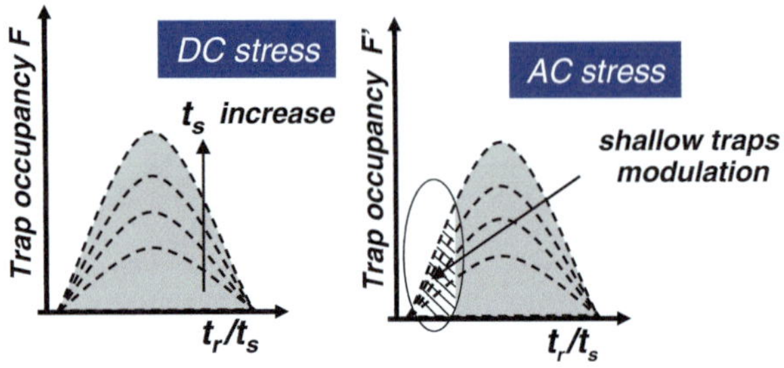

Fig. 21.11 Schematic comparison of the trap-filling process subject to DC and AC stresses. Notice that for AC stress case, occupancy modulation affects mostly the shallow traps

determined by the frequency and duty cycle and it mostly modulates the occupancy of shallow traps with relative short emission time t_r. In other words, for a given stress time, AC stress populates less shallow traps. On the other hand, deep traps are not as "responsive" to AC modulation. Thus, AC and DC stresses produce similar occupancy for deep traps. Therefore, at the beginning, AC relaxation always starts at a lower V_t shift value and shows a slower relaxation rate. Then, as relaxation proceeds, when shallow traps are mostly emptied and deep traps become more and more dominant, the AC relaxation merges with the DC relaxation.

21.3.2 Fast Transient Relaxation of PBTI in HKMG

Fast transient relaxation can be often observed after PBTI stress in HKMG NFETs [29–32]. The fast transient component builds up quickly and relaxes fast. The presence of fast transient relaxation can have important implications to PBTI measurement requirement, circuit level modeling, and, in extreme case, circuit yield and functionality at "time zero" [33, 34]. Here, a recent study of fast transient relaxation of PBTI in HKMG is reviewed.

Figure 21.12 shows a series of PBTI relaxation curves measured after a wide range of stress times. A fast transient relaxation can be clearly seen at the beginning of each relaxation curve. The fast component quickly decays and merges to a slow relaxation process. As the stress time t_s is reduced, the contribution from the slow traps becomes less while the fast component remains the same. When the stress time t_s is reduced to a minimum time of ~200 μsec, the fast transient relaxation becomes dominant and the overall relaxation approaches an empirical power law time dependence t^{α} with exponent $\alpha = 0.38$. Note that due to measurement limitation, the first relaxation measurement is done ~300 μs after the releasing of stress. An empirical equation as shown below is found to be able to model both the fast and slow relaxation:

$$\Delta V_t\,(t_r, t_s) = R(t_s)\left[1 + B(t_r/t_s)^{\beta}\right]^{-1} + A \times {t_r}^{-\alpha} \qquad (21.1)$$

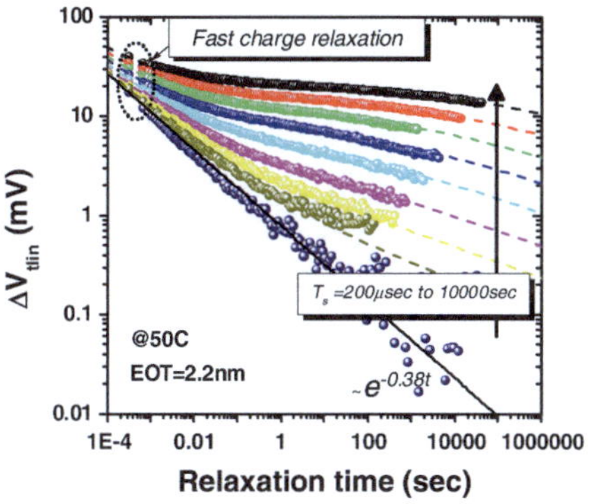

Fig. 21.12 Relaxation curves measured after different stress times. *Dash lines*, model fitting. *Solid black line*, extracted fast relaxation from model fitting, showing good agreement with the measured fast transient relaxation

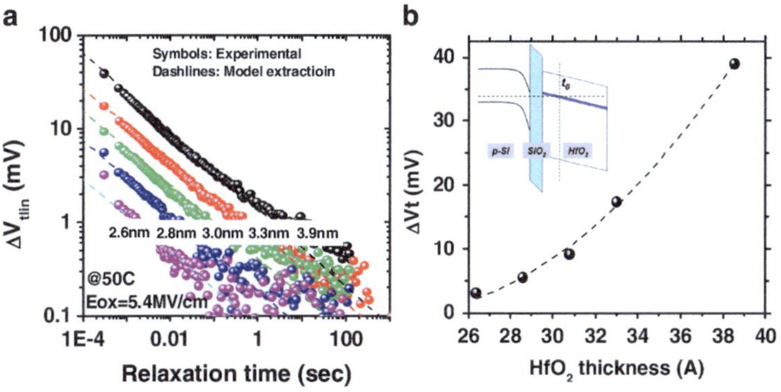

Fig. 21.13 (**a**) Fast transient relaxations measured from samples with different high-k thickness, showing the same power law dependence after short stress under constant stress field. The amplitude of the fast transient relaxation decreases as the high-k layer becomes thinner. (**b**) Parabolic high-k layer thickness dependence of fast transient V_t shift, suggesting the fast traps are located in a narrow energy band across the HK layer

where the first term on RHS describes the typical universal relaxation [29] and the second term is an empirical fitting of the fast transient component. As shown in Fig. 21.12, the model calculation using Eq. (21.1) (the solid line and the dash lines) fits the experimental data very well for both the fast and slow relaxation process over a wide range of stress times. Note that due to measurement limitation, the relaxation is measured ~300 μs after the releasing of stress; thus, the empirical power law dependence of the fast relaxation may not apply to shorter time scales. To understand the relaxation dynamics at shorter time, an ultrafast measurement technique is needed.

Both fast and slow V_t instabilities contribute to the overall V_t shift in HKMG nFETs. The fast transient component dominates the overall relaxation for PBTI stress with short stress time. Figure 21.13 shows that for samples with different HK thickness, the relaxation measured after short stress under constant field condition. The fast relaxation is strongly modulated by the HK thickness. As HK thickness

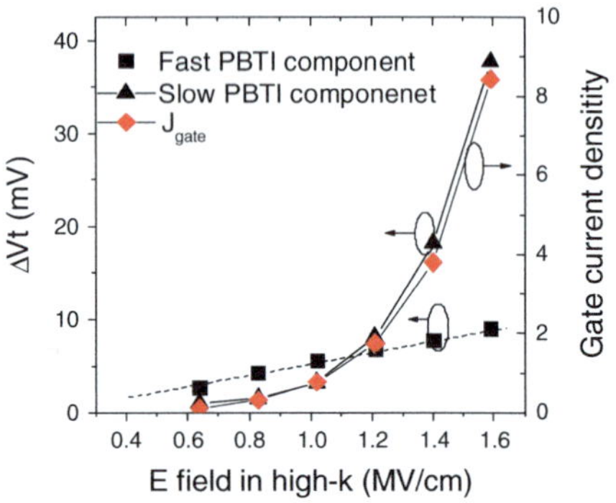

Fig. 21.14 Fast transient component follows linear field dependence and does not correlate with the gate leakage current. Increase of slow BTI component and the increase of the gate leakage current both follow nonlinear field dependence

decreases, the amplitude of fast transient V_t shift quickly diminishes to a negligible level, following a parabolic dependence. This suggests that the defects responsible to the fast transient component locate in a narrow energy band in high-k layer [35]:

$$\Delta Vt = n(E) \cdot \frac{\left(t_{HfO_2} - t_0\right)^2}{2 \cdot \varepsilon_0 \cdot \varepsilon_{HfO_2}} \tag{21.2}$$

Thanks to the strong HK thickness dependence, the fast transient V_t instability can be effectively suppressed as the HK thickness reduces with the general scaling trend.

Very different field dependence of the fast and slow PBTI component is shown in Fig. 21.14. It is interesting to notice that the increase of the slow PBTI component correlates the increase of gate leakage current and follow nonlinear field dependence. In [47, 48], it is reported that NBTI follows power law field dependence. On the other hand, the fast PBTI component does not correlate with gate leakage current and follows linear field dependence. This evidence again suggests that the fast transient component is likely associated with a group of defects in a narrow energy band near Fermi level. These defects have very large capture cross section, i.e., very short capture time constant, so that they can become fully occupied very quickly. As a result, the V_t shift caused by these defects are limited mostly by the total number of the available defect states under a given electrical field, i.e., $\Delta Vt = N(E)/C_{eff}$, rather than the current density.

21.3.3 AC Effect of BTI in HKMG

As a result of the relaxation phenomenon in NBTI and PBTI, a transistor under AC stress condition in general experiences less V_t shift than under DC stress condition. Figure 21.15 shows an example of PBTI under AC and DC stresses. Clearly under AC stress, V_t shift is much reduced at a given power-on time. In high-performance circuits, transistors are subject to AC switching during most of their

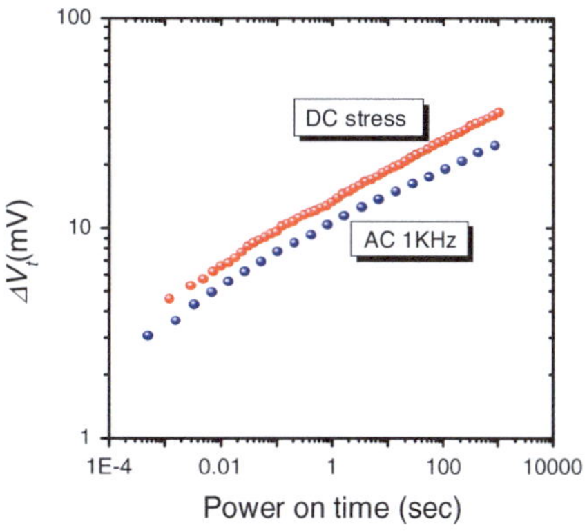

Fig. 21.15 PBTI under DC and AC stresses show clear reduction in V_t shift under AC stress at a given power-on time

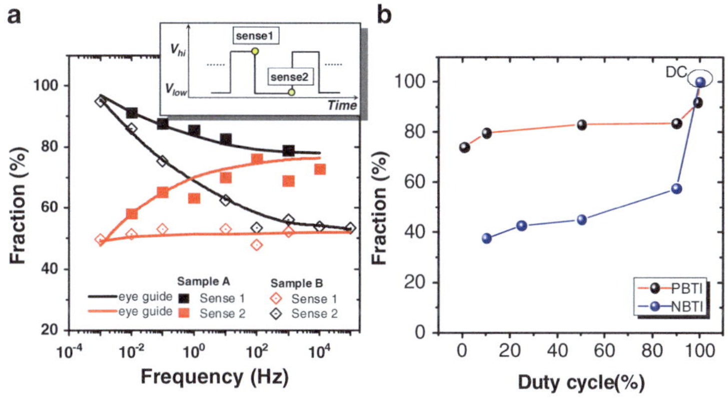

Fig. 21.16 (**a**) PBTI frequency dependence measured from two HKMG samples with different process conditions. Sense1 and sense 2 are measured after a V_{high} phase and a V_{low} phase, respectively. (**b**) Duty cycle dependence of NBTI and PBTI measured on HKMG PFET and NFET

lifetime. AC effect of NBTI and PBTI can provide significant relief to the overall circuit degradation over circuit lifetime. AC effect in both NBTI and PBTI is highly dependent on frequency and duty cycle of the AC bias.

Two examples of PBTI frequency dependence measured from different HKMG samples are shown in Fig. 21.16a. As shown in the inset, during AC stress, V_t shift is measured at sense 1 (end of a V_{high} phase) and sense 2 (end of a V_{low} phase). Fraction is defined as the ratio between V_t shift under AC stress and the reference DC stress at a given power-on time. The difference in fraction between sense 1 and sense 2 is caused by the relaxation occurred during the V_{low} phase. As frequency increases, the difference between sense 1 and sense 2 quickly reduces and fraction merges to a stable level and tends to become frequency independence in the tested frequency range. As can be seen, for the two samples with different gate stack processes, the fraction follows the similar general trend but merges at different levels. As will be discussed later, this could be understood from a single defect trapping/de-trapping point of view.

For both NBTI and PBTI in HKMG, the AC effects are also largely modulated by duty cycle. Figure 21.16b shows the typical duty cycle dependence of NBTI and PBTI measured on different HKMG samples. In the experiment, frequency is fixed at 100 Hz. As can be seen, in this case, for both NBTI and PBTI, the AC fraction exhibits "S" shape dependence on duty cycle, which is similar to what has been widely observed from other sources. The dependence is most sensitive at duty cycle range from 0 to 10% and 90 to 100%. Frequency dependence and duty cycle dependence are the two key properties that govern the overall AC effect and need to be captured in circuit level NBTI and PBTI model. As will be discussed next, AC effect and its frequency and duty cycle dependence can be largely understood from a single defect trapping/de-trapping point of view.

21.3.4 Understanding of Underlying Mechanism: Trapping/De-trapping of Single Defect

NBTI and PBTI in HKMG exhibit complex relaxation and AC dynamics. It is essentially important to capture these effects in NBTI and PBTI modeling in order to accurately predict any BTI-induced degradation under circuit operation environment. In recent years, substantial progress has been made towards the understanding of fundamental mechanism of NBTI and PBTI. With the availability of deeply scaled FETs, the properties of individual defects in gate dielectric have been studied through techniques such as Random Telegraph Noise (RTN) and Deep Level Transient Spectroscopy [11, 13, 36]. It has been demonstrated that the macroscopic NBTI dynamics can be reconstructed through summation of the microscopic trapping/de-trapping events of individual defects in PFETs with SiON gate stacks [11]. More recent study reveals that in HKMG, both NBTI and PBTI dynamics can also be well modeled with the similar approach [14]. This provides strong evidence that trapping/de-trapping of single defects is one of the fundamental mechanisms governing NBTI and PBTI dynamics in HKMG.

Trapping/de-trapping of a single defect in HKMG usually can be investigated through studying the random telegraph noise (RTN) measured from scaled FET. Figure 21.17 shows a series of RTN signal measured on a HKMG nFET with small gate area. In this experiment, the device under test is biased at different V_g and the RTN signal is measured by monitoring the linear drain current. When a trapping or de-trapping event occurs, the threshold voltage is disturbed and that translates to a sudden drain current decrease or increase. The capture time and emission time are the time interval between the current switches. As shown in Fig. 21.19, the capture time and emission time of a single defect follow exponential distribution:

$$P_c(t) = 1 - \exp\left(-\frac{t}{\tau_c}\right). \tag{21.3}$$

$$P_e(t) = 1 - \exp\left(-\frac{t}{\tau_e}\right). \tag{21.4}$$

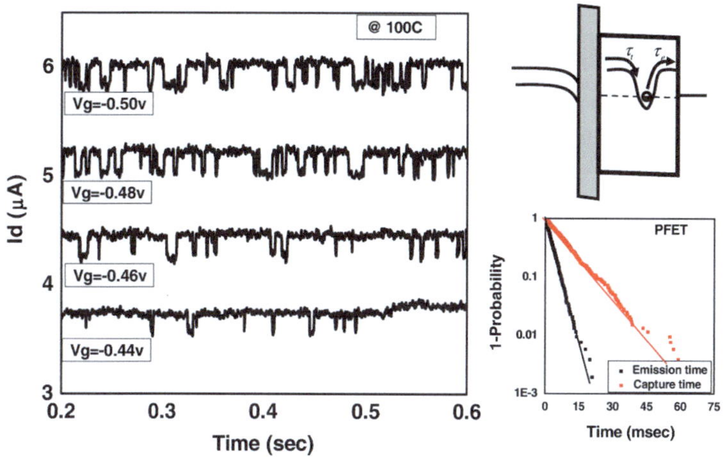

Fig. 21.17 Random Telegraph Noise (RTN) signal measured at different gate bias conditions on a 0.2 μm × 0.04 μm PFET. The capture time and emission time follow exponential distribution

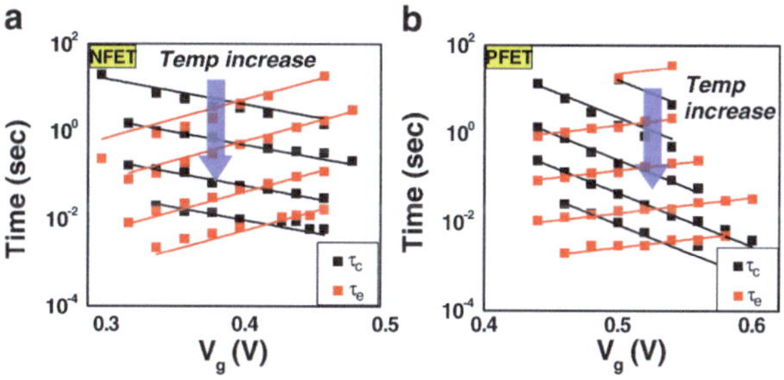

Fig. 21.18 For both HKMG NFET (**a**) and PFET (**b**), capture time and emission time of a single defect follow exponential dependence on V_g

By investigating a single defect trapping/de-trapping statistics at different gate bias and temperature conditions, an exponential voltage dependence within the bias range used in experiment and Arrhenius temperature dependence of capture time τ_c and emission time τ_e are observed as shown in Figs. 21.18 and 21.19. Comparisons are made side by side, showing very similar characteristics between HKMG NFETs and PFETs. The dependence of capture and emission time, τ_c, τ_e, on the gate voltage and temperature can be expressed as

$$\tau_c = \tau_{c0}\exp(-\beta_c \cdot V_g)\exp\left(-\frac{E_c}{kT}\right) \tag{21.5}$$

$$\tau_e = \tau_{e0}\exp(\beta_e \cdot V_g)\exp\left(-\frac{E_e}{kT}\right) \tag{21.6}$$

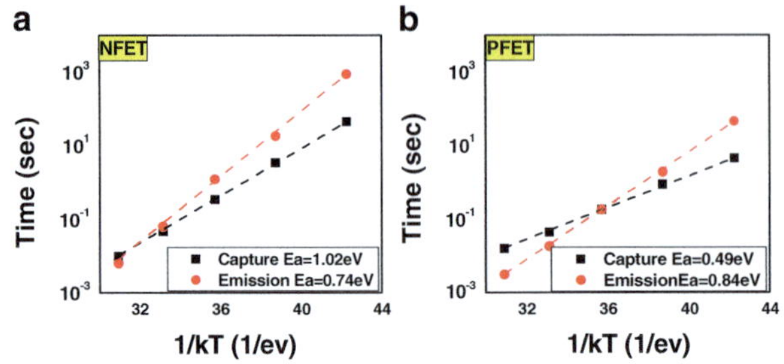

Fig. 21.19 Capture time and emission time of a single defect follow Arrhenius dependence on temperature in both HKMG NFET (**a**) and PFET (**b**)

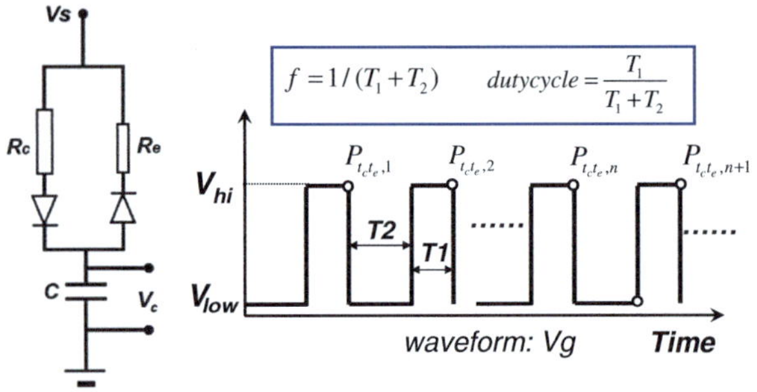

Fig. 21.20 Schematic diagram of the setup for modeling single defect occupancy under AC modulation. Note this can be generalized to DC case and AC with different frequency and duty cycle

where βc and βe are the voltage dependence factors and Ec and Ee are the activation energies. The strong dependence on temperature suggests that for the very thin gate stacks, direct tunneling process is unlikely governing the trapping/de-trapping process [37, 38]. On the other hand, the very similar characteristics between electron trapping in NFETs and hole trapping in PFETs indicate the underlying mechanism is likely related to the local atomic structure redistribution when capturing a charge carrier in gate dielectric [39–42], given that tunneling path for electrons and holes is very different in NFETs and PFETs.

With the insight that the trapping/de-trapping process follows exponential statistics, occupancy of a single defect in HKMG can be modeled as a RC element with different charging and discharging time constants [11, 46]. Based on this approach, a simple analytical model can be derived to understand how the occupancy of a single set of defects responds under AC stress. Since as discussed already, dynamics in NBTI and PBTI in HKMG is largely governed by the collective trapping/de-trapping response of many defects present in gate stack, understanding of the trapping/de-trapping response of a single defect can provide great insight to the general AC BTI dynamics.

Figure 21.20 shows the schematic diagram, illustrating the setup for single defect modeling under AC stress. A single set of defects with capture time of τ_c under

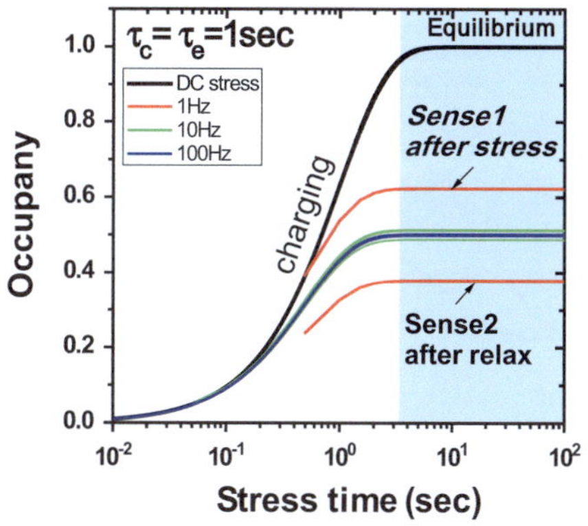

Fig. 21.21 Calculated occupancy of defects as a function of stress time under different stress modes. Sense 1 is the occupancy of the defects calculated right after a stress phase during an AC stress. Sense 2 is the occupancy of the defects calculated after the following relaxation phase. Note that an equilibrium state is reached at stress time $t_s \gg \tau_c$, where the occupancy level at both sense 1 and sense 2 does not grow anymore with stress time

stress condition and emission time of τ_e under relaxation condition are equivalent to a simple RC circuit with time constants $R_cC = \tau_c$ and $R_eC = \tau_e$ [3]. An AC stress signal Vs, with frequency equals 1/(T1 + T2) and duty cycle equals T1/(T1 + T2), is applied to the input terminal. The amplitude V_0 of the AC stress signal corresponds to the V_t shift when this single set of defects is all occupied. The occupancy level at a given moment can be monitored by calculating the charging state of the capacitor. As shown in the diagram, after n stress pulses, the occupancy level, $P_{t_ct_e,n}$, right after the stress phase and the occupancy level, $P'_{t_ct_e,n}$, after the relaxation phase can be derived analytically and are given by Eqs. (21.5) and (21.6):

$$P_{t_ct_e,n} = \left(\left(1-\exp\left(-\frac{T_1}{\tau_c}\right)\right)\left(1-\frac{1}{1-\exp\left(-\left(\frac{T_1}{\tau_c}+\frac{T_2}{\tau_e}\right)\right)}\right)\right) \times \exp\left(-(n-1)\left(\frac{T_1}{\tau_c}+\frac{T_2}{\tau_e}\right)\right) + \frac{\left(1-\exp\left(-\frac{T_1}{\tau_c}\right)\right)}{1-\exp\left(-\left(\frac{T_1}{\tau_c}+\frac{T_2}{\tau_e}\right)\right)} \tag{21.7}$$

$$P'_{\tau_c\tau_e,n} = P_{\tau_c\tau_e,n}\exp\left(-T_2/\tau_e\right) \tag{21.8}$$

Figure 21.21 shows for a single set of defects with $\tau_c = \tau_e = 1$ s (note here, τ_c is for under stress condition and τ_e is for under relaxation condition), how the occupancy changes as a function of time under different stress conditions. As wc can scc, under DC stress condition, the occupancy increases exponentially and approaches to 1 at stress time $t_s \gg \tau_c$. On the other hand, under AC stress, the occupancy only increases to a certain equilibrium level at long stress time. At the

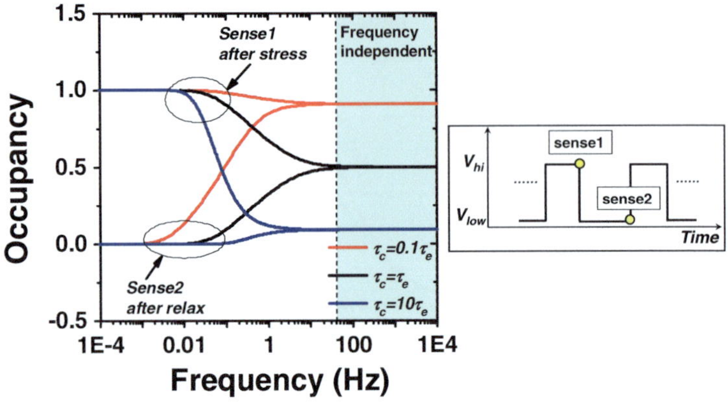

Fig. 21.22 Calculated frequency dependence of defects with different capture time τ_c and emission time τ_e. Note that it becomes frequency independent at frequency $f \gg [1/\tau_c, 1/\tau_e]$

equilibrium state, the occupancy jumps between two stable states, sense 1 state (right after the stress removal) and sense 2 state (after the following relaxation phase), showing that the capture probability during each stress phase equals the emission probability during the following relaxation phase. The expression for the occupancy at equilibrium is given by the second term in Eq. (21.7), showing that at equilibrium ($t_s \gg [\tau_c, \tau_e]$), the occupancy level is determined by the defect capture/emission time and the frequency and duty cycle of the AC stress signal. Note that the occupancy level at equilibrium under AC stress is always less than 1 and this explains the observation that AC BTI degradation is always lower than DC BTI degradation by a certain factor.

To further illustrate how the capture time and emission time impact the AC response, Fig. 21.22 plots, with duty cycle fixed at 50%, the calculated frequency dependence of defects with different capture and emission times. For a given set of defects, at a low-frequency region where f $\ll [1/\tau_c, 1/\tau_e]$, the occupancy of the defects at sense 1 and sense 2 switches between 1 and 0, indicating the defects are filled completely during a stress phase and then relaxed completely during the following relaxation phase. As the frequency increases, the occupancy levels at sense 1 and sense 2 gradually approach each other. At frequency f $\gg [1/\tau_c, 1/\tau_e]$, sense1 and sense 2 reach a common occupancy level and become frequency independent. From Eq. (21.7), the occupancy at high frequency can be expressed as

$$P_o\Big|_{t_s \gg \tau_c, f \gg [1/\tau_c, 1/\tau_e]} = \frac{1}{1 + \frac{\tau_c}{\tau_e}\left(\frac{1}{Dutycycle} - 1\right)} \tag{21.9}$$

As we can see, at equilibrium, for defects with very small τ_e/τ_c ratio, because the emission process during relax phase is much faster or more efficient than the capture process during stress phase, most defects remain emptied. On the other

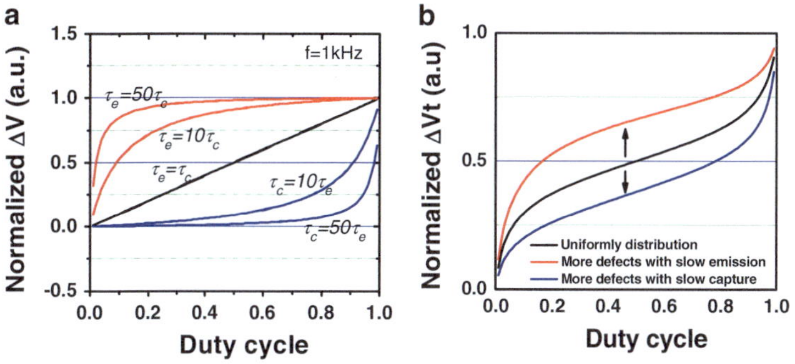

Fig. 21.23 (**a**) Duty cycle dependence of single sets of defects with different τ_e/τ_c ratios. (**b**) Duty cycle dependence of defects with different trap distributions

hand, for defects with large τ_e/τ_c ratio, the capture process during the stress phase becomes much more efficient and at equilibrium, most defects remain filled. More recent AC BTI data shows that complex frequency dependence may still exist at much higher-frequency region (>MHz) [43, 44]. This suggests a simple 2-state trapping/de-trapping model may not be sufficient here and a trap transformation model may need to be considered here [45].

Figure 21.23a shows the duty cycle dependence of defects with different τ_e/τ_c ratios. For defects with $\tau_e = \tau_c$, the occupancy at equilibrium shows a linear dependence on duty cycle. For defects with $\tau_e > \tau_c$, the duty cycle dependence is more sensitive at low duty cycle region. While for defects with $\tau_e < \tau_c$, it is more sensitive at high duty cycle region. Taking into account the defects distribution in terms of $\tilde{\tau}_e\tau_c$ ratio, the "S" shaped duty cycle dependence can be generated, as shown in Fig. 21.23b. For defects with symmetric distribution, the duty cycle dependence shows a symmetric behavior. On the other hand, for the defects with asymmetric distributions, the "S" curve moves up with more deep traps ($\tilde{\tau}_e\tau_c > 1$) or moves down with more shallow traps ($\tau_e/\tau_c < 1$). This suggests that in practical application, the shape of duty cycle dependence in general can be used as an indication of the trap distribution in terms of capture and emission times.

To expand from the single defect model to practically predict BTI behaviors under different dynamic stress conditions, the distribution of defects in terms of capture and emission times needs to be mapped out. The overall V_t shift can be calculated by the summation of V_t shift contribution from defects with different τ_c and τ_e, as shown in Eq. (21.10):

$$\Delta V_t(t) = \sum_{\tau_c}\sum_{\tau_e} P_{\tau_c\tau_e}(t)\Delta V_{t(\tau_c,\tau_e)} \tag{21.10}$$

where $P_{\tau_c\tau_e}(t)$ is the occupancy of defects with capture time τ_c and emission time τ_e and $\Delta V_{t(\tau_c,\tau_e)}$ is the V_t shift caused by the corresponding defects. A general method of mapping the defect distribution has been discussed in [11]. The basic idea of this method is to extract the capture time and emission time from a series

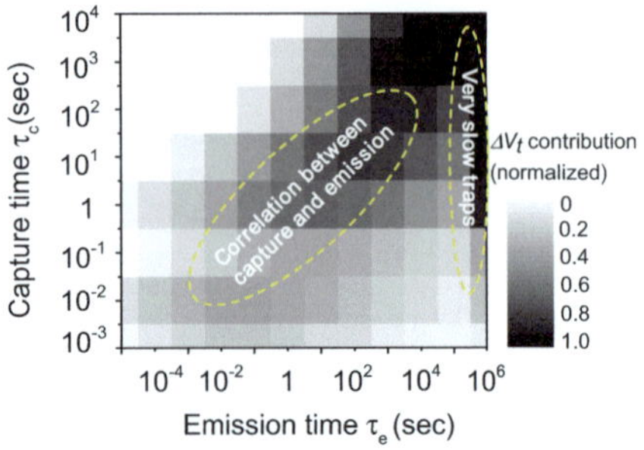

Fig. 21.24 Trap distribution of capture/emission time extracted from DC stress/relax traces [11]. Note a clear correlation between capture time and emission time is observed. Very slow traps ($\tau_e \gg \tau_c$) are found in the distribution

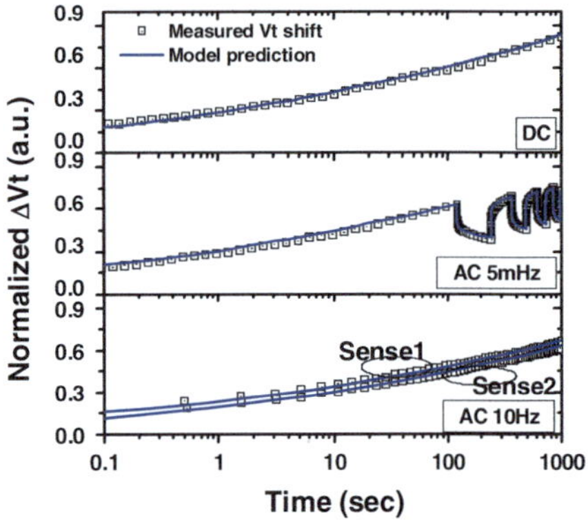

Fig. 21.25 PBTI degradation under different stress modes. *Symbols* are measured V_t shift. *Solid lines* are model prediction

of DC stress/relax measurements. The defect map is in general bias dependent. Figure 21.24 shows the defect map for the HKMG NFETs used in this work at stress voltage of 1.45 V and relax voltage of 0.3 V. Note that a clear correlation between capture time and emission time is observed.

Figure 21.25 shows the comparison between measured V_t shift and the corresponding model prediction under various stress conditions.

For the case of 10 Hz AC stress, V_t shift at sense 1 was measured right after stress (with a measurement delay of ~1 ms). V_t shift at sense 2 was measured after the relaxation phase. In the model calculation, the measurement delay time has been taken into account. As can be seen, the model prediction shows excellent agreement with the measured data for all cases. This further confirms the theory that in HKMG NFETs, PBTI dynamics is mostly governed by the distribution of defects and collective response (trapping/de-trapping) of individual defects to the applied stress.

The difference in relaxation dynamics between AC and DC PBTI and NBTI can also be well modeled from a trapping/de-trapping point of view. Figure 21.26a shows the experimental data and model predication of the relaxation dynamics after AC and DC PBTI stress. The AC relaxation shows much shallower relaxation rate at the beginning and then gradually merges to the DC relaxation. The model prediction shows excellent match to the experimental results. In Fig. 21.26b the occupancy of defects with different capture and emission times is plotted for both AC and

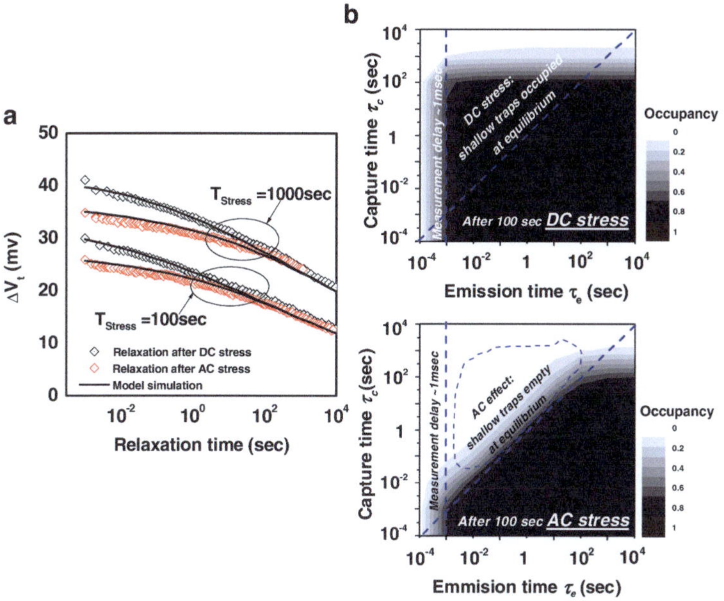

Fig. 21.26 (**a**) PBTI relaxation after DC and AC (100 Hz) stresses. *Symbols* are experimental data. *Solid lines* are model prediction. (**b**) Occupancy map calculated after AC and DC stresses. It is clearly demonstrated that the AC effect is essentially due to empty shallow traps at equilibrium. Note that the measurement delay time has been taken into account in the model calculation

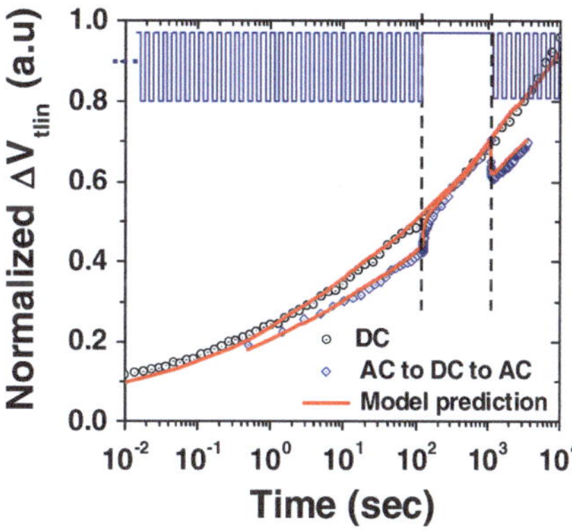

Fig. 21.27 Transition of PBTI between AC stress mode and DC stress mode. *Symbols* are experimental data. *Solid lines* are model prediction

DC PBTI (with the same net stress time). It is noticed that the defects with short emission time are mostly unoccupied because of the delay time (~1 ms) introduced in the measurement. It is clearly shown that the difference between AC PBTI and DC PBTI comes from the occupancy of shallow defects with $\tau_e < \tau_c$. For AC stress case, these shallow traps are mostly unoccupied at equilibrium.

In many circuit applications, the stress mode may change over time. Figure 21.27 shows an example where the stress mode is changed from AC to DC at ~100 s and then from DC back to AC at ~1,000 s. The degradation dynamics is compared to the continuous DC stress case. When the stress mode is changed from AC stress to DC stress, the V_t shift quickly approaches and merges to the DC degradation curve. Then, when the stress mode is changed from DC stress back to AC stress, the V_t shift relaxes towards the AC degradation curve and the degradation resumes after it merges with the AC degradation trend. The experiment was also simulated using the trapping/de-trapping model and the model prediction shows excellent agreement with the experimental observation. This simple example clearly demonstrates that the occupancy of each individual defect reaches certain equilibrium state under a given stress mode (DC, AC, frequency, duty cycle, etc.). When the stress mode is changed from one to another, the occupancy tends to transition to a new equilibrium condition appropriate to the stress condition.

21.3.5 Conclusion

In conclusion, experimental evidence suggests that although origin of defects is different between PBTI and NBTI in HKMG, those defects follow the similar statistical behavior. The capture and emission processes follow exponential statistics regardless whether it is electron trapping in HK layer or hole trapping in interfacial layer. By modeling the occupancy of single defects under AC stress, key insights can be summarized as the following: under AC stress, the occupancy level is determined by the equilibrium between charge capture during stress phase and charge emission during relaxation phase. τ_e/τ_c ratio is found to be the key parameter that determines the occupancy level at equilibrium. Excellent agreement between model simulation and the experimental data of PBTI under various AC stress conditions demonstrates clearly that the macroscopic BTI behaviors are largely governed by the microscopic response of each individual defects present in HKMG.

References

1. G. D. Wilk, R. M. Wallace and J. M. Anthony, App. Phys. Rev. 89, 10 (2001)
2. P. Packan, S. Akbar, M. Armstrong, D. Bergstrom, M. Brazier, H. Deshpande, K. Dev, G. Ding, T. Ghani, O. Golonzka, W. Han, J. He, R. Heussner, R. James, J. Jopling, C. Kenyon, S-H. Lee, M. Liu, S. Lodha, B. Mattis, A. Murthy, L. Neiberg, J. Neirynck, S. Pae, C. Parker, L. Pipes, J. Sebastian, J. Seiple, B. Sell, A. Sharma, S. Sivakumar, B. Song, A. St. Amour, K. Tone, T. Troeger, C. Weber, K. Zhang, Y. Luo, and S. Natarajan, IEEE IEDM (2009).
3. B. Greene, Q. Liang, K. Amarnath, Y. Wang, J. Schaeffer, M. Cai, Y. Liang, S. Saroop, J. Cheng, A. Rotondaro, S-J. Han, R. Mo, K. McStay, S. Ku, R. Pal, M. Kumar, B. Dirahoui, B. Yang, F. Tamweber, W.-H. Lee, M. Steigerwalt, H. Weijtmans, J. Holt, L. Black, S. Samavedam, M. Turner, K. Ramani, D. Lee, M. Belyansky, M. Chowdhury, D. Aime, B. Min, H. Van Meer, H. Yin, K. Chan, M. Angyal, M. Zaleski, O. Ogunsola, C. Child,

L. Zhuang, H. Yan, D. Permanaa, J. Sleight. D. Guo. S. Mittl, D. Ioannou, E. Wu, M. Chudzik, D.-G. Park, D. Brown, S. Luning, D. Mocuta, E. Maciejewski, K. Henson, E. Leobangung, VLSI Technology Symposium (2009).
4. H. Fukutome, D.H. Kim, S.M. Hwang, L.G. Jeong, S.C. Kim, J.C. Kim, I. Nakamatsu, M.K. Jung, W.C. Lee, Y. S. Kim, S.D. Kwon, G.H. Lyu, J.M. Youn, Y.M. Oh, M.H. Park, J.H. Ku, N. Lee, E.S. Jung, S. Paak, IEEE IEDM (2011).
5. G. Ribes J. Mitard, M. Denais, S. Bruyere, F. Monsieur, C. Parthasarathy, E. Vincent, G. Ghibaudo, IEEE Transactions on Device and Materials Reliability, 5, p. 5 (2005)
6. A.S. Oates, IEEE IEDM (2003)
7. K. Torii, H. Kitajima, T. Arikado, K. Shiraishi, S. Miyazaki, K. Yamabe. M. Boero, T. Chikyow, K. Yamada, IEEE IEDM (2004).
8. R. Degraeve, A. Kerber, P. Roussel, E. Cartier, T. Kauerauf, L. Pantisano, G. Groeseneken, IEEE IEDM (2003).
9. A. Kerber, E. Cartier, IEEE TDMR, 9, p. 147 (2009).
10. J.H. Stathis and S. Zafar, Microelectronics Reliability, 46, p. 270 (2006).
11. H. Reisinger, T. Grasser, W. Gustin and C. Schlünder, IEEE IRPS (2010)
12. B. Kaczer, S. Mahato, V. Valduga de Almeida Camargo, M. Toledao-Luque, P.J. Roussel, T. Grasser, F. Catthoor, P. Dobrovolnv, P. Zuber, G. Wirth, G. Groeseneken, IEEE IRPS (2011)
13. T. Grasser, H. Reisinger, P. -J. Wagner F. Schanovsky, W. Goes, B. Kaczer, IEEE IRPS (2010)
14. K. Zhao, J. H. Stathis, B. P. Linder, E. Cartier, A. Kerber, IEEE IRPS (2011)
15. E. Cartier, A. Kerber, T. Ando, M. Frank, K. Choi, S. Krishnan, B. Linder, K. Zhao, F. Monsieur, J. Stathis, V. Narayanan, IEEE IEDM (2011).
16. M. Aoulaiche, E. Simoen, C. Caillat, N. Collaert, G. Groeseneken, M. Jurczak, IEEE IEDM (2007).
17. S. Krishnan, M. Quevedo-Lopez, R. Choi, P. D. Kirsch, C. Young, R. Harris, J.J. Peterson, L. Hong-Jyh, L. Byoung Hun, J. C. Lee, IEEE Integrated Reliability Workshop Final Report (2005).
18. T. Grasser, P. J. Wagner, P. Hehenberger, W. Goes and B. Kaczer, IEEE Transactions on Device and Materials Reliability(2007)
19. K. Zhao, J. H. Stathis, A. Kerber and E. Cartier, IEEE IRPS (2010)
20. V. Huard, IEEE International IRPS (2010)
21. A. Kerber, E. Cartier, B. Linder, IEEE IEDM (2009)
22. S. Krishnan, U. Kwon, N. Moumen, M.W. Stoker, E. Harley, S. Bedell, D. Nair, B. Greene, K. Henson, M. Chowdhury, D. P. Prakash, E. Wu, D. Ioannou, E. Cartier, M. Na, S. Inumiya, K. McStay, L. Edge, R. Iijima. J. Cai, M, Frank, M. Hargrove, D. Guo, A. Kerber, H. Jagannathan, T. Ando, J. Shepard, S. Siddiqui, M. Dai, H. Bu, V. Narayanan, M. Chudzik, IEEE IEDM (2011).
23. L. Witters, J. Mitard, A. Veloso, A. Hikayy, J. Franco, T. Kauerauf, A. Steegen, N. Horiguchi, IEEE IEDM (2011).
24. S. Krishnan, E. Cartier, J. Stathis, M. Chudzik, IEEE IRPS (2011).
25. B. Kaczer, T. Grasser, P. J. Roussel, J. Martin-Martinez, R. O'Connor, B.J. O'Sullivan, G. Groeseneken, IEEE IRPS (2008)
26. T. Aichinger, M. Nelhiebel, T. Grasser, IEEE IRPS (2009)
27. D. Ielmini, M. Manigrasso, F. Gattel IEEE IRPS (2009)
28. S. Ramey, C. Prasad, M. Agostinelli, IEEE IRPS (2009)
29. T. Grasser, B. Kaczer, P. Hehenberger, W. Goes, R. O'Conner, H. Reisinger, W. Gustin, C. Schunder, IEEE IEDM 2007
30. D. Heh, C. Young, G. Bersuker, IEEE EDL, 29, 2, 2008
31. A. Kerber K. Maitra, A. Majumdar, M. Hargrove, R.J. Carter, E. Cartier, IEEE TED, 55, 11, 2008
32. K. Zhao, J. H. Stathis, E. Cartier, M. Wang, S. Zafar, IEEE IRPS (2012)
33. F. Adamu-Lema, C. Monzio Compagnoni, S.M. Amoroso, N. Castellani, L. Gerrer, s. Markov, A.S. Spinelli, A.L. Lacaita, A. Asenov, IEEE Transactions on Device and Materials Reliability (2013)

34. H. Reisinger, G. Steinlesberger, S. Jakschik, M. Gutsche, T. Hecht, M. Leonhard, U. Schroder, H. Seidl, D. Schumann, IEEE IEDM (2001)
35. A. Kerber, E. Cartier, IEEE TDMR, 9, 2, (2009)
36. S. S. Chung, C. M. Chang, Appl. Phys. Lett. (2008)
37. J. P. Campbell,J. Qin, K.P. Cheung, L.C. Yu, J.S. Suehle, A. Oates, K. Sheng, IEEE IRPS, (2009)
38. L. B. Freeman, W. E. Dahlke, Solid State Electron, 13, 11, (1970)
39. W. Goes, M. Karner, V. Sverdlov, T. Grasser, IEEE TDMR, (2008)
40. D. Veksler, G. Bersuker, S. Rumyyantsev, M. Shur, H. Park, C. Young, K.Y. Lim, W. Taylor, R. Jammy, IEEE International IRPS (2010)
41. P. Vashishta, R. K. Kalia, J. P. Rino, Phys. Rev. B, Condens. Matter, 41, 17, (1990)
42. G. Bersuker, D. Heh, C. Young, H. Park, P. Khanal, L. Larcher, A. Padovani, P. Lenahan, J. Ryan, B.H. Lee, H. Tseng, R. Jammy, IEEE IEDM, (2008)
43. T. Grasser, B. Kaczer, H. Reisinger, P.-J. Wagner, M. Toledano-Luque, IEEE, IRPS (2012)
44. Y-C Huang, T.-Y. Yew, W. Wang, Y.-H. Lee, J.R. Shih, K. Wu, IEEE, IRPS (2013)
45. T. Grasser, K. Rott, H. Reisinger, P. Wagner, W. Goes, F. Schanovsky, M. Waltl, M. Toledano-Luque, B. Kaczer, IEEE, IRPS (2013)
46. B. Kaczer, T. Grasser, J. Martin-Martinez, E. Simoen, M. Aoulaiche, P.J. Roussel, G. Groeseneken, IEEE, IRPS (2009)
47. H. Reisinger, R. Vollertsen, P. Wagner, T. Huttner, A. Martin, S, Aresu, W. Gustin T. Grasser, C. Schlünder, IEEE TRANSACTIONS ON DEVICE AND MATERIALS RELIABILITY (2009)
48. T. Grasser, B. Kaczer, W. Goes, Th. Aichinger, Ph. Hehenberger and M. Nelhiebel, IEEE IRPS (2009)
49. A. Kerber, E. Cartier, in *Bias Temperature Instability for Devices and Circuits*, ed. by T. Grasser (Springer, New York, 2013)

Chapter 22
PBTI in High-k Oxides

Chadwin D. Young and Gennadi Bersuker

Abstract Interaction of different materials in the multilayer high-k/metal gate stacks results in the formation of structural defects in the high-k dielectric and interfacial SiO_2 layer. This section discusses the impact that defects in *intrinsic* Hf-based dielectric layers have on the electron trapping process and concurrent defect generation occurring under positive bias stress in NMOS devices. A brief discussion on improvements to the intrinsic Hf-based films in regards to charge trapping is also discussed.

22.1 Introduction: Material Properties of High-k Dielectric Stacks

A multilayer structure and appreciably high density of as-grown structural defects in the transition metal oxides [1] complicates the evaluation of the electrical characteristics and reliability of the high-k gate stacks. Indeed, HfO_2 dielectrics were shown to have positively charged, mobile defects [2], consistent with characteristics of the oxygen vacancies in these materials as calculated by ab initio methods [3]. Oxygen vacancies in the high-k film induce, in turn, the oxygen vacancy generation in the underlying SiO_2 interfacial layer [4–6]. These preexisting defects, which act as electron traps, give rise to the fast transient charging (FTC) phenomenon [7], which can manifest itself in the observed threshold voltage instability and

C.D. Young (✉)
Department of Materials Science and Engineering, University of Texas at Dallas, Richardson, TX, USA
e-mail: Chadwin.Young@utdallas.edu

G. Bersuker
SEMATECH, Albany, NY, USA
e-mail: gennadi.bersuker@SEMATECH.org

T. Grasser (ed.), *Bias Temperature Instability for Devices and Circuits*,
DOI 10.1007/978-1-4614-7909-3_22, © Springer Science+Business Media New York 2014

mobility degradation [8]. The question then arises for separating contributions to the instability from the high-k and interfacial layers. Thus, valid assessments of the high-k gate stack properties require novel measurement techniques and analysis providing high spatial and time resolutions.

Extracting root causes of the instability in high-k gate stack structures requires deconvoluting the contributions from the preexisting electron traps and stress-generated traps in devices subjected to electrical stress [7, 9, 10], as well as delineating the interfacial SiO_2 layer and high-k layer contributions. Such tasks require testing sets of samples with different combinations of the gate stack materials (electrodes, high-k dielectrics, interfacial layer quality, substrate doping), processing conditions (anneal temperature, ambient, etc.), and thicknesses of individual layers, as well as adequate characterization techniques and data interpretations [11].

Here, we focus on the electrical measurements that allow the delineation of preexisting, bulk traps and generated, interfacial layer traps in high-k dielectric stack structures through the use of pulsed current–voltage (I–V) measurement methods [8, 12–14] and frequency-dependent charge pumping [11, 15–17] as a monitor during PBTI measurements [8, 12–14, 18–20].

22.2 Electron Trapping in High-k Dielectric Stacks

A schematic of the fast transient charging (FTC) process [21] is shown in Fig. 22.1. Substrate-injected electrons may tunnel through the interfacial layer (IL) and get trapped in the preexisting defects (electron traps) in the Hf-based film resulting in threshold voltage instability [9, 12, 22, 23] and degrading overall device performance [8, 24–27]. The inset shows the key time parameters of the fast transient charging effect (FTCE): the trapping occurs when the charging time, t_p, which is the sum of the pulse rise and width time, accedes the electron trapping time, which is a characteristic of the trap atomic structure and stress conditions [28] (see inset in Fig. 22.1). Therefore, to study the kinetics of trapping, the measurements must be faster than the characteristic trapping/detrapping times.

22.2.1 Fast Trapping Kinetics

Since the conventional DC measurement techniques are not fast enough to study the FTC kinetics, the pulsed I–V measurement was introduced [8, 12–14]. In this technique, the transistor under test is connected as part of an inverter circuit (inset, Fig. 22.2a). The pulsed I_d–V_g measurement output is shown in Fig. 22.2. The pulsed I_d–V_g curve exhibits a hysteresis—a shift between the rise (t_r) and fall (t_f) V_g

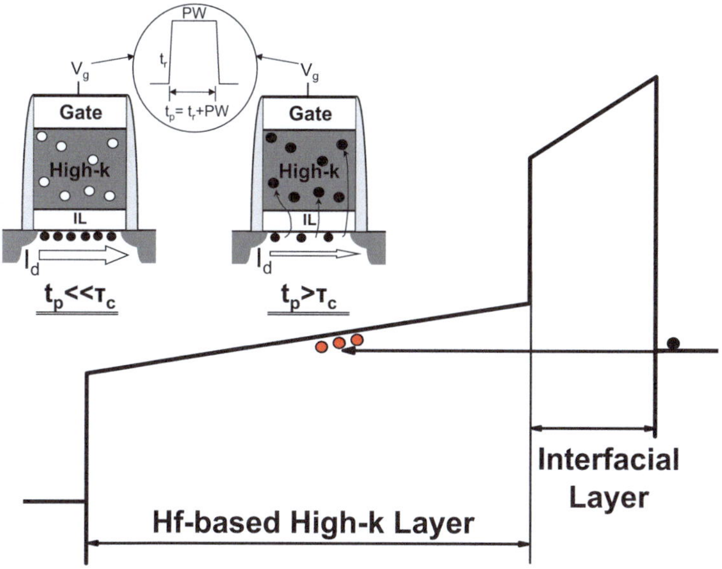

Fig. 22.1 Schematic of the charge trapping process and the corresponding energy band diagram. The substrate-injected electrons (nMOS in inversion) might get trapped by preexisting defects in the high-k layer if the total injection time t_p (the sum of t_r and pulse width—see inset) accedes the characteristic trapping time, τ_c

pulse swings (Fig. 22.2a)—while the I_d—time dependency shows that I_d decreases during the pulse width period. The width of the pulse represents an ultrafast constant voltage stress, and the I_d–V_g hysteresis and I_d—time degradation reflect the amount of charge trapped during this stress. These I_d changes can be translated into a threshold voltage shift (ΔV_t) allowing quantification of the trapped charge [8, 29]—which, in the non-optimized high-k example of Fig. 22.2, is 0.72 V.

22.2.2 *Fast Relaxation Kinetics*

Figure 22.3 illustrates the fast trapping and fast detrapping (i.e., relaxation) of the electrons when using the pulsed I_d—time approach for the given gate voltage pulse in the inset. This demonstrates that charges can be "detrapped" at a similar time scale as the fast trapping occurs. Therefore, fast trapping/detrapping can result in underestimating the charging effects when the conventional, DC-based "stress and sense" measurements are employed [16]. This is demonstrated in Fig. 22.4 showing significant detrapping during a very short relaxation period due to the delay between removal of the stress bias and sensing measurement.

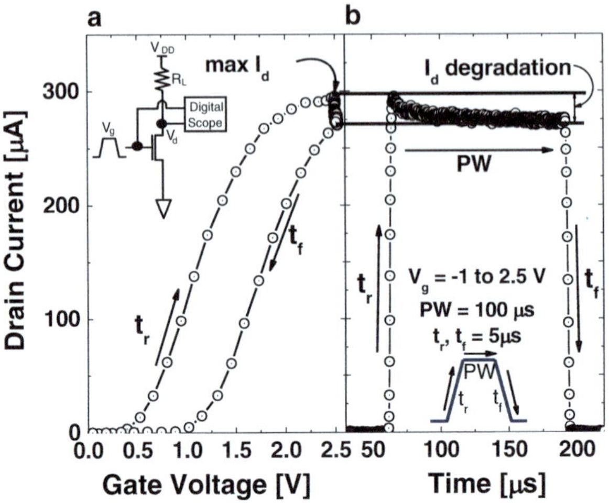

Fig. 22.2 "Single pulse" (**a**) I–V and (**b**) I-time measurements performed in the W/L = 10/0.5 μm transistor with a non-optimized 1 nm SiO_x/5 nm $HfSiO_x$/TiN stack

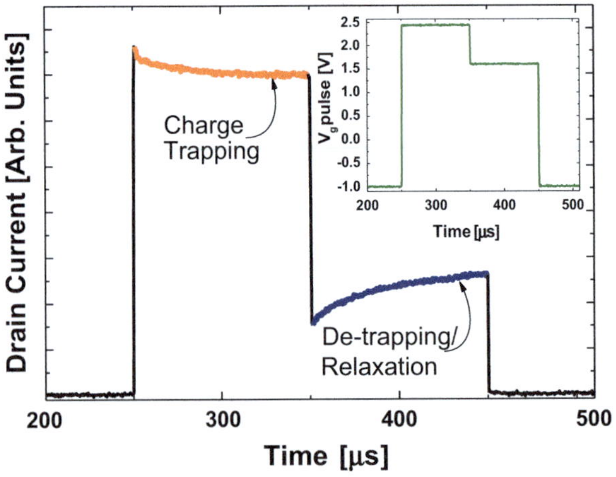

Fig. 22.3 An example of the fast transient charge trapping (I_d decay) and detrapping (I_d recovery) for the V_g pulse in the inset for a 1 nm SiO_x/3 nm HfO_2/TiN sample

22.3 Pulse-Based Bias Temperature Instability

Since fast relaxation can have a significant impact on the conventional DC-based "stress and sense" measurements used for the bias temperature instability (BTI) evaluation, more advanced measurement methodologies are required to evaluate the impact of stress on the instability (Fig. 22.5). An extremely useful approach uses pulsed I_d–V_g in conjunction with bias stress [30] with no measurement

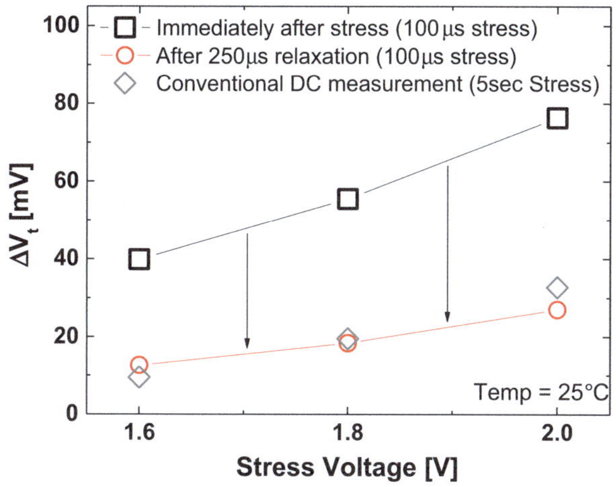

Fig. 22.4 V_t shifts immediately after the 100 μs stress and after 250 μs of post stress relaxation compared to conventional "stress and sense" approach demonstrating significant relaxation

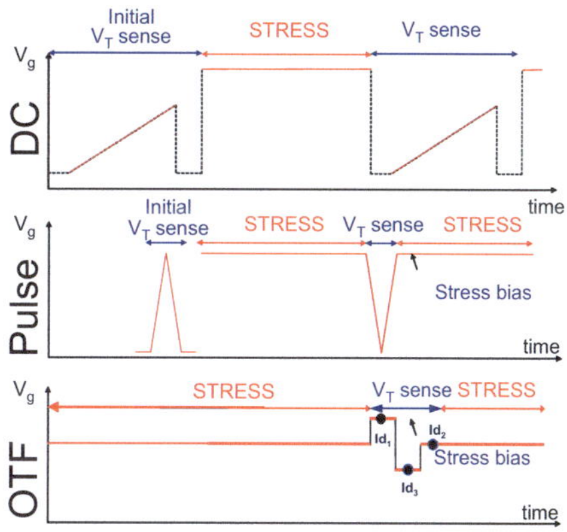

Fig. 22.5 Schematics of the various stress-sense schemes using the stress-interrupted DC I_d–V_g method, single pulse method, and on-the-fly method

interruption, thereby minimizing the effect of detrapping. Figure 22.6 shows data for the ΔV_{th} vs. stress time for various relaxation delay times before pulsed I_d sense (middle image: downward pulse, right at stress "removal") and the conventional DC technique (top image) to illustrate the impact of the delay. The ΔV_{th} values extracted from the pulsed measurements are usually higher because of less charge detrapping (relaxation). This data also demonstrates that the power law exponent value, typically used to evaluate device lifetime, is significantly impacted by the relaxation time [30].

To correlate ΔV_t measured using different approaches, one needs to understand the charge trapping kinetics in high-k nMOSFETs. The concept of a two-step process—comprised of fast (<1 s) and slow components—to explain the charge trapping in high-k devices under stress has been proposed [9]. The fast trapping kinetics, a characteristic time of which is about 100 μs, is shown to be temperature

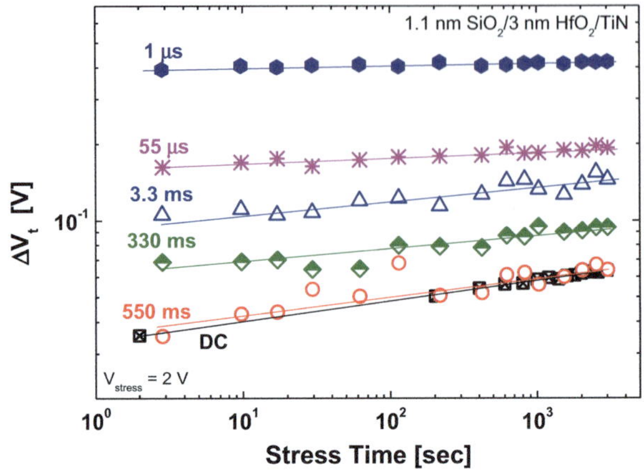

Fig. 22.6 The ΔV_{th} stress time dependencies as measured using DC and single pulse I_d–V_g methods. The slope of the curves decreases with shorter pulse time

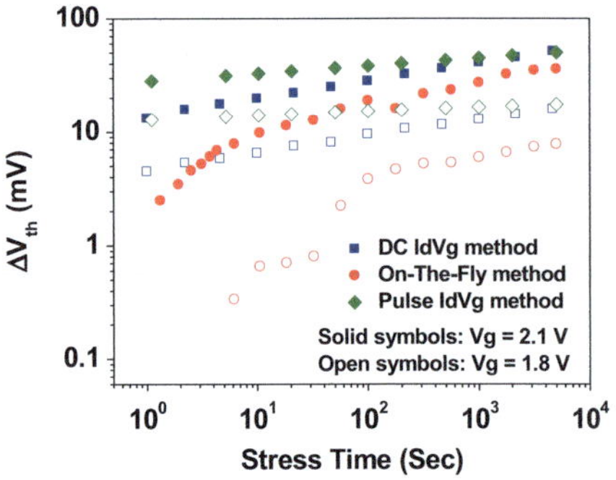

Fig. 22.7 The comparison of the stress time dependence of ΔV_{th} measured using the techniques (in Fig. 22.5) with different characteristic time delay. The stress time (the x-axis) does not include sense measurement time [33]

independent, suggesting the trapping of injected electrons at the preexisting defects occurring with a very small activation barrier [31]. Slow trapping can be related to either a different type of defects [32] or a different trapping mechanism, for instance, redistribution (assisted by temperature and applied field) of the trapped charges through the "charge migration" process to the nearby traps [9, 33]. The relaxation process also demonstrates a similar two-step (fast and slow) characteristic as shown by the pulse-based measurements [33].

The fast charging, which determines the initial ΔV_t values, strongly affects the slope of the $\Delta V_t(t)$ power law dependency and, thus, the PBTI life time estimate. Building on the above-discussed understanding, a practical approach to treating the $\Delta V_t(t)$ projection was proposed [33]. In this approach, the initial ΔV_t shift, $V_t(t = 1\ s)$, caused by fast transient charging, which is also sensitive to fast relaxation, is subtracted from all the subsequent $\Delta V_t(t)$ values: $\Delta V_t'(t) \equiv \Delta V_t(t) - \Delta V_t(t = 1\ s)$. The $\Delta V_t'(t)$ results in Fig. 22.7 generated by the different fast BTI techniques were reevaluated using this approach, Fig. 22.8. The $\Delta V_t'(t)$ curves obtained using the fast measurement methodologies with different characteristic relaxation

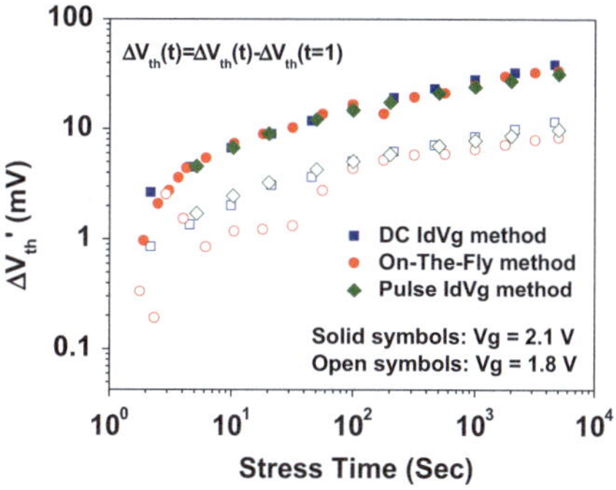

Fig. 22.8 The correlation of ΔV_{th} measured by different methods (Fig. 22.5) after correcting for the fast transient charging [33]

times exhibit an identical time dependency. This intrinsic time dependence, which is independent from the measurement approach, can be attributed solely to a slow charge trapping process [9, 33].

22.4 Factors That Aid in the Reduction of ΔV_t

From these pulse-based findings on the intrinsic fast trapping nature of Hf-based dielectric gate stacks, improvements in the reduction of the FTCE have occurred allowing high-k dielectric films to be in production. First, scaling the high-k layer has ultimately resulted in no detectible fast trapping [34, 35] which leads to improved device performance. Furthermore, this has obviously resulted in a significant reduction in the FTC contribution to ΔV_t. Another approach is the incorporation of silicon in the dielectric film. With as little as 20% of Si incorporation, there can be a significant reduction in the fast transient charging [36]. Yet, another technique is the use of a postdeposition anneal of the Hf-based film in ammonia (NH_3). This anneal incorporates nitrogen in the layer, thereby filling what would have been oxygen vacancies [35].

The incorporation of zirconium (Zr) has also shown drastic reduction of ΔV_t [37] and improved time-dependent dielectric breakdown times [38]. The lower observed PBTI degradation in the Zr:HfO_2 stack was shown to be caused by a smaller fast electron trapping component. Therefore, the Zr incorporation lowers the amount of preexisting defects present in the gate dielectric.

22.5 Defect Generation in Interfacial Oxide Layer Under Positive Bias Stress

A significant portion of the V_t shift caused by a positive bias stress can be recovered by applying a negative bias "discharge" step after the stress to empty the traps prior to the "sense" measurements (Fig. 22.9) [11, 15, 39]. Such V_t recovery was

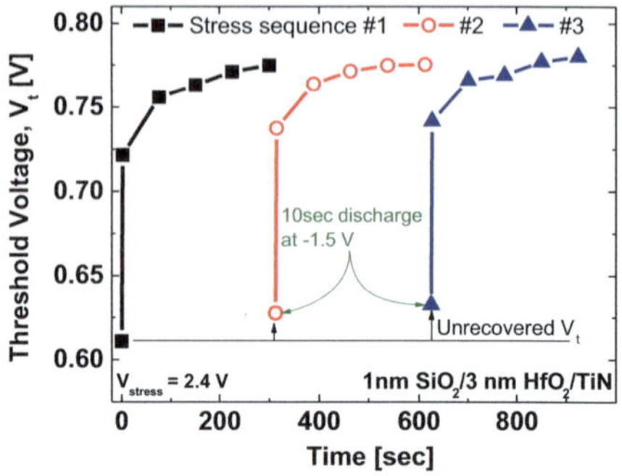

Fig. 22.9 Time evolution of the threshold voltage during stress demonstrating a repeatable and mostly reversible trend, with a propensity to saturate (near-exponential trend). A 10-s discharge at $V_g = -1.5$ V was performed before each stress sequence

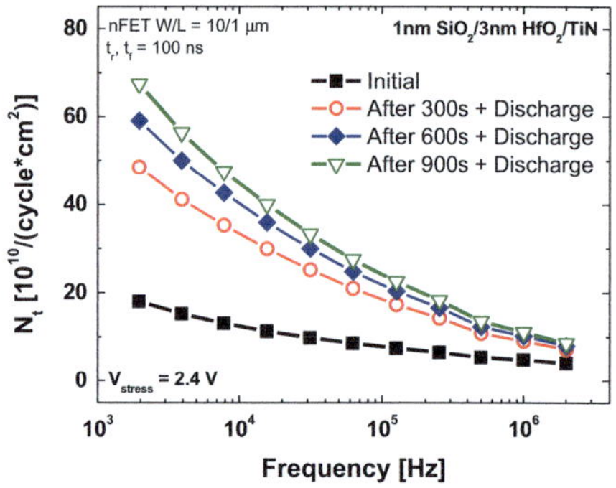

Fig. 22.10 Charge pumping N_t data for different frequencies before and after a stress sequence with $V_g = -1.5$ V discharge where trap generation is detected as frequency decreases

reported to be accompanied by stress-induced leakage current (SILC) recovery [40, 41]. The unrecoverable portion of the V_t shift has been attributed to trap generation during stress [11, 16]. Separating reversible, fast transient charge trapping from that caused by the generated traps can be done by implementing the discharge. Then, one can introduce various "sense" measurements which have different dielectric probing capabilities—both spatially and energetically. The charge pumping (CP) measurement is one of the techniques widely accepted for the trap profiling study [42, 43].

Following a discharge (by applying a negative bias to nMOS), a fixed-amplitude, fixed-base CP with a frequency sweep (typically in the range of 2 MHz to 2 kHz) method was used as a sense measurement during stress to study trap generation [11, 15, 17]. An example of these measurements is shown in Fig. 22.10 for a 300-s CVS/discharge/CP cycle that was repeated several times. A negative DC bias for 10 s ("discharge") was applied immediately prior to CP measurements during each stress interruption. The density of the traps sensed at lower frequencies are seen to increase after the stress, while only a very small increase is detected at higher frequencies (Fig. 22.10). This suggests an apparent trap generation farther away from the interface with the Si substrate [15–17, 39, 44, 45].

An interfacial layer gettering process, which significantly reduces the IL thickness and places the high-k layer in closer proximity to the substrate interface, allows

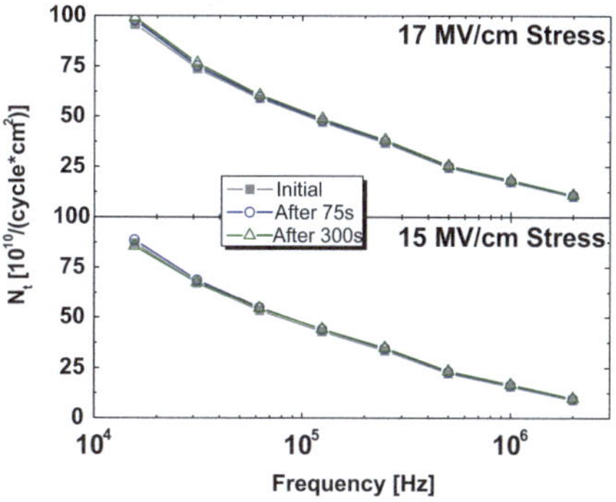

Fig. 22.11 Trap density as measured by CP before and after the 75-s and 300-s stresses of the gate stack with minimal IL. No detectable trap generation is observed

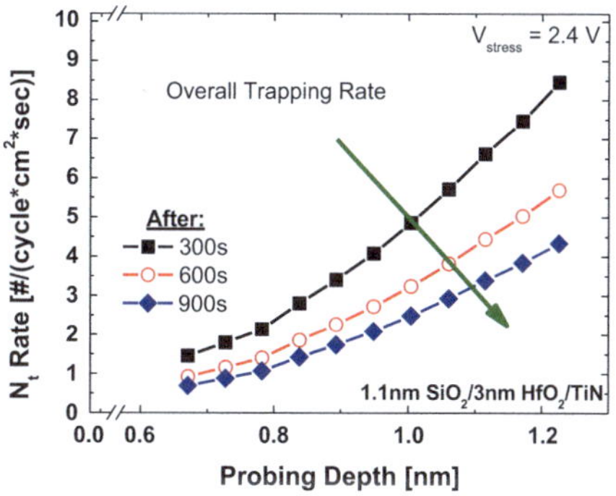

Fig. 22.12 Trap generation rate decreases with the stress time suggesting that the trap generation occurs primarily at "precursor" defects, suggesting that there is an initial density of these defects available for conversion to traps at the beginning of the stress, and then only the remaining defects are available for conversion with the next stress cycle and so on

access to the high-k layer during CP. This sample was subjected to the low-voltage stress/CP measurements [17]. There is no detectable stress-induced trap generation in this gate stack, Fig. 22.11; this is consistent with the proposed model of trap generation primarily within the interfacial layer since the bond formed by the d-electrons in the high-k materials is expected to be very stable [46].

Furthermore, the trap generation in the interfacial layer in close proximity to the overlying high-k layer has been shown to directly correlate to the increase in non-recoverable SILC data [11, 17, 47, 48], as shown in Figs. 22.12, 22.13, and 22.14. Figure 22.12 (N_t generation) and Fig. 22.13 (SILC) demonstrate a similar "rate of generation" where an initial density of "precursor" defects available for conversion to traps at the beginning of the stress is highest, and then, only the remaining defects are available for conversion with each subsequent stress cycle, resulting in a reduction in the generation rate with time. The direct correlation at various stress voltages is shown in Fig. 22.14. The excellent agreement between the low-frequency CP and SILC I_g–V_g curves corroborates the thought that SILC is controlled by defect generation in the IL. This degradation is determined to be the "weak link" or precursor to hard breakdown in time-dependent dielectric breakdown (TDDB) [47, 48].

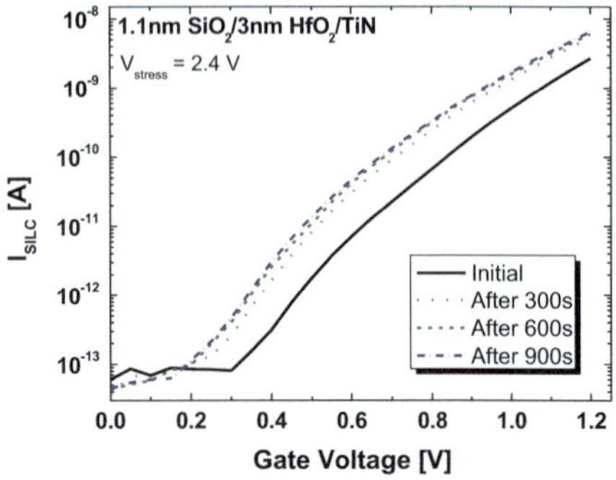

Fig. 22.13 SILC after several 300-s stress cycles. This data trends with the data in Fig. 22.10 in which the initial 300-s stress has the largest N_t increase with continually smaller increases thereafter

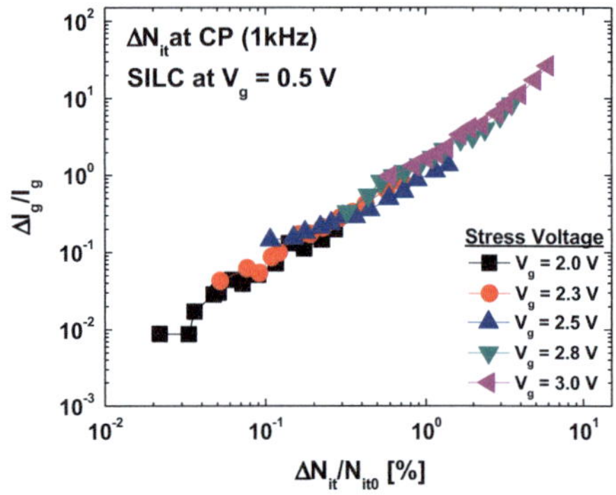

Fig. 22.14 Direct correlation of the low-voltage SILC and low-frequency N_{it} stress trend at different stress voltages

22.6 Summary

Positive bias stress coupled with various characterization techniques such as pulsed I–V and/or frequency-dependent charge pumping—along with robust analysis and meticulous data interpretation—has effectively enabled the separation of preexisting, fast transient charging and discharging defects from defects generated by stress. In addition, the ability to distinguish among electrically active bulk high-k traps and interfacial layer traps has also been demonstrated. In addition, a brief synopsis for the reduction of excessive ΔV_t in intrinsic HfO_2 was provided. Physical thickness scaling and Si or Zr incorporation are ways to minimize the fast trapping component of PBTI. The above results present the examples of the application of novel characterization approaches to effectively evaluate the properties high-k gate stack devices.

References

1. G. Bersuker, B. H. Lee, and H. R. Huff, *International Journal of High Speed Electronics and Systems*, vol. 16, pp. 221–239, 2006.
2. E. Hildebrandt, J. Kurian, M. M. Muller, T. Schroeder, H.-J. Kleebe, and L. Alff, *Applied Physics Letters*, vol. 99, p. 112902, 2011.

3. D. M. Ramo, J. L. Gavartin, A. L. Shluger, and G. Bersuker, *Physical Review B (Condensed Matter and Materials Physics)*, vol. 75, p. 205336, 2007.
4. G. Bersuker, C. S. Park, J. Barnett, P. S. Lysaght, P. D. Kirsch, C. D. Young, R. Choi, B. H. Lee, B. Foran, K. v. Benthem, S. J. Pennycook, P. M. Lenahan, and J. T. Ryan, *Journal of Applied Physics*, vol. 100, p. 094108, 2006.
5. J. T. Ryan, P. M. Lenahan, G. Bersuker, and P. Lysaght, *Applied Physics Letters*, vol. 90, p. 173513, 2007.
6. J. T. Ryan, P. M. Lenahan, J. Robertson, and G. Bersuker, *Applied Physics Letters*, vol. 92, p. 123506, 2008.
7. B. H. Lee, R. Choi, J. H. Sim, S. A. Krishnan, J. J. Peterson, G. A. Brown, and G. Bersuker, *IEEE Transactions on Device and Materials Reliability*, vol. 5, pp. 20–25, Mar 2005.
8. G. Bersuker, P. Zeitzoff, J. H. Sim, B. H. Lee, R. Choi, G. Brown, and C. D. Young, *Applied Physics Letters*, vol. 87, p. 042905, Jul 2005.
9. G. Bersuker, J. H. Sim, C. S. Park, C. D. Young, S. Nadkarni, R. Choi, and B. H. Lee, *IEEE International Reliability Physics Symposium*, 2006, pp. 179–183.
10. G. Bersuker, P. Zeitzoff, J. H. Sim, B. H. Lee, R. Choi, G. A. Brown, and C. D. Young, *IEEE Intl. Integrated Reliability Workshop Final Report*, 2004, pp. 141–144.
11. C. D. Young, D. Heh, S. V. Nadkarni, R. Choi, J. J. Peterson, J. Barnett, B. H. Lee, and G. Bersuker, *IEEE Transactions on Device and Materials Reliability*, vol. 6, pp. 123–131, 2006.
12. A. Kerber, E. Cartier, L. Pantisano, M. Rosmeulen, R. Degraeve, T. Kauerauf, G. Groeseneken, H. E. Maes, and U. Schwalke, *IEEE International Reliability Physics Symposium*, pp. 41–45, 2003.
13. C. Leroux, J. Mitard, G. Ghibaudo, X. Garros, G. Reimbold, B. Guillaumor, and F. Martin, *IEEE Intl. Electron Devices Meeting Tech. Digest*, 2004, pp. 737–740.
14. C. Shen, M. F. Li, X. P. Wang, Y. Yee-Chia, and D. L. Kwong, *IEEE Electron Device Letters*, vol. 27, pp. 55–57, 2006.
15. R. Degraeve, A. Kerber, P. Roussell, E. Cartier, T. Kauerauf, L. Pantisano, and G. Groeseneken, *IEEE Intl. Electron Devices Meeting Tech. Digest*, 2003, pp. 935–938.
16. C. D. Young, G. Bersuker, Y. G. Zhao, J. J. Peterson, J. Barnett, G. A. Brown, J. H. Sim, R. Choi, B. H. Lee, and P. Zeitzoff, *Microelectronics Reliability*, vol. 45, pp. 806–810, May-Jun 2005.
17. C. D. Young, S. Nadkarni, D. Heh, H. R. Harris, R. Choi, J. J. Peterson, J. H. Sim, S.A. Krishnan, J. Barnett, E. Vogel, B. H. Lee, P. Zeitzoff, G. A. Brown, and G. Bersuker, *IEEE International Reliability Physics Symposium*, 2006, pp. 169–173.
18. C. D. Young, Y. G. Zhao, M. Pendley, B. H. Lee, K. Matthews, J. H. Sim, R. Choi, G. A. Brown, R. W. Murto, and G. Bersuker, *Jap. J. of Applied Physics Part 1-Regular Papers Short Notes & Review Papers*, vol. 44, pp. 2437–2440, Apr 2005.
19. D. Heh, G. Bersuker, R. Choi, C. D. Young, and B. H. Lee, "A Novel Bias Temperature Instability Characterization Methodology for High-k MOSFETs," *ESSDERC*, 2006, pp. 387–390.
20. T. Yang, M. F. Li, C. Shen, C. H. Ang, Z. Chunxiang, Y. C. Yeo, G. Samudra, S. C. Rustagi, and M. B. Yu, *VLSI Symposium Technical Digest*, 2005, pp. 92–93.
21. B. H. Lee, C. D. Young, R. Choi, J. H. Sim, G. Bersuker, C. Y. Kang, R. Harris, G. A. Brown, K. Matthews, S. C. Song, N. Moumen, J. Barnett, P. Lysaght, K. S. Choi, H. C. Wen, C. Huffman, H. Alshareef, P. Majhi, S. Gopalan, J. J. Peterson, P. Kirsh, H.-J. Li, J. Gutt, M. Gardner, H. R. Huff, P. Zeitzoff, R. W. Murto, L. Larson, and C. Ramiller, *IEEE Intl. Electron Devices Meeting Tech. Digest*, 2004, pp. 859–862.
22. A. Kerber, E. Cartier, L. Pantisano, R. Degraeve, T. Kauerauf, Y. Kim, A. Hou, G. Groeseneken, H. E. Maes, and U. Schwalke, *IEEE Electron Device Letters*, vol. 24, pp. 87–89, Feb 2003.
23. S. Zafar, A. Kumar, E. Gusev, and E. Cartier, *IEEE Transactions on Device and Materials Reliability*, vol. 5, pp. 45–64, 2005.
24. G. Bersuker, J. H. Sim, C. D. Young, R. Choi, B. H. Lee, P. Lysaght, G. A. Brown, P. Zeitzoff, M. Gardner, R. W. Murto, and H. R. Huff, *Spring Meeting of the Material Research Society*, 2004, pp. 31–35.

25. L. Pantisano, E. Cartier, A. Kerber, R. Degraeve, M. Lorenzini, M. Rosmeulen, G. Groeseneken, and H. E. Maes, *VLSI Symposium Technical Digest*, 2003, pp. 163–164.
26. C. D. Young, G. Bersuker, G. A. Brown, P. Lysaght, P. Zeitzoff, R. W. Murto, and H. R. Huff, *IEEE International Reliability Physics Symposium*, 2004, pp. 597–598.
27. C. D. Young, A. Kerber, T. H. Hou, E. Cartier, G. A. Brown, G. Bersuker, Y. Kim, J. Gutt, P. Lysaght, J. Bennett, C. H. Lee, S. Gopalan, M. Gardner, P. M. Zeitzoff, G. Groeseneken, R. W. Murto, and H. R. Huff, *Fall Meeting of the Electrochemical Society, Physics and Technology of High-K Gate Dielectrics - II*, Orlando, FL, 2003, pp. 347–362.
28. L. Vandelli, A. Padovani, L. Larcher, R. G. Southwick, W. B. Knowlton, and G. Bersuker, *IEEE Transactions on Electron Devices*, vol. 58, pp. 2878–2887, 2011.
29. C. D. Young, R. Choi, J. H. Sim, B. H. Lee, P. Zeitzoff, Y. Zhao, K. Matthews, G. A. Brown, and G. Bersuker, *IEEE International Reliability Physics Symposium*, 2005, pp. 75–79.
30. D. Heh, R. Choi, C. D. Young, B. H. Lee, and G. Bersuker, *IEEE Electron Device Letters*, vol. 27, pp. 849–851, 2006.
31. D. Heh, C. D. Young, and G. Bersuker, *IEEE Electron Device Letters, vol. 29, pp. 180–182, 2008.*
32. L. Larcher and et al., *To be Published*, 2013.
33. D. Heh, R. Choi, and G. Bersuker, *IEEE Electron Device Letters*, vol. 28, pp. 245–247, 2007.
34. J. H. Sim, S. C. Song, P. D. Kirsch, C. D. Young, R. Choi, D. L. Kwong, B. H. Lee, and G. Bersuker, *Microelectronic Engineering*, vol. 80, pp. 218–221, Jun 2005.
35. P. D. Kirsch, M. A. Quevedo-Lopez, H.-J. Li, Y. Senzaki, J. J. Peterson, S. C. Song, S. A. Krishnan, N. Moumen, J. Barnett, G. Bersuker, P. Y. Hung, B. H. Lee, T. Lafford, Q. Wang, D. Gay, and J. G. Ekerdt, *Journal of Applied Physics*, vol. 99, p. 023508, 2006.
36. C. D. Young, D. Heh, A. Neugroschel, R. Choi, B. H. Lee, and G. Bersuker, *Microelectronics Reliability*, vol. 47, pp. 479–488, 2007.
37. S. Deora, G. Bersuker, C. D. Young, J. Huang, K. Matthews, K. W. Ang, T. Nagi, C. Hobbs, P. D. Kirsch, and R. Jammy, *International Symposium on VLSI Technology, Systems, and Applications (VLSI-TSA)*, 2012, pp. 1–2.
38. C. D. Young, G. Bersuker, M. Jo, K. Matthews, J. Huang, S. Deora, K. Ang, T. Ngai, C. Hobbs, P. D. Kirsch, A. Padovani, and L. Larcher, *IEEE International Reliability Physics Symposium (IRPS)*, 2012, pp. 5D.3.1-5D.3.5.
39. F. Crupi, R. Degraeve, A. Kerber, D. H. Kwak, and G. Groeseneken, *IEEE International Reliability Physics Symposium*, 2004, pp. 181–187.
40. E. Cartier and A. Kerber, *IEEE International Reliability Physics Symposium*, 2009, pp. 486–492.
41. A. Kerber and E. A. Cartier, *IEEE Transactions on Device and Materials Reliability*, vol. 9, pp. 147–162, 2009.
42. G. Groeseneken, H. E. Maes, N. Beltran, and R. F. De Keersmaecker, *IEEE Transactions on Electron Devices*, vol. 31, pp. 42–53, 1984.
43. D. Heh, E. Vogel, J. B. Bernstein, C. D. Young, G. A. Brown, P. Y. Hung, A. Diebold, and G. Bersuker, *IEEE Transactions on Electron Devices*, pp. 1338–1345, 2006.
44. D. Heh, C. D. Young, G. A. Brown, P. Y. Hung, A. Diebold, G. Bersuker, E. M. Vogel, and J. B. Bernstein, *Applied Physics Letters*, vol. 88, p. 152907, 2006.
45. C. D. Young, D. Heh, S. Nadkarni, R. Choi, J. J. Peterson, H. R. Harris, J. H. Sim, S. A. Krishnan, J. Barnett, E. Vogel, B. H. Lee, P. Zeitzoff, G. A. Brown, and G. Bersuker, *International Integrated Reliability Workshop Final Report*, 2005, pp. 79–83.
46. G. Bersuker, P. Zeitzoff, G. A. Brown, and H. R. Huff, *Materials Today*, pp. 26–33, 2004.
47. G. Bersuker, N. Chowdhury, C. Young, D. Heh, D. Misra, and R. Choi, *IEEE International Reliability Physics Symposium*, 2007, pp. 49–54.
48. G. Bersuker, D. Heh, C. Young, H. Park, P. Khanal, L. Larcher, A. Padovani, P. Lenahan, J. Ryan, B. H. Lee, H. Tseng, and R. Jammy, *IEEE International Electron Devices Meeting*, 2008, pp. 1–4.

Chapter 23
Characterization of Individual Traps in High-κ Oxides

M. Toledano-Luque and B. Kaczer

Abstract As the result of the vertical scaling of the CMOS technology, high-κ materials were introduced in the gate stack in order to reduce leakage current while keeping electrostatic control over the channel. Despite the high level of the bulk defects of these materials, only a handful of defects are present in the gate oxide due to the reduced lateral dimensions of the current CMOS technology. However, the relative impact of these traps on the device characteristics increases. In Chap. 17, it has been demonstrated for the conventional SiO_2/poly-Si stack that the properties of each Si/SiO_2 defect, such as its capture and emission times and its impact, are voltage and/or temperature dependent and widely distributed. In this chapter, we show that identical properties are followed by high-κ-based dielectrics. The stochastical nature of the behavior of the dielectric traps results in each of the nominally identical nm-scaled devices behaving very differently during operation and, therefore, increasing time-dependent variability (heteroskedasticity). Consequently, the lifetime of nm-sized high-κ devices cannot be predicted individually, but can be only described in terms of time (or workload)-dependent distributions.

23.1 Introduction

As the vertical scaling of the conventional poly-Si/SiO_2/Si field effect transistors (MOSFET) reached the nanometric scale, the high electric field over the gate oxide resulted in an intolerable gate leakage current. At that point, industry opted for replacing the conventional SiO_2/poly-Si structure with high-κ/metal gate stacks that tolerate physically thicker oxide while keeping or even increasing the oxide capacitance. Once the leakage current issue was circumvented, the so-called bias temperature instability (BTI) is becoming one of the most critical factors,

M. Toledano-Luque (✉) • B. Kaczer
Imec, Kapeldreef 75, Belgium
e-mail: toleda@imec.be; kaczer@imec.be

T. Grasser (ed.), *Bias Temperature Instability for Devices and Circuits*,
DOI 10.1007/978-1-4614-7909-3_23,

complicating the qualification of the future technology nodes [1–3]. Furthermore, the number of stochastically behaving gate oxide defects in each device decreases to a numerable level due to the lateral downscaling, while their relative impact on the device characteristics increases. For all these reasons, BTI lifetime cannot be described any longer by a unique average number, and *BTI lifetime distribution* has to be taken into consideration. As a consequence, even in the case of the *average BTI lifetime* meeting the ITRS [4] specifications, a fraction of nanoscaled devices will fail at low overdrives. In this chapter, the necessary physical understanding to predict the BTI lifetime distributions is developed for high-κ-based nFETs and pFETs.

We start by briefly reviewing the elementary definitions and experimental observations of BTI in large area and in nm-scaled high-κ devices. Specifically, in this chapter, stacks formed by 0.8 nm-SiO_2/1.8-nm-HfSiO/5 nm-TiN were studied and compared to SiO(N)/poly-Si FETs. As in the case of SiO_2 based devices, the continuous BTI relaxation curves observed on large area devices after bias temperature stress becomes quantized and giant discrete threshold voltage V_{TH} shifts are observed on nanoscaled high-κ devices [5]. Consequently, the properties of high-κ defects such as characteristic emission and capture times and V_{TH} impact can be directly extracted from BTI relaxation measurements in deeply scaled devices following identical methodology that was presented in Chap. 4 focused on SiO_2 devices [6]. Finally, we show how the understanding of gate oxide defect properties can be used to explain time dependent BTI variability in deeply scaled high-κ-based technologies.

23.2 Bias Temperature Instability BTI in Large and in Nanometer-Scaled High-κ Devices

During CMOS circuit operation, the devices typically undergo electrical stress at elevated temperature resulting in a shift of the device parameters such as its threshold voltage, channel mobility, transconductance, and subthreshold slope, instigating a decrease of the FET's drive current. Since these *instabilities* are strongly accelerated by *temperature T* and *gate bias V_G, they are known by the acronym BTI (Bias Temperature Instability).* These phenomena are mainly the consequence of charging of defects in the gate oxide and at its interface [7]. BTI in *n*-channel FET devices, which are typically biased at positive V_G in CMOS circuits, is referred to as positive BTI (PBTI), while negative BTI (NBTI) takes place in *p*-channel FETs.

Figure 23.1a illustrates the typical gradual shift of pFET threshold voltage ΔV_{TH} during accelerated stress at elevated temperature [8]. Data are typically measured at several V_Gs to obtain the maximum circuit operating voltage V_{DD} that the devices could withstand for 10 years, while the $|\Delta V_{TH}|$ is below a given value (typically 30 or 50 mV).

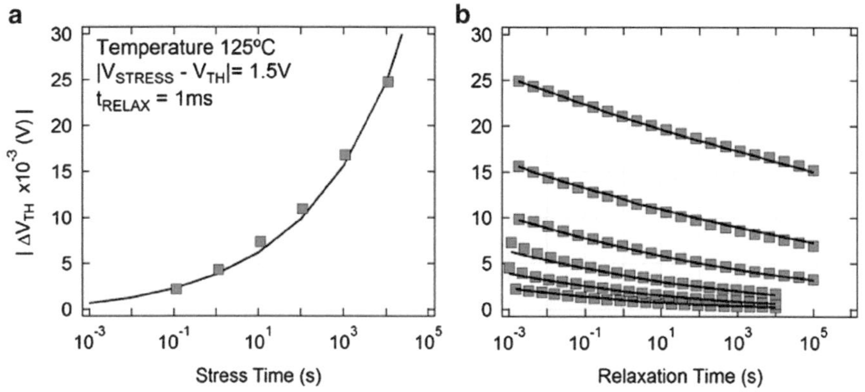

Fig. 23.1 (**a**) Threshold voltage shift ΔV_{TH} is observed during negative gate bias stress and high temperature (125 °C) in a $W \times L = 10 \times 0.5\ \mu m^2$ pFET formed by 0.8 nm-SiO_2/1.8-nm-HfSiO. (**b**) When the stress bias is removed, a recovery of the effect is observed

However, this extrapolation procedure is problematic due to the immediate ΔV_{TH} recovery after the stress bias is removed [5, 7], as illustrated in Fig. 23.1b. As we will discuss henceforth, this recovery or relaxation typically proceeds on many time scales, causing difficulties to extrapolate to both shorter and longer relaxation times and therefore to obtain the permanent degradation component [7, 9, 10]. This ΔV_{TH} relaxation is thus a crucial problem for BTI measurements, interpretation, and extrapolation. Understanding the recoverable component has been crucial to unraveling the BTI mechanism and has been greatly facilitated by means of the thorough study of deeply scaled devices.

Figure 23.2 displays the relaxation traces after a BTI stress obtained on $90 \times 35\ nm^2$ pFETs. Each trace reveals the combined response of multiple defects and every discrete drop is due to a single-carrier discharge event [5, 7]. The average relaxation resembles the curve taken on a large area device under equal stress condition, indicating that *identically behaving traps are responsible for BTI in both small and large area high-κ FETs* [8, 11].

The figure illustrates the wide variation in the behavior of individual nanoscale devices. We will show hereafter that this variation can be described analytically [12] by means of two parameters: the mean *total* ΔV_{TH} and the mean impact on V_{TH} per trap η, i.e., $\langle total\Delta V_{TH}\rangle = \eta \times N_T$, with N_T as the mean number of active traps channel percolation effects in small devices [13–15] as explained in the next sections.

23.3 BTI Nonsteady State Case of Random Telegraph Noise

From the quantized recovery behavior observed in nanoscaled FETs, it is straightforward to understand the recoverable component of BTI as the dynamic nonsteady state case of random telegraph noise (RTN) [15, 16]. As in the case of I_D-RTN,

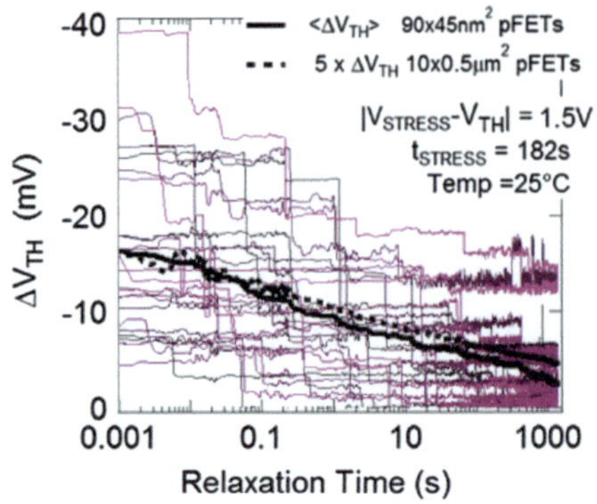

Fig. 23.2 Bias temperature instability (BTI) relaxation transients obtained on $W \times L_{eff} = 90 \times 35\ \text{nm}^2$ 0.8 nm-SiO_2/1.8 nm-HfSiO pFETs. Steps due to single-carrier discharge events are evident. The large dispersion is due to the stochastic distributions of the number of active traps after BTI stress and the impact of each trap. Note that the average relaxation resembles the curve taken on a large area device

the large quantized ΔV_{TH}s observed in Fig. 23.2 are explained by the nonuniform potential at the Si/SiO_2 interface caused by the random distributions of dopants in the channel and charged traps in the dielectric. The potential fluctuations produce variations of the inversion charge density and, consequently, preferential conduction paths from the source to the drain. The charging and discharging of single oxide traps over critical positions of the conduction paths can produce significant fluctuations of the drain current [13, 15]. The change of drain current can be in turn transformed into a V_{TH} shift, ΔV_{TH}, when taking the I_D–V_G curve of the fresh device as a reference [16, 17].

In the case of RTN, the emission and capture times are of the same order of magnitude causing random switching of the drain current at fixed V_G. In the case of BTI, the capture of charge is forced at high gate voltage (V_{STRESS}) and the emission at low voltage (V_{RELAX}). This allows studying states with dissimilar emission and capture times, reducing the prohibitive acquisition time of standard RTN experiments.

In this chapter, we present two approaches for the study of the discretized relaxation curves obtained on nm-scaled HfSiO-based devices: (1) repeatedly performing the same experiment on a single device or (2) conducting one single experiment on many devices.

From the former approach, the technique named *time-dependent defect spectroscopy* (TDDS) [6] presented in Chap. 4 allows the study of the kinetic properties of single high-κ defects as a function of stress/recovery bias conditions and temperature [4, 18–21]. These studies have revealed interesting facts about the charge trapping component that we summarize in the next section.

From the latter approach, we demonstrate the methodology to predict the ΔV_{TH} *distributions* after BTI stress through a detailed understanding of the *atomistic* impact of *individual* traps [12, 16, 22]. This approach has proven to be useful for reliability engineers [22] and circuit designers to predict time-dependent BTI variability [23] as shown in Chaps. 7 and 30.

23.4 Properties of Individual High-κ Traps: Impact of Single Charged Traps on the V_{TH}, Emission, and Capture Times

In this section, we first present a statistical comparison of the discrete threshold voltage shifts caused by charged traps in a larger number of SiO(N) FETs and, technologically more relevant, SiO_2/HfSiO devices after negative and positive bias stress. For pFETs, similar distributions are measured for SiO_2 and high-κ-based devices, reinforcing the idea that NBTI is mostly related to traps close to the oxide interface and the semiconductor. On the other hand, for nFETs, traps placed in the high-κ material play an important role [24, 25]. It is shown that even though the trap density in HfSiO is larger than in SiO_2, high-κ traps have a reduced impact on the total threshold voltage shift due to their larger separation from the channel. Next, a set of individual traps in a single SiO_2/HfSiO nFET is analyzed by the TDDS technique. Similarly to the SiO(N) traps, the emission and capture times of high-κ traps show strong thermal and bias dependences.

23.4.1 Statistical Comparison of Impact of Single Traps in SiO(N) and SiO_2/HfSiO FETs on V_{TH}

Figure 23.3 shows the typical relaxation transients obtained on (a) 2.1 nm EOT SiO(N) and (b) 1.4 nm EOT SiO_2/HfSiO 35 × 90 nm² nFETs after applying a gate

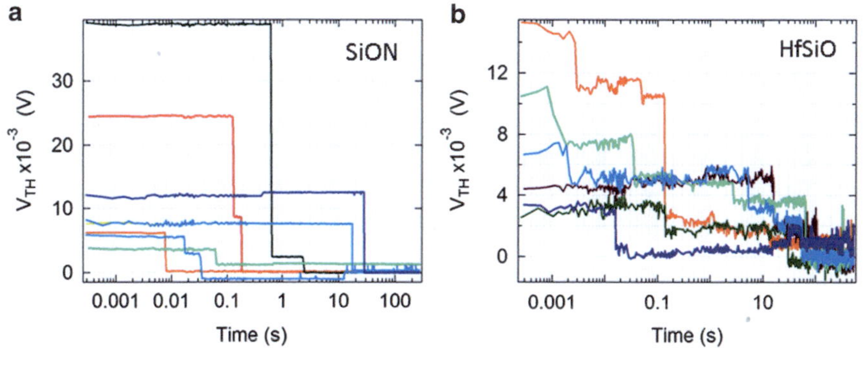

Fig. 23.3 Typical relaxation transients obtained in (**a**) SiO(N) and (**b**) SiO_2/HfSiO stacks with $W \times L_{eff} = 90 \times 35$ nm². Larger noise due to the higher trap density in the high-κ dielectric is observed in the HfSiO nFETs with respect to the "clean" SiO(N) traces

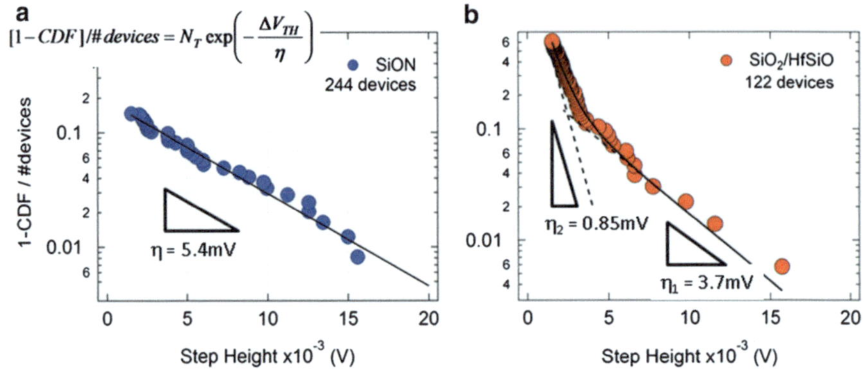

Fig. 23.4 Complementary cumulative distribution functions (CCDF) normalized to the number of devices of V_{TH} step heights larger than 1.5 mV for (**a**) 244 SiO(N) and (**b**) 122 SiO_2/HfSiO nFETs. Note that the number of traps (interceptions of fits with *y*-axes) is larger for SiO_2/HfSiO stacks than for SiO(N) devices. Complementary CDF for SiO(N) follows an exponential distribution with average value $\eta = 5.4$ mV. For the SiO_2/HfSiO stacks, the data can be fitted to a bimodal exponential distribution with $\eta_1 = 3.7$ mV and $\eta_2 = 0.85$ mV. The number of steps per device N_{T2} is ten times larger than N_{T1}

oxide electric field of 13 MV/cm for 240 ms. In both cases, the V_{TH} transients show a discrete behavior due to electron emission from individual traps [6, 13, 18]. As opposed to the "clean" V_{TH} relaxation traces obtained in SiO(N) nFETs, the high-κ stacks present a higher level of noise.

When the V_{TH} step heights are displayed in a complementary cumulative plot normalized to the number of traces (see Fig. 23.4), it is observed that the number of steps, i.e., traps, after identical stress conditions (the electric field and time) is larger for the HfSiO stacks than for the SiO(N) stacks. Indeed a significant number of SiO(N) devices did not show any step (not displayed in the graph).

The step heights for SiO(N) shown in Fig. 23.4 follow an exponential distribution [12]. The inverse of the slope of the distribution provides the average value η and the intercept of the fit with the *y*-axis the average of active traps N_T after stress when all the traces show a full recovery. For the case of SiON, Fig. 23.4a, the calculated average value is η of 5.4 mV and the trap density N_T is 0.19 trap/device.

For the high-κ nFETs, a clearly bimodal distribution can be observed in Fig. 23.4b. Each mode can be fitted by means of the maximum likelihood method using two exponential distributions with $\eta_1 = 3.7$ mV and $N_{T1} = 0.3$ trap/device and $\eta_2 = 0.85$ mV and $N_{T2} = 2.6$ trap/device, respectively. Distribution 1 (η_1) is similar to SiO(N), both in η and in magnitude N_T. Therefore, we can argue that this mode is due to defects in the SiO_2 layer and $\eta_1 = \eta_{IL}$ from now on. Since the η value is related to the centroid (x_0 distance from the gate) of the trapped electrons in the dielectric [4, 10], i.e., $\eta \propto x_0$, the higher value of η for the SiO(N) stacks is due to larger EOT. On the other hand, the low η value obtained for distribution 2 corresponds to the defects in the high-κ dielectric ($\eta_2 = \eta_{HK}$). Also it is worth noting that the number of steps per device N_{T2} is ten times (!) larger than N_{T1}.

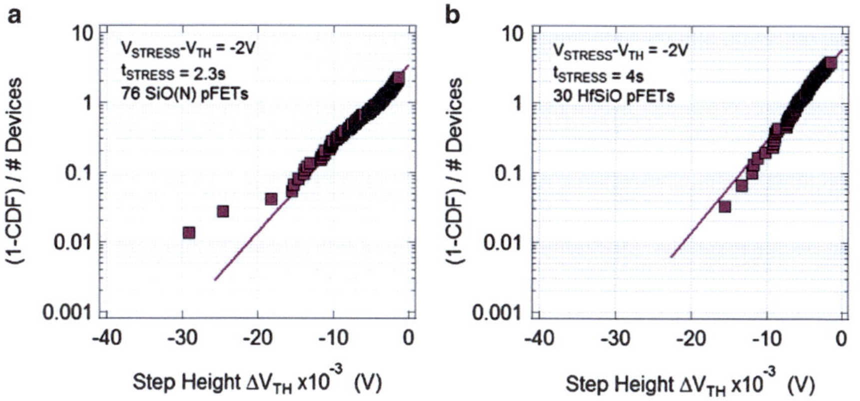

Fig. 23.5 Complementary cumulative distribution functions normalized to the number of devices of the single hole-discharged ΔV_{TH} for (**a**) SiO(N) and (**b**) HfSiO pFETs with similar EOT (~1.4 nm). Average step height $\langle \Delta V_{TH} \rangle = \eta$ is 3.6 mV for SiON devices and 3.4 mV for high-κ/metal gate pFETs

This indicates a *higher density of traps in the high-κ dielectric* with respect to SiO_2. However, the impact on the total V_{TH} is reduced since η_2 is significantly lower. Therefore, the large density of high-κ traps with a small impact on the V_{TH} explains the larger noise level observed in the high-κ stack (Fig. 23.3b).

In the case of pFETs (Fig. 23.5), a monomodal distribution of the single hole discharged ΔV_{TH} is measured after the NBTI relaxation curves for SiO(N) and HfSiO with similar EOT. Note that a single charged defect can cause up to tens of mV of ΔV_{TH}, for both cases with an average value η of 3.6 mV and 3.4 mV for SiON and HfSiO devices, respectively. These values are much larger than the value predicted by the simple charge sheet approximation $\eta_0 \sim C_{ox}/q \sim 1.7$ mV ($\eta/\eta_0 \approx 2.0$ in these cases). This is due to the amplifying effect of the random dopants in the FET channel as discussed in Sect. 3. The comparable η and N_T values obtained for both distributions strengthen the extended assumption that traps close to the Si/SiO_2 interface are mostly responsible of NBTI degradation [24].

23.4.2 *Kinetics of Individual Traps in a Single SiO_2/HfSiOnFET*

Figure 23.6 shows three typical *relaxation curves taken on a single SiO_2/HfSiOn FET*. Under the conditions of the experiment, the selected device has four active traps with V_{TH} step heights lower than 4 mV. From the bimodal distribution shown in Fig. 23.4b, the probability of observing a high-κ trap with a V_{TH} step height of 2 mV or lower is significantly larger than for SiO_2 traps. This suggests that there is a high probability that the traps observed in Fig. 23.6 are located in the high-κ material.

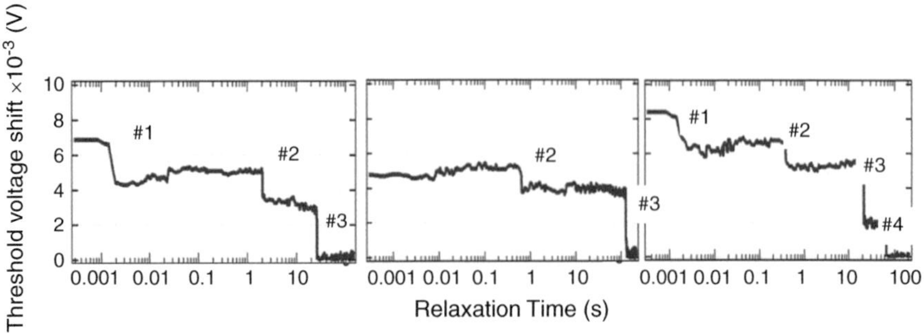

Fig. 23.6 Three typical ΔV_{TH} transients after applied $V_{STRESS} = 1.8$ V ($E_{OX} = 7$ MV/cm) for 189 ms at 25 °C to a single high-κ device. Up to four traps were active under the conditions of the experiment in this device. Note that their step heights are lower than 4 mV. Based on the bimodal distribution of Fig. 23.4b, we argue that the majority of these traps are in the high-κ layer

As in the case of SiO(N) devices [6], every high-κ trap has its characteristic emission time and V_{TH} shift, which form the "fingerprint" of the defect (see Fig. 23.7). However, the extracted clusters are not as compact as those observed in SiO(N) due to the higher level of noise present in the high-κ stack. Note that all four clusters in Fig. 23.7a shift to shorter emission times by about 1 order of the magnitude with an increasing temperature of only 25 °C (Fig. 23.7b). Trap #1 shifts out of the experimental window, while interestingly a new trap #5 that causes a negative V_{TH} shift appears. In the next paragraphs, we will analyze the emission and capture times of the trap #3 and the trap #5 as a function of stress time, stress voltage, and temperature.

Figure 23.8a shows the histogram of the emission times t_e of the trap #3 for two t_{STRESS} values. As expected for Markov processes [6, 18, 33], the emission times can be fitted to an exponential distribution in order to obtain the average characteristic emission time τ_e. Similarly to SiO(N) [6], the emission time τ_e is independent of the stress time. Figure 23.8b displays the Arrhenius plot of the characteristic emission time τ_e. It presents a strong thermal activation with $E_{ACT} = 0.48$ eV.

Figure 23.9 shows that the intensity of the cluster, probability of occupancy P_C, after stress, associated with trap #3 increases with stress time up to the saturation level determined by the characteristic emission (τ_e) and capture (τ_c) times at V_{STRESS} as expected from the equation of Fig. 23.9a [19, 23]. These characteristic times are strongly temperature (Fig. 23.9a) and voltage (Fig. 23.9b) dependent. The reduction of the occupancy with temperature is related to a higher activation energy of τ_e with respect to τ_c. The P_C increase with V_{STRESS} is due to the decrease of τ_c with increasing V_{STRESS} as (23.1).

$$P_c = \frac{\tau_e}{\tau_c + \tau_e} \exp\left[-\left(\frac{1}{\tau_e} + \frac{1}{\tau_c}\right) t_{stress}\right] \tag{23.1}$$

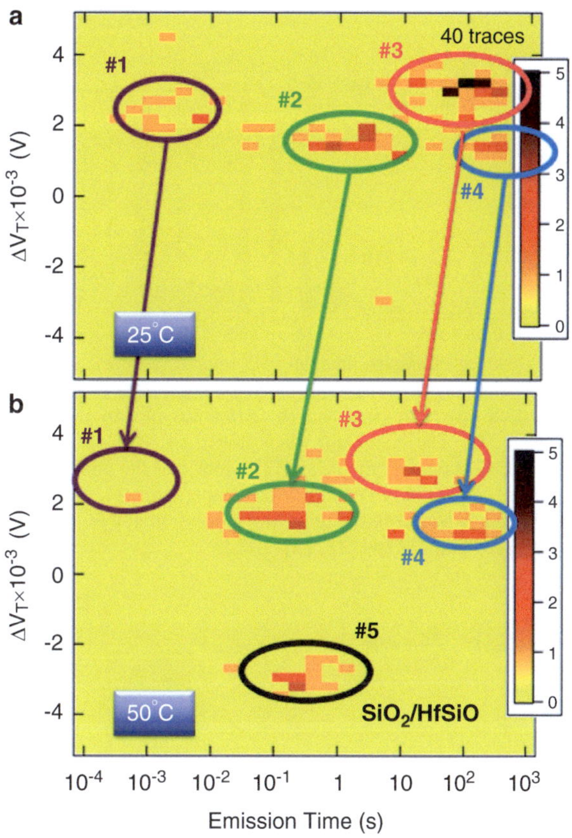

Fig. 23.7 TDDS spectra [6] of a single SiO_2/HfSiO stack at two temperatures extracted from 40 recovery traces under the bias and timing conditions of Fig. 23.6. At 25 °C (**a**), four homogenous clusters appear indicating the presence of four active traps under the conditions of the experiment. At 50 °C (**b**), a new cluster emerges (trap #5), which remarkably produces a negative V_{TH} shift. All the clusters shift to shorter emission times with temperature. Trap #1 even shifts out of the experimental window

As we already pointed out, in Fig. 23.7b, a new trap (trap #5) that *causes an unexpected negative* V_{TH} shift appears in the TDDS spectrum at high temperature. We hypothesize that this effect is due to electron discharge from the dielectric to the gate during stress [26]. Therefore, electron emission takes place during stress condition and electron capture occurs during relaxation. As we will see in the following, this trap follows analogous kinetics as the other four traps (#1–4). Figure 23.10a shows the intensity of cluster #5 as a function of the stress time for different temperatures. The emission probability increases with t_{STRESS} and can be described by the equation given in Fig. 23.9 after exchanging the capture and emission times, τ_c and τ_e, respectively. Again, this process is clearly thermally activated as shown in Fig. 23.10b. In the inset of Fig. 23.10b, the histogram of the capture times shows that the capture process can also be described by an exponential distribution. The activation energy obtained from the fit of the data to an Arrhenius law is 0.8 eV. The activation energies found in this study are close to the values obtained in SiO(N) pFETs after negative stress [6] and those of SiO(N) nFETs after positive stress [21].

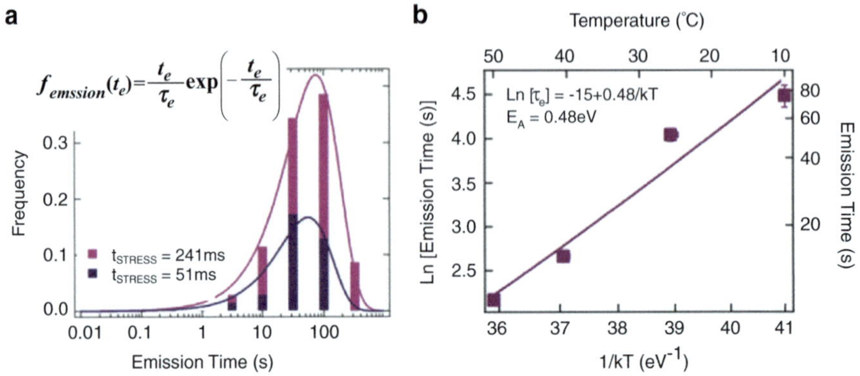

Fig. 23.8 (**a**) Histogram and (**b**) Arrhenius plot of the emission times t_e under the condition of Fig. 23.4. The histogram when plotted on the logarithmic scale matches with the theoretical expression shown in the *inset* [11]. Note that the emission times are independent of the stress time (**a**) but strongly dependent on temperature (**b**). This strong thermal dependence cannot be explained by a pure elastic tunneling process; therefore, non-radiative multiphonon processes have to be taken into account [18, 33]

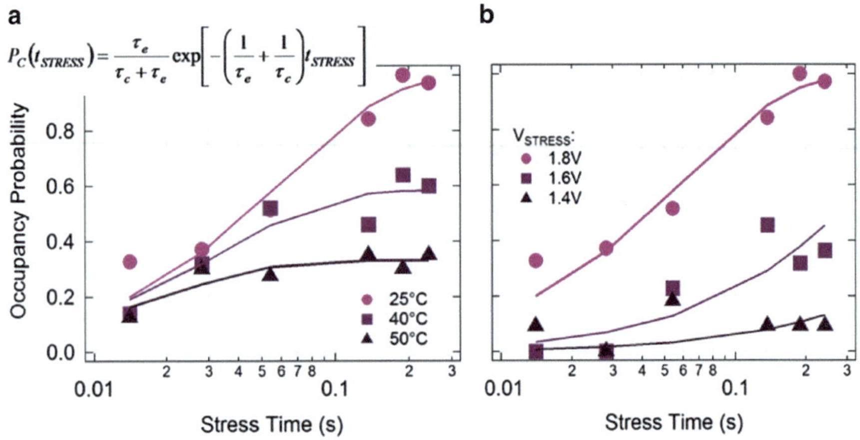

Fig. 23.9 (*Symbols*) Trap occupancy probability P_C of trap #3 vs. t_{STRESS} for different (**a**) temperatures and (**b**) V_{STRESS}. P_C increases with t_{STRESS} up to a saturation level dictated by the characteristic τ_e and τ_c times. These times depend strongly on temperature and V_{STRESS}. (*Lines*) Fit to the data according to the equation shown in the *inset* [19, 23]

We therefore conclude for all these cases that *both emission and capture in both electron and hole gate oxide traps are without any doubt thermally activated processes.* This experimental fact is incompatible with direct elastic tunneling theories widely used in different oxide trap characterization techniques and calculations. Consequently, a new model that takes into account this thermal dependence has to be considered. The most consistent explanation is provided by non-radiative multiphonon (NMP) theory [27] which has recently been applied to BTI data [18].

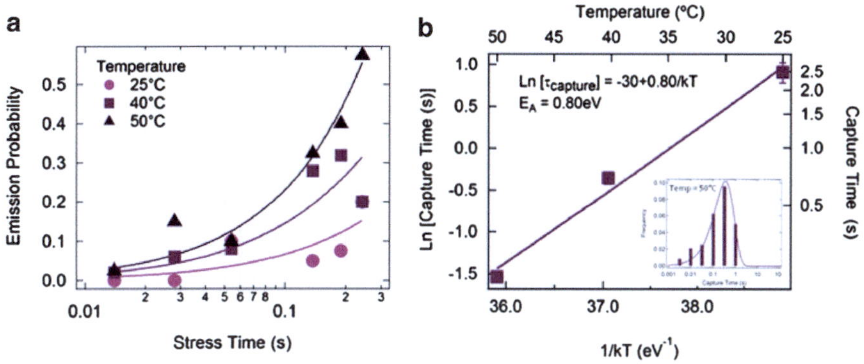

Fig. 23.10 (**a**) Emission probability P_E of trap #5 vs. t_{STRESS} for different temperatures. Emission increases with stress time and temperature. (*Lines*) Fit to the data according to the equation shown in the *inset* of Fig. 23.9a after exchanging τ_e and τ_c times. (**b**) Arrhenius plot of the electron capture time during relaxation. *Inset* shows that this process follows identical statistics as the emission times in Fig. 23.8

23.5 BTI Variability in Nanoscaled High-κ Devices

In large devices the random properties of many defects average out resulting in a well-defined lifetime as we showed in Sect. 2. However, in deeply scaled devices, the stochastic nature of a handful of defects becomes apparent. For this reason, the application of identical workload in such nanoscaled devices results in distributions of the parameter shifts [16, 28]. Therefore, the well-defined bias temperature instability (BTI) lifetime of large devices becomes widely distributed [12, 22, 29]. The atomistic understanding of the properties of individual defects and the demonstrated link between random telegraph noise and BTI presented in the previous section helped us to explain the large BTI variability during relaxation [16, 22, 30].

In the representative set of quantized NBTI relaxation transients presented in Fig. 23.2, the *total* ΔV_{TH} (ΔV_{TH} at given t_{RELAX}) strongly varies from device to device. Note that the *total* ΔV_{TH} ranges from a few mV up to 40 mV among devices under identical stress conditions. Figure 23.11 shows the complementary cumulative distribution (CCDF = 1 − CDF) of the individual step heights ΔV_{TH} normalized to the number of tested devices. Step heights follow an exponential distribution with an average step height $\langle \Delta V_{TH} \rangle = \eta$ equal to 3.4 mV, independent of stress conditions. The number of detected steps increases with stress time and stress voltage (Fig. 23.11).

The average number of traps per device N_T can be obtained from a maximum likelihood fit of the data with (23.5) in Table 23.1. Figure 23.12 shows that N_T follows a power-law voltage dependence and can be fitted with both power law and logarithmic time dependences.

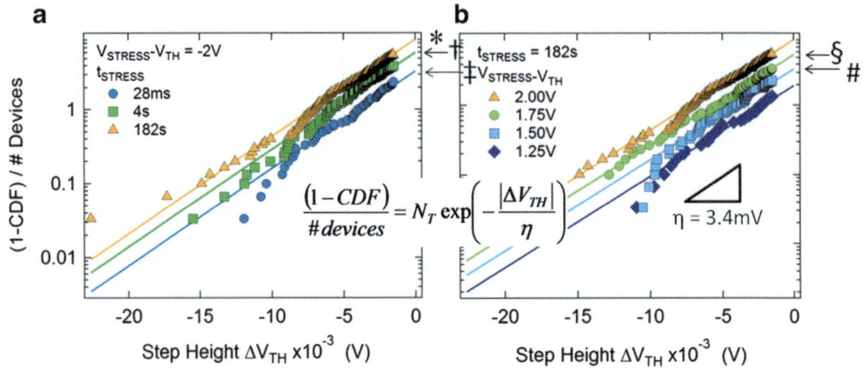

Fig. 23.11 Complementary cumulative distributions (CCDF = 1 − CDF) of step heights due to single oxide defects normalized to the number of tested pFETs after NBTI follow an exponential distribution ((23.5) in Table 23.1) with the average step height η. N_T values can be read from the intersection of the fit with the y-axis

As Fig. 23.13 shows, the number of steps causing ΔV_{TH} larger than 1.5 mV obtained from the relaxation curves following different stress times can be described by a Poisson distribution (23.8 in Table 23.1). The average value increases with increasing stress time and stress voltage as already emphasized by Fig. 23.12.

Figure 23.14 gives the *total* ΔV_{TH} for pFETs for different (a) t_{STRESS} and (b) V_{STRESS}. The *total* ΔV_{TH} distributions $H_{\eta,NT}$ (ΔV_{TH}) (23.9) [12, 31], a combination of exponential discrete ΔV_{TH} step distributions and the Poisson distributions with average N_T, are traced in Fig. 23.15 for $\eta = 3.4$ mV and different values of N_T. Note that the lines in Fig. 23.14 follow the experimental *total* ΔV_{TH} data, and the N_T values given by 23.9 excellently match those obtained *independently* in Fig. 23.11 (see symbols *, †, ‡, §, #), thus confirming the description derived in Table 23.1.

A 10-year lifetime CDF prediction of the *total* ΔV_{TH} is obtained by combining (23.9) with the N_T dependences on t_{STRESS} and V_{STRESS}. Figure 23.15 shows the predicted lifetimes for different conditions. We can conclude the following:

- For a fixed failure criteria of $\Delta V_{TH} = 30$, 50, and 100 mV at $t_{STRESS} = 10$ years, Fig. 23.15a allows one to readily read off the fraction of devices expected to exceed a given failure criterion.
- As already alluded to in Sect. 2, the predicted ΔV_{TH} distribution gets steeper ("tighter") with increasing device area A (Fig. 23.15b). Since the *average total* ΔV_{TH} is given by $N_T \times \eta$ and $N_T \propto A$, the median *total* $\langle \Delta V_{TH} \rangle$ is independent of A if $\eta \propto 1/A$ [32]. Therefore, Probit $(H_{\eta,N}) = 0$ determines the maximum overdrive for large, i.e., deterministic devices (vertical line in Fig. 23.15b). In contrast to that a considerable fraction of deeply scaled devices will exceed failure criteria even at low overdrives (see, e.g., circles in Fig. 23.15b).

Table 23.1 Flow to deduce the total ΔV_{TH} shift distribution [12, 22] presented in Chap. 7

Single defect: ΔV_{TH} exponentially distributed with $\eta = \langle \Delta V_{TH} \rangle \propto \frac{\langle x_0 \rangle \sqrt{N_A}}{WL}$ (23.2)		
$f_\eta\left(\Delta V_{TH}\right) = \frac{1}{\eta} e^{-\frac{\Delta V_{TH}}{\eta}}$ (23.3)	$F_\eta\left(\Delta V_{TH}\right) = 1 - e^{-\frac{\Delta V_{TH}}{\eta}}$ (23.4)	$\frac{1 - F_\eta(\Delta V_{TH})}{\#devices} = N_T e^{-\frac{\Delta V_{TH}}{\eta}}$ (23.5)
Devices: Total ΔV_{TH} convolution of n individual exponential distributions $= n$ traps		
$g_{\eta,n}\left(\Delta V_{TH}\right) = \frac{\Delta V_{TH}{}^{n-1}}{\eta^n (n-1)!} e^{-\frac{\Delta V_{TH}}{\eta}}$ (23.6)	$G_{\eta,n}\left(\Delta V_{TH}\right) = 1 - \frac{n}{n!} \Gamma\left(n, \Delta V_{TH}/\eta\right)$ (23.7)	
CHIP: Traps Poisson distributed with $\langle n \rangle = N_T$		
$P_{N_T}(n) = \frac{e^{-N} N_T^n}{n!}$ (23.8)		
Total ΔV_{TH} cumulative distribution in a chip is the sum up of $G_{\eta,n}$ weighted by P_{N_T}		
$H_{\eta,N_T}\left(\Delta V_{TH}\right) = \sum_{n=0}^{\infty} \frac{e^{-N} N_T^n}{n!} G_{\eta,n}\left(\Delta V_{TH}\right)$ (23.9)		

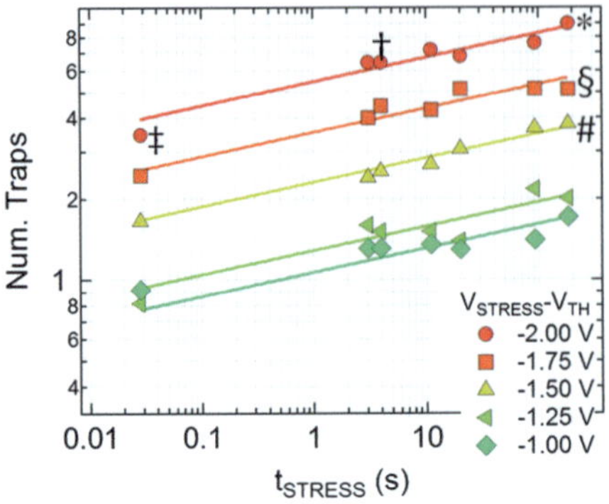

Fig. 23.12 The number of active traps per device N_T obtained from the fit of the CCDFs shown in Fig. 23.11 with (23.5) in Table 23.1 (intercept of CCDF with $\Delta V_{TH} = 0$). Note that N_T increases with stress time and voltage. Data can be fitted with both a power-law ($N_T = 0.83 \times t_{stress}{}^{0.094} \times (V_{stress} - V_{TH})^{2.53}$) and logarithmic time ($N_T = 0.19 \times \log(t_{stress}/2.63 \times 10^{-5}) \times (V_{stress} - V_{TH})^{2.53}$) dependences [22]

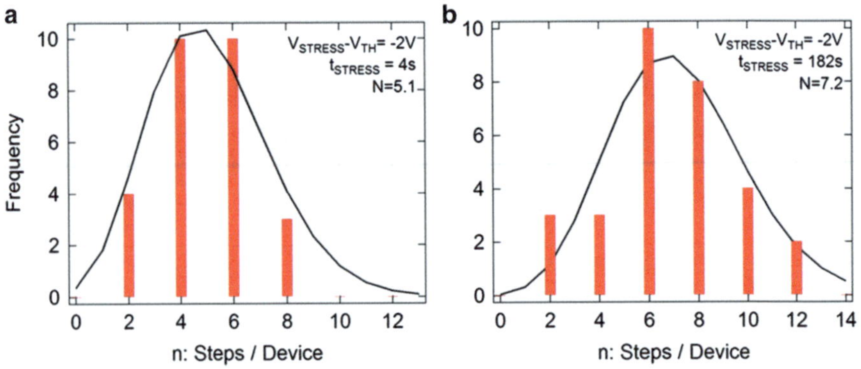

Fig. 23.13 Histogram of the number of steps per device detected ($\Delta V_{TH} > 1.5$ mV) from the relaxation curves for 30 pFETs following different stress times. Steps per device n are Poisson distributed (23.8). The average value N increases with increasing stress time

- Figure 23.15c shows that a reduction of the trap density N_T stretches out the overdrive (horizontal) axis, but the maximum fraction of working devices does not improve significantly at low overdrives.
- On the other hand, *a reduction of the* η *value shifts the lifetime prediction vertically*, boosting the number of working devices to high percentages over the whole overdrive range. *The largest gains in reliability can thus be achieved by moving to technologies with reduced dopant concentration* N_A *in the channel*; see (23.2) of Table 23.1 [3, 14, 22].

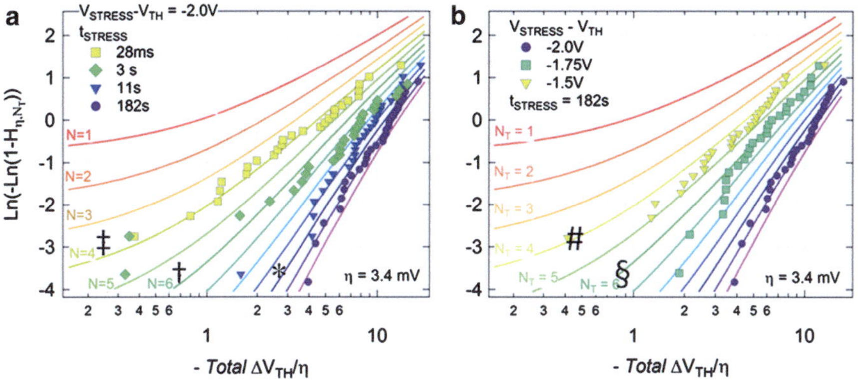

Fig. 23.14 (*Symbols*) Cumulative distributions of the total ΔV_{TH} normalized to $\eta = 3.4$ mV for 30 pFETs after stress (**a**) at different voltages and (**b**) for different times shown in Weibull plots. (*Lines*) Total ΔV_{TH} CDFs for different N_T values from (23.9) match excellently the experimental data

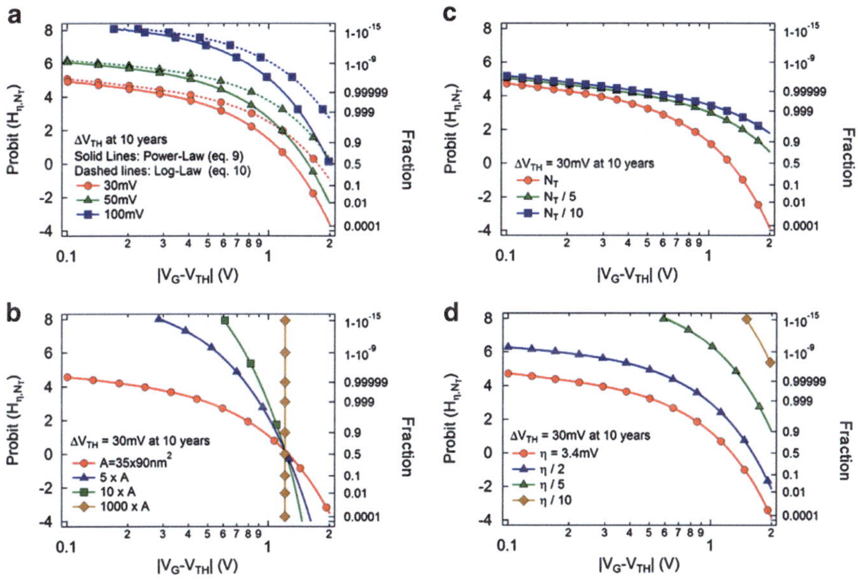

Fig. 23.15 Predicted 10-year-lifetime cumulative distributions of the total pFET ΔV_{TH} at $t_{RELAX} \sim 1$ ms. For different failure criteria (**a**), a slightly more optimistic prediction is given by a logarithmic time dependent law. For different device areas (**b**), it is observed that the median total ΔV_{TH} is independent of area. A significant fraction of deeply scaled devices exceeds failure criteria at lower overdrives. For different trap densities (**c**), CDF stretches out. For different η values (**d**), a significant boost of the fraction of working devices is obtained

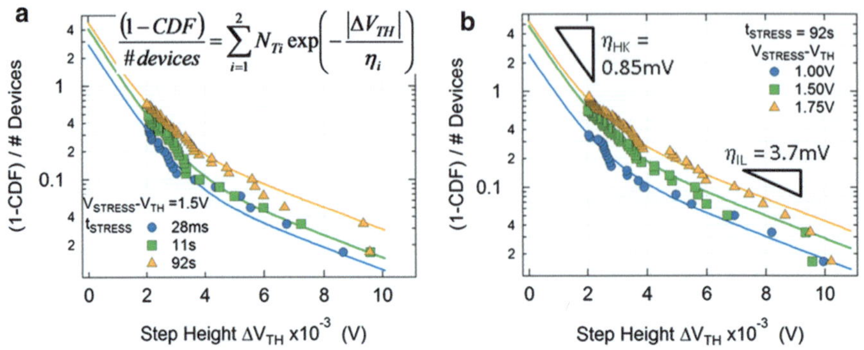

Fig. 23.16 (*Symbols*) CCDF of step heights normalized to the number of tested nFETs after positive stress. Data can be fitted with a bimodal distribution with $\eta_{IL} = 3.7$ mV and $\eta_{HK} = 0.9$ mV. Note that η_{IL} is similar to the η value obtained in pFETs

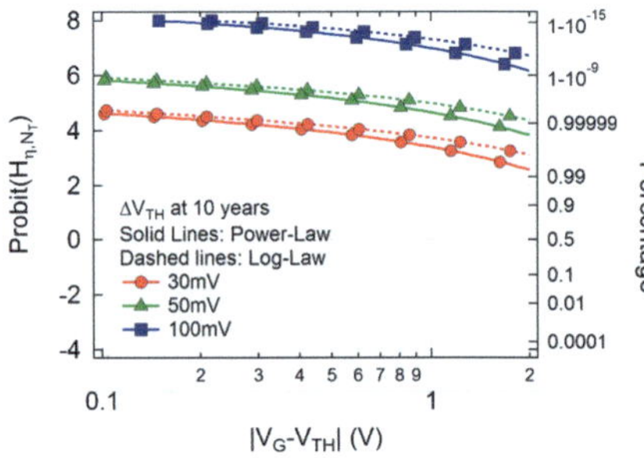

Fig. 23.17 Predicted 10-year lifetime CDF of the total nFET ΔV_{TH} indicates that PBTI is a less severe issue than NBTI (see Fig. 23.7a) in deeply scaled devices

An analogous analysis was performed on nFETs after positive gate bias stress. Figure 23.16 shows the CCDF for the step heights obtained from 60 nFETs. As already analyzed in Sect. 4.1 of this chapter, the CCDF follows a bimodal distribution with $\eta_{IL} = 3.7$ mV and $\eta_{HK} = 0.85$ mV. As shown in (23.2), the η value is related to the charge centroid $\langle x \rangle$: distance of the trapped charges in the dielectric from the gate [14, 18]. The lower value of η_{HK} corresponds to the defects in the high-κ dielectric. The trap density for η_{HK} is significantly higher, but it has a reduced impact on the total ΔV_{TH} shift due to the lower η_{HK} value. Note that η_{IL} value is close to the η value obtained from pFETs (Fig. 23.11), suggesting that it is related to border traps, however, its trap density is ten times lower w.r.t. pFETs under the same stress conditions. The ΔV_{TH} predictions for pFETs and nFETs (cf. Figs. 23.15a and 23.17) confirm that PBTI is a less severe issue than NBTI in deeply scaled devices.

From this study it is evident that a significant fraction of nm-scaled FETs will fail even at low overdrives assuming the classical failure criterion of maximum ΔV_{TH}. This conclusion was anticipated in the introduction, and it is obvious considering the link between RTN and BTI, since RTN is a phenomenon that causes giant V_{TH} oscillation in weak inversion, i.e., low overdrives. In future technological nodes, circuit design will become statistical (non deterministic). For this reason, the development of a circuit simulator that accounts for *heteroskedasticity* is compulsory to design reliable circuit with unreliable components [23].

23.6 Conclusions

In this chapter, we have summarized some recent insights into BTI achieved from the comprehensive study of deeply scaled devices. Among the most relevant, it is the close link between RTN and the recoverable component of BTI, indicating that identically behaving traps are responsible for both effects. Useful information about the kinetic properties of individual traps has been straightforwardly extracted from the recently developed TDDS technique. These insights helped to understand the charge exchange mechanisms between silicon substrate and gate oxide traps. Based on detailed understanding of the behavior and statistics of individual defects, we have demonstrated a new methodology to predict the BTI lifetime distributions of deeply scaled high-κ-based nFETs and pFETs.

Acknowledgments The authors would like to acknowledge the stimulating discussion with the imec DRE group (imec) and Prof. Tibor Grasser's group (Technical University of Vienna). This work was performed as part of imec's Core Partner Program. It has been in part supported by the European Commission under the seventh Framework Programme (Collaborative project MORDRED, contract No. 261868).

References

1. E. Cartier, A. Kerber, T. Ando, M. M. Frank, K. Choi, S. Krishnan, B. Linder, K. Zhao, F. Monsieur, J. Stathis and V. Narayanan,Tech. Dig. Int. Electron Devices Meet.2011, 441.
2. M. Cho, M. Aoulaiche, R. Degraeve, B. Kaczer, J. Franco, T. Kauerauf, Ph. J. Roussel, L. Å. Ragnarsson, J. Tseng, T.Y. Hoffmann and G. Groeseneken, IEEE Int. Reliab. Phys.Symp. Proc. 1095 (2010).
3. J. Franco, B. Kaczer, G. Eneman, J. Mitard, A. Stesmans, V. Afanas'ev, T. Kauerauf, Ph.J. Roussel, M. Toledano-Luque, M. Cho, R. Degraeve, T. Grasser, L.-Å. Ragnarsson, L. Witters, J. Tseng, S. Takeoka, W.-E. Wang, T.Y. Hoffmann and G. Groeseneken, Tech. Dig. Int. Electron Devices Meet.2010, 70.
4. International Technology Roadmap for Semiconductors available at http://public.itrs.net.
5. B. Kaczer, T. Grasser, J. Martin-Martinez, E. Simoen, M. Aoulaiche, Ph.J. Roussel and G. Groeseneken, IEEE Int. Reliab. Phys. Symp. Proc. 55 (2009).
6. T. Grasser, H. Reisinger, P. Wagner, F. Schanovsky, W. Goes and B. Kaczer, IEEE Int. Reliab. Phys. Symp. Proc. 16 (2010).
7. V. Huard, M. Denais and C. Parthasarathy, Microelectron.Reliab.**46**, 1 (2006).
8. M. Toledano-Luque, B. Kaczer, T. Grasser, Ph. J. Roussel, J. Franco and G. Groeseneken, J. Vac. Sci. Technol. B **31**, 01A114 (2013).
9. T. Grasser, B. Kaczer, P. Hehenberger, W. Goes, R. O'Connor, H. Reisinger, W. Gustinand C. Schlunder, Tech. Dig. Int. Electron Devices Meet 2007, 801.

10. T. Grasser, Th. Aichinger, G. Pobegen, H. Reisinger, P.-J. Wagner, J. Franco, M. Nelhiebel and B. Kaczer, , IEEE Int. Reliab. Phys.Symp. Proc.605 (2011).
11. H. Reisinger, T. Grasser, W. Gustin and C. Schlünder, IEEE Int. Reliab. Phys.Symp. Proc. 7 (2010).
12. B. Kaczer, Ph. J. Roussel, T. Grasser and G. Groeseneken, IEEE Electron Device Lett.**31**, 411 (2010).
13. A. Asenov, R. Balasubramaniam, A.R. Brown and J.H. Davies, IEEE Trans. Electron Devices **50,** 839 (2003).
14. M.F.Bukhori, S. Roy and A. Asenov, IEEE Trans. Electron Devices **57**, 795 (2010).
15. A. Ghetti, C.M. Compagnoni, A.S. Spinelli and A. Visconti, IEEE Trans. Electron Devices **56**, 1746 (2009).
16. B. Kaczer, T. Grasser, Ph. J. Roussel, J. Franco, R. Degraeve, L.-A. Ragnarsson, E. Simoen, G. Groeseneken and H. Reisinger, IEEE Int. Reliab. Phys. Symp. Proc. 26 (2010).
17. B. Kaczer, T. Grasser, P.J. Roussel, J. Martin-Martinez, R. O'Connor, B.J. O'Sullivan and G. Groeseneken, IEEE Int. Reliab. Phys. Symp. Proc. 20 (2008).
18. T. Grasser, H. Reisinger, P.-J. Wagner and B. Kaczer, Phy. Rev. B **82**, 245318 (2010).
19. M. Toledano-Luque, B. Kaczer, Ph.J Roussel, T. Grasser, G.I. Wirth, J. Franco, C. Vrancken, N. Horiguchi and G. Groeseneken, IEEE Int. Reliab. Phys. Symp. Proc 364 (2011).
20. M. Toledano-Luque, B. Kaczer, E. Simoen, Ph. J. Roussel, A. Veloso, T. Grasser and G. Groeseneken, Microelectron. Reliab.**88**, 1243 (2011).
21. M. Toledano-Luque, B. Kaczer, Ph. Roussel, M.J. Cho, T. Grasser and G. Groeseneken, J. Vac. Sci. Technol. B **29**, 01AA04 (2011).
22. M. Toledano-Luque, B. Kaczer, J. Franco, Ph. J. Roussel, T. Grasser, T.Y. Hoffmann and G. Groeseneken, Symposium on VLSI Technology Digest of Technical Papers2011, 152.
23. B. Kaczer, S. Mahato, V. Valduga de Almeida Camargo, M. Toledano Luque, Ph.J. Roussel, T. Grasser, F. Catthoor, P. Dobrovolny, P. Zuber, G. Wirth and G. Groeseneken,IEEE Int. Reliab. Phys. Symp. Proc. 915 (2011).
24. M. Cho, J.-D. Lee, M. Aoulaiche, B. Kaczer,Ph. J. Roussel,T. Kauerauf, R. Degraeve, J. Franco, L. Ragnarsson, G. Groeseneken, IEEE Trans. Electron Devices**59**, 2042 (2012).
25. A. Kerber, E. Cartier, L. Pantisano, R. Degraeve, T. Kauerauf, Y. Kim, A. Hou, G. Groeseneken, H.E. Maes and U. Schwalke, IEEE Electron Device Lett. **24**, 87 (2003).
26. S. Lee, H.-J. Cho, Y. Son, D.S. Lee and H. Shin, Tech. Dig. - Int. Electron Devices Meet. 2009, 759.
27. M. Uren, M. Kirton and S. Collins, Phy. Rev. B **37,** 8346 (1988).
28. V. Huard, C. Parthasarathy, C. Guerin, T. Valentin, E. Pion, M. Mammasse, N. Planes and L. Camus, IEEE Int. Reliab. Phys. Symp. Proc.289 (2008).
29. T. Grasser, P.-J. Wagner, H. Reisinger, Th. Aichinger, G. Pobegen, M. Nelhiebeland B. Kaczer, Tech. Dig. - Int. Electron Devices Meet. (2011), 6618.
30. T. Grasser, B. Kaczer, W. Goes, H. Reisinger, Th. Aichinger, Ph. Hehenberger, P.-J. Wagner, F. Schanovsky, J. Franco, Ph. Roussel and M. Nelhiebel, Tech. Dig. - Int. Electron Devices Meet. (2010). 82.
31. K. Takeuchi, T. Nagumo, S. Yokogawa, K. Imai and Y. Hayashi, Symposium on VLSI Technology Digest of Technical Papers 54 (2009).
32. J.Franco, B. Kaczer, M. Toledano-Luque, Ph.J. Roussel, J. Mitard, L.-Å. Ragnarsson, L. Witters, T. Chiarella, M. Togo, N. Horiguchi, G. Groeseneken, M. F. Bukhori, T. Grasser and A. Asenov, IEEE Electron Device Lett.**33**, 779 (2012).
33. T. Grasser, Microel. Reliab. **52**, 39–70 (2012).

Chapter 24
NBTI in (Si)Ge Channel Devices

Jacopo Franco and Ben Kaczer

Abstract This chapter focuses on the negative bias temperature instability (NBTI) of the novel Ge-based high-mobility channel pMOS technology, with Si passivation scheme and SiO_2/HfO_2 dielectric stack. We observe that this technology offers a remarkable reliability improvement. In particular, a significantly reduced NBTI is obtained by optimizing the gate stack with a high Ge fraction in the channel, a sufficiently thick channel quantum well, and a Si passivation layer of reduced thickness. By means of such optimization, sufficiently reliable ultrathin EOT SiGe pMOSFETs with a 10-year lifetime at operating conditions are demonstrated in both gate-first and gate-last process flows. Furthermore, the reliability improvement is observed to be process independent and architecture independent, proving to be an intrinsic property of the studied MOS system consisting of a Ge-based channel and a SiO_2/HfO_2 dielectric stack.

We ascribe this superior reliability chiefly to a reduced interaction between channel inversion holes and dielectric defects, thanks to a favorable energy alignment of the Fermi level in the (Si)Ge channel. This beneficial effect considerably alleviates also the time-dependent variability which arises as devices scale toward atomistic dimensions.

The extensive experimental results here reviewed strongly support (Si)Ge technology as a clear front-runner for future CMOS technology nodes, offering a solution to the reliability issues for ultrathin EOT nanoscale pMOS devices.

24.1 Introduction

Negative bias temperature instability (NBTI) is considered as the most severe reliability issue for scaled CMOS technologies [1]. The quasi-constant supply

J. Franco (✉) • B. Kaczer
IMEC, Kapeldreef 75, B-3001 Leuven, Belgium
e-mail: francoj@imec.be; kaczer@imec.be

T. Grasser (ed.), *Bias Temperature Instability for Devices and Circuits*,
DOI 10.1007/978-1-4614-7909-3_24, © Springer Science+Business Media New York 2014

voltage scaling proposed by the International Technology Road Map [2] for the recent technology nodes enhances NBTI due to the ever increasing interfacial oxide electric field (E_{ox}). As a consequence, although several groups have already demonstrated well-behaving CMOS devices with aggressively scaled EOT down to 0.5 nm [3, 4], a 10-year lifetime cannot be guaranteed for the expected operating voltages [5, 6]. Hence, the device reliability is becoming an impending showstopper for further scaling.

Meanwhile, the use of high-mobility materials such as SiGe or *pure* Ge for p-type channels and III-V compounds for n-type channel is being considered for further enhancement of the CMOS performance [7–10]. The main benefits promised by the Ge-based technology can be briefly summarized as (1) enhanced mobility which can alleviate the mobility reduction caused by the defective high-k layer coming closer to the channel due to the scaling of the SiO_2 interfacial layer (IL) and (2) pMOS threshold voltage tuning toward the roadmap target.

In this chapter, we discuss the NBTI reliability of Ge-based quantum well (QW) pMOSFETs. Already in 2009 we have observed that the incorporation of Ge into the channel significantly improves the NBTI robustness [11, 12]. Later on we have observed this reliability improvement to be process independent and architecture independent and therefore intrinsically related to the incorporation of Ge in the channel layer.

Here, we review the extensive experimental results collected over the recent years, and we discuss the physical model we have proposed to understand the superior NBTI reliability of this technology. It is made clear how incorporation of Ge into the pMOSFET channel opens a new degree of freedom for optimizing the NBTI reliability of ultrathin EOT devices. In particular, a reliability-oriented gate-stack optimization with a high Ge fraction, a thick QW, and a thin Si passivation layer is shown to boost the allowed gate voltage overdrive for 10 years operation above the expected operating V_{DD} for devices with ultrathin EOT (down to ~0.6 nm EOT) [13].

In the first part of the chapter, we discuss experimental results obtained on large area ($W \times L = 10 \times 1\ \mu m^2$) test devices, as customary for standard NBTI testing. On such large area devices, the random properties of the many defects in the gate oxide (e.g., for a defect density of $\sim 10^{11}\ cm^{-2}$, one device contains $\sim 10^4$ defects) average out, yielding the same, well-defined, BTI degradation curve on each device [14, 15]. Conversely, recent works have shown that as the device geometries scale toward atomistic dimensions, the number of dopant atoms, as well as the number of dielectric defects in each transistor, is reduced to numerable levels (e.g., for a defect density of $\sim 10^{11}\ cm^{-2}$, a $90 \times 35\ nm^2$ device would include an average of only ~3 defects per device). As an implication, both the fresh device parameters and the parameter shifts during operation become statistically distributed [14–17, 42]. In other words, both a time-zero (i.e., as-fabricated) variability and a considerable time-dependent variability (i.e., reduced reliability) arise. As a consequence, the deterministic "average" lifetime which is normally assessed on large area devices should be replaced by lifetime distributions [15, 18]. This time-dependent variability (i.e., nanoscale NBTI reliability) can be studied in terms of statistical measurements of the charging and discharging of individual defects [19–21]. We and others have recently shown that the properties of individual charged-gate oxide defects can be directly observed and measured by looking at the individual discharge events visible in NBTI ΔV_{th} relaxation transients recorded on nanoscaled devices [14, 15, 17, 41]. This approach recently led to the introduction of a novel defect characterization technique, the time-dependent defect spectroscopy (TDDS [22]).

For these reasons, in the second part of the chapter we focus on the NBTI of nanoscale SiGe pFETs and compare with the results obtained on large area devices. The use of a SiGe channel is shown to offer a considerable reliability improvement also for deeply scaled devices, which is expected to significantly alleviate the time-dependent variability issue.

The extensive experimental results collected on a variety of processed wafers and reviewed here strongly support SiGe channel technology as a promising candidate for future CMOS technology nodes, offering a significant relief to the reliability issue for ultrathin EOT nanoscale pMOS devices.

24.2 Device and Experimental Methodology

The experiments here discussed were performed on SiGe channel pFETs fabricated at *imec* on 300 mm Si wafers. A sketch of the device gate stack and its band diagram in inversion are depicted in Fig. 24.1. The channel consists of an epitaxially grown compressively strained thin $Si_{1-x}Ge_x$ layer, with thickness varying between 3 and 7 nm. Ge fractions up to $x = 0.55$ were used. *Pure* Ge (i.e., $x = 1$) unstrained channel devices were also used.

On top of the (Si)Ge layer, a thin undoped Si cap was grown epitaxially. The physical thicknesses of this thin Si cap varied between 0.65 and 2 nm, as estimated from capacitance–voltage (C–V) curves and TEM pictures of the final device.

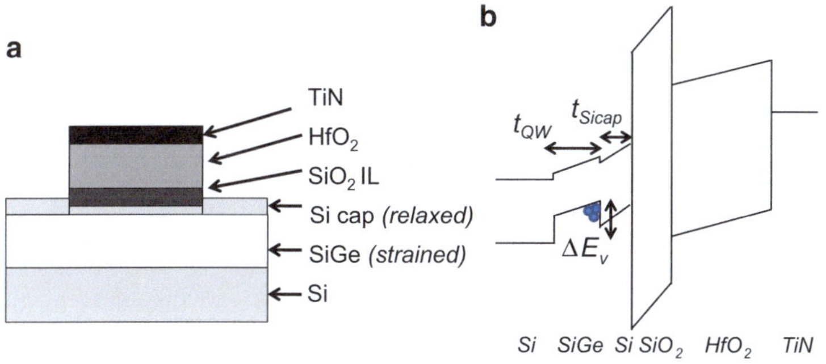

Fig. 24.1 (**a**) Gate-stack sketch of the (Si)Ge devices used in this work. (**b**) Band diagram in inversion. Channel holes are confined into the (Si)Ge quantum well due to the valence band offset (ΔE_v) between the channel and the Si cap. The Si cap thickness (t_{Sicap}) therefore contributes to the T_{inv} of the gate stack

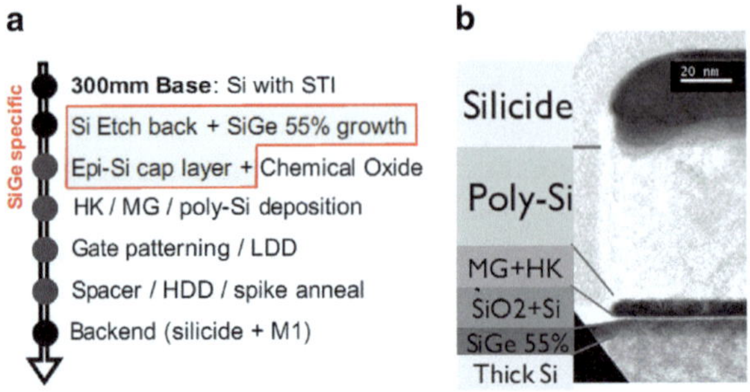

Fig. 24.2 (**a**) SiGe channel device process flow schematic. With respect to a standard Si flow, two epitaxy steps are introduced for the (Si)Ge channel and the Si cap growths. (**b**) TEM picture of the (Si)Ge channel device gate stack [8]

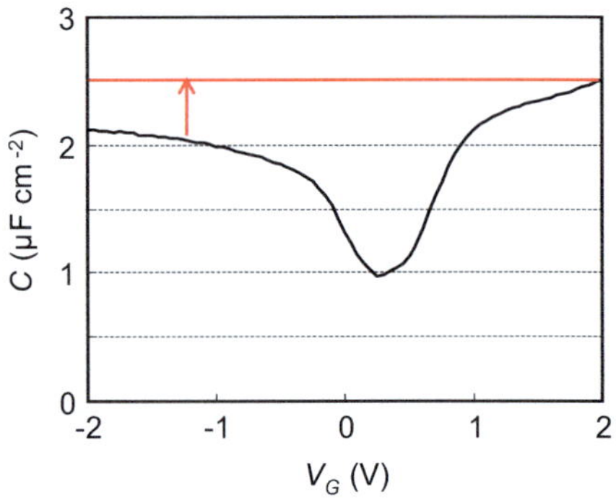

Fig. 24.3 Full C–V curve (i.e., "gate to all") measured on a pure Ge channel device. Due to the band alignment of the (Si)Ge channel toward the Si cap, the latter acts as an additional displacement for inversion holes only, i.e., an asymmetry between inversion and accumulation oxide capacitance is observed [11]

Gate-stack fabrication started with a wet chemical oxidation (*imec clean* [23]) of the Si cap. On top of this SiO_2 interfacial layer (IL), ~1.8 nm of HfO_2 was deposited using an atomic layer deposition (ALD). Finally, a PVD TiN metal gate was deposited. The metal gate thickness controlled the final IL thickness by means of the oxygen-scavenging technique, as discussed in [3]. A schematic of the process flow is depicted in Fig. 24.2 together with a representative TEM picture of the final devices. With respect to a standard Si device process flow, only the two epitaxy steps for the growth of the thin (Si)Ge channel layer and Si cap were added [24]. For comparison purposes, standard Si channel devices with an identical gate stack were also used.

Due to the valence band offset between the (Si)Ge layer and the Si cap (see Fig. 24.1b), inversion channel holes are confined in the (Si)Ge layer, which therefore acts as a quantum well. This causes the Si cap thickness to lower the inversion capacitance as compared to the accumulation capacitance, as documented by Fig. 24.3. For a fair benchmarking of these devices, it is therefore necessary to consider the capacitance-equivalent thickness in inversion (T_{inv}, evaluated at $V_G = V_{th} - 0.66$ V) which includes the contribution of the Si caps of varying thicknesses.

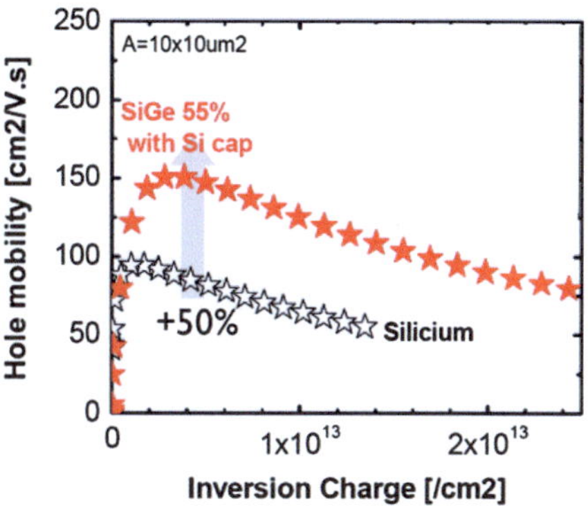

Fig. 24.4 Split-CV mobility curves for SiGe devices with EOT = 0.85 nm compared to the universal Si hole mobility. A significant mobility boost is observed even at such scaled EOT [8]

The effective mobility enhancement factor of SiGe devices with respect to Si control (Fig. 24.4) ranged between 1.5× and 2.4×, depending on the process parameters [7, 8]. Three major process parameters of the (Si)Ge pMOSFETs, i.e., the Ge fraction, the quantum well thickness, and the Si cap thickness, were varied separately in order to assess their individual impact on the device reliability.

NBTI stress experiments were performed using the extended measure-stress-measure technique [25]. The devices were stressed at $T = 125°C$ with several gate overdrives, while the sensing bias was $V_G = V_{th0}$. To minimize NBTI relaxation effects for the device lifetime predictions, ΔV_{th} was evaluated at $t_{relax} = 1$ ms, i.e., the minimum delay of the used setup (Keithley 2602 Fast Source Meter Units). This delay was fixed in the experiments to allow cross-comparison. For each gate voltage, the stress time needed to reach a failure criterion, assumed at 30 mV threshold voltage shift, was extracted (i.e., the time to failure). The maximum operating overdrive for a 10-year lifetime (V_{op}, i.e., the maximum $|V_G - V_{th0}|$) was then extrapolated by fitting a power law to the time-to-failure vs. gate overdrive dataset (see, e.g., Fig. 24.5).

24.3 (Si)Ge Gate-Stack Optimization for Reliability

In this section we review the impact of the three main gate-stack parameters on the NBTI reliability, namely, the Ge fraction in the channel, the quantum well thickness, and the Si cap thickness. In this preliminary set of experiments, the EOT was not aggressively scaled (EOT ~ 1.2 nm). For comparison purposes, a set of standard Si channel devices with an identical high-k/metal gate stack was also used as a reference.

As shown in Fig. 24.5a, the introduction of Ge in the channel significantly improved the NBTI reliability. The extrapolated maximum operating overdrive voltage for 10-year lifetime (*maximum* $|V_G - V_{th0}| \equiv V_{op}$) increased from 0.46 V for the Si reference up to 0.8 V for 45% Ge fraction device with a SiGe layer

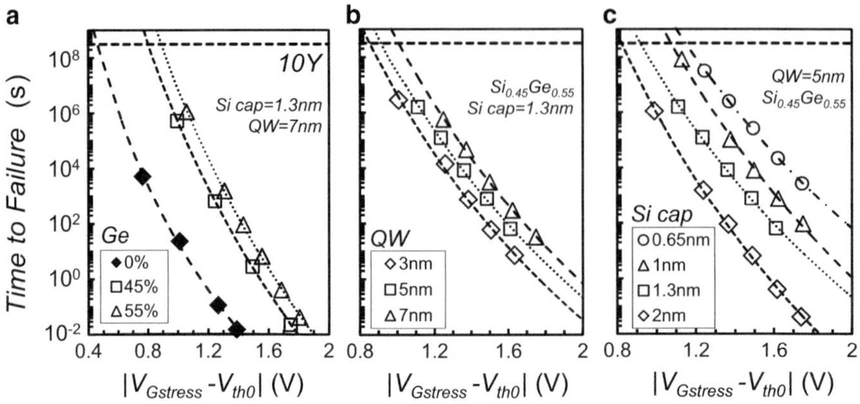

Fig. 24.5 Extrapolated lifetimes as a function of the gate overdrive voltage for devices: (**a**) with varying Ge content, a higher Ge fraction boosts the NBTI robustness; (**b**) with varying quantum well (QW) thickness, a thicker quantum well boosts the NBTI robustness; and (**c**) with varying Si cap thickness (0.65–2 nm), a reduced thickness of the Si cap boosts the NBTI robustness while enabling T_{inv} reduction

thickness of 7 nm and a Si cap thickness of 1.3 nm. Increasing the Ge fraction to 55% while fixing the other parameters boosted the operating overdrive voltage even more, reaching 0.9 V.

Increasing the thickness of the SiGe quantum well resulted in an additional improvement of the NBTI reliability (Fig. 24.5b): V_{op} increased from 0.85 V up to 1.01 V when moving from a 3 nm-thick SiGe layer to a 7 nm one. This observation was made while fixing the Si cap thickness to 1.3 nm on devices with 55% Ge fraction.

The most significant impact on the NBTI reliability was observed when varying the Si cap thickness (Fig. 24.5c). Interestingly, a *reduced* thickness of this layer clearly improved the NBTI robustness. Naively, one would expect the thinner Si cap to act as a reduced tunneling barrier for holes, but conversely V_{op} increased from 0.82 to 1.14 V when the Si cap thickness was decreased from 2 to 0.65 nm. This counterintuitive observation is crucial for understanding the superior reliability of SiGe devices, as we will discuss later in this chapter. Moreover, the observation is particularly relevant since a reduced Si cap thickness, while improving the NBTI reliability, also reduces the device T_{inv} (thanks to reduced hole displacement; see Fig. 24.1b) and therefore enhances the current drive performance.

The estimated V_{op} values for different Si cap thicknesses are shown in Fig. 24.6 in a benchmark plot vs. the T_{inv} values and compared with reference data measured on Si channel high-k/metal gate pMOSFETs. It is clear that reducing the Si cap thickness yields a significant V_{op} boost together with a T_{inv} reduction. Such a V_{op} boost for reduced Si cap thickness was observed consistently at several T_{inv} ranges obtained by SiO_2 IL scaling [3]. This remarkable trend is clearly different from the data collected on Si channel devices where a T_{inv} reduction is always associated with a reliability reduction (see Fig. 24.6 diamonds).

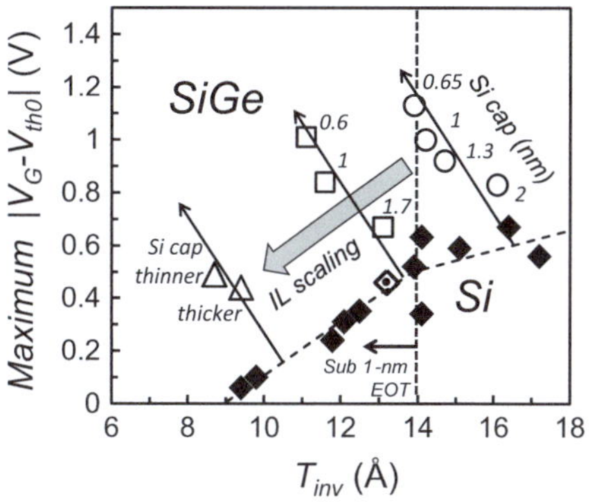

Fig. 24.6 Maximum operating gate overdrive (V_{op}) for 10-year lifetime under NBTI stress ($T = 125°C$, failure criterion $\Delta V_{th} = 30$ mV) vs. T_{inv}. SiGe devices (*open symbols*) with a thin Si cap offer improved NBTI reliability, i.e., higher maximum operating gate overdrive. The observation is consistently reproduced while scaling the SiO_2 IL

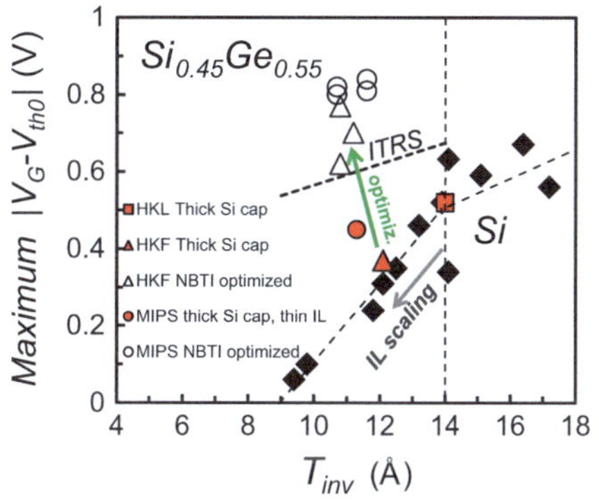

Fig. 24.7 A high Ge fraction (55%) in a 6.5 nm-thick SiGe quantum well, combined with a Si cap of reduced thickness (0.8 nm) boost NBTI lifetime to meet the target V_{DD} at ultrathin EOT in a MIPS process flow (*open circles*, as compared to *solid circle*). The optimization was also implemented in a RMG process flow (*open triangles*, as compared to *solid triangle* and *square*). The results were reproduced for several thermal budgets

The remarkable impact of the individual SiGe gate-stack parameters discussed above can be proficiently used to optimize the NBTI robustness and restore the reliability of ultrathin EOT devices with aggressively scaled IL. Figure 24.7 shows that the beneficial effects of a *high Ge fraction, a thick quantum well, and a Si cap of reduced thickness* can be combined to boost the NBTI lifetime above the ITRS [2] target V_{DD} for ultrathin EOT devices (10-year continuous operation at $|V_G - V_{th0}| \approx 0.6$ V at $T_{inv} \approx 1$ nm, EOT ≈ 0.6 nm). This result was first obtained in a metal inserted poly-Si (MIPS) process flow and reproduced for several thermal budgets (open circles in Fig. 24.7). Furthermore, the optimization was also implemented in a replacement metal gate (RMG) process flow: high-k last SiGe sample with a thick Si cap (square) shows poor NBTI robustness; an IL reduction by means of O-scavenging in a high-k first process flow further reduces NBTI robustness as discussed above (solid triangle); however, the SiGe gate-stack optimization (open triangles) boosts the V_{op} above the ITRS target. Also for the RMG process flow, the result was reproduced for several process thermal budgets.

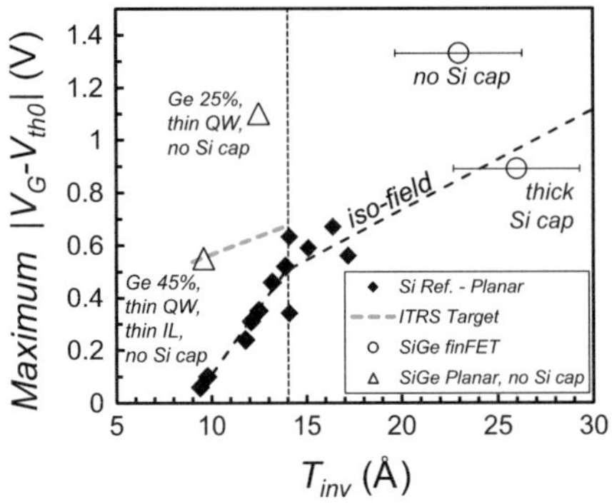

Fig. 24.8 SiGe channel bulk pFinFETs *without* a Si cap show improved NBTI reliability w.r.t. to the same devices with a thick Si cap and w.r.t. Si planar pFETs. The dashed trend line for $T_{inv} > 1.4$ nm demarcates planar Si pFET constant field scaling (*iso-field*). Uncertainty in T_{inv} is related to the finFET dimensions. Note: Planar SiGe devices without a Si cap also show improved NBTI reliability (e.g., two different planar gate stacks are shown, triangles, $Si_{0.75}Ge_{0.25}$ 3 nm-thick QW, and $Si_{0.55}Ge_{0.45}$ 3 nm-thick QW with reduced IL thickness by O-scavenging)

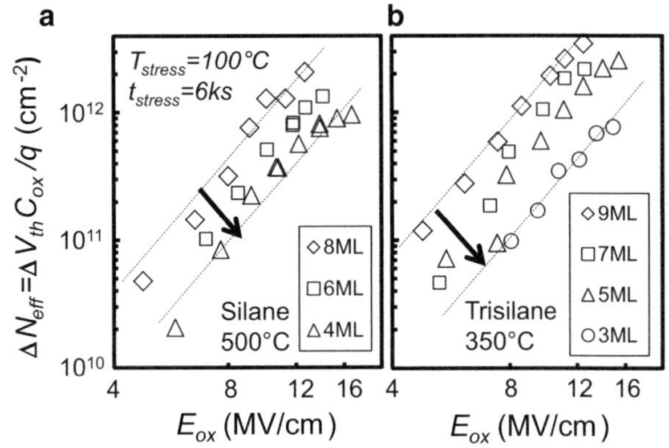

Fig. 24.9 A decreased thickness of the Si passivation layer improves the NBTI robustness of Ge pMOSFETs, independently of the Si cap epi-process used, (**a**) silane 500°C or (**b**) trisilane 350°C

These process-independent results suggest the reliability improvement to be an intrinsic property of the Ge-based channel/SiO_2/HfO_2 MOS system. Moreover, the improved reliability was observed to be also architecture independent: preliminary results on novel SiGe wrapped-channel bulk pFinFETs [26] showed improved NBTI lifetime w.r.t. Si planar ref. when removing the Si cap (Fig. 24.8).

Furthermore, the experiment with varying Si cap thickness was repeated on *pure* Ge channel pMOSFETs with 4, 6, and 8 Si monolayers (ML) epi-grown from silane precursor at 500°C. A reduced thickness of the Si layer again resulted in a reduced NBTI at fixed stress conditions (electric field, stress time, stress temperature, sensing delay): four MLs devices degraded about four times less than eight MLs devices (Fig. 24.9a). The same trend was also observed for Si caps grown using a 350°C epi-growth from trisilane precursor: a $\sim 8\times$ reduction of the degradation was observed when reducing the Si from nine to three MLs (Fig. 24.9b).

Independent confirmations of improved NBTI reliability in Ge-based pMOSFETs have been reported lately by several other groups [9, 27, 28]. All these process- and architecture-independent results suggest that *the reduced NBTI is an intrinsic property of the pMOS system consisting of a Ge-based channel and a SiO_2/HfO_2 dielectric stack*, further emphasizing the use of a (Si)Ge channel as a promising candidate for future CMOS technology nodes. In the next sections, the physical mechanisms behind this observed property are discussed, and a model for the improved NBTI reliability is proposed.

24.4 NBTI Kinetics on (Si)Ge

A reduction of the Si cap thickness on SiGe was shown to yield the most significant reliability boost. It is then worth discussing this remarkable experimental result in more detail. Figure 24.10 shows the typical NBTI ΔV_{th} evolution vs. the stress time for the Si ref. and the SiGe devices with different Si cap thickness. As one can see, the SiGe devices show a significantly reduced ΔV_{th}, especially for the samples with reduced Si cap thickness. The NBTI ΔV_{th} evolution is often described as a power law of the stress time:

$$\Delta V_{th} = At^{n}_{stress}. \tag{24.1}$$

The power-law pre-factor A is dependent on the stress bias, while the apparent exponent n is typically reported to be in the range of 0.1–0.25 [1], depending on the relaxation allowed by the measurement delay [25]. Figure 24.11 documents the extracted power-law pre-factors for all the devices here considered. A dramatic NBTI reduction for the SiGe devices is apparent, in particular for a reduced Si cap thickness. Moreover, it is evident that the pre-factors show a significantly stronger E_{ox} acceleration for SiGe with respect to the Si reference device, yielding further improvement at the lower operating fields.

Figure 24.12 reports NBTI-induced ΔV_{th} at different stress temperatures on a Si ref. device, SiGe devices with a 2 nm and with a 0.65 nm-thick Si caps. No

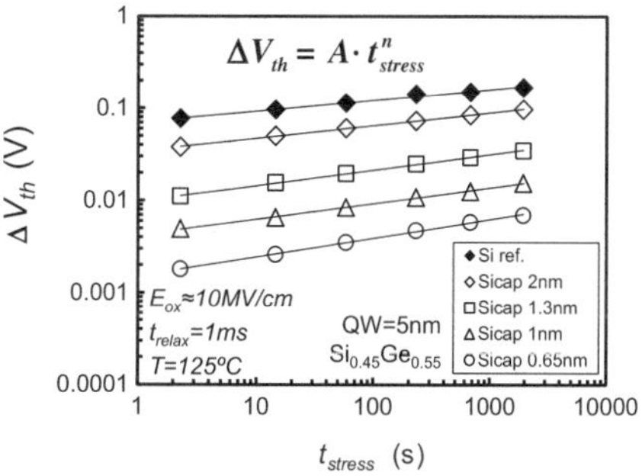

Fig. 24.10 Measured ΔV_{th} during NBTI stress at fixed stress conditions on a Si ref. and on SiGe devices with different Si cap thicknesses. The SiGe device with the thinnest Si cap shows most reduced V_{th} instability

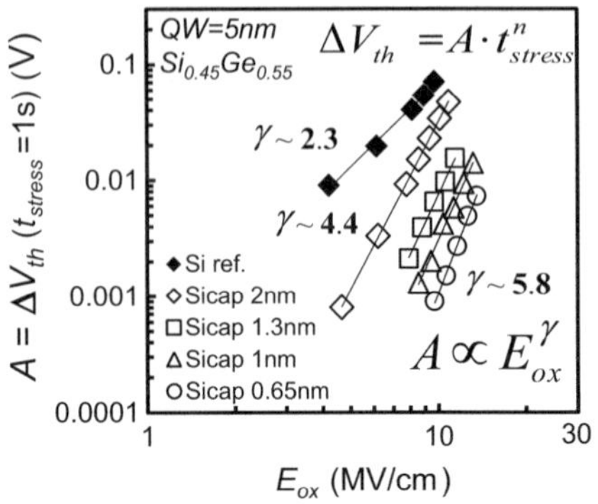

Fig. 24.11 Extracted power-law pre-factors: a significant reduction for the SiGe devices is observed, especially with a reduced Si cap thickness. A stronger E_{ox} acceleration for SiGe with respect to the Si ref. device is also observed

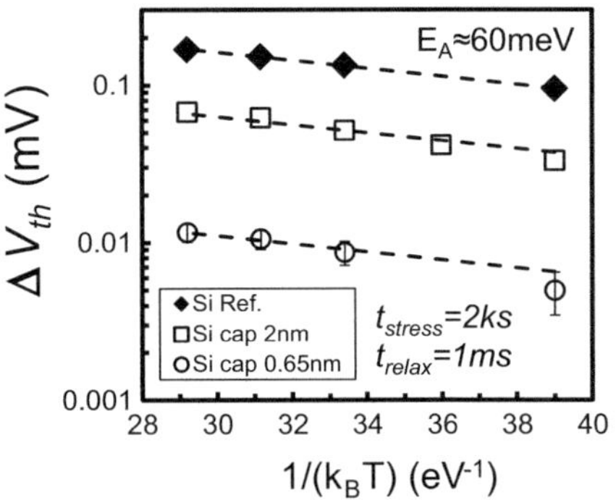

Fig. 24.12 NBTI-induced ΔV_{th} measured at different stress temperatures on the Si ref. devices and on SiGe devices with two different Si cap thicknesses (2 and 0.65 nm). No clear difference in the apparent ΔV_{th}-activation energy is observed (extracted $E_A \approx 60$ meV)

clear difference in the activation energies is observed for the different devices. The extracted apparent ΔV_{th}-activation energy is ~60 meV, in the typically reported range [1].

NBTI degradation is often described as the combination of two different components [1, 29, 30]: a recoverable component (R) related to hole trapping in preexisting bulk oxide defects (ΔN_{ot}) and a so-called permanent component (P) typically associated with the creation of new interfaces states (ΔN_{it}). To get a deeper insight into the measured NBTI trends, the charge pumping (CP) technique [31] was used to monitor the interface state creation during the NBTI stress. While ΔN_{it} was monitored by CP, ΔN_{ot} was calculated by subtracting the ΔN_{it} contribution from the total ΔV_{th} measured. Figure 24.13 reports ΔN_{it} and ΔN_{ot} evolutions measured on SiGe devices with two different Si cap thicknesses and on a Si reference device for fixed stress conditions ($E_{ox} = 10$ MV/cm, $T = 125$°C). The SiGe device with a thin Si cap shows both reduced R and P, with the reduction in R being of higher relevance on the total ΔV_{th}, since R contributes most to the degradation at short t_{stress}. Interestingly, ΔN_{it} follows a power law with stress time with the same exponent of ~0.25 for the Si and for SiGe devices with different Si caps, suggesting the same interface bond breaking process. This conclusion is also supported by the P component showing the same dependence on the stress electric field for the SiGe and Si reference devices, as illustrated in Fig. 24.14.

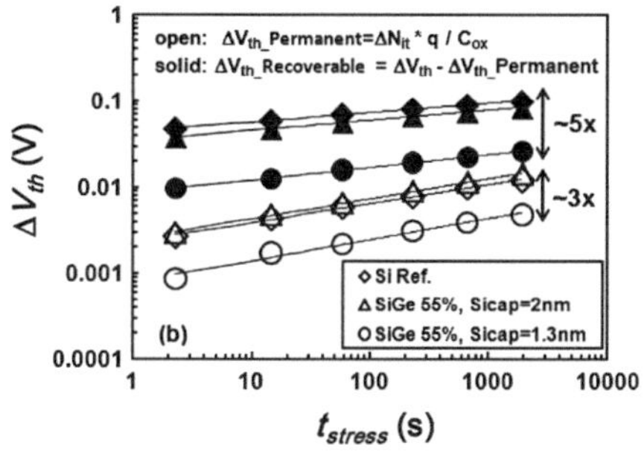

Fig. 24.13 Total ΔV_{th} split into the so-called permanent (*P*) ΔV_{th}, assumed to be caused by ΔN_{it}, and the recoverable (*R*) ΔV_{th}, assumed to be caused by filling of preexisting oxide traps (N_{ot}). ΔN_{it} measured with charge pumping during NBTI stress were converted to $\Delta V_{th\_Permanent}$ ($=\Delta N_{it} \cdot q/C_{ox}$) in order to decouple their contribution from the total measured ΔV_{th}. ΔN_{it} follows a power law on the stress time with the same exponent (~0.25) on all three samples. However, SiGe devices with thinner Si cap show both reduced *P* and *R*, with the reduction of *R* having a higher impact on the total ΔV_{th}

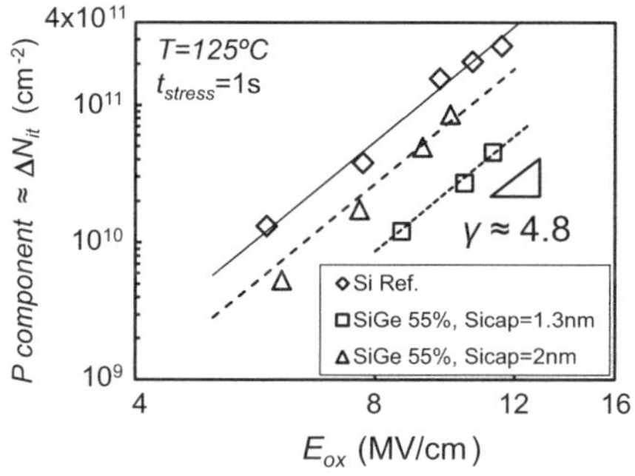

Fig. 24.14 *P* component (~ΔN_{it}) extracted with the universal relaxation methodology [25, 29] and plotted vs. the stress electric field (E_{ox}) for the Si reference device and SiGe devices with two different Si cap thicknesses (2 and 1.3 nm). A reduced *P* component is observed for the SiGe device with a thinner Si cap, while the similar E_{ox} dependence suggests the same interface bond breaking process taking place as in the Si device

The above-discussed experimental observations of the NBTI kinetics in optimized SiGe devices with a reduced Si cap thickness as compared to Si reference devices can be summarized as follows:

1. Reduced overall ΔV_{th}, i.e., lower power-law pre-factor
2. Similar temperature activation (apparent $E_A \approx 60$ meV)
3. Significantly reduced ΔN_{it} and ΔN_{ot}, with the latter reduction being of greater relevance
4. Similar ΔN_{it} time exponent (~0.25) and field dependence (~4.8), suggesting same interface bond breaking process
5. Stronger E_{ox} acceleration of the overall ΔV_{th} (~5.8 vs. ~2.3), to be attributed to the recoverable charge trapping component

24.5 Model

As discussed in the previous section, SiGe devices with reduced Si cap thickness show both reduced *P* and *R* NBTI components. In the following subsections, models for reduced *P* and *R* are proposed.

24.5.1 Reduced *P* ($\sim\Delta N_{it}$)

N_{it} creation during NBTI stress is commonly attributed to de-passivation of H-passivated Si dangling bonds (P_{b0}) at the Si/SiO_2 interface. Electron spin resonance spectroscopy (ESR) [32] measured on a Ge substrate with a thick Si cap revealed a high P_{b0} density ($\sim 1 \times 10^{12}$ cm^{-2}), while it could not detect these defects ($<10^{11}$ cm^{-2}) for a very thin Si cap (Fig. 24.15). This suggests that the higher Ge segregation at the Si/SiO_2 interface reported for thin Si caps [33] can reduce the N_{it} precursor defect density and therefore reduce the creation of interface states (ΔN_{it}) during NBTI stress.

As a confirmation of this, we observed on pure Ge channel devices that the NBTI reliability showed a dependence on the growth temperature of the epitaxial process of the Si passivation layer. In particular, an increased NBTI was observed in devices with a Si cap grown at 350°C from trisilane precursor with respect to the more common 500°C epi from silane precursor (Fig. 24.16). This experimental result confirmed that a higher Ge segregation at the interface, caused by a higher temperature process [34], might reduce the interface state creation during stress thanks to a lower precursor defect availability.

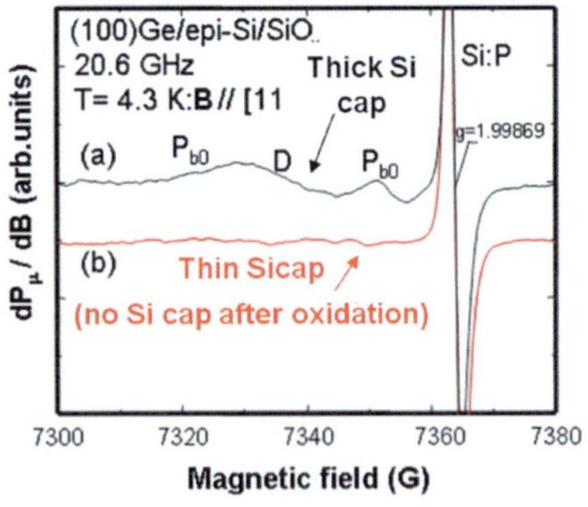

Fig. 24.15 ESR measurements performed on Ge (100) single crystal p-type substrates, passivated with epitaxial Si caps subsequently UV oxidized at room *T*. In the sample with a thick cap, a high P_{b0} density was found ($\sim 10^{12}$ cm^{-2}), while these defects could not be detected ($<10^{11}$ cm^{-2}) for a very thin and almost completely oxidized Si cap. Data courtesy of Profs. A. Stesmans and V. Afanas'ev, Dept. of Physics and Astronomy, KU Leuven. Similar results were obtained by the same group on samples prepared with *imec clean* oxidation combined with a HfO_2 layer

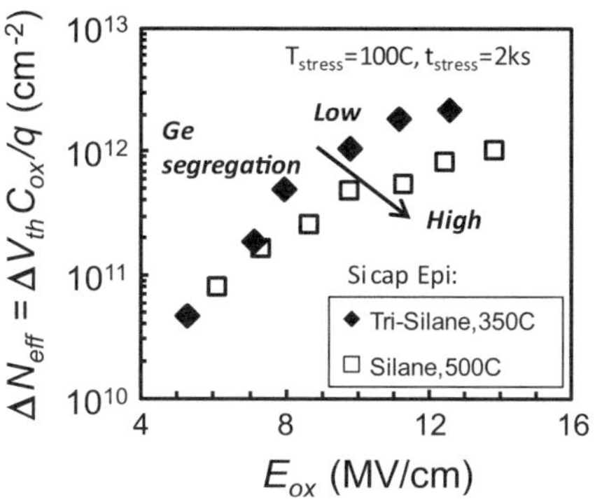

Fig. 24.16 NBTI-induced threshold voltage shift for pure Ge channel devices with similar Si cap thickness ($\sim$1 nm) but different epitaxial growth temperature. The higher temperature epi-process yields reduced NBTI at high equivalent oxide stress field (i.e., possibly in a ΔN_{it}-dominated degradation regime), possibly owing to a reduced precursor defect density linked to Ge segregation at the interface

The reduced creation of interface states undoubtedly plays a role in the improved NBTI reliability of (Si)Ge channel devices (see Fig. 24.13). In particular, since the P component has a higher time exponent ($n \sim 0.25$) with respect to R ($n \sim 0.05$–0.1) [25], it is expected to dominate at the device end of life—a reduced P component extends the predicted device lifetime.

However, the reduced P component alone cannot explain the dramatically reduced overall NBTI degradation in optimized SiGe devices ($\sim$0.01$\times$ as compared to Si reference, see Fig. 24.10) already observed at short stress times, i.e., when R is unarguably the dominating component. Furthermore, in ultrathin EOT devices with aggressively scaled SiO_2 IL, the enhanced hole trapping in high-k defects increases the contribution of R to the total degradation [6]. It is therefore crucial to understand the origin of the reduction in the R component in (Si)Ge devices, as discussed next.

24.5.2 *Reduced R ($\sim\Delta N_{ot}$): A Model for the Superior NBTI Reliability of (Si)Ge Channel pMOSFETs*

We have proposed that the reduction in R is related to a favorable alignment shift of the Fermi level E_F in the SiGe QW with respect to the preexisting bulk oxide defect energy levels (Fig. 24.17). Larger misalignment yields a reduced interaction of carriers with oxide traps (N_{ot}) since fewer defect levels are energetically favorable for charging by channel holes.

To model this effect, we assumed the existence of a defect band both in the SiO_2 IL and in the high-k layer; we note that interacting defects have to be located in both the dielectric layers since the same NBTI trends on SiGe with different Si caps were

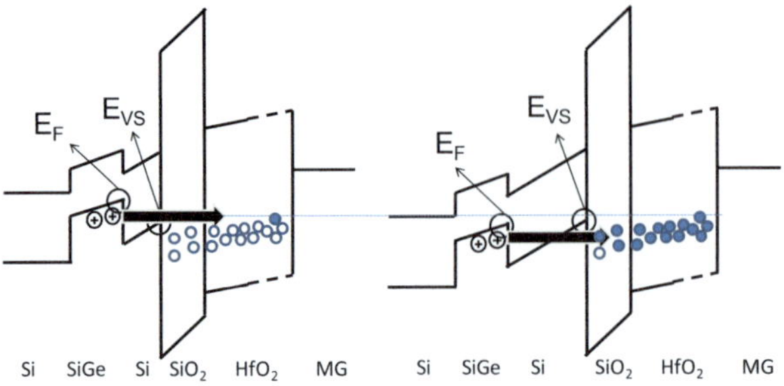

Fig. 24.17 Sketch of the proposed model based on a favorable alignment shift of the Fermi level E_F in the (Si)Ge quantum well. A significant fraction of the preexisting bulk oxide defects might be inaccessible (i.e., energetically unfavorable) for channel holes in the SiGe channel. This beneficial effect is lost for a thicker Si cap, since the additional voltage drop on it (at constant electric field) "pushes down" the channel energy, making all defects accessible as in the standard Si channel devices

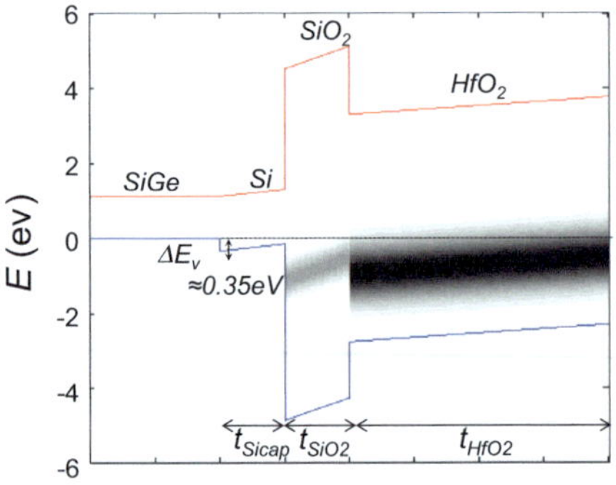

Fig. 24.18 A model including defect bands centered at 0.95 and 1.4 eV below the Si valence band in the SiO_2 IL and in the HfO_2, respectively. The channel Fermi level determines which part of the defect bands is accessible to channel holes. The defect band is modeled as a Gaussian distribution over energy. Charged defects at different spatial positions contribute differently to the total ΔV_{th} due to electrostatic

consistently observed when scaling the IL (see Fig. 24.6). As depicted in Fig. 24.17, the Fermi level in the channel determines which part of the defect band is accessible to channel holes. The defect bands were modeled as Gaussian distributions over the dielectric energy bandgap:

$$\mathrm{N_{ot}}(\mathrm{E}, \mu, \sigma, \mathrm{x}) = N_{ot0} \cdot \frac{e^{-\frac{(\mathrm{E}-\mu(\mathrm{x}))^2}{2\sigma^2}}}{\sigma\sqrt{2\pi}}, \tag{24.2}$$

where E is the energy within the dielectric bandgap; μ and σ are the mean and the standard deviations of the Gaussian distributions, respectively; and x is the spatial position inside the dielectric layer (Fig. 24.18). This proposed representation of the defect levels is a mere assumption serving the sole purpose of simplifying the mathematical treatment of the model. Different energy distributions of the defects in the dielectric layer might exist in reality. Nevertheless, a similar beneficial effect by shifting up the Fermi-level energy in the channel would be obtained independently of the chosen defect-level representation [e.g., even for a uniform energy distribution of defect levels, a fraction of defects would become unfavorable for holes at the (Si)Ge channel Fermi level].

While the exact nature of the defects causing NBTI is still not clear, oxygen vacancy-related defects are typically blamed due to their ubiquitous presence in dielectric layers [35]. We chose therefore to pin the mean values of the distributions at the theoretically calculated values of common oxygen vacancies, i.e., at 0.95 eV below the Si valence band for the IL (corresponding to the E'γ[$E_{0/+}$] center in SiO_2 [36]) and at 1.4 eV below the Si valence band for the high-k (corresponding to the neutral oxygen vacancy [O^o] level in HfO_2 [37]). Notice that the use of a particular defect representation is purely illustrative and does not affect the general concept of the model proposed here. As a function of the applied gate voltage, all the defects located above the channel Fermi level are considered occupied by trapped holes, while all the defects below are considered neutral (note: no trapping/de-trapping kinetics is included in this calculation, i.e., thermodynamic equilibrium condition [38]).

The model was first calibrated using the *R* component NBTI data obtained on the Si reference devices: the standard deviations of the Gaussian distributions were used as fitting parameters (obtained values, ~0.3 eV in the SiO_2 IL and ~0.5 eV in the HfO_2) in order to capture the experimentally observed electric field dependence, while the defect densities were fitted in order to match the observed ΔV_{th} magnitude. The varying contribution to ΔV_{th} of defects located at varying depths due to their electrostatic effect was also included. Once the defect band parameters were obtained to reproduce the experimental data of standard Si channel devices, the expected ΔV_{th} was calculated for SiGe channel devices, by including the valence band offset of +0.35 eV in the channel and the varying voltage drop on Si cap with varying thicknesses.

As one can see in Fig. 24.19, the simple model matches excellently the experimental data relative to the recoverable component. Specifically, the model readily captures *both the reduced NBTI and the stronger field dependence* observed for SiGe devices with reduced Si cap thicknesses. The model explains also the other experimental observations previously made concerning the Ge fraction and the quantum well thickness. In order to minimize the percentage of accessible defects, i.e., in order to "*push up*" the Fermi level in the channel with respect to the defect band, the valence band offset between SiGe and Si has to be maximized: higher Ge fractions (reduced bandgap and higher ΔE_v) and thick quantum wells (to reduce quantization) are therefore beneficial.

The observation of a distinct relation between the fresh device V_{th0} and the NBTI observed in SiGe devices with various process parameters (Fig. 24.20a) further supports the proposed model: as calculated with MEDICI for, e.g., a Si cap thickness

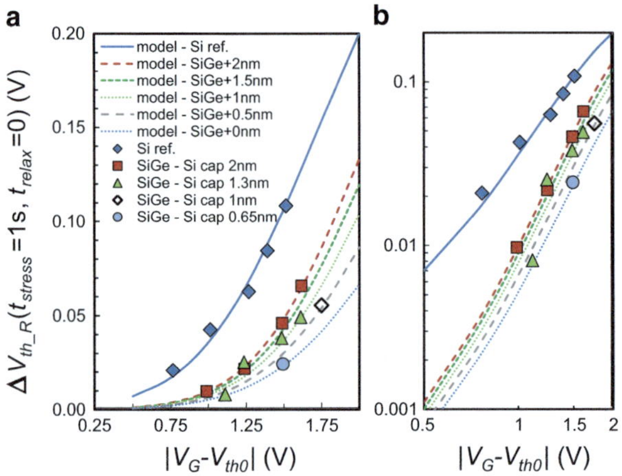

Fig. 24.19 Calculated ΔV_{th} vs. experimental data of the recoverable component. The model was first calibrated on the Si ref. data, then the same defect band parameters were used to calculate the expected ΔV_{th} for SiGe devices (including the valence band offset between the SiGe and the Si cap and the voltage drop on different Si cap thickness). The simple model matches the experimental data remarkably well on (**a**) a lin–lin scale and (**b**) on a log–log scale

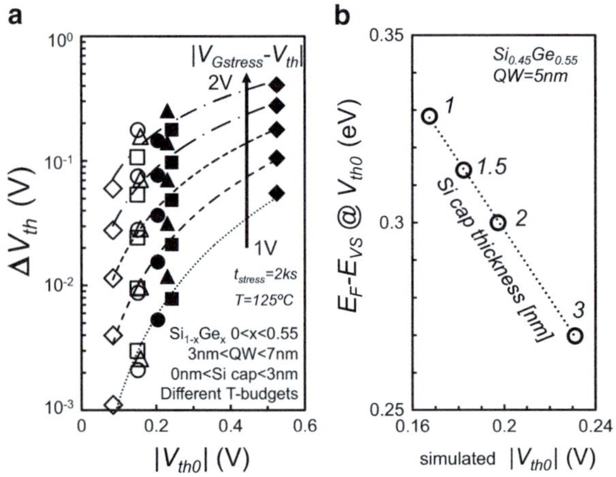

Fig. 24.20 (**a**) A clear correlation between the initial V_{th0} and NBTI-induced ΔV_{th} is consistently observed on SiGe devices with a variety of gate-stack parameters: devices with lower initial V_{th0} always show a reduced V_{th} instability, at any given stress condition ($|V_{Gstress} - V_{th0}|$). This was not observed for Si devices. (**b**) A lower V_{th0} corresponds to a higher channel Fermi-level energy (E_F) with respect to the Si valence band (E_{VS}), as shown with MEDICI simulations. The higher Fermi level in the channel is beneficial for reducing the carrier-trap interaction, according to the model proposed here (cf. Fig. 24.17)

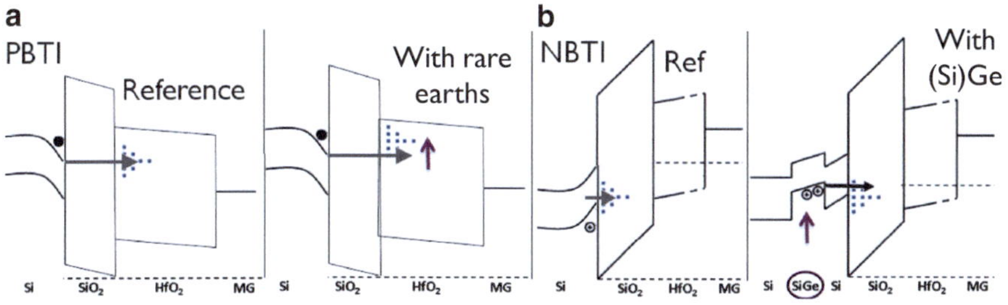

Fig. 24.21 Unified picture for BTI mechanism in high-k-based MOS gate stacks. (**a**) nMOSFETs: improved positive-BTI (PBTI) reliability has been obtained by rare earth doping [39] which pushes up the electron trap energy level [40]. (**b**) Equivalently, for pMOSFETs a significantly improved reliability is obtained by pushing up the channel hole energy level by Ge incorporation

split, gate stacks with lower $|V_{th0}|$ (i.e., less negative V_{th0}) have higher channel Fermi-level energy (Fig. 24.20b) and therefore benefit from reduced interaction between holes and oxide defects.

Further on, the model also predicts improved NBTI reliability for SiGe channel devices even *without* any Si cap (i.e., *no voltage drop on the cap* $\rightarrow$ *maximum* Fermi *energy shift*, cf. Fig. 24.17) as observed experimentally for planar devices and finFETs (see Fig. 24.8). It should be also noted that an alternative explanation for the reduced NBTI reliability with thicker Si caps could be related to a spillover of inversion holes into the cap at high oxide fields: according to the proposed model, holes at the valence band of the Si cap would be favorably trapped in the dielectric defects, and therefore the benefit of using a Ge-based channel would be partially lost.

Finally, the model we have proposed for improved NBTI in Ge-based channel pMOSFETs yields a unified picture (Fig. 24.21) of the understanding of BTI mechanism in thin EOT high-k-based MOS gate stacks, which appears to be mainly controlled by the energy alignment between channel carriers and oxide defects. For nMOSFETs, an improved reliability has been obtained by doping the high-k layer with rare earth metals (La, Gd, Dy) [39]. This beneficial effect has been attributed to a favorable shift of the energy level of electron traps toward the high-k conduction band [40]. Equivalently, for pMOSFET a significantly improved reliability is obtained by shifting the carrier energy level by Ge incorporation in the channel, as shown here.

24.6 Performance vs. Reliability

In the previous sections, we have shown that a reduced Si cap thickness is the key for improved NBTI robustness on SiGe devices. However, previous work reported reduced channel hole mobility for SiGe devices with a reduced Si cap thickness [33].

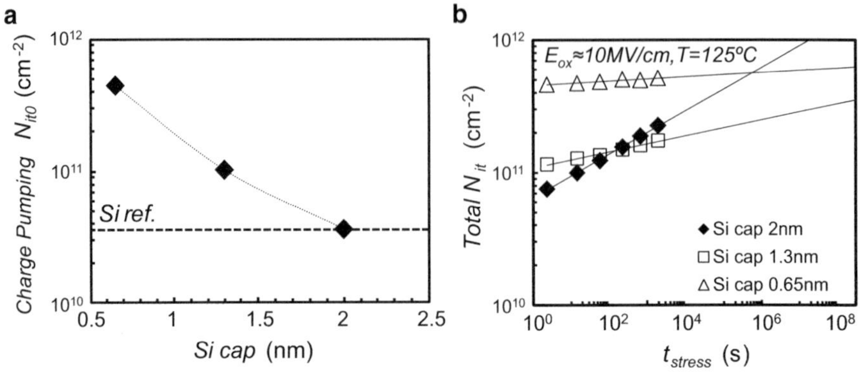

Fig. 24.22 (**a**) Initial interface state density (N_{it0}) measured with CP on Si and SiGe devices. A thick Si cap yields an excellent Si/SiO_2 interface passivation, comparable with the Si reference devices. A reduced Si cap thickness yields increased interface state density due to enhanced Ge segregation from the channel. (**b**) *Total* N_{it} monitored by CP during the same NBTI stress on SiGe devices with different Si cap thicknesses. The higher ΔN_{it} observed during a typical NBTI stress for a 2 nm-thick Si cap sample (see Fig. 24.13) causes the *total* N_{it} to soon overtake the value measured on a medium-thick Si cap sample

This mobility loss was ascribed to poorer interface passivation: for a thinner Si cap, the Ge segregation from the channel toward the interface is enhanced [34], causing a higher density of preexisting interface states. This can be seen in Fig. 24.22a, where N_{it0} values extracted from CP measurements are reported for three different Si cap thickness. However, it is worth emphasizing that the higher ΔN_{it} observed during NBTI stress for devices with thicker Si caps (see Fig. 24.13) quickly causes the interface quality of these samples to become worse than the one of the devices with reduced Si cap thickness, as documented by Fig. 24.22b.

Moreover, it is worth noting that a thinner Si cap, while causing a detrimental mobility reduction, beneficially increases the gate stack C_{ox} thanks to reduced hole displacement (reduced T_{inv} , see Fig. 24.1b), as shown in Fig. 24.23a. When looking at the I_{ON} [i.e., $I_D(V_G = V_D = V_{DD})$] performance of the SiGe devices for different Si cap thickness, the best performance is observed for a medium-thick Si cap of ~1.2 nm, where an optimal trade-off between higher C_{ox} and reduced mobility is obtained (Fig. 24.23b). However, the I_{ON} stays within a ±5% range for the whole Si cap thickness range considered here. Looking at the subthreshold swing of the devices (which ultimately determines the I_{OFF} figure of merit, when combined with the device V_{th0}), a negligible increase is observed for the thinnest Si cap (~3%, Fig. 24.23b), thanks to the higher C_{ox} reducing the effect of a poorer interface passivation.

In conclusion, the Si cap shows an overall limited impact on the I_{ON}/I_{OFF} device metrics, while it has a dramatic impact on the device reliability. Therefore, when implementing a SiGe channel process it is advisable to perform a Si cap thickness optimization based on performance/leakage metrics first and then reduce this thickness as slightly as needed to meet the NBTI reliability specifications.

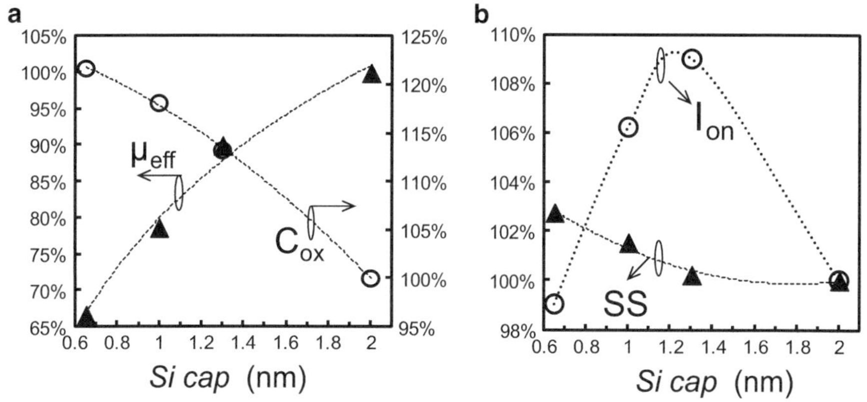

Fig. 24.23 (**a**) Thinner Si cap samples show reduced mobility due to poorer interface passivation but also increased C_{ox} thanks to reduced hole displacement; see Fig. 24.1b (Data courtesy of J. Mitard). (**b**) The trade-off between reduced mobility and increased C_{ox} yields an optimum I_{ON} performance for a medium Si cap thickness. On the other hand, the subthreshold swing is observed to be only marginally dependent on the Si cap thickness, thanks to the higher C_{ox} reducing the detrimental effect of a poorer interface passivation for thin Si caps

24.7 NBTI of Nanoscale SiGe Devices

As discussed in the introduction of this chapter, as the device area scales toward atomistic dimensions, the stochastic properties of the handful of gate oxide defects included in each device become apparent [41]. BTI reliability is then perceived as time-dependent device-to-device variability, and deterministic degradation kinetics has to be replaced by a statistical description [14, 15, 17]. In order to illustrate this, a representative set of typical NBTI relaxation transients recorded on nanoscale SiGe devices is shown in Fig. 24.24.a Several observations can be made:

1. As previously reported for Si devices [14, 17, 42], the total ΔV_{th} observed after the same NBTI stress strongly varies from device to device.
2. Single discharge events are visible, each causing a different ΔV_{th} step.
3. Each device shows a different number of charging/discharging events (i.e., a different number of active oxide traps, $\langle N_T \rangle$).
4. The ΔV_{th} step heights appear to be approximately exponentially distributed (Fig. 24.24b), with an average value η (i.e., the inverse of the slope of the exponential distribution) being ~3.9 mV but with some single-charged oxide defects easily causing gigantic ΔV_{th} [14] as large as ~20 mV (probability of ~1 in ~240 observed defects) [42, 43].

This value is about ~9 times larger than expected from a simple charge sheet approximation [35]. For comparison, it is worth recalling that the typical BTI failure criteria considered for process qualification range between 30 and 50 mV of V_{th} shift. In other words, two such charged defects might already jeopardize

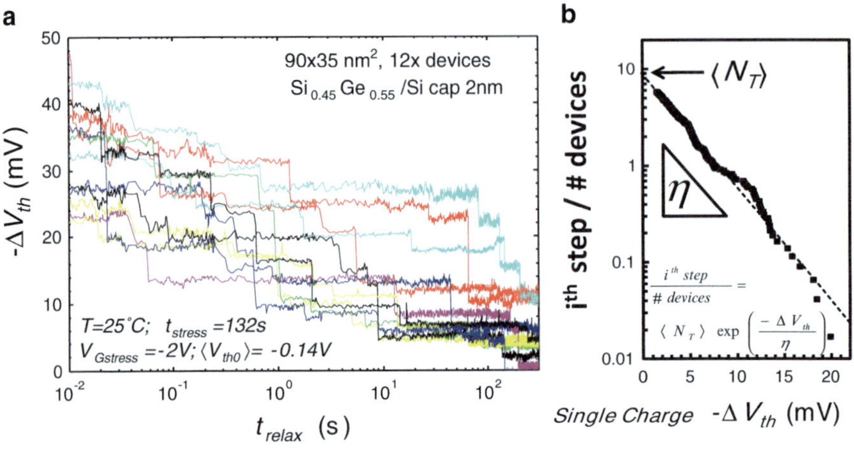

Fig. 24.24 (**a**) NBTI relaxation transients recorded on different nanoscale SiGe devices. For each device, multiple single-defect discharge events are visible. (**b**) Weighted complementary cumulative distribution function (Weighted *CCDF*) plot of the individual ΔV_{th} step heights observed on multiple (41) devices. The ΔV_{th} step heights appear to be exponentially distributed, with an average value $\eta \approx 3.9$ mV. The average number of defects per device, $\langle N_T \rangle$, can be easily read off in this plot as the intersection of the distribution with the y-axis [18]

the device reliable operation. Such anomalous large ΔV_{th} are commonly ascribed to the percolative nature of the current in nanoscale devices associated with channel potential nonuniformity induced by variability sources (e.g., random dopant distribution, line edge roughness, metal gate granularity, etc.) [16]. In the unlucky case of a gate oxide defect spatially located close to (i.e., directly on top) the critical point of a channel current percolation path, a single charge/discharge event can result in a significant change in the device current (i.e., observed as a large ΔV_{th} step in the NBTI relaxation trace).

The time-dependent device-to-device variability has been described by means of the average number of active (i.e., charging/discharging) oxide defects, $\langle N_T \rangle$, and the average ΔV_{th} impact per defect, η [14, 18, 42]. In order to estimate $\langle N_T \rangle$ and η for each gate stack studied here, the same sequence including a charging phase and a relaxation transient was repeated on a large set of nominally identical devices. The device set size was chosen to be large enough to capture the signatures of some hundred active defects for each gate stack (i.e., typically a few tens of devices but up to 160 devices for SiGe devices with a reduced Si cap thickness). The ΔV_{th} steps observed in the relaxation traces were then collected into weighted complementary cumulative distribution function (Weighted CCDF; see Fig. 24.24b) plots, and a maximum likelihood fit was performed in order to estimate the exponential distribution parameters η and $\langle N_T \rangle$.

As discussed in Sect. 24.4, for large area devices, we found that a reduced Si cap thickness on SiGe pFETs results in a significantly reduced ΔV_{th} at fixed stress conditions (see Fig. 24.10). In a similar way, a thinner Si cap yields a $\sim$10$\times$ reduction of the average number of ΔV_{th} steps (i.e., average number of active defects

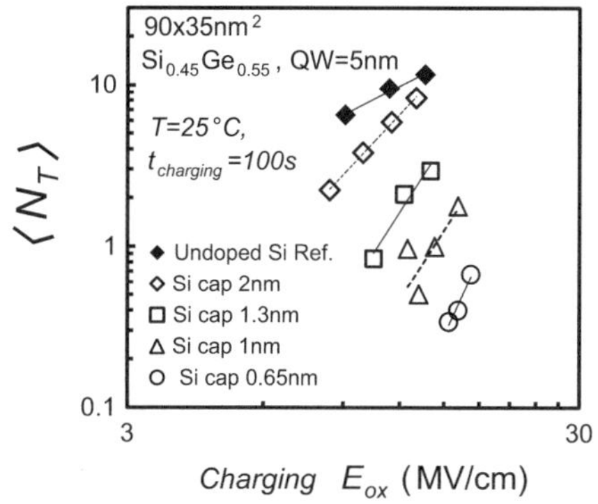

Fig. 24.25 Consistently with large area device (cf. Fig. 24.11), nanoscaled SiGe channel pMOSFETs with a reduced Si cap thickness show reduced average number of charging/discharging defects per device $\langle N_T \rangle$ and a stronger field acceleration. Note: Very high equivalent oxide fields were needed for the charging phase in order to be able to observe active defects in SiGe devices with the thinnest Si cap ($\langle N_T \rangle$ as low as ~0.33 at 15 MV/cm, i.e., one defect observed for every three measured devices)

$\langle N_T \rangle$) observed in nanoscale devices (Fig. 24.25). Note also the stronger dependence of $\langle N_T \rangle$ on the charging oxide electric field, consistently with the observation made on large area devices (cf. Fig. 24.11). It is worth noting that very low $\langle N_T \rangle$ values (as low as ~0.33 for the short charging time used here, i.e., only one defect observed per three measured devices) are observed for SiGe channel devices with the thinnest Si cap. Such low $\langle N_T \rangle$ values complicate the experiment, requiring larger sample set (up to ~160 devices were used for this particular gate stack) to observe a sufficient number of charging/discharging events. However, as discussed later, such low $\langle N_T \rangle$ values might still jeopardize the reliability of a fraction of devices in a realistic device population (billions of devices) [18, 42].

By looking at the weighted *CCDF* plots of the ΔV_{th} step heights observed on SiGe devices with two different Si cap thicknesses, a reduced η is also found for the thinnest Si cap (Fig. 24.26a, $\eta \approx 3.9$ mV for a 2 nm Si cap and $\eta \approx 1.8$ mV for a 0.65 nm Si cap). In order to benchmark more correctly the η values estimated on different gate stacks, a normalization of η for the expected ΔV_{th} per single charge calculated according to the electrostatic charge sheet approximation (i.e., $\eta_0 = q/C_{ox}$) is proposed. Moreover, it is important to note that, since the device-to-device variability is known to depend on the doping level in the channel [16, 44, 45], a beneficial effect is expected for SiGe devices, owing to their undoped epitaxially grown channel. Therefore, for correct benchmarking, a particular Si channel reference gate stack was considered, including a ~8 nm-thick undoped Si channel layer grown epitaxially. The normalized η/η_0 values for SiGe devices with different Si cap thicknesses and with two different thickness of the SiO_2 interfacial layer are shown in Fig. 24.26b and benchmarked against the reference value measured on the undoped Si channel devices. A ~2× reduction in η is found for the thin Si cap devices.

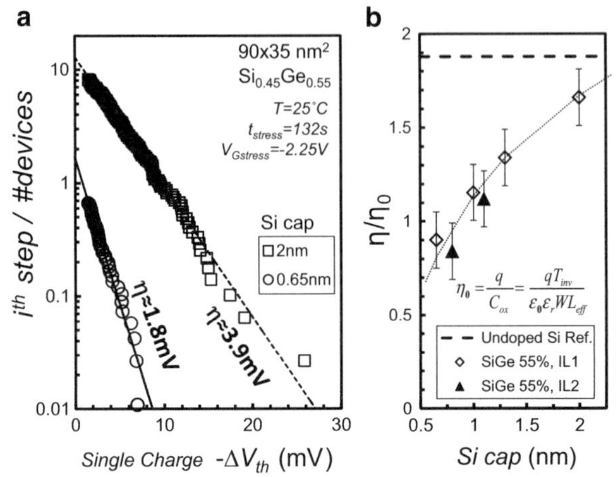

Fig. 24.26 (**a**) Weighted *CCDF* plot of the ΔV_{th} step heights observed on SiGe devices with two different Si cap thicknesses. The average ΔV_{th} step height η is significantly reduced for the devices with the thinnest Si cap. Note also the reduced $\langle N_T \rangle$ (lower y-axis intercept). (**b**) Extracted average ΔV_{th} step heights η for SiGe devices with different Si cap and for undoped Si channel devices after a pre-charging phase at $E_{ox} \approx 12$ MV/cm. SiGe devices with the thinnest Si cap show a significantly lower η ($\sim 2\times$). The observation is confirmed on SiGe devices with two different SiO_2 interfacial layer thicknesses. The *dashed line* demarcates the benchmark value experimentally estimated on undoped Si channel ref. devices. The error bars on the estimated η values are related to the lower $\langle N_T \rangle$ observed for SiGe, while the *dotted line* is a guide to the eye

The above-discussed experimental observations on nanoscale SiGe devices with reduced Si cap thickness can be summarized as follows:

1. Reduced average number of active oxide defects, $\langle N_T \rangle$
2. Stronger dependence of $\langle N_T \rangle$ on the electric field
3. Reduced average ΔV_{th} impact per charged defect, η

These experimental observations are readily explained by the model discussed in Sect. 24.5.2 (cf. Fig. 24.17): fewer oxide defects are energetically favorable for SiGe channel holes (i.e., lower $\langle N_T \rangle$), with the energetically favorable defects preferentially located on the gate side of the dielectric, resulting in a reduced electrostatic effect on the channel (i.e., lower η) [46].

As proposed by our group [14, 18, 42], the fraction of a realistic population (i.e., billions of devices) expected to be still functional after 10 years of continuous operation can be estimated from the $\langle N_T \rangle$ and η values extracted on individual devices. The calculation is based on the convolution of a Poisson-distributed number of defects with the mean value $\langle N_T \rangle$, with an exponential distribution of impact per single-charged defects on the device threshold voltage with mean value η. The mathematical details can be found in [42]. Figure 24.27 illustrates the projected fraction of device still functional after 10 years of continuous operation at varying gate overdrive voltages, based on the experimental estimation of $\langle N_T \rangle$ and η for

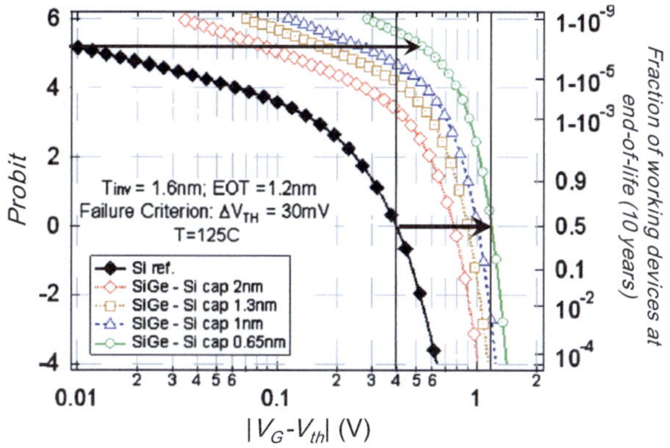

Fig. 24.27 Calculated fractions of working devices after 10-year continuous operation at varying operating voltages for the different gate stacks studied here. A dramatic improvement of the distributions for SiGe devices with reduced Si cap thickness is apparent. Note: Large area devices would appear in this plot as a *vertical dashed line* (whole population fails above maximum allowed operating voltage, while it passes for lower voltages) with same median value (probit = 0) of the respective nanoscale device distribution. The reliability improvement previously observed in large area SiGe devices (demarcated by the *solid arrow*, distance at probit = 0) is magnified at high percentiles (demarcated by the *dotted arrow*, at ~1 ppb)

Si and SiGe devices with varying Si cap thicknesses. A dramatic improvement of the distribution for optimized SiGe devices is apparent, particularly at the high percentiles (e.g., ~one failure per billion devices).

24.8 Conclusions

We have reviewed a broad range of experimental NBTI data of novel Ge-based high-mobility channel pMOSFETs, with Si passivation scheme and SiO2/HfO2 dielectric stack, showing that this technology offers a *significantly improved NBTI reliability.* A (Si)Ge *gate-stack optimization,* including a *high Ge fraction* in the channel, a sufficiently *thick quantum well,* and a *reduced Si cap thickness*, has been identified to *maximize the NBTI reliability.* By implementing this optimization, *ultrathin EOT SiGe devices with 10-year NBTI reliability at operating* V_{DD} *were demonstrated* both in gate-first (MIPS) and gate-last (RMG) process flows. The improved NBTI reliability was shown to be process independent (i.e., consistently observed for different process flows, different epi-precursors, different epi-growth temperatures, and different process thermal budgets) and architecture independent (SiGe channel planar and finFET architectures, pure Ge channel planar), and therefore it is to be considered *an intrinsic property of the pMOS system consisting of a Ge-based channel and a* SiO_2/HfO_2 *dielectric stack.*

While typical NBTI temperature and time dependencies were observed on (Si)Ge devices, suggesting the same degradation mechanism as in the standard Si channel counterparts, a stronger electric field acceleration was highlighted and attributed to the charge trapping component. This stronger field dependence projects to further improvement at lower operating voltages.

Although a reduced creation of interface states was observed in (Si)Ge devices and attributed to a lower precursor defect density, the *superior NBTI reliability was ascribed chiefly to a favorable alignment shift of the Fermi level* in the small bandgap (Si)Ge channel *with respect to preexisting defect energy levels* in the dielectric layers, which effectively reduced the carrier-defect interaction. A mathematical representation of the proposed model was shown to excellently reproduce the experimental observations.

The NBTI of nanoscale SiGe devices was also discussed. As for Si counterparts, individual discharge events were observed in the NBTI ΔV_{th} relaxation traces, with exponentially distributed step heights. *The use of a Si cap of reduced thickness on SiGe* (i.e., the use of the proposed reliability-optimized gate stack) *was found to yield a significant reduction of the average number of active oxide defects* $\langle N_T \rangle$ ($\sim$10$\times$) causing the observed charge/discharge events *and of the average* ΔV_{th} *impact per charged defect* η ($\sim$2$\times$). The proposed model based on energy decoupling between channel holes and preexisting dielectric defect energy levels readily explains also these additional experimental observations, suggesting that *fewer defects preferentially located further from the channel are energetically favorable* for channel holes in optimized SiGe devices. Thanks to this effect, a significant improvement of the time-dependent variability of a realistic population of nanoscale devices has been illustrated.

The extensive experimental results here reviewed strongly support (Si)Ge technology as a clear front-runner for future CMOS technology nodes, offering a solution to the NBTI reliability issue for ultrathin EOT nanoscale devices.

Acknowledgments The authors gratefully acknowledge Philippe Roussel and M. Toledano-Luque for continuous discussion and collaboration; J. Mitard, L. Witters, A. Hikavyy, and G. Eneman for device process development and characterization; Prof. A. Stesmans and Prof. V. Afanas'ev for providing the ESR spectra shown in Fig. 24.15 and for discussion; and Prof. M. Heyns, Prof. Tibor Grasser, and Prof. Guido Groeseneken for discussion and support.

This work was performed as part of imec's Core Partner Program. It has been in part supported by the European Commission under the 7th Framework Programme (Collaborative project MORDRED, Contract No. 261868).

References

1. V. Huard, M. Denais, C. Parthasarathy, "NBTI degradation: from physical mechanism to modeling", in Microelectronic Reliability, Vol. 46, No. 1, pp. 1–23, 2006;
2. International Technology Roadmap for Semiconductors available at http://public.itrs.net;
3. L.-Å. Ragnarsson, Z. Li, J. Tseng, T. Schram, E. Rohr, M. Cho, T. Kauerauf, T. Conrad, Y. Okuno, B. Parvais, P. Absil, S. Biesemans, T.Y. Hoffmann, "Ultra low-EOT (5Å) gate-first and gate-last high performance CMOS achieved by gate-electrode optimization", in IEEE *Proc.* International Electron Device Meeting (IEDM), pp. 663–666, 2009;

4. T. Ando, M.M. Frank, K. Choi, C. Choi, J. Bruley, M. Hopstaken, M. Copel, E. Cartier, A. Kerber, A. Callegari, D. Lacey, S. Brown, Q. Yang, V. Narayanan, "Understanding mobility mechanisms in extremely scaled HfO_2 (EOT 0.42nm) using remote interfacial layer scavenging technique and V_t-tuning dipoles with gate-first process", in IEEE *Proc.* IEDM, pp. 423–426, 2009;
5. E. Cartier, A. Kerber, T. Ando, M.M. Frank, K. Choi, S. Krishnan, B. Linder, K. Zhao, F. Monsieur, J. Stathis, V. Narayanan, "Fundamental Aspects of HfO_2-based High-k Metal Gate Stack Reliability and Implication on tinv-Scaling", in IEEE *Proc.* IEDM, pp. 441–444, 2011;
6. M. Cho, J.-D. Lee, M. Aoulaiche, B. Kaczer, Ph. J. Roussel, T. Kauerauf, R. Degraeve, J. Franco, L.-Å. Ragnarsson, G. Groeseneken, "Insight into Negative and Positive Bias Temperature Instability (N/PBTI) mechanism in sub-nanometer EOT devices", in IEEE Trans. Electron Devices, Vol. 59, no. 8, pp. 2042–2048, 2012;
7. L. Witters, S. Takeoka, S. Yamaguchi, A. Hikavyy, D. Shamiryan, M. Cho, T. Chiarella, L.-Å. Ragnarsson, R. Loo, C. Kerner, Y. Crabbe, J. Franco, J. Tseng, W.E. Wang, R. Rohr, T. Schram, O. Richard, H. Bender, S. Biesemans, P. Absil, T.Y. Hoffman, "8Å Tinv gate-first dual channel technology achieving low-V_t high performance CMOS", in *Proc.* Symp. on VLSI Technology, pp. 181–182, 2010;
8. J. Mitard, L. Witters, M.G. Bardon, P. Christie, J. Franco, A. Mercha, P. Magnone, M. Alioto, F. Crupi, L.-Å. Ragnarsson, A. Hikavyy, B. Vincent, T. Chiarella, R. Loo, J. Tseng, S. Yamaguchi, S. Takeoka, W.E. Wang, P. Absil, T.Y. Hoffmann, "High-mobility 0.85nm-EOT $Si_{0.45}Ge_{0.55}$-pFETs: Delivering high performance at scaled V_{DD}", in IEEE *Proc.* IEDM, pp. 249–252, 2010;
9. S. Krishnan, U. Kwon, N. Moumen, M.W. Stoker, E.C.T. Harley, S. Bedell, D. Nair, B. Greene, W. Henson, M. Chowdhury, D.P. Prakash, E. Wu, D. Ioannou, E. Cartier, M.-H. Na, S. Inumiya, K. McStay, L. Edge, R. Iijima, J. Cai, M. Frank, M. Hargrove, D. Guo, A. Kerber, H. Jagannathan, T. Ando, J. Shepard, S. Siddiqui, M. Dai, H. Bu, J. Schaeffer, D. Jaeger, K. Barla, T. Wallner, S. Uchimura, Y. Lee, G. Karve, S. Zafar, D. Schepis, Y. Wang, R. Donaton, S. Saroop, P. Montanini, Y. Liang, J. Stathis, R. Carter, R. Pal, V. Paruchuri, H. Yamasaki, J.-H. Lee, M. Ostermayr, J.-P. Han, Y. Hu, M. Gribelyuk, D.-G. Park, X. Chen, S. Samavedam, S. Narasimha, P. Agnello, M. Khare, R. Divakaruni, V. Narayanan, M. Chudzik, "A manufacturable dual channel (Si and SiGe) high-k metal gate CMOS technology with multiple oxides for high performance and low power applications", in IEEE *Proc.* IEDM, pp. 634–637, 2011;
10. K.J. Kuhn, "Considerations for Ultimate CMOS Scaling", in IEEE Trans. Electron Devices, vol.59, no. 7, pp. 1813–1828, 2012;
11. B. Kaczer, J. Franco, Ph. J. Roussel, A. Veloso, G. Groeseneken, "Improvements in NBTI Reliability of Si-passivated Ge/high-k/metal-gate pFETs", in Microelectronic Engineering, vol. 86, no. 7–9, pp. 1582–1584, 2009;
12. J. Franco, B. Kaczer, M. Cho, G. Eneman, T. Grasser, G. Groeseneken, "Improvements of NBTI reliability in SiGe p-FETs", in IEEE *Proc.* International Reliability Physics Symposium (IRPS), pp. 1082–1085, 2010;
13. J. Franco, B. Kaczer, G. Eneman, J. Mitard, A. Stesmans, V. Afanas'ev, T. Kauerauf, Ph. J. Roussel, M. Toledano-Luque, M. Cho, R. Degraeve, T. Grasser, L.-Å. Ragnarsson, L. Witters, J. Tseng, S. Takeoka, W.E. Wang, T.Y. Hoffmann, G. Groeseneken, "6Å EOT $Si_{0.45}Ge_{0.55}$ pMOSFET with Optimized Reliability (V_{DD}=1V): Meeting the NBTI Lifetime Target at Ultra-Thin EOT , in IEEE *Proc.* IEDM, pp. 70–73, 2010;
14. B. Kaczer, T. Grasser, Ph. J. Roussel, J. Franco, R. Degraeve, L.-Å. Ragnarsson, E. Simoen, G. Groeseneken, H. Reisinger, "Origin of NBTI Variability in Deeply Scaled pFETs", in IEEE *Proc.* IRPS, pp. 26–32, 2010;
15. T. Grasser, B. Kaczer, W. Gös, H. Reisinger, T. Aichinger, P. Hehenberger, P.-J. Wagner, F. Schanovsky, J. Franco, Ph. J. Roussel, M. Nelhiebel, "Recent Advances in Understanding the Bias Temperature Instability", in IEEE *Proc.* IEDM, pp. 82–85, 2010;

16. A. Asenov, R. Balasubramaniam, A. R. Brown, and J. H. Davies, "RTS Amplitude in Decananometer MOSFETs: 3-D Simulation Study", in IEEE Trans. Electron Devices, Vol. 50, no. 3, pp. 839–845, 2003;
17. V. Huard, C. Parthasarathy, C. Guerin, T. Valentin, E. Pion, M. Mammasse, N. Planes, L. Camus, "NBTI Degradation: from Transistor to SRAM Arrays", in IEEE *Proc.* IRPS, pp. 289–300, 2008;
18. M. Toledano-Luque, B. Kaczer, J. Franco, Ph. J. Roussel, T. Grasser, T.Y. Hoffmann, G. Groeseneken, "From Mean Values to Distributions of BTI Lifetime of Deeply Scaled FETs through Atomistic Understanding of the Degradation", in *Proc.* VLSI Symp., pp. 152–153, 2011;
19. V. Huard, F. Cacho, Y. Mamy Randriamihaja, A. Bravaix, "From Defects Creation to Circuit Reliability", in Microelectronic Engineering, Vol. 88, no. 7, pp. 1396–1407, 2011;
20. M. Nafria, R. Rodriguez, M. Porti, J. Martin-Martinez, M. Lanza, X. Aymerich, "Time-dependent Variability of high-k based MOS devices: nanoscale characterization and inclusion in circuit simulators", in IEEE *Proc.* IEDM, pp. 127–130, 2011;
21. B. Kaczer, S. Mahato, V.V. de Almeida Camargo, M. Toledano-Luque, Ph. J. Roussel, T. Grasser, F. Catthoor, P. Dobrovolny, P. Zuber, G. Wirth, G. Groeseneken, "Atomistic Approach to Variability of Bias-Temperature Instability in Circuit Simulations", in IEEE *Proc.* IRPS, pp. 915–919, 2011;
22. T. Grasser, H. Reisinger, P.-J. Wagner, B. Kaczer, Phys. Rev. B, 82(24), 245318, 2010.
23. M. Meuris, P. Mertens, A. Opdebeeck, H. Schmidt, M. Depas, G. Vereecke, M. Heyns, A. Philipossian, "The IMEC clean: A new concept for particle and metal removal on Si surfaces", in Solid State Technology, 38(7), pp. 109–113, 1995;
24. A. Hikavyy, R. Loo, L. Witters, S. Takeoka, J. Geypen, B. Brijs, C. Merckling, M. Caymax, J. Dekoster, "SiGe SEG Growth For Buried Channel p-MOS Devices", in ECS Transactions, Vol. 25, No. 7, pp. 201–210, 2009;
25. B. Kaczer, T. Grasser, Ph. J. Roussel, J. Martin-Martinez, R. O'Connor, B.J. O'Sullivan, G. Groeseneken, "Ubiquitous Relaxation in BTI Stressing–New Evaluation and Insights", in *Proc.* IRPS, pp. 20–27, 2008;
26. T. Chiarella, L. Witters, A. Mercha, C. Kerner, M. Rakowski, C. Ortolland, L.-Å. Ragnarsson, B. Parvais, A. De Keersgieter, S. Kubicek, A. Redolfi, C. Vrancken, S. Brus, A. Lauwers, P. Absil, S. Biesemans, T.Y. Hoffmann, "Benchmarking SOI and bulk FinFET alternatives for PLANAR CMOS scaling succession", in Solid-State El., Vol. 54, No. 9, pp. 855–860, 2010;
27. M.M. Frank, E. Cartier, T. Ando, S.W. Bedell, J. Bruley, Y. Zhu, V. Narayanan, "Aggressive SiGe Channel Gate Stack Scaling by Remote Oxygen Scavenging: pFET Performance and Reliability", in *Proc.* ECS Fall meeting in ECS Trans., vol. 50, 2012;
28. X. Gong, S. Shaojian, B. Liu, L. Wang, W. Wang, Y. Yang, R. Cheng, E. Kong, B. Cheng, G. Han, Y.-C. Yeo, "Negative Bias Temperature Instability Study on $Ge_{0.97}$ $Sn_{0.03}$ p-MOSFETs with Si_2H_6 Passivation, HfO_2 High-k Dielectric and TaN Metal Gate", in *Proc.* ECS Fall meeting in ECS Trans., vol. 50, no. 9, 2012;
29. T. Grasser and B. Kaczer, "Negative Bias Temperature Instability: recoverable versus permanent degradation", in *Proc.* ESSDERC, pp. 127–130, 2007;
30. T. Grasser, B. Kaczer, P. Hehenberger, W. Gös, R. O'Connor, H. Reisinger, W. Gustin, C. Schunder, "Simultaneous extraction of recoverable and permanent components contributing to Bias-Temperature Instability", in IEEE *Proc.* IEDM, pp. 801–804, 2007;
31. G. Groeseneken, H.E. Maes, N. Beltran, R.F. De Keersmaecker, "A reliable approach to Charge Pumping measurements in MOS transistors", in IEEE Trans. Electron Devices, vol. 31, no. 1, pp. 42–53, 1984;
32. A. Stesmans and V. Afanas'ev, "ESR of interfaces and nanolayers in semiconductor heterostructures", in Characterization of Semiconductor Heterostructures and Nanostructures, Elsevier, pp. 435–489, 2008;
33. J. Mitard, K. Martens, B. De Jaeger, J. Franco, C. Shea, C. Plourde, F.E. Leys, R. Loo, G. Hellings, G. Eneman, W.E. Wang, J.C. Lin, B. Kaczer, K. De Meyer, T.Y. Hoffmann, S. Degendt, M. Caymax, M. Meuris, M. Heyns, "Impact of Epi-Si Growth Temperature on Ge-pFET Performance", in *Proc.* ESSDERC 2009, pp. 411–414;

34. M. Caymax, F. Leys, J. Mitard, K. Martens, L. Yang, G. Pourtois, W. Vandervorst, M. Meuris, R. Loo, "The influence of the epitaxial growth process parameters on layer characteristics and device performance in Si-passivated Ge pMOSFETs", in J. Electrochem. Soc., vol. 156, no. 12, pp. H979–985, 2009;
35. T. Grasser, B. Kaczer, W. Gös, H. Reisinger, T. Aichinger, P. Hehenberger, P.-J. Wagner, F. Schanovsky, J. Franco, M. Toledano-Luque, M. Nelhiebel, "The Paradigm Shift in Understanding the Bias Temperature Instability: from Reaction–diffusion to Switching Oxide Traps", in *IEEE* Trans. Electron Devices, vol. 58, no. 11, pp. 3652–3666, 2011;
36. W. Gös, "Hole Trapping and the Negative Bias Temperature Instability", Ph.D. dissertation, T.U. Wien, 2011, available at http://www.iue.tuwien.ac.at/ phd/goes/dissse19.html;
37. A. S. Foster, F. Lopez Gejo, A. L. Shluger, and R. M. Nieminen, Phys. Rev. B 65, 174117, 2002;
38. T. Grasser, "Stochastic Charge Trapping in Oxides: From Random Telegraph Noise to Bias Temperature Instabilities", in Microelectronic Reliability, vol. 52, no. 1, pp. 39–70, 2012;
39. B. Kaczer, A. Veloso, M. Aoulaiche, G. Groeseneken, "Significant reduction of Positive Bias Temperature Instability in high-k/metal-gate nFETs by incorporation of rare earth metals", in Microelectronic Engineering, vol. 86, no. 7–9, pp. 1894–1896, 2009;
40. D. Liu and J. Robertson, "Passivation of oxygen vacancy states and suppression of Fermi pinning in HfO2 by La addition", in Appl. Phys. Lett., vol. 94, pp. 042904.1-4, 2009;
41. M. Toledano-Luque, B. Kaczer, Characterization of individual traps in high-κ oxides, in *Bias Temperature Instability for Devices and Circuits*, ed. by T. Grasser (Springer, Heidelberg, 2013).
42. B. Kaczer, M. Toledano-Luque, J. Franco, P. Weckx, Statistical distribution of defect parameters, in *Bias Temperature Instability for Devices and Circuits*, ed. by T. Grasser (Springer, Heidelberg, 2013).
43. S.M. Amoroso, L. Gerrer, F. Adamu-Lema, S. Markov, A. Asenov, Statistical study of bias temperaure instabilities by means of 3D 'atomistic' simulation, in *Bias Temperature Instability for Devices and Circuits*, ed. by T. Grasser (Springer, Heidelberg, 2013).
44. A. Ghetti, C.M. Compagnoni, A.S. Spinelli, A. Visconti, "Comprehensive Analysis of Random Telegraph Noise Instability and Its Scaling in Deca-Nanometer Flash Memories", in IEEE Trans. Electron Devices, Vol. 56, no. 8, pp. 1746–1752, 2009;
45. J. Franco, B. Kaczer, M. Toledano-Luque, Ph. J. Roussel, B. Schwarz, M. Bina, M. Waltl, P.-J. Wagner, T. Grasser, G. Groeseneken, "Reduction of the BTI Time-Dependent Variability in Nanoscaled MOSFETs by Body Bias", in IEEE *Proc.* IRPS, pp. 2D.3.1-6, 2013;
46. J. Franco, B. Kaczer, M. Toledano-Luque, Ph. J. Roussel, P. Hehenberger, T. Grasser, J. Mitard, G. Eneman, L. Witters, T.Y. Hoffmann, G. Groeseneken, "On the impact of the Si passivation layer thickness on the NBTI of nanoscaled $Si_{0.45}Ge_{0.55}$ pMOSFETs", in Microelectronic Engineering, Vol. 88, No. 7, pp. 1388–1391, 2011.

Chapter 25
Characteristics of NBTI in Multi-gate FETs for Highly Scaled CMOS Technology

Ru Huang, Runsheng Wang, and Ming Li

Abstract The multi-gate devices have been paid much attention nowadays. The multi-gate FinFET has been used in manufacturing recently, and the gate-all-around (GAA) silicon nanowire transistor (SNWT) is a promising device structure for ultimate CMOS applications near the end of the technology roadmap. This chapter briefly reviews the reliability of multi-gate FETs, with focuses on the negative bias temperature instability (NBTI) behavior in GAA SNWTs, which exhibits some new characteristics due to its unique structural nature of quasi-1D channel and 3D surrounding gate.

25.1 Introduction

For the last five decades, IC technology has been driven by device scaling to continuously enhance performance, as well as reduce cost and maintain low-power consumption. However, as entering sub-100 nm region, performance gain of Si MOSFETs has become increasingly difficult by conventional scaling. Many critical issues, such as complex process integration, increased leakage current, short-channel effects, high-field effects, reliability, variability, noise, and parasitic effects, pose more obstructions for highly scaled CMOS devices. Therefore, a paradigm shift has been occurring in the academe and industry, where material and device structure innovation is becoming the primary enabler for performance enhancement in CMOS technology [1–4]. New materials have been studied for many years and have already been adopted in the state-of-the-art Si CMOS products. For example, embedded SiGe source/drain for strained-Si channel and high-k gate dielectrics has been introduced into products since 90 nm node and 45 nm node [4], respectively. The new device structure, especially the multi-gate FET (MuGFET), is attracting

R. Huang (✉) • R. Wang • M. Li
Institute of Microelectronics, Peking University, Beijing 100871, China
e-mail: ruhuang@pku.edu.cn

T. Grasser (ed.), *Bias Temperature Instability for Devices and Circuits*, DOI 10.1007/978-1-4614-7909-3_25, © Springer Science+Business Media New York 2014

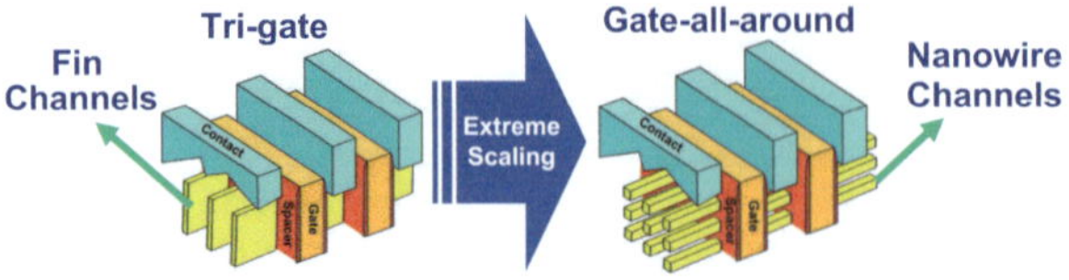

Fig. 25.1 From tri-gate FinFET to gate-all-around nanowire FETs

Fig. 25.2 Schematic view of a GAA SNWT with cylindrical channel [7]

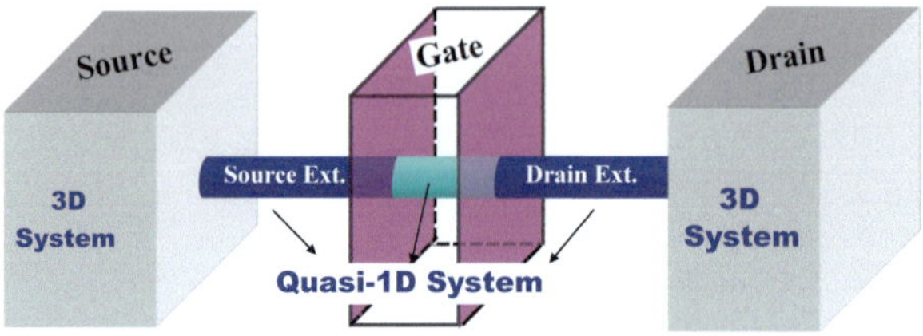

more and more attentions recently for ultimately scaled CMOS technology, due to their superior gate control capability than conventional planar devices [1–5]. Very recently, Intel has adopted tri-gate device architecture with Fin-shaped fully depleted channel (i.e., Fin-shaped FET, or FinFET for short) as one of the key enablers for its 22 nm products. Other companies, such as TSMC, have also announced their technology roadmaps to use multi-gate FinFET device structure for 16/14 nm node and beyond. Therefore, the multi-gate FinFET will be the mainstream device technology until ultimate scaling.

At the ultimate scale, as shown in Fig. 25.1, one can improve the gate architecture from tri-gate to gate-all-around (GAA) for getting the strongest gate controllability. This enhancement can also greatly relax the stringent process requirements for more flexible device design, e.g., the Si channel thickness can be comparable to (or even double of) the gate length (L_G) in GAA structure, rather than about 1/3–2/3 L_G in FinFETs [5]. And with the feature size continuously shrinking, the Fin-channel thickness should be reduced accordingly and eventually will go to nanowire-like geometry, which is called as the Si nanowire transistor (SNWT). In addition, the GAA SNWTs with quasi-1D nanowire channel can achieve improved transport properties from volume inversion and quasi-ballistic transport [6]. Therefore, the GAA SNWT with top-down approach has been considered as one of the most promising structures for ultimately scaled device at the end of the roadmap.

As shown in Fig. 25.2, this kind of device has the unique structural nature [6, 7]: for the nanowire channel, it is (quasi-) one-dimensional and strongly confined, while within the surrounding gate stack, it is three-dimensional and has multiple crystallographic interface orientations; for the source/drain region, there exists a sharp transition from large (3D) source/drain to the (1D) nanowire. Therefore, the above-mentioned critical issues may be even more complicated that would give rise to new challenges in device engineering of SNWTs, and need careful characterization and analysis [7]. On the other hand, these nanowire devices also provide a unique opportunity to investigate the device physics and performance in quasi-one dimensions.

Reliability is one of the critical factors which should be comprehensively evaluated before the new device goes into practical applications, especially for nanoscale devices. For the GAA SNWT, the above unique device structure of quasi-1D channel and surrounding gate of multiple crystallographic interface orientations can result in some special reliability behaviors [8]. The small intrinsic area and confinement effects from the device architectures, combined with the special trap behavior may cause some new phenomena. Therefore, this chapter will discuss the reliability in GAA SNWTs as well as in multi-gate FinFETs.

In the following parts of this chapter, a short overview of the process integration of multi-gate devices will be given first, followed by a brief review of the results of reliability in multi-gate FinFETs. Then the recent results of negative bias temperature instability (NBTI) in GAA SNWTs and the related characterization will be focused. Finally, the summary will be given.

25.2 Short Overview of Process Integration for Multi-gate Devices

As the MuGFET has been discussed extensively as the alternative to conventional planar MOSFET, how to fabricate it becomes a realistic problem. Device reliability behavior can also be affected by the fabrication process, particularly for the nonplanar MuGFET. Due to the three-dimensional nature, the fabrication technology of MuGFET is very different from that of planar MOSFET. As known until now, the following process challenges exist in the fabrication of a MuGFET: (1) three-dimensional active region patterning, (2) sidewall channel surface roughness, (3) channel strain engineering, (4) source/drain parasitic resistance, (5) threshold voltage adjustment, and (6) process variation. The following context will give a brief introduction to the state-of-the-art fabrication technologies of both multi-gate FinFETs and GAA SNWTs.

For a scaled FinFET technology, the Fin patterning is the first technical challenge. By using the advanced optical lithography process with double patterning technique, Fin width can achieve 8 nm [9]. As feature size continuously scales down, the sidewall spacer transferring technique [10, 11] can help to pattern smaller and denser fins, as shown in Fig. 25.3. This technology can improve not only the Fin resolution but also the Fin edge roughness due to highly uniform deposition process.

Another issue for FinFET fabrication is the roughness of sidewall surface which can cause mobility and reliability degradation, resulting from the (110) orientation of sidewall on the extensively used (100) wafer and the dry etching plasma damage [12]. By hydrogen annealing and halogen passivation, sidewall surface roughness can be effectively reduced for channel mobility and gate dielectric reliability improvement due to increased surface atom migration [13]. As to the strain technologies for FinFETs, the process-induced uniaxial stress such as embedded SiGe and Si:C is the most effective way. On the other hand, reduction of source/drain parasitic resistance is especially important for MuGFET technology since the source

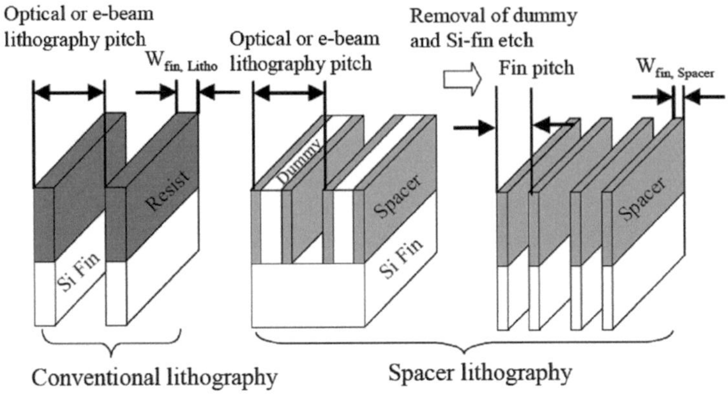

Fig. 25.3 Comparison of Fin width and pitch resolution between conventional lithography and spacer pattern transferring technology. The spacer pattern transferring technology can produce narrower and denser Fin arrays than a conventional lithography [10]

and drain fan-out region becomes smaller for this kind of structure. A traditional method to reduce source and drain resistance is to selectively grow SiGe or Si on source/drain [14]. Conformal doping technique is also needed to form highly doped and steep lateral gradient junction to decrease the spread resistance.

Threshold voltage adjustment is another challenge for FinFETs. The channel ion implantation is not effective to adjust the threshold voltage in FinFET due to the fully depleted channel. Work function engineering has to be adopted to obtain multiple threshold voltage. Combined with channel length modulation [15], the dual metal gates can provide enough threshold voltage tuning range for SoC application [16].

Although FinFET has a fully depleted thin body to control short-channel effect, it still has the bottom leakage when built on bulk substrate. To solve this problem, an anti-punch-through implantation is required. Some researchers also proposed to use a localized isolation beneath the channel to cut off the bottom leakage path [17], which can integrate the isolation merit of SOI and thermal dissipation merit of bulk Si.

As feature size is scaled down to sub-10 nm, MuGFETs will evolve into GAA nanowire transistor type from FinFET as discussed above. Due to the surrounding gate structure, the most difficult step in fabricating a GAA SNWT is to form a suspending nanowire channel on the substrate. By using wet etching of oxide, it is relatively easy to form a nanowire on SOI substrate. On bulk Si substrate, however, this process becomes extremely difficult. One method was reported to adopt sacrificial layer method to form silicon nanowire on bulk substrate [18]. This method grows SiGe/Si layers on the bulk Si substrate firstly, and the SiGe under channel is then selectively removed to release the nanowire. Another method was proposed to make use of the saturation characteristics of stress/temperature-dependent oxidation (a kind of self-limiting oxidation) to form the self-limited nanowire on the prepared silicon pillar [19]. A kind of functional circuit was also

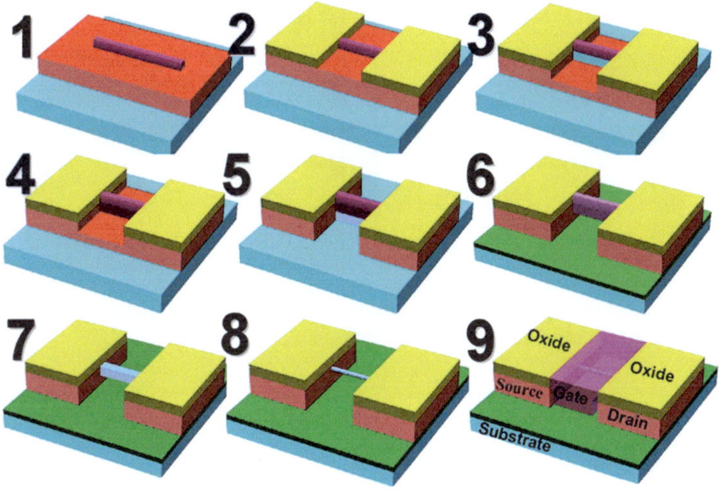

Fig. 25.4 Schematic process sequence of GAA SNWT by self-limiting oxidation [19]

demonstrated based on this method [20]. As shown in Fig. 25.4, a vertical silicon Fin is firstly formed, and the bottom of the Fin is then isotropically etched with Fin sidewall protected by nitride spacer. After the nitride spacer is striped, a self-limiting oxidation process is then performed to trim and round the Fin body. Due to retardation effect of the oxidation, the profile of the Fin will be changed into triangle firstly and then cylinder [19, 21]. Compared to the sacrificial layer method, the latter method shows higher compatibility to conventional CMOS process and can effectively avoid the junction leakage issue resulting from the interface of the remaining SiGe and Si substrate in the sacrificial method.

For GAA SNWTs, in addition to new strain technology and threshold voltage tuning methods specially needed, the parasitic resistance and process variation control becomes even more severe than FinFETs due to its structural features, which have attracted more and more attention recently.

Despite of the above-mentioned challenges in fabrication, MuGFET technology is rapidly progressing as the solution of ultimate scaling due to its incomparable scalability.

25.3 Reliability Behaviors of Multi-gate FinFETs

In general, the main reliability challenge in multi-gate FinFETs comes from the edges and corners existing in this kind of nonplanar device structure, which is known as the corner effect. For example, as shown in Fig. 25.5, large electric field in the gate oxide occurs at the sharp corner/edge of the Fin-channel [22], which would easily cause the oxide breakdown as demonstrated in Fig. 25.6a. Therefore,

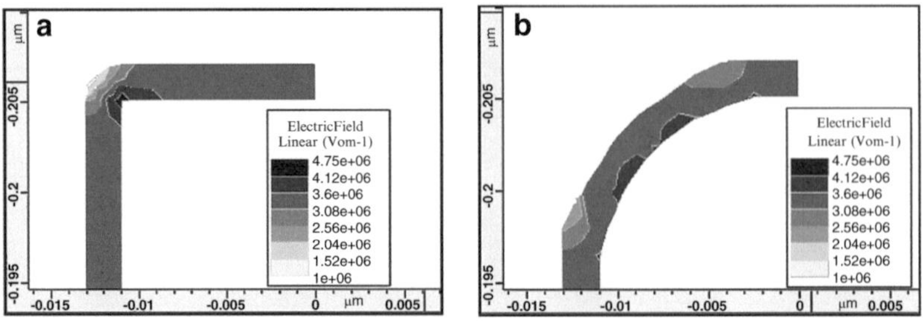

Fig. 25.5 (**a**) Large oxide electric fields occur at the sharp corner/edge of the Fin-channel. (**b**) Reduced oxide electric fields for corner rounding-shaped Fin-channel [22]

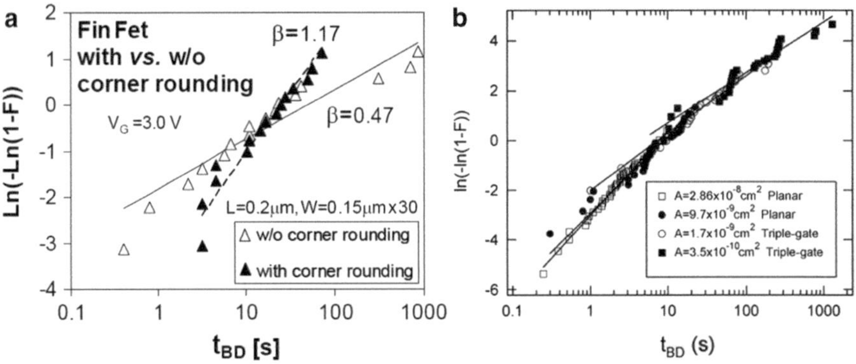

Fig. 25.6 (**a**) The TDDB for FinFET with and without corner rounding [22]. (**b**) TDDB in Tri-gate FinFETs with good corner rounding is similar to planar devices [23]

the corners should be carefully rounded for multi-gate FinFETs to reduce the corner/edge electric fields. Figure 25.6b shows that the time-dependent dielectric breakdown (TDDB) of multi-gate FinFETs with good corner rounding is similar to that of planar devices. The perfect area scaling with corner-rounded multi-gate devices indicates the same TDDB mechanism as for planar devices [23].

Another reliability challenge for FinFETs is the different surface orientations of the sidewall channel rather than conventional (100) channel surface of planar devices. NBTI was reported to be slightly worse for FinFETs due to (110) sidewall orientation [24]. Therefore, special process optimization should be used, for example, F surface passivation and surface NH_3 nitridation have been proposed for improving NBTI in multi-gate FinFETs [25]. Other reports show that, in spite of the different surface orientations, similar values of the effective interface trap density for tri-gate and planar FETs can be obtained [23], as shown in Fig. 25.7.

Therefore, in principle, there are no serious reliability problems in multi-gate FinFET with well-rounded Fin-structure design and careful process optimizations. Intel's data show that the reliability issues in their 22 nm Tri-gate FinFETs have been well controlled to achieve good TDDB, hot carrier, and BTI reliability results [9, 16], as shown in Fig. 25.8.

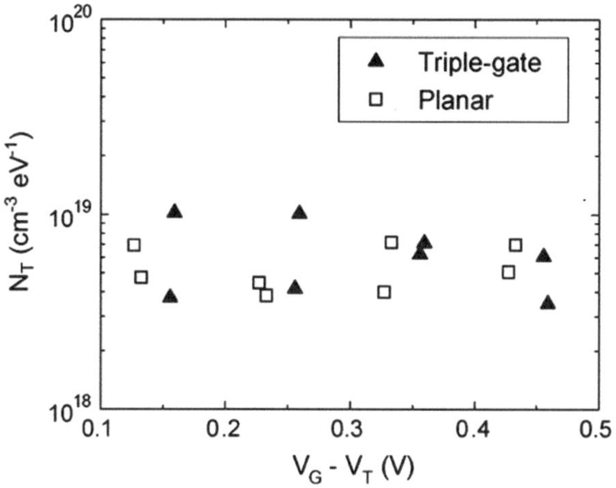

Fig. 25.7 The interface trap density of tri-gate and planar FETs, extracted from flicker noise [23]

25.4 NBTI in GAA SNWTs

As shown above, the reliability behavior of FinFETs (with well-designed corner rounding) is generally having similar degradation mechanisms with the conventional planar devices, due to the fact that the FinFET is just the vertical version of a planar FET. However, as mentioned in the introduction of this chapter, different from FinFETs, the SNWT actually has a much distinct structural nature when compared with planar FETs. Thus, the reliability behavior, especially the NBTI, could be exhibiting new characteristics in SNWTs. The following will discuss the NBTI and the related characterization of SNWTs with rounded channel (i.e., cylindrical nanowire channel for avoiding corner effects).

25.4.1 Intrinsic (Average) NBTI Behavior in SNWTs

NBTI in p-type SNWTs exhibits new characteristics both in stress and recovery stages, resulting from the unique device structure.

In the degradation stage, fast initial threshold voltage (V_t) shift (large power exponential factor) and quick saturation (less than 1,000 s) of NBTI are observed, as shown in Fig. 25.9. And higher stress voltages result in faster initial degradation and longer saturation time [26, 27]. These are due to the structural nature of nanowire devices. The GAA structure results in remarkable enhancement of the electrical field near the channel surface, due to the large curvature of the concentric cylinder capacitance, which accelerates the oxide hole trapping effect. It can be further enhanced by the strain in gate oxide induced during the self-limiting oxidation [28, 29] for nanowire thinning and shaping. And the cross-sectional geometry effect of

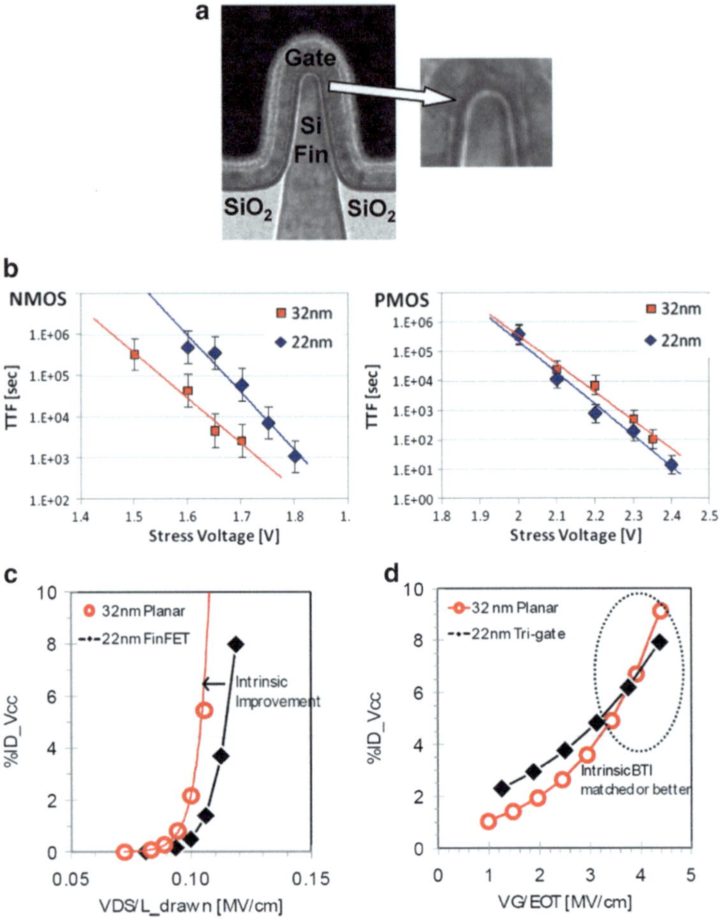

Fig. 25.8 (**a**) Intel's 22 nm Tri-gate FinFET device shows good corner rounding structure, which results in comparable or even better (**b**) TDDB, (**c**) HCI, and (**d**) NBTI reliability compared to its 32 nm planar devices [9, 16] (Note: %ID_Vcc in (**c**) and (**d**) means the on-current degradation amount)

the cylinder wire can also cause faster trap generation at the initial stages. In the SNWT with small gate area, the thin gate oxide only includes a small number of oxide defects, which can result in quick saturation. Rangan et al. [30] and Grasser et al. [31] have also found similar varying degradation slopes in large planar devices, but with slower gradual changing due to larger gate area than SNWTs.

In the recovery stage, both effects of oxide hole trap detrapping and interface trap passivation are observed in SNWTs. We know that the subthreshold swing shift is mainly caused by the interface trap, while both the interface trap and oxide trap contribute to the threshold voltage shift. From Fig. 25.10, it is found that the interface trap recovery is relatively small and is inert to the gate voltage, while the oxide hole trap relaxation is sensitive to the recovery voltage.

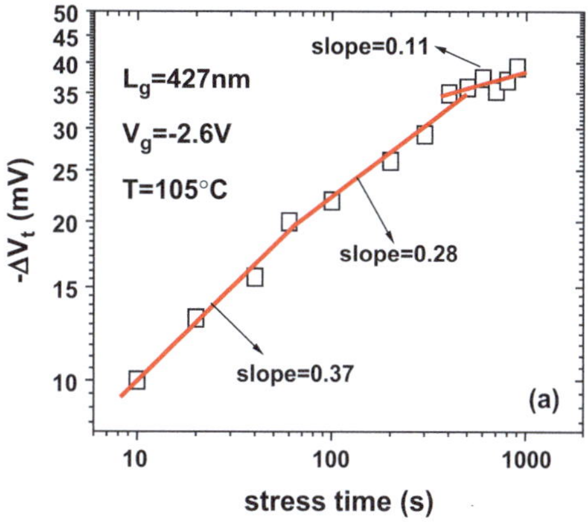

Fig. 25.9 The typical results of V_t degradation of NBTI in p-SNWTs [26]

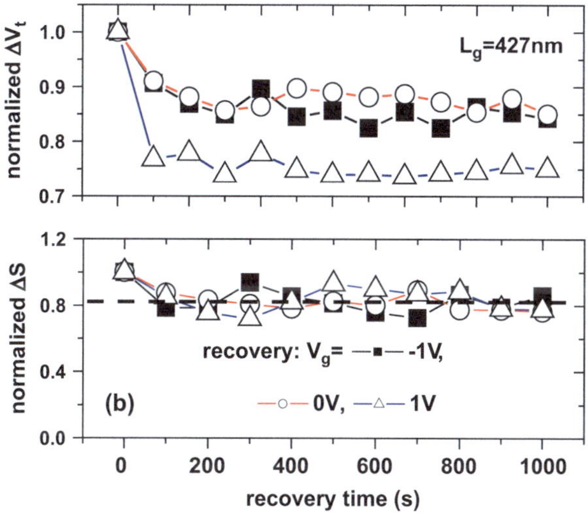

Fig. 25.10 The typical results of NBTI recovery characteristics in p-SNWTs [26]

In addition, in some SNWTs the impact of electron trapping-detrapping on the NBTI behavior is observed, which is usually neglected in the traditional devices [32]. Over 100% recovery of both V_t and gate current (I_g) after NBTI stress is observed in some extreme cases, as shown in Fig. 25.11a and b. For stress behavior, we can find abnormally non-monotonic variation of the monitored I_g during the stress time, as shown in Fig. 25.11c. I_g increases in the first few seconds and then reduces rapidly. Based on the different trapping rate of electrons and holes [34, 35], we deduce that the initial I_g increase mainly results from the injected electrons from the TiN metal gate and captured by the as-grown electron traps, which can distort the band diagram and enhance the gate current. As the stress time increases, hole trapping induced by NBT stress dominates and is responsible

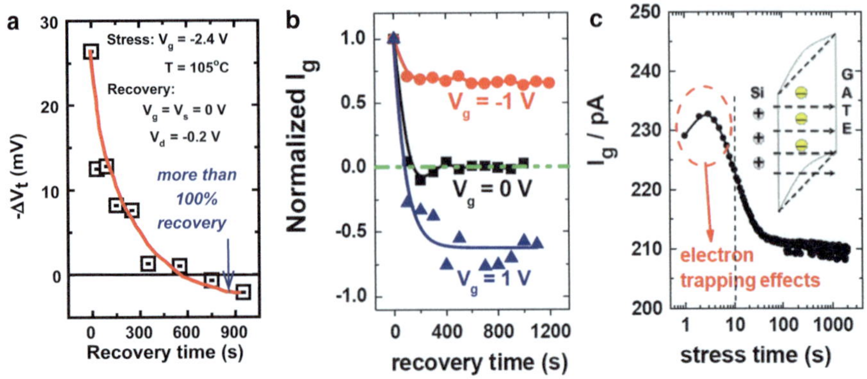

Fig. 25.11 Over 100% recovery of (**a**) V_t and (**b**) I_g can be ascribed to the un-detrapped electrons. (**c**) The typical monitored I_g during stress [32]

for the accordingly reduced gate current. Since electrons are easier to be trapped and more difficult to be detrapped compared with holes [34, 35], in the stress stage, both trapped electrons and trapped holes exist in the gate dielectrics, while during the positive V_g or/and negative V_d recovery stage, holes are detrapped, with some electrons still being trapped, resulting in more than 100% recovery. The non-negligible impacts of electron traps in SNWTs are mainly caused by the cylinder 1D channel with multiple surface crystal orientations and strain induced by the self-limiting oxidation for more as-grown defects. For the devices with gate trimming process (see Fig. 25.12a), which can induce large amount of electron traps in the oxide [32], the electron trapping-detrapping impact can be enlarged, as shown in Fig. 25.12b, c. Therefore, for SNWTs, both electron traps and hole traps should be involved in NBTI mechanisms, which is quite different from traditional theory with hole trap action dominating.

25.4.2 Statistical NBTI Behavior in SNWTs

25.4.2.1 NBTI Fluctuations in Short-Channel SNWTs

Besides the above intrinsic NBTI characteristics, short-channel SNWTs suffer additional NBTI fluctuation effects. Experiments on sub-100 nm SNWTs showed that the V_t and I_D degradation appears to be randomly fluctuated under NBT stress (Fig. 25.13) [26]. This NBTI fluctuation mainly originates from the ultrasmall gate area of short-channel SNWTs and the statistical nature of randomly trapped charges. The gate area of short SNWTs is ultrasmall, thus only few trapped charges in the oxide. The random dynamic variations in the number and spatial distribution of the trapped charges can affect the drain current, depending on the local current density

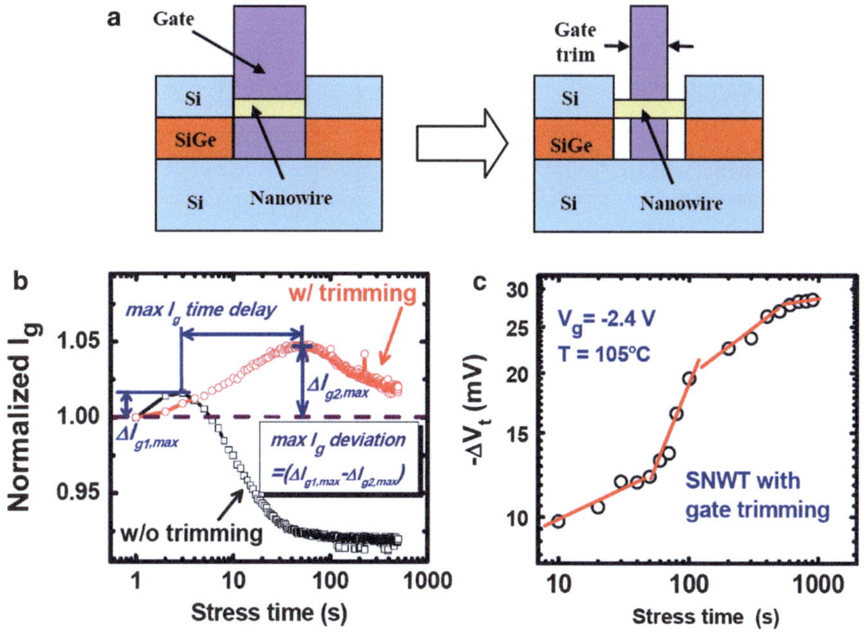

Fig. 25.12 (**a**) The schematic illustration of the gate trimming process for SNWTs [33]. (**b**) Typical monitored gate current under NBT stress for SNWTs with and without gate trimming process [32]. (**c**) The typical NBTI degradation in SNWTs with gate trimming process [32]

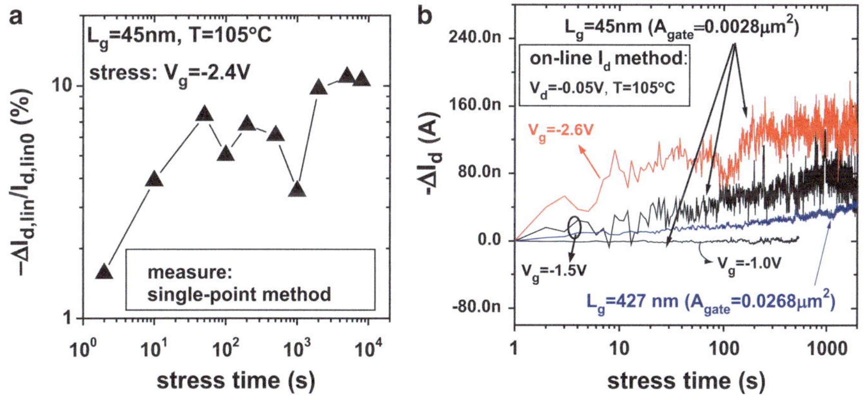

Fig. 25.13 Typical results of NBTI degradation in short-channel SNWTs [26]

in its percolation path through the channel. While, in large-area planar devices, this random nature is hidden by average effects. Therefore, how to characterize the NBTI in ultra-scaled SNWTs is challenging. Traditionally, people usually talk about NBTI only in large-area devices. However, in ultra-scaled nanowire devices with ultrasmall gate area, people have to deal with NBTI with large fluctuations.

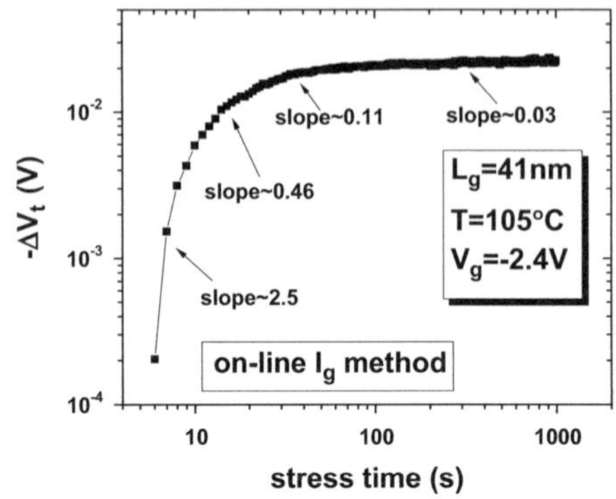

Fig. 25.14 Demonstration of fluctuation-immunity measurement of NBTI in an ultrasmall SNWT [26] using the developed on-line I_g method [38]. Even faster saturation behavior of the NBTI degradation in short-channel SNWTs can be observed

In order to study NBTI in SNWTs, long-channel devices should be used to suppress the statistical uncertainty from the above analysis. However, investigation of ultrashort SNWTs is important for practical technology qualification. The results above show that, the conventional I_D-based methods (whether sweep mode or on-the-fly mode) are not suitable. This problem can be alleviated in three ways. First, multi-wire structure can be used to obtain a large gate area, yet it depends on the requirements of the designed circuits. Second, advanced models can be developed to predict the median NBTI degradation. The fluctuation can then be calculated using statistical models based on the percolation theory [36, 37]. Third, new characterization methods can be developed, which can deal with this kind of fluctuation. Based on the principle that the trapped charge at the Si/SiO_2 interface and in the oxide bulk can modify the oxide electric field and thus the gate direct tunneling current, an on-line I_g method [38] is developed and demonstrated, which can effectively suppress the NBTI fluctuation of short-channel SNWTs (Fig. 25.14). Another strategy is to live with the fluctuations rather than suppress it. That is, to do the measurement statistically, as demonstrated very recently [39–44]. Since it is essentially the process of "stimulating/stressing the trap and recording its response statistically", we named it "statistical trap response" (STR) method [43]. This method can be used to directly study the impacts of single/few trap in SNWTs during NBT stress, as will be discussed in Sect. 25.4.2.2.

It is worth noticing that this kind of stochastic effect is not only relevant in SNWTs, but it also affects planar FETs with very small area [39–44]. However, this effect would be more serious in SNWTs even with large gate area. That is to say, the impact of single/few trap in SNWTs is enhanced, as will be shown next.

25.4.2.2 Enhanced Single/Few Trap Behavior in SNWTs

In nanoscale devices, a single trap can induce non-negligible degradation in the device, due to the following two reasons. One is a kind of simple size effect, i.e., the ΔV_{TH} ($\sim q/C_{OX}$) is increased with scaled gate area. Another is the localization effect due to the channel percolation mechanism [37, 45]. In SNWTs or other kinds

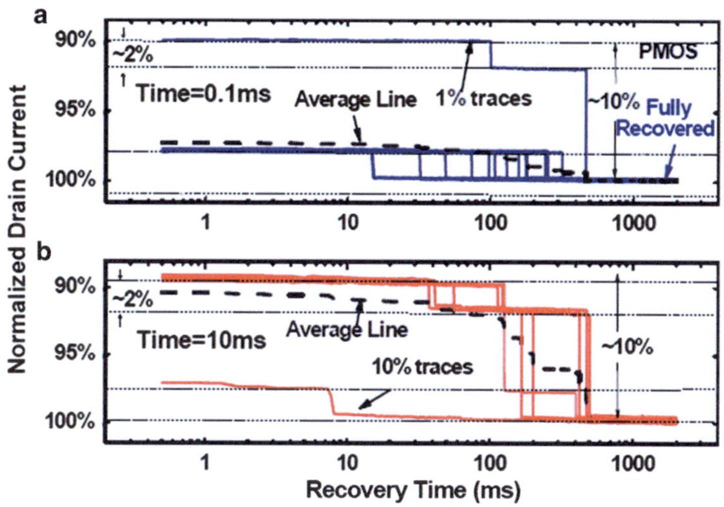

Fig. 25.15 The typical STR results in SNWTs under different stress time [46]

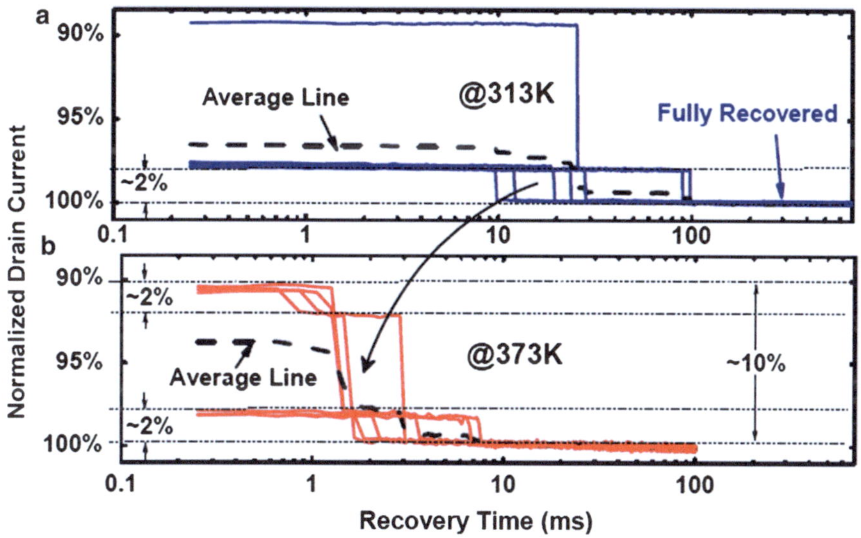

Fig. 25.16 The typical STR results in SNWTs under different stress temperature [46]

of thin-body devices, the single/few trap-induced degradation can be serious if the oxide trap is located above the critical channel conduction path and may even lead to device failure directly.

To study this single/few trap behavior in SNWTs, the STR method is adopted [46]. With 100 STR measurement cycles, after different stress time (Fig. 25.15) and under different stress temperature (Fig. 25.16), four discrete levels/states in drain

Table 25.1 The comparison of the single trap-induced degradation in planar devices and SNWTs

Device structure	Planar bulk FET [45]	GAA SNWT
Gate area (μm^2)	0.003	0.36
Expected single trap-induced ΔV_{TH} ($\sim q/C_{OX}$)	$\sim$2 mV	$\sim$0.02 mV
Measured single trap-induced ΔV_{TH} (mean value)	4.75 mV	$\sim$10 mV
Measured single trap-induced ΔV_{TH} (extreme case)	$\sim$30 mV	$\sim$50 mV

current (induced by two traps) can be observed, which has the probability of about 50 mV apparent threshold voltage shift, induced by the single trap in p-SNWTs. On the contrary, considering a simple charge sheet approximation, number fluctuation of a single carrier can only induce less than 1 mV of the threshold voltage shift. Table 25.1 further compares the amplitude of single trap-induced degradation in planar device and SNWT with comparable gate area (to exclude the strong impacts of quantum confinement on the switching behaviors of oxide traps in SNWTs [46], large-diameter nanowires are adopted here). The enhanced single/few trap-induced degradation in SNWTs is partly due to limited percolation path number in fully depleted undoped channels.

Moreover, as shown in the figures, higher temperature increases the trap capture probability and reduce the emission time, while longer stress time only increases capture probability. And the experimental results show that the influence of single electron trap in n-SNWTs is much smaller compared with single hole trap in p-SNWT devices.

25.4.2.3 Other Open Issues

The above shows the NBTI fluctuation in SNWTs and the related characterization method. But how would this "new" reliability behavior impact on SNWT-based circuits? One of the answers is that it will affect the NBTI-induced dynamic variability. As we have demonstrated first on scaled planar FETs, the statistical NBTI behavior in small gate-area devices adds the cycle-to-cycle variation effects into the conventional time-dependent device-to-device variation during the circuit aging [43, 44]. For SNWTs, since the trap-induced degradation is enhanced, this dynamic variability issue would be more announced in SNWT circuits, which should be further studied. And future investigation on the modeling and characterization for accurate lifetime prediction in SNWTs and trap-aware circuit design are also needed.

Another issue is that the experiments have shown the link between RTN and NBTI fluctuations in SNWTs, which is consistent with the recent discussions in ultra-scaled bulk MOSFETs [41, 47, 48]. As pointed out by Grasser et al., the link is that the RTN and the recoverable component of BTI are caused by the same oxide defects [41, 47, 48]. That is why some people would like to consider RTN as a kind of "microscopic BTI". It is worth noticing that RTN is near-equilibrium behavior of traps, while BTI is nonstationary response of traps. Some experimental results have shown that the trap capture statistics exhibits Weibull distribution with Weibull slope $\beta < 1$ under high stress biases [43, 44, 49], which is different from the expected exponential distribution (i.e., the Weibull slope $\beta = 1$) in RTN theory. But some other reports showed that it is still exponential distribution ($\beta \approx 1$) under stress voltages [50, 51]. However, this issue has not been fully understood by now. Nevertheless, to some extent, BTI can be regarded as the nonstationary RTN. More studies are still needed to get full understanding of the trap behavior under different conditions. And further questions for SNWTs are how RTN and BTI are interacting or enhanced by the nanowire structure and how do they influence the reliability and variability in nanowire-based circuits, which need further investigations as well.

25.5 Summary

The multi-gate FinFET structure has been used in manufacturing since 22 nm technology node. The reliability of FinFETs can be improved by well-rounded Fin-structure design (to overcome corner effects) and careful process optimizations (for sidewall channel surface treatment). Therefore, there is no serious reliability problem for multi-gate FinFET devices. When approaching the end of CMOS technology roadmap, the GAA SNWT is a promising architecture for ultimate CMOS applications. The main part of this chapter briefly reviews the recent understanding of the NBTI reliability behavior in GAA SNWTs. The experimentally characterized reliability behaviors exhibit some new features unique to this structure with quasi-1D channel and 3D surrounding gate, which should be carefully considered in their circuit applications. The intrinsic (or average) NBTI characteristics of GAA SNWTs show fast initial degradation, quick degradation saturation, and bias dependent of recovery with more oxide hole detrapping at $V_G > 0$. And additional electron trapping-detrapping impacts are also observed especially for gate-trimmed SNWT devices. The statistical NBTI characteristics are studied in ultra-scaled SNWTs, which exhibit NBTI fluctuation during degradation with enhanced single/few trap behavior. In addition to further in-depth analysis needed, special device-circuit co-design methodology or trap-aware design is highly expected to maximize the inherent advantages of this kind of emerging device.

Acknowledgments The work in this chapter was supported in part by 973 Projects (2011CBA 00601), NSFC (61106085, 60625403), National S&T Major Project (2009ZX02035-001) and the Collaborative Project between Peking University and Samsung Electronics. The authors would like to thank the staff of the National Key Laboratory of Micro/Nano Fabrication Technology, Peking University, for their assistance in device fabrication and Prof. T. Grasser for the helpful discussions.

References

1. International Technology Roadmap for Semiconductors (ITRS), http://public.itrs.net/
2. R. Huang, H. M. Wu, J. F. Kang, D. Y. Xiao, X. L. Shi, X. An, Y. Tian, R. S. Wang, L. L. Zhang, X. Zhang and Y. Wang, Sci. China Ser. F, Inf. Sci., 52, 1491 (2009).
3. Y. Y. Wang, Sci China Inf Sci, 54, 915 (2011).
4. M. Li, Sci China-Phys Mech Astron, 55, 2316 (2012)
5. J.-P. Colinge (eds.), FinFETs and Other Multi-Gate Transistors, Springer (2008).
6. R. Huang and R. S. Wang, Front. Phys. China, 5, 414 (2010).
7. R. Huang, R. Wang, J. Zhuge, C. Liu, T. Yu, L. Zhang, X. Huang, Y. Ai, J. Zou, Y. Liu, J. Fan, H. Liao and Y. Wang, IEEE Custom Integrated Circuits Conference (CICC), 2-1 (2011).
8. R. Huang, R. Wang, C. Liu, L. Zhang, J. Zhuge, Y. Tao, J. Zou, Y. Liu, Y. Wang, Microelectronics Reliability, 51, 1515 (2011).
9. C. Auth, C. Allen, A. Blattner, D. Bergstrom, M. Brazier, M. Bost, M. Buehler, et al., VLSI, 131 (2012).
10. Y.-K. Choi, N. Lindert, P. Xuan, S. Tang, D. Ha, et al., in IEDM Tech. Dig., 421 (2001).
11. A. Kaneko, A. Yagishita, K. Yahashi, T. Kubota, M. Omura, et al., in IEDM Tech. Dig., 844 (2005).
12. C. J. Petti, J.P. McVittc and J. D. Plammer, in IEDM Tech. Dig., 88 (1988).
13. Y.-K. Choi, L. Chang, P. Ranade, J.-S. Lee, D. Ha, et al., in IEDM Tech. Dig., 259 (2002).
14. A. Kaneko, A. Yagishita, K.Yahashi et al., in IEDM Tech.Dig., 863 (2005).
15. S.Chouksey, J. G. Fossum, A.Behnam, S.Agrawal and L. Mathew, IEEE Trans. on Electron Dev., 56, 10, 2348 (2009).
16. C.-H. Jan, U. Bhattacharya, R. Brain, S .- J. Choi, G. Curello, G. Gupta, et al., IEDM Tech. Dig., 44 (2012).
17. X.Xu, R.Wang, R.Huang, J. Zhuge, G. Chen, X. Zhang and Y. Wang, IEEE Trans. on Electron Dev., 55, 11, p.3246 (2008).
18. S. D. Suk, S.-Y. Lee, S.-M.Kim, E.-J.Yoon, M.-S. Kim, et al., in IEDM Tech. Dig., 717 (2005).
19. Y. Tian, R. Huang, Y. Wang, J. Zhuge, R. Wang, J. Liu, X. Zhang and Y. Wang, IEDM Tech. Dig., 895 (2007).
20. R. Huang, J. Zou, R. Wang, C. Fan, Y. Ai, J. Zhuge and Y. Wang, IEEE Trans. Electron Devices, 58, 3639 (2011).
21. Y. Ai, R. Huang, Z. Hao, C. Fan, R. Wang, S. Pu, and Y. Wang, Physica E, 43, 102 (2010).
22. A. Shickova, N. Collaert, R. Rooyackers, A. De Keersgieter, T. Kauerauf, M. Jurczak, B. Kaczer and G. Groeseneken, Proceedings of the ULIS conference, 141 (2006).
23. F. Crupi, B. Kaczer, R. Degraeve, V. Subramanian, P. Srinivasan, E. Simoen, A. Dixit, M. Jurczak and G. Groeseneken, IEEE Transaction on Electron Devices, 53, 2351 (2006).
24. S. Maeda, J.-A. Choi, J.-H. Yang, Y.-S. Jin, S.-K. Bae, Y.-W. Kim,and K.-P. Suh, IRPS, 8 (2004).
25. A. Shickova, N. Collaert, P. Zimmerman, M. Demand, E. Simoen, G. Pourtois, A. De Keersgieter, L. Trojman, I. Ferain, F. Leys, W. Boullart, A. Franquet, B. Kaczer, M. Jurczak, H. Maes and G. Groeseneken, VLSI Symp., 112 (2007).
26. R. Wang, R. Huang, D. W. Kim, Y. He, Z. Wang, G. Jia, D. Park and Y. Wang, IEDM Tech. Dig., 821 (2007).
27. C. Liu, T. Yu, R. Wang, L. Zhang, R. Huang, D.-W. Kim, D. Park and Y. Wang, IEEE Trans. Electron Devices, 57, 3442 (2010).
28. T. Irisawa, T. Numata, E. Toyoda, N. Hirashita, T. Tezuka, N. Sugiyama and S. Takagi, IEEE Trans. Electron Devices, 55, 3159 (2008).
29. H. Ohta, T. Watanabe and I. Ohdomari, Jpn. J. Appl. Phys., 46, 3277 (2007).
30. S. Rangan, N. Mielke, E. C.C. Yeh, IEDM Tech. Dig., 14.3.1 (2003)
31. T. Grasser, and B. Kaczer, IEEE T-ED, 56, 1056 (2009)
32. L. Zhang, R. Wang, J. Zhuge, R. Huang, D. W. Kim, D. Park and Y. Wang, EDM Tech. Dig., 123 (2008).

33. K. H. Yeo, S. D. Suk, M. Li, et al., in IEDM Tech. Dig., 20.2 (2006).
34. K. P. Cheung, Proc. IEEE Int. Symp. P2ID, 181 (1997).
35. J. P. Campbell, K.P. Cheung, J.S. Suehle and A. Oates, VLSI, 76 (2008).
36. D. Ielmini, C. M. Compagnoni, A. S. Spinelli, A. L. Lacaita and C. Gerardi, IRPS, 515 (2004).
37. A. Asenov, R. Balasubramaniam, A.R. Brown, J.H. Davies., IEEE Trans Electron Dev., 50, 839 (2003).
38. G. Jia, and M. Xu, ICSICT, 1144 (2006).
39. V. Huard, C. Parthasarathy, C. Guerin, T. Valentin, E. Pion, M. Mammasse and N. Planes, L. Camus, IRPS, 289 (2008).
40. H. Reisinger, T. Grasser, K. Hofmann, W. Gustin, and C. Schlünder, IIRW, 30 (2009).
41. T. Grasser, H. Reisinger, W. Goes, Th. Aichinger, Ph. Hehenberger, P.-J. Wagner, M. Nelhiebel, J. Franco, and B. Kaczer, IEDM, 729 (2009).
42. B. Kaczer, T. Grasser, P. J. Roussel, J. Franco, R. Degraeve, L. Ragnarsson, E. Simoen, G. Groeseneken, H. Reisinger, IRPS, 26 (2010).
43. C. Liu, J. Zou, R. Wang, R. Huang, X. Xu, J. Liu, H. Wu, and Y. Wang, IEDM Tech. Dig., 571 (2011).
44. C. Liu, P. Ren, R. Wang, R. Huang, J. Ou, J. Wang, J. Wu, S. Yu, S.-W. Lee, and Y. Wang, IEDM Tech. Dig., 466 (2012).
45. B. Kaczer, Ph. J. Roussel, T. Grasser and G. Groeseneken, IEEE EDL, 31, 411 (2010).
46. C. Liu, R. Wang, J. Zou, R. Huang, C. Fan, L. Zhang, J. Fan, Y. Ai, Y. Wang, IEDM Tech. Dig., 521 (2011).
47. T. Grasser, B. Kaczer, W. Goes, H. Reisinger, T. Aichinger, P. Hehenberger, P.-J. Wagner, F. Schanovsky, J. Franco, M. Toledano Luque, and M. Nelhiebel, IEEE T-ED, 58, 3652 (2011).
48. T. Grasser, Microelectronics Reliability, 52, 39 (2012).
49. P. Ren, C. Liu, R. Wang, N. Gong, J. Liu, H. Wu, and R. Huang, ICSICT, 339 (2012).
50. T. Grasser, H. Reisinger, P.-J. Wagner, B. Kaczer, Physical Review B, 82, 245318 (2010).
51. T. Grasser, H. Reisinger, K. Rott, M. Toledano-Luque, B. Kaczer, IEDM Tech. Dig., 470 (2012).

Chapter 26
Bias-Temperature Instabilities in Silicon Carbide MOS Devices

D.M. Fleetwood, E.X. Zhang, X. Shen, C.X. Zhang, R.D. Schrimpf, and S.T. Pantelides

We have investigated bias-temperature instabilities (BTIs) in 4H-SiC transistors and capacitors under a range of stress conditions. The threshold voltage V_{TH} of *n*MOS transistors decreases for elevated temperature stress under negative bias, when the surface is accumulated. Devices stressed with the surface inverted do not exhibit significant V_{TH} shifts. Similar results are observed for *n*MOS and *p*MOS capacitors stressed in accumulation (measurable shift) or inversion (no significant shift). V_{TH} shifts due to BTI stressed under negative bias correlate strongly with the additional ionization of deep dopants in SiC at elevated temperatures. The charge that leads to BTI lies in interface traps that are more than 0.6 eV below the SiC conduction band, nitrogen-related defects, and O vacancies in the SiO_2.

D.M. Fleetwood (✉) • S.T. Pantelides
Electrical Engineering and Computer Science Department, Vanderbilt University, Nashville, TN 37235, USA

Department of Physics and Astronomy, Vanderbilt University, Nashville, TN 37235, USA
e-mail: dan.fleetwood@vanderbilt.edu; pantelides@vanderbilt.edu

E.X. Zhang • C.X. Zhang • R.D. Schrimpf
Electrical Engineering and Computer Science Department, Vanderbilt University, Nashville, TN 37235, USA
e-mail: enxia.zhang@vanderbilt.edu; xuan.zhang@vanderbilt.edu; ron.schrimpf@vanderbilt.edu

X. Shen
Department of Physics and Astronomy, Vanderbilt University, Nashville, TN 37235, USA
e-mail: xiao.shen@vanderbilt.edu

T. Grasser (ed.), *Bias Temperature Instability for Devices and Circuits*,
DOI 10.1007/978-1-4614-7909-3_26,

26.1 Introduction

Silicon carbide is attractive for high-power and high-temperature applications because of its wide band gap (∼3.26 eV for 4H-SiC), high breakdown field strength, high saturation electron drift velocity, high thermal conductivity, and compatibility with Si processing [1–4]. Although bias-temperature instability (BTI) is a critical reliability issue for SiC devices and circuits [5–11], only limited information on underlying mechanisms is available. As shown for Si MOS devices in other chapters of this volume, especially at low applied electric fields, the depassivation under BTI stress of interfacial Si dangling bonds that were previously passivated with hydrogen can lead to interface-trap buildup and oxide-trap charge [12]. Especially at higher stressing fields, charge trapping in O vacancies can dominate the observed BTI for Si MOS devices [13, 14]. Hydrogen processing is less effective in reducing SiC-SiO_2 interface-trap density than in Si MOS devices [14, 15]. Moreover, as a result of their thicker oxides, applied electric fields tend to be lower for SiC MOS than for Si MOS devices. Because of these reasons, the wide bandgap for SiC and differences in processing compared to Si MOS devices, the mechanisms of BTI in SiC MOS devices can differ significantly from those in Si MOS devices.

We have investigated bias-temperature instabilities (BTIs) for 4H-SiC-based *n*MOSFETs and *n*MOS and *p*MOS capacitors with similar oxide processing. Steady-state and switched-bias stresses were performed at positive and negative bias. Significant shifts are observed for steady-state stresses at elevated temperatures when the surface is in accumulation; negligible shifts are observed for steady-state stresses at elevated temperatures when the surface is in inversion. We describe the dynamics of the resulting charge trapping and recovery in detail.

26.2 Experimental Details

nMOSFETs and *n*MOS and *p*MOS capacitors were fabricated on an aluminum-doped (*p*-type) 4H-SiC epitaxial layer with a 55 nm, NO-nitrided gate oxide. A schematic diagram is shown in Fig. 26.1 [16]. All devices received a post-oxidation NO anneal at 1,175°C for 2 h to reduce their interface-trap densities [15, 17–21]. Steady-state and switched-bias stress experiments were performed at gate bias of ±17.5 V with the other terminals grounded at temperatures that ranged from 25 to 300°C. Electric fields were well below the thresholds for Fowler–Nordheim charge injection [22, 23]. For the transistors, drain current versus gate voltage (I_D–V_G) measurements were performed at room temperature with a HP4156A parameter analyzer. For the capacitors, high-frequency (100 kHz) capacitance–voltage (C–V) measurements were performed at room temperature with a HP4284A precision LCR meter, after the device was allowed to equilibrate, at a ramp rate of ∼0.1 V/s. No significant C–V hysteresis was observed for these devices and experimental conditions. At least three devices were measured for each case shown; the results represent the average response of the tested devices.

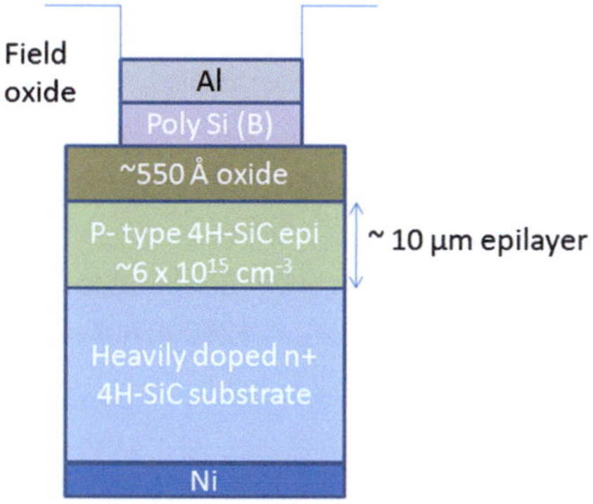

Fig. 26.1 Schematic illustration of p-substrate capacitor. The wafer is n^+ 4H-SiC, the epitaxial layer is p-type 4H-SiC with doping $\sim 5 \times 10^{15}$ cm^{-3} (Al), the gate oxide thickness t_{ox} is 55 nm, and the gate is p-type poly-crystalline Si, capped by Al. For n-substrate capacitors (not shown), the wafer is also p^+ 4H-SiC, the epitaxial layer is n-type 4H-SiC with doping $\sim 5 \times 10^{15}$ cm^{-3} (N), t_{ox} is 67.5 nm, and the gate is also p-type poly-crystalline Si, capped by Al (After [16], © IEEE, 2012, reprinted with permission)

26.3 Experimental Results and Discussion

26.3.1 Time Dependence

Figure 26.2 shows I_D–V_G curves measured at room temperature as a function of time for nMOSFETs stressed in inversion at an applied gate bias of 17.5 V at a temperature of 150°C [24]. The curves do not exhibit any significant shift with stressing time; the threshold voltage shift ΔV_{TH} is less than 1 mV. Even if we increase the stress temperature to 250°C, the magnitude of ΔV_{TH} is only $\sim$4 mV after 20-h stress, as shown in Fig. 26.3. This shows the relative stability of these 4H-SiC devices under positive bias (inversion) at elevated temperatures.

Figure 26.4 shows I_D–V_G curves measured at room temperature as a function of time for nMOSFETs stressed at 150°C under negative bias at an applied gate bias of -17.5 V with the other terminals grounded. For these stress conditions, the curves shift monotonically negatively with stress time due to hole trapping. The corresponding threshold voltage shift is shown as a function of stressing time in Fig. 26.5. The magnitude of the midgap voltage increases rapidly at early stress times, with a significant decrease in degradation rate at longer times. These negative shifts are strongly correlated with the ionization of deeper acceptor levels, as we discuss further below [25–28].

Figure 26.6 shows C–V curves measured at room temperature before and after stress for n- and p-substrate capacitors processed under conditions similar to the nMOSFETs of Figs. 26.2, 26.3, 26.4, and 26.5. The C–V curves shift positively after 20 min of stress at 20 V applied to the gate for the n-substrate capacitor and negatively for the p-substrate capacitor stressed for 20 min with -17 V applied

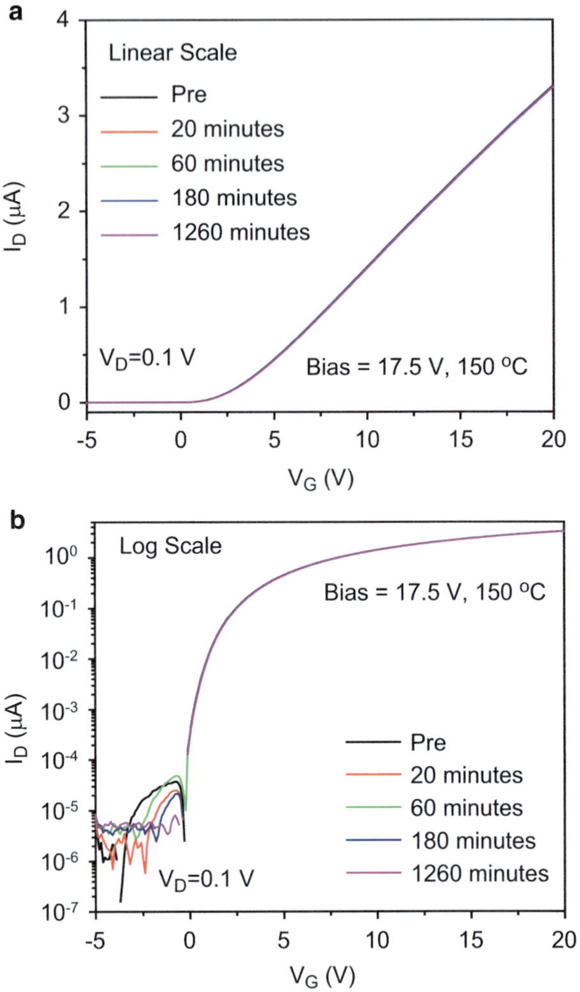

Fig. 26.2 Drain current versus gate voltage (I_D–V_G) and stress time for 4H-SiC nMOSFETs, for (**a**) linear and (**b**) logarithmic scale. The applied gate bias is 17.5 V, with other pins grounded, and the temperature is 150°C (After [24], © ECS, 2011, reprinted with permission)

to the gate. For *n*-substrate capacitors in Fig. 26.6a stressed under positive bias, negative oxide-trapped charge and electrons in deeper interface traps (energies more than 0.6 eV below the conduction band edge) can each lead to a positive midgap voltage shifts in SiC MOS capacitors [29]. Interface traps with shallower levels would lead instead to stretch-out in the C–V curve; no significant stretch-out is observed in these devices under these stressing conditions. Some of the defects that cause the midgap voltage shifts in Fig. 26.6 likely are process-induced; others may be created during the BT stressing. In Fig. 26.6b, for *p*-substrate capacitors stressed at negative bias, negative midgap voltage shifts are observed. These shifts are consistent with positive oxide-trapped charge and/or holes in deeper interface traps (energies more than 0.6 eV above the valence band edge) [30–35]. Again, no significant increase in C–V stretch-out is observed.

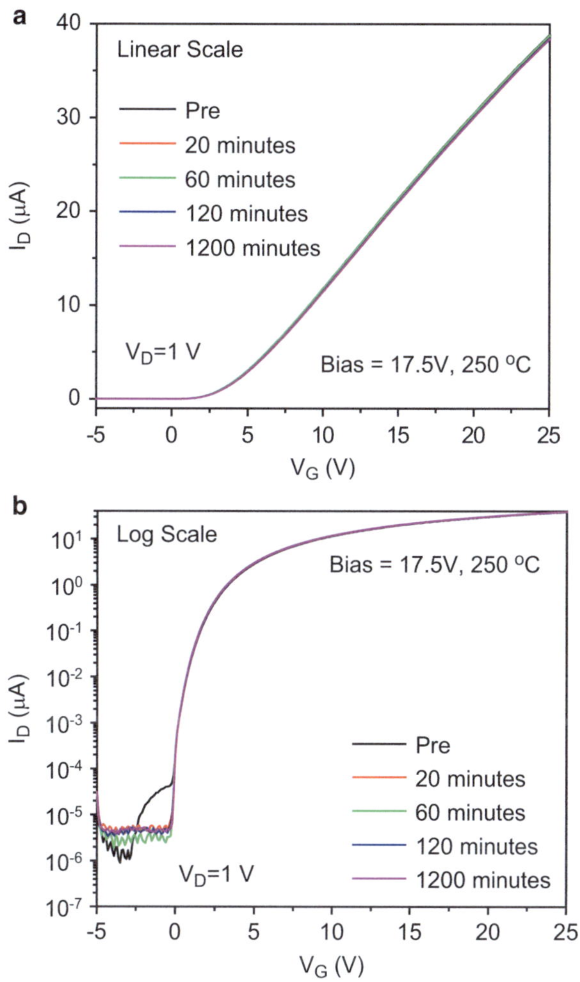

Fig. 26.3 Drain current as a function of gate voltage (I_D–V_G) and stress time for 4H-SiC nMOSFETs, for (**a**) linear scale and (**b**) log scale. The applied gate bias is 17.5 V with the other terminals grounded, and the temperature is 250°C (After [23], © ECS, 2011, reprinted with permission)

Figure 26.7 shows C–V curves as a function of stressing time for a *p*-substrate capacitor; the gate voltage is −17 V and the stress temperature is 150°C. The curves shift monotonically to more negative values, again consistent with hole trapping. Figure 26.8 shows the midgap voltage shift ΔV_{mg} versus stress time for *n*-substrate (Fig. 26.8a) and *p*-substrate (Fig. 26.8b) capacitors at biases of ±17 V on the gate at a stress temperature of 150°C. In each case, the trapping shows a monotonic increase in the magnitude of the midgap voltage shifts at early stress times, with saturation at longer times. This saturation represents the filling of process and stress-induced interface and near-interfacial oxide (border) traps in these devices [19, 33].

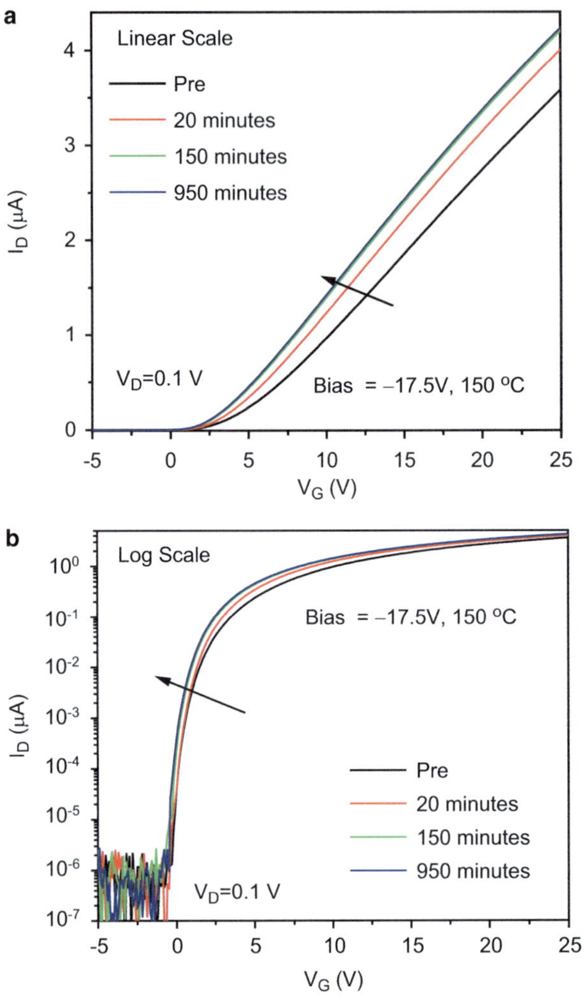

Fig. 26.4 Drain current as a function of gate voltage (I_D–V_G) and stress time for 4H-SiC nMOSFETs, for (**a**) linear scale and (**b**) log scale. The applied gate bias is -17.5 V with the other terminals grounded, and the temperature is 150°C (After [24], © ECS, 2011, reprinted with permission)

26.3.2 Temperature Dependence

Figure 26.9 shows the midgap voltage shifts as a function of stressing temperature for *n*- and *p*-substrate SiC MOS capacitors stressed for 20 min at $+20$ V and -17 V on the gates, respectively. The best fit, effective activation energies of the resulting BTI are 0.12 ± 0.02 eV (for positive threshold voltage shifts), although at higher temperatures the value may increase by as much as $\sim$50% (as shown by the dotted

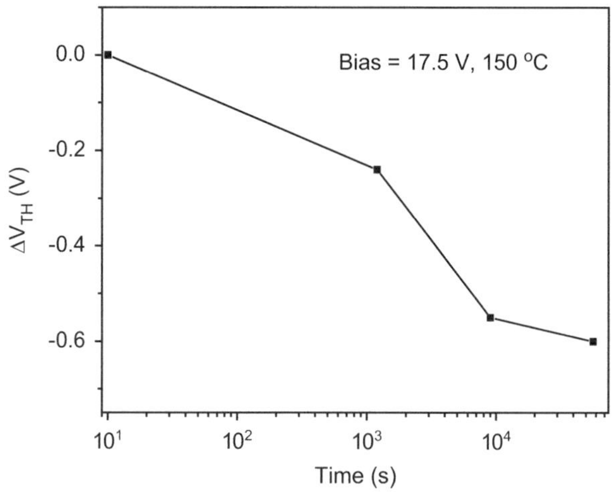

Fig. 26.5 Threshold voltage shifts as a function of stress time for the 4H-SiC nMOSFETs and stress conditions of Fig. 26.4 (After [24], © ECS, 2011, reprinted with permission)

blue line in Fig. 26.9), and 0.23 ± 0.02 eV (for negative threshold voltage shifts) for the *n*- and *p*-substrate capacitors [28]. The activation energies for the *n*-substrate SiC capacitors in Fig. 26.9 are somewhat lower than those typically measured for SiO_2 on Si devices, but the *p*-substrate devices show effective activation energies similar to those reported for SiO_2 on Si [30, 35–37].

Calculations show that, for the Al acceptors in the *p*-substrate capacitors, assuming an ionization energy of ~0.22 (0.20) eV, only ~37% (49%) are ionized at room temperature [28]. At the elevated temperatures of the BT stress, additional carriers are released from the dopants under these accumulation stress-bias conditions. The excess carriers are available to tunnel into border traps [37] and increase the midgap voltage shift. This mechanism appears to be more significant here than in Si because SiC has more defects and fewer carriers than Si. We also note that no significant PBTI is observed for transistors biased in inversion in Fig. 26.3, under conditions that are higher in temperature and longer in time than those shown for devices in accumulation in Figs. 26.4, 26.5, and 26.6. This is because carrier densities are much lower in inversion than accumulation for these devices [38–40]. Moreover, interface-trap levels that are away from the band edges are quite slow to populate in SiC MOS devices [40, 41]. Hence, carrier availability can affect BTI due to both interface and oxide-trap charge effects in SiC MOS devices much more significantly than in Si MOS devices.

For an ionization energy of 0.125 eV (0.105 eV), ~92% (~96%) of the N-donors in *n*-type SiC are ionized at 25°C. However, Chatterjee et al. have suggested that the NO anneal may lead to a significant increase in the effective N-doping of the SiC channel [42]. This may decrease the fraction of ionized N-dopants at 25°C, consistent with the possibility that dopant ionization is also rate limiting for BTI in *n*-type SiC over a similar range of temperatures [28].

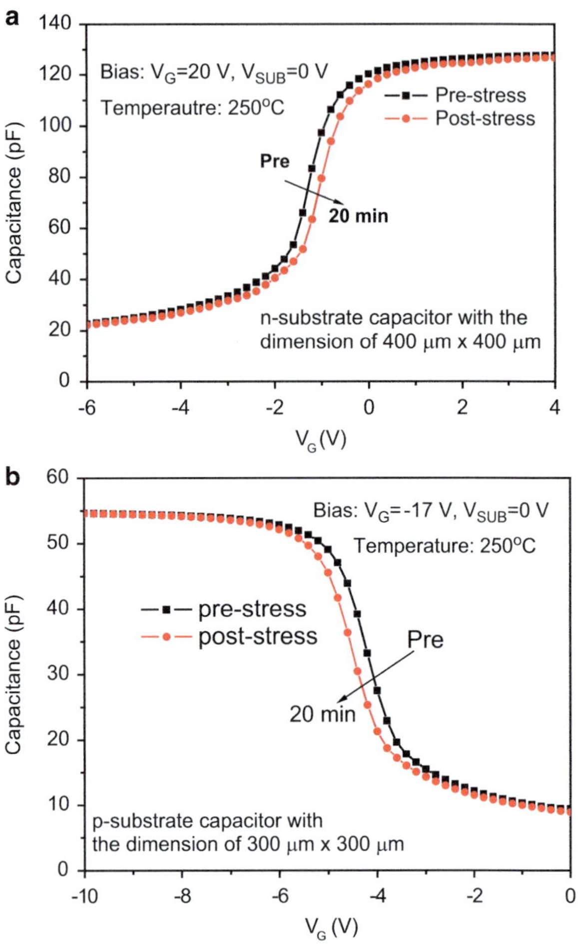

Fig. 26.6 Capacitance as a function of gate voltage measured at room temperature for (**a**) an *n*-substrate/4H-SiC capacitor, stressed with a gate voltage of 20 V at a temperature of 250°C, and (**b**) a *p*-substrate/4H-SiC capacitor, stressed at a gate voltage of −17 V at a temperature of 250°C (After [16], © IEEE, 2012, reprinted with permission)

There are other differences between Si MOS and SiC MOS devices that must be considered. The gate oxides on SiC are thicker and grown at higher temperatures than the ultrathin nitrided SiO_2 and/or high-K gate dielectrics used in Si-based integrated circuit technologies [18–21, 35–43]. For Si MOS technologies, hydrogen is known to play an important role in both the passivation of process-induced interface traps and the creation of defects during bias-temperature stress [12, 44]. Hydrogen reactions at the SiC interface are typically not as effective in passivating interface traps as at the Si/SiO_2 interface [15, 17–21, 30, 45, 47], so the role of hydrogen in BTI on SiC is less clear. Excess nitrogen associated with the NO nitridation treatments used to reduce interface-trap densities in these SiC devices [1, 19, 21, 46] may also introduce a relatively small but finite density of N-related trap levels in the near-interfacial SiO_2. N-related centers are observed to function as both electron traps and hole traps [47–49]. Moreover, nitrogen has been demonstrated to increase NBTI in Si/SiO_2 structures [45, 48, 49]. On the other hand, NO processing

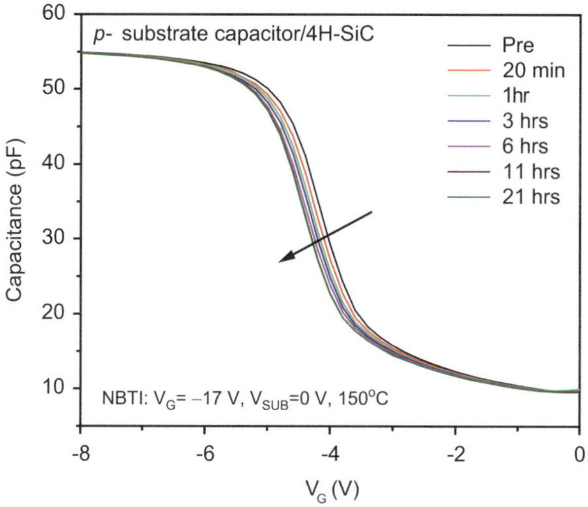

Fig. 26.7 Capacitance as a function of gate voltage and stress time for a 300 μm × 300 μm *p*-substrate 4H-SiC capacitor. The stressing bias is −17 V on the gate and the temperature is 150°C (After [16], © IEEE, 2012, reprinted with permission)

has been found to reduce the density of interface defects [11, 19, 20, 50]. Thus, if N-related defects play a significant role in the observed response, it is the excess nitrogen concentration above and beyond the levels required to passivate the process-induced defects that is likely contributing to the BTI. Further, holes trapped by N-related defects typically are re-emitted when the temperature is raised past 125°C [19]. So it seems more likely that O vacancies play a key role in the hole trapping in these devices [28], similar to what is found for Si MOS devices stressed at similar temperatures but higher electric fields (e.g., >5 MV/cm) [13, 14, 37, 51, 52].

If devices are stressed for longer times and higher temperatures than we have employed in these studies, it is sometimes observed that a potentially different trap formation mechanism is observed [30, 53]. Particularly for SiC MOS devices that are intended for use in high-temperature electronics, these kinds of studies will be particularly important to continue and extend in the future. Moreover, changes in interface-trap occupancy with temperature can shift the threshold voltage, mobility, and drive current of SiC MOS devices much more as the device operating temperature changes [29, 54–56] than the BTIs that we observe when devices are measured at a fixed temperature. This is illustrated in Fig. 26.10, where values of V_{TH} extracted from I_D–V_G characteristics are shown as a function of T and compared with TCAD simulations [54]. The slope of the V_{TH} versus T curve was reproduced well in TCAD simulation with an effective interface-trap density that reaches a maximum of $\sim 2 \times 10^{13}$ cm^{-2} eV^{-1}, decreasing approximately exponentially in density as the distance from the band edge increases [56]. The inset of Fig. 26.10 schematically illustrates the occupancy of interface traps as a function of T. For increasing T, the Fermi level moves from the conduction band toward midgap, decreasing the

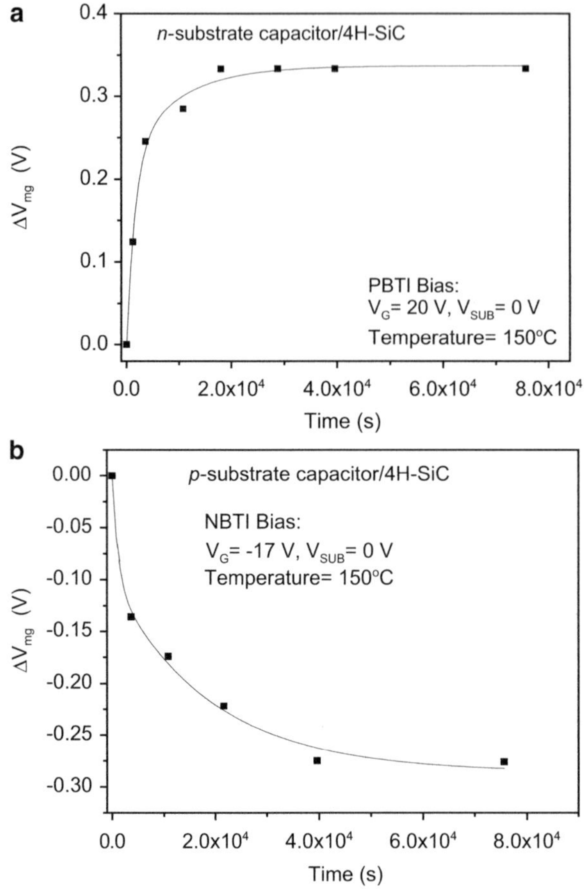

Fig. 26.8 Midgap voltage shift (ΔV_{mg}) versus stress time for (**a**) *n*-substrate SiC capacitors stressed at 150°C and gate bias of ±20 V and (**b**) *p*-substrate 4H-SiC capacitors stressed at 150°C and gate bias of ±17 V. The *lines* are guides to the eye (After [16], © IEEE, 2012, reprinted with permission)

percentage of interface traps occupied by electrons [54–56]. Note that V_{TH} changes by more than 1 V as the device is heated from room temperature to ~150°C (423 K). So this is another important consideration in the practical implementation of Si MOS electronics over a wide temperature range.

26.3.3 Switched-Bias Stress

Figure 26.11 shows the values of ΔV_{TH} as a function of switched stressing biases for 4H-SiC-based nMOSFETs. The characterization was performed at room temperature before and after BT stressing at 150°C. The V_{TH} shifts negatively

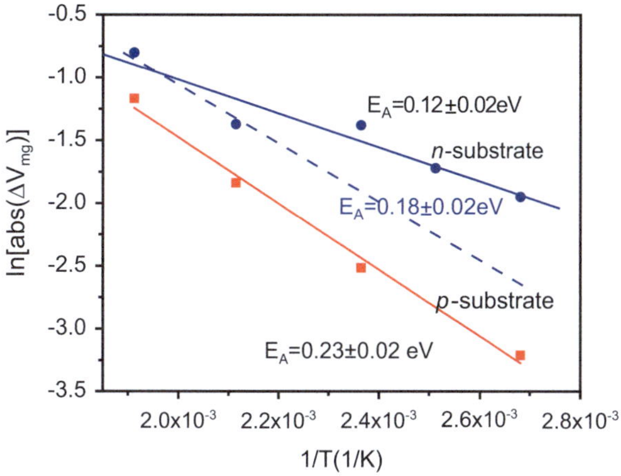

Fig. 26.9 Midgap voltage shift as a function of stressing temperature for *n*- and *p*-substrate/4H-SiC capacitors stressed in accumulation for 20 min at 20 V (*n*-substrate capacitors) and -17 V (*p*-substrate capacitors) on the gates. The effective activation energies are 0.23 ± 0.02 eV and from 0.12 ± 0.02 eV to 0.18 ± 0.02 eV for *p*- and *n*-substrate capacitors, respectively. The standard deviation in Arrhenius slope is ± 0.02 eV in each case (After [16] and [28], © AIP and IEEE, 2012, reprinted with permission)

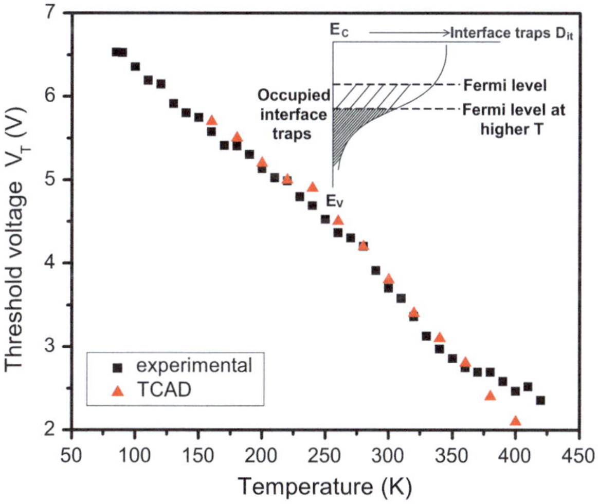

Fig. 26.10 Experimental and calculated threshold voltage V_T as a function of temperature T; the *inset* illustrates the change in occupancy of interface traps with T (After [56], © IEEE, 2013, reprinted with permission)

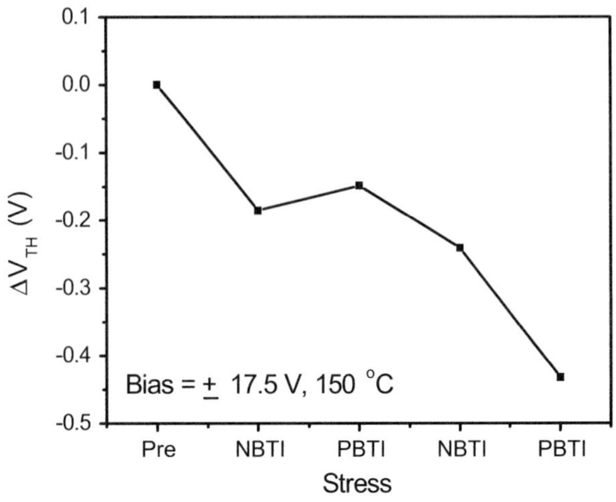

Fig. 26.11 ΔV_{TH} as a function of switched-bias stress for 4H-SiC nMOSFETs. The applied gate bias during the stress was ± 17.5 V; the temperature was 150°C; the stress times were 20 min for the first pair of stresses, and 60 min for the second pair (After [24], © ECS 2011, reprinted with permission)

after the first 20-min negative-bias stress and then recovers somewhat during the next 20-min positive-bias stress. In contrast, an additional increase in V_{TH} shift is observed in the second period (60 min) of positive-bias stress that follows the second negative-bias stress (60 min). Smaller shifts but significantly more reversibility is typically observed for Si MOS capacitors subjected to similar switched-bias stressing conditions [57, 58]. For the SiC MOS devices of Fig. 26.11, the majority of trapped holes remain in interface and/or oxide traps despite the reverse-bias stress, showing that hole traps are more stable than electron traps in these devices [28]. The significant increase in the magnitude of the midgap voltage shift during the second period of PBT stress may result from the bias-induced motion of trapped positive charge in the SiO_2 layer from trapping sites within the oxide to sites at or closer to the SiC/SiO_2 interface. The trapped charge is likely to comprise mostly trapped holes. In oxides in Si, many of these holes would be neutralized via electron compensation near the interface [52, 53], but charge generation rates in SiC are significantly less than in Si [33, 59], so fewer electrons are available to compensate the trapped holes in SiC MOS capacitors than Si MOS capacitors. It is also possible that some of the trapped positive charge is in the form of H^+ [33, 60]. If that is the case in these oxides on SiC, then the results of Fig. 26.10 suggest that the H^+ in the SiO_2 can move closer to the SiO_2/SiC interface during the application of PBT stress. However, there must not be a significant amount of interface-trap formation, or the midgap voltage shift would be positive instead of negative, under positive bias. Trapping of H^+ at defects in the near-interfacial SiO_2 regions of oxides with high O vacancy densities has been observed in Si MOS devices [61], so this interpretation of the results is also consistent with Fig. 26.10.

26.4 Conclusion

We have performed a detailed experimental study of bias-temperature instabilities in 4H-SiC-based MOS capacitors. We find that hole traps are more stable in these materials than electron traps. The effective activation energy between room temperature and $\sim$250°C for NBTI is 0.12 ± 0.02 eV for the n-substrate capacitors and 0.23 ± 0.02 eV for PBTI for the p-substrate capacitors. The positive midgap voltage shifts for n-substrate/4H-SiC capacitors under positive bias are due to net electron trapping at or near the SiC-SiO_2 interface, which can be enhanced at elevated temperatures by the ionization of deeper donors, and the negative midgap voltage shifts for p-substrate/4H-SiC capacitors under negative bias are due to net hole trapping at or near the SiC-SiO_2 interface, which can be enhanced at elevated temperatures by the ionization of deeper acceptors. The charge trapping in these devices likely results primarily from near-interfacial oxygen vacancies, deep interface traps, and/or N-related defects in the near-interfacial SiO_2. All of these kinds of defects can capture either electrons or holes, consistent with the results of this study. The low effective activation energies are comparable to results observed for Si MOS capacitors, but the mechanisms responsible for the BTIs in SiC MOS devices appear to be different in origin from those in Si MOS devices. The magnitudes and the time dependencies of the BTI responses observed in these SiC-based MOS structures differ significantly from what are typically observed for Si-based MOS devices with ultrathin gate oxides. Shifts are larger for SiC-based devices than for Si-based devices, and do not follow a simple power-law time dependence, as commonly observed in Si-based MOS devices. Switched-bias stress experiments reveal a complex interplay among charge trapping and redistribution effects that warrant follow-on studies.

Acknowledgments This chapter is based upon material that was previously published in [16, 24, 28, 56]. We would like to thank S. Dhar and S. H. Ryu for providing the devices for this study and C. Scozzie, B. Geil, and A. J. Lelis at the US Army Research Laboratory, Adelphi, MD, for supporting the fabrication of the SiC MOS capacitors at Cree, Inc. This work was supported in part by the Air Force Office of Scientific Research through a MURI program, by the Defense Threat Reduction Agency's Basic and Applied Sciences Program, by National Science Foundation grant DMR-09-07385, and the McMinn Endowment at Vanderbilt University.

References

1. A. Agarwal and S.H. Ryu, *Proc.CS MANTECH Conference*, Vancouver, British Columbia, Canada (2006), pp. 215–218.
2. M. Bhatnagar and B. J. Baliga, *IEEE Trans. Electron Dev.* **40**, 645 (1993).
3. A. A. Orouji and H. Elahipanah, *IEEE Trans. Dev. Mater. Reliab.* **10**, 92 (2010).
4. T. Okayama, S. D. Arthur, J. L. Garrett, and M. V. Rao, *Solid-State Electron.* **52**, 164 (2008).
5. J. A. Cooper, Jr., M. R. Melloch, R. Singh, A. Agarwal, and J. W. Palmour, *IEEE Trans. Electron Dev.* **49**, 658 (2002).

6. A. V. Kuchuk, M. Guziewicz, R. Ratajczak, M. Wzorek, V.P. Kladko, and A. Piotrowska, *Microelectron. Engineering* **85**, 2142 (2008).
7. R. Arora, J. Rozen, D. M. Fleetwood, K. F. Galloway, C. X. Zhang, J. Han, S. Dimitrijev, F. Kong, L. C. Feldman, S. T. Pantelides, and R. D. Schrimpf, *IEEE Trans. Nucl. Sci.* **56**, 3185 (2009).
8. K.Y. Cheong, J. H. Moon, H. J. Kim, W. Bahng, and N. K. Kim, *Thin Solid Films* **518**, 3255 (2010) .
9. K. Kawahara, M. Krieger, J. Suda, and T. Kimoto, *J. Appl. Phys.* **108**, 023706 (2010)
10. M. Gurfinkel, H. D. Xiong, K. P. Cheung, J. S. Suehle, J. B. Bernstein, Y. Shapira, A. J. Lelis, D. Habersat, and N. Goldsman, *IEEE Trans. Electron Dev.* **55**, 2004 (2008).
11. A. J. Lelis, D. Habersat, G. Lopez, J. M. McGarrity, F. B. McLean, and N. Goldsman, *Mater. Sci. Forum* **527–529**, 1317 (2006).
12. L. Tsetseris, X. J. Zhou, D. M. Fleetwood, R. D. Schrimpf, L. Tsetseris, and S. T. Pantelides, *Appl. Phys. Lett.* **86**, 142103 (2005).
13. T. Grasser, W. Gos, and B. Kaczer, *IEEE Trans. Dev. Mater. Reliab.* **8**, 79 (2008).
14. T. Grasser, B. Kaczer, W. Goes, H. Reisinger, T. Aichinger, P. Hehenberger, P. J. Wagner, F. Schanovsky, J. Franco, M. T. Luque, and M. Nelhiebel, *IEEE Trans. Electron Dev.* **58**, 3652 (2011).
15. J. Rozen, S. Dhar, S. T. Pantelides, L. C. Feldman, S. Wang, J. R. Williams, and V. V. Afanas'ev, *Appl. Phys. Lett.* **91**, 153503 (2007).
16. E. X. Zhang, C. X. Zhang, D. M. Fleetwood, R. D. Schrimpf, S. Dhar, S. H. Ryu, X. Shen, and S. T. Pantelides, *IEEE Trans. Dev. Mater. Reliab.* **12**, 391 (2012).
17. S. T. Pantelides, S. Wang, A. Franceschetti, R. Buczko, M. Di Ventra, S. N. Rashkeev, L. Tsetseris, M. H. Evans, I. G. Batyrev, L. C. Feldman, S. Dhar, K. McDonald, R. A. Weller, R. D. Schrimpf, D. M. Fleetwood, X. J. Zhou, J. R. Williams, C. C. Tin, G. Y. Chung, T. Isaacs-Smith, S. R. Wang, S. J. Pennycook, G. Duscher, K. Van Benthem, and L. M. Porter, *Mater. Sci. Forum* **527**, 935 (2006).
18. S. Wang, S. Dhar, S.-R. Wang, A. C. Ahyi, A. Franceschetti, J. R. Williams, L. C. Feldman, and S. T. Pantelides, *Phys. Rev. Lett.* **98**, 026101 (2007).
19. J. Rozen, S. Dhar, S. K. Dixit, V. V. Afanas'ev, F. O. Roberts, H. L. Dang, S. Wang, S. T. Pantelides, J. R. Williams, and L. C. Feldman, *J. Appl. Phys.* **103**, 124513 (2008).
20. J. Rozen, S. Dhar, M. E. Zvanut, J. R. Williams, and L. C. Feldman, *J. Appl. Phys.* **105**, 124506 (2009).
21. G. Y. Chung, C. C. Tin, J. R. Williams, K. McDonald, M. D. Ventra, S. T. Pantelides, L. C. Feldman, and R. A. Weller, *Appl. Phys. Lett.* **76**, 1713 (2000).
22. H. Li, S. Dimitrijev, H. B. Harrison, and D. Sweatman, *Appl. Phys. Lett.* **70**, 2028 (1997).
23. H.-F. Li, S. Dimitrijev, D. Sweatman, and H. B. Harrison, *Microelectron. Reliab.* **40**, 283 (2000).
24. E. X. Zhang, C. X. Zhang, D. M. Fleetwood, R. D. Schrimpf, S. Dhar, S. H. Ryu, X. Shen, and S. T. Pantelides, *ECS Trans.* **35**, No. 4, 369 (2011).
25. A. K. Agarwal, S.Seshadri, and L. B. Rowland, *IEEE Electron Dev. Lett.* **18**, 592 (1997).
26. A. V. Los and M. S. Mazzola, J. Electron. Mater. **30**, 235 (2001).
27. I. G. Ivanov, A. Magnusson, and E. Janzen, *Phys. Rev. B***67**, 165212 (2003).
28. X. Shen, E. X. Zhang, C. X. Zhang, D. M. Fleetwood, R. D. Schrimpf, S. Dhar, S. H. Ryu, and S. T. Pantelides, *Appl. Phys. Lett.* **98**, 063507 (2011).
29. J. N. Shenoy, G. L. Chindalore, M. R. Melloch, J. A. Cooper, J. W. Palmour, and K. G. Irvine, *J. Electron. Mater.* **24**, 303 (1995).
30. S. Dhar, L. C. Feldman, S. Wang, T. Isaacs-Smith, and J. R. Williams, *J. Appl. Phys.* **98**, 014902 (2005).
31. M. J. Marinella, D. K. Schroder, T. Isaacs-Smith, A. C. Ahyi, J. R. Williams, G. Y. Chung, J. W. Wan, and M. J. Loboda, *Appl. Phys. Lett.* **90**, 253508 (2007).
32. D. M. Fleetwood, S. L. Kosier, R. N. Nowlin, R. D. Schrimpf, R. A. Reber, Jr., M. DeLau, P. S. Winokur, A. Wei, W. E. Combs, and R. L. Pease, *IEEE Trans. Nucl. Sci.* **41**, 1871 (1994).

33. D. M. Fleetwood, W.L. Warren, J. R. Schwank, P. S. Winokur, M. R. Shaneyfelt, and L. C. Riewe, *IEEE Trans. Nucl. Sci.* **42**, 1698 (1995).
34. P. S. Winokur, J. R. Schwank, P. J. McWhorter, P. V. Dressendorfer, and D. C. Turpin, *IEEE Trans. Nucl. Sci.* **31**, 1453 (1984).
35. M. A. Alam and S. Mahapatra, *Microelectron. Reliab.*, **45**, 71, 2005.
36. X. J. Zhou, L. Tsetseris, S. N. Rashkeev, D. M. Fleetwood, R. D. Schrimpf, S. T. Pantelides, J. A. Felix, E. P. Gusev, and C. D'Emic, *Appl. Phys. Lett.* **84**, 4394 (2004).
37. J. H. Stathis and S. Zafar, *Microelectron. Reliab.* **46**, 270 (2006).
38. D. M. Fleetwood and N. S. Saks, *J. Appl. Phys.* **79**, 1583 (1996).
39. I. G. Ivanov, A. Henry, and E. Janzen, *Phys. Rev. B***98**, 241201(R) (2005).
40. P. Neudeck, S. Kang, J. Petit, and M. Tabibazar, J. Appl. Phys. **75**, 7949 (1994).
41. J. N. Shenoy, G. L. Chindalore, M. R. Melloch, J. A. Cooper, J. W. Palmour, and K. G. Irvine, *J. Electron. Mater.* **24**, 303 (1995).
42. A. Chatterjee, K. Matocha, V. Tilak, J. A. Fronheiser, and H. Piao, *Mater. Sci. Forum***645–648**, 478 (2009).
43. Z. Q. Fang, B. Claflin, D. C. Look, L. Polenta, and W. C. Mitchel, *J. Electron. Mater.* **34**, 336 (2005).
44. D. M. Fleetwood, *Microelectron. Reliab.* **42**, 523 (2002).
45. J. P. Campbell, P. M. Lenahan, C. J. Cochrane, A. T. Krishnan, and S. Krishnan, *IEEE Trans. Dev. Mater. Reliab.* **7**, 540 (2007).
46. D. T. Krick, P. M. Lenahan, and J. Kanicki, *J.Appl. Phys.* **64**, 3558 (1988).
47. D. J. DiMaria and J. H. Stathis, *J. Appl. Phys.* **70**, 1500 (1991).
48. S. S. Tan, T. P. Chen, J. M. Soon, K. P. Loh, C. H. Ang, and L. Chan, *Appl. Phys. Lett.* **82**, 1881 (2003).
49. P. M. Lenahan, *Proc. IRPS*, doi:10.1109/IRPS.2010.5488669, p. XT11 (2010).
50. A. J. Lelis, D. Haberstat, R. Green, A. Ogunniyi, M. Gurfinkel, J. Suehle, and N. Goldsman, *IEEE Trans. Electron Dev.* **55**, 1835 (2008).
51. D. K. Schroder and J. A. Babcock, *J. Appl. Phys.* **94**, 1 (2003).
52. J. P. Campbell, P. M. Lenahan, C. J. Cochrane, A. T. Krishnan, and S. Krishnan, *IEEE Trans. Dev. Mater. Reliab.* **7**, 540 (2007).
53. S. DasGupta, R. Brock, R. Kaplar, M. Marinella, M. Smith, and S. Atcitty, *Appl. Phys. Lett.* **99**, 023503 (2011).
54. E. Arnold, *IEEE Trans. Electron Dev.* **46**, 497 (1999).
55. C.-Y. Lu, J. A. Cooper, T. Tsuji, G. Chung, J. R. Williams, K. McDonald, and L. C. Feldman, *IEEE Trans. Electron Dev.*, **58**, 1582 (2003).
56. C. X. Zhang, E. X. Zhang, D. M. Fleetwood, R. D. Schrimpf, S. Dhar, S.-H. Ryu, X. Shen, and S. T. Pantelides, *IEEE Electron Dev. Lett.*, **34**, 117 (2013).
57. X. J. Zhou, D. M. Fleetwood, J. A. Felix, E. P. Gusev, and C. D'Emic, *IEEE Trans. Nucl. Sci.* **52**, 2231 (2005).
58. X. J. Zhou, D. M. Fleetwood, L. Tsetseris, R. D. Schrimpf, and S. T. Pantelides, IEEE Trans. Nucl. Sci. **53**, 3636 (2006).
59. A. J. Lelis, T. R. Oldham, H. E. Boesch, Jr., and F. B. McLean, *IEEE Trans. Nucl. Sci.* **36**, 1808 (1989).
60. S. T. Pantelides, S. N. Rashkeev, R. Buczko, D. M. Fleetwood, and R. D. Schrimpf, *IEEE Trans. Nucl. Sci.* **47**, 2262 (2000).
61. D. M. Fleetwood, M. J. Johnson, T. L. Meisenheimer, P. S. Winokur, W. L. Warren, and S. C. Witczak, *IEEE Trans. Nucl. Sci.* **44**, 1810 (1997).

Part IV

Chapter 27
On-Chip Silicon Odometers for Circuit Aging Characterization

John Keane, Xiaofei Wang, Pulkit Jain, and Chris H. Kim

27.1 Introduction

The parametric shifts or circuit failures caused by Bias Temperature Instability (BTI) and other aging mechanisms in CMOS transistors have become more severe with shrinking device sizes and voltage margins. These mechanisms must be studied in order to develop accurate reliability models which are used to design robust circuits. Another option for addressing aging effects is to use on-chip reliability monitors that can trigger real-time adjustments to compensate for reduced performance or device failures. The need for efficient technology characterization and aging compensation is exacerbated by the rapid introduction of process changes, such as high-k/metal gate stacks and new transistor architectures.

Much of the device aging data gathered for process characterization is obtained through individual probing experiments. However, several on-chip aging measurement systems have been implemented recently to circumvent the shortcomings of expensive probing stations and to compliment the characterization efforts undertaken with that type of test hardware. Performing reliability experiments with on-chip circuits provides several advantages in addition to enabling the use of simpler test equipment.

J. Keane (✉) • P. Jain
Department of Electrical and Computer Engineering, University of Minnesota, 200 Union Street SE, Minneapolis, MN 55455, USA

Advanced Design, Logic Technology Development, Intel Corporation, RA3-256, 2501 NW 229th Avenue, Hillsboro, OR 97124, USA
e-mail: john.keane@intel.com; pulkit.jain@intel.com

X. Wang • C.H. Kim
Department of Electrical and Computer Engineering, University of Minnesota, 200 Union Street SE, Minneapolis, MN 55455, USA
e-mail: xfwang@umn.edu; chriskim@umn.edu

T. Grasser (ed.), *Bias Temperature Instability for Devices and Circuits*,
DOI 10.1007/978-1-4614-7909-3_27,

First, using on-chip logic to control the measurements enables better timing resolution. This is critical when interrupting stress to record BTI measurements, as BTI degradation is known to partially recover within microseconds or less. Next, fast and sensitive embedded circuits facilitate high measurement resolution. A digital beat frequency detection system presented in this chapter enables the measurement of ring oscillator (ROSC) frequency shifts with resolution ranging down to a theoretical limit of less than 0.01 %. In addition, standard digital logic can be used to control experiments on many devices in parallel, resulting in a large test time speedup when monitoring statistical processes.

This chapter provides an overview of several on-chip systems used to measure transistor or circuit aging. These systems are often referred to as "Silicon Odometers," because they measure the degradation of circuits much like a car's odometer tells one how many miles it has covered as an indication of wear and tear. Utilizing the benefits of Silicon Odometers to obtain accurate CMOS aging information will allow manufacturers to avoid wasteful overdesign and frequency guard banding based on pessimistic degradation projections and hence more fully realize the benefits of CMOS scaling.

27.2 Summary of Advanced Reliability Monitoring Techniques

This section will review a sampling of the large number of device and circuit aging measurement techniques invented over the past decade—using both on- and off-chip infrastructure—as design margins have shrunk and interest in transistor degradation mechanisms has grown.

Denais presented an "on-the-fly" BTI measurement technique in order to avoid the recovery inherent in previously disclosed characterization setups when stress conditions are removed for measurements [1, 2]. In this method the gate stress or recovery voltage on a single transistor under test is kept quasi-constant, and the linear drain current ($I_{D,lin}$) is periodically measured with off-chip equipment to monitor degradation or recovery. However, the on-the-fly method relies on a translation of $\Delta I_{D,lin}$ into ΔV_{th} (i.e., the threshold voltage shift) which requires approximations, and other researchers state that this method underestimates the total degradation due to a slow initial "prestress" measurement at the stress voltage which causes unrecorded degradation [3]. Additionally, the time required for each measurement is typically in the range of milliseconds, and it is difficult to get an accurate reading of $\Delta I_{D,lin}$ at the stress voltage level, all of which could make on-the-fly results less reliable [4].

Next, Shen used a 100 ns I–V sweep technique to monitor NBTI degradation and demonstrated the fundamental differences in NBTI that are observed with ultrahigh-speed measurements which minimize unwanted recovery [3]. This technique may experience drawbacks associated with high-speed off-chip device probing though, such as losses and cross talk.

Fischer presented an array-based BTI characterization system for SRAM devices in a configuration that matches a product-like SRAM array layout as closely as possible while facilitating individual device current measurements [5]. This design relies on a long stress time of 10,000 s and a relatively long recovery time of 500 s before recording drain current measurements, after which time no further recovery in device threshold shift was observed. It does not attempt to record fast measurements before unwanted BTI recovery takes place when stress conditions are removed.

Karl designed two separate compact circuits for monitoring NBTI and TDDB (time-dependent dielectric breakdown), with the goal of facilitating real-time characterization on products in the field [6]. First he measured the frequency shift of a ROSC with a PMOS header that is placed under NBTI stress and then biased in subthreshold during measurements for high V_{th} sensitivity. This work relies on a mathematical model to map temperature and V_{th} variations to the measured ROSC frequencies after calibration. The TDDB aging results were provided in the form of a frequency shift of a Schmitt trigger oscillator which is modified by the increasing gate leakage through a pair of stressed PMOS transistors.

Singh introduced another in situ monitor for providing an early indication of the onset of TDDB [7]. This design takes advantage of the prevalence of PMOS headers used for power gating in certain applications to provide a low-overhead reliability monitor. The circuits gated by the PMOS headers are periodically taken offline, and the gate bias of a weak PMOS added in parallel to the supply switch is swept while recording the virtual supply rail voltage between the header and the circuit under test with an on-chip ADC. The resulting V_{bias} versus V_{rail} characteristic is strongly nonlinear in the fresh circuit, but becomes more linear as breakdown paths are formed through gate stacks in the power-gated circuits under test. The authors state that the differences in this curve can be used as highly sensitive indications of early TDDB degradation. Next, Hofmann described a product-level BTI and HCI (hot carrier injection) aging monitoring system where measurement times of 80 μs were achieved with 0.2% resolution, although no details were given about how frequencies were recorded in this single-ended system [8]. This system stressed critical paths replicating those in real products by raising the supply voltage and controlling the number of switching events in an effort to bridge the gap between device-level aging data and the degradation of meaningful circuits.

Saneyoshi presented a fast method to detect NBTI degradation in delay lines by monitoring the number of stages an input edge could travel through before and after stress using edge capture logic so that the timing resolution is set by one stage delay rather than a longer ROSC period [9]. A similar US patent was filed in 2008 [10]. Chen used another time to digital converter to sample degrading circuit path delays which allows one to detect the asymmetric aging of rising versus falling edges that can be induced by clock or power gating [11]. The delay averaging of rising and falling edges inherent in ROSC measurements is avoided here by monitoring the delay of a single edge through the open-loop circuit under test.

Da Silva implemented an array of devices that facilitates parallel BTI stressing and uses on-chip control to ensure that all devices experience the same stress and relaxation time [12]. Large experimental sample sets are now required for BTI characterization as variations in the number of defects lead to a larger range of threshold voltage shifts in aggressively scaled devices and leveraging on-chip circuits to facilitate that can lead to faster and more simple testing.

Finally, several recent publications have focused on methods to both measure circuit aging and take corrective actions to maintain proper functionality. For example, Dadgour demonstrated a time-borrowing scheme in which a more severely aged pipe stage can borrow time from a less degraded neighbor [13].

The remaining sections will cover several other on-chip Silicon Odometer designs that have been implemented with the goal of achieving high timing and measurement resolution, reducing test time, and considering different aspects of circuit aging with simple digital test interfaces.

27.3 The Original Silicon Odometer Based on Beat Frequency Detection

The Silicon Odometer beat frequency detection system measures frequency changes in stressed ROSCs with the concept presented in Fig. 27.1 [14]. This illustration of the beat frequency between pairs of signals uses low speeds for clarity. The faster signal catches up to the slower one, they overlap, and then the faster one pulls ahead. This repeats, and the time between the overlapping points is the period of the beat frequency. When the 63 Hz signal is superimposed on the 64 Hz signal, the beat period is (64 – 63 Hz) = 1 Hz (note the one second horizontal axis). With 62 and 64 Hz, the period is (64 – 62 Hz) = 2 Hz. Therefore by measuring the beat frequency, one can ascertain the difference between two faster frequencies.

The beat frequency is measured using the circuit shown in Fig. 27.2 [15, 16]. During the short measurement periods, a phase comparator uses a fresh reference ROSC to sample the output of an identical stressed ROSC. The output signal of this phase comparator exhibits the beat frequency: $f_{PC} = f_{ref} - f_{stress}$. A counter is used to measure the beat frequency by counting the number of reference ROSC periods during one period of the phase comparator output signal. This count is recorded after each stress period to calculate the reduction in the stressed ROSC frequency.

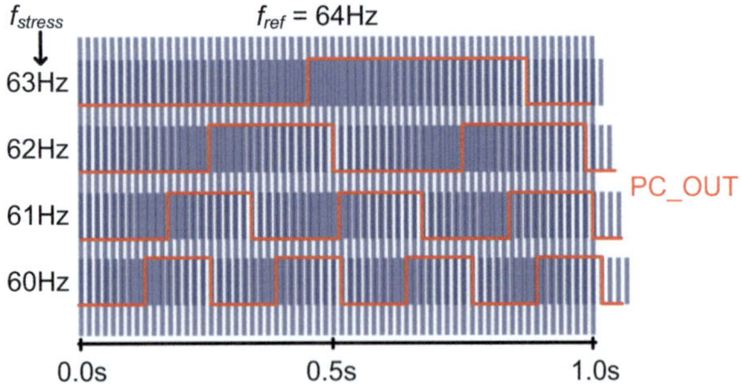

Fig. 27.1 The beat phenomenon between two ROSCs switching at different speeds is illustrated here with low-frequency signals. PC_OUT is the output of the phase comparator in Fig. 27.2 (*Figure credit: David Schneider of IEEE Spectrum*)

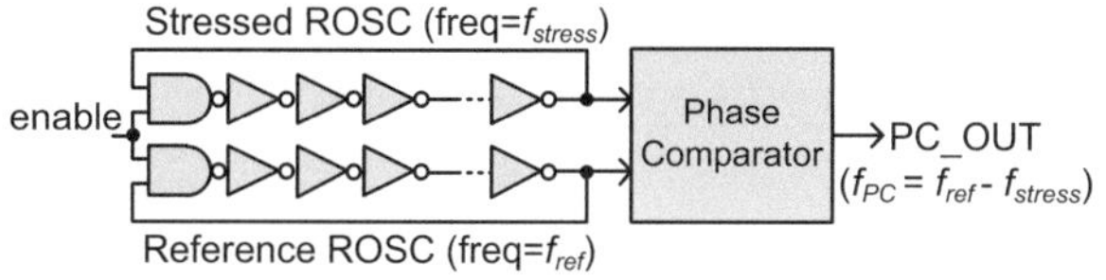

Fig. 27.2 Basic setup for beat frequency detection between a stressed and an unstressed ROSC. This system achieves picosecond-order frequency shift resolution for the stressed ROSC, as well as sub-μs measurement times

If the initial frequency of the reference ROSC is called f_{ref}, that of the fresh ROSC to be stressed is f_{stress}, and the initial Odometer output count is N_1, then assuming f_{ref} is higher than f_{stress}, the following relationship holds:

$$\left(\frac{1}{f_{ref}}\right) \bullet N_1 = \left(\frac{1}{f_{stress}}\right) \bullet (N_1 - 1) \tag{27.1}$$

The $(N_1 - 1)$ term arises from the fact that the stressed ROSC with the lower frequency, f_{stress}, will take one less period to cycle back to the same point in the reference ROSC period while both are oscillating. After a stress period ends, f_{ref} will remain unchanged because the reference ROSC is powered down during stress, but f_{stress} will decrease due to aging and the new value is named f'_{stress}. The resulting equation with the new output count N_2 is

$$\left(\frac{1}{f_{ref}}\right) \bullet N_2 = \left(\frac{1}{f'_{stress}}\right) \bullet (N_2 - 1) \tag{27.2}$$

These two equations are combined to calculate the frequency shift during stress as follows:

$$\frac{f'_{stress}}{f_{stress}} - 1 = \frac{N_1 \bullet (N_2 - 1)}{N_2 \bullet (N_1 - 1)} - 1 = \frac{(N_2 - N_1)}{N_2 \bullet (N_1 - 1)} \tag{27.3}$$

Those simple calculations show that if f_{ref} is only slightly higher than f_{stress}, the output count is high. For example, the count is 100 for a 1% difference, and this slight difference is ensured with trimming capacitors. The subsequent small decreases in f_{stress} due to aging cause a large change in this count. For instance, a 2% difference between the ROSC frequencies gives a count of 50, so a 1% shift to that point is translated into a decreased count of 50. Therefore, with high frequency ROSCs, the beat frequency detection system achieves picosecond-order frequency shift measurement resolution.

The Odometer output count relationship with the difference between the reference ROSC (REF_ROSC) and stressed ROSC (STR_ROSC) frequencies is illustrated in Fig. 27.3a. This figure shows that the Odometer operates correctly with a reference ROSC frequency that is either slower or faster than the stressed ROSC.

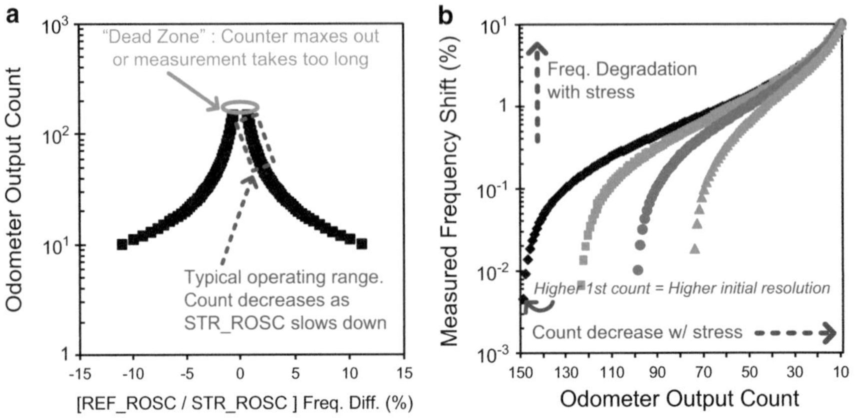

Fig. 27.3 (**a**) Silicon Odometer output count versus the frequency difference between the reference and stressed ROSCs. (**b**) Output count versus frequency shift during a stress experiment. *Curves* are shown for varied initial counts, where a higher count corresponds to a smaller frequency difference between the two ROSCs

In the former case, the output count will increase with stress, while it decreases in the latter. A slower reference frequency is accounted for in Eqs. (27.1) through (27.3) by changing the $(N_{\#} - 1)$ terms to $(N_{\#} + 1)$, because the faster stressed ROSC in this case goes through *one more* period than the slow reference during the beat frequency measurement, rather than *one less*. Additionally, it is possible for the reference frequency to transition from being slower to faster than the stressed ROSC, but this involves moving through a "dead zone" where either the output count will equal the counter max value or, if the counter is large enough, the measurement time will become excessively long as the difference between the two ROSC frequencies becomes extremely small.

Experiments are generally started with a reference frequency that is slightly faster than the stressed ROSC frequency in order to obtain a monotonic decrease in the output counts with stress. This maximizes the frequency measurement resolution in the early phases of stress and avoids the dead zone. Figure 27.3b shows measurement result characteristics with monotonic count decreases and four different initial counts. Note again that a smaller difference between the two ROSC periods leads to a higher initial count and therefore a higher initial frequency resolution, while slightly lengthening the measurement time. Maximum starting counts of $\sim$125 can often be achieved in measurements, which corresponds to initial frequency shift measurements theoretically ranging down to 0.0065%, although this will be limited by practical issues such as noise in the measurement lab setup. The resolution decreases with time, but we are primarily concerned with the small initial degradation steps that can be obtained with stress that is closer to real operating conditions.

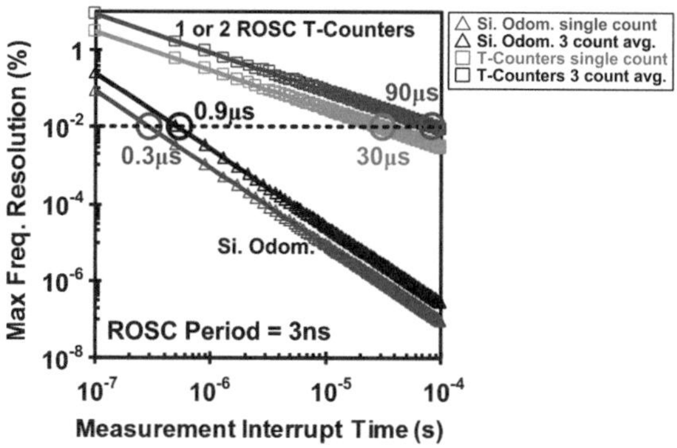

Fig. 27.4 Maximum frequency measurement resolution versus the total stress interruption time for measurements

The plot in Fig. 27.4 shows the theoretical maximum frequency measurement resolution for three measurement setups during a fixed time. In the "1 ROSC T-Counter" system (where T stands for period), a single ROSC's degradation is recorded with a single period counter during an externally controlled measurement time. The "2 ROSC T-Counter" measures the degradation in one stressed ROSC by counting the number of periods it cycles through while a set number of periods in a fresh reference ROSC are counted (see Fig. 27.6). Since the resolution of these period counters is simply the measurement time divided by the ROSC period, while that of the Odometer can be derived from Eq. (27.3), the latter reaches a maximum resolution of 0.01% within only 0.3 μs in the ideal case with a single measurement recorded, while the former systems require 100× more time. A large improvement is still seen when three counts are recorded during each Odometer measurement period for averaging or to eliminate unpredictable initial counts (see Fig. 27.7). The longer measurement times in the standard period counter systems would result in unwanted BTI recovery.

In addition to the high frequency resolution, the Odometer benefits from a high immunity to voltage or temperature variations due to its differential nature. Given that the reference and stressed ROSCs are identical structures placed near each other, both will see essentially identical temporal variations, so their frequencies should be affected by roughly the same amount. The simulation results shown in Fig. 27.5 illustrate this noise immunity and compare the Odometer behavior with that of the ROSC T-Counter setups using an industrial 65 nm process. In these simulations the stressed ROSC started out 0.64% slower than the reference when measured at nominal VCC (1.2 V) and temperature (25°C), and the former structure is slowed by 0.38% due to aging (in the 1 ROSC T-Counter, we only consider the 0.38% shift since there is no reference ROSC). However, if the post-stress measurement takes place under a different temperature or voltage condition, it will lead to some deviation from 0.38% in the recorded value. Figure 27.5 shows a clear benefit for the differential Odometer system. Also note that since the measurement time is limited to the ideal required by the Odometer system here, the T-Counters suffer from low-frequency resolution, which results in further rounding errors.

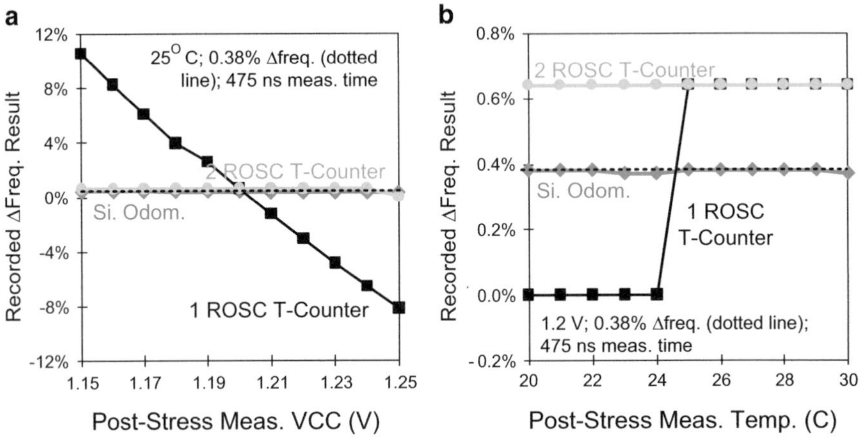

Fig. 27.5 Simulated effects of (**a**) voltage and (**b**) temperature variations on the Silicon Odometer and both 1 and 2 ROSC period counter (T-Counter) systems after the initial fresh measurement prior to stress was taken at 1.2 V, 25°C

System	1 ROSC T-Counter	2 ROSC T-Counter	Silicon Odometer
Block Diagram	Stress ROSC → Counter = Variable N2	T1 = constant: Ref. ROSC → Counter = Constant N1; T2 = degrades: Stress ROSC → Counter = Variable N2	T1 = constant: Ref. ROSC; T2 = degrades: Stress ROSC → Phase Comp. → PC_OUT ($f_{PC} = f_{ref} - f_{stress}$)
Function	Count Stress ROSC periods during externally controlled meas. time	Count Stress ROSC periods during N1 periods of Ref. ROSC	Count Ref. ROSC periods during one period of PC_OUT
Features	Simple; compact	Simple; immune to common mode variations	High resolution w/ short meas. time; immune to common mode variations
Issues	Voltage and temp. varations; meas. time vs. resolution tradeoff; requires absolute timing reference (e.g. oscilloscope)	Meas. time vs. resolution tradeoff	Requires extra circuits (e.g., Phase Comp., edge detector, etc...)
Meas. time for 1% max resolution *	30 μs	30 μs	0.3 μs
Meas. error wrt. common mode variations **	+10.18% / -8.57%	+0.26% / -0.38%	+0.06% / -0.07%

*ROSC period = 3 ns ** simulated with +/- 0.4% ΔVCC; 0.38% stress shift; 475 ns measurement time; error = (measured %) – (0.38%)

Fig. 27.6 Comparison of simple ROSC period counting systems with the Silicon Odometer

Figure 27.6 compares the three frequency measurement systems that have been discussed. While the Odometer requires additional circuits for the beat frequency detection, it achieves a significantly higher frequency measurement resolution in a shorter measurement time and is immune to environmental variations that are common to the reference and stressed ROSCs.

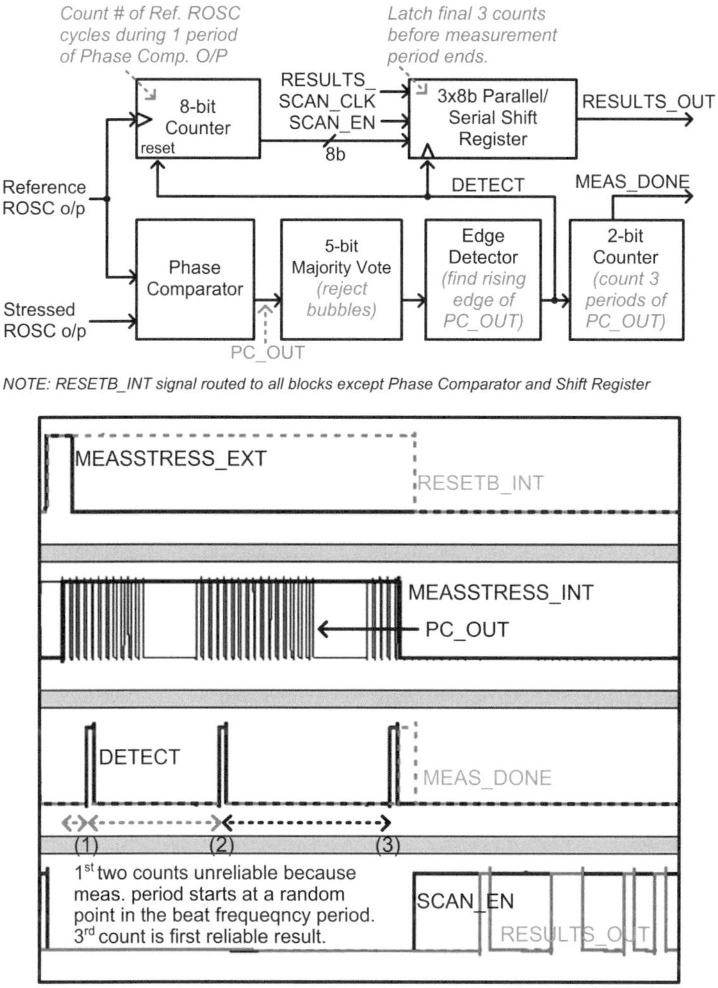

Fig. 27.7 Block diagram of the beat frequency detection circuit and simulation results illustrating the operation of this system

A block diagram of the full beat frequency detection system is presented in Fig. 27.7. It contains logic which automatically sends the ROSC under test back into stress after three results are recorded in order to achieve measurement times of $\leq 1\ \mu s$. The completion of a measurement period is flagged by the MEAS_DONE signal when the three rising edges from the phase comparator are counted, meaning three 8 bit count results have been recorded. In this automated scheme, the first two counts are generally smaller than the true result due to the unpredictable starting location of the measurement at some midpoint in the phase comparator period, so they are discarded. The validity of the third count is verified during calibration by using an externally controlled longer measurement period in which the initial smaller counts are overwritten by subsequent results. In this case, all counts should be roughly identical and equal to the third result recorded during the shorter automated measurements. Moving on, the MEAS_DONE flag is sent to a Finite State Machine (FSM), which restarts stress. Using on-chip logic to control this timing allows us to avoid generating very short, accurate measurement pulses externally.

The majority voting circuit rejects a lone "1" signal in a series of "0"s, or vice versa. These "bubbles" could be caused by noise impacting the phase comparator, for example. The edge detector is used to find the beginning of each period of the phase comparator. Its output (DETECT) is used to sample the counter output and then to reset the counter for a new period.

Figure 27.7 also contains simulation waveforms illustrating the operation of this system. After the MEASSTRESS_EXT signal is asserted by the tester, its internal counterpart is driven high by the FSM, which connects switching signals from the ROSCs to the phase comparator, and starts the measurement. After three high PC_OUT periods, the MEAS_DONE signal goes high. This causes the FSM to end the measurement period and switch the parallel/serial shift registers to scan mode. An external clock is then used to scan out the results. The registers will be put back into parallel mode when MEASSTRESS_EXT is next asserted.

27.4 All-in-One Odometer for Separating BTI and HCI Effects

HCI and TDDB are two critical reliability mechanisms impacting CMOS devices along with BTI. Although some of the underlying physical explanations for these mechanisms are similar, each of them has unique sensitivities to operating conditions and process changes and can be more critical in certain circuit topologies.

In this work, we endeavor to monitor all three of these aging mechanisms with a pair of ROSCs which are representative of standard circuits [16]. A "backdrive" concept is employed in which one ROSC drives the transitions in both structures during stress, such that the driving oscillator ages due to both BTI and HCI, while the other only suffers from BTI. The latter ROSC is gated off from the supplies during stress so that no current is driven through the channels of its transistors, and therefore, the charge carriers cannot become "hot." In addition, long-term or high-voltage experiments facilitate TDDB measurements. Since BTI degradation has been observed to start recovering within microseconds, the beat frequency detection method is used to enable sub-μs measurements and minimize unwanted device recovery during stress interruptions. Frequency measurement resolution on the order of picoseconds is achieved for finely tuned HCI and BTI readings, and experiments are automated through a straightforward digital interface. This design facilitates the examination of frequency, temperature, and voltage dependencies of the stress mechanisms. In addition, it can be used to monitor both sustained stress and recovery characteristics and to observe the effects of increased load capacitance on the frequency shift induced by aging.

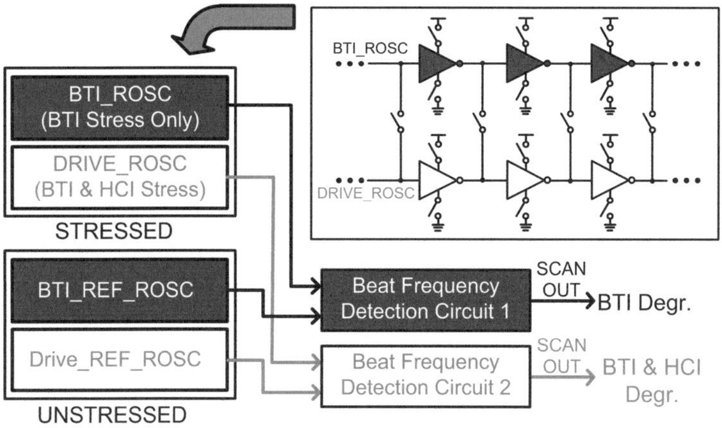

Fig. 27.8 High-level diagram of the All-in-One Silicon Odometer

A block diagram of the reliability monitor for separating the effects of HCI and BTI is shown in Fig. 27.8. This circuit contains four ROSCs in total: two stressed and two unstressed to maintain fresh reference points. Each of the stressed oscillators is paired with its identical, fresh reference during measurements, and its frequency degradation is monitored with the Silicon Odometer beat frequency detection circuit.

27.4.1 Illustration of the Backdrive Concept

Figure 27.9 presents the pair of stressed ROSCs in both stress and measurement modes. During stress, the BTI_ROSC stages are gated off from the power supplies, while the DRIVE_ROSC maintains a standard inverter configuration with the supply set at VSTRESS. Both ROSC loops are opened, and the input of the DRIVE_ROSC is driven by a stress clock generated by an on-chip voltage controlled oscillator (VCO) whose output is level-shifted up to VSTRESS. The switches between these two ROSCs are closed so the DRIVE_ROSC can drive the internal node transitions for both structures.

Simulated voltage and current waveforms are shown in Fig. 27.9c. The internal nodes of the BTI_ROSC switch between the supply level (VSTRESS) and 0 V, as would be the case in standard operation. However, the peak drain current through the "on" devices in this structure is only 3–5% of that in the DRIVE_ROSC, since their sources are gated off from the supplies. Note that the sources of these "on" devices in the stressed BTI_ROSC are held at their respective supply levels due to the backdriving action of the DRIVE_ROSC. Therefore, the BTI_ROSC will age due only to BTI stress, while the DRIVE_ROSC suffers both BTI and HCI. One can extract the contribution of HCI to the latter ROSC's frequency degradation with

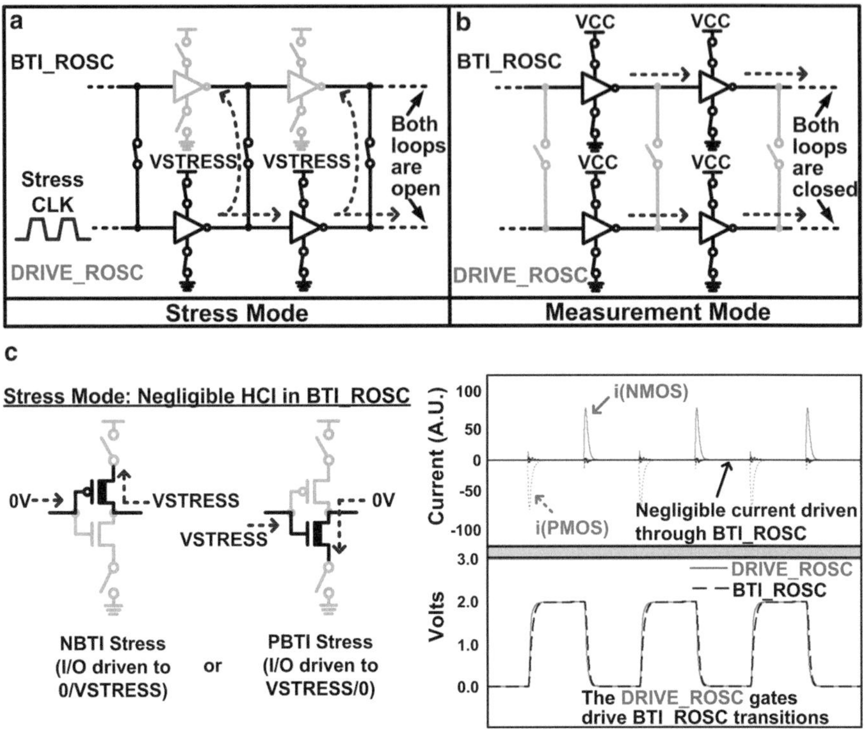

Fig. 27.9 ROSC configuration during (**a**) stress and (**b**) measurement modes. (**c**) The BTI_ROSC transistors suffer the same amount of BTI as the DRIVE_ROSC transistors during stress, but with negligible HCI degradation, since very little current is driven through the channels of the devices under test the former structure

the equation $HCI_{DEG} = DRIVE_{DEG} - BTI_{DEG}$, where DEG stands for degradation. During measurement periods, both ROSCs are connected to the digital logic power supply (VCC) and the switches between them are opened, so they each operate independently in a standard closed-loop configuration.

27.4.2 ROSC Design Details for Backdrive

A detailed schematic of one stage of the paired ROSCs is shown in Fig. 27.10. The thick oxide I/O devices should not age appreciably during stress experiments aimed at the thin oxide core transistors. All core devices are either stressed devices under test (DUTs) or have no voltage drops across any pair of terminals during stress, so they will not age. The header and footer transistors in each inverter pin the source nodes of those gates to the supply levels when closed. The M/S signal here is used to start and end measurement periods. This signal is timed and driven

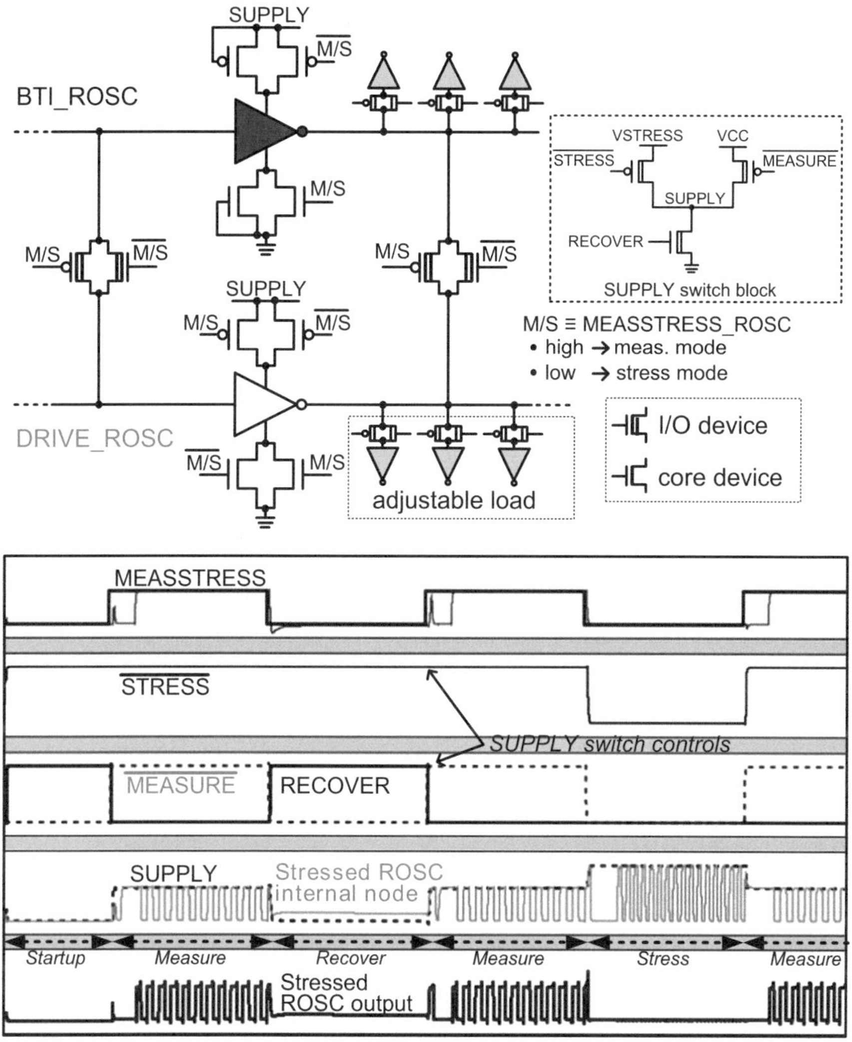

Fig. 27.10 Schematic of one stage of the paired ROSCs and simulation waveforms from a stressed ROSC during measurement, stress, and recovery periods. Note that any initial lone pulses seen at the stressed ROSC output are rejected by the beat frequency detection logic

by the on-chip FSM after the external MEASSTRESS_EXT signal is asserted. Both ROSCs contain three levels of adjustable fanout which facilitate an examination of the effects of additional load capacitance on aging.

Figure 27.10 also contains waveforms from a stressed ROSC during measurement, stress, and recovery periods. After MEASSTRESS_INT is driven high by the FSM, there is a short delay before MEASSTRESS_ROSC goes high, which then causes the tapped output from the stressed ROSC to be connected to the input of the

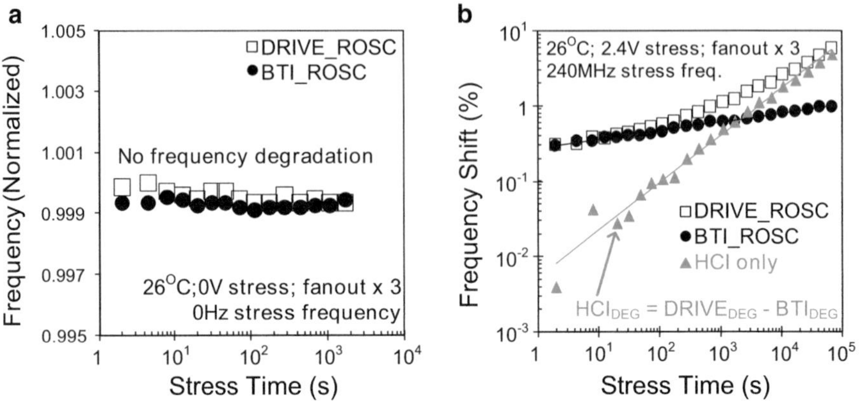

Fig. 27.11 (**a**) Results from a no-stress experiment. (**b**) Example measurements with AC stress conditions

Odometer measurement system. This delay allows the SUPPLY node to settle after being switched to the standard operating supply of VCC and having the ROSC loop closed. An external control signal is set high any time recovery mode is desired between measurements, but will not take effect until the end of the subsequent measurement period.

27.4.3 All-in-One Odometer Test Chip Measurements

A test circuit was implemented in a 65 nm bulk CMOS process for concept verification. Measurements were automated with LabVIEW™ software through a National Instruments data acquisition board. Trimming capacitors were used in each 33 stage ROSC to ensure that the frequencies of the stressed structures began slightly slower than the reference frequencies (see Sect. 27.3). Trimming was also utilized to push apart the oscillating frequencies of the two sets of paired ROSCs to prevent injection locking. The DUTs were 1.5 μm/60 nm NMOS and 3 μm/60 nm PMOS transistors in the inverter stages of the stressed ROSCs. All automated measurement times were under 1 μs, but varied according to the exact beat frequency count results.

The results of a 0 V no-stress experiment were first checked, meaning the SUPPLY node of both normally stressed ROSCs was dropped to 0 V between measurement periods by keeping the RECOVER_EXT signal high, so no aging should have taken place. The plots in Fig. 27.11a confirm this outcome, indicating that frequency shifts shown in later results are not due to aging elsewhere in this system or other circuit effects.

Figure 27.11b presents example measurement results for both ROSCs under 2.4 V stress, as well as the calculated degradation due to HCI (HCI_{DEG}). Both BTI

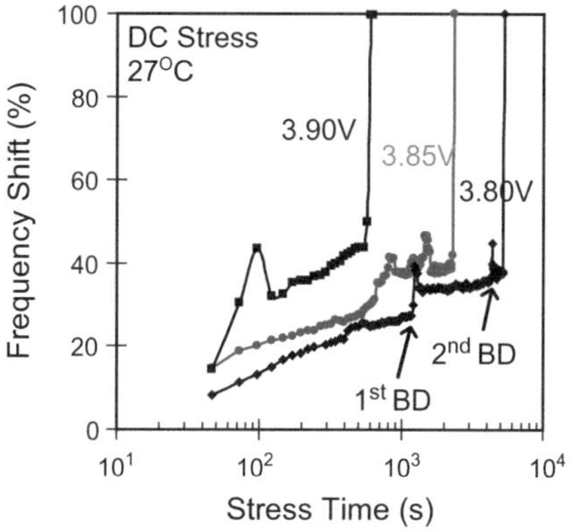

Fig. 27.12 ROSC frequency jumps attributed to TDDB before final circuit failure

and HCI degradation follow a power law behavior, although the latter is seen to saturate at long stress times. The power law exponent for BTI in this case was 0.12 while that of HCI was larger as expected [17], measuring in at 0.63 in the range fitted on this plot. Since device drive current, and therefore ROSC frequency, is linearly proportional to V_{th}, these results can be compared with trends found in studies of V_{th} degradation, and frequency degradation results can be used to deduce the V_{th} shift of devices in the ROSCs with modeling or simulation [18, 19].

Figure 27.12 presents three examples of high-voltage stress experiment results in which sudden jumps in ROSC frequency are interpreted as gate breakdown events. In HCI and BTI experiments, we typically ignore TDDB because it acts on a much longer timescale at lower stress voltages. In these experiments involving large frequency shifts, the beat frequency detection framework was not used since it is aimed at high-resolution measurements for smaller shifts. Instead, the frequency was directly read off-chip with an oscilloscope. Note that longer-term experiments, or those done in future technology generations where soft breakdowns are more prevalent, will be able to make use of the Odometer system. Figure 27.12 indicates that ROSCs continue to function after one or more breakdowns, which only lead to reduced output swing and lower frequencies, as long as subsequent logic stages in the ROSC can restore full-rail swing [20].

27.5 Statistical Odometer for Collecting BTI Variability Data

Variations in the number and characteristics of charges or traps contributing to transistor degradation lead to a distribution of device "ages" at any given time. This section presents an overview of a measurement system that facilitates efficient statistical aging measurements involving BTI and HCI in an array of ring oscillators. Microsecond measurements for minimal BTI recovery, as well as frequency shift measurement resolution ranging down to the error floor of 0.07%, are achieved with three beat frequency detection systems working in tandem.

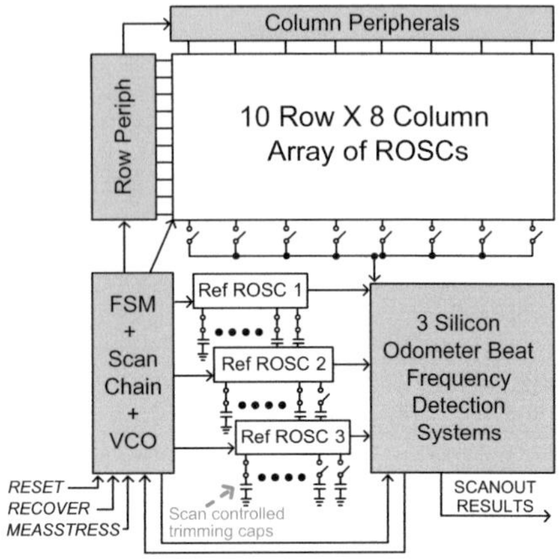

Fig. 27.13 Top level system diagram of the Statistical Odometer

This Odometer consists of a 10×8 array of cells containing ROSCs to be stressed, an FSM, a scan chain, and three Silicon Odometers with their reference ROSCs (Fig. 27.13) [21]. This implementation was limited to 80 ROSC cells due to the available silicon area, but that number could be increased on future test chips. During tests the whole array of ROSCs, or any one rectangular group of them, are stressed in parallel and selected one by one for measurements. Alternatively, all of the ROSCs under test can be put into a recovery state (i.e., 0 V supply) along with any cells that are not selected. During stress the ROSC loops are opened so that their frequencies can be controlled by an on-chip VCO. When each oscillator is selected for a measurement with the *MEASSTRESS* signal from the controlling software, its supply is set to the standard digital level of 1.2 V, the loop is closed, and its frequency shift is measured by the three Odometer systems. Reference ROSCs are put into a 0 V no-stress state in between measurements, and their supply is set at 1.2 V during those brief events so they should not measurably age. Even if short periods under the nominal supply are sufficient to cause any aging, this effect is cancelled out by the differential measurement setup.

27.5.1 Statistical Odometer Ring Oscillator Cell Design

Each ROSC cell contains its own supply switch that sets the local virtual supply (CSUPPLY) at the stress level (VSTRESS), 1.2 V (VCC), or 0 V as appropriate (Fig. 27.14). The ROSC cells also include selection logic to switch a cell into measurement mode if its *row[n]* and *col[m]* signals are both high by closing the loop and then connecting one tapped output node to the bitline after the virtual supply has had time to settle to VCC. This timing is indicated in Fig. 27.14 by the order of the *meas* ("measure") and *stress* signals. The former go high during measurement periods, and the latter are driven high during stress or recovery.

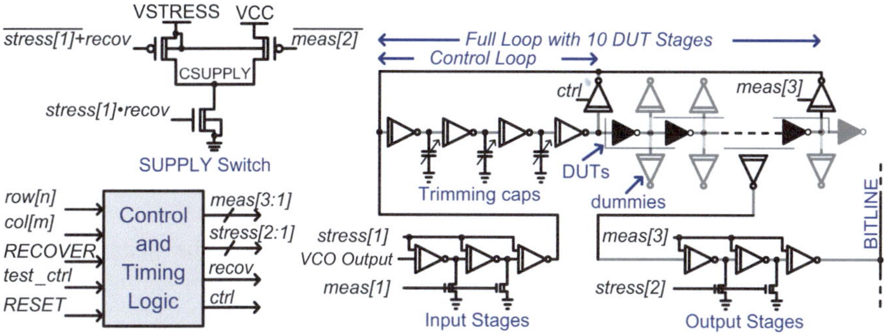

Fig. 27.14 ROSC cell design. The thin oxide logic stages under test are colored *black*, and all other transistors are thick oxide I/O devices (indicated by *double lines*)

Fig. 27.15 Waveforms illustrating the basic operation of a ROSC cell. Only two cells are included in this simulation to demonstrate the functionality as cell[0] goes into and out of stress periods

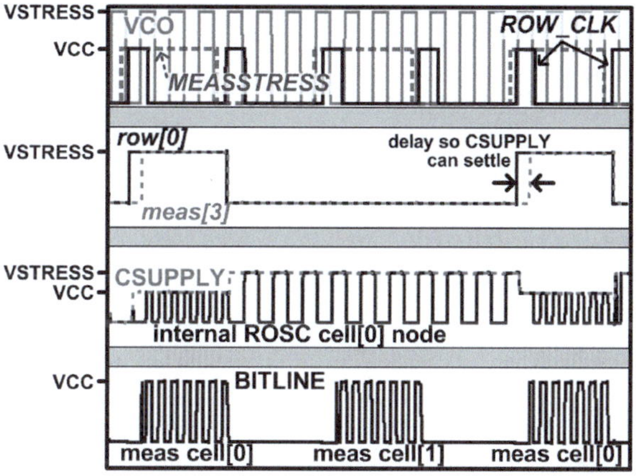

When a measurement is started, *meas[1]/stress[1]* first turn off the input stages and the VSTRESS switch. Next *meas[2]* turns the VCC switch on, and finally *meas[3]/stress[2]* close the loop and connect it to the bitline.

When a cell is sent back to stress, or recovery if the *RECOVER* signal is high, this ordering is roughly reversed. The ROSC is first disconnected from the bitline to prevent any unwanted stress in other parts of the array. At nearly the same time, the ROSC loop is opened, the VCC switch is turned off, and the input path from the VCO is turned on. Finally, the VSTRESS or GND supply switch is closed to start stress or recovery, respectively. Several tri-state inverters and pull-down transistors are placed between the VCO input and the ROSC, as well as the ROSC and the bitline, in order to prevent any coupling when those connection paths are shut off. The *RESET* signal asynchronously sends the whole array into recovery mode.

Basic cell operation is illustrated in Fig. 27.15. Only two cells are included in this simulation to demonstrate the functionality of one of them as it is measured multiple times. The external *MEASSTRESS* signal controls the internal *ROW_CLK* signal, which starts and then stops each measurement with two consecutive pulses, as will be explained in Sect. 27.5.3. The delay between CSUPPLY dropping to VCC and *meas[3]* closing the loop prevents unstable oscillations. Ten inverters in each ROSC are constructed with 1.2 V thin oxide logic DUTs. These stages will age during stress experiments while the rest of the gates, composed of 2.5 V thick oxide I/O transistors, will not degrade significantly. However, the thick oxide control stages contribute to the full loop delay, and this must be accounted for when calculating the stressed stages' delay shift due to aging.

Therefore, in addition to the full loop that is selected with the *meas[3]* signal, a replica control path is selected with *ctrl* which is asserted at the appropriate time if the scan bit *test_ctrl* is high. The delay of the replica path is roughly equivalent to that of the control logic in the full loop. So by first measuring the control loop frequency, and then that of the full loop during automated circuit calibration, one can calculate the percentage of the fresh full loop delay accounted for by the DUTs (i.e., $1-f_{full}/f_{ctrl}$). The total frequency shift of each stressed full loop is divided by the percentage of the fresh delay taken by the DUTs in order to calculate the degradation in those thin oxide stages. All DUT stages have identical loads and layouts due to the use of dummy cells.

27.5.2 Multiple Silicon Odometer Beat Frequency Detection Setup

As described in Sect. 27.3, the Silicon Odometer beat frequency detection system provides high-resolution frequency shift measurements when the speeds of the ROSC under test and reference are close. A small initial difference between f_{ref} and f_{stress} is ensured with trimming capacitors, and in other Odometer test circuits, each capacitor on both oscillators was individually controlled with scan chain bits. However, in the present circuit where many ROSCs are stressed in parallel and selected one by one for measurements, controlling the capacitors in each stressed oscillator would be very time and area consuming. Therefore, nine of fifteen capacitors were instead hardwired "on" in each of those ROSCs, while individually controlling all fifteen in the three references.

The Odometers associated with those references all record counts corresponding to the beat frequency for each ROSC measurement. During post-processing, the highest-resolution degradation characteristic is selected from that set of three for each oscillator that was stressed. Figure 27.16a presents an example distribution of 80 fresh full loop frequencies, along with the range covered by the three reference ROSCs under all trimming conditions. Turning on each trimming capacitor slowed a reference ROSC by roughly 900 kHz, or 0.57% of the mean fresh full loop frequency under nominal operating conditions. During calibration, the reference ROSCs are trimmed to positions within the fresh array distribution such that we maximize the resolution of the group of degradation characteristics gathered from each full stress experiment.

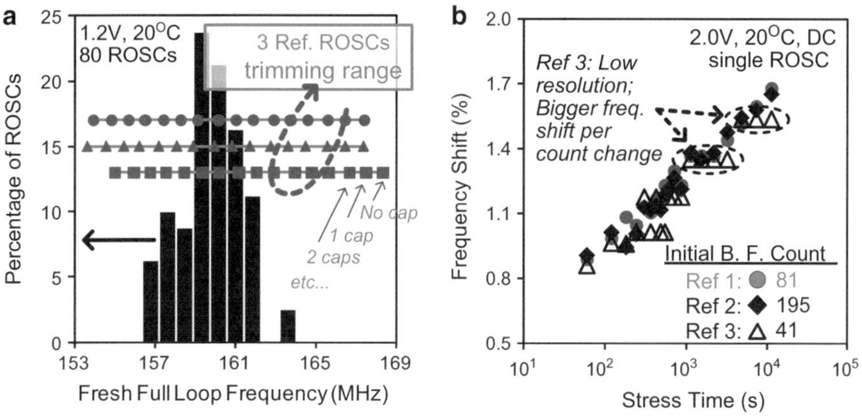

Fig. 27.16 (**a**) Example fresh full loop frequency distribution for an 80-cell array, with the corresponding reference ROSC trimming range. (**b**) Measured results from all three Odometers for one ROSC under test

Figure 27.16b shows a group of three degradation characteristics gathered by the references from one ROSC under test. Starting measurements with the reference and test ROSC frequencies close together, but the latter slightly slower, leads to a high-resolution measurement with a monotonic decrease in the output counts (see Sect. 27.3). The odometer output count is equal to the number of reference ROSC periods (N_{ref}) counted throughout one period of the beat frequency, during which time one less cycle is observed in the stressed ROSC ($N_{stressed} = N_{ref} - 1$). Therefore, according to Eq. (27.1) from Sect. 27.3, with $N_{ref} = 41$ reference ROSC 3 started out 2.44% faster than the ROSC under test in the current example (i.e., $1 - N_{stressed}/N_{ref} = 1 - 40/41$). The resulting low resolution is apparent from the highly quantized outputs of the corresponding Odometer. However, reference 2 was initially only 0.513% faster than the ROSC under test, so this high-resolution result is saved instead. Finally, as stated earlier, measurement results from a reference ROSC that starts out *slower* than the ROSC under test can also be used, and the corresponding output count will *increase* with stress. Therefore, the distribution of fresh frequencies in the array under test can be more easily covered with only three appropriately trimmed references.

27.5.3 Test Interface and Procedure

Calibration and measurements are automated through a simple digital interface. During calibration the fresh control and full loop frequencies from each ROSC in the array are recorded by reading those values with an oscilloscope. The error in this step is minimized by averaging thirty results for each loop. Next, we sweep through the trimming range in the three reference ROSCs, again averaging

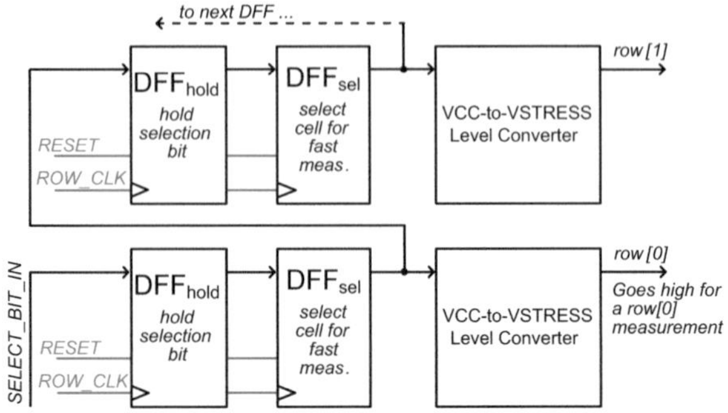

Fig. 27.17 Row selection logic ("Row Periph" from Fig. 27.13). Two flops are included for each row so that as soon as a measurement is finished on row[n], the selection bit can move to row[$n+1$] without starting a measurement there until the system is ready

the measured frequency results from thirty samples. After that point the optimal trimming configurations are selected in order to cover the distribution of frequencies of the ROSCs to be tested.

A *RESET* signal is asserted before stress conditions are set which prevents any cells from being selected, and puts them all into recovery mode. During experiments, ROSC cells are cycled through without the need to send or decode cell addresses, in order to simplify the logic and attain faster measurement times. The first cell is selected with an initialization sequence, and *MEASSTRESS* is asserted each time that the controlling software is ready for a new measurement. The row selection signal is incremented with each measurement, and the column selection shifts after all of the cells in a column have been selected. Any cells not selected for stress are kept in a 0 V no-stress state by asserting the *RECOVER* signal appropriately.

The logic used to store the row selection signal has to minimize the time when a ROSC is taken out of stress in order to prevent unwanted BTI recovery. Although the Odometer logic enables measurement times of down to 1 μs, the time required to scan out the results must be accounted for. Therefore, two DFFs were used for each row, as shown in Fig. 27.17. *SELECT_BIT_IN* is clocked into the first DFF_{hold} with *ROW_CLK* during initialization. The next *ROW_CLK* pulse starts a measurement on *row[0]* by moving the select bit to the DFF_{sel} on the first rising edge of *MEASSTRESS*. That selection bit is then sent to the DFF_{hold} in *row[1]* by another pulse of *ROW_CLK* on the falling edge of *MEASSTRESS* and is held there while the results are scanned out. The next assertion of *MEASSTRESS* from the controlling software starts a measurement on *row[1]*, and this process is repeated through the rows and columns as necessary.

27.5.4 Statistical Odometer Test Chip Measurements

A test circuit was implemented in 65 nm bulk CMOS. PBTI (Positive Bias Temperature Instability) in NMOS is not significant in this technology, so only half of the PMOS devices and none of the NMOS undergo substantial BTI aging

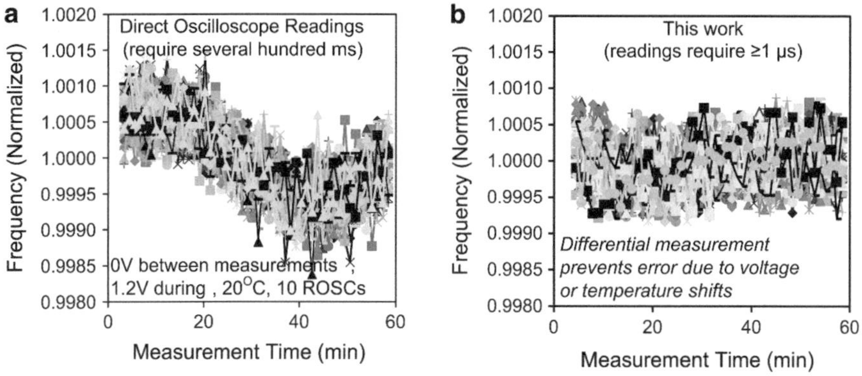

Fig. 27.18 Error (i.e., deviation from 1.0) in (**a**) oscilloscope and (**b**) faster Odometer measurements from 40 ROSCs during no-stress experiments

in between voltage transitions. Under AC stress all PMOS are stressed with NBTI conditions for half of the time between transitions, and HCI degradation occurs in both types of transistors during the switching events.

In this system where three reference ROSCs are used rather than one, measurement times were manually set to 2.5 μs for most experiments, which allowed the majority of beat frequency counts to complete correctly. As discussed in Sect. 27.3, the initial two counts reported by the beat frequency Odometer are unpredictable because they begin at random midpoints in the beat frequency period. Therefore, the last three count results that are latched before the measurement period is stopped are recorded, and the final one is saved for post-processing. If the time required to latch these first three counts is greater than the chosen measurement time due to a long beat frequency period (i.e., high Odometer output count), or long reference and stressed ROSC periods, then the desired results cannot be recorded. Shorter measurements were possible, but in that case the higher count results which did not have time to complete were discarded, and the next highest-resolution output was selected for post-processing. In addition to allowing more beat frequency counts to complete, 2.5 μs interrupts were maintained because measurement results showed that the difference between the frequency degradation measured with this value and 1 μs was negligible [21].

Figure 27.18 illustrates frequency measurements from 0 V no-stress experiments, so ideally there should be no shift (i.e., the normalized frequency should remain at 1.0). The characteristics of forty ROSCs are displayed and are representative of results seen from the entire arrays measured on multiple chips. Figure 27.18a was directly recorded by a 100 MHz, 1.25GS/s oscilloscope after the frequencies were divided down by 1,024 on-chip. A worst case error of 0.18% was observed, and a drift is apparent in the measured values due to some slight change in operating conditions. Figure 27.18b shows a smaller worst case error of 0.07% in the frequency calculated by the Silicon Odometer, along with the fact that this

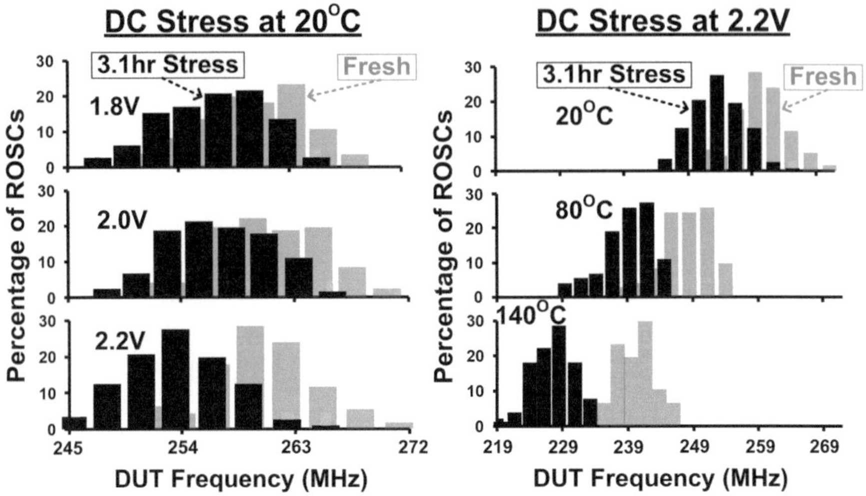

Fig. 27.19 Frequency distributions after 3.1 h stress (*black bars*), along with the fresh distributions (*gray bars*). Each 20°C distribution was gathered from 120 ROSCs, while those at higher temperatures came from 80 due to a limited amount of dies

differential system eliminates the effects of variations common to the reference and stressed ROSCs. Similar error floors were found for these systems during repeated tests, setting the lower bound on the range of frequency shifts that they can accurately measure. Finally, note that the automated oscilloscope readings required over 500 ms, while the Odometer measurements take down to 1 μs.

Histograms of fresh DUT frequencies are shown in Fig. 27.19 with the resulting distributions after 3.1 h of DC stress (11,200 s). The distributions at 20°C were obtained from 120 ROSCs each, while the higher temperature measurements involved 80 ROSCs. Smaller numbers were used when necessary due to the limited number of available dies. In order to prevent any particular systematic inter-die process shifts from fully impacting one set of the results, each experiment that spanned more than 40 ROSCs came from multiple dies. The primary degradation mechanism at work in these DC experiments was NBTI, and Fig. 27.20a indicates that there was no correlation between the fresh ROSC frequency and the stress-induced shift. This lines up with previous findings that the stress-induced V_{th} mismatch in PMOS pairs was uncorrelated to the initial mismatch [22] and that the initial spread in the V_{th} is not correlated to that caused by aging [23–25]. Figure 27.20b shows the average (μ) frequency shifts and the standard deviation (σ) of the shifts versus stress time. The σ increases with stress [23, 26–28], roughly following a power law with an exponent (n) of just under 1/2 that of the μ shift. Therefore, the σ/μ ratio of the shift decreases with stress time [29].

The σ of the frequency itself (rather than the frequency shift) did not show a clear trend with stress time. This value was poorly fitted by the power law (R-squared values of only ~0.03–0.30), with the exponent of this fit ranging from −0.002 to

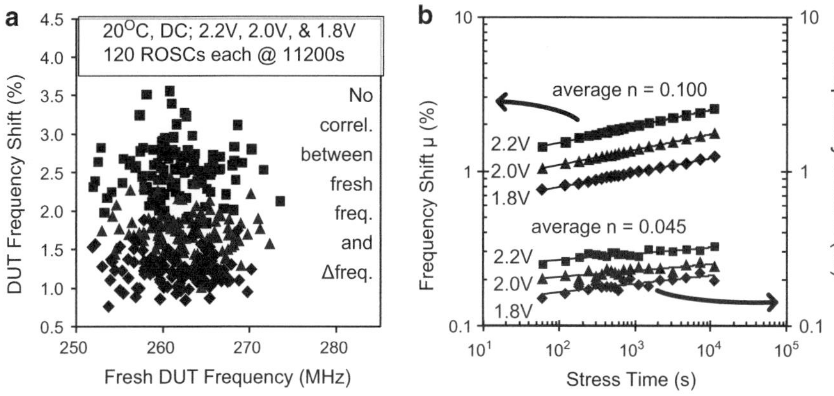

Fig. 27.20 (**a**) The fresh ROSC frequencies were uncorrelated with the stress-induced shifts. (**b**) The mean and standard deviation of the frequency shifts both increase with stress time

0.028, meaning the σ value remained generally flat during stress. That behavior is expected because the spread in the fresh frequency is larger than that of the spread in the aging-induced shifts, and the increase in the latter value is modest during stress, as seen in Fig. 27.20b.

27.6 Interconnect Odometer for Measuring the Impact of Wire Length on BTI and HCI

Interconnect plays a major role in the performance and reliability of critical circuits such as clock networks, signal buses, networks-on-chip, memory wordlines and bitlines, and high-speed I/Os. Devices in these circuits with large interconnect loads experience unique time-varying voltage stresses which result in performance degradation due to BTI and HCI and should be studied to enable reliable system designs.

Although BTI is frequently cited as the primary reliability concern in sub-100 nm processes, HCI remains an important mechanism in applications demanding high current levels [30, 31]. HCI degradation is also exacerbated in drivers with large loads and high activity factors such as clock buffers, where the transistors are exposed to stronger hot carrier stress due to the longer active switching phases [32].

BTI and HCI have different sensitivities to operating conditions, which vary with specific circuit paths. Sheet resistance and the parasitic capacitance of long wires have not been scaling favorably in advanced processes, which could lead to interconnect-dominated paths having drastically different aging behavior compared to logic-dominated paths [33]. Although previous work has shown the impact of gate fanout load on circuit aging [34, 35], little attention has been paid to the aging behavior in logic gates with long wire loads. Understanding the impact of

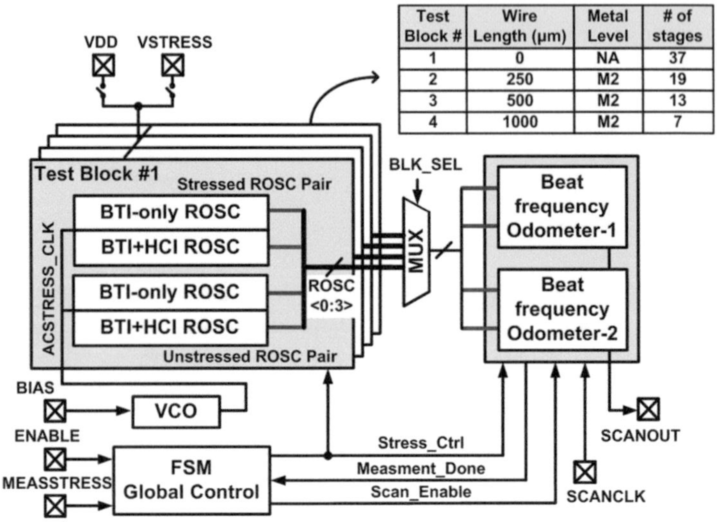

Fig. 27.21 Interconnect odometer test chip diagram. Four ROSCs (stressed pair and unstressed pair) are used for each wire configuration to separately monitor BTI- and HCI-induced frequency shifts

interconnect length on circuit degradation will enable a more complete picture of system-level aging.

This work presents measurement results highlighting the dependence of BTI- and HCI-induced aging on wire length for the first time [36]. The All-in-One Silicon Odometer beat frequency detection framework was adopted to separate the BTI and HCI contributions with picosecond-order resolution and microsecond-order stress interruptions for measurement (see Sect. 27.4). Measurement data from a 65 nm test chip shows that BTI-induced degradation decreases monotonically with longer interconnect length, while HCI exhibits a non-monotonic behavior. Models for BTI- and HCI-induced degradation in long interconnect drivers are developed and show good agreement with the measured data.

27.6.1 *Interconnect Odometer Design*

A top level block diagram of the interconnect odometer test chip is shown in Fig. 27.21. This system includes four ROSC configurations with different interconnect lengths ranging from 0 to 1,000 μm. In each of the four configurations, one ROSC suffers from BTI stress exclusively, another undergoes both BTI and HCI stress, and the final two are unstressed references. Every stressed oscillator is paired up with its unstressed counterpart, and their outputs are fed into the beat frequency detection system through multiplexers. The transistor dimensions of each ROSC stage are $(W/L)_{PMOS} = 6$ μm/0.06 μm and $(W/L)_{NMOS} = 3$ μm/0.06 μm. On-chip

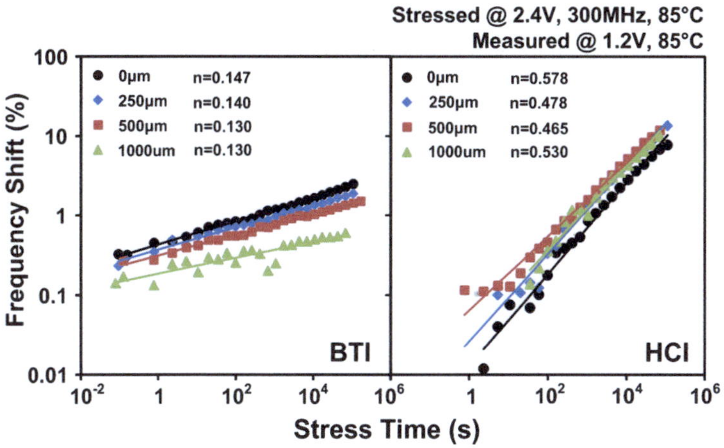

Fig. 27.22 Measured frequency degradation induced by BTI and HCI for different interconnect lengths

power gates provide fast local stress voltage switching, while a VCO generates an AC stress frequency. Note that electromigration (EM) is negligible in the 65 nm process used for the test chip presented here when only AC stress conditions are applied, and no excessively high stress temperatures or current densities are employed.

27.6.2 Interconnect Odometer Test Chip Results

Figure 27.22 presents BTI- and HCI-induced frequency degradation versus stress time for different interconnect lengths. Both BTI and HCI degradation can be fitted by a power law function of stress time with the exponents marked on the plots. BTI is the primary contributor to aging at early stress times while HCI surpasses BTI at longer stress times due to its larger power law exponent. Under identical stress conditions, the ROSC without any interconnect suffers the most BTI degradation, while HCI is most severe for the ROSC with 500 μm wire length.

BTI-induced frequency shifts after 19 h of stress at 2.4 V are shown in Fig. 27.23a for different interconnect lengths. The amount of BTI aging decreases monotonically with longer interconnects for all three stress conditions. This can be explained by the longer transition time observed in longer wires, which translates into a shorter amount of time the PMOS transistor is exposed to a full static BTI stress bias, or the "BTI duty cycle." The distributed interconnect RC reduces the slew rate for each signal transition as shown in Fig. 27.23b. HSPICE simulation results in that same figure confirm a 20× longer transition time ($t_T = t_R + t_F$) as the wire length is increased from 0 to 1,000 μm. Note that PBTI in NMOS is negligible

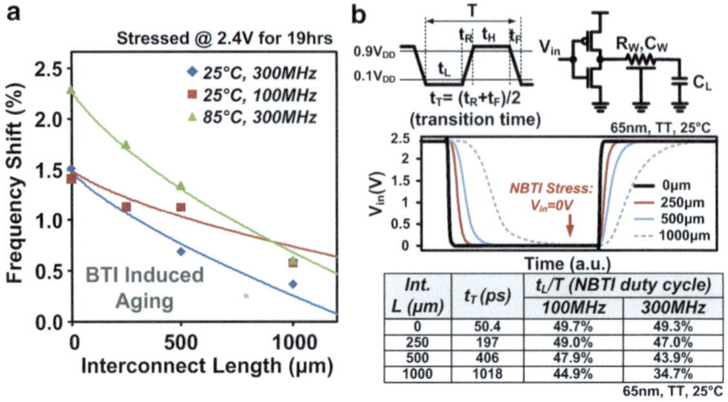

Int. L (μm)	t_T (ps)	t_L/T (NBTI duty cycle) 100MHz	300MHz
0	50.4	49.7%	49.3%
250	197	49.0%	47.0%
500	406	47.9%	43.9%
1000	1018	44.9%	34.7%

Fig. 27.23 (**a**) Measured data (markers) and the aging model presented in Sect. 27.6.3.2 (*curves*) for BTI-induced frequency degradation. (**b**) Effective stress time (t_L) decreases in longer interconnects resulting in a smaller BTI degradation as shown in (**a**)

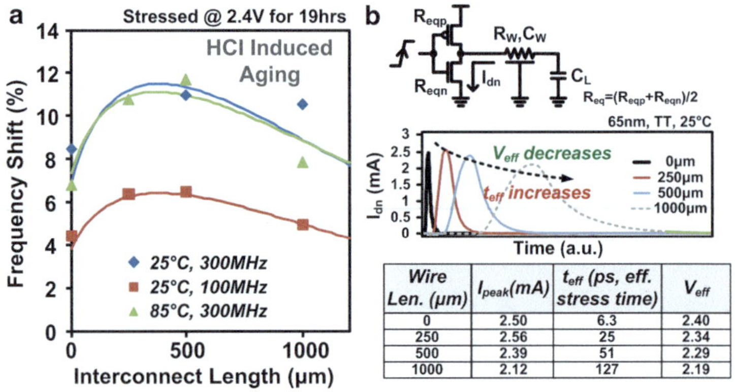

Wire Len. (μm)	I_{peak}(mA)	t_{eff} (ps, eff. stress time)	V_{eff}
0	2.50	6.3	2.40
250	2.56	25	2.34
500	2.39	51	2.29
1000	2.12	127	2.19

Fig. 27.24 (**a**) Measured data (markers) and the aging model presented in Sect. 27.6.3.3 (*curves*) for HCI-induced frequency degradation. (**b**) The effective stress time increases in longer interconnects while the effective stress voltage decreases, resulting in a non-monotonic HCI degradation characteristic

in this 65 nm process as it does not employ high-k/metal-gate devices. However, the general trend will not change in the presence of PBTI as the duty cycle for the NMOS is also reduced for longer interconnects. Figure 27.23a also reveals that slower frequencies result in a lower sensitivity of BTI to interconnect length because the devices spend less time in transition states and hence more time under the full BTI stress bias.

Figure 27.24a shows that HCI degradation has a non-monotonic relationship with wire length. This phenomenon is explained by two competing factors: a reduction of the effective stress voltage and the increase in current pulse duration, both with longer wire loads.

A driver with a longer wire load has a smaller peak current due to the voltage division between the wire resistance and the driver's equivalent resistance as shown in Fig. 27.24b. Simulation results modeling ROSC stages driving a distributed RC network confirm that the maximum discharging current through the NMOS decreases with longer interconnect loading for this reason. The reduction of the peak current has a similar effect to lowering the effective stress voltage and therefore leads to a smaller frequency shift. Note that the peak current also drops for wires shorter than 200 μm due to the fast input slew rate that causes the NMOS to turn off before it enters the saturation mode.

The second factor contributing to this non-monotonic characteristic is the wider current pulse caused by the increased RC loading of longer wires. A longer current pulse width equates to longer HCI stress time, which leads to more severe degradation for longer interconnects at an equivalent effective stress voltage [37].

27.6.3 BTI and HCI Aging Models for Interconnect Drivers

As demonstrated in the previous sections, the amount of circuit aging caused by BTI and HCI depends on transition times and bias conditions, which in turn are modulated by the interconnect load. In this section, an analytical BTI and HCI model applicable to global interconnect drivers is proposed and shown to closely match experimental results. The general approach for modeling the frequency degradation in drivers with long interconnect loads follows a two-step approach.

First, the frequency degradation of an interconnect-dominated path is less sensitive to the device aging compared to a logic-dominated path due to the invariant interconnect RC delay components in the former. This difference is captured by introducing a Sensitivity Factor in Sect. 27.6.3.1. Second, the amount of BTI and HCI aging depends on the stress time and voltage, and those vary with interconnect length. In the following sections, existing BTI and HCI models are used with modified stress parameters to derive the final models.

27.6.3.1 Sensitivity Factor

Wire RC dominates the delay of drivers with long interconnect loads, so performance is less sensitive to an aging-induced V_{th} shift compared with a logic-dominated path. This effect can be accounted for in our models by introducing the Sensitivity Factor α, defined as the ratio between the percentage frequency degradation of an interconnect-dominated path and that of a logic-dominated path for the same amount of device aging:

$$\left(\frac{\Delta f}{f}\right)_{interconnect} = \alpha\left(\frac{\Delta f}{f}\right)_{logic} \tag{27.4}$$

For a given interconnect resistance (R_W), interconnect capacitance (C_W), load capacitance (C_L), and equivalent driver resistance before stress (R_{eq}) and after stress (R_{eq}'), the frequency degradation can be expressed as

$$\left(\frac{\Delta f}{f}\right)_{interconnect} = \frac{\Delta R_{eq}(C_W + C_L)}{R'_{eq}(C_W + C_L) + R_W\left(\frac{C_W}{2} + C_L\right)} \tag{27.5}$$

The frequency degradation for a logic-dominated path without long wire loads can be written as

$$\left(\frac{\Delta f}{f}\right)_{logic} = \frac{\Delta R_{eq} C_L}{R'_{eq} C_L} = \frac{\Delta R_{eq}}{R'_{eq}} \tag{27.6}$$

R_W and C_W of the wire load can be calculated from the sheet resistance and metal capacitance parameters. Here, $\Delta R_{eq} = R_{eq}' - R_{eq}$, which is the change in equivalent driver resistance due to stress. The expression for α can be derived using Eqs. (27.5) and (27.6):

$$\alpha = \frac{\left(\frac{\Delta f}{f}\right)_{interconnect}}{\left(\frac{\Delta f}{f}\right)_{logic}} = \frac{R'_{eq}(C_W + C_L)}{R'_{eq}(C_W + C_L) + R_W\left(\frac{C_W}{2} + C_L\right)} \tag{27.7}$$

27.6.3.2 BTI Aging Model for Interconnect Drivers

BTI aging is determined by the time a device is biased in a strong inversion with a drain-to-source voltage drop of zero. Hence, it can be expressed using the cycle time parameter t_L/T (for the case of NBTI specifically) where t_L and T are defined in Fig. 27.23b. Employing the methodology presented by Fernandez [38], the deviation of BTI from the ideal 50% duty cycle case can be expressed using $(50\% - t_L/T)^k$, where k is determined empirically. The overall BTI-induced frequency shift can be expressed as

$$\begin{aligned}\left(\frac{\Delta f}{f}\right)_{BTI} &= \left(\frac{\Delta f}{f}\right)_{@50\%}\left[1 - B\left(50\% - \frac{t_L}{T}\right)^k\right] \\ &= Aexp(\gamma \mathbf{V_{str}})t^n\left[1 - B\left(\frac{2t_T}{T}\right)^k\right]\end{aligned} \tag{27.8}$$

Here, t_L is the low duty cycle of the input signal, T is the AC stress cycle, γ is the voltage acceleration factor, V_{str} is the stress voltage, t is the BTI stress time of a logic-only path, and n is the NBTI time exponent. B and k are empirical parameters found to be 0.01 and 0.7 at 25°C and 0.003 and 0.75 at 85°C in the 65 nm technology used for this work. Parameter A follows Arrhenius behavior with temperature, or

$\exp(E_a/kT)$, where E_a is the temperature activation energy. Both E_a and γ values are experimentally determined constants, which can be found based on the type of CMOS device.

The transition time t_T is interconnect RC dependent, which can be denoted as

$$t_T = R_{eq}(C_W + C_L) + R_W\left(\frac{C_W}{2} + C_L\right) \tag{27.9}$$

Using the Sensitivity Factor, the overall BTI frequency degradation for long interconnects can be derived as

$$\left(\frac{\Delta f}{f}\right)_{BTI_{interconnect}} = \alpha\left(\frac{\Delta f}{f}\right)_{BTI} = \frac{R'_{eq}(C_W+C_L)}{R'_{eq}(C_W+C_L)+R_W\left(\frac{C_W}{2}+C_L\right)} \times \left\{A\exp(\gamma \mathbf{V}_{\mathbf{str}})t^n\left[1 - B\left(\frac{2R_{eq}(C_W+C_L)+2R_W\left(\frac{C_W}{2}+C_L\right)}{T}\right)^k\right]\right\} \tag{27.10}$$

27.6.3.3 HCI Aging Model for Interconnect Drivers

The degradation of frequency with HCI can be approximated as presented by Hu [39]:

$$\left(\frac{\Delta f}{f}\right)_{HCI} = C\exp\left(-\frac{D}{V_{eff}}\right)t_{eff}^m \tag{27.11}$$

where C, D, and m are empirical process parameters, t_{eff} is the effective HCI stress time which is directly related to the transition time, and V_{eff} is the effective drain-to-source voltage during stress. The experimental and simulation results in Sect. 27.6.2 show that the effective voltage and stress time depend on the interconnect RC load.

The wire resistance R_W divides the stress voltage applied on transistor drain while charging and discharging. So the effective HCI stress voltage can be estimated as

$$V_{eff} = \frac{R_{eq}}{R_{eq} + R_W}V_{str} \tag{27.12}$$

where R_{eq} is the equivalent driver resistance, R_W is the interconnect resistance, and V_{str} is the HCI stress voltage in a path without significant interconnect loading.

Under the assumption that the HCI stress time is proportional to the transition time, the effective stress time considering interconnect impact can be expressed as

$$t_{eff} = \frac{R_{eq}(C_W + C_L) + R_W\left(\frac{C_W}{2} + C_L\right)}{R_{eq}C_L}t \tag{27.13}$$

Here, t is the time a device with a fanout of one in a logic-dominated path is under HCI stress. Finally, the HCI-induced frequency degradation can be derived by incorporating our Sensitivity Factor:

$$\left(\frac{\Delta f}{f}\right)_{HCI_{interconnect}} = \alpha\left(\frac{\Delta f}{f}\right)_{HCI} = \frac{R'_{eq}(C_W+C_L)}{R'_{eq}(C_W+C_L)+R_W\left(\frac{C_W}{2}+C_L\right)} \times Cexp\left[-D/\left(\frac{R_{eq}}{R_{eq}+R_W}V_{str}\right)\right]\left(\frac{R_{eq}(C_W+C_L)+R_W\left(\frac{C_W}{2}+C_L\right)}{R_{eq}C_L}t\right)^m \quad (27.14)$$

The models in Eqs. (27.10) and (27.14) are plotted in Figs. 27.23 and 27.24, where they show close agreement with measured silicon data.

27.7 SRAM Odometer for Recovery-Free Evaluation of NBTI and PBTI

Bias Temperature Instability is a primary reliability concern in sub-32 nm SRAMs [5, 40–42]. NBTI and PBTI under the DC stress condition that dominate in SRAM bit cells lead to an increase in read V_{MIN} and a decrease in write V_{MIN} as illustrated in Fig. 27.25.

While there is a pressing need to do in situ statistical characterization of BTI on large memory arrays, this task is complicated by the phenomenon of fast BTI recovery, which can lead to inaccurate results if the measurement time, T_{MEAS}, is not on the order of microseconds (Fig. 27.26) [21, 43]. In simple test circuits such as ROSCs, it is possible to gate stress on or off in small blocks, which helps

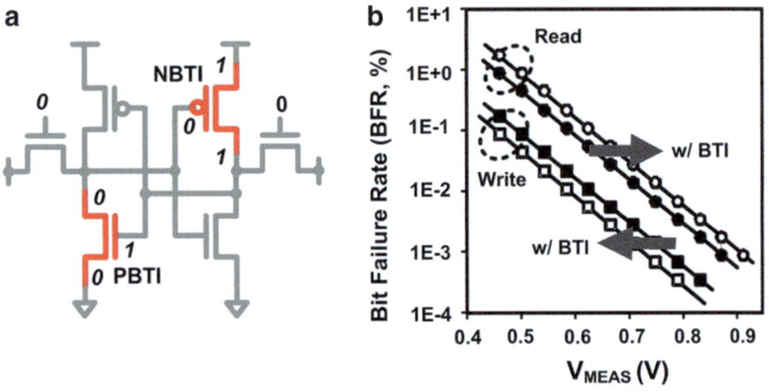

Fig. 27.25 (**a**) SRAM static stress conditions promote BTI stress in the two highlighted MOSFETs. (**b**) SRAM read V_{MIN} degrades while write V_{MIN} improves under the influence of BTI stress

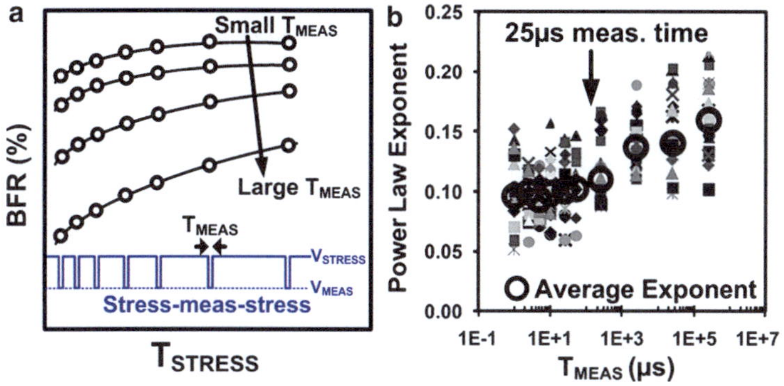

Fig. 27.26 (**a**) Longer T_{MEAS} results in optimistic BTI data (i.e., lower bit cell failure rate) due to the unwanted fast recovery. (**b**) Power law exponents measured at different T_{MEAS} indicates a recovery time constant of at most ∼25 μs [21]

to facilitate faster stress interruptions for measurements. However, this approach cannot be directly extended to SRAM arrays where the supply rail is generally shared across all rows, and a large amount of stored data is processed in parallel. Moreover, all of the stored data generally needs to be sent off-chip during a read operation for experiments like this where each bit's result should be recorded, as on-chip storage would be too costly in terms of area. Considering a typical tester's data acquisition frequency of few megahertz, the requirement for fast BTI measurements becomes problematic and has been the main limitation of existing approaches.

Kim et al. used off-chip control of the supply during measurements to obtain the SRAM V_{MIN}, which takes a few seconds and hence leads to extensive unwanted BTI recovery [42]. Drapatz et al. presented a BFR (Bit Fail Rate) tracking approach with local data storage similar to this work for fast measurements [44]. However, the overall approach was not scalable to full SRAM arrays and could not be used for progressive evaluation of BTI. This system only produced an end-of-life degradation estimation which has limited use for reliability modeling. The test structure presented in this section is the first to facilitate recovery-free evaluation of the progression of NBTI- and PBTI-induced degradation in an SRAM macro [45]. Results from a 32 nm high-k/metal gate SOI test chip show a 35 mV improvement in read V_{MIN} measurement accuracy and a 10−100× improvement in the accuracy of BFR estimation using a T_{MEAS} of 3 μs, which is better timing resolution than the closest related works by three to four orders of magnitude [5, 42, 44].

27.7.1 SRAM Reliability Macro Design

Figure 27.27 illustrates the SRAM reliability characterization macro. All critical components are designed to be representative of a product subarray to the extent possible, while still facilitating fast BTI measurements. Single-Ended Sensing

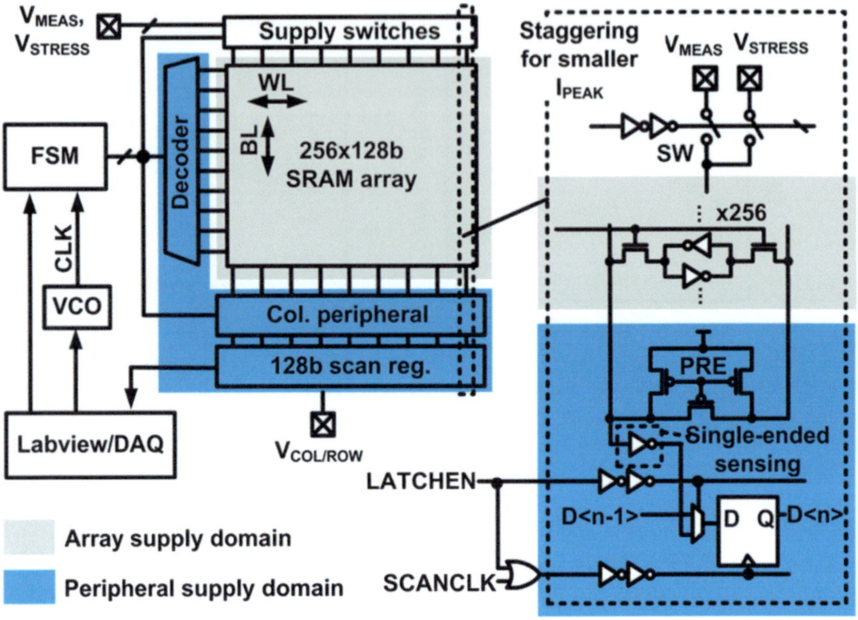

Fig. 27.27 SRAM Odometer macro architecture

(SES) with a slow scan-based readout is used rather than differential sense amplifiers and column multiplexing in order to simplify testing and reduce the pin count. A marker row with alternate hardwired 1 s and 0 s is used to verify the address decoding and other peripheral circuitry during dynamic operation. An on-chip FSM handles the timing and execution of sensitive functions such as controlling the supply switches for measurement and stress modes, modulating measurement times, pulse width control, asserting read/write commands, and address sequencing. A VCO is used to generate high-speed signals, and slower functions like BFR readout are run by software controlling the tester. On-chip per-column supply switches are used with delayed firing of signals to reduce current spikes during supply switching and to optimize the overall switching time.

27.7.2 Read Timing Sequence: Pseudo-reads with Deferred Stressed Readout (PR-SR)

Figure 27.28 shows example timing diagrams of the conventional [42] and recovery-free methods. Prior to applying V_{STRESS}, all bit cells are initialized through a blanket write 0 with the supply set to the nominal supply level (V_{NOM}). Next, the peripheral supply is lowered to V_{MEAS}, a level corresponding to a target read BFR. This

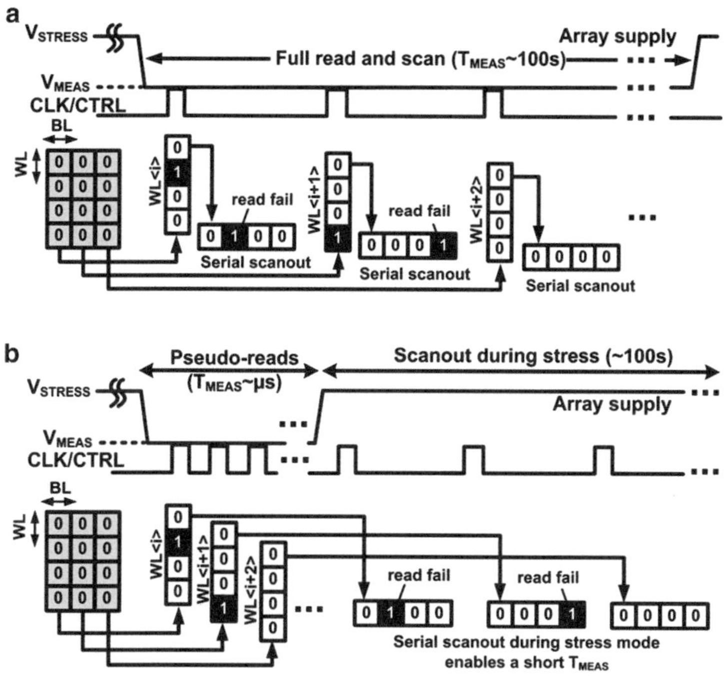

Fig. 27.28 PR-SR sequence for SNM failures in an array initialized to zero [45]. (**a**) In the conventional method, the supply is lowered to V_{MEAS} followed by a full read and slow scan-out which results in a long T_{MEAS}. (**b**) The recovery-free approach consists of a pseudo-read (i.e., sequential WL perturbations) which stores pass/fail info in the array under test. The array is immediately put back into stress mode to prevent unwanted recovery, followed by a full reliable read and scan-out at the V_{STRESS} supply level

completes the initialization step. After that, stress is applied in a stress-measure-stress routine with exponentially increasing stress intervals using an array supply of V_{STRESS}. In the short measurement windows, the array supply is lowered to V_{MEAS} using on-chip switches, with 20% of T_{MEAS} dedicated to supply switching.

A pseudo-read burst consisting of up to 256 sequential WL (wordline) perturbations follows next. Sufficiently "weak" cells on affected rows experience a data flip, while "strong" cells retain their original values. Thus pass/fail information corresponding to this fast measurement interrupt is stored locally in each bit cell. After this, the array supply is switched back to V_{STRESS} to prevent unwanted BTI recovery. The full read and off-chip data acquisition is deferred to the stress period. Due to the relatively long duration of stress periods, this data readout can be done slowly without interrupting the overall test procedure. Note that since the array operates at a high stress voltage during this read and scan operation, the chance of read failures occurring is remote. After all read data has been captured and the stress cycle is complete, the entire measurement and readout procedure is repeated. An extension of this approach can be used to track V_{MIN} as illustrated

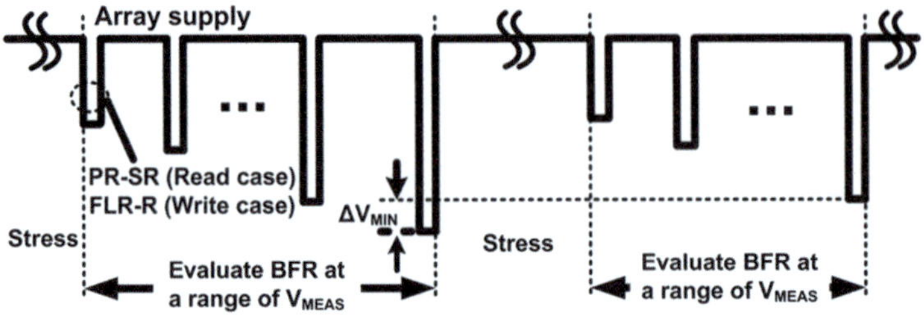

Fig. 27.29 Extension of the read BFR test sequence in Fig. 27.28 for read V_{MIN} measurements with microsecond range T_{MEAS} [45]. Here, V_{MEAS} is stepped down until a target BFR is reached

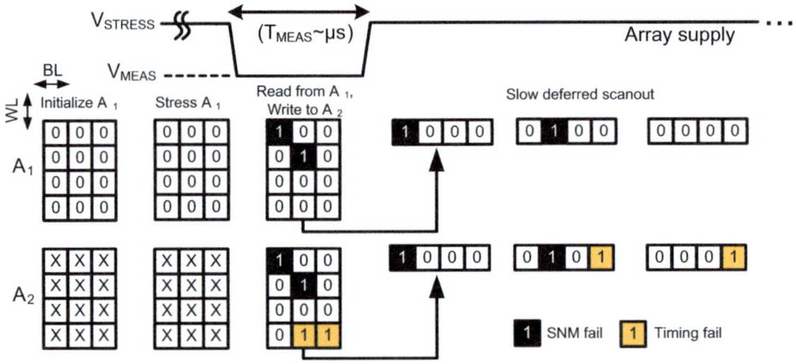

Fig. 27.30 PR-SR sequence to capture and separate SNM and access time failures exacerbated by the stress in the SRAM cell

in Fig. 27.29. Here, V_{MEAS} is stepped down until a target BFR is reached. Note that the milliseconds of stress in between increasingly lower V_{MEAS} value tests are negligible compared with the long stress cycles before and after each of these V_{MIN} searches.

In addition to recording SNM-related read failures, this design can be extended to examine the access time degradation due to the reduced drive strength of aging bit cell transistors with the approach outlined in Fig. 27.30. In this scheme (which was not implemented on the test chip), two subarrays are used: "A1" and "A2." The supply for A1 is switched between V_{MEAS} and V_{STRESS}, while A2's supply is fixed at $V_{NOM} = 0.9$ V for reliable operation. A1 is first initialized to a known data pattern and then stressed. Next, its supply is relaxed for fast, unreliable reads at V_{MEAS}, and the data is written reliably into A2. The flips stored in A1 indicate SNM flips while those seen only in A2 indicate access time fails occurring during reads from A1. The data in A1 is then read off-chip with a slow deferred scan-out operation during the next stress cycle. Using on-chip storage for a full subarray's failure data is costly in terms of area, but the overall cost is reduced by holding part of that information directly in the subarray under test (A1).

An approach similar to the one described above would not work for a typical write case. A successful write to a cell would mean the cell data flips. Consequently,

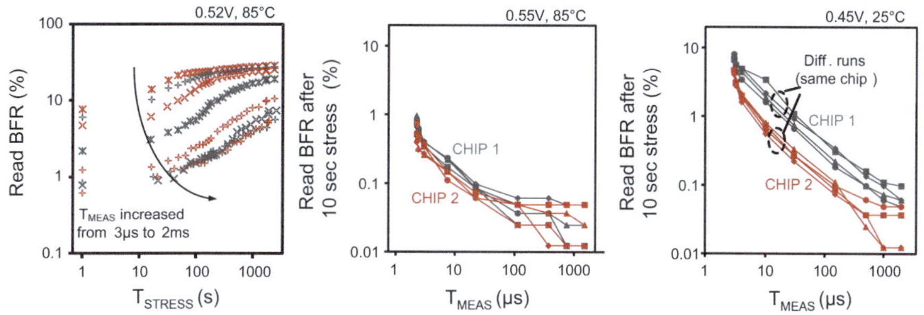

Fig. 27.31 Read BFR degradation with different T_{STRESS} and T_{MEAS} (V_{STRESS} must be kept confidential). The minimum T_{MEAS} possible in order to cover the whole array at $T_{CYCLE} = 10$ ns is 3 μs in the current test setup. A high BFR range (e.g. >0.1%) was chosen to obtain a smooth curve

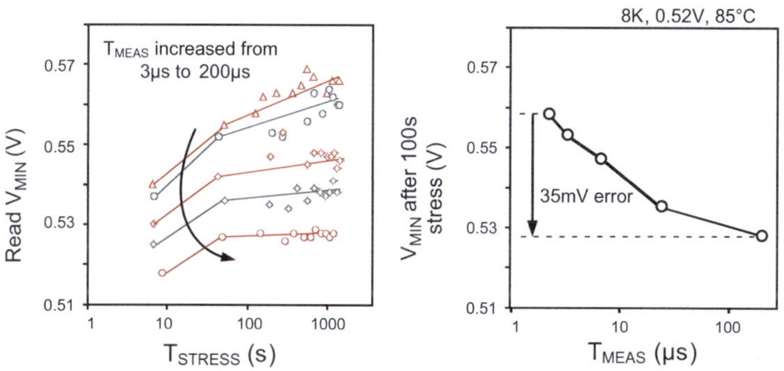

Fig. 27.32 Read V_{MIN} degradation with different T_{STRESS} and T_{MEAS}

BTI due to the prior DC stress would start to recover unless an immediate second flip (or write-back) to the original state is done. The details of the modified write test approach are given in an earlier publication [45].

27.7.3 Read Failure Test Chip Measurements

Figure 27.31 presents read BFR versus stress time at different T_{MEAS}, showing expected degradation trends. The BFR rises by around 3–10× over a 2,000 s stress period with $T_{MEAS} = 3$ μs. T_{MEAS} would be more than few milliseconds without the PR-SR technique, causing BFR errors of as much as 10–100×. Several measurements from multiple chips were performed to verify the consistency of these results.

Figure 27.32 shows the effect of BTI on V_{MIN}, which increases by 13–26 mV during a stress period of 2,000 s with $T_{MEAS} = 3$ μs. The PR-SR technique measures a V_{MIN} that is 35 mV higher than a conventional method requiring hundreds of milliseconds, because the latter experiences unwanted BTI recovery during long stress interruptions. Note that measurements of V_{MIN} required external supply changes as shown in Fig. 27.29 leading to larger time between measurement samples. This time discrepancy was calibrated out during post-processing.

27.8 Conclusions

This chapter provided an overview of a number of unique circuits that demonstrate the benefits of utilizing on-chip logic and a simple test interface to automate circuit or transistor aging characterization. In addition to avoiding the use of expensive probing equipment, implementing on-chip logic to control the measurements enables a better combination of measurement and timing resolutions. This is critical when interrupting stress to record BTI measurements, as that mechanism is known to recover within microseconds or less. The high frequency resolution of the beat frequency detection Odometer facilitates the acquisition of aging data at low stress conditions, close to or matching those of normal operation. Next, the silicon area needed to collect statistical data is significantly reduced compared with traditional device probing, as multiple test structures can share the same readout circuitry and I/O pads. Finally, compact all-digital systems could enable circuit aging prognostics and enhanced real-time adaptation in products.

In conclusion, Silicon Odometers can play a critical role in understanding the aging behavior of nanoscale devices and circuits. This capability will allow chip manufacturers to develop techniques to avoid wasteful overdesign and frequency guard banding based on pessimistic degradation projections and hence more fully realize the benefits of CMOS scaling while ensuring that products remain fully operational for their intended lifetimes.

References

1. M. Denais, C. Parthasarathy, G. Ribes, Y. Rey-Tauriac, N. Revil, A. Bravaix, V. Huard, and F. Perrier. On-the-fly Characterization of NBTI in Ultra-Thin Gate Oxide PMOSFETs. *IEEE International Electron Devices Meeting*, pages 109-112, 2004.
2. M. Denais, A. Bravaix, V. Huard, C. Parthasarathy, C. Guerin, G. Ribes, F. Perrier, M. Mairy, and D. Roy. Paradigm Shift for NBTI Characterization in Ultra-Scaled CMOS Technologies. *IEEE International Reliability Physics Symposium*, pages 735-736, 2006.
3. C. Shen, M.-F. Li, C. Foo, T. Yang, D. Huang, A. Yap, G. Samudra, and Y.-C. Yeo. Characterization and Physical Origin of Fast Vth Transient in NBTI of pMOSFETs with SiON Dielectrics. *IEEE Electron Devices Meeting*, pages 1-4, 2006.
4. T. Grasser, W. Gös, V. Sverdlov, and B. Kaczer. The Universality of NBTI Relaxation and Its Implications for Modeling and Characterization. *IEEE International Reliability Physics Symposium*, pages 268-280, 2007.

5. T. Fischer, E. Amirante, K. Hofmann, M. Ostermayr, P. Huber, and D. Schmitt-Landsiedel. A 65 nm Test Structure for the Analysis of NBTI Induced Statistical Variation in SRAM Transistors. *IEEE European Solid-State Device Research Conference*, pages 51-54, 2008.
6. E. Karl, P. Singh, D. Blaauw, and D. Sylvester. Compact In-Situ Sensors for Monitoring Negative-Bias-Temperature-Instability Effect and Oxide Degradation. *IEEE International Solid-State Circuits Conference*, pages 410-411, 2008.
7. P. Singh, Z. Foo, M. Wieckowski, S. Hanson, M. Fojtik, D. Blaauw, and D. Sylvester. Early Detection of Oxide Breakdown through In Situ Degradation Sensing. *IEEE International Solid-State Circuits Conference*, pages 190-191, 2010.
8. K. Hofmann, H. Reisinger, K. Ermisch, C. Schlunder, W. Gusting, T. Pompl, G. Georgakos, K. v. Arnim, J. Hatsch, T. Kodytek, T. Baumann, and C. Pacha. Highly Accurate Product-Level Aging Monitoring in 40 nm CMOS. *IEEE Symposium on VLSI Technology*, pages 27-28, 2010.
9. E. Saneyoshi, K. Nose, and M. Mizuno. A Precise-Tracking NBTI-Degradation Monitor Independent of NBTI Recovery Effect. *IEEE International Solid-State Circuits Conference*, pages 192-193, 2010.
10. F. Gebara, J. Hayes, J. Keane, S. Nassif, and J. Schaub. Delay-Based Bias Temperature Instability Recovery Measurements for Characterizing Stress Degradation and Recovery. U. S. Patent Application 12/142,294, Filed June 19, 2008.
11. M. Chen, V. Reddy, J. Carulli, S. Krishnan, V. Rentala, and V. Srinivasan. A TDC-based Test Platform for Dynamic Circuit Aging Characterization. *IEEE International Reliability Physics Symposium*, pages 2B.2.1-2B.2.5, 2011.
12. M.B. da Silva, B. Kaczer, G. Van der Plas, G.I. Wirth, and G. Groeseneken. On-Chip Circuit for Massively Parallel BTI Characterization. *IEEE International Integrated Reliability Workshop*, pages 90-93, 2011.
13. H. F. Dadgour and K. Banerjee. A Built-in Aging Detection and Compensation Technique for Improving Reliability of Nanoscale CMOS Designs. *IEEE International Integrated Reliability Workshop*, pages CR.1.1-CR.1.4, 2010.
14. J. Keane and C. H. Kim. On-Chip Silicon Odometers and their Potential Use in Medical Electronics. *IEEE International Reliability Physics Symposium*, pages 4C.1.1-4C.1.8, 2012.
15. T. H. Kim, R. Persaud, and C. H. Kim. Silicon Odometer: An On-Chip Reliability Monitor for Measuring Frequency Degradation of Digital Circuits. *IEEE Journal of Solid-State Circuits*, vol. 43, no. 4, pages 874-880, 2008.
16. J. Keane, D. Persaud, and C. H. Kim. An All-In-One Silicon Odometer for Separately Monitoring HCI, BTI, and TDDB. *IEEE VLSI Circuits Symposium*, pages 108-109, 2009.
17. J. Velamala, V. Reddy, R. Zheng, and Y. Cao. On the Bias Dependence of Time Exponent in NBTI and CHC Effects. *IEEE International Reliability Physics Symposium*, pages 5E.2.1-5E.2.5, 2010.
18. K. Hofmann, H. Reisinger, K. Ermisch, C. Schlunder, W. Gusting, T. Pompl, G. Georgakos, K. v. Arnim, J. Hatsch, T. Kodytek, T. Baumann, and C. Pacha. Highly Accurate Product-Level Aging Monitoring in 40 nm CMOS. *IEEE Symposium on VLSI Technology*, pages 27-28, 2010.
19. H. Reisinger, T. Grasser, K. Hofmann, W. Gustin, and C. Schlunder. The Impact of Recovery on BTI Reliability Assessments. *IEEE International Integrated Reliability Workshop*, pages 12-16, 2010.
20. B. Kaczer, R. Degraeve, M. Rasras, K. Van de Mieroop, P. Roussel, and G. Groeseneken. Impact of MOSFET Gate Oxide Breakdown on Digital Circuit Operation and Reliability. *IEEE Transactions on Electron Devices*, pages 500-506, vol. 49, no. 3, 2002.
21. J. Keane, W. Zhang, and C. H. Kim. An Array-Based Odometer System for Statistically Significant Circuit Aging Characterization. *IEEE Journal of Solid-State Circuits*, vol. 46, no. 10, pages 2374-2385, 2011.
22. S. Rauch. The Statistics of NBTI-Induced VT and β Mismatch Shifts in pMOSFETs. *IEEE Transactions on Device and Materials Reliability*, vol. 2, no. 4, pages 89-93, 2002.
23. S. Pae, J. Maiz, C. Prasad, and B. Woolery. Effect of BTI Degradation on Transistor Variability in Advanced Semiconductor Technologies. *IEEE Transactions on Device Materials and Reliability*, vol. 8, no. 3, pages 519-525, 2008.

24. B. Kaczer, T. Grasser, P. Roussel, J. Franco, R. Degraeve, L. Ragnarsson, E. Simoen, G. Groeseneken, and H. Reisinger. Origin of NBTI Variability in Deeply Scaled pFETs. *IEEE International Reliability Physics Symposium*, pages 26-32, 2010.
25. T. Fischer, E. Amirante, K. Hofmann, M. Ostermayr, P. Huber, and D. Schmitt-Landsiedel. A 65nm Test Structure for the Analysis of NBTI Induced Statistical Variation in SRAM Transistors. *IEEE European Solid-State Device Research Conference*, pages 51-54, 2008.
26. M. Agostinelli, S. Pae, W. Yang, C. Prasad, D. Kencke, S. Ramey, E. Snyder, S. Kashyap, and M. Jones. Random Charge Effects for PMOS NBTI in Ultra-Small Gate Area Devices. *IEEE International Reliability Physics Symposium*, pages 529-532, 2005.
27. V. Huard, C. Parthasarathy, C. Guerin, T. Valentin, E. Pion, M. Mammasse, N. Planes, and L. Camus. NBTI Degradation: From Transistor to SRAM Arrays. *IEEE International Reliability Physics Symposium*, pages 289-300, 2008.
28. K. Kang, S. Park, K. Roy, and M. Alam. Estimation of Statistical Variation in Temporal NBTI Degradation and its Impact on Lifetime Circuit Performance. *IEEE/ACM Int. Conference on Computer-Aided Design*, pages 730-734, 2007.
29. S. Rauch. Review and Reexamination of Reliability Effects Related to NBTI-Induced Statistical Variations. *IEEE Transactions on Device Materials and Reliability*, vol. 7, no. 4, pages 524-530, 2007.
30. C. Schlunder, S. Aresu, G. Georgakos, W. Kanert, H. Reisinger, K. Hofmann and W. Gustin. HCI vs. BTI? – Neither one's out. *IEEE International Reliability Physics Symposium*, pages 2F.4.1-2F.4.6, 2012.
31. P. Moens, G. Van den bosch, and G. Croeseneken. Hot-carrier degradation phenomena in lateral and vertical DMOS transistors. *IEEE Transactions Electron Devices*, vol. 51, no. 4, pages 623-628, 2004.
32. Y. Leblebici. Design considerations for CMOS digital circuits with improved hot-carrier reliability. *IEEE Journal of Solid-State Circuits*, vol.31, pages 1014-1024, 1996.
33. N. S. Nagaraj, W. R. Hunter, P. R. Chidambaram, T. Y. Garibay, U. Narasimha, A. Hill and H. Shichijo. Impact of interconnect technology scaling on SoC design methodologies. *IEEE Interconnect Technology Conference*, pages 71-73, 2005.
34. W. Weber, H. M. Brox, T. Kunemund, M Muhlhoff, and D. Schmitt-Landsiedel. Dynamic degradation in MOSFET's—Part II: Application in the Circuit Environment. *IEEE Transactions Electron Devices*, vol. 38, no. 8, pages 1859-1867, 1991.
35. W. Jiang, H. Le, J. Chung, T. Kopley, P. Marcoux, and C. Dai. Assessing Circuit-Level Hot-Carrier Reliability. *IEEE International Reliability Physics Symposium*, pages 173-179, 1998.
36. X. Wang, P. Jain, D. Jiao, and C. H. Kim. Impact of Interconnect Length on BTI and HCI Induced Frequency Degradation. *IEEE International Reliability Physics Symposium*, pages 2F.5.1-2F.5.6, 2012
37. K. N. Quader, E. R. Minami, W. J. Ko, P. K. Ko and C. Hu. Hot-carrier-reliability design guidelines for CMOS logic circuits. *IEEE Journal of Solid State Circuits*, vol. 29, pages 253-262, 1994.
38. R. Fernandez, B. Kaczer, A. Nackaerts, S. Demuynck, R. Rodriguez, M. Nafria and G. Groeseneken. AC NBTI studied in the 1 hz – 2 GHz range on dedicated on-chip CMOS circuits. *IEEE International Electron Devices Meeting*, pages 1-4, 2006.
39. C. Hu, S. C. Tam, F. C. Hsu, P. K. Ko, T. Y. Chan and K. W. Terrill. Hot-electron induced MOSFET degradation – model, monitor and improvement. *IEEE Transactions Electron Devices*, vol. 32, no. 2, pages 375-385, 1985.
40. A. T. Krishnan, V. Reddy, D. Aldrich, J. Raval, K. Christensen, J. Rosal, C. O'Brien, R. Khamankar, A. Marshall, W. Loh, R. McKee, and S. Krishnan. SRAM Cell Static Noise Margin and VMIN Sensitivity to Transistor Degradation. *IEEE International Reliability Physics Symposium*, pages 1-4, 2006.
41. V. Huard, R. Chevallier, C. Parthasarathy, A. Mishra, N. Ruiz-Amador, F. Persin, V. Robert, V, A. Chimeno, E. Pion, N. Planes, D. Ney, F. Cacho, N. Kapoor, V. Kulshrestha, S. Chopra, and N. Vialle. Managing SRAM reliability from bitcell to library level. *IEEE International Reliability Physics Symposium*, pages 655-664, 2010.

42. T. Kim, W. Zhang, and C. H. Kim. An SRAM Reliability Test Macro for Fully Automated Statistical Measurements of Degradation. *IEEE Custom Integrated Circuits Conference*, pages 231-234, 2009.
43. T. Grasser, H. Reisinger, W. Goes, T. Aichinger, P. Hehenberger, P-J. Wagner, M. Nelheibel, J. Franco, and B. Kaczer. Switching oxide traps as the missing link between negative bias temperature instability and random telegraph noise. *IEEE International Electron Devices Meeting*, pages 1-4, 2009.
44. S. Drapatz, K. Hofmann, G. Georgakos, and D. Schmitt-Landsiedel. Impact of fast-recovering NBTI degradation on stability of large-scale SRAM arrays. *European Solid-State Device Research Conference*, pages 146-149, 2010.
45. P. Jain, A. Paul, X. Wang, and C. H. Kim. A 32nm SRAM Reliability Macro for Recovery Free Evaluation of NBTI and PBTI Induced Bit Failures. *IEEE International Electron Devices Meeting*, pages 9.7.1-9.7.4, 2012.

Chapter 28
Multilevel Reliability Simulation for IC Design

Ketul B. Sutaria, Jyothi B. Velamala, Venkatesa Ravi, Gilson Wirth, Takashi Sato, and Yu Cao

Abstract With CMOS technology scaling, design for reliability becomes a vitally important part in today's design cycle. Aging mechanisms, such as NBTI and CHC, degrade the performance of a circuit over time, eventually causing system functional failure. NBTI predominantly affects digital circuits, inducing delay shift in logic paths and data instability in memory cells, while CHC impacts the mismatch, gain, and offset in analog/mixed signal (AMS) circuits. Accurate long-term modeling of these aging effects is key to circuit failure analysis. In addition, simulation efficiency is critical to reliability diagnosis in a modern design at the scale of multi-million gates. This chapter presents a new simulation flow that integrates long-term aging models, which are sensitive to dynamic voltage scaling and switching activities, with aging-aware standard cell library, predicting the degradation for both digital and AMS circuits. Different from conventional reliability tools that rely on extrapolation for long-term aging prediction, the new methodology continuously monitors the shift in operating conditions and circuit performance metrics toward the end of the lifetime. As implemented into representative VLSI design tools, the newly developed aging models and tools dramatically improve simulation efficiency and accuracy, supporting design practice for reliability with scaled CMOS technology.

28.1 Introduction

The evolution of electronics will continually be driven by the scaling of CMOS technology and the increase in the total number of devices per chip. Moore's law predicts that the total number of transistors placed on a chip will be approximately

K.B. Sutaria • J.B. Velamala • V. Ravi • Y. Cao (✉)
e-mail: Yu.Cao@asu.edu

G. Wirth
UFRGS - Electrical Eng Department, Porto Alegre, Brazil

T. Sato
Graduate School of Informatics, Kyoto University, Japan

T. Grasser (ed.), *Bias Temperature Instability for Devices and Circuits*,
DOI 10.1007/978-1-4614-7909-3_28, © Springer Science+Business Media New York 2014

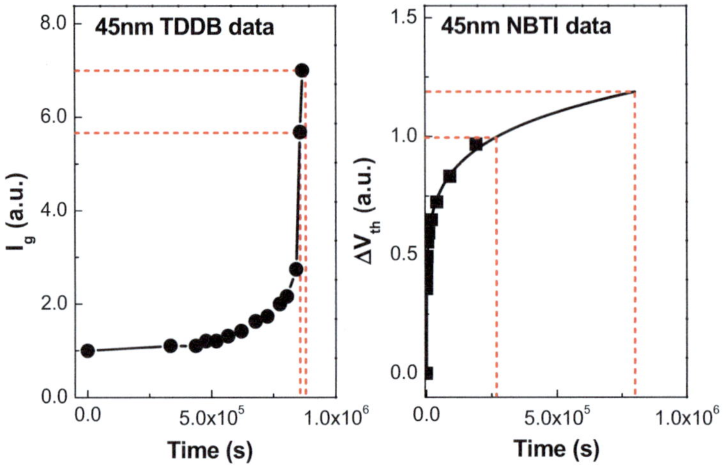

Fig. 28.1 Traditional definition of reliability is only appropriate for sudden failures (*left*): for TDDB, the exact threshold value has little impact on the prediction of lifetime. But such a definition is not applicable to gradual shift (*right*): the prediction of lifetime is highly sensitive to the threshold value, as observed in NBTI

doubled every 18 months [1, 2]. The increased transistor count has directly led to improved capabilities of integrated analog and digital circuits on a single platform called system on chip (SoC). Present SoCs pack close to 800 million transistors incorporating analog and mixed signal designs such as data converters (A/D, D/A) and PLLs, extracting high performance at a lower cost.

Aggressive scaling of CMOS technology brings forth multiple variability and reliability issues such as negative bias temperature instability (NBTI), channel hot carrier (CHC), and time-dependent dielectric breakdown (TDDB) [3–14]. As an aftermath of mentioned reliability issues, circuit performance degrades over time, which is called circuit aging. Circuit aging becomes more pronounced in the nanoscale regime due to process scaling techniques that are introduced to improve device and circuit performance. Scaling of supply voltage and threshold voltage does not go hand in hand with scaling of device feature size. This greatly increases vertical and lateral electrical field causing the exacerbation of NBTI and CHC.

Previous reliability concerns, such as oxide breakdown and electromigration (EM), are evaluated by an empirical threshold of performance shift, as shown in Fig. 28.1 [10]. Since these effects usually induce sudden failures, the exact value of reliability threshold only has a marginal impact on the lifetime (Fig. 28.1, left). But for NBTI and CHC, their effect is gradual (Fig. 28.1, right). A small difference in reliability threshold may result in a dramatic shift in determining the lifetime. In this case, a better approach is to provide parametric prediction of the lifetime, such that designers are able to examine the detailed trade-offs among speed, power, cost, and reliability. This new trend requires the development of accurate aging models and simulation tools that correctly capture the physics and efficiently support aging diagnosis during the design stage.

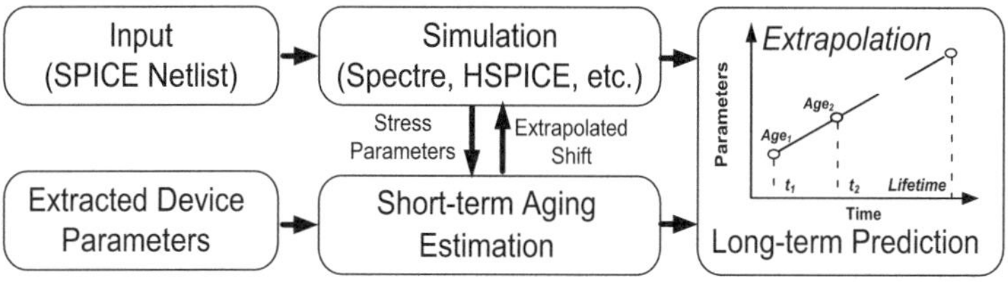

Fig. 28.2 The simulation flow employed by conventional reliability tools. The extrapolation method is used for long-term lifetime prediction

28.1.1 Circuit Reliability Simulation: Challenges and Needs

To date, research work on aging mechanisms has been active only within the communities of device and reliability physics [15–19]. This is partially due to its complexity and emerging status. The lack of design knowledge and CAD tools further creates the barrier for managing impact of device degradation on circuit performance. Leading industrial companies develop their own reliability models and tools. These tools, however, are usually proprietary and customized to a specific technology, not available for general usage.

Commercially available aging tools [20–23] suffer from issues of inaccuracy in aging prediction mainly due to their extrapolation method. One example is conventional lifetime prediction tools based on Berkeley reliability simulation framework [20]. Figure 28.2 presents a typical flow of such tools. In this flow, several reliability parameters are needed at the device level. These device parameters are extracted from the silicon data collected by stressing devices at high temperature and voltage to accelerate the aging process. In addition to device parameters, reliability simulators require design schematic or netlist files, as well as their input stimulus. Simulation of the input files reveals their operating voltages and thereby dynamic stress conditions. Based on these conditions and device parameters, which are extracted from short-term measurements and simulations, the aging rate and the lifetime are predicted using the extrapolation method (Fig. 28.2).

The tracking of stressed parameters through SPICE simulations makes these tools computationally expensive, as it consumes a large portion of the memory. While these tools can calculate the degradation of circuits with a limited number of transistors, performance evaluation of large-scale designs with millions of gates is impractical. To overcome these problems, a generic simulation tool that efficiently predicts the degradation would be extremely useful. A good circuit-aging simulator requires capabilities such as high capacity, high speed, and high accuracy. The simulation of aging in large logic designs is difficult, since circuit degradation rate depends on both process and operation conditions such as V_{dd}, temperature (T) and input signal duty cycle (α) [24]. These parameters are not spatially or temporally uniform, but vary significantly from gate to gate and from time to time due to the uncertainty in circuit topologies and operations. A simple static analysis may provide an extremely pessimistic estimate and, consequently, result in over-margining. To estimate the degradation bound under various α's, a rudimentary approach resorts to exhaustive simulations. Yet such method is inhibitive in computation cost, especially for circuits with a large number of inputs.

Lifetime prediction in AMS design is even more challenging than in digital logic circuits [25]. While aging-induced V_{th} shift does not change the operating conditions in logic gates, parameters in AMS designs, such as the bias condition, offset, and gain, are more vulnerable to V_{th} shift. The extrapolation method based on prestressed model parameters does not account for the changing operating conditions during aging which may lead to overly optimistic results. Furthermore, small AC signals affect the device degradation which is not accounted in current aging models. Hence, commercial tools inaccurately estimate the aging in AMS designs.

In summary, it is necessary to develop new models and simulation methodology in order to improve circuit reliability prediction in both VLSI and AMS design under dynamic operations. This chapter presents a cross-layer approach toward this goal, from device-level modeling of reliability mechanisms to circuit-level long-term aging models that are customized for digital and AMS design and to large-scale reliability simulation methods. The results are demonstrated with design examples that experience severe reliability threat.

28.2 Reliability Physics: Device-Level Modeling

The primary impact of NBTI and CHC at the device level is the gradual increase in transistor V_{th}, whereas the degradation of other device parameters is less pronounced. Since the threshold voltage directly affects the delay of a digital gate, circuit operating frequency of a logic path decreases temporally. Similarly, in AMS circuits, shift in V_{th} degrades the gain and other performance metrics. To estimate circuit-aging rate, the fundamental step is to model device V_{th} shift.

28.2.1 CHC: Power Law Model

The power law model (t^n model, where t is the stress time) explains the time dependence of gradual shift in the threshold voltage due to CHC, since CHC can be microscopically described as the generation of interface or oxide traps (charges) at the Si-SiO_2 interface. In fact, until recently, such a reaction–diffusion (RD) modeling framework was also used to explain the NBTI effect. This model assumes that when a gate voltage is applied, it initiates a field-dependent reaction at the semiconductor–oxide interface [26–34]. There are two critical phases described in this modeling framework: the reaction phase and the diffusion phase. In the reaction phase, some Si–H or Si–O bonds at the substrate/gate oxide interface are broken under electrical stress [29]. The species that trigger such reactions can be positive holes in NBTI and hot electrons in CHC [32]. While NBTI happens uniformly in the channel, CHC impacts primarily the drain end.

Channel hot carrier is mainly observed in NMOS transistors. The main source of the hot carriers is the heating inside the channel of the MOSFET during dynamic circuit operation. These energetic carriers can lead to impact ionization within the substrate and the generated electrons or holes inside the channel, or the heated carriers themselves can be injected in to the gate oxide. During this process, the injected carriers can generate interface or bulk oxide defects, and as a result, the MOSFET characteristics, like threshold voltage, transconductance, etc., degrade over time. The degradation of V_{th} caused by CHC is given by [32]:

$$\Delta V_{th} = \frac{q}{C_{ox}} K_2 \sqrt{Q_i} \exp\left(\frac{E_{ox}}{E_{o2}}\right) \exp\left(-\frac{\varphi_{it}}{q\lambda E_m}\right) t^n \qquad (28.1)$$

where E_{ox} is the vertical electric field and E_m is the maximum lateral electric field. λ is interpreted as the hot-electron mean-free path and φ_{it} the minimum energy in electron volts that a hot electron must have in order to create an impact ionization. $\varphi_{it}/(qE_m)$ is the distance that an electron must travel in the electric field E_m to gain energy φ_{it}, and $\exp(-\varphi_{it}/(qE_m))$ is the probability of an electron travelling a sufficient distance to gain energy φ_{it} or more without suffering a collision. The temporal degradation rate is governed by the time exponent, n, which is about 0.45. Channel hot carrier does not recover as seen in NBTI.

Limitation of RD in NBTI: The RD theory may partially explain the degradation behavior during the stress phase of NBTI and predict the aging under various bias voltages and temperature. However, one of the limitations of the RD model is that it is deterministic and cannot account for stochastic aging variability. In addition, the RD model predicts that the recovery upon stress removal is independent of the electric field and the temperature during the recovery phase. Recent publications [35–39] show that it is dependent on the temperature applied during the recovery, contradicting to the classical RD theory. Similarly, [40] shows that the recovery is also a function of the electric field. These limitations question the physical background of RD in explaining NBTI and incline more toward the trapping/de-trapping (TD) theory that is explained in Sect. 2.2.

28.2.2 NBTI: Trapping/De-Trapping Model

The RD theory has several limitations as mentioned in previous section. Charge trapping and de-trapping at localized states (charge traps) at the gate oxide interface resolve the limitations of the RD theory. Several works have presented the evidence of TD mechanism through the discrete V_{th} shifts, especially during the recovery phase in NBTI observed with the fast measurement techniques [34, 40]. Figure 28.3

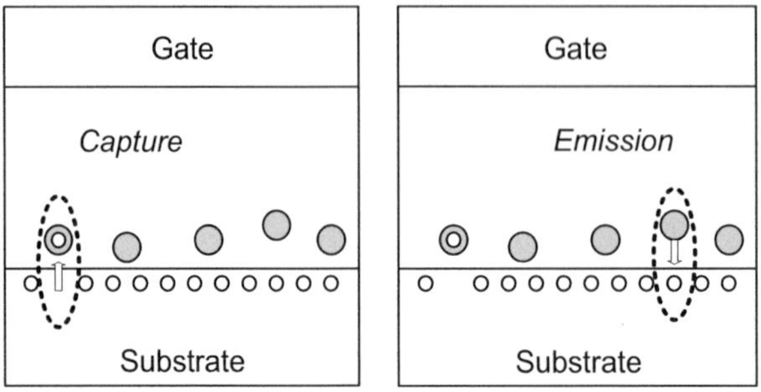

Fig. 28.3 Statistical trapping/de-trapping events lead to the generation of interface charges during the aging process of NBTI

illustrates the physical picture of TD: when a negative bias voltage is applied to the gate of a PMOS device, the trap energy (relative to the Fermi energy level) is modulated. If the trap gains sufficient energy, it may capture a charge carrier, thus reducing the number of available carriers in the channel [41]. The charged trap state modulates the local V_{th} and acts as a scattering source, reducing the effective mobility [42, 43]. In faster traps (with shorter time constants) having a higher probability of capturing carriers, the occupation probability increases with voltage and temperature. Trapping and de-trapping events are stochastic in nature and hence a compact model is based on the statistics of trap properties.

The basic assumptions in the modeling effort based on TD theory are the same as the ones used in modeling of low-frequency noise, since the charge trapping dynamics (capture and emission time statistics) that contribute to the degradation of device performance over time is similar to that causing low-frequency noise [25]. The three main assumptions of the trap properties are:

- The number of traps follows a Poisson distribution, which is common for a discrete process.
- Capture and emission time constants are uniformly distributed on logarithmic scale. This microscopic assumption is critical to derive the logarithmic time evolution at the macro scale.
- The distribution of trap energy follows a U shape, which is verified by silicon measurement and key to the voltage and temperature dependence on aging.

The location of traps is close to the interface (appropriate when $T_{ox} < 1$ nm). Based on these assumptions, TD-based static and dynamic models are presented in the next subsection. More details are available in Sect. 4.2.4 of this book.

28.2.2.1 Static NBTI Model

Based on the TD theory, the V_{th} shift at a given stress time is the result of a number of traps ($n(t)$) occupied by the channel carriers [36]. The probability of a particular trap, initially empty (state 0) to be occupied (state 1) after an elapsed time *t*, is given by $P_{01}(t)$. This occupation probability can be calculated by observing that

$$P_{01}(t+dt) = P_{01}(t)p_{11}(dt) + P_{00}(t)p_{01}(dt) \tag{28.2}$$

where $p_{01}(dt) = 1/\tau_c$ and $p_{11}(dt) = 1 - p_{10}(dt) = 1/\tau_e$. Integrating it from t_0 to t:

$$P_{01}(t+t_0) = \frac{\tau_{eq}}{\tau_c}\left(1 - e^{-t/\tau_{eq}}\right) + P_{01}(t_0)e^{-t/\tau_{eq}} \tag{28.3}$$

where $1/\tau_{eq} = 1/\tau_c + 1/\tau_e$. τ_c, τ_e are random in nature, representing capture and emission time constants, respectively, and dependent on bias point and temperature. The values are determined by [43]

$$\tau_c = 10^p(1+e^{-q}) \tag{28.4}$$

$$\tau_e = 10^p(1+e^{+q}) \tag{28.5}$$

where $p \in [p_{min}, p_{max}]$, where p_{min} and p_{max} define the time constants for fastest and slowest traps, respectively ($p_{min} \sim 1$ and $p_{max} > 10$). This assumption of the existence of defects with wide distribution of time constants is in line with recent NBTI data [46, 47]. Since p is assumed to be uniformly distributed, the characteristic time constants are uniformly distributed on logarithmic scale. The parameter q is given by $(E_T - E_F)/kT$, where E_T is the trap energy and E_F is the Fermi energy level. The trap energy (relative to Fermi energy) is a function of applied electric field. Consequently, τ_c and τ_e are dependent on voltage and temperature.

The occupation probability of the trap at time t, assuming that it is under constant stress from time $t_0 = 0$, is obtained by substituting $P_{01}(0) = 1 - P_{01}(0) = 0$ in Eq. (28.4), integrating P_{01}, and multiplying with the number of available traps, the average number of occupied traps obtained by substituting the logarithmic distribution of time constants and the U-shaped distribution of trap energies:

$$n(t) = \frac{N}{\ln 10(p_{\max} - p_{\min})} \int_0^{ET\max} \frac{g(E_T)dE_T}{1+\exp\left(-\frac{E_T - E_F}{kT}\right)} \cdot \int_{10^{-p_{\min}}t}^{10^{-p_{\max}}t} \frac{e^{-u}-1}{u}du \tag{28.6}$$

where $g(E_T)$ is the trap energy distribution and p_{min} and p_{max} represent fast and slow traps, respectively. The trap energy, E_T, changes as a function of electric field (E_{ox}). Assuming $p_{min} \sim 1$ and $p_{max} > 10$, and $E_T \sim 1/E_{ox}$,

$$n(t) = \frac{N}{\ln 10(p_{\max} - p_{\min})} \exp\left(\frac{\beta V_g}{T_{ox}kT}\right) \exp\left(\frac{-E_0}{kT}\right) \left[A + B\log 10^{-p_{\max}}t\right] \tag{28.7}$$

Equation (28.7) describes the aging under a constant stress voltage and temperature. Similar as previous RD model [32], it is an exponential function of the stress

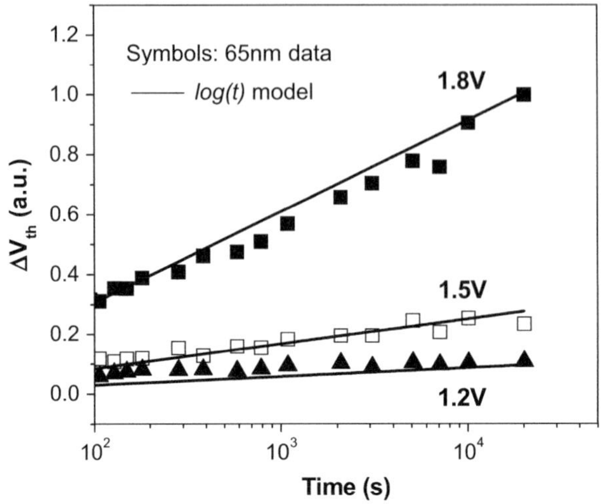

Fig. 28.4 The TD-based compact model matches the logarithmic time dependence and exponential voltage dependence

voltage, temperature, and T_{ox}. Furthermore, it has a statistical nature with N, an index for the number of traps per device. For the simplicity of model derivation and data analysis, Eq. (28.7) in compact form is written as

$$\Delta V_{th}(t) = \phi\,[A + B\log(1 + Ct)] \tag{28.8}$$

Equation (28.8) shows the logarithmic relation of degradation with stress time in contrary to the power law behavior. TD-based model also predicts the exponential dependence of voltage and temperature. Figure 28.4 shows the model prediction matches with 65 nm silicon data. Such a time evolution has a far-reaching impact on the aging behavior. More modeling details are presented in Sect. 4.2.4.

28.2.2.2 Dynamic NBTI Model

Today's circuits typically have a reduced activity factor (or duty cycle) through dynamic voltage scaling (DVS) to reduce power consumption. Therefore, a significant portion of the operation is under lower supply voltage, resulting in large recovery. Since the degradation is highly sensitive to the stress voltage, DVS leads to different amounts of circuit aging. To handle such a voltage transition, using a nonzero time, t_0, to calculate the occupation probability at time t (time elapsed after t_0) using Eq. (28.3),

$$P_{01}(t + t_0) = \frac{\tau_{eq2}}{\tau_{c2}}\left(1 - e^{-t/\tau_{eq2}}\right) + P_{01}(t_0)e^{-t/\tau_{eq2}} \tag{28.9}$$

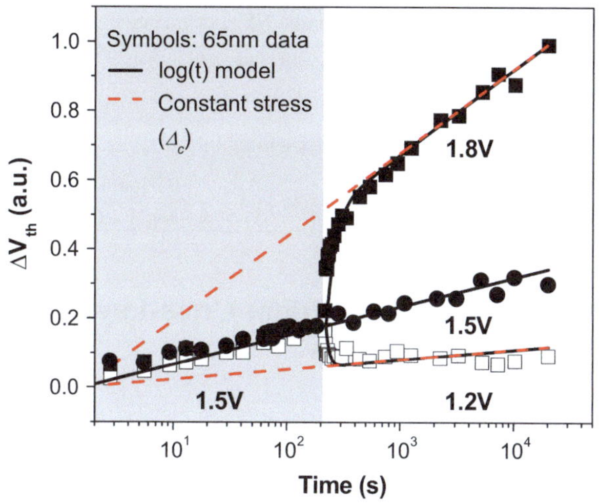

Fig. 28.5 The *log*(*t*) model predicts the dynamic behavior under voltage tuning, such as the transition period, and the convergence to the constant stress condition

where τ_{eq2}, τ_{c2} represent the time constants under voltage V_2. Using Eqs. (28.4) and (28.5), $\tau_{eq1} = \tau_{eq2}$, since τ_{eq} depends only on parameter p, which is independent of the voltage. Substituting this property and $P_{01}(t_0)$, we get

$$P_{01}(t+t_0) = \frac{\tau_{eq}}{\tau_{c1}}\left(1 - e^{-t/\tau_{eq}}\right) - \frac{\tau_{eq}}{\tau_{c2}}\left(e^{-t/\tau_{eq}} - e^{-(t+t_0)/\tau_{eq}}\right) \qquad (28.10)$$

where τ_{c1} and τ_{c2} correspond to voltages V_1 and V_2. Following similar steps as in static model derivation, we arrive at a closed-form solution

$$\Delta V_{th}(t) = \phi_2\left[A + B\log(1+Ct)\right] + \phi_1.B\left[\log\left(\frac{1+C(t+t_0)}{1+Ct}\right)\right] \qquad (28.11)$$

where ϕ_1 corresponds to the voltage V_1 and ϕ_2 corresponds to V_2. The degradation in Eq. (28.11) is physically interpreted as a sum of two components, Δ_1 and Δ_2 which are proportional to ϕ_1 and ϕ_2, respectively. When the voltage is changed to a lower voltage, traps emit some of the charge carriers, and the number of occupied traps reaches a new equilibrium. Δ_2 dominates initially, which contributes to the recovery. If the operation under V_2 continues for a longer time, Δ_1 eventually takes over and ΔV_{th} increases. Such a non-monotonic behavior is correctly predicted from Eq. (28.11). When the voltage is changed to a higher voltage, the degradation rate increases at the point of voltage change. Figure 28.5 validates the dynamic model. Non-monotonic behavior when voltage transition to a higher and lower value is correctly predicted by the model. Table 28.1 presents the summary of static and dynamic models.

Table 28.1 Summary of TD-based NBTI degradation models

Constant stress	$\Delta V_{th}(t) = \phi \cdot [A + B\log(1 + Ct)]$
Cycle to cycle	$\Delta V_{th}(t + t_0) = \Delta_1 + \Delta_2$ $\Delta_1 = \varphi\,(A + B\log(1 + Ct))$, $\Delta_2 = \Delta V_{th}(t_0)\left(1 - \dfrac{k + \log(1 + Ct)}{k + \log(1 + C(t + t_0))}\right)$

28.3 Circuit Reliability Prediction: Cross-Layer Modeling and Simulation Solutions

The accurate aging models at the device level are critical to predict the circuit aging. The timing paths which meet the timing requirements in the fresh circuit may turn critical over time due to aging, leading to a timing violation. **Sy**stem-level **R**eliability **A**nalyzer (**SyRA**) tool is developed to estimate the impact of aging in both digital and AMS circuits. Device-level compact aging models that predict the V_{th} shift of the transistor under various operating conditions are modified for digital and AMS designs. Both the power law model and the TD model are integrated into this framework for accurate and efficient lifetime prediction. For digital logic, the gate-level models that use device models to predict gate delay shift are developed. The new tool is integrated with standard Static Timing Analysis (STA) flow and is capable of computing delay shift in VLSI design with thousands of gates. For AMS circuits, the aging models are modified to predict long-term aging by incorporating AC dependence as well as varying bias conditions due to aging. The entire SyRA approach transfers the microscopic understanding of device aging physics into system-level reliability for digital and AMS designs.

With technology scaling, the relative importance of different aging mechanisms changes for different circuits as seen in Fig. 28.6. Digital circuits operating on full power supply are more affected by NBTI. Impact of CHC on digital circuits becomes less relevant as transistors degrade only during the switching phase. In AMS designs, transistors are usually biased in the saturation region, and sometimes long-channel devices are used. Due to these reasons, CHC is still pronounced in AMS designs.

28.3.1 SyRA for Digital Circuits

28.3.1.1 Long-Term Aging Model Under Dynamic Digital Operations

Under the extensive usage of DVS, it is preferred to have a long-term model that directly estimates aging at the end of a given operation time, without tracking the stress recovery over many cycles. This long-term model predicts a tight upper bound under multi-cycle operations under DVS. Based on the multi-cycle model in the

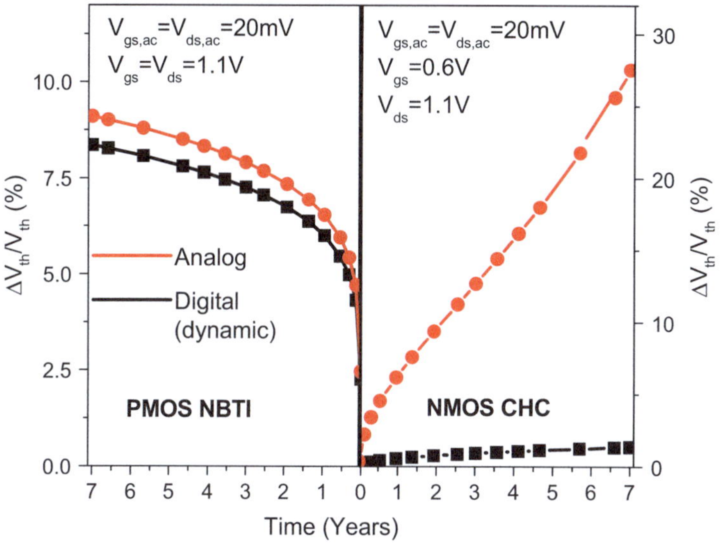

Fig. 28.6 NBTI dominates the aging of digital circuits, while both NBTI and CHC are important for AMS circuit aging

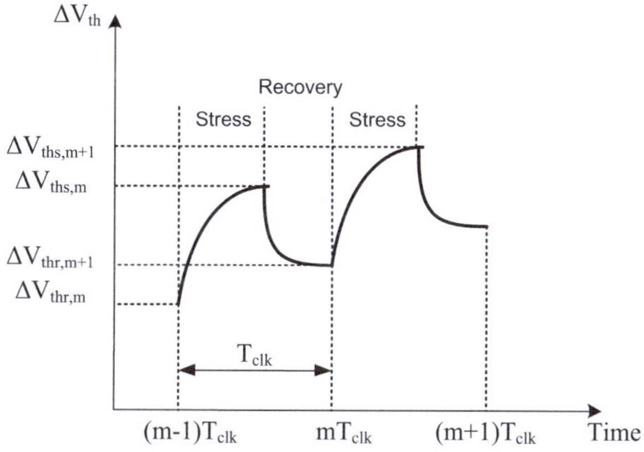

Fig. 28.7 The V_{th} shift during the stress and recovery cycles, representing the parameters used in the multi-cycle model

previous section, stress ($\Delta V_{ths,m}$) and recovery ($\Delta V_{ths,m+1}$) as shown in Fig. 28.7 are connected by

$$\Delta V_{ths,m+1} = \phi_1 \left[A + B\log\left(1 + C\alpha T_{Clk}\right)\right] + \phi_2 \left[A + B\log\left(1 + C(1-\alpha)T_{Clk}\right)\right]\beta_{1,m} + \Delta_{Vths,m}\left(1 - \beta_{1,m}\right)\left(1 - \beta_{2,m}\right) \tag{28.12}$$

Using Eq. (28.12) and repeatedly replacing the $\Delta V_{ths,m+1}$ by $\Delta V_{ths,i}$ for $i=m$, ….,1, we get

$$\Delta V_{ths,m} = \phi_1\left[A+B\log\left(1+C\alpha T_{Clk}\right)\right]\left(1+\sum_{i=1}^{m}\prod_{j=m-i+1}^{m}\beta_{1,.j}.\beta_{2,.j}\right)$$
$$+\phi_2\left[A+B\log\left(1+C\left(1-\alpha\right)T_{Clk}\right)\right]\beta_{1,m}\left(1+\sum_{i=1}^{m}\prod_{j=m-i+1}^{m}\beta_{1,.j-1}.\beta_{2,.j}\right) \quad (28.13)$$

Since obtaining a closed-form solution for Eq. (28.13) is not straightforward, we use the property $\beta_{1,m-1} < \beta_{1,m}$ and $\beta_{2,m-1} < \beta_{2,m}$:

$$\Delta V_{ths,m+1} \le \phi_1\left[A+B\log\left(1+C\alpha T_{Clk}\right)\right]\left(1+\beta_{1,.m}.\beta_{2,.m}+\left(\beta_{1,.m}.\beta_{2,.m}\right)^2+\ldots\right)$$
$$+\phi_2\beta_{1,m}\left[A+B\log\left(1+C\left(1-\alpha\right)T_{Clk}\right)\right]\left(1+\beta_{1,.m}.\beta_{2,.m}+\left(\beta_{1,.m}.\beta_{2,.m}\right)^2+\ldots\right) \quad (28.14)$$

Equation (28.14) is a geometric series and the upper bound of degradation is

$$\Delta V_{ths,m} = \phi_1\left[A+B\log\left(1+C\alpha T_{Clk}\right)\right]\frac{1}{1-\beta_{1,.m}.\beta_{2,.m}}$$
$$+\phi_2\left[A+B\log\left(1+C\left(1-\alpha\right)T_{Clk}\right)\right]\frac{\beta_{1,.m}}{1-\beta_{1,.m}.\beta_{2,.m}} \quad (28.15)$$

Equation (28.15) is sensitive to the duty cycle, α (ratio of time under V_1 to time under V_2), time period (sum of operation times under V_1 and V_2 for a single cycle), and the stress voltages. 40 cycles under 1.8 V, 1.2 V stress, and the cycle-to-cycle model capture the dynamic V_{th} shift. Figure 28.8 shows the cycle-to-cycle model matching with Monte Carlo simulations and experimental data; the long-term model directly captures the upper bound of multiple cycles. The long-term model captures the tight upper bound of cycle-to-cycle prediction, as illustrated in Fig. 28.8. Furthermore, Figure 28.9 presents the aging behavior under a wide range of duty cycles as validated with the silicon data. The degradation rate changes rapidly when $\alpha \sim 0$ or 1 but gradual for intermediate duty cycles. This behavior is due to the sudden change in degradation at the beginning of the stress and recovery phase, due to the voltage-dependent time constants. It is well predicted by a single equation in the long-term model. These V_{th} shift can result in delay shifts in digital logic circuits. With millions of gates, evaluation using each transistor becomes impractical. To overcome this constraint, gate delay models are derived.

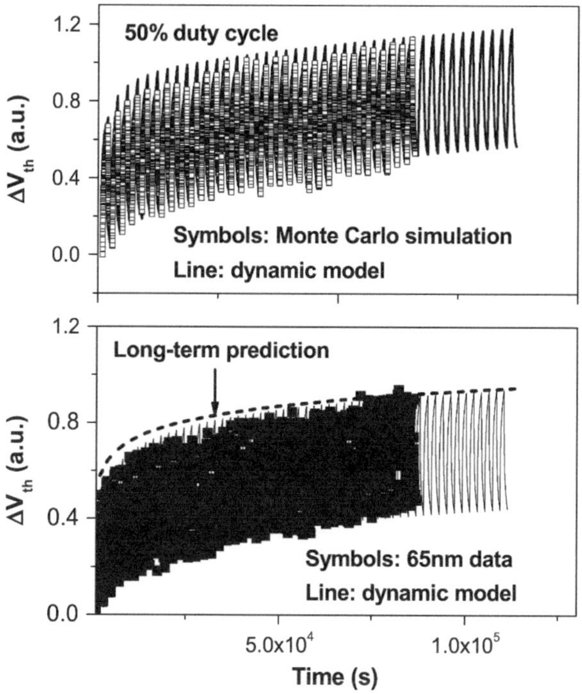

Fig. 28.8 The *log*(t)-based long-term model predicts a tight upper bound of aging under voltage tuning, as validated by TCAD, Si data, and the cycle-to-cycle model

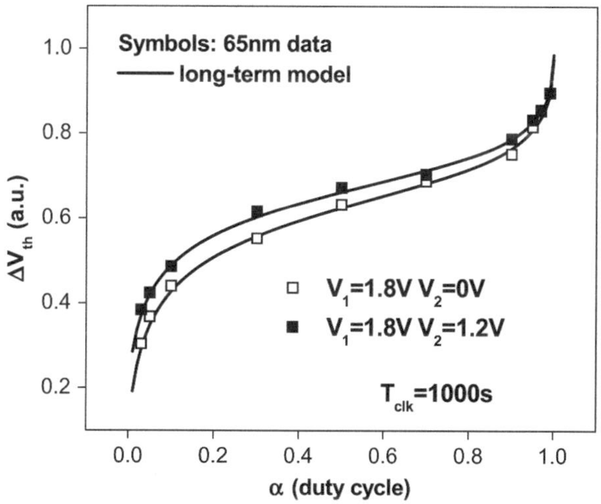

Fig. 28.9 Degradation behavior under a wide range of α (0.01–0.99), as predicted by the long-term model and validated with 65 nm silicon measurement

28.3.1.2 Gate Delay Models Under the Aging Effect

Previous works use complicated models and methodology to predict delay shifts in digital gates due to NBTI [48–53]. A simple gate delay model is shown in this chapter that calculates the delay change due to ΔV_{th} from V_{dd} information in the cell library. This model facilitates designers to estimate the delay degradation with time directly from the standard cell library, without having to rely on time-consuming circuit simulations to re-characterize the library under several aging conditions. Since the primary impact of NBTI at the device level is the increase in V_{th}, the primary effect at the circuit level is the increase in gate delay under larger V_{th}. Based on the drain current of a short channel device, the delay of a digital gate (t_d) is expressed by [54, 55]

$$t_d \propto \frac{CV_{dd}}{V_{dd} - V_{th}} \tag{28.16}$$

where C is the output capacitance of the gate. The change in gate delay when both V_{dd} and V_{th} are subject to change is

$$\frac{\Delta t_d}{t_d} = \frac{\Delta V_{dd}}{V_{dd}} - \frac{\Delta V_{dd} - \Delta V_{th}}{V_{dd} - V_{th}} \tag{28.17}$$

The delay change occurs when only V_{th} is changed ($\Delta t_d V_{th}$) or only when V_{dd} is subject to change ($\Delta t_d V_{dd}$) and is given by

$$\frac{\Delta t_{dVth}}{t_d} = \frac{\Delta V_{th}}{V_{dd} - V_{th}} \tag{28.18}$$

$$\frac{\Delta t_{dVdd}}{t_d} = \frac{-V_{th}\Delta V_{dd}}{V_{dd}\,(V_{dd} - V_{th})} \tag{28.19}$$

The above two equations can be combined to relate $\Delta t_d V_{dd}$ and $\Delta t_d V_{th}$:

$$\Delta t_{dVth} = -\frac{V_{dd}}{V_{th}}\left(\frac{\Delta V_{th}}{\Delta V_{dd}}\right)\Delta t_{dVdd} \tag{28.20}$$

This model calculates the delay shift due to V_{th} shift that is calculated by the long-term aging model. Figures 28.10 and 28.11 present the model validation with simulation results under wide range of output capacitance (C_L) and input slew rate (T_r).

The new model predicts the shift in gate delay in case of inverter and NAND gates where a single PMOS exists between switching input and output and also between V_{dd} and output. However, the situation is different for a gate like NOR, where there are multiple transistors between the switching input and output. Figure 28.12 shows two switching cases in a 2-input NOR gate. In Case 1, two PMOS transistors are present between the switching input (in1) and output, whereas a single transistor is present between switching input (in2) and output in Case 2. The gate delay model

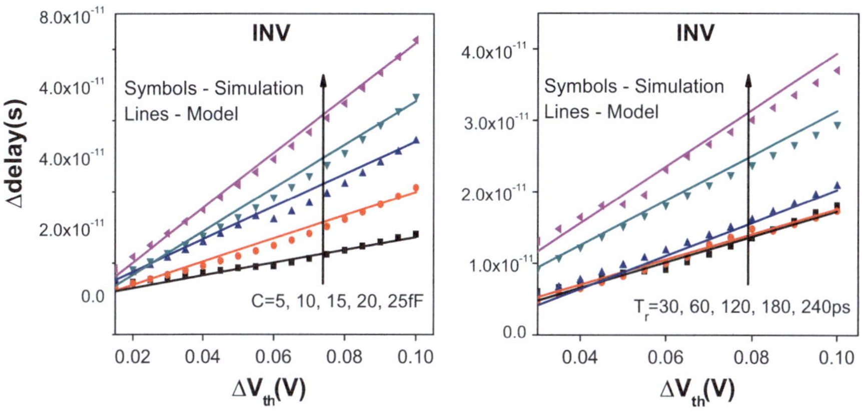

Fig. 28.10 Validation of delay model in an inverter under wide output capacitance and slew rate range

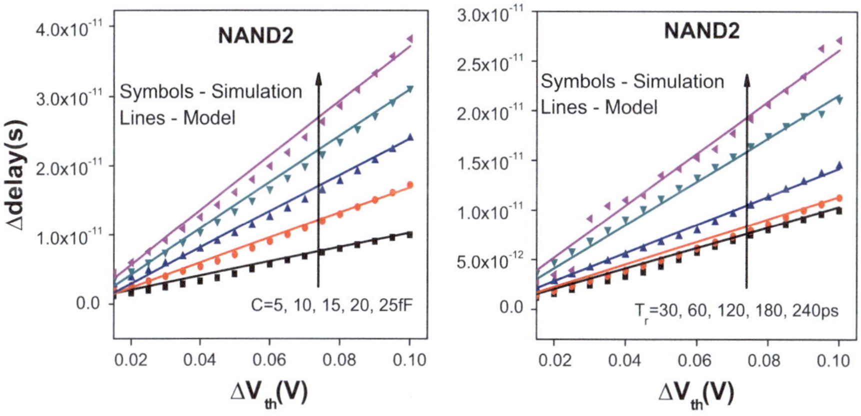

Fig. 28.11 Validation of delay model in NAND2 gate

for the inverter handles the situation in Case 1, since the same number (two) of transistors exists between input and output and between supply voltage and output (analogous to inverter). Hence, it can use the delay sensitivity to supply voltage and predict change in delay due V_{th} shift. However, in Case 2, the contribution of two PMOS devices toward the gate delay is different. The V_{th} shift in M2 has a larger impact on gate delay than M1, since M2 is in the path between switching input and output and also in the path between V_{dd} and output. The shift in delay due to V_{th} shift in this case is modeled by

$$\Delta t_{dVth} = -\frac{V_{dd}}{V_{th}} \left(\frac{k\Delta V_{th1} + \Delta V_{th2}}{2\Delta V_{dd}} \right) \Delta t_{dVdd} \qquad (28.21)$$

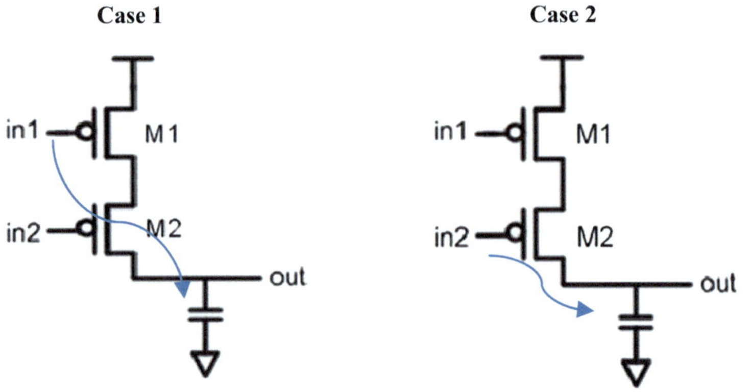

Fig. 28.12 Two switching cases in a 2-input NOR gate

where k (~0.25) denotes the contribution of V_{thM1} to the delay shift compared to that from V_{thM2} in Case 2. For a NOR gate with N inputs, the delay shift is

$$\Delta t_{dVth} = -\frac{V_{dd}}{V_{th}} \left(\frac{k \sum_{i=0}^{N-m} \Delta V_{thi} + \sum_{j=0}^{m} \Delta V_{thj}}{N \Delta V_{dd}} \right) \Delta t_{dVdd} \tag{28.22}$$

where m transistors exist in the path of the switching input and output that have more contribution toward delay. The delay model is validated in both the switching cases of NOR2 gate (Fig. 28.13). The prediction of delay shift is simple and accurate, enabling reliable failure assessment under aging effect.

28.3.1.3 Simulation Framework: Digital Circuits

Aging analysis in digital circuits can be implemented at the SPICE level where each transistor in the circuit is replaced by the sub-circuit model of the aged device [56]. For larger circuits, replacing every transistor in the circuit is not a practical approach, and a gate-level timing analysis is required. Figure 28.14 presents the experimental setup and static timing analysis framework implemented in SyRA. The aging-aware library is used to calculate the delay shift in digital gates. Our framework uses delay information under different V_{dd} in standard library and predicts the delay shift due to change in V_{th} using a simple gate delay model.

For a given digital circuit, we begin with Static Timing Analysis (STA) which generates a fresh timing report with timing information of all the paths in the circuit, without considering the NBTI effect. Logic analysis is performed on the circuit to obtain activity factors (α) in case of AC stress and node voltages in case of static stress. Based on the stress condition, PMOS V_{th} shift is calculated by the

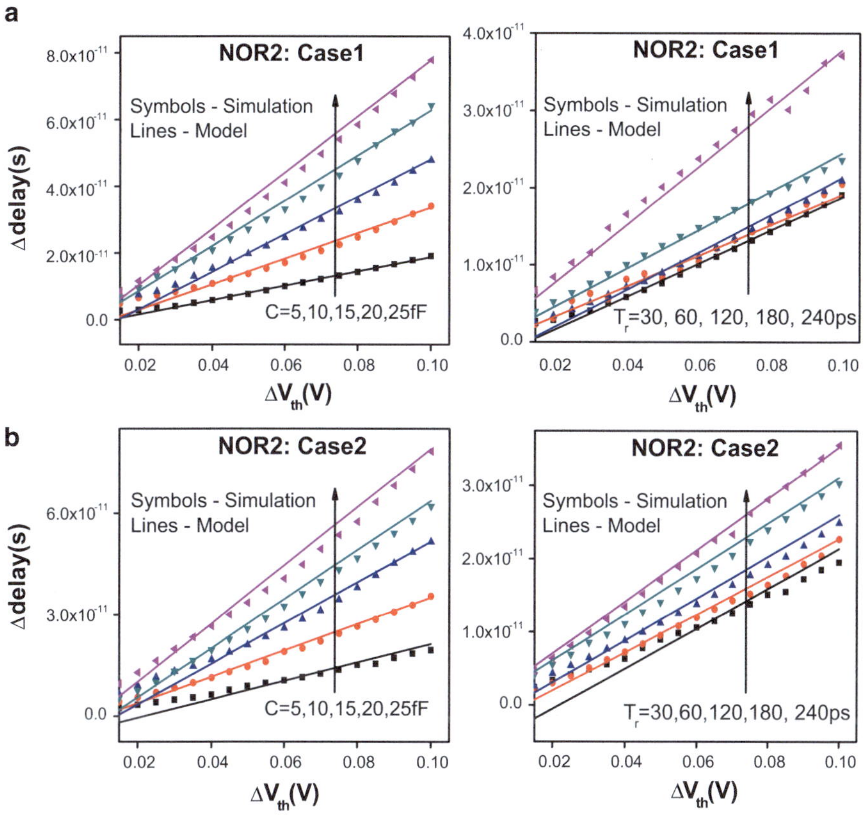

Fig. 28.13 Validation of two switching cases in a NOR2 gate: the nominal capacitance ~ 5fF and the nominal slew = 30 ps

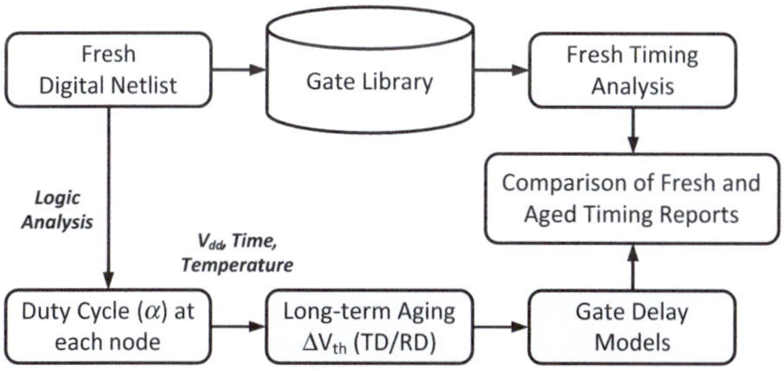

Fig. 28.14 The SyRA framework for failure diagnosis in digital circuits

long-term NBTI model, and gate delay shifts are computed using delay information from standard cell library under different slew rates and load capacitances. An aged timing report is then obtained by updating gate and path delay shifts in the fresh timing report, further helping identify the paths with timing violation.

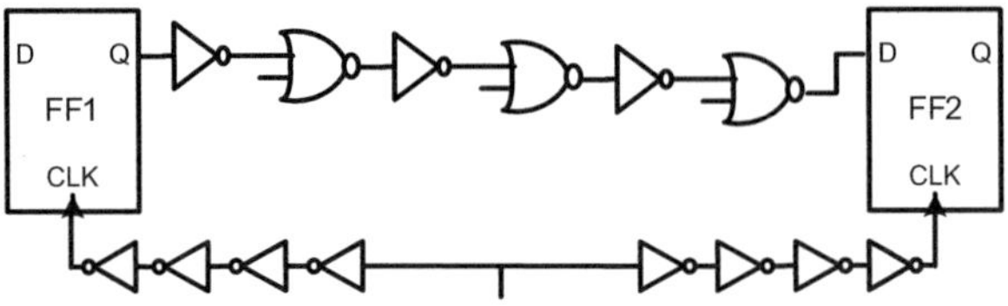

Fig. 28.15 Logic path in an ISCAS89 circuit; NBTI does not result in hold violation due to symmetric clock tree at two FFs

The SyRA framework is general and can be extended to other aging mechanisms such as positive bias temperature instability (PBTI). The strength of the implementation is that it is integrated into the standard STA flow. The sizing and complexity of the circuit on which the aging analysis can be run using this method directly depends on the STA capability being used. The implementation of this aging flow is performed using PrimeTime, a commercial STA tool from Synopsys. The library used in following aging analysis is a 45 nm Nangate standard cell library characterized with the Predictive Technology Model (PTM) [57]. The aging-aware delay model captures the shift in the rising delays of each gate in the circuit. The noncritical paths in the fresh circuit may turn critical over time due to aging, depending on the size of the paths and types of gates.

28.3.1.4 Design Example: Asymmetric Aging

The aging timing analysis illustrated above has to be performed at the critical instants in the operation. Aging at these critical moments is prominent and has maximum impact on circuit performance. NBTI increases the path delay of a logic path which depends on the type, size, and number of gates in the path. The increase in path delay causes the decrease of setup slack, leading to possible timing violation and logic failures. When a digital circuit is operated in alternate standby and normal modes, it is under alternate static and dynamic stress phases, respectively. The degradation in these phases is estimated using long-term models. The degradation at the end of a particular standby mode depends on the switching activity of the previous dynamic phase. NBTI-induced delay shifts are present in the clock tree as shown in the Fig. 28.15 (logic path from s5378, an ISCAS89 circuit). Since the clock tree is symmetrical to minimize skew, clock ticks at the same moment in both sequential elements even under aging (ignoring variations in degradation).When the STA tool is invoked, fresh timing report is generated which consists delays for all the paths in the circuit. STA tool is invoked to generate fresh timing report, and 20% paths with maximum accumulated delay are identified for setup analysis. Similarly, paths with minimum delay are identified for hold analysis; PrimeTime also supports logic analysis, which generates intermediate node voltages or α-value depending on the type of circuit operation. ΔV_{th} is predicted using long-term model based on the stress conditions. Gate delay shifts are computed by delay models in previous

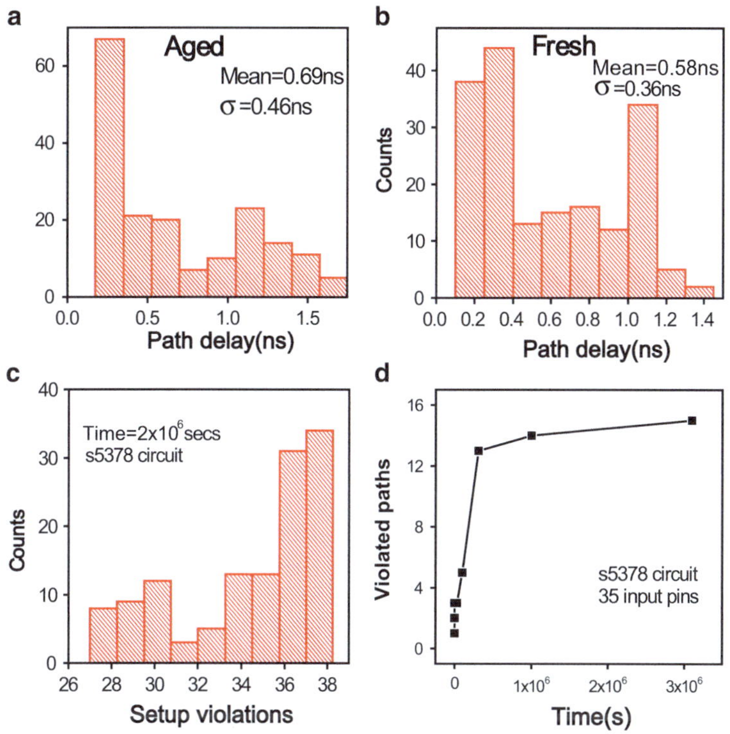

Fig. 28.16 Delay distribution in (**a**) fresh and (**b**) aged s5378 circuit, distribution of (**c**) setup violations under different input patterns, and (**d**) timing violations increase with time

section, and aged timing report is generated. Fig. 28.16a shows the distribution of shift in path delays when the proposed framework is implemented in s5378 circuit with 179 paths. The distribution of path delays has a mean (μ) of 0.58 ns and variance (σ) of 0.36 ns as shown in Fig. 28.16a. Such a wide distribution is due to the variety of gate types, path structures, and large number of gates in the circuit. Under aging, the mean increases by approximately 20% when the circuit is stressed for 2×10^6s. Since both the minimum and maximum path delays are shifted by same percentage, standard deviation increases due to aging and shifts by 25% in s5378 circuit (Fig. 28.16b). Though the fresh circuit meets the timing requirements, increase in path delays leads to timing violations under the aging effect.

Furthermore, the internal node voltages depend upon the input voltages when the circuit is in the standby mode. The input of any PMOS transistor can be either 0 (GND) or 1 (V_{dd}) depending on the input voltage pattern. If the input is 0, the PMOS V_{th} shifts, resulting in gate delay increase, and the delay is predominant in case of gates such as NOR4. The input pattern dependence is shown in Fig. 28.16c, showing the distribution of setup violations due to aging with 27 different input patterns. The distribution is wide, indicating the importance of input voltages on aging in failure

Table 28.2 Setup violations in ISCAS89 circuits

Design	Clock period (ns)	t = 1year	t = 5years	t = 10years
S27	0.48	1	1	1
S382	0.9	1	2	2
S420	0.87	1	1	2
S444	1.05	0	1	1
S510	0.95	1	1	1
S641	2.76	3	4	4
S713	2.91	4	4	5
S820	2.3	1	2	2
S832	2.4	2	2	2

diagnosis under NBTI. Figure 28.16d shows the timing violation of paths in s5378 circuit with stress time. Timing violations occur even when the stress time is low and the increase in number of violations gradually decreases for longer stress times. This behavior is similar to the ΔV_{th} shift with stress time.

Reliability simulation methodology is comprehensively demonstrated in different ISCAS89 circuits and is summarized in Table 28.2. As the number of gates and complexity of the circuit increases, there is an increase in the number of timing violations. The framework can be applied to any large-scale circuit using commercially available STA tools.

28.3.2 SyRA for Analog and Mixed Signal Circuits

Reliability simulation in AMS circuits is fundamentally more challenging than in digital circuits mainly due to different operating conditions encountered in AMS designs. Moreover, their design quality is much more sensitive to V_{th} of devices making them more vulnerable to aging effects. Here, a new approach of lifetime prediction for analog and mixed signal circuits is proposed.

28.3.2.1 Long-Term Aging Models for AMS Circuits

Operation Patterns of AMS Designs

Analog and mixed signal circuit operation is not confined to the saturation region. The operation range of AMS circuits extends to linear as well as sub-threshold regions in current low-power regime. Due to the wide range of operating voltages, a single degradation model for a specific stress condition is not able to predict long-term degradation for all AMS circuits. Hence, the first step to derive a long-term model is to classify AMS operations into three representative categories. Then long-term aging models are developed for each category in order to accurately predict the degradation and the lifetime of a given circuit.

Table 28.3 Classification of AMS circuits for aging prediction based on operation condition of NMOS

Circuit classification	Operating condition of NMOS	Input signal	Output signal
Type 1	Saturation	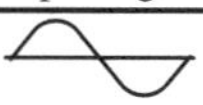	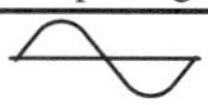
Type 2	Linear/Saturation		
Type 3	Saturation	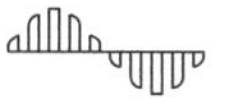	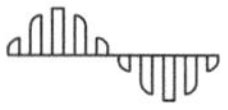

CHC aging mechanism in NMOS has a strong dependence on both gate-source voltage (V_{gs}) and drain-source voltage (V_{ds}), while NBTI in PMOS is only a strong function of V_{gs} but relatively independent on V_{ds}. Therefore, the classification, which is based on the operation characteristics of a transistor, mainly targets the CHC effect as follows:

(a) *Linear circuits* (type 1) are biased in the saturation region for their entire operation, such as the amplifier design. Both gate and drain terminals have a fixed DC bias with small AC signal as shown in Table 28.3. Since the transistors are always in saturation, the degradation is continuous, and V_{th} shift accumulates through varying circuit conditions over the lifetime.
(b) *Nonlinear circuits* (type 2) typically employ positive feedback in an effort to get a large output swing. Comparators are the example, which compare input signal with a reference signal and gives a rail to rail output. Due to the large swing at the drain terminal of NMOS, the operation can be either in the saturation or the linear region. As a result, CHC degradation which occurs only in the saturation region becomes a function of both input and reference voltage signals.
(c) *Dynamic circuits* (type 3) include amplifiers used in the sample and hold unit. Though these circuits are biased in the saturation region, AC signals on gate and drain terminals are not continuous. In one operation phase, the amplifier is in the reset mode due to which CHC degradation is not impacted by AC signal. Thus, the degradation is a function of both DC and AC signal in one phase, while a function of DC signals only in other. This leads to different degradation rate in each cycle.

Long-Term CHC Models

Charge generation is localized to drain region rendering CHC to be a strong function of drain-source voltage. The short-term degradation model for CHC is described by [32]

$$d\Delta {V_{th}}^{1/n'} = \left[\frac{q}{C_{ox}} K_2 \sqrt{C_{ox}\left(V_{gs} - V_{th}\right)} e^{\frac{V_{gs}-V_{th}}{t_{ox}E_{o2}} - \frac{\varphi_{it} l_p}{q\lambda_f\left(V_{ds}-V_{Dsat}\right)}} \right]^{1/n'} dt \qquad (28.23)$$

Table 28.4 Summary of long-term CHC models for AMS circuits

Type 1	$\Delta V_{th} = \frac{q}{C_{ox}} K_2 \sqrt{C_{ox}(V_{gs,dc} - V_{th}) e^{\frac{V_{gs,dc}-V_{th}}{t_{ox}E_{o2}}}}\, e^{-\frac{\phi_{it} l_p}{q\lambda_f (V_{ds,dc}-V_{dsat})}} [F_2(t)]^{n'}$
	$F_2(t) = t + \left[(b^2 - 2b)\frac{c^2}{2} + abc + \frac{a^2}{2}\right]\frac{t}{2}$
	$a = \frac{V_{gs,ac}}{n' t_{ox} E_{o2}}, b = \frac{\phi_{it} l_p}{n' q \lambda_f (V_{ds,dc} - V_{dsat})}, c = \frac{V_{ds,ac}}{V_{ds,dc} - V_{dsat}}$
Type 2	$\Delta V_{th} = \frac{q}{C_{ox}} K_2 \sqrt{C_{ox}(V_{gs,dc} - V_{th}) e^{\frac{V_{gs,dc}-V_{th}}{t_{ox}E_{o2}}}}\, \beta e^{-\frac{\phi_{it} l_p}{q\lambda_f (V_{ds,dc}-V_{dsat})}} [F(t)]^{n'}$
	$F(t) = \left[t + \frac{1}{2}\left(\frac{4V_{gs,ac}}{E_{o1} t_{ox}}\right)\frac{t}{2} + \frac{1}{24}\left(\frac{4V_{gs,ac}}{E_{o1} t_{ox}}\right)^4 \frac{3}{8} t\right]$
Type 3	$\Delta V_{th1} = \frac{q}{C_{ox}} K_2 \sqrt{C_{ox}(V_{gs,dc} - V_{th}) e^{\frac{V_{gs,dc}-V_{th}}{t_{ox}E_{o2}}}}\, e^{-\frac{\phi_{it} l_p}{q\lambda_f (V_{ds,dc}-V_{dsat})}} [t_1]^{n'}$
	$\Delta V_{th2} = \frac{q}{C_{ox}} K_2 \sqrt{C_{ox}(V_{gs,dc} - V_{th}) e^{\frac{V_{gs,dc}-V_{th}}{t_{ox}E_{o2}}}}\, e^{-\frac{\phi_{it} l_p}{q\lambda_f (V_{ds,dc}-V_{dsat})}} [F_2(t_2)]^{n'}$
	$\Delta V_{th} = \Delta V_{th1} + \Delta V_{th2}$

where $n' \sim 0.45$. For type 1 circuits, since V_{gs} and V_{ds} have both DC and AC components, they can be expressed as $V_{gs} = V_{gs,dc} + V_{gs,ac}\sin\omega t$ and $V_{ds} = V_{ds,dc} + V_{ds,ac} \sin\omega t$. Substituting this into Eq. (28.23), the direct integration leads to Bessel functions of the first type. To simplify the solution, we expand the exponential term in Eq. (28.23) using the Taylor series:

$$\Delta V_{th}{}^{1/n'} \approx \left[\frac{q}{C_{ox}} K_2 \sqrt{C_{ox}\left(V_{gs,dc} - V_{th}\right) \mathrm{e}^{\frac{V_{gs,dc}-V_{th}}{t_{ox}E_{o2}} - \frac{\varphi_{it} l_p}{q\lambda_f \left(V_{ds,dc}-V_{Dsat}\right)}}}\right]^{1/n'} .F_2(t) \quad (28.24)$$

$$F_2(t) \approx t + \left[\left(b^2 - 2b\right)\frac{c^2}{2} + abc + \frac{a^2}{2}\right]\frac{t}{2} \quad (28.25)$$

and a, b, and c are dimensionless intermediate parameters:

$$a = \frac{V_{gs,ac}}{n' t_{ox} E_{o2}}, b = \frac{\varphi_{it} l_p}{n' q \lambda_f \left(V_{ds,dc} - V_{Dsat}\right)}, c = \frac{V_{ds,ac}}{V_{ds,dc} - V_{Dsat}} \quad (28.26)$$

In general, Table 28.4 summarizes the long-term CHC models for three types of AMS designs incorporating DC and AC dependence. Similarly, the long-term NBTI model can be derived. Leveraging these equations, a new framework is implemented for lifetime prediction.

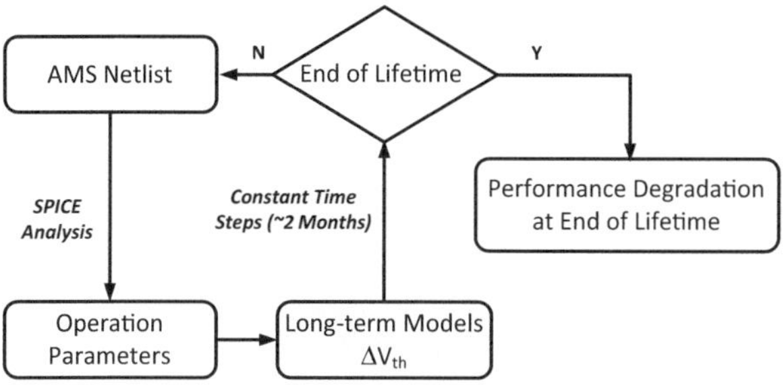

Fig. 28.17 SyRA simulation framework for AMS circuits

28.3.2.2 Simulation Framework: AMS Circuits

Besides having accurate long-term models for AMS designs, it is essential to have an efficient simulation methodology to predict aging of various circuit parameters and to estimate lifetime at the design stage. Commercial reliability tools [20–23] perform aging prediction by the extrapolation method (Fig. 28.2). The extrapolation points are calculated based on short-term stressed model parameters. Thus, they do not account for the changing operating conditions due to V_{th} degradation. In reality, V_{th} change results in higher drain-source voltage which further accelerates V_{th} shift. This may even lead to positive feedback between V_{th} shift and V_{ds}, causing an exponential degradation over the stress time instead of the expected power law relation. Hence, commercial tools may significantly underestimate aging in AMS circuits. In addition, these tools require a large amount of memory for short-term SPICE simulations in order to generate pre-aged models and collect enough sampling points for the extrapolation.

To overcome these limits, an efficient simulation methodology needs to be developed. Figure 28.17 presents the SyRA flow for AMS design. As a cornerstone, the long-term degradation models are used for accurate and efficient aging prediction of AMS circuits. We start with a test bench and input conditions which are fed to the SPICE simulator to obtain the operation voltages based on DC and transient analysis. The long-term aging models use the extracted node voltages and input conditions to predict threshold voltage shift for all transistors in the circuit.

The long-term models can predict device-level degradation for any given time, if the operation conditions remain constant. However, as V_{th} shift affects the bias conditions in AMS circuits, the aging prediction is performed for constant time step, followed by the update of the operation conditions. The time step is chosen to keep the prediction error low. After one time step, the long-term models use updated node voltages to predict the degradation for the next time step. This process is repeated for N = lifetime/step size times. Many works suggest the use of logarithmic time steps in order to improve the simulation speed, assuming very less change in degradation after certain period of time [58–60]. However, for such cases where the degradation is exponentially increased, a logarithmic time step will present a dramatic amount

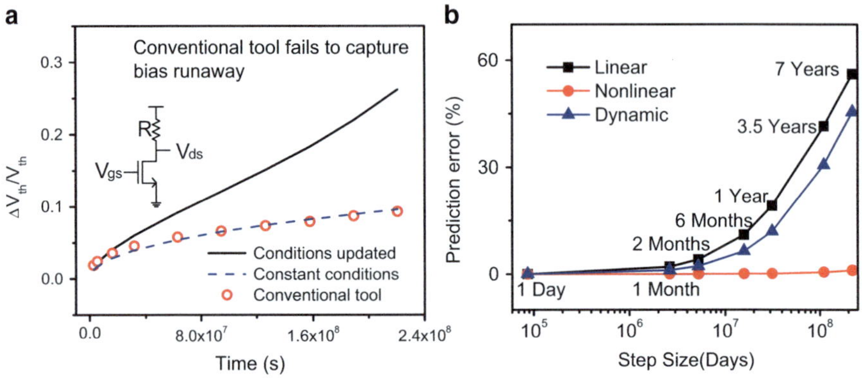

Fig. 28.18 (**a**) Validation of proposed long-term model and SyRA with a conventional method [21] and (**b**) prediction error with step size

of error, justifying the need of constant time steps for accurate analysis, as in SyRA. Since SyRA does not involve extracting pre-aged parameters with short-term SPICE simulations, it consumes much lower memory and is faster compared to commercial tools. Due to high efficiency and accuracy, this methodology promises reliable aging analysis in AMS designs.

28.3.2.3 Tool Evaluation with Benchmark Circuits

As the initial validation of SyRA, a simple resistive load common source circuit is used. The aging prediction by SyRA and conventional tool is presented in Fig. 28.18a [21]. The new analysis method predicts the same V_{th} shift when the parameters are kept constant. However, in actual circuit environment, V_{th} shift changes the operating condition, thereby changing the degradation rate (solid curve in Fig. 28.18a).

The time step for updating circuit parameters plays an important role in accurate aging analysis. Larger time steps would cause high error while finer steps reduce the efficiency. Figure 28.18b shows the prediction error for different types of circuits. Based on the error evaluation, the optimal time step is determined to be 2 months for all further analysis. The simulation methodology is demonstrated in 65 nm benchmark circuits in this section [57], with significant improvement in simulation efficiency and accuracy compared to commercial tools [21].

Differential Operational Amplifier: Figure 28.19a presents the schematic of a differential amplifier (type 1) used for reliability analysis. Since the transistors always operate in the saturation region, V_{th} shift is large due to continuous stress. Current mirror-based operational amplifier is used intentionally in the example to induce loading mismatch. Due to loading mismatch of the input transistors, they experience different degradation rate. This further increases the mismatch between

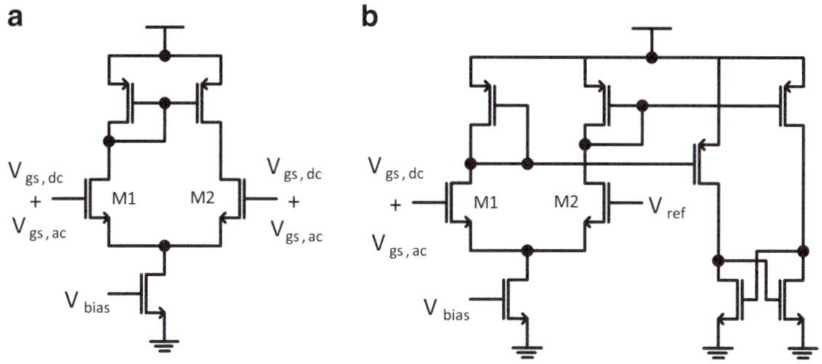

Fig. 28.19 Schematics of 65 nm benchmark circuits: (**a**) Amplifier with load mismatch. (**b**) Comparator with a preamplifier and a latch

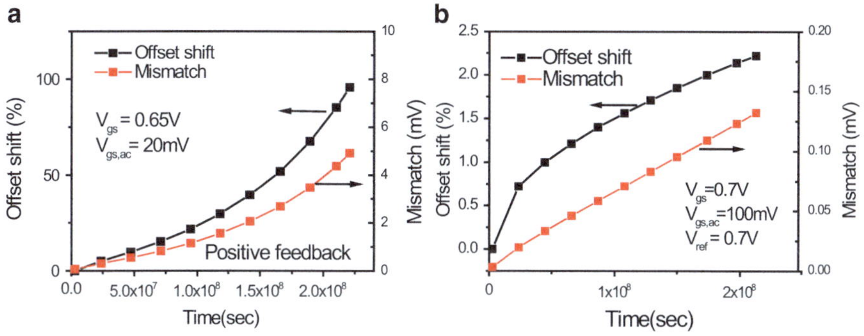

Fig. 28.20 (**a**) Offset shift and mismatch increase due to different loads in amplifier and (**b**) aging prediction of the comparator

input NMOS devices. Due to this reason, the offset of the given circuit increases at a much higher rate, eventually leading to circuit failure. This behavior is well captured by SyRA. Figure 28.20a presents the increase in offset and the mismatch of input transistors of the amplifier. Over a lifetime of 7 years, the offset increases by more than 90%.

Comparator: Comparators are nonlinear circuits (type 2) where the transistors drive out of the saturation region due to large voltage swing. These circuits are usually cascade stages of linear amplifiers with the latch at the end for generating rail to rail voltage signal. Figure 28.19b presents the schematic of a comparator with such configuration. The preamplifier belongs to type 1 circuits and the latch stage falls into the category of type 2. Appropriate long-term aging models are applied to correctly predict the degradation. Figure 28.20b shows the shift in the offset and the mismatch of input transistors. Due to symmetric loading, the mismatch is much lower compared to the previous case. Similarly, the offset shift is much less compared to the amplifier.

Table 28.5 Efficiency comparison with a commercial reliability tool

	3 Years			
Lifetime	Memory Usage (MB)		Simulation Time (ms)	
Amplifier	5.10	0.20	4.45E+04	5.88E+03
Comparator	5.25	0.35	5.25E+04	4.28E+04
SH-Amplifier	5.10	0.20	4.45E+04	6.30E+03
	7 Years			
Lifetime	Memory Usage (MB)		Simulation Time (ms)	
Amplifier	5.13	0.20	8.57E+04	1.26E+04
Comparator	5.26	0.35	1.08E+05	8.57E+04
SH-Amplifier	5.13	0.20	8.65E+04	1.43E+04

□ Commercial tool [21]
▒ SyRA for AMS

Sample and Hold Amplifier: The sample and hold (SH) amplifier has a similar structure as shown in Fig. 28.19a. Since such an amplifier operates in sample and reset phases, the long-term model for type 3 circuits is used. AC signals are not present in the sample mode as the amplifier resets itself, and hence, the degradation is comparatively less than that in type 1 circuits. The operation conditions of these circuits are also updated.

Efficiency Comparison: In summary, Table 28.5 compares the memory usage and simulation time of benchmark circuits compared to a commercial tool [21]. SyRA achieves up to 10× speed up in the prediction and 30× reduction in memory usage compared to the commercial tool.

28.3.2.4 Design Example: Bias Runaway

As an emerging threat, bias runaway represents an intriguing challenge to reliability analysis. Since constant current sources used in AMS circuits are difficult and expensive to design, an AMS design typically only uses a couple of them to bias the entire design. Figure 28.21a presents such current mirroring circuit where a NMOS transistor copies the current from the reference I_{bias} generating the bias voltage for a folded cascode amplifier. In order to accurately copy the current, gate length of the mirroring transistor is usually longer than the minimum length, leading to higher V_{gs} (V_{bias}). Under CHC, higher V_{bias} causes higher degradation in threshold voltage which further increases V_{bias}. Depending on the initial design value, there exists a relation between device degradation and the operating condition which becomes unstable after a critical condition. The positive feedback of such condition leads to an exponential shift in V_{bias}, termed as "bias runway" in AMS circuits. Due this phenomenon, the gain decreases at a much faster rate than expected, while SyRA is able to track the rapid change with constant time step as seen in Fig. 28.21b. It is very important to identify such critical design conditions, especially in IO circuits, and mitigate them at early design stage.

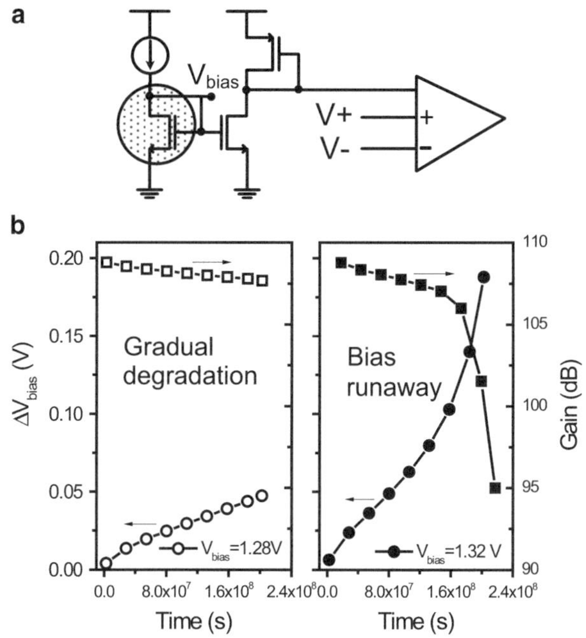

Fig. 28.21 (**a**) Biasing circuit for a folded cascode amplifier; (**b**) improper bias design may lead to bias runaway and much faster degradation

28.4 Summary

A cross-layer approach of circuit reliability prediction is presented in this chapter. Figure 28.22 illustrates the approach from the device-level to system-level aging analysis. Physical mechanisms contributing to aging are investigated at the device level. These device-level models are accordingly customized for digital and AMS circuit operations to enable accurate and efficient long-term prediction.

Based on long-term aging models, a new **Sy**stem-level **R**eliability **A**nalyzer (**SyRA**) tool is developed for digital and AMS circuits. SyRA for digital circuit is integrated with standard Static Timing Analysis (STA) flow and is capable of aging computation in circuits with thousands of gates. The new tool supports the analysis of timing errors in critical paths and provides the necessary information for guard banding and adaptive design protection.

For AMS designs, aging analysis is fundamentally more difficult than in digital circuit due to the varying nature of operation conditions. To handle the diversity in bias conditions, SyRA identifies representative operation characteristics to develop long-term AMS aging models; it further employs finite-step analysis to achieve high accuracy, especially in the case where positive feedback may occur in circuit aging.

With the significant advantage of SyRA in simulation efficiency and accuracy, the new tool better equips IC designers in design practice for reliability.

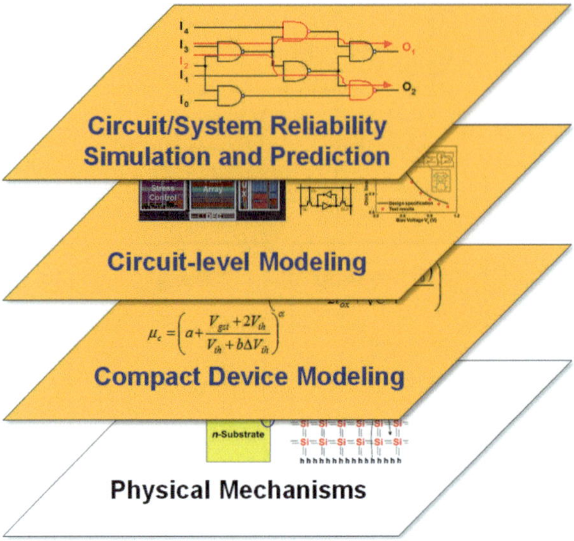

Fig. 28.22 A cross-layer approach of circuit reliability analysis

References

1. C. Constantinescu, "Trends and challenges in vlsi circuit reliability", *IEEE Computer Society*, pages 14–19, 2003.
2. Semiconductor Industry Associate, *International Technology Roadmap for Semiconductors*, 2010. (Available at http://public.itrs.net).
3. D. K. Schroder and J. A. Babcock, "Negative bias temperature instability: Road to cross in deep submicron silicon semiconductor manufacturing," *Journal of Applied Physics*, vol. 94, no. 1, pp. 1–18, 2003.
4. A. R. Brown, V. Huard and A. Asenov, "Statistical simulation of progressive NBTI degradation in a 45-nm technology pMOSFET," *IEEE Transactions on Electron Devices*, vol. 57, no. 9, pp. 2320–2323, Sept. 2010.
5. T. Austin, V. Bertacco, S. Mahlke, Y. Cao, "Reliable systems on unreliable fabrics," *IEEE Design & Test of Computers*, vol. 25, no. 4, pp. 322–332, July/August 2008.
6. B. H. Calhoun, Y. Cao, X. Li, K. Mai, L. T. Pileggi, R. A. Rutenbar, and K. L. Shepard, "Digital circuit design challenges and opportunities in the era of nanoscale CMOS," *Proceedings of the IEEE*, vol. 96, no. 2, pp. 343–365, Feb. 2008.
7. S. Borkar, T. Karnik, S. Narendra, J. Tschanz, A. Keshavarzi, V. De, "Parameter variations and impact on circuits and microarchitecture," *Design Automation Conference*, pp. 338–342, 2003.
8. F. Arnaud, L. Pinzelli, C. Gallon, M. Rafik, P. Mora, F. Boeuf, "Challenges and opportunity in performance, variability and reliability in sub-45 nm CMOS technologies," *Microelectronics Reliability*, vol. 51, no. 9–11, pp. 1508–1514, Sept.-Nov. 2011.
9. N. Kimizuka, T. Yamamoto, T. Mogami, K. Yamaguchi, K. Imai, and T. Horiuchi, "The impact of bias temperature instability for direct-tunneling ultra-thin gate oxide on MOSFET scaling," *VLSI Symposium on Technology*, pp. 73–74, 1999.
10. A. S. Goda, G. Kapila, "Design for degradation: CAD tools for managing transistor degradation mechanisms," *International Symposium on Quality Electronics and Design*, pp. 416–420, 2005.

11. J. Hicks, D. Bergstrom, M. Hattendorf, J. Jopling, J. Maiz, S. Pae, et al., "45 nm transistor reliability," *Intel Technology Journal*, vol. 12, no. 02, pp. 131–144, Jun. 2008.
12. V. Huard, N. Ruiz, F. Cacho, E. Pion, "A bottom-up approach for system-on-chip reliability," *Microelectronics Reliability*, vol. 51, no. 9–11, pp. 1425–1439, Sept.-Nov. 2011.
13. V. Huard, C. R. Parthasarathy, A. Bravaix, T. Hugel, C. Guerin, E. Vincent, "Design-in-reliability approach for NBTI and hot-carrier degradations in advanced nodes," *IEEE Transactions on Device and Materials Reliability*, vol. 7, no. 4, pp. 558–570, Dec. 2007.
14. S. Mitra, K. Brelsford, Y. Kim, K. Lee, Y. Li, "Robust system design to overcome CMOS reliability challenges," *IEEE Journal on Emerging and Selected Topics in Circuits and Systems*, vol. 1, no. 1, pp. 30–41, Mar. 2011.
15. J. Lee, I. Chen and C. Hu, "Modeling and characterization of oxide reliability," *IEEE Transactions on Electron Devices*, vol. 35, no. 12, pp. 2268–2278, Dec. 1988.
16. C. Hu, S. C. Tam, F.-C. Hsu, P.-K. Ko, T.-Y. Chan, K. W. Terrill, "Hot-electron-induced MOSFET degradation – model, monitor, and improvement," *IEEE Transactions on Electron Devices*, vol. 32, no. 2, pp. 375–385, Feb. 1985.
17. J. B. Bernstein, M. Gurfinkel, X. Li, J. Walters, Y. Shapira, M. Talmor, "Electronic circuit reliability modeling," *Microelectronics Reliability*, vol. 46, no. 12, pp. 1957–1979, Dec. 2006.
18. J. J. Clement, "Electromigration modeling for integrated circuit interconnect reliability analysis," *IEEE Transactions on Device and Materials Reliability*, vol. 1, no. 1, pp. 33–42, Mar. 2001.
19. E. Rosenbaum, "Interconnect thermal modeling for accurate simulation of circuit timing and reliability," *IEEE Transactions on Computer-Aided Design of Integrated Circuits and Systems*, vol. 19, no. 2, pp. 197–205, Feb. 2000.
20. R. Tu, E. Rosenbaum, W. Chan, C. Li, E. Minami, K. Quader, et al., "Berkeley reliability tools – BERT," *IEEE Transactions on Computer-Aided Design of Integrate Circuits and Systems*, vol. 12, no. 10, pp. 1524–1534, Oct. 1993.
21. *Reliability Simulation in Integrated Circuit Design*, Cadence, 2003.
22. *MOS Device Aging Analysis with HSPICE and CustomSim*, Synopsys, 2011.
23. *ELDO User's Manual*, Mentor Graphics 2005.
24. S. Ogawa and N. Shiono, *Physical Review B*, vol. 51, no. 7, pp. 4218–4230, Feb. 1995.
25. E. Maricau and G. Gielen, "Computer-aided analog circuit design for reliability in nanometer CMOS," *IEEE Transactions on Emerging and Selected Topics in Circuits and Systems*, vol. 1, no. 1, pp. 50–58, Mar. 2011.
26. K. O. Jeppson and C. M. Svensson, "Negative bias stress of MOS devices at high electric fields and degradation of MNOS devices," *Journal of Applied Physics*, vol. 48, no. 5, pp. 2004–2014, 1977.
27. G. Chen, K. Y. Chuah, M. F. Li, D. S. H. Chan, C. H. Ang, J. Z. Zheng, Y. Jin, et al., "Dynamic NBTI of PMOS transistors and its impact on device lifetime," *International Reliability Physics Symposium*, pp. 196–202, 2003.
28. S. Chakravarthi, A. T. Krishnan, V. Reddy, C. F. Machala and S. Krishnan, "A comprehensive framework for predictive modeling of negative bias temperature instability," *International Reliability Physics Symposium*, pp. 273–282, 2004.
29. M. A. Alam, S. Mahapatra, "A comprehensive model of PMOS NBTI degradation," *Microelectronics Reliability*, vol. 45, pp. 71–81, 2005.
30. V. Huard, C. R. Parthasarathy, C. Guerin, M. Denais, "Physical modeling of negative bias temperature instabilities for predictive extrapolation," *International Reliability Physics Symposium*, pp. 733–734, 2006.
31. S. V. Kumar, C. H. Kim, and S. S. Sapatnekar, "An analytical model for Negative Bias Temperature Instability (NBTI)," *International Conference for Computer Aided Design*, pp. 493–496, 2006.

32. W. Wang, V. Reddy, A. Krishnan, R. Vattikonda, S. Krishnan and Y. Cao, "Compact modeling and simulation of circuit reliability for 65 nm CMOS technology," *IEEE Transactions on Device and Materials Reliability*, vol. 7, no. 4, pp. 509–517, Dec. 2007.
33. M. Denais, C. R. Parthasarathy, G. Ribes, Y. Rey-Tauriac, N. Revil, A. Bravaix, V. Huard, F. Perrier, "On-the-fly characterization of NBTI in ultra-thin gate oxide PMOSFET's," *International Electron Devices Meeting*, pp. 109–112, 2004.
34. H. Reisinger, O. Blank, W. Heinrigs, A. Muhlhoff, W. Gustin, C. Schlunder, "Analysis of NBTI degradation- and recovery-behavior based on ultra fast V_T-measurements," *International Reliability Physics Symposium*, 2006.
35. V. Huard, M. Denais, "Hole trapping effect on methodology for DC and AC negative bias temperature instability measurements in pMOS transistors," *International Reliability Physics Symposium*, pp. 40–45, 2004.
36. T. Grasser, B. Kaczer, W. Goes, H. Reisinger, T. Aichinger, P. Hehenberger, et al., "The paradigm shift in understanding the bias temperature instability: from reaction–diffusion to switching oxide traps," *IEEE Transactions on Electron Devices*, vol. 58, no. 11, pp. 3652–3666, Nov. 2011.
37. G. I. Wirth, R. da Silva and B. Kaczer, "Statistical model for MOSFET bias temperature instability component due to charge trapping," *IEEE Transactions on Electron Devices*, vol. 58, no. 8, pp. 2743–2751, Aug. 2011.
38. J. Velamala, K. Sutaria, T. Sato and Y. Cao, "Aging statistics based on trapping/detrapping: silicon evidence, modeling and long-term prediction," to appear at *International Reliability Physics Symposium*, 2012.
39. J. Keane, W. Zhang and C. Kim, "An on-chip monitor for statistically significant circuit aging characterization," *IEEE International Electron Devices Meeting*, pp. 4.2.1-4.2.4, 2010.
40. T. Grasser, H. Reisinger, W. Goes, T. Aichinger, P. Hehenberger, P.-J. Wagner, M. Nelhiebel, J. Franco, B. Kaczer, "Switching oxide traps as the missing link between negative bias temperature instability and random telegraph noise," *Electron Devices Meeting (IEDM), 2009 IEEE International*, pp.1-4, 2009.
41. G. I. Wirth, J. Koh, R. da Silva, R. Thewes, and Ralf Brederlow, "Modeling of statistical low-frequency noise of deep-submicron MOSFETs," *Trans. on Electron Dev.*, vol. 52, pp. 1576–1588, 2005.
42. A. P. van der Wel, E. A. M. Klumperink, J. S. Kolhatkar, E. Hoekstra, M. S. Snoeij, C. Salm, H. Wallinga, B. Nauta, "Low-Frequency Noise Phenomena in Switched MOSFETs," *Solid-State Circuits, IEEE Journal of* , vol. 42, no. 3, pp.540-550, March 2007.
43. M. J. Kirton and M. J. Uren, "Noise in solid-state microstructures: A new perspective on individual defects, interface states and sow-frequency (1/f) noise", *Advances in Physics*, vol. 38, p. 367–468, 1989.
44. G. Wirth, R. da Silva and R. Brederlow, "Statistical model for the circuit bandwidth dependence of low-frequency noise in deep-submicrometer MOSFETs," *Trans. on Electron Dev*, vol. 54, pp. 340-345, Feb. 2007.
45. G. Wirth, R. da Silva, P. Srinivasan, J. Krick and R. Brederlow. "Statistical model for MOSFET low-frequency noise under cyclo-stationary conditions," *Electron Devices Meeting (IEDM)*, p.30.5.1-4, 2009.
46. B. Kaczer, T. Grasser, Ph. J. Rousse, J. Martin-Martinez, R. O'Connor, B. J. O'Sullivan, and G. Groeseneken, "Ubiquitous relaxation in BTI stressing—new evaluation and insights," *IRPS*, pp. 20–27, 2008.
47. T. Grasser, H. Reisinger, P.-J. Wagner, F. Schanovsky, W. Goes and B. Kaczer, "The time dependent defect spectroscopy (TDDS) for the characterization of the bias temperature instability," *IRPS*, pp. 16–25, 2010.
48. V. Reddy, A. Krishnan, A. Marshall, J. Rodriguez, S. Natarajan, T. Rost, S. Krishnan, "Impact of negative bias temperature instability on digital circuit reliability," *IEEE International Reliability Physics Symposium*, pp. 248–254, 2002.

49. B. C. Paul, K. Kang, H. Kufluoglu, M. A. Alam, and K. Roy, "Impact of NBTI on the temporal performance degradation of digital circuits," *IEEE Electron Device Letters*, vol. 26, no. 8, pp. 560–562, Aug. 2003.
50. K. Hang, S. Gangwal, S. P. Park, and K. Roy, "NBTI induced performance degradation in logic and memory circuits: how effectively can we approach a reliability solution?" *Asia and South Pacific Design Automation Conference*, pp. 726–731, 2008.
51. T. Nigam, "Impact of transistor level degradation on product reliability," *Custom Integrated Circuits Conference*, pp. 431–438, 2009.
52. W. Wang, S. Yang, S. Bhardwaj, R. Vattikonda, S. Vrudhula, F. Liu, Y. Cao, "The impact of NBTI effect on combinational circuit: modeling, simulation, and analysis," *IEEE Transactions on VLSI Systems*, vol. 18, no. 2, pp. 173–183, Feb. 2010.
53. H. Sangwoo, K. Juho, "NBTI-aware statistical timing analysis framework," *IEEE International SOC Conference*, pp.158-163, 2010.
54. T. Sakurai and A. R. Newton, "Alpha-Power Law Mosfet Model and its Application to CMOS Inverter Delay and Other Formulas," *IEEE J. Solid-State Circuits*, vol. 25, no. 2, pp. 584–594, Apr. 1990.
55. J. Velamala, V. Ravi, Y. Cao, "Failure diagnosis of asymmetric aging under NBTI," *International Conference on Computer Aided Design*, pp. 428–433, 2011.
56. R. Vattikonda, Y. Luo, A. Gyure, X. Qi, S. Lo, M. Shahram, Y. Cao, K. Singhal, and D. Toffolon, "A New Simulation Method for NBTI Analysis in SPICE Environment," *Int. Symposium on Quality Electronic Design*, pp. 41–46, 2007.
57. W. Zhao, Y. Cao, "New generation of predictive technology model for sub-45nm early design exploration," *IEEE Transactions on Electron Devices*, vol. 53, no. 11, pp. 2816–2823, Nov. 2006. (Available at http://ptm.asu.edu).
58. E. Maricau and G. Gielen, "Reliability simulation of analog ICs under time varying stress in nanoscale CMOS," *IEEE Workshop on Design for Reliability and Variability*, 2008.
59. E. Maricau, P. D. Wit, and G. Gielen, "An analytical model for hot carrier degradation in nanoscale CMOS suitable for the simulation of degradation in analog IC applications," *Microelectronics Reliability*, vol. 48, no. 8–9, pp. 1576–1580, Aug/Sept 2008.
60. N. Jha, P. Reddy, D. Sharma, R. Rao, "NBTI degradation and its impact for analog circuit reliability", *IEEE Transactions on Electron Devices*, pp. 2609–2615, 2005.

Chapter 29
Charge Trapping in MOSFETS: BTI and RTN Modeling for Circuits

Gilson Wirth, Yu Cao, Jyothi B. Velamala, Ketul B. Sutaria, and Takashi Sato

Abstract This chapter presents experimental investigation and statistical modeling of charge trapping in the context of random telegraph noise (RTN) and bias temperature instability (BTI). The goal is to develop circuit (electrical) level models to support circuit designers. The developed modeling approach is based on discrete device physics quantities, which are shown to cause statistical variability in the electrical behavior of MOSFETs. Besides evaluating the average behavior, the modeling approach here proposed allows the derivation of statistically relevant parameters. It allows the derivation of an analytical formulation for the both noise (RTN) and aging (BTI) behavior. Monte Carlo simulations are also discussed and presented. Good agreement between experimental data, Monte Carlo simulations, and model is found.

29.1 Introduction

Charge trapping and de-trapping at localized states (charge traps) at the interface or in the gate dielectric is a significant reliability issue for CMOS applications. It is known to be playing a significant role in bias temperature instability (BTI), besides being the major source of low-frequency noise in MOS devices. MOSFET low-frequency (LF) noise is dominated by charge capture and emission by defects

G. Wirth (✉)
UFRGS – Electrical Eng Department, Porto Alegre, Brazil
e-mail: wirth@inf.ufrgs.br

Y. Cao • J.B. Velamala • K.B. Sutaria
Arizona State University, Tempe, AZ, USA
e-mail: yu.cao@asu.edu; jvelamal@asu.edu; kbsutari@asu.edu

T. Sato
Graduate School of Informatics, Kyoto University, Kyoto, Japan
e-mail: takashi@i.kyoto-u.ac.jp

T. Grasser (ed.), *Bias Temperature Instability for Devices and Circuits*,
DOI 10.1007/978-1-4614-7909-3_29,

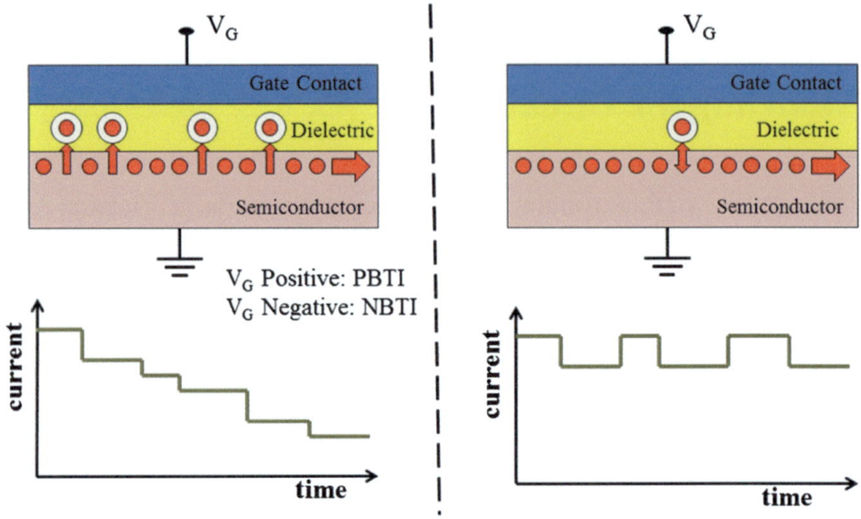

Fig. 29.1 (*Left side*) Traps that contribute to BTI are the ones that stay occupied after a capture event occurs. This leads to a degradation of transistor on-current over time. (*Right side*) Traps that contribute to noise are the ones that keep switching their state over time, exchanging charge carriers with the channel inversion layer. This leads to random telegraph noise in the device current. Current is assumed to flow along the horizontal direction. The *red circles* depict the charge carriers, while the *white* (*larger*) *circles* depict the traps. In NMOSFET the applied gate bias (V_G) is positive (in relation to the body terminal), leading to PBTI, while in PMOSFET it is negative, leading to NBTI

(traps) close to the Si–SiO_2 interface. Standard LF noise models used today (e.g., BSIM and PSP) do not properly model noise behavior under large signal excitation. A circuit level modeling and simulation approach, valid at both DC and large signal (AC) biasing, is presented. The role of charge trapping and de-trapping in BTI (Bias Temperature Instability) is also discussed and modeled.

The major goal of this chapter is to compile and critically discuss recent work performed by the authors on modeling of charge trapping and de-trapping phenomena in nanometer scale CMOS devices [1–6, 18–21]. The role of charge trapping and de-trapping in both low-frequency noise and BTI is discussed. We start by discussing the impact of charge trapping events on the current carried by a MOSFET. The capture or emission of a charge carrier by a trap changes the conductance of the channel of a MOSFET. A charge trapping event changes the number of carriers available to conduct current and also affects the mobility of channel charge carriers. This leads to discrete steps in the current flowing through the channel, as depicted in Fig. 29.1. Figure 29.1 depicts the mechanisms leading to both BTI and low-frequency noise.

Traps that contribute to noise are the ones that keep switching their state between occupied and empty, as depicted in Fig. 29.2. These are the traps with occupation probability close to 50%, which means that their capture and emission times are similar. These traps show significant activity, by capture and subsequent emission of charge carriers from the channel region.

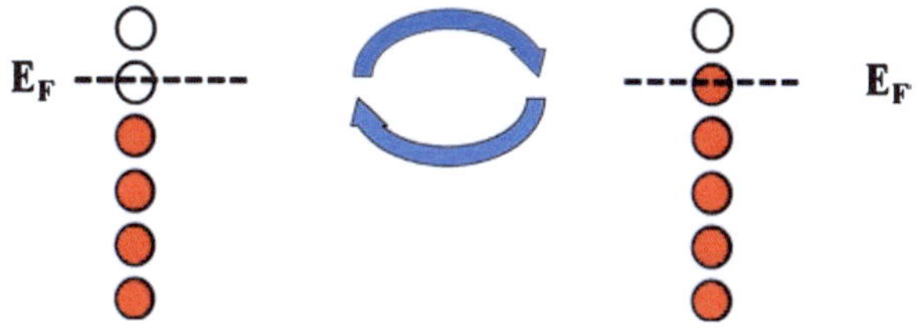

Fig. 29.2 Traps within a few kT from the Fermi level contribute to noise. These traps keep switching its state between empty and occupied. A *filled* (*red*) *circle* represents a trap occupied by a charge carrier, while an empty (white) *circle* represents a trap that is not occupied by a charge carrier

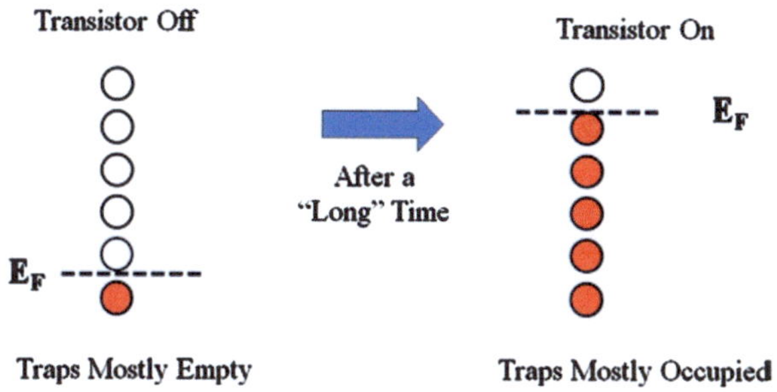

Fig. 29.3 If a transistor is abruptly turned on, the surface potential (or Fermi level) abruptly changes and trap occupation probability follows the change of the electric potential level. Charge trapping/de-trapping is not an instantaneous event. It is governed by characteristic time constants, and the trap occupation reflects the new biasing condition (surface potential) after an elapsed time. The picture reflects the situation for an NMOSFET, where electron traps below the Fermi level (E_F) are expected to become occupied and traps above the Fermi level are expected to remain empty

Traps that contribute to BTI are the ones that have a high probability to stay occupied after a charge trapping event. These are the traps with occupation probability close to 100%, which means that their capture time is much shorter than the emission time. For these traps, charge capture is by far the most likely event.

For instance, if a transistor is turned on by applying a voltage at its gate terminal, the surface potential (or Fermi level, E_F) changes in such a way that trap occupation probability is increased, as shown in Fig. 29.3. The rate at which charge carriers are captured abruptly becomes larger than the rate at which carriers are emitted, and the number of trapped charge increases over time. Traps change their occupation state according to their characteristic time constant, meaning that the number of trapped charge does not instantaneously reflect the new occupation probability. The faster traps (the ones with shorter capture time constants) become filled first, while the slowest traps take longer to become filled. Each trap that becomes occupied degrades the channel conductivity, decreasing the device current. This current decrease is seen to occur in discrete steps, each step being related to the capture of a single channel carrier. Since the dynamics of this occupation depends on the bias point and temperature, it may lead to bias temperature instability (BTI).

The first sections of this chapter cover the modeling of low-frequency noise, while the last sections cover the modeling of BTI.

29.2 Random Telegraph Noise

In nanometer scale MOSFETs the alternate capture and emission of carriers at individual defect sites (traps) generates discrete fluctuations in the device conductance. These fluctuations, also called random telegraph noise (RTN), are the main source of low-frequency noise in deep-submicron MOSFETs. This work covers analysis and modeling of these fluctuations, targeting the development of models and methodologies for electrical (circuit level) simulations.

The low-frequency noise model here presented is based on device physics parameters which cause statistical variation in low-frequency noise behavior of individual devices. It includes detailed consideration of statistical effects for distribution of number of traps per device, the trap energy distribution, trap location, and its device bias-dependent noise contributions [1, 2]. Microscopic discrete quantities are used in model derivation, and analytical equations for the statistical parameters are provided.

In many practical applications, the MOS device is not biased at steady state, but periodically switched. This operation regime is called cyclo-stationary excitation. The modeling approach adopted allows analysis of noise both at steady state operation as well as under cyclo-stationary excitation.

We start by studying the noise power generated by a single trap. First the autocorrelation of the RTN signal generated by a single trap is calculated, and then the Wiener-Khinchin formula applied, leading to an analytical formulation for the RTN spectrum due to a single trap.

After the evaluation of the power spectrum due to a single trap, the noise behavior resulting from the combined effect of all traps found in a device is derived.

We discuss and model both DC and AC (large signal) biasing conditions. The modeling approach is valid for steady state (DC) biasing as well as for any periodic excitation signal and allows the derivation of relevant statistical parameters. Square wave excitation is used as a case study, to explore noise behavior in detail.

It is shown that since RTN amplitude depends on the bias point strong variations of noise performance may appear not only between devices but also for a single device operated under different bias conditions.

If the trap energy distribution in the bandgap is a convex curve (e.g., "U"-shaped), the model here presented yields a reduction in LF-noise under cyclo-stationary excitation when the device is biased between strong and weak inversion or even slight accumulation, which is in agreement to experimental results found in the literature [3, 7–11]. However, while the average noise power is seen to decrease, the variability (normalized standard deviation) of noise power increases, as shown by the mathematical derivation and experimental observations discussed in this chapter.

29.3 Power Spectrum of the RTN Noise Due to a Single Trap

The origin of RTN noise is the alternate capture and emission of charge carriers at discrete trap levels near the interface between the semiconductor and the dielectric. Figure 29.4 depicts the cross section of an n-channel MOSFET through the location of the charge trap. The influence of the traps on the electrical current flowing through the channel is twofold. On the one hand, the occupation of a trap changes the number of free carriers in the inversion layer. On the other hand, a charged trap state has an influence on the local mobility near to its position due to Coulomb scattering. If the MOSFET biasing is kept constant, a stationary RTN is observed at the terminals of the device as a discrete fluctuation in electrical current, δI_d being the amplitude of the current fluctuation, as shown in the inset of Fig. 29.5. The average high current time corresponds to the electron capture time constant (τ_c), and the average low current time corresponds to the emission time constant (τ_e).

The power spectrum of a RTN fluctuation can be evaluated by calculating the auto-covariance of the signal and then applying the Wiener-Khinchin formula [12]. The power spectrum is then the Fourier transform of the auto-covariance and is a Lorentzian, as given by Eq. (29.1) and depicted in Fig. 29.5:

$$S(\omega) = \frac{\delta^2}{\pi} \cdot \frac{\beta}{(1+\beta)^2} \cdot \frac{1}{\omega_0} \cdot \frac{1}{1+(\omega/\omega_0)^2} \tag{29.1}$$

where ω_0 is the angular corner frequency (given by τ_c and τ_e as discussed below) and β is equal to τ_c/τ_e.

Equation (29.1) assumes that the capture and emission time constants (τ_c and τ_e) are time-independent, constant values. This is true for constant (time invariant) biasing. However, under cyclo-stationary excitation, the applied bias voltage is a periodic function of time, and $\tau_c(t)$ and $\tau_e(t)$ become periodic functions of time. Figure 29.6 depicts the situation under cyclo-stationary excitation.

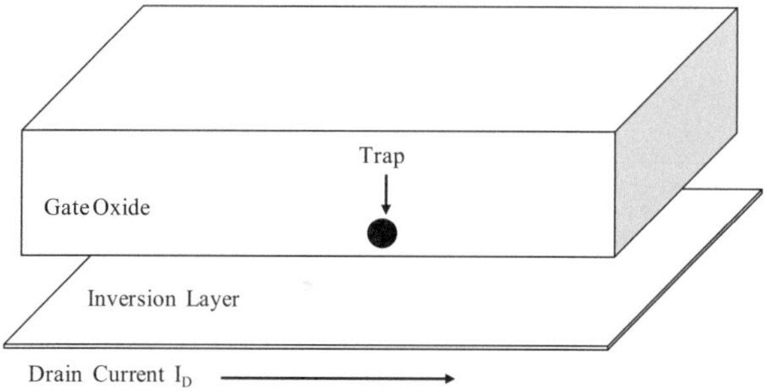

Fig. 29.4 Schematic cross section of the inversion layer of a MOS transistor through the location of an interface trap. If the trap is electrically charged, the inversion layer is disturbed by the trap, affecting the drain current I_D. The trap not only affects the number of free carriers in the inversion layer but is also a source of electrical charge carrier scattering

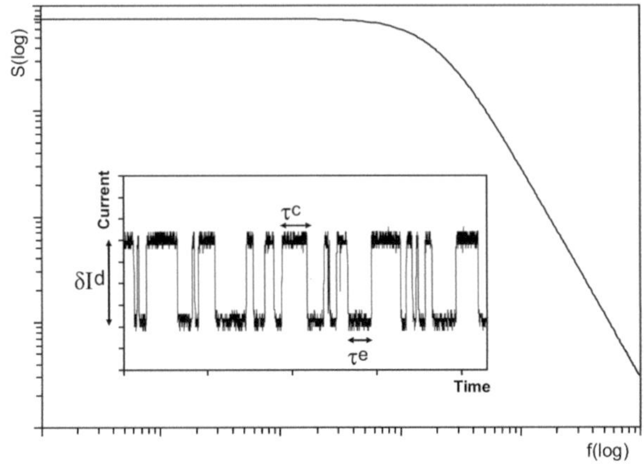

Fig. 29.5 Time and frequency domain representation of a stationary random telegraph noise (RTN). In frequency domain, the power spectrum of a RTN is a Lorentzian. In time domain discrete fluctuations are observed in the drain current, where τ_c is the average time in the high current state, which corresponds to the state where the trap is electrically neutral (empty). τ_e is the average time in the low current state, which corresponds to the state where the trap is electrically charged. δI_d is the amplitude of the current fluctuation

Fig. 29.6 Energy band diagram of MOSFET with noise relevant traps during two different phases (*dashed and dotted*) of square wave gate voltage biasing as shown on *top* of the figure. E_T stands for a trap energy level, while E_f stands for the Fermi level. $g_t(E)$ shows a U-shaped trap density

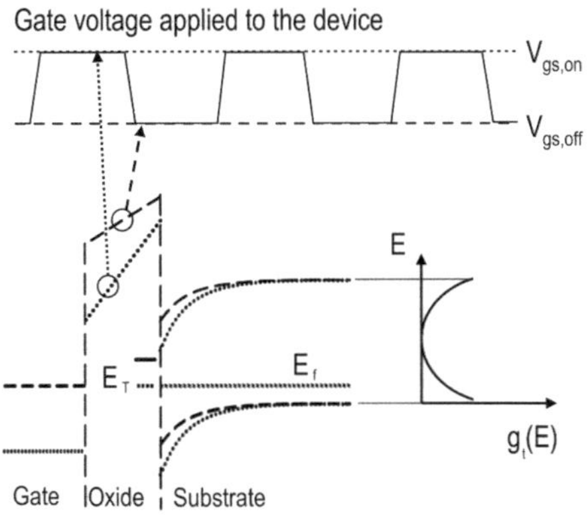

In order to derive the power spectrum under cyclo-stationary excitation, we did follow the methodology originally proposed by Machlup [12] for stationary RTN. A RTN is considered to be a purely random signal, which may be in one of two states, called 1 and 0. If the signal is in state 1, the probability of making a transition to 0 in a short time *dt* is assumed to be $dt/\tau_c(t)$. If the signal is in state 0, the probability of making a transition to 1 is assumed to be $dt/\tau_e(t)$. In this form, the state 1 is related to the empty trap, i.e., high current state in Fig. 29.5, while the state 0 is related to the occupied trap, i.e., low current state.

In order to derive the low-frequency noise spectrum of cyclo-stationary RTN, we first calculate the autocorrelation of the RTN and then apply the Wiener-Khinchin formula to obtain the spectrum. Let the RTN signal be $x(t)$.

The autocorrelation is then given by

$$A(s) = \langle x(t).x(t+s)\rangle_{\text{average}} = P(x(t)=1).P_{11}(s) \tag{29.2}$$

where $P_{11}(s)$ is the probability of an even number of transitions in time s, given we start in state 1. P(x(t)=1), the probability of being in state 1 at time t, is given by

$$P(x(t)=1) = \frac{\frac{1}{T}\int_0^T \frac{1}{\tau_e(t)}dt}{\frac{1}{T}\int_0^T \frac{1}{\tau_e(t)}dt + \frac{1}{T}\int_0^T \frac{1}{\tau_c(t)}dt} = \frac{\left\langle \frac{1}{\tau_e(t)} \right\rangle}{\left\langle \frac{1}{\tau_e(t)} \right\rangle + \left\langle \frac{1}{\tau_c(t)} \right\rangle} \tag{29.3}$$

where T is the period of the cyclo-stationary excitation and symbol $\langle\bullet\rangle$ is an abbreviation for $(1/T)\int_0^T \bullet dt$. In other words, $\langle\bullet\rangle$ is the time average value.

The autocorrelation can then be calculated as

$$A(s) = \frac{\left\langle \frac{1}{\tau_e(t)} \right\rangle}{\left\langle \frac{1}{\tau_e(t)} \right\rangle + \left\langle \frac{1}{\tau_c(t)} \right\rangle}.$$

$$\cdot\left(1 + \int_0^s e^{\int_0^x \left(\left\langle \frac{1}{\tau_c(y)} \right\rangle + \frac{1}{\tau_e(y)}\right)dy} \frac{1}{\tau_e(x)} dx\right) e^{-\int_0^s \left(\left\langle \frac{1}{\tau_c(y)} \right\rangle + \frac{1}{\tau_e(y)}\right)dy}. \tag{29.4}$$

This formulation for the autocorrelation is a generalization of the Machlup formula and is valid for any kind of periodic excitation and any frequency. If τ_c and τ_e become constant (independent of time), Eq. (29.4) becomes equal to Eq. (7) in [12].

29.3.1 Approximation for Excitation Frequencies Higher than the Noise Frequency

The case of interest for most practical applications is for the noise at frequencies below the frequency of the cyclo-stationary excitation signal.

If the probability of a trap to switch state during one period T of the cyclo-stationary excitation signal is very small, a simplification may be done in the calculation of the autocorrelation. This case corresponds to the limit where $\tau_c(t)$ and $\tau_e(t)$ are much larger than the period T, leading to small transition probabilities $T/\tau_c(t)$ and $T/\tau_e(t)$.

In this case we can write, without loss of generality, that $s{=}nT$, where n is a positive integer (1, 2, 3,…). In this situation we have

$$\int_0^S \left(\frac{1}{\tau_e(y)}+\frac{1}{\tau_c(y)}\right)dy = \sum_{i=0}^{n-1}\int_{iT}^{(i+1)T}\frac{1}{\tau_e(y)}dy + \sum_{i=0}^{n-1}\int_{iT}^{(i+1)}\frac{1}{\tau_c(y)}dy$$
$$= nT\left(\left\langle\frac{1}{\tau_e(t)}\right\rangle + \left\langle\frac{1}{\tau_c(t)}\right\rangle\right) \tag{29.5}$$

leading to

$$A(s) = \frac{\left\langle\frac{1}{\tau_e(t)}\right\rangle}{\left\langle\frac{1}{\tau_e(t)}\right\rangle+\left\langle\frac{1}{\tau_c(t)}\right\rangle}\left[1\text{-}\frac{T\left\langle\frac{1}{\tau_e(t)}\right\rangle}{e^{T\left(\left\langle\frac{1}{\tau_e(t)}\right\rangle+\left\langle\frac{1}{\tau_c(t)}\right\rangle\right)}\text{-}1}\right]e^{-S\left(\left\langle\frac{1}{\tau_e(t)}\right\rangle+\left\langle\frac{1}{\tau_c(t)}\right\rangle\right)}$$
$$+\frac{\left\langle\frac{1}{\tau_e(t)}\right\rangle}{\left\langle\frac{1}{\tau_e(t)}\right\rangle+\left\langle\frac{1}{\tau_c(t)}\right\rangle}\frac{T\left\langle\frac{1}{\tau_e(t)}\right\rangle}{e^{T\left(\left\langle\frac{1}{\tau_e(t)}\right\rangle+\left\langle\frac{1}{\tau_c(t)}\right\rangle\right)}\text{-}1} \tag{29.6}$$

For small values of T, a Taylor expansion may be employed, leading to

$$A(s) = \frac{\left\langle\frac{1}{\tau_e(t)}\right\rangle}{\left\langle\frac{1}{\tau_e(t)}\right\rangle+\left\langle\frac{1}{\tau_c(t)}\right\rangle}\left[1\text{-}\frac{\left\langle\frac{1}{\tau_e(t)}\right\rangle}{\left\langle\frac{1}{\tau_e(t)}\right\rangle+\left\langle\frac{1}{\tau_c(t)}\right\rangle}\right]e^{-S\left(\left\langle\frac{1}{\tau_e(t)}\right\rangle+\left\langle\frac{1}{\tau_c(t)}\right\rangle\right)}$$
$$+\frac{\left\langle\frac{1}{\tau_e(t)}\right\rangle^2}{\left(\left\langle\frac{1}{\tau_e(t)}\right\rangle+\left\langle\frac{1}{\tau_c(t)}\right\rangle\right)^2} \tag{29.7}$$

This means that if $\tau_c(t)$ and $\tau_e(t)$ are much larger than the period T of the excitation signal, the values of $1/\tau_c(t)$ and $1/\tau_e(t)$ in the integrals of Eq. (29.4) are equivalent to their time averages, written as $\langle 1/\tau_c(t)\rangle$ and $\langle 1/\tau_e(t)\rangle$.

The power spectrum $S_i(\omega)$, due to a single trap (the i-th trap), is then calculated as the Fourier transform of the autocorrelation, leading to

$$S_i(\omega) = \frac{1}{2\pi}\int_{-\infty}^{\infty} A(s)e^{i\omega s}\,ds \tag{29.8}$$

which is evaluated as being

$$S_i(\omega) = \frac{\delta_i^2}{\pi}\cdot\frac{\beta_{eq}}{(1+\beta_{eq})^2}\cdot\frac{1}{\omega_i}\cdot\frac{1}{1+(\omega/\omega_i)^2} \tag{29.9}$$

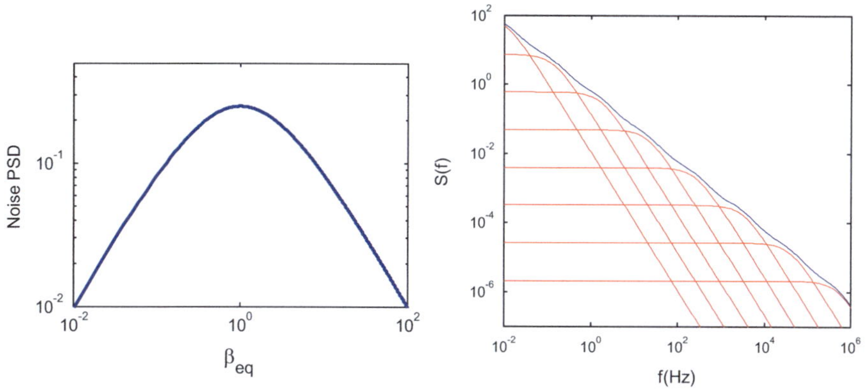

Fig. 29.7 *Left hand side*: Noise power spectral density as a function of β_{eq}. For $\beta_{eq}=1$, i.e., $\langle 1/\tau_e(t)\rangle = \langle 1/\tau_c(t)\rangle$, the noise power reaches its maximum. *Right hand side*: If the corner frequencies of the Lorentzians (*red lines*) corresponding to different traps are equally spaced on a log scale, the summation of the power spectrum due to all traps lead to 1/f noise (*blue line*)

Here δ_i determines the trap's voltage amplitude and ω_i the angular frequency, and

$$\beta_{eq} = \langle 1/\tau_e(t)\rangle / \langle 1/\tau_c(t)\rangle \tag{29.10}$$

Figure 29.7 depicts the behavior of noise power as a function of β_{eq}.

The cyclo-stationary noise spectrum is still Lorentzian, with angular corner frequency ω_i given by

$$\omega_i = \left\langle \frac{1}{\tau_c(t)} \right\rangle + \left\langle \frac{1}{\tau_e(t)} \right\rangle \tag{29.11}$$

This takes us to the conclusion that making a RTN signal cyclo-stationary leads to a Lorentzian spectrum with corner frequency equal to the sum of the inverse time average values of the capture and emission times. Please note that for stationary RTN, the corner frequency is equal to the sum of the inverse values of the constant capture and emission times. Please refer to Eq. (9) in [12].

The result for this limit is valid for any kind of periodic excitation. This is the limit studied in [11] and [13]. However, for this limit, we obtain the same result in a much simpler derivation than in [11] and [13] and without making any further assumption or simplification. The single assumption is that the transition probabilities $dt/\tau_e(t)$ and $dt/\tau_c(t)$ are much smaller than the excitation period T.

Equations (29.7) and (29.9) are a generalization of the Machlup formulation for cyclo-stationary RTN with excitation frequency higher than the noise frequency. The Machlup equations for the autocorrelation and power spectrum are recovered if we consider $\langle 1/\tau_e(t)\rangle = 1/\tau_e$ and $\langle 1/\tau_c(t)\rangle = 1/\tau_c$, i.e., constant, not time-dependent values.

29.4 Average Power Spectrum of the RTN Due to the Ensemble of Traps

The noise behavior of a device results from the combined effect of all traps found in the device. The noise power spectrum may then be written as the summation of the contribution of each one of the N_{tr} traps found in the device. The observation that traps are Poisson distributed results in an average noise spectral density given by

$$\langle S\rangle = \sum_{N_{tr}=0}^{\infty} \sum_{i=1}^{N_{tr}} \langle S_i\rangle \frac{e^{-N} N_{tr}^{N}}{N_{tr}!} = N\langle S_i\rangle \tag{29.12}$$

Here $\langle S_i\rangle$ is the average noise contribution of a trap, N_{tr} is the actual number of traps in a particular device, and N is the average number of traps in an ensemble of devices.

Equation (29.12) can be interpreted as being a sum of Lorentzians, where each Lorentzian represents the power spectrum of a single trap. If the corner frequencies of the Lorentizians are assumed to be uniformly distributed in a log scale, the resultant power spectrum will be 1/f, as depicted in Fig. 29.7.

The equation above is valid for constant bias, as well as for excitation by any periodic signal. In order to explore the noise behavior in detail and allow comparison to experimental results, a case of particular interest for practical applications will be studied. It is square wave excitation.

29.4.1 Square Wave Excitation

Square wave excitation is chosen as case study, because of its interest in practical applications and because of the availability of experimental data. Comparison of model results to relevant experimental data from the literature is performed below.

Under square wave excitation, the bias voltage abruptly alternates between two states, called *on* and *off*, as depicted in Fig. 29.6. Please note that the state names *on* and *off* do not imply that the device has necessarily to be periodically turned on and off. The names refer to two distinct states, with different gate bias. With periodically changing gate bias, the Fermi level becomes a periodic function of time $E_F(t)$. This implies that the Fermi level alternates between two levels: E_{on} being the Fermi level during the *on* state and E_{off} being the Fermi level during the *off* state. The duty cycle α is the fraction of the period T in which the device is in the *on* state. The capture and emission time constants of a trap are affected by the Fermi level.

For square wave cyclo-stationary excitation with duty cycle α, the time-averaged capture and emission time constants may be written as

$$\langle 1/\tau_c\rangle = \left(\alpha/\tau_{c,on} + (1-\alpha)/\tau_{c,off}\right) \tag{29.13}$$

$$\langle 1/\tau_e\rangle = \left(\alpha/\tau_{e,on} + (1-\alpha)/\tau_{e,off}\right) \tag{29.14}$$

The duty cycle α is the fraction of time spent in the *on* state, with $0 \leq \alpha \leq 1$. Using these values and Eq. (29.10), β_{eq} becomes

$$\beta_{eq} = \psi(E_{on}, E_{off}, \alpha)e^{2E_t/k_BT} \tag{29.15}$$

with

$$\psi(E_{on}, E_{off}, \alpha) = \frac{\alpha e^{-E_{on}/k_BT} + (1-\alpha)e^{-E_{off}/k_BT}}{\alpha e^{E_{on}/k_BT} + (1-\alpha)e^{E_{off}/k_BT}} \tag{29.16}$$

Assuming statistical independence of the random variables, the average noise of a transistor $\langle S \rangle$ is then given by

$$\langle S \rangle = \frac{N_{dec}\langle \delta^2 \rangle}{\pi W L \omega} \int_{E_v}^{E_c} \frac{\psi e^{2E_t/k_BT}}{(1+\psi e^{2E_t/k_BT})^2} g(E_t) dE_t \tag{29.17}$$

where W is the channel width, L the channel length and N_{dec} is the trap density per unit area and frequency decade. Please see [1] for a detailed explanation of these parameters. This equation again looks similar to the DC noise behavior (see [1]). In case the cyclo-stationary period T is short compared to the trap's time constants, their time averages and the trap numbers close to the two Fermi levels determine the noise level. At frequencies significantly higher than the excitation frequency, the noise is given by DC theory.

Equation (29.17) above clearly relates the noise reduction to the distribution of traps over energy. From the above equation, it follows that if the distribution of traps in energy over the bandgap is uniform, $\langle S \rangle$ is expected to remain approximately constant under cyclo-stationary excitation. In this case, the peak of the noise contribution of a trap always probes the same trap density, since the trap density around to the Fermi level is always the same. Note that the peak of the noise contribution of a trap occurs for $E_T = E_F$. If the trap density is uniform over the bandgap, $g(E_t)$ corresponding to the peak of β_{eq} is always the same.

The behavior is different if a U-shaped trap density is assumed (see Fig. 29.8). In this situation, $\langle S \rangle$ is expected to be reduced under cyclo-stationary excitation, as depicted in Fig. 29.8. This behavior will be discussed in more detail below, where we also discuss the experimental observations of the noise reduction.

To further investigate the behavior predicted by the analytical equations just presented, Monte Carlo (MC) simulations are performed in time domain. The RTN of each trap is evaluated, and then the power spectral density is calculated. The Monte Carlo simulation show very good agreement to the analytical results, as well as to relevant experimental results from the literature, as discussed below.

In Fig. 29.9 the results of Eq. (29.17) show noise reduction as a function of the Fermi levels in the "*on*" and "*off*" state at frequencies lower than the excitation frequency. In good agreement to [11], the trap energy distribution $g(E_t)$ is key for

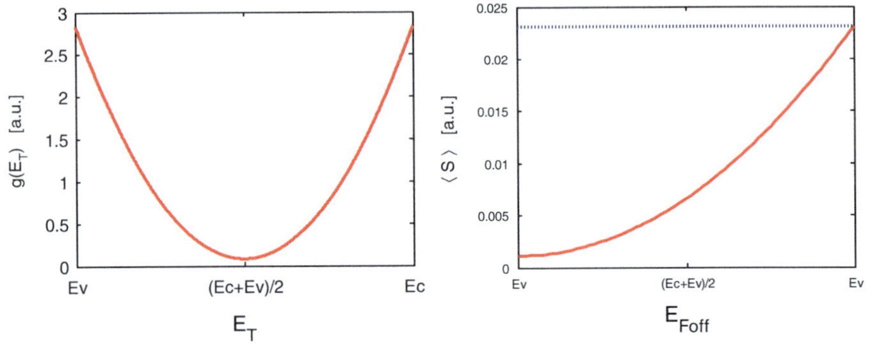

Fig. 29.8 *Left hand side*: For a U-shaped trap density, the trap density is higher close to the valance and conduction bands and lower in the *center* of the bandgap. *Right hand side*: Noise power as a function of the Fermi level in the *off* phase. In the *on* phase of the device, the Fermi level is assumed to be fixed, close to the conduction band. The *blue dotted line* shows the behavior as predicted by standard models used today (e.g., BSIM). The *red solid line* shows the behavior as predicted by the model here presented

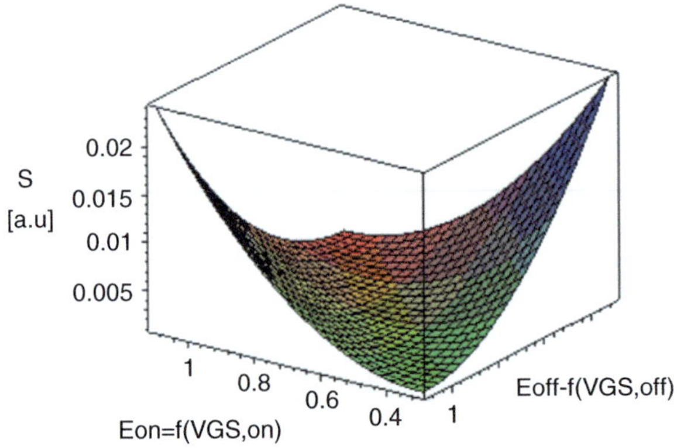

Fig. 29.9 Noise reduction as a function of the Fermi level in the "on" and "off" states. The noise power S is evaluated according to Eq. (29.17), considering the U-shaped trap density given by Eq. (29.18). The same parameter $a = 11$ is used in all figures and evaluations performed in this work. Note that the noise reduction is larger if the biasing levels are symmetrical in relation to the center of the U-shaped trap density

explaining experimentally observed findings. A parabolic U-shaped trap density function is assumed here:

$$g(E_t) = aE_t^2 - a(E_c - E_v)E_t + k \tag{29.18}$$

where a is a fitting parameter. The integral of $g(E_t)$ from E_v to E_c is normalized to one with

$$k = \frac{1}{E_c - E_v}\left(\frac{a}{6}(E_c^3 - E_v^3) + \frac{a}{2}(E_c^2 E_v - E_v^2 E_c) + 1\right) \tag{29.19}$$

It can be seen that a higher noise reduction can be expected if the biasing drives the Fermi level to maximum and minimum energy values that are symmetrical in relation to the center of the U-shaped trap density. In this case, the traps that contribute most to the noise power are the ones close to the center of the bandgap, where the density is lowest.

29.5 Variability in the Power Spectrum of the RTN Noise Due to the Ensemble of Traps

Since the microscopic approach maintains the statistically relevant parameters, the model proposed here also allows the modeling of the statistically relevant parameters.

The standard deviation of noise power is given by

$$\sigma_S = \sqrt{\langle S^2 \rangle - \langle S \rangle^2} \tag{29.20}$$

As derived in [2], the normalized standard deviation of noise performance under constant biasing is

$$\frac{\sigma_{\mathrm{np}}}{\langle \mathrm{np_{BW}} \rangle} = \frac{2}{\pi} \frac{1}{\sqrt{\mathrm{N_{dec}WL}}} \sqrt{\frac{\langle A^4 \rangle}{\langle A^2 \rangle^2} \frac{\mathrm{b}}{\left(\frac{\mathrm{f_H}}{\mathrm{f_L}}\right)^{\mathrm{c}}}} \tag{29.21}$$

Here f_H and f_L are the lower and upper boundaries of the bandwidth of interest in a given circuit design, respectively, and b and c are constants, with $b = 0.74$ and $c = 0.05$ [2].

A similar formulation can also be derived for cyclo-stationary operation. Hence, in addition to $\langle S \rangle^2$ based on Eq. (29.17) above, it is necessary to evaluate $\langle S^2 \rangle$. Again following a similar approach to [12], the resulting normalized standard deviation under square wave excitation is given by

$$\frac{\sigma_S}{\langle S \rangle} = \frac{\sqrt{\frac{\langle \delta^4 \rangle}{\langle \delta^2 \rangle^2}}}{\sqrt{N_{dec} WL}} \frac{\left(\int_{E_v}^{E_c} \frac{\psi^2 e^{4E_t/k_B T}}{(1+\psi e^{2E_t/k_B T})^4} g(E_t) dE_t \right)^{\frac{1}{2}}}{\int_{E_v}^{E_c} \frac{\psi e^{2E_t/k_B T}}{(1+\psi e^{2E_t/k_B T})^2} g(E_t) dE_t} \tag{29.22}$$

Figure 29.10 shows the normalized standard deviation as a function of the Fermi levels in the *on* and *off* states for the U-shaped trap density given by (29.18). The normalized standard deviation becomes higher when the traps that contribute most to noise are the ones close to the center of the bandgap. In this case the normalized standard deviation increases even though the average noise power decreases.

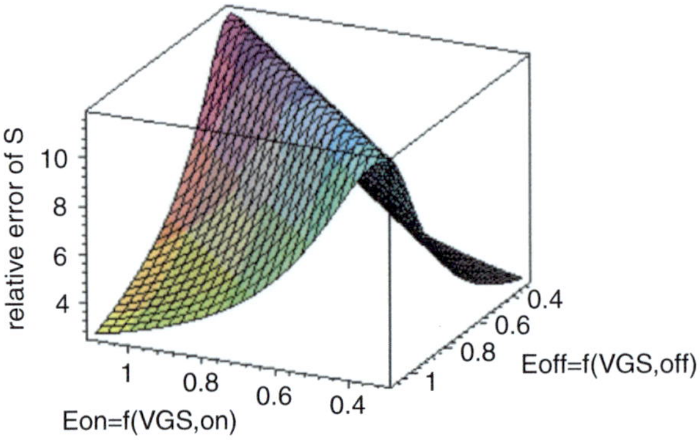

Fig. 29.10 Normalized standard deviation of noise performance as given by Eq. (29.22). An increase in normalized standard deviation is seen under cyclo-stationary excitation, especially for conditions where the average cyclo-stationary noise reduction is high

29.6 Compact Modeling of RTN for Circuit (SPICE) Simulation

For DC behavior, Eqs. (29.1) and (29.21) above properly describe the average noise behavior and its standard deviation, respectively. And for cyclo-stationary excitation, the equations that describe the average noise behavior and its standard deviation are (29.17) and (29.22), respectively. Considering circuit simulation purposes, the equations that describe the behavior under cyclo-stationary excitation have the drawback of demanding evaluations for calculating the Quasi Fermi levels during both the *on* and *off* semicycles.

It is of great practical interest to provide a simpler equation. An empirical approach, considering an equivalent Fermi level $E_{F,eff}$, would help compact modeling. In this section we will introduce such an approach, writing an equivalent Fermi level $E_{F,eff}$ as a function of duty cycle and using it to fit the experimental data.

Standard low-frequency (LF) noise models used today (e.g., BSIM and PSP) do not properly model noise behavior under large signal excitation. In these models, noise power depends solely on the Fermi level at the current bias point (i.e., there is no dependency on the "history" of the Fermi level). This means that simple modulation theory is applied to the noise source. Eventually, the circuit simulator may use techniques like harmonic balance to account for the distortion in nonlinear circuits. However, these techniques usually account solely for nonlinearities in the circuit transfer function, and do not address the impact of large signal excitation on the noise source itself. This also brings us to the most serious limitation of a SPICE like simulator for such type of problem. The correct noise evaluation for a device needs to know the state of the device in the past period. Not all types of SPICE simulations can provide this information—however periodic steady state or transient analysis can do so. For cyclo-stationary noise as well as for periodic steady state analysis, a periodic change in gate voltage is assumed. Here the necessary bias information is present and the practical benefit of noise model enhancement is highest as this simulation type is used to check circuit phase noise performance. Therefore the authors recommend implementing the proposed noise models for this simulation type first.

The problem with the usual noise modeling and simulation approaches is that they evaluate the noise power assuming that at each time instant the traps could be modeled by their behavior at the corresponding DC bias value. However, under large signal excitation, this is not correct, since the trapping and emission process are not instantaneous: its statistic is governed by the evolution of the device bias over the whole excitation period, as discussed above. The use of equivalent time constants, or alternatively an equivalent Fermi level, is the correct approach.

In this approach, $E_{F,eff}$ is written as the time average of the Fermi level, as already proposed in the context of the charge trapping component of bias temperature instability [21]. For the case of square wave excitation with duty cycle α, we have [21]

$$E_{F,eff}(\alpha) = \alpha E_{on} + (1-\alpha)E_{off} \tag{29.23}$$

With this empirical approach, the same compact model equations can model both steady (DC) bias and cyclo-stationary excitation. These equations are (29.1) for a single trap and (29.12) for the ensemble of traps. With this additional information, any modern compact model can make use of the modeling approach presented here.

Modern compact models include the modeling of the low-frequency noise at DC biasing. For instance, compact models like BSIM and PSP use formulations that account for the trap density and correlated carrier number and mobility fluctuations, leading to the following equation for the input-referred noise [17]:

$$S_{VG}(E_f) = \frac{q^2kT}{\gamma WLC_{ox}^2 f}[1+\alpha\mu_0C_{ox}(V_G-V_T)]^2N_t(E_f) \tag{29.24}$$

γ is the attenuation coefficient of the electron wave function in the oxide. α is the Coulomb scattering parameter (a measure of the mobility fluctuations induced by the trapped charge). $N_t(E_f)$ is the trap density at the Fermi level E_f. The channel inversion carrier density is assumed to be proportional to $C_{ox}(V_G - V_T)$. Comparing this equation to Eq. (29.17) above, we see that the term $\frac{q^2[1+\alpha\mu_0C_{ox}(V_G-V_T)]^2}{C_{ox}^2}$ models the noise contribution δ^2 of the traps.

The major contribution to the integral in (29.17) will be from traps for which $\beta_{eq}(E_t) \approx 1$. For DC biasing, these are the traps with energy levels within a few kT from E_f. In case of a varying gate voltage, this is still true but the Fermi level constantly changes. Still, the noise power is proportional to $N_t(E_f,t)$. As a conclusion, Eqs. (29.17) and (29.24) are of the same form. Both equations model the noise power to be proportional to the trap density at the Fermi level of interest multiplied by a second term that accounts for the amplitude of the noise contribution of the traps.

This is also true for the flicker noise equations used in the standard noise simulators like BSIM and PSP. They use an equivalent oxide trap density $N^*_t(E_f)$, which for simulation purposes is approximated by a three-parameter empirical function of the channel carrier density:

$$N^*_t\left(E_f\right) = NOIA + NOIB \cdot \mathrm{N} + NOIC \cdot \mathrm{N}^2 \qquad (29.25)$$

where *NOIA*, *NOIB*, and *NOIC* are technology-dependent SPICE model parameters and N is the channel inversion carrier density. Since N is a function of the Fermi level, N^*_t is also a function of the Fermi level. The noise level at a given bias is modeled to be proportional to N^*_t.

Therefore the same (BSIM and PSP) compact model equations can be used, substituting E_f in Eq. (29.25) by $E_{F,eff}$, i.e., instead of using the channel inversion carrier density corresponding to the Fermi level at DC bias, the inversion carrier density corresponding to the equivalent Fermi level $E_{F,eff}$ at cyclo-operation should be used.

The gap between the modeling approach here proposed and standard models can be seen if we consider the noise behavior of a MOSFET under periodic switching, with the bias point in the off state being changed, while the bias point in the on state is kept fixed.

According to the standard models, the noise at a given point in time depends only on the biasing at that time point. Since the device contributes to circuit noise only in the *on* state, and the bias point in the *on* state is fixed, the noise should not depend on the bias point in the *off* state (according to the standard models). However, according to our theories and experimental data, the opposite is true. The noise does depend on the bias point in the *off* state. Even more relevant for practical applications, it is possible to reduce the noise by proper choice of the bias in the off state.

Figure 29.8 depicts the prediction of our model for this behavior. The noise power as a function of the Fermi level in the *off* phase is shown. In the *on* phase of the device, the Fermi level is assumed to be fixed, close to the conduction band. This means that $V_{G,ON}$ is always the same, while $V_{G,OFF}$ is varied. The device is assumed to contribute to noise during the *on* phase only (i.e., it is assumed that no current flows through the device during the *off* phase).

The dotted line shows the behavior as predicted by standard models used today (e.g., BSIM and PSP). In these models, noise power depends solely on the Fermi at the current bias point (i.e., no dependency on the "history" of the Fermi level). Since the bias point where the device contributes to noise ($V_{GS,ON}$) is fixed, the predicted noise power is a constant, not depending on E_{off}.

The solid line shows the behavior as predicted by the model. Although the transistor contributes to noise during the *on* phase only, the generated noise is a function of the Fermi level in both *on* and *off* phases. In this case, a reduction in noise power is predicted, in agreement to experimental data. The dependence of the noise power on $V_{GS,OFF}$ is experimentally observed, for instance, in [9, 15]. In [15] forward substrate bias during the off state is used to significantly decrease the phase noise in a VCO design, and in [9] noise is seen to decrease when the transistor is cycled from inversion to accumulation.

Another relevant experiment showing noise reduction is also found in [7], where a circuit which enables the switching of device pairs between two gate-to-source voltage values was used. There is always one of the devices of the pair turned on, while the other one is turned off. Using this technique, significant noise reduction was achieved, if compared to DC biasing.

29.7 The Charge Trapping Component of Bias Temperature Instability

It is widely accepted that charge trapping plays a role in the threshold voltage shifts (ΔV_T) produced by bias temperature stress. It was reported that a significant fraction of the threshold voltage change is recovered spontaneously once the bias temperature stress is removed [23–27]. Although first observed decades ago, the phenomenon still remains controversial in both experimental and theoretical terms.

In PMOS transistor, it is called negative bias temperature instability (NBTI), while in NMOS transistors it is called positive bias temperature instability (PBTI).

There is a vast literature on negative bias temperature instability (NBTI) in PMOSFETs, where most models are based on reaction–diffusion theory. Reaction–diffusion models involve breaking of hydrogen–silicon bonds at the silicon-gate dielectric interface, related to the trapping of inversion layer holes, with the release and diffusion of a hydrogenic species. Recovery (relaxation) is assumed to occur with re-bonding of the hydrogen–silicon bonds, i.e., interface trap annealing out [23, 26]. Although reaction diffusion models have been very useful and successful, some aspects of NBTI are difficult to be fully explained in a reaction–diffusion framework, as, for instance, the fast recovery which occurs if bias stress is removed [23, 28]. Other phenomena, such as charge trapping, needed to be considered in the NBTI mechanism. Besides our work, several other works showed evidence of the role of charge trapping and de-trapping in BTI [22–32]. Clear steps caused by single trapping or de-trapping events were seen in experimental works, showing the discrete nature of V_T shifts [24, 27, 31, 32].

The focus of this work is not on the elucidation of the origin and nature of the charge traps. Detailing lattice dynamics and tunneling mechanisms is not the focus of this work. This work focuses instead on the charge trapping statistics (stochastic capture and emission events) that contribute to degradation of transistor on-current over time (BTI). The basic assumptions made in the modeling effort presented in this work are the same ones as in our work on modeling of RTN (low-frequency) noise: (1) charge trapping and de-trapping are stochastic events, governed by characteristic time constants, which are uniformly distributed on a log scale; (2) the number of traps is assumed to be Poisson distributed, and the parameter of the Poisson distribution is assumed to be constant over the time interval of interest; (3) trap energy distribution is assumed to be U-shaped (this last assumption is key to explain the AC behavior); and (4) the amplitude of the fluctuation induced by a single trap is a random variable.

In this section, we develop a theoretical analysis to describe the density of occupied traps as a function of bias, temperature, and time, aiming to understand and model the charge trapping component of the aging (degradation) process that occurs in MOSFETs.

The same equation (model) applies to the degradation process, i.e., increase in number of occupied traps, as well as to the recovery process, i.e., decrease in number of occupied traps.

The capture and emission of charge carriers by a trap are described as simple Poisson processes governed by rates τ_c and τ_e, where the capture occurs with probability $p_{01}(dt) = dt/\tau_c$ and emission occurs with probability $p_{10}(dt) = dt/\tau_e$. State 1 stands for the occupied trap, while state 0 stands for the empty trap. τ_c and τ_e are then the average residence time in states 0 and 1, respectively [12].

We start by evaluating the device degradation process, which is described by the average number of occupied traps at time t, which we denote $\langle n(t) \rangle$. First we write the equation for the probability of a particular trap, which is initially empty (state 0), to remain in the same state after an elapsed time t. We denote this probability as $P_{00}(t)$. This probability can be calculated observing that

$$P_{01}(t+dt) = P_{01}(t)p_{11}(dt) + P_{00}(t)p_{01}(dt) \tag{29.26}$$

where $p_{11}(dt) = 1 - p_{10}(dt) = 1 - dt/\tau_e$ and $P_{00}(t) = 1 - P_{01}(t)$. This leads to a simple differential equation. The solution of this differential equation is [16]

$$P_{01}(t) = \left[1 - \exp(-t/\tau_{eq})\right] \tau_e/(\tau_c + \tau_e) \tag{29.27}$$

where $1/\tau_{eq} = 1/\tau_c + 1/\tau_e$. A similar evaluation can be performed for $P_{11}(t)$, leading to

$$P_{11}(t) = \left[\tau_e + \tau_c \exp(-t/\tau_{eq})\right] / (\tau_c + \tau_e) \tag{29.28}$$

For a device which has N_{tr} traps, we can write the number of occupied traps at time t, $n(t)$, as being $n(t) = \sum_{i=1}^{Ntr} \theta_i(t)$, where $\theta_i(t)$ can assume the values 0 or 1. We can then evaluate the average number of traps occupied at time t as being

$$\langle n(t) \rangle = \sum_{Ntr=0}^{\infty} \frac{N^{Ntr} e^{-N}}{N_{tr}!} \sum_{k=0}^{Ntr} k \, P(k,t) \tag{29.29}$$

where $N^{Ntr}e^{-N}/N_{tr}!$ is the probability that N_{tr} traps are found in a device, i.e., the number of traps is assumed to be Poisson distributed, with parameter $\langle N_{tr} \rangle = N$, while P(k, t) is the probability that k traps are occupied at time t, with $k = 0 \ldots N_{tr}$. The averaging must be performed over the successive stochastic capture and emission events of a trap and over the number of traps of the ensemble of devices.

The average threshold voltage shift is then obtained by multiplying the average number of occupied traps by $\langle\delta\rangle$, which is the average fluctuation due to a single trap. This leads to

$$\langle \Delta V_T(t)\rangle = \langle\delta\rangle\,\langle n(t)\rangle \tag{29.30}$$

Since traps have different time constants, which are statistically independent and identically distributed random variables, for considering P(k, t) in Eq. (29.29), we evaluate the average probability over different traps by writing a binomial. In this case

$$\overline{\mathrm{P(k,t)}} = \binom{\mathrm{Ntr}}{\mathrm{k}} \overline{\mathrm{P_{01}(\tau_c,\tau_e,t)}}^{\mathrm{k}}\, \overline{\mathrm{P_{00}(\tau_c,\tau_e,t)}}^{\mathrm{Ntr-k}} \tag{29.31}$$

where $\overline{\mathrm{P_{01}(\tau_c,\tau_e,t)}} = \iint d\tau_c d\tau_e \mathrm{P_{01}(\tau_c,\tau_e,t)f(\tau_c,\tau_e)}$ and $\overline{\mathrm{P_{00}(\tau_c,\tau_e,t)}} = \iint d\tau_c d\tau_e\, \mathrm{P_{00}(\tau_c,\tau_e,t)f(\tau_c,\tau_e)}$, with $\mathrm{f}(\tau_c,\ \tau_e)$ being the joint probability function of time constants τ_c and τ_e over all traps.

τ_c and τ_e are random variables, whose values depend on temperature and bias point and which we assume to follow [3, 14]

$$\tau_c = 10^{\mathrm{p}}\left[1+\exp(-\mathrm{q})\right] \tag{29.32}$$

$$\tau_e = 10^{\mathrm{p}}\left[1+\exp(+\mathrm{q})\right] \tag{29.33}$$

where $p \in [\mathrm{p_{min}},\ \mathrm{p_{max}}]$. Please note that $\mathrm{p_{min}}$ and $\mathrm{p_{max}}$ define the times constants of the fastest and the slowest trap, respectively. This also limits the time interval in which the model here proposed is valid, in a similar way as in the analysis of low-frequency noise [1–6]. Since p is assumed to be uniformly distributed, the charge trap characteristic time constants are uniformly distributed on a log scale. Again, this is the same assumption done in low-frequency noise analysis [1–6]. The variable q is given by $q = (\mathrm{E_T}-\mathrm{E_F})/\mathrm{k_BT} \in [(\mathrm{E_V}-\mathrm{E_F})/\mathrm{k_BT},\ (\mathrm{E_C}-\mathrm{E_F})/\mathrm{k_BT}]$, where $\mathrm{E_C}$ is the conduction band edge, while $\mathrm{E_V}$ is the valence band edge. $\mathrm{E_T}$ is the energy level of the trap. Consequently, τ_c and τ_e are temperature dependent. Also note that the assumption that time constants are uniformly distributed on a log scale is in line with low-frequency noise data. The assumption of the existence of individual defects with a very wide distribution of time constants is also in line with recent NBTI data [21, 28, 32]. The assumption of a single p, governing both τ_c and τ_e, may be restrictive and deserves deeper investigation in future works, both theoretical and experimental. Nevertheless, we did run numerical analysis (Monte Carlo simulations) assuming independent p values for capture and emission. The BTI behavior is observed to stay essentially the same (no qualitative deviation on the BTI behavior is observed), as long as the p values remain uniformly distributed.

For the evaluation of Eq. (29.29), we consider Eqs. (29.31), (29.32), and (29.33). In evaluating the average over the different random variables, we separate the average over number of traps

$$\langle(\cdot)\rangle = \sum_{Ntr=0}^{\infty} (\cdot)\frac{N^{Ntr}e^{-N}}{N_{tr}!}$$

and time constants

$$\overline{(\cdot)} = \iint d\tau_c d\tau_e (\cdot) \mathrm{f}(\tau_c, \tau_e)$$

Please remember that the time constants τ_c and τ_e depend on Fermi level, trap energy level, and temperature, as given by Eqs. (29.32) and (29.33). Evaluation of the averages in this manner facilitates the analytical analysis.

So we have [16]

$$\overline{\langle n(t)\rangle} = \overline{\mathrm{P}_{01}(\tau_c, \tau_e, \mathrm{t})} \sum_{Ntr=0}^{\infty} \frac{N^{Ntr}e^{-N}}{N_{tr}!} N_{tr} = N\overline{\mathrm{P}_{01}(\tau_c, \tau_e, \mathrm{t})}$$

$$= \frac{N}{ln10\ (p_{max} - p_{min})} \left(\int_{Ev}^{Ec} \frac{g(E_T)\, dE_T}{1 + e^{-(E_T - E_F)/k_B T}} \right) \left(\int_{10^{-p_{min}}t}^{10^{-p_{max}}t} \frac{(e^{-u} - 1)}{u} du \right) \tag{29.34}$$

where $g(E_T)$ describes the trap energy distribution in the bandgap, and in the second integral, a change of variable was made, p = −log (u/t), $dp = -\, du\, /\, (u \ln 10)$. N is the average number of traps found in a device.

In Eq. (29.34), the first integral contains the Fermi level and temperature dependence, while the second integral contains the time dependence. This means that this equation has the time dependence separated from the Fermi level (i.e., bias point or oxide electric field, since the Fermi level is defined by them) and temperature dependence. Hence, the model predicts that for measurements carried out at different temperatures, there are scaling factors that can be used as a multiplicative coefficients for the threshold voltage shift, making the curves for different temperatures to overlap (at all measured times). The same applies for measurements carried out at different Fermi levels, i.e., different stress voltages (bias points). Note that the Fermi level is a function of the applied voltage. Since Eq. (29.34) is valid for both stress and recovery phases, it implies that the temperature and voltage dependence (scaling factor) during stress and during recovery is the same. This is very relevant and in agreement to experimental data, as discussed below. Equation (29.34) is evaluated numerically, leading to

$$\langle n(t)\rangle \sim \varphi(T, E_F)(A + B\, log(t)) \tag{29.35}$$

where A and B are constants, and the last term clearly shows that the time evolution of number of occupied traps shows a log(t) behavior. The term $\varphi(T, E_F)$ describes the temperature and Fermi level dependence. Please note that Eq. (29.35) above is of the same form as Eq. (4) in [30], which was empirically written, as an approximation for the experimentally observed behavior. The actual form of the term $\varphi(T, E_F)$ depends on the trap energy distribution in the bandgap $g(E_T)$. $g(E_T)$ is usually found to be a convex curve (e.g., U-shaped); see, for instance, [3]. In this work, a U-shaped trap density function is investigated. As in the case of low-frequency noise analysis for abruptly changing gate bias, the trap energy distribution $g(E_T)$ is key for explaining experimentally observed findings [3].

Equation (29.34) models both stress and recovery phases. If the density of initially occupied traps is lower than the value expected for the bias point, the number of occupied traps increases logarithmically. This corresponds to stress phase of BTI. On the other hand, if the density of initially occupied traps is higher than the value expected for the bias point, the density of occupied states decreases logarithmically. This corresponds to the recovery phase of BTI. Besides analytical analysis and evaluation, we did run Monte Carlo simulations to confirm this behavior (Monte Carlo simulations were performed starting from different numbers of initially occupied traps) [20]. Hence, both stress and recovery phases of BTI are described by the same Eq. (29.34). This is corroborated by the experimental results shown below.

The model here presented considers that the number of traps is constant over time, i.e., there is no trap generation or annihilation. What changes over time is trap occupation, but not number of traps. For stress or recovery over long time intervals, there may be generation of traps during the stress phase, or annihilation of traps (annealing) during the recovery phase. In this case, the average number of traps $\langle N_{tr} \rangle$ in Eq. (29.29) becomes a function of time, $\langle N_{tr}(t) \rangle$. If the number or traps increases with time during the stress phase, the evolution of number of occupied traps may become faster than log(t) as time evolves, as experimentally observed in many works, where the time dependence is found to follow a power law for long stress times [22–26]. If a proper equation for $\langle N_{tr}(t) \rangle$ is available, the modeling approach here presented can be used to evaluate this behavior. Modeling N_{tr} as a function of time is out of the scope of this work.

The results of numerical analysis and Monte Carlo simulations confirm that the time evolution of number of occupied shows a logarithmic behavior. For numerical analysis and Monte Carlo simulations performed to confirm the logarithmic behavior, $(p_{max} - p_{min})$ was chosen to be 7, i.e., it is assumed that p is uniformly distributed over 7 decades. The behavior does not depend on p_{max} and p_{min}, as long as the time being considered is much longer than the shortest time constant and much shorter than longest time constant [20].

Note that the logarithmic behavior is related to the trap model used here, which assumes trap characteristic time constants are uniformly distributed on a log scale, as given by Eqs. (29.32) and (29.33).

29.8 Compact Modeling of BTI for Circuit (SPICE) Simulation

This section addresses the circuit level modeling of BTI. The modeling is based on the underlying physics, as presented in the previous sections of this chapter, and derives a compact aging model, thereby enhancing the robustness of BTI modeling practice. The main contributions of this section include (1) compact aging model for circuit simulation, based on the statistical trapping/de-trapping (T–D) assumptions, illustrating the logarithmic time dependence of degradation; (2) accurate prediction of aging variability due to stochastic nature of number of traps and their parameters, such as their time constants, amplitude, and energy distribution; and (3) comprehensive validation of proposed aging models with statistical device data at 65 nm under various voltage tuning.

The T–D-based models for the time evolution of BTI is derived from first principles and can handle the non-monotonic behavior during voltage transitions. It avoids overly pessimistic predictions and hence helps to reduce the margin in reliable design.

To derive the compact model equations we start from Eq. (29.34). We assume that the trap energy (E_T) changes as a function of electric field in the dielectric (E_{ox}) and that the electric field is constant over the dielectric. Hence, E_T will be inversely proportional to E_{ox}, i.e., $E_T \sim 1/E_{ox}$. Assuming that trap time constants spawn over at least ten time decades, i.e., assuming $p_{min} \sim 1$ and $p_{max} > 10$ in Eq. (29.34), leads to

$$n(t) = \frac{Ntr}{\ln 10(p_{\max} - p_{\min})} \exp\left(\frac{\beta V_g}{T_{ox} kT}\right) \exp\left(\frac{-E_0}{kT}\right) \left[A + B\log 10^{-p_{\max}} t\right] \quad (29.36)$$

Equation (29.36) describes the aging under a constant stress voltage and temperature. Similar as the R-D model [19], it is an exponential function of the stress voltage, temperature, and T_{ox}. Furthermore, it has a statistical nature with N_{tr}, an index for the number of traps per device. For the simplicity of model derivation and data analysis, Eq. (29.36) is compacted as

$$\Delta V_{th}(t) = \phi\left[A + B\log(1 + Ct)\right] \quad (29.37)$$

Equation (29.37) shows logarithmic relation of degradation with stress time in contrary to the power law behavior predicted by R-D models.

Along with technology specifications, the degradation rate under BTI depends on dynamic circuit operating conditions. Traditional reliability tests and aging prediction methods are conducted at constant stress conditions, e.g., fixed supply voltage, temperature, and activity factor (α) [33, 34]. Although such an approach simplifies test and modeling procedures, it does not match a realistic circuit operation. The situation is compounded with low-power circuit operations, where techniques such as dynamic voltage scaling (DVS) are employed in today's design, in order to aggressively reduce power consumption. Such techniques further complicate aging

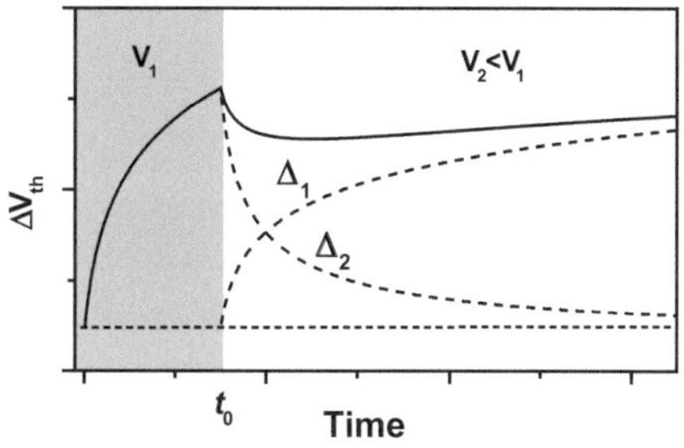

Fig. 29.11 The V_{th} shift under DVS is non-monotonic; when the stress voltage is changed from V_1 to V_2 ($V_2 < V_1$), the degradation is partitioned into two components; the device experiences initial recovery and degradation eventually catches up

prediction as stress and recovery phases are mixed. For instance, when the stress voltage is lowered, the degradation partially recovers at the beginning, before the stress eventually catches up. These phenomenon demands a simple, yet coherent, explanation in order to support accurate lifetime prediction in design for reliability. Further, if the device is stressed under two voltages for multiple cycles, long-term prediction over a given time facilitates designers to eliminate the need of tracking the degradation at every cycle. Since the degradation is highly sensitive to the applied voltage, dynamic voltage scaling leads to different amounts of circuit aging. Aiming at providing simple equations suitable for compact modeling, the voltage tuning is categorized into two cases.

In case 1, stress voltage is changed from V_1 to a higher voltage V_2, and in case 2, V_2 is lower than V_1 at time t_0. Please see Fig. 29.11, which illustrates case 2. To handle such a voltage transition and using a nonzero time t_0 to calculate the occupation probability at time t (time elapsed after t_0), we arrive at a closed form solution for the degradation:

$$\Delta V_{th}(t) = \phi_2 \left[A + B\log(1+Ct)\right] + \phi_1.B\left[\log\left(\frac{1+C(t+t_0)}{1+Ct}\right)\right] \tag{29.38}$$

where ϕ_1 corresponds to the voltage V_1 and ϕ_2 corresponds to V_2. The parameters ϕ_1 and ϕ_2 are a linear function of the number of traps initially occupied and an exponential function of stress voltage and temperature. The degradation in Eq. (29.38) is physically interpreted as a sum of two components, Δ_1 and Δ_2, which are proportional to ϕ_1 and ϕ_2 respectively, as shown in Fig. 29.11. When the voltage is changed to a lower voltage, traps emit some of the charge carriers, and the number of occupied traps reaches a new equilibrium. This behavior is shown in Fig. 29.11, where the V_{th} shift initially recovers and the degradation eventually catches up. Δ_2 dominates initially, which contributes to the recovery. If operation under V_2 continues for a longer time, Δ_1 eventually takes over and ΔV_{th} increases. Such a non-monotonic behavior is correctly predicted by Eq. (29.38), which is a sum of two components. For t values not much larger than t_0, the second component dominates and recovery is observed; and for $t >> t_0$, the second component saturates to a constant value and the first component increases logarithmically with stress

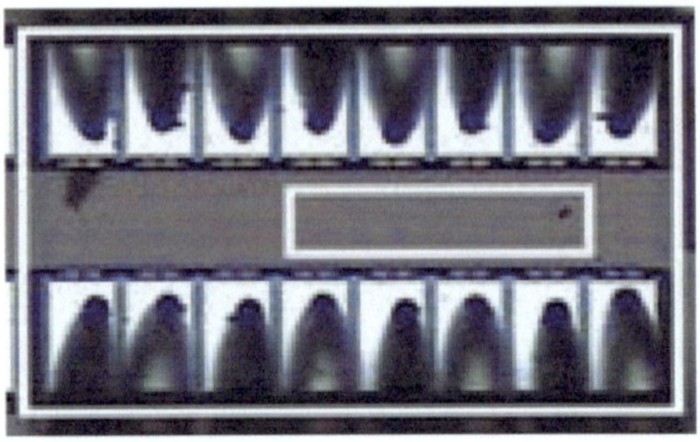

Fig. 29.12 Microphotograph of a 65 nm test chip ($489 \times 332\ \mu m^2$) with an 11 metal layer CMOS; 128 PMOS transistors of four different aspect ratios

time. When the voltage is changed to a higher voltage, the degradation rate increases at the point of transition.

To validate the compact model here presented, statistical aging data is collected from a 65 nm test chip. Various sequences of voltages are applied to probe both stress and recovery phases, and the models are validated across these voltages.

The measurement delay plays a crucial role in NBTI test since even a small measurement time leads to large recovery, resulting in inaccurate aging data. Hence, obtaining degradation data by removing stress from all devices leads to a large measurement error. One solution is to place multiple DUTs on a chip so that stress periods and threshold voltage measurements can be conducted in parallel. This approach is very expensive and needs a larger area. Contrary to parallel measurement method, a parallelized stress period in a pipeline manner can be implemented, and V_{th} measurements for the DUTs are conducted sequentially. The data presented in this work was obtained in this way [18].

Figure 29.12 shows the microphotograph of test chip implemented in 65 nm and 11 metal layer CMOS process. The area of the inner rectangle of the test chip is $489 \times 332\ \mu m^2$, and 128 PMOS devices of four different aspect ratios are implemented as devices under test (DUTs). Initially, aging measurements are conducted when all the devices are stressed at 1.8 V and a temperature of 125°C for 200 ks. A DUT in the transistor array is changed from the stress phase to the recovery phase using pass gates as switches. These measurements are intended to analyze device to device statistical aging behavior under constant stress and dynamic operations. Further, measurements are conducted when all the devices are stressed under multiple V_{DD} to realize the aging in DVS operation. Figure 29.13 presents the test structure and measurement principle implemented in our work. Except for the device in measurement, all other devices are stressed and V_{th} is measured using the constant current method with a resolution of 0.2 mV. Measurement time for each DUT is less than 0.4 ms, minimizing the recovery.

Statistical aging analysis is performed on the data collected from the test chip. Initial data is collected by stressing all the PMOS transistors at 1.8 V for 200 ks. From this stress data, model parameters for the *log*(t) model are extracted. Figure 29.14 compares aging statistics with model prediction: the variation is attributed mainly due to the randomness in parameter ϕ, proportional to the fluctuation of trap numbers. The *log*(t) model reliably captures the increase of both

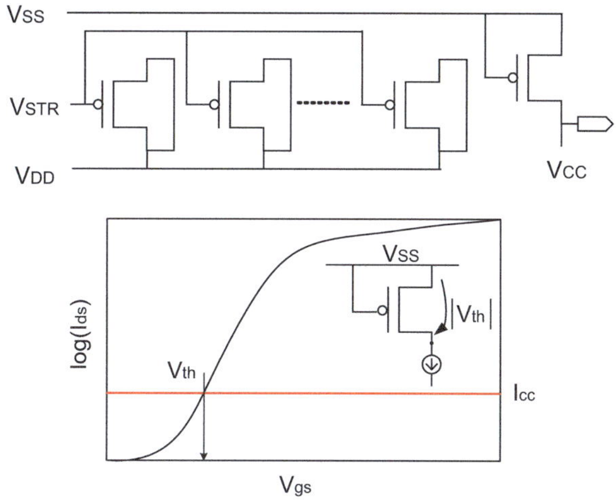

Fig. 29.13 Measurement setup: (**a**) All devices are under stress except the device under measurement (**b**) V_{th} measurement using constant current method

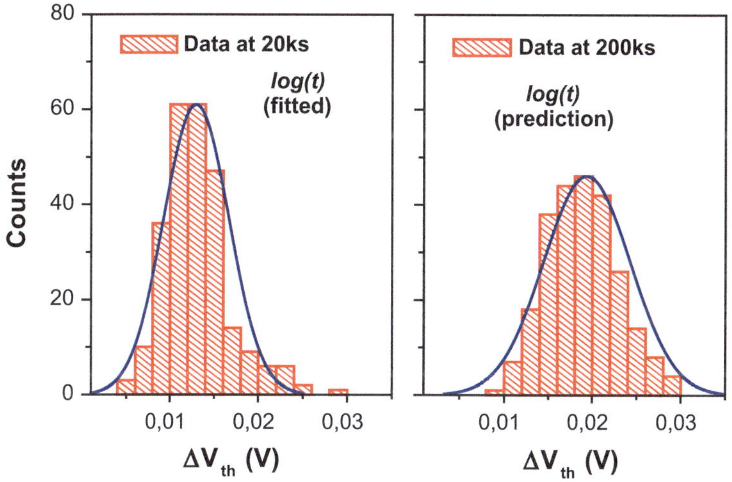

Fig. 29.14 The *log*(*t*) compact model fitted from data < 20 ks under 1.8 V, 125°C, well predicts the long-term statistics at 200 ks. The shift of V_{th} follows the normal distribution

mean and variance with longer stress time [6]. Other model parameters do not suffer a high amount of variations. The variation in parameter ϕ has only linear impact on the V_{th} shift. The modeling framework enables a one-time extraction to predict aging under various voltage tuning scenarios. The randomness in parameter ϕ extracted from the static measurement can be used to predict aging under DVS. The model prediction under different stress voltages using the current extraction was validated as described below.

To further test the accuracy of model prediction under static stress, model parameters are initially extracted from stress data for stress times lower than 20 ks. These model parameters are then used to predict aging to 200 ks by extrapolation. The experimental results showed that the *log*(t) model correctly captures the shift in mean and variance of the ΔV_{th} in long term. As indicated in Eq. (29.36), the degradation follows a logarithmic dependence on stress time. The experimental characterization was performed for different constant stress voltages, and the *log*(t) model did properly fit the experimental data. The experimental data also confirmed that the degradation has an exponential relation with stress voltage. Such an exponential dependence on voltage is similar to that predicted by the t^n, i.e., power law, model [5]. The time dependence of degradation is the major difference between R–D-based t^n and T–D-based *log*(t) compact models.

The T–D mechanism expects fast and discrete recovery events from individual trapped charges. In this work, various sequences of stress voltages are applied, in order to calibrate the model prediction by the logarithmic model. After a stress period of 200 s under 1.5 V, the stress voltage changes to different values, such as 1.8, 1.5, or 1.2 V. Depending on the second voltage, the degradation rate may increase or decrease. More interestingly, if the voltage is lowered the device may experience a transition period, before the stress goes back to the equilibrium condition. Eventually, the degradation rate goes to the same value as the constant stress under the second voltage. This behavior is predicted from Eq. (29.38), where the second component dominates initially, resulting in the recovery. After $t \gg 200$ s, the second component decays down and first component dominates, leading to the stress under second voltage. Experimental results from the test chip are shown in Fig. 29.15 and well validate these non-monotonic behaviors, supporting further study on aging under DVS. The two components in Eq. (29.38) play an important role in long-term prediction under multiple cycles.

Figure 29.16 evaluates the model prediction, with different periods under the same voltage. In this study, the device is initially stressed under 1.8 V, for 50 s or 200 s; then the voltage is switched to 1.65 V. As the stress voltage is lowered, a temporary recovery behavior is observed due to the emission of excessive amount of trapped charges. The *log*(t) model captures such behavior in both the cases. The stress profile for the case where the device is first stressed under 1.8 V for 200 s is higher compared to the case where the initial stress is 50 s only. However, since in both the cases, the device is later stressed for a much longer time ($\sim$10 ks) at 1.65 V, the degradation converges to the constant stress condition at 1.65 V. This validation helps to predict the aging under different biasing conditions and switching activities (α).

Figure 29.17 illustrates two cases where a PMOS device is stressed under different voltages sequentially. In case 1 (left), the device is stressed at 1.5 V, 1.8 V, 1.5 V for 5,000 s during each period. When the stress voltage is changed from 1.5 to 1.8 V, the degradation rate is increased. The number of occupied traps increases and V_{th} shift increases as shown in Fig. 29.17a and predicted from *log*(t) model. When the voltage is changed back to 1.5 V from 1.8 V, the degradation undergoes recovery due to traps emitting the excess charge carriers (Fig. 29.17a). This recovery

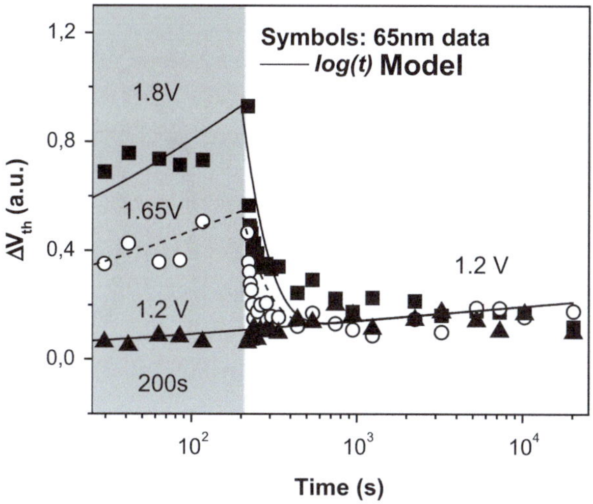

Fig. 29.15 The *log*(*t*) model predicts the degradation under voltage tuning; when the voltage is lowered to 1.2 V, the degradation reaches a new equilibrium, approaching that under 1.2 V

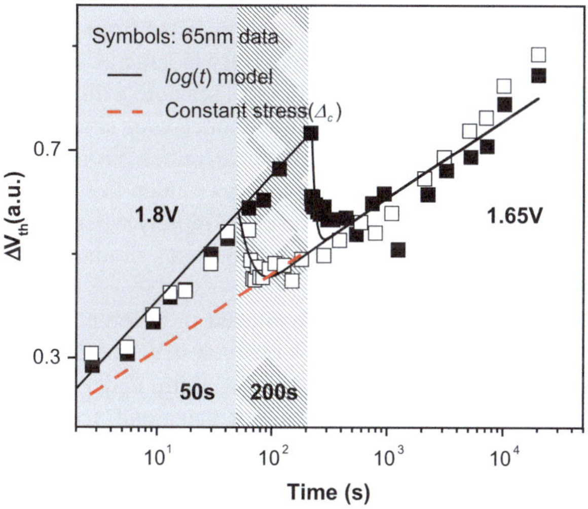

Fig. 29.16 V_{th} shift under voltage tuning from 1.8 to 1.65 V. The shift eventually converges to the log(t) curve of the final voltage, independent on previous stress history

behavior is captured by trapping/de-trapping model, as discussed above. In case 2, the device is stressed at 1.8, 1.2, 1.5 V for 5,000 s during each period. The recovery is furthermore significant when operated under a much lower voltage of 1.2 V as presented in Fig. 29.17b.

Figure 29.18 presents the measurement under different patterns of voltages and duty cycles. Figure 29.18a, b show the multiple cycle prediction stressed at 1.8 V, 1.2 V, and 1.8 V, 0 V at $\alpha = 0.5$. When voltage is lowered from 1.8 V, a recovery is

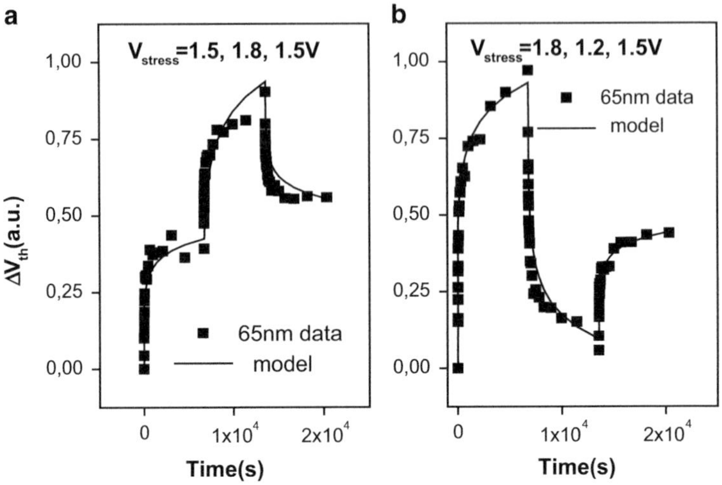

Fig. 29.17 Threshold voltage shift under different V_{DD} sequences; T-D model accurately captures changing stress and recovery behaviors

observed from the silicon data, which is well predicted from the model. Further, if the voltage is lowered down to 0 V, the recovery is large as shown in Fig. 29.18b. The compact model expressed by Eq. (29.38) is scalable with different duty cycles. Figs. 29.18c, d show the validation of our model with silicon data at very low ($\alpha = 0.03$) and very high ($\alpha = 0.97$) duty cycle values. From the figures, voltage transition even for a short duration will result in a sudden shift in V_{th}, due to the fast trapping/de-trapping. The degradation increases rapidly when voltage is increased momentarily to 1.8 V as illustrated in Fig. 29.18c and a similar behavior is observed in recovery as shown in Fig. 29.18d.

Further, statistical measurement is conducted in different PMOS devices with stress condition at 1.8 V altering with recovery at 0 V. The parameter ϕ extracted from the static measurement is used to predict aging under this stress condition. Figure 29.19 presents the model prediction of mean and upper bounds using the randomness in parameter ϕ. The model well matches the statistical silicon data reconfirming that one-time extraction from static measurement is sufficient to predict the aging under DVS.

Device level aging due to trapping/de-trapping exhibit large variations due to randomness in number of available traps. The variability at ($t>0$) is also a function of transistor sizing similar to process induced variations ($t = 0$) [13]. Aging variability increases with downsizing the devices and hence, it becomes significant with CMOS technology scaling. Further, it is important to understand and estimate the translation of device aging to circuit level degradation [27, 28].

Figure 29.20 explores the dependence of the aging effect on the choice of V_{dd} values, as well as the control of α. Since the amount of degradation is an exponential function of voltage, lower stress voltage helps reduce the degradation rate. The degradation significantly decreases when the ratio of V_{low}/V_{high} drops from 1.

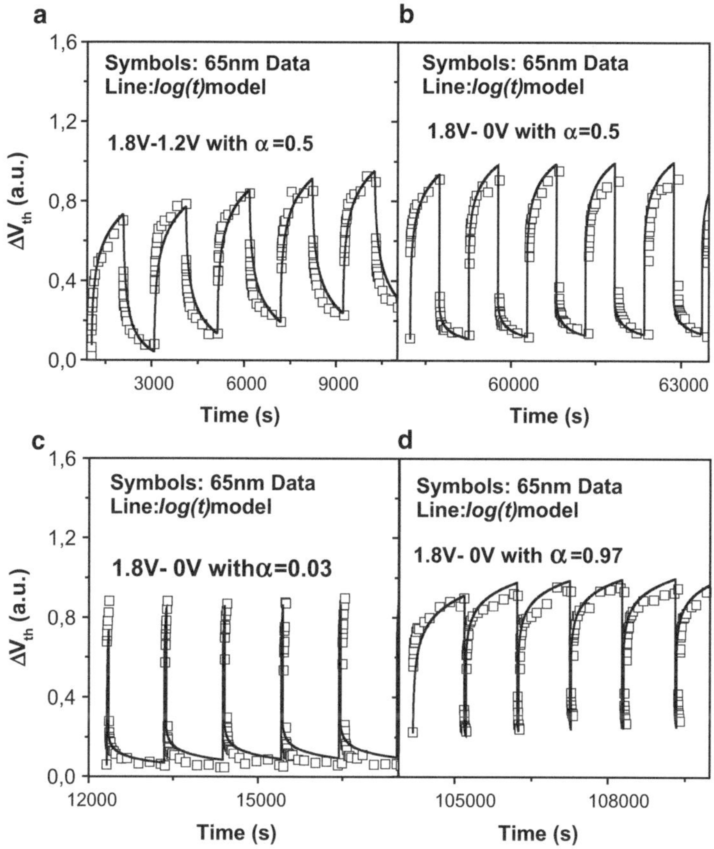

Fig. 29.18 Validation of the compact model under different voltages and duty cycles

However, the reduction rate becomes less as the V_{low} is further lowered, since the operation period with V_{high} dominates the entire degradation and therefore, further scaling of V_{low} does not help minimizing the aging. Lower α, i.e., the shorter period in V_{high} operation, is still effective in reducing the aging, as predicted by the proposed model.

29.9 Conclusion

A statistical model for the charge trapping phenomena in MOS devices is presented. The model is based on discrete microscopic device physics parameters. Besides evaluating the average behavior, the model here proposed allows the derivation of statistically relevant parameters. The model is applied to study the random telegraph noise (RTN) and bias temperature instability (BTI).

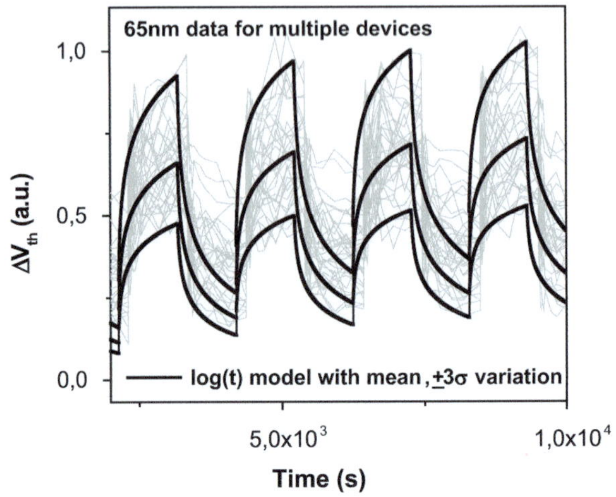

Fig. 29.19 Statistical measurement under DVS operation; models with ϕ values, in which the mean and 3σ are extracted from static stress data, predict statistical aging under dynamic circuit conditions

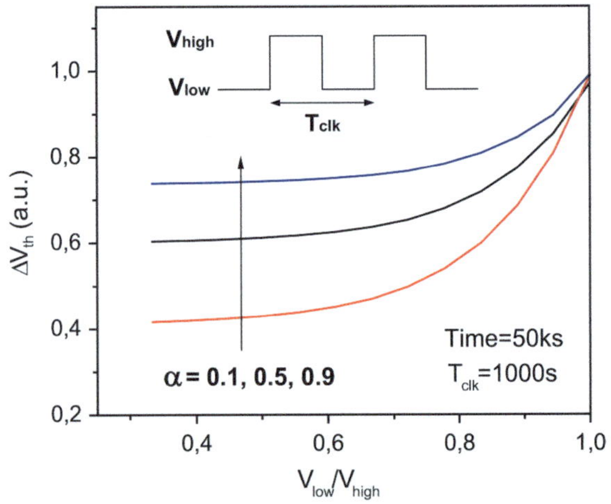

Fig. 29.20 Aging prediction from the compact model varying the lower operating voltage in DVS

The impact of charge trapping is modeled as momentary changes in threshold voltage. In analog circuits, this may lead to low-frequency noise and be a source of jitter in oscillators. In digital circuits, it may lead to aging effects and transient effects, since circuit behavior may change due to transient V_T fluctuations between two logic operations of a digital circuit.

The modeling approach focuses on operation conditions relevant for digital and analog design, including large signal AC operation.

Regarding BTI, an analytical model for both stress and recovery phases of BTI is presented. Furthermore, the model properly describes device behavior under different switching, including dynamic voltage scaling (DVS). It is shown that a universal logarithmic law describes the time dependence of charge trapping in both stress and recovery phases and that the time dependence may be separated from the temperature and bias point dependence.

The proposed models facilitate the circuit aging prediction under device level variations and dynamic operation conditions. The proposed approach transfers the latest understanding of the trapping/de-trapping mechanism into compact models, and provides a simple and practical solution for the circuit design community.

It is shown that if the density of traps in the bandgap is a convex curve ("U"-shaped), a reduction in noise power may be achieved under cyclo-stationary excitation. However, the variability in noise behavior (normalized standard deviation of noise power) is shown to increase. This increase in variability will be another challenge to analog and RF circuit designs in deep-submicron CMOS technologies. Assuming a U-shaped trap density is also key for properly modeling BTI under different bias conditions.

References

1. G Wirth et al. "Modeling of Statistical Low-Frequency Noise of Deep-Submicron MOSFETs", *IEEE Trans. Electron Dev.*, **52**, p. 1576 (2005).
2. G Wirth, R da Silva and R Brederlow. "Statistical Model for the Circuit Bandwidth Dependence of Low-Frequency Noise in Deep-Submicrometer MOSFETs", *IEEE Trans. Electron Dev.*, **54**, p. 340–345 (2007).
3. G Wirth et al. "Statistical model for MOSFET low-frequency noise under cyclo-stationary conditions" *Int Electron Dev Meeting - IEDM 2009*, p. 30.5.1 (2009).
4. G Wirth and R da Silva, "Low-Frequency Noise Spectrum of Cyclo-Stationary Random Telegraph Signals", *Electrical Eng.*, **90**, p. 435 (2008).
5. R da Silva, G Wirth and L Brusamarello. "An appropriate model for the noise power spectrum produced by traps at the Si SiO interface: a study of the influence of a time-dependent Fermi level. Journal of Statistical Mechanics" *J. of Statistical Mechanics. Theory and Experiment*, **2008**, p. P10015 (2008).
6. R Brederlow et al. "Low Frequency Noise Considerations for CMOS Analog Circuit Design" *Proceedings of the 2005 Int. Conf. on Noise and Fluctuations* (*ICNF*). p. 703 (2005).
7. R Brederlow, J Koh and R Thewes. "A physics-based low frequency noise model for MOSFETs under periodic large signal excitation" *Solid-State El.*, **50**, p. 668 (2006).
8. I Bloom and Y Nemirovsky. "1/f noise reduction of metal-oxide-semiconductor transistors by cycling from inversion to accumulation" *Appl Phys Lett*, **58**, p.1664 (1991).
9. B Dierickx and E Simoen. "The decrease of "random telegraph signal" noise in metal-oxide-semiconductor field-effect transistors when cycled from inversion to accumulation". *J Appl Phys*, **71**, p. 2028 (1992).
10. M Ertürk, T Xia and W Clark "Gate voltage dependence of mosfet 1/f noise statistics" *IEEE Elec Dev Let*, **28** p. 812 (2007).
11. A van der Wel et al. "Relating random telegraph signal noise in metal-oxide-semiconductor transistors to interface trap energy distribution" *Appl Phys Lett.*, **87**, p. 183507 (2005).

12. S Machlup, "Noise in semiconductors: spectrum of a twoparameter random signal" *J Appl Phys*, **35**, p. 341 (1954).
13. A Roy and C Enz, "Analytical modeling of large-signal cyclo-stationary low-frequency noise with arbitrary periodic input", *IEEE Trans. Electron Dev.*, **54**, p.2537 (2007).
14. M Kirton and M Uren "Noise in solid-state microstructures: a new perspective on individual defects, interface states and low-frequency (1/f) noise" *Adv. in Phys.*, **38**, p. 367 (1989).
15. S Ekbote et al., "45nm low-power CMOS SoC technology with aggressive reduction of random variation for SRAM and analog transistors", *VLSI Tech. Symp.*, p. 160 (2008).
16. V Camargo et al. "Impact of rdf and rts on the performance of sram cells" *J. of Comput. Electronics*, p. 122 (2010).
17. K.K. Hung et al. "A physics-based mosfet noise model for circuit simulators" *IEEE Trans. Electron Dev.*, **37** p. 1323 (1990).
18. T. Sato, et al., "A device array for efficient bias temperature instability measurements," ESSDERC, pp. 143–146, 2011.
19. W. Wang et al. "Compact modeling and simulation of circuit reliability for 65 nm CMOS technology," *IEEE Trans. Dev. Mat. Reliab.*, **7**, p. 509 (2007).
20. R da SILVA and G Wirth. "Logarithmic behavior of the degradation dynamics of metal oxide semiconductor devices", *Journal of Statistical Mechanics*, v. P04025, p. 01 (2010).
21. G Wirth, R da Silva, and B Kaczer "Statistical model for mosfet bias temperature instability component due to charge trapping" *IEEE Trans. on Electron Dev.*, **58**, p.2743 (2011).
22. T. L. Tewksbury and H.-S. Lee, "Characterization, modeling and minimization of transient threshold voltage shifts in MOSFETs" *IEEE J. Solid-State Circ.*, **29**, p. 239 (1994).
23. D. K. Schroder, "Negative bias temperature instability: What do we understand?" *Microel. Reliab.*, **47**, p. 841 (2007).
24. V. Huard et al. "NBTI degradation: From Transistor to SRAM Arrays," *Proc. Int. Rel. Phys. Symp.*, p. 289 (2008).
25. S. E. Rauch, "Review and reexamination of reliability effects related to NBTI statistical variations" *IEEE Trans. Dev. Mat. Rel.*, **7**, p. 524 (2007).
26. M.A. Alam et al. "A comprehensive model for pmos nbti degradation: recent progress" *Microel. Reliab.* p. 853 (2007).
27. B. Kaczer et al. "NBTI from the perspective of defect states with widely distributed times," *Proc. Int. Rel. Phys. Symp.*, p. 55 (2009).
28. Reisinger et al. "The statistical analysis of individual defects constituting NBTI and its implications for modeling DC- and AC-stress" *Proc. Int. Rel. Phys. Symp.*, p.7 (2010).
29. V. Huard et al. "New characterization and modeling approach for NBTI degradation from transistor to product level". *Int Electron Dev Meeting*, p. 797 (2007).
30. T Grasser and B Kaczer, "Evidence That Two Tightly Coupled Mechanisms Are Responsible for Negative Bias Temperature Instability in Oxynitride MOSFETs", *IEEE Trans. on Electron Dev.*, **56**, p. 1056 (2009).
31. T. Grasser et al. "Switching oxide traps as the missing link between negative bias temperature instability and random telegraph noise", *Int. Electron Dev. Meeting*, p. 729 (2009).
32. B. Kaczer et al. "Ubiquitous Relaxation in BTI stressing—New Evaluation and Insights", *Proc. Int. Reliab. Phys. Symp.*, p. 20 (2008).
33. Z. Liu, B. W. McGaughy, and J. Z. Ma, "Design tools for reliability analysis," *DAC – Desing Aut. Conf.*, p. 182 (2006).
34. M. Denais et al. "On-the-fly characterization of nbti in ultra-thin gate oxide pmosfets" *Int. Elec. Dev. Meet.*, p. 109 (2004).

Chapter 30
Simulation of BTI-Related Time-Dependent Variability in CMOS Circuits

Javier Martin-Martinez, Rosana Rodriguez, and Montse Nafria

Abstract The correct evaluation of BTI impact on the circuit performance and reliability is a major concern in current technologies. Since BTI in ultrascaled devices is a stochastic mechanism and aging must be evaluated under the actual operation conditions of devices in the circuit, SPICE and Monte Carlo simulations are customary combined with this purpose. The key point in these simulations is the correct description of the BTI effects in the device and their inclusion in circuit simulators. In this subchapter, the different adopted approaches are presented, pointing out their pros and cons, and illustrated with examples of BTI effects on several analog and digital circuits.

30.1 Introduction

Among the different aging mechanisms that can affect the MOSFET performance, i.e., channel hot carrier degradation, bias temperature instability (BTI), and time-dependent dielectric breakdown, BTI is assumed to be the most severe in determining the circuit's end of life [1–3]. The different fabrication approaches that can be adopted to reduce BTI aging can only partially mitigate the problem [4–11]. Therefore, acting during the design phase of the circuit becomes mandatory, in what has been called Design for Reliability (DfR) [12–14]. In the DfR scenario, several techniques, at different architectural levels, to reduce the BTI detrimental effects have been proposed [15–23]. However, some important problems still need to be urgently solved. According to the ITRS [24], due to the strong BTI dependencies

J. Martin-Martinez (✉) • R. Rodriguez • M. Nafria
Dept. Enginyeria Electrònica, Universitat Autònoma de Barcelona (UAB),
Escola d'Enginyeria, Edifici Q, 08193 Bellaterra, Spain
e-mail: javier.martin.martinez@uab.es; rosana.rodriguez@uab.es; montse.nafria@uab.es

T. Grasser (ed.), *Bias Temperature Instability for Devices and Circuits*,
DOI 10.1007/978-1-4614-7909-3_30, © Springer Science+Business Media New York 2014

on stress and relaxation times, voltage, and temperature, new tools that correctly evaluate the BTI degradation of transistors at their operation conditions within the circuit are urgently required [25–30]. This implies that physics-based and accurate BTI models must be developed and incorporated in these tools [31–38]. Moreover, in ultrascaled devices BTI becomes a stochastic phenomenon [39–47], because few defects can provoke large V_T shifts. Consequently, statistically distributed V_T shifts have to be taken into account to correctly evaluate the device lifetime [48]. The BTI aging at device level will have to be properly considered during circuit simulation, to evaluate its impact on the circuit performance and reliability. With this purpose, a methodology that combines SPICE, to consider the actual operation of devices within the circuit, and Monte Carlo simulations, to account for the statistically distributed V_T shifts (because of the fabrication related variability and/or the BTI aging itself), has been proposed [49–51]. The key point of this methodology is how the device aging is evaluated and considered during the circuit simulations. In this subchapter, the most common adopted approaches are described. The most straightforward is to include the BTI degradation effects in the SPICE model of the MOSFETs, to describe the change in their electrical performance. However, the most extended approach is to connect to the MOSFET terminals a suitable electrical circuit, based on the BTI underlying physics, to account for the BTI effects in the device. The advantages and inconveniences of both approaches will be discussed. Finally, to be useful to circuit designers, the whole procedure (calculation of the device BTI effects in the device, their inclusion in the circuit simulator, and statistical simulation of the circuit) should be integrated in a reliability circuit simulation tool. RELAB, an example of such tool, will be described.

30.2 Combined SPICE and Monte Carlo Simulation Methodology

The first approach to include the V_T shift related to BTI degradation in circuit simulators consists in the variation of the adequate parameters of the SPICE model that describes the electrical behavior of the aged MOSFETs. To include also the device variability, which could be associated to the fabrication process [52–56] or to the BTI mechanism itself [39–45], these parameters will be statistically distributed. To account for the resulting device time-dependent variability during circuit simulation, a methodology that combines SPICE and Monte Carlo (MC) simulations has been proposed [51]. The validity of this simulation technique will be checked with experimental data obtained from isolated devices, i.e., test structures. As an example of applicability, the effect of BTI degradation and fabrication process variability on the performance of three different configurations of differential amplifiers will be studied.

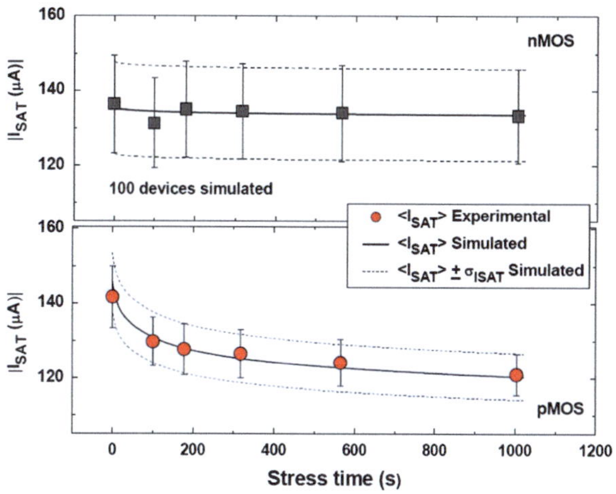

Fig. 30.1 Time evolution of I_{SAT} for (*top*) NMOS and (*bottom*) PMOS devices. *Dots* are the experimental average values and *bars* show the σ of the distribution. *Continuous lines* correspond to the mean value and *dashed lines* to the dispersion of I_{SAT} obtained in MC simulations [51]

30.2.1 Device Characterization and Simulation Methodology

The simulation of the device BTI aging effects on the circuit performance will have to take into account the fabrication technology since it will strongly affect the device degradation [4–11]. Therefore, first of all, an experimental characterization of the NMOS and PMOS BTI aging is required [57–61]. Since variability could be present, either related to the fabrication process or to the BTI mechanism itself, a statistical characterization will be required. As an example of BTI aging results, Fig. 30.1 shows the evolution with time of the saturation current measured in NMOS (top) and PMOS (bottom) transistors subjected to BTI stress. NMOS and PMOS with SiON as gate dielectric (EOT = 2.2 nm), same length (0.13 μm), and different channel width (0.3 μm for NMOS and 0.6 μm for PMOS) have been considered. The nominal supply voltage was 1.2 V. Transistors were degraded at room temperature with a conventional measurement–stress–measurement (MSM) technique. The voltage applied to the gate was 2.9 V for NMOS and −2.9 V for PMOS, and the other terminals were grounded. Therefore, PBTI and NBTI are the aging mechanisms for the NMOS and the PMOS transistors, respectively. During the stress interruption, the I_D–V_{GS} and I_D–V_{DS} characteristics were registered after a relaxation time determined by the measurement instrument. From these curves, the threshold voltage (V_T) and the saturation current at operation voltage (I_{SAT}) were extracted. To get statistical data, several transistors of each type were stressed. As expected, for both NMOS and PMOS transistors, due to BTI effects, the gate-dielectric damage leads to an increase of the threshold voltage (in absolute value) and a decrease of the saturation current (I_{SAT}). The saturation current decrease is associated to the V_T increase but also to the mobility degradation caused when high-voltage stresses are applied [62]. As previously observed for SiON devices, a much-larger degradation is observed in PMOS transistors (17% I_{SAT} decay after 1,000 s of stress), degraded by NBTI, compared to NMOS devices, degraded by PBTI. Concerning to the I_{SAT} variability, the error bars shown in Fig. 30.1 indicate the standard deviation of the I_{SAT} distribution (σ_{ISAT}) for NMOS and PMOS transistors. Any or negligible variation of σ_{ISAT} with the stress time is observed in both transistor types, which indicates that the initial device variability (related to the fabrication process) dominates to the possible variability induced by the electrical stress applied.

A three-step simulation methodology has been designed to include the observed variability and gate oxide degradation effects in a circuit simulator, whose simulation flow is shown in Fig. 30.2. The first two phases correspond to the extraction from the experimental data (Fig. 30.1) of the SPICE model parameters that better describe the experimentally observed electrical behavior of the devices (fresh and stressed) and the third to the circuit simulation itself.

In the first step (Phase 1 in Fig. 30.2), the complete set of BSIM4 SPICE model parameters [63] of the fresh transistors (before any stress is applied) are obtained from the measured I_D–V_{GS} and I_D–V_{DS} curves using AURORA software tool [64]. In the second step of the proposed simulation methodology (Phase 2 in Fig. 30.2), the BSIM4 model parameters of the stressed devices are obtained from the I_D–V_{DS} and I_D–V_{GS} transistor curves measured when the stress was interrupted. However, we have observed that the I_D–V_{DS} and I_D–V_{GS} curves of the degraded transistor at different stress times can be reproduced by varying only two physically meaningful parameters of the BSIM4 parameter set: VTH0 (related to V_T) and U0 (related to the transistor mobility). Therefore, the aim of the second phase is to extract VTH0 and U0 for each device, at different stress times. The extraction procedure is as follows, which is repeated for all the devices. First, VTH0 is varied to fit the measured degraded I_D–V_{GS} characteristics. For this fitting, only the medium-voltage range of the curve ($V_T \pm \sim 0.2$ V) is considered, since, for larger voltages, the current is affected by mobility degradation, and for lower voltages, subthreshold slope is affected by the stress-generated interfacial traps (changes in the subthreshold slope could be taken into account by adding an extra BSIM4 parameter, CIT). Next, the U0 BSIM4 parameter is varied to fit the stressed I_D–V_{GS} curve (at voltages larger than $V_T + 0.2$ V) and the stressed I_D–V_{DS} characteristics. This process is repeated until fitting convergence. When all the degraded device curves have been fitted for a given stress time, the VTH0 and U0 mean values and their standard deviations (σ_{VTH0} and σ_{U0}) are calculated. Afterwards, the next stress time is considered. Figure 30.3 shows the evolution with the stress time of the extracted mean value variations of VTH0 (ΔVTH0, top) and U0 (ΔU0, bottom) parameters. Both parameters obey a power law with the stress time (note that the time exponent in the ΔVTH0 evolution is close to that usually observed for NBTI induced V_T change, ~0.25 at these

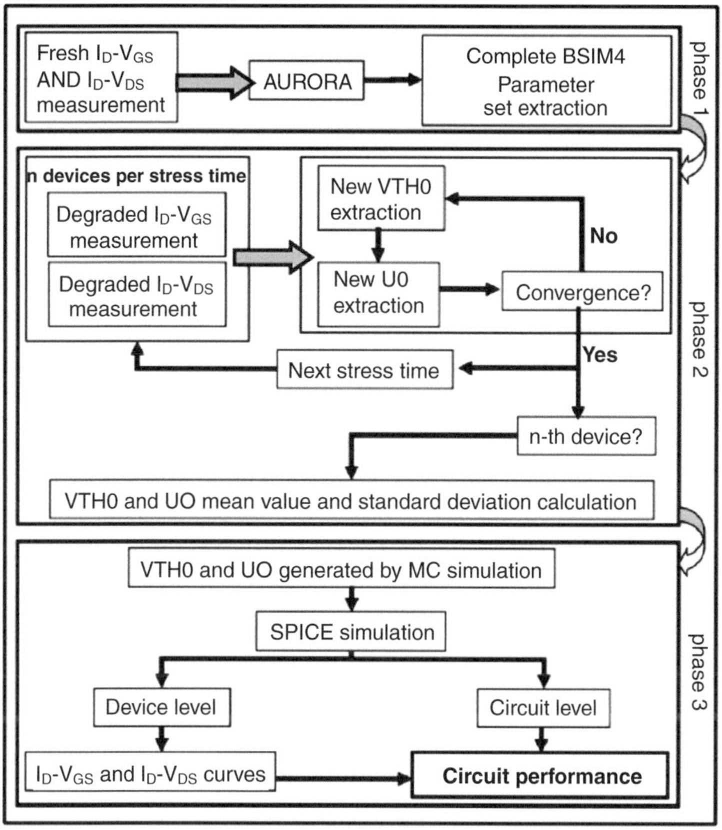

Fig. 30.2 Simulation flow of the proposed methodology to simultaneously include process variability and BTI effects in a circuit simulator

measurements conditions [65]). These temporal dependencies will allow estimating the VTH0 and U0 parameters in the stressed devices from the parameter values given for the fresh ones, for any stress time.

In the last step of the procedure (Phase 3 in Fig. 30.2), the circuit and/or device response after stress is evaluated, combining MC and SPICE simulations. The experimental mean values and standard deviations of VTH0 and U0 distributions derived in Phase 2 and the time dependence of the mean values (σ does not change with time, Fig. 30.1) are the inputs to a MC simulator, which generates a set of new VTH0 and U0 parameters. The output VTH0–U0 pairs of the MC simulator (together with the rest of the nonvaried BSIM4 parameters obtained from fresh devices in Phase 1) are the inputs to a circuit simulator, which will calculate the I_D–V_{DS} and I_D–V_{GS} curves of the device and the circuit performance as a function of the stress time. From the results at device and/or circuit levels, the effects of variability and wear-out will be evaluated.

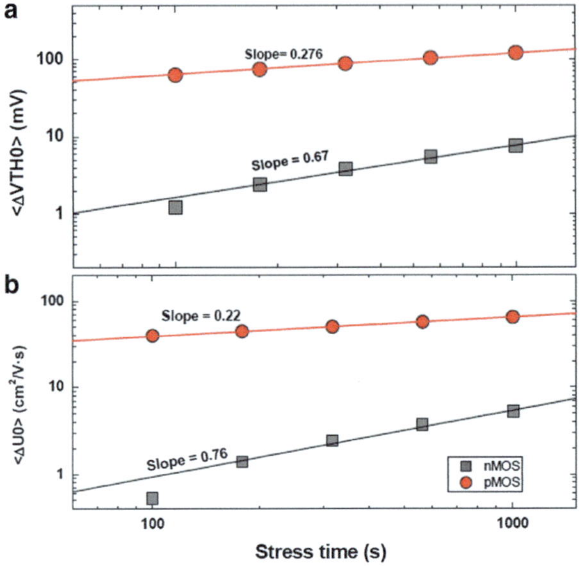

Fig. 30.3 BSIM4 VTH0 (*top*) and U0 (*bottom*) parameter variation with the stress time extracted from the experimental data. The extracted values have been fitted to a power law. The time exponent obtained for the PMOS is close to the value usually observed for the V_T shift related to NBTI degradation

The proposed simulation methodology has been used to reproduce the experimental results observed in the stressed transistors (device simulation). For each of the considered stress times, 100 I_D–V_{DS} characteristics have been simulated, using ADS circuit simulator [66]. From the simulated curves, the I_{SAT} mean value and standard deviation have been calculated, which are shown in Fig. 30.1. In this figure, continuous lines correspond to the mean I_{SAT} value obtained from the simulations, and the dashed lines represent the $\langle I_{SAT}\rangle \pm \sigma_{ISAT}$ range. The good match between experimental data and simulation in Fig. 30.1 demonstrates that it is only necessary to modify the average of two physically meaningful SPICE parameters (VTH0 and U0) to correctly reproduce the experimentally observed I_{SAT} evolution and its variability. Moreover, it also validates the initial assumption of a normal distribution of these parameters, which has been found to be described by a time-independent standard deviation. In this approach, a power law has been used to model the VTH0 and U0 time dependence, since it provides a simple description of the experimental observations. However, the proposed simulation technique is independent of the considered BTI model, and the used power law can be replaced by more complete models that account for other BTI concerns, such as voltage dependence for lifetime estimation and temperature dependence.

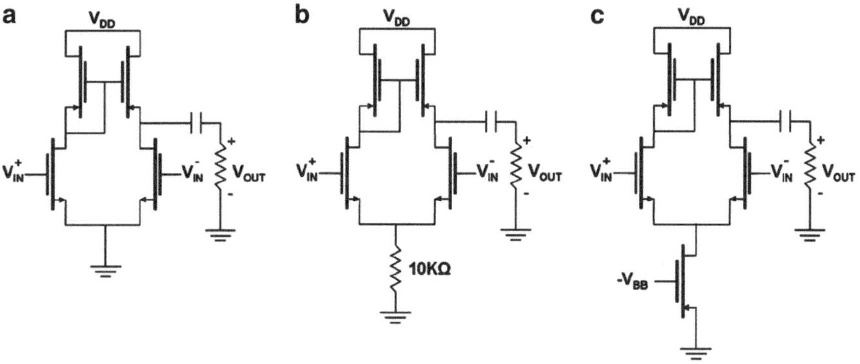

Fig. 30.4 Three amplifier configurations (based in an NMOS differential pair biased with a PMOS current mirror) have been considered to simultaneously study the variability and stress impact on circuit performance. The configurations differ in the current-limiting mechanism used: (**a**) none, (**b**) a 10-KΩ resistor, and (**c**) a PMOS transistor

30.2.2 Device Degradation and Process Variability Impact on Differential Amplifiers Performance

To analyze the influence on the circuit performance of the time-dependent variability related to the BTI gate oxide degradation, NMOS differential pairs biased with a PMOS current mirror (Fig. 30.4) have been simulated using the proposed methodology. Three different configurations have been considered, which differ in the current-limiting mechanism used: none, Fig. 30.4a; a 10-KΩ resistor, Fig. 30.4b; or a PMOS transistor, Fig. 30.4c. The amplifier gain and bandwidth (BW) have been taken as figures of merit of their performance. As the first approach, we have assumed that all the devices within the circuit are subjected to the same stress conditions; but according to Fig. 30.3, their degradation rates are different.

To simulate the circuit performance, fresh and stressed MOSFET electrical characteristics are considered, which have been built from the MC-generated VTH0 and U0 parameters. Device-level variability and the aging effects in its characteristics are included by considering different VTH0–U0 pairs for each transistor within the circuit and the time dependence of their mean values, respectively. First, the Bode diagram of the amplifier has been simulated for different stress times, without taking into account the variability of the gate oxide wear-out-related transistor parameters, i.e., $\sigma_{U0} = \sigma_{VTH0} = 0$, and considering the time dependence of the average values of U0 of VTH0. As a consequence of the dielectric wear-out, the gain slightly decreases (e.g., for the circuit in Fig. 30.4a, a small reduction of 0.68 dB was obtained after 10^5 s of stress). Second, 500 Bode diagrams have been generated, taking into account the device variability. In this case, the experimental average of U0 and VTH0 (Fig. 30.3) have been considered, but σ_{U0} and σ_{VTH0} have been chosen so as to obtain a 5% (a standard value) V_T variability (σ_{VT}) for the NMOS and PMOS transistors. As an example, Fig. 30.5 shows 20 of the Bode diagrams

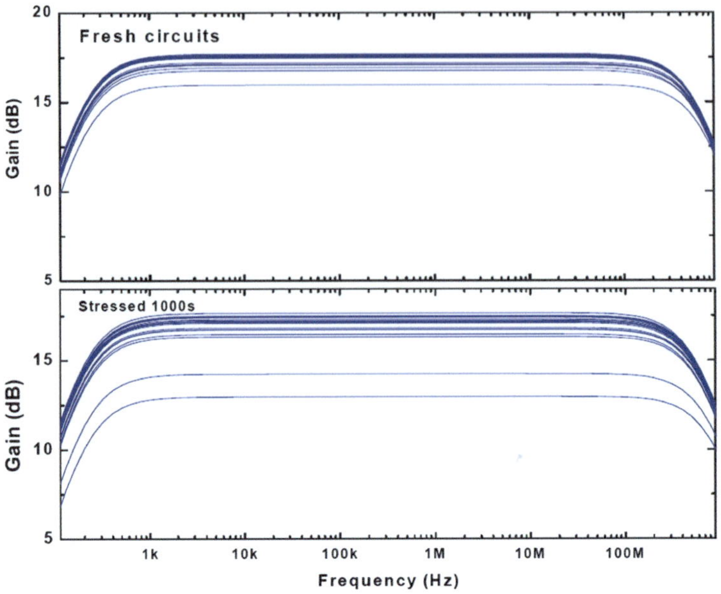

Fig. 30.5 Bode diagrams of 20 (*top*) fresh and (*bottom*) after 1,000 s of stress simulated differential amplifiers as those shown in Fig. 30.4a. $\sigma_{VT} = 5\%$ has been considered in the simulations. The stress has a small impact in the BW and gain mean values but increases their standard deviation

obtained with this simulation technique. The graph on the top corresponds to circuits simulated using the fresh BSIM4 parameters for the transistors and the graph on the bottom after 1,000 s of stress. After the stress, a small number of amplifiers show a strong gain reduction.

To analyze more carefully this point, the Bode diagrams have been simulated at different stress times, again considering $\sigma_{VT} = 5\%$ for the three circuit configurations shown in Fig. 30.4. Results are shown in Fig. 30.6. The graph on the top shows the amplifier average gain in dB, and the graph on the bottom shows its standard deviation. For all the circuit configurations, a large gain reduction is obtained if V_T variability is considered. As an example, for the circuit shown in Fig. 30.4a, after 10^5 s of stress, the average gain decreases around 2 dB, in contrast to the observed 0.68-dB reduction when V_T variability was not accounted for. These results suggest that the V_T variability plays an important role in the circuit performance, not only at $t = 0$ but also after long operation times. Moreover, the gain standard deviation continuously increases with the stress time. Another interesting point is the dependence of the results on the circuit configuration. Although Fig. 30.6 shows the same general trend for all the circuit configurations (i.e., average decrease and standard deviation increase), when a current-limiting mechanism is included, the gain decay is slower and its standard deviation is lower, suggesting that the circuit reliability can be improved if the circuit configuration is properly chosen.

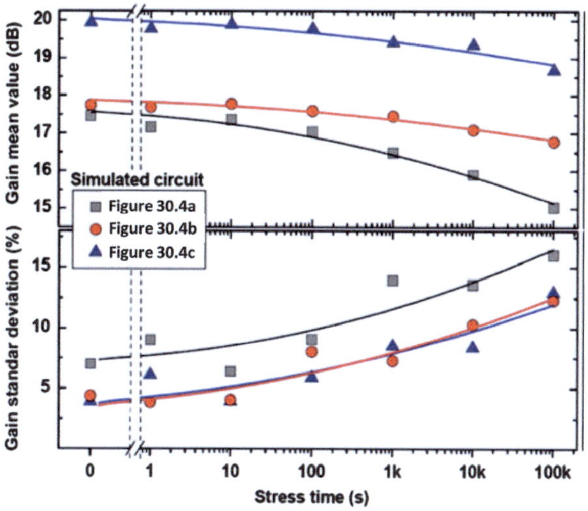

Fig. 30.6 (*Top*) Average gain and (*bottom*) its standard deviation as a function of the stress time, for the amplifiers shown in Fig. 30.4, obtained by combining SPICE and MC simulations ($\sigma_{VT} = 5\%$). The circuit configuration strongly influences the circuit variability and performance

The amplifiers BW can also be studied from the simulations performed. Similar simulations show that the BW mean value does not change significantly as a consequence of the stress, but the standard deviation considerably increases. As an example, for the circuit shown in Fig. 30.4a, BW ≈ 660 MHz during the analyzed time interval, but its standard deviation increases from 5.1% in the fresh amplifiers to 27.8% after 10^5 s of stress. To analyze the time-dependent variability impact in circuit performance and reliability in more detail, a study of the BTI effects in the amplifier gain has been done as a function of stress time and σ_{VT}. Seven different stress times (ranging from 0, i.e., fresh device, up to 10^5 s) and 12 V_T standard deviations (from 0 to 12%) have been considered, and for each stress time σ_{VT} pair, 500 amplifiers have been simulated (giving a total of 4.2×10^4 circuits). We will focus on the circuit configuration in Fig. 30.4c, since the effects of oxide degradation on its performance are the lowest. To determine the circuit reliability, the next failure criterion has been chosen: the circuit fails if the gain decays 3 dB from the obtained value in the fresh amplifier, without device V_T variability. Figure 30.7 shows the 3D plot of the cumulative failure distribution F, plotted versus σ_{VT} and t. As can be observed for low σ_{VT} (below 4–5%), F is low during the whole simulated stress time interval (neither yield nor reliability problem). For large enough σ_{VT} (>10%), F is so high at $t_s = 0$ (without stress) that it is only slightly influenced by the electrical stress (yield problem). However, for intermediate σ_{VT} values, F is low at the beginning of the stress but increases quickly with t (reliability problem). This result indicates that the V_T variability resulting from the fabrication process has a strong influence not only in the circuit yield but also on its reliability.

In this first section, a simulation methodology based on combined SPICE and MC simulations has been proposed for the evaluation of the impact of process

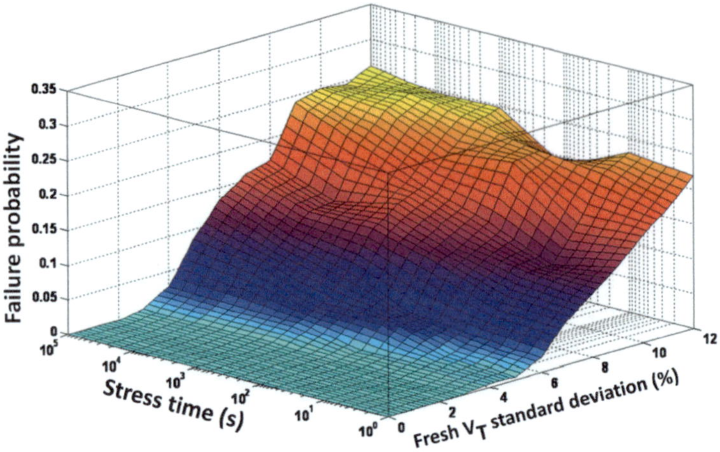

Fig. 30.7 Cumulative failure distribution of amplifiers shown in Fig. 30.4c as a function of the stress time and the initial V_T standard deviation. As failure criterion, a 3-dB decrease of the gain has been set. The number of circuits simulated for each t–σ_{VT} pair (500) allows to estimate the failure distribution with a 0.002 of precision [51]

variations and oxide wear-out on the performance of devices and circuits. Device BTI aging has been considered by changing the SPICE model parameters that describe the device behavior. At device level, we have shown that the methodology is able to reproduce the experimental observations, by only changing the averages of two of the physically meaningful parameters of the BSIM4 model parameter set obtained in fresh devices (VTH0 and U0). At circuit level, the simulation technique has been used to evaluate the gain and BW of differential amplifiers based on NMOS transistors biased with a PMOS current mirror, taking into account the stress effects and the variability observed at device level. As the stress proceeds, as a general trend, degradation of the average gain and larger spread are observed, but their values and temporal evolutions are strongly dependent on the circuit configuration. In addition, the initial threshold-voltage variability strongly affects the circuit lifetime: intermediate values of σ will reduce the reliability of circuits that showed an acceptable yield. Therefore, these results show that the device mismatch related to the variability of the fabrication process has a strong influence, not only in the circuit yield but also on its reliability.

30.3 Equivalent Circuit for the BTI Aging

In the previous section, the effects of BTI and fabrication process variability have been included in circuit simulators by varying the SPICE model parameters of MOSFETs. However, in this approach, the required changes of these parameters have been derived from experimental data on test structures, so that the obtained laws for the parameter shifts could not be easily and/or meaningfully extrapolated to other operation conditions and/or device geometries. In addition, the behavior of the device during the transients in the inputs cannot be considered. Moreover, a particular SPICE model has been chosen for the description of the MOSFETs, so that changing the MOSFET model would mean determining the parameters of the alternative model that would have to be modified and repeating phases 1 and 2 of the process in Fig. 30.2. An approach that overcomes these limitations is required, which should have into account the physics of the aging phenomenon.

In this section, it will be shown that the aging effects can be modeled in the form of an equivalent circuit, which, when connected to the device terminals, accounts for the change of the electrical properties of the device. The electrical circuit topology (built with capacitors and diodes) is based on the physics of the BTI aging and is suitable to be included in circuit simulators. It will be shown that the circuit model accounts for the BTI stress and relaxation time dependencies, and, then, the stress history of the device is considered for during the BTI degradation evaluation. The circuit is tested under DC and AC stress conditions, and the dependencies of the circuit output with frequency and duty factor (duty cycle) will be described. Finally, as an example, the circuit will be used to evaluate the BTI degradation effects in the delay time of inverters.

30.3.1 Construction of the BTI Electrical Circuit

BTI aging is attributed to charge trapping at the interface or bulk gate oxide at high voltages, charges that can be detrapped at low voltages. The probability that one trap is occupied or empty depends on its characteristic emission (τ_e) and capture (τ_c) times [67–71], which are voltage and temperature dependent [72–74] and determined by the energy level of the traps and the structure band diagram [70]. At a given condition of voltage and temperature, the occupation probability of a single trap can be electrically described by the use of resistor–capacitor (Fig. 30.8a [70]) or diode–capacitor (Fig. 30.8b [75]) elements, where the voltage drop at the capacitor is equivalent to the occupation probability of the trap. In both circuits, two diodes control the charge and discharge at the capacitor. However, in the first case, the diodes are used as ideal switches; consequently, the voltage of the capacitor in Fig. 30.8a is linear with stress voltage and has no dependence with the temperature. In the second case, the Shockley ideal model can be used to describe the diodes, so that the current through the diode, I_D, is given by Eq. (30.1):

$$I_D = I_s \cdot exp\left(\frac{nq(V_{stress} - V_c)}{KT} - 1\right) \quad (30.1)$$

being the saturation current (I_S) and the ideality factor (n) the parameters that characterize the diode. Note that I_D given by Eq. (30.1) is dependent on the stress voltage (V_{stress}) and temperature (T), so that the operation conditions of the device

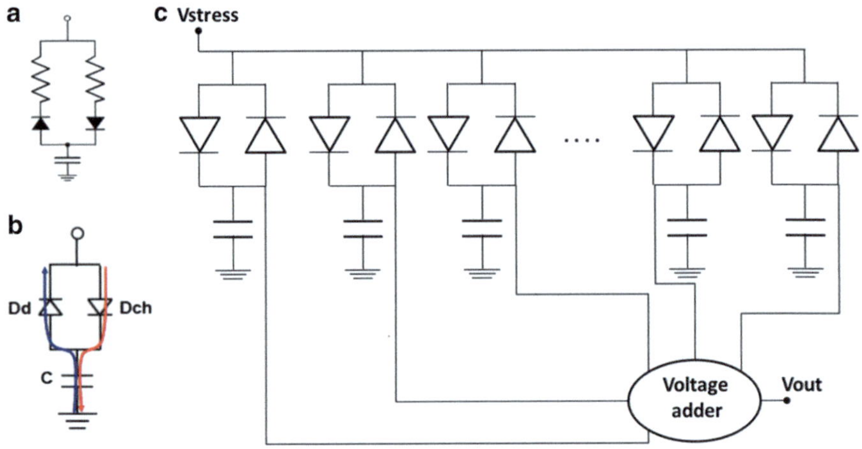

Fig. 30.8 The occupation probability of a trap can be described by the circuit in figure (**a**), where the diodes represent ideal switches. Other option is the use of the circuit (**b**), where the resistors and ideal switches of (**a**) are replaced by two diodes described by the Shockley ideal model (BTI cell). (**c**) The BTI degradation in a device is reproduced by connecting a network of different BTI cells, which represent different traps in the transistor (BTI circuit) [75]

are directly included. The circuit in Fig. 30.8b, which describes the occupation probability of a BTI trap, will be referred to as the *BTI cell*. The parameters of the BTI cell, that is, the value of the capacitor (C) and I_S and n of the charge and discharge diodes, must be chosen to obtain charge and discharge times equal to τ_c and τ_e, respectively. In a device several defects coexist, whose parameters τ_c and τ_e are widely distributed in a logarithmic scale. Therefore, to account for the BTI degradation in a device, one BTI cell for each trap should be included, leading to the *BTI circuit* (Fig. 30.8c). The output of this circuit is the sum of the voltage drop at the capacitors of each BTI cell.

30.3.2 Stress and Relaxation Under DC and AC Conditions

First, the ability of the model to reproduce the BTI degradation when a transistor is DC stressed is analyzed. The BTI circuit in Fig. 30.8c was excited with a DC source (1 V) for 10^5 s and the circuit output as a function of the stress time was simulated. The results are shown in Fig. 30.9 (black squares). As expected, for BTI degradation, the simulated V_T shows a linear dependence in a semi-log scale. After the stress, the input was grounded in order to analyze the circuit output during the capacitors discharge, i.e., BTI relaxation (Fig. 30.9, open circles), showing that the V_T also decreases linearly in a semi-log scale. These behaviors are observed over ten time decades, reproducing one of the basic properties of the BTI recovery [76].

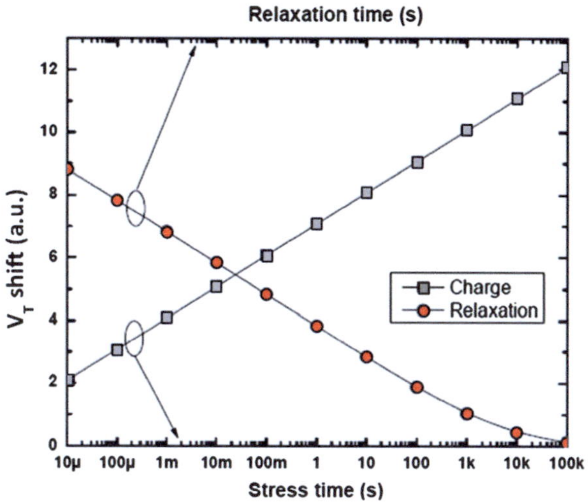

Fig. 30.9 Simulation of the V_T shift using the proposed BTI equivalent circuit (Fig. 30.8c) for stress (*filled squares*) and relaxation (*open circles*) stages. For both processes, V_T presents a linear evolution in a semi-log plot on ten time decades, in agreement with experimental observations [77]

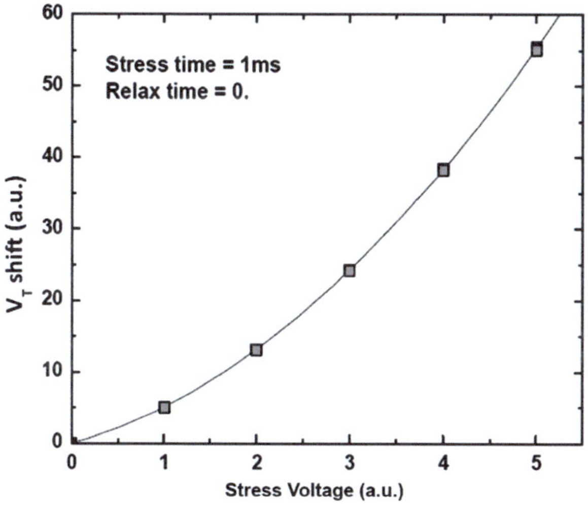

Fig. 30.10 The V_T shift follows a power law with the stress voltage applied

The suitability of the model to reproduce the stress voltage dependence was tested as well. To do so, the BTI equivalent circuit is DC stressed at different stress voltages. The results are shown in Fig. 30.10. Since nonlinear elements (diodes) control the charge of the capacitors, a superlinear relation between the V_T shift and the stress voltage can be observed, in accordance with experimental observations [77].

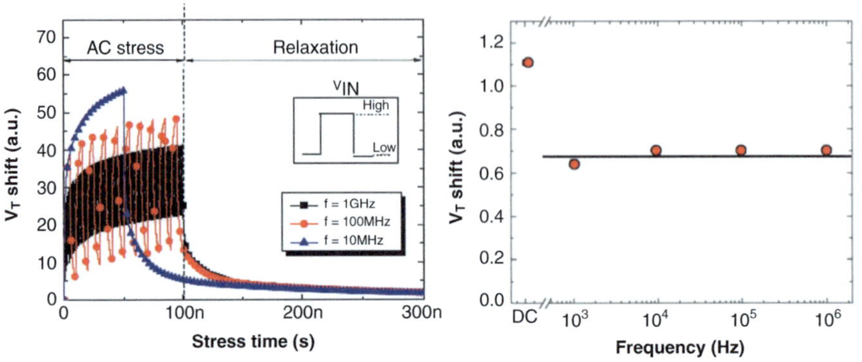

Fig. 30.11 *Left*: V_T shift evolution with time when the circuit is AC stressed with square waveforms of three different frequencies. *Right*: V_T shift dependence on frequency after 1 ms of relaxation time. DC case has been included for comparison. The results are in agreement with the experiments in [77]

The BTI equivalent circuit response to an AC input signal has been also simulated. First, a simulation of the V_T shift evolution with time was done, when the circuit is excited with unipolar square waveforms of 1 V amplitude and three different frequencies (10 MHz, 100 MHz and 1 GHz) during a short stress time (100 ns). Afterwards, the input was grounded to study the frequency dependence after a certain relaxation time. The results are shown in Fig. 30.11 (left), where a clear difference during the stress can be observed when the circuit is excited at different frequencies. After a short relaxation time (~200 ns), however, no differences in the relaxation component are observed. This is in agreement with experimental results, which showed that the V_T variation after a short relaxation time (the instrument response time) is not frequency dependent in the 1 Hz–2 GHz range [77]. This is confirmed in Fig. 30.11 (right), where the simulated V_T shift after 0.1 s of DC and AC stresses at different frequencies and a relaxation time of 1 ms is shown. In the simulations no dependence on frequency is observed in the same frequency range.

The V_T shift component after 1 ms of relaxation time was also studied for different stress times and duty factors (DF). As shown in Fig. 30.12 (left), the model is able to describe the peculiar nonlinear dependence on DF, as it was experimentally found in [76, 77]. Note that the peak at DF = 100 is more pronounced for longer stress times. In addition, the relaxation of V_T when the equivalent circuit is excited with a unipolar square waveform considering different duty factors has also been studied. Figure 30.12 (right) shows the V_T shift as a function of DF after a stress of 100 s and different relaxation times. BTI relaxation is faster when DF is close to 100, as experimentally observed [76].

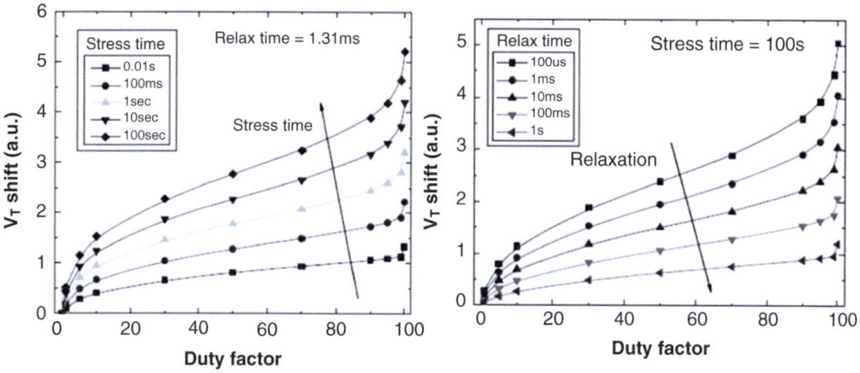

Fig. 30.12 *Left*: V_T shift as function of the duty factor after different stress times. *Right*: V_T shift as function of DF for different relaxation times after 100 s of stress. The stress frequency was 10 kHz

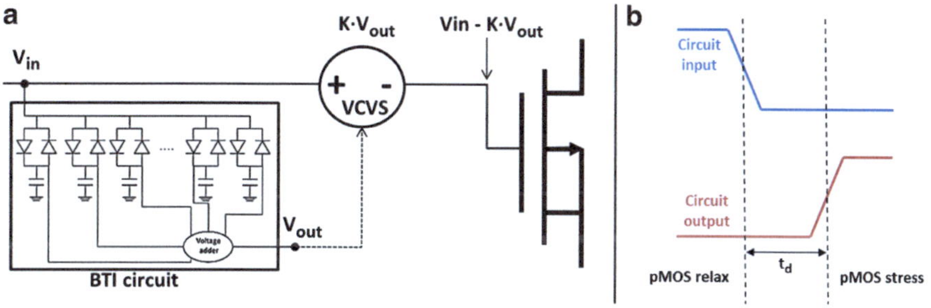

Fig. 30.13 (**a**) Connection of the BTI circuit in circuit simulators to account for the BTI-related V_T shift in devices during the circuit operation. (**b**) Schematic view of the inverter input and output during a transition in the input from logic state "1" to "0"

30.3.3 BTI Degradation in CMOS Inverters

The proposed equivalent circuit model was used to evaluate the BTI effects in a CMOS inverter. Only V_T variations in the PMOS transistor were considered. To do this, as shown in Fig. 30.13a, a voltage-controlled voltage source (VCVS) is connected to the transistor gate. The value of the VCVS is proportional to the output of the BTI circuit, whose input is the actual gate voltage applied to the gate of the device. The value of the proportionally constant, K, is related with the number of traps of the device, N, and the mean value of the VT shift caused by the traps when filled $\langle\eta\rangle$, following Eq. (30.2):

$$K = N \cdot \langle\eta\rangle \tag{30.2}$$

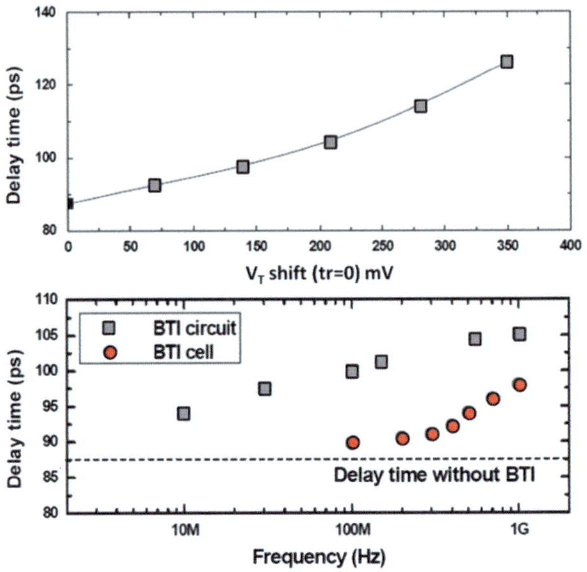

Fig. 30.14 *Top*: Delay time in a CMOS inverter as a function of the V_T shifts of the pMOS transistor (relaxation phase) calculated using the proposed equivalent BTI circuit. *Bottom*: Delay time as a function of frequency simulated using the BTI circuit and only one BTI cell. The *dashed line* indicates the delay time value if BTI effects are not considered

The inverter was excited with a 1 V square input, and, from the output waveform, the delay time (td), defined as the time interval elapsed between the time at which Vin = 0.5 V and Vout = 0.5 V after a pulse transition from "1" to "0" logic states in the circuit input (which implies that PMOS changes from "off" to "on," i.e., from relaxation phase to stress phase), was calculated (Fig. 30.13b). The delay time for different values of V_T shift of the pMOS transistor is shown in Fig. 30.14 (top) for a 1 GHz stress. The results show an increase of the delay time for larger V_T [78]. The delay time as a function of the frequency of the stress waveform was also calculated, and the results are shown in Fig. 30.14 (bottom—black squares). An increase of the delay time with frequency is observed. To understand this result, it should be taken into account that the delay time is evaluated during the transition between states, changing the pMOS from the relaxation to the stress phase (Fig. 30.13b). Therefore, when the circuit is operating at high frequency, the pMOS transistor has less time to relax, and, consequently, higher V_T shift and t_d are obtained.

In this section an equivalent circuit for the BTI degradation has been presented, which can be easily introduced in a circuit simulator to evaluate the effects of BTI on circuit performance. The BTI circuit response accurately describes the experimentally observed dependencies of BTI degradation on the operation conditions in MOSFETs. Under DC stress conditions, the proposed circuit is able to describe the V_T shift for the stress and relaxation phases and shows a nonlinear relation with the applied voltage. It is also able to reproduce the frequency and

DF dependencies of BTI under AC stress conditions. The delay time in a CMOS inverter has been studied using this electrical model when BTI aging is considered in the PMOS transistors. Simulations show that the delay time increases with the V_T value, which can limit the circuit operation frequency.

30.4 RELAB: A Tool to Include Time-Dependent Variability in SPICE Simulators

The BTI circuit model described in the previous section is able to correctly estimate the V_T shift in a MOSFET related to BTI aging. However, the connection of such kind of circuit (with multiple circuit elements) to the gate of all the transistors in a circuit containing a large number of devices could considerably increase the complexity of the circuit that finally has to be simulated (in an iterative and statistical process), so that the simulation time could be unaffordable. An alternative that could reduce the simulation time would be the use of a much more simplified aging circuit, whose value could be calculated by a purposely developed software module [25–29]. Actually, it has been proposed that V_T shift in a device can be considered during a circuit simulation by adding a voltage source to the gate of the MOSFET, which would reproduce the observed reduced current drive capability [75, 79].

In addition, simulation tools for circuit reliability evaluation should be thought to be used by circuit designers, who probably will not be experts in the reliability field. Therefore, the reliability simulation process should be as transparent to the user as possible.

In this last section, a new reliability circuit simulation tool, RELAB (*R*eliability *E*valuation Too*L* of *U*niversitat *A*utonoma de *B*arcelona) is presented which, by combining Monte Carlo and SPICE simulations (i.e., simulation methodology in Sect. 30.1), evaluates the impact of process variability and/or aging mechanisms in the response of CMOS circuits. Though RELAB is able to consider BTI and BD effects [25, 26], the RELAB features that will be shown here will focus mainly in BTI degradation. In RELAB, variability and aging are introduced in the circuit simulations by connecting circuit elements (as suggested in Sect. 30.2), which depend on the phenomena under study, to the MOSFET terminals and whose statistically distributed values are computed taking into account the physics of the phenomenon, the operation conditions, the geometry, and channel type of the devices within the circuit, so that the procedure becomes independent of the transistor compact model. For BTI, a voltage source is connected to the gate of the MOSFET to consider the associated V_T shift. In addition to the evaluation of the impact of aging on circuit performance and reliability, the tool allows performing sensitivity analysis, to detect the "weak" transistors that have larger influence. In this section, the tool is used to evaluate the process variability and BTI impact on the performance of a digital block and SRAM.

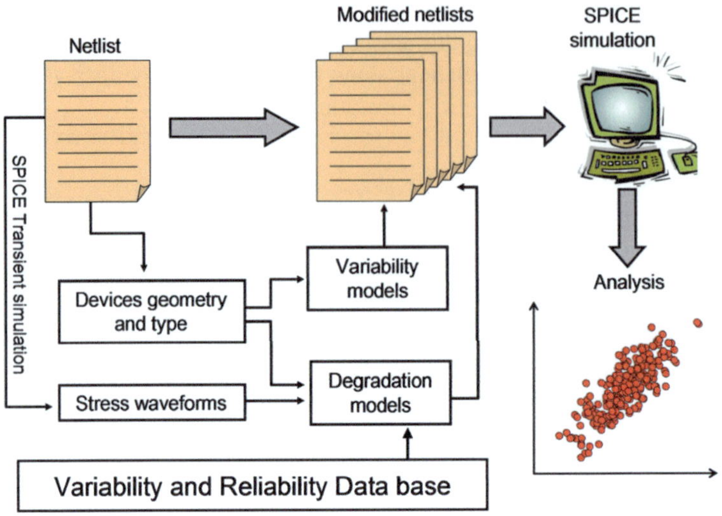

Fig. 30.15 Simulation flow of RELAB. A netlist which contains the circuit under study is the input. Variability and degradation effects on each transistor within the circuit are evaluated according to the physical models implemented in RELAB and the operation conditions of the devices within the circuits. The effect of variability and aging mechanisms is included in new netlists by connecting circuit elements to the transistor terminals. Finally, SPICE simulations of the modified netlists are launched to evaluate the circuit performance after device aging

30.4.1 RELAB Simulation Flow

RELAB simulation flow is depicted in Fig. 30.15. The input of the tool is a netlist that contains the description of the circuit under study. RELAB analyzes this netlist to obtain the information on the characteristics of the transistors which is relevant for the variability and degradation evaluation, such as transistor channel width (W), length (L), and substrate type (nMOS or pMOS). Additionally, for degradation studies, the knowledge of the voltages applied at the transistor terminals and their evolution with time are essential, in order to correctly evaluate the damage induced in the devices. To obtain this information, RELAB launches a SPICE transient simulation of the circuit. These data are used to calculate the degradation that each transistor suffers within the circuit, with the aid of the physical models included in the tool and a variability/reliability data base measured on test structures (representative of the fabrication technology) from which the model parameters are extracted. New netlists are created in which the initial circuit is changed by adding external circuit elements connected to the MOSFET terminals that account for the variability/degradation effects predicted by the models. The modification of the netlist does not significantly increase the simulation time of the circuit. RELAB launches a SPICE simulation each time that the netlist is modified.

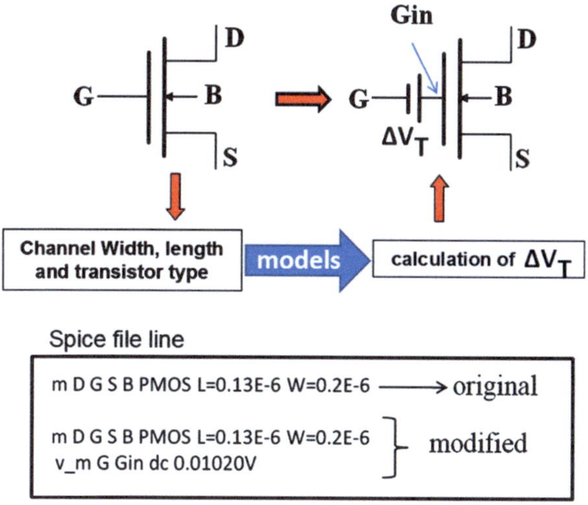

Fig. 30.16 To consider transistor threshold-voltage shifts during circuit simulation, RELAB adds a voltage source, with value ΔV_T, to the gate of each transistor. RELAB also reads the transistor parameters needed to calculate the ΔV_T value. With this purpose, the RELAB includes models for the different phenomena (variability and/or aging) under study, whose parameters have been experimentally determined

To consider the V_T shift of the transistors during the circuit simulation, which can be due to process variation or BTI aging, RELAB modifies the SPICE circuit description by adding a constant voltage source of value ΔV_T to the gate of each transistor (Fig. 30.16) in the circuit. From the transistor data (width, length, channel type, and voltages at the terminals) and the implemented variability/aging models, the corresponding ΔV_T values are calculated and fed into the new SPICE script. To make this calculation, the model parameters have been previously obtained from experimental data on test structures.

The key point of the simulations is the physical model for the device BTI aging. RELAB evaluates the BTI-related V_T shift using the proposed probabilistic occupancy model (PDO) [36], which allows computing the corresponding V_T shift in a very efficient way. To do so, several traps are included in each transistor in the circuit. The V_T shift in the device will be dependent on the number of occupied traps. As previously described in this book [46], the number of traps (N) and their characteristic times (τ_c and τ_e) are statistically distributed [36, 79]. The voltage and temperature dependencies of τ_c and τ_e are included, considering the laws obtained from the experimental characterization [73, 74]. The stress waveform applied to each device, provided by SPICE from the transient simulation, is used to evaluate the occupancy probability of each trap within the circuit. For an empty trap, the occupation probability (F_c) at time t is given by Eq. (30.3):

$$F_c = 1 - e^{-t/\tau_c(V_G(t),T)} \qquad (30.3)$$

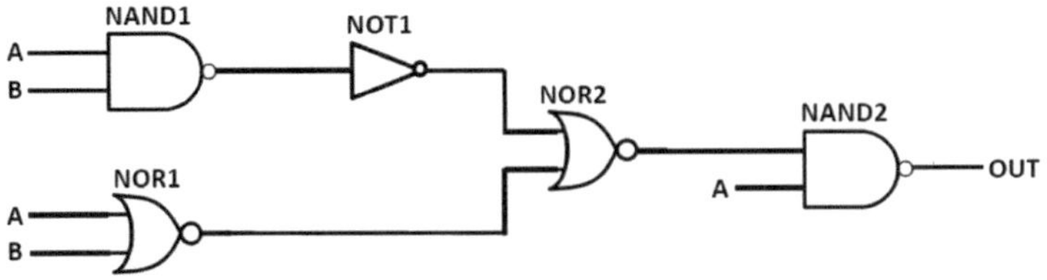

Fig. 30.17 Logic CMOS circuit used as an example to show the main features of RELAB. The circuit has 18 (9 nMOS and 9 pMOS) devices and its transistors have been designed with different geometries (ranging from W × L = 0.15 μm × 0.13 μm to 1 μm × 0.13 μm)

Similarly, if a trap is occupied, the detrapping cumulative probability (F_e) is given by Eq. (30.4):

$$F_e = 1 - e^{-t/\tau_e(V_G(t),T)} \tag{30.4}$$

Then, from Monte Carlo simulations, taking into account the distributions given by Eqs. (30.3) and (30.4), the occupation state of the trap (empty or occupied) is evaluated. If the trap is occupied, a V_T shift of value η is produced in the MOSFET. In a set of traps, η is exponentially distributed [39, 43]. Finally, ΔV_T is calculated as the contribution of all the defects in the device following Eq. (30.5):

$$\Delta V_T|_{NBTI} = \sum_i^N k_i(\tau_c, \tau_e) \cdot \eta_i \tag{30.5}$$

where i is an index that denotes each trap, η_i is the V_T shift caused by the i-th trap when charged, and k_i is a coefficient that takes the values of 1 or 0 if the trap is occupied or empty, respectively.

30.4.2 Evaluation of BTI Degradation and Variability Effects on a Digital Band SRAM Cells

Using RELAB, the BTI degradation impact in the performance of the circuit of Fig. 30.17 has been evaluated. The area of the devices is small enough to observe the statistical nature of BTI aging.

Figure 30.18 shows an example of the stress voltages and V_T shift obtained in three transistors of the circuit in Fig. 30.17. Black lines (left axis) correspond to the voltage at the gate provided by the transient simulation, or in other words, the stress history of each transistor. Several simulations are launched, considering a statistical distribution of defects in the devices, to induce the variability in V_T associated to BTI. The dots (right axis) indicate the average of the V_T shift caused by the BTI stress on each transistor obtained from 200 simulations. As can be observed,

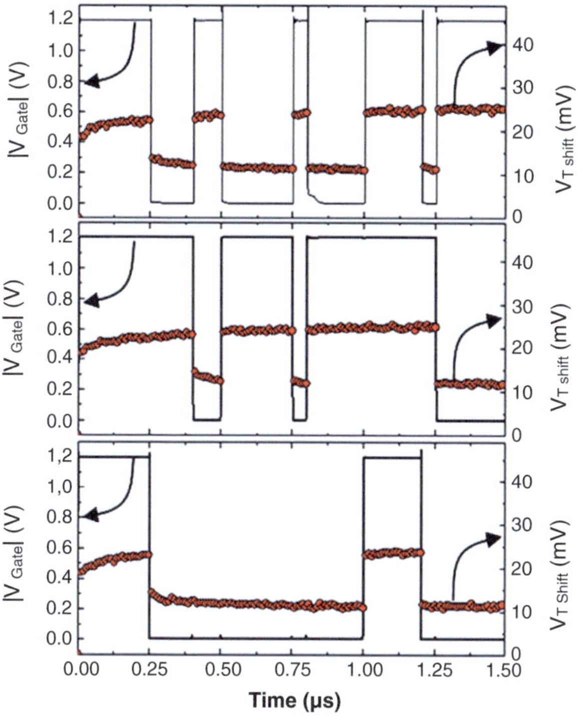

Fig. 30.18 ΔV_T caused by BTI stress calculated in three different transistors of the circuit in Fig. 30.17. The actual stress waveforms applied to each transistor (*lines*, *left axis*) are used to calculate the BTI-induced V_T shift (*symbols*, *right axis*)

RELAB considers the particular operation conditions of the devices, so that the impact of BTI on the circuit performance considering the actual damage suffered by each transistor is evaluated.

Figure 30.19 shows that the cumulative probability of the circuit delay time has been simulated for four operation times (symbols). These distributions have been used to extrapolate the V_T shift to 1 year of operation time (continuous line), by assuming that the voltage input is periodic. Note that not only the average value increases but also, due to the stochastic BTI nature, the delay time variability.

Another useful feature of RELAB is the detection of "weak transistors" in the circuit, that is, those MOSFETs that have a larger influence on the circuit output, by studying the relation between the device ΔV_T and the circuit performance. As an example, in Fig. 30.20, the delay of the circuit in Fig. 30.17 is plotted versus the ΔV_T values in two of the transistors. In this case, ΔV_T has been assumed to be caused only by the process variations, so that normal distributions with standard deviations given by the Pelgrom's rule (since process variability depends on the device area [79]) have been considered to obtain the V_T shift values. Figure 30.20 shows that the circuit delay is not affected by changes in the V_T of transistor #4 (left). However, a

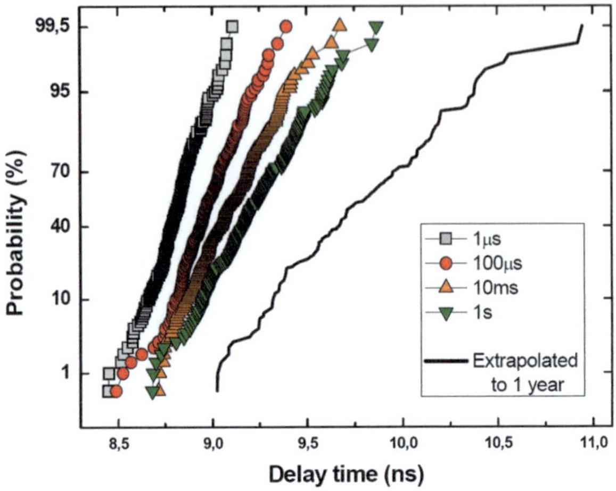

Fig. 30.19 Simulated delay time statistical distributions of the circuit in Fig. 30.17 for different BTI stress times. The results take into account the actual stress that each transistor has received during the circuit operation. The *line* corresponds to the extrapolation of the delay time to 1 year of operation calculated from the simulation results and assuming periodic voltage waveforms at the device terminals

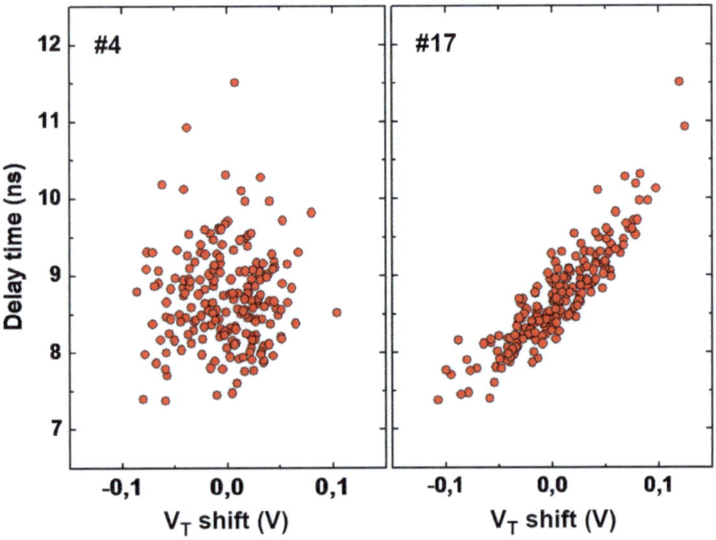

Fig. 30.20 Delay time of the circuit in Fig. 30.17 versus ΔV_T of two transistors, when only their process variability is considered. No influence of the V_T deviation in transistor #4 (*left*) is observed in the circuit response. However, the circuit delay time is clearly related to the ΔV_T of transistor #17 (*right*). This demonstrates that RELAB can be used to detect weak points in the circuits caused by V_T shifts

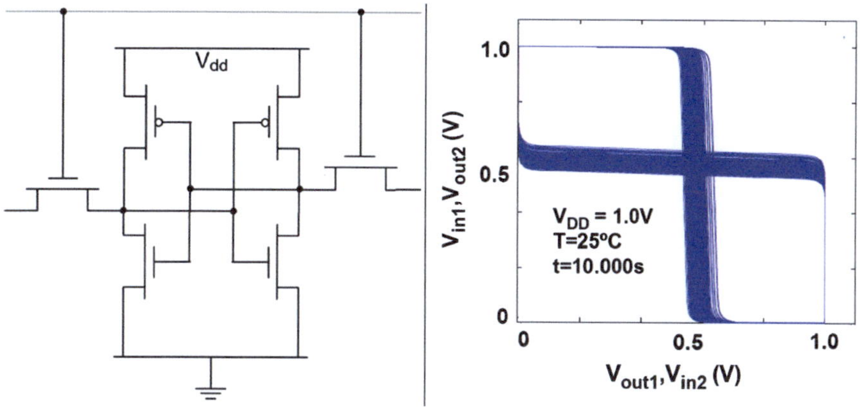

Fig. 30.21 *Left*: RELAB has been used to evaluate the BTI impact at different operation conditions on SRAM cells. *Right*: simulated butterfly plots of the SRAM cells at t = 10,000 s, T = 125°C, and $V_{DD} = 1$ V

clear correlation between delay and V_T shift of transistor #17 is observed. This result indicates that RELAB can be used to perform sensitivity analysis, whose information could be used to optimize the circuit design.

Another important issue for the circuit reliability evaluation is that BTI is strongly dependent on the operation conditions (voltage, temperature, operation time). RELAB is able to consider different conditions by changing (internally) the model parameters. To do so, RELAB calculates τ_e and τ_c of all the traps within the devices in the circuit at each particular operation condition using the PDO model and the dependencies of the model parameters with temperature and voltage. As an example, RELAB has been used to compute the BTI effects in SRAM cells (Fig. 30.21, left) at different operation conditions. Again, the devices are small enough to observe the BTI-related V_T variability. Figure 30.21 (right) shows 200 simulations of the typical butterfly curves of the SRAM obtained when the logic states change between "0" and "1" every 0.1 s. A temperature of 125°C and supply voltage of $V_{DD} = 1$ V and operation time (t) of 10,000 s have been considered.

We can clearly observe the dispersion of the transfer characteristics introduced by the BTI aging. To evaluate its dependence with voltage, temperature, and time, the static noise margin (SNM) of the SRAM cells has been calculated at three different operation conditions. The SNM distributions are presented in Fig. 30.22. At $V_{DD} = 0.6$ V, T = 25°C, and t = 1 s (Fig. 30.22, top), a low spread of the SNM distribution is observed because only few traps in the device are potentially capable to trap/detrap charge. Actually, typical RTN signals are observed for the V_T shits [74]. For larger V_{DD} (Fig. 30.22, middle), the number of active traps increases and, consequently, a larger SNM spread is observed. In the last case (Fig. 30.22, bottom), for a higher temperature and operation time, several defects that were empty in cases top and bottom are now occupied (leading to a net shift of V_T, typical of BTI), and, consequently, a shift of the distribution to lower SNM values and an increase

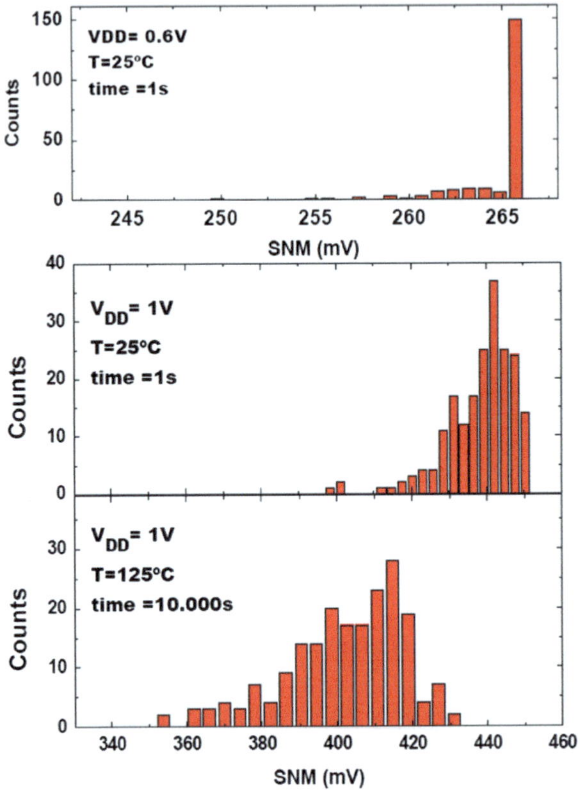

Fig. 30.22 SNM distributions obtained for SRAM cells at different operation conditions

of the spread are obtained. Finally, these simulations show that if the behavior of individual defects are properly modeled and characterized, the impact of BTI and its variability on the circuit performance can be evaluated, taking into account the operation conditions of the transistors within the circuit.

In this last section, RELAB circuit reliability simulation tool has been presented, which allows including process variability and aging mechanisms into SPICE simulators. Here, RELAB has been used to evaluate the effects of BTI (and its inherent variability) on the circuit electrical properties, from the BTI aging physics-based implemented models. V_T shifts are included in the SPICE simulator by automatically connecting a voltage source to the device terminals, which accounts for the aging of the devices. Iterative Monte Carlo simulations allow evaluating the statistical evolution with time and dependence on the operation conditions of the circuit performance. RELAB can perform sensitivity analysis that can be used to detect the weakest devices in the circuits and optimize the circuit design. In summary, RELAB provides a user-friendly framework for circuit designers to evaluate the IC reliability.

30.5 Conclusions

In this subchapter, different alternatives to translate the effects of BTI degradation on the electrical properties of devices into circuit performance and reliability have been addressed. With this purpose, a simulation methodology, based on the combination of SPICE and Monte Carlo (MC) simulations, has been developed, which allows considering the variability of the electrical parameters of MOSFETs, related to the fabrication process or to the device wear-out, during circuit simulation, so that their impact on the performance and reliability of circuits can be evaluated. To account for the variability and aging of the devices during the circuit simulation, different approaches can be adopted. In the first one, some parameters of the MOSFET SPICE model are changed, according to the empirical laws obtained from the experimental data. Using differential amplifiers as circuit example, it has been shown that the circuit topology can strongly improve the reliability. Moreover, the simulations show that fabrication process variability can have a very detrimental effect in the circuit lifetime. In the second section, circuit elements (whose topology will depend on the phenomena under analysis) are connected to the device terminals, which account for the variability and aging on the device electrical properties. One of these circuits has been described in detail: a physics-based equivalent circuit (built with diodes and capacitors) that reproduces threshold-voltage shifts related to BTI aging. A much-simpler circuit could be also used, which improves the computing efficiency at the expense of accuracy: a voltage source whose value is computed during the simulations from the physics of the phenomenon. Finally, the main features of a reliability circuit simulation tool, called RELAB (*R*eliability *E*valuation Too*L* of Universitat *A*utonoma de *B*arcelona), have been presented. By means of RELAB, in a process which is transparent to the user, the impact of BTI degradation and its inherent variability on the circuit performance is rapidly analyzed, by considering physical models of the phenomena and the device operation conditions of the devices within the circuit. The tool can be very useful to circuit designers to choose the best design option from the reliability point of view.

Acknowledgments This work was partially supported by the Spanish MINECO (TEC2010-16126) and by the Generalitat de Catalunya (2009 SGR-783).

References

1. J. Hicks *et al.*, *Intel Technology Journal***12**, 131 (2008).
2. A. Haggag, et. Al., IEEE Int. Reliab. Phys.Symp. (IRPS) Proc. 93 (2006).
3. Dieter K. Schroder and Jeff A. Babcock, J. Appl. Phys. **94**, 1 (2003)
4. E. Takeda, E. Murakami, K. Torii, Y. Okuyama, E. Ebe, K. Hinode, and S.Kimura, Microelectron. Rcliab. **42**, 493 (2002).
5. M. M. Albert and N. H. Tolk, Phys. Rev. B **63**, 035308 (2001).
6. C. H. Lin, M. H. Lee, and C. W. Liu, Appl. Phys. Lett. **78**, 637 (2001).
7. C. H. Liu et al., Tech. Dig. Int. Electron Devices Meet, 861 (2001).

8. M. Aoulaiche, et. al., IEEE Int. Reliab. Phys.Symp. (IRPS) Proc., 358 (2008).
9. S. Mahapatra and V.D. Maheta, Proceedings of Solid-State and Integrated-Circuit Technology (ICSICT) Proc. 616 (2008).
10. S. Krishnan et. al., IEEE Int. Reliab. Phys.Symp. (IRPS) Proc. 5.A.1.1 (2012).
11. S. Mahapatra, FEOL and BEOL process dependence of NBTI, in *Bias Temperature Instability for Devices and Circuits*, ed. by T. Grasser (Springer, Heidelberg, 2013).
12. M. Silverman and A. Kleyner, Proceedings Reliability and Maintainability Symposium (RAMS) Proc. 1 (2012).
13. S. Mitra, K. Brelsford, Y. M. Kim, H.-H. Lee and Y. Li, Emerging and Selected Topics in Circuits and Systems, **1**, 30 (2011).
14. E. Maricau and G. Gielen. Emerging and Selected Topics in Circuits and Systems **1**, 50 (2011).
15. R. Joshi, R. Kanj, C. Adams and J.Warnock. IEEE Int. Reliab. Phys.Symp. (IRPS) Proc. 3A.6.1(2013).
16. U. Y. Ogras, R. Marculescu, D. Marculescu, IEEE Design Automation Conference Proc. 614 (2008).
17. D. Sylvester, D. Blaauw and E. Karl, IEEE Design & Test of Computers Proc., 484 (2006).
18. J. Tschanz et al., Solid-State Circuits Conference, 2007. ISSCC 2007. Digest of Technical Papers. IEEE International 292 (2007).
19. J. Tschanz et al., Symposium on VLSI Technology Digest of Technical Papers 112 (2009).
20. E. Karl, D. Blaaw, D. Sylvester and T. Mudge IEEE Trans. on VLSI **16**, 476 (2008).
21. K. Mihic, T. Simuni and G. De Micheli, Digital System Design (DSD) Euromicro Symposium on 5 (2004)
22. J. Srinivasan, S. V. Adve, P. Bose and J. A. Rivers, IEEE *Micro,***25**, 70 (2005).
23. E. Mintarno, V. Chandra, D. Pietromonaco, R. Aitken and R. W. Dutto, IEEE Int. Reliab. Phys.Symp. (IRPS) Proc. p. 3A1.1 (2013)
24. International Technology Roadmap for Semiconductors available at http://public.itrs.net.
25. M. Nafria, R. Rodriguez, M. Porti, J. Martin-Martinez, M. Lanza, and X. Aymerich, Int. Electron Devices Meeting Tech. Dig., 6.3.1 (2011).
26. J. Martin-Martinez, N. Ayala, R. Rodriguez, M. Nafria and X. Aymerich, Synthesis, Modeling, Analysis and Simulation Methods and Applications to Circuit Design (SMACD), Conference on. 249 (2012).
27. L. Gerrer et. al., IEEE Int. Reliab. Phys.Symp. (IRPS) Proc. 3A.2 (2013).
28. B. Kaczer et. al., IRPS Int. Reliab. Phys.Symp. (IRPS) Proc. XT.3.1 (2011).
29. P. Weckx, B. Kaczer, M. Toledano-Luque, T. Grasser, P.J. Roussel, et. al. Int. Reliab. Phys. Symp. (IRPS) Proc. 3A4.1 (2013).
30. L. Brusamarello, G. I. Wirth, Ph. Roussel, M. Miranda, Microel. Reliab. ***51****,2341 (2011).*
31. M. A. Alam and S. Mahapatra, *Microel. Reliab.***45**, 71(2005).
32. M. A. Alam, H. Kufluoglu, D. Varghese, and S. Mahapatra, *Microel. Reliab.***47**, 853 (2007).
33. V. Huard, M. Denais, and C. Parthasarathy, *Microel. Reliab.***46** 1(2006).
34. B. Kaczer, V. Arkhipov, M. Jurczak, and G. Groeseneken, *Microel.Eng.***80**, 122 (2005).
35. T. Grasser, B. Kaczer, W. Goes, H. Reisinger, T. Aichinger, P. Hehenberger, P.-J. Wagner, F. Schanovsky, J. Franco, P. J. Roussel, and M. Nelhiebel. Int. Electron Devices Meeting Tech. Dig., 4.4.1 (2010)
36. J. Martin-Martinez, B. Kaczer, M. Toledano-Luque, R. Rodriguez, M. Nafria, X. Aymerich, G. Groeseneken, IEEE Int. Reliab. Phys. Symp. (IRPS) Proc., XT4.1 (2011)
37. S. Mahapatra, A comprehensive modeling framework for DC and AC NBTI, in *Bias Temperature Instability for Devices and Circuits*, ed. by T. Grasser (Springer, Heidelberg, 2013).
38. T. Grasser, The capture/emission time map approach to the bias temperature instability, in *Bias Temperature Instability for Devices and Circuits*, ed. by T. Grasser (Springer, Heidelberg, 2013).
39. B. Kaczer, T. Grasser, Ph. J. Roussel, J. Franco, R. Degraeve, L.-A. Ragnarsson, E. Simoen, G. Groeseneken and H. Reisinger, IEEE Int. Reliab. Phys. Symp. (IRPS) Proc. 26 (2010).
40. K. Zhao, J. H. Stathis, B. P. Linder, E. Carties and A. Kerber, IEEE Int. Reliab. Phys. Symp. (IRPS) Proc. 4A.3.1 (2011)

41. B. Kaczer, J. Franco, M. Toledano-Luque, Ph. J. Roussel, M. F. Bukhori, A. Asenov, B. Schwarz, M. Bina, T. Grasser, G. Groeseneken, IEEE Int. Reliab. Phys. Symp. Proc. 5A.2.1 (2012)
42. S. E. Rauch, IEEE T. Dev. Mat. Rel. 7, 524 (2007)
43. J. Franco, B. Kaczer, M. Toledano-Luque, Ph. J. Roussel, J. Mitard, L.-Å. Ragnarsson, L. Witters, T. Chiarella, M. Togo, N. Horiguchi, G. Groeseneken, M. F. Bukhori, T. Grasser and A. Asenov, IEEE Int. Reliab. Phys. Symp. Proc. 5A.4.1 (2012).
44. M. Toledano-Luque, B. Kaczer, Ph.J. Roussel, J. Franco, T. Grasser, C. Vrancken, N.Horiguchi, and G. Groeseneken, Proc. IEEE Int. Reliab. Phys. Symp., 4A.2.1 (2011).
45. H. Reisinger, T. Grasser, K. Ermisch, H. Nielen, W. Gustin, and C. Schlünder, Proc. IEEE Int. Reliab. Phys. Symp., 6A.1.1 (2011)
46. S. E. Rauch, III, BTI induced statistical variations, in *Bias Temperature Instability for Devices and Circuits*, ed. by T. Grasser (Springer, Heidelberg, 2013).
47. B. Kaczer, M. Toledano-Luque, J. Franco, P. Weckx, Statistical distribution of defect parameters, in *Bias Temperature Instability for Devices and Circuits*, ed. by T. Grasser (Springer, Heidelberg, 2013).
48. M. Toledano-Luque, B. Kaczer, J. Franco, Ph. J. Roussel, T. Grasser, T. Y. Hoffmann, G. Groeseneken, Proc. VLSI Symp. 152 (2011)
49. P. Asenov, N.A. Kamsani, D. Reid, C. Millar, S. Roy, A. Asenov, in Proc. of the European Solid-State Device Research Conference (ESSDERC), 130. (2010)
50. B. Cheng, D. Dideban, N. Moezi, C. Millar, G. Roy, X. Wang, S. Roy and A. Asenov, Design, Automation & Test in Europe (DATE), 650, (2010).
51. J. Martin-Martinez, R. Rodriguez, M. Nafria and X. Aymerich, IEEE T. Dev. Mat. Rel., **9**, 305 (2009).
52. M. Pelgrom, A. Duinmaijer and A. Welbers, IEEE J. Solid-State Circuits, **24**, 1433 (1989).
53. T. Mizutani, A. Kumar and T. Hiramoto, Int. Electron Devices Meeting Tech. Dig., 25.2.1.(2011)
54. A. Asenov, Proc. VLSI Symp., 86 (2007).
55. S.K. Saha, Design & Test of Computers, IEEE, **27**, 8 (2010).
56. A. Tajalli and Y. Leblebici, Circuit and Systems I: Regular Papers, IEEE Transactions on, **58**, 2189 (2011).
57. V. Huard, C. Parthasarathy, G. Ribes, F. Perrier, N. Revil and A. Bravaix, 265 (2004).
58. A. Kerber, K. Maitra, A. Majumdar, M. Hargrove, R. J. Carter and E. A.Cartier, IEEE T. Electron Dev., **55**, 3175 (2008).
59. H. Reisinger, O. Blank, W. Heinrigs, A. Mühlhoff, W. Gustin, and C. Schlünder, IEEE Int. Reliab. Phys. Symp. (IRPS) Proc., 448 (2006).
60. B. Kaczer, T. Grasser, Ph. J. Roussel, J. Martin-Martinez, R. O'Connor, B. J. O'Sullivan, G. Groeseneken, IEEE Int. Reliab. Phys. Symp. (IRPS) Proc., 20 (2008).
61. A. Kerber, E. Cartier, Bias temperature instability characterization methods, in *Bias Temperature Instability for Devices and Circuits*, ed. by T. Grasser (Springer, Heidelberg, 2013).
62. Y. Z. Hu, D. S. Ang, ans Z. Q. Teo, IEEE T. Electron Dev., 57, 2027, (2010).
63. BSIM4. Available: www.device.eecs.berkeley.edu/bsim/Files/BSIM4/BSIM460/doc/BSIM460_Manual.pdf
64. AURORA User's manual. www.synopsys.com
65. J. H- Stathis and S. Zafar. Microelectron. Reliab., **46**, 270 (2006)
66. Agilent technologies. Advanced Design System.
67. T. Grasser, H. Reisinger, P.-J. Wagner, F. Schanovsky, W. Goes, and B. Kaczer, Proc. IEEE Int. Reliab. Phys. Symp., 16 (2010)
68. S. M. Amoroso, L. Gerrer, S. Markov, F. Adamu-Lema and A. Asenov., Proc. of the European Solid-State Device Research Conference (ESSDERC), 109 (2012).
69. B. Kaczer, Ph.J. Roussel, T. Grasser and G. Groeseneken, IEEE Electron Dev. Letters, **31**, 411(2010).
70. H. Reisinger, T. Grasser, W. Gustin and C. Schlünder, IEEE Int. Reliab. Phys. Symp. (IRPS) Proc., 7, (2010)

71. M. Toledano-Luque, B. Kaczer, Characterization of individual traps in high-κ oxides, in *Bias Temperature Instability for Devices and Circuits*, ed. by T. Grasser (Springer, Heidelberg, 2013).
72. M. Toledano-Luque, B. Kaczer, T. Grasser, Ph. J. Roussel, J. Franco and G. Groeseneken, J. Vac. Sci. Technol. B 31, 01A114 (2013).
73. T. Grasser, P.-J. Wagner, H. Reisinger, Th. Aichinger, G. Pobegen, M. Nelhiebel, and B. Kaczer, Int. Electron Devices Meeting Tech. Dig., 27.4.1 (2011).
74. N. Ayala, J. Martin-Martinez, R. Rodriguez, M. Nafria and X. Aymerich, Proc. of the European Solid-State Device Research Conference (ESSDERC), 266 (2012).
75. J. Martin-Martinez, R. Rodriguez, M. Nafria, X. Aymerich, B. Kaczer and G. Groeseneken., Proc. of the European Solid-State Device Research Conference (ESSDERC), 55 (2008)
76. B. Kazcer, T. Grasser, P. J. Roussel, J. Martin-Martinez, R. O'Connor, B. J. O'Sullivan and G. Groeseneken, IEEE Int. Reliab. Phys. Symp. (IRPS) Proc., 20 (2008).
77. R. Fernandez, B. Kazcer, A. Nackaerts, S. Demuynck, R. Rodriguez, M. Nafria and G. Groeseneken, Int. Electron Devices Meeting Tech. Dig., 1 (2004).
78. Ring Oscillator Based Test Structure for NBTI Analysis M.B. Ketchen, M. Bhushan, R. Bolam, Microelectronic Test Structures, ICMTS '07. IEEE International Conference on, 42 (2007).
79. B. Kaczer, T. Grasser, J. Martin-Martinez, E. Simoen, M. Aoulaiche, Ph.J. Roussel and G. Groeseneken, Proc. IEEE Int. Reliab. Phys. Symp. (IRPS) Proc. , 55 (2009)